COMPLETE GUIDE TO PRESCRIPTION & NONPRESCRIPTION

DRUGS

By H. WINTER GRIFFITH, M.D.

Revised and Updated by Stephen W. Moore, M.D.

Technical Consultants:
Kevin Boesen, Pharm.D.
Cindy Boesen, Pharm.D.

Over 6000 Brand Names
Over 1000 Generic Names

A Perigee Book

PERIGEE
An imprint of Penguin Random House LLC
375 Hudson Street, New York, New York 10014

COMPLETE GUIDE TO PRESCRIPTION AND NONPRESCRIPTION DRUGS

ISBN: 978-0-399-17573-2
ISSN: 1082-2585

PUBLISHING HISTORY
2016–2017 Perigee trade paperback edition / November 2015

PRINTED IN THE UNITED STATES OF AMERICA

10 9 8 7 6 5 4 3 2

Neither the publisher nor the author is engaged in rendering professional advice or services to the individual reader. The ideas, procedures, and suggestions contained in this book are not intended as a substitute for consulting with your physician. All matters regarding your health require medical supervision. Neither the author nor the publisher shall be liable or responsible for any loss or damage allegedly arising from any information or suggestion in this book.

While the author has made every effort to provide accurate telephone numbers and Internet addresses at the time of publication, neither the publisher nor the author assumes any responsibility for errors, or for changes that occur after publication. Further, publisher does not have any control over and does not assume any responsibility for author or third-party websites or their content.

Most Perigee books are available at special quantity discounts for bulk purchases for sales promotions, premiums, fund-raising, or educational use. Special books, or book excerpts, can also be created to fit specific needs. For details, write: SpecialMarkets@penguinrandomhouse.com.

Contents

About the Author

H. Winter Griffith, M.D., authored 25 medical books, including the *Complete Guide to Symptoms, Illness & Surgery* and *Complete Guide to Sports Injuries*, each published by The Body Press/Perigee Books. Dr. Griffith received his medical degree from Emory University in Atlanta, Georgia. After 20 years in private practice, he established and was the first director of a basic medical science program at Florida State University. He then became an associate professor of Family and Community Medicine at the University of Arizona College of Medicine. Until his death in 1993, Dr. Griffith lived in Tucson, Arizona.

Editor

Stephen Moore, M.D.
Family physician, Tucson, Arizona

Technical Consultants

Kevin Boesen, Pharm.D.
Clinical Assistant Professor, College of Pharmacy, University of Arizona

Cindy Boesen, Pharm.D.
Clinical Instructor, College of Pharmacy, University of Arizona

Technical Editor

Jo A. Griffith

Drugs and You

What is in This Book

The purpose of this book is to give you information about the most widely used drugs (prescription and nonprescription). The information is derived from many authoritative sources and represents the consensus of many experts. Every effort has been made to ensure accuracy and completeness. However, because drug information is constantly changing, you should always talk to your doctor or pharmacist if you have any questions or concerns.

The information applies to generic drugs in both the United States and Canada. Generic names do not vary in these countries, but brand names do. Each year, new drug charts are added and existing charts are updated when appropriate. For the most part, drugs that are injected by a medical professional, used mainly in a hospital (or medical clinic) or have rare usage are not included.

A drug cannot "cure." It aids the body's natural defenses to promote recovery. Likewise, a manufacturer or doctor cannot guarantee a drug will help every person. The complexity of the human body, individual responses in different people and in the same person under different circumstances, past and present health, age and gender impact how well a drug works.

All effective drugs produce desirable changes in the body, but can also cause undesirable adverse reactions or side effects. Before you decide whether to take a drug, you or your doctor must decide, "Will the benefits outweigh the risks?"

In the United States, it is the responsibility of the Food and Drug Administration (FDA) to ensure that drugs are safe and effective. For more information, you may contact the FDA at 1-888-INFO-FDA or visit the website: www.fda.gov.

Your Role

Learn the generic names and brand names of all your medicines. For example, acetaminophen is the generic name for the brand Tylenol. Write them down to help you remember. If a drug is a combination, learn the names of its generic ingredients.

Filling a Prescription

Once a prescription is written you may purchase the medication from various sources. Pharmacies are usually located in a drug or grocery store. You may need to consider your options: Does your health insurance limit where prescriptions can be filled? Is the location convenient? Does the pharmacy maintain patient records and are the employees helpful and willing to answer drug related questions?

Insurance companies or an HMO (Health Maintenance Organization) may specify certain pharmacies. Some insurance companies have chosen a mail-order pharmacy. Normally a prescription is sent to the mail-order pharmacy or phoned in by the physician. Mail order is best used for maintenance (long-term medications). Short-term medications such as antibiotics should be purchased at a local pharmacy.

Once a pharmacy has been chosen it is best to stay with that one so an accurate drug history can be maintained. The pharmacist can more easily check for drug interactions that may be potentially harmful to the patient or decrease the efficacy of one or more of the medications.

You can phone the pharmacy for a refill. Provide the prescription number, name of medication, and name of the patient.

Taking A Drug

Read the instructions provided with the drug and follow all directions for taking or using it.

Never take medicine in the dark! Recheck the label before each use. You could be taking the wrong drug!

Tell your doctor about any unexpected new symptoms you have while taking or using a drug. You may need to change drugs or have a dose adjustment.

Storage

Keep all medicines out of children's reach and in childproof containers. Store drugs in a cool, dry place, such as a kitchen cabinet or bedroom. Avoid medicine cabinets in bathrooms. They get too moist and warm at times.

Keep medicine in its original container, tightly closed. Don't remove the label! If directions call for refrigeration, keep the medicine cool, but don't freeze it.

Discarding

Don't save leftover medicine to use later. Discard it on or before the expiration date shown on the container. Dispose safely to protect children and pets. See page xx.

Alertness

Many of the medicines used to treat disorders may alter your alertness. If you drive, work around machinery, or must avoid sedation, discuss the problem with your doctor; usually there are ways (e.g., the time of day you take the medicine) to manage the problem.

Alcohol & Medications

Alcohol and drugs of abuse defeat the purpose of many medications. For example, alcohol causes depression; if you drink and are depressed, antidepressants will not relieve the depression. If you have a problem with drinking or drugs, discuss it with your doctor. There are ways to help.

Learn About Drugs

Study the information in this book's charts regarding your medications. Read each chart completely. Because of space limitations, most information that fits more than one category appears only once. Any time you are prescribed a new medication, read the information on the chart for that drug, then take the time to review the charts on other medications you already take. Read any instruction sheets or printed warnings provided by your doctor or pharmacist.

Drug Advertising

Ads can cause confusion. Be sure and get sufficient information about any drug you think may help you. Ask your doctor or pharmacist.

Be Safe! Tell Your Doctor

Some suggestions for wise drug use apply to all drugs. Always give your doctor, dentist, or healthcare provider complete information about the drugs and supplements you take, including your medical history, your medical plans and your progress while under medication.

Medical History

Tell the important facts of your medical history including illness and previous experience with drugs. Include allergic or adverse reactions you have had to any medicine or other substance in the past. Describe the allergic symptoms you have, such as hay fever, asthma, eye watering and itching, throat irritation and reactions to food. People who have allergies to common substances are more likely to develop drug allergies.

List all drugs you take. Don't forget vitamin and mineral supplements; skin, rectal or vaginal medicines; eyedrops and eardrops; antacids; antihistamines; cold and cough remedies; inhalants and nasal sprays; aspirin, aspirin combinations or other pain relievers; motion sickness remedies; weight-loss aids; salt and sugar substitutes; caffeine; oral contraceptives; sleeping pills; laxatives; "tonics" or herbal preparations.

Future Medical Plans

Discuss plans for elective surgery (including dental surgery), pregnancy and breastfeeding. These conditions may require discontinuing or modifying the dosages of medicines you may be taking.

Questions

Don't hesitate to ask questions about a drug. Your doctor or pharmacist will be able to provide helpful information if they are familiar with you and your medical history.

Guide to Drug Charts

The drug information in this book is organized in condensed, easy-to-read charts. Each drug is described in a two-page format, as shown in the sample chart below and opposite. Charts are arranged alphabetically by drug generic names, such as ACETAMINOPHEN or by drug class name, such as ANTIHISTAMINES.

A generic name is the official chemical name for a drug. A brand name is a drug manufacturer's registered trademark for a generic drug. Brand names listed on the charts include those from the United States

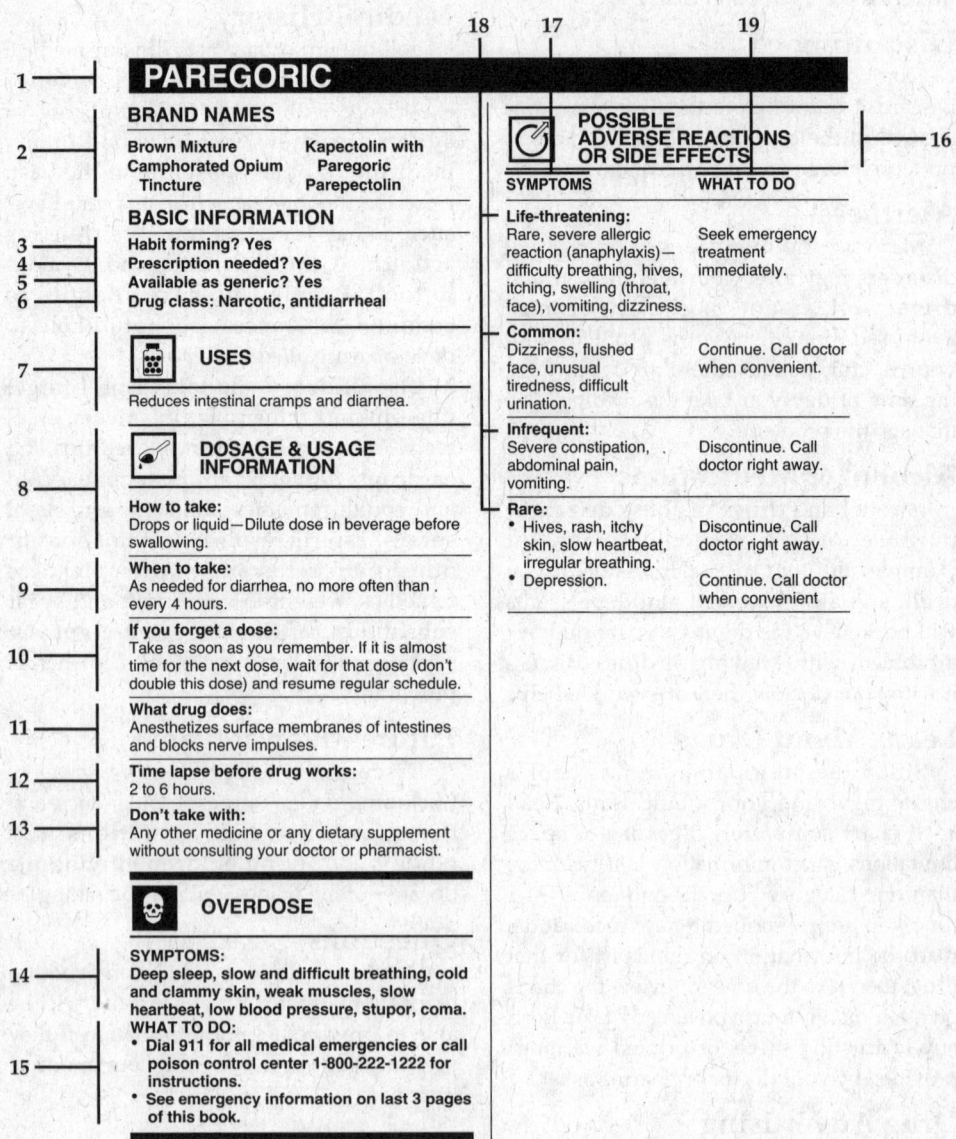

18 17 19

1 **PAREGORIC**

BRAND NAMES

2 Brown Mixture
Camphorated Opium Tincture
Kapectolin with Paregoric
Parepectolin

BASIC INFORMATION

3 Habit forming? Yes
4 Prescription needed? Yes
5 Available as generic? Yes
6 Drug class: Narcotic, antidiarrheal

7 **USES**

Reduces intestinal cramps and diarrhea.

DOSAGE & USAGE INFORMATION

8 **How to take:**
Drops or liquid—Dilute dose in beverage before swallowing.

9 **When to take:**
As needed for diarrhea, no more often than every 4 hours.

10 **If you forget a dose:**
Take as soon as you remember. If it is almost time for the next dose, wait for that dose (don't double this dose) and resume regular schedule.

11 **What drug does:**
Anesthetizes surface membranes of intestines and blocks nerve impulses.

12 **Time lapse before drug works:**
2 to 6 hours.

13 **Don't take with:**
Any other medicine or any dietary supplement without consulting your doctor or pharmacist.

OVERDOSE

14 **SYMPTOMS:**
Deep sleep, slow and difficult breathing, cold and clammy skin, weak muscles, slow heartbeat, low blood pressure, stupor, coma.

15 **WHAT TO DO:**
- **Dial 911 for all medical emergencies or call poison control center 1-800-222-1222 for instructions.**
- **See emergency information on last 3 pages of this book.**

POSSIBLE ADVERSE REACTIONS OR SIDE EFFECTS 16

SYMPTOMS	WHAT TO DO
Life-threatening:	
Rare, severe allergic reaction (anaphylaxis)—difficulty breathing, hives, itching, swelling (throat, face), vomiting, dizziness.	Seek emergency treatment immediately.
Common:	
Dizziness, flushed face, unusual tiredness, difficult urination.	Continue. Call doctor when convenient.
Infrequent:	
Severe constipation, abdominal pain, vomiting.	Discontinue. Call doctor right away.
Rare:	
• Hives, rash, itchy skin, slow heartbeat, irregular breathing.	Discontinue. Call doctor right away.
• Depression.	Continue. Call doctor when convenient

and Canada. A generic drug may have one, a few, or many brand names.

To find information about a generic drug, look it up in the index. To learn about a brand name, check the index, where each brand name is followed by the name(s) of its generic ingredients and their chart page number(s).

The chart design is the same for every drug. When you are familiar with the chart, you can quickly find information you want to know about a drug.

On the next few pages, each of the numbered chart sections below is explained. This information will guide you in reading and understanding the charts that begin on page 2.

PAREGORIC

20 **WARNINGS & PRECAUTIONS**

21 **Don't take if:**
You are allergic to any narcotic.*

22 **Before you start, consult your doctor if:**
You have impaired liver or kidney function.

23 **Over age 60:**
Unknown effect.

24 **Pregnancy:**
Decide with your doctor if drug benefits justify risk to unborn baby. Risk category C (see page xviii).

25 **Breastfeeding; Lactation; Nursing Mothers:**
This drug is generally not recommended with breastfeeding. Discuss risks and benefits with your doctor.

26 **Infants & children up to age 18:**
Follow instructions provided by your child's doctor.

27 **Prolonged use:**
Causes psychological and physical dependence.

28 **Skin & sunlight:**
No problems expected.

29 **Driving, piloting or hazardous work:**
Don't drive or pilot aircraft until you learn how medicine affects you. Don't work around dangerous machinery. Don't climb ladders or work in high places. Danger increases if you drink alcohol or take medicine affecting alertness and reflexes, such as antihistamines, tranquilizers, sedatives, pain medicine, narcotics and mind-altering drugs.

30 **Discontinuing:**
May be unnecessary to finish medicine. Follow doctor's instructions.

31 **Others:**
Advise any doctor, dentist or pharmacist whom you consult that you take this medicine.

POSSIBLE INTERACTION WITH OTHER DRUGS 32

GENERIC NAME OR DRUG CLASS	COMBINED EFFECT
Analgesics*	Increased analgesic effect.
Anticholinergics*	Increased risk of constipation.
Antidepressants*	Increased sedation.
Antidiarrheal preparations*	Increased sedative effect. Avoid.
Antihistamines*	Increased sedation.
Central nervous system (CNS) depressants*	Increased central nerve system depression.
Naloxone	Decreased paregoric effect.
Naltrexone	Decreased paregoric effect.
Narcotics,* other	Increased narcotic effect.

POSSIBLE INTERACTION WITH OTHER SUBSTANCES 33

INTERACTS WITH	COMBINED EFFECT
Alcohol:	Increases alcohol's intoxicating effect. Avoid.
Beverages:	None expected.
Cocaine:	Increased risk of side effects. Avoid.
Foods:	None expected.
Marijuana:	Increased risk of side effects. Avoid.
Tobacco:	None expected.

1—Generic or Class Name

Each drug chart is titled by generic name or by the name of the drug class, such as DIGITALIS PREPARATIONS.

All drugs have a generic name. These generic names are the same worldwide. Sometimes a drug is known by more than one generic name. The chart is titled by the most common one. Less common generic names appear in parentheses. For example, vitamin C is also known as ascorbic acid. Its chart title is VITAMIN C (Ascorbic Acid). The index will include both names.

Your drug container may show a generic name, a brand name or both. If you have only a brand name, use the index to find the drug's generic name(s) and chart page number(s).

If your drug container shows no name, ask your doctor or pharmacist for the name and write it on the container.

2—Brand Names

A brand name is usually shorter and easier to remember than the generic name. The brand name is selected by the drug manufacturer

The brand names listed for each generic drug in this book do not include all brands available in the United States and Canada. The more common names are listed. New brands appear on the market, and brands are sometimes removed from the market. No list can reflect every change. In the instances in which the drug chart is titled with a drug class name instead of a generic name, the generic and brand names all appear under the heading GENERIC AND BRAND NAMES. The GENERIC NAMES are in capital letters and the BRAND NAMES are in lower-case letters.

Inclusion of a brand name does not imply recommendation or endorsement. Exclusion does not imply that a missing brand name is less effective or less safe than the ones listed. Some drug charts have too many generic and brand names to list on the page. A complete list is on the page indicated on the chart.

Lists of brand names don't differentiate between prescription and nonprescription drugs. The active ingredients are the same.

If you buy a nonprescription drug, look for generic names of the active ingredients on the container. Common nonprescription drugs are described in this book under their generic components. Many are also listed in the index by brand name.

Most drugs contain inert, or inactive, ingredients that are fillers, dyes or solvents for the active ingredients. Manufacturers choose inert ingredients that preserve the drug without interfering with the action of the active ingredients.

Inert substances are listed on labels of nonprescription drugs. They do not appear on prescription drugs. Your pharmacist can tell you all active and inert ingredients in a prescription drug.

A tablet, capsule or liquid may contain small amounts of sodium, sugar or potassium. If you are on a diet that severely restricts any of these, ask your pharmacist or doctor to suggest another form.

Some liquid medications contain alcohol. Avoid them if you are susceptible to the adverse effects of alcohol consumption.

BASIC INFORMATION

3—Habit Forming?

Yes—means the drug is capable of leading to physical and/or psychological dependence.

Physical dependence includes tolerance (requiring larger dosages or repeated use) and withdrawal symptoms (mental and physical) when it is stopped.

Psychological dependence involves repeated use of a drug to bring about effects that are pleasurable or satisfying, or it reduces undesirable feelings.

4—Prescription Needed?

Yes—means a doctor must prescribe the drug for you. "No" means you can buy the drug without prescription. Sometimes low strengths of a drug are available without prescription, while higher strengths require prescription.

The information about the drug applies whether it requires prescription or not. If the generic ingredients are the same, nonprescription drugs have the same dangers, warnings, precautions and interactions as prescription drugs. A nonprescription (over-the-counter) drug has dosing and other instructions printed on the container label. Always read them carefully before you take the drug. The information and warnings on containers for nonprescription drugs may not be as complete as the information in this book. Check both sources.

5—Available as Generic?

Some drugs have patent restrictions that protect the manufacturer or distributor of that drug. These drugs may be purchased only by a specific brand name.

Drugs purchased by generic name are usually less expensive than brand names. Once the patent expires, other drug companies can sell that particular drug. They will choose their own brand name.

Some states allow pharmacists to fill a prescription by brand name or generic name (if available). This allows patients to buy the least expensive form of a drug.

A doctor may specify a brand name because he or she trusts a known source more than an unknown manufacturer of generic drugs. You and your doctor should decide together whether you should buy a medicine by generic name or brand name.

Generic drugs manufactured in other countries are not subject to regulation by the U.S. Food and Drug Administration. All drugs manufactured in the United States are subject to regulation.

6—Drug Class

Drugs that possess similar chemical structures or similar therapeutic effects are grouped into classes. Most drugs within a class produce similar benefits, side effects, adverse reactions and interactions with other drugs and substances. For example, all the generic drugs in the narcotic drug class will have similar effects on the body.

Some information on the charts applies to all drugs in a class. The index lists the classes (such as narcotics) and lists the drug charts in that class.

Names for classes of drugs are not standardized; class names listed in other references may vary from the class names in this book.

7— Uses

This section lists the disease or disorder for which a drug is prescribed.

Most uses listed are approved by the U.S. Food and Drug Administration. Some uses are listed if experiments and clinical trials indicate effectiveness and safety. Still, other uses are included that may not be officially sanctioned, but for which doctors may prescribe the drug.

The use for which your doctor prescribes the drug may not appear. You and your doctor should discuss the reason for any prescription medicine you take. You alone will probably decide whether to take a nonprescription drug. This section may help you make a wise decision.

DOSAGE & USAGE INFORMATION

8—How To Take

Drugs are available in forms of tablets, capsules, chewables, liquids, powders, thin film, specialty tablets, suppositories, injections, transdermal patches (used on the skin), aerosol inhalants and topical forms such as drops, sprays, creams, gels, ointments and lotions. This section gives brief instructions for taking or using each form.

The information here supplements the drug label information. If your doctor's instructions differ from these suggestions, follow your doctor's instructions.

Instructions are left out for how much to take. Dose amounts can't be generalized. Dosages of prescription drugs must be individualized for you by your doctor. Be sure the dosage instructions are on the label. Advice to "take as directed" is not helpful if you forget the doctor's instructions or didn't understand them. Nonprescription drugs have instructions on the labels regarding how much to take.

9—When To Take

Dose schedules vary for medicines and for patients.

Drugs prescribed on a schedule should usually be taken at approximately the same times each day. Some must be taken at regular intervals to maintain a steady level of the drug in the body. If the schedule is a problem for you, consult your doctor.

Instructions to take on an empty stomach mean the drug is absorbed best in the body this way. Some drugs must be taken with food because they irritate the stomach.

Instructions for other dose schedules are usually on the label. Variations in standard dose schedules may apply because some medicines interact with others if you take them at the same time.

10—If You Forget a Dose

Suggestions in this section vary from drug to drug. Most tell you when to resume taking the medicine if you forget a scheduled dose.

Establish good habits so you won't forget doses. Forgotten doses decrease a drug's therapeutic effect.

11—What Drug Does

This is a simple description of the drug's action in the body. The wording is generalized and may not be a complete explanation of the complex chemical process that takes place. For some drugs, the method of action is unknown.

12—Time Lapse Before Drug Works

The times given are approximations. Times vary a great deal from person to person, and from time to time in the same person. The figures give you some idea of when to expect improvement or side effects.

13—Don't Take With

Some drugs create problems when taken in combination with other substances. Most problems are detailed in the Interaction section of each chart.

Occasionally, an interaction is singled out if the combination is particularly harmful.

OVERDOSE

14—Symptoms

The symptoms listed are most likely to develop with accidental or purposeful overdose. Overdosage may not cause all symptoms listed. Sometimes symptoms are identical to ones listed as side effects. The difference is intensity and severity. You will have to judge. Consult a doctor or poison control center if you have any doubt.

15—What To Do

If you suspect an overdose, whether symptoms are apparent or not, get medical help or advice, and if needed, follow instructions in this section. Additional instructions for emergency treatment for overdose are at the end of the book.

16—Possible Adverse Reactions or Side Effects

Adverse reactions or side effects are symptoms that may occur when you take a drug. They are effects on the body other than the desired therapeutic effect.

The term side effects implies expected and usually unavoidable effects of a drug. Side effects have nothing to do with the drug's intended use.

For example, the generic drug paregoric reduces intestinal cramps and vomiting. It also often causes a flushed face. The flushing is a side effect that is harmless and does not affect the drug's therapeutic potential. Many side effects disappear in a short time without treatment.

The term adverse reaction is more significant. For example, paregoric can cause a serious adverse allergic reaction in some people. This reaction can include hives, rash and severe itch.

Some adverse reactions can be prevented, which is one reason this information is included in the book. Most adverse reactions are minor and last only a short time. With many drugs, adverse reactions that might occur will frequently diminish in intensity as your body adjusts to the medicine.

The majority of drugs, used properly for valid reasons, offer benefits that outweigh potential hazards.

17—Symptoms

Symptoms of commonly known side effects and adverse reactions are listed. Other drug responses may be listed under "Prolonged Use," "Skin & Sunlight" or "Others." You may experience a symptom that is not listed. It may be a side effect or adverse reaction to the drug, or it may be an additional symptom of the illness. If you are unsure, call your doctor.

18—Frequency

This is an estimation of how often symptoms occur in persons who take the drug. The four most common categories of frequency can be found under the SYMPTOMS heading and are as follows: Life-threatening means exactly what it says; seek emergency treatment immediately. Common means these symptoms are expected and sometimes inevitable. Infrequent means the symptoms occur in approximately 1% to 10% of patients. Rare means symptoms occur in fewer than 1%.

19—What To Do

Follow the guidelines provided opposite the symptoms that apply to you. These are general instructions. If you are concerned or confused, always call your doctor.

20— Warnings & Precautions

Read these entries to determine special information that applies to you.

21—Don't Take If

This section lists circumstances when drug use is not safe. On some drug labels and in formal medical literature, these circumstances are called contraindications.

22—Before You Start, Consult Your Doctor If

This section lists conditions, especially disease conditions, under which a drug should be used only with caution and medical supervision.

23—Over Age 60

As a person ages, physical changes occur that require special considerations. Liver and kidney functions decrease, metabolism slows and the prostate gland enlarges in men.

Most drugs are metabolized or excreted at a rate dependent on kidney and liver functions. Smaller doses or longer intervals between doses may be necessary to prevent unhealthy concentration of a drug. Toxic effects and adverse reactions occur more frequently and cause more serious problems in older people.

24—Pregnancy

This section will advise a pregnant woman to discuss all possible drug uses with her doctor.

The definitions of the pregnancy risk categories of drugs used by the Food and Drug Administration (FDA) are listed on page xviii. The page also contains other information about pregnancy and drugs.

The best rule to follow during pregnancy is to avoid all drugs, including tobacco and alcohol. Any medicine—prescription or nonprescription—requires medical advice and supervision.

25—Breastfeeding; Lactation; Nursing Mothers

Drugs are listed as *generally acceptable* or *generally not recommended* in breastfeeding. The decision to continue or stop breastfeeding during drug therapy should take into account the risk of drug exposure to the infant and the benefits of treatment to the mother. A nursing mother should consult her doctor before using any drug. Read the drug's label for more information.

26—Infants & Children Under Age 18

Information is provided about drug's use in this age group. Talk to your child's doctor about prescription drug use. Read labels carefully for nonprescription drugs.

27—Prolonged Use

With the exception of immediate allergic reactions, most drugs produce no ill effects during short periods of treatment. However, relatively safe drugs taken for long periods may produce unwanted effects. These are listed. Drugs should be taken in the smallest doses and for the shortest time possible. Nevertheless, some diseases and conditions require a prolonged or even lifelong period of treatment. Therefore, follow-up medical exams and lab tests may be recommended. Your doctor may want to change drugs or your treatment plan to minimize problems.

The words "functional dependence" may appear in this section. Sometimes a body function ceases to work naturally because it has been replaced or interfered with by the drug. The body then becomes dependent on the drug to continue the function.

28—Skin & Sunlight

Many drugs cause photosensitivity, which means increased skin sensitivity to ultraviolet rays from sunlight or artificial rays from a sunlamp. This section will alert you to this potential problem.

29—Driving, Piloting or Hazardous Work

Any drug that alters moods or that decreases alertness, muscular coordination or reflexes may make these activities particularly hazardous. The effects may not appear in all people, or they may disappear after a short exposure to the drug. If this section contains a warning, use caution until you determine how a new drug affects you.

30—Discontinuing

This section gives special warnings about discontinuing a drug.

Some patients stop taking a drug when symptoms begin to go away, although complete recovery may require longer treatment.

Other patients continue taking a drug when it is no longer needed.

Some drugs cause symptoms days or weeks after they have been discontinued.

31—Others

Warnings and precautions appear here if they don't fit into the other categories. This section includes special instructions, reminders, storage instructions, warnings to persons with chronic illness and other information.

32—Possible Interaction With Other Drugs

People often must take two or more drugs at the same time. Many of these drug combinations have the potential to interact adversely. Fortunately, this adverse reaction occurs in only a small proportion of people who take the interacting combinations. Drugs interact in your body with other drugs, whether prescription or non-prescription. Interactions affect absorption, metabolism, elimination or distribution of either drug. Other factors that can influence drug interactions are the patient's age, state of health and the way the drugs are administered: Time taken, how taken, dosage, dosage forms and duration of treatment. The chart lists interactions by generic name, drug class or drug-induced effect. An asterisk (*) in this column reminds you to "See Glossary" in the back of the book, where that entry is further explained.

If a drug class appears, the drug you are looking up may interact with any drug in that class. Drugs in each class that are included in the book are listed in the index. Occasionally drugs that are not included in this book appear in the Interaction column.

Interactions are sometimes beneficial. You may not be able to determine from the chart which interactions are good and which are bad. Don't guess. Consult your doctor or pharmacist if you take drugs that interact. Some combinations can be fatal.

Some drugs have too many interactions to list on one chart. The additional interactions appear on the continuation page indicated at the bottom of the list.

Testing has not been done on all possible drug combinations. It is important to let your doctor or pharmacist know about any drugs you take, both prescription and nonprescription.

33— Possible Interaction With Other Substances

The substances listed here are repeated on every drug chart. All people eat food and drink beverages. Many adults consume alcohol. Some people use cocaine and/or smoke tobacco or marijuana. This section shows possible interactions between these substances and each drug.

Checklist for Safer Drug Use

- Tell your doctor about any drug you take (even aspirin, allergy pills, cough and cold preparations, antacids, laxatives, herbal preparations, vitamins, etc.) before you take any new drug.
- Learn all you can about drugs you may take *before* you take them. Information sources are your doctor, your nurse, your pharmacist, this book, other books in your public library and the Internet.
- Keep an up-to-date list of all the medicines you take in a wallet or purse. Include name, dose and frequency.
- Don't take drugs prescribed for someone else—even if your symptoms are the same.
- Keep your prescription drugs to yourself. Your drugs may be harmful to someone else.
- Tell your doctor about any symptoms you believe are caused by a drug—prescription or nonprescription—that you take.
- Take only medicines that are *necessary*. Avoid taking nonprescription drugs while taking prescription drugs for a medical problem.
- Before your doctor prescribes for you, tell him about your previous experiences with any drug—beneficial results, side effects, adverse reactions or allergies.
- Take medicine in good light after you have identified it. If you wear glasses to read, put them on to check drug labels. It is easy to take the wrong drug or take a drug at the wrong time.

- Don't keep any drugs that change mood, alertness or judgment—such as sedatives, narcotics or tranquilizers—by your bedside. These can cause accidental deaths. You may unknowingly repeat a dose when you are half asleep or confused.
- Know the names of your medicines. These include the generic name, the brand name and the generic names of all ingredients in a combination drug. Your doctor, nurse or pharmacist can give you this information.
- Study the labels on all nonprescription drugs. If the information is incomplete or if you have questions, ask the pharmacist for more details.
- If you must deviate from your prescribed dose schedule, tell your doctor.
- Shake liquid medicines before taking (if directed).
- Store all medicines away from moisture and heat. Bathroom medicine cabinets are usually unsuitable.
- If a drug needs refrigeration, don't freeze.
- Obtain a standard measuring spoon from your pharmacy for liquid medicines. Kitchen teaspoons and tablespoons are not accurate enough.
- Follow diet instructions when you take medicines. Some work better on a full stomach, others on an empty stomach. Some drugs are more useful with special diets. For example, medicine for high blood pressure may be more effective if accompanied by a sodium-restricted diet.
- Tell your doctor about any allergies you have to any substance (e.g., food) or adverse reactions to medicines you've had in the past. A previous allergy to a drug may make it dangerous to prescribe again. People with other allergies, such as eczema, hay fever, asthma, bronchitis and food allergies, are more likely to be allergic to drugs.
- Prior to surgery, tell your doctor, anesthesiologist or dentist about any drug you have taken in the past few weeks. Advise them of any cortisone drugs you have taken within two years.
- If you become pregnant while taking any medicine, including birth control pills, tell your doctor immediately.
- Avoid *all* drugs while you are pregnant, if possible. If you must take drugs during pregnancy, record names, amounts, dates and reasons.
- If you see more than one doctor, tell each one about drugs others have prescribed.
- When you use nonprescription drugs, report it so the information is on your medical record.
- Store all drugs away from the reach of children.
- Note the expiration date on each drug label. Discard outdated ones safely. If no expiration date appears and it has been at least one year since taking the medication, it may be best to discard it. Follow any specific disposal instructions on the drug label or patient information that came with the drug. See page xx for instructions about disposing of drugs.
- Pay attention to the information in the drug charts about safety while driving, piloting or working in dangerous places.
- Alcohol, cocaine, marijuana or other mood-altering drugs, as well as tobacco—mixed with some drugs—can cause a life-threatening interaction, prevent your medicine from being effective or delay your return to health. Be sure that you avoid them during illness.
- Some medications are subject to theft. For example, a repair person in your home who is abusing drugs may ask to use your bathroom, and while there "checkout" your medicine cabinet. Sedatives, stimulants and analgesics are especially likely to be stolen, but almost any medication is subject to theft.

- If possible, use the same pharmacy for all your medications. Every pharmacy keeps a "drug profile," and if it is complete, the pharmacist may stop medications that are likely to cause serious interactions. Also, having a record of all your medications in one place helps your doctor or an emergency room doctor get a complete picture in case of an emergency.
- If you have a complicated medical history or a condition that might render you unable to communicate (e.g., diabetes or epilepsy), wear some type of medical identification (small tag, bracelet, neck chain or other) to identify your condition. (The MedicAlert Corporation provides this type of product. Call 888-633-4298 or go online www.medicalert.org for information.) Some people carry a wallet card with the information. A newer type of medical alert ID is the USB tag. It is basically a USB flash drive that contains a person's medical emergency information and can store much more information if desired. Emergency medical people can access the information using any available computer.
- If you are giving medicine to children, read all instructions carefully. Use the specific dosing device (dropper, cup, etc.) that comes with the product. Don't exceed recommended dose. It does not help and can cause health risks.
- Ask your pharmacist if they will provide an extra, prescription-labeled small bottle or container to use for one or two doses of your medication. It can be used when going out for a meal, during short trips or for a student in school. This avoids carrying all your medications with you.

Compliance with Doctors' Instructions

For medical purposes, compliance is defined as the extent to which a patient follows the instructions of a doctor and includes taking medications on schedule, keeping appointments and following directions for changes in lifestyle, such as changing one's diet or exercise.

Although the cost of obtaining medical advice and medication is one of the largest items in a family budget, many people defeat the health-care process by departing from the doctor's recommendations. This failure to carry out the doctor's instructions is the single most common cause of treatment failure. Perhaps the instructions were not presented clearly, or you may not have understood them or realized their importance and benefits.

Factors That Can Cause Problems With Compliance:

- Treatment recommendations that combine two or more actions (such as instructions to take medication, see a therapist and join a support group).
- Recommendations that require lifestyle changes (such as dieting).
- Recommendations that involve long-term treatment (such as taking a medication for life).
- Recommendations for very young patients or for the elderly (when another person has to be responsible for following the instructions).
- Other factors include traveling and time changes, busy work schedules and lack of organization.

Examples of Noncompliance:

- Medications are forgotten or discontinued too soon. Forgetting to take a medication is the most common of all shortcomings, especially if a medication must be taken more than once a day. If you need to take a medication several times a day, set out a week's supply in an inexpensive pill box that you can carry with you.

- Side effects of medications are a common problem. Almost all medications have some unpleasant side effects. Often these disappear after a few days, but if they don't, let your doctor know right away. Side effects can often be controlled by changing dosage, switching to a similar medication or by adding medications that control the side effects.
- Not taking a drug because it is unpleasant (e.g., bad tasting). Ask your doctor about options.
- Cost is another reason why there are treatment failures. Because of a tight budget, a person may take a medication less frequently than prescribed or just not purchase it. If you can't afford a medication, perhaps a less costly one can be prescribed or your doctor can find other ways to provide it.
- Laboratory tests, x-rays or other recommended medical studies are not obtained, perhaps due to concerns about costs or fear of the tests themselves.
- Recommendations about behavioral changes such as diet or exercise are ignored (old habits are difficult for anyone to change).
- Suggested immunizations are not obtained, sometimes due to fear of needles.
- Follow-up visits to the doctor are not made, or appointments are canceled, perhaps due to problems finding transportation or long waiting times in the doctor's office.

Communicating With Your Doctor:

- If you don't understand something, ask.
- If there are reasons why you cannot follow a recommendation, speak up.
- If you have reservations or fears about treatment, discuss them.

Remember, it is your health and your money that are at issue. You and your doctor are—or should be—working together to make you well and keep you healthy.

Cough and Cold Medicines

There are hundreds of nonprescription (over-the-counter or OTC) drugs available to treat symptoms of the common cold and other minor respiratory illnesses.

Some of these drugs contain a single active ingredient that relieves one specific symptom (such as a cough). Other drugs contain combinations of two or more ingredients intended to relieve a number of different symptoms (such as a cough, stuffy nose and pain).

Each of the active ingredients in a drug has a generic (or chemical) name. Generics are names such as acetaminophen, ibuprofen, aspirin, pseudoephedrine, guaifenesin, diphenhydramine and others.

Cough and cold drugs can be sorted into five categories (or classes). They include:

Antihistamines dry up secretions of the respiratory tract and help control allergy symptoms. Antihistamines may cause drowsiness, slow reflexes and decrease ability to concentrate. Therefore, don't drive vehicles or pilot aircraft until you learn how this medicine affects you. Don't work around dangerous machinery. Don't climb ladders or work in high places. The danger increases if you drink alcohol or take other medicines that affect alertness and reflexes. Antihistamines can cause other side effects or adverse reactions.

Decongestants relieve symptoms of nasal or bronchial congestion. Decongestants may cause nervousness, irregular heartbeats in some people, dizziness, confusion and other side effects. Conduct your daily activities with these effects in mind.

Antitussives (Cough suppressants) reduce the frequency and severity of a cough. These may be either narcotic (e.g., codeine and hydrocodone) or non-narcotic (e.g., dextromethorphan or carbetapentane). Narcotic cough suppressants are habit-forming and may cause some of the same mental changes that can take place with antihistamines. The most common non-narcotic antitussive—dextromethorphan—is not habit-forming but has other side effects. Carbetapentane is a similar drug (but not covered in this book).

Expectorants loosen secretions to make them easier to cough up. The most common expectorant is guaifenesin, which has very few side effects.

Analgesics relieve aches, pains and fever. Common analgesics in cough and cold medicines are aspirin, acetaminophen and ibuprofen.

These nonprescription drugs are generally recognized as safe, but they do have side effects, possible interactions with other drugs, and need to be used with caution in certain individuals.

There is no drug that will cure a cold (including antibiotics). If you decide to take a drug for specific symptoms, a single ingredient one is usually a better choice, unless each of the drug's ingredients is necessary to relieve your particular symptoms. Before taking a drug, read the label and follow all the directions. Also, read the information in this book about the drug's generic ingredients. To find information about a drug:

- Look on the label for the generic ingredients that are in the drug product.
- Consult the index for each generic name and find the page number for its drug chart. Then look up each drug chart. Read all the information, especially regarding the adverse reactions and side effects, precautions, and interactions with other drugs you may be taking.

Always consult a doctor if any cold symptoms cause you concern.

Practice prevention—wash your hands often and keep your hands away from your face.

Nonprescription cough and cold drugs in children

- Read the label to be sure it is approved for your child's age. Labels advise you not to use these medicines in children under age 4. Do not give children medications labeled only for adults.

- Choose nonprescription cough and cold medicines with child-resistant safety caps, when available. After using, make sure to close the cap tightly and store the medicine out of the sight and reach of children.

- Check the "active ingredients" section of the DRUG FACTS label of the medicines that you choose. This will help you know what symptoms the "active ingredients" in the medicine are intended to treat.

- Be careful about giving more than one medicine to a child. If you do, make sure they do not have the same type of "active ingredients." If you use two medicines that have the same or similar active ingredients, a child could get too much of an ingredient and that may hurt your child.

- Carefully follow the directions for how to use the medicine in the DRUG FACTS part of the label. The directions tell you how much medicine to give and how often you can give it. If you have a question about how to use the medicine, ask your pharmacist or doctor. Overuse or misuse of these products can lead to serious and potentially life threatening side effects.

- Only use measuring devices that come with the medicine or those specially made for measuring drugs. Do not use common household spoons to measure medicines for children because household spoons come in different sizes and are not meant for measuring medicines.

- Understand that using nonprescription cough and cold medicines does not cure the cold or cough. These medicines only treat the symptom(s) such as runny nose, congestion, fever and aches. They do not shorten the length of time your child is sick.

Pregnancy

Pregnancy Section on the Drug Charts

Caution must always be taken with drug use in pregnancy. The recommendation in this section is that a woman and her doctor decide together about the benefits and risks of any drugs used during pregnancy. This recommendation can help remind and encourage a woman to seek medical advice.

The pregnancy risk category (see below) assigned to the drug is listed in this section.

Space limitations on the drug charts prevent a detailed report about the drug's benefits and risks during pregnancy. Your doctor knows your specific situation and is always the best source for this information.

Pregnancy Risk Category

Pregnancy risk categories as defined by the Food and Drug Administration (FDA) are assigned to prescription drugs and appear on the drug's label (prescribing information). Each lettered category A, B, C, D and X describes potential risks for the drug to cause harmful effects to an unborn baby (fetus).

Doctors, pharmacists and other health - care providers use the risk category as one of their resources when advising pregnant women about drugs.

Category letters and definitions include:

- **Category A**: *No evidence of risk exists.* Adequate and well-controlled studies in pregnant women have not shown an increased risk of fetal abnormalities.
- **Category B**: *The risk of fetal harm is possible but remote.* Animal studies have revealed no evidence of harm to the fetus; however, there are no adequate and well-controlled studies in pregnant women, or animal studies have shown an adverse effect, but adequate and well-controlled studies in pregnant women have failed to show a risk to the fetus.
- **Category C**: *Fetal risk can't be ruled out.* Animal studies have shown an adverse effect, and there are no adequate and well-controlled studies in pregnant women, or no animal studies have been conducted, and there are no adequate and well-controlled studies in pregnant women.
- **Category D**: *Positive evidence of fetal risk exists, but, potential benefits from the drug may outweigh the risk. For example, the drug may be acceptable in a life-threatening situation or serious disease if safer drugs can't be used or are ineffective.* Adequate well-controlled or observational studies in pregnant women have demonstrated a risk to the fetus.
- **Category X**: *Contraindicated during pregnancy.* Adequate well-controlled or observational studies in animals or pregnant women have demonstrated positive evidence of fetal abnormalities. The use of the product is contraindicated in women who are or who may become pregnant.
- **Category NR**: *Not rated.*

Pregnancy & FDA Drug Labeling

The FDA issued new guidelines in 2014 that require details on prescription drug labels about the drug's risks and benefits in pregnancy (and in breast-feeding). This additional detail will eventually replace the current letters A, B, C, D or X used on drug labels to categorize risk.

The new guidelines took effect in 2015 for new drugs, but drug manufacturers of existing drugs (FDA-approved since 2001) have several years to revise their labels.

During this drug label transition period (2015 to about 2020), doctors, pharmacists, other healthcare providers and pregnant women who are discussing the use of a drug may find:

1. Label has the current pregnancy risk category A, B, C, D or X (see previous column for definition).

2. New type of label with detailed information (as explained in next column).

3. Label on a drug approved prior to 2001 may have the current pregnancy risk category definition, but it will not include the associated letter—A, B, C, D or X.

4. Nonprescription drugs are not included in the guidelines. Their pregnancy labeling often states "ask a health professional before use." Some may also list a specific warning.

New Label Details

The Pregnancy portion of the revised drug label will have these subsections:

• **Pregnancy Exposure Registry**: This is shown if a registry (see below) exists for the drug. It includes the telephone number of the registry and how to enroll.

• **Fetal Risk Summary**: This section provides information about the likelihood of fetal problems attributed to drug use and if this information is based on animal or human (or both) studies.

• **Clinical Considerations**: This section provides information on prescribing (e.g., dosage adjustment), education (e.g., about risks to mother and fetus from disease being treated) and counseling.

• **Data**: This section will provide background on how the information in Fetal Risk Summary and Clinical Considerations was determined.

Pregnancy Exposure Registry

Pregnancy Exposure Registries are research studies that collect information from women who take prescription drugs or vaccines during pregnancy. Pregnancy registries help women and their doctors learn more about which drugs are safe to take during pregnancy.

The FDA does not run pregnancy studies, but it keeps a list of all registries. To see if there is a registry for your drug. go to www.fda.gov/pregnancyregistries.

Enrolling in a registry involves a few interviews, both during and after your pregnancy. The information collected in these interviews, as well as the fact of your participation, will be kept confidential. Some registries may ask that you enroll through your doctor or healthcare provider.

Ask your doctor about a Pregnancy Exposure Registry if you are pregnant and are taking any prescription drugs.

Buying Prescription Drugs Online

The Food and Drug Administration (FDA) offers these suggestions. When it comes to buying medicine online, it is important to be very careful. Some websites sell medicine that may not be safe to use and could put your health at risk.

• Some websites that sell medicine:

1. Aren't U.S. state-licensed pharmacies or aren't pharmacies at all.
2. May give a diagnosis that is not correct and sell medicine that is not right for you or your condition.
3. Won't protect your personal information.

• Some medicines sold online:

1. Are fake (counterfeit or "copycat" medicines).
2. Are too strong or too weak.
3. Have dangerous ingredients.
4. Have expired (are out-of-date).
5. Aren't FDA-approved (haven't been checked for safety and effectiveness).
6. Aren't made using safe standards.
7. Aren't safe to use with other medicine or products you use.
8. Aren't labeled, stored, or shipped correctly.

What you should do:

1. Talk with your doctor and have a physical exam before you get any new medicine for the first time.
2. Use only medicine that has been prescribed by your doctor or another trusted professional who is licensed in the U.S. to write prescriptions for medicine.
3. Ask your doctor if there are any special steps you need to take to fill your prescription.

These tips will help protect you if you buy medicines online:

• Know your source to make sure it's safe.

• Make sure a website is a state-licensed pharmacy that is located in the United States. Pharmacies and pharmacists in the United States are licensed by a state's board of pharmacy. Your state board of pharmacy can tell you if a website is a state-licensed pharmacy, is in good standing, and is located in the United States. Find a list of state boards of pharmacy on the National Association of Boards of Pharmacy (NABP) website at www.nabp.info.

• The NABP has a program to help you find some of the pharmacies that are licensed to sell medicine online. Internet websites that display the seal of this program have been checked to make sure they meet state and federal rules. For more on this program and a list of pharmacies that display the Verified Internet Pharmacy Practice Sites™ Seal, (VIPPS® Seal), go to www.vipps.info.

• Look for websites with practices that protect you. A safe website should:
1. Be located in the United States and licensed by the state board of pharmacy where the website is operating.
2. Have a licensed pharmacist to answer your questions.
3. Require a prescription from your doctor or other healthcare professional who is licensed in the United States to write prescriptions for medicine.
4. Have a way for you to talk to a person if you have problems.

• Be sure your privacy is protected.
1. Look for privacy and security policies that are easy-to-find and easy-to-understand.
2. Don't give any personal information (such as social security number, credit card, or medical or health history), unless you are sure the website will keep your information safe and private.
3. Make sure that the site will not sell your information, unless you agree.

Report websites you are not sure of, or if you want to file a complaint.

Go to www.fda.gov/buyonline and click on "Report a Problem"

Guidelines for Disposing of Drugs

Follow any specific disposal instructions on the prescription drug labeling or patient information that accompanies the medicine. Do not flush medicines down the sink or toilet unless this information specifically instructs you to do so.

Take advantage of community drug take-back programs that allow the public to bring unused drugs to a central location for proper disposal. Call your city or county government's household trash and recycling service to see if a take-back program is available in your community. If no disposal instructions are given on the prescription drug labeling and no take-back program is available in your area, throw the drugs in the household trash following these steps.

1. Remove them from their original containers and mix them with an undesirable substance, such as used coffee grounds or kitty litter (this makes the drug less appealing to children and pets, and unrecognizable to people who may intentionally go through the trash seeking drugs). 2. Place the mixture in a sealable bag, empty can, or other container to prevent the drug from leaking or breaking out of a garbage bag.

Before throwing out a medicine container, scratch out all identifying information on the prescription label to make it unreadable. This will help protect your identity and the privacy of your personal health information.

Do not give your medicine to friends. Doctors prescribe medicines based on a person's specific symptoms and medical history. A medicine that works for you could be dangerous for someone else.

When in doubt about proper disposal, talk to your pharmacist.

The same disposal methods for prescription drugs can apply to nonprescription drugs as well.

Why Some Drugs are Flushed Away

Prescription drugs such as powerful narcotic pain relievers and other controlled substances carry instructions for flushing to reduce the danger of unintentional use or overdose and illegal abuse.

For example, the fentanyl patch, an adhesive patch that delivers a potent pain medicine through the skin, comes with instructions to flush used or leftover patches. Too much fentanyl can cause severe breathing problems and lead to death in babies, children, pets, and even adults, especially those who have not been prescribed the medicine.

Even after a patch is used, a lot of the medicine remains in the patch so don't throw something in the trash that contains a powerful and potentially dangerous narcotic that could harm others.

Disposal of Inhaler Products

Inhalers are used by people who have asthma or other breathing problems, such as chronic obstructive pulmonary disease. Read handling instructions on the labeling of inhalers and aerosol products because they could be dangerous if punctured or thrown into a fire or incinerator. To ensure safe disposal that complies with local regulations and laws, contact your local trash and recycling facility.

Getting Help From Your Pharmacist

One of the most important services a pharmacist can offer is to talk to you about your medicines. A pharmacist can help you understand how and when to take your medicines, what side effects you might expect, or what interactions may occur. A pharmacist can answer your questions privately in the pharmacy or over the telephone.

Here are some other ways your pharmacist can help:

Many pharmacists keep track of medicines on their computer. If you buy your medicines at one store and tell your pharmacist all the nonprescription and prescription drugs or dietary supplements you take, your pharmacist can help make sure your medicines don't interact harmfully with one another.

Ask your pharmacist to place your prescription medicines in easy-to-open containers if you have a hard time taking off child-proof caps and do not have young children living in or visiting your home. Remember to keep all medicines out of the sight and reach of children.

Your pharmacist may be able to print labels on prescription medicine containers in larger type, if reading the medicine label is hard for you.

Your pharmacist may be able to give you written information to help you learn more about your drugs. This information may be available in large type or in a language other than English. (Information from the U.S. Food & Drug Administration.)

Cutting Medicine Costs

Medicines are an important part of treating an illness because they often allow people to remain active and independent. But medicine can be expensive. Here are some ideas to help lower costs:

Tell your doctor if you are worried about the cost of your drug. Your doctor may not know the cost of a prescription, but may be able to tell you about another less expensive medicine, such as a generic drug or OTC product.

Ask for a senior citizen's discount.

Shop around. Look at prices at different stores or pharmacies. Lower medicine prices may not be a bargain if you need other services, such as home delivery, patient medicine profiles, or pharmacist consultation, or if you cannot get a senior citizen discount.

Ask for medicine samples. If your doctor gives you a prescription for a new medicine, ask your doctor for samples you can try before filling the prescription.

Buy bulk. If you need to take medicine for a long period of time and your medicine does not expire quickly, you may be able to buy a larger amount of the medicine for less money.

Try mail order. Mail-order pharmacies can provide drugs at lower prices. It is a good idea to talk with your doctor before using such a service. Make sure to find a backup pharmacy in case there is a problem with the mail service.

Buy nonprescription medicines when they are on sale. Check the expiration dates and use them before they expire. If you need help choosing a nonprescription medicine, ask the pharmacist for help.

If you decide to buy drugs on the Internet, first read Buying Prescription Drugs Online on page xix.

(Information from the U.S. Food & Drug Administration.)

Aging and Taking Medicines

As you get older you may be faced with more health conditions that you need to treat on a regular basis. It is important to be aware that more use of medicines and normal aging body changes can increase the chance of unwanted or maybe even harmful drug interactions. The more you know about your drugs and the more you talk with your healthcare providers, the easier it is to avoid problems.

Body changes can affect the way medicines are absorbed and used. For example, changes in the digestive system can affect how fast drugs enter the bloodstream. Changes in body weight can influence the amount of medicine you need to take and how long it stays in your body. The circulatory system may slow down, which can affect how fast drugs get to the liver and kidneys. The liver and kidneys also may work more slowly, affecting the way a drug breaks down and is removed from the body.

Because of these body changes, there is also a bigger risk of drug interactions among older adults. It's important to know about drug interactions.

Side effects are unplanned symptoms or feelings you have when taking a medicine. Most side effects are not serious and go away on their own; others can be more bothersome and even serious. To help prevent possible problems with medicines, seniors must know about the medicine they take and how it makes them feel.

Keep track of side effects to help your doctor know how your body is responding to a medicine. New symptoms or mood changes may not be a result of getting older, but could be from the medicine you're taking or another factor, such as a change in diet or routine. If you have an unwanted side effect, call your doctor.

Information about Substances of Abuse

Each of the drug charts list the interactions of alcohol, marijuana and cocaine. These three drugs are singled out because of their wide use and abuse.

Drugs of abuse include those that are addictive and harmful.

Common drugs of abuse:

Tobacco (nicotine)

What it does: Tobacco smoke contains noxious, addictive and cancer-producing ingredients. They include nicotine, carbon monoxide, ammonia and tars. Carcinogens probably come from the tars. Tars are present in chewing tobacco, snuff, cigarettes, cigars and pipes.
Short-term effects of average amount: Relaxation of mood if you are a steady smoker. Constriction of blood vessels.
Short-term effects of large amount inhaled: Headache, appetite loss, nausea.
Long-term effects: Risk of lung cancer and other cancers. Breathing problems and chronic lung disease, heart and blood vessel disease, risk of abortion and reduced birth weight of baby born to women who smoke during pregnancy.

Alcohol

What it does:

• *Central Nervous System*
It depresses normal mental activity and normal muscle function. Alcoholism is associated with accidents of all types. Marital, family, work, legal and social problems also occur. Abuse of alcohol may result in nerve damage and cause various types of brain disorders.

• *Gastrointestinal System*
Increases stomach acid, poisons liver function. Chronic alcoholism frequently leads to permanent damage to the liver.

• *Heart and Blood Vessels*
Decreased normal function, leading to heart diseases such as cardiomyopathy and disorders of the blood vessels and kidney, such as high blood pressure. Bleeding from the esophagus and stomach frequently accompany liver disease.

• *Unborn Fetus (teratogenicity)*
Alcohol abuse in the mother carrying a fetus causes *fetal alcohol spectrum disorder (FASD). It* includes mental deficiency, facial abnormalities, slow growth and other major and minor problems in a newborn.

Signs of Use:
Early signs: Smell of alcohol on the breath, behavior changes (aggressive; passive; lack of sexual inhibition; poor judgment; uncontrolled emotion, such as rage).
Intoxication signs: Unsteady gait, slurred speech, poor performance of any brain or muscle function, stupor or coma in *severe* alcoholic intoxication.

Long-term effects:
Addiction: Compulsive alcohol use. Persons addicted to alcohol have severe withdrawal symptoms if alcohol is unavailable.
Liver disease: Usually cirrhosis.
Loss of sexual function: Impotence, erectile dysfunction, loss of libido.
Increased incidence of cancer: Lung and other types. Interferes with medications.

Marijuana (cannabis, hashish)

What it does: Heightens perception, causes mood swings, relaxes mind and body.
Signs of use: Red eyes, lethargy, uncoordinated body movements.
Long-term effects: Decreased motivation. Possible brain, heart, lung and reproductive system damage; schizophrenia.

Amphetamines (including ecstasy)

What they do: Speed up physical and mental processes to cause a false sense of energy and excitement. The moods are temporary and unreal.
Signs of use: Dilated pupils, insomnia, trembling.

Long-term effects or overdose: Violent behavior, paranoia, inflammation of blood vessels, renal failure, possible death from overdose.

Anabolic Steroids

What they do: Enhance strength, increase muscle mass.
Signs of use: Significant mood swings, aggressiveness.
Long-term effects or overdose: Possible heart problems, paranoia and mania, liver damage, male infertility and impotence, male characteristics in females.

Barbiturates

What they do: Produce drowsiness and lethargy.
Signs of use: Confused speech, lack of coordination and balance.
Long-term effects or overdose: Disrupt normal sleep pattern. Possible death from overdose, especially with alcohol abuse.

Sedative-hypnotics (benzodiazepines, "party drugs" that include gammahydroxybutyrate and rohypnol)

What they do: Produce drowsiness and lethargy.
Signs of use: Slow breathing, low blood pressure, vomiting, delirium, amnesia, possible coma.
Long-term effects or overdose: Disrupt normal sleep pattern. Possible death from overdose, especially in combination with alcohol.

Cocaine

What it does: Stimulates the nervous system, heightens sensations and may produce hallucinations.
Signs of use: Trembling, intoxication, dilated pupils, constant sniffling.
Long-term effects or overdose: Soreness of nasal passages. Itching all over body, some with open sores. Possible brain damage or heart rhythm disturbance. Possible death from overdose.

Crystal Methamphetamine

What it does: Stimulates the nervous system, heightens sensations and has long-lasting euphoric effects.
Signs of use: Obsessively picking at the face or body, hallucinations, teeth grinding, extreme energy and no sleep for 2-3 days, major loss of weight.
Long-term effects or overdose: Rapid heart rate, high blood pressure, damaged blood vessels in the brain, fever, convulsions and death.

Opiates (codeine, heroin, morphine, methadone, opium)

What they do: Relieve pain, create temporary and false sense of well-being.
Signs of use: Small pupils, mood swings, slurred speech, sore eyes, lethargy, weight loss, sweating.
Long-term effects or overdose: Loss of appetite, infections, need to increase drug amount to produce same effects, death.

Phencyclidine (PCP, angeldust)

What they do: Produce euphoria along with a feeling of numbness.
Signs of use: Violent behavior, dizziness, loss of motor skills, disorientation.
Long-term effects or overdose: Seizures, high or low blood pressure, rigid muscles. Possible death from overdose.

Psychedelic Drugs (LSD, mescaline)

What they do: Produce hallucinations, either pleasant or frightening.
Signs of use: Dilated pupils, sweating, trembling, fever, chills.
Long-term effects or overdose: Lack of motivation, unpredictable behavior, hallucinations, death from overdose.

Volatile Substances (glue, solvents, nitrous oxide, other volatile compounds)

What they do: Produce hallucinations, false sense of well-being.
Signs of use: Dilated pupils, flushed face, confusion, respiratory failure, coma.
Long-term effects or overdose: Permanent brain, liver, kidney damage; death.

Medical Conditions and Their Commonly Used Drugs

This list contains the names of many medical problems and the names of drugs that may be used for their treatment. The drugs are listed either as a generic name (e.g., Acetaminophen) or class name (e.g., Antihistamines). Specific brand or trade names of drugs are not shown. This list of drugs is intended only as a guide and is not meant to be 100% complete. Use it for a general reference.

The inclusion of a drug name does not mean it is necessarily an appropriate treatment for you. Also, your doctor may prescribe a drug for you that is not listed, but is quite appropriate for treatment. Your doctor knows your medical history and can prescribe the drug that should work best for you.

You can find information about the drugs listed by looking up the name in the General Index and referring to the page listed. Do not be concerned if the drug chart does not list your specific illness in the USES section. For example, that section may state that a drug is used for bacterial infections and not list specific bacterial disorders (such as a vaginal infection or urinary tract infection).

Acid Indigestion & Upset Stomach
- Antacids
- Bismuth Salts
- Histamine H2 Receptor Antagonists
- Hyoscyamine
- Proton Pump Inhibitors
- Simethicone
- Sodium Bicarbonate

Acne
- Antiacne Cleansing (Topical)
- Antibacterials for Acne (Topical)
- Azelaic Acid
- Benzoyl Peroxide
- Erythromycins
- Isotretinoin
- Keratolytics
- Nitroimidazoles
- Retinoids (Topical)
- Tetracyclines

Actinic Keratoses
- Fluorouracil (Topical)

Acute Myocardial Infarction
- Angiotensin-Converting Enzyme (ACE) Inhibitors

Addison's Disease
- Adrenocorticoids (Systemic)

Aging
- Dehydroepiandrosterone (DHEA)

AIDS – See HIV Infection

Alcohol Withdrawal
- Acamprosate
- Benzodiazepines
- Beta-Adrenergic Blocking Agents
- Carbamazepine
- Disulfiram
- Hydroxyzine
- Lithium
- Naltrexone
- Thiamine

Allergies & Allergic Reactions
- Adrenocorticoids (Nasal Inhalation)
- Adrenocorticoids (Oral Inhalation)
- Adrenocorticoids (Systemic)
- Antihistamines
- Antihistamines (Nasal)
- Antihistamines, Nonsedating

Antihistamines, Phenothiazine-Derivative
Cromolyn
Decongestants (Ophthalmic)
Ephedrine
Hydroxyzine
Leukotriene Modifiers
Pollen Allergen Extracts

Alopecia
5-Alpha Reductase Inhibitors
Minoxidil

Altitude Illness
Carbonic Anhydrase Inhibitors

Alzheimer's Disease
Cholinesterase Inhibitors
Memantine

Amebiasis
Chloroquine
Iodoquinol
Nitroimidazoles

Amenorrhea
Bromocriptine
Progestins

Amyotrophic Lateral Sclerosis (ALS)
Riluzole

Anal Fissures
Adrenocorticoids (Rectal)

Anemia
Adrenocorticoids (Systemic)
Androgens
Cyclosporine
Folic Acid
Iron Supplements
Leucovorin
Vitamin B-12

Angina
Antithyroid Drugs
Beta-Adrenergic Blocking Agents
Calcium Channel Blockers
Dipyridamole
Nitrates
Ranolazine

Ankylosing Spondylitis
Adrenocorticoids (Systemic)
Anti-inflammatory Drugs, Nonsteroidal (NSAIDs)
Anti-inflammatory Drugs, Nonsteroidal (NSAIDs) COX-2 inhibitors
Aspirin
Methotrexate
Sulfasalazine
Tumor Necrosis Factor Blockers

Anorexia
Antidepressants, Tricyclic
Progestins
Selective Serotonin Reuptake Inhibitors (SSRIs)

Anticoagulants – See Blood Clots

Anxiety
Antidepressants, Tricyclic
Barbiturates
Benzodiazepines
Beta-Adrenergic Blocking Agents
Buspirone
Haloperidol
Hydroxyzine
Loxapine
Meprobamate
Phenothiazines
Selective Serotonin Reuptake Inhibitors (SSRIs)
Serotonin & Norepinephrine Reuptake Inhibitors (SNRIs)
Thiothixene

Appetite Stimulant
Antihistamines
Dronabinol

Appetite Suppressant
Appetite Suppressants

Arrhythmias – See Heart Rhythm Disorders

Arthritis
Acetaminophen
Adrenocorticoids (Systemic)
Anakinra
Anti-Inflammatory Drugs, Nonsteroidal (NSAIDs)
Anti-Inflammatory Drugs, Nonsteroidal (NSAIDs) COX-2 Inhibitors
Apremilast

Aspirin
Azathioprine
Capsaicin
Chloroquine
Cyclosporine
Diclofenac (Topical)
Gold Compounds
Hydroxychloroquine
Leflunomide
Meloxicam
Methotrexate
Salicylates
Tofacitinib
Tumor Necrosis Factor Blockers

Asthma
Adrenocorticoids (Nasal Inhalation)
Adrenocorticoids (Oral Inhalation)
Adrenocorticoids (Systemic)
Bronchodilators, Adrenergic (Long-Acting)
Bronchodilators, Adrenergic (Short-Acting)
Bronchodilators, Xanthine
Cromolyn
Ephedrine
Ipratropium
Leukotriene Modifiers

Athlete's Foot
Antifungals (Topical)

Attention Deficit Hyperactivity Disorder (ADHD) & ADD
Amphetamines
Atomoxetine
Central Alpha Agonists
Stimulant Medications
Stimulants, Amphetamine-Related

Autism
Aripiprazole
Haloperidol
Olanzapine
Quetiapine
Selective Serotonin Reuptake Inhibitors (SSRIs)
Serotonin-Dopamine Antagonists
Stimulant Medications

Bacterial Infections
Acetohydroxamic Acid (AHA)
Cephalosporins
Clindamycin
Erythromycins
Fluoroquinolones
Kanamycin
Lincomycin
Linezolid
Macrolide Antibiotics
Neomycin (Oral)
Nitrofurantoin
Nitroimidazoles
Penicillins
Penicillins & Beta-Lactamase Inhibitors
Rifamycins
Sulfonamides
Telithromycin
Tetracyclines
Trimethoprim
Vancomycin

Baldness – See Hair Loss

Bedwetting (Enuresis)
Antidepressants, Tricyclic
Desmopressin

Bipolar Disorder
Aripiprazole
Asenapine
Carbamazepine
Lithium
Divalproex
Quetiapine
Serotonin-Dopamine Antagonists
Valproic Acid
Ziprasidone

Birth Control – See Contraception

Bites & Stings
Adrenocorticoids (Topical)
Anesthetics (Topical)

Bladder Infection – See Cystitis

Bladder, Overactive – See Overactive Bladder

Bladder Spasms
Clidinium
Propantheline

Bleeding
Antifibrinolytic Agents
Vitamin K

Blood Circulation
Intermittent Claudication Agents
Vitamin E

Blood Clots
Anticoagulants (Oral)
Dabigatran
Dipyridamole
Factor Xa Inhibitors
Platelet Inhibitors
Ticagrelor
Vorapaxar

Bronchial Spasms
Anticholinergics
Bronchodilators, Adrenergic (Long-Acting)
Bronchodilators, Adrenergic (Short-Acting)

Bronchitis
Bronchodilators, Adrenergic (Long-Acting)
Bronchodilators, Adrenergic (Short-Acting)
Bronchodilators, Xanthine
Cephalosporins
Dextromethorphan
Fluoroquinolones
Ipratropium
Macrolide Antibiotics
Sulfonamides
Tetracyclines

Bulimia
Antidepressants, Tricyclic
Lithium
Selective Serotonin Reuptake Inhibitors (SSRIs)

Burns
Anesthetics (Topical)
Zinc Supplements

Bursitis
Adrenocorticoids (Systemic)
Anti-Inflammatory Drugs, Nonsteroidal (NSAIDs)
Aspirin
Salicylates

Cancer
Adrenocorticoids (Systemic)
Aminoglutethimide
Androgens
Antiandrogens, Nonsteroidal
Antifungals, Azoles
Busulfan
Capecitabine
Chlorambucil
Cyclophosphamide
Estramustine
Estrogens
Etoposide
Hydroxyurea
Imatinib
Lomustine
Melphalan
Mercaptopurine
Methotrexate
Paclitaxel
Procarbazine
Progestins
Raloxifene
Tamoxifen
Thyroid Hormones
Toremifene

Cancer of the Skin
Fluorouracil (Topical)
Mechlorethamine (Topical)

Canker Sores
Amlexanox
Anesthetics (Mucosal-Local)

Chickenpox
Acetaminophen
Antihistamines
Antivirals for Herpes Virus

Cholesterol, High
Cholestyramine
Colestipol
Ezetimibe
Gemfibrozil
HMG-CoA Reductase Inhibitors
Neomycin (Oral)
Niacin
Raloxifene

Chronic Obstructive Pulmonary Disease (COPD)
Adrenocorticoids (Systemic)
Bronchodilators, Adrenergic (Long-Acting)

Bronchodilators, Adrenergic (Short-Acting)
Bronchodilators, Anticholinergic
Bronchodilators, Xanthine
Ipratropium
Roflumilast

Cirrhosis
Colchicine
Cyclosporine
Thiamine (Vitamin B-1)

Colds & Cough
Acetaminophen
Anticholinergics
Antihistamines
Antihistamines, Nonsedating
Anti-Inflammatory Drugs, Nonsteroidal (NSAIDs)
Aspirin
Dextromethorphan
Ephedrine
Guaifenesin
Oxymetazoline
Phenylephrine
Phenylephrine (Ophthalmic)
Pseudoephedrine

Colic
Hyoscyamine
Simethicone

Colitis – See Inflammatory Bowel Disease

Congestion, Chest
Bronchodilators, Adrenergic (Long-Acting)
Bronchodilators, Adrenergic (Short-Acting)
Ephedrine
Oxymetazoline
Phenylephrine
Pseudoephedrine
Xylometazoline

Congestive Heart Failure
Angiotensin-Converting Enzyme (ACE) Inhibitors
Beta-Adrenergic Blocking Agents
Beta-Adrenergic Blocking Agents & Thiazide Diuretics
Digitalis Preparations
Diuretics, Loop
Diuretics, Potassium-Sparing
Diuretics, Potassium-Sparing & Hydrochlorothiazide
Diuretics, Thiazide
Nitrates

Conjunctivitis (Pink Eye)
Antibacterials (Ophthalmic)
Antivirals (Ophthalmic)

Conjunctivitis, Seasonal Allergic
Anti-Inflammatory Drugs, Nonsteroidal (Ophthalmic)
Antiallergic Agents (Ophthalmic)

Constipation
Laxatives, Bulk-Forming
Laxatives, Osmotic
Laxatives, Softener/Lubricant
Laxatives, Stimulant
Linaclotide
Lubiprostone
Tegaserod

Contraception
Contraceptives, Oral & Skin
Contraceptives, Vaginal
Selective Progesterone Receptor Modulators

Convulsions (Epilepsy; Seizures)
Anticonvulsants, Hydantoin
Anticonvulsants, Succinimide
Barbiturates
Benzodiazepines
Carbamazepine
Divalproex
Felbamate
Gabapentin
Lacosamide
Lamotrigine
Levetiracetam
Oxcarbazepine
Primidone
Tiagabine
Topiramate
Valproic Acid
Zonisamide

Corneal Ulcers
Antibacterials (Ophthalmic)

Crohn's Disease – See Inflammatory Bowel Disease

Cushing's Disease
- Adrenocorticoids (Systemic)
- Aminoglutethimide
- Antifungals, Azoles
- Metyrapone

Cystitis
- Phenazopyridine
- See also – Bacterial Infections

Dandruff
- Antifungals (Topical)
- Antiseborrheics (Topical)
- Coal Tar

Deep Vein Thrombosis – See Blood Clots

Dementia
- Buspirone
- Cholinesterase Inhibitors
- Ergoloid Mesylates
- Haloperidol

Depression
- Antidepressants, Tricyclic
- Aripiprazole
- Bupropion
- Ergoloid Mesylates
- Loxapine
- Maprotiline
- Mirtazapine
- Monoamine Oxidase (MAO) Inhibitors
- Monoamine Oxidase Type B (MAO-B) Inhibitors
- Nefazodone
- Selective Serotonin Reuptake Inhibitors (SSRIs)
- Serotonin & Norepinephrine Reuptake Inhibitors (SNRIs)
- Stimulant Medications
- Trazodone
- Vilazodone
- Vortioxetine

Dermatitis
- Adrenocorticoids (Systemic)
- Adrenocorticoids (Topical)
- Anesthetics (Topical)
- Antiseborrheics (Topical)
- Coal Tar
- Colchicine
- Dapsone
- Keratolytics

Dermatomyositis
- Adrenocorticoids (Systemic)
- Methotrexate

Diabetes
- Acarbose
- Bromocriptine
- Colesevelam
- DPP-4 Inhibitors
- GLP-1 Receptor Agonists
- Insulin
- Insulin Analogs
- Meglitinides
- Metformin
- Miglitol
- Pramlintide
- SGLT2 Inhibitors
- Sulfonylureas
- Thiazolidinediones

Diarrhea
- Attapulgite
- Bismuth Salts
- Charcoal, Activated
- Difenoxin & Atropine
- Diphenoxylate & Atropine
- Kaolin & Pectin
- Loperamide
- Nitazoxanide
- Paregoric
- Rifaximin

Dietary Supplement
- Calcium Supplements
- Iron Supplements
- Niacin
- Vitamin A
- Vitamin B-12 (Cyanocobalamin)
- Vitamin C (Ascorbic Acid)
- Vitamin D
- Vitamin E
- Vitamin K
- Zinc Supplements

Digestive Spasms
- Clidinium
- Dicyclomine
- Difenoxin & Atropine
- Hyoscyamine
- Propantheline

Diverticulitis
- Cephalosporins
- Clindamycin
- Fluoroquinolones
- Nitroimidazoles
- Penicillins

Drowsiness
- Caffeine

Dry Eyes
- Protectant (Ophthalmic)

Dry Mouth
- Pilocarpine (Oral)

Dysmenorrhea – See Menstrual Cramps

Ear Infections or Problems
- Antibacterials (Otic)
- Anti-Inflammatory Drugs, Steroidal (Otic)
- Antipyrine & Benzocaine (Otic)
- Phenylephrine
- See also – Bacterial Infections

Ear Wax
- Antipyrine & Benzocaine (Otic)

Eczema
- Adrenocorticoids (Topical)
- Coal Tar
- Doxepin (Topical)
- Keratolytics

Edema – See Fluid Retention

Emphysema
- Adrenocorticoids (Systemic)
- Bronchodilators, Adrenergic (Long-Acting)
- Bronchodilators, Adrenergic (Short-Acting)
- Bronchodilators, Anticholinergic
- Bronchodilators, Xanthine
- Ipratropium

Endometriosis
- Danazol
- Nafarelin

Epilepsy – See Convulsions

Erectile Dysfunction
- Alprostadil
- Erectile Dysfunction Agents
- Papaverine

Esophagitis
- Histamine H_2 Receptor Antagonists
- Metoclopramide
- Proton Pump Inhibitors

Estrogen Deficiency
- Estrogens

Eye Allergies
- Antiallergic Agents (Ophthalmic)

Eye Conditions
- Antibacterials (Ophthalmic)
- Cromolyn
- Cycloplegic, Mydriatic (Ophthalmic)
- Cyclopentolate (Ophthalmic)
- Decongestants (Ophthalmic)
- Natamycin (Ophthalmic)
- Phenylephrine (Ophthalmic)

Fatigue
- Caffeine

Fever
- Acetaminophen
- Anti-Inflammatory Drugs, Nonsteroidal (NSAIDs)
- Aspirin
- Salicylates

Fibrocystic Breast Disease
- Danazol
- Vitamin E

Fibromyalgia
- Pregabalin

Flu – See Influenza

Fluid Retention
- Carbonic Anhydrase Inhibitors
- Diuretics, Loop
- Diuretics, Potassium-Sparing
- Diuretics, Thiazide
- Guanethidine & Hydrochlorothiazide
- Indapamide

Fungal Infections
- Antifungals, Azoles
- Antifungals (Topical)
- Antifungals (Vaginal)

Griseofulvin
Nystatin
Terbinafine (Oral)

Gallstones
Ursodiol

Gastroesophageal Reflux
Histamine H_2 Receptor Antagonists
Proton Pump Inhibitors
Sucralfate

Genital Warts
Condyloma Acuminatum Agents

Giardiasis
Nitroimidazoles

Gingivitis & Gum Disease
Chlorhexidine
Erythromycins
Penicillins
Tetracyclines

Glaucoma
Antiglaucoma, Adrenergic Agonists
Antiglaucoma, Anticholinesterases
Antiglaucoma, Beta Blockers
Antiglaucoma, Carbonic Anhydrase Inhibitors
Antiglaucoma, Cholinergic Agonists
Antiglaucoma, Prostaglandins
Carbonic Anhydrase Inhibitors

Gonorrhea
Cephalosporins
Erythromycins
Fluoroquinolones
Macrolide Antibiotics
Penicillins
Tetracyclines

Gout
Adrenocorticoids (Systemic)
Antigout Drugs
Anti-Inflammatory Drugs, Nonsteroidal (NSAIDs)
Colchicine
Meloxicam
Probenecid

Hair, Excess Facial
Eflornithine

Hair Loss
5-Alpha Reductase Inhibitors
Anthralin (Topical)
Minoxidil (Topical)

Hay Fever
Antiallergic Agents (Ophthalmic)
Antihistamines
Antihistamines, Nonsedating
Antihistamines, Phenothiazine-Derivative
Antihistamines, Piperazine
Ephedrine
Guaifenesin
Hydroxyzine
Phenylephrine (Ophthalmic)

Headache (Cluster, Migraine, Sinus, Tension, Vascular)
Acetaminophen
Antidepressants, Tricyclic
Antihistamines
Anti-Inflammatory Drugs, Nonsteroidal (NSAIDs)
Aspirin
Beta-Adrenergic Blocking Agents
Buspirone
Butorphanol
Caffeine
Calcium Channel Blockers
Central Alpha Agonists
Divalproex
Ergot Derivatives
Lithium
Monoamine Oxidase Inhibitors
Topiramate
Triptans

Heart Rhythm Disorders
Antiarrhythmics, Benzofuran-Type
Beta-Adrenergic Blocking Agents
Calcium Channel Blockers
Digitalis Preparations
Disopyramide
Dofetilide
Flecainide Acetate
Mexiletine
Propafenone
Quinidine

Heartburn
Antacids
Histamine H_2 Receptor Antagonists

Proton Pump Inhibitors
Sodium Bicarbonate

Hemorrhoids
Adrenocorticoids (Rectal)
Adrenocorticoids (Topical)
Anesthetics (Rectal)

Herpes
Antivirals (Topical)
Antivirals for Herpes Virus

Hepatitis B
Nucleotide Reverse Transcriptase Inhibitors

High Blood Pressure – See Hypertension

HIV Infection
Fusion Inhibitors
Integrase Inhibitors
Maraviroc
Non-Nucleoside Reverse Transcriptase Inhibitors
Nucleoside Reverse Transcriptase Inhibitors
Nucleotide Reverse Transcriptase Inhibitors
Protease Inhibitors

Hives (Urticaria)
Antihistamines
Antihistamines, Nonsedating
Antihistamines, Phenothiazine-Derivative
Hydroxyzine

Huntington's
Haloperidol

Hypercalcemia
Colesevelam
Colestipol
Dextrothyroxine
HMG-CoA Reductase Inhibitors

Hypertension
Alpha Adrenergic Receptor Blockers
Angiotensin II Receptor Antagonists
Angiotensin-Converting Enzyme (ACE) Inhibitors
Beta-Adrenergic Blocking Agents
Beta-Adrenergic Blocking Agents & Thiazide Diuretics
Calcium Channel Blockers
Central Alpha Agonists
Diuretics, Loop
Diuretics, Potassium-Sparing
Diuretics, Potassium-Sparing & Hydrochlorothiazide
Diuretics, Thiazide
Eplerenone
Guanethidine
Guanethidine & Hydrochlorothiazide
Hydralazine
Indapamide
Minoxidil

Hyperthyroidism
Antithyroid Drugs

Hypertriglyceridemia
Fibrates
Gemfibrozil
HMG-CoA Reductase Inhibitors
Omega-3-Fatty Acids

Hypoglycemia
Glucagon

Hypothyroidism
Thyroid Hormones

Impotence - See Erectile Dysfunction

Incontinence
Antidepressants, Tricyclic
Estrogens
Muscarinic Receptor Antagonists

Indigestion – See Heartburn

Infertility
Bromocriptine
Clomiphene
Danazol
Progestins

Inflammation
Anti-Inflammatory Drugs, Nonsteroidal (NSAIDs)
Anti-Inflammatory Drugs, Nonsteroidal (NSAIDs) COX-2 Inhibitors
Aspirin
Mesalamine
Salicylates

Inflammatory Bowel Disease
- Adrenocorticoids (Systemic)
- Cyclosporine
- Infliximab
- Mesalamine
- Nitroimidazoles
- Olsalazine

Influenza
- Antivirals for Influenza
- Antivirals for Influenza, Neuraminidase Inhibitors
- Ribavirin

Insomnia
- Barbiturates
- Benzodiazepines
- Eszopiclone
- Melatonin
- Meprobamate
- Orexin Receptor Antagonists
- Ramelteon
- Trazodone
- Triazolam
- Zaleplon
- Zolpidem

Intermittent Claudication
- Intermittent Claudication Agents
- Pentoxifylline

Irregular Heartbeat – See Heart Rhythm Disorders

Irritable Bowel Syndrome
- Linaclotide
- Tegaserod

Itching
- Adrenocorticoids (Topical)
- Doxepin (Topical)

Jet Lag
- Melatonin

Jock Itch
- Antifungals (Topical)

Joint Pain
- Anti-Inflammatory Drugs, Nonsteroidal (NSAIDs)
- Anti-Inflammatory Drugs, Nonsteroidal (NSAIDs) COX-2 Inhibitors
- Aspirin

Kidney Stones
- Antigout Drugs
- Citrates
- Diuretics, Thiazide
- Penicillamine
- Sodium Bicarbonate
- Tiopronin

Labyrinthitis
- Antihistamines, Phenothiazine-Derivative
- Antihistamines, Piperazine
- Benzodiazepines

Leg Pain or Cramps
- Intermittent Claudication Agents
- Pentoxifylline
- Quinine

Leukemia
- Imatinib

Lice
- Pediculicides

Lupus (Skin & Systemic)
- Adrenocorticoids (Systemic)
- Adrenocorticoids (Topical)
- Anti-Inflammatory Drugs, Nonsteroidal (NSAIDs)
- Anti-Inflammatory Drugs, Nonsteroidal (NSAIDs) COX-2 Inhibitors
- Hydroxychloroquine
- Methotrexate

Lyme Disease
- Cephalosporins
- Erythromycins
- Macrolide Antibiotics
- Penicillins
- Tetracyclines

Malabsorption
- Vitamin K

Malaria
- Antimalarial
- Atovaquone
- Chloroquine
- Hydroxychloroquine
- Primaquine
- Proguanil
- Quinidine
- Quinine

Sulfadoxine & Pyrimethamine
Tetracyclines

Male Hormone Deficiency
Androgens

Melanoma
Hydroxyurea
Melphalan

Meniere's Disease
Antihistamines
Antihistamines, Piperazine
Benzodiazepines
Scopolamine (Hyoscine)

Menopause Symptoms
Estrogens
Ospemifene
Progestins
Selective Serotonin Reuptake Inhibitors (SSRIs)

Menorrhagia – See Menstruation, Excessive

Menstrual Cramps
Anti-Inflammatory Drugs, Nonsteroidal (NSAIDs)
Anti-Inflammatory Drugs, Nonsteroidal (NSAIDs) COX-2 Inhibitors
Contraceptives, Oral & Skin

Menstruation, Excessive (Menorrhagia)
Antifibrinolytic Agents
Contraceptives, Oral & Skin
Danazol
Estrogens
Progestins

Mental & Emotional Disturbances
Loxapine
Serotonin-Dopamine Antagonists

Motion Sickness
Antihistamines
Antihistamines, Nonsedating
Antihistamines, Phenothiazine-Derivative
Antihistamines, Piperazine
Clotrimazole
Diphenidol
Scopolamine

Multiple Sclerosis
Adrenocorticoids (Systemic)
Baclofen
Tizanidine

Muscle Cramp, Spasm, Strain
Baclofen
Cyclobenzaprine
Dantrolene
Muscle Relaxants, Skeletal
Quinine
Tizanidine

Myasthenia Gravis
Adrenocorticoids (Systemic)
Antimyasthenics
Azathioprine
Cyclosporine

Narcolepsy
Amphetamines
Pemoline
Stimulant Medications
Stimulants, Amphetamine-Related

Narcotic Withdrawal
Central Alpha Agonists
Naloxone
Naltrexone

Nasal Allergy
Adrenocorticoids (Nasal Inhalation)
Antihistamines (Nasal)

Nausea & Vomiting
Antihistamines, Phenothiazine-Derivative
Bismuth Salts
Diphenidol
Dronabinol
Hydroxyzine
Metoclopramide
Nabilone
Phenothiazines
Scopolamine
Trimethobenzamide

Nerve Disorders
Pregabalin

Neural Tube Defects (prevention)
Folic Acid

Obesity
- Appetite Suppressants
- Lorcaserin
- Orlistat
- Selective Serotonin Reuptake Inhibitors (SSRIs)

Obsessive Compulsive Disorder
- Antidepressants, Tricyclic
- Selective Serotonin Reuptake Inhibitors (SSRIs)

Ocular Hypertension
- Antiglaucoma, Carbonic Anhydrase Inhibitors
- Beta-Adrenergic Blocking Agents (Ophthalmic)

Osteoarthritis – See Arthritis

Osteoporosis
- Bisphosphonates
- Bone Formation Agents
- Calcitonin
- Calcium Supplements
- Estrogens
- Raloxifene
- Sodium Fluoride
- Vitamin D

Otitis Media – See Ear Infection

Overactive Bladder
- Mirabegron
- Muscarinic Receptor Antagonists

Paget's Disease
- Bisphosphonates
- Colchicine

Pain
- Acetaminophen
- Anti-Inflammatory Drugs, Nonsteroidal (NSAIDs)
- Anti-Inflammatory Drugs, Nonsteroidal (NSAIDs) COX-2 Inhibitors
- Aspirin
- Butorphanol
- Carbamazepine
- Narcotic Analgesics
- Salicylates
- Tapentadol
- Tramadol
- Trazodone

Pain In Mouth
- Anesthetics (Mucosal-Local)

Panic Disorder
- Antidepressants, Tricyclic
- Benzodiazepines
- Monoamine Oxidase Inhibitors

Parasites
- Anthelmintics
- Pentamidine

Parkinson's Disease
- Antidyskinetics
- Antihistamines
- Antivirals for Influenza
- Bromocriptine
- Carbidopa & Levodopa
- COMT Inhibitors
- Dopamine Agonists, Nonergot
- Levodopa
- Monoamine Oxidase Type B (MAO-B) Inhibitors

Parkinson's Tremors
- Antihistamines
- Niacin

Peripheral Neuropathy
- Serotonin & Norepinephrine Reuptake Inhibitors (SNRIs)

Pneumocystis Jiroveci
- Atovaquone
- Dapsone
- Primaquine
- Trimethoprim

Pneumonia
- Fluoroquinolones
- Ribavirin
- Sulfonamides

Pneumonia, Community Acquired
- Telithromycin

Poisoning
- Charcoal, Activated

Potassium Deficiency
- Potassium Supplements

Premenstrual Syndrome (PMS)
- Antidepressants, Tricyclic

Anti-Inflammatory Drugs, Nonsteroidal (NSAIDs)
Buspirone
Calcium Supplements
Contraceptives, Oral & Skin
Danazol
Pyridoxine (Vitamin B-6)
Selective Serotonin Reuptake Inhibitors (SSRIs)
Vitamin E

Pressure Sores
Benzoyl Peroxide

Proctitis
Adrenocorticoids (Rectal)

Prostate Hyperplasia, Benign
5-Alpha Reductase Inhibitors
Alpha Adrenergic Receptor Blockers

Psoriasis
Adrenocorticoids (Topical)
Anthralin (Topical)
Apremilast
Biologics for Psoriasis
Coal Tar
Cyclosporine
Keratolytics
Methotrexate
Psoralens
Retinoids (Oral)
Retinoids (Topical)
Tumor Necrosis Factor Blockers
Vitamin D (Topical)

Psychotic Disorders
Aripiprazole
Asenapine
Carbamazepine
Clozapine
Haloperidol
Loxapine
Lurasidone
Olanzapine
Phenothiazines
Quetiapine
Serotonin-Dopamine Antagonists
Thiothixene
Ziprasidone

Pulmonary Arterial Hypertension (PAH)
Endothelin Receptor Antagonists

Rashes – See Skin Disorders

Rectal Fissures
Anesthetics (Rectal)

Respiratory Syncytial Virus (RSV)
Ribavirin

Restless Legs Syndrome
Antidyskinetics
Benzodiazepines
Carbidopa & Levodopa
Dopamine Agonists, Nonergot
Narcotic Analgesics

Rheumatoid Arthritis – See Arthritis

Rickets
Vitamin D

Ringworm – See Fungal Infections

Rosacea
Antibacterials (Topical)
Antibacterials for Acne (Topical)
Azelaic Acid
Benzoyl Peroxide
Brimonidine
Nitroimidazoles
Pediculicides
Tetracyclines

Scabies
Pediculicides

Schizophrenia
Aripiprazole
Carbamazepine
Clozapine
Haloperidol
Olanzapine
Phenothiazines
Quetiapine
Serotonin-Dopamine Antagonists

Seasonal Affective Disorder (SAD)
Bupropion
Selective Serotonin Reuptake Inhibitors (SSRIs)

Seizures – See Convulsions

Shingles
Antivirals (Topical)
Antivirals for Herpes Virus
Capsaicin

Sickle Cell Disease
Hydroxyurea

Sinusitis
Cephalosporins
Erythromycins
Macrolide Antibiotics
Penicillins
Penicillins & Beta-Lactamase Inhibitors
Sulfonamides
Tetracyclines
Trimethoprim
Xylometazoline

Skin Cancer – See Cancer of the Skin

Skin Disorders
Anesthetics (Topical)
Antibacterials (Topical)
Cyclophosphamide
Condyloma Acuminatum Agents
Fluorouracil (Topical)
Isotretinoin
Retinoids (Topical)

Skin Lines & Wrinkles
Botulinum Toxin Type A

Smoking Cessation
Bupropion
Central Alpha Agonists
Nicotine
Varenicline

Sore Throat
Anesthetics (Mucosal-Local)
See also – Bacterial Infections

Stroke Prevention
Platelet Inhibitors

Sunburn
Adrenocorticoids (Topical)
Anesthetics (Topical)

Swelling – See Fluid Retention

Thyroid Disorders – See Hyperthyroidism; Hypothyroidism

Tonsillitis
Cephalosporins
Macrolide Antibiotics

Tourette's Syndrome
Antidyskinetics
Haloperidol

Toxoplasmosis
Atovaquone

Transplantation, Organ (Antirejection)
Azathioprine
Cyclosporine
Immunosuppressive Agents

Tremors
Benzodiazepines
Beta-Adrenergic Blocking Agents

Trichomoniasis
Nitroimidazoles

Trigeminal Neuralgia
Baclofen
Carbamazepine

Tuberculosis
Cycloserine
Ethionamide
Isoniazid
Nitroimidazoles
Rifamycins

Ulcers
Antacids
Anticholinergics
Bismuth Salts
Glycopyrrolate
Histamine H_2 Receptor Antagonists
Nitroimidazoles
Proton Pump Inhibitors
Sodium Bicarbonate
Sucralfate
Tetracyclines

Ulcers, Skin
Becaplermin

Ulcerative Colitis
Adrenocorticoids (Rectal)
Olsalazine
Sulfasalazine

Urethra Spasms
Clidinium
Propantheline

Urethritis
Erythromycins
Fluoroquinolones
Macrolide Antibiotics
Phenazopyridine
Tetracyclines

Urinary Frequency
Muscarinic Receptor Antagonists

Urinary Retention
Antimyasthenics
Bethanechol

Urinary Tract Infection
Acetohydroxamic Acid (AHA)
Atropine, Hyoscyamine, Methenamine
Cephalosporins
Cycloserine
Fluoroquinolones
Methenamine
Penicillins
Penicillins & Beta-Lactamase Inhibitors
Phenazopyridine
Sulfonamides
Tetracyclines
Trimethoprim

Urine Acidity
Citrates
Vitamin C

Uveitis
Anti-Inflammatory Drugs, Steroidal (Ophthalmic)

Vaginal Infections or Irritation
Clindamycin
Estrogens
Nitroimidazoles
Progestins

Vaginal Yeast Infections
Antifungals (Vaginal)

Vertigo
Antihistamines, Piperazine
Niacin

Virus Infections of the Eye
Antivirals (Ophthalmic)

Vitamin Deficiency
Pantothenic Acid
Riboflavin
Vitamin A
Vitamin B-12
Vitamin C
Vitamin D
Vitamin E
Vitamin K

Vitiligo
Psoralens

Vomiting – See Nausea & Vomiting

Warts
Keratolytics
Retinoids (Topical)

Wilson's Disease
Penicillamine
Zinc Supplements

Worms
Anthelmintics

Zinc Deficiency
Zinc Supplements

Drug Charts

5-ALPHA REDUCTASE INHIBITORS

GENERIC AND BRAND NAMES

DUTASTERIDE
Avodart
Duagen
Jalyn

FINASTERIDE
Propecia
Proscar

BASIC INFORMATION

Habit forming? No
Prescription needed? Yes
Available as generic? Yes, for some
Drug class: Dihydrotestosterone inhibitor

USES

- Treats symptoms of benign prostatic hypertrophy (BPH) in men with enlarged prostate.
- Treatment of male pattern hair loss in men.

DOSAGE & USAGE INFORMATION

How to take:
Tablet or capsule—Swallow with liquid. May take with or without food. If you can't swallow whole, ask your doctor or pharmacist for advice.

When to take:
Once a day or as directed on prescription.

If you forget a dose:
Take as soon as you remember. If it is almost time for the next dose, wait for that dose (don't double this dose) and resume regular schedule.

What drug does:
Inhibits the enzyme needed for the conversion of testosterone to dihydrotestosterone. Dihydrotestosterone is required for the development of benign prostatic hypertrophy.

Continued next column

OVERDOSE

SYMPTOMS:
Effects unknown.
WHAT TO DO:
If person uses much larger amount than prescribed or if accidentally swallowed, call poison control center 1-800-222-1222 for instructions or dial 911 (emergency) for help.

Time lapse before drug works:
- May require up to 6 months for full therapeutic effect for BPH.
- For treatment of hair loss, may not see any benefit for 3 months or more.

Don't take with:
Any other medicine or dietary supplement without consulting your doctor or pharmacist, especially nonprescription decongestants.

POSSIBLE ADVERSE REACTIONS OR SIDE EFFECTS

SYMPTOMS	WHAT TO DO
Life-threatening:	
None expected.	
Common:	
None expected.	
Infrequent:	
Decreased volume of ejaculation, headache, dizziness.	Continue. Call doctor when convenient.
Rare:	
Impotence, decreased libido, breast changes (enlargement, tenderness, lumps, pain, discharge from nipple), depressed mood, rash.	Continue. Call doctor when convenient.

WARNINGS & PRECAUTIONS

Don't take if:
- You are allergic to dutasteride or finasteride. Allergic reactions are rare, but may occur.
- You are a female, are pregnant or a child.

Before you start, consult your doctor if:
- You have had prostate cancer or not had a blood test to check for prostate cancer.
- Your sexual partner is pregnant or may become pregnant.
- You have a liver disorder.
- You have reduced urinary flow.
- You have large residual urinary volume.

Over age 60:
No special problems expected.

Pregnancy:
- Not recommended for women.
- Pregnant women should not handle the crushed tablets.
- Ask your doctor if you should avoid exposure to mate's semen if he takes these drugs.
- Risk category X (see page xviii).

Breastfeeding; Lactation; Nursing Mothers:
These drugs are not recommended for women.

Infants & children up to age 18:
Not recommended for this age group.

Prolonged use:
Talk to your doctor about the need for follow-up medical examinations or laboratory studies to check the effectiveness of the treatment.

Skin & sunlight:
No special problems expected.

Driving, piloting or hazardous work:
No special problems expected.

Discontinuing:
- Don't discontinue without medical advice.
- Consult doctor if some adverse effects such as problems with libido, ejaculation and orgasm continue after stopping the drug.

Others:
- Advise any doctor, dentist or pharmacist whom you consult that you take this medicine.
- May affect results of some medical tests (especially prostate specific antigen [PSA] which is a test for prostate cancer).
- For those who respond well, the drug must be continued indefinitely.
- Taking these drugs may increase risk for a prostate cancer called high grade prostate cancer. Consult your doctor about your risks.
- Drug use may lead to male infertility and/or poor semen quality. Once drug is stopped, these effects improve or return to normal.
- Don't donate blood until 6 months have passed after you stop taking the drug.

POSSIBLE INTERACTION WITH OTHER DRUGS

GENERIC NAME OR DRUG CLASS	COMBINED EFFECT
Enzyme inhibitors*	Unknown effect with dutasteride. Consult doctor.

POSSIBLE INTERACTION WITH OTHER SUBSTANCES

INTERACTS WITH	COMBINED EFFECT
Alcohol:	None expected.
Beverages:	
Grapefruit juice.	Ask your doctor or pharmacist.
Cocaine:	Unknown. Avoid.
Foods:	
Grapefruit.	Ask your doctor or pharmacist.
Marijuana:	Seek medical advice.
Tobacco:	None expected.

*See Glossary

ACAMPROSATE

BRAND NAMES

Campral

BASIC INFORMATION

Habit forming? No
Prescription needed? Yes
Available as generic? Yes
Drug class: Alcohol-abuse deterrent

USES

Treatment for alcohol dependence. It is used after alcohol withdrawal to help maintain abstinence by reducing cravings. The drug should be part of a complete treatment plan (e.g., counseling, behavior therapy and support of family and friends). The drug does not diminish or eliminate withdrawal symptoms.

DOSAGE & USAGE INFORMATION

How to take:
Delayed-release tablet—Swallow whole with liquid. May be taken with or without food. Do not chew, crush or split tablet.

When to take:
Usually three times a day at the same times each day (such as at mealtimes).

If you forget a dose:
Take as soon as you remember. If it is almost time for the next dose, wait for that dose (don't double this dose) and resume regular schedule.

What drug does:
Action is not fully understood. It appears to affect chemicals (called neurotransmitters) in the brain and reduces the craving for alcohol.

Time lapse before drug works:
About one week.

Don't take with:
Any other medicine or any dietary supplement without consulting your doctor or pharmacist.

OVERDOSE

SYMPTOMS:
Diarrhea and possibly other symptoms.
WHAT TO DO:
If person uses much larger amount than prescribed or if accidentally swallowed, call poison control center 1-800-222-1222 for instructions or dial 911 (emergency) for help.

POSSIBLE ADVERSE REACTIONS OR SIDE EFFECTS

SYMPTOMS	WHAT TO DO
Life-threatening: None expected.	
Common: Diarrhea, dizziness, itching, nausea, gas, bloating, vomiting, abdominal pain, headache, insomnia.	Continue. Call doctor when convenient.
Infrequent: Depression, anxiety, increase or decrease in sexual desire, rash or other skin reaction, heart palpitations, fainting, muscle pain, swelling (face, feet or ankles).	Continue, but call doctor right away.
Rare:	
• Thoughts of suicide or patient talks of suicide.	Discontinue. Get emergency help.
• Low blood pressure, increased heart rate, other symptoms that may occur (can be due to the drug or are a result of alcohol abuse or alcohol withdrawal symptoms).	Continue. Call doctor when convenient.

WARNINGS & PRECAUTIONS

Don't take if:
You are allergic to acamprosate. Allergic reactions are rare, but may occur.

Before you start, consult your doctor if:
- You have kidney (renal) disease or severe kidney impairment.
- You suffer from depression, have thoughts about suicide or have history of drug abuse.

Over age 60:
No special problems expected.

Pregnancy:
Decide with your doctor if drug benefits justify risk to unborn baby. Risk category C (see page xviii).

Breastfeeding; Lactation; Nursing Mothers:
This drug is generally acceptable with breastfeeding. Discuss risks and benefits with your doctor.

Infants & children up to age 18:
Safety and effectiveness in this age group have not been established. Consult doctor.

Prolonged use:
Talk to your doctor about the need for follow-up medical examinations or laboratory studies to determine drug's effectiveness and to monitor for symptoms and suicidal thoughts.

Skin & sunlight:
No problems expected.

Driving, piloting or hazardous work:
- Avoid if the drug makes you feel dizzy; otherwise no problems expected.
- If you start drinking alcohol again, avoid driving, piloting or hazardous work.

Discontinuing:
Do not stop taking acamprosate without talking to your doctor. Continue taking acamprosate even if you start drinking alcohol again.

Others:
- Advise any doctor, dentist or pharmacist whom you consult that you take this medicine.
- Drug may affect the accuracy of some medical tests.
- Contact your doctor if you develop symptoms of depression, have suicidal thoughts or start drinking alcohol again.

POSSIBLE INTERACTION WITH OTHER DRUGS

GENERIC NAME OR DRUG CLASS	COMBINED EFFECT
None expected.	

POSSIBLE INTERACTION WITH OTHER SUBSTANCES

INTERACTS WITH	COMBINED EFFECT
Alcohol:	None expected, but all alcohol should be avoided.
Beverages:	None expected.
Cocaine:	Effects unknown. Avoid.
Foods:	None expected. Avoid foods that contain alcohol.
Marijuana:	Seek medical advice.
Tobacco:	None expected.

ACARBOSE

BRAND NAMES

Precose

BASIC INFORMATION

Habit forming? No
Prescription needed? Yes
Available as generic? Yes
Drug class: Antihyperglycemic, antidiabetic

USES

Treatment for hyperglycemia (excess sugar in the blood) that cannot be controlled by diet alone in patients with type 2 diabetes. It may be used alone or in combination with other antidiabetic drugs.

DOSAGE & USAGE INFORMATION

How to take:
Tablet—Swallow whole with liquid. Take at the very beginning of a meal.

When to take:
Usually 3 times a day or as directed by doctor. Dosage may be increased at 4- to 8-week intervals until maximum benefits are achieved.

Continued next column

OVERDOSE

SYMPTOMS:

- **Symptoms of lactic acidosis (acid in the blood)—chills, diarrhea, severe muscle pain, sleepiness, slow heartbeat, breathing difficulty, unusual weakness.**
- **Symptoms of hypoglycemia (not a problem with acarbose used alone)—stomach pain, anxious feeling, cold sweats, chills, confusion, convulsions, cool pale skin, excessive hunger, nausea or vomiting, rapid heartbeat, nervousness, shakiness, unsteady walk, unusual weakness or tiredness, vision changes, unconsciousness.**

WHAT TO DO:

- **For mild low blood sugar symptoms, drink or eat something containing sugar right away.**
- **For more severe symptoms, dial 911 (emergency) for medical help or call poison control center 1-800-222-1222 for instructions.**
- **See emergency information on last 3 pages of this book.**

If you forget a dose:
Take as soon as you remember. If it is almost time for the next dose, wait for that dose (don't double this dose) and resume regular schedule.

What drug does:
Impedes the digestion and absorption of carbohydrates and their subsequent conversion into glucose, improving control of blood glucose, and may reduce the complications of diabetes. However, acarbose does not cure diabetes.

Time lapse before drug works:
May take several weeks for full effectiveness.

Don't take with:
Any other medicine or any dietary supplement without consulting your doctor or pharmacist.

POSSIBLE ADVERSE REACTIONS OR SIDE EFFECTS

SYMPTOMS	WHAT TO DO
Life-threatening:	
In case of overdose or low blood sugar, see previous column.	
Common:	
Diarrhea, stomach cramps, gas, bloating, feeling of fullness in stomach, nausea.	Continue. Call doctor when convenient.
Infrequent:	
None expected.	
Rare:	
• Lactic acidosis or severe low blood sugar (see symptoms under Overdose).	Discontinue. Call doctor right away or seek emergency help.
• Jaundice (yellow skin and eyes), severe stomach pain, easy bruising.	Discontinue. Call doctor right away.

WARNINGS & PRECAUTIONS

Don't take if:
You are allergic to acarbose. Allergic reactions are rare, but may occur.

Before you start, consult your doctor if:
- You have any kidney or liver disease or any heart or blood vessel disorder.
- You have any chronic health problem.
- You have an infection, illness or any condition that can cause low blood sugar.
- You have a history of acid in the blood (metabolic acidosis or ketoacidosis).
- You have inflammatory bowel disease or any other intestinal disorder.
- You are allergic to any medication, food or other substance.

Over age 60:
No special problems expected. A lower starting dosage may be recommended by your doctor.

Pregnancy:
Decide with your doctor if drug benefits justify risks to unborn baby. Risk category B (see page xviii).

Breastfeeding; Lactation; Nursing Mothers:
This drug is generally acceptable with breastfeeding. Discuss risks and benefits with your doctor.

Infants & children up to age 18:
Safety and effectiveness in this age group have not been established. Consult doctor.

Prolonged use:
- Schedule regular doctor visits to determine if the drug is continuing to be effective in controlling the diabetes and to check for any problems in kidney function.
- You will most likely require an antidiabetic medicine for the rest of your life.
- You will need to test your blood glucose levels several times a day; or for some, once to several times a week.
- Acarbose may reduce absorption of iron, causing anemia. Discuss with your doctor.

Skin & sunlight:
No special problems expected.

Driving, piloting or hazardous work:
No special problems expected.

Discontinuing:
Don't discontinue without consulting your doctor even if you feel well. You can have diabetes without feeling any symptoms. Untreated diabetes can cause serious problems.

Others:
- Advise any doctor, dentist or pharmacist whom you consult that you take this medicine.
- Drug may interfere with the accuracy of some medical tests.
- Follow any special diet your doctor may prescribe. It can help control diabetes.
- Consult doctor if you become ill with vomiting or diarrhea.
- Use caution when exercising. Ask your doctor about an appropriate exercise program.
- Wear or carry medical identification stating that you have diabetes and take this drug.
- Learn to recognize the symptoms of low blood sugar. You and your family need to know what to do if these symptoms occur.
- Have a glucagon kit and syringe in the event severe low blood sugar occurs.
- High blood sugar (hyperglycemia) may occur with diabetes. Ask your doctor about symptoms to watch for and treatment steps to take.
- Educate yourself about diabetes.

POSSIBLE INTERACTION WITH OTHER DRUGS

GENERIC NAME OR DRUG CLASS	COMBINED EFFECT
Amylase (Pancreatic enzyme)	Decreased acarbose effect.
Charcoal, activated	Decreased acarbose effect.
Hyperglycemia-causing medications*	Increased risk of hyperglycemia.
Insulin	Increased risk of low blood sugar.
Pancreatin (Pancreatic enzyme)	Decreased acarbose effect.
Pramlintide	Decreased absorption of nutrients.
Sulfonylureas	Increased risk of low blood sugar.

POSSIBLE INTERACTION WITH OTHER SUBSTANCES

INTERACTS WITH	COMBINED EFFECT
Alcohol:	Avoid alcohol. It can lower blood sugar.
Beverages:	None expected.
Cocaine:	Unknown. Avoid.
Foods:	None expected.
Marijuana:	Seek medical advice.
Tobacco:	None expected.

***See Glossary**

ACETAMINOPHEN

BRAND NAMES

See full list of brand names in the *Generic and Brand Name Directory*, page 829.

BASIC INFORMATION

Habit forming? No
Prescription needed? No, for most
Available as generic? Yes
Drug class: Analgesic; fever reducer

USES

Treatment of mild to moderate pain and fever. Acetaminophen does not relieve redness, stiffness or swelling of joints or tissue inflammation. Use other drugs for inflammation.

DOSAGE & USAGE INFORMATION

How to take:

- Note—Acetaminophen comes in several dosage forms and is also an ingredient in other painkillers and many cold, cough and flu remedies. Read labels carefully. Do not use, or give, more than the dosages listed for adults or children. Too much acetaminophen can result in liver or kidney damage.
- Tablet or capsule—Swallow with liquid. Do not open, crush or chew tablet or capsule.
- Chewable tablet—Chew and swallow.
- Liquid; liquid drops—Follow label instructions.
- Powder—Mix as instructed with water or other liquid and swallow. May be mixed with small amount of food and eaten.
- Granules—Mix with a small amount of applesauce, ice cream or jam and eat.

Continued next column

- Dissolving tablet—Place tablet on tongue and let dissolve (or chew) before swallowing.
- Extended-release or controlled-release tablet—Swallow whole. Don't crush or chew.
- Squeeze packet (for child)—one dose per packet. Follow instructions on product.
- Rectal suppository—Remove wrapper and moisten suppository with water. Gently insert into rectum. Push well into rectum with finger.

When to take:
As needed, depending on the dosage. May range from 4 to 8 hours between dosages.

If you forget a dose:
Take as soon as you remember. Wait 4 to 8 hours for next dose (depends on dosage used).

What drug does:
Exact mechanism is not fully known. It blocks pain impulses in central nervous system. Reduces fever by its action on the brain's heat regulating center.

Time lapse before drug works:
15 to 30 minutes. May last 4 or more hours.

Don't take with:

- Other drugs that contain acetaminophen without consulting your doctor or pharmacist.
- Any other medicine or any dietary supplement without consulting your doctor or pharmacist.

OVERDOSE

SYMPTOMS:
May take up to 12 hours after an overdose for symptoms to occur. Sweating, diarrhea, nausea, vomiting, stomach upset and cramping or pain, irritability, loss of appetite, yellow skin or eyes, seizures, coma.
WHAT TO DO:

- **If you suspect overdose, even if not sure, call poison control center 1-800-222-1222 for help. Symptoms may not appear until damage has occurred.**
- **Dial 911 (emergency) if symptoms occur.**
- **See emergency information on last 3 pages of this book.**

POSSIBLE ADVERSE REACTIONS OR SIDE EFFECTS

SYMPTOMS	WHAT TO DO
Life-threatening:	
Rare, severe allergic reaction (anaphylaxis)—difficulty breathing, hives, itching, swelling (throat, face), vomiting, dizziness.	Seek emergency treatment immediately.
Common:	
None expected.	
Infrequent:	
None expected.	
Rare:	
Extreme fatigue, skin rash or itch or hives, sore throat and fever, unexplained bleeding or bruising, blood in urine, painful or decreased or frequent urination, yellow skin or eyes, black or tarry stools, illness symptoms not present before taking drug.	Discontinue. Call doctor right away.

WARNINGS & PRECAUTIONS

Don't take if:
You are allergic to acetaminophen or have had a serious skin reaction while using the drug.

Before you start, consult your doctor if:
- You have kidney or liver disease.
- You drink alcohol.

Over age 60:
No special problems expected as long as proper dosage is taken.

Pregnancy:
Decide with your doctor if drug benefits outweigh risks to unborn child. Risk category B (see page xviii). Category can change with combination drug products.

Breastfeeding; Lactation; Nursing Mothers:
This drug is generally acceptable with breastfeeding. Discuss risks and benefits with your doctor.

Infants & children up to age 18:
Read the label on the product to see if it is approved for your child's age. Always follow the directions on product's label about how to use. Use proper measuring device or dropper provided. If unsure, consult doctor or pharmacist.

Prolonged use:
Talk to your doctor about the need for follow-up medical exams or lab studies to check liver and kidney functions.

Skin & sunlight:
No problems expected.

Driving, piloting or hazardous work:
No problems expected with acetaminophen-only product.

Discontinuing:
- Follow instructions on label for nonprescription product.
- Consult doctor before discontinuing use of prescription acetaminophen.

Others:
- Advise any doctor, dentist or pharmacist whom you consult that you take this medicine (if you take it regularly).
- There is a risk for severe liver injury if person takes more than recommended dose, takes high doses on regular basis or takes with other drug containing acetaminophen.
- This drug has been associated with a risk of rare but serious skin reactions (that could be fatal). Stop taking drug and seek medical help if symptoms develop (e.g., blisters, serious rash, reddening or peeling skin).
- Diabetic patients may get false blood glucose results while taking this drug. Check with the doctor if any changes occur.
- May interfere with the accuracy of some medical tests.
- If the sore throat or fever symptoms don't improve after 2 to 3 days use or if pain symptoms continue more than 5 days in children or 10 days for adults, call your doctor.

POSSIBLE INTERACTION WITH OTHER DRUGS

GENERIC NAME OR DRUG CLASS	COMBINED EFFECT
Acetaminophen-containing drugs	May increase risk of liver injury.
Anticoagulants, oral*	May increase anticoagulant effect. Consult doctor.
Anti-inflammatory drugs, nonsteroidal (NSAIDs)*	Increased risk of adverse effects of both drugs.
Aspirin and other salicylates*	Increased risk of adverse effects of both drugs.
Cholestyramine	May decrease effect of acetaminophen.
Enzyme inducers*	May increase risk of side effects.
Hepatotoxic medications*	Increased risk of liver damage.

POSSIBLE INTERACTION WITH OTHER SUBSTANCES

INTERACTS WITH	COMBINED EFFECT
Alcohol:	Increased risk of liver damage. Avoid.
Beverages:	None expected.
Cocaine:	Unknown. Avoid.
Foods:	None expected.
Marijuana:	May increase pain relief. However, marijuana may slow body's recovery. Consult doctor.
Tobacco:	May increase risk of liver damage. Consult doctor.

***See Glossary**

ACETOHYDROXAMIC ACID (AHA)

BRAND NAMES

Lithostat

BASIC INFORMATION

Habit forming? No
Prescription needed? Yes
Available as generic? No
Drug class: Antibacterial (antibiotic), antiurolithic

USES

- Treatment for chronic urinary tract infections.
- Prevents formation of urinary tract stones. Will not dissolve stones already present.

DOSAGE & USAGE INFORMATION

How to take:
Tablet—Swallow with liquid. Take on an empty stomach (1 hour before or 2 hours after a meal). If you can't swallow whole, crumble tablet and take with liquid or food.

When to take:
At the same times each day, according to instructions on prescription label.

If you forget a dose:
Take as soon as you remember. If it is almost time for the next dose, wait for that dose (don't double this dose) and resume regular schedule.

What drug does:
Stops enzyme action that makes urine too alkaline. Alkaline urine favors bacterial growth and stone formation and growth.

Time lapse before drug works:
1 to 3 weeks.

Don't take with:
- Alcohol or iron supplement.
- Any other medicine or any dietary supplement without consulting your doctor or pharmacist.

OVERDOSE

SYMPTOMS:
Appetite loss, tremor, nausea, vomiting, anxiety, feeling sluggish, lack of well-being.
WHAT TO DO:
If person uses much larger amount than prescribed or if accidentally swallowed, call poison control center 1-800-222-1222 for instructions or dial 911 (emergency) for help.

POSSIBLE ADVERSE REACTIONS OR SIDE EFFECTS

SYMPTOMS	WHAT TO DO
Life-threatening:	
Rare, severe allergic reaction (anaphylaxis)—hives, itching, swelling (face, throat), wheezing, vomiting, dizziness.	Seek emergency treatment immediately.
Common:	
Appetite loss, nausea, vomiting, anxiety, depression, mild headache, unusual tiredness.	Continue. Call doctor when convenient.
Infrequent:	
• Loss of coordination, slurred speech, severe headache, sudden change in vision, shortness of breath, pain over a blood vessel, sudden chest pain, leg pain in calf (deep vein blood clot).	Discontinue. Seek emergency treatment.
• Rash on arms and face.	Continue. Call doctor when convenient.
Rare:	
• Sore throat, fever, unusual bleeding, bruising.	Discontinue. Call doctor right away.
• Hair loss.	Continue. Call doctor when convenient.

WARNINGS & PRECAUTIONS

Don't take if:
You are allergic to AHA.

Before you start, consult your doctor if:
- You are anemic or have blood disorder.
- You have kidney disease or liver problems.
- You are a female of reproductive age.
- You have or have had blood clots (e.g., phlebitis or thrombophlebitis).

Over age 60:
Adverse reactions and side effects may be more frequent and severe than in younger persons.

Pregnancy:
Consult doctor. Drug should not be used during pregnancy. Can cause harm to unborn baby. Risk category X (see page xviii).

Breastfeeding; Lactation; Nursing Mothers:
This drug is generally not recommended with breastfeeding. Discuss risks and benefits with your doctor.

Infants & children up to age 18:
Follow instructions provided by your child's doctor.

Prolonged use:
Talk to your doctor about the need for follow-up medical examinations or laboratory studies to check blood pressure, liver function, kidney function, urinary pH.

Skin & sunlight:
No problems expected.

Driving, piloting or hazardous work:
Don't drive or pilot aircraft until you learn how medicine affects you. Don't work around dangerous machinery. Don't climb ladders or work in high places. Danger increases if you drink alcohol or take medicine affecting alertness and reflexes, such as antihistamines, tranquilizers, sedatives, pain medicines, narcotics and mind-altering drugs.

Discontinuing:
Don't discontinue without consulting doctor. Dose may require gradual reduction if you have taken drug for a long time. Doses of other drugs may also require adjustment.

Others:
Advise any doctor, dentist or pharmacist whom you consult that you take this medicine.

POSSIBLE INTERACTION WITH OTHER DRUGS

GENERIC NAME OR DRUG CLASS	COMBINED EFFECT
Iron	Decreased effects of both drugs.

POSSIBLE INTERACTION WITH OTHER SUBSTANCES

INTERACTS WITH	COMBINED EFFECT
Alcohol:	Severe skin rash common in many patients within 30 to 45 minutes after drinking alcohol.
Beverages:	None expected.
Cocaine:	Unknown. Avoid.
Foods:	None expected.
Marijuana:	Consult doctor.
Tobacco:	None expected.

ADRENOCORTICOIDS (Nasal Inhalation)

GENERIC AND BRAND NAMES

BECLOMETHASONE (nasal)
- Beconase
- Beconase AQ
- Qnasl
- Vancenase
- Vancenase AQ

BUDESONIDE (nasal)
- Rhinocort Aqua
- Rhinocort Nasal Inhaler
- Rhinocort Turbuhaler

CICLESONIDE (nasal)
- Omnaris
- Zetonna

FLUTICASONE (nasal)
- Dymista
- Flonase
- Flonase Allergy Relief
- Veramyst

MOMETASONE (nasal)
- Nasonex
- Nasonex Aqueous Nasal Spray

TRIAMCINOLONE (nasal)
- AllerNaze
- Nasacort Allergy 24HR
- Nasacort AQ
- Nasacort HFA
- Tri-Nasal

BASIC INFORMATION

Habit forming? No
Prescription needed? Yes, for most
Available as generic? Yes, for some
Drug class: Adrenocorticoid (nasal); anti-inflammatory (steroidal), nasal

USES

- Treats allergic conditions such as hay fever (seasonal rhinitis).
- Treats nasal polyps and noninfectious-inflammatory nasal conditions.

OVERDOSE

SYMPTOMS:
May have upset stomach if swallowed by mouth.
WHAT TO DO:
If person uses much larger amount than prescribed or if accidentally swallowed, call poison control center 1-800-222-1222 for instructions or dial 911 (emergency) for help.

DOSAGE & USAGE INFORMATION

How to take:
Spray—Read patient instruction sheet supplied with your prescription. Usually 1 or 2 sprays into each nostril every 12 hours. Save container for possible refills.

When to take:
At the same time each day, according to instructions on prescription label.

If you forget a dose:
Use as soon as you remember. If it is almost time for the next dose, wait for that dose (don't double this dose) and resume regular schedule.

What drug does:
- Subdues inflammation by decreasing secretion of prostaglandins in cells of the lining of the nose and by inhibiting release of histamine.
- Very little, if any, of the nasal adrenocorticoid gets absorbed into the bloodstream.

Time lapse before drug works:
Usually 5 to 7 days, but may be as long as 2 to 3 weeks.

Don't take with:
Any other medicine or any dietary supplement without consulting your doctor or pharmacist.

POSSIBLE ADVERSE REACTIONS OR SIDE EFFECTS

SYMPTOMS	WHAT TO DO
Life-threatening: None expected.	
Common: Burning or dryness of nose, sneezing.	Continue. Call doctor when convenient.
Infrequent: Crusting inside the nose, nosebleed, sore throat, sores in nose, cough, dizziness, headache, hoarseness, nausea, runny nose, bloody mucus.	Discontinue. Call doctor right away.
Rare: White patches in nose or throat, wheezing, any changes in vision, other unexpected symptoms (may occur if drug absorbed in body).	Discontinue. Call doctor right away.

WARNINGS & PRECAUTIONS

Don't take if:
You are allergic to cortisone or any cortisone-like medication.

Before you start, consult your doctor if:
- You are allergic to any of the propellants in the spray. These include benzalkonium chloride, disodium acetate, phenylethanol, fluorocarbons and propylene glycol.
- You have sores in the nose, have had recent surgery or injury involving the nose.
- You have amebiasis, asthma, type 2 diabetes, glaucoma, herpes eye infection, liver disease, tuberculosis, underactive thyroid, any heart condition or any infection.

Over age 60:
No special problems expected.

Pregnancy:
Decide with your doctor if drug benefits justify risk to unborn baby. Risk category C (budesonide is category B) (see page xviii).

Breastfeeding; Lactation; Nursing Mothers:
Drugs in this group are generally acceptable with breastfeeding. Discuss risks and benefits with your doctor.

Infants & children up to age 18:
- Follow instructions provided by your child's doctor. Approval for use in children varies for these drugs.
- Adrenocorticoids taken by mouth may slow or decrease growth rate or cause reduced adrenal gland function. The nasal form is generally considered safer than the oral form. Be sure you and your child's doctor discuss all benefits and risks of the drug.

Prolonged use:
Not recommended.

Skin & sunlight:
No special problems expected.

Driving, piloting or hazardous work:
Don't drive or pilot aircraft until you learn how medicine affects you. Don't work around dangerous machinery. Don't climb ladders or work in high places. Danger increases if you drink alcohol or take medicine affecting alertness and reflexes.

Discontinuing:
No special problems expected.

Others:
Advise any doctor, dentist or pharmacist whom you consult that you use this medicine.

POSSIBLE INTERACTION WITH OTHER DRUGS

GENERIC NAME OR DRUG CLASS	COMBINED EFFECT
Ephedrine	Decreased effect of nasal adrenocorticoid.
Phenobarbital	Decreased effect of nasal adrenocorticoid.
Rifampin	Decreased effect of nasal adrenocorticoid.
Ritonavir	Increased effect of fluticasone.

POSSIBLE INTERACTION WITH OTHER SUBSTANCES

INTERACTS WITH	COMBINED EFFECT
Alcohol:	None expected.
Beverages:	None expected.
Cocaine:	Unknown. Avoid.
Foods:	None expected.
Marijuana:	Consult doctor.
Tobacco:	None expected.

*See Glossary

ADRENOCORTICOIDS (Oral Inhalation)

GENERIC AND BRAND NAMES

BECLOMETHASONE (oral inhalation)
Beclodisk
Becloforte
Qvar
Vanceril
BUDESONIDE (oral inhalation)
Pulmicort Flexhaler
Pulmicort Respules
Symbicort
CICLESONIDE (oral inhalation)
Alvesco Inhalation
FLUTICASONE (oral inhalation)
Advair Diskus
Advair HFA
Arnuity Ellipta
Breo Ellipta
Flovent
Flovent HFA
MOMETASONE (oral inhalation)
Asmanex HFA
Asmanex Twisthaler
Dulera

BASIC INFORMATION

Habit forming? No
Prescription needed? Yes
Available as generic? Yes, for some
Drug class: Anti-inflammatory (inhalation), antiasthmatic

USES

- Treatment for prevention of symptoms of chronic bronchial asthma. Does not relieve the symptoms of an acute asthma attack.
- Treatment for chronic obstructive pulmonary disease (COPD).

DOSAGE & USAGE INFORMATION

How to take:
Oral inhaler—Follow instructions that come with your prescription or from your doctor. If you don't understand the instructions or have any questions, consult your doctor or pharmacist. Most effective if taken regularly. More effective if taken with a spacer.

Continued next column

When to take:
Your doctor will determine the dosage amount and schedule that will help control the asthma symptoms and lessen risks of side effects. Usually 1 to 2 inhaled puffs 3 to 4 times a day is sufficient.

If you forget a dose:
Use as soon as you remember. If it is almost time for the next dose, wait for that dose (don't double this dose) and resume regular schedule.

What drug does:
Helps prevent inflammation in the lungs and breathing passages. May decrease progression of severe disease.

Time lapse before drug works:
1 to 4 weeks for the initial response and up to several months for full benefits.

Don't take with:
Any other medicine or any dietary supplement without consulting your doctor or pharmacist.

OVERDOSE

SYMPTOMS:
Unknown.
WHAT TO DO:
If person uses much larger amount than prescribed or if accidentally swallowed, call poison control center 1-800-222-1222 for instructions or dial 911 (emergency) for help.

POSSIBLE ADVERSE REACTIONS OR SIDE EFFECTS

SYMPTOMS	WHAT TO DO
Life-threatening: None expected.	
Common: Dry mouth, cough, throat irritation, hoarseness or other voice changes.	Continue. Call doctor when convenient.
Infrequent: Dry throat, headache, nausea, skin bruising, unpleasant taste, white curd-like patches in mouth or throat, pain when eating or swallowing.	Continue. Call doctor when convenient.
Rare: Increased wheezing, difficulty in breathing, pain or tightness or burning in chest, behavior changes if using budesonide (restlessness or nervousness or depression).	Continue, but call doctor right away.

ADRENOCORTICOIDS (Oral Inhalation)

WARNINGS & PRECAUTIONS

Don't use if:
You are allergic to any corticosteroids.*

Before you start, consult your doctor if:
- You have osteoporosis.
- You have or have had tuberculosis.
- You are taking oral corticosteroid drugs.

Over age 60:
No special problems expected.

Pregnancy:
Decide with your doctor if drug benefits justify risk to unborn baby. Risk category C (see page xviii).

Breastfeeding; Lactation; Nursing Mothers:
Drugs in this group are generally acceptable with breastfeeding. Discuss risks and benefits with your doctor.

Infants & children up to age 18:
- Should be safe with regular low-dosage regimen. These drugs may slow or decrease growth rate or cause reduced adrenal gland function. Be sure you and your child's doctor discuss all benefits and risks of the drug.
- Children using large doses of this drug are more susceptible to infectious disease (chicken pox, measles). Avoid exposure to infected people and keep all immunizations up to date.

Prolonged use:
- Talk to your doctor about the need for follow-up medical examinations or laboratory studies to check adrenal function, growth and development in children, pulmonary function and inhalation technique.
- The drug may lose its effectiveness. If this occurs, consult your doctor.

Skin & sunlight:
No special problems expected.

Driving, piloting or hazardous work:
No special problems expected.

Discontinuing:
Don't discontinue this drug after prolonged use without consulting doctor. Dosage may require a gradual reduction before stopping to avoid any withdrawal symptoms.

Others:
- Advise any doctor, dentist or pharmacist whom you consult that you use this medicine.
- Carry or wear identification to state that you use this medicine.
- Call your doctor if you have any injury, infection or other stress to your body.
- Take medicine only as directed. Do not increase or reduce dosage without doctor's approval.

*See Glossary

POSSIBLE INTERACTION WITH OTHER DRUGS

GENERIC NAME OR DRUG CLASS	COMBINED EFFECT
None significant.	

POSSIBLE INTERACTION WITH OTHER SUBSTANCES

INTERACTS WITH	COMBINED EFFECT
Alcohol:	None expected.
Beverages:	None expected.
Cocaine:	Unknown. Avoid.
Foods:	None expected.
Marijuana:	Consult doctor.
Tobacco:	None expected, but asthma patients should not smoke.

ADRENOCORTICOIDS (Rectal)

GENERIC AND BRAND NAMES

BUDESONIDE (Rectal)
- **Uceris Foam**

HYDROCORTISONE (Rectal)
- **Anusol-HC**
- **Cortifoam**
- **Cortizone 10 Anti-Itch Anal Cream**

HYDROCORTISONE (Rectal) (Con't)
- **Cortizone 10 Anti-Itch Anal Ointment**
- **Proctofoam**
- **Xyralid RC**

BASIC INFORMATION

Habit forming? No
Prescription needed? Yes, for some
Available as generic? Yes, for most
Drug class: Anti-inflammatory, steroidal (rectal)

USES

Used in or around the anal and rectal areas to relieve inflammation (swelling), itching or pain caused by conditions such as hemorrhoids, anal fissures, ulcerative colitis and proctitis.

DOSAGE & USAGE INFORMATION

How to use:
- Rectal cream or ointment—Always wash hands and clean anal area before using. Apply to surface of anal area with fingers. Rub in gently. For use inside the anus, read and follow instructions provided with product.
- Suppository or aerosol foam—Read and follow instructions provided with each product.

When to use:
Follow instructions provided with product. It is usually applied or used 1 to 4 times a day.

Continued next column

OVERDOSE

SYMPTOMS:
Unknown. May have symptoms as listed under Rare side effects (see next column).
WHAT TO DO:
If person uses much larger amount than prescribed or if accidentally swallowed, call poison control center 1-800-222-1222 for instructions or dial 911 (emergency) for help.

If you forget a dose:
If used on a regular basis, use as soon you remember. If it is almost time for the next dose, wait for next scheduled dose (don't double this dose).

What drug does:
Reduces inflammation in the anal and rectal areas which helps relieve the pain and itching.

Time lapse before drug works:
5 to 15 minutes; may take several weeks for full effect.

Don't use with:
Other rectal medicines without consulting your doctor or pharmacist.

POSSIBLE ADVERSE REACTIONS OR SIDE EFFECTS

SYMPTOMS	WHAT TO DO
Life-threatening: Rare, severe allergic reaction (anaphylaxis)—difficulty breathing, hives, itching, swelling (throat, face), vomiting, dizziness.	Seek emergency treatment immediately.
Common: None expected.	
Infrequent: Mild skin symptoms in treated area (burning, itching, dryness, color change, scaly skin), nausea.	Continue. Call doctor if symptoms persist.
Rare: Rectal bleeding or infection or severe irritation. Very rarely, too much of the drug is absorbed into the bloodstream (such as with long-term use) and may cause symptoms (such as bruising, acne, rounding of face, swelling of feet or ankles, extreme tiredness, headache, weight loss, excess thirst or urination, vision problems or other unexpected symptoms).	Discontinue. Call doctor right away.

WARNINGS & PRECAUTIONS

Don't use if:
You are allergic to any topical adrenocorticoid.

Before you start, consult your doctor if:
- You have skin infection or bleeding in area to be treated, an abscess, intestinal obstruction or perforation, or anal fistula.
- You have diabetes, an untreated infection or tuberculosis.
- You are taking ant cortisone" medications.
- You have a medical problem that affects the liver, heart, stomach, intestines or eyes.

Over age 60:
No problems expected.

Pregnancy:
Decide with your doctor if drug benefits justify risk to unborn baby. Risk category C (see page xviii).

Breastfeeding; Lactation; Nursing Mothers:
Drugs in this group may or may not be recommended during breastfeeding. Discuss risks and benefits with your doctor.

Infants & children up to age 18:
Follow instructions provided by your child's doctor. Children are more likely to absorb more of the drug into the bloodstream which can lead to adverse effects.

Prolonged use:
Don't use longer than stated on instructions provided with product. Long-term use is generally not recommended, but may be prescribed for certain patients.

Skin & sunlight:
No problems expected.

Driving, piloting or hazardous work:
No problems expected.

Discontinuing:
No problems expected.

Others:
- Don't use a bandage or wrapping or other covering on the treated area unless instructed to do so.
- May have increased risk of infections. Consult doctor if you develop fever, chills, aches, fatigue, nausea or vomiting.
- Advise any doctor, dentist or pharmacist whom you consult that you use this drug.

POSSIBLE INTERACTION WITH OTHER DRUGS

GENERIC NAME OR DRUG CLASS	COMBINED EFFECT
Enzyme inhibitors*	May increase effect of budesonide.

POSSIBLE INTERACTION WITH OTHER SUBSTANCES

INTERACTS WITH	COMBINED EFFECT
Alcohol:	None expected.
Beverages: Grapefruit juice.	May increase effect of budesonide.
Cocaine:	Unknown. Avoid.
Foods:	None expected.
Marijuana:	None expected.
Tobacco:	None expected.

*See Glossary

ADRENOCORTICOIDS (Systemic)

GENERIC AND BRAND NAMES

See full list of generic and brand names in the *Generic and Brand Name Directory*, page 832.

BASIC INFORMATION

Habit forming? No
Prescription needed? Yes
Available as generic? Yes, for most
Drug class: Anti-inflammatory (steroidal), corticosteroid, immunosuppressant

USES

- Used for their anti-inflammatory and immunosuppressive effect in the treatment of many medical disorders.
- Treats allergies, asthma, arthritis, Addison's disease, skin problems, ulcerative colitis, some cancers and numerous other conditions.

DOSAGE & USAGE INFORMATION

How to take:
- Tablet or capsule or extended-release tablet or extended-release capsule—Swallow whole with liquid. Take with food to lessen stomach irritation. If you have trouble swallowing drug, ask doctor or pharmacist for advice.
- Oral disintegrating tablet—Let tablet dissolve in mouth. Do not chew or swallow tablet.
- Oral suspension, syrup or enema—Follow instructions on label.

When to take:
At the same time(s) each day. Take once-a-day or once-every-other-day doses in mornings.

If you forget a dose:
Take as soon as you remember. If it is almost time for the next dose, wait for that dose (don't double this dose) and resume regular schedule.

Continued next column

What drug does:
Decreases inflammatory responses. Suppresses immune response. Stimulates bone marrow.

Time lapse before drug works:
Starts working within hours, but may take days to weeks for full benefit.

Don't take with:
Any other medicine or any dietary supplement without consulting your doctor or pharmacist.

OVERDOSE

SYMPTOMS:
May have psychological changes, heart rhythm problems or allergic symptoms (e.g., shortness of breath).
WHAT TO DO:
- **Dial 911 for all medical emergencies or call poison control center 1-800-222-1222 for instructions.**
- **See emergency information on last 3 pages of this book.**

POSSIBLE ADVERSE REACTIONS OR SIDE EFFECTS

SYMPTOMS	WHAT TO DO
Life-threatening:	
Rare, severe allergic reaction (anaphylaxis)—difficulty breathing, hives, itching, swelling (throat, face), vomiting, dizziness.	Seek emergency treatment immediately.
Common:	
Increased appetite, indigestion, stomach irritation, insomnia, mood changes, feeling nervous or restless; with long term use—high blood pressure or puffy face or weak bones or fractures.	Continue. Call doctor when convenient.
Infrequent:	
• Infections, blurred or decreased vision, frequent urination, increased thirst, unusual mental or emotional changes.	Continue. Call doctor right away.
• Feeling dizzy or light-headed, changes in skin color, flushing, hiccups, sweating, spinning sensation, acne, hair loss.	Continue. Call doctor when convenient.
Rare:	
Long term use or high doses—pain (stomach, bone, joint, other), black or tarry stools, eye pain or redness or sensitive to light, muscle cramps, menstrual problems, headaches, irregular heartbeat, stretch marks or thin skin, nausea or vomiting, child's growth stunted, sleep problems, unusual hair growth, bruising, tiredness or weakness, rapid weight gain, slow wound healing, feet or ankle swelling.	Continue. Call doctor when convenient.

WARNINGS & PRECAUTIONS

Don't take if:
- You are allergic to any cortisone* drug.
- You have an active case of tuberculosis, systemic fungal infection, herpes infection of eyes or peptic ulcer disease.

Before you start, consult your doctor if:
You have, or have had, heart disease, congestive heart failure, diabetes, AIDS, HIV infection, glaucoma, underactive or overactive thyroid, high blood pressure, myasthenia gravis, blood clots in legs or lungs, peptic ulcer disease, tuberculosis, recent or current chickenpox or measles, kidney or liver disease, esophagitis, cold sores, osteoporosis, systemic lupus erythematosus or hyperlipidemia.

Over age 60:
With long term use, the adverse reactions and side effects may be more frequent or severe.

Pregnancy:
Risk factors vary for drugs in this group. See category list on page xviii and consult doctor.

Breastfeeding; Lactation; Nursing Mothers:
Drugs in this group are generally acceptable with breastfeeding. Discuss risks and benefits with your doctor.

Infants & children up to age 18:
- Follow instructions provided by your child's doctor. These drugs may slow or decrease growth rate or cause reduced adrenal gland function. Be sure you and your child's doctor discuss all benefits and risks of the drug.
- Children using large doses of this drug are more susceptible to infectious disease.

Prolonged use:
- Greatly increases the risk of adverse effects.
- Talk to your doctor about the need for follow-up medical exams or laboratory studies.

Skin & sunlight:
No problems expected.

Driving, piloting or hazardous work:
No problems expected (unless you feel dizzy).

Discontinuing:
- Don't discontinue without doctor's advice until you complete prescribed dose, even though symptoms diminish or disappear.
- Don't stop drug suddenly. Drug dose usually needs to be gradually reduced (tapered). Consult doctor if withdrawal symptoms occur (e.g., fatigue, weakness, stomach pain, nausea or vomiting, diarrhea or low blood pressure).
- Drug can affect your response to surgery, illness, injury or stress for 2 years after discontinuing. Tell anyone who takes medical care of you within 2 years about use of this drug.

Others:
- Consult your doctor before receiving any type of vaccination. Some vaccines may be less effective in persons taking this drug.
- Resistance to infection is less while taking this medicine. Consult doctor if infection occurs.
- Call doctor about swelling or rapid weight gain.
- Advise any doctor, dentist or pharmacist whom you consult that you take this medicine.
- May cause recurrence of tuberculosis.
- Can interfere with the accuracy of some medical tests.
- Wear or carry medical identification that indicates use of this drug (if using long term).

POSSIBLE INTERACTION WITH OTHER DRUGS

GENERIC NAME OR DRUG CLASS	COMBINED EFFECT
Antacids*	May decrease effect of adrenocorticoid.
Anticholinergics*	Risk of glaucoma.
Anticoagulants,* oral	May increase or decrease anti-coagulant effect.
Antidepressants, tricyclic*	Increased risk of side effects.
Antidiabetics,* oral	Decreased anti-diabetic effect.
Antifungals, azole	Increased effect of adrenocorticoid.
Anti-inflammatory drugs, nonsteroidal (NSAIDs)*	Increased risk of side effects.

Continued on page 885

POSSIBLE INTERACTION WITH OTHER SUBSTANCES

INTERACTS WITH	COMBINED EFFECT
Alcohol:	Stomach ulcer risk.
Beverages: Grapefruit juice.	Ask your doctor or pharmacist.
Cocaine:	Unknown. Avoid.
Foods: Grapefruit.	Ask your doctor or pharmacist.
Marijuana:	Consult doctor.
Tobacco:	May decrease effect of adrenocorticoid.

*See Glossary

ADRENOCORTICOIDS (Topical)

GENERIC AND BRAND NAMES

See full list of generic and brand names in the *Generic and Brand Name Directory*, page 833.

BASIC INFORMATION

Habit forming? No
Prescription needed? Yes, for some
Available as generic? Yes, for most
Drug class: Adrenocorticoid (topical)

USES

Relieves redness, swelling, itching, skin discomfort of hemorrhoids; insect bites; poison ivy, oak, sumac; soaps, cosmetics; jewelry; burns; sunburn; numerous skin rashes; eczema; discoid lupus erythematosus; swimmer's ear; sun poisoning; hair loss; scars; pemphigus; psoriasis; pityriasis rosea.

DOSAGE & USAGE INFORMATION

How to use:

- Cream, lotion, ointment, gel—Apply small amount and rub in gently.
- Foam—Follow directions on container. Don't breathe vapors.
- Other forms—Follow directions on container.

When to use:
When needed or as directed. Don't use more often than directions allow.

If you forget an application:
Use as soon as you remember. If it is almost time for the next dose, wait for that dose (don't double this dose) and resume regular schedule.

What drug does:
Reduces inflammation by affecting enzymes that produce inflammation.

Continued next column

OVERDOSE

SYMPTOMS:
Unknown.
WHAT TO DO:
If person uses much larger amount than prescribed or if accidentally swallowed, call poison control center 1-800-222-1222 for instructions or dial 911 (emergency) for help.

Time lapse before drug works:
15 to 20 minutes.

Don't use with:
Any other topical medicine without consulting your doctor or pharmacist.

POSSIBLE ADVERSE REACTIONS OR SIDE EFFECTS

SYMPTOMS	WHAT TO DO
Life-threatening:	
None expected.	
Common:	
None expected.	
Infrequent:	
Infection on skin (pain, redness, blisters, pus), skin irritation (burning, itching, blistering or peeling), acne-like skin eruptions.	Continue. Call doctor when convenient.
Rare:	
None expected.	

Note: Side effects are unlikely if topical adrenocorticoids are used in low doses for short periods of time. High doses for long periods can possibly cause the adverse reactions of cortisone, listed under ADRENOCORTICOIDS (Systemic).

WARNINGS & PRECAUTIONS

Don't take if:
You are allergic to any topical adrenocorticoid (cortisone) preparation. Allergic reactions are rare, but may occur.

Before you start, consult your doctor if:
- You have diabetes.
- You have infection at treatment site.
- You have stomach ulcer.
- You have tuberculosis.

Over age 60:
Adverse reactions and side effects may be more frequent and severe than in younger persons, especially thinning of the skin.

Pregnancy:
Decide with your doctor if drug benefits justify risk to unborn child. Risk category C (see page xviii).

Breastfeeding; Lactation; Nursing Mothers:
Drugs in this group are generally acceptable with breastfeeding. Discuss risks and benefits with your doctor.

Infants & children up to age 18:
- Use with caution. Read label to be sure product is approved for your child's age. Consult doctor or pharmacist if unsure.
- If too much of the drug is used for too long, it can be absorbed into blood stream through skin and cause adverse reactions.

Prolonged use:
- Increases chance of absorption into blood stream which may lead to the side effects of systemic adrenocorticoid (cortisone) drugs.
- May thin skin where used.
- Talk to your doctor about the need for follow-up medical examinations or laboratory studies.

Skin & sunlight:
Desoximetasone may cause rash or intensify sunburn in areas exposed to sun or ultraviolet light (photosensitivity reaction). Avoid over-exposure. Notify doctor if reaction occurs.

Driving, piloting or hazardous work:
No problems expected.

Discontinuing:
May be unnecessary to finish medicine. Follow doctor's instructions.

Others:
- Don't use a plastic dressing longer than 2 weeks.
- Aerosol spray—Store in cool place. Don't use near heat or open flame or while smoking. Don't puncture, break or burn container.
- Don't use for acne or gingivitis.

POSSIBLE INTERACTION WITH OTHER DRUGS

GENERIC NAME OR DRUG CLASS	COMBINED EFFECT
Antibacterials* (topical)	May decrease antibiotic effect.
Antifungals* (topical)	may decrease antifungal effect.

POSSIBLE INTERACTION WITH OTHER SUBSTANCES

INTERACTS WITH	COMBINED EFFECT
Alcohol:	None expected.
Beverages:	None expected.
Cocaine:	Unknown. Avoid.
Foods:	None expected.
Marijuana:	Consult doctor.
Tobacco:	None expected.

***See Glossary**

ALPHA ADRENERGIC RECEPTOR BLOCKERS

GENERIC AND BRAND NAMES

ALFUZOSIN
UroXatral
DOXAZOSIN
Cardura
Cardura XL
PRAZOSIN
Minipress
Minizide
SILODOSIN
Rapaflo
TAMSULOSIN
Flomax
Jalyn
TERAZOSIN
Hytrin

BASIC INFORMATION

Habit forming? No
Prescription needed? Yes
Available as generic? Yes, for some
Drug class: Antihypertensive

USES

- Treatment for high blood pressure.
- May improve congestive heart failure.
- Treatment for Raynaud's disease.
- Treatment for benign prostatic hyperplasia.

DOSAGE & USAGE INFORMATION

How to take:
- Tablet (doxazosin)—Swallow with liquid. If you can't swallow whole, crush or crumble tablet and take with liquid or food.
- Tablet or capsule or extended-release tablet—Swallow whole with liquid. Do not crumble or crush or chew tablet or open capsule.

When to take:
At the same times each day.

Continued next column

OVERDOSE

SYMPTOMS:
Extreme weakness; rapid or irregular heartbeat; loss of consciousness; cold, sweaty skin; weak, rapid pulse; coma.
WHAT TO DO:
- **Dial 911 for all medical emergencies or call poison control center 1-800-222-1222 for instructions.**
- **See emergency information on last 3 pages of this book.**

If you forget a dose:
Take as soon as you remember. If it is almost time for the next dose, wait for that dose (don't double this dose) and resume regular schedule.

What drug does:
Expands and relaxes blood vessel walls to lower blood pressure.

Time lapse before drug works:
30 minutes; may take days or weeks for full effect.

Don't take with:
Any other medicine or any dietary supplement without consulting your doctor or pharmacist.

POSSIBLE ADVERSE REACTIONS OR SIDE EFFECTS

SYMPTOMS	WHAT TO DO
Life-threatening: In case of overdose, see previous column.	
Common: Headache, dizziness.	Continue. Call doctor when convenient.
Infrequent: • Rash or itchy skin, blurred vision, shortness of breath, difficulty breathing, chest pain, rapid heartbeat.	Discontinue. Call doctor right away.
• Appetite loss, constipation or diarrhea, abdominal pain, nausea, vomiting, fluid retention, joint or muscle aches, tiredness, weakness and faintness when arising from bed or chair, little or no semen when ejaculating, headache, irritability, depression, dry mouth, stuffy nose, increased urination, drowsiness.	Continue. Call doctor when convenient.
Rare: Decreased sexual function, numbness or tingling in hands or feet.	Continue. Call doctor when convenient.

ALPHA ADRENERGIC RECEPTOR BLOCKERS

WARNINGS & PRECAUTIONS

Don't take if:
You are allergic to alpha adrenergic receptor blockers. Allergic reactions are rare, but may occur.

Before you start, consult your doctor if:
- You experience lightheadedness or fainting with other antihypertensive drugs.
- You suffer from depression.
- You have impaired brain circulation or have had a stroke.
- You will have surgery within 2 months, including eye or dental surgery.
- You have coronary heart disease (with or without angina).
- You have kidney disease or impaired liver function.

Over age 60:
The starting dosage may be increased gradually by your doctor. Sudden changes in position may cause falls. Sit or lie down promptly if you feel dizzy. If you have impaired brain circulation or coronary heart disease, excessive lowering of blood pressure should be avoided. Report problems to your doctor right away.

Pregnancy:
Decide with your doctor if drug benefits justify risk to unborn baby. Risk category B or category C (see page xviii).

Breastfeeding; Lactation; Nursing Mothers:
Drugs in this group are generally acceptable with breastfeeding. Discuss risks and benefits with your doctor.

Infants & children up to age 18:
Not recommended for this age group.

Prolonged use:
Talk to your doctor about the need for follow-up medical examinations or laboratory studies.

Skin & sunlight:
No problems expected.

Driving, piloting or hazardous work:
Don't drive or pilot aircraft until you learn how medicine affects you. Don't work around dangerous machinery. Don't climb ladders or work in high places.

Discontinuing:
Don't discontinue without doctor's advice until you complete prescribed dose, even though symptoms diminish or disappear.

Others:
- First dose likely to cause dizziness or lightheadedness. Take drug at night and get out of bed slowly next morning.
- Advise any doctor, dentist or pharmacist whom you consult that you take this medicine.
- May affect the results in some medical tests.

POSSIBLE INTERACTION WITH OTHER DRUGS

GENERIC NAME OR DRUG CLASS	COMBINED EFFECT
Antihypertensives,* other	Increased anti-hypertensive effect. Dosages may require adjustments.
Anti-inflammatory drugs, nonsteroidal (NSAIDs)*	Decreased effect of alpha adrenergic blocker.
Enzyme inhibitors*	Increased effect of alpha adrenergic blocker.
Estrogen	Decreased effect of alpha adrenergic blocker.
Ritonavir	Increased effect of alfuzosin.
Sympathomimetics*	Decreased effect of alpha adrenergic blocker.

POSSIBLE INTERACTION WITH OTHER SUBSTANCES

INTERACTS WITH	COMBINED EFFECT
Alcohol:	Excessive blood pressure drop.
Beverages: Grapefruit juice.	May increase effect of alpha adrenergic blocker.
Cocaine:	Increased risk of heart block and high blood pressure. Avoid.
Foods:	None expected.
Marijuana:	Fainting. Avoid.
Tobacco:	Possible spasm of coronary arteries. Avoid.

***See Glossary**

ALPROSTADIL

BRAND NAMES

Caverject
Edex
Muse

BASIC INFORMATION

Habit forming? No
Prescription needed? Yes
Available as generic? No
Drug class: Impotence therapy

USES

Treatment for impotence in some men who have erectile dysfunction due to neurologic, vascular, psychological or mixed causes.

DOSAGE & USAGE INFORMATION

How to use:

- Injection—The first injection will be given in the doctor's office to determine proper dosage and to train you in preparing and self-injecting the drug. When using it at home, follow the instructions provided with the prescription or use as directed by your doctor to inject drug into the penis.
- Intraurethral—Use as a single dose suppository 10 to 30 minutes prior to intercourse.

When to take:
Usually 10 to 30 minutes prior to sexual intercourse. Do not use injection more than 3 times in one week and do not use more than once in a 24-hour period. Do not use more than 2 suppositories in one 24-hour period.

If you forget a dose:
Not used on a scheduled basis.

What drug does:
Increases the blood flow into the penis and decreases the blood flow from the penis. The change in blood flow causes the penis to swell and elongate.

Continued next column

Time lapse before drug works:
5 to 20 minutes. Erections may last up to 60 minutes.

Don't take with:
Any other medicine or any dietary supplement without consulting your doctor or pharmacist.

OVERDOSE

SYMPTOMS:
Prolonged penile erection.
WHAT TO DO:
If person uses much larger amount than prescribed or if accidentally swallowed, call poison control center 1-800-222-1222 for instructions or dial 911 (emergency) for help.

POSSIBLE ADVERSE REACTIONS OR SIDE EFFECTS

SYMPTOMS	WHAT TO DO
Life-threatening:	
Rare, severe allergic reaction (anaphylaxis)—difficulty breathing, hives, itching, swelling (throat, face), vomiting, dizziness.	Seek emergency treatment immediately.
Common:	
Bleeding (for short time) or pain at injection site, pain with erection, intraurethral form may cause urethra pain or mild bleeding.	Discontinue. Call doctor when convenient.
Infrequent:	
None expected.	
Rare:	
• Erection lasting more than 4 hours.	Call doctor right away or seek emergency care.
• Curving of the erect penis, testicle pain, injection site has bruising or clotted blood. If too much drug absorbed by body may have dizziness, fainting, flu-like symptoms or pelvic pain. Female partners may have mild vaginal itching or burning.	Discontinue. Call doctor when convenient.

WARNINGS & PRECAUTIONS

Don't take if:
You are allergic to alprostadil or you have been advised not to have sex.

Before you start, consult your doctor if:
- You have liver disease.
- You have sickle cell anemia or trait.
- You have multiple myeloma or leukemia.
- You have a penile implant or any type of penile malformation.
- You are allergic to any other medications.
- You have a history of priapism (prolonged penile erection).

Over age 60:
Effects on this age group are variable. Consult doctor.

Pregnancy:
Not used by females. Men should not use the product to have sexual intercourse with a pregnant woman unless the couple uses a condom barrier.

Breastfeeding; Lactation; Nursing Mothers:
Not used by females.

Infants & children up to age 18:
Not used in this age group.

Prolonged use:
Have regular checkups with your doctor while using this drug to determine the effectiveness of the treatment and to check for any penile problems.

Skin & sunlight:
No special problems expected.

Driving, piloting or hazardous work:
No special problems expected.

Discontinuing:
No special problems expected.

Others:
- Don't increase dosage or frequency of use without your doctor's approval.
- Follow label instructions and dispose of all needles properly after use. Do not reuse or share needles.
- The injection of this drug provides no protection from sexually transmitted diseases. Other protective measures, such as condoms, should be used when necessary to prevent the spread of sexually transmitted diseases.
- Slight bleeding may occur at injection site. Apply pressure if this occurs. If bleeding persists, consult doctor.

POSSIBLE INTERACTION WITH OTHER DRUGS

GENERIC NAME OR DRUG CLASS	COMBINED EFFECT
None significant.	

POSSIBLE INTERACTION WITH OTHER SUBSTANCES

INTERACTS WITH	COMBINED EFFECT
Alcohol:	None expected.
Beverages:	None expected.
Cocaine:	Unknown. Avoid.
Foods:	None expected.
Marijuana:	Consult doctor.
Tobacco:	None expected.

*See Glossary

AMINOGLUTETHIMIDE

BRAND NAMES

Cytadren

BASIC INFORMATION

Habit forming? No
Prescription needed? Yes
Available as generic? No
Drug class: Antiadrenal, antineoplastic

USES

- Treats Cushing's syndrome.
- Treats breast and prostate cancer.

DOSAGE & USAGE INFORMATION

How to take:
Tablet—Swallow with liquid. If you can't swallow whole, crumble tablet and take with liquid or food. Instructions to take on empty stomach mean 1 hour before or 2 hours after eating.

When to take:
Follow doctor's instructions exactly.

If you forget a dose:
Take as soon as you remember. If it is almost time for the next dose, wait for that dose (don't double this dose) and resume regular schedule.

What drug does:
Suppresses adrenal cortex.

Time lapse before drug works:
1 to 2 hours.

Don't take with:
Any other medicines (including nonprescription drugs such as cough and cold medicines, laxatives, antacids, diet pills, caffeine, nose drops, vitamins or other diet supplements) without consulting your doctor or pharmacist.

OVERDOSE

SYMPTOMS:
May be more severe form of adverse effects.
WHAT TO DO:
- **Dial 911 for all medical emergencies or call poison control center 1-800-222-1222 for instructions.**
- **See emergency information on last 3 pages of this book.**

POSSIBLE ADVERSE REACTIONS OR SIDE EFFECTS

SYMPTOMS	WHAT TO DO
Life-threatening: None expected.	
Common:	
• Skin rash on face and hands, feeling drowsy.	Continue, but call doctor right away.
• Nausea, loss of appetite.	Continue. Call doctor when convenient.
Infrequent: Dizziness, clumsiness, unusual tiredness or weakness, unusual eye movements, depression, fast heartbeat, feeling shaky, slurred speech.	Continue, but call doctor right away.
Rare:	
• Unusual bleeding or bruising, black or tarry stools, fever and chills along with cough or low back pain or painful urination, shortness of breath, pinpoint spots on skin, blood in urine.	Discontinue. Call doctor right away.
• Tenderness or swelling of the neck, headache, vomiting, menstrual changes, hair growth or deeper voice in females, muscle pain.	Continue. Call doctor when convenient.

WARNINGS & PRECAUTIONS

Don't take if:
You are allergic to aminoglutethimide or glutethimide. Allergic reactions are rare, but may occur.

Before you start, consult your doctor if:
- You have recently been exposed to chicken pox.
- You have shingles (herpes zoster).
- You have decreased thyroid function (hypothyroidism).
- You have any liver or kidney disorder.
- You have any form of infection.

Over age 60:
Adverse reactions and side effects may be more frequent and severe than in younger persons.

Pregnancy:
Consult doctor. Use of the drug in pregnancy is not recommended. It could cause harm to the unborn baby. Risk category D (see page xviii).

Breastfeeding; Lactation; Nursing Mothers:
This drug is generally not recommended with breastfeeding. Discuss risks and benefits with your doctor.

Infants & children up to age 18:
Safety and effectiveness in this age group have not been established. Consult doctor.

Prolonged use:
Talk to your doctor about the need for follow-up medical examinations or laboratory studies to check thyroid function, liver function, serum electrolytes (sodium, potassium, chloride) and blood pressure.

Skin & sunlight:
No problems expected.

Driving, piloting or hazardous work:
No problems expected.

Discontinuing:
No special problems expected.

Others:
- Advise any doctor, dentist or pharmacist whom you consult that you take this medicine.
- May affect results in some medical tests.
- May cause decreased thyroid function.

POSSIBLE INTERACTION WITH OTHER DRUGS

GENERIC NAME OR DRUG CLASS	COMBINED EFFECT
Anticoagulants*	Decreased anticoagulant effect.
Central nervous system (CNS) depressants*	Increased risk of drowsiness.
Cortisone-like drugs*	Decreased cortisone effects.
Dexamethasone	Decreased dexamethasone effect.

POSSIBLE INTERACTION WITH OTHER SUBSTANCES

INTERACTS WITH	COMBINED EFFECT
Alcohol:	Increased stomach irritation.
Beverages: Coffee, tea, cocoa.	Increased stomach irritation.
Cocaine:	Unknown. Avoid.
Foods:	None expected.
Marijuana:	Consult doctor.
Tobacco:	None expected.

*See Glossary

AMLEXANOX

BRAND NAMES

Aphthasol
OraDisc

BASIC INFORMATION

Habit forming? No
Prescription needed? Yes
Available as generic? No
Drug class: Antiaphthous ulcer agent

USES

Treatment for severe canker sores (aphthous ulcers) in the mouth.

DOSAGE & USAGE INFORMATION

How to use:

- Oral paste—Use fingertips to apply paste directly to each canker sore following oral hygiene.
- Patch—Apply to affected area in the mouth. The drug will dissolve slowly.

When to use:
Use as soon as symptoms of a canker sore appear. Apply four times a day—after meals and before bedtime. Wash hands after application.

If you forget a dose:
Use as soon as you remember. If it is almost time for the next dose, wait for that dose (don't double this dose) and resume regular schedule.

What drug does:
Exact healing mechanism is unknown. Appears to stop the inflammatory process and hypersensitivity reaction.

Time lapse before drug works:
Pain relief may occur within hours or up to 24 hours. Complete healing time will take several days.

Don't use with:
Any other medicine for mouth ulcers without consulting your doctor or pharmacist.

OVERDOSE

SYMPTOMS:
Unknown.
WHAT TO DO:
If person uses much larger amount than prescribed or if accidentally swallowed, call poison control center 1-800-222-1222 for instructions or dial 911 (emergency) for help.

POSSIBLE ADVERSE REACTIONS OR SIDE EFFECTS

SYMPTOMS	WHAT TO DO
Life-threatening: None expected.	
Common: None expected.	
Infrequent: Slight pain, stinging or burning at site of application.	No action necessary.
Rare: Diarrhea, nausea, rash.	Discontinue. Call doctor when convenient.

WARNINGS & PRECAUTIONS

Don't take if:
You are allergic to amlexanox. Allergic reactions are rare, but may occur.

Before you start, consult your doctor if:
- You are allergic to any medication, food or other substance.
- You have a weak immune system due to drugs or illness.

Over age 60:
No special problems expected.

Pregnancy:
Decide with your doctor if drug benefits outweigh risks to unborn child. Risk category B (see page xviii).

Breastfeeding; Lactation; Nursing Mothers:
This drug is generally acceptable with breastfeeding. Discuss risks and benefits with your doctor.

Infants & children up to age 18:
Safety and effectiveness in this age group have not been established. Consult doctor.

Prolonged use:
Normally only used for up to 10 days of treatment.

Skin & sunlight:
No problems expected.

Driving, piloting or hazardous work:
No problems expected.

Discontinuing:
May be unnecessary to finish medicine. Discontinue when canker sores heal.

Others:
- If canker sores do not heal after 10 days, consult your dentist or health care provider.
- Advise any doctor, dentist or pharmacist whom you consult that you take this medicine.

POSSIBLE INTERACTION WITH OTHER DRUGS

GENERIC NAME OR DRUG CLASS	COMBINED EFFECT
None expected.	

POSSIBLE INTERACTION WITH OTHER SUBSTANCES

INTERACTS WITH	COMBINED EFFECT
Alcohol:	None expected.
Beverages:	None expected.
Cocaine:	Unknown. Avoid.
Foods:	None expected.
Marijuana:	Consult doctor.
Tobacco:	None expected.

AMPHETAMINES

GENERIC AND BRAND NAMES

AMPHETAMINE & DEXTRO-AMPHETAMINE
Adderall
Adderall XR
DEXTROAMPHETAMINE
Dexedrine Spansule
LISDEXAMFETAMINE
Vyvanse
METHAMPHETAMINE
Desoxyn
Desoxyn Gradumet

BASIC INFORMATION

Habit forming? Yes
Prescription needed? Yes
Available as generic? Yes
Drug class: Central nervous system stimulant

USES

- Treats narcolepsy (sleep attacks).
- Treats attention deficit hyperactivity disorder in adults, adolescents and children.
- Treatment for other conditions as determined by your doctor.

DOSAGE & USAGE INFORMATION

How to take:
- Tablet—Swallow with liquid.
- Extended-release capsule or tablet—Swallow each dose whole with liquid; do not crush.
- Solution—Follow instructions on prescription.

When to take:
- Short-acting form—Don't take later than 6 hours before bedtime.
- Long-acting form—Take on awakening.

If you forget a dose:
Take as soon as you remember. If it is almost time for the next dose, wait for that dose (don't double this dose) and resume regular schedule.

Continued next column

OVERDOSE

SYMPTOMS:
Rapid heartbeat, hyperactivity, high fever, hallucinations, suicidal or homicidal feelings, convulsions, coma.
WHAT TO DO:
- **Dial 911 for all medical emergencies or call poison control center 1-800-222-1222 for instructions.**
- **See emergency information on last 3 pages of this book.**

What drug does:
- Hyperactivity—Decreases motor restlessness and increases ability to pay attention.
- Narcolepsy—Increases motor activity and mental alertness; diminishes drowsiness.

Time lapse before drug works:
Takes several weeks to see if drug is effective.

Don't take with:
Any other medicine or any diet supplement without consulting your doctor or pharmacist.

POSSIBLE ADVERSE REACTIONS OR SIDE EFFECTS

SYMPTOMS	WHAT TO DO
Life-threatening:	
In case of overdose, see previous column.	
Common:	
• Fast, pounding heartbeat.	Discontinue. Call doctor right away.
• Irritability, insomnia, euphoria, nervousness, dry mouth, signs of addiction.*	Continue. Call doctor when convenient.
Infrequent:	
• Dizziness, reduced alertness, blurred vision, unusual sweating.	Discontinue. Call doctor right away.
• Headache, diarrhea or constipation, appetite loss, stomach pain, nausea, vomiting, weight loss, diminished sex drive, impotence.	Continue. Call doctor when convenient.
Rare:	
• Rash, hives, chest pain, irregular heartbeat, trouble breathing, fainting, hallucinations, becoming suspicious, manic behavior, uncontrollable movements (head, neck, arms, legs).	Discontinue. Call doctor right away.
• Mood changes, swollen breasts.	Continue. Call doctor when convenient.

WARNINGS & PRECAUTIONS

Don't take if:
- You are allergic to any amphetamine. Allergic reactions are rare, but may occur.
- Patient has certain underlying heart defects.

Before you start, consult your doctor if:
- You plan to become pregnant.
- You have glaucoma, diabetes, overactive thyroid, anxiety or suffer from stress.

- You have a history of substance abuse.
- You have any heart or blood vessel disorder, high blood pressure or tic disorder (e.g., Tourette syndrome).
- Adult or child patient has a mental illness.

Over age 60:
Adverse reactions and side effects may be more frequent and severe than in younger persons.

Pregnancy:
Decide with your doctor if drug benefits justify risk to unborn baby. Risk category C (see page xviii).

Breastfeeding; Lactation; Nursing Mothers:
Drugs in this group are generally not recommended with breastfeeding. Discuss risks and benefits with your doctor.

Infants & children up to age 18:
You and your child's doctor should discuss all risks and benefits before using this drug.

Prolonged use:
- Drug can be habit forming. Ask your doctor about the risks involved.
- Talk to your doctor about the need for follow-up medical examinations or laboratory studies to check blood pressure, growth charts in children and need for continued treatment.

Skin & sunlight:
No problems expected.

Driving, piloting or hazardous work:
Don't drive or pilot aircraft until you learn how medicine affects you. Don't work around dangerous machinery. Don't climb ladders or work in high places. Danger increases if you drink alcohol or take medicine affecting alertness and reflexes.

Discontinuing:
- May be unnecessary to finish medicine, but don't suddenly stop. Follow doctor's instructions.
- During a withdrawal phase, may cause prolonged sleep of several days.

Others:
- Use of this drug must be closely supervised by healthcare provider.
- Advise any doctor, dentist or pharmacist whom you consult that you take this medicine.
- Don't use drug for fatigue or to replace rest. Don't use for appetite control or depression.
- Drug may rarely cause serious heart and psychiatric (mental) problems, including sudden death. Read warning information provided with prescription. Call doctor right away if symptoms develop (e.g., chest pain, shortness of breath, fainting or hallucinations).
- Drug may rarely cause blood vessel problems in fingers or toes; they feel cold or numb and very rarely, can develop wounds. Contact doctor right away if symptoms occur.

POSSIBLE INTERACTION WITH OTHER DRUGS

GENERIC NAME OR DRUG CLASS	COMBINED EFFECT
Antidepressants, tricyclic*	Decreased amphetamine effect.
Antihypertensives*	Decreased anti-hypertensive effect.
Beta-adrenergic blocking agents*	High blood pressure, slow heartbeat.
Carbonic anhydrase inhibitors*	Increased amphetamine effect.
Central nervous system (CNS) stimulants,* other	Excessive CNS stimulation.
Doxazosin	Decreased doxazosin effect.
Furazolidone	Sudden and severe high blood pressure.
Haloperidol	Decreased amphetamine effect.
Monoamine oxidase (MAO) inhibitors*	Severe increase in blood pressure.
Phenothiazines*	Decreased amphetamine effect.
Prazosin	Decreased prazosin effect.
Sodium bicarbonate	Increased amphetamine effect.
Sympathomimetics*	Seizure risk.
Thyroid hormones*	Irregular heartbeat.

POSSIBLE INTERACTION WITH OTHER SUBSTANCES

INTERACTS WITH	COMBINED EFFECT
Alcohol:	Decreased amphetamine effect. Avoid.
Beverages: Caffeine drinks.	Consult doctor.
Cocaine:	Dangerous risk to body's nervous system. Avoid.
Foods:	None expected.
Marijuana:	Increased risk of side effects. Avoid.
Tobacco:	None expected.

***See Glossary**

ANAGRELIDE

BRAND NAMES

Agrylin

BASIC INFORMATION

Habit forming? No
Prescription needed? Yes
Available as generic? Yes
Drug class: Platelet count-reducing agent; antithrombocythemia

USES

Reduces elevated platelet counts and the risk of thrombosis (formation of a blood clot); also makes symptoms more tolerable in patients with essential thrombocythemia.

DOSAGE & USAGE INFORMATION

How to take:
Capsule—Swallow with liquid. Take with or without food.

When to take:
At the same time each day. Dose may be adjusted to maintain proper platelet count.

If you forget a dose:
Take as soon as you remember. If it is almost time for the next dose, wait for that dose (don't double this dose) and resume regular schedule.

What drug does:
Exact mechanism is unknown.

Time lapse before drug works:
One to two weeks.

Don't take with:
Any other medicine or any dietary supplement without consulting your doctor or pharmacist.

OVERDOSE

SYMPTOMS:
None expected immediately. May lower platelet count, leading to increased bleeding.
WHAT TO DO:
If person uses much larger amount than prescribed or if accidentally swallowed, call poison control center 1-800-222-1222 for instructions or dial 911 (emergency) for help.

POSSIBLE ADVERSE REACTIONS OR SIDE EFFECTS

SYMPTOMS	WHAT TO DO
Life-threatening:	
Severe headache or weakness; pain or pressure in chest, jaw, neck, back or arms; swelling of feet or legs; severe tiredness or weakness; increased heart rate; difficulty breathing or shortness of breath.	Seek emergency treatment immediately.
Common:	
• Abdominal pain, weakness, dizziness, palpitations, shortness of breath.	Continue, but call doctor right away.
• Diarrhea, heartburn, gas, bloating, headache, loss of appetite, general feeling of discomfort or illness, nausea.	Continue. Call doctor when convenient.
Infrequent:	
• Blurred or double vision, painful or difficult urination, blood in urine, tingling in hands or feet, unusual bruising or bleeding, flushing, faintness.	Discontinue. Call doctor right away.
• Canker sore, joint pain, back pain, confusion, fever or chills, insomnia, constipation, leg cramps, runny nose, depression, nervousness, ringing in ears, skin rash, sensitivity to light, itching, sleepiness, vomiting.	Continue. Call doctor when convenient.
Rare:	
Hair loss.	Continue. Call doctor when convenient.

WARNINGS & PRECAUTIONS

Don't take if:
You are allergic to anagrelide. Allergic reactions are rare, but may occur.

Before you start, consult your doctor if:
- You have any kidney or liver disease or any heart or blood vessel disorder.
- You have any chronic health problem.
- You are pregnant or nursing.

Over age 60:
No special problems expected.

Pregnancy:
Decide with your doctor if drug benefits justify risk to unborn child. Risk category C (see page xviii).

Breastfeeding; Lactation; Nursing Mothers:
This drug is generally acceptable with breastfeeding. Discuss risks and benefits with your doctor.

Infants & children up to age 18:
Safety and efficacy have not been established in patients under age 16. Consult doctor.

Prolonged use:
Schedule regular visits with your doctor for laboratory examinations to monitor the continued effectiveness of the medication.

Skin & sunlight:
No problems expected.

Driving, piloting or hazardous work:
No problems expected.

Discontinuing:
Don't discontinue without consulting your doctor even if you feel well.

Others:
- Close medical supervision, including frequent platelet counts, required at start of therapy with this drug.
- Advise any doctor, dentist or pharmacist whom you consult that you use this medicine.

POSSIBLE INTERACTION WITH OTHER DRUGS

GENERIC NAME OR DRUG CLASS	COMBINED EFFECT
Sucralfate	May interfere with anagrelide absorption. Don't take at the same time.

POSSIBLE INTERACTION WITH OTHER SUBSTANCES

INTERACTS WITH	COMBINED EFFECT
Alcohol:	None expected.
Beverages:	None expected.
Cocaine:	Unknown. Avoid.
Foods:	None expected.
Marijuana:	Consult doctor.
Tobacco:	None expected.

ANAKINRA

BRAND NAMES

Kineret

BASIC INFORMATION

Habit forming? No
Prescription needed? Yes
Available as generic? No
Drug class: Antirheumatic; biological response modifier

USES

Treatment of moderately to severely active rheumatoid arthritis. Used for patients who have not responded to one or more disease modifying antirheumatic drugs (DMARDs). May be used alone or in combination with certain other arthritis drugs.

DOSAGE & USAGE INFORMATION

How to take:
Injection—The drug is self-injected under the skin (subcutaneously). Follow your doctor's instructions and the directions provided with the prescription on how and where to inject. Do not use the medication unless you are sure about the proper method for injection. Store medication in the refrigerator (do not freeze) until you plan to use it. After each use, throw away the syringe and any medicine left in it (ask your pharmacist about disposal methods). Never reuse the needle or syringe.

When to take:
Inject every day at the same time each day.

If you forget a dose:
Inject as soon as you remember. If it is almost time for the next dose, wait for that dose (don't double this dose) and resume regular schedule.

Continued next column

OVERDOSE

SYMPTOMS:
Unknown.
WHAT TO DO:
If person uses much larger amount than prescribed or if accidentally swallowed, call poison control center 1-800-222-1222 for instructions or dial 911 (emergency) for help.

What drug does:
Reduces the actions of chemicals in the body that cause inflammatory and immune responses. It helps prevent progressive joint destruction.

Time lapse before drug works:
It will take several weeks before full benefits of the drug are noticeable.

Don't use with:
Any other medicine or any dietary supplement without consulting your doctor or pharmacist.

POSSIBLE ADVERSE REACTIONS OR SIDE EFFECTS

SYMPTOMS	WHAT TO DO
Life-threatening:	
Rare, severe allergic reaction (anaphylaxis)—difficulty breathing, hives, itching, swelling (throat, face), vomiting, dizziness.	Seek emergency treatment immediately.
Common:	
Diarrhea, headache, nausea, mild stomach pain.	Continue. Call doctor when convenient.
Infrequent:	
Reaction at injection site (pain, purple discoloration, inflammation), pain in bones or joints, fever or chills, chest pain, cold or flu symptoms (sneezing, cough, sore throat, muscle aches, runny or stuffy nose), skin symptoms (itching, redness, swelling, hot), trouble with breathing, unusual tiredness, insomnia, vomiting, other signs or symptoms of an infection.	Discontinue. Call doctor right away.
Rare:	
Rash, hives, unusual bruising, dizziness, puffiness of face, fast heartbeat.	Discontinue. Call doctor right away.

WARNINGS & PRECAUTIONS

Don't take if:
You are allergic to anakinra or the components of the drug (including proteins made from bacterial cells such as *E coli*). Allergic reactions are rare, but may occur.

Before you start, consult your doctor if:
- You have a chronic disorder or infection or are immunosuppressed.
- You have asthma (increases risk of infections).
- You have a kidney disorder.
- You have an active infection.
- You are allergic or sensitive to latex.

Over age 60:
Used with caution in elderly patients, since infections are more common in this age group.

Pregnancy:
Decide with your doctor if drug benefits justify risk to unborn child. Risk category B (see page xviii).

Breastfeeding; Lactation; Nursing Mothers:
This drug is generally acceptable with breastfeeding. Discuss risks and benefits with your doctor.

Infants & children up to age 18:
Safety and effectiveness in this age group have not been established. Consult doctor.

Prolonged use:
- No specific problems expected.
- Talk to your doctor about the need for follow-up medical examinations or laboratory studies to check effectiveness of the drug and to monitor for infections.

Skin & sunlight:
No problems expected.

Driving, piloting or hazardous work:
No problems expected.

Discontinuing:
No problems expected. Consult doctor.

Others:
- Advise any doctor, dentist or pharmacist whom you consult that you take this medicine.
- Using this drug increases the risk of infections. Consult your doctor if any signs or symptoms of infection occur.
- Avoid immunizations unless approved by your doctor.

POSSIBLE INTERACTION WITH OTHER DRUGS

GENERIC NAME OR DRUG CLASS	COMBINED EFFECT
Tumor necrosis factor blockers	Increased risk of infections. Use only with close medical supervision.
Vaccines, live	Vaccines may not be effective.

POSSIBLE INTERACTION WITH OTHER SUBSTANCES

INTERACTS WITH	COMBINED EFFECT
Alcohol:	None expected.
Beverages:	None expected.
Cocaine:	Unknown. Avoid.
Foods:	None expected.
Marijuana:	Consult doctor.
Tobacco:	None expected.

ANDROGENS

GENERIC AND BRAND NAMES

See full list of generic and brand names in the *Generic and Brand Name Directory*, page 835.

BASIC INFORMATION

Habit forming? No
Prescription needed? Yes
Available as generic? Yes, for some
Drug class: Androgen

USES

- Corrects male hormone deficiency. Reduces "male menopause" symptoms (loss of sex drive, depression, anxiety).
- Decreases calcium loss of osteoporosis (softened bones).
- Blocks breast cancer cell growth in women.
- Stimulates start of puberty in certain boys.
- Augments treatment of aplastic anemia.
- Stimulates weight gain after illness, injury or for chronically underweight persons.
- Stimulates growth in treatment of dwarfism.

DOSAGE & USAGE INFORMATION

How to take:

- Tablet or capsule—Take with food to lessen stomach irritation.
- Topical (gel, cream, solution), transdermal patch, buccal system, nasal gel—Follow instructions provided with prescription or by your doctor.
- Injection—Given as needed for the condition by medical professional. May be self-injected as prescribed by your doctor.

When to take:
Follow directions provided with your prescription.

If you forget a dose:
Take or use as soon as you remember. If it is almost time for the next dose, wait for the next scheduled dose (don't double this dose). For gel or patch, usually wait for next scheduled time.

Continued next column

OVERDOSE

SYMPTOMS:
Unknown.
WHAT TO DO:
If person uses much larger amount than prescribed or if accidentally swallowed, call poison control center 1-800-222-1222 for instructions or dial 911 (emergency) for help.

What drug does:

- Stimulates cells that produce male sex characteristics.
- Replaces hormone deficiencies.
- Stimulates red-blood-cell production.
- Suppresses production of estrogen (female sex hormone).

Time lapse before drug works:
Varies with problems treated. May require 2 or 3 months of regular use for desired effects.

Don't take with:
Any other medicine or any dietary supplement without consulting your doctor or pharmacist.

POSSIBLE ADVERSE REACTIONS OR SIDE EFFECTS

SYMPTOMS	WHAT TO DO
Life-threatening:	
Rare, severe allergic reaction (anaphylaxis)—hives, itching, swelling (face, throat), wheezing, vomiting, dizziness.	Seek emergency treatment immediately.
Common:	
Acne or oily skin, deep voice, enlarged clitoris in females, frequent or continuing erections, sore or swollen breasts in men, pain or sores under patch.	Discontinue. Call doctor right away.
Infrequent:	
• Yellow skin or eyes, depression or confusion, other changes in moods, flushed face, rash or itch, nausea, vomiting, diarrhea, swollen feet or legs, headache, rapid weight gain, difficult or frequent urination, unusual bleeding, scrotum pain.	Discontinue. Call doctor right away.
• Sore mouth, higher or lower sex drive, decreased testicle size, impotence in men, mild redness or itching at site of patch, hair loss, pubic hair growth.	Continue. Call doctor when convenient

Rare:

Black stools, symptoms continue at patch site, vomiting up blood, fever or chills or hives.	Discontinue. Call doctor right away.

WARNINGS & PRECAUTIONS

Don't take if:
You are allergic to any male hormone.

Before you start, consult your doctor if:
- You might be pregnant.
- You have cancer of- or enlarged prostate.
- You have heart disease or arteriosclerosis.
- You have kidney or liver disease.
- You have breast cancer (male or female).
- You have high blood pressure.
- You have diabetes (drug can affect blood sugar).

Over age 60:
Your doctor will discuss the risks and benefits of taking the drug. While on the drug, visit your doctor on a regular basis to determine continued effectiveness and for adverse effects.

Pregnancy:
Consult doctor. Drug should not be used during pregnancy. Can cause harm to unborn baby. Risk category X (see page xviii).

Breastfeeding; Lactation; Nursing Mothers:
Drugs in this group are generally acceptable with breastfeeding. Discuss risks and benefits with your doctor.

Infants & children up to age 18:
Follow instructions provided by your child's doctor. See also information under Others.

Prolonged use:
- May cause liver cancer, possible kidney stones. In women may cause unnatural hair growth and deep voice.
- Talk to your doctor about the need for follow-up medical examinations or laboratory studies to check effects of the drug.

Skin & sunlight:
No problems expected.

Driving, piloting or hazardous work:
No problems expected.

Discontinuing:
No problems expected.

Others:
- Reduces sperm count and volume of semen. This effect is usually temporary.
- With patch or gel, drug can pass to sexual partner. Consult doctor if partner starts getting any of the side effects listed.
- Wash hands carefully after applying gel and cover area where gel applied.
- Children and women need to avoid contact with areas of skin where men have applied the gel. It can cause serious side effects, especially in children. Consult doctor if symptoms occur.
- Use of testosterone can increase risk of blood clots in the veins (venous thromboembolism or VTE). Call doctor if symptoms occur (pain, swelling, warmth and redness in leg or may have shortness of breath).
- Advise any doctor, dentist or pharmacist whom you consult that you take this medicine.
- In women, may cause changes such as deepened voice, increased hair growth, enlarged clitoris. Some changes may not go away after drug is discontinued.

POSSIBLE INTERACTION WITH OTHER DRUGS

GENERIC NAME OR DRUG CLASS	COMBINED EFFECT
Anticoagulants*	Increased anticoagulant effect.
Antidiabetic agents*	Increased antidiabetic effect.
Chlorzoxazone	Decreased androgen effect.
Cyclosporine	Increased cyclosporine effect.
Hepatotoxic drugs* (other)	Increased liver toxicity.
Insulin	Increased antidiabetic effect.
Oxyphenbutazone	Decreased androgen effect.
Phenobarbital	Decreased androgen effect.
Phenylbutazone	Decreased androgen effect.

POSSIBLE INTERACTION WITH OTHER SUBSTANCES

INTERACTS WITH	COMBINED EFFECT
Alcohol:	None expected.
Beverages:	None expected.
Cocaine:	Unknown. Avoid.
Foods:	None expected.
Marijuana:	Consult doctor.
Tobacco:	None expected.

***See Glossary**

ANESTHETICS (Mucosal-Local)

GENERIC AND BRAND NAMES

BENZOCAINE
- Anbesol Baby Gel
- Anbesol Maximum Strength Gel
- Anbesol Maximum Strength Liquid
- Baby Anbesol
- Baby Orabase
- Baby Oragel
- Baby Oragel Nighttime Formula
- Benzodent
- Children's Chloraseptic Lozenges
- Dentapaine
- Dentocaine
- Dent-Zel-Ite
- Hurricaine
- Numzident
- Num-Zit Gel
- Num-Zit Lotion
- Orabase-B with Benzocaine
- Orajel Extra Strength
- Orajel Liquid
- Orajel Maximum Strength
- Oratect Gel
- Rid-A-Pain
- SensoGARD Canker Sore Relief
- Spec-T Sore Throat Anesthetic
- Topicaine

BENZOCAINE & MENTHOL
- Chloraseptic Lozenges Cherry Flavor

BENZOCAINE & PHENOL
- Anbesol Gel
- Anbesol Liquid
- Anbesol Regular Strength Gel
- Anbesol Regular Strength Liquid

DYCLONINE
- Children's Sucrets
- Sucrets Maximum Strength
- Sucrets Regular Strength

LIDOCAINE
- Xylocaine
- Xylocaine Viscous
- Zilactin-L

TETRACAINE
- Supracaine

BASIC INFORMATION

Habit forming? No
Prescription needed? Yes, for some
Available as generic? Yes, for some
Drug class: Anesthetic (mucosal-local)

OVERDOSE

SYMPTOMS:
Overabsorption by body—Dizziness, blurred vision, seizures, drowsiness.
WHAT TO DO:
If person uses much larger amount than prescribed or if accidentally swallowed, call poison control center 1-800-222-1222 for instructions or dial 911 (emergency) for help.

USES

Relieves pain or irritation in mouth caused by toothache, teething, mouth sores, dentures, braces, dental appliances. Also relieves pain of sore throat for short periods of time.

DOSAGE & USAGE INFORMATION

How to use:
- For mouth problems—Apply to sore places with cotton-tipped applicator. Don't swallow.
- For throat—Gargle, but don't swallow.
- For aerosol spray—Don't inhale.

When to use:
As directed by doctor or label on package.

If you forget a dose:
Use as soon as you remember. If it is almost time for the next dose, wait for that dose (don't double this dose) and resume regular schedule.

What drug does:
Blocks pain impulses to the brain.

Time lapse before drug works:
Immediately.

Don't use with:
Any other medicine for your mouth without consulting your doctor or pharmacist.

POSSIBLE ADVERSE REACTIONS OR SIDE EFFECTS

SYMPTOMS	WHAT TO DO
Life-threatening: None expected.	
Common: None expected.	
Infrequent: Redness, irritation, sores not present before treatment, rash, itchy skin, hives.	Discontinue. Call doctor right away.
Rare: None expected.	

WARNINGS & PRECAUTIONS

Don't take if:
You are allergic to any of the products listed. Allergic reactions are rare, but may occur.

Before you start, consult your doctor if:
- You have infection, canker sores or other sores in your mouth.
- You take medicine for myasthenia gravis, eye drops for glaucoma or any sulfa medicine.

Over age 60:
No special problems expected.

Pregnancy:
Decide with your doctor if drug benefits justify risk to unborn baby. Risk category B or C (see page xviii).

Breastfeeding; Lactation; Nursing Mothers:
Drugs in this group are generally acceptable with breastfeeding. Discuss risks and benefits with your doctor.

Infants & children up to age 18:
- Read label on product to be sure it is approved for your child's age.
- Do not use viscous lidocaine products for teething pain in babies. It can cause serious harm (may be life-threatening). Don't use other topical nonprescription products for teething pain without talking to your child's doctor.

Prolonged use:
Not intended for prolonged use.

Skin & sunlight:
No problems expected.

Driving, piloting or hazardous work:
Wait to see if causes dizziness, sweating, drowsiness or blurred vision. If not, no problems expected.

Discontinuing:
No problems expected.

Others:
- Keep cool, but don't freeze product.
- Don't puncture, break or burn aerosol containers.
- Don't eat, drink or chew gum for 1 hour after use.
- Heat and moisture in bathroom medicine cabinet can cause breakdown of medicine. Store someplace else.
- Before anesthesia, tell dentist about any medicines you take or use.

POSSIBLE INTERACTION WITH OTHER DRUGS

GENERIC NAME OR DRUG CLASS	COMBINED EFFECT
None expected.	

POSSIBLE INTERACTION WITH OTHER SUBSTANCES

INTERACTS WITH	COMBINED EFFECT
Alcohol:	Adverse reactions more common.
Beverages:	None expected.
Cocaine:	May cause too much nervousness and trembling. Avoid.
Foods:	None expected.
Marijuana:	None expected.
Tobacco:	Avoid. Tobacco makes mouth problems worse.

ANESTHETICS (Rectal)

GENERIC AND BRAND NAMES

BENZOCAINE
Americaine Hemorrhoidal
Ethyl Aminobenzoate
DIBUCAINE
Nupercainal
LIDOCAINE
Anamantle HC Cream Kit
Peranex HC Cream
Recticare Anorectal
Xyralid RC
PRAMOXINE
Fleet Relief
ProCort
Proctofoam
Tronolane
Tronothane
TETRACAINE
Pontocaine Cream
TETRACAINE & MENTHOL
Pontocaine Ointment

BASIC INFORMATION

Habit forming? No
Prescription needed? Yes, for some
Available as generic? Yes
Drug class: Anesthetic (rectal)

USES

- Relieves pain, itching and swelling of hemorrhoids (piles).
- Relieves pain of rectal fissures (breaks in lining membrane of the anus).

DOSAGE & USAGE INFORMATION

How to use:

- Rectal cream or ointment—Apply to surface of rectum with fingers. Insert applicator into rectum no farther than 1/2 and apply inside. Wash applicator carefully or discard.
- Aerosol foam—Read patient instructions. Don't insert into rectum. Use the special applicator and wash carefully after using.

Continued next column

OVERDOSE

SYMPTOMS:
Unknown.
WHAT TO DO:
If person uses much larger amount than prescribed or if accidentally swallowed, call poison control center 1-800-222-1222 for instructions or dial 911 (emergency) for help.

- Suppository—Remove wrapper and moisten with water. Lie on side. Push blunt end of suppository into rectum with finger. If suppository is too soft, run cold water over wrapper or put in refrigerator for 15 to 45 minutes before using.
- Pads—For external use only. Follow instructions on label. Do not use for more than 1 week without doctor's approval.

When to use:
As directed on the product's label.

If you forget a dose:
Use as soon as you remember. If it is almost time for the next dose, wait for that dose (don't double this dose) and resume regular schedule.

What drug does:
Deadens nerve endings to pain and touch.

Time lapse before drug works:
5 to 15 minutes.

Don't use with:
Any other topical rectal medicine without consulting your doctor or pharmacist.

POSSIBLE ADVERSE REACTIONS OR SIDE EFFECTS

SYMPTOMS	WHAT TO DO
Life-threatening: None expected.	
Common: None expected.	
Infrequent: Skin where applied has new symptoms (itching, redness, stinging, hives, burning, tenderness).	Continue. Call doctor when convenient.
Rare: If too much of drug is absorbed by body may have anxiety, nervousness, slow or irregular heartbeat, trembling, ringing in ears, sweating, paleness.	Discontinue. Call doctor right away.

WARNINGS & PRECAUTIONS

Don't use if:
You are allergic to any topical anesthetic. Allergic reactions are rare, but may occur.

Before you start, consult your doctor if:
- You have skin infection at site of treatment.
- You have had severe or extensive skin disorders such as eczema or psoriasis.
- You have bleeding hemorrhoids.

Over age 60:
No special problems expected.

Pregnancy:
Decide with your doctor if drug benefits justify risk to unborn baby. Risk category B or C or unclassified (see page xviii).

Breastfeeding; Lactation; Nursing Mothers:
Drugs in this group are generally acceptable with breastfeeding. Discuss risks and benefits with your doctor.

Infants & children up to age 18:
Use caution. More likely to be absorbed through skin and cause adverse reactions.

Prolonged use:
Possible excess absorption. Don't use longer than 3 days for any one problem.

Skin & sunlight:
No problems expected.

Driving, piloting or hazardous work:
No problems expected.

Discontinuing:
May be unnecessary to finish medicine. Follow doctor's instructions.

Others:
- Report any rectal bleeding to your doctor.
- Store medicine per label instructions.

POSSIBLE INTERACTION WITH OTHER DRUGS

GENERIC NAME OR DRUG CLASS	COMBINED EFFECT
Sulfa drugs*	Decreased anti-infective effect of sulfa drugs.

POSSIBLE INTERACTION WITH OTHER SUBSTANCES

INTERACTS WITH	COMBINED EFFECT
Alcohol:	None expected.
Beverages:	None expected.
Cocaine:	Unknown. Avoid.
Foods:	None expected.
Marijuana:	Consult doctor.
Tobacco:	None expected.

*See Glossary

ANESTHETICS (Topical)

GENERIC AND BRAND NAMES

See full list of generic and brand names in the *Generic and Brand Name Directory*, page 836.

BASIC INFORMATION

Habit forming? No
Prescription needed?
High strength: Yes
Low strength: No
Available as generic? Yes
Drug class: Anesthetic (topical)

USES

Relieves pain and itch of sunburn, insect bites, scratches and other minor skin irritations.

DOSAGE & USAGE INFORMATION

How to use:
All forms—Use only enough to cover irritated area. Follow instructions on label or use as directed by doctor. Avoid using on large areas of skin.

When to use:
When needed for discomfort, no more often than every hour.

If you forget an application:
Use as needed.

What drug does:
Blocks pain impulses from skin to brain.

Time lapse before drug works:
3 to 15 minutes.

Don't take with:
Any other topical medicine without consulting your doctor or pharmacist.

OVERDOSE

SYMPTOMS:
Unknown.
WHAT TO DO:
If person uses much larger amount than prescribed or if accidentally swallowed, call poison control center 1-800-222-1222 for instructions or dial 911 (emergency) for help.

POSSIBLE ADVERSE REACTIONS OR SIDE EFFECTS

SYMPTOMS	WHAT TO DO
Life-threatening: None expected.	
Common: None expected.	
Infrequent: Hive-like swellings on skin or in mouth or throat, skin problems not present before treatment (rash, burning, stinging, tenderness, redness).	Discontinue. Call doctor right away.
Rare: If too much of drug absorbed into body (very rare)—Nervousness, slow heartbeat, dizziness, blurred or double vision, confusion, convulsions, noises in ears, feeling hot or cold, numbness, trembling, anxiety, paleness, tiredness or weakness.	Discontinue. Call doctor right away.

WARNINGS & PRECAUTIONS

Don't use if:
You are allergic to any topical anesthetic. Allergic reactions are rare, but may occur.

Before you start, consult your doctor if:
- You have skin infection at site of treatment.
- You have had severe or extensive skin disorders such as eczema or psoriasis.
- You have bleeding hemorrhoids.

Over age 60:
No special problems expected.

Pregnancy:
Decide with your doctor if drug benefits justify risk to unborn baby. Risk category B or C or unclassified (see page xviii).

Breastfeeding; Lactation; Nursing Mothers:
Drugs in this group are generally acceptable with breastfeeding. Discuss risks and benefits with your doctor.

Infants & children up to age 18:
Use caution. More likely to be absorbed through skin and cause adverse reactions.

Prolonged use:
Possible excess absorption.

Skin & sunlight:
May cause rash or intensify sunburn in areas exposed to sun or ultraviolet light (photosensitivity reaction). Avoid overexposure. Notify doctor if reaction occurs.

Driving, piloting or hazardous work:
No problems expected.

Discontinuing:
May be unnecessary to finish medicine. Follow doctor's instructions.

Others:
- Contact doctor if condition being treated doesn't improve within a week. Call sooner if new symptoms develop or pain worsens.
- Wash hands carefully after use.

POSSIBLE INTERACTION WITH OTHER DRUGS

GENERIC NAME OR DRUG CLASS	COMBINED EFFECT
Sulfa drugs*	Decreased effect of sulfa drugs for infection.

POSSIBLE INTERACTION WITH OTHER SUBSTANCES

INTERACTS WITH	COMBINED EFFECT
Alcohol:	None expected.
Beverages:	None expected.
Cocaine:	Unknown. Avoid.
Foods:	None expected.
Marijuana:	Consult doctor.
Tobacco:	None expected.

*See Glossary

ANGIOTENSIN II RECEPTOR ANTAGONISTS

GENERIC AND BRAND NAMES

AZILSARTAN
Edarbi
Edarbyclor
CANDESARTAN
Atacand
Atacand Plus
EPROSARTAN
Teveten
Teveten HCT
IRBESARTAN
Avalide
Avapro
LOSARTAN
Cozaar
Hyzaar
OLMESARTAN
Azor
Benicar
Benicar HCT
Tribenzor
TELMISARTAN
Micardis
Micardis HCT
Micardis Plus
Twynsta
VALSARTAN
Diovan
Diovan HCT
Diovan Oral
Exforge
Exforge HCT
Valturna

BASIC INFORMATION

Habit forming? No
Prescription needed? Yes
Available as generic? Yes, for some
Drug class: Antihypertensive, angiotensin II receptor antagonist

USES

- Treatment for hypertension (high blood pressure) and heart failure. May be used alone or in combination with other anti-hypertensive medications.
- Reduces risk of heart attack or stroke in certain patients.

DOSAGE & USAGE INFORMATION

How to take:
Tablet—Swallow with liquid. May be taken with or without food.

Continued next column

When to take:
Once or twice daily as directed.

If you forget a dose:
Take as soon as you remember. If it is almost time for the next dose, wait for that dose (don't double this dose) and resume regular schedule.

What drug does:
Lowers blood pressure by relaxing the blood vessels to allow improved blood flow in the body.

Time lapse before drug works:
May take several weeks for full effectiveness.

Don't take with:
Any other medicine or any dietary supplement without consulting your doctor or pharmacist.

OVERDOSE

SYMPTOMS:
Slow or irregular heartbeat, faintness, dizziness, lightheadedness.
WHAT TO DO:
- **Dial 911 for all medical emergencies or call poison control center 1-800-222-1222 for instructions.**
- **See emergency information on last 3 pages of this book.**

POSSIBLE ADVERSE REACTIONS OR SIDE EFFECTS

SYMPTOMS	WHAT TO DO
Life-threatening:	
Rare, severe allergic reaction (anaphylaxis)—hives, itching, swelling (face, throat), wheezing, vomiting, dizziness.	Seek emergency treatment immediately.
Common:	
Headache.	Continue. Call doctor when convenient.
Infrequent:	
Dizziness, faintness when rising, diarrhea, cold symptoms (such as stuffy nose).	Continue. Call doctor when convenient.
Rare:	
• Confusion, fast or irregular heartbest, severe vomiting or severe diarrhea, swelling of legs or feet or other body area.	Discontinue. Call doctor right away.
• Dry cough, trouble sleeping, muscle cramps, leg pain, weakness, stomach pain, tiredness.	Continue. Call doctor when convenient.

WARNINGS & PRECAUTIONS

Don't take if:
You are allergic to angiotensin II receptor antagonists.

Before you start, consult your doctor if:
- You have any kidney or liver disease.
- You are pregnant or planning pregnancy.
- You are allergic to any medication, food or other substance.

Over age 60:
No special problems expected.

Pregnancy:
- Decide with your doctor if drug benefits justify risks to unborn baby. Risk category C for first trimester and category D (could harm unborn baby) for second/third trimesters (see page xviii).
- Contact doctor and stop taking losartan right away if pregnancy occurs.

Breastfeeding; Lactation; Nursing Mothers:
Drugs in this group are generally not recommended with breastfeeding. Discuss risks and benefits with your doctor.

Infants & children up to age 18:
Olmesartan and valsartan approved to treat high blood pressure in ages over 6. For other angiotensin II receptor antagonists, consult doctor.

Prolonged use:
- No special problems expected. Hypertension usually requires life-long treatment.
- Schedule regular doctor visits to determine if drug is continuing to be effective in controlling the hypertension and to check for any kidney problems.

Skin & sunlight:
No special problems expected.

Driving, piloting or hazardous work:
Don't drive or pilot aircraft until you learn how medicine affects you. Don't work around dangerous machinery. Don't climb ladders or work in high places. Danger increases if you drink alcohol or take other medicines affecting alertness and reflexes.

Discontinuing:
Don't discontinue without consulting your doctor, even if you feel well. You can have hypertension without feeling any symptoms. Untreated high blood pressure can cause serious problems.

Others:
- Advise any doctor, dentist or pharmacist whom you consult that you take this medicine.
- May inter-fere with the accuracy of some medical tests.
- Follow any special diet your doctor may prescribe. It can help control hypertension.
- Consult doctor if you become ill with vomiting or diarrhea.
- Olmesartan can cause intestinal problems (sprue-like enteropathy). May occur after months or years of use. Symptoms include severe, chronic diarrhea and weight loss. Call doctor if symptoms develop.
- Use caution when exercising or performing activities in hot weather and with excessive sweating. You may experience dizziness, lightheadedness or faintness.

POSSIBLE INTERACTION WITH OTHER DRUGS

GENERIC NAME OR DRUG CLASS	COMBINED EFFECT
Anti-inflammatory drugs, nonsteroidal (NSAIDs)*	Decreased anti-hypertensive effect.
Cyclosporine	Excess potassium levels in the body.
Diuretics*	Increased anti-hypertensive effect.
Diuretics, potassium-sparing*	Excess potassium levels in the body.
Hypotension-causing drugs,* other	Increased anti-hypertensive effect.
Indomethacin	Decreased anti-hypertensive effect.
Potassium-containing medications	Excess potassium levels in the body.
Potassium supplements*	Excess potassium levels in the body.
Sympathomimetics*	Decreased anti-hypertensive effect.

POSSIBLE INTERACTION WITH OTHER SUBSTANCES

INTERACTS WITH	COMBINED EFFECT
Alcohol:	Unknown effect. Consult doctor.
Beverages: Grapefruit juice.	Decreased effect of losartan.
Cocaine:	Unknown. Avoid.
Foods:	None expected.
Marijuana:	Sedation. Avoid.
Tobacco:	None expected.

ANGIOTENSIN-CONVERTING ENZYME (ACE) INHIBITORS

GENERIC AND BRAND NAMES

See full list of generic and brand names in the *Generic and Brand Name Directory*, page 837.

BASIC INFORMATION

Habit forming? No
Prescription needed? Yes
Available as generic? Yes, for some
Drug class: Antihypertensive

USES

- Treatment for hypertension (high blood pressure) and congestive heart failure.
- Used for kidney disease in diabetic patients.
- Helps prevent complications in patients with stable coronary artery disease.
- Treatment for acute myocardial infarction.

DOSAGE & USAGE INFORMATION

How to take:
Capsule, tablet, long-acting tablet, liquid—Follow directions provided with your prescription.

When to take:
At the same times each day, usually 2-3 times daily. Captopril should be taken on an empty stomach 1 hour before or 2 hours after eating.

If you forget a dose:
Take as soon as you remember. If it is almost time for the next dose, wait for that dose (don't double this dose) and resume regular schedule.

What drug does:
Relaxes artery walls and lowers blood pressure.

Time lapse before drug works:
60 to 90 minutes.

Don't take with:
Any other medicine or any dietary supplement without consulting your doctor or pharmacist.

OVERDOSE

SYMPTOMS:
Low blood pressure, fever, chills, sore throat, fainting, convulsions, coma.
WHAT TO DO:
- **Dial 911 for all medical emergencies or call poison control center 1-800-222-1222 for instructions.**
- **See emergency information on last 3 pages of this book.**

POSSIBLE ADVERSE REACTIONS OR SIDE EFFECTS

SYMPTOMS	WHAT TO DO
Life-threatening	
Rare, severe allergic reaction (anaphylaxis)—difficulty breathing, hives, itching, swelling (throat, face), vomiting, dizziness.	Seek emergency treatment immediately.
Common:	
Rash, loss of taste.	Discontinue. Call doctor right away.
Infrequent:	
• Swelling of mouth, face, hands or feet.	Discontinue. Seek emergency treatment.
• Dizziness, fainting, chest pain, fast or irregular heartbeat, confusion, nervousness, numbness and tingling in hands or feet.	Discontinue. Call doctor right away.
• Diarrhea, headache, tiredness, cough.	Continue. Call doctor when convenient.
Rare:	
• Sore throat, cloudy urine, fever, chills.	Discontinue. Call doctor right away.
• Nausea, vomiting, indigestion, abdominal pain.	Continue. Call doctor when convenient.

WARNINGS & PRECAUTIONS

Don't take if:
- You are allergic to any ACE inhibitor.*
- You are receiving blood from a blood bank.
- You will have surgery within 2 months, including dental surgery, requiring general or spinal anesthesia.

Before you start, consult your doctor if:
- You have had a stroke.
- You have angina or heart or blood vessel disease.
- You have any autoimmune disease, including AIDS or lupus.
- You have high level of potassium in blood.
- You have kidney or liver disease.
- You are on severe salt-restricted diet.
- You have a bone marrow disorder.

Over age 60:
Your doctor may start drug with a lower dosage that will gradually be increased.

ANGIOTENSIN-CONVERTING ENZYME (ACE) INHIBITORS

Pregnancy:
Consult doctor. Use of the drug in pregnancy is not recommended. It could cause harm to the unborn baby. Risk category D (see page xviii).

Breastfeeding; Lactation; Nursing Mothers:
Drugs in this group are generally acceptable with breastfeeding. Discuss risks and benefits with your doctor.

Infants & children up to age 18:
Follow instructions provided by your child's doctor.

Prolonged use:
Request periodic laboratory blood counts and urine tests.

Skin & sunlight:
One or more drugs in this group may cause rash or intensify sunburn in areas exposed to sun or ultraviolet light (photosensitivity reaction). Avoid overexposure. Notify doctor if reaction occurs.

Driving, piloting or hazardous work:
Avoid if you become dizzy or faint. Otherwise, no problems expected.

Discontinuing:
Don't discontinue without consulting doctor. Dose may require gradual reduction if you have taken drug for a long time. Doses of other drugs may also require adjustment.

Others:
- Use caution when exercising in hot weather.
- Advise any doctor, dentist or pharmacist whom you consult that you take this medicine.

POSSIBLE INTERACTION WITH OTHER DRUGS

GENERIC NAME OR DRUG CLASS	COMBINED EFFECT
Amiloride	Possible excessive potassium in blood.
Antihypertensives, other*	Increased anti-hypertensive effect. Dosage of each may require adjustment.
Anti-inflammatory drugs nonsteroidal (NSAIDs), cox-2 inhibitors	May decrease ACE inhibitor effect.
Beta-adrenergic blocking agents*	Increased anti-hypertensive effect. Dosage of each may require adjustment.
Carteolol	Increased effect of both drugs. Dosages may require adjustment.
Chloramphenicol	Possible blood disorders.
Diuretics*	Possible severe blood pressure drop with first dose.
Diclofenac	May decrease ACE inhibitor effect.
Guanfacine	Increased effect of both drugs.
Meloxicam	Decreased effect of ACE inhibitor.
Nicardipine	Possible excessive potassium in blood. Dosages may require adjustment.
Nimodipine	Possible excessive potassium in blood. Dangerous blood pressure drop.
Nitrates*	Possible excessive blood pressure drop.
Pentamidine	May increase bone marrow depression or make kidney damage more likely.
Pentoxifylline	Increased anti-hypertensive effect.
Potassium	May raise potassium levels in blood to toxic levels.

Continued on page 885

POSSIBLE INTERACTION WITH OTHER SUBSTANCES

INTERACTS WITH	COMBINED EFFECT
Alcohol:	Possible excessive blood pressure drop.
Beverages: Low-salt milk.	Possible excessive potassium in blood.
Cocaine	Risk of heart block and high blood pressure. Avoid.
Foods: Salt substitutes.	Possible excessive potassium.
Marijuana:	Consult doctor.
Tobacco:	May decrease ACE inhibitor effect.

*See Glossary

ANTACIDS

GENERIC AND BRAND NAMES

See full list of generic and brand names in the *Generic and Brand Name Directory*, page 837.

BASIC INFORMATION

Habit forming? No
Prescription needed? No
Available as generic? Yes, for some
Drug class: Antacid

USES

Treatment for hyperacidity in upper gastrointestinal tract, including stomach and esophagus. Symptoms may be heartburn or acid indigestion. Diseases include peptic ulcer, gastritis, esophagitis, hiatal hernia.

DOSAGE & USAGE INFORMATION

How to take:
Follow package instructions.

When to take:
1 to 3 hours after meals unless directed otherwise by your doctor.

If you forget a dose:
Take as soon as you remember, but not simultaneously with any other medicine.

What drug does:
- Neutralizes some of the hydrochloric acid in the stomach.
- Reduces action of pepsin, a digestive enzyme.

Time lapse before drug works:
15 minutes for antacid effect.

Continued next column

OVERDOSE

SYMPTOMS:
Diarrhea or constipation, nausea, excess thirst, stomach pain, other symptoms.
WHAT TO DO:
If person uses much larger amount than prescribed or if accidentally swallowed, call poison control center 1-800-222-1222 for instructions or dial 911 (emergency) for help.

Don't take with:
- Any other medicine or any dietary supplement without consulting your doctor or pharmacist.
- Other drugs at the same time. Decreases absorption of that drug. Wait 2 hours.

POSSIBLE ADVERSE REACTIONS OR SIDE EFFECTS

SYMPTOMS	WHAT TO DO
Life-threatening: None expected.	
Common: Chalky taste.	Continue. Call doctor when convenient.
Infrequent: Mild constipation, increased thirst, laxative effect, bad taste in mouth, stomach cramps, stool color changes.	Continue. Call doctor when convenient.
Rare: Bone pain, frequent or urgent urination, muscle weakness or pain, nausea, weight gain, dizziness, severe constipation, headache, appetite loss, mood changes, vomiting, nervousness, swollen feet and ankles, tiredness or weakness.	Discontinue. Call doctor right away.

Note: Side effects are rare unless too much medicine is taken for a long time.

WARNINGS & PRECAUTIONS

Don't take if:
- You are allergic to any antacid. Allergic reactions are rare, but may occur.
- You have a high blood-calcium level.

Before you start, consult your doctor if:
You have kidney disease, chronic constipation, colitis, diarrhea, symptoms of appendicitis, stomach or intestinal bleeding, irregular heartbeat.

Over age 60:
Usually no problems with occasional use. Advise your doctor if you use antacids regularly.

Pregnancy:
Generally acceptable with pregnancy. Discuss risks and benefits with your doctor before using.

Breastfeeding; Lactation; Nursing Mothers:
Drugs in this group are generally acceptable with breastfeeding. Discuss risks and benefits with your doctor.

Infants & children up to age 18:
Read product's label to be sure it is approved for your child's age. Follow label instructions on dosage. Consult doctor or pharmacist if unsure.

Prolonged use:
- High blood level of calcium (if your antacid contains calcium) which can disturb electrolyte balance.
- Kidney stones, impaired kidney function.
- Talk to your doctor about the need for follow-up medical examinations or laboratory studies.

Skin & sunlight:
No problems expected.

Driving, piloting or hazardous work:
No problems expected.

Discontinuing:
May be unnecessary to finish medicine. Follow doctor's instructions.

Others:
- Don't take longer than 2 weeks unless under medical supervision.
- Advise any doctor, dentist or pharmacist whom you consult that you take this medicine. May affect results in some medical tests.

POSSIBLE INTERACTION WITH OTHER DRUGS

GENERIC NAME OR DRUG CLASS	COMBINED EFFECT
Alendronate	Decreased alendronate effect. Take antacid 30 minutes after alendronate.
Antifungals, azoles	Decreased azole absorption.
Anti-inflammatory drugs nonsteroidal (NSAIDs), COX-2 inhibitors	Decreased pain relief.
Capecitabine	Increased risk of capecitabine toxicity.
Chlorpromazine	Decreased chlorpromazine effect.
Ciprofloxacin	May cause kidney dysfunction.
Dexamethasone	Decreased dexa-methasone effect.
Digitalis preparations*	Decreased digitalis effect.
Iron supplements*	Decreased iron effect.
Isoniazid	Decreased isoniazid effect.
Levodopa	Increased levodopa effect.
Mecamylamine	Increased mecamylamine effect.
Meperidine	Increased meperidine effect.
Methenamine	Reduced methenamine effect.
Nalidixic acid	Decreased nalidixic acid effect.
Nicardipine	Possible decreased nicardipine effect.
Nizatidine	Decreased nizatidine absorption.
Ofloxacin	Decreased ofloxacin effect.
Oxyphenbutazone	Decreased oxyphenbutazone effect.
Para-aminosalicylic acid (PAS)	Decreased PAS effect.
Penicillins*	Decreased penicillin effect.
Prednisone	Decreased prednisone effect.
Pseudoephedrine	Increased pseudo-ephedrine effect.
Salicylates*	Increased salicylate effect.
Tetracyclines	Decreased tetracycline effect.
Ticlopidine	Decreased ticlopidine effect.

POSSIBLE INTERACTION WITH OTHER SUBSTANCES

INTERACTS WITH	COMBINED EFFECT
Alcohol:	Decreased antacid effect.
Beverages:	No proven problems.
Cocaine:	Unknown. Avoid.
Foods:	Decreased antacid effect. Wait 1 hour after eating.
Marijuana:	Consult doctor.
Tobacco:	None expected.

***See Glossary**

ANTHELMINTICS

GENERIC AND BRAND NAMES

ALBENDAZOLE
- Albenza

IVERMECTIN
- Stromectol

PYRANTEL
- Antiminth
- Aut
- Cobantril
- Helmex
- Lombriareu
- Reese's Pinworm Medicine
- Trilombrin

THIABENDAZOLE
- Foldan
- Mintezol
- Mintezol Topical
- Minzolum
- Triasox

BASIC INFORMATION

Habit forming? No
Prescription needed? Yes
Available as generic? Yes
Drug class: Anthelmintics, antiparasitic

USES

- Treatment of roundworms, pinworms, whipworms, hookworms and other intestinal parasites.
- Treatment of hydatid disease, neurocysticercosis, strongyloidiasis and onchocerciasis.

DOSAGE & USAGE INFORMATION

How to take or apply:
- Tablet—Swallow with liquid or food to lessen stomach irritation.
- Topical suspension—Apply to end of each tunnel or burrow made by worm.
- Chewable tablet—Chew thoroughly before swallowing.
- Oral suspension—Follow package instructions.

Continued next column

OVERDOSE

SYMPTOMS:
Increased severity of adverse reactions and side effects.
WHAT TO DO:
If person uses much larger amount than prescribed or if accidentally swallowed, call poison control center 1-800-222-1222 for instructions or dial 911 (emergency) for help.

When to take:
Morning and evening with food to increase uptake.

If you forget a dose:
Skip dose and begin treatment again. Often only one or two doses are needed to complete treatment.

What drug does:
Kills or paralyzes the parasites. They then pass out of the body in the feces. Usually the type of worm parasite must be identified so the appropriate drug can be prescribed.

Time lapse before drug works:
Some take only hours; others, 1 to 3 days.

Don't take with:
Any other medicine or any dietary supplement without consulting your doctor or pharmacist.

POSSIBLE ADVERSE REACTIONS OR SIDE EFFECTS

SYMPTOMS	WHAT TO DO
Life-threatening: None expected.	
Common: None expected.	
Infrequent: Abdominal pain, diarrhea, dizziness, nausea, rectal itching, headache.	Continue. Call doctor when convenient.
Rare:	
• Skin rash, itching, sore throat, fever, unusual weakness or tiredness.	Discontinue. Call doctor right away.
• Hair loss.	Continue. Call doctor when convenient.

WARNINGS & PRECAUTIONS

Don't take if:
You are allergic to any anthelmintic. Allergic reactions are rare, but may occur.

Before you start, consult your doctor if:
- You have liver disease.
- You have Crohn's disease.
- You have ulcerative colitis.

Over age 60:
Adverse reactions and side effects may be more frequent and severe than in younger persons.

Pregnancy:
Decide with your doctor if drug benefits justify risk to unborn baby. Risk category C (see page xviii).

Breastfeeding; Lactation; Nursing Mothers:
Drugs in this group are generally acceptable with breastfeeding. Discuss risks and benefits with your doctor.

Infants & children up to age 18:
Follow instructions provided by your child's doctor.

Prolonged use:
- Not intended for long-term use.
- Talk to your doctor about the need for follow-up medical examinations or laboratory studies to check stools, cellophane tape swabs pressed against rectal area to check for parasite eggs, complete blood counts (white blood cell count, platelet count, red blood cell count, hemoglobin, hematocrit).

Skin & sunlight:
Thiabendazole may cause rash or intensify sunburn in areas exposed to sun or ultraviolet light (photosensitivity reaction). Avoid overexposure. Notify doctor if reaction occurs.

Driving, piloting or hazardous work:
Use caution if the medicine causes you to feel dizzy or weak. Otherwise, no problems expected.

Discontinuing:
No problems expected.

Others:
- Take full course of treatment. Repeat course may be necessary if follow-up examinations reveal persistent infection.
- Advise any doctor, dentist or pharmacist whom you consult that you take this medicine.
- Wash all bedding after treatment to prevent re-infection.

POSSIBLE INTERACTION WITH OTHER DRUGS

GENERIC NAME OR DRUG CLASS	COMBINED EFFECT
Cimetidine	Increased effect of albendazole.
Corticosteroids*	Increased effect of albendazole.
Praziquantel	Increased effect of albendazole.
Xanthines*	Dosage of xanthine drug may need adjusting.

POSSIBLE INTERACTION WITH OTHER SUBSTANCES

INTERACTS WITH	COMBINED EFFECT
Alcohol:	None expected.
Beverages: Grapefruit juice.	May increase effect of anthelmintic.
Cocaine:	Unknown. Avoid.
Foods:	None expected.
Marijuana:	Consult doctor.
Tobacco:	None expected.

***See Glossary**

ANTHRALIN (Topical)

BRAND NAMES

Anthra-Derm
Anthraforte
Anthranol
Anthrascalp
Dithranol
Drithocreme
Drithocreme HP
Dritho-Scalp
Lasan
Lasan HP
Lasan Pomade
Lasan Unguent
Micanol

BASIC INFORMATION

Habit forming? No
Prescription needed? Yes
Available as generic? No
Drug class: Antipsoriatic, hair growth stimulant

USES

- Treats quiescent or chronic psoriasis.
- Stimulates hair growth in some people (not an approved use by the FDA).

DOSAGE & USAGE INFORMATION

How to use:

- Wear plastic gloves for all applications.
- If directed, apply at night.
- Cream, lotion, ointment—Bathe and dry area before use. Apply small amount and rub gently.
- If for short contact, same as above for cream.
- Leave on 20 to 30 minutes. Then remove medicine by bathing or shampooing.
- If for scalp overnight—Shampoo before use to remove scales or medicine. Dry hair. Part hair several times and apply to scalp. Wear plastic cap on head. Clean off next morning with petroleum jelly, then shampoo.

When to use:
As directed.

Continued next column

OVERDOSE

SYMPTOMS:
Unknown.
WHAT TO DO:
If person uses much larger amount than prescribed or if accidentally swallowed, call poison control center 1-800-222-1222 for instructions or dial 911 (emergency) for help.

If you forget a dose:
Use as soon as you remember. If it is almost time for the next dose, wait for that dose (don't double this dose) and resume regular schedule.

What drug does:
Reduces growth activity within abnormal cells by inhibiting enzymes.

Time lapse before drug works:
May require several weeks or more.

Don't use with:
Any other medicine or any dietary supplement without consulting your doctor or pharmacist.

POSSIBLE ADVERSE REACTIONS OR SIDE EFFECTS

SYMPTOMS	WHAT TO DO
Life-threatening: None expected.	
Common: None expected.	
Infrequent: Redness or irritation of skin not present before application, rash.	Discontinue. Call doctor when convenient.
Rare: None expected.	

WARNINGS & PRECAUTIONS

Don't use if:
- You are allergic to anthralin. Allergic reactions are rare, but may occur.
- You have infected skin.

Before you start, consult your doctor if:
- You have chronic kidney disease.
- You are allergic to anything.

Over age 60:
No problems expected, but check with doctor.

Pregnancy:
Decide with your doctor if drug benefits justify risk to unborn baby. Risk category C (see page xviii).

Breastfeeding; Lactation; Nursing Mothers:
This drug is generally acceptable with breastfeeding. Discuss risks and benefits with your doctor.

Infants & children up to age 18:
Safety and effectiveness in this age group have not been established. Consult doctor.

Prolonged use:
No problems expected, but check with doctor.

Skin & sunlight:
May cause rash or intensify sunburn in areas exposed to sun or ultraviolet light (photosensitivity reaction). Avoid overexposure. Notify doctor if reaction occurs.

Driving, piloting or hazardous work:
No problems expected.

Discontinuing:
No problems expected.

Others:
- Keep cool, but don't freeze.
- Apply petroleum jelly to normal skin or scalp to protect areas not being treated.
- Will stain hair, clothing, shower, bathtub or sheets. Wash as soon as possible.
- Advise any doctor, dentist or pharmacist whom you consult that you take this medicine.
- Heat and moisture in bathroom medicine cabinet can cause breakdown of medicine. Store someplace else.

POSSIBLE INTERACTION WITH OTHER DRUGS

GENERIC NAME OR DRUG CLASS	COMBINED EFFECT
Antidiabetic agents*	Increased sensitivity to sun exposure.
Coal tar preparations*	Increased sensitivity to sun exposure.
Diuretics, thiazide*	Increased sensitivity to sun exposure.
Griseofulvin	Increased sensitivity to sun exposure.
Methoxsalen	Increased sensitivity to sun exposure.
Nalidixic acid	Increased sensitivity to sun exposure.
Phenothiazines*	Increased sensitivity to sun exposure.
Sulfa drugs*	Increased sensitivity to sun exposure.
Tetracyclines*	Increased sensitivity to sun exposure.
Trioxsalen	Increased sensitivity to sun exposure.

POSSIBLE INTERACTION WITH OTHER SUBSTANCES

INTERACTS WITH	COMBINED EFFECT
Alcohol:	None expected.
Beverages:	None expected.
Cocaine:	Unknown. Avoid.
Foods:	None expected.
Marijuana:	Consult doctor.
Tobacco:	None expected.

*See Glossary

ANTIACNE, CLEANSING (Topical)

GENERIC AND BRAND NAMES

ALCOHOL & ACETONE
Seba-Nil
ALCOHOL & SULFUR
Liquimat
Postacne
SULFURATED LIME
Vlemasque
Vleminckx Solution

BASIC INFORMATION

Habit forming? No
Prescription needed? No
Available as generic? Yes
Drug class: Antiacne agent, cleansing agent

USES

Treats acne or oily skin.

DOSAGE & USAGE INFORMATION

How to use:

- Lotion, gel or pledget—Start with small amount and wipe over face to remove dirt and surface oil. Don't apply to wounds or burns. Don't rinse with water and avoid contact with eyes. Skin may be more sensitive in dry or cold climates.
- Plaster—Follow package instructions.

When to use:
As directed. May increase frequency up to 3 or more times daily as tolerated. Warm, humid weather may allow more frequent use.

If you forget a dose:
Use as soon as you remember. If it is almost time for the next dose, wait for that dose (don't double this dose) and resume regular schedule.

Continued next column

OVERDOSE

SYMPTOMS:
Unknown.
WHAT TO DO:
If person uses much larger amount than prescribed or if accidentally swallowed, call poison control center 1-800-222-1222 for instructions or dial 911 (emergency) for help.

What drug does:
Helps remove oil from skin's surface.

Time lapse before drug works:
Works immediately.

Don't use with:
Other topical acne treatments unless directed by doctor.

POSSIBLE ADVERSE REACTIONS OR SIDE EFFECTS

SYMPTOMS	WHAT TO DO
Life-threatening: None expected.	
Common: None expected.	
Infrequent: None expected.	
Rare: Skin infection, redness, peeling. dryness.	Discontinue. Call doctor if symptoms persist

WARNINGS & PRECAUTIONS

Don't use if:
You have to apply over a wounded or burned area.

Before you start, consult your doctor if:
You use benzoyl peroxide, resorcinol, salicylic acid, sulfur or tretinoin (vitamin A acid).

Over age 60:
No special problems expected.

Pregnancy:
Pregnancy risk category not assigned to this drug group. Consult doctor about use.

Breastfeeding; Lactation; Nursing Mothers:
Drugs in this group are generally acceptable with breastfeeding. Discuss risks and benefits with your doctor.

Infants & children up to age 18:
Read product's label to be sure it is approved for your child's age. Follow label instructions on dosage. Consult doctor or pharmacist if unsure.

Prolonged use:
Excessive drying of skin.

Skin & sunlight:
No special problems expected.

Driving, piloting or hazardous work:
No problems expected, but check with doctor.

Discontinuing:
No problems expected, but check with doctor.

Others:
Some antiacne agents are flammable. Don't use near fire or while smoking.

POSSIBLE INTERACTION WITH OTHER DRUGS

GENERIC NAME OR DRUG CLASS	COMBINED EFFECT
Abrasive or medicated soaps	Irritation or too much drying.
After-shave lotions	Irritation or too much drying.
Antiacne topical preparations (other)	Irritation or too much drying.
"Cover-up" cosmetics	Irritation or too much drying.
Drying cosmetic soaps	Irritation or too much drying.
Isotretinoin	Irritation or too much drying.
Mercury compounds	May stain skin black and smell bad.
Perfumed toilet water	Irritation or too much drying.
Preparations containing skin-peeling agents such as benzoyl peroxide, resorcinol, salicylic acid, sulfur, tretinoin	Irritation or too much drying.

POSSIBLE INTERACTION WITH OTHER SUBSTANCES

INTERACTS WITH	COMBINED EFFECT
Alcohol:	None expected.
Beverages:	None expected.
Cocaine:	None expected.
Foods:	None expected.
Marijuana:	None expected.
Tobacco:	None expected.

*See Glossary

ANTIALLERGIC AGENTS (Ophthalmic)

GENERIC AND BRAND NAMES

ALCAFTADINE
Lastacaft
AZELASTINE (ophthalmic)
Optivar
BEPOTASTINE
Bepreve
EMEDASTINE
Emadine
EPINASTINE
Elestat
KETOTIFEN
Alaway
Claritin Eye
Refresh Eye Itch Relief
Zaditor
Zyrtec Eye Drops
LEVOCABASTINE
Livostin
LODOXAMIDE
Alomide
NEDOCROMIL
Alocril
OLOPATADINE
Pataday
Patanol
Pazeo
PEMIROLAST
Alamast

BASIC INFORMATION

Habit forming? No
Prescription needed? Yes, for some
Available as generic? Yes, for some
Drug class: Ophthalmic antiallergic agents, antihistaminic

USES

Prevention and treatment of seasonal allergic (hay fever) eye disorders. May be referred to as seasonal conjunctivitis, vernal conjunctivitis, vernal keratitis or vernal keratoconjunctivitis.

DOSAGE & USAGE INFORMATION

How to use:
Eye solution
- Wash hands.
- Apply pressure to inside corner of eye with middle finger.
- Continue pressure for 1 minute after placing medicine in eye.

Continued next column

- Tilt head backward. Pull lower lid away from eye with index finger of the same hand.
- Drop eye drops into pouch and close eye. Don't blink.
- Keep eyes closed for 1 to 2 minutes.

When to use:
1 to 2 drops 4 times a day or as directed by doctor or instructions on product.

If you forget a dose:
Use as soon as you remember. If it is almost time for the next dose, wait for that dose (don't double this dose) and resume regular schedule.

What drug does:
Prevents a hypersensitivity reaction to certain allergens such as pollen.

Time lapse before drug works:
Relief of symptoms may begin immediately, but full benefit might take a few days.

Don't use with:
Any other eye medications without consulting your doctor or pharmacist.

OVERDOSE

SYMPTOMS:
Unknown.
WHAT TO DO:
If person uses much larger amount than prescribed or if accidentally swallowed, call poison control center 1-800-222-1222 for instructions or dial 911 (emergency) for help.

POSSIBLE ADVERSE REACTIONS OR SIDE EFFECTS

SYMPTOMS	WHAT TO DO
Life-threatening: None expected.	
Common: Brief and mild burning or stinging when drops are administered.	No action necessary.
Infrequent: Blurred vision, feeling that something is in the the eye, redness of eye, eye irritation not present before, eye tearing or discharge.	Discontinue. Call doctor right away.
Rare: • Pain or swelling of eye, sensitivity to light, headache, dizziness, skin rash.	Discontinue. Call doctor right away.
• Crusting in corner of eye or eyelid, other eye symptoms.	Continue. Call doctor when convenient.

WARNINGS & PRECAUTIONS

Don't use if:
You are allergic to any of the antiallergic ophthalmic drugs.

Before you start, consult your doctor if:
- You wear soft contact lenses.
- You are allergic to any other medications, foods or other substances.

Over age 60:
No special problems expected.

Pregnancy:
Risk category B for azelastine, emedastine, levocabastine, lodoxamide, nedocromil and pemirolast. Risk category C for epinastine, olopatadine and ketotifen. See page xviii for category information. Consult doctor about use.

Breastfeeding; Lactation; Nursing Mothers:
Drugs in this group are generally acceptable with breastfeeding. Discuss risks and benefits with your doctor.

Infants & children up to age 18:
Follow instructions provided by doctor for prescription product. Read label on non-prescription product to be sure it's approved for your child's age.

Prolonged use:
No special problems expected.

Skin & sunlight:
No special problems expected.

Driving, piloting or hazardous work:
Avoid if you feel dizzy or side effects cause vision problems.

Discontinuing:
No special problems expected.

Others:
- Don't use leftover medicine for other eye problems without your doctor's approval.
- If symptoms don't improve after a few days of use, call your doctor.

POSSIBLE INTERACTION WITH OTHER DRUGS

GENERIC NAME OR DRUG CLASS	COMBINED EFFECT
None significant.	

POSSIBLE INTERACTION WITH OTHER SUBSTANCES

INTERACTS WITH	COMBINED EFFECT
Alcohol:	None expected.
Beverages:	None expected.
Cocaine	None expected.
Foods:	None expected.
Marijuana:	None expected.
Tobacco:	None expected.

*See Glossary

ANTIANDROGENS, NONSTEROIDAL

GENERIC AND BRAND NAMES

BICALUTAMIDE
Casodex
FLUTAMIDE
Euflex
Eulexin
NILUTAMIDE
Amandron
Nilandron

BASIC INFORMATION

Habit forming? No
Prescription needed? Yes
Available as generic? No
Drug class: Antineoplastic

USES

Treatment for prostate cancer. Used in combination with a testosterone lowering measure such as surgery (removal of the testicles) or use of a special monthly injection of luteinizing hormone-releasing hormone (LHRH).

DOSAGE & USAGE INFORMATION

How to take:
Tablet or capsule—Swallow with liquid. May be taken with or without food.

When to take:
According to doctor's instructions. Normally at the same times each day.

If you forget a dose:
Take as soon as you remember. If it is almost time for the next dose, wait for that dose (don't double this dose) and resume regular schedule.

What drug does:
Interferes with utilization of androgen (male hormone) testosterone by body cells. Prostate cancer cells require testosterone in order to grow and reproduce.

Continued next column

OVERDOSE

SYMPTOMS:
Diarrhea, nausea, vomiting, tiredness, headache, dizziness, breast tenderness.
WHAT TO DO:
If person uses much larger amount than prescribed or if accidentally swallowed, call poison control center 1-800-222-1222 for instructions or dial 911 (emergency) for help.

Time lapse before drug works:
Starts working within two hours, but may take several weeks to be effective.

Don't take with:
Any other medicine or any dietary supplement without consulting your doctor or pharmacist.

POSSIBLE ADVERSE REACTIONS OR SIDE EFFECTS

SYMPTOMS	WHAT TO DO
Life-threatening:	
Rare, severe allergic reaction (anaphylaxis)—difficulty breathing, hives, itching, swelling (throat, face), vomiting, dizziness.	Seek emergency treatment immediately.
Common:	
• Decreased sex drive, diarrhea, appetite loss, nausea, vomiting, cough or hoarseness, fever, runny nose, sneezing, sore throat, tightness in chest or wheezing, constipation, insomnia.	Continue. Call doctor when convenient.
• Hot flashes with mild sweating.	No action necessary.
Infrequent:	
• Hands and feet tingling or numb, painful or swollen breasts, swollen feet and legs, chest pain, shortness of breath.	Continue, but call doctor right away.
• Bloody or black tarry stools, itching, back or side pain, depression, muscle weakness, unusual tiredness, skin rash, bloated feeling, confusion, dry mouth, nervousness, color vision changes (with nilutamide).	Continue. Call doctor when convenient.
Rare:	
Jaundice (yellow eyes and skin), pain or tenderness in the stomach, bluish colored lips or skin or nails, dark urine, dizziness or fainting, unusual bleeding or bruising.	Continue, but call doctor right away.

Note: Adverse effects may also be due to use of LHRH or prostate cancer symptoms.

ANTIANDROGENS, NONSTEROIDAL

WARNINGS & PRECAUTIONS

Don't take if:
You are allergic to any of the antiandrogens.

Before you start, consult your doctor if:
- You have liver disease.
- You use tobacco.
- You have lung disease or other breathing problems.
- You have glucose-6-phosphate dehydrogenase (G6PD) deficiency or hemoglobin M disease.
- You are planning on starting a family. May decrease sperm count.

Over age 60:
Adverse reactions and side effects may be more frequent and severe than in younger persons.

Pregnancy:
Consult doctor. These drugs are not intended for use in women.

Breastfeeding; Lactation; Nursing Mothers:
Not intended for use in women.

Infants & children up to age 18:
Not intended for use in infants and children.

Prolonged use:
Talk to your doctor about the need for follow-up medical examinations or laboratory studies to check liver and pulmonary functions, PSA levels, chest x-rays, and other tests as recommended.

Skin & sunlight:
May cause rash or intensify sunburn in areas exposed to sun or ultraviolet light (photosensitivity reaction). Avoid overexposure. Notify doctor if reaction occurs.

Driving, piloting or hazardous work:
You may experience vision problems when going from a dark area to a lighted area and vice versa (such as driving in and out of tunnels). Use caution.

Discontinuing:
No special problems expected. Don't discontinue drug without doctor's approval.

Others:
- Advise any doctor, dentist or pharmacist whom you consult that you take this medicine.
- May affect results in some medical tests.
- May decrease sperm count.

POSSIBLE INTERACTION WITH OTHER DRUGS

GENERIC NAME OR DRUG CLASS	COMBINED EFFECT
Anticoagulants*	Increased effect of anticoagulant.
Phenytoin	Increased effect of phenytoin with nilutamide.
Theophylline	Increased effect of theophylline.

POSSIBLE INTERACTION WITH OTHER SUBSTANCES

INTERACTS WITH	COMBINED EFFECT
Alcohol:	Nilutamide may cause alcohol intolerance reaction. Avoid alcohol while on this drug.
Beverages:	None expected.
Cocaine:	Unknown. Avoid.
Foods:	None expected.
Marijuana:	Consult doctor.
Tobacco:	Increased risk of toxicity. Avoid.

*See Glossary

ANTIARRHYTHMICS, BENZOFURAN-TYPE

GENERIC AND BRAND NAMES

AMIODARONE
Cordarone

DRONEDARONE
Multaq

BASIC INFORMATION

Habit forming? No
Prescription needed? Yes
Available as generic? Yes, for amiodarone
Drug class: Antiarrhythmic

USES

Prevents and treats certain types of life-threatening irregular heart rhythms.

DOSAGE & USAGE INFORMATION

How to take:
Tablet—Swallow whole with liquid or food to lessen stomach irritation. If you can't swallow tablet whole, ask doctor or pharmacist for advice.

When to take:
Amiodarone is usually taken once a day. Dronedarone is usually taken twice daily (with morning and evening meals).

If you forget a dose:
Take as soon as you remember. If it is almost time for the next dose, wait for that dose (don't double this dose) and resume regular schedule.

What drug does:
Antiarrhythmic drugs slow the electrical impulses in the heart to help restore, maintain or control normal heart rhythm.

Time lapse before drug works:
May take up to 2 weeks for therapeutic effect.

Don't take with:
Any other medicine or any dietary supplement without consulting your doctor or pharmacist.

OVERDOSE

SYMPTOMS:
Weakness, slow heart rate, lightheadedness, or fainting.

WHAT TO DO:
- **Dial 911 for all medical emergencies or call poison control center 1-800-222-1222 for instructions.**
- **See emergency information on last 3 pages of this book.**

POSSIBLE ADVERSE REACTIONS OR SIDE EFFECTS

SYMPTOMS	WHAT TO DO
Life-threatening:	
Rare, severe allergic reaction (anaphylaxis)—difficulty breathing, hives, itching, swelling (throat, face), vomiting, dizziness.	Seek emergency treatment immediately.
Common:	
• Painful breathing, cough, shortness of breath, coughing up blood.	Continue, but call doctor right away or seek emergency. treatment.
• Dizziness, fainting, lightheadedness, low fever, tingling or numbness in fingers or toes, hands tremble or shake, trouble walking, uncontrolled body movements.	Continue, but call doctor right away.
• Diarrhea, weakness, nausea or vomiting, tiredness, stomach pain, headache, loss of appetite, constipation.	Continue. Call doctor when convenient.
Infrequent:	
• Skin color change to blue-gray, unusual tiredness, difficulty breathing (can occur when lying down or sleeping), eye symptoms (blurred vision or less clarity, dry eyes, sensitive to light, seeing halos, other vision changes), scrotum swelling or pain, swollen feet or ankles or hands, fast or slow or irregular heartbeat, chest pain, weight gain or loss, sweating, sensitive to heat, feeling hot or cold.	Continue, but call doctor right away.
• Odd taste or smell, decreased libido, insomnia, flushed face, nervousness.	Continue. Call doctor when convenient.
Rare:	
• Yellow skin or eyes, severe stomach pain.	Continue, but call doctor right away.
• Skin symptoms (rash, itchy, red or puffy), heartburn, other new symptoms.	Continue. Call doctor when convenient.

ANTIARRHYTHMICS, BENZOFURAN-TYPE

WARNINGS & PRECAUTIONS

Don't take if:
You are allergic to amiodarone or dronedarone.

Before you start, consult your doctor if:
- You have diabetes; liver, kidney, thyroid, lung, or electrolyte (e.g., low potassium) disorder.
- You are pregnant or plan to become pregnant.
- You take herbal or vitamin supplements.
- You have any heart problem or heart disorder (e.g., slow heartbeat, congestive heart failure, high blood pressure, or stroke).

Over age 60:
No special problems expected.

Pregnancy:
Consult doctor. Drug could harm unborn baby. Use of amiodarone (risk category D) is not recommended and dronedarone (risk category X) should not be used in pregnancy. See page xviii.

Breastfeeding; Lactation; Nursing Mothers:
Drugs in this group are generally not recommended with breastfeeding. Discuss risks and benefits with your doctor.

Infants & children up to age 18:
Safety and effectiveness in this age group have not been established. Consult doctor.

Prolonged use:
- Blue-gray discoloration of skin may occur with amiodarone.
- Talk to your doctor about the need for follow-up medical exams or lab tests.

Skin & sunlight:
May cause rash or intensify sunburn in areas exposed to sun or ultraviolet light (called photosensitivity reaction). Avoid overexposure. Notify doctor if reaction occurs.

Driving, piloting or hazardous work:
Don't drive or pilot aircraft until you learn how medicine affects you. Don't work around dangerous machinery. Don't climb ladders or work in high places. Danger increases if you drink alcohol or take other medicines affecting alertness and reflexes.

Discontinuing:
- Don't discontinue without doctor's advice. Dose may require gradual reduction if you have taken drug for a long time. Doses of other drugs may also require adjustment.
- Notify doctor if cough, fever, breathing difficulty or other symptoms occur after discontinuing the drug.

Others:
- Call doctor right away if you have symptoms of lung toxicity (e.g., cough or painful breathing) or liver toxicity (e.g., yellow skin or eyes, swelling of feet and ankles).
- Advise any doctor, dentist or pharmacist whom you consult that you take this medicine.
- Dronedarone has been linked to increased risk of death, stroke, and heart failure in some patients. Ask your doctor about your risks.
- May interfere with the accuracy of some medical tests.
- It is important that you have regular eye exams before and during treatment.
- Carry or wear medical identification stating that you are taking this drug.

POSSIBLE INTERACTION WITH OTHER DRUGS

GENERIC NAME OR DRUG CLASS	COMBINED EFFECT
Antiarrhythmics, other*	Increased risk of irregular heartbeat.
Anticoagulants*	May increase anticoagulant effect.
Beta-adrenergic blocking agents*	Increased risk of slow heartbeat.
Calcium channel blockers*	Increased risk of slow heartbeat.
Cholestyramine	May decrease amiodarone effect.
Clopidogrel	May decrease effect of clopidogrel.
Cholestyramine	May decrease effect of amiodarone.

Continued on page 886

POSSIBLE INTERACTION WITH OTHER SUBSTANCES

INTERACTS WITH	COMBINED EFFECT
Alcohol:	None expected.
Beverages: Grapefruit juice.	Increased risk of benzofuran-type antiarrhythmic toxicity. Avoid.
Cocaine:	Unknown. Avoid.
Foods: Grapefruit.	Increased risk of benzofuran-type antiarrhythmic toxicity. Avoid.
Marijuana:	Consult doctor.
Tobacco:	None expected.

*See Glossary

ANTIBACTERIALS FOR ACNE (Topical)

GENERIC AND BRAND NAMES

CLINDAMYCIN (topical)
- Acanya
- Benzaclin
- Cleocin
- Clinda-Derm
- ClindaReach
- Dalacin T Topical Solution
- DUAC Topical Gel
- Evoclin Foam
- Onexton
- Veltin Gel
- Ziana Gel

ERYTHROMYCIN (topical)
- Akne-Mycin
- Benzamycin
- EryGel
- Erythro-statin
- Sans-Acne
- Theramycin Z

BASIC INFORMATION

Habit forming? No
Prescription needed? Yes
Available as generic? Yes
Drug class: Antibacterial (topical)

USES

Treats acne by killing skin bacteria that may be part of the cause of acne.

DOSAGE & USAGE INFORMATION

How to use:
- Pledgets and solutions are flammable. Use away from flame or heat.
- Apply drug to entire area, not just to pimples.
- If you use other acne medicines on skin, wait an hour after using erythromycin before applying other medicine.
- Cream, lotion, ointment—Wash and dry area Then apply small amount and rub gently.

When to use:
2 times a day, morning and evening, or as directed by your doctor.

Continued next column

OVERDOSE

SYMPTOMS:
Unknown.
WHAT TO DO:
If person uses much larger amount than prescribed or if accidentally swallowed, call poison control center 1-800-222-1222 for instructions or dial 911 (emergency) for help.

If you forget a dose:
Use as soon as you remember. If it is almost time for the next dose, wait for that dose (don't double this dose) and resume regular schedule.

What drug does:
Kills bacteria on skin, skin glands or in hair follicles.

Time lapse before drug works:
3 to 4 weeks to begin improvement.

Don't use with:
Other skin medicine without consulting your doctor or pharmacist.

POSSIBLE ADVERSE REACTIONS OR SIDE EFFECTS

SYMPTOMS	WHAT TO DO
Life-threatening:	
None expected.	
Common:	
Stinging or burning of skin for a few minutes after application.	No action needed.
Infrequent:	
Red, peeling, itching, irritated or dry skin.	Continue. Call doctor when convenient.
Rare (extremely):	
Symptoms of excess medicine absorbed by body—Abdominal pain, diarrhea, fever, nausea, vomiting, bloating, thirst, weakness, weight loss.	Discontinue. Call doctor right away.

WARNINGS & PRECAUTIONS

Don't use if:
You are allergic to clindamycin or erythromycin. Allergic reactions are rare, but may occur.

Before you start, consult your doctor if:
- You are allergic to any substance that touches your skin.
- You use benzoyl peroxide, resorcinol, salicylic acid, sulfur or tretinoin (vitamin A acid).

Over age 60:
No special problems expected.

Pregnancy:
Decide with your doctor if drug benefits justify risk to unborn baby. Risk category B (see page xviii).

Breastfeeding; Lactation; Nursing Mothers:
Drugs in this group are generally acceptable with breastfeeding. Discuss risks and benefits with your doctor.

Infants & children up to age 18:
Read product's label to be sure it is approved for your child's age. Follow label instructions on dosage. Consult doctor or pharmacist if unsure.

Prolonged use:
Excess irritation to skin.

Skin & sunlight:
No problems expected.

Driving, piloting or hazardous work:
No problems expected.

Discontinuing:
No problems expected, but check with doctor.

Others:
- Use water-base cosmetics.
- Keep medicine away from mouth or eyes.
- If accidentally gets into eyes, flush immediately with clear water.
- Keep medicine away from heat or flame.
- Keep medicine cool, but don't freeze.

POSSIBLE INTERACTION WITH OTHER DRUGS

GENERIC NAME OR DRUG CLASS	COMBINED EFFECT
Abrasive or medicated soaps	Irritation or too much drying.
After-shave lotions	Irritation or too much drying.
Antiacne topical preparations (other)	Irritation or too much drying.
"Cover-up" cosmetics	Irritation or too much drying.
Drying cosmetic soaps	Irritation or too much drying.
Isotretinoin	Irritation or too much drying.
Preparations containing skin peeling agents such as benzoyl peroxide, resorcinol, salicylic acid, sulfur, tretinoin	Irritation or too much drying.

POSSIBLE INTERACTION WITH OTHER SUBSTANCES

INTERACTS WITH	COMBINED EFFECT
Alcohol:	None expected.
Beverages:	None expected.
Cocaine:	None expected.
Foods:	None expected.
Marijuana:	None expected.
Tobacco:	None expected.

ANTIBACTERIALS (Ophthalmic)

GENERIC AND BRAND NAMES

See full list of generic and brand names in the *Generic and Brand Name Directory*, page 839.

BASIC INFORMATION

Habit forming? No
Prescription needed? Yes
Available as generic? Yes, for some
Drug class: Antibacterial (ophthalmic)

USES

- Helps body overcome eye infections on surface tissues of the eye.
- Treatment for corneal ulcers.

DOSAGE & USAGE INFORMATION

How to use:
Eye drops
- Wash hands.
- Apply pressure to inside corner of eye with middle finger.
- Continue pressure for 1 minute after placing medicine in eye.
- Tilt head backward. Pull lower lid away from eye with index finger of the same hand.
- Drop eye drops into pouch and close eye. Don't blink.
- Keep eyes closed for 1 to 2 minutes.

Eye ointment
- Wash hands.
- Pull lower lid down from eye to form a pouch.
- Squeeze tube to apply thin strip of ointment into pouch.
- Close eye for 1 to 2 minutes.
- Don't touch applicator tip to any surface (including the eye). If you accidentally touch tip, clean with warm soap and water.
- Keep container tightly closed.
- Wash hands immediately after using.

Continued next column

OVERDOSE

SYMPTOMS:
Unknown.
WHAT TO DO:
If person uses much larger amount than prescribed or if accidentally swallowed, call poison control center 1-800-222-1222 for instructions or dial 911 (emergency) for help.

When to use:
As directed. Don't miss doses.

If you forget a dose:
Use as soon as you remember. If it is almost time for the next dose, wait for that dose (don't double this dose) and resume regular schedule.

What drug does:
Penetrates bacterial cell membrane and prevents cells from multiplying.

Time lapse before drug works:
Begins in 1 hour. May require 7 to 10 days to control infection.

Don't use with:
Any other eye drops or ointment without checking with your doctor or pharmacist.

POSSIBLE ADVERSE REACTIONS OR SIDE EFFECTS

SYMPTOMS	WHAT TO DO
Life-threatening: None expected.	
Common: Ointments cause blurred vision for a few minutes.	Continue. Call doctor when convenient.
Infrequent:	
• Signs of irritation not present before drug use.	Discontinue. Call doctor right away.
• Burning or stinging of the eye.	Continue. Call doctor when convenient.
Rare: None expected.	

WARNINGS & PRECAUTIONS

Don't use if:
You are allergic to any antibiotic used in the eyes or on skin, ears, vagina and rectum.

Before you start, consult your doctor if:
You have had an allergic reaction to any medicine, food or other substances.

Over age 60:
No special problems expected.

Pregnancy:
Risk factors vary for drugs in this group. See category list on page xviii and consult doctor.

Breastfeeding; Lactation; Nursing Mothers:
Drugs in this group are generally acceptable with breastfeeding. Discuss risks and benefits with your doctor.

Infants & children up to age 18:
Read product's label to be sure it is approved for your child's age. Follow label instructions on dosage. Consult doctor or pharmacist if unsure.

Prolonged use:
Sensitivity reaction may develop.

Skin & sunlight:
No problems expected.

Driving, piloting or hazardous work:
No problems expected.

Discontinuing:
No problems expected.

Others:
- Notify doctor if symptoms fail to improve in 2 to 4 days.
- Keep medicine cool, but don't freeze.

POSSIBLE INTERACTION WITH OTHER DRUGS

GENERIC NAME OR DRUG CLASS	COMBINED EFFECT

Clinically significant interactions with oral or injected medicines unlikely.

POSSIBLE INTERACTION WITH OTHER SUBSTANCES

INTERACTS WITH	COMBINED EFFECT
Alcohol:	None expected.
Beverages:	None expected.
Cocaine:	None expected.
Foods:	None expected.
Marijuana:	None expected.
Tobacco:	None expected.

***See Glossary**

ANTIBACTERIALS (Otic)

GENERIC AND BRAND NAMES

ACETIC ACID
VoSol
ACETIC ACID & ALUMINUM ACETATE
ACETIC ACID & HYDROCORTISONE
Acetasol HC
VoSol HC
CIPROFLOXACIN (otic)
Cetraxal
CIPROFLOXACIN & DEXAMETHASONE
Ciprodex
CIPROFLOXACIN & HYDROCORTI-SONE
Cipro HC
FINAFLOXACIN
Xtoro
NEOMYCIN, COLISTIN & HYDROCORTI-SONE
Coly-Mycin S
NEOMYCIN, POLYMIXIN B & HYDROCORTI-SONE (otic)
Cortatrigen
Cort-Biotic
Cortisporin Otic
Drotic
Ear-Eze
Masporin Otic
Octigen
Oticair
Otimar
Otocidin
Otocor
Pediotic
OFLOXACIN (otic)
Floxin Otic

BASIC INFORMATION

Habit forming? No
Prescription needed? Yes
Available as generic? Yes, for some
Drug class: Antibacterial (otic)

USES

- Treatment for outer ear infection (called swimmer's ear or otitis externa).
- Treats certain acute or chronic middle ear infections (e.g., otitis media).
- Most of these drugs contain an antibacterial to fight the infection and an anti-inflammatory to help provide relief from redness, irritation and discomfort.

OVERDOSE

SYMPTOMS:
Unknown.
WHAT TO DO:
If person uses much larger amount than prescribed or if accidentally swallowed, call poison control center 1-800-222-1222 for instructions or dial 911 (emergency) for help.

DOSAGE & USAGE INFORMATION

How to use:
As directed by your doctor or pharmacist. The following are general instructions.
How to use ear drops:
- Wash and dry hands.
- Warm drops by holding container in your hand for a few minutes.
- Lie down with affected ear up.
- Adults—Pull ear lobe back and up.
- Children—Pull ear lobe down and back.
- Put the correct number of drops into the ear. Do not allow dropper to touch the ear.
- Wipe away any spilled drops.
- Stay lying down for 2 to 5 minutes.

When to use:
As directed on label. The number of daily doses will vary depending on the specific drug.

If you forget a dose:
Use as soon as you remember. If it is almost time for the next dose, wait for that dose (don't double this dose) and resume regular schedule.

What drug does:
Antibacterials destroy the bacteria causing the infection. Anti-inflammatories reduce symptoms of inflammation.

Time lapse before drug works:
Symptoms should improve within a few days. Complete healing of the infection will take longer.

Don't use with:
Other ear medications unless directed by your doctor or pharmacist.

POSSIBLE ADVERSE REACTIONS OR SIDE EFFECTS

SYMPTOMS	WHAT TO DO
Life-threatening:	
None expected.	
Common:	
None expected.	
Infrequent:	
Burning or stinging of the ear.	Continue. Call doctor if symptoms persist.
Rare:	
• Itching, redness, hives, swelling (allergic reaction), hearing changes, bleeding from ear.	Discontinue. Call doctor right away.
• Headache, fever, sore throat, runny or stuffy nose, taste changes, ear ringing.	Continue. Call doctor when convenient.

WARNINGS & PRECAUTIONS

Don't use if:
You are allergic to any of the drugs listed or other antibiotics, fluoroquinolones or steroid medications.

Before you start, consult your doctor if:
- Your eardrum is punctured.
- You have tendonitis (if ofloxacin prescribed).
- You have a viral infection such as chickenpox (varicella) or herpes simplex.

Over age 60:
No special problems expected.

Pregnancy:
Decide with your doctor if drug benefits justify risk to unborn child. Risk category C for most of these drugs (see page xviii).

Breastfeeding; Lactation; Nursing Mothers:
Drugs in this group are generally acceptable with breastfeeding. Discuss risks and benefits with your doctor.

Infants & children up to age 18:
Follow instructions provided by your doctor. Use the correct dosage for your infant or child's age and weight.

Prolonged use:
- Not intended for prolonged use. Don't use longer than prescribed by your doctor.
- Overuse or unnecessary use of the drug can lead to its decreased effectiveness in fighting infections.

Skin & sunlight:
No problems expected.

Driving, piloting or hazardous work:
Avoid if you experience dizziness or balance problems caused by the ear infection.

Discontinuing:
Don't discontinue without doctor's advice until you complete prescribed dosage.

Others:
- Follow your doctor's instructions for additional ear care at home.
- Call your doctor if ear symptoms worsen or don't improve after a few days of treatment.
- There is a slight risk of a secondary infection (one that occurs during or after treatment of another infection). Consult doctor if new, unexpected symptoms develop.
- Rarely, some of these drugs may increase the risk of hearing or balance problems. Discuss the drug's benefits and risks with your doctor.
- Advise any doctor, dentist or pharmacist whom you consult that you are using this drug.

POSSIBLE INTERACTION WITH OTHER DRUGS

GENERIC NAME OR DRUG CLASS	COMBINED EFFECT
None expected.	

POSSIBLE INTERACTION WITH OTHER SUBSTANCES

INTERACTS WITH	COMBINED EFFECT
Alcohol:	None expected.
Beverages:	None expected.
Cocaine:	None expected.
Foods:	None expected.
Marijuana:	None expected.
Tobacco:	None expected.

***See Glossary**

ANTIBACTERIALS (Topical)

GENERIC AND BRAND NAMES

MUPIROCIN
- Bactroban
- Bactroban Nasal

NEOMYCIN (Topical)
- Myciguent

NEOMYCIN, POLYMIXIN B & BACITRACIN
- Cortisporin Cream
- Cortisporin Ointment
- Triple Antibiotic

RETAPAMULIN
- Altabax

SILVER SULFADIAZINE
- Flamazine
- Silvadene
- SSD
- SSD AF
- Thermazene

BASIC INFORMATION

Habit forming? No
Prescription needed? Yes, for some
Available as generic? Yes, for some
Drug class: Antibacterial (topical)

USES

Treats skin infections that may accompany burns, superficial boils, insect bites or stings, skin ulcers, impetigo, minor surgical wounds.

DOSAGE & USAGE INFORMATION

How to use:
- Cream, lotion, ointment—First bathe and dry skin area. Apply small amount and rub gently. Cover with gauze or bandage if desired.
- Nasal ointment—Follow instructions provided.

When to use:
3 or 4 times daily, or as directed by doctor.

If you forget a dose:
Use as soon as you remember. If it is almost time for the next dose, wait for that dose (don't double this dose) and resume regular schedule.

What drug does:
Kills susceptible bacteria by interfering with bacterial DNA and RNA.

Continued next column

OVERDOSE

SYMPTOMS:
Unknown.
WHAT TO DO:
If person uses much larger amount than prescribed or if accidentally swallowed, call poison control center 1-800-222-1222 for instructions or dial 911 (emergency) for help.

Time lapse before drug works:
Begins first day. May require treatment for a week or longer to cure infection.

Don't use with:
Any other topical medicine without consulting your doctor or pharmacist.

POSSIBLE ADVERSE REACTIONS OR SIDE EFFECTS

SYMPTOMS	WHAT TO DO
Life-threatening:	
None expected.	
Common:	
None expected.	
Infrequent:	
• Itching or swollen or red skin or rash.	Discontinue. Call doctor right away.
• Nasal ointment may cause symptoms (stinging or burning in nose, headache, taste changes, sore throat, stuffy or runny nose).	Continue. Call doctor if symptoms don't go away.
Rare:	
Any loss of hearing, blistering or pain at application site.	Discontinue. Call doctor right away.

WARNINGS & PRECAUTIONS

Don't use if:
You are allergic to any topical antibacterial drug. Allergic reactions are rare, but may occur.

Before you start, consult your doctor if:
Any of the lesions on the skin are open sores.

Over age 60:
No special problems expected.

Pregnancy:
Decide with your doctor if drug benefits justify risk to unborn baby. Risk category B or C or unassigned (see page xviii).

Breastfeeding; Lactation; Nursing Mothers:
Drugs in this group are generally acceptable with breastfeeding. Discuss risks and benefits with your doctor.

Infants & children up to age 18:
Read product's label to be sure it is approved for your child's age. Follow label instructions on dosage. Consult doctor or pharmacist if unsure.

Prolonged use:
No problems expected, but check with doctor.

Skin & sunlight:
No problems expected.

Driving, piloting or hazardous work:
No problems expected.

Discontinuing:
No problems expected

Others:
- Heat and moisture in bathroom medicine cabinet can cause breakdown of medicine. Store someplace else.
- Keep medicine cool, but don't freeze.

POSSIBLE INTERACTION WITH OTHER DRUGS

GENERIC NAME OR DRUG CLASS	COMBINED EFFECT
Any other topical medication	Hypersensitivity* reactions more likely to occur.

POSSIBLE INTERACTION WITH OTHER SUBSTANCES

INTERACTS WITH	COMBINED EFFECT
Alcohol:	None expected.
Beverages:	None expected.
Cocaine:	None expected.
Foods:	None expected.
Marijuana:	None expected.
Tobacco:	None expected.

***See Glossary**

ANTICHOLINERGICS

GENERIC AND BRAND NAMES

HOMATROPINE
Hydromet
Tussigon
MEPENZOLATE
Cantil
METHSCOPOLAMINE
AH-Chew
Pamine
Pamine Forte
Prehist D

BASIC INFORMATION

Habit forming? No
Prescription needed?
Low strength: No
High strength: Yes
Available as generic? Yes
Drug class: Antispasmodic, anticholinergic

USES

- Reduces spasms of digestive system, bladder and urethra.
- Treatment of bronchial spasms.
- Used as a component in some cough and cold preparations.
- Treatment of peptic ulcers.

DOSAGE & USAGE INFORMATION

How to take:
Tablet—Swallow with liquid or food to lessen stomach irritation.

When to take:
30 minutes before meals (unless directed otherwise by doctor).

If you forget a dose:
Take as soon as you remember. If it is almost time for the next dose, wait for that dose (don't double this dose) and resume regular schedule.

Continued next column

OVERDOSE

SYMPTOMS:
Dilated pupils, rapid pulse and breathing, dizziness, fever, hallucinations, confusion, slurred speech, agitation, flushed face, convulsions, coma.
WHAT TO DO:
- **Dial 911 for all medical emergencies or call poison control center 1-800-222-1222 for instructions.**
- **See emergency information on last 3 pages of this book.**

If you forget a dose:
Take as soon as you remember. If it is almost time for the next dose, wait for that dose (don't double this dose) and resume regular schedule.

What drug does:
Blocks nerve impulses at parasympathetic nerve endings, preventing muscle contractions and gland secretions of organs involved.

Time lapse before drug works:
15 to 30 minutes.

Don't take with:
- Antacids* or antidiarrheals.*
- Any other medicine or any dietary supplement without consulting your doctor or pharmacist.

POSSIBLE ADVERSE REACTIONS OR SIDE EFFECTS

SYMPTOMS	WHAT TO DO
Life-threatening: In case of overdose, see previous column.	
Common:	
• Confusion, delirium, rapid heartbeat.	Discontinue. Call doctor right away.
• Nausea, vomiting, decreased sweating, constipation.	Continue. Call doctor when convenient.
• Dryness in ears, nose, throat, mouth.	No action necessary.
Infrequent:	
• Lightheadedness.	Discontinue. Call doctor right away.
• Headache, difficult or painful urination, nasal congestion, altered taste, increased sensitivity to light.	Continue. Call doctor when convenient.
Rare:	
Rash or hives, eye pain, blurred vision, fever.	Discontinue. Call doctor right away.

WARNINGS & PRECAUTIONS

Don't take if:
- You are allergic to any anticholinergic. Allergic reactions are rare, but may occur.
- You have trouble with stomach bloating.
- You have difficulty emptying your bladder completely.
- You have narrow-angle glaucoma.
- You have severe ulcerative colitis.

Before you start, consult your doctor if:
- You have open-angle glaucoma.
- You have angina or any heart disease or heart rhythm problem.
- You have chronic bronchitis or asthma.
- You have liver, kidney or thyroid disease.
- You have hiatal hernia or esophagitis.
- You have enlarged prostate or urinary retention.
- You have myasthenia gravis.
- You have peptic ulcer.
- You will have surgery within 2 months, including dental surgery, requiring general or spinal anesthesia.

Over age 60:
Adverse reactions and side effects may be more frequent and severe than in younger persons.

Pregnancy:
Decide with your doctor if drug benefits justify risk to unborn baby. Risk category C for homatropine and methscopolamine; category B for mepenzolate (see page xviii).

Breastfeeding; Lactation; Nursing Mothers:
Drugs in this group are generally acceptable with breastfeeding. Discuss risks and benefits with your doctor.

Infants & children up to age 18:
Safety and effectiveness in this age group have not been established. Consult doctor.

Prolonged use:
Chronic constipation, possible fecal impaction. Consult doctor immediately.

Skin & sunlight:
No special problems expected.

Driving, piloting or hazardous work:
Use disqualifies you for piloting aircraft. Otherwise, no problems expected.

Discontinuing:
Follow doctor's instructions.

Others:
Advise any doctor, dentist or pharmacist whom you consult that you take this medicine.

POSSIBLE INTERACTION WITH OTHER DRUGS

GENERIC NAME OR DRUG CLASS	COMBINED EFFECT
Adrenocorticoids, systemic	Possible glaucoma.
Amantadine	Increased anticholinergic effect.
Antacids*	Space doses of drugs 2-3 hours apart.
Anticholinergics, other*	Increased anticholinergic effect.
Antidepressants, tricyclic*	Increased anticholinergic effect. Increased sedation.
Antifungals, azoles	Decreased azole absorption.
Antihistamines*	Increased anticholinergic effect.
Attapulgite	Decreased anticholinergic effect.
Haloperidol	Increased internal eye pressure.
Methylphenidate	Increased anticholinergic effect.
Molindone	Increased anticholinergic effect.
Monoamine oxidase (MAO) inhibitors*	Increased anticholinergic effect.
Narcotics*	Increased risk of severe constipation.
Orphenadrine	Increased anticholinergic effect.
Phenothiazines*	Increased anticholinergic effect.
Potassium supplements*	Possible intestinal ulcers with oral potassium tablets.
Quinidine	Increased anticholinergic effect.

POSSIBLE INTERACTION WITH OTHER SUBSTANCES

INTERACTS WITH	COMBINED EFFECT
Alcohol:	None expected.
Beverages:	None expected.
Cocaine:	Increased risk of side effects. Avoid.
Foods:	None expected.
Marijuana:	Drowsiness and dry mouth. Avoid.
Tobacco:	None expected.

***See Glossary**

ANTICOAGULANTS (Oral)

GENERIC AND BRAND NAMES

WARFARIN
Coumadin
Jantoven

BASIC INFORMATION

Habit forming? No
Prescription needed? Yes
Available as generic? Yes
Drug class: Anticoagulant

USES

It is used to help prevent harmful blood clots from forming (or growing larger) in the blood vessels or heart. Preventing clots helps reduce the risk of stroke, heart attack or pulmonary embolism.

DOSAGE & USAGE INFORMATION

How to take:
Tablet—Swallow with liquid. It may be taken with or without food. If you can't swallow whole, crumble tablet and take with liquid or food.

When to take:
Once a day at the same time each day.

If you forget a dose:
Take as soon as you remember. If it is almost time for the next dose, wait for that dose (don't double this dose) and resume regular schedule.

What drug does:
It is in a class of drugs called anticoagulants (blood thinners). It works by decreasing the clotting ability of the blood.

Time lapse before drug works:
It will begin to work within 24 hours, but the full effect may take 3 to 5 days.

Don't take with:
Any other medicine or any dietary supplement without consulting your doctor or pharmacist. Many interactions are possible with warfarin.

OVERDOSE

SYMPTOMS:
Bloody vomit, coughing blood, bloody or black stools, red urine.
WHAT TO DO:
- **Dial 911 for all medical emergencies or call poison control center 1-800-222-1222 for instructions.**
- **See emergency information on last 3 pages of this book.**

POSSIBLE ADVERSE REACTIONS OR SIDE EFFECTS

SYMPTOMS	WHAT TO DO
Life-threatening:	
Rare, severe allergic reaction (anaphylaxis)—difficulty breathing, hives, itching, swelling (throat, face), vomiting, dizziness.	Seek emergency treatment immediately.
Common:	
None expected.	
Infrequent:	
• Black or red or tarry stools, nosebleeds, bleeding gums, red or brown urine, coughing up blood, vomiting blood or coffee-ground material, easy bruising, heavy menstrual flow, vaginal or rectal bleeding, cuts that won't stop bleeding, purple or red spots under the skin, discomfort or pain or swelling in any part of the body, headaches, dizziness, weakness.	Continue, but call doctor right away.
• Diarrhea, cramps, nausea or vomiting, bloating, rash, itch.	Continue. Call doctor when convenient.
Rare:	
• Purple or dark color of toes or foot, skin tissue death (pain, change in color or temperature in any body area), any bleeding that will not stop.	Seek emergency treatment.
• Yellow skin or eyes.	Continue, but call doctor right away.
• Changes in taste, tiredness, pale skin, loss of hair, feeling cold or having chills, fever, low blood pressure, any unusual symptoms that cause concern.	Continue. Call doctor when convenient.

WARNINGS & PRECAUTIONS

Don't take if:
You are allergic to any oral anticoagulant.

ANTICOAGULANTS (Oral)

Before you start, consult your doctor if:

- You have a bleeding disorder.
- You have an active ulcer.
- You have high blood pressure (hypertension).
- You have a stomach or intestinal infection.
- You have congestive heart failure or have had deep venous thrombosis (DVT) or a stroke.
- You drink alcohol or have problems with alcohol abuse.
- You have diabetes.
- You fall often.
- You are a female of reproductive age.
- You have a bladder catheter.
- You have protein C or protein S deficiency.
- You have liver or kidney disease.
- You have memory problems and may not be able to take drug as prescribed.
- You will have surgery in the near future (including eye or dental surgery).

Over age 60:
May have increased risk of side effects.

Pregnancy:
Consult doctor. Drug should not be used during pregnancy. Can cause harm to unborn baby. Risk category X (risk category D for women with mechanical heart valves). See page xviii.

Breastfeeding; Lactation; Nursing Mothers:
This drug is generally acceptable with breastfeeding. Discuss risks and benefits with your doctor.

Infants & children up to age 18:
Safety and effectiveness in this age group have not been established. Consult doctor.

Prolonged use:
Your doctor will schedule regular blood tests to monitor drug levels.

Skin & sunlight:
No problems expected.

Driving, piloting or hazardous work:
Avoid hazardous activities that could cause injury. Don't drive if you experience dizziness.

Discontinuing:
Don't discontinue drug without doctor's approval. Stopping the drug can increase the risk of blood clots and their complications.

Others:

- Advise any doctor, dentist or pharmacist whom you consult that you take this drug.
- Get regular blood tests to check your response to the drug.
- Your doctor or dentist may tell you to stop taking warfarin or change the dosage before surgery or a medical procedure.
- Carry or wear medical ID to state that you take this drug.
- Your genetic makeup may affect response to warfarin. Ask doctor about genetic testing.
- Use of this drug may cause severe bleeding (can be life-threatening).

POSSIBLE INTERACTION WITH OTHER DRUGS

GENERIC NAME OR DRUG CLASS	COMBINED EFFECT
Acetaminophen	Increased risk of bleeding.
Antibiotics*	Increased risk of bleeding.
Anticoagulants,* other	Increased risk of bleeding.
Antifungals*	Increased risk of bleeding.
Anti-inflammatory drugs, nonsteroidal (NSAIDs)*	Increased risk of bleeding.
Antiplatelet drugs*	Increased risk of bleeding.
Aspirin	Increased risk of bleeding.
Capecitabine	Increased risk of bleeding.
Cholestyramine	Decreased effect of warfarin.
Dietary supplements or herbal products	Increased risk of bleeding or decrease in warfarin effect.
Enzyme inducers*	Decreased effect of warfarin.
Enzyme inhibitors*	Increased effect of warfarin.
Selective serotonin reuptake inhibitors	Increased risk of bleeding.
Tamoxifen	Increased risk of bleeding.

POSSIBLE INTERACTION WITH OTHER SUBSTANCES

INTERACTS WITH	COMBINED EFFECT
Alcohol:	Increase or decrease warfarin effect. Avoid.
Beverages:	None expected.
Cocaine:	Bleeding risk Avoid.
Foods: High in vitamin K—green, leafy vegetables, Brussels sprouts, others.	Decreased anti-coagulant effect.
Marijuana:	Bleeding risk Avoid.
Tobacco:	Unclear effect.

*See Glossary

ANTICONVULSANTS, HYDANTOIN

GENERIC AND BRAND NAMES

ETHOTOIN
- Peganone

PHENYTOIN
- Dilantin
- Dilantin 30
- Dilantin 125
- Dilantin Infatabs
- Dilantin Kapseals
- Diphenylan

BASIC INFORMATION

Habit forming? No
Prescription needed? Yes
Available as generic? Yes
Drug class: Anticonvulsant (hydantoin)

USES

- Prevents some forms of epileptic seizures.
- Stabilizes irregular heartbeat.

DOSAGE & USAGE INFORMATION

How to take:
- Tablet—Swallow with liquid.
- Chewable tablet—Chew well before swallowing.
- Suspension—Shake solution well before taking with liquid.

When to take:
At the same time each day.

If you forget a dose:
- If drug taken 1 time per day—Take as soon as you remember. If it is almost time for the next dose, wait for the next scheduled dose (don't double this dose).
- If taken several times per day—Take as soon as possible, then return to regular schedule.

Continued next column

OVERDOSE

SYMPTOMS:
Jerky eye movements; stagger; slurred speech; imbalance; drowsiness; blood pressure drop; slow, shallow breathing; coma.

WHAT TO DO:
- **Dial 911 for all medical emergencies or call poison control center 1-800-222-1222 for instructions.**
- **See emergency information on last 3 pages of this book.**

What drug does:
Stabilizes electrical activity in the brain.

Time lapse before drug works:
7 to 10 days continual use.

Don't take with:
Any other medicine or any dietary supplement without consulting your doctor or pharmacist.

POSSIBLE ADVERSE REACTIONS OR SIDE EFFECTS

SYMPTOMS	WHAT TO DO
Life-threatening:	
Rare, severe allergic reaction (anaphylaxis)—difficulty breathing, hives, itching, swelling (throat, face), vomiting, dizziness.	Seek emergency treatment immediately.
Common:	
• Bleeding, swollen or tender gums.	Continue, but call doctor right away.
• Mild dizziness or drowsiness, constipation.	Continue. Call doctor when convenient.
Infrequent:	
• Hallucinations, confusion, stagger, fever, uncontrolled eye movements, increase in seizures, rash, change in vision, agitation, sore throat, diarrhea, slurred speech, muscle twitching.	Continue, but call doctor right away.
• Increased body and facial hair, breast swelling, insomnia, enlargement of facial features.	Continue. Call doctor when convenient.
Rare:	
Nausea, vomiting, unusual bleeding or bruising, swollen lymph nodes, stomach pain, yellow skin or eyes, joint pain, light color stools, loss of appetite, weight loss, trouble breathing, uncontrolled movements (arms, legs, hands, lips, tongue or cheeks), slowed growth, learning problems.	Continue, but call doctor right away.

WARNINGS & PRECAUTIONS

Don't take if:
You are allergic to any hydantoin anticonvulsant.

ANTICONVULSANTS, HYDANTOIN

Before you start, consult your doctor if:
- You have had impaired liver function or disease.
- You will have surgery within 2 months, including dental surgery, requiring general or spinal anesthesia.
- You have diabetes.
- You have a blood disorder.

Over age 60:
Adverse reactions and side effects may be more frequent and severe than in younger persons.

Pregnancy:
Decide with your doctor if drug benefits justify risk to unborn baby. Risk category C (see page xviii).

Breastfeeding; Lactation; Nursing Mothers:
Drugs in this group are generally acceptable with breastfeeding. Discuss risks and benefits with your doctor.

Infants & children up to age 18:
Follow instructions provided by your child's doctor.

Prolonged use:
- Weakened bones.
- Lymph gland enlargement.
- Possible liver damage.
- Numbness and tingling of hands and feet.
- Continual back-and-forth eye movements.
- Talk to your doctor about the need for follow-up medical exams or lab studies.

Skin & sunlight:
One or more drugs in this group may cause rash or intensify sunburn in areas exposed to sun or ultraviolet light (photosensitivity reaction). Avoid overexposure. Notify doctor if reaction occurs.

Driving, piloting or hazardous work:
Don't drive or pilot aircraft until you learn how medicine affects you. Don't work around dangerous machinery. Don't climb ladders or work in high places. Danger increases if you drink alcohol or take medicine affecting alertness and reflexes.

Discontinuing:
Don't discontinue without consulting doctor. Dose may require gradual reduction if you have taken drug for a long time. Doses of other drugs may also require adjustment.

Others:
- May cause learning disability.
- Good dental care is important while using this medicine.
- Advise any doctor, dentist or pharmacist whom you consult about the use of this drug.
- Rarely, antiepileptic drugs may lead to suicidal thoughts and behaviors. Call doctor right away if suicidal symptoms or unusual behaviors occur.

POSSIBLE INTERACTION WITH OTHER DRUGS

GENERIC NAME OR DRUG CLASS	COMBINED EFFECT
Adrenocorticoids, systemic	Decreased adreno-corticoid effect.
Amiodarone	Increased anticonvulsant effect.
Antacids*	Decreased anticonvulsant effect.
Antiandrogens, nonsteroidal	Increased effect of phenytoin.
Anticoagulants*	Increased effect of both drugs.
Antidepressants, tricyclic*	May need to adjust anticonvulsant dose.
Antifungals, azoles	Increased anticoagulant effect.
Antivirals, HIV/AIDS*	Increased risk of peripheral neuropathy with phenytoin.
Barbiturates*	Changed seizure pattern.
Calcium	Decreased effects of both drugs.
Carbamazepine	Possible increased anticonvulsant metabolism.
Carbonic anhydrase inhibitors*	Increased chance of bone disease.

Continued on page 886

POSSIBLE INTERACTION WITH OTHER SUBSTANCES

INTERACTS WITH	COMBINED EFFECT
Alcohol:	Possible decreased anticonvulsant effect. Use with caution.
Beverages:	None expected.
Cocaine:	Unknown. Avoid.
Foods:	None expected.
Marijuana:	Drowsiness, unsteadiness. Avoid.
Tobacco:	None expected.

***See Glossary**

ANTICONVULSANTS, SUCCINIMIDE

GENERIC AND BRAND NAMES

ETHOSUXIMIDE	**METHSUXIMIDE**
Zarontin	**Celontin**

BASIC INFORMATION

Habit forming? No
Prescription needed? Yes
Available as generic? Yes, for some
Drug class: Anticonvulsant (succinimide)

USES

Controls seizures in treatment of some forms of epilepsy.

DOSAGE & USAGE INFORMATION

How to take:
Capsule or syrup—Swallow with liquid or food to lessen stomach irritation.

When to take:
Every day in regularly spaced doses, according to prescription.

If you forget a dose:
Take as soon as you remember. If it is almost time for the next dose, wait for that dose (don't double this dose) and resume regular schedule.

What drug does:
Depresses nerve transmissions in part of brain that controls muscles.

Time lapse before drug works:
3 hours.

Don't take with:
Any other medicine or any dietary supplement without consulting your doctor or pharmacist.

OVERDOSE

SYMPTOMS:
Severe drowsiness, slow or irregular breathing, coma.
WHAT TO DO:
- **Dial 911 for all medical emergencies or call poison control center 1-800-222-1222 for instructions.**
- **See emergency information on last 3 pages of this book.**

POSSIBLE ADVERSE REACTIONS OR SIDE EFFECTS

SYMPTOMS	WHAT TO DO
Life-threatening:	
Rare, severe allergic reaction (anaphylaxis)—difficulty breathing, hives, itching, swelling (throat, face), vomiting, dizziness.	Seek emergency treatment immediately.
Common:	
• Muscle pain, skin rash or itching, swollen glands, sore throat, fever.	Continue, but call doctor right away.
• Nausea, vomiting, appetite loss, dizziness, drowsiness, hiccups, stomach pain, headache, loss of appetite.	Continue. Call doctor when convenient.
• Change in urine color (pink, red, red-brown).	No action necessary.
Infrequent:	
Nightmares, irritability, mood changes, tiredness, difficulty concentrating.	Continue, but call doctor right away.
Rare:	
Unusual bleeding or bruising, depression, swollen glands, chills, increased seizures, shortness of breath, wheezing, chest pain, sores in mouth or on lips.	Continue, but call doctor right away.

WARNINGS & PRECAUTIONS

Don't take if:
You are allergic to any succinimide anticonvulsant.

Before you start, consult your doctor if:
- You plan to become pregnant within medication period.
- You take other anticonvulsants.
- You have blood disease.
- You have kidney or liver disease.

Over age 60:
Adverse reactions and side effects may be more frequent and severe than in younger persons.

Pregnancy:
Decide with your doctor if drug benefits justify risk to unborn baby. Risk category C (see page xviii).

Breastfeeding; Lactation; Nursing Mothers:
Drugs in this group are generally not recommended with breastfeeding. Discuss risks and benefits with your doctor.

Infants & children up to age 18:
Follow instructions provided by your child's doctor.

Prolonged use:
Talk to your doctor about the need for follow-up medical examinations or laboratory studies to check complete blood counts (white blood cell count, platelet count, red blood cell count, hemoglobin, hematocrit), liver function, kidney function, urine.

Skin & sunlight:
No problems expected.

Driving, piloting or hazardous work:
Don't drive or pilot aircraft until you learn how medicine affects you. Don't work around dangerous machinery. Don't climb ladders or work in high places. Danger increases if you drink alcohol or take medicine affecting alertness and reflexes, such as antihistamines, tranquilizers, sedatives, pain medicine, narcotics and mind-altering drugs.

Discontinuing:
Don't discontinue without doctor's advice until you complete prescribed dose, even though symptoms diminish or disappear.

Others:
- Your response to medicine should be checked regularly by your doctor. Dose and schedule may have to be altered frequently to fit individual needs.
- Periodic blood cell counts, kidney and liver function studies recommended.
- May discolor urine pink to red-brown. No action necessary.
- Advise any doctor, dentist or pharmacist whom you consult that you use this medicine.
- Rarely, antiepileptic drugs may lead to suicidal thoughts and behaviors. Call doctor right away if suicidal symptoms or unusual behaviors occur.

POSSIBLE INTERACTION WITH OTHER DRUGS

GENERIC NAME OR DRUG CLASS	COMBINED EFFECT
Anticonvulsants, other*	Increased effect of both drugs.
Antidepressants, tricyclic*	May provoke seizures.
Antipsychotics*	May provoke seizures.
Central nervous system (CNS) depressants*	Decreased anticonvulsant effect.
Haloperidol	Decreased haloperidol effect; changed seizure pattern.
Phenytoin	Increased phenytoin effect.

POSSIBLE INTERACTION WITH OTHER SUBSTANCES

INTERACTS WITH	COMBINED EFFECT
Alcohol:	Increased risk of side effects.
Beverages:	None expected.
Cocaine:	Consult doctor.
Foods:	None expected.
Marijuana:	Increased risk of side effects.
Tobacco:	None expected.

***See Glossary**

ANTIDEPRESSANTS, TRICYCLIC

GENERIC AND BRAND NAMES

See full list of generic and brand names in the *Generic and Brand Name Directory*, page 839.

BASIC INFORMATION

Habit forming? No
Prescription needed? Yes
Available as generic? Yes
Drug class: Antidepressant (tricyclic)

USES

- Treatment for several types of depression.
- Treatment for obsessive compulsive disorder and bedwetting.
- May be used to treat narcolepsy, bulimia, anxiety, panic attacks, cocaine withdrawal, attention-deficit disorder, chronic pain, migraines, restless leg syndrome and other disorders. The brand name Silenor treats insomnia.

DOSAGE & USAGE INFORMATION

How to take:
Tablet, capsule or syrup—Swallow with liquid.

When to take:
At the same time each day, usually at bedtime.

If you forget a dose:
Bedtime dose—If you forget your once-a-day bedtime dose, don't take it more than 3 hours late. If more than 3 hours, wait for next scheduled dose. Don't double this dose.

What drug does:
Probably affects part of brain that controls messages between nerve cells.

Continued next column

OVERDOSE

SYMPTOMS:
Hallucinations, drowsiness, enlarged pupils, respiratory failure, fever, cardiac arrhythmias, convulsions, coma.
WHAT TO DO:
- **Dial 911 for all medical emergencies or call poison control center 1-800-222-1222 for instructions.**
- **See emergency information on last 3 pages of this book.**

Time lapse before drug works:
2 to 4 weeks. May require 4 to 6 weeks for maximum benefit.

Don't take with:
Any other medicine or any dietary supplement without consulting your doctor or pharmacist.

POSSIBLE ADVERSE REACTIONS OR SIDE EFFECTS

SYMPTOMS	WHAT TO DO
Life-threatening: In case of overdose, see previous column.	
Common:	
• Tremor.	Discontinue. Call doctor right away.
• Headache, dry mouth or unpleasant taste, constipation or diarrhea, nausea, indigestion, fatigue, weakness, drowsiness, nervousness, anxiety, excessive sweating, insomnia.	Continue. Call doctor when convenient.
Infrequent:	
• Convulsions.	Discontinue. Seek emergency treatment.
• Hallucinations, shakiness, dizziness, fainting, blurred vision, eye pain, vomiting, irregular heartbeat or slow pulse, inflamed tongue, abdominal pain, jaundice, hair loss, rash, fever, chills, joint pain, palpitations, visual changes.	Discontinue. Call doctor right away.
• Difficult or frequent urination, decreased sex drive, muscle aches, abnormal dreams, weakness and faintness when arising from bed or chair, back pain, nasal congestion.	Continue. Call doctor when convenient.
Rare:	
Itchy skin, sore throat, involuntary movements (jaw, lips and tongue), nightmares, confusion, swollen breasts, swollen testicles.	Discontinue. Call doctor right away.

WARNINGS & PRECAUTIONS

Don't take if:
- You are allergic to any tricyclic antidepressant. Allergic reactions are rare, but may occur.
- You drink alcohol in excess.
- You have had a heart attack in past 6 weeks.
- You have taken MAO inhibitors* within 2 weeks.
- Patient is younger than age 12.

Before you start, consult your doctor if:
- You will have surgery within 2 months, including dental surgery, requiring anesthesia.
- You have an enlarged prostate or glaucoma.
- You have high blood pressure, heart disease or stomach or intestinal problems.
- You have an overactive thyroid.
- You have asthma or liver disease.

Over age 60:
May have increased risk for side effects.

Pregnancy:
Risk factors vary for drugs in this group. See category list on page xviii and consult doctor.

Breastfeeding; Lactation; Nursing Mothers:
Most drugs in this group are generally acceptable with breastfeeding. Discuss risks and benefits with your doctor.

Infants & children up to age 18:
- Not recommended for ages 12 and under.
- Carefully read information provided with prescription. Contact doctor right away if depression symptoms get worse or there is any talk of suicide or suicide behaviors. Also, read information under Others.

Prolonged use:
Talk to your doctor about the need for follow-up medical examinations or laboratory studies to check complete blood counts (white blood cell count, platelet count, red blood cell count, hemoglobin, hematocrit), blood pressure, eyes, teeth.

Skin & sunlight:
One or more drugs in this group may cause rash or intensify sunburn in areas exposed to sun or ultraviolet light (photosensitivity reaction). Avoid overexposure and use sunscreen. Notify doctor if reaction occurs.

Driving, piloting or hazardous work:
Don't drive or pilot aircraft until you learn how medicine affects you. Don't work around dangerous machinery. Don't climb ladders or work in high places. Danger increases if you drink alcohol or take medicine affecting alertness and reflexes.

Discontinuing:
- Don't discontinue without consulting doctor. Dose may require gradual reduction if you have taken drug for a long time. Doses of other drugs may also require adjustment.
- Physical or emotional withdrawal symptoms may occur once you stop drug. Contact your doctor if any symptoms cause concern.

Others:
- Adults and children taking antidepressants may experience a worsening of the depression symptoms and may have increased suicidal thoughts or behaviors. Call doctor right away if these symptoms or behaviors occur.
- Advise any doctor, dentist or pharmacist whom you consult that you take this medicine.

POSSIBLE INTERACTION WITH OTHER DRUGS

GENERIC NAME OR DRUG CLASS	COMBINED EFFECT
Adrenocorticoids, systemic	Increased risk of mental side effects.
Anticoagulants,* oral	Possible increased anticoagulant effect.
Anticholinergics*	Increased anticholinergic effect.
Antifungals, azoles	Increased effect of antidepressant.
Antiglaucoma agents*	Decreased ocular hypertensive effect.
Antihistamines*	Increased antihistamine effect.
Barbiturates*	Decreased anti-depressant effect. Increased sedation.

Continued on page 888

POSSIBLE INTERACTION WITH OTHER SUBSTANCES

INTERACTS WITH	COMBINED EFFECT
Alcohol:	Sedation. Avoid.
Beverages:	Grapefruit juice may increase effect of clomipramine.
Cocaine:	Drowsiness. Risk of side effects. Avoid.
Foods:	None expected.
Marijuana:	Drowsiness. Risk of side effects. Avoid.
Tobacco:	Consult doctor.

***See Glossary**

ANTIDYSKINETICS

GENERIC AND BRAND NAMES

See full list of generic and brand names in the *Generic and Brand Name Directory*, page 840.

BASIC INFORMATION

Habit forming? No
Prescription needed? Yes
Available as generic? Yes
Drug class: Antidyskinetic, antiparkinsonism, dopamine agonists

USES

- Treatment of Parkinson's disease.
- Treatment of adverse effects of certain central nervous system drugs.
- Treatment of moderate to severe restless leg syndrome.
- Treatment for Tourette syndrome.

DOSAGE & USAGE INFORMATION

How to take:
- Tablet—Swallow with liquid. If you can't swallow whole, ask doctor or pharmacist for advice.
- Extended-release capsule or tablet—Swallow whole with liquid.
- Elixir—Follow directions on prescription label.
- All forms—Take with, or right after, a meal to lessen stomach irritation (unless otherwise directed by doctor).

When to take:
At the same time(s) each day.

If you forget a dose:
Take as soon as you remember. If it is almost time for the next dose, wait for that dose (don't double this dose) and resume regular schedule.

Continued next column

OVERDOSE

SYMPTOMS:
Agitation, dilated pupils, hallucinations, dry mouth, rapid heartbeat, sleepiness.
WHAT TO DO:
- **Dial 911 for all medical emergencies or call poison control center 1-800-222-1222 for instructions.**
- **See emergency information on last 3 pages of this book.**

What drug does:
- Balances chemical reactions necessary to send nerve impulses within base of brain.
- Improves muscle control and reduces stiffness.

Time lapse before drug works:
1 to 2 hours. Full effect may take 2 to 3 days.

Don't take with:
Any other medicine or any dietary supplement without consulting your doctor or pharmacist.

POSSIBLE ADVERSE REACTIONS OR SIDE EFFECTS

SYMPTOMS	WHAT TO DO
Life-threatening:	
In case of overdose, see previous column.	
Common:	
• Blurred vision, light sensitivity, unusual body movements, painful or difficult or frequent urination, vomiting, hallucinations.	Continue, but call doctor right away.
• Dry mouth, tiredness, weakness, insomnia or drowsiness, constipation, nausea, lightheadedness.	Continue. Call doctor when convenient.
Infrequent:	
Headache, memory loss, abdominal pain, weakness and faintness when rising from bed or chair, nervousness, impotence, sore throat, cough or wheezing, viral infection, appetite loss, restlessness.	Continue. Call doctor when convenient.
Rare:	
• Rash, hives, eye pain, delusions, amnesia, paranoia, fever, swollen neck glands, vision changes, chest pain, swallowing or breathing difficulty, numbness or tingling or swelling in hands or feet, urine bloody or cloudy, ear buzzing, irregular heartbeat.	Continue, but call doctor right away.
• Confusion, dizziness, sore mouth or tongue, muscle cramps or weakness, depression.	Continue. Call doctor when convenient.

Note: Many symptoms caused by side effects either disappear or decrease when dose is reduced. Consult doctor.

WARNINGS & PRECAUTIONS

Don't take if:
You are allergic to any antidyskinetic. Allergic reactions are rare, but may occur.

Before you start, consult your doctor if:
- You have glaucoma or retinal problems.
- You have had high blood pressure, heart disease, impaired liver function.
- You have hypotension or orthostatic hypotension.*
- You have had tardive dyskinesia.*
- You have had kidney disease, urination difficulty, prostatic hypertrophy or intestinal obstruction.
- You have myasthenia gravis.

Over age 60:
May be more sensitive to the drug and have increased risk of adverse effects.

Pregnancy:
Decide with your doctor if drug benefits justify risk to unborn baby. Risk category C (see page xviii).

Breastfeeding; Lactation; Nursing Mothers:
Drugs in this group are generally not recommended with breastfeeding. Discuss risks and benefits with your doctor.

Infants & children up to age 18:
Follow instructions provided by your child's doctor.

Prolonged use:
- Possible glaucoma.
- Talk to your doctor about the need for follow-up medical examinations to assess drug's effectiveness and examination to check eye pressure.

Skin & sunlight:
No special problems expected.

Driving, piloting or hazardous work:
Don't drive or pilot aircraft until you learn how medicine affects you. Don't work around dangerous machinery. Don't climb ladders or work in high places. Danger increases if you drink alcohol or take medicine affecting alertness and reflexes, such as antihistamines, tranquilizers, sedatives, pain medicine, narcotics and mind-altering drugs.

Discontinuing:
- Don't discontinue without consulting doctor. Dose may require gradual reduction if you have taken drug for a long time. Doses of other drugs may also require adjustment.
- After discontinuing, if you experience extrapyramidal reaction* recurrence or worsening, orthostatic hypotension, fast heartbeat, or trouble sleeping, consult doctor.

Others:
- Internal eye pressure should be measured regularly.
- Avoid becoming overheated.
- Use caution when arising from a sitting or lying position.
- Advise any doctor, dentist or pharmacist whom you consult that you take this medicine.

POSSIBLE INTERACTION WITH OTHER DRUGS

GENERIC NAME OR DRUG CLASS	COMBINED EFFECT
Antacids*	Possible decreased absorption.
Anticholinergics, others*	Increased anti-cholinergic effect.
Antidepressants, tricyclic*	Increased anti-dyskinetic effect.
Antihistamines*	Increased antidyskinetic effect.
Carbidopa	Increased effect of carbidopa.
Central nervous system (CNS) depressants*	May add to any sedative effect.
Chlorpromazine	Decreased effect of chlorpromazine.
Ciprofloxacin	Increased effect of ropinirole.
Dopamine antagonists*	Decreased effect of pramipexole and ropinirole.
Estrogens	Increased effect of ropinirole.

Continued on page 889

POSSIBLE INTERACTION WITH OTHER SUBSTANCES

INTERACTS WITH	COMBINED EFFECT
Alcohol:	Oversedation. Avoid.
Beverages:	None expected.
Cocaine:	Decreased anti-dyskinetic effect. Avoid.
Foods:	None expected.
Marijuana:	Consult doctor.
Tobacco:	Decreased effect of ropinirole.

*See Glossary

ANTIFIBRINOLYTIC AGENTS

GENERIC AND BRAND NAMES

AMINOCAPROIC ACID	TRANEXAMIC ACID
Amicar	Cyklokapron
	Lysteda

BASIC INFORMATION

Habit forming? No
Prescription needed? Yes
Available as generic? Yes, for some
Drug class: Antifibrinolytic, antihemorrhagic

USES

- Treats serious bleeding, especially that occurring after surgery, dental or otherwise.
- Treatment of women with menorrhagia (heavy menstrual bleeding).
- May be used before surgery to help prevent risk of excessive bleeding in patients with disorders that increase the chance of serious bleeding.

DOSAGE & USAGE INFORMATION

How to take:
- Tablet—Swallow with liquid or food to lessen stomach irritation. If you can't swallow whole, crumble tablet and take with liquid or food.
- Syrup—Take as directed on label.

When to take:
As directed by your doctor.

If you forget a dose:
Take as soon as you remember. If it is almost time for the next dose, wait for that dose (don't double this dose) and resume regular schedule.

What drug does:
Inhibits activation of plasminogen to cause blood clots to disintegrate.

Continued next column

OVERDOSE

SYMPTOMS:
Unknown.
WHAT TO DO:
- **Dial 911 for all medical emergencies or call poison control center 1-800-222-1222 for instructions.**
- **See emergency information on last 3 pages of this book.**

Time lapse before drug works:
Within 2 hours.

Don't take with:
- Thrombolytic chemicals such as streptokinase or urokinase.
- Any other medicine or any dietary supplement without consulting your doctor or pharmacist.

POSSIBLE ADVERSE REACTIONS OR SIDE EFFECTS

SYMPTOMS	WHAT TO DO
Life-threatening:	
Rare, severe allergic reaction (anaphylaxis)—difficulty breathing, hives, itching, swelling (throat, face), vomiting, dizziness.	Seek emergency treatment immediately.
Common:	
Diarrhea, nausea, vomiting, severe menstrual cramps.	Continue. Call doctor when convenient.
Infrequent:	
Dizziness, headache, muscular pain and weakness, red eyes, ringing in ears, skin rash, abdominal pain, stuffy nose, decreased urine, swelling (of feet, face, legs), rapid weight gain, unusual tiredness.	Continue. Call doctor when convenient.
Rare:	
Signs of thrombosis (headache that is sudden and severe, pain in chest or groin or legs, loss of coordination, shortness of breath, slurred speech, vision changes, weakness or numbness in arms or leg).	Seek emergency treatment.

WARNINGS & PRECAUTIONS

Don't take if:
- You are allergic to aminocaproic acid or tranexamic acid.
- You have a diagnosis of disseminated intravascular coagulation (DIC).

Before you start, consult your doctor if:
- You have heart disease.
- You have bleeding from the kidney.
- You have had impaired liver function.
- You have had kidney disease or urination difficulty.
- You have blood clots in parts of the body.

Over age 60:
No special problems expected.

Pregnancy:
Decide with your doctor if drug benefits justify risk to unborn baby. Risk category C for aminocaproic acid and risk category B for tranexamic acid (see page xviii).

Breastfeeding; Lactation; Nursing Mothers:
Drugs in this group are generally not recommended with breastfeeding. Discuss risks and benefits with your doctor.

Infants & children up to age 18:
Safety and effectiveness in this age group have not been established. Consult doctor.

Prolonged use:
Talk to your doctor about the need for follow-up medical examinations or laboratory studies.

Skin & sunlight:
No problems expected.

Driving, piloting or hazardous work:
Don't drive or pilot aircraft until you learn how medicine affects you. Don't work around dangerous machinery. Don't climb ladders or work in high places. Danger increases if you drink alcohol or take medicine affecting alertness and reflexes, such as antihistamines, tranquilizers, sedatives, pain medicine, narcotics and mind-altering drugs.

Discontinuing:
Don't discontinue without consulting doctor. Dose may require gradual reduction if you have taken drug for a long time. Doses of other drugs may also require adjustment.

Others:
- Should not be used in patients with disseminated intravascular coagulation.
- Advise any doctor, dentist or pharmacist whom you consult that you take this medicine.
- Have eyes checked frequently.

POSSIBLE INTERACTION WITH OTHER DRUGS

GENERIC NAME OR DRUG CLASS	COMBINED EFFECT
Contraceptives, oral*	Increased possibility of blood clotting.
Estrogens*	Increased possibility of blood clotting.
Thrombolytic agents* (alteplase, streptokinase, urokinase)	Decreased effects of both drugs.

POSSIBLE INTERACTION WITH OTHER SUBSTANCES

INTERACTS WITH	COMBINED EFFECT
Alcohol:	Decreases drug effectiveness. Avoid.
Beverages:	No problems expected.
Cocaine:	Unknown. Avoid.
Foods:	None expected.
Marijuana:	Consult doctor.
Tobacco:	None expected.

***See Glossary**

ANTIFUNGALS, AZOLES

GENERIC AND BRAND NAMES

FLUCONAZOLE
Diflucan
ITRACONAZOLE
Sporanox
KETOCONAZOLE
Nizoral
POSACONAZOLE
Noxafil
VORICONAZOLE
Vfend

BASIC INFORMATION

Habit forming? No
Prescription needed? Yes, for some
Available as generic? Yes, for some
Drug class: Antifungal

USES

- Treatment for fungal infections.
- Treatment for certain types of meningitis.
- Treatment for prostate cancer.

DOSAGE & USAGE INFORMATION

How to take:
- Capsule or tablet—Swallow with liquid. If you can't swallow whole, crumble tablet or open capsule and take with liquid or food.
- Oral suspension—Shake well before using; follow instructions supplied with medication.

When to take:
At the same time(s) each day.

If you forget a dose:
Take as soon as you remember. If it is almost time for the next dose, wait for that dose (don't double this dose) and resume regular schedule.

What drug does:
- Prevents fungi from growing and reproducing.
- For prostate cancer, ketoconazole decreases male hormone (testosterone) levels.

Continued next column

OVERDOSE

SYMPTOMS:
Mostly unknown. May have hallucinations, paranoid behavior, vision problems, loss of balance, shortness of breath, drooling, enlarged pupils, seizures.
WHAT TO DO:
- **Dial 911 for all medical emergencies or call poison control center 1-800-222-1222 for instructions.**
- **See emergency information on last 3 pages of this book.**

Time lapse before drug works:
Several weeks or months for full benefit.

Don't take with:
Any other medicine or any dietary supplement without consulting your doctor or pharmacist.

POSSIBLE ADVERSE REACTIONS OR SIDE EFFECTS

SYMPTOMS	WHAT TO DO
Life-threatening:	
Rare, severe allergic reaction (anaphylaxis)—difficulty breathing, hives, itching, swelling (throat, face), vomiting, dizziness.	Seek emergency treatment immediately.
Common:	
None expected.	
Infrequent:	
• Skin rash.	Discontinue. Call doctor right away.
• Diarrhea, nausea, vomiting, appetite loss, constipation, headache, stomach pain.	Continue. Call doctor when convenient.
Rare:	
• Pale stools, yellow skin or eyes, dark or amber urine, unusual tiredness or weakness.	Discontinue. Call doctor right away.
• Diminished sex drive in males, swollen breasts in males, increased sensitivity to light, drowsiness, dizziness, insomnia.	Continue. Call doctor when convenient.

WARNINGS & PRECAUTIONS

Don't take if:
You are allergic to any azole antifungal.

Before you start, consult your doctor if:
- You have impaired kidney or liver function.
- You have congestive heart failure.
- You have been diagnosed with reduced stomach acidity.

Over age 60:
No special problems expected.

Pregnancy:
Risk factors vary for drugs in this group. See category list on page xviii and consult doctor.

Breastfeeding; Lactation; Nursing Mothers:
Drugs in this group are generally acceptable with breastfeeding. Discuss risks and benefits with your doctor.

Infants & children up to age 18:
Follow instructions provided by your child's doctor.

Prolonged use:
Request periodic liver function studies.

Skin & sunlight:
No problems expected.

Driving, piloting or hazardous work:
Don't drive or pilot aircraft until you learn how drug affects you. Don't work around dangerous machinery. Don't climb ladders or work in high places. Danger increases if you drink alcohol or take medicine affecting alertness and reflexes.

Discontinuing:
Don't discontinue without consulting doctor. Dose may require gradual reduction if you have taken drug for a long time. Doses of other drugs may also require adjustment.

Others:
- Advise any doctor, dentist or pharmacist whom you consult that you take this medicine.
- Ketoconazole taken orally increases risk of liver damage (can be life threatening) and adrenal gland problems. Its use is restricted. Itraconazole should not be taken if you have congestive heart failure. Discuss all risks and benefits of these drugs with your doctor.

POSSIBLE INTERACTION WITH OTHER DRUGS

GENERIC NAME OR DRUG CLASS	COMBINED EFFECT
Adrenocorticoids, systemic	Decreased azole antifungal effect.
Antacids*	Decreased azole antifungal effect.
Anticholinergics*	Decreased azole antifungal effect.
Anticoagulants, oral*	Increased effect of anticoagulant.
Antidepressants, tricyclic*	Increased effect of antidepressant.
Antidiabetics, oral*	Increased risk of hypoglycemia.
Antivirals, HIV/AIDS*	Reduced effect of both drugs. Risk of pancreatitis.
Atropine	Decreased azole antifungal effect.
Carbamazepine	Decreased azole antifungal effect.
Cimetidine	Decreased azole antifungal effect.
Clidinium	Decreased azole antifungal effect.
Contraceptives, oral*	Decreased effect of contraceptive.
Cyclosporine	Increased risk of toxicity to kidney.
Digoxin	Possible toxic levels of digoxin.
Ergot preparations*	Can cause serious or life-threatening problems with blood circulation. Avoid.
Enzyme inducers*	Decreased effect of azole antifungal.
Enzyme inhibitors*	Increased effect of azole antifungal.
Eszopiclone	Decreased effect of ketoconazole.
Famotidine	Reduced azole antifungal effect. Take 2 hours apart.
Glycopyrrolate	Decreased azole antifungal effect.
Hepatotoxic medications*	Increased risk of toxicity to kidney.
Histamine H_2 receptor antagonists	Decreased azole antifungal effect.
HMG-CoA reductase inhibitors	Risk of muscle toxicity.
Hyoscyamine	Decreased azole antifungal effect.
Hypoglycemics, oral	Increased effect of oral hypoglycemics.
Isoniazid	Decreased azole antifungal effect.

Continued on page 889

POSSIBLE INTERACTION WITH OTHER SUBSTANCES

INTERACTS WITH	COMBINED EFFECT
Alcohol:	Liver damage risk or disulfiram reaction.*
Beverages:	None expected.
Cocaine:	Unknown. Avoid.
Foods:	None expected.
Marijuana:	Consult doctor.
Tobacco:	None expected.

***See Glossary**

ANTIFUNGALS (Topical)

GENERIC AND BRAND NAMES

See full list of generic and brand names in the *Generic and Brand Name Directory*, page 840.

BASIC INFORMATION

Habit forming? No
Prescription needed? Yes, for some
Available as generic? Yes, for some
Drug class: Antifungal (topical)

USES

- Treats skin fungus infections such as ringworm of the scalp (tinea capitis), athlete's foot (tinea pedis), jock itch (tinea cruris), nail fungus (onychomycosis), sun fungus (tinea versicolor) and others.
- Treatment of fungus (yeast) infection of the mouth and throat (also called oral thrush).
- Treatment of seborrheic dermatitis (dandruff, cradle cap).

DOSAGE & USAGE INFORMATION

How to use:

- Solution—Follow package instructions when applying to affected area.
- Cream, lotion, ointment, gel—Bathe and dry area before use. Apply small amount and rub gently.
- Powder—Apply lightly to skin.
- Shampoo—Follow package instructions.
- Buccal tablet—Follow package instructions.
- Don't bandage or cover treated areas with plastic wrap. Follow other instructions listed on label.

When to use:
Follow instructions provided with the product or use as directed by your doctor.

Continued next column

OVERDOSE

SYMPTOMS:
Unknown.
WHAT TO DO:
If person uses much larger amount than prescribed or if accidentally swallowed, call poison control center 1-800-222-1222 for instructions or dial 911 (emergency) for help.

If you forget a dose:
Use as soon as you remember. If it is almost time for the next dose, wait for that dose (don't double this dose) and resume regular schedule.

What drug does:
Kills fungi by damaging the fungal cell wall.

Time lapse before drug works:
May require 6 to 8 weeks or longer for cure.

Don't use with:
Other skin medicines without consulting your doctor or pharmacist.

POSSIBLE ADVERSE REACTIONS OR SIDE EFFECTS

SYMPTOMS	WHAT TO DO
Life-threatening: None expected.	
Common: None expected.	
Infrequent: Itching, redness, swelling of treated skin not present before treatment.	Continue. Call doctor if symptoms persist.
Rare: With buccal form—diarrhea, headache, taste changes, nausea or vomiting, stomach pain.	Continue. Call doctor if symptoms persist.

WARNINGS & PRECAUTIONS

Don't use if:
You are allergic to any topical antifungal medicine listed. Allergic reactions are rare, but may occur.

Before you start, consult your doctor if:
- You are allergic to anything that touches your skin.
- You are using buccal tablet form of drug and have liver disease or milk protein allergy.

Over age 60:
No special problems expected.

Pregnancy:
Risk factors vary for drugs in this group. See category list on page xviii and consult doctor.

Breastfeeding; Lactation; Nursing Mothers:
Drugs in this group are generally acceptable with breastfeeding. Discuss risks and benefits with your doctor.

Infants & children up to age 18:
Read product's label to be sure it is approved for your child's age. Follow label instructions on dosage. Consult doctor or pharmacist if unsure.

Prolonged use:
No problems expected, but check with doctor.

Skin & sunlight:
No special problems expected.

Driving, piloting or hazardous work:
No problems expected, but check with doctor.

Discontinuing:
Follow instructions on product's label.

Others:
- Avoid contact with eyes.
- Heat and moisture in bathroom medicine cabinet can cause breakdown of medicine. Store someplace else.
- Keep medicine cool, but don't freeze.
- Store away from heat or sunlight.
- Don't use on other members of the family without consulting your doctor.
- If using for jock itch, avoid wearing tight underwear.
- If using for athlete's foot, dry feet carefully after bathing, wear clean cotton socks with sandals or well-ventilated shoes.

POSSIBLE INTERACTION WITH OTHER DRUGS

GENERIC NAME OR DRUG CLASS	COMBINED EFFECT
Other drugs	Consult doctor about interactions if using buccal tablet form.

POSSIBLE INTERACTION WITH OTHER SUBSTANCES

INTERACTS WITH	COMBINED EFFECT
Alcohol:	None expected.
Beverages:	None expected.
Cocaine:	None expected.
Foods:	None expected.
Marijuana:	None expected.
Tobacco:	None expected.

ANTIFUNGALS (Vaginal)

GENERIC AND BRAND NAMES

See full list of generic and brand names in the *Generic and Brand Name Directory*, page 841.

BASIC INFORMATION

Habit forming? No
Prescription needed? Yes, for some
Available as generic? Yes, for some
Drug class: Antifungal (vaginal)

USES

Treats fungus infections of the vagina.

DOSAGE & USAGE INFORMATION

How to use:

- Vaginal cream—Insert into vagina with applicator as illustrated in patient instructions that come with prescription.
- Vaginal tablet—Insert with applicator as illustrated in instructions.
- Vaginal suppository—Insert as illustrated in instructions.

When to use:
According to instructions. Usually once or twice daily.

If you forget a dose:
Use as soon as you remember. If it is almost time for the next dose, wait for that dose (don't double this dose) and resume regular schedule.

What drug does:
Destroys fungus cell membrane causing loss of essential elements to sustain fungus cell life.

Time lapse before drug works:
Begins immediately. May require 2 weeks of treatment to cure vaginal fungus infections. Recurrence common.

Don't use with:
Other vaginal preparations or douches unless otherwise instructed by your doctor.

OVERDOSE

SYMPTOMS:
Unknown.
WHAT TO DO:
If person uses much larger amount than prescribed or if accidentally swallowed, call poison control center 1-800-222-1222 for instructions or dial 911 (emergency) for help.

POSSIBLE ADVERSE REACTIONS OR SIDE EFFECTS

SYMPTOMS	WHAT TO DO
Life-threatening: None expected.	
Common: None expected.	
Infrequent: Vaginal burning, itching, irritation, swelling of labia, redness, increased discharge (not present before starting medicine).	Discontinue. Call doctor right away.
Rare: Skin rash, hives, irritation of sex partner's penis.	Discontinue. Call doctor right away.

WARNINGS & PRECAUTIONS

Don't use if:
- You are allergic to any of the products listed. Allergic reactions are rare, but may occur.
- You have pre-existing liver disease.

Before you start, consult your doctor if:
You are pregnant.

Over age 60:
No special problems expected.

Pregnancy:
Risk factors vary for drugs in this group. See category list on page xviii and consult doctor.

Breastfeeding; Lactation; Nursing Mothers:
Drugs in this group are generally acceptable with breastfeeding. Discuss risks and benefits with your doctor.

Infants & children up to age 18:
Follow instructions provided by your child's doctor.

Prolonged use:
No problems expected.

Skin & sunlight:
No problems expected.

Driving, piloting or hazardous work:
No problems expected.

Discontinuing:
Recurrence likely if you stop before time suggested.

Others:
- Gentian Violet and some of the other products can stain clothing. Sanitary napkins may protect against staining.
- Keep the genital area clean. Use plain unscented soap.
- Take showers rather than tub baths.
- Wear cotton underpants or pantyhose with a cotton crotch. Avoid underpants made from non-ventilating materials. Wear freshly laundered underpants.
- Don't sit around in wet clothing—especially a wet bathing suit.
- After urination or bowel movements, cleanse by wiping or washing from front to back (vagina to anus).
- Don't douche unless your doctor recommends it.
- If urinating causes burning, urinate through a tubular device, such as a toilet-paper roll or plastic cup with the end cut out.

POSSIBLE INTERACTION WITH OTHER DRUGS

GENERIC NAME OR DRUG CLASS	COMBINED EFFECT
Warfarin	May cause bleeding with miconazole vaginal cream.

POSSIBLE INTERACTION WITH OTHER SUBSTANCES

INTERACTS WITH	COMBINED EFFECT
Alcohol:	None expected.
Beverages:	None expected.
Cocaine:	None expected.
Foods:	None expected.
Marijuana:	None expected.
Tobacco:	None expected.

ANTIGLAUCOMA, ADRENERGIC AGONISTS

GENERIC AND BRAND NAMES

APRACLONIDINE
- Iopidine

BRIMONIDINE
- Alphagan
- Alphagan P
- Combigan
- Simbrinza

DIPIVEFRIN
- AKPro
- Ophtho-Dipivefrin
- Propine
- Propine C Cap

EPINEPHRINE
- Epifren
- Epinal
- Eppy/N
- Glaucon

BASIC INFORMATION

Habit forming? No
Prescription needed? Yes
Available as generic? Yes, for some
Drug class: Antiglaucoma

USES

Treats open-angle glaucoma, secondary glaucoma and ocular hypertension. May be used with eye surgery.

DOSAGE & USAGE INFORMATION

How to use:
Eye drops
- Wash hands.
- Apply pressure to inside corner of eye with middle finger.
- Continue pressure for 1 minute after placing medicine in eye.
- Tilt head backward. Pull lower lid away from eye with index finger of the same hand.
- Drop eye drops into pouch and close eye. Don't blink.
- Keep eyes closed for 1 to 2 minutes.
- If using more than one eye solution, wait at least 10 minutes between instillations to avoid a "wash-out" effect.

Continued next column

When to use:
As directed on label.

If you forget a dose:
Apply as soon as you remember. If it is almost time for the next dose, wait for that dose (don't double this dose) and resume regular schedule.

What drug does:
Inactivates enzyme and facilitates movement of fluid (aqueous humor) into and out of the eye.

Time lapse before drug works:
30 minutes to 4 hours.

Don't use with:
Any other eye medicine without consulting your doctor or pharmacist.

OVERDOSE

SYMPTOMS:
If accidental overdose in eye, flush with water. Other symptoms are unknown.
WHAT TO DO:
If person uses much larger amount than prescribed or if accidentally swallowed, call poison control center 1-800-222-1222 for instructions or dial 911 (emergency) for help.

POSSIBLE ADVERSE REACTIONS OR SIDE EFFECTS

SYMPTOMS	WHAT TO DO
Life-threatening: None expected.	
Common:	
• Allergic reaction (itching, redness, tearing of eye).	Discontinue. Call doctor right away.
• Headache, eye discomfort, dry mouth.	Continue. Call doctor when convenient.
Infrequent:	
Eye symptoms: pain, changes in vision, blurred vision, discharge or swelling, color change in white of eye, feeling of something in the eye, stinging, burning, watering, light sensitivity, crusting on eyelid, paleness of eye or inner eyelid.	Continue, but call doctor right away.
Rare:	
• Symptoms of too much drug absorbed in body: faintness, skin paleness, chest pain, increased or fast or irregular heartbeat, swelling (face, hands, or feet), dizziness, numbness or tingling in fingers or toes, wheezing, troubled breathing.	Discontinue. Call doctor right away.
• Other symptoms of drug absorbed in body: sore throat, muscle aches, nausea, smell or taste changes, anxiety, nervousness, depression, constipation, insomnia or drowsiness.	Continue. Call doctor when convenient.

WARNINGS & PRECAUTIONS

Don't use if:
You are allergic to any of the adrenergic agonist antiglaucoma drugs.

Before you start, consult your doctor if:
- You suffer from depression.
- You have any eye disease.
- You have heart problems, high or low blood pressure or thromboangiitis obliterans.
- You have Raynaud's disease.
- You have a history of vasovagal attacks* (if using apraclonidine).
- You have liver or kidney problems.

Over age 60:
No special problems expected.

Pregnancy:
Decide with your doctor if drug benefits justify risk to unborn baby. Risk category B or C (see page xviii).

Breastfeeding; Lactation; Nursing Mothers:
Drugs in this group are generally acceptable with breastfeeding. Discuss risks and benefits with your doctor.

Infants & children up to age 18:
Follow instructions provided by your child's doctor.

Prolonged use:
May be necessary.

Skin & sunlight:
No problems expected.

Driving, piloting or hazardous work:
Your vision may be blurred or there may be a change in your near or far vision or night vision for a short time after drug use. Don't drive or pilot aircraft until you learn how medicine affects you. Don't work around dangerous machinery. Don't climb ladders or work in high places.

Discontinuing:
Don't discontinue without consulting your doctor. Dose may require gradual reduction if you have used drug for a long time. Doses of other drugs may also require adjustment.

Others:
- Advise any doctor, dentist or pharmacist whom you consult that you use this medicine.
- Drugs may cause your eyes to become more sensitive to light. Wear sunglasses and avoid too much exposure to bright light.
- Brimonidine contains a preservative that could be absorbed by soft contact lenses. Wait at least 15 minutes after putting eye drops in before you put in your soft contact lenses.
- If you have any eye infection, injury or wound, consult doctor before using this medicine.
- Keep appointments for regular eye examinations to measure pressure in the eye.

POSSIBLE INTERACTION WITH OTHER DRUGS

GENERIC NAME OR DRUG CLASS	COMBINED EFFECT
Antidepressants, tricyclic	May decrease ocular hypertensive effect. Dipivefrin may cause heart rhythm problem, high blood pressure.
Antiglaucoma, beta blockers	May help decrease eye pressure.
Antihypertensives*	May decrease blood pressure.
Central nervous system (CNS) depressants*	May increase CNS depressant effect.
Digitalis preparations*	Increased risk of heart problems.
Maprotiline	Heart rhythm problems, high blood pressure.
Monoamine oxidase (MAO) inhibitors*	Separate use by at least 14 days.

POSSIBLE INTERACTION WITH OTHER SUBSTANCES

INTERACTS WITH	COMBINED EFFECT
Alcohol:	None expected.
Beverages:	None expected.
Cocaine:	Unknown. Avoid.
Foods:	None expected.
Marijuana:	Unknown.Consult doctor.
Tobacco:	None expected.

***See Glossary**

GENERIC AND BRAND NAMES

DEMECARIUM
Humorsol

ECHOTHIOPHATE
Phospholine Iodide

BASIC INFORMATION

Habit forming? No
Prescription needed? Yes
Available as generic? No
Drug class: Antiglaucoma

USES

- Treatment for certain types of glaucoma.
- Used for diagnosis and treatment for other eye conditions.

DOSAGE & USAGE INFORMATION

How to use:
Eye drops

- Wash hands. Tilt head back.
- Press finger gently on the skin right under the lower eyelid; pull the eyelid away from the eye to make a space or small pocket.
- Drop the medicine into this pocket, then let go of the skin and gently close the eyes; don't blink.
- Keep the eyes closed and apply pressure to the inner corner of the eye with your finger for 1 to 2 minutes.
- Wash hands again after using the drops.
- To keep the solution germ-free, do not allow the applicator tip to touch the skin or eye.
- If using more than one eye solution, wait at least 10 minutes between instillations to avoid a "wash-out" effect.

Continued next column

OVERDOSE

SYMPTOMS:
Fast heartbeat, diarrhea, heavy sweating, breathing difficulty, unable to control bladder, shock. If accidental overdose in eye, flush with water.

WHAT TO DO:

- **Dial 911 for all medical emergencies or call poison control center 1-800-222-1222 for instructions.**
- **See emergency information on last 3 pages of this book.**

When to use:
As directed on label.

If you forget a dose:
Apply as soon as you remember. If it is almost time for the next dose, wait for that dose (don't double this dose) and resume regular schedule.

What drug does:
Inactivates an enzyme to reduce pressure inside the eye.

Time lapse before drug works:
5 to 60 minutes.

Don't use with:
Any other eye medicine without consulting your doctor or pharmacist.

POSSIBLE ADVERSE REACTIONS OR SIDE EFFECTS

SYMPTOMS	WHAT TO DO
Life-threatening:	
None expected.	
Common:	
Stinging, burning watery eyes.	Continue. Call doctor when convenient.
Infrequent:	
Blurred vision, change in vision, change in night vision, eyelids twitch, headache, ache in brow area.	Continue. Call doctor when convenient.
Rare:	
Decreased vision with veil or curtain appearing in part of vision, eye redness, symptoms of too much of drug absorbed in body (loss of bladder control, slow heartbeat, increased sweating, weakness, difficult breathing, vomiting, nausea, diarrhea, stomach pain or cramping).	Discontinue. Call doctor right away.

ANTIGLAUCOMA, ANTICHOLINESTERASES

WARNINGS & PRECAUTIONS

Don't use if:
You are allergic to demecarium or echothiophate.

Before you start, consult your doctor if:
- You have eye infection or other eye disease.
- You have ulcers in stomach or duodenum or other stomach disorder.
- You have myasthenia gravis, overactive thyroid or urinary tract blockage.
- You have asthma, epilepsy, Down syndrome, heart disease, high or low blood pressure, Parkinson's disease.

Over age 60:
No special problems expected, but visit your doctor on a regular basis while using this drug.

Pregnancy:
- Demecarium—Consult doctor. Drug should not be used during pregnancy. Can cause harm to unborn baby. Risk category X (see page xviii).
- Echothiophate—Decide with your doctor if drug benefits justify risk to unborn baby. Risk category C (see page xviii).

Breastfeeding; Lactation; Nursing Mothers:
Drugs in this group are generally acceptable with breastfeeding. Discuss risks and benefits with your doctor.

Infants & children up to age 18:
Follow instructions provided by your child's doctor.

Prolonged use:
Cataracts or other eye problems may occur. Be sure to see your doctor for regular eye examinations.

Skin & sunlight:
No problems expected.

Driving, piloting or hazardous work:
Your vision may be blurred or there may be a change in your near or far vision or night vision for a short time after drug use. Don't drive or pilot aircraft until you learn how medicine affects you. Don't work around dangerous machinery. Don't climb ladders or work in high places.

Discontinuing:
Don't discontinue without consulting your doctor. Dose may require gradual reduction if you have used drug for a long time. Doses of other drugs may also require adjustment.

Others:
- Advise any doctor, dentist or pharmacist whom you consult that you use this medicine.
- Keep appointments for regular eye examinations to measure pressure in the eye.
- If you have any eye infection, injury or wound, consult doctor before using this medicine.

POSSIBLE INTERACTION WITH OTHER DRUGS

GENERIC NAME OR DRUG CLASS	COMBINED EFFECT
Anticholinergics*	Increased risk of toxicity.
Antimyasthenics*	Increased risk of side effects.
Cholinesterase inhibitors*	Increased risk of toxicity.
Insecticides or pesticides with organic phosphates	Increased toxic absorption of pesticides.
Topical anesthetics	Increased risk of toxic effects of antiglaucoma eye medicines.

POSSIBLE INTERACTION WITH OTHER SUBSTANCES

INTERACTS WITH	COMBINED EFFECT
Alcohol:	None expected.
Beverages:	None expected.
Cocaine:	Unknown. Avoid.
Foods:	None expected.
Marijuana:	Unknown.Consult doctor.
Tobacco:	None expected.

***See Glossary**

ANTIGLAUCOMA, BETA BLOCKERS

GENERIC AND BRAND NAMES

BETAXOLOL (ophthalmic)
- **Betoptic**
- **Betoptic S**

CARTEOLOL (ophthalmic)
- **Ocupress**

LEVOBUNOLOL (ophthalmic)
- **AKBeta**
- **Betagen C Cap B.I.D.**
- **Betagen C Cap Q.D.**
- **Betagen Standard Cap**

METIPRANOLOL (ophthalmic)
- **OptiPranolol**

TIMOLOL (ophthalmic)
- **Apo-Timop**
- **Beta-Tim**
- **Betimol**
- **Combigan**
- **Cosopt**
- **Cosopt PF**
- **Gen-Timolo**
- **Istalol**
- **Med Timolol**
- **Novo-Timolol**
- **Nu-Timolol**
- **Timodal**
- **Timoptic**
- **Timoptic in Ocudose**
- **Timoptic-XE**
- **Xalcom**

BASIC INFORMATION

Habit forming? No
Prescription needed? Yes
Available as generic? Yes, for some
Drug class: Antiglaucoma

USES

Treatment for glaucoma and ocular hypertension. May be used in eye surgery.

DOSAGE & USAGE INFORMATION

How to take:
Eye drops—Follow directions on prescription.

Continued next column

When to take:
At the same time each day, usually in the morning. Follow your doctor's instructions.

If you forget a dose:
- Once-a-day dose—Apply as soon as you remember. If almost time for next dose, wait and apply at regular time (don't double this dose).
- More than once-a-day dose—Apply as soon as you remember. If close to time for next dose, wait and apply at regular time (don't double this dose).

What drug does:
Appears to reduce production of aqueous humor (fluid inside eye), thereby reducing pressure inside eye.

Time lapse before drug works:
30 minutes to 1 hour.

Don't take with:
Any other eye medicine without consulting your doctor or pharmacist.

OVERDOSE

SYMPTOMS:
Slow heartbeat, low blood pressure, bronchospasm, heart failure (these symptoms are what might be expected if similar drugs were taken orally). If accidental overdose in eye, flush with water.
WHAT TO DO:
- **Dial 911 for all medical emergencies or call poison control center 1-800-222-1222 for instructions.**
- **See emergency information on last 3 pages of this book.**

POSSIBLE ADVERSE REACTIONS OR SIDE EFFECTS

SYMPTOMS	WHAT TO DO
Life-threatening:	
In case of overdose, see previous column.	
Common:	
• Redness of eyes or inside of eyelids.	Continue. Call doctor right away.
• Temporary blurred vision, night vision decreased, eye irritation or discomfort when drug is used.	Continue. Call doctor when convenient.
Infrequent:	
Ongoing blurred vision, other vision changes, different size pupils, eyeball discolored, droopy eyelid, eye pain, or swelling or irritation.	Continue. Call doctor when convenient.
Rare:	
• Increased sensitivity to light, sensation of foreign body in eye, crusty eyelids.	Continue. Call doctor when convenient.
• Symptoms of body absorbing too much of drug include problems with heart, stomach, lungs (breathing difficulties), skin, nervous system, hair loss, and others.	Discontinue. Call doctor right away.

WARNINGS & PRECAUTIONS

Don't take if:
You are allergic to any beta-adrenergic blocking agent taken orally or used in the eye.

Before you start, consult your doctor if:
- You have asthma, a bronchial disorder or pulmonary disease.
- You have any heart disease or heart problem.
- You suffer from depression.
- You have diabetes or low blood sugar, overactive thyroid or myasthenia gravis.

Over age 60:
No special problems expected, but visit your doctor on a regular basis while using this drug.

Pregnancy:
Decide with your doctor if drug benefits justify risk to unborn baby. Risk category C (see page xviii).

Breastfeeding; Lactation; Nursing Mothers:
Drugs in this group are generally acceptable with breastfeeding. Discuss risks and benefits with your doctor.

Infants & children up to age 18:
Follow instructions provided by your child's doctor.

Prolonged use:
Talk to your doctor about the need for follow-up medical examinations to check pressure inside eye.

Skin & sunlight:
No special problems expected.

Driving, piloting or hazardous work:
Your vision may be blurred or there may be a change in your near or far vision or night vision for a short time after drug use. Don't drive or pilot aircraft until you learn how medicine affects you. Don't work around dangerous machinery. Don't climb ladders or work in high places.

Discontinuing:
- Don't discontinue without doctor's approval.
- May need to discontinue drug temporarily before major surgery. Your doctor will provide instructions.

Others:
- Advise any doctor, dentist or pharmacist whom you consult that you take this medicine.
- These drugs may affect blood sugar levels in diabetic patients.
- Keep appointments for regular eye examinations to measure pressure in the eye.
- If you have any eye infection, injury or wound, consult doctor before using this medicine.

POSSIBLE INTERACTION WITH OTHER DRUGS

GENERIC NAME OR DRUG CLASS	COMBINED EFFECT

Drug interactions are unlikely unless a significant amount of the eye medication is absorbed into the system. Potential interactions that may occur are similar to those listed in Possible Interactions With Other Drugs under Beta-Adrenergic Blocking Agents.

POSSIBLE INTERACTION WITH OTHER SUBSTANCES

INTERACTS WITH	COMBINED EFFECT
Alcohol:	None expected.
Beverages:	None expected.
Cocaine:	Unknown. Avoid.
Foods:	None expected.
Marijuana:	Unknown. Consult doctor.
Tobacco:	None expected.

ANTIGLAUCOMA, CARBONIC ANHYDRASE INHIBITORS

GENERIC AND BRAND NAMES

BRINZOLAMIDE	DORZOLAMIDE
Azopt	Cosopt
Simbrinza	Cosopt PF
	Trusopt

BASIC INFORMATION

Habit forming? No
Prescription needed? Yes
Available as generic? Yes, for some
Drug class: Antiglaucoma

USES

- Treatment for open-angle glaucoma (increased pressure in the eye).
- Treatment for ocular hypertension.

DOSAGE & USAGE INFORMATION

How to use:
Eye drops
- Wash hands. Tilt head back.
- Press finger gently on the skin right under the lower eyelid; pull the eyelid away from the eye to make a space or small pocket.
- Drop the medicine into this pocket, then let go of the skin and gently close eyes; don't blink.
- Keep the eyes closed and apply pressure to the inner corner of the eye with your finger for 1 to 2 minutes.
- Wash hands again after using the drops.
- To keep the solution germ-free, do not allow the applicator tip to touch the skin or eye.
- If using more than one eye solution, wait at least 10 minutes between instillations to avoid a "wash-out" effect.

When to use:
Normally used 3 times a day (about 8 hours apart). Always use as directed by your doctor.

Continued next column

If you forget a dose:
Use as soon as you remember. If it is almost time for the next dose, wait for that dose (don't double this dose) and resume regular schedule.

What drug does:
This medicine is a topically applied carbonic anhydrase inhibitor that helps decrease production of aqueous humor (the fluid in the eye) and lowers the pressure inside the eye.

Time lapse before drug works:
30 to 60 minutes.

Don't take with:
Any other eye medicine without consulting your doctor or pharmacist.

OVERDOSE

SYMPTOMS:
If accidental overdose in eye, flush with water. Other symptoms are unknown.
WHAT TO DO:
If person uses much larger amount than prescribed or if accidentally swallowed, call poison control center 1-800-222-1222 for instructions or dial 911 (emergency) for help.

POSSIBLE ADVERSE REACTIONS OR SIDE EFFECTS

SYMPTOMS	WHAT TO DO
Life-threatening:	
None expected.	
Common:	
• Allergic reaction (redness, itching or swelling of eye or eyelid), feeling of something in the eye, continued or severe sensitivity to light.	Discontinue. Call doctor right away.
• Bitter taste, burning, stinging or discomfort when medicine is used, mild sensitivity to light.	Continue. Call doctor when convenient.
Infrequent:	
Blurred vision, dryness or mild tearing of eyes, tiredness or weakness, headache, hair loss.	Continue. Call doctor when convenient.
Rare:	
Blood in urine; continued nausea or vomiting, hives, pain (in chest, back, side or abdomen), eye pain, severe or continued tearing, seeing double, skin rash, shortness of breath.	Discontinue. Call doctor right away.

ANTIGLAUCOMA, CARBONIC ANHYDRASE INHIBITORS

WARNINGS & PRECAUTIONS

Don't use if:
You are allergic to ophthalmic carbonic anhydrase inhibitors or any sulfonamide* medications.

Before you start, consult your doctor if:
- You have kidney or liver disease.
- You are allergic to any medication, food, preservatives or other substances.

Over age 60:
No special problems expected.

Pregnancy:
Decide with your doctor if drug benefits justify risk to unborn baby. Risk category C (see page xviii).

Breastfeeding; Lactation; Nursing Mothers:
Drugs in this group are generally acceptable with breastfeeding. Discuss risks and benefits with your doctor.

Infants & children up to age 18:
Follow instructions provided by your child's doctor.

Prolonged use:
Schedule regular appointments with your eye doctor for eye examinations to be sure the medication is controlling the glaucoma.

Skin & sunlight:
No special problems expected.

Driving, piloting or hazardous work:
Your vision may be blurred or there may be a change in your near or far vision or night vision for a short time after drug use. Don't drive or pilot aircraft until you learn how medicine affects you. Don't work around dangerous machinery. Don't climb ladders or work in high places.

Discontinuing:
Don't discontinue without consulting your doctor.

Others:
- If you have any eye infection, injury or wound, consult doctor before using this medicine.
- Advise any doctor, dentist or pharmacist whom you consult that you use this medicine.
- Keep appointments for regular eye examinations to measure pressure in the eye.
- Wear sunglasses when outside in sunlight.

POSSIBLE INTERACTION WITH OTHER DRUGS

GENERIC NAME OR DRUG CLASS	COMBINED EFFECT
Amphetamines	Increased risk of side effects.
Carbonic anhydrase inhibitors (oral)	Increased effect of both drugs. Avoid.
Mecamylamine	Increased risk of side effects.
Quinidine	Increased risk of side effects.
Salicylates (High doses)	Increased risk of adverse effects.

POSSIBLE INTERACTION WITH OTHER SUBSTANCES

INTERACTS WITH	COMBINED EFFECT
Alcohol:	None expected.
Beverages:	None expected.
Cocaine:	Unknown. Avoid.
Foods:	None expected.
Marijuana:	Unknown. Consult doctor.
Tobacco:	None expected.

*See Glossary

ANTIGLAUCOMA, CHOLINERGIC AGONISTS

GENERIC AND BRAND NAMES

CARBACHOL
- **Carbastat**
- **Carboptic**
- **Miostat**

PILOCARPINE
- **Adsorbocarpine**
- **Akarpine**
- **Almocarpine**
- **Carpine**
- **Isopto Carpine**
- **Minims**
- **Miocarpine**
- **Ocu-Carpine**
- **Ocusert Pilo**
- **Pilocarpine Pilocar**
- **Piloptic**
- **Pilostat**
- **P.V. Carpine Liquifilm**

BASIC INFORMATION

Habit forming? No
Prescription needed? Yes
Available as generic? Yes, for some
Drug class: Antiglaucoma

USES

Treatment for glaucoma and other eye conditions. May be used in eye surgery.

DOSAGE & USAGE INFORMATION

How to take:
- Drops—Apply to eyes. Close eyes for 1 or 2 minutes to absorb medicine.
- Eye insert system—Follow label directions.
- Gel—Follow label directions.

When to use:
As directed on label.

Continued next column

OVERDOSE

SYMPTOMS:
If accidental overdose in eye, flush with water. If swallowed—nausea, vomiting, diarrhea, sweating.
WHAT TO DO:
If person uses much larger amount than prescribed or if accidentally swallowed, call poison control center 1-800-222-1222 for instructions or dial 911 (emergency) for help.

If you forget a dose:
- For eye drops or gel, use as soon as possible. If it is almost time for your next dose, skip the missed dose and return to regular schedule (don't double this dose).
- For eye insert, replace it as soon as possible. Then return to your regular schedule.

What drug does:
Reduces internal eye pressure.

Time lapse before drug works:
75 minutes to 4 hours.

Don't take with:
Any other eye medicine without consulting your doctor or pharmacist.

POSSIBLE ADVERSE REACTIONS OR SIDE EFFECTS

SYMPTOMS	WHAT TO DO
Life-threatening: None expected.	
Common: Blurred or altered vision (near or distant vision), eye stinging or burning.	Continue. Call doctor when convenient.
Infrequent: Headache, eye irritation, redness, of eye, eyelid twitching.	Continue. Call doctor when convenient.
Rare: Eye pain, a veil or curtain appears across part of vision, symptoms of too much of drug absorbed in the body (increased sweating, muscle trembling, nausea, vomiting, diarrhea, troubled breathing or wheezing, mouth watering, stomach cramps, fainting, flushing or redness of face, urge to urinate).	Discontinue. Call doctor right away.

ANTIGLAUCOMA, CHOLINERGIC AGONISTS

WARNINGS & PRECAUTIONS

Don't take if:
You are allergic to carbachol or pilocarpine.

Before you start, consult your doctor if:
- You have other eye problems.
- You have heart disease, overactive thyroid, Parkinson's disease, ulcers, or urinary blockage problems.
- You have asthma.

Over age 60:
No special problems expected.

Pregnancy:
Decide with your doctor if drug benefits justify risk to unborn baby. Risk category C (see page xviii).

Breastfeeding; Lactation; Nursing Mothers:
Drugs in this group are generally acceptable with breastfeeding. Discuss risks and benefits with your doctor.

Infants & children up to age 18:
Not recommended.

Prolonged use:
- You may develop tolerance* for drug, making it ineffective. Your doctor may switch antiglaucoma drugs for a period of time to return effectiveness.
- Talk to your doctor about the need for follow-up medical examinations to check eye pressure.

Skin & sunlight:
Safety and effectiveness in this age group have not been established. Consult doctor.

Driving, piloting or hazardous work:
Your vision may be blurred or there may be a change in your near or far vision or night vision for a short time after drug use. Don't drive or pilot aircraft until you learn how medicine affects you. Don't work around dangerous machinery. Don't climb ladders or work in high places.

Discontinuing:
Don't discontinue without consulting your doctor.

Others:
- Advise any doctor, dentist or pharmacist whom you consult that you use this medicine.
- Keep appointments for regular eye examinations to measure pressure in the eye.
- If you have any eye infection, injury or wound, consult doctor before using this medicine.

POSSIBLE INTERACTION WITH OTHER DRUGS

GENERIC NAME OR DRUG CLASS	COMBINED EFFECT
Belladonna (ophthalmic)	Decreased antiglaucoma effect.
Cyclopentolate	Decreased antiglaucoma effect.
Flurbiprofen (ophthalmic)	Decreased antiglaucoma effect.

POSSIBLE INTERACTION WITH OTHER SUBSTANCES

INTERACTS WITH	COMBINED EFFECT
Alcohol:	None expected.
Beverages:	None expected.
Cocaine:	Unknown. Avoid.
Foods:	None expected.
Marijuana:	Unknown. Consult doctor.
Tobacco:	None expected.

***See Glossary**

ANTIGLAUCOMA, PROSTAGLANDINS

GENERIC AND BRAND NAMES

BIMATOPROST	**LATANOPROST**
Latisse	**Xalatan**
Lumigan	**TAFLUPROST**
Xalcom	**Zioptan**
ISOPROPYL UNOPROSTONE	**TRAVOPROST**
Rescula	**Izba**
	Travatan
	Travatan Z

BASIC INFORMATION

Habit forming? No
Prescription needed? Yes
Available as generic? Yes, for some
Drug class: Antiglaucoma

USES

- Treats diseases of the eye like glaucoma and hypertension of the eye.
- Treats hypotrichosis (reduced amount of hair) of the eyelashes.

DOSAGE & USAGE INFORMATION

How to use:
Eye drops
- Wash hands.
- Apply pressure on the skin just beneath the lower eyelid. Pull the lower eyelid away from the eye to make a space.
- Drop the medicine into this space.
- Release eyelid and gently close eyes.
- Don't blink.
- Keep eyes closed for 1 to 2 minutes.
- Remove excess solution from around the eye with a clean tissue, being careful not to touch the eye.
- Don't touch applicator tip to any surface (including the eye). If you accidentally touch tip, clean with warm water and soap.
- Keep container tightly closed.
- Wash hands immediately after using.

Continued next column

OVERDOSE

SYMPTOMS:
Unknown. If overdose in eye, flush with water.
WHAT TO DO:
If person uses much larger amount than prescribed or if accidentally swallowed, call doctor or poison control center 1-800-222-1222 for help.

When to use:
As directed on label.

If you forget a dose:
Use as soon as you remember. If it is almost time for the next dose, wait for that dose (don't double this dose) and resume regular schedule.

What drug does:
Helps lower intraocular eye pressure by increasing drainage of fluid (aqueous humor) out of the eyeball.

Time lapse before drug works:
10 to 30 minutes.

Don't use with:
Any other eye medicine without consulting your doctor or pharmacist.

POSSIBLE ADVERSE REACTIONS OR SIDE EFFECTS

SYMPTOMS	WHAT TO DO
Life-threatening: None expected.	
Common: Eye symptoms: itching, discomfort, mild pain, redness, feeling of something in eye, vision decreased.	Continue. Call doctor when convenient.
Infrequent: Eye tearing or dry, crusting on eyelid, eyes more sensitive to light, eye discharge, color vision or other vision changes, hair growth increased.	Continue. Call doctor when convenient.
Rare: • Faintness, increased sweating, irregular or fast heartbeat, chest pain or tightness, shortness of breath, wheezing, unusual tiredness, paleness, heartburn, indigestion, coughing up mucus, fainting, chills or fever, dizziness, pain and stiffness in muscles or joints, headache, urination problems, cold symptoms, back pain, mental and mood changes.	Discontinue. Call doctor right away.

• May cause changes in the treated eye only (color of the iris and eyelid). It may change eyelashes (thicker, longer, color). Changes may take months or years and may be permanent.	Continue. Call doctor when convenient.

WARNINGS & PRECAUTIONS

Don't take if:
You are allergic to any prostaglandin eye medicine.

Before you start, consult your doctor if:
- You plan to have eye or dental surgery.
- You have any eye disease.
- You have heart problems or high blood pressure.
- You have liver or kidney problems.

Over age 60:
No special problems expected, but visit your doctor on a regular basis while using this drug.

Pregnancy:
Decide with your doctor if drug benefits justify risk to unborn baby. Risk category C (see page xviii).

Breastfeeding; Lactation; Nursing Mothers:
Drugs in this group are generally acceptable with breastfeeding. Discuss risks and benefits with your doctor.

Infants & children up to age 18:
Safety and effectiveness in this age group have not been established. Consult doctor.

Prolonged use:
May be necessary.

Skin & sunlight:
No problems expected.

Driving, piloting or hazardous work:
Your vision may be blurred or there may be a change in your near or far vision or night vision for a short time after drug use. Don't drive or pilot aircraft until you learn how medicine affects you. Don't work around dangerous machinery. Don't climb ladders or work in high places.

Discontinuing:
Don't discontinue without consulting your doctor. Dose may require gradual reduction if you have used drug for a long time. Doses of other drugs may also require adjustment.

Others:
- If you have any eye infection, injury or wound, consult doctor before using this medicine.
- Advise any doctor, dentist or pharmacist whom you consult that you use this medicine.
- Keep appointments for regular eye examinations to measure pressure in the eye.

POSSIBLE INTERACTION WITH OTHER DRUGS

GENERIC NAME OR DRUG CLASS	COMBINED EFFECT
None specific	Other drugs may increase or decrease antiglaucoma effect. Consult doctor.

POSSIBLE INTERACTION WITH OTHER SUBSTANCES

INTERACTS WITH	COMBINED EFFECT
Alcohol:	None expected.
Beverages:	None expected.
Cocaine:	Unknown. Avoid.
Foods:	None expected.
Marijuana:	Unknown. Consult doctor.
Tobacco:	None expected.

***See Glossary**

ANTIGOUT DRUGS

GENERIC AND BRAND NAMES

ALLOPURINOL	**FEBUXOSTAT**
Alloprin	**Uloric**
Apo-Allopurinol	
Lopurin	
Novopural	
Purinol	
Zyloprim	

BASIC INFORMATION

Habit forming? No
Prescription needed? Yes
Available as generic? Yes, for some
Drug class: Antigout

USES

- Treats symptoms of chronic gout.
- Treatment for increased levels of uric acid (hyperuricemia) in the body.
- Prevention of kidney stones caused by uric acid.

DOSAGE & USAGE INFORMATION

How to take:
Tablet—Swallow with liquid or food to lessen stomach irritation. Drink extra fluids each day.

When to take:
At the same time(s) each day.

If you forget a dose:
Take as soon as you remember. If it is almost time for the next dose, wait for that dose (don't double this dose) and resume regular schedule.

What drug does:
Slows formation of uric acid by inhibiting enzyme (xanthine oxidase) activity.

Continued next column

OVERDOSE

SYMPTOMS:
May have nausea, vomiting and diarrhea.
WHAT TO DO:
If person uses much larger amount than prescribed or if accidentally swallowed, call poison control center 1-800-222-1222 for instructions or dial 911 (emergency) for help.

Time lapse before drug works:
Reduces blood uric acid in 1 to 3 weeks. May require 6 months to prevent acute gout attacks.

Don't take with:
Any other medicine or any dietary supplement without consulting your doctor or pharmacist.

POSSIBLE ADVERSE REACTIONS OR SIDE EFFECTS

SYMPTOMS	WHAT TO DO
Life-threatening:	
None expected.	
Common:	
Rash, hives, itch.	Discontinue. Call doctor right away.
Infrequent:	
Drowsiness, diarrhea, stomach ache or pain, nausea or vomiting without other symptoms, headache.	Continue. Call doctor when convenient.
Rare:	
• Sore throat, fever, unusual bleeding or bruising, black or tarry stools, mouth or lips sores, blood in urine or stools, skin problems (redness, peeling, burning, tenderness, scaly, thickened), red or itchy eyes, wheezing or shortness of breath or trouble breathing or swelling (hands, feet, lower legs, fingers), fast weight gain, unusual tiredness or weakness, yellow skin or eyes, abnormal heartbeat, chest pain, dizziness, vision problems, hearing loss, slurred speech, numbness or tingling in face or arm or legs.	Discontinue. Call doctor right away.
• Loose fingernails, pain in lower back or side, unexpected nosebleeds, hair loss.	Continue. Call doctor when convenient.

WARNINGS & PRECAUTIONS

Don't take if:
You are allergic to allopurinol or febuxostat. Allergic reactions are rare, but may occur.

Before you start, consult your doctor if:
- You have liver or kidney problems.
- You have history of heart disease or stroke.
- You have diabetes.
- You have hypertension (high blood pressure).
- You are having an acute gout attack.

Over age 60:
No special problems expected.

Pregnancy:
Decide with your doctor if drug benefits justify risk to unborn baby. Risk category C (see page xviii).

Breastfeeding; Lactation; Nursing Mothers:
Drugs in this group are generally acceptable with breastfeeding. Discuss risks and benefits with your doctor.

Infants & children up to age 18:
Usually not recommended for ages under 18. Allopurinol may be used to treat certain rare conditions in children.

Prolonged use:
Talk to your doctor about the need for follow-up medical examinations or laboratory studies to check liver function, kidney function and serum uric-acid levels.

Skin & sunlight:
No problems expected.

Driving, piloting or hazardous work:
Avoid if you feel drowsy. Use of this drug may disqualify you for piloting aircraft.

Discontinuing:
Don't discontinue without doctor's advice, even though symptoms diminish or disappear. These drugs are usually prescribed for long-term use to prevent gout attacks.

Others:
- Acute gout attacks may increase during first weeks of use. If so, consult doctor about your symptoms.
- Consult your doctor before taking vitamin C supplement as it can increase the risk for kidney stones.
- Febuxostat may rarely increase the risk for heart attack or stroke.
- Advise any doctor, dentist or pharmacist whom you consult that you use this medicine.

POSSIBLE INTERACTION WITH OTHER DRUGS

GENERIC NAME OR DRUG CLASS	COMBINED EFFECT
Amoxicillin	Risk of skin rash (with allopurinol).
Ampicillin	Risk of skin rash (with allopurinol).
Anticoagulants, oral*	May increase anticoagulant effect (with allopurinol).
Antineoplastics*	Usage needs to be carefully monitored (with allopurinol).
Azathioprine	Increased effect of azathioprine. Don't use with febuxostat.
Chlorpropamide	Increased effect of chlorpropamide (with allopurinol).
Chlorthalidone	Decreased allopurinol effect.
Cyclosporine	Increased cyclosporine effect (with allopurinol).
Diuretics, thiazide*	Risk of adverse effects (with allopurinol).
Mercaptopurine	Increased mercaptopurine effect. Don't use with febuxostat.
Probenecid	Increased allopurinol effect.
Theophylline	Increased effect of theophylline. Don't use with febuxostat.

POSSIBLE INTERACTION WITH OTHER SUBSTANCES

INTERACTS WITH	COMBINED EFFECT
Alcohol:	May increase uric acid. Avoid.
Beverages:	None expected.
Cocaine:	Unknown. Avoid.
Foods:	None expected.
Marijuana:	Consult doctor.
Tobacco:	None expected.

*See Glossary

ANTIHISTAMINES

GENERIC AND BRAND NAMES

See full list of generic and brand names in the *Generic and Brand Name Directory*, page 842.

BASIC INFORMATION

Habit forming? No
Prescription needed? No, for most
Available as generic? Yes
Drug class: Antihistamine

USES

- Treats symptoms such as itchy, watery eyes; sneezing; itchy nose or throat; runny nose; rash or hives caused by allergies, hay fever and the common cold. The drug product may contain only the antihistamine or the antihistamine may be an ingredient in a combination drug product that treats multiple allergy/cold symptoms.
- Helps prevent motion sickness, dizziness, nausea and vomiting.
- Used as a sleep aid for insomnia.
- Other uses as determined by your doctor.

DOSAGE & USAGE INFORMATION

How to take:
Tablet, capsule, chewable tablet, extended-release forms, liquid or syrup—Follow label directions.

When to take:
Varies with dosage form and your symptoms. Antihistamines may be taken daily or only when you have symptoms or as a preventive.

If you forget a dose:
Take as soon as you remember. If it is almost time for the next dose, wait for that dose (don't double this dose) and resume regular schedule.

Continued next column

OVERDOSE

SYMPTOMS:
Excess drowsiness, blurred vision, ringing in ears, tremors, dizziness, hallucinations, seizures, coma; a child may become overexcited and have a fever.
WHAT TO DO:
- **Dial 911 for all medical emergencies or call poison control center 1-800-222-1222 for instructions.**
- **See emergency information on last 3 pages of this book.**

What drug does:
- Blocks action of histamine after an allergic response triggers histamine release in sensitive cells. Histamines cause itching, sneezing and the other symptoms.
- Appears to work in the vomiting center of the brain to control nausea and vomiting and help prevent motion sickness.

Time lapse before drug works:
15 minutes to 1 hour.

Don't take with:
Any other medicine or any dietary supplement without consulting your doctor or pharmacist.

POSSIBLE ADVERSE REACTIONS OR SIDE EFFECTS

SYMPTOMS	WHAT TO DO
Life-threatening:	
Rare, severe allergic reaction (anaphylaxis)—difficulty breathing, hives, itching, swelling (throat, face), vomiting, dizziness.	Seek emergency treatment immediately.
Common:	
Drowsiness, dryness of mouth or nose or throat, constipation.	Continue. Call doctor when convenient.
Infrequent:	
• Changes in vision, clumsiness, rash, painful or difficult urination.	Discontinue. Call doctor right away.
• Appetite loss, upset stomach.	Continue. Call doctor when convenient.
Rare:	
Nightmares, agitation, irritability, sore throat, fever, rapid heartbeat, unusual bleeding or bruising, fatigue, weakness, confusion, fainting, seizures.	Discontinue. Call doctor right away.

Note: Other side effects that occur may be due to other ingredients in the drug product.

WARNINGS & PRECAUTIONS

Don't take if:
You are allergic to any antihistamine.

Before you start, consult your doctor if:
- You have glaucoma.
- You have enlarged prostate gland.
- You have lung disorder (e.g., asthma or COPD).
- You have intestinal or bladder obstruction or have difficulty with urination.
- You have heart disease or high blood pressure or overactive thyroid.

Over age 60:
Don't exceed recommended dose. Adverse reactions and side effects may be more frequent and severe than in younger persons, especially urination difficulty, diminished alertness and other brain and nervous-system symptoms.

Pregnancy:
Risk factors vary for drugs in this group. See category list on page xviii and consult doctor.

Breastfeeding; Lactation; Nursing Mothers:
Drugs in this group are generally acceptable with breastfeeding. Discuss risks and benefits with your doctor.

Infants & children up to age 18:
- Read the label on the product to see if it is approved for your child's age. Always follow the directions on product's label about how to use. If unsure, ask your doctor or pharmacist.
- Do not use antihistamines for the purpose of making a child sleepy.

Prolonged use:
Follow guidelines on label or consult your doctor.

Skin & sunlight:
May cause rash or intensify sunburn in areas exposed to sun or sunlamp.

Driving, piloting or hazardous work:
Don't drive or pilot aircraft until you learn how medicine affects you. Don't work around dangerous machinery. Don't climb ladders or work in high places. Danger increases if you drink alcohol or take medicine affecting alertness and reflexes, such as other antihistamines, tranquilizers, sedatives, pain medicine, narcotics and mind-altering drugs.

Discontinuing:
No problems expected. Consult doctor if the drug was prescribed for you.

Others:
- Advise any doctor, dentist or pharmacist whom you consult that you take this medicine.
- Check labels of other nonprescription drugs you take to be sure an antihistamine is not included in the product.
- If the antihistamine is in a combination drug product, read the charts in this book for the other drug(s) for complete information.
- Consult doctor or pharmacist if your allergy or cold symptoms do not improve after a week of treatment or sleep problem (insomnia) does not improve after 2 weeks of treatment.

POSSIBLE INTERACTION WITH OTHER DRUGS

GENERIC NAME OR DRUG CLASS	COMBINED EFFECT
Anticholinergics*	Increased anti-cholinergic effect.
Antidepressants*	Increased risk of side effects.
Antihistamines, other	Increased risk of side effects.
Central nervous system (CNS) depressants*	Increased risk of side effects.
Hypnotics*	Increased risk of side effects.
Mind-altering drugs*	Increased risk of side effects.
Narcotics*	Increased risk of side effects.
Sedatives*	Increased risk of side effects.
Sleep inducers*	Increased risk of side effects.
Tranquilizers*	Increased risk of side effects.

POSSIBLE INTERACTION WITH OTHER SUBSTANCES

INTERACTS WITH	COMBINED EFFECT
Alcohol:	Excess sedation. Avoid.
Beverages:	None expected.
Cocaine:	Increased risk of side effects. Avoid.
Foods:	None expected.
Marijuana:	Increased risk of side effects. Avoid.
Tobacco:	None expected.

***See Glossary**

ANTIHISTAMINES (Nasal)

GENERIC AND BRAND NAMES

AZELASTINE
Astelin
Astepro
Dymista

OLOPATADINE
Patanase

BASIC INFORMATION

Habit forming? No
Prescription needed? Yes
Available as generic? No
Drug class: Antihistamine

USES

- Reduces allergic symptoms caused by hay fever (seasonal allergic rhinitis), such as sneezing, itching, runny nose and other nasal symptoms of allergies.
- Azelastine is used to treat vasomotor rhinitis (also called nonallergic rhinitis) which is not caused by allergic reactions, but has similar symptoms.

DOSAGE & USAGE INFORMATION

How to use:
Nasal spray—Gently blow nose before using. Prime the pump per package instructions. Use 2 sprays per nostril. Use 1 spray of azelastine in children ages 5 through 11. Avoid eyes.

When to take:
Usually twice a day or according to doctor's instructions. Effects of spray last for 12 hours.

If you forget a dose:
Use as soon as you remember. If it is almost time for the next dose, wait for that dose (don't double this dose) and resume regular schedule.

Continued next column

What drug does:
- Blocks action of histamines which are released in the body during an allergic reaction (such as to seasonal pollens). Histamines cause itching, swollen tissues, sneezing, runny nose and eyes and other symptoms.
- Azelastine also blocks the action of other inflammatory chemicals that cause stuffy and runny nose.

Time lapse before drug works:
Thirty minutes to 1 to 3 hours.

Don't take with:
Any other medicine or any dietary supplement without consulting your doctor or pharmacist.

OVERDOSE

SYMPTOMS:
May include extreme drowsiness, and feeling restless or agitated, but an overdose with this dosage form is unlikely to occur.
WHAT TO DO:
If person uses much larger amount than prescribed or if accidentally swallowed, call poison control center 1-800-222-1222 for instructions or dial 911 (emergency) for help.

POSSIBLE ADVERSE REACTIONS OR SIDE EFFECTS

SYMPTOMS	WHAT TO DO
Life-threatening:	
None expected.	
Common:	
• Mild drowsiness, bitter taste.	Usually no action needed. If symptoms continue, call doctor.
• Mild nosebleeds, irritation/soreness in the nose, runny nose, headache.	Discontinue. Call doctor when convenient.
Infrequent:	
Dry mouth, sore or painful throat, cough, nausea.	Discontinue. Call doctor when convenient.
Rare:	
• Extreme drowsiness, severe or frequent nosebleeds, rapid or forceful heartbeat, nasal perforation (pain and swelling).	Discontinue. Call doctor right away.
• Fatigue, flu or cold-like symptoms, post nasal drip, sinus infection, dizziness, weight gain, burning or pain when urinating, changes in urination, other unexplained symptoms.	Discontinue. Call doctor when convenient.

WARNINGS & PRECAUTIONS

Don't take if:
You are allergic to azelastine or olopatadine. Allergic reactions are rare, but may occur.

Before you start, consult your doctor if:
- You have kidney problems.
- You have any disorder or injury involving the nose (such as a deviated septum).
- You are allergic to any medication, food or other substance.

Over age 60:
No special problems expected.

Pregnancy:
Decide with your doctor if drug benefits justify risk to unborn baby. Risk category C (see page xviii).

Breastfeeding; Lactation; Nursing Mothers:
Drugs in this group are generally acceptable with breastfeeding. Discuss risks and benefits with your doctor.

Infants & children up to age 18:
- Azelastine is approved in children 12 years and older for vasomotor rhinitis and in children 5 years and older for allergic rhinitis.
- Olopatadine is approved in children 12 years and older for allergic rhinitis.

Prolonged use:
- Antihistamines are normally used during the hay fever season. They are not intended for long-term uninterrupted use.
- Consult your doctor about long-term use of azelastine for vasomotor rhinitis symptoms.

Skin & sunlight:
No problems expected.

Driving, piloting or hazardous work:
Don't drive or pilot aircraft until you learn how medicine affects you. Don't work around dangerous machinery. Don't climb ladders or work in high places. Danger increases if you drink alcohol or take other medicines affecting alertness and reflexes such as antihistamines, tranquilizers, sedatives, pain medicine, narcotics and mind-altering drugs.

Discontinuing:
No problems expected. Consult your doctor if you have been using the spray for a long time.

Others:
- Don't exceed recommended dose. It could increase the risk of adverse reactions.
- Advise any doctor, dentist or pharmacist whom you consult that you take this medicine.
- Consult doctor if new nasal problems occur.
- Avoid getting the spray in your eyes. Rinse eyes with water if this occurs.

POSSIBLE INTERACTION WITH OTHER DRUGS

GENERIC NAME OR DRUG CLASS	COMBINED EFFECT
Central nervous system (CNS) depressants*	May add to any sedative effect.
Cimetidine	Increased azelastine effect. May result in increased sedation.

POSSIBLE INTERACTION WITH OTHER SUBSTANCES

INTERACTS WITH	COMBINED EFFECT
Alcohol:	May cause excessive sedation. Avoid.
Beverages:	None expected.
Cocaine:	Increased risk of side effects. Avoid.
Foods:	None expected.
Marijuana:	May cause sedation. Avoid.
Tobacco:	None expected.

*See Glossary

ANTIHISTAMINES, NONSEDATING

GENERIC AND BRAND NAMES

See full list of generic and brand names in the *Generic and Brand Name Directory*, page 846.

BASIC INFORMATION

Habit forming? No
Prescription needed? Yes, for some
Available as generic? Yes, for some
Drug class: Antihistamine

USES

- Reduces allergic symptoms caused by hay fever (seasonal allergic rhinitis) and perennial rhinitis, such as sneezing, runny nose, itchy nose or throat, itchy and watery eyes.
- Treatment for urticaria (hives).
- Used to help relieve some asthma symptoms.
- Other uses as recommended by your doctor.

DOSAGE & USAGE INFORMATION

How to take:

- Capsule, tablet, suspension, syrup—Swallow with liquid. Most may be taken with food or milk to lessen stomach irritation.
- Chewable tablet—Chew the tablets well before swallowing.
- Oral disintegrating tablet—Let tablet dissolve in mouth. Don't chew. No need to drink fluid.

When to take:
Varies with form and brand. Follow label directions.

If you forget a dose:
Take as soon as you remember. If it is almost time for the next dose, wait for that dose (don't double this dose) and resume regular schedule.

Continued next column

OVERDOSE

SYMPTOMS:
Serious irregular heartbeat, convulsions, being clumsy or unsteady, drowsiness, nausea, severe headache, hallucinations.

WHAT TO DO:

- **Dial 911 for all medical emergencies or call poison control center 1-800-222-1222 for instructions.**
- **See emergency information on last 3 pages of this book.**

What drug does:
Blocks action of histamine after an allergic response triggers histamine release in sensitive cells. Histamines cause itching, sneezing, runny nose and eyes and other symptoms.

Time lapse before drug works:
1 to 2 hours.

Don't take with:
Any other medicine or any dietary supplement without consulting your doctor or pharmacist.

POSSIBLE ADVERSE REACTIONS OR SIDE EFFECTS

SYMPTOMS	WHAT TO DO
Life-threatening:	
None expected.	
Common:	
Dryness of mouth, nose or throat.	Continue. Call doctor when convenient.
Infrequent:	
Increased appetite, weight gain, mild stomach or intestinal problems, cold or flu-like symptoms.	Continue. Call doctor when convenient.
Rare:	
• Heart rhythm disturbances, fainting.	Discontinue. Call doctor right away or get emergency care.
• Allergic reaction such as mild skin rash, headache, nausea, dizziness, nervousness, fatigue, muscle aches. Drowsiness may occur even though these drugs are nonsedating.	Discontinue. Call doctor when convenient.

WARNINGS & PRECAUTIONS

Don't take if:
You are allergic to any antihistamine. Allergic reactions are rare, but may occur.

Before you start, consult your doctor if:
- You have any type of heart disorder.
- You have glaucoma.
- You have enlarged prostate or urinary retention problems.
- You have asthma or a respiratory disease.
- You have liver or kidney disease.
- You have peptic ulcer.
- You have electrolyte abnormality, such as low potassium (hypokalemia).

Over age 60:
No special problems expected.

Pregnancy:
Decide with your doctor if drug benefits justify risk to unborn baby. Risk category B or C (see page xviii).

Breastfeeding; Lactation; Nursing Mothers:
Drugs in this group are generally acceptable with breastfeeding. Discuss risks and benefits with your doctor.

Infants & children up to age 18:
Read the label on the product to see if it is approved for your child's age. Always follow the directions on product's label about how to use. If unsure, ask your doctor or pharmacist.

Prolonged use:
Antihistamines are normally taken during the hay fever season. Longer use may be recommended by your doctor depending on the disorder being treated.

Skin & sunlight:
Rarely, may cause rash or intensify sunburn in areas exposed to sun or ultraviolet light (photosensitivity reaction). Avoid overexposure. Notify doctor if reaction occurs.

Driving, piloting or hazardous work:
No problems expected.

Discontinuing:
No problems expected.

Others:
- Don't exceed recommended dose. This can increase the risk of adverse reactions.
- Advise any doctor, dentist or pharmacist whom you consult that you take this medicine.

POSSIBLE INTERACTION WITH OTHER DRUGS

GENERIC NAME OR DRUG CLASS	COMBINED EFFECT
Anticholinergics	Increased anti-cholinergic effect.
Central nervous system (CNS) depressants*	May add to any sedative effect.
Erythromycins*	Heart rhythm problems. Avoid.
Fluvoxamine	Increased antihistamine effect.
Leukotriene modifiers	Effects unknown. Consult doctor.
Macrolide antibiotics	Increased risk of adverse reactions.
Monoamine oxidase (MAO) inhibitors	Increased sedation. Avoid.

POSSIBLE INTERACTION WITH OTHER SUBSTANCES

INTERACTS WITH	COMBINED EFFECT
Alcohol:	May cause sedation. Avoid.
Beverages:	Grapefruit juice may increase the effect of fexofenadine.
Cocaine:	Increased risk of side effects. Avoid.
Foods:	None expected.
Marijuana:	May cause sedation. Avoid.
Tobacco:	None expected.

***See Glossary**

ANTIHISTAMINES, PHENOTHIAZINE-DERIVATIVE

GENERIC AND BRAND NAMES

See full list of generic and brand names in the *Generic and Brand Name Directory*, page 846.

BASIC INFORMATION

Habit forming? No
Prescription needed? Yes
Available as generic? Yes, for most
Drug class: Tranquilizer (phenothiazine), antihistamine

USES

- Relieves itching of hives, skin allergies, chickenpox.
- Treatment for hay fever, motion sickness, vertigo.
- Treatment for nausea and vomiting.

DOSAGE & USAGE INFORMATION

How to take:
- Tablet or syrup—Swallow with liquid or food to lessen stomach irritation.
- Extended-release capsule—Swallow each dose whole. If you take regular tablets, you may chew or crush them.

When to take:
At the same times each day.

If you forget a dose:
Take as soon as you remember. If it is almost time for the next dose, wait for that dose (don't double this dose) and resume regular schedule.

Continued next column

OVERDOSE

SYMPTOMS:
Fast heartbeat, flushed face, shortness of breath, clumsiness, drowsiness, muscle spasms, jerking movements of head and face.
WHAT TO DO:
- **Dial 911 for all medical emergencies or call poison control center 1-800-222-1222 for instructions.**
- **See emergency information on last 3 pages of this book.**

What drug does:
- Blocks action of histamine after an allergic response triggers histamine release in sensitive cells. Histamines cause itching, sneezing, runny nose and eyes and other symptoms.
- Appears to work in the vomiting center of the brain to control nausea and vomiting and help prevent motion sickness.

Time lapse before drug works:
15 minutes to 1 hour.

Don't take with:
- Antacid or medicine for diarrhea.
- Nonprescription drug for cough, cold or allergy.
- Any other medicine or any dietary supplement without consulting your doctor or pharmacist.

POSSIBLE ADVERSE REACTIONS OR SIDE EFFECTS

SYMPTOMS	WHAT TO DO
Life-threatening: None expected.	
Common: Drowsiness, dryness (of mouth, nose or throat), stuffy nose.	Continue. Call doctor when convenient.
Infrequent: • Difficult urination, blurred or changed vision, dizziness, ringing in ears, skin rash, uncontrolled, jerky movements (with high doses), slow or unusual movement of arms, spasm of neck muscles, stiffening of tongue, eyes rolling upward.	Discontinue. Call doctor right away.
• Nightmares, unusual excitement, nervousness, irritability, loss of appetite, sweating.	Continue. Call doctor when convenient.
Rare: Sore throat, fever, confusion, yellow skin or eyes, fast heartbeat, feeling faint, unusual tiredness or weakness, unusual bleeding or bruising.	Discontinue. Call doctor right away.

ANTIHISTAMINES, PHENOTHIAZINE-DERIVATIVE

WARNINGS & PRECAUTIONS

Don't take if:
- You are allergic to any phenothiazine. Allergic reactions are rare, but may occur.
- You have a blood or bone marrow disease.

Before you start, consult your doctor if:
- You will have surgery within 2 months, including dental surgery, requiring anesthesia.
- You have asthma, emphysema or other lung disorder.
- You take nonprescription ulcer medicine, asthma medicine or amphetamines.

Over age 60:
Adverse reactions and side effects may be more frequent and severe than in younger persons.

Pregnancy:
Decide with your doctor if drug benefits justify risk to unborn baby. Risk category C (see page xviii).

Breastfeeding; Lactation; Nursing Mothers:
Drugs in this group are generally acceptable with breastfeeding. Discuss risks and benefits with your doctor.

Infants & children up to age 18:
Read the label on the product to see if it is approved for your child's age. Always follow the directions on product's label about how to use. If unsure, ask your doctor or pharmacist.

Prolonged use:
- May lead to tardive dyskinesia (involuntary movement of jaws, lips, tongue, chewing).
- Talk to your doctor about the need for follow-up medical examinations or laboratory studies to check complete blood counts, liver function, and eyes.

Skin & sunlight:
One or more drugs in this group may cause rash or intensify sunburn in areas exposed to sun or ultraviolet light (photosensitivity reaction). Avoid overexposure. Notify doctor if reaction occurs.

Driving, piloting or hazardous work:
Don't drive or pilot aircraft until you learn how medicine affects you. Don't work around dangerous machinery. Don't climb ladders or work in high places. Danger increases if you drink alcohol or take medicine affecting alertness and reflexes.

Discontinuing:
May be unnecessary to finish medicine. Follow doctor's instructions.

Others:
- Advise any doctor, dentist or pharmacist whom you consult that you take this medicine.
- May affect results in some medical tests.

POSSIBLE INTERACTION WITH OTHER DRUGS

GENERIC NAME OR DRUG CLASS	COMBINED EFFECT
Antacids*	Decreased antihistamine effect.
Anticholinergics*	Increased anticholinergic effect.
Anticonvulsants, hydantoin*	Increased anticonvulsant effect.
Antidepressants, tricyclic*	Increased antihistamine effect.
Antihistamines,* other	Increased antihistamine effect.
Antithyroid drugs*	Increased risk of bone marrow depression.
Appetite suppressants*	Decreased appetite suppressant effect.
Barbiturates*	Oversedation.
Carteolol	Decreased antihistamine effect.
Central nervous system (CNS) depressants*	Dangerous degree of sedation.
Cisapride	Decreased antihistamine effect.

Continued on page 890

POSSIBLE INTERACTION WITH OTHER SUBSTANCES

INTERACTS WITH	COMBINED EFFECT
Alcohol:	Dangerous oversedation.
Beverages:	None expected.
Cocaine:	Increased risk of side effects. Avoid.
Foods:	None expected.
Marijuana:	Drowsiness. Avoid.
Tobacco:	None expected.

***See Glossary**

ANTIHISTAMINES, PIPERAZINE (Antinausea)

GENERIC AND BRAND NAMES

CYCLIZINE
Marezine

MECLIZINE
Antivert
Antivert/25
Antivert/50
Bonamine
Bonine
Dramamine II
D-Vert 15
D-Vert 30
Meclicot
Medivert
Zentrip

BASIC INFORMATION

Habit forming? No
Prescription needed? Yes, for some
Available as generic? Yes
Drug class: Antihistamine, antiemetic, anti-motion sickness

USES

- Prevention and treatment of motion sickness.
- Treats nausea and vomiting after operations or radiation treatment.
- Treatment for vertigo.

DOSAGE & USAGE INFORMATION

How to take:

- Tablet or capsule—Swallow with liquid or food to lessen stomach irritation. If you can't swallow whole, crumble tablet or open capsule and take with liquid or food.
- Chewable tablet—May be chewed, swallowed whole or mixed with food.
- Quick-dissolving strips—Let strip dissolve on the tongue. Water or fluid is not needed.

Continued next column

OVERDOSE

SYMPTOMS:
Drowsiness, confusion, incoordination, stupor, coma, weak pulse, blurred vision, shallow breathing, hallucinations, seizures.
WHAT TO DO:

- **Dial 911 for all medical emergencies or call poison control center 1-800-222-1222 for instructions.**
- **See emergency information on last 3 pages of this book.**

When to take:
30 minutes to 1 hour before traveling, or as directed by doctor, or follow label instructions.

If you forget a dose:
Take as soon as you remember. If it is almost time for the next dose, wait for that dose (don't double this dose) and resume regular schedule.

What drug does:
It is not known just how drug works. Appears to reduce sensitivity of nerve endings in inner ear, blocking messages to brain's vomiting center.

Time lapse before drug works:
30 to 60 minutes.

Don't take with:
Any other medicine or any dietary supplement without consulting your doctor or pharmacist.

POSSIBLE ADVERSE REACTIONS OR SIDE EFFECTS

SYMPTOMS	WHAT TO DO
Life-threatening: In case of overdose, see previous column.	
Common: Drowsiness.	Continue. Call doctor when convenient.
Infrequent: Headache, diarrhea or constipation, upset stomach, dry mouth or nose or throat.	Continue. Call doctor when convenient.
Rare: Restlessness, insomnia, blurred vision, frequent or difficult urination, hallucinations, dizziness, increased heartbeat, loss of appetite.	Discontinue. Call doctor if symptoms continue or are severe.

ANTIHISTAMINES, PIPERAZINE (Antinausea)

WARNINGS & PRECAUTIONS

Don't take if:
You are allergic to meclizine or cyclizine. Allergic reactions are rare, but may occur.

Before you start, consult your doctor if:
- You have glaucoma.
- You have an intestinal or bladder blockage.
- You have heart problems.
- You have chronic obstructive pulmonary disease (COPD).
- You have prostate enlargement.

Over age 60:
Adverse reactions and side effects may be more frequent and severe than in younger persons.

Pregnancy:
Decide with your doctor if drug benefits justify risk to unborn baby. Risk category B (see page xviii).

Breastfeeding; Lactation; Nursing Mothers:
Drugs in this group are generally acceptable with breastfeeding. Discuss risks and benefits with your doctor.

Infants & children up to age 18:
Follow doctor's instructions or read product's label to be sure drug is approved for your child's age. Follow label instructions on dosage. Consult doctor or pharmacist if unsure.

Prolonged use:
No problems expected.

Skin & sunlight:
No problems expected.

Driving, piloting or hazardous work:
Don't drive or pilot aircraft until you learn how medicine affects you. Don't work around dangerous machinery. Don't climb ladders or work in high places. Danger increases if you drink alcohol or take medicine affecting alertness and reflexes.

Discontinuing:
No problems expected.

Others:
- Advise any doctor, dentist or pharmacist whom you consult that you take this medicine (if taking it on a regular basis).
- May interfere with some skin allergy tests.
- If you have dry mouth from using this drug, try sugarless candy or gum or small pieces of ice. If dry mouth continues longer than 2 weeks, consult doctor.

POSSIBLE INTERACTION WITH OTHER DRUGS

GENERIC NAME OR DRUG CLASS	COMBINED EFFECT
Anticholinergics*	Increased effect of both drugs.
Central nervous system (CNS) depressants*	Increased depressive effects of both drugs.

POSSIBLE INTERACTION WITH OTHER SUBSTANCES

INTERACTS WITH	COMBINED EFFECT
Alcohol:	Increased sedation. Avoid.
Beverages:	None expected.
Cocaine:	Increased risk of side effects. Avoid.
Foods:	None expected.
Marijuana:	Increased drowsiness, dry mouth. Avoid.
Tobacco:	None expected.

*See Glossary

ANTI-INFLAMMATORY DRUGS, NONSTEROIDAL (NSAIDs)

GENERIC AND BRAND NAMES

See full list of generic and brand names in the *Generic and Brand Name Directory*, page 847.

BASIC INFORMATION

Habit forming? No
Prescription needed? Yes, for some.
Available as generic? Yes, for some.
Drug class: Anti-inflammatory (nonsteroidal), analgesic, antigout agent, fever-reducer

USES

- Treatment for pain, fever and inflammation of a variety of disorders, illnesses and injuries.
- Treatment for joint pain, rheumatoid arthritis, gout, menstrual cramps, osteoarthritis, juvenile idiopathic arthritis and others.

DOSAGE & USAGE INFORMATION

How to take:
- Tablet, capsule, extended-release tablet, delayed-release tablet or extended-release capsule—Swallow with liquid and take with food. If you can't swallow whole, ask doctor or pharmacist for advice.
- Oral suspension, rectal suppository, chewable tablet—Take as directed on label.
- Nasal spray—Spray into each (or one) nostril every 6 hours (for no more than 5 days).

When to take:
At the same times each day (if taken on a regular schedule). NSAIDs are often taken as needed.

If you forget a dose:
Take as soon as you remember. If it is almost time for the next dose, wait for that dose (don't double this dose) and resume regular schedule.

Continued next column

OVERDOSE

SYMPTOMS:
May include nausea, vomiting, drowsiness, confusion, agitation, severe headache, convulsions, coughing up blood, bloody stools, shallow breathing, fainting, coma.
WHAT TO DO:
- **Dial 911 for all medical emergencies or call poison control center 1-800-222-1222 for instructions.**
- **See emergency information on last 3 pages of this book.**

What drug does:
They decrease production of prostaglandins which are substances in the body that cause inflammation, pain and fever.

Time lapse before drug works:
30 minutes to 24 hours (depends on drug). May take 3 weeks regular use for maximum benefit.

Don't take with:
Any other medicine or any dietary supplement without consulting your doctor or pharmacist.

POSSIBLE ADVERSE REACTIONS OR SIDE EFFECTS

SYMPTOMS	WHAT TO DO
Life-threatening:	
Rare, severe allergic reaction (anaphylaxis)—difficulty breathing, hives, itching, swelling (throat, face), vomiting, dizziness.	Seek emergency treatment immediately.
Common:	
Stomach cramps, mild headache, dizziness, drowsiness, nausea, vomiting, heartburn, diarrhea.	Continue. Call doctor when convenient.
Infrequent:	
• Urination changes (difficult, painful, more frequent, burning, has odor, bloody, odd color, cloudy, more or less urine), cuts or sores bleed longer, eyes are dry or painful or red, cough, vision or hearing changes, thirstiness, skin symptoms (e.g., rash or spots), mental or mood changes (confused, depressed, forgetful), ringing in ears, muscle weakness or cramps or pain, unusual vaginal bleeding, high blood pressure, severe back or side pain, yellow eyes or skin, severe headache, swallowing or speaking difficulty, numbness or tingling in hands or feet, heart-beat is fast or pounding.	Discontinue. Call doctor right away.
• Taste changes, feeling bloated, appetite loss, flushing, nose pain with spray use, general ill feeing, eyes sensitive	Continue. Call doctor when convenient.

to light, insomnia, being anxious or nervous, unexpected weight loss, increased sweating, unusual tiredness, rectal bleeding (with suppository drug).

Rare:	
Convulsions, black or tarry or bloody or light-color stools, ongoing stomach pain or severe indigestion, mouth or lip sores, chest pain, fever, chills, vomiting or spitting up blood, unusual bleeding or bruising.	Discontinue. Call doctor right away.

WARNINGS & PRECAUTIONS

Don't take if:
You are allergic to or intolerant of any nonsteroidal anti-inflammatory drug or aspirin.

Before you start, consult your doctor if:
You have epilepsy, Parkinson's disease, ulcers, gastritis, enteritis, ileitis, ulcerative colitis, heart disease, asthma, high blood pressure, bleeding problems, impaired kidney or liver function, fluid retention, lupus erythematosus, alcohol abuse, hemorrhoids, diabetes, anemia, mental illness, recent tobacco use, porphyria, temporal arteritis, polymyalgia rheumatica or mouth sores.

Over age 60:
Adverse reactions and side effects may be more frequent and severe than in younger persons.

Pregnancy:
Always consult doctor. Risk categories vary for drugs in this group. See category list (page xviii).

Breastfeeding; Lactation; Nursing Mothers:
Drugs in this group are generally acceptable with breastfeeding. Discuss risks and benefits with your doctor.

Infants & children up to age 18:
- Use only as directed by your child's doctor.
- For nonprescription drugs, read labels and use only those approved for your child's age.

Prolonged use:
- Stomach (gastrointestinal) bleeding, ulcers, and may raise risk of heart attack or stroke.
- Talk to your doctor about the need for follow-up medical exams or lab studies.

Skin & sunlight:
One or more drugs in this group may cause rash or intensify sunburn in areas exposed to sun or ultraviolet light (photosensitivity reaction). Avoid overexposure. Notify doctor if reaction occurs.

Driving, piloting or hazardous work:
Don't drive or pilot aircraft until you learn how medicine affects you. Don't work around dangerous machinery. Don't climb ladders or work in high places. Danger increases if you drink alcohol or take medicine affecting alertness and reflexes.

Discontinuing:
No problems expected. If drug has been taken for a long time, consult doctor before stopping.

Others:
- Advise any doctor, dentist or pharmacist whom you consult that you take this drug. May increase risk of bleeding if surgery is required.
- Consult doctor if symptoms don't improve in 10 days or a fever lasts more than 3 days.

POSSIBLE INTERACTION WITH OTHER DRUGS

GENERIC NAME OR DRUG CLASS	COMBINED EFFECT
Antacids*	May decrease NSAID effect.
Anticoagulants, oral*	Increased risk of bleeding.
Antihypertensives*	May decrease effect of antihypertensive.
Anti-inflammatory pain relievers (any combination of)	Increased risk of side effects.
Antiplatelet drugs*	Increased risk of bleeding.

Continued on page 890

POSSIBLE INTERACTION WITH OTHER SUBSTANCES

INTERACTS WITH	COMBINED EFFECT
Alcohol:	Risk of stomach problems.
Beverages:	None expected.
Cocaine:	Unknown. Avoid.
Foods:	None expected.
Marijuana:	Consult doctor.
Tobacco:	None expected.

***See Glossary**

ANTI-INFLAMMATORY DRUGS, NONSTEROIDAL (NSAIDs) COX-2 INHIBITORS

GENERIC AND BRAND NAMES

CELECOXIB
Celebrex

BASIC INFORMATION

Habit forming? No
Prescription needed? Yes
Available as generic? Yes
Drug class: Anti-inflammatory (nonsteroidal), analgesic, antigout agent, fever-reducer

USES

- Treatment for joint pain, stiffness, inflammation and swelling of rheumatoid arthritis, osteoarthritis, ankylosing spondylitis and gout.
- Treatment for pain, fever and inflammation.
- Treatment for dysmenorrhea (painful or difficult menstruation).
- Treatment for colorectal polyps.

DOSAGE & USAGE INFORMATION

How to take:
- Tablet or capsule—Swallow with liquid. If you can't swallow whole, open capsule or crumble tablet and take with liquid or food.
- Suspension—Shake the bottle well before use. Measure dose with measuring spoon or cup.

When to take:
At the same times each day.

If you forget a dose:
Take as soon as you remember. If it is almost time for the next dose, wait for that dose (don't double this dose) and resume regular schedule.

Continued next column

OVERDOSE

SYMPTOMS:
Breathing problems, chest tightness, thirst, swelling, decreased urine , bloody or black stools, tiredness or weakness, stomach pain, dizziness, headache, nausea or vomiting.
WHAT TO DO:
- **Dial 911 for all medical emergencies or call poison control center 1-800-222-1222 for instructions.**
- **See emergency information on last 3 pages of this book.**

What drug does:
Reduces tissue concentration of prostaglandins (hormones which produce inflammation and pain).

Time lapse before drug works:
Begins in 2 to 3 hours. May require 3 weeks of regular use for maximum benefit.

Don't take with:
Any other medicine or any dietary supplement without consulting your doctor or pharmacist.

POSSIBLE ADVERSE REACTIONS OR SIDE EFFECTS

SYMPTOMS	WHAT TO DO
Life-threatening:	
Rare, severe allergic reaction (anaphylaxis)—difficulty breathing, hives, itching, swelling (throat, face), vomiting, dizziness.	Seek emergency treatment immediately.
Common:	
• Cough, fever, skin rash, swelling of face, fingers, feet and/or lower legs.	Discontinue. Call doctor right away.
• Back pain, dizziness, gas, headache, nausea, heartburn, burning in throat, sleeplessness, stuffy or runny nose.	Continue. Call doctor when convenient.
Infrequent:	
• Bloody or tarry stools, chills, congestion, diarrhea, fatigue, loss of appetite, muscle pains, blood in urine, pale skin, shortness of breath, severe stomach pain, weight gain, vomiting.	Discontinue. Call doctor right away.
• Anxiety, vision changes, noises in ears, changes in sense of taste, difficulty swallowing, dry mouth, constipation, depression, rapid heartbeat, increased sweating, numbness in fingers or toes, sleepiness.	Continue. Call doctor when convenient.
Rare:	
None expected.	

ANTI-INFLAMMATORY DRUGS, NONSTEROIDAL (NSAIDs) COX-2 INHIBITORS

WARNINGS & PRECAUTIONS

Don't take if:
- You are allergic to aspirin or any nonsteroidal, anti-inflammatory drug.
- You are in the third trimester of your pregnancy.
- You have had recent heart surgery.
- You have not read special note under Uses.

Before you start, consult your doctor if:
- You have a history of alcohol abuse.
- You have bleeding problems or ulcers.
- You have used tobacco recently.
- You have impaired kidney or liver function.
- You have anemia, asthma, dehydration or fluid retention.
- You have high blood pressure or heart disease.

Over age 60:
Adverse reactions and side effects may be more frequent and severe than in younger persons.

Pregnancy:
Decide with your doctor if drug benefits justify risk to unborn baby. Risk category C (see page xviii).

Breastfeeding; Lactation; Nursing Mothers:
This drug is generally acceptable with breastfeeding. Discuss risks and benefits with your doctor.

Infants & children up to age 18:
Celecoxib is used to treat juvenile idiopathic arthritis (JIA) in patients age 2 years and older.

Prolonged use:
- Eye damage; reduced hearing.
- Sore throat, fever.
- Weight gain.
- Talk to your doctor about the need for follow-up medical exams or laboratory studies to check complete blood counts, liver function, stools for blood, eyes.

Skin & sunlight:
No problems expected.

Driving, piloting or hazardous work:
Don't drive or pilot aircraft until you learn how medicine affects you. Don't work around dangerous machinery. Don't climb ladders or work in high places. Danger increases if you drink alcohol or take medicine affecting alertness and reflexes, such as antihistamines, tranquilizers, sedatives, pain medicine, narcotics and mind-altering drugs.

Discontinuing:
No problems expected. If drug has been taken for a long time, consult doctor before discontinuing.

Others:
- May affect results in some medical tests.
- This class of drugs has been linked to an increased risk of heart attack and stroke. Consult doctor before using.
- Advise any doctor, dentist or pharmacist whom you consult that you take this medicine.

POSSIBLE INTERACTION WITH OTHER DRUGS

GENERIC NAME OR DRUG CLASS	COMBINED EFFECT
Angiotensin-converting enzyme (ACE) inhibitors*	May decrease ACE inhibitor effect.
Antacids*	Decreased pain relief.
Antifungals, azole	Increased risk of side effects.
Anti-inflammatory drugs, nonsteroidal (NSAIDs)* other	Increased risk of side effects.
Aspirin	Increased risk of stomach ulcer.
Dextromethorphan	Increased effect of dextromethorphan.
Diuretics	Decreased diuretic effect.
Lithium	Increased lithium effect.
Methotrexate	Increased methotrexate effect.
Phenytoin	Decreased pain relief.
Rifampin	Decreased rifampin effect.
Warfarin	Increased risk of bleeding problems.

POSSIBLE INTERACTION WITH OTHER SUBSTANCES

INTERACTS WITH	COMBINED EFFECT
Alcohol:	Possible stomach ulcer or bleeding.
Beverages:	None expected.
Cocaine:	Unknown. Avoid.
Foods:	None expected.
Marijuana:	Consult doctor.
Tobacco:	None expected.

***See Glossary**

ANTI-INFLAMMATORY DRUGS, NONSTEROIDAL (NSAIDs) (Ophthalmic)

GENERIC AND BRAND NAMES

BROMFENAC
- Prolensa
- Xibrom

DICLOFENAC
- Voltaren Ophtha
- Voltaren Ophthalmic

FLURBIPROFEN
- Ocufen

INDOMETHACIN
- Indocid

KETOROLAC
- Acular
- Acular LS

NEPAFENAC
- Ilevro
- Nevanac

SUPROFEN
- Profenal

BASIC INFORMATION

Habit forming? No
Prescription needed? Yes
Available as generic? Yes, for some
Drug class: Ophthalmic anti-inflammatory agents, nonsteroidal

USES

- Used to prevent problems during and following eye surgery, such as cataract removal.
- Treatment for eye itching caused by seasonal allergic conjunctivitis.

DOSAGE & USAGE INFORMATION

How to use:
Eye solution
- Wash hands.
- Apply pressure to inside corner of eye with middle finger.
- Continue pressure for 1 minute after placing medicine in eye.
- Tilt head backward. Pull lower lid away from eye with index finger of the same hand.
- Drop eye drops into pouch and close eye. Don't blink.
- Keep eyes closed for 1 to 2 minutes.

Continued next column

When to use:
As directed by your doctor or on the label. Your doctor or nurse may instill the drug before an eye operation.

If you forget a dose:
Use as soon as you remember. If it is almost time for the next dose, wait for that dose (don't double this dose) and resume regular schedule.

What drug does:
Blocks prostaglandin production. Prostaglandins cause inflammatory responses and constriction of the pupil.

Time lapse before drug works:
Immediately.

Don't use with:
Any other medicine or any dietary supplement without consulting your doctor or pharmacist.

OVERDOSE

SYMPTOMS:
Unknown.
WHAT TO DO:
If person uses much larger amount than prescribed or if accidentally swallowed, call poison control center 1-800-222-1222 for instructions or dial 911 (emergency) for help.

POSSIBLE ADVERSE REACTIONS OR SIDE EFFECTS

SYMPTOMS	WHAT TO DO
Life-threatening:	
None expected.	
Common:	
Brief and mild burning or stinging when drops are administered.	No action necessary.
Infrequent:	
None expected.	
Rare:	
Allergic reaction (itching, tearing), redness or swelling or bleeding in eye not present before, eye pain, sensitivity to light.	Discontinue. Call doctor right away.

ANTI-INFLAMMATORY DRUGS, NONSTEROIDAL (NSAIDs) (Ophthalmic)

WARNINGS & PRECAUTIONS

Don't use if:
- You are allergic to any eye medication.
- You are allergic to any nonsteroidal anti-inflammatory drugs taken orally, e.g., aspirin.

Before you start, consult your doctor if:
- You have any bleeding disorder such as hemophilia.
- You have or have had herpes simplex keratitis (an inflammation of the cornea).
- You have allergies to any medications, foods or other substances.

Over age 60:
No special problems expected.

Pregnancy:
Decide with your doctor if drug benefits justify risk to unborn baby. Risk category C (see page xviii).

Breastfeeding; Lactation; Nursing Mothers:
Drugs in this group are generally acceptable with breastfeeding. Discuss risks and benefits with your doctor.

Infants & children up to age 18:
Safety and effectiveness in this age group have not been established. Consult doctor.

Prolonged use:
Not intended for long-term use.

Skin & sunlight:
No special problems expected.

Driving, piloting or hazardous work:
No special problems expected.

Discontinuing:
No special problems expected.

Others:
Don't use leftover medicine for other eye problems without your doctor's approval. Some eye infections could be made worse.

POSSIBLE INTERACTION WITH OTHER DRUGS

GENERIC NAME OR DRUG CLASS	COMBINED EFFECT
Anticoagulants,* oral	May increase bleeding tendency.
Antiglaucoma drugs*	May decrease antiglaucoma effect (with flurbiprofen).
Carbachol	Decreased carbachol effect.

POSSIBLE INTERACTION WITH OTHER SUBSTANCES

INTERACTS WITH	COMBINED EFFECT
Alcohol:	None expected.
Beverages:	None expected.
Cocaine:	None expected.
Foods:	None expected.
Marijuana:	None expected.
Tobacco:	None expected.

*See Glossary

ANTI-INFLAMMATORY DRUGS, STEROIDAL (Ophthalmic)

GENERIC AND BRAND NAMES

See full list of generic and brand names in the *Generic and Brand Name Directory*, page 848.

BASIC INFORMATION

Habit forming? No
Prescription needed? Yes
Available as generic? Yes
Drug class: Adrenocorticoid (ophthalmic); anti-inflammatory, steroidal (ophthalmic)

USES

- Relieves redness and irritation due to allergies or other irritants.
- Prevents damage to eye.
- Treatment for anterior uveitis (an eye infection).

DOSAGE & USAGE INFORMATION

How to use:
Eye drops
- Wash hands.
- Apply pressure to inside corner of eye with middle finger.
- Continue pressure for 1 minute after placing medicine in eye.
- Tilt head backward. Pull lower lid away from eye with index finger of the same hand.
- Drop eye drops into pouch and close eye. Don't blink.
- Keep eyes closed for 1 to 2 minutes.

Eye ointment
- Wash hands.
- Pull lower lid down from eye to form a pouch.
- Squeeze tube to apply thin strip of ointment into pouch.
- Close eye for 1 to 2 minutes.
- Don't touch applicator tip to any surface (including the eye). If you accidentally touch tip, clean with warm water and soap.
- Keep container tightly closed.

Continued next column

- Keep cool, but don't freeze.
- Wash hands immediately after using.

When to use:
As directed.

If you forget a dose:
Use as soon as you remember. If it is almost time for the next dose, wait for that dose (don't double this dose) and resume regular schedule.

What drug does:
Affects cell membranes and decreases response to irritating substances.

Time lapse before drug works:
Immediately.

Don't use with:
Medicines for abdominal cramps or glaucoma without first consulting your doctor.

OVERDOSE

SYMPTOMS:
Unknown.
WHAT TO DO:
If person uses much larger amount than prescribed or if accidentally swallowed, call poison control center 1-800-222-1222 for instructions or dial 911 (emergency) for help.

POSSIBLE ADVERSE REACTIONS OR SIDE EFFECTS

SYMPTOMS	WHAT TO DO
Life-threatening:	
None expected.	
Common:	
None expected.	
Infrequent:	
Watery, stinging, burning eyes.	Continue. Call doctor when convenient.
Rare:	
• Eye pain, blurred vision, drooping eyelid, seeing halos around lights, enlarged pupils, seeig flashes of light.	Discontinue. Call doctor right away.
• Eye symptoms (discharge, dryness, irritation, tearing, sensation of foreign body), sore throat, runny or stuffy nose.	Continue. Call doctor when convenient.

ANTI-INFLAMMATORY DRUGS, STEROIDAL (Ophthalmic)

WARNINGS & PRECAUTIONS

Don't use if:
You are allergic to any cortisone medicine.

Before you start, consult your doctor if:
- You have or ever have had any eye infection, glaucoma, virus (herpes) or fungus infection of the eye or tuberculosis of the eye.
- You wear contact lenses (may need to discontinue wearing temporarily).

Over age 60:
No special problems expected.

Pregnancy:
Risk factors vary for drugs in this group. See category list on page xviii and consult doctor.

Breastfeeding; Lactation; Nursing Mothers:
Drugs in this group are generally acceptable with breastfeeding. Discuss risks and benefits with your doctor.

Infants & children up to age 18:
Follow instructions provided by your child's doctor.

Prolonged use:
- Recheck with eye doctor at regular intervals.
- May develop glaucoma, hypertension of the eye, damage to optic nerve, vision changes, cataracts or infections due to suppressive effects of drug.

Skin & sunlight:
No problems expected.

Driving, piloting or hazardous work:
No problems expected.

Discontinuing:
No problems expected.

Others:
- Cortisone eye medicines should not be used for bacterial, viral, fungal or tubercular infections.
- Keep cool, but don't freeze.
- Notify doctor if condition doesn't improve within 3 days.
- Contact lens wearers have increased risk of infection.

POSSIBLE INTERACTION WITH OTHER DRUGS

GENERIC NAME OR DRUG CLASS	COMBINED EFFECT
Antiglaucoma drugs,* long- and short-acting	Decreased antiglaucoma effect.

POSSIBLE INTERACTION WITH OTHER SUBSTANCES

INTERACTS WITH	COMBINED EFFECT
Alcohol:	None expected.
Beverages:	None expected.
Cocaine:	None expected.
Foods:	None expected.
Marijuana:	None expected.
Tobacco:	None expected.

***See Glossary**

ANTI-INFLAMMATORY DRUGS, STEROIDAL (Otic)

GENERIC AND BRAND NAMES

BETAMETHASONE (otic)	**DEXAMETHASONE (otic)**
Betnesol	**Decadron**

BASIC INFORMATION

Habit forming? No
Prescription needed? Yes
Available as generic? Yes
Drug class: Anti-inflammatory, steroidal (otic); adrenocorticoid (otic)

USES

- Treats inflammation symptoms (redness, swelling, itching) of the ear due to allergies or other disorders.
- Treats seborrheic and eczematoid dermatitis involving the ear.

DOSAGE & USAGE INFORMATION

How to use:
As directed by your doctor or pharmacist. The following are general instructions.

How to use ear drops:
- Wash and dry hands.
- Warm drops by holding container in your hand for a few minutes.
- Lie down with affected ear up.
- Adults—Pull ear lobe back and up.
- Children—Pull ear lobe down and back.
- Put the correct number of drops into the ear. Do not allow dropper to touch the ear.
- Wipe away any spilled drops.
- Stay lying down for 2 to 5 minutes.

When to use:
As directed by your doctor.

Continued next column

If you forget a dose:
Use as soon as you remember. If it is almost time for the next dose, wait for that dose (don't double this dose) and resume regular schedule.

What drug does:
Decreases inflammation.

Time lapse before drug works:
Symptoms should improve within 5 to 7 days. Complete healing may take up to several weeks.

Don't use with:
Other ear medications unless directed by your doctor.

OVERDOSE

SYMPTOMS:
Unknown.
WHAT TO DO:
If person uses much larger amount than prescribed or if accidentally swallowed, call poison control center 1-800-222-1222 for instructions or dial 911 (emergency) for help.

POSSIBLE ADVERSE REACTIONS OR SIDE EFFECTS

SYMPTOMS	WHAT TO DO
Life-threatening:	
None expected.	
Common:	
None expected.	
Infrequent:	
Burning or stinging of the ear.	Continue. Call doctor if symptoms persist.
Rare:	
New or unusual symptoms occur (may be due to drug being absorbed into the body).	Discontinue. Call doctor right away.

ANTI-INFLAMMATORY DRUGS, STEROIDAL (Otic)

WARNINGS & PRECAUTIONS

Don't use if:
You are allergic to any corticosteroid* drugs.

Before you start, consult your doctor if:
- Eardrum is punctured.
- You have diabetes, heart disease, epilepsy, glaucoma, high blood pressure, osteoporosis, or tuberculosis.
- You have a chronic ear infection or other ear problem.
- You have a viral or fungal infection.

Over age 60:
No special problems expected.

Pregnancy:
Decide with your doctor if drug benefits justify risk to unborn baby. Risk category C (see page xviii).

Breastfeeding; Lactation; Nursing Mothers:
Drugs in this group are generally acceptable with breastfeeding. Discuss risks and benefits with your doctor.

Infants & children up to age 18:
Follow instructions provided by your child's doctor.

Prolonged use:
Not intended for prolonged use.

Skin & sunlight:
No problems expected.

Driving, piloting or hazardous work:
No problems expected.

Discontinuing:
No problems expected.

Others:
- Do not increase or decrease dosage without doctor's approval.
- Follow your doctor's instructions for additional ear care at home.
- Call your doctor if ear symptoms worsen or don't improve after a few days of treatment.
- Advise any doctor, dentist or pharmacist whom you consult that you are using this drug.

POSSIBLE INTERACTION WITH OTHER DRUGS

GENERIC NAME OR DRUG CLASS	COMBINED EFFECT
Phenytoin	May decrease effect of anti-inflammatory.

POSSIBLE INTERACTION WITH OTHER SUBSTANCES

INTERACTS WITH	COMBINED EFFECT
Alcohol:	None expected.
Beverages:	None expected.
Cocaine:	None expected.
Foods:	None expected.
Marijuana:	None expected.
Tobacco:	None expected.

***See Glossary**

ANTIMALARIAL

GENERIC NAMES

HALOFANTRINE **MEFLOQUINE**

BASIC INFORMATION

Habit forming? No
Prescription needed? Yes
Available as generic? Yes, for some
Drug class: Antiprotozoal, antimalarial, antiparasitic

USES

- Treats malaria caused by *Plasmodium falciparum* (either chloroquine-sensitive or chloroquine-resistant).
- Treats malaria caused by *Plasmodium vivax.*
- Mefloquine helps prevent malaria in people traveling into areas where malaria is prevalent.

DOSAGE & USAGE INFORMATION

How to take:
- Mefloquine tablet—Swallow with food, milk or water to lessen stomach irritation.
- Halofantrine tablet—Take on an empty stomach.
- Halofantrine suspension—Varies by age and weight; follow physician's directions.

When to take:
- Mefloquine treatment is usually given as 5 tablets in a single dose, while prevention with mefloquine should start a week prior to travel.
- Halofantrine tablets and suspension are taken every 6 hours, 3 times a day for one day on an empty stomach, 1 hour before or 2 hours after a meal.

Continued next column

OVERDOSE

SYMPTOMS:
Seizures, heart rhythm disturbances.
WHAT TO DO:
- **Dial 911 for all medical emergencies or call poison control center 1-800-222-1222 for instructions.**
- **See emergency information on last 3 pages of this book.**

If you forget a dose:
Take as soon as you remember. If it is almost time for the next dose, wait for that dose (don't double this dose) and resume regular schedule.

What drug does:
Exact mechanism unknown. Mefloquine kills parasite in one of its developmental stages, while halofantrine treats malaria in its acute stage.

Time lapse before drug works:
6 to 24 hours.

Don't take with:
- Sulfadoxine and pyrimethamine combination (Fansidar).
- Any other medicines (including over-the-counter drugs such as cough and cold medicines, laxatives, antacids, diet pills, caffeine, nose drops or vitamins) without consulting your doctor or pharmacist.

POSSIBLE ADVERSE REACTIONS OR SIDE EFFECTS

SYMPTOMS	WHAT TO DO
Life-threatening:	
None expected.	
Common:	
• Dizziness, headache, lightheadedness, abdominal pain, diarrhea, nausea or vomiting, rash, visual disturbances.	Discontinue. Call doctor right away.
• Insomnia, appetite loss.	Continue. Call doctor when convenient.
Infrequent:	
None expected.	
Rare:	
Change in heart rate, confusion, anxiety, depression, seizure, hallucinations, psychosis, black urine or decrease in urine amount, chest or lower back pain, rapid breathing.	Discontinue. Call doctor right away.

WARNINGS & PRECAUTIONS

Don't take if:
You are allergic to mefloquine, halofantrine, quinine, quinidine or related medications. Allergic reactions are rare, but may occur.

Before you start, consult your doctor if:
- You plan to become pregnant within the medication period or 2 months after.
- You have heart trouble, especially heart block.
- You have depression or other emotional problems.
- You are giving this to a child under 40 pounds of body weight.
- You have epilepsy or a seizure disorder.

Over age 60:
Adverse reactions and side effects may be more frequent and severe than in younger persons.

Pregnancy:
Decide with your doctor if drug benefits justify risk to unborn baby. Risk category C (see page xviii).

Breastfeeding; Lactation; Nursing Mothers:
Drugs in this group are generally acceptable with breastfeeding. Discuss risks and benefits with your doctor.

Infants & children up to age 18:
- For halofantrine, consult your child's doctor.
- Mefloquine is approved for over age 2.

Prolonged use:
Not recommended.

Skin & sunlight:
No problems expected.

Driving, piloting or hazardous work:
Don't drive or pilot aircraft until you learn how medicine affects you. Don't work around dangerous machinery. Don't climb ladders or work in high places. Danger increases if you drink alcohol or take medicine affecting alertness and reflexes.

Discontinuing:
Don't discontinue without doctor's advice until you complete the prescribed dosage. Consult doctor if side effect symptoms continue after stopping the drug.

Others:
- Periodic physical (including eye) examinations and blood studies recommended.
- Use of mefloquine can increase the risk for serious neurological side effects.(e.g., dizziness and loss of balance) that could be permanent. The drug can also increase the risk of psychiatric side effects (e.g., anxiety, depression, hallucinations). Discuss risks and benefits with your doctor.
- Resistance to one or more of these drugs by some strains of malaria has been reported, so prevention and treatment of malaria may not be uniformly effective.

POSSIBLE INTERACTION WITH OTHER DRUGS

GENERIC NAME OR DRUG CLASS	COMBINED EFFECT
Anticonvulsants*	Possible lowered seizure control.
Beta-adrenergic blocking agents*	Heartbeat irregularities or cardiac arrest. Avoid.
Calcium channel blockers*	Heartbeat irregularities.
Chloroquine	Increased chance of seizures. Avoid.
Propranolol	Heartbeat irregularities.
Quinidine	Increased chance of seizures and heart rhythm disturbances.
Quinine	Increased chance of seizures and heart rhythm disturbances.
Typhoid vaccine (oral)	Concurrent use may decrease effectiveness of vaccine.

POSSIBLE INTERACTION WITH OTHER SUBSTANCES

INTERACTS WITH	COMBINED EFFECT
Alcohol:	Possible liver toxicity. Avoid.
Beverages: Any alcoholic beverage.	Possible liver toxicity. Avoid.
Cocaine:	Unknown. Avoid.
Foods:	None expected.
Marijuana:	Consult doctor.
Tobacco:	None expected.

*See Glossary

ANTIMYASTHENICS

GENERIC AND BRAND NAMES

NEOSTIGMINE
Prostigmin

PYRIDOSTIGMINE
Mestinon
Mestinon Timespans
Regonol

BASIC INFORMATION

Habit forming? No
Prescription needed? Yes
Available as generic? Yes, for some
Drug class: Cholinergic, antimyasthenic

USES

- Diagnosis and treatment of myasthenia gravis.
- Treatment of urinary retention and abdominal distension.
- Antidote to adverse effects of muscle relaxants used in surgery.

DOSAGE & USAGE INFORMATION

How to take:
- Tablet or syrup—Swallow with liquid or food to lessen stomach irritation.
- Extended-release tablet—Swallow each dose whole. If you take regular tablets, you may chew or crush them.

When to take:
As directed, usually 3 or 4 times a day.

If you forget a dose:
Take as soon as you remember. If it is almost time for the next dose, wait for that dose (don't double this dose) and resume regular schedule.

Continued next column

OVERDOSE

SYMPTOMS:
Muscle weakness or paralysis, cramps, twitching or clumsiness; severe diarrhea, nausea, vomiting, stomach cramps or pain; breathing difficulty; confusion, irritability, nervousness, restlessness, fear; unusually slow heartbeat; seizures; blurred vision; extreme fatigue.

WHAT TO DO:
- **Dial 911 for all medical emergencies or call poison control center 1-800-222-1222 for instructions.**
- **See emergency information on last 3 pages of this book.**

What drug does:
Inhibits the chemical activity of an enzyme (cholinesterase) so nerve impulses can cross the junction of nerves and muscles.

Time lapse before drug works:
Usually takes 10 to 14 days to determine if drug helps relieve symptoms.

Don't take with:
Any other medicine or any dietary supplement without consulting your doctor or pharmacist.

POSSIBLE ADVERSE REACTIONS OR SIDE EFFECTS

SYMPTOMS	WHAT TO DO
Life-threatening:	
In case of overdose, see previous column.	
Common:	
Excess saliva, unusual sweating, mild diarrhea, nausea, vomiting, stomach cramps or pain.	Continue. Call doctor when convenient.
Infrequent:	
Constricted pupils, watery eyes, lung congestion, frequent urge to urinate, confusion, slurred speech.	Continue, but call doctor right away.
Rare:	
Other symptoms.	Continue. Call doctor when convenient.

WARNINGS & PRECAUTIONS

Don't take if:
You are allergic to any cholinergic* or bromide. Allergic reactions are rare, but may occur.

Before you start, consult your doctor if:
- You have bronchial asthma.
- You have heartbeat irregularities.
- You have urinary obstruction or urinary tract infection.

Over age 60:
Adverse reactions and side effects may be more frequent and severe than in younger persons.

Pregnancy:
Decide with your doctor if drug benefits justify risk to unborn baby. Risk category C (see page xviii).

Breastfeeding; Lactation; Nursing Mothers:
Drugs in this group are generally acceptable with breastfeeding. Discuss risks and benefits with your doctor.

Infants & children up to age 18:
Safety and effectiveness in this age group have not been established. Consult doctor.

Prolonged use:
Medication may lose effectiveness. Ask your doctor about discontinuing drug for a few days to possibly help restore effect.

Skin & sunlight:
No problems expected.

Driving, piloting or hazardous work:
Don't drive or pilot aircraft until you learn how medicine affects you. Don't work around dangerous machinery. Don't climb ladders or work in high places. Danger increases if you drink alcohol or take medicine affecting alertness and reflexes, such as antihistamines, tranquilizers, sedatives, pain medicine, narcotics and mind-altering drugs.

Discontinuing:
Don't discontinue without doctor's advice until you complete prescribed dose, even though symptoms diminish or disappear.

Others:
- Advise any doctor, dentist or pharmacist whom you consult that you take this medicine.
- Be cautious about participating in hot weather activities since drug may cause excessive sweating.

POSSIBLE INTERACTION WITH OTHER DRUGS

GENERIC NAME OR DRUG CLASS	COMBINED EFFECT
Anesthetics, local or general*	Decreased effect of antimyasthenic.
Antiarrhythmics*	Decreased effect of antimyasthenic.
Anticholinergics*	May mask severe side effects.
Cholinergics,* other	Possible brain and nervous system toxicity.
Guanadrel	Decreased effect of antimyasthenic.
Guanethidine	Decreased effect of antimyasthenic.
Mecamylamine	Decreased effect of antimyasthenic.
Procainamide	Decreased effect of antimyasthenic.
Quinidine	Decreased effect of antimyasthenic.

POSSIBLE INTERACTION WITH OTHER SUBSTANCES

INTERACTS WITH	COMBINED EFFECT
Alcohol:	No proven problems with small doses.
Beverages:	None expected.
Cocaine:	Unknown. Avoid.
Foods:	None expected.
Marijuana:	Consult doctor.
Tobacco:	None expected.

***See Glossary**

ANTIPYRINE & BENZOCAINE (Otic)

BRAND NAMES

A/B Otic	**Aurodex**
Allergen	**Dolotic**
Analgesic Ear Drops	**Earache Drops**
Antiben	**Earocol**
	Otiprin
Auralgan	**Otocalm**

BASIC INFORMATION

Habit forming? No
Prescription needed?
Yes (in US;
No in Canada)
Available as generic? Yes
Drug class: Analgesic (otic); anesthetic

USES

- Relieves symptoms of middle ear infections (otitis media). It does not treat the infection itself.
- Used to soften earwax so it can be removed.

DOSAGE & USAGE INFORMATION

How to use ear drops:
- Wash and dry hands.
- Warm drops by holding container in your hand for a few minutes.
- Lie down with affected ear up.
- Adults—Pull ear lobe back and up.
- Children—Pull ear lobe down and back.
- Put the correct number of drops into the ear. Do not allow dropper to touch the ear.
- Wipe away any spilled drops.
- Stay lying down for 2 to 5 minutes.
- For ear wax removal, follow doctor's instructions.

When to use:
Every 1 to 2 hours for 4 hours, then 4 times a day when needed for pain.

Continued next column

OVERDOSE

SYMPTOMS:
Unknown.
WHAT TO DO:
If person uses much larger amount than prescribed or if accidentally swallowed, call poison control center 1-800-222-1222 for instructions or dial 911 (emergency) for help.

If you forget a dose:
Use as soon as you remember. If it is almost time for the next dose, wait for that dose (don't double this dose) and resume regular schedule.

What drug does:
Helps relieve the pain, congestion and swelling of ear infections. It does not cure the infection.

Time lapse before drug works:
10 minutes.

Don't use with:
Any other ear medicine without consulting your doctor or pharmacist.

POSSIBLE ADVERSE REACTIONS OR SIDE EFFECTS

SYMPTOMS	WHAT TO DO
Life-threatening: None expected.	
Common: None expected.	
Infrequent: Itching or burning in ear (probably represents allergic reaction).	Discontinue. Call doctor right away.
Rare: None expected.	

ANTIPYRINE & BENZOCAINE (Otic)

WARNINGS & PRECAUTIONS

Don't use if:
You are allergic to any local anesthetic (name usually ends with "caine"). Allergic reactions are rare, but may occur.

Before you start, consult your doctor if:
Eardrum is ruptured.

Over age 60:
No special problems expected.

Pregnancy:
Decide with your doctor if drug benefits justify risk to unborn baby. Risk category C (see page xviii).

Breastfeeding; Lactation; Nursing Mothers:
Drugs in this group are generally acceptable with breastfeeding. Discuss risks and benefits with your doctor.

Infants & children up to age 18:
Read product's label to be sure it is approved for your child's age. Follow label instructions on dosage. Consult doctor or pharmacist if unsure.

Prolonged use:
Not intended for prolonged use.

Skin & sunlight:
No problems expected.

Driving, piloting or hazardous work:
No problems expected.

Discontinuing:
No problems expected.

Others:
- Keep cool, but don't freeze.
- Don't touch tip of dropper to any other surface.
- Don't rinse the dropper. Wipe with clean cloth and close tightly.

POSSIBLE INTERACTION WITH OTHER DRUGS

GENERIC NAME OR DRUG CLASS	COMBINED EFFECT
None expected.	

POSSIBLE INTERACTION WITH OTHER SUBSTANCES

INTERACTS WITH	COMBINED EFFECT
Alcohol:	None expected.
Beverages:	None expected.
Cocaine:	None expected.
Foods:	None expected.
Marijuana:	None expected.
Tobacco:	None expected.

ANTISEBORRHEICS (Topical)

GENERIC AND BRAND NAMES

See full list of generic and brand names in the *Generic and Brand Name Directory,* page 848.

BASIC INFORMATION

Habit forming? No
Prescription needed? Yes
Available as generic? Yes
Drug class: Antiseborrheic

USES

Treats dandruff or seborrheic dermatitis of scalp.

DOSAGE & USAGE INFORMATION

How to use:

- Wet hair and scalp.
- Apply enough medicine to form lather.
- Rub in well. Keep away from eyes.
- Allow to remain on scalp 3 to 5 minutes, then rinse.
- Repeat above steps once.

When to use:
As directed by doctor. Twice a week for shampoo is average.

If you forget a dose:
Use as soon as you remember. If it is almost time for the next dose, wait for that dose (don't double this dose) and resume regular schedule.

What drug does:
Slows cell growth in scales on scalp.

Time lapse before drug works:
Varies a great deal. If no improvement in 2 weeks, notify doctor.

Don't use with:
Other hair or scalp preparations without consulting your doctor or pharmacist.

OVERDOSE

SYMPTOMS:
Unknown.
WHAT TO DO:
If person uses much larger amount than prescribed or if accidentally swallowed, call poison control center 1-800-222-1222 for instructions or dial 911 (emergency) for help.

POSSIBLE ADVERSE REACTIONS OR SIDE EFFECTS

SYMPTOMS	WHAT TO DO
Life-threatening: None expected.	
Common: None expected.	
Infrequent: Irritation not present before using, rash, dry or itching scalp.	Continue. Call doctor or discontinue if symptoms persist.
Rare: None expected.	

WARNINGS & PRECAUTIONS

Don't use if:
- You are allergic reaction to chloroxine, clioquinol (iodochlorhydroxyquin), iodoquinol (diiodohydroxyquin) or sodium edate. Allergic reactions are rare, but may occur.
- Scalp is blistered or infected with oozing or raw areas.

Before you start, consult your doctor if:
You are allergic to anything.

Over age 60:
No special problems expected.

Pregnancy:
Risk factors vary for drugs in this group. See category list on page xviii and consult doctor.

Breastfeeding; Lactation; Nursing Mothers:
Drugs in this group are generally acceptable with breastfeeding. Discuss risks and benefits with your doctor.

Infants & children up to age 18:
Read product's label to be sure it is approved for your child's age. Follow label instructions on dosage. Consult doctor or pharmacist if unsure.

Prolonged use:
No problems expected, but check with doctor.

Skin & sunlight:
No special problems expected.

Driving, piloting or hazardous work:
No problems expected, but check with doctor.

Discontinuing:
No problems expected, but check with doctor.

Others:
- If medicine accidentally gets into eyes, flush them immediately with cool water.
- Heat and moisture in bathroom medicine cabinet can cause breakdown of medicine. Store someplace else.
- Keep cool, but don't freeze.

POSSIBLE INTERACTION WITH OTHER DRUGS

GENERIC NAME OR DRUG CLASS	COMBINED EFFECT
Other medicated shampoos	May increase adverse reactions of each medicine.

POSSIBLE INTERACTION WITH OTHER SUBSTANCES

INTERACTS WITH	COMBINED EFFECT
Alcohol:	None expected.
Beverages:	None expected.
Cocaine:	None expected.
Foods:	None expected.
Marijuana:	None expected.
Tobacco:	None expected.

*See Glossary

ANTITHYROID DRUGS

GENERIC AND BRAND NAMES

METHIMAZOLE	PROPYLTHIOURACIL
Tapazole	Propyl-Thyracil
Thiamazole	

BASIC INFORMATION

Habit forming? No
Prescription needed? Yes
Available as generic? Yes
Drug class: Antihyperthyroid

USES

- Treatment of overactive thyroid (hyperthyroidism).
- Treatment of angina in patients who have overactive thyroid.

DOSAGE & USAGE INFORMATION

How to take:
Tablet—Swallow with liquid or food to lessen stomach irritation. If you can't swallow whole, crumble tablet and take with liquid or food.

When to take:
At the same times each day.

If you forget a dose:
Take as soon as you remember. If it is almost time for the next dose, wait for that dose (don't double this dose) and resume regular schedule.

What drug does:
Prevents thyroid gland from producing excess thyroid hormone.

Time lapse before drug works:
10 to 20 days.

Don't take with:
Any other medicine or any dietary supplement without consulting your doctor or pharmacist.

OVERDOSE

SYMPTOMS:
Bleeding, spots on skin, jaundice (yellow eyes and skin), loss of consciousness.
WHAT TO DO:
- **Dial 911 for all medical emergencies or call poison control center 1-800-222-1222 for instructions.**
- **See emergency information on last 3 pages of this book.**

POSSIBLE ADVERSE REACTIONS OR SIDE EFFECTS

SYMPTOMS	WHAT TO DO
Life-threatening: None expected.	
Common: Skin symptoms (rash, itching).	Continue, but call doctor right away.
Infrequent:	
• Dizziness, sore, throat. chills, fever, abdominal pain.	Continue, but call doctor right away.
• Taste loss, constipation, diarrhea.	Continue. Call doctor when convenient.
Rare: Headache, enlarged lymph glands, irregular or rapid heartbeat, unusual bruising or bleeding, backache, numbness or tingling (in toes, fingers or face), joint pain, muscle aches, menstrual irregularities, jaundice (yellow eyes or skin), tiredness or weakness, listless, swollen eyes or feet, black stools, excessive cold feeling, puffy skin, irritability. Propylthiouracil may cause severe liver problems (low fever, itching, nausea, stomach pain, loss of appetite, dark urine, clay-colored stools, yellow skin or eyes).	Continue, but call doctor right away.

WARNINGS & PRECAUTIONS

Don't take if:
You are allergic to antithyroid medicines. Allergic reactions are rare, but may occur.

Before you start, consult your doctor if:
- You have liver disease or a blood disease.
- You have an infection.
- You take anticoagulants.

Over age 60:
Side effects may be more frequent and severe.

Pregnancy:
Consult doctor. Use of the drug in pregnancy is not recommended. It could cause harm to the unborn baby. Risk category D (see page xviii).

Breastfeeding; Lactation; Nursing Mothers:
Drugs in this group are generally acceptable with breastfeeding. Discuss risks and benefits with your doctor.

Infants & children up to age 18:
Follow instructions provided by your child's doctor.

Prolonged use:
- Adverse reactions and side effects more common.
- Talk to your doctor about the need for follow-up medical examinations or laboratory studies to check thyroid function, complete blood counts (white blood cell count, platelet count, red blood cell count, hemoglobin, hematocrit).

Skin & sunlight:
No problems expected.

Driving, piloting or hazardous work:
Don't drive or pilot aircraft until you learn how medicine affects you. Don't work around dangerous machinery. Don't climb ladders or work in high places. Danger increases if you drink alcohol or take medicine affecting alertness and reflexes, such as antihistamines, tranquilizers, sedatives, pain medicine, narcotics and mind-altering drugs.

Discontinuing:
Don't discontinue without consulting doctor. Dose may require gradual reduction if you have taken drug for a long time. Doses of other drugs may also require adjustment.

Others:
- Advise any doctor, dentist or pharmacist whom you consult that you take this medicine.
- Ask your doctor about the symptoms of overactive or underactive thyroid and what to do if they occur.

POSSIBLE INTERACTION WITH OTHER DRUGS

GENERIC NAME OR DRUG CLASS	COMBINED EFFECT
Amiodarone	Decreased antithyroid effect.
Anticoagulants*	Increased effect of anticoagulants.
Antineoplastic drugs*	Increased chance to suppress bone marrow.
Chloramphenicol	Increased chance to suppress bone marrow.
Clozapine	Toxic effect on bone marrow.
Digitalis preparations*	Increased digitalis effect.
Iodine	Decreased antithyroid effect.
Levamisole	Increased risk of bone marrow depression.
Lithium	Decreased thyroid activity.
Potassium iodide	Decreased antithyroid effect.
Tiopronin	Increased risk of toxicity to bone marrow.

POSSIBLE INTERACTION WITH OTHER SUBSTANCES

INTERACTS WITH	COMBINED EFFECT
Alcohol:	Increased possibility of liver toxicity. Avoid.
Beverages:	None expected.
Cocaine:	Unknown. Avoid.
Foods:	None expected.
Marijuana:	Consult doctor.
Tobacco:	None expected.

*See Glossary

ANTIVIRALS FOR HERPES VIRUS

GENERIC AND BRAND NAMES

ACYCLOVIR
- **Alti-Acyclovir**
- **Avirax**
- **Sitavig**
- **Zovirax**

FAMCICLOVIR
- **Famvir**

VALACYCLOVIR
- **Valtrex**

BASIC INFORMATION

Habit forming? No
Prescription needed? Yes
Available as generic? Yes, for some
Drug class: Antiviral

USES

- Treatment for symptoms of herpes virus infections (does not cure the disorders). These infections include herpes simplex, genital herpes and herpes zoster (also known as shingles). Herpes infections may occur on the lips and mouth, genitals, skin and the brain.
- May be used to treat chickenpox and other viral infections as prescribed by your doctor.

DOSAGE & USAGE INFORMATION

How to take:
- Tablet, capsule or oral suspension—Swallow with liquid. If you can't swallow whole, open capsule or crumble tablet and take with liquid or food. Drugs may be taken with or without food. Measure oral suspension with specially marked measuring device.
- Buccal tablet—Follow the instructions provided with the drug.

When to take:
At the same times each day and night. Drugs work best if started within 48 hours of diagnosis (or when symptoms first appear).

Continued next column

OVERDOSE

SYMPTOMS:
Tiredness, seizures, agitation, reduced or no urine output, coma.
WHAT TO DO:
- **Dial 911 for all medical emergencies or call poison control center 1-800-222-1222 for instructions.**
- **See emergency information on last 3 pages of this book.**

If you forget a dose:
Take as soon as you remember. If it is almost time for the next dose, wait for that dose (don't double this dose) and resume regular schedule.

What drug does:
Inhibits the growth and spread of the virus thereby decreasing length of infection and lessening the severity of the symptoms.

Time lapse before drug works:
Begins the first day, but may take several days for symptoms (pain, burning and blisters) to improve.

Don't take with:
Any other medicine or any dietary supplement without consulting your doctor or pharmacist.

POSSIBLE ADVERSE REACTIONS OR SIDE EFFECTS

SYMPTOMS	WHAT TO DO
Life-threatening:	
None expected.	
Common:	
General feeling of illness or discomfort.	Continue. Call doctor when convenient.
Infrequent:	
Headache, nausea, diarrhea, vomiting, tiredness, dizziness.	Continue. Call doctor when convenient.
Rare:	
Vision changes, eyes irritated, swelling of different parts of the body, agitation, fever, chills, confusion, hallucinations, sore throat or mouth, muscle cramps, skin symptoms (rash, blister, itch, peel).	Discontinue. Call doctor right away.

WARNINGS & PRECAUTIONS

Don't take if:
You are allergic to antiviral agents. Allergic reactions are rare, but may occur.

Before you start, consult your doctor if:
- You have kidney disease, neurological problems or any disease of the blood.
- You are allergic to any medication, food or other substance.

Over age 60:
No special problems expected.

Pregnancy:
Decide with your doctor if drug benefits justify risk to unborn baby. Risk category B (see page xviii).

Breastfeeding; Lactation; Nursing Mothers:
Drugs in this group are generally acceptable with breastfeeding. Discuss risks and benefits with your doctor.

Infants & children up to age 18:
Follow instructions provided by your child's doctor.

Prolonged use:
See your doctor for regular visits to check effectiveness of the drug and to check for any blood problems.

Skin & sunlight:
No special problems expected.

Driving, piloting or hazardous work:
Avoid if you experience dizziness after taking drug, otherwise no problems expected.

Discontinuing:
Don't discontinue without doctor's advice until you complete prescribed dose, even though symptoms diminish or disappear.

Others:
- Advise any doctor, dentist or pharmacist whom you consult that you take this medicine.
- If symptoms don't improve within a few days, or if they worsen, consult doctor.
- See your eye doctor regularly if you are taking the drug for eye infection.
- If drug is prescribed for genital herpes, be sure you use proper precautions to prevent spreading the disorder to your sexual partner. If unsure, ask your doctor for information.
- Keep affected skin area clean and dry.
- Decrease blister irritation by wearing loose-fitting clothing.

POSSIBLE INTERACTION WITH OTHER DRUGS

GENERIC NAME OR DRUG CLASS	COMBINED EFFECT
Cimetidine	Increased antiviral effect.
Probenecid	Increased effect of antivirals.
Nephrotoxics*	Increased risk of kidney problems.
Nucleotide reverse transcriptase inhibitors	Increased effect of nucleotide reverse transcriptase inhibitor.
Zidovudine	Increased risk of adverse effects.

POSSIBLE INTERACTION WITH OTHER SUBSTANCES

INTERACTS WITH	COMBINED EFFECT
Alcohol:	None expected.
Beverages:	None expected.
Cocaine:	Unknown. Avoid.
Foods:	None expected.
Marijuana:	Consult doctor.
Tobacco:	None expected.

***See Glossary**

ANTIVIRALS FOR INFLUENZA

GENERIC AND BRAND NAMES

AMANTADINE	RIMANTADINE
Symadine	Flumadine
Symmetrel	

BASIC INFORMATION

Habit forming? No
Prescription needed? Yes
Available as generic? Yes
Drug class: Antiviral, antiparkinsonism

USES

- Prevention and treatment for Type-A flu infections.
- Relief for symptoms of Parkinson's disease (amantadine).

DOSAGE & USAGE INFORMATION

How to take:
- Capsule—Swallow with liquid or food to lessen stomach irritation.
- Syrup—Dilute dose in beverage before swallowing.

When to take:
At the same times each day. For Type-A flu it is especially important to take regular doses as prescribed.

If you forget a dose:
Take as soon as you remember. If it is almost time for the next dose, wait for that dose (don't double this dose) and resume regular schedule.

What drug does:
- Type-A flu—May block penetration of tissue cells by infectious material from virus cells.
- Parkinson's disease and drug-induced extrapyramidal* reactions—Improves muscular condition and coordination.

Continued next column

OVERDOSE

SYMPTOMS:
Heart rhythm problems, blood pressure drop, convulsions, hallucinations, violent behavior, confusion, slurred speech, rolling eyes, breathing difficulty.
WHAT TO DO:
- **Dial 911 for all medical emergencies or call poison control center 1-800-222-1222 for instructions.**
- **See emergency information on last 3 pages of this book.**

Time lapse before drug works:
- Type-A flu—48 hours.
- Parkinson's disease—2 days to 2 weeks.

Don't take with:
- Alcohol.
- Any other medicine or any dietary supplement without consulting your doctor or pharmacist.

POSSIBLE ADVERSE REACTIONS OR SIDE EFFECTS

SYMPTOMS	WHAT TO DO
Life-threatening:	
In case of overdose, see previous column.	
Common:	
Headache, difficulty in concentrating, dizziness or lightheadedness, insomnia, irritability, nervousness, nightmares (these side effects infrequent with rimantadine).	Continue. Call doctor when convenient.
Infrequent:	
• With amantadine—Blurred or changed vision, confusion, difficult urination, hallucinations, fainting.	Discontinue. Call doctor right away.
• Constipation, dry mouth or nose or throat, vomiting, appetite loss, nausea.	Continue. Call doctor when convenient.
Rare:	
• With amantadine—swelling or irritated eyes, depression, swelling (of hands, legs or feet), skin rash.	Discontinue. Call doctor right away.
• Seizures (may occur in persons with a history of seizures).	Discontinue. Seek emergency help.

WARNINGS & PRECAUTIONS

Don't take if:
You are allergic to amantadine or rimantadine. Allergic reactions are rare, but may occur.

Before you start, consult your doctor if:
- You have had epilepsy or other seizures.
- You have had heart disease or heart failure.
- You have had liver or kidney disease.
- You have had peptic ulcers.
- You have had eczema or skin rashes.
- You have had emotional or mental disorders or taken drugs for them.

Over age 60:
Adverse reactions and side effects may be more frequent and severe than in younger persons.

Pregnancy:
Decide with your doctor if drug benefits justify risk to unborn baby. Risk category C (see page xviii).

Breastfeeding; Lactation; Nursing Mothers:
Drugs in this group are generally not recommended with breastfeeding. Discuss risks and benefits with your doctor.

Infants & children up to age 18:
Follow instructions provided by your child's doctor.

Prolonged use:
Skin splotches, feet swelling, rapid weight gain, shortness of breath. Consult doctor.

Skin & sunlight:
One or more drugs in this group may cause rash or intensify sunburn in areas exposed to sun or ultraviolet light (photosensitivity reaction). Avoid overexposure. Notify doctor if reaction occurs.

Driving, piloting or hazardous work:
Don't drive or pilot aircraft until you learn how medicine affects you. Don't work around dangerous machinery. Don't climb ladders or work in high places. Danger increases if you drink alcohol or take medicine affecting alertness and reflexes.

Discontinuing:
- Parkinson's disease—Don't discontinue without doctor's advice until you complete prescribed dose, even though symptoms diminish or disappear.
- Type-A flu—Discontinue 48 hours after symptoms disappear.

Others:
- Parkinson's disease—May lose effectiveness in 3 to 6 months. Consult doctor.
- These drugs are not effective for influenza-B virus.
- Drug-resistant strains of the virus may occur within the same household.

POSSIBLE INTERACTION WITH OTHER DRUGS

GENERIC NAME OR DRUG CLASS	COMBINED EFFECT
Acetaminophen	With rimantadine—Decreased antiviral effect.
Anticholinergics*	With amantadine—Increased risk of side effects.
Antidepressants, tricyclic*	With amantadine—Increased risk of side effects.
Antidyskinetics*	With amantadine—Increased risk of side effects.
Antihistamines*	With amantadine—Increased risk of side effects.
Aspirin	With rimantadine—Decreased antiviral effect.
Carbidopa & Levodopa	With amantadine—Increased effect of carbidopa & levodopa.
Central nervous system (CNS) stimulants*	With amantadine—Increased risk of adverse reactions.
Memantine	With amantadine—Adverse effects of either drug.

POSSIBLE INTERACTION WITH OTHER SUBSTANCES

INTERACTS WITH	COMBINED EFFECT
Alcohol:	Increased alcohol effect. Possible fainting.
Beverages:	None expected.
Cocaine:	Unknown. Avoid.
Foods:	None expected.
Marijuana:	Consult doctor.
Tobacco:	None expected.

***See Glossary**

ANTIVIRALS FOR INFLUENZA, NEURAMINIDASE INHIBITORS

GENERIC AND BRAND NAMES

OSELTAMIVIR
Tamiflu
PERAMIVIR
Rapivab
ZANAMIVIR
Relenza

BASIC INFORMATION

Habit forming? No
Prescription needed? Yes
Available as generic? No
Drug class: Anti-influenza

USES

- Shortens the duration of influenza types A and B. It is best to start using this medicine within 2 days of onset of symptoms.
- Used to help prevent influenza types A and B.

DOSAGE & USAGE INFORMATION

How to take:

- Capsule—Swallow with liquid. If you can't swallow whole, open capsule and take with liquid or food.
- Oral solution—Take as directed on label.
- Powder—This medication is to be used with a device called a Diskhaler. Read and carefully follow the instructions provided with the device. If you are still unsure, consult your pharmacist for detailed instructions.
- Injection—Given by a health care provider.

When to take:

- For oseltamivir, take two doses daily at the same times.
- For zanamivir, take two doses on the first day separated by 2 hours and then two doses daily for 5 days separated by 12 hours.
- For peramivir, given in one injection.

Continued next column

OVERDOSE

SYMPTOMS:
May be similar to adverse reactions.
WHAT TO DO:
If person uses much larger amount than prescribed or if accidentally swallowed, call poison control center 1-800-222-1222 for instructions or dial 911 (emergency) for help.

If you forget a dose:
Take as soon as you remember. If it is almost time for the next dose, wait for that dose (don't double this dose) and resume regular schedule.

What drug does:
Inhibits the spread of the virus by preventing release of the cells within the respiratory tract.

Time lapse before drug works:
Begins in 2 to 3 days.

Don't take with:
Any other medicine or any dietary supplement without consulting your doctor or pharmacist.

POSSIBLE ADVERSE REACTIONS OR SIDE EFFECTS

SYMPTOMS	WHAT TO DO
Life-threatening:	
Rare, severe allergic reaction (anaphylaxis)—difficulty breathing, hives, itching, swelling (throat, face), vomiting, dizziness.	Seek emergency treatment immediately.
Common:	
• Cough, fever, skin rash, swelling (of face, fingers, feet and/or lower legs).	Discontinue. Call doctor right away.
• Back pain, dizziness, gas, headache, nausea, heartburn, burning in throat, sleeplessness, stuffy or runny nose.	Continue. Call doctor when convenient.
Infrequent:	
• Bloody or tarry stools, chills, congestion, diarrhea, fatigue, loss of appetite, muscle pains, shortness of breath, severe stomach pain, weight gain, vomiting.	Discontinue. Call doctor right away.
• Anxiety, vision changes, noises in ears, change in sense of taste, difficulty swallowing, dry mouth, constipation, depression, rapid heartbeat, increased sweating, numbness in fingers or toes, sleepiness.	Continue. Call doctor if symptoms persist.
Rare:	
Abnormal behaviors.	Discontinue. Call doctor right away.

ANTIVIRALS FOR INFLUENZA, NEUROMINIDASE INHIBITORS

WARNINGS & PRECAUTIONS

Don't take if:
You are allergic to neuraminidase inhibitors.

Before you start, consult your doctor if:
- You have a history of asthma or chronic obstructive pulmonary disease (zanamivir).
- You have kidney disease (oseltamivir).

Over age 60:
Adverse reactions and side effects may be more frequent and severe than in younger persons.

Pregnancy:
Decide with your doctor whether drug benefits justify risk to unborn child. Risk category B for zanamivir and risk category C for oseltamivir (see page xviii).

Breastfeeding; Lactation; Nursing Mothers:
Drugs in this group are generally acceptable with breastfeeding. Discuss risks and benefits with your doctor.

Infants & children up to age 18:
- Oseltamivir is approved for children over age 2 weeks for influenza treatment and over age 1 for influenza prevention.
- Zanamivir is approved for children over age 7 for influenza treatment and over age 5 for influenza prevention.
- Peramivir not approved in ages under 18.

Prolonged use:
Not intended for long-term use.

Skin & sunlight:
No problems expected.

Driving, piloting or hazardous work:
Don't drive or pilot aircraft until you learn how medicine affects you. Don't work around dangerous machinery. Don't climb ladders or work in high places. Danger increases if you drink alcohol or take medicine affecting alertness and reflexes, such as antihistamines, tranquilizers, sedatives, pain medicine, narcotics and mind-altering drugs.

Discontinuing:
Do not discontinue until you finish all of your medicine; otherwise, symptoms may return.

Others:
- You should continue receiving an annual flu shot according to guidelines on immunization practices or as recommended by your doctor.
- People with the flu (including children and adolescents) may be at increased risk shortly after taking oseltamivir of self-injury, confusion, hallucinations, delirium and abnormal behavior leading to injury which in some cases may have fatal outcomes. Watch patient's behavior closely and call doctor if symptoms develop.
- Use of these drugs does not reduce the risk of transmitting influenza to others.
- Advise any doctor, dentist or pharmacist whom you consult that you take this medicine.

POSSIBLE INTERACTION WITH OTHER DRUGS

GENERIC NAME OR DRUG CLASS	COMBINED EFFECT
None expected.	

POSSIBLE INTERACTION WITH OTHER SUBSTANCES

INTERACTS WITH	COMBINED EFFECT
Alcohol:	None expected.
Beverages:	None expected.
Cocaine:	Unknown. Avoid.
Foods:	None expected.
Marijuana:	Consult doctor.
Tobacco:	None expected.

ANTIVIRALS (Ophthalmic)

GENERIC AND BRAND NAMES

IDOXURIDINE
- Herplex Eye Drops
- Stoxil Eye Ointment

GANCICLOVIR (ophthalmic)
- Zirgan

TRIFLURIDINE
- Trifluorothymidine
- Viroptic

BASIC INFORMATION

Habit forming? No
Prescription needed? Yes
Available as generic? No
Drug class: Antiviral (ophthalmic)

USES

Treats virus infections of the eye (e.g., herpes simplex virus) and acute herpetic keratitis (dendritic ulcers).

DOSAGE & USAGE INFORMATION

How to use:
Eye drops
- Wash hands.
- Apply pressure to inside corner of eye with middle finger.
- Continue pressure for 1 minute after placing medicine in eye.
- Tilt head backward. Pull lower lid away from eye with index finger of the same hand.
- Drop eye drops into pouch and close eye. Don't blink.
- Keep eyes closed for 1 to 2 minutes.

Eye ointment
- Wash hands.
- Pull lower lid down from eye to form a pouch.
- Squeeze tube to apply thin strip of ointment into pouch.
- Close eye for 1 to 2 minutes.
- Don't touch applicator tip to any surface (including the eye). If you accidentally touch tip, clean with warm water and soap.

Continued next column

OVERDOSE

SYMPTOMS:
Unknown.
WHAT TO DO:
If person uses much larger amount than prescribed or if accidentally swallowed, call poison control center 1-800-222-1222 for instructions or dial 911 (emergency) for help.

- Keep container tightly closed.
- Keep cool, but don't freeze.
- Wash hands immediately after using.

When to use:
As directed. Usually 1 drop every 2 hours up to maximum of 9 drops daily.

If you forget a dose:
Use as soon as you remember. If it is almost time for the next dose, wait for that dose (don't double this dose) and resume regular schedule.

What drug does:
Destroys reproductive capacity of virus.

Time lapse before drug works:
Begins to work immediately. Usual course of treatment is 7 days.

Don't use with:
Other eye drugs or products, boric acid or ointment without consulting doctor or pharmacist.

POSSIBLE ADVERSE REACTIONS OR SIDE EFFECTS

SYMPTOMS	WHAT TO DO
Life-threatening:	
None expected.	
Common:	
Stinging or burning eyes.	Continue. Call doctor when convenient.
Infrequent:	
Blurred vision for a few minutes (with ointment).	No action necessary.
Rare:	
Itchy and red eyes, swollen eyelid or eye, excess flow of tears, dimming or blurred vision.	Discontinue. Call doctor right away.

WARNINGS & PRECAUTIONS

Don't use if:
You are allergic to trifluridine, idoxuridine or ganciclovir.

Before you start, consult your doctor if:
- You have had any other eye problems.
- You use eye drops for glaucoma.

Over age 60:
No special problems expected.

Pregnancy:
Decide with your doctor if drug benefits justify risk to unborn baby. Risk category C (see page xviii).

Breastfeeding; Lactation; Nursing Mothers:
Drugs in this group are generally acceptable with breastfeeding. Discuss risks and benefits with your doctor.

Infants & children up to age 18:
Follow instructions provided by your child's doctor.

Prolonged use:
Avoid unless directed by your eye doctor.

Skin & sunlight:
No problems expected.

Driving, piloting or hazardous work:
No problems expected.

Discontinuing:
Don't discontinue without consulting doctor.

Others:
- Don't use more often or longer than prescribed.
- Keep cool, but don't freeze.
- If problem doesn't improve within a week, notify your doctor.

POSSIBLE INTERACTION WITH OTHER DRUGS

GENERIC NAME OR DRUG CLASS	COMBINED EFFECT
Eye products containing boric acid	Increased risk of toxicity to eye.

POSSIBLE INTERACTION WITH OTHER SUBSTANCES

INTERACTS WITH	COMBINED EFFECT
Alcohol:	None expected.
Beverages:	None expected.
Cocaine:	None expected.
Foods:	None expected.
Marijuana:	None expected.
Tobacco:	None expected.

ANTIVIRALS (Topical)

GENERIC AND BRAND NAMES

ACYCLOVIR (topical)
Lipsovir
Xerese Cream
Zovirax Ointment
DOCOSANOL (topical)
Abreva
PENCICLOVIR (topical)
Denavir

BASIC INFORMATION

Habit forming? No
Prescription needed? Yes
Available as generic? Yes, for some
Drug class: Antiviral

USES

- Treatment of symptoms of herpes infections of the skin, mucous membranes, lips, mouth and genitals.
- May be used for other skin disorders as prescribed by your doctor.

DOSAGE & USAGE INFORMATION

How to take:

- Acyclovir ointment—Apply to skin and mucous membranes every 3 hours (6 times a day) for 7 days. Use rubber glove when applying. Apply 1/2-inch strip to each sore or blister. Wash before using.
- Penciclovir cream—Use only on lips and face. Avoid eye area. Use every 2 hours, while awake, for 4 days.
- Docosanol—Apply directly to the affected area at the first sign of a cold sore. It should be used five times daily until the cold sore or fever blister is completely healed.

When to use:
Use as soon as symptoms begin to appear (burning, pain or blisters).

If you forget a dose:
Apply as soon as you remember. If it is almost time for the next dose, wait for that dose (don't double this dose) and resume regular schedule.

Continued next column

OVERDOSE

Unknown.
WHAT TO DO:
If person uses much larger amount than prescribed or if accidentally swallowed, call poison control center 1-800-222-1222 for instructions or dial 911 (emergency) for help.

What drug does:

- Inhibits reproduction of virus in cells without killing normal cells.
- Does not cure. Herpes breakout often recurs.

Time lapse before drug works:
2 hours.

Don't take with:
Any other topical medicine without consulting your doctor or pharmacist.

POSSIBLE ADVERSE REACTIONS OR SIDE EFFECTS

SYMPTOMS	WHAT TO DO
Life-threatening:	
None expected.	
Common:	
May cause mild pain, burning, itching or stinging.	Continue. Call doctor if symptoms persist.
Infrequent:	
None expected.	
Rare:	
Skin rash.	Continue. Call doctor if symptoms persist.

WARNINGS & PRECAUTIONS

Don't take if:
You are allergic to topical acyclovir, docosanol or penciclovir. Allergic reactions are rare, but may occur.

Before you start, consult your doctor if:
You are allergic to any medication, food or other substance.

Over age 60:
No special problems expected.

Pregnancy:
Decide with your doctor whether drug benefits justify risk to unborn child. Risk category C (see page xviii). Docosanol has no risk category.

Breastfeeding; Lactation; Nursing Mothers:
Drugs in this group are generally acceptable with breastfeeding. Discuss risks and benefits with your doctor.

Infants & children up to age 18:
Safety and effectiveness in this age group have not been established. Consult doctor.

Prolonged use:
Don't use longer than prescribed time.

Skin & sunlight:
No problems expected.

Driving, piloting or hazardous work:
No problems expected.

Discontinuing:
May be unnecessary to finish medicine. Follow doctor's instructions.

Others:
- Women: Get pap smear every 6 months because those with herpes infections are possibly at increased risk to develop cancer of the cervix. Avoid sexual activity until all blisters or sores heal.
- Don't get topical medicine in eyes.
- Check with doctor if no improvement in 1 week.

POSSIBLE INTERACTION WITH OTHER DRUGS

GENERIC NAME OR DRUG CLASS	COMBINED EFFECT
None expected.	

POSSIBLE INTERACTION WITH OTHER SUBSTANCES

INTERACTS WITH	COMBINED EFFECT
Alcohol:	None expected.
Beverages:	None expected.
Cocaine:	None expected.
Foods:	None expected.
Marijuana:	None expected.
Tobacco:	None expected.

*See Glossary

APPETITE SUPPRESSANTS

GENERIC AND BRAND NAMES

BENZPHETAMINE
Didrex
DIETHYLPROPION
Tenuate
Tenuate Dospan
PHENDIMETRAZINE
Bontril PDM
Bontril Slow Release
PHENTERMINE
Adipex-P
Qsymia
Suprenza

BASIC INFORMATION

Habit forming? Yes
Prescription needed? Yes
Available as generic? Yes, for most
Drug class: Appetite suppressant

USES

Suppresses appetite. Temporary treatment for overweight and obesity.

DOSAGE & USAGE INFORMATION

How to take:
- Tablet or capsule—Swallow with liquid. You may chew or crush tablet.
- Extended-release tablet or capsule—Swallow each dose whole with liquid; do not chew, crush or open.
- Oral disintegrating tablet—Let dissolve on top of tongue; then swallow with or without water.

When to take:
- Long-acting forms—10 to 14 hours before bedtime.
- Short-acting forms—1 hour before meals. Last dose no later than 4 to 6 hours before bedtime.

Continued next column

OVERDOSE

SYMPTOMS:
Irritability, overactivity, trembling, insomnia, mood changes, fever, rapid heartbeat, confusion, disorientation, hallucinations, convulsions, coma.
WHAT TO DO:
- **Dial 911 for all medical emergencies or call poison control center 1-800-222-1222 for instructions.**
- **See emergency information on last 3 pages of this book.**

If you forget a dose:
Take as soon as you remember. If it is almost time for the next dose, wait for that dose (don't double this dose) and resume regular schedule.

What drug does:
Apparently stimulates brain's appetite control center.

Time lapse before drug works:
Begins in 1 hour. Short-acting form lasts 4 hours. Long-acting form lasts 14 hours.

Don't take with:
Any other medicine or any dietary supplement without consulting your doctor or pharmacist.

POSSIBLE ADVERSE REACTIONS OR SIDE EFFECTS

SYMPTOMS	WHAT TO DO
Life-threatening:	
In case of overdose, see previous column.	
Common:	
Irritability, nervousness, insomnia, false sense of well-being.	Continue. Call doctor when convenient.
Infrequent:	
• Irregular or pounding heartbeat, urgent or difficult urination.	Discontinue. Call doctor right away.
• Blurred vision, unpleasant taste or dry mouth, constipation or diarrhea, nausea, vomiting, cramps, changes in sex drive, increased sweating, headache, nightmares, weakness.	Continue. Call doctor when convenient.
Rare:	
• Rash or hives, breathing difficulty.	Discontinue. Call doctor right away.
• Hair loss.	Continue. Call doctor when convenient.

WARNINGS & PRECAUTIONS

Don't take if:
You are allergic to benzphetamine, diethylpropion, mazindol, phendimetrazine or phentermine. Allergic reactions are rare, but may occur.

Before you start, consult your doctor if:
- You have high blood pressure or heart disease.
- You have an overactive thyroid, nervous tension or anxiety.
- You have epilepsy.
- You will have surgery within 2 months, including dental surgery, requiring general or spinal anesthesia.
- You have glaucoma.
- You have taken MAO inhibitors within 2 weeks.
- You plan to become pregnant within medication period.
- You have a history of drug or alcohol abuse.
- You have irregular or rapid heartbeat.

Over age 60:
Adverse reactions and side effects may be more frequent and severe than in younger persons.

Pregnancy:
Risk factors vary for drugs in this group. See category list on page xviii and consult doctor.

Breastfeeding; Lactation; Nursing Mothers:
Drugs in this group are generally not recommended with breastfeeding. Discuss risks and benefits with your doctor.

Infants & children up to age 18:
Follow instructions provided by your child's doctor.

Prolonged use:
- Loses effectiveness. Avoid.
- Talk to your doctor about the need for follow-up medical examinations or laboratory studies.

Skin & sunlight:
No problems expected.

Driving, piloting or hazardous work:
Don't drive or pilot aircraft until you learn how medicine affects you. Don't work around dangerous machinery. Don't climb ladders or work in high places. Danger increases if you drink alcohol or take medicine affecting alertness and reflexes, such as antihistamines, tranquilizers, sedatives, pain medicine, narcotics and mind-altering drugs.

Discontinuing:
- Don't discontinue without consulting doctor. Dose may require gradual reduction if you have taken drug for a long time. Doses of other drugs may also require adjustment.
- Consult doctor if following symptoms occur after stopping the drug—depression, nausea and vomiting, stomach cramps, insomnia, nightmares, extreme tiredness or weakness.

Others:
- Don't increase dose without doctor's approval.
- Advise any doctor, dentist or pharmacist whom you consult that you take this medicine.

POSSIBLE INTERACTION WITH OTHER DRUGS

GENERIC NAME OR DRUG CLASS	COMBINED EFFECT
Antidiabetic agents,* oral or insulin	May require dosage adjustment of antidiabetic agent.
Antihypertensives*	Decreased antihypertensive effect.
Appetite suppressants other*	Dangerous overstimulation.
Caffeine	Increased stimulant effect.
Central nervous system (CNS) depressants*	Increased depressive effects of both drugs.
Central nervous system (CNS) stimulants*	Increased stimulant effects of both drugs.
Guanethidine	Decreased guanethidine effect.
Methyldopa	Decreased methyldopa effect.
Monoamine oxidase (MAO) inhibitors*	Dangerous blood pressure rise.
Phenothiazines*	Decreased appetite suppressant effect.
Rauwolfia alkaloids*	Decreased effect of rauwolfia alkaloids.

POSSIBLE INTERACTION WITH OTHER SUBSTANCES

INTERACTS WITH	COMBINED EFFECT
Alcohol:	Increased sedation.
Beverages: Caffeine drinks.	Excessive stimulation.
Cocaine:	Increased risk of side effects. Avoid.
Foods:	None expected.
Marijuana:	Increased risk of side effects. Avoid.
Tobacco:	None expected.

*See Glossary

APREMILAST

BRAND NAMES

Otezla

BASIC INFORMATION

Habit forming? No
Prescription needed? Yes
Available as generic? No
Drug class: Phosphodiesterase-4 (PDE4) inhibitor

USES

Treatment for psoriatic arthritis and plaque psoriasis in adults.

DOSAGE & USAGE INFORMATION

How to take:
Tablet—Swallow whole with liquid. Do not crush, break or chew tablet. Take with or without food.

When to take:
Treatment begins with a starter pack. The dose is gradually increased for 5 days until the recommended twice-a-day dose is reached.

If you forget a dose:
Take as soon as you remember. If it is almost time for the next dose, wait for the next scheduled dose (don't double this dose).

What drug does:
It works by reducing production of an enzyme known as phosphodiesterase 4 (PDE4). This action helps reduce inflammation in the body. The exact mechanism in treating psoriatic arthritis and plaque psoriasis is not known.

Time lapse before drug works:
The initial effect may take several weeks. The full benefits of the drug may take several months.

Don't take with:
Any other medicine or any dietary supplement without consulting your doctor or pharmacist.

OVERDOSE

SYMPTOMS:
Unknown.
WHAT TO DO:
If person takes much larger amount than prescribed or if accidentally swallowed, call poison control center 1-800-222-1222 for instructions or dial 911 (emergency) for help.

POSSIBLE ADVERSE REACTIONS OR SIDE EFFECTS

SYMPTOMS	WHAT TO DO
Life-threatening:	
Rare, severe allergic reaction (anaphylaxis)—difficulty breathing, hives, itching, swelling (throat, face), vomiting, dizziness.	Seek emergency treatment immediately.
Common:	
Nausea, headache, diarrhea, weight loss.	Continue. Call doctor when convenient.
Infrequent:	
• Depression or other mood changes.	Discontinue. Call doctor right away.
• Cold-like symptoms (runny or stuffy nose, sore throat, cough), muscle or other body aches, stomach pain, vomiting.	Continue. Call doctor when convenient.
Rare:	
• Thoughts of suicide or suicidal behavior.	Discontinue. Call doctor right away or seek emergency treatment.
• Indigestion or heartburn, belching, loss of appetite.	Continue. Call doctor when convenient.

WARNINGS & PRECAUTIONS

Don't take if:
You are allergic to apremilast.

Before you start, consult your doctor if:
- You have any kidney disorder.
- You have a history of anxiety, depression or mood or psychiatric disorders.
- You have a history of suicidal thoughts or behavior.

Over age 60:
No problems expected.

Pregnancy:
Decide with your doctor if drug benefits justify risk to unborn baby. Risk category C (see page xviii).

Breastfeeding; Lactation; Nursing Mothers:
This drug should be used with caution in breastfeeding. Discuss risks and benefits with your doctor.

Infants & children up to age 18:
Safety and effectiveness in this age group have not been established. Consult doctor.

Prolonged use:
- Consult with your doctor on a regular basis while taking this drug to monitor your progress, to check for side effects and to get recommended lab tests.
- The drug may cause weight loss in some patients. Check your weight on a regular basis. If excess weight loss occurs, consult your doctor.

Skin & sunlight:
No problems expected.

Driving, piloting or hazardous work:
No problems expected.

Discontinuing:
No problems expected, but consult your doctor before stopping the drug.

Others:
- Advise any doctor, dentist or pharmacist whom you consult that you take this drug.
- Caregivers and families should be alert to unusual changes in the patient's mood or behavior while taking this drug. If new symptoms occur or milder symptoms worsen, consult doctor.

POSSIBLE INTERACTION WITH OTHER DRUGS

GENERIC NAME OR DRUG CLASS	COMBINED EFFECT
Enzyme inducers*	Decreased effect of apremilast. Avoid.

POSSIBLE INTERACTION WITH OTHER SUBSTANCES

INTERACTS WITH	COMBINED EFFECT
Alcohol:	None expected.
Beverages:	None expected.
Cocaine:	Unknown. Avoid.
Foods:	None expected.
Marijuana:	Consult doctor.
Tobacco:	None expected.

*See Glossary

ARIPIPRAZOLE

BRAND NAMES

Abilify
Abilify DiscMelt
Abilify Maintena
Abilify Oral Solution

BASIC INFORMATION

Habit forming? Not expected to be
Prescription needed? Yes
Available as generic? Yes
Drug class: Antipsychotic

USES

- Treats nervous, mental and emotional conditions. Helps in managing the signs and symptoms of schizophrenia (treats positive symptoms such as hearing voices and negative symptoms such as social withdrawal).
- Treatment for bipolar disorder.
- Treatment for major depressive disorder (MDD).
- Treatment of irritability associated with autistic disorder in children ages 6 to 17 years.
- May be used for treatment of other disorders as determined by your doctor.

DOSAGE & USAGE INFORMATION

How to take:

- Tablet or capsule—Swallow with liquid. May be taken with or without food.
- Oral disintegrating tablet—Let tablet dissolve in mouth. Do not chew or swallow.
- Oral solution—Use the oral dosing cup that came with the bottle. Store open bottle in a refrigerator (for up to 6 months after opening).
- An injectable form is given by a health care provider.

When to take:
Usually once a day at the same time each day for oral forms. Injectable form dosage schedule is determined by doctor.

Continued next column

If you forget a dose:
Take as soon as you remember. If it is almost time for the next dose, wait for the next scheduled dose (don't double this dose). Consult doctor if you miss appointment for injection.

What drug does:
The exact mechanism is unknown. It appears to block certain nerve impulses between nerve cells.

Time lapse before drug works:
1 to 4 weeks. A further increase in the dosage amount may be necessary to relieve symptoms for some patients. Do not increase dosage without your doctor's approval.

Don't take with:
Any other medicine or any dietary supplement without consulting your doctor or pharmacist.

OVERDOSE

SYMPTOMS:
Vomiting, tremors, sleepiness, confusion, slow heartbeat, seizures, unconsciousness.

WHAT TO DO:

- **Dial 911 for all medical emergencies or call poison control center 1-800-222-1222 for instructions.**
- **See emergency information on last 3 pages of this book.**

POSSIBLE ADVERSE REACTIONS OR SIDE EFFECTS

SYMPTOMS	WHAT TO DO
Life-threatening:	
High fever, rapid pulse, profuse sweating, muscle rigidity, confusion and irritability, seizures (rare neuroleptic malignant syndrome).	Discontinue. Seek emergency treatment.
Common:	
Anxiety, insomnia, headache, stomach upset, vomiting, constipation, light-headedness, restlessness (akathisia), weight loss.	Continue. Call doctor when convenient.
Infrequent:	
Runny nose, cough, skin rash, dry mouth, weight gain, drowsiness.	Continue. Call doctor when convenient.
Rare:	
• Tardive dyskinesia (involuntary movements, especially of the face, lips, jaw and tongue, sometimes involves twitching of hands or feet), high blood sugar (thirstiness, frequent urination, increased hunger, weakness).	Discontinue. Call doctor right away.
• Other symptoms causing concern, not listed above.	Continue. Call doctor when convenient.

WARNINGS & PRECAUTIONS

Don't take if:
You are allergic to aripiprazole. Allergic reactions are rare, but may occur.

Before you start, consult your doctor if:
- You have or have had liver or kidney disease, heart disease or stroke.
- You have irregular heartbeat, problems with blood pressure or a blood disorder.
- You have depression, or a history of alcohol or drug abuse.
- You have a family history of, are at risk for, or have diabetes.
- You have difficulty in swallowing.
- You are allergic to any other medications.
- You have a history of seizures.

Over age 60:
- Adverse reactions and side effects may be more likely than in younger persons.
- Use of antipsychotic drugs in elderly patients with dementia-related psychosis may increase risk of death. Consult doctor.

Pregnancy:
Decide with your doctor if drug benefits justify risk to unborn baby. Risk category C (see page xviii).

Breastfeeding; Lactation; Nursing Mothers:
This drug is generally not recommended with breastfeeding. Discuss risks and benefits with your doctor.

Infants & children up to age 18:
- Used in the treatment of schizophrenia in adolescents age 13 and over.
- Used for treatment of manic and mixed episodes associated with bipolar disorder in age 10 and over.

Prolonged use:
Consult with your doctor on a regular basis while taking this drug to check your progress or to discuss any increase or changes in side effects, blood sugar levels, and the need for continued treatment.

Skin & sunlight:
No problems expected.

Driving, piloting or hazardous work:
Don't drive or pilot aircraft until you learn how medicine affects you. Don't work around dangerous machinery. Don't climb ladders or work in high places. Danger increases if you drink alcohol or take medicine affecting alertness and reflexes.

Discontinuing:
Don't discontinue this drug without consulting doctor even if you feel well. Dosage may require a gradual reduction before stopping.

Others:
- Get up slowly from a sitting or lying position to avoid any dizziness, faintness or lightheadedness.
- The drug may reduce the body's ability to reduce body temperature. Avoid getting overheated or dehydrated.
- Advise any doctor, dentist or pharmacist whom you consult that you take this medicine.
- Take medicine as directed. Do not increase or reduce dosage without doctor's approval.
- Consult doctor if weight loss is a problem.

POSSIBLE INTERACTION WITH OTHER DRUGS

GENERIC NAME OR DRUG CLASS	COMBINED EFFECT
Enzyme inducers*	Decreased effect of aripiprazole.
Enzyme inhibitors*	Increased effect of aripiprazole.

POSSIBLE INTERACTION WITH OTHER SUBSTANCES

INTERACTS WITH	COMBINED EFFECT
Alcohol:	Increased sedative affect. Avoid.
Beverages:	None expected.
Cocaine:	Increased risk of side effects. Avoid.
Foods:	None expected.
Marijuana:	Increased risk of side effects. Avoid.
Tobacco:	None expected.

*See Glossary

ASENAPINE

BRAND NAMES

Saphris

BASIC INFORMATION

Habit forming? No
Prescription needed? Yes
Available as generic? No
Drug class: Antipsychotic

USES

- Treatment for schizophrenia.
- Treatment for acute manic or mixed episodes associated with bipolar disorder (manic-depression).

DOSAGE & USAGE INFORMATION

How to take:
Sublingual tablet—Follow prescription instructions to remove tablet from pack. Place tablet under the tongue and allow it to dissolve slowly (takes a few seconds). Don't swallow, crush or chew tablet. Don't eat or drink for 10 minutes after taking drug.

When to take:
Usually twice a day at the same times each day.

If you forget a dose:
Take as soon as you remember. If it is almost time for the next dose, wait for that dose (don't double this dose) and resume regular schedule.

What drug does:
The exact way the drug works is unknown. It appears to suppress excess levels of the brain chemicals dopamine and serotonin.

Time lapse before drug works:
Starts working within hours, but can take up to several weeks for full effect.

Don't take with:
Any other medicine or any dietary supplement without consulting your doctor or pharmacist.

OVERDOSE

SYMPTOMS:
Agitation, confusion, severe dizziness, or uncontrolled muscle movements.
WHAT TO DO:
- **Dial 911 for all medical emergencies or call poison control center 1-800-222-1222 for instructions.**
- **See emergency information on last 3 pages of this book.**

POSSIBLE ADVERSE REACTIONS OR SIDE EFFECTS

SYMPTOMS	WHAT TO DO
Life-threatening:	
Rare, severe allergic reaction (anaphylaxis)—difficulty breathing, hives, itching, swelling (throat, face), vomiting, dizziness.	Seek emergency treatment immediately.
Common:	
Dizziness, headache, drowsiness, unable to sit still, loss of feeling around the mouth, weight gain, restlessness.	Continue. Call doctor when convenient.
Infrequent:	
Dry mouth, tiredness, taste changes, increased appetite, joint aches, toothache, being agitated or anxious, indigestion, trouble sleeping, depression.	Continue. Call doctor when convenient.
Rare:	
• More or less urine output, fainting, loss of consciousness, trouble swallowing, sore throat, fever or chills, muscle pain or weakness or stiffness, fast or slow or irregular heartbeat, mental or mood changes, seizures, uncontrolled body movements (of arms, legs, tongue, jaw or cheeks), tremors, twitching, unusual sweating, high blood sugar symptoms (excess thirst and urination, dry mouth, fatigue), mouth or lip sores, white patches in mouth, sudden and severe headache, problem with speech or vision or balance, sudden numbness or weakness, unusual bleeding or bruising, suicidal thoughts or actions, fainting.	Discontinue. Call doctor right away.
• Other new symptoms that cause concern.	Continue. Call doctor when convenient.

WARNINGS & PRECAUTIONS

Don't take if:
You are allergic to asenapine.

Before you start, consult your doctor if:
- You have or have had liver disease, heart disease, heart rhythm problems, stroke, QT prolongation, heart failure, blood vessel problems, low blood pressure, recent heart attack, breast cancer, Parkinson disease, low blood cell counts, trouble swallowing, hyperprolactinemia or seizures.
- You have or have had suicidal thoughts or attempts or alcohol abuse or dependence.
- Patient has Alzheimer's or dementia.
- You have a family history of, or have diabetes.
- You have tardive dyskinesia.
- You have hypokalemia (low potassium) or hypomagnesemia (low magnesium).
- You have neuroleptic malignant syndrome (serious or fatal problems may occur).

Over age 60:
- Adverse reactions and side effects may be more severe than in younger persons. A lower starting dosage is usually recommended.
- Use of antipsychotic drugs in elderly patients with dementia-related psychosis may increase risk of death. Consult doctor.

Pregnancy:
Decide with your doctor if drug benefits justify risk to unborn baby. Risk category C (see page xviii).

Breastfeeding; Lactation; Nursing Mothers:
This drug is generally acceptable with breastfeeding. Discuss risks and benefits with your doctor.

Infants & children up to age 18:
Treatment for bipolar disorder in ages over 10.

Prolonged use:
Consult with your doctor on a regular basis while taking this drug to monitor your progress, check for side effects and for recommended lab tests.

Skin & sunlight:
Drug may affect body's ability to maintain normal temperature. Use caution with strenuous exercising or exposure to extreme heat. Drink plenty of water to avoid dehydration.

Driving, piloting or hazardous work:
Don't drive or pilot aircraft until you learn how medicine affects you. Don't work around dangerous machinery. Don't climb ladders or work in high places. Danger increases if you drink alcohol or take medicine affecting alertness and reflexes.

Discontinuing:
Don't discontinue this drug without consulting doctor. Dosage may require a gradual reduction before stopping.

Others:
- Get up slowly from a sitting or lying position to avoid dizziness, faintness or lightheadedness.
- Advise any doctor, dentist or pharmacist whom you consult that you take this medicine.
- Take medicine only as directed. Do not change the dosage without doctor's approval.

POSSIBLE INTERACTION WITH OTHER DRUGS

GENERIC NAME OR DRUG CLASS	COMBINED EFFECT
Anticholinergics*	Increased risk of side effects of asenapine.
Antihypertensives*	Increased risk of low blood pressure.
Antiparkinsonism drugs*	Decreased effect of antiparkinsonism drug.
Central nervous system (CNS) depressants*	Increased sedative effect.
Enzyme inhibitors*	May increase effect of asenapine and/or enzyme inhibitor.
Metoclopramide	Increased risk of side effects of asenapine.
Fluvoxamine	Increased risk of side effects of asenapine.
QT interval prolongation-causing drugs*	Heart rhythm problems. Avoid.

POSSIBLE INTERACTION WITH OTHER SUBSTANCES

INTERACTS WITH	COMBINED EFFECT
Alcohol:	Increased sedative affect. Avoid.
Beverages: Grapefruit juice.	May increase effect of drug.
Cocaine:	Unknown. Avoid.
Foods: Grapefruit.	May increase effect of drug.
Marijuana:	Sedation. Avoid.
Tobacco:	None expected.

*See Glossary

ASPIRIN

BRAND NAMES

See full list of brand names in the *Generic and Brand Name Directory*, page 849.

BASIC INFORMATION

Habit forming? No
Prescription needed? No
Available as generic? Yes
Drug class: Anti-inflammatory (nonsteroidal); analgesic; platelet aggregation inhibitor

USES

- Reduces pain, fever and inflammation of many different disorders.
- Relieves swelling, stiffness and joint pain of arthritis and other rheumatoid disorders.
- Used for its antiplatelet action to reduce the risk of heart attack and/or stroke.

DOSAGE & USAGE INFORMATION

How to take:
- Tablet or capsule—Swallow with liquid or food to lessen stomach irritation.
- Extended-release tablet or capsule—Swallow each dose whole.
- Effervescent tablet—Dissolve in water.
- Chewing gum tablet—Chew completely. Don't swallow whole.
- Dispersible tablet—Dissolve in the mouth before swallowing.
- Chewable tablet—Chew before swallowing or dissolve in liquid before swallowing.

Continued next column

OVERDOSE

SYMPTOMS:
- **Mild overdose—Confusion, severe diarrhea, stomach pain, increased thirst, ringing or buzzing in ears, vision problems, severe headache, dizziness, lightheadedness.**
- **Severe overdose—Bloody urine, severe nervousness or excitement or confusion, hallucinations, convulsions, coma.**
- **In some children—The only symptoms may be behavior changes, severe drowsiness or tiredness, fast or deep breathing.**

WHAT TO DO:
- **Dial 911 for all medical emergencies or call poison control center 1-800-222-1222 for instructions.**
- **See emergency information on last 3 pages of this book.**

- Crystals—Let them dissolve on tongue.
- Suppository—Remove wrapper, moisten suppository with water and gently insert into rectum.

When to take:
- Pain, fever, inflammation—As needed; no more often than every 4 to 6 hours (per instructions on product).
- Antiplatelet use—Take daily dose at same time each day.

If you forget a dose:
Take as soon as you remember. If it is almost time for the next dose, wait for that dose (don't double this dose) and resume regular schedule.

What drug does:
- Blocks certain chemicals in the body that trigger pain, inflammation and fever.
- Prevents blood cells called platelets from clumping together to form clots. Clots can block or restrict blood flow.

Time lapse before drug works:
30 to 60 minutes for pain or fever.

Don't take with:
Any other medicine or any dietary supplement without consulting your doctor or pharmacist.

POSSIBLE ADVERSE REACTIONS OR SIDE EFFECTS

SYMPTOMS	WHAT TO DO
Life-threatening:	
Rare, severe allergic reaction (anaphylaxis)—difficulty breathing, hives, itching, swelling (throat, face), vomiting, dizziness.	Seek emergency treatment immediately.
Common:	
Heartburn, indigestion, mild nausea or vomiting or headache, drowsiness.	Continue. Call doctor when convenient.
Infrequent:	
Rectal irritation (with suppository).	Continue. Call doctor when convenient.
Rare:	
Severe headache, convulsions, extreme drowsiness, flushing or change in skin color, any loss of hearing, severe vomiting, swelling of face, vision problems, bloody or black stools, ringing in ears, severe or ongoing stomach cramps or pain (all symptoms are more likely with repeated doses for long periods).	Discontinue. Call doctor right away or seek emergency treatment for severe. symptoms

WARNINGS & PRECAUTIONS

Don't take if:
- You are allergic to aspirin or salicylates or nonsteroidal anti-inflammatory drugs.
- Aspirin has decomposed (vinegar-like odor).

Before you start, consult your doctor if:
You have stomach or duodenal ulcers, gout, asthma, nasal polyps, a bleeding or blood clotting disorder, overactive thyroid, anemia, heart disease, high blood pressure, kidney or liver disease, hemophilia, glucose-6-phosphate dehydrogenase (G6PD) deficiency or are sensitive to aspirin (non-immune drug reaction).

Over age 60:
- Adverse reactions and side effects may be more frequent and severe.
- More likely to cause hidden bleeding in stomach or intestines. Watch for dark stools.

Pregnancy:
Decide with your doctor if drug benefits justify risk to unborn baby. Risk category C 1st and 2nd trimester; category D (drug can cause harm to unborn baby) 3rd trimester (see page xviii).

Breastfeeding; Lactation; Nursing Mothers:
This drug is generally not recommended with breastfeeding. Discuss risks and benefits with your doctor.

Infants & children up to age 18:
- Follow doctor's or product's label instructions.
- Do not give aspirin to persons under age 18 who have fever and discomfort of viral illness, especially chicken pox and influenza. It can increase risk of Reye's syndrome.*

Prolonged use:
May increase risk of kidney or liver problems, ulcers or bleeding problems. Ask doctor about having follow-up medical exams or lab tests.

Skin & sunlight:
No problems expected.

Driving, piloting or hazardous work:
No restrictions unless you feel drowsy.

Discontinuing:
For a chronic illness, don't discontinue without doctor's advice, even though symptoms diminish or disappear.

Others:
- Consult doctor before taking aspirin routinely to treat or prevent heart conditions or stroke.
- Advise any doctor, dentist or pharmacist whom you consult about the use of this drug.
- Consult doctor if you are taking aspirin for pain that lasts longer than 10 days (5 days for children), a fever that lasts more than 3 days, a sore throat lasts for over 2 days, or new or more severe symptoms develop.
- Aspirin is available in many combination drug products (e.g., cold remedies). Be sure to read drug charts on other ingredients.

POSSIBLE INTERACTION WITH OTHER DRUGS

GENERIC NAME OR DRUG CLASS	COMBINED EFFECT
Acetazolamide	Increased risk of adverse effects of acetazolamide.
Adrenocorticoids, systemic	May increase risk of stomach irritation.
Alendronate	Increased risk of stomach irritation.
Angiotensin-converting enzyme (ACE) inhibitors*	Decreased ACE inhibitor effect.
Antacids*	Decreased aspirin effect.
Anticoagulants*	Increased risk for bleeding.
Antidiabetic agents, oral*	Increased effect of antidiabetic agent
Anti-inflammatory drugs nonsteroidal (NSAIDs)*	Risk of stomach bleeding, ulcers and adverse effects.
Beta-adrenergic blocking agents*	Decreased effect of beta adrenergic blocker.
Diuretics*	May decrease effect of diuretic.
Ibuprofen	May decrease anti-platelet effect of aspirin.

Continued on page 891

POSSIBLE INTERACTION WITH OTHER SUBSTANCES

INTERACTS WITH	COMBINED EFFECT
Alcohol:	Possible stomach irritation and bleeding. Avoid.
Beverages:	None expected.
Cocaine:	Unknown. Avoid.
Foods:	None expected.
Marijuana:	Consult doctor.
Tobacco:	None expected.

***See Glossary**

ATOMOXETINE

BRAND NAMES

Strattera

BASIC INFORMATION

Habit forming? Not expected
Prescription needed? Yes
Available as generic? Yes
Drug class: Selective norepinephrine reuptake inhibitor

USES

Treatment of symptoms (e.g., inattention, hyperactivity and impulsiveness) of attention deficit hyperactivity disorder (ADHD) in children, adolescents and adults. The drug is used as part of a total treatment program for ADHD that may include other therapy measures (psychological, educational and social) for patients.

DOSAGE & USAGE INFORMATION

How to take:
Capsule—Swallow whole with liquid. It may be taken with or without food. Do not open capsule.

When to take:
- Once-a-day dose at the same time each day usually in the morning.
- Twice-a-day dose at the same times each day usually in the morning and then late afternoon or early evening.

If you forget a dose:
Take as soon as you remember. If it is almost time for the next dose, wait for that dose (don't double this dose) and resume regular schedule.

Continued next column

OVERDOSE

SYMPTOMS:
Sleepiness, agitation, being hyperactive, abnormal behavior, stomach symptoms, blurred vision, fast heartbeat or dry mouth.
WHAT TO DO:
- **Dial 911 for all medical emergencies or call poison control center 1-800-222-1222 for instructions.**
- **See emergency information on last 3 pages of this book.**

What drug does:
The exact mechanism by which the drug works is unknown. It appears to block a chemical (neurotransmitter) in the brain having to do with attention and activity.

Time lapse before drug works:
1 to 2 hours, but full benefits may take up to 4 weeks with possible dosage increases. Don't increase dosage without doctor's approval.

Don't take with:
Any other medicine or any dietary supplement without consulting your doctor or pharmacist.

POSSIBLE ADVERSE REACTIONS OR SIDE EFFECTS

SYMPTOMS	WHAT TO DO
Life-threatening:	
None expected.	
Common:	
In children—appetite loss, mood swings, nausea, vomiting, upset stomach, dizziness, tiredness, weight loss, runny nose, eyes tearing.	Continue. Call doctor when convenient.
Infrequent:	
In adults—insomnia, dry mouth, stomach pain, constipation, weight loss, loss of libido, impotence, ejaculatory difficulty, problems in urinating, menstrual pain; in adults and children—may feel lightheaded when getting up after sitting or lying down.	Continue. Call doctor when convenient.
Rare:	
Mental problems (hallucinations, manic behavior, becoming suspicious), liver problems (dark urine, itching, yellow skin or eyes, tender abdomen, flu-like symptoms), suicidal thoughts or behaviors.	Discontinue. Call doctor right away.

WARNINGS & PRECAUTIONS

Don't take if:
You are allergic to atomoxetine. Allergic reactions are rare, but may occur.

Before you start, consult your doctor if:
- You have liver or kidney disease.
- You have narrow angle glaucoma.
- You take MAO inhibitors* or have taken them in the last 14 days.
- You have heart problems, high or low blood pressure, or blood vessel disorder.
- You have history of mental disorders.
- You have history of alcohol or drug abuse.

Over age 60:
Drug has not been studied in this age group. Consult doctor.

Pregnancy:
Decide with your doctor if drug benefits justify risk to unborn baby. Risk category C (see page xviii).

Breastfeeding; Lactation; Nursing Mothers:
This drug is generally not recommended with breastfeeding. Discuss risks and benefits with your doctor.

Infants & children up to age 18:
- Approved for children age 6 or older.
- Regular doctor visits are important to monitor drug's effectiveness and side effects.
- Read the warnings under Others.

Prolonged use:
It is unknown about the long term effects of taking this drug. Talk to your doctor about the need for follow-up medical examinations to check weight and height changes in children and adolescents, and to check the effectiveness of the drug in treating ADHD.

Skin & sunlight:
No problems expected.

Driving, piloting or hazardous work:
Don't drive or pilot aircraft until you learn how medicine affects you. Don't work around dangerous machinery. Don't climb ladders or work in high places. Danger increases if you drink alcohol or take other medicines affecting alertness and reflexes such as antihistamines, tranquilizers, sedatives, pain medicine, narcotics and mind-altering drugs.

Discontinuing:
No problems expected, but talk to your doctor before discontinuing the drug.

Others:
- Consult doctor if your child is not gaining weight or growing at an expected or satisfactory level.
- Will commonly cause a slight increase in blood pressure. Your doctor will monitor your blood pressure on a regular basis.
- In children and teenagers, the drug may increase suicidal thoughts or behaviors. Watch for any new or increased suicidal thoughts, mood or behavioral changes (e.g., becoming irritable or anxious). Call doctor right away if any symptoms occur.
- Rarely, the drug can cause liver damage. Call doctor if you have dark urine, itching, yellowing of skin or eyes, abdominal tenderness or flu-like symptoms.
- Rarely, use of this drug in children, teenagers and adults can lead to serious heart problems, including sudden unexplained death. Adults may have stroke or heart attack. Patients with heart defects or serious heart problems are more at risk. Consult doctor.
- Consult doctor if new mental symptoms occur (e.g., abnormal thoughts or behaviors).
- Follow your doctor's recommendation for any additional measures for treating ADHD.
- Advise any doctor, dentist or pharmacist whom you consult about the use of this drug.

POSSIBLE INTERACTION WITH OTHER DRUGS

GENERIC NAME OR DRUG CLASS	COMBINED EFFECT
Beta agonists*	Increased risk of heart problems.
Enzyme inhibitors*	May increase effect of atomoxetine.
Monoamine oxidase (MAO) inhibitors*	Serious reactions (potentially fatal). Take at least 2 weeks apart.
Vasoconstrictors* (vasopressors)	May increase heart rate and/or blood pressure.

POSSIBLE INTERACTION WITH OTHER SUBSTANCES

INTERACTS WITH	COMBINED EFFECT
Alcohol:	None expected. Best to avoid.
Beverages:	None expected.
Cocaine:	Unknown. Avoid.
Foods:	None expected.
Marijuana:	Consult doctor.
Tobacco:	None expected. Best to avoid.

*See Glossary

ATOVAQUONE

BRAND NAMES

Malarone
Mepron

BASIC INFORMATION

Habit forming? No
Prescription needed? Yes
Available as generic? Yes
Drug class: Antiprotozoal, antimalarial.

USES

- Treats mild to moderate pneumocystis jiroveci pneumonia.
- May be effective for other parasitic infections such as toxoplasmosis and malaria.

DOSAGE & USAGE INFORMATION

How to take:
- Tablet—Swallow with liquid. Take with a meal. If you cannot swallow whole, crumble tablet and take with liquid or food.
- Suspension—Follow instructions provided with product.

When to take:
At the same times each day. Take with meals that are high in fat (eggs, cheese, butter, milk, meat, pizza, nuts) to increase absorption.

If you forget a dose:
Take as soon as you remember. If it is almost time for the next dose, wait for that dose (don't double this dose) and resume regular schedule.

What drug does:
Stops harmful growth of susceptible organisms.

Time lapse before drug works:
3 weeks.

Don't take with:
Any other medicine or any dietary supplement without consulting your doctor or pharmacist.

OVERDOSE

SYMPTOMS:
Stomach upset, vomiting, sores in mouth, bruising or bleeding, peeling of skin on hands or feet.
WHAT TO DO:
If person uses much larger amount than prescribed or if accidentally swallowed, call poison control center 1-800-222-1222 for instructions or dial 911 (emergency) for help.

POSSIBLE ADVERSE REACTIONS OR SIDE EFFECTS

SYMPTOMS	WHAT TO DO
Life-threatening:	
None expected.	
Common:	
• Fever, skin rash.	Discontinue. Call doctor right away.
• Nausea or vomiting, diarrhea, headache, cough, trouble sleeping.	Continue. Call doctor when convenient.
Infrequent:	
None expected.	
Rare:	
None expected.	

WARNINGS & PRECAUTIONS

Don't take if:
You are allergic to atovaquone. Allergic reactions are rare, but may occur.

Before you start, consult your doctor if:
You have any gastrointestinal disorder.

Over age 60:
No special problems expected.

Pregnancy:
Decide with your doctor if drug benefits justify risk to unborn baby. Risk category C (see page xviii).

Breastfeeding; Lactation; Nursing Mothers:
This drug is generally acceptable with breastfeeding. Discuss risks and benefits with your doctor.

Infants & children up to age 18:
Approved for children over age 13. Follow instructions provided by your child's doctor.

Prolonged use:
Not intended for long-term use.

Skin & sunlight:
No problems expected.

Driving, piloting or hazardous work:
No problems expected.

Discontinuing:
No problems expected.

Others:
- Advise any doctor, dentist or pharmacist whom you consult that you take this medicine.
- May affect the results in some medical tests.
- Talk to your doctor about the need for follow-up medical examinations or laboratory studies to check blood counts and liver function.

POSSIBLE INTERACTION WITH OTHER DRUGS

GENERIC NAME OR DRUG CLASS	COMBINED EFFECT
None significant.	

POSSIBLE INTERACTION WITH OTHER SUBSTANCES

INTERACTS WITH	COMBINED EFFECT
Alcohol:	None expected.
Beverages:	None expected.
Cocaine:	Unknown. Avoid.
Foods:	None expected.
Marijuana:	Consult doctor.
Tobacco:	None expected.

***See Glossary**

ATROPINE, HYOSCYAMINE, METHENAMINE, METHYLENE BLUE, PHENYLSALICYLATE & BENZOIC ACID

BRAND NAMES

Atrosept
Dolsed
Hexalol
Prosed/DS
Trac Tabs 2X
UAA
Uribel
Uridon Modified
Urimed
Urinary Antiseptic No. 2
Urised
Uritab
Uritab
Uritin
Urogesic-Blue
Uro-Ves

BASIC INFORMATION

Habit forming? No
Prescription needed? Yes
Available as generic? No
Drug class: Analgesic (urinary), antispasmodic, anti-infective (urinary)

USES

A combination drug to control infection, spasms and pain caused by urinary tract infections.

DOSAGE & USAGE INFORMATION

How to take:
Tablet—Swallow with liquid or food to lessen stomach irritation.

When to take:
30 minutes before meals (unless directed otherwise by doctor).

Continued next column

OVERDOSE

SYMPTOMS:
Dilated pupils, rapid pulse and breathing, dizziness, fever, hallucinations, confusion, slurred speech, agitation, flushed face, convulsions, coma.
WHAT TO DO:
- **Dial 911 for all medical emergencies or call poison control center 1-800-222-1222 for instructions.**
- **See emergency information on last 3 pages of this book.**

If you forget a dose:
Take as soon as you remember. If it is almost time for the next dose, wait for that dose (don't double this dose) and resume regular schedule.

What drug does:
Makes urine acid. Blocks nerve impulses at para-sympathetic nerve endings, preventing muscle contractions and gland secretions of organs involved. Methenamine destroys some germs.

Time lapse before drug works:
15 to 30 minutes.

Don't take with:
- Antacids* or antidiarrheals* at the same time.
- Any other medicine or any dietary supplement without consulting your doctor or pharmacist.

POSSIBLE ADVERSE REACTIONS OR SIDE EFFECTS

SYMPTOMS	WHAT TO DO
Life-threatening:	
Rare, severe allergic reaction (anaphylaxis)—difficulty breathing, hives, itching, swelling (throat, face), vomiting, dizziness.	Seek emergency treatment immediately.
Common:	
Dryness (throat, ears, nose).	Continue. Call doctor when convenient.
Infrequent:	
• Flushed face, drowsiness, difficult urination, nausea and vomiting, abdominal pain, ringing or buzzing in ears, severe drowsiness, back pain, lightheadedness.	Discontinue. Call doctor right away.
• Headache, nasal congestion, altered taste.	Continue. Call doctor when convenient.
Rare:	
Blurred vision, pain in eyes, skin rash or hives.	Discontinue. Seek emergency treatment.

WARNINGS & PRECAUTIONS

Don't take if:
- You are allergic to any of the ingredients or aspirin.
- There is brain damage in a child.
- You have glaucoma.

ATROPINE, HYOSCYAMINE, METHENAMINE, METHYLENE BLUE, PHENYLSALICYLATE & BENZOIC ACID

Before you start, consult your doctor if:
- You are on any special diet such as low-sodium.
- You have had a hiatal hernia, bronchitis, liver disease, asthma, stomach or duodenal ulcers.
- You have asthma, nasal polyps, bleeding disorder, glaucoma or enlarged prostate.
- You will have any surgery within 2 months.
- You have heart disease.

Over age 60:
- Adverse reactions and side effects may be more frequent and severe than in younger persons.
- More likely to cause hidden bleeding in stomach or intestines. Watch for dark stools.

Pregnancy:
Risk factors vary for drugs in this group. See category list on page xviii and consult doctor.

Breastfeeding; Lactation; Nursing Mothers:
This drug is generally acceptable with breastfeeding. Discuss risks and benefits with your doctor.

Infants & children up to age 18:
Follow instructions provided by your child's doctor.

Prolonged use:
May lead to constipation or kidney damage. Request lab studies to monitor effects of prolonged use.

Skin & sunlight:
No problems expected.

Driving, piloting or hazardous work:
May disqualify for piloting aircraft during time you take medicine.

Discontinuing:
May be unnecessary to finish medicine. Follow your symptoms and doctor's advice.

Others:
- Salicylates can complicate surgery, pregnancy, labor and delivery, and illness.
- Urine tests for blood sugar may be inaccurate.
- The generic ingredients in the brand names listed may vary. Read labels.
- Ask doctor about drinking cranberry juice to help make urine more acid.

POSSIBLE INTERACTION WITH OTHER DRUGS

GENERIC NAME OR DRUG CLASS	COMBINED EFFECT
Allopurinol	Decreased allopurinol effect.
Amantadine	Increased atropine and belladonna effect.
Antacids*	Decreased salicylate and methenamine effect.
Anticoagulants*	Increased anti-coagulant effect. Abnormal bleeding.
Anticholinergics, other*	Increased atropine and belladonna effect.
Antidepressants, other*	Increased sedation.
Antidiabetics, oral*	Low blood sugar.
Antifungals, azoles	Reduced azole effect.
Antihistamines*	Increased atropine and hyoscyamine effect.
Anti-inflammatory drugs, nonsteroidal (NSAIDs)*	Risk of stomach bleeding and ulcers.
Aspirin	Likely salicylate toxicity.
Beta-adrenergic blocking agents*	Decreased anti-hypertensive effect.
Carbonic anhydrase inhibitors*	Decreased methenamine effect.

Continued on page 891

POSSIBLE INTERACTION WITH OTHER SUBSTANCES

INTERACTS WITH	COMBINED EFFECT
Alcohol:	Excessive sedation. Possible stomach irritation and bleeding. Avoid.
Beverages:	None expected.
Cocaine:	Unknown. Avoid.
Foods:	None expected.
Marijuana:	Consult doctor.
Tobacco:	Dry mouth.

ATTAPULGITE

BRAND NAMES

- **Diar-Aid**
- **Diasorb**
- **Fowlers Diarrhea Tablets**
- **Kaopectate**
- **Kaopectate Advanced Formula**
- **Kaopectate Maximum Strength**
- **Rheaban**
- **St. Joseph Antidiarrheal**

BASIC INFORMATION

Habit forming? No
Prescription needed? No
Available as generic? Yes
Drug class: Antidiarrheal

USES

Treats diarrhea. Used in conjunction with fluids, appropriate diet and rest. Treats symptoms only. Does not cure any disorder that causes diarrhea.

DOSAGE & USAGE INFORMATION

How to take:

- Tablet—Swallow with liquid. If you can't swallow whole, crumble tablet and take with liquid or food. Instructions to take on empty stomach mean 1 hour before or 2 hours after eating.
- Chewable tablet—Chew well before swallowing.
- Oral suspension—Follow label instructions.

When to take:
2 hours before or 3 hours after taking any other oral medications. Outside of this restriction, take a dose after each loose bowel movement until diarrhea is controlled.

If you forget a dose:
Take as soon as you remember. If it is almost time for the next dose, wait for that dose (don't double this dose) and resume regular schedule.

Continued next column

OVERDOSE

SYMPTOMS:
Unknown.
WHAT TO DO:
If person uses much larger amount than prescribed or if accidentally swallowed, call poison control center 1-800-222-1222 for instructions or dial 911 (emergency) for help.

What drug does:
Absorbs bacteria and toxins and reduces water loss. Attapulgite does not get absorbed into the body.

Time lapse before drug works:
5 to 8 hours.

Don't take with:
Any other medicine or any dietary supplement without consulting your doctor or pharmacist.

POSSIBLE ADVERSE REACTIONS OR SIDE EFFECTS

SYMPTOMS	WHAT TO DO
Life-threatening:	
None expected.	
Common:	
None expected.	
Infrequent:	
Constipation (usually mild and of short duration).	Continue. Call doctor when convenient.
Rare:	
None expected.	

WARNINGS & PRECAUTIONS

Don't take if:
- You are allergic to attapulgite. Allergic reactions are rare, but may occur.
- You or your doctor suspects intestinal obstruction.

Before you start, consult your doctor if:
You are dehydrated (signs are a dry mouth, loose skin, sunken eyes and parched lips).

Over age 60:
No special problems expected, but may be more at risk for dehydration from diarrhea.

Pregnancy:
Risk category not designated. See list on page xviii and consult doctor.

Breastfeeding; Lactation; Nursing Mothers:
This drug is generally acceptable with breastfeeding. Discuss risks and benefits with your doctor.

Infants & children up to age 18:
Follow instructions provided by your child's doctor.

Prolonged use:
Not intended for prolonged use.

Skin & sunlight:
No special problems expected.

Driving, piloting or hazardous work:
No special problems expected.

Discontinuing:
May be unnecessary to finish medicine. Follow doctor's instructions.

Others:
No special problems expected.

POSSIBLE INTERACTION WITH OTHER DRUGS

GENERIC NAME OR DRUG CLASS	COMBINED EFFECT
Digitalis	May decrease effectiveness of digitalis.
Lincomycins*	May decrease effectiveness of lincomycins.
Any other medicine taken by mouth	When taken at the same time, neither drug may be as effective. Take other medicines 2 hours before or 3 hours after attapulgite.

POSSIBLE INTERACTION WITH OTHER SUBSTANCES

INTERACTS WITH	COMBINED EFFECT
Alcohol:	None expected.
Beverages:	None expected.
Cocaine:	Unknown. Avoid.
Foods: Prunes, prune juice and other fruits or foods that may cause diarrhea.	Decreased effect of attapulgite.
Marijuana:	Consult doctor.
Tobacco:	None expected.

***See Glossary**

AZATHIOPRINE

BRAND NAMES

Azasan
Imuran

BASIC INFORMATION

Habit forming? No
Prescription needed? Yes
Available as generic? Yes
Drug class: Immunosuppressant, antirheumatic

USES

- Protects against rejection of transplanted organs (e.g., kidney, heart).
- Treats severe active rheumatoid arthritis and other immunologic diseases if simpler treatment plans have been ineffective.

DOSAGE & USAGE INFORMATION

How to take:
Tablet—Swallow with liquid. If you can't swallow whole, crumble tablet and take with liquid or food. Instructions to take on empty stomach mean 1 hour before or 2 hours after eating.

When to take:
Follow your doctor's instructions. Usually once a day.

If you forget a dose:
Take as soon as you remember. If it is almost time for the next dose, wait for that dose (don't double this dose) and resume regular schedule.

What drug does:
Unknown; probably inhibits synthesis of DNA and RNA.

Continued next column

OVERDOSE

SYMPTOMS:
Nausea, vomiting, diarrhea, stomach pain, fever, chills, bleeding.
WHAT TO DO:
- **Dial 911 for all medical emergencies or call poison control center 1-800-222-1222 for instructions.**
- **See emergency information on last 3 pages of this book.**

Time lapse before drug works:
6 to 8 weeks.

Don't take with:
Any other medicines (including over-the-counter drugs such as cough and cold medicines, laxatives, antacids, diet pills, caffeine, nose drops or vitamins) without consulting your doctor.

POSSIBLE ADVERSE REACTIONS OR SIDE EFFECTS

SYMPTOMS	WHAT TO DO
Life-threatening: None expected.	
Common:	
• Infection or low blood count symptoms (fever and chills, back pain, cough, painful urination), anemia (tiredness or weakness), nausea or vomiting.	Discontinue. Call doctor right away.
• Appetite loss.	Continue. Call doctor when convenient.
Infrequent:	
Jaundice (yellow eyes, skin), skin rash.	Discontinue. Call doctor right away.
Rare:	
Low platelet count causing bleeding or bruising, tarry or black stools, bloody urine, red spots under skin, severe abdominal pain, mouth sores.	Discontinue. Call doctor right away.

WARNINGS & PRECAUTIONS

Don't take if:
- You are allergic to azathioprine. Allergic reactions are rare, but may occur.
- You have chicken pox.
- You have shingles (herpes zoster).

Before you start, consult your doctor if:
- You have gout.
- You have liver or kidney disease.
- You have an infection.

Over age 60:
Adverse reactions and side effects may be more frequent and severe than in younger persons.

Pregnancy:
Consult doctor. Use of the drug in pregnancy is not recommended. It could cause harm to the unborn baby. Risk category D (see page xviii).

Breastfeeding; Lactation; Nursing Mothers:
This drug is generally acceptable with breastfeeding. Discuss risks and benefits with your doctor.

Infants & children up to age 18:
Safety and effectiveness in this age group have not been established. Consult doctor.

Prolonged use:
- May increase likelihood of problems upon discontinuing.
- Talk to your doctor about the need for follow-up medical examinations or laboratory studies. May need every week during first two months, then once a month.

Skin & sunlight:
No special problems expected.

Driving, piloting or hazardous work:
Avoid if you feel confused, drowsy or dizzy.

Discontinuing:
May still experience symptoms of bone marrow depression, such as blood in stools, fever or chills, blood spots under the skin, back pain, hoarseness, bloody urine. If any of these occur, call your doctor right away.

Others:
- Advise any doctor, dentist or pharmacist whom you consult that you take this medicine.
- May affect results in some medical tests.

POSSIBLE INTERACTION WITH OTHER DRUGS

GENERIC NAME OR DRUG CLASS	COMBINED EFFECT
Allopurinol	Greatly increased azathioprine activity.
Antivirals, HIV/AIDS*	Increased risk of pancreatitis.
Clozapine	Toxic effect on bone marrow.
Immunosuppressants, other*	Higher risk of developing infection or malignancies.
Levamisole	Increased risk of bone marrow depression.
Tiopronin	Increased risk of toxicity to bone marrow.
Vaccines	May decrease effectiveness or cause disease itself.

POSSIBLE INTERACTION WITH OTHER SUBSTANCES

INTERACTS WITH	COMBINED EFFECT
Alcohol:	Consult doctor about any limits.
Beverages:	None expected.
Cocaine:	Increased risk of adverse reactions. Avoid.
Foods:	None expected.
Marijuana:	Increased risk of adverse reactions. Avoid.
Tobacco:	None expected.

*See Glossary

AZELAIC ACID

BRAND NAMES

Azelex
Finacea
Finevin

BASIC INFORMATION

Habit forming? No
Prescription needed? Yes
Available as generic? No
Drug class: Antiacne agent, hypopigmentation agent

USES

- Topical treatment for mild to moderate acne vulgaris.
- May be used for treatment of melasma (chloasma), a skin condition in which brownish patches of pigmentation appear on the face.

DOSAGE & USAGE INFORMATION

How to use:
Cream—Wash the affected skin area and then apply the prescribed amount of cream and rub it into the skin. Rub it in thoroughly, but gently, to avoid irritation. Wash hands after applying.

When to use:
Usually twice a day (morning and evening).

If you forget a dose:
Use as soon as you remember. If it is almost time for the next dose, wait for that dose (don't double this dose) and resume regular schedule.

What drug does:
The drug helps prevents the development of new acne lesions (whiteheads), but the exact mechanism is unknown. It appears to have some antibacterial and anti-inflammatory effect, and also helps in skin renewal.

Continued next column

OVERDOSE

SYMPTOMS:
May be similar to symptoms of adverse reactions.
WHAT TO DO:
If person uses much larger amount than prescribed or if accidentally swallowed, call poison control center 1-800-222-1222 for instructions or dial 911 (emergency) for help.

Time lapse before drug works:
Results should be visible in about 4 weeks, but full benefits may take months.

Don't use with:
Other topical medications without consulting your doctor or pharmacist.

POSSIBLE ADVERSE REACTIONS OR SIDE EFFECTS

SYMPTOMS	WHAT TO DO
Life-threatening: None expected.	
Common: Peeling, itching, redness or dryness of skin; tingling, burning or stinging may occur when medicine first used.	Continue. Call doctor if symptoms persist.
Infrequent: None expected.	
Rare: Lightening of skin or white spots in persons with darker complexions.	Discontinue. Call doctor when. convenient.

WARNINGS & PRECAUTIONS

Don't use if:
You are sensitive to azelaic acid. Allergic reactions are rare, but may occur.

Before you start, consult your doctor if:
- You are allergic to any medicine, food or other substance, or have a family history of allergies.
- You have a dark complexion.

Over age 60:
No special problems expected, however the drug has not been tested extensively in this age group.

Pregnancy:
Decide with your doctor if drug benefits justify risk to unborn baby. Risk category B (see page xviii).

Breastfeeding; Lactation; Nursing Mothers:
This drug is generally acceptable with breastfeeding. Discuss risks and benefits with your doctor.

Infants & children up to age 18:
Safety and effectiveness in children under age 12 have not been established. Consult doctor.

Prolonged use:
No problems expected.

Skin & sunlight:
No problems expected.

Driving, piloting or hazardous work:
No problems expected.

Discontinuing:
No problems expected.

Others:
- Use as directed. Don't increase or decrease dosage without doctor's approval. Using more of the cream or using it more frequently than prescribed won't improve results and may cause excessive skin irritation.
- The side effects involving skin irritation usually go away with continued use. If they continue beyond 4 weeks, or are severe, consult doctor about reducing the dosage to once a day.
- You may use water-based cosmetics while undergoing treatment with this drug.

POSSIBLE INTERACTION WITH OTHER DRUGS

GENERIC NAME OR DRUG CLASS	COMBINED EFFECT
None expected.	

POSSIBLE INTERACTION WITH OTHER SUBSTANCES

INTERACTS WITH	COMBINED EFFECT
Alcohol:	None expected.
Beverages:	None expected.
Cocaine:	None expected.
Foods:	None expected.
Marijuana:	None expected.
Tobacco:	None expected.

***See Glossary**

BACLOFEN

BRAND NAMES

Kemstro
Lioresal

BASIC INFORMATION

Habit forming? No
Prescription needed? Yes
Available as generic? Yes
Drug class: Muscle relaxant

USES

- Relieves spasms, cramps and spasticity of muscles caused by medical problems, including multiple sclerosis and spine injuries.
- Reduces number and severity of trigeminal neuralgia attacks.

DOSAGE & USAGE INFORMATION

How to take:
- Tablet—Swallow with liquid or food to lessen stomach irritation.
- Oral disintegrating tablet—Dissolves in the mouth.

When to take:
3 or 4 times daily as directed.

If you forget a dose:
Take as soon as you remember. If it is almost time for the next dose, wait for that dose (don't double this dose) and resume regular schedule.

What drug does:
Blocks body's pain and reflex messages to brain.

Time lapse before drug works:
Variable. Few hours to weeks.

Don't take with:
Any other medicine or any dietary supplement without consulting your doctor or pharmacist.

OVERDOSE

SYMPTOMS:
Blurred vision, blindness, difficult breathing, vomiting, drowsiness, muscle weakness, seizures.
WHAT TO DO:
- **Dial 911 for all medical emergencies or call poison control center 1-800-222-1222 for instructions.**
- **See emergency information on last 3 pages of this book.**

POSSIBLE ADVERSE REACTIONS OR SIDE EFFECTS

SYMPTOMS	WHAT TO DO
Life-threatening: In case of overdose, see previous column.	
Common: Dizziness, nausea, lightheadedness, drowsiness, confusion.	Continue. Call doctor when convenient.
Infrequent:	
• Rash with itching, numbness or tingling in hands or feet.	Discontinue. Call doctor right away.
• Headache, stomach pain, diarrhea or constipation, loss of appetite, difficult or painful urination, sexual problems, stuffy nose, insomnia, clumsiness.	Continue. Call doctor when convenient.
Rare:	
• Fainting, weakness, hallucinations, depression, chest or muscle pain, pounding heartbeat.	Discontinue. Call doctor right away.
• Ringing in ears, lowered blood pressure, dry mouth, weight gain, taste change, overexcitement.	Continue. Call doctor when convenient.

WARNINGS & PRECAUTIONS

Don't take if:
- You are allergic to any muscle relaxant. Allergic reactions are rare, but may occur.
- Muscle spasm is due to strain or sprain.

Before you start, consult your doctor if:
- You have Parkinson's disease.
- You have cerebral palsy.
- You have had a recent stroke or head injury.
- You have arthritis, diabetes or epilepsy.
- You have psychosis.
- You have kidney disease.
- You will have surgery within 2 months, including dental surgery, requiring general or spinal anesthesia.

Over age 60:
Adverse reactions and side effects may be more frequent and severe than in younger persons.

Pregnancy:
Consult doctor. Use of the drug during pregnancy is not recommended. It could cause harm to the unborn baby. Risk category D (see page xviii).

Breastfeeding; Lactation; Nursing Mothers:
This drug is generally acceptable with breastfeeding. Discuss risks and benefits with your doctor.

Infants & children up to age 18:
Approved for children age 12 and over. Follow instructions provided by your child's doctor.

Prolonged use:
Epileptic patients should be monitored with EEGs. Diabetic patients should more closely monitor blood sugar levels. Obtain periodic liver function tests.

Skin & sunlight:
No problems expected.

Driving, piloting or hazardous work:
Don't drive or pilot aircraft until you learn how medicine affects you. Don't work around dangerous machinery. Don't climb ladders or work in high places. Danger increases if you drink alcohol or take medicine affecting alertness and reflexes, such as antihistamines, tranquilizers, sedatives, pain medicine, narcotics and mind-altering drugs.

Discontinuing:
Don't discontinue without consulting doctor. Dose may require gradual reduction if you have taken drug for a long time. Doses of other drugs may also require adjustment.

Others:
Advise any doctor, dentist or pharmacist whom you consult that you take this medicine.

POSSIBLE INTERACTION WITH OTHER DRUGS

GENERIC NAME OR DRUG CLASS	COMBINED EFFECT
Anesthetics, general*	Increased sedation. Low blood pressure. Avoid.
Antidiabetic drugs,* insulin or oral	Need to adjust diabetes medicine dosage.
Central nervous system (CNS) depressants*	Increased sedation. Low blood pressure. Avoid.
Clozapine	Toxic effect on the central nervous system.
Ethinamate	Dangerous increased effects of ethinamate. Avoid combining.
Fluoxetine	Increased depressant effects of both drugs.
Guanfacine	May increase depressant effects of either drug.
Leucovorin	High alcohol content of leucovorin may cause adverse effects.
Methyprylon	Increased sedative effect, perhaps to dangerous level. Avoid.
Nabilone	Greater depression of the central nervous system.
Sertraline	Increased depressive effects of both drugs.

POSSIBLE INTERACTION WITH OTHER DRUGS

INTERACTS WITH	COMBINED EFFECT
Alcohol:	Increased sedation. Low blood pressure. Avoid.
Beverages:	None expected.
Cocaine:	Increased risk of side effects. Avoid.
Foods:	None expected.
Marijuana:	Increased risk of side effects. Avoid.
Tobacco:	None expected.

***See Glossary**

BARBITURATES

GENERIC AND BRAND NAMES

See full list of generic and brand names in the *Generic and Brand Name Directory*, page 850.

BASIC INFORMATION

Habit forming? Yes
Prescription needed? Yes
Available as generic? Yes, for some
Drug class: Sedative-hypnotic agent, anticonvulsant

USES

- Reduces likelihood of seizures (tonic-clonic seizure pattern and simple partial) in epilepsy.
- Preventive treatment for febrile seizures.
- Reduces anxiety or nervous tension.
- Used in combination drugs to treat gastro-intestinal disorders, headaches and asthma.
- Aids sleep at night (on a short-term basis).

DOSAGE & USAGE INFORMATION

How to take:

- Capsule—Swallow with liquid. If you can't swallow whole, open capsule and take with liquid or food. Take on empty stomach (1 hour before or 2 hours after eating).
- Elixir—Swallow with liquid.
- Rectal suppository—Remove wrapper and moisten suppository with water. Gently insert into rectum, pointed end first. If suppository is too soft, chill first in refrigerator or cool water.
- Tablet—Swallow with liquid or food to lessen stomach irritation. If you can't swallow whole, crumble tablet and take with liquid or food.

When to take:
At the same times each day.

Continued next column

OVERDOSE

SYMPTOMS:
Difficulty with thinking, impaired judgment, trouble breathing, slow or slurred speech, weak pulse, staggering, deep sleep, coma.
WHAT TO DO:

- **Dial 911 for all medical emergencies or call poison control center 1-800-222-1222 for instructions.**
- **See emergency information on last 3 pages of this book.**

If you forget a dose:
Take as soon as you remember. If it is almost time for the next dose, wait for that dose (don't double this dose) and resume regular schedule.

What drug does:
Blocks the transmission of nerve impulses from the brain to other parts of the body.

Time lapse before drug works:
60 minutes, but will take several weeks for maximum antiepilepsy effect.

Don't take with:
Any other medicine or any dietary supplement without consulting your doctor or pharmacist.

POSSIBLE ADVERSE REACTIONS OR SIDE EFFECTS

SYMPTOMS	WHAT TO DO
Life-threatening:	
Rare, severe allergic reaction (anaphylaxis)—difficulty breathing, hives, itching, swelling (throat, face), vomiting, dizziness.	Seek emergency treatment immediately.
Common:	
Dizziness, drowsiness, clumsiness, unsteadiness, signs of addiction.*	Continue. Call doctor when convenient.
Infrequent:	
Confusion, headache, irritability, feeling faint, nausea, vomiting, depression, nightmares, trouble sleeping.	Continue, but call doctor right away.
Rare:	
Agitation, slow heartbeat, difficult breathing, bleeding sores on lips, fever, chest pain, unexplained bleeding or bruising, muscle or joint pain, skin rash or hives, thickened or scaly skin, white spots in mouth, tightness in chest, face swelling, sore throat, yellow eyes or skin, hallucinations, unusual tiredness or weakness, sleep-related behaviors.*	Continue, but call doctor right away.

WARNINGS & PRECAUTIONS

Don't take if:

- You are allergic to any barbiturate.
- You have porphyria.

Before you start, consult your doctor if:

- You have epilepsy, kidney or liver problems, asthma, anemia or chronic pain.
- You will have surgery within 2 months, including dental surgery, requiring anesthesia.

Over age 60:
Adverse reactions and side effects may be more frequent and severe. Use small doses.

Pregnancy:
Consult doctor. Use of the drug during pregnancy is not recommended. It could cause harm to the unborn baby. Risk category D (see page xviii).

Breastfeeding; Lactation; Nursing Mothers:
Drugs in this group are generally not recommended with breastfeeding. Discuss risks and benefits with your doctor.

Infants & children up to age 18:
Follow instructions provided by your child's doctor.

Prolonged use:

- May cause addiction, anemia, chronic intoxication. Unlikely to occur with the usual anticonvulsant or sedative dosage levels.
- May lower body temperature, making exposure to cold temperatures hazardous.
- Talk to your doctor about the need for follow-up medical exams or lab studies to check your progress while taking this drug.

Skin & sunlight:
No problems expected.

Driving, piloting or hazardous work:
Don't drive or pilot aircraft until you learn how medicine affects you. Don't work around dangerous machinery. Don't climb ladders or work in high places. Danger increases if you drink alcohol or take medicine affecting alertness and reflexes.

Discontinuing:
Don't stop taking barbiturates suddenly. Seek medical help for safe withdrawal.

Others:

- May affect results in some medical tests.
- Barbiturate addiction is common. Withdrawal effects may be fatal.
- You may become physically or mentally dependent on the drug. Ask doctor about risks.
- Advise any doctor, dentist or pharmacist whom you consult that you take this medicine.

POSSIBLE INTERACTION WITH OTHER DRUGS

GENERIC NAME OR DRUG CLASS	COMBINED EFFECT
Adrenocorticoids, systemic	Decreased effect of prednisone.
Anticoagulants, oral*	Decreased effect of anticoagulant.
Anticonvulsants*	Changed seizure patterns.
Antidepressants, tricyclic*	Decreased anti-depressant effect. Possible dangerous oversedation.
Antidiabetic, agents, oral*	Increased effect of barbiturate.
Antihistamines*	Dangerous sedation. Avoid.
Aspirin	Decreased aspirin effect.
Beta-adrenergic blocking agents*	Decreased effect of beta-adrenergic blocker.
Carbamazepine	Decreased carbamazepine effect.
Carteolol	Increased barbiturate effect. Dangerous sedation.
Clozapine	Toxic effect on the central nervous system.
Contraceptives, oral*	Decreased contraceptive effect.
Dextrothyroxine	Decreased barbiturate effect.
Doxycycline	Decreased doxycycline effect.
Griseofulvin	Decreased griseofulvin effect.
Lamotrigine	Decreased lamotrigine effect.
Leukotriene modifiers	Decreased montelukast effect.

Continued on page 892

POSSIBLE INTERACTION WITH OTHER DRUGS

INTERACTS WITH	COMBINED EFFECT
Alcohol:	Possible serious oversedation. Avoid.
Beverages:	None expected.
Cocaine:	Excessive sedation.
Foods:	None expected.
Marijuana:	Excessive sedation. Avoid.
Tobacco:	None expected.

*See Glossary

BECAPLERMIN

BRAND NAMES

Regranex

BASIC INFORMATION

Habit forming? No
Prescription needed? Yes
Available as generic? No
Drug class: Platelet-derived growth factor

USES

Treatment of skin ulcers in patients with diabetes.

DOSAGE & USAGE INFORMATION

How to use:
Gel—Apply to the affected area. Follow all instructions provided with the prescription. Dosage may change as wound heals.

When to use:
At the same time each day. Change the wound dressing between applications of the medication.

If you forget a dose:
Apply as soon as you remember. If it is almost time for the next dose, wait for that dose (don't double this dose) and resume regular schedule.

What drug does:
Stimulates growth of cells involved in wound repair.

Time lapse before drug works:
Up to six months.

Don't use with:
Any other medicine or any dietary supplement without consulting your doctor or pharmacist.

OVERDOSE

SYMPTOMS:
Unknown.
WHAT TO DO:
If person uses much larger amount than prescribed or if accidentally swallowed, call poison control center 1-800-222-1222 for instructions or dial 911 (emergency) for help.

POSSIBLE ADVERSE REACTIONS OR SIDE EFFECTS

SYMPTOMS	WHAT TO DO
Life-threatening:	
None expected.	
Common:	
None expected.	
Infrequent:	
Rash in area of skin ulcer.	Discontinue. Call doctor right away.
Rare:	
None expected.	

WARNINGS & PRECAUTIONS

Don't take if:
- You are allergic to becaplermin, parabens or metacresol. Allergic reactions are rare, but may occur.
- You have any new growths or wounds in the application area.

Before you start, consult your doctor if:
- You have any other medical problem.
- You are allergic to any other substances, such as food preservatives or dyes.
- You have a malignancy or cancer.

Over age 60:
No special problems expected.

Pregnancy:
Decide with your doctor if drug benefits justify risk to unborn baby. Risk category C (see page xviii).

Breastfeeding; Lactation; Nursing Mothers:
It is unknown if this drug is acceptable with breastfeeding. Discuss risks and benefits with your doctor.

Infants & children up to age 18:
Safety and efficacy in children under age 16 have not been established. Consult doctor.

Prolonged use:
No problems expected. Your doctor should periodically evaluate your response to the drug and adjust the dose according to the rate of change in the width and length of the diabetic ulcer.

Skin & sunlight:
No problems expected.

Driving, piloting or hazardous work:
No problems expected.

Discontinuing:
Don't discontinue without consulting doctor.

Others:
- There is an increased risk of cancer death in patients who use 3 or more tubes of this drug. Consult your doctor about your risks.
- Do not place tip of tube onto ulcer or any other object; it may contaminate the medication.
- Be sure you follow application instructions carefully.
- Avoid bearing weight on the affected extremity.
- Wash hands carefully before preparing your dose.
- Keep this medication in refrigerator; do not freeze.
- Advise any doctor, dentist or pharmacist whom you consult that you take this medicine.

POSSIBLE INTERACTION WITH OTHER DRUGS

GENERIC NAME OR DRUG CLASS	COMBINED EFFECT
None expected.	

POSSIBLE INTERACTION WITH OTHER DRUGS

INTERACTS WITH	COMBINED EFFECT
Alcohol:	None expected.
Beverages:	None expected.
Cocaine:	Unknown. Avoid.
Foods:	None expected.
Marijuana:	Consult doctor.
Tobacco:	None expected.

*See Glossary

BENZODIAZEPINES

GENERIC AND BRAND NAMES

See full list of generic and brand names in the *Generic and Brand Name Directory*, page 850.

BASIC INFORMATION

Habit forming? Yes
Prescription needed? Yes
Available as generic? Yes, for most
Drug class: Tranquilizer (benzodiazepine), anticonvulsant

USES

- Treatment for anxiety disorders and panic disorders.
- Treatment for muscle spasm.
- Treatment for seizure disorders.
- Treatment for alcohol withdrawal.
- Treatment for insomnia (short-term).

DOSAGE & USAGE INFORMATION

How to take:
- Tablet or capsule—Swallow with liquid. If you can't swallow whole, crumble tablet or open capsule and take with liquid or a bite of food.
- Extended-release capsule—Swallow capsule whole. Do not open or chew.
- Oral suspension—Dilute dose in water, soda or sodalike beverage or small amount of food such as applesauce or pudding.
- Sublingual tablet—Do not chew or swallow. Place under tongue until dissolved.
- Disintegrating tablet—Let dissolve on tongue.
- Rectal gel—Follow instructions provided with prescription or as directed by the doctor.

When to take:
At the same time each day, according to instructions on prescription label.

Continued next column

OVERDOSE

SYMPTOMS:
Drowsiness, weakness, tremor, dizziness, confusion, anxiety, blurred vision, agitation, stupor, coma.
WHAT TO DO:
- **Dial 911 for all medical emergencies or call poison control center 1-800-222-1222 for instructions.**
- **See emergency information on last 3 pages of this book.**

If you forget a dose:
Take as soon as you remember. If it is almost time for the next dose, wait for that dose (don't double this dose) and resume regular schedule.

What drug does:
Affects limbic system of brain, the part that controls emotions.

Time lapse before drug works:
May take 6 weeks for full benefit; depends on drug when treating anxiety.

Don't take with:
Any other medicine or any dietary supplement without consulting your doctor or pharmacist.

POSSIBLE ADVERSE REACTIONS OR SIDE EFFECTS

SYMPTOMS	WHAT TO DO
Life-threatening:	
In case of overdose, see previous column.	
Common:	
Clumsiness, light-headed, dizziness, unsteadiness, drowsiness, slurred speech.	Continue. Call doctor when convenient.
Infrequent:	
• Memory loss, fast heartbeat, anxiety, depression.	Discontinue. Call doctor right away.
• Constipation or diarrhea, nausea, vomiting, urination problems, stomach pain, headache, mouth is dry or watering, muscle spasm, changes in sexual function.	Continue. Call doctor when convenient.
Rare:	
Behavior changes (may be bizarre), delusions, outbursts of anger, loss of reality, infection symptoms (fever, chills), unusual tiredness or weakness, unusual bleeding or bruising, skin rash or itching or peeling or blisters, uncontrolled body movements, sores in mouth, hallucinations, yellow skin or eyes, sleep-related behaviors.*	Discontinue. Call doctor right away.

WARNINGS & PRECAUTIONS

Don't take if:
You are allergic to any benzodiazepine. Allergic reactions are rare, but may occur.

Before you start, consult your doctor if:
- You have myasthenia gravis.
- You have a history of drug or alcohol abuse, or severe depression or mental disorder.
- You have liver, kidney or lung disease.
- You have diabetes, seizure disorder, a swallowing problem (in children), or porphyria.
- You have sleep apnea.
- You have glaucoma.

Over age 60:
Adverse reactions and side effects may be more frequent and severe than in younger persons.

Pregnancy:
Risk factors vary for drugs in this group. Always consult doctor. See category list on page xviii.

Breastfeeding; Lactation; Nursing Mothers:
Drugs in this group are generally not recommended with breastfeeding. Discuss risks and benefits with your doctor.

Infants & children up to age 18:
Follow instructions provided by your child's doctor.

Prolonged use:
Risk of physical or psychological dependence.

Skin & sunlight:
One or more drugs in this group may cause rash or intensify sunburn in areas exposed to sun or ultraviolet light (photosensitivity reaction). Avoid overexposure and use sunscreen. Notify doctor if reaction occurs.

Driving, piloting or hazardous work:
Don't drive or pilot aircraft until you learn how medicine affects you. Don't work around dangerous machinery. Don't climb ladders or work in high places. Danger increases if you drink alcohol or take medicine affecting alertness and reflexes.

Discontinuing:
- Don't discontinue without consulting doctor. Adverse effects can occur (may be life-threatening) if the drug has been taken for longer periods. Dose may require gradual reduction. Doses of other drugs may also require adjustment.
- If withdrawal symptoms (emotional or physical) occur after stopping the drug, call doctor.

Others:
- Don't use for insomnia more than 4-7 days.
- Hot weather, heavy exercise and sweating may increase risk of heat stroke.
- Advise any doctor, dentist or pharmacist whom you consult that you take this drug.
- Use of clobazam may rarely cause severe, life threatening skin reactions. Consult doctor right away if skin rash, blisters or peeling occur.
- Rarely, anticonvulsant (antiseizure) drugs may lead to suicidal thoughts and behaviors. Call doctor right away if suicidal symptoms or unusual behaviors occur.

POSSIBLE INTERACTION WITH OTHER DRUGS

GENERIC NAME OR DRUG CLASS	COMBINED EFFECT
Antidepressants, tricyclic*	Increased sedative effect of both drugs.
Carbamazepine	Decreased effect of benzodiazepine.
Central nervous system (CNS) depressants*	Increased sedative effect.
Cimetidine	Increased effect of benzodiazepine.
Clozapine	Low blood pressure.
Contraceptives, oral*	Increased effect of benzodiazepine.
Enzyme inhibitors*	Increased effect of benzodiazepine.
Erythromycins*	Increased effect of benzodiazepine.
Fluoxetine	Increased effect of benzodiazepine.
Fluvoxamine	Increased effect of benzodiazepine.
Isoniazid	Increased effect of benzodiazepine.

Continued on page 892

POSSIBLE INTERACTION WITH OTHER DRUGS

INTERACTS WITH	COMBINED EFFECT
Alcohol:	Sedation. Avoid.
Beverages: Grapefruit juice.	Increased effect of benzodiazepine.
Cocaine:	Unknown. Avoid.
Foods:	None expected.
Marijuana:	Sedation. Avoid.
Tobacco:	None expected.

***See Glossary**

BENZOYL PEROXIDE

BRAND NAMES

See full list of brand names in the *Generic and Brand Name Directory*, page 851.

BASIC INFORMATION

Habit forming? No
Prescription needed? No
Available as generic? Yes
Drug class: Antiacne (topical)

USES

- Treatment for acne.
- Treats pressure sores.

DOSAGE & USAGE INFORMATION

How to use:
Cream, gel, pads, sticks, lotion, cleansing bar, foam or facial mask—Wash affected area with plain soap and water. Dry gently with towel. Apply product as directed into affected areas. Keep away from eyes, nose, mouth. Wash hands after using.

When to use:
Apply as directed on product.

If you forget an application:
Use as soon as you remember. If it is almost time for the next dose, wait for that dose (don't double this dose) and resume regular schedule.

What drug does:
Slowly releases oxygen from skin, which controls some skin bacteria. Also causes peeling and drying, helping control blackheads and whiteheads.

Time lapse before drug works:
1 to 2 weeks.

Don't use with:
Any other topical medicine without consulting your doctor or pharmacist.

OVERDOSE

SYMPTOMS:
Unknown.
WHAT TO DO:
If person uses much larger amount than prescribed or if accidentally swallowed, call poison control center 1-800-222-1222 for instructions or dial 911 (emergency) for help.

POSSIBLE ADVERSE REACTIONS OR SIDE EFFECTS

SYMPTOMS	WHAT TO DO
Life-threatening:	
Rare, severe allergic reaction (anaphylaxis)—difficulty breathing, hives, itching, swelling (throat, face), vomiting, dizziness.	Seek emergency treatment immediately.
Common:	
Mild redness and chapping of skin during first few weeks of use.	No action necessary.
Infrequent:	
Rash, excessive dryness or peeling or irritation of skin, acne gets worse in first few weeks of use.	Continue. Call doctor or discontinue using if symptoms persist.
Rare:	
None expected.	

WARNINGS & PRECAUTIONS

Don't take if:
You are allergic to benzoyl peroxide.

Before you start, consult your doctor if:
- You plan to become pregnant within medication period.
- You take oral contraceptives.
- You are using any other prescription or nonprescription medicine for acne.
- You are using abrasive skin cleansers or medicated cosmetics.

Over age 60:
No problems expected.

Pregnancy:
Decide with your doctor if drug benefits justify risk to unborn baby. Risk category C (see page xviii).

Breastfeeding; Lactation; Nursing Mothers:
This drug is generally acceptable with breastfeeding. Discuss risks and benefits with your doctor.

Infants & children up to age 18:
Approved for age 12 and over. Follow instructions provided with product.

Prolonged use:
Permanent rash or scarring.

Skin & sunlight:
May cause rash or intensify sunburn in areas exposed to sun or ultraviolet light (photosensitivity reaction). Avoid overexposure. Notify doctor if reaction occurs.

Driving, piloting or hazardous work:
No problems expected.

Discontinuing:
- May be unnecessary to finish medicine. Discontinue when acne improves.
- If acne doesn't improve in 2 weeks, call doctor.

Others:
- Drug may bleach hair or dyed fabrics, including clothing or carpet.
- Store away from heat in cool, dry place.
- Avoid contact with eyes, lips, nose and sensitive areas of the neck.

POSSIBLE INTERACTION WITH OTHER DRUGS

GENERIC NAME OR DRUG CLASS	COMBINED EFFECT
Antiacne topical preparations, other	Excessive skin irritation.
Skin-peeling agents (salicylic acid, sulfur, resorcinol, tretinoin)	Excessive skin irritation.

POSSIBLE INTERACTION WITH OTHER SUBSTANCES

INTERACTS WITH	COMBINED EFFECT
Alcohol:	None expected.
Beverages:	None expected.
Cocaine:	None expected.
Foods: Cinnamon, foods with benzoic acid.	Skin rash.
Marijuana:	None expected.
Tobacco:	None expected.

BETA CAROTENE

BRAND NAMES

Solatene
Numerous multiple vitamin and mineral supplements. Check labels.

BASIC INFORMATION

Habit forming? No
Prescription needed? No
Available as generic? Yes
Drug class: Nutritional supplement

USES

- Used as a nutritional supplement.
- Used as an adjunct to the treatment of steatorrhea, chronic fever, obstructive jaundice, pancreatic insufficiency, protein deficiency, total parenteral nutrition and photosensitivity in photo porphyria.

DOSAGE & USAGE INFORMATION

How to take:
Tablet or capsule—Swallow with liquid. If you can't swallow whole, crumble tablet or open capsule and take with liquid or food.

When to take:
At the same time each day, according to directions on package or prescription label.

If you forget a dose:
Take as soon as you remember. If it is almost time for the next dose, wait for that dose (don't double this dose) and resume regular schedule.

What drug does:
Enables the body to manufacture vitamin A, which is essential for the normal functioning of the retina, normal growth and development and normal testicular and ovarian function.

Time lapse before drug works:
Total effect may take several weeks.

Don't take with:
No restrictions unless advised by doctor.

OVERDOSE

SYMPTOMS:
Yellow skin of hands and feet, diarrhea.
WHAT TO DO:
If person uses much larger amount than prescribed or if accidentally swallowed, call poison control center 1-800-222-1222 for instructions or dial 911 (emergency) for help.

POSSIBLE ADVERSE REACTIONS OR SIDE EFFECTS

SYMPTOMS	WHAT TO DO
Life-threatening:	
None expected.	
Common:	
Yellow palms, hands, soles of feet.	Continue. Call doctor if concerned.
Infrequent:	
None expected.	
Rare:	
• Joint pain, unusual bleeding or bruising.	Discontinue. Call doctor right away.
• Diarrhea, dizziness.	Continue. Call doctor when convenient.

WARNINGS & PRECAUTIONS

Don't take if:
You are allergic to beta carotene. Allergic reactions are rare, but may occur.

Before you start, consult your doctor if:
- You have liver or kidney disease.
- You have hypervitaminosis.*

Over age 60:
No problems expected.

Pregnancy:
Decide with your doctor if drug benefits justify risk to unborn baby. Risk category C (see page xviii).

Breastfeeding; Lactation; Nursing Mothers:
This drug is generally acceptable with breastfeeding. Discuss risks and benefits with your doctor.

Infants & children up to age 18:
Read product's label to be sure it is approved for your child's age. Follow label instructions on dosage. Consult doctor or pharmacist if unsure.

Prolonged use:
No problems expected.

Skin & sunlight:
No special problems expected.

Driving, piloting or hazardous work:
No special problems expected.

Discontinuing:
No special problems expected.

Others:
- Some researchers claim that beta carotene may reduce the occurrence of some cancers. There is insufficient data to substantiate this claim.
- Advise any doctor, dentist or pharmacist whom you consult that you take this medicine.
- May affect results of some medical tests.

POSSIBLE INTERACTION WITH OTHER DRUGS

GENERIC NAME OR DRUG CLASS	COMBINED EFFECT
Cholestyramine	Decreased absorption of beta carotene.
Colestipol	Decreased absorption of beta carotene.
Mineral oil	Decreased absorption of beta carotene.
Neomycin	Decreased absorption of beta carotene.

POSSIBLE INTERACTION WITH OTHER DRUGS

INTERACTS WITH	COMBINED EFFECT
Alcohol:	None expected.
Beverages:	None expected.
Cocaine:	None expected.
Foods:	None expected.
Marijuana:	None expected.
Tobacco:	None expected.

***See Glossary**

BETA-ADRENERGIC BLOCKING AGENTS

GENERIC AND BRAND NAMES

See full list of generic and brand names in the *Generic and Brand Name Directory*, page 852.

BASIC INFORMATION

Habit forming? No
Prescription needed? Yes
Available as generic? Yes, for some.
Drug class: Antiadrenergic, antianginal, antiarrhythmic, antihypertensive

USES

- Treats high blood pressure (hypertension).
- Some beta-blockers are used to relieve angina (chest pain).
- May be used to treat irregular heartbeat.
- May be used to treat anxiety disorders and other conditions as determined by your doctor.
- Treats tremors (some types).
- Reduces frequency of vascular headaches (does not relieve headache pain).

DOSAGE & USAGE INFORMATION

How to take:
- Tablet, capsule, extended-release capsule or extended-release tablet—Swallow whole with liquid. If you can't swallow whole, ask doctor or pharmacist for advice.
- Oral solution—Measure dose with proper device (e.g., oral syringe). Don't use a regular teaspoon or tablespoon (may not be accurate).

When to take:
With meals or immediately after.

If you forget a dose:
Take as soon as you remember. If it is almost time for the next dose, wait for that dose (don't double this dose) and resume regular schedule.

Continued next column

OVERDOSE

SYMPTOMS:
Weakness, slow or weak pulse, blood pressure drop, fainting, difficulty breathing, convulsions, cold and sweaty skin.
WHAT TO DO:
- **Dial 911 for all medical emergencies or call poison control center 1-800-222-1222 for instructions.**
- **See emergency information on last 3 pages of this book.**

What drug does:
- Blocks certain actions of sympathetic nervous system.
- Lowers heart's oxygen requirements.
- Slows nerve impulses through heart.
- Reduces blood vessel contraction in heart, scalp and other body parts.

Time lapse before drug works:
1 to 4 hours.

Don't take with:
Any other medicine or any dietary supplement without consulting your doctor or pharmacist.

POSSIBLE ADVERSE REACTIONS OR SIDE EFFECTS

SYMPTOMS	WHAT TO DO
Life-threatening:	
None expected.	
Common:	
Drowsiness, unusual tiredness or weakness, less sexual ability, trouble with sleeping.	Continue. Call doctor when convenient.
Infrequent:	
• Difficult breathing, dizziness, swelling of ankles or feet or lower legs, cold hands or feet, depression.	Discontinue. Call doctor right away.
• Nervousness or anxiety, diarrhea or constipation, stuffy nose, nausea, vomiting, stomach discomfort.	Continue. Call doctor when convenient.
Rare:	
• Rash, pain in back or chest or joints, yellow eyes or skin, dark urine, confusion, hallucinations, unusual bleeding or bruising, skin is red and crusty, fever, sore throat, lightheaded when getting up from sitting or lying, irregular heartbeat.	Discontinue. Call or doctor right away.
• Frequent urination, burning or dry eyes, numbness and tingling of fingers or toes, nightmares or vivid dreams, changes in taste, itching skin.	Continue. Call doctor when convenient.

BETA-ADRENERGIC BLOCKING AGENTS

WARNINGS & PRECAUTIONS

Don't take if:
- You are allergic to beta-adrenergic blockers. Allergic reactions are rare, but may occur.
- You have taken a monoamine oxidase (MAO) inhibitor* in the past 2 weeks.

Before you start, consult your doctor if:
- You have heart disease or poor circulation to the extremities.
- You have hay fever, asthma, chronic bronchitis, emphysema.
- You have overactive thyroid function.
- You have impaired liver or kidney function.
- You will have surgery within 2 months, including dental surgery, requiring general or spinal anesthesia.
- You have diabetes or hypoglycemia.

Over age 60:
Adverse reactions and side effects may be more frequent and severe than in younger persons.

Pregnancy:
Risk factors vary for drugs in this group. Always consult doctor. See category list on page xviii.

Breastfeeding; Lactation; Nursing Mothers:
Drugs in this group are generally acceptable with breastfeeding. Discuss risks and benefits with your doctor.

Infants & children up to age 18:
Safety and effectiveness in this age group have not been established. Consult doctor.

Prolonged use:
Talk to your doctor about the need for follow-up medical examinations or laboratory studies.

Skin & sunlight:
No problems expected.

Driving, piloting or hazardous work:
Don't drive or pilot aircraft until you learn how medicine affects you. Don't work around dangerous machinery. Don't climb ladders or work in high places. Danger increases if you drink alcohol or take medicine affecting alertness and reflexes.

Discontinuing
- Don't discontinue without consulting doctor. Dose may require gradual reduction if you have taken drug for a long time. Doses of other drugs may also require adjustment.
- Advise doctor of symptoms that occur after discontinuing (e.g., changes in heartbeat, feeling ill, weakness, headache, shortness of breath, sweating).

Others:
- May mask diabetic hypoglycemia symptoms.
- May affect results in some medical tests.
- Advise any doctor, dentist or pharmacist whom you consult that you take this medicine.

POSSIBLE INTERACTION WITH OTHER DRUGS

GENERIC NAME OR DRUG CLASS	COMBINED EFFECT
Angiotensin-converting (ACE) inhibitors*	Increased anti-hypertensive effects of both drugs. Dosages may require adjustment.
Antidiabetics*	Increased anti-diabetic effect.
Antihistamines*	Decreased antihistamine effect.
Antihypertensives*	Increased anti-hypertensive effect.
Anti-inflammatory drugs, nonsteroidal (NSAIDs)*	Decreased anti-hypertensive effect of beta blocker.
Betaxolol eyedrops	Possible increased beta blocker effect.
Calcium channel blockers*	Additional blood pressure drop.
Clonidine	Additional blood pressure drop. High blood pressure if clonidine stopped abruptly.
Dextrothyroxine	Possible decreased beta blocker effect.

Continued on page 893

POSSIBLE INTERACTION WITH OTHER SUBSTANCES

INTERACTS WITH	COMBINED EFFECT
Alcohol:	Blood pressure drop. Avoid.
Beverages:	None expected.
Cocaine:	Unknown. Avoid.
Foods:	None expected.
Marijuana:	Consult doctor.
Tobacco:	None expected.

***See Glossary**

BETA-ADRENERGIC BLOCKING AGENTS & THIAZIDE DIURETICS

GENERIC AND BRAND NAMES

See full list of generic and brand names in the *Generic and Brand Name Directory*, page 853.

BASIC INFORMATION

Habit forming? No
Prescription needed? Yes
Available as generic? Yes
Drug class: Beta-adrenergic blocker, diuretic (thiazide)

USES

- Controls (doesn't cure) high blood pressure.
- Reduces fluid retention (edema).
- Reduces angina attacks.
- Stabilizes irregular heartbeat.
- Reduces frequency of migraine headaches. (Does not relieve headache pain.)
- Other uses as determined by your doctor.

DOSAGE & USAGE INFORMATION

How to take:
- Tablet—Swallow with liquid.
- Extended-release capsule or tablet—Swallow whole with liquid. Do not open, chew or crush.

When to take:
At the same time each day.

If you forget a dose:
Take as soon as you remember. If it is almost time for the next dose, wait for that dose (don't double this dose) and resume regular schedule.

What drug does:
It reduces the amount of work of the heart. It helps the heart beat more regularly and slows down the heart rate. It reduces excess fluid in the body which helps lower blood pressure.

Continued next column

OVERDOSE

SYMPTOMS:
Irregular heartbeat (often, too slow), seizures, confusion, fainting, bluish nails, coma.
WHAT TO DO:
- **Dial 911 for all medical emergencies or call poison control center 1-800-222-1222 for instructions.**
- **See emergency information on last 3 pages of this book.**

Time lapse before drug works:
- 1 to 4 hours for beta-blocker effect.
- May take a few weeks to lower blood pressure.

Don't take with:
Any other medicines, even over-the-counter drugs such as cough/cold medicines, diet pills, nose drops, or caffeine, without consulting your doctor or pharmacist.

POSSIBLE ADVERSE REACTIONS OR SIDE EFFECTS

SYMPTOMS	WHAT TO DO
Life-threatening:	
None expected.	
Common:	
Drowsiness, unusual tiredness or weakness, less sexual ability, trouble with sleeping.	Continue. Call doctor when convenient.
Infrequent:	
• Difficult breathing, dizziness, swelling of ankles or feet or lower legs, cold hands or feet, slow heartbeat, depression.	Discontinue. Call doctor right away.
• Nervousness or anxiety, diarrhea or constipation, stuffy nose, nausea, vomiting, stomach discomfort.	Continue. Call doctor when convenient.
Rare:	
• Rash, pain in back or chest or joints, yellow eyes or skin, confusion, chills, fever, sore throat, hallucinations, unusual bleeding or bruising, skin is red and crusty, lightheaded when getting up from sitting or lying, irregular heartbeat, black or tarry stools, bloody or dark urine, seizures, mood or mental changes, muscle pain or cramps.	Discontinue. Call doctor right away.
• Frequent urination, burning or dry eyes, numbness and tingling of fingers or toes, nightmares or vivid dreams, changes in taste, itching skin.	Continue. Call doctor when convenient.

BETA-ADRENERGIC BLOCKING AGENTS & THIAZIDE DIURETICS

WARNINGS & PRECAUTIONS

Don't take if:
- You are allergic to any beta-adrenergic blocker or any thiazide diuretic drug. Allergic reactions are rare, but may occur.
- You have taken MAO inhibitors in past two weeks.

Before you start, consult your doctor if:
- You have heart disease or poor circulation to the extremities.
- You have hay fever, asthma, chronic bronchitis, emphysema, overactive thyroid function, impaired liver or kidney function, gout, diabetes, hypoglycemia, pancreas disorder, systemic lupus erythematosus.
- You are allergic to any sulfa drug or tartrazine dye.
- You will have surgery within 2 months, including dental surgery, requiring general or spinal anesthesia.

Over age 60:
Adverse reactions and side effects may be more frequent and severe than in younger persons, especially dizziness and excessive potassium loss.

Pregnancy:
Risk factors vary for drugs in this group. Always consult doctor. See category list on page xviii.

Breastfeeding; Lactation; Nursing Mothers:
Drugs in this group are generally acceptable with breastfeeding. Discuss risks and benefits with your doctor.

Infants & children up to age 18:
Safety and effectiveness in this age group have not been established. Consult doctor.

Prolonged use:
- Weakens heart muscle contractions.
- You may need medicine to treat high blood pressure for the rest of your life.
- Talk to your doctor about the need for follow-up medical examinations or laboratory studies.

Skin & sunlight:
One or more drugs in this group may cause rash or intensify sunburn in areas exposed to sun or ultraviolet light (photosensitivity reaction). Avoid overexposure. Notify doctor if reaction occurs.

Driving, piloting or hazardous work:
Don't drive or pilot aircraft until you learn how medicine affects you. Don't work around dangerous machinery. Don't climb ladders or work in high places. Danger increases if you drink alcohol or take medicine affecting alertness and reflexes.

Discontinuing:
Don't discontinue without consulting doctor. Dose may require gradual reduction if you have taken drug for a long time. Doses of other drugs may also require adjustment.

Others:
- May mask hypoglycemia symptoms.
- Hot weather and fever may cause dehydration and drop in blood pressure. Dose may require temporary adjustment. Weigh daily and report any unexpected weight decreases to your doctor.
- May cause rise in uric acid, leading to gout.
- May cause blood sugar rise in diabetics.

POSSIBLE INTERACTION WITH OTHER DRUGS

GENERIC NAME OR DRUG CLASS	COMBINED EFFECT
Allopurinol	Decreased allopurinol effect.
Aminophylline	Decreased effectiveness of both.
Antidepressants, tricyclic*	Dangerous drop in blood pressure. Avoid combination unless under medical supervision.

Continued on page 894

POSSIBLE INTERACTION WITH OTHER SUBSTANCES

INTERACTS WITH	COMBINED EFFECT
Alcohol:	Dangerous blood pressure drop. Avoid.
Beverages:	None expected.
Cocaine:	Irregular heartbeat, decreased beta blocker effect. Avoid.
Foods: Licorice.	Excessive potassium loss that causes dangerous heart rhythms.
Marijuana:	Consult doctor.
Tobacco:	May increase blood pressure and make heart work harder. Avoid.

***See Glossary**

BETHANECHOL

BRAND NAMES

Duvoid
Urabeth
Urecholine

BASIC INFORMATION

Habit forming? No
Prescription needed? Yes
Available as generic? Yes
Drug class: Cholinergic

USES

- Helps initiate urination following surgery, or for persons with urinary infections or enlarged prostate.
- Treats reflux esophagitis.

DOSAGE & USAGE INFORMATION

How to take:
Tablet—Swallow with liquid, 1 hour before or 2 hours after eating.

When to take:
At the same times each day.

If you forget a dose:
Take as soon as you remember. If it is almost time for the next dose, wait for that dose (don't double this dose) and resume regular schedule.

What drug does:
Affects chemical reactions in the body that strengthen bladder muscles.

Time lapse before drug works:
30 to 90 minutes.

Don't take with:
Any other medicine or any dietary supplement without consulting your doctor or pharmacist.

OVERDOSE

SYMPTOMS:
Upset stomach, excess saliva, flushing, feeling of warmth, sweating, nausea, vomiting.
WHAT TO DO:
If person uses much larger amount than prescribed or if accidentally swallowed, call poison control center 1-800-222-1222 for instructions or dial 911 (emergency) for help.

POSSIBLE ADVERSE REACTIONS OR SIDE EFFECTS

SYMPTOMS	WHAT TO DO
Life-threatening: None expected.	
Common: None expected.	
Infrequent: Dizziness, headache, faintness, blurred or changed vision, diarrhea, nausea, vomiting, stomach discomfort, belching, excessive urge to urinate.	Continue. Call doctor when convenient.
Rare: Shortness of breath, wheezing, tightness in chest.	Discontinue. Call doctor right away.

WARNINGS & PRECAUTIONS

Don't take if:
You are allergic to any cholinergic. Allergic reactions are rare, but may occur.

Before you start, consult your doctor if:
- You plan to become pregnant within medication period.
- You have asthma.
- You have epilepsy.
- You have heart or blood vessel disease.
- You have high or low blood pressure.
- You have overactive thyroid.
- You have intestinal blockage.
- You have Parkinson's disease.
- You have stomach problems (including ulcer).
- You have had bladder or intestinal surgery within 1 month.

Over age 60:
Adverse reactions and side effects may be more frequent and severe than in younger persons.

Pregnancy:
Decide with your doctor if drug benefits justify risk to unborn baby. Risk category C (see page xviii).

Breastfeeding; Lactation; Nursing Mothers:
This drug is generally not recommended with breastfeeding. Discuss risks and benefits with your doctor.

Infants & children up to age 18:
Follow instructions provided by your child's doctor.

Prolonged use:
No problems expected.

Skin & sunlight:
No problems expected.

Driving, piloting or hazardous work:
Don't drive or pilot aircraft until you learn how medicine effects you. Don't work around dangerous machinery. Don't climb ladders or work in high places. Danger increases if you drink alcohol or take medicine affecting alertness and reflexes, such as antihistamines, tranquilizers, sedatives, pain medicine, narcotics and mind-altering drugs.

Discontinuing:
May be unnecessary to finish medicine. Follow doctor's instructions.

Others:
- Be cautious about standing up suddenly.
- Advise any doctor, dentist or pharmacist whom you consult that you take this medicine.
- May interfere with laboratory studies of liver and pancreas function.

POSSIBLE INTERACTION WITH OTHER DRUGS

GENERIC NAME OR DRUG CLASS	COMBINED EFFECT
Cholinergics,* other	Increased effect of both drugs. Possible toxicity.
Ganglionic blockers*	Decreased blood pressure.
Nitrates*	Decreased bethanechol effect.
Procainamide	Decreased bethanechol effect.
Quinidine	Decreased bethanechol effect.

POSSIBLE INTERACTION WITH OTHER SUBSTANCES

INTERACTS WITH	COMBINED EFFECT
Alcohol:	None expected.
Beverages:	None expected.
Cocaine:	Unknown. Avoid.
Foods:	None expected.
Marijuana:	Consult doctor.
Tobacco:	None expected.

***See Glossary**

BIOLOGICS FOR PSORIASIS

GENERIC AND BRAND NAMES

SECUKINUMAB
Cosentyx

USTEKINUMAB
Stelara

BASIC INFORMATION

Habit forming? No
Prescription needed? Yes
Available as generic? No
Drug class: Immunosuppressant; antipsoriatic

USES

- Treatment for adult patients with moderate to severe, chronic, plaque psoriasis. Drug helps treat the cause of psoriasis as well as treating the symptoms.
- Treats adult patients with psoriatic arthritis.

DOSAGE & USAGE INFORMATION

How to take:
- Injection—Injected by a health care provider.
- Self-injected—You will be trained on technique. Follow instructions provided by your doctor.

When to take:
Your doctor will determine the schedule.

If you forget a dose:
Take as soon as you remember. If it is almost time for the next dose, wait for that dose (don't double this dose) and resume regular schedule.

What drug does:
The exact mechanism is unknown. The drug blocks certain cells in the body's immune system to help prevent skin inflammation that leads to psoriasis.

Time lapse before drug works:
Improvement may be seen in 4 weeks, but it may take 3 months for maximum benefits.

Don't take with:
Any other medicine or any dietary supplement without consulting your doctor or pharmacist.

OVERDOSE

SYMPTOMS:
Unknown.
WHAT TO DO:
If person uses much larger amount than prescribed, call poison control center 1-800-222-1222 for instructions or dial 911 (emergency) for help.

POSSIBLE ADVERSE REACTIONS OR SIDE EFFECTS

SYMPTOMS	WHAT TO DO
Life-threatening:	
Rare, severe allergic reaction (anaphylaxis)—difficulty breathing, hives, itching, swelling (throat, face), vomiting, dizziness.	Seek emergency treatment immediately.
Common:	
• Chills, fever, cough, urination painful or difficult, lower back or side pain, hoarseness.	Discontinue. Call doctor right away.
• Injection site problems (pain, swelling, rash, bleeding, lumps), tiredness.	Continue. Call doctor when convenient.
Infrequent:	
• Congestion, dry or sore throat, body aches or pain, runny nose, swollen or tender neck glands, signs of infection, swallowing difficulty, voice changes.	Discontinue. Call doctor right away.
• Dizziness, itching skin, painful or swollen joints, muscle aches or stiffness, difficulty in moving, headache.	Continue. Call doctor when convenient.
Rare:	
Chest symptoms (pain, heaviness, tightness, discomfort), arm or jaw pain, fast or irregular heartbeat, shortness of breath, nausea or vomiting, sweating, bloating, dark urine, tiredness or weakness, light color stools, loss of appetite, yellow eyes or skin, flu-like symptoms, neurological disorder (headache, seizures, confusion, vision changes).	Discontinue. Call doctor right away.

WARNINGS & PRECAUTIONS

Don't take if:
You are allergic to ustekinumab.

Before you start, consult your doctor if:
- You have kidney or liver problems.
- You have heart or blood vessel disorders.
- You have or have had cancer.
- You have any type of infection or have recurrent or chronic infections.
- You have a weak (suppressed) immune system due to illness or drugs.
- You have or have had tuberculosis.
- You have diverticulitis.
- You are getting phototherapy treatment.
- You are allergic to any medication, food or other substance.

Over age 60:
Unknown effect. Adverse reactions and side effects may be more frequent and severe.

Pregnancy:
Decide with your doctor if drug benefits justify risk to unborn baby. Risk category B (see page xviii).

Breastfeeding; Lactation; Nursing Mothers:
This drug is generally acceptable with breastfeeding. Discuss risks and benefits with your doctor.

Infants & children up to age 18:
Not recommended for ages under 18.

Prolonged use:
Visit the doctor regularly to see if the drug continues to be effective and to monitor your blood and platelet counts. If they get too low, the drug may be stopped on a temporary or permanent basis.

Skin & sunlight:
No special problems expected.

Driving, piloting or hazardous work:
Avoid if you feel dizzy, otherwise no special problems expected.

Discontinuing:
Consult doctor about discontinuing.

Others:
- Advise any doctor, dentist or pharmacist whom you consult about the use of this drug.
- The drug may increase the risk of developing cancer. Consult doctor about risks.
- Because the drug affects the immune system, you are at risk for new infections or reactivation of a chronic infection that has not been active. These include bacterial, viral and fungal infections that can be serious, possibly fatal. Call your doctor right away if symptoms of an infection develop.
- Avoid people with infections and people who have recently had a live virus vaccine.

POSSIBLE INTERACTION WITH OTHER DRUGS

GENERIC NAME OR DRUG CLASS	COMBINED EFFECT
Immunosuppressants,* other	Increased risk of infections or cancer.
Other drugs	Unknown. Consult doctor or pharmacist.
Vaccines, live virus	Unknown. May decrease effect of vaccine or may be harmful.

POSSIBLE INTERACTION WITH OTHER SUBSTANCES

INTERACTS WITH	COMBINED EFFECT
Alcohol:	None expected. Best to avoid.
Beverages:	None expected.
Cocaine:	Unknown. Avoid.
Foods:	None expected.
Marijuana:	Consult doctor.
Tobacco:	None expected.

***See Glossary**

BISMUTH SALTS

GENERIC AND BRAND NAMES

BISKALCITRATE
Pylera

BISMUTH SUBSALICYLATE
Bismatrol
Helidac
Maalox Total Stomach Relief
Pepto-Bismol

BASIC INFORMATION

Habit forming? No
Prescription needed? No
Available as generic? Yes
Drug class: Antidiarrheal; antacid

USES

- Treats symptoms of diarrhea, heartburn, nausea, acid indigestion.
- Helps prevent traveler's diarrhea.
- Treats ulcers.
- Used with other medications to treat a stomach infected by the bacteria *H. pylori*.

DOSAGE & USAGE INFORMATION

How to take:
- Tablet—Swallow with water.
- Chewable tablet—Chew well before swallowing.
- Liquid—Take as directed on label.

When to take:
As directed on label or by your doctor.

If you forget a dose:
Take as soon as you remember. If it is almost time for the next dose, wait for that dose (don't double this dose) and resume regular schedule.

Continued next column

OVERDOSE

SYMPTOMS:
Hearing loss, ringing or buzzing in the ears, severe drowsiness or tiredness, severe excitement or nervousness, fast or deep breathing, unconsciousness.
WHAT TO DO:
- **Dial 911 for all medical emergencies or call poison control center 1-800-222-1222 for instructions.**
- **See emergency information on last 3 pages of this book.**

What drug does:
- Decreases inflammation and increased motility of the intestinal muscles and lining.
- In combination with other drugs, it works to destroy certain bacterial infections.

Time lapse before drug works:
30 minutes to 1 hour.

Don't take with:
Any other medicine or any dietary supplement without consulting your doctor or pharmacist.

POSSIBLE ADVERSE REACTIONS OR SIDE EFFECTS

SYMPTOMS	WHAT TO DO
Life-threatening:	
In case of overdose, see previous column.	
Common:	
Black stools, dark tongue (symptoms like this are normal).	No action necessary.
Infrequent:	
None expected.	
Rare:	
Abdominal pain, increased sweating, muscle weakness, drowsiness, anxiety, trembling, hearing loss, ringing or buzzing in ears, confusion, dizziness, headache, increased thirst, vision problems, severe constipation, continuing diarrhea, trouble breathing (all more likely to occur with high doses or chronic use).	Discontinue. Call doctor right away.

WARNINGS & PRECAUTIONS

Don't take if:
- You are allergic to aspirin, salicylates or other nonsteroidal anti-inflammatory drugs. Allergic reactions are rare, but may occur.
- You have stomach ulcers that have ever bled.
- The patient is a child with fever.

Before you start, consult your doctor if:
- You are on a low-sodium, low-sugar or other special diet.
- You have had diarrhea for more than 24 hours. This is especially applicable to infants, children and those over 60.
- You have had kidney disease.

Over age 60:
Consult doctor before using.Older patients are more at risk for fluid and electrolyte loss from diarrhea.

Pregnancy:
Consult doctor. Use of the drug during pregnancy is not recommended. It could cause harm to the unborn baby. Risk category C/D (see page xviii).

Breastfeeding; Lactation; Nursing Mothers:
Drugs in this group are generally acceptable with breastfeeding. Discuss risks and benefits with your doctor.

Infants & children up to age 18:
Read product's label to be sure it is approved for your child's age. Follow label instructions on dosage. Consult doctor or pharmacist if unsure.

Prolonged use:
May cause constipation.

Skin & sunlight:
No problems expected.

Driving, piloting or hazardous work:
Don't drive or pilot aircraft if you take high or prolonged dose until you learn how medicine affects you. Don't work around dangerous machinery. Don't climb ladders or work in high places. Danger increases if you drink alcohol or take medicine affecting alertness and reflexes, such as antihistamines, tranquilizers, sedatives, pain medicine, narcotics and mind-altering drugs.

Discontinuing:
No problems expected.

Others:
- Pepto-Bismol contains salicylates. When given to children with flu or chicken pox, salicylates may cause a serious illness called Reye's syndrome.* An overdose in children can cause the same problems as aspirin poisoning.
- May cause false urine glucose tests.
- Dehydration can develop if too much body fluid has been lost. Consult doctor if any of the following symptoms occur: decreased urination, dizziness or lightheadedness, dryness of mouth, increased thirst, wrinkled skin.
- Don't store tablet form of drug in bathroom or near kitchen sink. Heat and moisture can cause it to break down.
- Consult doctor if diarrhea doesn't improve within 2 days.
- Read labels of any other drugs being used, such as for pain or inflammation. They may contain salicylates* and can lead to increased risk of side effects and overdose.

POSSIBLE INTERACTION WITH OTHER DRUGS

GENERIC NAME OR DRUG CLASS	COMBINED EFFECT
Anticoagulants*	Increased risk of bleeding.
Insulin or oral antidiabetic drugs*	Increased insulin effect. May require dosage adjustment.
Probenecid	Decreased effect of probenecid.
Salicylates,* other	Increased risk of salicylate toxicity.
Sulfinpyrazone	Decreased effect of sulfinpyrazone.
Tetracyclines*	Decreased absorption of tetracycline.
Thrombolytic agents*	Increased risk of bleeding.

POSSIBLE INTERACTION WITH OTHER SUBSTANCES

INTERACTS WITH	COMBINED EFFECT
Alcohol:	None expected.
Beverages:	None expected.
Cocaine:	Decreased bismuth subsalicylate effect. Avoid.
Foods:	None expected.
Marijuana:	Consult doctor.
Tobacco:	None expected.

***See Glossary**

BISPHOSPHONATES

GENERIC AND BRAND NAMES

See full list of generic and brand names in the *Generic and Brand Name Directory*, page 853.

BASIC INFORMATION

Habit forming? No
Prescription needed? Yes
Available as generic? Yes, for some
Drug class: Osteoporosis therapy, bisphosphonate; osteopenia therapy

USES

- Prevention and treatment of postmenopausal osteopenia and osteoporosis (thinning of bones). Treatment for osteoporosis in men.
- Treats osteoporosis caused by certain drugs.
- May be used to treat other bone disease or bone cancer as determined by your doctor.
- Treatment for Paget's disease of bone.
- Treatment for hypercalcemia (high calcium).

DOSAGE & USAGE INFORMATION

How to take:

- Tablet or extended-release tablet (alendronate or risedronate)—Swallow with a full glass of water (6 to 8 oz.). To help the medicine reach your stomach faster and to prevent throat irritation, stay upright for 30 minutes after you take it. Don't lie down.
- Tablet (etidronate or tiludronate)—Take morning, midday or evening 2 hours before or after any food.
- Tablet (ibandronate)—Take before first meal of the day. Swallow whole with water (6 to 8 oz). Do not lie down for 60 minutes.
- Effervescent tablet (alendronate)—Follow instructions provided with prescription.
- Injection—Etidronate, pamidronate or zoledronic acid are given by a medical person.

Continued next column

OVERDOSE

SYMPTOMS:
Breathing difficulty, confusion, tingling and numbness (mouth, feet, hands), irregular heartbeat, muscle cramps, convulsions.
WHAT TO DO:

- **Dial 911 for all medical emergencies or call poison control center 1-800-222-1222 for instructions.**
- **See emergency information on last 3 pages of this book.**

When to take:

- Daily dose (alendronate or risedronate), take first thing in the morning at least 30 to 60 minutes before eating, drinking or taking any other medications.
- Daily dose (etidronate or tiludronate), take anytime during the day 2 hours before or 2 hours after eating.
- Once-a-week dose, take on the same day each week. Follow instructions as daily dose.
- Once a month dose (ibandronate), take on the same date each month.

If you forget a dose:

- Daily dose taken first thing in morning: skip the missed dose entirely, then resume schedule the next day. Do not double this dose.
- For weekly dose: take the next morning and then return to your regular weekly schedule.
- Daily dose taken anytime: take as soon as you remember. If it is almost time for the next dose, wait for the next scheduled dose (don't double this dose).

What drug does:
Slows down the loss of bone tissue and increases bone mass. Osteoporosis and osteopenia are progressive diseases in which bone breakdown occurs faster than bone formation.

Time lapse before drug works:
Up to 6 months or longer.

Don't take with:

- Any other medicine or any dietary supplement without consulting your doctor or pharmacist.
- Any other medication at the same time as the bisphosphonate. Follow doctor's instructions.

POSSIBLE ADVERSE REACTIONS OR SIDE EFFECTS

SYMPTOMS	WHAT TO DO
Life-threatening: None expected.	
Common: Stomach pain.	Continue. Call doctor when convenient.
Infrequent: Mild bone or muscle pain, nausea, diarrhea, constipation, gas, leg cramps, bloated feeling, anxiety, depression, throat pain or irritation, mild heartburn, swallowing difficulty, headache, weak muscles.	Continue. Call doctor when convenient.

Rare:

• Chest pain, severe heartburn or throat pain, leg or groin pain, severe muscle pain.	Discontinue. Call doctor right away.
• Skin rash, ankle or leg swelling, eye or dental problems, cold or flu symptoms.	Continue. Call doctor when convenient.

WARNINGS & PRECAUTIONS

Don't take if:
You are allergic to any bisphosphonate. Allergic reactions are rare, but may occur.

Before you start, consult your doctor if:
- You currently have a gastrointestinal problem or serious esophageal disease.
- You have low blood levels of calcium (hypocalcemia) or a vitamin D deficiency.
- You have asthma or heart disease.
- You have dental disease or plan dental surgery. Consult your dentist also.
- You have a kidney (renal) disorder.

Over age 60:
No special problems expected.

Pregnancy:
Risk factors vary for drugs in this group. Always consult doctor. See category list on page xviii.

Breastfeeding; Lactation; Nursing Mothers:
Drugs in this group are generally acceptable with breastfeeding. Discuss risks and benefits with your doctor.

Infants & children up to age 18:
Not recommended for this age group.

Prolonged use:
Visit your doctor regularly to determine if the drug is continuing to control bone loss.

Skin & sunlight:
No special problems expected.

Driving, piloting or hazardous work:
No special problems expected.

Discontinuing:
No problems expected, but don't discontinue without your doctor's approval. After stopping the drug, it still remains in the body bound to the bone for as long as 10 years in some patients.

Others:
- Bisphosphonates may rarely increase the risk of a femoral (thigh bone) fracture.
- May affect the results of some medical tests.
- To avoid throat (esophagus) problems, carefully follow directions for taking the drug.
- Using zoledronic acid increases risk of kidney failure in certain patients. Consult your doctor.
- Advise any doctor and especially any dentist whom you consult that you take this medicine.
- In addition to the drug, your doctor may recommend exercises, diet changes, and calcium and vitamin D supplements.
- Though rare, bisphosphonates can increase the risk for osteonecrosis (bone destruction), especially of the jaw. Dental disease, oral surgery or tooth removal add to the risk. Talk to your doctor and dentist about your risks.

POSSIBLE INTERACTION WITH OTHER DRUGS

GENERIC NAME OR DRUG CLASS	COMBINED EFFECT
Antacids*	Decreased effect of bisphosphonate. Take 30 minutes after bisphosphonate.
Aspirin-containing products	Increased risk of stomach irritation.
Calcium supplements*	Decreased effect of bisphosphonate. Take 30 minutes after bisphosphonate.
Mineral or vitamin supplements	Decreased effect of bisphosphonate. Take 30 minutes after bisphosphonate.

POSSIBLE INTERACTION WITH OTHER SUBSTANCES

INTERACTS WITH	COMBINED EFFECT
Alcohol:	None expected. Alcohol will increase risk for osteoporosis. Try to avoid.
Beverages: Any beverage other than plain water.	Decreased effect of drug. Wait 30 minutes after you take drug.
Cocaine:	Unknown. Avoid.
Foods: Any food.	Decreased effect of drug. Wait 30 minutes to 2 hours after you take drug.
Marijuana:	Consult doctor.
Tobacco:	None expected. Smoking increases risk for osteoporosis. Try to avoid.

BONE FORMATION AGENTS

GENERIC AND BRAND NAMES

TERIPARATIDE
Forteo

BASIC INFORMATION

Habit forming? No
Prescription needed? Yes
Available as generic? No
Drug class: Osteoporosis therapy

USES

- Treatment of advanced postmenopausal osteoporosis (thinning of bones) in females. Osteoporosis is a cause of bone fractures.
- Treatment of osteoporosis associated with sustained, systemic glucocorticoid therapy in patients at high risk of fracture.
- Treatment for hypogonadal osteoporosis in men at high risk for fracture.

DOSAGE & USAGE INFORMATION

How to take:
Injection—Inject under the skin (subcutaneously) with the pen device provided. Follow your doctor's instructions and the directions provided with the prescription on how, when and where to inject. Do not use the medication unless you are sure about the proper method for injection. Store medication in the refrigerator (do not freeze) until you plan to use it.

When to take:
At the same time each day.

If you forget a dose:
Inject as soon as you remember. If it is almost time for the next dose, wait for that dose (don't double this dose) and resume regular schedule.

Continued next column

OVERDOSE

SYMPTOMS:
Nausea, weakness, low blood pressure.
WHAT TO DO:
If person uses much larger amount than prescribed or if accidentally swallowed, call poison control center 1-800-222-1222 for instructions or dial 911 (emergency) for help.

What drug does:
Increases the action of osteoblasts, the body's bone building cells. The bones become more dense and more resistant to fractures.

Time lapse before drug works:
Up to 3 months or longer.

Don't take with:
Any other medicine or any dietary supplement without consulting your doctor or pharmacist.

POSSIBLE ADVERSE REACTIONS OR SIDE EFFECTS

SYMPTOMS	WHAT TO DO
Life-threatening: None expected.	
Common: None expected.	
Infrequent: Injection site discomfort or redness, nausea, headache, stomach cramps, dizziness.	Continue. Call doctor when convenient.
Rare: Leg cramps, light-headedness when rising after sitting or lying down, syncope (fainting), vertigo.	Continue. Call doctor when convenient.

WARNINGS & PRECAUTIONS

Don't take if:
You are allergic to teriparatide or its components. Allergic reactions are rare, but may occur.

Before you start, consult your doctor if:
- You have had radiation treatment on the skeleton (bones).
- You have excess calcium in blood (hypercalcemia) or urine (hypercalcuria).
- You have or have had bone cancer/disease.
- You have Paget's disease.
- You have urolithiasis.
- You are allergic to any medication, food or other substance or latex.

Over age 60:
No special problems expected, but caution should be used in the elderly.

Pregnancy:
Normally not used in premenopausal women. Consult doctor. Risk category C (see page xviii).

Breastfeeding; Lactation; Nursing Mothers:
Normally not used in premenopausal women.

Infants & children up to age 18:
Not recommended for this age group.

Prolonged use:
- Long-term use after 2 years has not been established. Discuss with your doctor.
- Visit your doctor regularly to determine if the drug is continuing to be effective and to monitor your calcium levels.

Skin & sunlight:
No special problems expected.

Driving, piloting or hazardous work:
No special problems expected.

Discontinuing:
No problems expected. Consult doctor.

Others:
- In medical studies on rats injected with teriparatide, a few developed bone cancer (osteosarcoma). Risk in humans is unknown.
- Too much calcium in the blood (hypercalcemia) occurs in some patients using this drug. Consult your doctor about blood tests.
- Other therapies and nondrug routines (e.g., weight-bearing exercise), for treating osteoporosis may be recommended by your doctor. Teriparatide works effectively with certain other drugs for osteoporosis because the drugs work by different mechanisms.
- Advise any doctor, dentist or pharmacist whom you consult that you take this medicine.
- Smoking and alcohol consumption are risk factors for osteoporosis and should be discontinued.

POSSIBLE INTERACTION WITH OTHER DRUGS

GENERIC NAME OR DRUG CLASS	COMBINED EFFECT
Other drugs	Consult doctor or pharmacist.

POSSIBLE INTERACTION WITH OTHER SUBSTANCES

INTERACTS WITH	COMBINED EFFECT
Alcohol:	None expected, but alcohol is a risk factor for osteoporosis.
Beverages:	No special problems expected.
Cocaine:	Unknown. Avoid.
Foods:	None expected.
Marijuana:	Consult doctor.
Tobacco:	None expected, but smoking is a risk factor for osteoporosis.

*See Glossary

BOTULINUM TOXIN TYPE A

BRAND NAMES

Botox
Dysport
Myobloc
Xeomin

BASIC INFORMATION

Habit forming? No
Prescription needed? Yes
Available as generic? No
Drug class: Neuromuscular blocking agent

USES

- Provides temporary improvement in appearance in the frown lines between the eyebrows (glabellar lines). May also be used for lines and wrinkles in the forehead, around the eyes, in the lower face area and the neck.
- Treats strabismus (lazy eye) and blepharospasm (uncontrolled eye blinking).
- Treats certain facial nerve disorders and cervical dystonia (neck and shoulder tightness).
- Treatment for migraine, upper limb spasticity, excessive sweating, and urinary incontinence.
- May be used for writer's cramp, tremor, muscle-related disorders, pain, effects of a stroke, and other disorders.

DOSAGE & USAGE INFORMATION

How to take:
Injection—The medicine is administered by your doctor or a healthcare provider. It is injected into the muscle in or around the area being treated.

When to take:
As directed by your doctor.

If you forget a dose:
Injection is done only by scheduled appointment.

What drug does:
It paralyzes, weakens or relaxes the injected muscle by blocking the release of a chemical that normally signals the muscle to contract or tighten. The effect is temporary and most patients will require repeat treatments.

Continued next column

OVERDOSE

SYMPTOMS:
Unknown. May cause body weakness.
WHAT TO DO:
If person uses much larger amount than prescribed or if accidentally swallowed, call poison control center 1-800-222-1222 for instructions or dial 911 (emergency) for help.

Time lapse before drug works:
Improvement may be seen in 1-3 days and lasts up to 3-6 months. The degree of improvement will vary from person to person and will depend on the disorder being treated.

Don't take with:
Any other medicine or any dietary supplement without consulting your doctor or pharmacist.

POSSIBLE ADVERSE REACTIONS OR SIDE EFFECTS

SYMPTOMS	WHAT TO DO
Life-threatening:	
Rare, severe allergic reaction (anaphylaxis)—difficulty breathing, hives, itching, swelling (throat, face), vomiting, dizziness.	Seek emergency treatment immediately.
Common:	
• With blepharospasm (dry eyes, eyelid does not close completely).	Call doctor right away.
• With blepharospasm or strabismus (eye irritation or watering, eyelid drooping or bruised, light sensitivity).	Call doctor if you are concerned or symptoms continue.
Infrequent:	
• With blepharospasm or strabismus (blinking decreased, cornea irritation, eyelid edge turns in or out, skin rash, eyelid swelling, vision changes, eye pointing up or down).	Call doctor right away.
• With lines/wrinkles (injection site numb, burning or swelling).	Call doctor if you are concerned or symptoms continue.
Rare:	
• Any problems with speech, breathing or swallowing or heart symptoms occur or allergic reaction occurs.	Call doctor right away or seek emergency treatment if symptoms severe.
• With lines/wrinkles (drooping eyelids, redness or bruising at injection site, facial pain, skin rash or itching, headache, nausea, flu or cold symptoms).	Call doctor if you are concerned or symptoms continue.

• Other side effects or adverse reactions may occur depending on the disorder being treated. Drugs injected into muscles can be absorbed by the body and cause symptoms.	Call doctor if you are concerned or symptoms continue or they are severe.

WARNINGS & PRECAUTIONS

Don't take if:
You are allergic to botulinum toxin type A.

Before you start, consult your doctor if:
- You have heart problems.
- You have a nerve or muscle disorder, or a problem with swallowing.
- You have inflammation in the muscle area to be treated.
- You are allergic to any medication, food or other substance.
- You have a history of infection involving botulism poisoning.

Over age 60:
No special problems expected. Currently, for wrinkle treatment, the drug is approved for people between ages 18 and 65.

Pregnancy:
Decide with your doctor if drug benefits justify risk to unborn baby. Risk category C (see page xviii). Consult doctor if you become pregnant and have had a botulinum injection.

Breastfeeding; Lactation; Nursing Mothers:
This drug is generally acceptable with breastfeeding. Discuss risks and benefits with your doctor.

Infants & children up to age 18:
Approved for children age 12 and over for strabismus or blepharospasm treatment. Follow instructions provided by your child's doctor.

Prolonged use:
- Long term effects are unknown. Discuss with you doctor about long term use. Benefits and risks will differ depending on the problem being treated.
- Benefits may decrease with continued use.
- For facial lines and wrinkles, the injections should be at least 3 months apart.

Skin & sunlight:
No special problems expected.

Driving, piloting or hazardous work:
Since this medicine may be used for treatment of a variety of disorders (including eye muscle disorders, muscle contraction problems and muscle spasms), always consult your doctor about your individual circumstances.

Discontinuing:
Symptoms and signs of the problem being treated will most likely return.

Others:
- This treatment is given in a medical office and the risks and benefits will be explained to you. The information provided in this topic does not replace the information or special instructions provided by your doctor.
- Patients who have been inactive (sedentary) should resume activities gradually after receiving an injection.
- Very rarely, botulinum toxin may affect areas of the body away from the injection site and cause symptoms of botulism (a serious condition). It can happen hours to weeks after an injection. Symptoms of botulism include: loss of strength and muscle weakness all over the body, double vision, blurred vision and drooping eyelids, hoarseness or change or loss of voice, trouble saying words clearly, loss of bladder control, trouble breathing and swallowing. Call doctor or seek emergency care.
- Advise any doctor, dentist or pharmacist whom you consult (within the few months after the injection) that you have used this drug.

POSSIBLE INTERACTION WITH OTHER DRUGS

GENERIC NAME OR DRUG CLASS	COMBINED EFFECT
Aminoglycosides*	Increased effect of botulinum toxin type A.

POSSIBLE INTERACTION WITH OTHER SUBSTANCES

INTERACTS WITH	COMBINED EFFECT
Alcohol:	None expected.
Beverages:	None expected.
Cocaine:	None. Avoid.
Foods:	None expected.
Marijuana:	Consult doctor.
Tobacco:	None expected. May contribute to facial lines. Avoid.

*See Glossary

BRIMONIDINE (Topical)

BRAND NAMES

Mirvaso

BASIC INFORMATION

Habit forming? No
Prescription needed? Yes
Available as generic? No
Drug class: Vasoconstrictor (topical)

USES

Treatment for persistent rosacea in adults. It is used on the skin (topical) to treat facial redness (erythema is the medical term).

DOSAGE & USAGE INFORMATION

How to take:
Gel—Use a pea-sized amount for each of five areas of the face—forehead, nose, chin and each cheek. Spread the gel smoothly and evenly in a thin layer. Avoid contact with eyes and lips. Don't use on irritated skin or open wounds. Wash and dry hands before and right after applying the drug.

When to take:
Once a day at the same time each day.

If you forget a dose:
Use as soon as you remember. If it is almost time for the next dose, wait for that dose (don't double this dose) and resume regular schedule.

Continued next column

OVERDOSE

SYMPTOMS:

- **If accidentally swallowed, may have slow heartbeat, breathing difficulty, tiredness, sweating, muscle spasms, confusion.**
- **For topical overdose, symptoms are unknown.**

WHAT TO DO:

- **If swallowed—Dial 911 (emergency) for medical help. See emergency information on last 3 pages of this book.**
- **If person uses much larger amount on skin than prescribed, call doctor or poison control center 1-800-222-1222 for help.**

What drug does:
The exact cause of rosacea is unknown, but redness symptoms may be due in part to dilated (widened) blood vessels in the skin. The drug constricts (narrows) the dilated blood vessels which helps reduce the redness.

Time lapse before drug works:
Starts working in about 30 minutes and reduces its effect after 12 hours. Maximum effect is somewhere between 6 and 12 hours. It is a temporary treatment for rosacea.

Don't use with:
Any other topical facial medicine without consulting your doctor or pharmacist.

POSSIBLE ADVERSE REACTIONS OR SIDE EFFECTS

SYMPTOMS	WHAT TO DO
Life-threatening: None expected.	
Common: Redness, flushing, itching or burning feeling of facial skin.	Discontinue. Call doctor when convenient.
Infrequent: Skin irritation (e.g., soreness, flaking, crusting, blemishes, dryness, swelling, severe redness, acne).	Discontinue. Call doctor when convenient.
Rare: Headache, cold symptoms, blurred vision.	Discontinue. Call doctor when convenient.

WARNINGS & PRECAUTIONS

Don't take if:
You are allergic to topical brimonidine. Allergic reactions are rare, but may occur.

Before you start, consult your doctor if:
- You suffer from depression.
- You have any blood vessel disorder.
- You have heart disease.
- You have Raynaud's phenomenon.
- You have orthostatic hypotension.
- You have thromboangiitis obliterans, Sjögren's syndrome or scleroderma.

Over age 60:
No special problems expected.

Pregnancy:
Decide with your doctor if drug benefits justify risk to unborn baby. Risk category B (see page xviii).

Breastfeeding; Lactation; Nursing Mothers:
This drug is generally not recommended with breastfeeding. Discuss risks and benefits with your doctor.

Infants & children up to age 18:
Not recommended for ages under 18.

Prolonged use:
No problems expected.

Skin & sunlight:
No problems expected with drug use. Sun exposure can worsen rosacea. Follow doctor's advice about sun protection.

Driving, piloting or hazardous work:
No problems expected.

Discontinuing:
No problems expected, but call doctor if any symptoms occur after stopping drug that cause concern.

Others:
- Advise any doctor, dentist or pharmacist whom you consult that you use this drug.
- Keep drug in a safe place out of the reach of children.
- Be cautious with use of other skin care products that may irritate the skin.
- Consult your doctor if skin redness doesn't improve or it worsens while using this drug.

POSSIBLE INTERACTION WITH OTHER DRUGS

GENERIC NAME OR DRUG CLASS	COMBINED EFFECT
None expected.	

POSSIBLE INTERACTION WITH OTHER SUBSTANCES

INTERACTS WITH	COMBINED EFFECT
Alcohol:	None expected.
Beverages:	None expected.
Cocaine:	None expected.
Foods:	None expected.
Marijuana:	None expected.
Tobacco:	None expected.

BROMOCRIPTINE

BRAND NAMES

Alti-Bromocriptine
Apo-Bromocriptine
Cycloset
Parlodel
Parlodel Snaptabs

BASIC INFORMATION

Habit forming? No
Prescription needed? Yes
Available as generic? Yes
Drug class: Antiparkinsonism; antidiabetic

USES

- Controls Parkinson's disease symptoms such as rigidity, tremors and unsteady gait.
- Treats male and female infertility.
- Treats acromegaly (an overproduction of growth hormone).
- Treatment for diabetes type 2 (along with diet and exercise).
- Treats some pituitary tumors.

DOSAGE & USAGE INFORMATION

How to take:
Tablet or capsule—Swallow with liquid or food to lessen stomach irritation. If you can't swallow whole, crumble tablet or open capsule and take with liquid or food.

When to take:
At the same times each day. Take brand name Cycloset with food within 2 hours after waking up in the morning.

Continued next column

OVERDOSE

SYMPTOMS:
Vomiting, sleepiness, low blood pressure, lightheadedness, tiredness, paleness, paranoia, hallucinations, aggressive behavior.
WHAT TO DO:
- **Dial 911 for all medical emergencies or call poison control center 1-800-222-1222 for instructions.**
- **See emergency information on last 3 pages of this book.**

If you forget a dose:
Take as soon as you remember. If it is almost time for the next dose, wait for that dose (don't double this dose) and resume regular schedule.

What drug does:
- Restores chemical balance necessary for normal nerve impulses.
- It is unknown how it works to lower blood sugar in diabetic patients.

Time lapse before drug works:
2 to 3 weeks to improve; several months or longer for maximum benefit.

Don't take with:
Any other medicine or any dietary supplement without consulting your doctor or pharmacist.

POSSIBLE ADVERSE REACTIONS OR SIDE EFFECTS

SYMPTOMS	WHAT TO DO
Life-threatening: In case of overdose, see previous column.	
Common: Dizziness, mild nausea, lightheadedness when getting up, headache.	Continue. Call doctor when convenient.
Infrequent: Constipation, diarrhea, tiredness, drowsiness, dry mouth, depression, tingling and numbness of hands and feet.	Continue. Call doctor when convenient.
Rare:	
• Severe nausea and vomiting (may be bloody), vision changes, nervousness, sudden weakness, unusual headache, excess sweating, seizures, fainting, chest pain, black or tarry stools, uncontrollable body movements.	Discontinue. Call doctor right away.
• Stomach or back pain, runny nose, urinary frequency.	Continue. Call doctor when convenient.

WARNINGS & PRECAUTIONS

Don't take if:
- You are allergic to bromocriptine or ergotamine. Allergic reactions are rare, but may occur.
- You have taken a monoamine oxidase (MAO) inhibitor* in the past 2 weeks.
- You have glaucoma (narrow-angle type).

Before you start, consult your doctor if:
- You have diabetes or epilepsy.
- You have had high blood pressure, heart or lung disease or have a peptic ulcer.
- You have had liver or kidney disease.
- You have a history of mental problems.
- You will have surgery within 2 months, requiring general or spinal anesthesia.

Over age 60:
Adverse reactions and side effects may be more frequent and severe than in younger persons.

Pregnancy:
Decide with your doctor if drug benefits justify risk to unborn baby. Risk category C (see page xviii).

Breastfeeding; Lactation; Nursing Mothers:
This drug is generally not recommended with breastfeeding. Discuss risks and benefits with your doctor.

Infants & children up to age 18:
Not recommended if under age 15.

Prolonged use:
- May lead to uncontrolled movements of head, face, mouth, tongue, arms or legs.
- Changes in lung tissue and excess fluid in chest cavity may occur.
- Talk to your doctor about the need for follow-up medical examinations or laboratory studies to check blood pressure, x-rays, growth hormone levels, or blood sugar levels.

Skin & sunlight:
No problems expected.

Driving, piloting or hazardous work:
Don't drive or pilot aircraft until you learn how medicine effects you. Don't work around dangerous machinery. Don't climb ladders or work in high places. Danger increases if you drink alcohol or take medicine affecting alertness and reflexes, such as antihistamines, tranquilizers, sedatives, pain drugs or narcotics.

Discontinuing:
Don't discontinue without doctor's advice until you complete prescribed dose, even though symptoms diminish or disappear.

Others:
- May start treatment with small doses and increase gradually to lessen frequency and severity of adverse reactions.
- For diabetes type 2 patients: you and your family should educate yourselves about diabetes; learn to recognize hypoglycemia and treat it with sugar or glucagon.
- Advise any doctor, dentist or pharmacist whom you consult that you take this medicine.

POSSIBLE INTERACTION WITH OTHER DRUGS

GENERIC NAME OR DRUG CLASS	COMBINED EFFECT
Antihypertensives*	May decrease blood pressure.
Antiparkinsonism drugs, other*	Increased bromocriptine effect.
Ergot alkaloids, other	Increased risk of high blood pressure.
Erythromycin	Increased bromocriptine effect.
Haloperidol	Decreased bromocriptine effect.
Levodopa	Decreased antiparkinson effect.
Methyldopa	Decreased bromocriptine effect.
Papaverine	Decreased bromocriptine effect.
Phenothiazines*	Decreased bromocriptine effect.
Risperidone	Increased bromocriptine effect.
Ritonavir	Increased bromocriptine effect.

POSSIBLE INTERACTION WITH OTHER SUBSTANCES

INTERACTS WITH	COMBINED EFFECT
Alcohol:	Decreased alcohol tolerance. Avoid.
Beverages:	None expected.
Cocaine:	Unknown. Avoid.
Foods:	None expected.
Marijuana:	Consult doctor.
Tobacco:	None expected.

***See Glossary**

BRONCHODILATORS, ADRENERGIC (Long-Acting)

GENERIC AND BRAND NAMES

ARFORMOTEROL
Brovana
FORMOTEROL
Dulera
Foradil Aerolizer
Perforomist
Symbicort
INDACATEROL
Arcapta
OLODATEROL
Striverdi Respimat
SALMETEROL
Advair Diskus
Advair HFA
Serevent
VILANTEROL
Anoro Ellipta
Breo Ellipta

BASIC INFORMATION

Habit forming? No
Prescription needed? Yes
Available as generic? No
Drug class: Long-acting beta-agonist (LABA)

USES

- Used (in combination with an inhaled corticosteroid) for long-term control of asthma symptoms and to help prevent exercise-induced asthma. The drugs are not used to treat an asthma attack (bronchospasm).
- Long-term treatment (not for flare-ups) for chronic obstructive pulmonary disease (COPD).

DOSAGE & USAGE INFORMATION

How to take:
Powder, aerosol or solution—Inhale by mouth. Use with inhaler device or a nebulizer. Follow the directions on the prescription. If unsure how to use inhaler or nebulizer, ask your doctor or pharmacist about correct technique.

When to take:
Usually twice a day; longer acting type is used once a day. Use at the same time(s) each day.

Continued next column

OVERDOSE

SYMPTOMS:
Chest pain, fast and irregular heartbeat, severe dizziness, fainting, severe anxiety, severe muscle cramps, seizures.
WHAT TO DO:
- **Dial 911 for all medical emergencies or call poison control center 1-800-222-1222 for instructions.**
- **See emergency information on last 3 pages of this book.**

If you forget a dose:
Use as soon as you remember. If it is almost time for the next dose, wait for the next scheduled dose (don't double this dose).

What drug does:
In asthma and COPD patients, the airways (bronchi) of the lungs are narrowed causing breathing difficulties. The drugs help the airway muscles to relax, widening the airways, which leads to easier breathing.

Time lapse before drug works:
About 20 minutes if inhaled; 10 minutes if nebulized.

Don't take with:
- Nonprescription drugs for cough, cold, allergy or asthma without consulting doctor.
- Any other medicine or any dietary supplement without consulting your doctor or pharmacist.

POSSIBLE ADVERSE REACTIONS OR SIDE EFFECTS

SYMPTOMS	WHAT TO DO
Life-threatening:	
Rare, severe allergic reaction (anaphylaxis)—difficulty breathing, hives, itching, swelling (throat, face), vomiting, dizziness.	Seek emergency treatment immediately.
Common:	
Cough, runny or stuffy nose, sore throat, chills, fever, sneezing, headache, tremor, infection in mouth or throat, nervousness, anxiety, insomnia.	Continue. Call doctor when convenient.
Infrequent:	
Body aches or pain, voice changes, mild breathing problems, nausea, tiredness, dizziness, dry mouth, stomach pain, rash, diarrhea, high or low blood pressure.	Continue. Call doctor when convenient.
Rare:	
• Difficulty breathing that worsens quickly, chest pain, fainting, seizures, severe dizziness.	Seek emergency treatment.
• Muscle weakness or cramping, increased thirst or urination, fast/pounding or irregular heartbeat.	Discontinue. Call doctor right away.

WARNINGS & PRECAUTIONS

Don't take if:
- You are allergic to long-acting adrenergic bronchodilators.
- You are an asthma patient who is not also using an inhaled corticosteroid.

Before you start, consult your doctor if:
- You have high blood pressure or any heart or blood vessel problems.
- You have diabetes, overactive thyroid, seizure disorder, lactose intolerance, milk allergy, glaucoma, cataracts, low potassium level or adrenal gland problem.

Over age 60:
May have increased risk of side effects.

Pregnancy:
Decide with your doctor if drug benefits justify risk to unborn baby. Risk category C (see page xviii).

Breastfeeding; Lactation; Nursing Mothers:
Drugs in this group are generally acceptable with breastfeeding. Discuss risks and benefits with your doctor.

Infants & children up to age 18:
- Salmeterol used for over age 4; formoterol for over age 5; other drugs not used under age 18.
- Use of these drugs in children with asthma will depend on the child's age, symptoms, activities, asthma triggers, child's ability to use an inhaler and response to the drug. Carefully follow instructions provided by your child's doctor.
- Drug use may increase risk of asthma-related hospital admissions in patients under age 18.

Prolonged use:
It's important to consult your doctor on a regular basis while using this drug to monitor your progress and check for any unwanted effects.

Skin & sunlight:
No problems expected.

Driving, piloting or hazardous work:
No problems expected.

Discontinuing:
- Don't stop using the drug for asthma or COPD without consulting your doctor.
- Once your asthma is well controlled and maintained, your doctor may discontinue this drug.

Others:
- Don't exceed prescribed dose. Consult doctor if drug does is not helping your asthma or COPD symptoms or you're using more rescue inhalers due to increased symptoms.
- These drugs may increase the risk of severe asthma episodes, and death when an episode occurs. Ask your doctor about your risks.
- These drugs will not relieve an asthma attack or COPD flare-up that has already started. Your doctor will prescribe other drugs for quick relief.
- Advise any doctor, dentist or pharmacist whom you consult that you use this drug.
- Wear or carry a medical ID that states you have asthma or COPD and the drugs you use.
- For brand names that combine an inhaled corticosteroid, also read the drug chart—ADRENOCOTICOIDS (Oral Inhalation).

POSSIBLE INTERACTION WITH OTHER DRUGS

GENERIC NAME OR DRUG CLASS	COMBINED EFFECT
Adrenocorticoids, systemic	May decrease levels of potassium.
Beta-adrenergic blocking agents*	Decreased effect of long-acting adrenergic bronchodilator. Increased risk of bronchospasm.
Diuretics*	May decrease levels of potassium.
Enzyme inhibitors*	Increased effect of some long-acting adrenergic bronchodilators. Ask doctor.
Monoamine oxidase (MAO) inhibitors*	Increased risk of heart problems.
QT interval prolongation-causing drugs*	Risk of heart rhythm problems.
Sympathomimetics*	Increased risk of side effects.

Continued on page 895

POSSIBLE INTERACTION WITH OTHER SUBSTANCES

INTERACTS WITH	COMBINED EFFECT
Alcohol:	None expected.
Beverages:	None expected.
Cocaine:	Unknown. Avoid
Foods:	None expected.
Marijuana:	Consult doctor.
Tobacco:	No interaction. You should not smoke.

*See Glossary

BRONCHODILATORS, ADRENERGIC (Short-Acting)

GENERIC AND BRAND NAMES

See full list of generic and brand names in the *Generic and Brand Name Directory*, page 853.

BASIC INFORMATION

Habit forming? No
Prescription needed? Yes
Available as generic? Yes, for some
Drug class: Short-acting beta-agonist (SABA)

USES

Used to treat or prevent asthma attack symptoms (wheezing, shortness of breath, cough) caused by bronchospasm. Used to treat a flare-up of chronic obstructive pulmonary disease (COPD).

DOSAGE & USAGE INFORMATION

How to take:
- Aerosol, solution, powder—Inhale by mouth. Use with inhaler device or a nebulizer. Follow the instructions on the prescription. If unsure how to use inhaler or nebulizer, ask your doctor or pharmacist about correct technique.
- Tablet—Swallow whole with liquid. Do not crush or chew extended-release tablet.
- Syrup—Measure correct dosage and swallow.

When to take:
- Use inhaler or nebulizer as soon as symptoms develop. Repeat as needed according to your prescription.
- Use inhaler or nebulizer prior to exercising to prevent exercise-induced bronchospasm (EIB). Use before being around known triggers that bring on an asthma attack or COPD flare-up.
- Tablet or syrup usually taken one or more times daily at the same times each day.

Continued next column

If you forget a dose:
Generally used as needed. If drug is taken on a daily schedule, take/use as soon as you remember. If it is almost time for the next dose, wait for the next scheduled dose (don't double this dose).

What drug does:
In asthma and COPD patients, the airways (bronchi) of the lungs are narrowed causing breathing difficulties. The drugs help the airway muscles to relax, widening the airways, which leads to easier breathing.

Time lapse before drug works:
- Inhaled: 3 to 5 minutes (lasts 2 to 8 hours).
- Tablet or syrup: 30 minutes (lasts 4 to 12 hours).

Don't take with:
Any other drug (including nonprescription drugs for cough, cold, allergy) or dietary supplements without consulting your doctor or pharmacist.

OVERDOSE

SYMPTOMS:
Chest pain, fast and irregular heartbeat, severe dizziness, fainting, severe anxiety, severe muscle cramps, seizures.
WHAT TO DO:
- **Dial 911 for all medical emergencies or call poison control center 1-800-222-1222 for instructions.**
- **See emergency information on last 3 pages of this book.**

POSSIBLE ADVERSE REACTIONS OR SIDE EFFECTS

SYMPTOMS	WHAT TO DO
Life-threatening:	
Rare, severe allergic reaction (anaphylaxis)—difficulty breathing, hives, itching, swelling (throat, face), vomiting, dizziness.	Seek emergency treatment immediately.
Common:	
Cough, runny or stuffy nose, sore throat, chills, fever, sneezing, headache, tremor, nervousness, infection in mouth or throat, anxiety, taste changes.	Continue. Call doctor when convenient.
Infrequent:	
Body aches or pain, voice changes, mild breathing problems, nausea, trouble sleeping, tiredness, dizziness, dry mouth, diarrhea, rash, high or low blood pressure.	Continue. Call doctor when convenient.
Rare:	
• Difficulty breathing that worsens quickly, chest pain, fainting, seizures, severe dizziness.	Seek emergency treatment.
• Muscle weakness or cramping, increased thirst or urination, fast/pounding heartbeat.	Discontinue. Call doctor right away.

BRONCHODILATORS, ADRENERGIC (Short-Acting)

WARNINGS & PRECAUTIONS

Don't take if:
You are allergic to short-acting adrenergic bronchodilators.

Before you start, consult your doctor if:
- You have high blood pressure or any heart or blood vessel problems.
- You have diabetes, overactive thyroid, seizure disorder, low potassium level or adrenal gland problem.

Over age 60:
May have increased risk of side effects.

Pregnancy:
Decide with your doctor if drug benefits justify risk to unborn baby. Risk category C for all except terbutaline is category B (see page xviii).

Breastfeeding; Lactation; Nursing Mothers:
Drugs in this group are generally acceptable with breastfeeding. Discuss risks and benefits with your doctor.

Infants & children up to age 18:
- Use of these drugs in children with asthma will depend on the child's age, symptoms, activities, asthma triggers, child's ability to use an inhaler and response to the drug. Carefully follow instructions provided by your child's doctor.
- Be sure the drug is always available for your child (e.g., at school or sports activities).

Prolonged use:
It's important to consult your doctor on a regular basis while using this drug to monitor your progress and check for any unwanted effects.

Skin & sunlight:
No problems expected.

Driving, piloting or hazardous work:
No problems expected.

Discontinuing:
Asthma and COPD patients should not stop using the drug without consulting their doctor.

Others:
- Advise any doctor, dentist or pharmacist whom you consult that you use this drug.
- Consult your doctor if the drug does not seem to help your asthma or COPD symptoms or you are using the drug more often than usual.
- These drugs are not effective as maintenance treatment for asthma or COPD. Your doctor may prescribe other drugs (such as a long-acting adrenergic bronchodilator) to be used on a daily schedule.
- Wear or carry some type of medical ID that states you have asthma or COPD and the drugs you use.

POSSIBLE INTERACTION WITH OTHER DRUGS

GENERIC NAME OR DRUG CLASS	COMBINED EFFECT
Adrenocorticoids, systemic	May decrease levels of potassium.
Beta-adrenergic blocking agents*	Decreased effect of short-acting adrenergic bronchodilator. Increased risk of bronchospasm.
Digoxin	May decrease levels of digoxin.
Diuretics*	May decrease levels of potassium.
Monoamine oxidase (MAO) inhibitors*	Increased risk of heart problems.
QT interval prolongation-causing drugs*	Risk of heart rhythm problems.
Sympathomimetics*	Increased risk of side effects.
Xanthines*	May decrease levels of potassium. Risk of heart problems.

POSSIBLE INTERACTION WITH OTHER SUBSTANCES

INTERACTS WITH	COMBINED EFFECT
Alcohol:	None expected.
Beverages:	None expected.
Cocaine:	Unknown. Avoid
Foods:	None expected.
Marijuana:	Consult doctor.
Tobacco:	No interaction, but asthma and COPD patients should not smoke.

***See Glossary**

BRONCHODILATORS, ANTICHOLINERGIC

GENERIC AND BRAND NAMES

ACLIDINIUM
Tudorza Pressair
TIOTROPIUM
Spiriva
Spiriva Respimat
UMECLIDINIUM
Anoro Ellipta
Incruse Ellipta

BASIC INFORMATION

Habit forming? No
Prescription needed? Yes
Available as generic? No
Drug class: Anticholinergic; bronchodilator

USES

Maintenance treatment for bronchospasms that occur with chronic obstructive pulmonary disease (COPD), including chronic bronchitis and emphysema. These drugs are not used to relieve an acute attack of breathing difficulty.

DOSAGE & USAGE INFORMATION

How to take:
Inhalation powder or inhalation spray—Carefully follow the instructions provided with each product. Review them often; don't depend on memory. Don't swallow capsule.

When to take:
Once or twice a day per prescription instructions.

If you forget a dose:
Take as soon as you remember. If it is almost time for the next dose, wait for that dose (don't double this dose) and resume regular schedule.

Continued next column

OVERDOSE

SYMPTOMS:
Inhaled overdose is unlikely. Dry mouth, conjunctivitis, tremors, stomach pain, confusion may occur. Swallowed capsule is unlikely to cause symptoms as capsule is not well-absorbed by the gastrointestinal tract.
WHAT TO DO:
If person inhales much larger amount than prescribed or someone accidentally swallows a capsule, dial 911 (emergency) for medical help or call poison control center 1-800-222-1222 for instructions.

What drug does:
Relaxes the muscles around narrowed airways in the lungs and helps to keep them open and make breathing easier.

Time lapse before drug works:
Begins working right away, but will take about 2 to 3 weeks for full maintenance benefits.

Don't take with:
Any other medicine or any dietary supplement without consulting your doctor or pharmacist.

POSSIBLE ADVERSE REACTIONS OR SIDE EFFECTS

SYMPTOMS	WHAT TO DO
Life-threatening: None expected.	
Common: Dry mouth, cold or sinus infection symptoms (stuffy nose, sneezing, cough, sore throat), headache.	Continue. Call doctor when convenient.
Infrequent:	
• Painful or frequent urination, urinary retention.	Discontinue. Call doctor right away.
• Constipation, upset stomach, diarrhea.	Continue. Call doctor when convenient.
Rare:	
• Increased heart rate, chest pain, bloody nose, breathing problem, induced broncho-spasm (wheezing or cough), eye symptoms (pain, redness, blurred vision, seeing halos or colors around lights).	Discontinue. Call doctor right away.
• Heartburn, rash, vomiting, muscle or bone aches, dizziness.	Continue. Call doctor when convenient.

WARNINGS & PRECAUTIONS

Don't take if:
You are allergic to aclidinium, tiotropium, umeclidinium or atropine (or similar drugs). Allergic reactions are rare, but may occur.

Before you start, consult your doctor if:
- You have a kidney disorder.
- You have bladder problems, or urinary blockage or trouble urinating.
- You have enlarged prostate gland (prostatic hypertrophy).
- You have eye problems (e.g., glaucoma).
- You are allergic to milk proteins.
- You are allergic to any medications.

Over age 60:
No special problems expected.

Pregnancy:
Decide with your doctor if drug benefits justify risk to unborn baby. Risk category C (see page xviii).

Breastfeeding; Lactation; Nursing Mothers:
Drugs in this group are generally acceptable with breastfeeding. Discuss risks and benefits with your doctor.

Infants & children up to age 18:
Not recommended for this age group. COPD does not normally occur in children.

Prolonged use:
See your doctor for regular visits to make sure the drug is working properly and to check for unwanted effects.

Skin & sunlight:
No problems expected.

Driving, piloting or hazardous work:
Don't drive or pilot aircraft until you learn how medicine affects you. Don't work around dangerous machinery. Don't climb ladders or work in high places.

Discontinuing:
No problems expected. Consult your doctor before discontinuing.

Others:
- Advise any doctor, dentist or pharmacist whom you consult that you take this drug.
- Do not increase the dose or use drug more often than prescribed without checking with your doctor.
- The drug should not be used for sudden breathing problems. It is not a rescue inhaler.
- For dry mouth, suck sugarless hard candy or chew sugarless gum. If dry mouth continues, consult your doctor or dentist.

POSSIBLE INTERACTION WITH OTHER DRUGS

GENERIC NAME OR DRUG CLASS	COMBINED EFFECT
Anticholinergics,* other	May increase anticholinergic effect. Consult doctor.

POSSIBLE INTERACTION WITH OTHER SUBSTANCES

INTERACTS WITH	COMBINED EFFECT
Alcohol:	None expected.
Beverages:	None expected.
Cocaine:	Effect unknown. Avoid.
Foods:	None expected.
Marijuana:	Consult doctor.
Tobacco:	May decrease effect of drug. People with COPD should not smoke.

***See Glossary**

BRONCHODILATORS, XANTHINE

GENERIC AND BRAND NAMES

DYPHYLLINE
Lufyllin
OXTRIPHYLLINE
Choledyl SA
THEOPHYLLINE
Elixophyllin
Theo-24
Theochron
Theolair
Uniphyl

BASIC INFORMATION

Habit forming? No
Prescription needed? Yes
Available as generic? Yes, for some
Drug class: Bronchodilator (xanthine)

USES

Long-term treatment for patients with asthma or chronic obstructive pulmonary disease (COPD). These drugs are not used to treat an asthma attack (bronchospasm) or a COPD flare-up.

DOSAGE & USAGE INFORMATION

How to take:
- Tablet, extended-release tablet or extended-release capsule—Swallow whole with liquid. Do not chew or crush. If you can't swallow whole, ask doctor or pharmacist for advice.
- Elixir—Use a medicine measuring device for prescribed dosage amount and swallow.

When to take:
- Will depend on the dosage type and individual patient. May take once or several times a day.
- Most effective taken on empty stomach 1 hour before or 2 hours after eating. However, may take with food to lessen stomach upset.

Continued next column

OVERDOSE

SYMPTOMS:
Agitation, confusion, fast and irregular heartbeat, stomach pain, dizziness, diarrhea, vomiting (may be bloody), seizures, collapse.
WHAT TO DO:
- **Dial 911 for all medical emergencies or call poison control center 1-800-222-1222 for instructions.**
- **See emergency information on last 3 pages of this book.**

If you forget a dose:
Take as soon as you remember. If it is almost time for the next dose, wait for that dose (don't double this dose) and resume regular schedule.

What drug does:
Appears to improve lung function by relaxing the smooth muscles lining the airways of the lungs, decreasing inflammation in the airways and reducing mucus production.

Time lapse before drug works:
Starts working in 15 minutes to 2 hours (depends on dosage type); takes a few days for full benefit.

Don't take with:
Any other medicine or any dietary supplement without consulting your doctor or pharmacist.

POSSIBLE ADVERSE REACTIONS OR SIDE EFFECTS

SYMPTOMS	WHAT TO DO
Life-threatening:	
In case of overdose, see previous column.	
Common:	
Nervousness, nausea.	Continue. Call doctor when convenient.
Infrequent:	
Heartburn, vomiting, insomnia, trembling, headache, diarrhea, frequent urination, irritability.	Continue. Call doctor when convenient.
Rare:	
Fast and irregular heartbeat, seizures, dizziness, bloody vomit, stomach pain, confusion, decreased urine.	Discontinue. Call doctor right away or seek emergency care.

WARNINGS & PRECAUTIONS

Don't take if:
You are allergic to xanthine bronchodilators. Allergic reactions are rare, but may occur.

Before you start, consult your doctor if:
- You have a seizure disorder.
- You have liver or thyroid problems.
- You have stomach ulcer.
- You have heart failure or other heart disorder.
- You are a tobacco or marijuana smoker.

Over age 60:
Adverse reactions and side effects may be more frequent and severe than in younger persons.

Pregnancy:
Decide with your doctor if drug benefits justify risk to unborn baby. Risk category C (see page xviii).

Breastfeeding; Lactation; Nursing Mothers:
Drugs in this group are generally acceptable with breastfeeding. Discuss risks and benefits with your doctor.

Infants & children up to age 18:
Use of these drugs in children with asthma will depend on the child's age, weight, symptoms, activities, asthma triggers and child's response to the drug. Follow instructions provided by your child's doctor.

Prolonged use:
- Consult your doctor on a regular basis to check for continued effectiveness of drug, any unwanted side effects and to get blood tests to measure the level of the drug in your body.
- Stomach irritation may occur.

Skin & sunlight:
No problems expected.

Driving, piloting or hazardous work:
No problems expected.

Discontinuing:
Asthma and COPD patients should consult their doctor before stopping use of the drug.

Others:
- Advise any doctor, dentist or pharmacist whom you consult that you take this drug.
- A blood test will be done soon after starting the drug to be sure dosage is correct.
- Do not increase the dose or take drug more often than prescribed without consulting your doctor. An overdose can occur suddenly or over time if you take too much.
- The drug should not be used for sudden breathing problems. Other drugs are used for asthma attacks or COPD flare-ups.

POSSIBLE INTERACTION WITH OTHER DRUGS

GENERIC NAME OR DRUG CLASS	COMBINED EFFECT
Benzodiazepines*	Decreased effect of benzodiazepine.
Enzyme inducers*	Decreased xanthine bronchodilator effect.
Enzyme inhibitors*	Increased xanthine bronchodilator effect.
Telithromycin	Increased effect of telithromycin.
Ticlopidine	Increased theophylline effect.
Troleandomycin	Increased bronchodilator effect.
Zafirlukast	May increase effect of zafirlukast.
Zileuton	Increased theophylline effect.

POSSIBLE INTERACTION WITH OTHER SUBSTANCES

INTERACTS WITH	COMBINED EFFECT
Alcohol:	Increased risk of side effects.
Beverages: Caffeine drinks.	Nervousness and insomnia.
Cocaine:	Excess stimulation. Avoid.
Foods:	None expected.
Marijuana:	Decreased xanthine bronchodilator effect. Avoid.
Tobacco:	Decreased xanthine bronchodilator effect. Avoid

*See Glossary

BUPROPION

BRAND NAMES

Aplenzin
Contrave
Forfivo XL
Wellbutrin
Wellbutrin SR
Wellbutrin XL
Zyban

BASIC INFORMATION

Habit forming? No
Prescription needed? Yes
Available as generic? Yes
Drug class: Antidepressant

USES

- Treatment for major depressive disorder (MDD).
- May be used in combination with other therapy for smoking cessation.
- Prevention of major depressive episodes in those with seasonal affective disorder (SAD).
- Used in a combination drug for weight loss in obese and overweight adults.
- Treatment for other disorders as determined by your doctor.

DOSAGE & USAGE INFORMATION

How to take:

- Tablet—Swallow with liquid. May take with food to lessen stomach irritation. If you can't swallow whole, ask doctor or pharmacist for advice.
- Sustained-release tablet or extended release tablet—Swallow with liquid. Do not crush or crumble tablet.

When to take:
At the same time(s) each day, according to instructions on prescription label.

If you forget a dose:
Take as soon as you remember. If it is almost time for the next dose, wait for that dose (don't double this dose) and resume regular schedule.

Continued next column

OVERDOSE

SYMPTOMS:
Confusion, agitation, hallucinations, loss of consciousness, fast heartbeat, seizures, coma.
WHAT TO DO:

- **Dial 911 for all medical emergencies or call poison control center 1-800-222-1222 for instructions.**
- **See emergency information on last 3 pages of this book.**

What drug does:
It is unclear how the drug works. It appears to affect a number of brain chemicals.

Time lapse before drug works:
For depression symptoms, about 3 to 4 weeks. For smoking cessation, may take 7 to 12 weeks.

Don't take with:
Any other medicine or any dietary supplement without consulting your doctor or pharmacist.

POSSIBLE ADVERSE REACTIONS OR SIDE EFFECTS

SYMPTOMS	WHAT TO DO
Life-threatening:	
In case of overdose, see previous column.	
Common:	
• Agitation, anxiety.	Discontinue. Call doctor right away.
• Dry mouth, dizziness, insomnia, loss of appetite, constipation, stomach or muscle pain, nausea or vomiting, sore throat, unusual weight loss, trembling, increased sweating.	Continue. Call doctor when convenient.
Infrequent:	
• Skin rash or hives or itching, severe headache, hearing noises in ears.	Discontinue. Call doctor right away.
• Drowsiness, change in taste, blurred vision, frequent urination, euphoria.	Continue. Call doctor when convenient.
Rare:	
Fainting, seizures, confusion, delusions, hallucinations, feeling paranoid, unable to concentrate, fast or irregular heartbeat, higher blood pressure, behavior changes, thoughts/talk of suicide.	Discontinue. Call doctor right away.

WARNINGS & PRECAUTIONS

Don't take if:
You are allergic to bupropion. Allergic reactions are rare, but may occur.

Before you start, consult your doctor if:

- You have or have had any mental illness, a seizure disorder, anorexia nervosa or bulimia.
- You have a history of alcohol or drug abuse.
- You have liver, kidney or heart disease.

- You have had a recent head injury.
- You have a brain or spinal cord tumor.

Over age 60:
May be more sensitive to drug's effects. You may need smaller doses for shorter periods of time.

Pregnancy:
Decide with your doctor if drug benefits justify risk to unborn baby. Risk category C (see page xviii).

Breastfeeding; Lactation; Nursing Mothers:
This drug is generally acceptable with breastfeeding. Discuss risks and benefits with your doctor.

Infants & children up to age 18:
If prescribed for age under 18, follow instructions provided by your child's doctor. Contact doctor right away if depression symptoms get worse or there is any talk of suicide or suicide behaviors. Also, read information under Others.

Prolonged use:
Talk to your doctor about the need for follow-up medical examinations or laboratory studies to check kidney function, liver function, blood pressure and levels of bupropion in your blood.

Skin & sunlight:
No problems expected.

Driving, piloting or hazardous work:
Don't drive or pilot aircraft until you learn how medicine affects you. Don't work around dangerous machinery. Don't climb ladders or work in high places. Danger increases if you drink alcohol or take drugs affecting alertness and reflexes.

Discontinuing:
Don't discontinue without doctor's advice. Dose may require gradual reduction.

Others:
- If drug is taken to help stop smoking, be sure to follow all medical instructions.
- Use of this drug may lead to serious mental health events. They include worsening or new depression symptoms, changes in behavior (hostility, agitation) and may have increased suicidal thoughts or behaviors. Call doctor right away if these symptoms or behaviors occur.
- Advise any doctor, dentist or pharmacist whom you consult that you take this medicine.

POSSIBLE INTERACTION WITH OTHER DRUGS

GENERIC NAME OR DRUG CLASS	COMBINED EFFECT
Adrenocorticoids (systemic)	Increased risk of seizures.
Antidepressants, tricyclic*	Increased risk of seizures.
Clozapine	Increased risk of seizures.
Enzyme inducers*	Decreased effect of bupropion.
Enzyme inhibitors*	Increased effect of bupropion.
Fluoxetine	Increased risk of seizures.
Haloperidol	Increased risk of seizures.
Levodopa	Increased risk of side effects.
Lithium	Increased risk of seizures.
Loxapine	Increased risk of seizures.
Maprotiline	Increased risk of seizures.
Molindone	Increased risk of seizures.
Monoamine oxidase (MAO) inhibitors*	Increased risk of toxicity. Take at least 14 days apart.
Nicotine	May increase blood pressure.
Phenothiazines*	Increased risk of seizures.
Phenytoin	Increased phenytoin effect and risk of seizures.
Ritonivir	Increased risk of seizures.
Thioxanthenes*	Increased risk of seizures.
Trazodone	Increased risk of seizures

POSSIBLE INTERACTION WITH OTHER SUBSTANCES

INTERACTS WITH	COMBINED EFFECT
Alcohol:	Seizure risk. Avoid.
Beverages: Grapefruit juice.	Toxicity risk. Avoid.
Cocaine:	Unknown. Avoid.
Foods:	None expected.
Marijuana:	Consult doctor.
Tobacco:	None expected.

***See Glossary**

BUSPIRONE

BRAND NAMES

BuSpar

BASIC INFORMATION

Habit forming? Probably not
Prescription needed? Yes
Available as generic? Yes
Drug class: Antianxiety agent

USES

- Treats chronic anxiety disorders with nervousness or tension. Not intended for treatment of ordinary stress of daily living. Causes less sedation than some antianxiety drugs. Not useful for acute anxiety.
- Useful in agitation associated with dementia.
- Reduces aggression and irritability in patients with dementia, brain injury, mental retardation.
- Used for anxiety in alcoholics or substance abusers (buspirone has low abuse potential).
- Reduces frequency of vascular headaches (does not relieve headache pain).
- Not useful in withdrawal from sedatives.

DOSAGE & USAGE INFORMATION

How to take:
Tablet—Swallow with water. Take with or without food, but be consistent and always take it the same way.

When to take:
As directed. Usually 3 times daily. Food increases absorption.

If you forget a dose:
Take as soon as you remember. If it is almost time for the next dose, wait for that dose (don't double this dose) and resume regular schedule.

Continued next column

OVERDOSE

SYMPTOMS:
Severe drowsiness, dizziness, nausea, vomiting, small eye pupils, stomach upset.
WHAT TO DO:

- **Dial 911 for all medical emergencies or call poison control center 1-800-222-1222 for instructions.**
- **See emergency information on last 3 pages of this book.**

What drug does:
Affects certain chemicals in the brain and causes a calming effect.

Time lapse before drug works:
1 to 2 weeks before beneficial effects may be observed.

Don't take with:

- Alcohol, other tranquilizers, antihistamines, muscle relaxants, sedatives or narcotics.
- Any other medicine or any dietary supplement without consulting your doctor or pharmacist.

POSSIBLE ADVERSE REACTIONS OR SIDE EFFECTS

SYMPTOMS	WHAT TO DO
Life-threatening:	
None expected.	
Common:	
Lightheadedness, headache, nausea, restlessness, dizziness.	Continue. Call doctor when convenient.
Infrequent:	
Drowsiness, dry mouth, ringing in ears, nightmares or vivid dreams, unusual fatigue.	Continue. Call doctor when convenient.
Rare:	
Numbness or tingling in feet or hands, sore throat, fever, depression, confusion, uncontrollable movements (of tongue, lips, arms and legs), slurred speech, blurred vision, chest pain, fast or pounding heartbeat, muscle weakness, skin rash or hives, incoordination.	Discontinue. Call doctor right away.

WARNINGS & PRECAUTIONS

Don't take if:
You are allergic to buspirone. Allergic reactions are rare, but may occur.

Before you start, consult your doctor if:
- You have ever been addicted to any substance.
- You have chronic kidney or liver disease.
- You are already taking any medicine.

Over age 60:
No special problems expected.

Pregnancy:
Decide with your doctor if drug benefits justify risk to unborn baby. Risk category B (see page xviii).

Breastfeeding; Lactation; Nursing Mothers:
This drug is generally not recommended with breastfeeding. Discuss risks and benefits with your doctor.

Infants & children up to age 18:
Safety and effectiveness in this age group have not been established. Consult doctor.

Prolonged use:
- Not recommended for prolonged use. Adverse side effects more likely.
- Request follow-up studies to check kidney function, blood counts, and platelet counts.

Skin & sunlight:
No problems expected.

Driving, piloting or hazardous work:
Don't drive or pilot aircraft until you learn how medicine affects you. Don't work around dangerous machinery. Don't climb ladders or work in high places. Danger increases if you drink alcohol or take medicine affecting alertness and reflexes, such as antihistamines, tranquilizers, sedatives, pain medicine, narcotics and mind-altering drugs.

Discontinuing:
No problems expected.

Others:
- Before elective surgery requiring local or general anesthesia, tell your dentist, surgeon or anesthesiologist that you take buspirone.
- Advise any doctor, dentist or pharmacist whom you consult that you take this medicine.

POSSIBLE INTERACTION WITH OTHER DRUGS

GENERIC NAME OR DRUG CLASS	COMBINED EFFECT
Antihistamines*	Increased sedative effect of both drugs.
Barbiturates*	Excessive sedation. Sedative effect of both drugs may be increased.
Benzodiazepines*	Recent use of benzodiazepines may lessen effect of buspirone.
Central nervous system (CNS) depressants*	Increased sedative effect.
Monoamine oxidase MAO inhibitors*	May increase blood pressure.
Narcotics*	Excessive sedation. Sedative effect of both drugs may be increased.

POSSIBLE INTERACTION WITH OTHER SUBSTANCES

INTERACTS WITH	COMBINED EFFECT
Alcohol:	Excess sedation. Use caution.
Beverages:	
Caffeine-containing drinks.	Avoid. Decreased antianxiety effect of buspirone.
Grapefruit juice.	Toxicity risk. Avoid.
Cocaine:	Increased risk of side effects. Avoid.
Foods:	None expected.
Marijuana:	Consult doctor.
Tobacco:	None expected.

***See Glossary**

BUSULFAN

BRAND NAMES

Myleran

BASIC INFORMATION

Habit forming? No
Prescription needed? Yes
Available as generic? No
Drug class: Antineoplastic, immunosuppressant

USES

- Treatment for chronic myelogenous leukemia.
- Suppresses immune response after transplant and in immune disorders.

DOSAGE & USAGE INFORMATION

How to take:
Tablet—Swallow with liquid after light meal. Don't drink fluids with meals. Drink extra fluids between meals. Avoid sweet or fatty foods.

When to take:
At the same time each day.

If you forget a dose:
Take as soon as you remember. If it is almost time for the next dose, wait for that dose (don't double this dose) and resume regular schedule.

What drug does:
Inhibits abnormal cell reproduction. May suppress immune system.

Time lapse before drug works:
Up to 6 weeks for full effect.

Don't take with:
Any other medicine or any dietary supplement without consulting your doctor or pharmacist.

OVERDOSE

SYMPTOMS:
Symptoms may develop over time rather than sudden onset. Bleeding, easy bruising, infection (chills, fever), fatigue, weakness, shortness of breath.
WHAT TO DO:
- **Dial 911 for all medical emergencies or call poison control center 1-800-222-1222 for instructions.**
- **See emergency information on last 3 pages of this book.**

POSSIBLE ADVERSE REACTIONS OR SIDE EFFECTS

SYMPTOMS	WHAT TO DO
Life-threatening:	
In case of overdose, see previous column.	
Common:	
• Unusual bleeding or bruising, mouth sores, sore throat, chills and fever, black stools, lip sores, bloody urine, red spots on skin, painful or difficult urination.	Discontinue. Call doctor right away.
• Stomach pain, anxiety, diarrhea, menstrual changes, tiredness, weakness, nausea, vomiting, loss of appetite, sudden weight loss, muscle pain, trouble sleeping, headache.	Continue. Call doctor when convenient.
Infrequent:	
• Fast or irregular heartbeat, chest pain, shortness of breath, low blood pressure, excess sweating, swelling (hands, arms, feet, legs or face), mental confusion.	Discontinue. Call doctor right away.
• Cough, depression, constipation, bloody or stuffy or runny nose, dry mouth, itching.	Continue. Call doctor when convenient
Rare:	
Blurred vision, severe stomach and back pain, problem with swallowing, vomiting blood.	Discontinue. Call doctor right away.

WARNINGS & PRECAUTIONS

Don't take if:
- You are allergic to busulfan. Allergic reactions are rare, but may occur.
- Your physician has not explained the serious nature of your medical problem and risks of taking this medicine.

Before you start, consult your doctor if:
- You have gout.
- You have had kidney stones.
- You have active infection.
- You have impaired kidney or liver function.
- You have taken other antineoplastic drugs or had radiation treatment in last 3 weeks.

Over age 60:
Adverse reactions and side effects may be more frequent and severe than in younger persons.

Pregnancy:
Consult doctor. Use of the drug during pregnancy is not recommended. It could cause harm to the unborn baby. Risk category D (see page xviii).

Breastfeeding; Lactation; Nursing Mothers:
This drug is generally not recommended with breastfeeding. Discuss risks and benefits with your doctor.

Infants & children up to age 18:
Follow instructions provided by your child's doctor.

Prolonged use:
- Adverse reactions more likely the longer drug is required.
- Talk to your doctor about the need for follow-up medical examinations or laboratory studies.

Skin & sunlight:
No problems expected.

Driving, piloting or hazardous work:
No problems expected.

Discontinuing:
Don't discontinue without doctor's advice until you complete prescribed dose, even though symptoms diminish or disappear. Some side effects may follow discontinuing. Report to doctor blurred vision, convulsions, confusion, persistent headache or other new symptoms.

Others:
- Advise any doctor, dentist or pharmacist whom you consult that you take this medicine.
- May increase chance of developing lung or blood problems. Consult doctor about risks.

POSSIBLE INTERACTION WITH OTHER DRUGS

GENERIC NAME OR DRUG CLASS	COMBINED EFFECT
Antigout drugs*	Decreased antigout effect.
Antineoplastic drugs, other*	Increased effect of all drugs (may be beneficial).
Bone marrow depressants,* other	Increased risk of bone marrow depression.
Enzyme inducers*	Decreased effect of busulfan.
Enzyme inhibitors*	Increased effect of busulfan.
Vaccines, live or killed	Increased risk of toxicity or reduced effectiveness of vaccine.

POSSIBLE INTERACTION WITH OTHER SUBSTANCES

INTERACTS WITH	COMBINED EFFECT
Alcohol:	Increased risk of adverse reactions.
Beverages:	None expected.
Cocaine:	Increases risk of toxicity. Avoid
Foods:	None expected.
Marijuana:	Consult doctor.
Tobacco:	None expected.

*See Glossary

BUTORPHANOL

BRAND NAMES

Butorphanol Tartrate Nasal Spray
Stadol NS

BASIC INFORMATION

Habit forming? Yes
Prescription needed? Yes
Available as generic? No
Drug class: Narcotic analgesic

USES

- Treatment for migraine headache pain and postoperative pain.
- Treatment for other types of pain for which a narcotic analgesic is appropriate.

DOSAGE & USAGE INFORMATION

How to take:
Nasal spray—Spray in one nostril using the metered-dose pump.

When to take:
For pain as directed by your doctor. Usual treatment consists of one dose in one nostril followed by a second dose in 60 to 90 minutes if pain persists. Your doctor may direct that the initial 2-dose sequence may be repeated in 3 to 4 hours as needed.

If you forget a dose:
Unlikely to be a problem since the drug is taken for pain and not routinely.

What drug does:
Blocks the pain impulses at specific sites in the brain and spinal cord.

Time lapse before drug works:
Within 15 minutes of the first dose.

Don't take with:
Any other medicine or any dietary supplement without consulting your doctor or pharmacist.

OVERDOSE

SYMPTOMS:
Severe drowsiness, confusion, pinpoint pupils, heartbeat irregularities, breathing difficulty, cold and clammy skin, coma.
WHAT TO DO:
- **Dial 911 for all medical emergencies or call poison control center 1-800-222-1222 for instructions.**
- **See emergency information on last 3 pages of this book.**

POSSIBLE ADVERSE REACTIONS OR SIDE EFFECTS

SYMPTOMS	WHAT TO DO
Life-threatening:	
In case of overdose, see previous column.	
Common:	
Drowsiness, dizziness, nausea or vomiting, nasal congestion or irritation.	Discontinue. Call doctor when convenient.
Infrequent:	
Constipation (with continued use), faintness, high or low blood pressure.	Discontinue. Call doctor when convenient.
Rare:	
• Difficult breathing, heart palpitations.	Discontinue. Call doctor right away.
• Taste changes, ear ringing, dry mouth.	No action necessary.

WARNINGS & PRECAUTIONS

Don't take if:
You are allergic to butorphanol or the preservative benzethonium chloride,* which is used in the manufacture of the drug. Allergic reactions are rare, but may occur.

Before you start, consult your doctor if:
- You have a respiratory disorder or a central nervous system disease.
- You have had adverse reactions to other narcotics.*
- You have a history of emotional problems.
- You have heart, kidney or liver disease.

Over age 60:
Adverse reactions and side effects (particularly dizziness) may be more frequent and severe than in younger persons.

Pregnancy:
Decide with your doctor if drug benefits justify risk to unborn baby. Risk category C (see page xviii).

Breastfeeding; Lactation; Nursing Mothers:
This drug is generally not recommended with breastfeeding. Discuss risks and benefits with your doctor.

Infants & children up to age 18:
Safety and effectiveness in this age group have not been established. Consult doctor.

Prolonged use:
Long-term use effects are unknown. Probably habit forming. Consult with your doctor on a regular basis while using this drug.

Skin & sunlight:
No special problems expected.

Driving, piloting or hazardous work:
Don't drive or pilot aircraft until you learn how medicine affects you. Don't work around dangerous machinery. Don't climb ladders or work in high places. Danger increases if you drink alcohol or take medicine affecting alertness and reflexes.

Discontinuing:
Don't discontinue this drug after prolonged use without consulting doctor. Dosage may require a gradual reduction before stopping to avoid any withdrawal symptoms.

Others:
- When first using this drug, get up slowly from a sitting or lying position to avoid any dizziness, faintness or lightheadedness.
- Advise any doctor, dentist or pharmacist whom you consult that you take this medicine.
- Take medicine only as directed. Do not increase or reduce dosage without doctor's approval.

POSSIBLE INTERACTION WITH OTHER DRUGS

GENERIC NAME OR DRUG CLASS	COMBINED EFFECT
Central nervous system (CNS) depressants,* other	Increased sedative effect.
Oxymetazoline	Delays start of butorphanol effect.

POSSIBLE INTERACTION WITH OTHER SUBSTANCES

INTERACTS WITH	COMBINED EFFECT
Alcohol:	Increased sedative affect. Avoid.
Beverages:	None expected.
Cocaine:	Increased risk of side effects. Avoid.
Foods:	None expected.
Marijuana:	Increased risk of side effects. Avoid.
Tobacco:	None expected.

***See Glossary**

CAFFEINE

BRAND NAMES

See full list of brand names in the *Generic and Brand Name Directory*, page 854.

BASIC INFORMATION

Habit forming? Yes
Prescription needed? No
Available as generic? Yes
Drug class: Stimulant (xanthine), vasoconstrictor

USES

- Treatment for drowsiness and fatigue (occasional use only).
- Treatment for migraine and other vascular headaches in combination with ergot.

DOSAGE & USAGE INFORMATION

How to take:
- Tablet or liquid—Swallow with liquid or food to lessen stomach irritation. If you can't swallow whole, crumble tablet and take with liquid or food.
- Extended-release capsule—Swallow whole with liquid.
- Powder—Stir powder into water or other liquid. The powder may also be placed on the tongue and then followed by liquid.

When to take:
At the same times each day.

If you forget a dose:
Take as soon as you remember. If it is almost time for the next dose, wait for that dose (don't double this dose) and resume regular schedule.

What drug does:
- Constricts blood vessel walls.
- Stimulates central nervous system.

Continued next column

OVERDOSE

SYMPTOMS:
Excitement, insomnia, rapid heartbeat, fever, confusion, hallucinations, dizziness, thirst, convulsions, muscle twitching, breathing difficulty.
WHAT TO DO:
- **Dial 911 for all medical emergencies or call poison control center 1-800-222-1222 for instructions.**
- **See emergency information on last 3 pages of this book.**

Time lapse before drug works:
30 minutes.

Don't take with:
Any other medicine or any dietary supplement without consulting your doctor or pharmacist.

POSSIBLE ADVERSE REACTIONS OR SIDE EFFECTS

SYMPTOMS	WHAT TO DO
Life-threatening:	
In case of overdose, see previous column.	
Common:	
Nervousness, insomnia.	Continue. Call doctor if symptoms persist.
Infrequent:	
Indigestion, burning feeling in stomach. overexcitement.	Continue. Call doctor if symptoms persist.
Rare:	
• Rapid heartbeat, low blood sugar (hunger, anxiety, cold sweats, rapid pulse, tremor).	Discontinue. Call doctor right away.
• Confusion, nausea, irritability, agitation.	Continue. Call doctor if symptoms persist.

WARNINGS & PRECAUTIONS

Don't take if:
You are allergic to caffeine. Allergic reactions are rare, but may occur.

Before you start, consult your doctor if:
- You have irregular heartbeat.
- You have hypoglycemia (low blood sugar).
- You have heart disease.
- You have active peptic ulcer of stomach or duodenum.
- You have epilepsy.
- You have a seizure disorder.
- You have high blood pressure.
- You have insomnia.

Over age 60:
Adverse reactions and side effects may be more frequent and severe than in younger persons.

Pregnancy:
Decide with your doctor if drug benefits justify risk to unborn baby. Risk category not designated (see page xviii).

Breastfeeding; Lactation; Nursing Mothers:
This drug is generally acceptable with breastfeeding. Discuss risks and benefits with your doctor.

Infants & children up to age 18:
- Follow instructions provided by your child's doctor for prescription product.
- Read nonprescription product's label to be sure it is approved for your child's age. Follow label instructions on dosage. Consult doctor or pharmacist if unsure.

Prolonged use:
Can affect people in different ways. Consult your doctor if you are concerned.

Skin & sunlight:
No problems expected.

Driving, piloting or hazardous work:
No problems expected.

Discontinuing:
Will cause withdrawal symptoms of headache, irritability, drowsiness. Discontinue gradually if you use caffeine for a month or more.

Others:
Consult your doctor if drowsiness or fatigue continues, recurs or is not relieved by caffeine.

POSSIBLE INTERACTION WITH OTHER DRUGS

GENERIC NAME OR DRUG CLASS	COMBINED EFFECT
Caffeine-containing drugs, other	Increased risk of overstimulation.
Central nervous system (CNS) stimulants*	Increased risk of overstimulation.
Cimetidine	Increased caffeine effect.
Contraceptives, oral*	Increased caffeine effect.
Isoniazid	Increased caffeine effect.
Monoamine oxidase (MAO) inhibitors*	Dangerous blood pressure rise.
Sympathomimetics*	Overstimulation.
Xanthines*	Increased risk of overstimulation.

POSSIBLE INTERACTION WITH OTHER SUBSTANCES

INTERACTS WITH	COMBINED EFFECT
Alcohol:	Decreased alcohol effect.
Beverages: Caffeine drinks (coffee, tea or soft drinks).	Increased caffeine effect. Use caution.
Cocaine:	Unknown. Avoid.
Foods:	None expected.
Marijuana:	Consult doctor.
Tobacco:	None expected.

***See Glossary**

CALCITONIN

BRAND NAMES

Fortical
Miacalcin

BASIC INFORMATION

Habit forming? No
Prescription needed? Yes
Available as generic? No
Drug class: Osteoporosis therapy

USES

Treatment for postmenopausal osteoporosis (thinning of bones) in females. Osteoporosis is a major cause of bone fractures.

DOSAGE & USAGE INFORMATION

How to take:
Nasal spray—One spray per day in a nostril, alternating nostrils daily. Follow directions on label about activating and using the pump supplied with the medication.

When to take:
At the same time each day.

If you forget a dose:
Take as soon as you remember. If it is almost time for the next dose, wait for that dose (don't double this dose) and resume regular schedule.

What drug does:
The exact mechanism is not fully understood. It slows down the loss of bone tissue and increases bone mass in women with osteoporosis.

Time lapse before drug works:
Up to 6 months or longer.

Don't take with:
Any other medicine or any dietary supplement without consulting your doctor or pharmacist.

OVERDOSE

SYMPTOMS:
May have nausea and vomiting.
WHAT TO DO:
If person uses much larger amount than prescribed or if accidentally swallowed, call poison control center 1-800-222-1222 for instructions or dial 911 (emergency) for help.

POSSIBLE ADVERSE REACTIONS OR SIDE EFFECTS

SYMPTOMS	WHAT TO DO
Life-threatening:	
Rare, severe allergic reaction (anaphylaxis)—difficulty breathing, hives, itching, swelling (throat, face), vomiting, dizziness.	Seek emergency treatment immediately.
Common:	
Nasal symptoms (inflammation, dryness, crusting, sores, itching, irritation, redness, swollen, runny, stuffy, slight bleeding, tender, discomfort).	Continue. Call doctor when convenient.
Infrequent:	
Back pain, joint pain, headache, flushing, nausea, sinus infection.	Continue. Call doctor when convenient.
Rare:	
None expected.	

WARNINGS & PRECAUTIONS

Don't take if:
You are allergic to calcitonin.

Before you start, consult your doctor if:
You are allergic to any medication, food or other substance. A skin test may be performed before beginning treatment with calcitonin.

Over age 60:
No special problems expected.

Pregnancy:
Drug not normally used in premenopausal women. Consult doctor. Risk category C (see page xviii).

Breastfeeding; Lactation; Nursing Mothers:
Not normally used in premenopausal women.

Infants & children up to age 18:
Safety and effectiveness in this age group have not been established. Consult doctor.

Prolonged use:
- No special problems expected.
- Visit your doctor regularly to determine if the drug is continuing to control bone loss and to have periodic nasal examinations to check for ulceration or irritation.

Skin & sunlight:
No special problems expected.

Driving, piloting or hazardous work:
No special problems expected.

Discontinuing:
Don't discontinue without your doctor's approval.

Others:
- In addition to taking the drug, weight-bearing exercise and adequate dietary intake of calcium and vitamin D are essential in preventing bone loss. Dietary supplements of 1000 mg elemental calcium and 400 I.U. vitamin D daily may be recommended by your doctor.
- Advise any doctor, dentist or pharmacist whom you consult that you take this medicine.
- May affect the results of some medical tests.

POSSIBLE INTERACTION WITH OTHER DRUGS

GENERIC NAME OR DRUG CLASS	COMBINED EFFECT
None significant.	

POSSIBLE INTERACTION WITH OTHER SUBSTANCES

INTERACTS WITH	COMBINED EFFECT
Alcohol:	None expected.
Beverages:	None expected.
Cocaine:	Unknown. Avoid.
Foods:	None expected.
Marijuana:	Consult doctor.
Tobacco:	None expected.

*See Glossary

CALCIUM CHANNEL BLOCKERS

GENERIC AND BRAND NAMES

See full list of generic and brand names in *Generic and Brand Name Directory*, page 854.

BASIC INFORMATION

Habit forming? No
Prescription needed? Yes
Available as generic? Yes, for most
Drug class: Calcium channel blocker, antiarrhythmic, antianginal

USES

- Used for high blood pressure (hypertension) angina attacks and irregular heartbeat.
- Treats migraines and Raynaud's disease.

DOSAGE & USAGE INFORMATION

How to take:
- Tablet or capsule—Swallow with liquid. You may chew or crush tablet.
- Extended-release tablet or capsule—Swallow each dose whole with liquid; do not crush tablet or open capsule.

When to take:
At the same times each day. Take verapamil with food.

If you forget a dose:
Take as soon as you remember. If it is almost time for the next dose, wait for that dose (don't double this dose) and resume regular schedule.

What drug does:
- Reduces work that heart must perform.
- Reduces normal artery pressure.
- Increases oxygen to heart muscle.

Continued next column

OVERDOSE

SYMPTOMS:
Fast or slow heartbeat, drowsiness, nausea, dizziness, weakness, constipation, shortness of breath, slurred speech.
WHAT TO DO:
- **Dial 911 for all medical emergencies or call poison control center 1-800-222-1222 for instructions.**
- **See emergency information on last 3 pages of this book.**

Time lapse before drug works:
1 to 2 hours.

Don't take with:
Any other medicine or any dietary supplement without consulting your doctor or pharmacist.

POSSIBLE ADVERSE REACTIONS OR SIDE EFFECTS

SYMPTOMS	WHAT TO DO
Life-threatening:	
In case of overdose, see previous column.	
Common:	
Tiredness.	Continue. Call doctor when convenient.
Infrequent:	
• Unusually fast or slow heartbeat, wheezing, cough, shortness of breath, dizziness, numbness or tingling in hands and feet, swollen feet or ankles or legs, difficult urination.	Discontinue. Call doctor right away.
• Nausea, constipation.	Continue. Call doctor when convenient.
Rare:	
• Fainting, depression, unusual mental changes, rash, yellow skin or eyes.	Discontinue. Call doctor right away.
• Headache, insomnia, vivid dreams, hair loss.	Continue. Call doctor when convenient.

WARNINGS & PRECAUTIONS

Don't take if:
- You are allergic to calcium channel blockers. Allergic reactions are rare, but may occur.
- You have very low blood pressure.

Before you start, consult your doctor if:
- You have kidney or liver disease.
- You have low blood pressure, aortic stenosis, bowel blockage or certain rare hereditary problems.
- You have heart disease or heart attack.

Over age 60:
No special problems expected.

Pregnancy:
Decide with your doctor if drug benefits justify risk to unborn baby. Risk category C (see page xviii).

Breastfeeding; Lactation; Nursing Mothers:
Drugs in this group are generally acceptable with breastfeeding. Discuss risks and benefits with your doctor.

Infants & children up to age 18:
Safety and effectiveness in this age group have not been established. Consult doctor.

Prolonged use:
Talk to your doctor about the need for follow-up medical exams or lab studies to check drug's effectiveness.

Skin & sunlight:
One or more drugs in this group may cause rash or intensify sunburn in areas exposed to sun or ultraviolet light (photosensitivity reaction). Avoid overexposure. Notify doctor if reaction occurs.

Driving, piloting or hazardous work:
Avoid if you feel dizzy. Otherwise, no problems expected.

Discontinuing:
Don't discontinue without doctor's advice until you complete prescribed dose, even though symptoms diminish or disappear.

Others:
- Learn to check your own pulse rate. If it drops to 50 beats per minute or lower, don't take drug until you consult your doctor.
- Advise any doctor, dentist or pharmacist whom you consult that you take this medicine.

POSSIBLE INTERACTION WITH OTHER DRUGS

GENERIC NAME OR DRUG CLASS	COMBINED EFFECT
Angiotensin-converting enzyme (ACE) inhibitors*	Possible excessive potassium in blood. Dosages may require adjustment.
Antiarrhythmics*	Possible increased effect and toxicity of each drug.
Anticoagulants, oral*	Possible increased anticoagulant effect.
Anticonvulsants, hydantoin*	Increased anticonvulsant effect.
Antihypertensives*	Blood pressure drop. Dosages may require adjustment.
Beta-adrenergic blocking agents*	Possible irregular heartbeat and congestive heart failure.
Calcium (large doses)	Possible decreased effect of calcium channel blocker.
Carbamazepine	May increase carbamazepine effect and toxicity.
Cimetidine	Possible increased effect of calcium channel blocker.
Cyclosporine	Increased cyclosporine toxicity.
Digitalis preparations*	Increased digitalis effect. May need to reduce dose.
Disopyramide	May cause dangerously slow, fast or irregular heartbeat.
Diuretics*	Dangerous blood pressure drop. Dosages may require adjustment.
Dofetilide	Increased risk of heart problems.
Encainide	Increased effect of toxicity on heart muscle.
Fluvoxamine	Slow heartbeat (with diltiazem).
HMG-CoA reductase inhibitors	Increased effect of HMG-CoA reductase inhibitor.
Hypokalemia-causing medications*	Increased anti-hypertensive effect.
Leukotriene modifiers	Increased effect of calcium channel blocker.

Continued on page 895

POSSIBLE INTERACTION WITH OTHER SUBSTANCES

INTERACTS WITH	COMBINED EFFECT
Alcohol:	Very low blood pressure. Avoid.
Beverages: Grapefruit juice.	Possible increased drug effect.
Cocaine:	Unknown. Avoid.
Foods: Grapefruit.	Possible increased drug effect.
Marijuana:	Consult doctor.
Tobacco:	Possible rapid heartbeat. Avoid.

*See Glossary

CALCIUM SUPPLEMENTS

GENERIC AND BRAND NAMES

See full list of generic and brand names in the *Generic and Brand Name Directory*, page 855.

BASIC INFORMATION

Habit forming? No
Prescription needed? For some
Available as generic? Yes
Drug class: Antihypocalcemic, dietary replacement

USES

- Treats or prevents osteoporosis (thin, porous, easily fractured bones).
- Helps heart, muscle and nervous system to work properly.
- Dietary supplement when calcium ingestion is insufficient or there is a deficiency such as osteomalacia or rickets.

DOSAGE & USAGE INFORMATION

How to take:
- Tablet or capsule—Swallow whole with liquid.
- Chewable tablet—Chew tablet well before swallowing.
- Syrup—Take before meals.
- Powder—Follow directions on label.
- Suspension—Swallow with liquid.

When to take:
Take 1 to 1 1/2 hours after meals (except for syrup) in 3 to 4 daily doses. Try to avoid taking oral drugs within 1 to 2 hours of taking calcium.

If you forget a dose:
Use as soon as you remember. If it is almost time for the next dose, wait for that dose (don't double this dose) and resume regular schedule.

Continued next column

OVERDOSE

SYMPTOMS:
Confusion, irregular heartbeat, stomach pain, depression, headache, muscle twitching, nausea, vomiting.
WHAT TO DO:
- **Dial 911 for all medical emergencies or call poison control center 1-800-222-1222 for instructions.**
- **See emergency information on last 3 pages of this book.**

What drug does:
Calcium helps maintain strong bones and teeth. It also helps heart function, muscle contraction, blood clotting and nerve transmission.

Time lapse before drug works:
Starts within 15 to 30 minutes, but may take months or years to improve certain conditions.

Don't take with:
- Any other oral medicine until 1 to 2 hours have passed since taking calcium.
- Any other medicine or any dietary supplement without consulting your doctor or pharmacist.

POSSIBLE ADVERSE REACTIONS OR SIDE EFFECTS

SYMPTOMS	WHAT TO DO
Life-threatening: None expected.	
Common: None expected.	
Infrequent: Constipation, upset stomach.	Continue. Call doctor if symptoms persist.
Rare: Urination changes (difficult, more frequent, painful), increased thirst, nausea, vomiting, weight loss, mood changes, muscle pain, unusual weakness or tiredness.	Discontinue. Call doctor right away.

WARNINGS & PRECAUTIONS

Don't take if:
- You are allergic to calcium.
- You have a high blood calcium level.

Before you start, consult your doctor if:
You have diarrhea, heart disease, kidney stones, kidney disease, sarcoidosis or malabsorption.

Over age 60:
No special problems expected.

Pregnancy:
Decide with your doctor if drug benefits justify risk to unborn baby. Risk category not designated (see page xviii).

Breastfeeding; Lactation; Nursing Mothers:
This drug is generally acceptable with breastfeeding. Discuss risks and benefits with your doctor.

Infants & children up to age 18:
Read nonprescription product's label to be sure it is approved for your child's age. Follow label instructions on dosage. Consult doctor or pharmacist if unsure.

Prolonged use:
Talk to your doctor about the need for any follow-up laboratory studies to check calcium levels in the body.

Skin & sunlight:
No problems expected.

Driving, piloting or hazardous work:
No problems expected.

Discontinuing:
No problems expected.

Others:
- Exercise, along with vitamin D from sunshine and calcium, helps prevent osteoporosis.
- Advise any doctor, dentist or pharmacist whom you consult that you take this medicine.

POSSIBLE INTERACTION WITH OTHER DRUGS

GENERIC NAME OR DRUG CLASS	COMBINED EFFECT
Alendronate	Decreased effect of alendronate. Take calcium 30 minutes after alendronate.
Anticoagulants, oral*	Decreased anticoagulant effect.
Calcitonin	Decreased calcitonin effect.
Calcium-containing medicines, other	Increased calcium effect.
Chlorpromazine	Decreased effect of chlorpromazine.
Contraceptives, oral*	May increase absorption of calcium—frequently a desirable combined effect.
Corticosteroids*	Decreased calcium absorption and effect.
Digitalis preparations*	Decreased digitalis effect.
Diuretics, thiazide*	Increased calcium in blood.
Estrogens*	May increase absorption of calcium.
Etidronate	Decreased effect of etidronate. Take drugs 2 hours apart.
Iron supplements*	Decreased iron effect.
Meperidine	Increased meperidine effect.
Mexiletine	May slow elimination of mexiletine and cause need to adjust dosage.
Nalidixic acid	Decreased effect of nalidixic acid.
Nicardipine	Possible decreased nicardipine effect.
Nimodipine	Possible decreased nimodipine effect.
Oxyphenbutazone	Decreased oxyphenbutazone effect.
Para-aminosalicylic acid (PAS)	Decreased PAS effect.
Penicillins*	Decreased penicillin effect.
Pentobarbital	Decreased pentobarbital effect.
Phenylbutazone	Decreased effect of phenylbutazone.
Phenytoin	Decreased phenytoin effect.
Propafenone	Increased effects of both drugs and increased risk of toxicity.
Pseudoephedrine	Increased pseudoephedrine effect.

Continued on page 895

POSSIBLE INTERACTION WITH OTHER SUBSTANCES

INTERACTS WITH	COMBINED EFFECT
Alcohol:	Decreased absorption of calcium.
Beverages:	None expected.
Cocaine:	Unknown. Avoid.
Foods:	None expected.
Marijuana:	Consult doctor.
Tobacco:	None expected.

***See Glossary**

CAPECITABINE

BRAND NAMES

Xeloda

BASIC INFORMATION

Habit forming? No
Prescription needed? Yes
Available as generic? Yes
Drug class: Antineoplastic

USES

Treatment of cancer of the colon after surgery and colorectal or breast cancer that has spread to other parts of the body (metastatic cancer).

DOSAGE & USAGE INFORMATION

How to take:
Tablet—Swallow with water. If you can't swallow whole, crumble tablet and take with liquid or food.

When to take:
Take in two divided doses every 12 hours after a meal and with water.

If you forget a dose:
Take as soon as you remember. If it is almost time for the next dose, wait for that dose (don't double this dose) and resume regular schedule.

What drug does:
Capecitabine is converted in the body to the substance 5-fluorouracil. In some patients, this substance kills cancer cells and decreases the size of the tumor.

Time lapse before drug works:
Results may not show for several months.

Don't take with:
Any other medicine or any dietary supplement without consulting your doctor or pharmacist.

OVERDOSE

SYMPTOMS:
Nausea, vomiting, diarrhea, gastrointestinal irritation, bleeding, fatigue, infection.
WHAT TO DO:
If person uses much larger amount than prescribed or if accidentally swallowed, call poison control center 1-800-222-1222 for instructions or dial 911 (emergency) for help.

POSSIBLE ADVERSE REACTIONS OR SIDE EFFECTS

SYMPTOMS	WHAT TO DO
Life-threatening: None expected.	
Common:	
• Diarrhea (if you have more than 4 bowel movements in a day or any diarrhea at night), vomiting (more than once a day), nausea, loss of appetite, stomatitis (pain, redness or swelling in mouth), hand and foot syndrome (pain, redness or swelling in hands or feet), fever.	Discontinue. Call doctor right away.
• Nausea, vomiting, rash, dry or itchy skin, tiredness, headache, weakness, dizziness.	Continue. Call doctor when convenient.
Infrequent:	
Jaundice (yellow skin and eyes).	Continue, but call doctor right away.
Rare: None expected.	

WARNINGS & PRECAUTIONS

Don't take if:
You are allergic to capecitabine or 5-fluorouracil. Allergic reactions are rare, but may occur.

Before you start, consult your doctor if:
- You have an infection or any other medical problem.
- You have heart problems.
- You are taking folic acid.
- You are pregnant or if you plan to become pregnant.
- You have kidney or liver problems.

Over age 60:
Adverse reactions and side effects, especially gastrointestinal, may be more severe and frequent than in younger patients.

Pregnancy:
Consult doctor. Use of the drug during pregnancy is not recommended. It could cause harm to the unborn baby. Risk category D (see page xviii).

Breastfeeding; Lactation; Nursing Mothers:
This drug is generally not recommended with breastfeeding. Discuss risks and benefits with your doctor.

Infants & children up to age 18:
Safety and effectiveness in this age group have not been established. Consult doctor.

Prolonged use:
Talk to your doctor about the need for follow-up medical examinations or laboratory studies.

Skin & sunlight:
No problems expected.

Driving, piloting or hazardous work:
Avoid if you feel side effects such as nausea and vomiting.

Discontinuing:
Your doctor will determine the schedule.

Others:
Advise any doctor, dentist or pharmacist whom you consult that you take this medicine.

POSSIBLE INTERACTION WITH OTHER DRUGS

GENERIC NAME OR DRUG CLASS	COMBINED EFFECT
Antacids	Increased risk of capecitabine toxicity.
Leucovorin	Increased risk of capecitabine toxicity.
Warfarin	Increased risk of bleeding. Consult doctor.

POSSIBLE INTERACTION WITH OTHER SUBSTANCES

INTERACTS WITH	COMBINED EFFECT
Alcohol:	None expected.
Beverages:	None expected.
Cocaine:	Unknown. Avoid.
Foods:	None expected.
Marijuana:	Consult doctor.
Tobacco:	None expected.

CAPSAICIN

BRAND NAMES

ArthriCare
ARTH-RX
Axsain
Capsagel
Dura-Patch
Dura Patch Joint
Methacin
Qutenza
Sinol Nasal Spray
Sinus Buster
WellPatch Capsaicin Pain Patch
Zostrix
Zostrix-HP
Zostrix Neuropathy Cream

BASIC INFORMATION

Habit forming? No
Prescription needed? No
Available as generic? Yes
Drug class: Analgesic (topical)

USES

- Treats neuralgias, such as pain that occurs following shingles (herpes zoster) or neuropathy of the feet, ankles and fingers (common in diabetes).
- Nasal spray helps relieves sinus, allergy, and headache symptoms, including migraines.
- Treats discomfort caused by arthritis.

OVERDOSE

SYMPTOMS:
Unknown.
WHAT TO DO:
If person uses much larger amount than prescribed or if accidentally swallowed, call poison control center 1-800-222-1222 for instructions or dial 911 (emergency) for help.

DOSAGE & USAGE INFORMATION

How to use:
- Cream—Apply a small amount and rub carefully on the affected areas. Use every day. Wash hands after applying. Don't apply to irritated skin. Don't bandage over treated areas.
- Nasal spray—Use will depend on condition being treated. Follow package instructions.
- Patch—Follow package instructions for proper application.

When to use:
Apply 3 or 4 times a day.

If you forget a dose:
Use as soon as you remember. If it is almost time for the next dose, wait for that dose (don't double this dose) and resume regular schedule.

What drug does:
It is made from red chili peppers. When applied to the skin, it appears to decrease a chemical that causes pain. The nasal spray product can help reduce nasal inflammation and desensitize nasal passages to allergens.

Time lapse before drug works:
Begins to work immediately. Frequently takes 2 to 3 weeks for full benefit, but may take up to 6 or 8 weeks.

Don't use with:
No problems expected.

POSSIBLE ADVERSE REACTIONS OR SIDE EFFECTS

SYMPTOMS	WHAT TO DO
Life-threatening:	
None expected.	
Common:	
Stinging or burning sensation at application site.	Nothing. It usually improves in 2 to 3 days or becomes less severe the longer you use the drug.
Infrequent:	
None expected.	
Rare:	
Serious burns or blistering on the skin where product applied.	Discontinue. Call doctor right away.

WARNINGS & PRECAUTIONS

Don't use if:
You are allergic to capsaicin or to the fruit of capsaicin plants (for example, hot peppers).

Before you start, consult your doctor if:
You have any allergies.

Over age 60:
No problems expected.

Pregnancy:
Decide with your doctor if drug benefits justify risk to unborn baby. Risk category B (see page xviii).

Breastfeeding; Lactation; Nursing Mothers:
This drug is generally acceptable with breastfeeding. Discuss risks and benefits with your doctor.

Infants & children up to age 18:
Read nonprescription product's label to be sure it is approved for your child's age. Follow label instructions on dosage. Consult doctor or pharmacist if unsure.

Prolonged use:
No problems expected.

Skin & sunlight:
No problems expected.

Driving, piloting or hazardous work:
No problems expected.

Discontinuing:
Discontinue if there are no signs of improvement within a month.

Others:
- Capsaicin is not a local anesthetic.*
- Although capsaicin may help relieve the pain of neuropathy, it does not cure any disorder.
- If you accidentally get some capsaicin in your eye, flush with water.

POSSIBLE INTERACTION WITH OTHER DRUGS

GENERIC NAME OR DRUG CLASS	COMBINED EFFECT
None expected.	

POSSIBLE INTERACTION WITH OTHER SUBSTANCES

INTERACTS WITH	COMBINED EFFECT
Alcohol:	None expected.
Beverages:	None expected.
Cocaine:	None expected.
Foods:	None expected.
Marijuana:	None expected.
Tobacco:	None expected.

*See Glossary

CARBAMAZEPINE

BRAND NAMES

Apo-Carbamazepine
Carbatrol
Epitol
Equetro
Mazepine
Novocarbamaz
Nu-Carbamazepine
PMS Carbamazepine
Taro-Carbamazepine
Tegretol
Tegretol Chewtabs
Tegretol CR
Tegretol XR

BASIC INFORMATION

Habit forming? No
Prescription needed? Yes
Available as generic? Yes
Drug class: Analgesic, anticonvulsant, antimanic agent

USES

- Treatment for trigeminal neuralgia.
- Treats bipolar (manic-depressive) disorder.
- Used to control certain types of seizures.
- Used for pain relief, restless leg syndrome, alcohol/drug withdrawal and other disorders.

DOSAGE & USAGE INFORMATION

How to take:

- Tablet—Swallow with liquid. Take with food to lessen stomach upset. For chewable tablet, chew well before swallowing.
- Extended-release capsule or tablet or controlled release tablet—Swallow with liquid. May be taken with or without food. Capsule may be opened and sprinkled on food (e.g., teaspoon of applesauce). Do not chew or crush tablet or capsule.
- Oral suspension—Follow label instructions. Do not mix it with other liquid drugs.

Continued next column

OVERDOSE

SYMPTOMS:
Abnormal movements, slurred speech, nausea and vomiting, drowsiness, high or low blood pressure, dilated pupils, flushed skin, irregular heartbeat and breathing, decreased urine, seizures, tremor, fainting, coma.
WHAT TO DO:

- **Dial 911 for all medical emergencies or call poison control center 1-800-222-1222 for instructions.**
- **See emergency information on last 3 pages of this book.**

When to take:
At the same times each day as directed.

What drug does:
It works by decreasing impulses in nerves that cause seizures and pain.

If you forget a dose:
Take as soon as you remember. If it is almost time for the next dose, wait for that dose (don't double this dose) and resume regular schedule.

Time lapse before drug works:

- Tic douloureux—24 to 72 hours.
- Bipolar disorder—7 to 10 days.
- Seizures—Hours to days in different patients.

Don't take with:
Any other medicine or any dietary supplement without consulting your doctor or pharmacist. May inactivate other medications, such as birth control pills.

POSSIBLE ADVERSE REACTIONS OR SIDE EFFECTS

SYMPTOMS	WHAT TO DO
Life-threatening:	
In case of overdose, see previous column.	
Common:	
• Back-and-forth eye movements, double or blurred vision.	Discontinue. Call doctor right away.
• Being clumsy or unsteady, mildly dizzy or lightheaded, mild nausea or vomiting.	Continue. Call doctor when convenient.
Infrequent:	
• Behavior changes in a child or in the elderly (e.g., confusion or agitation), ongoing headache, severe diarrhea or nausea or vomiting or drowsiness, increase in seizures, skin rash or hives.	Discontinue. Call doctor right away.
• Mild diarrhea, dry mouth, constipation, impotence, muscle aches, sore tongue or mouth, hair loss, upset stomach, appetite loss, increased sweating.	Continue. Call doctor when convenient.
Rare:	
Breathing difficulty, irregular or pounding or slow heartbea, chest pain, uncontrollable body movements, numbness, weakness or tingling in hands and feet, swollen legs or feet or face,	Discontinue. Call doctor right away.

unusual bleeding or bruising, stool changes (black, tarry, bloody or pale), urine changes (more frequent, decreased amount, bloody, dark or painful), unusual pain, yellow eyes or skin, infection (fever, chills), rapid weight gain, noises in ears, slurred speech, hallucinations, unusual tiredness or weakness.

WARNINGS & PRECAUTIONS

Don't take if:
- You are allergic to carbamazepine or any tricyclic antidepressant.* Allergic reactions are rare, but may occur.
- You have taken a monoamine oxidase (MAO) inhibitor* in the past 2 weeks.

Before you start, consult your doctor if:
- You have high blood pressure, heart block, thrombophlebitis or heart disease.
- You have glaucoma.
- You have emotional or mental problems.
- You have diabetes, liver or kidney disease.
- You have a blood disorder (e.g., anemia) or bone marrow depression or disease.
- You have had reactions to other drugs.
- You are of Asian ancestry (a genetic blood test can tell if you are at increased risk of developing a rare, but serious, skin reaction).
- You drink more than 2 alcoholic drinks a day.

Over age 60:
Adverse reactions and side effects may be more frequent and severe than in younger persons.

Pregnancy:
Consult doctor. Use of the drug during pregnancy is not recommended. It could cause harm to the unborn baby. Risk category D (see page xviii).

Breastfeeding; Lactation; Nursing Mothers:
This drug is generally acceptable with breastfeeding. Discuss risks and benefits with your doctor.

Infants & children up to age 18:
Follow instructions provided by your child's doctor.

Prolonged use:
Talk to your doctor about the need for follow-up medical examinations, blood and urine studies, eye and dental exams, liver and kidney function tests, bone density tests, and others as needed.

Skin & sunlight:
May cause rash or intensify sunburn in areas exposed to sun or ultraviolet light (photosensitivity reaction). Avoid overexposure and use sunscreen. Notify doctor if reaction occurs.

Driving, piloting or hazardous work:
Don't drive or pilot aircraft until you learn how drug effects you. Don't work around dangerous machinery. Don't climb ladders or work in high places. Danger increases if you drink alcohol or take drugs affecting alertness and reflexes.

Discontinuing:
Don't discontinue without doctor's advice until you complete prescribed dose, even though symptoms diminish or disappear.

Others:
- Wear or carry medical identification that states that you take this drug.
- Periodic blood tests are needed.
- Advise any doctor, dentist or pharmacist whom you consult that you take this medicine.
- Rarely, anticonvulsant (antiepileptic) drugs may lead to suicidal thoughts and behaviors. Call doctor right away if suicidal symptoms or unusual behaviors occur.

POSSIBLE INTERACTION WITH OTHER DRUGS

GENERIC NAME OR DRUG CLASS	COMBINED EFFECT
Acetaminophen	Increased risk of liver problems.
Adrenocorticoids, systemic	Decreased adrenocorticoid effect.
Antibiotics, macrolide	Increased effect of carbamazepine.
Anticoagulants, oral*	Decreased anticoagulant effect.
Anticonvulsants, hydantoin* or succinimide*	Decreased effect of both drugs.

Continued on page 895

POSSIBLE INTERACTION WITH OTHER SUBSTANCES

INTERACTS WITH	COMBINED EFFECT
Alcohol:	Increased sedative effect. Avoid.
Beverages: Grapefruit juice.	Risk of toxicity. Avoid.
Cocaine:	Increased adverse effects. Avoid.
Foods:	None expected.
Marijuana:	Increased adverse effects. Avoid.
Tobacco:	None expected.

***See Glossary**

CARBIDOPA & LEVODOPA

BRAND NAMES

Parcopa	Sinemet CR
Rytary	Stalevo
Sinemet	

BASIC INFORMATION

Habit forming? No
Prescription needed? Yes
Available as generic? Yes
Drug class: Antiparkinsonism

USES

Controls Parkinson's disease symptoms such as rigidity, tremor and unsteady gait.

DOSAGE & USAGE INFORMATION

How to take:

- Tablet—Swallow with liquid or food to lessen stomach irritation. If you can't swallow whole, crumble tablet and take with liquid or food. Do not crush, break or chew Stalevo tablet.
- Extended-release tablet—Swallow each dose whole; do not crumble.

When to take:
At the same times each day.

If you forget a dose:
Take as soon as you remember. If it is almost time for the next dose, wait for that dose (don't double this dose) and resume regular schedule.

What drug does:
Restores chemical balance necessary for normal nerve impulses.

Continued next column

OVERDOSE

SYMPTOMS:
Muscle twitch, spastic eyelid closure, nausea, vomiting, diarrhea, irregular and rapid pulse, weakness, fainting, confusion, agitation, hallucination, coma.
WHAT TO DO:

- **Dial 911 for all medical emergencies or call poison control center 1-800-222-1222 for instructions.**
- **See emergency information on last 3 pages of this book.**

Time lapse before drug works:
2 to 3 weeks to improve; 6 weeks or longer for maximum benefit.

Don't take with:
Any other medicine or any dietary supplement without consulting your doctor or pharmacist.

POSSIBLE ADVERSE REACTIONS OR SIDE EFFECTS

SYMPTOMS	WHAT TO DO
Life-threatening:	
In case of overdose, see previous column.	
Common:	
• Mood changes, uncontrollable body movements, diarrhea.	Continue. Call doctor when convenient.
• Dry mouth, body odor.	No action necessary.
Infrequent:	
• Fainting, severe dizziness, headache, insomnia, nightmares, rash, itch, nausea, vomiting, irregular heartbeat.	Discontinue. Call doctor right away.
• Flushed face, blurred vision, muscle twitching, discolored or dark urine, difficult urination, tiredness, constipation.	Continue. Call doctor when convenient.
Rare:	
• High blood pressure.	Discontinue. Call doctor right away.
• Upper abdominal pain, anemia.	Continue. Call doctor when convenient.

WARNINGS & PRECAUTIONS

Don't take if:

- You are allergic to levodopa or carbidopa. Allergic reactions are rare, but may occur.
- You have taken MAO inhibitors in past 2 weeks.
- You have glaucoma (narrow-angle type).

Before you start, consult your doctor if:

- You have diabetes or epilepsy.
- You have had high blood pressure, heart or lung disease.
- You have had liver or kidney disease.
- You have a peptic ulcer.
- You have malignant melanoma.
- You will have surgery within 2 months, including dental surgery, requiring general or spinal anesthesia.

Over age 60:
No special problems expected.

Pregnancy:
Decide with your doctor if drug benefits justify risk to unborn baby. Risk category C (see page xviii).

Breastfeeding; Lactation; Nursing Mothers:
This drug is generally acceptable with breastfeeding. Discuss risks and benefits with your doctor.

Infants & children up to age 18:
Safety and effectiveness in this age group have not been established. Consult doctor.

Prolonged use:
- May lead to uncontrolled movements of head, face, mouth, tongue, arms or legs.
- Talk to your doctor about the need for follow-up medical examinations or laboratory studies.

Skin & sunlight:
No problems expected.

Driving, piloting or hazardous work:
Don't drive or pilot aircraft until you learn how medicine affects you. Don't work around dangerous machinery. Don't climb ladders or work in high places. Danger increases if you drink alcohol or take medicine affecting alertness and reflexes, such as antihistamines, tranquilizers, sedatives, pain medicine, narcotics and mind-altering drugs.

Discontinuing:
Don't discontinue without doctor's advice until you complete prescribed dose, even though symptoms diminish or disappear.

Others:
- Expect to start with small dose and increase gradually to lessen frequency and severity of adverse reactions.
- Advise any doctor, dentist or pharmacist whom you consult that you take this medicine.

POSSIBLE INTERACTION WITH OTHER DRUGS

GENERIC NAME OR DRUG CLASS	COMBINED EFFECT
Anticonvulsants,* hydantoin	Decreased effect of carbidopa and levodopa.
Antidepressants*	Weakness or faintness when arising from bed or chair.
Antihypertensives*	Decreased blood pressure and effect of carbidopa and levodopa.
Antiparkinsonism drugs, other*	Increased effect of carbidopa and levodopa.
Bupropion	Increased levodopa effect.
Haloperidol	Decreased effect of carbidopa and levodopa.
Methyldopa	Decreased effect of carbidopa and levodopa.
Monoamine oxidase (MAO) inhibitors*	Dangerous rise in blood pressure.
Papaverine	Decreased effect of carbidopa and levodopa.
Phenothiazines*	Decreased effect of carbidopa and levodopa.
Phenytoin	Decreased effect of carbidopa and levodopa.
Pyridoxine (Vitamin B-6)	Decreased effect of carbidopa and levodopa.
Rauwolfia alkaloids*	Decreased effect of carbidopa and levodopa.
Selegiline	May require adjustment in dosage of carbidopa and levodopa.

POSSIBLE INTERACTION WITH OTHER SUBSTANCES

INTERACTS WITH	COMBINED EFFECT
Alcohol:	None expected.
Beverages:	None expected.
Cocaine:	Unknown. Avoid.
Foods:	None expected.
Marijuana:	Consult doctor.
Tobacco:	None expected.

***See Glossary**

CARBONIC ANHYDRASE INHIBITORS

GENERIC AND BRAND NAMES

ACETAZOLAMIDE
- **Acetazolam**
- **Ak-Zol**
- **Apo Acetazolamide**
- **Dazamide**
- **Diamox**
- **Storzolamide**

METHAZOLAMIDE
- **Neptazane**
- **MZM**

BASIC INFORMATION

Habit forming? No
Prescription needed? Yes
Available as generic? Yes
Drug class: Carbonic anhydrase inhibitor

USES

- Treatment of glaucoma.
- Treatment of epileptic seizures.
- Treatment of body fluid retention.
- Treatment for shortness of breath, insomnia and fatigue at high altitudes.
- Treatment for prevention of altitude illness.

DOSAGE & USAGE INFORMATION

How to take:
- Sustained-release tablet—Swallow whole with liquid or food to lessen stomach irritation.
- Extended-release capsule—Swallow whole with liquid.
- Topical—1 drop in the affected eye 3 times a day.

When to take:
- 1 dose per day—At the same time each morning.
- More than 1 dose per day—Take last dose several hours before bedtime.

Continued next column

OVERDOSE

SYMPTOMS:
Drowsiness, confusion, nausea, vomiting, numbness in hands and feet, tingling, ringing in the ears.

WHAT TO DO:
- **Dial 911 for all medical emergencies or call poison control center 1-800-222-1222 for instructions.**
- **See emergency information on last 3 pages of this book.**

If you forget a dose:
Take as soon as you remember. If it is almost time for the next dose, wait for that dose (don't double this dose) and resume regular schedule.

What drug does:
- Inhibits action of carbonic anhydrase, an enzyme. This lowers the internal eye pressure by decreasing fluid formation in the eye.
- Forces sodium and water excretion, reducing body fluid.

Time lapse before drug works:
2 hours.

Don't take with:
Any other medicine or any dietary supplement without consulting your doctor or pharmacist.

POSSIBLE ADVERSE REACTIONS OR SIDE EFFECTS

SYMPTOMS	WHAT TO DO
Life-threatening:	
None expected.	
Common:	
Diarrhea, loss of appetite, nausea or vomiting, weight loss, metallic taste, general ill feeling, tingling or burning in feet or hands.	Continue. Call doctor when convenient.
Infrequent:	
Unusual tiredness or weakness, urination changes (bloody, difficult, painful or decreased amount), depression, back pain, yellow eyes or skin.	Continue, but call doctor right away.
Rare:	
• Stools are bloody or tarry or pale, fever, clumsiness, skin symptoms (hives, itch, rash, sores), ringing in ears, sore throat, easy bleeding or bruising, trembling, seizures, dark urine, potassium loss (dry mouth, thirstiness, muscle cramps, weak pulse, irregular heartbeat, mood changes).	Continue, but call doctor right away.
• Dizziness, loss of taste or smell, headache, lump in throat feeling, constipation.	Continue. Call doctor when convenient.

WARNINGS & PRECAUTIONS

Don't take if:
You are allergic to any carbonic anhydrase inhibitor. Allergic reactions are rare, but may occur.

Before you start, consult your doctor if:
- You have gout or lupus.
- You have liver or kidney disease or Addison's disease (adrenal gland failure).
- You have diabetes.
- You will have surgery within 2 months, including dental surgery, requiring general or spinal anesthesia.

Over age 60:
No special problems expected.

Pregnancy:
Decide with your doctor if drug benefits justify risk to unborn baby. Risk category C (see page xviii).

Breastfeeding; Lactation; Nursing Mothers:
Drugs in this group are generally acceptable with breastfeeding. Discuss risks and benefits with your doctor.

Infants & children up to age 18:
Not recommended for children younger than 12. Follow instructions provided by your child's doctor.

Prolonged use:
May cause kidney stones, vision change, loss of taste and smell, jaundice or weight loss.

Skin & sunlight:
One or more drugs in this group may cause rash or intensify sunburn in areas exposed to sun or ultraviolet light (photosensitivity reaction). Avoid overexposure. Notify doctor if reaction occurs.

Driving, piloting or hazardous work:
Avoid if you feel drowsy or dizzy. Otherwise, no problems expected.

Discontinuing:
Don't discontinue without medical advice.

Others:
- Drug may increase blood sugar levels. Diabetics may need insulin adjustment.
- Advise any doctor, dentist or pharmacist whom you consult that you take this medicine.

POSSIBLE INTERACTION WITH OTHER DRUGS

GENERIC NAME OR DRUG CLASS	COMBINED EFFECT
Adrenocorticoids, systemic	Increased loss of calcium & potassium.
Amphetamines*	Increased amphetamine effect.
Anticonvulsants*	Increased loss of bone minerals.
Antidiabetics, oral*	May need dosage adjustment.
Antiglaucoma, carbonic anhydrase inhibitors	Increased effect of both drugs. Avoid.
Ciprofloxacin	May cause kidney dysfunction.
Digitalis preparations*	Possible digitalis toxicity.
Diuretics*	Increased potassium loss.
Lithium	Decreased lithium effect.
Mecamylamine	Increased mecamylamine effect.
Memantine	Increased effect of memantine.
Methenamine	Decreased methenamine effect.
Mexiletine	May slow elimination of mexiletine and cause need to adjust dosage.
Quinidine	Increased quinidine effect.
Salicylates*	Salicylate toxicity.
Sympathomimetics*	Increased sympathomimetic effect.

POSSIBLE INTERACTION WITH OTHER SUBSTANCES

INTERACTS WITH	COMBINED EFFECT
Alcohol:	None expected.
Beverages:	None expected.
Cocaine:	Unknown. Avoid
Foods: Potassium-rich foods.*	Eat these to decrease potassium loss.
Marijuana:	Consult doctor.
Tobacco:	May decrease effect of carbonic anhydrase inhibitors.

*See Glossary

CENTRAL ALPHA AGONISTS

GENERIC AND BRAND NAMES

CLONIDINE
- Catapres
- Catapres-TTS
- Dixarit
- Jenloga
- Kapvay
- Nexiclon XR

GUANABENZ
- Wytensin

GUANFACINE
- Intuniv
- Tenex

METHYLDOPA
- Aldoclor
- Aldomet
- Aldoril
- Apo-Methyldopa
- Dopamet
- Novodoparil
- Novomedopa
- Nu-Medopa
- PMS Dopazide
- Supres

BASIC INFORMATION

Habit forming? No
Prescription needed? Yes
Available as generic? Yes
Drug class: Antihypertensive

USES

- Control of hypertension (high blood pressure).
- May be used for symptoms of narcotic, alcohol or nicotine withdrawal.
- Prevention of migraine headaches.
- Treatment for menstrual cramps or hot flashes due to menopause.
- Treats attention deficit hyperactivity disorder.

DOSAGE & USAGE INFORMATION

How to take:
- Tablet—Swallow with liquid with or without food.
- Transdermal patch (attaches to skin)—Apply to clean, dry, hairless skin on arm or trunk. Follow all prescription instructions carefully.
- Extended-release tablet—Swallow whole with small amount of liquid. Don't crush or chew.
- Solution—Follow label instructions.

Continued next column

OVERDOSE

SYMPTOMS:
Vomiting, slow heartbeat, weakness, low blood pressure, lightheadedness, cold feeling, extreme tiredness, breathing difficulty drowsiness.

WHAT TO DO:
- **Dial 911 for all medical emergencies or call poison control center 1-800-222-1222 for instructions.**
- **See emergency information on last 3 pages of this book.**

When to take:
- Oral form—One to four times a day according to the instructions on your prescription.
- Transdermal patch—Replace as directed, usually once a week.

If you forget a dose:
- Take tablet as soon as you remember. If it is almost time for the next dose, wait for the next scheduled dose (don't double this dose). If you miss 2 doses in a row, call your doctor.
- If the once-a-week patch change is 3 days late, call your doctor for advice.

What drug does:
- Hypertension—Lowers blood pressure by relaxing and dilating (widening) blood vessels. This helps increase flow of blood in the body.
- Withdrawal symptoms—Thought to block nerve impulses in the area of the brain responsible for the symptoms.
- Attention deficit hyperactivity disorder—The way it works is unclear.

Time lapse before drug works:
2 to 3 weeks for full benefit.

Don't take with:
Any other medicine or any dietary supplement without consulting your doctor or pharmacist.

POSSIBLE ADVERSE REACTIONS OR SIDE EFFECTS

SYMPTOMS	WHAT TO DO
Life-threatening:	
None expected.	
Common:	
Dizziness, constipation, drowsiness, irritated skin (with skin patch), dry mouth, tiredness, headache, insomnia, weakness, low blood pressure.	Continue. Call doctor when convenient.
Infrequent:	
• Swollen feet and legs, slow or fast heartbeat, fever.	Continue, but call doctor right away.
• Darkened skin (with skin patch), lightheadedness, dry or burning eyes, anxiety, decreased sex function, nausea, vomiting, appetite loss, nervousness, depression.	Continue. Call doctor when convenient.

Rare:

• Cold fingers and toes, dark urine, chills, breathing difficulty, yellow eyes or skin, stomach pain.	Continue, but call doctor right away.
• Joint pain, pale stools, skin rash or itching, nightmares, confusion.	Continue. Call doctor when convenient.

WARNINGS & PRECAUTIONS

Don't take if:
You are allergic to any central alpha agonist. Allergic reactions are rare, but may occur.

Before you start, consult your doctor if:
- You will have surgery within 2 months requiring general or spinal anesthesia.
- You have heart, liver or kidney disease.
- You had a recent stroke or heart attack.
- You have a peripheral circulation disorder (intermittent claudication, Raynaud's syndrome, Buerger's disease).
- You have Parkinson's disease.
- You have a history of depression.
- You have a disorder affecting the skin or any skin irritation (with use of transdermal patch).

Over age 60:
Adverse reactions and side effects may be more frequent and severe than in younger persons.

Pregnancy:
Decide with your doctor if drug benefits justify risk to unborn baby. Risk category B for guanfacine and methyldopa; risk category C for clonidine and guanabenz (see page xviii).

Breastfeeding; Lactation; Nursing Mothers:
Drugs in this group are generally acceptable with breastfeeding. Discuss risks and benefits with your doctor.

Infants & children up to age 18:
One or more of these drugs may be prescribed for attention deficit hyperactivity disorder for ages 6 and over. Follow instructions provided by your child's doctor.

Prolonged use:
Talk to your doctor about the need for follow-up medical or eye exams or laboratory tests.

Skin & sunlight:
None expected.

Driving, piloting or hazardous work:
Don't drive or pilot aircraft until you learn how medicine affects you. Don't work around dangerous machinery. Don't climb ladders or work in high places. Danger increases if you drink alcohol or take medicine affecting alertness and reflexes.

Discontinuing:
- Don't discontinue abruptly. May cause a withdrawal syndrome (anxiety, chest pain, headache, nausea, insomnia, irregular heartbeat, flushed face, sweating). Consult doctor if any symptoms occur after stopping the drug.
- Dose may require gradual reduction if you have taken drug for a long time. Doses of other drugs may also require adjustment.

Others:
- Advise any doctor, dentist or pharmacist whom you consult that you take this medicine.
- For dry mouth, suck sugarless hard candy or chew sugarless gum. If dry mouth continues, consult your dentist.

POSSIBLE INTERACTION WITH OTHER DRUGS

GENERIC NAME OR DRUG CLASS	COMBINED EFFECT
Antidepressants, tricyclic*	Decreased effect of central alpha agonist.
Antihypertensives,* other	Excessive lowering of blood pressure.
Beta-adrenergic blocking agents*	Increased risk of adverse reactions and excessive low blood pressure.

Continued on page 897

POSSIBLE INTERACTION WITH OTHER SUBSTANCES

INTERACTS WITH	COMBINED EFFECT
Alcohol:	Increased sedative effect of alcohol and very low blood pressure. Avoid.
Beverages:	None expected.
Cocaine:	Increased risk of heart problems and high blood pressure. Avoid.
Foods:	None expected.
Marijuana:	Consult doctor.
Tobacco:	None expected. Persons with high blood pressure should not smoke.

***See Glossary**

CEPHALOSPORINS

GENERIC AND BRAND NAMES

CEFACLOR
- Ceclor
- Ceclor CD
- Raniclor

CEFADROXIL
- Duricef
- Ultracef

CEFDINIR
- Omnicef

CEFDITOREN
- Spectracef

CEFIXIME
- Suprax

CEFOTETAN
- Cefotan

CEFPODOXIME
- Vantin

CEFPROZIL
- Cefzil

CEFTIBUTEN
- Cedax

CEFUROXIME
- Ceftin

CEPHALEXIN
- Apo-Cephalex
- Cefanex
- Ceporex
- C-Lexin
- Keflex
- Keftab
- Novolexin
- Nu-Cephalex

CEPHRADINE
- Anspor
- Velosef

BASIC INFORMATION

Habit forming? No
Prescription needed? Yes
Available as generic? Yes
Drug class: Antibacterial

USES

Treatment of bacterial infections. Will not cure viral infections such as cold and flu.

DOSAGE & USAGE INFORMATION

How to take:
- Tablet or capsule—Swallow with liquid. If you can't swallow whole, crumble tablet or open capsule and take with liquid or food.
- Extended-release tablet—Swallow with liquid. Do not crush or chew tablet.

Continued next column

OVERDOSE

SYMPTOMS:
Abdominal cramps, nausea, vomiting, severe diarrhea with mucus or blood in stool, convulsions.
WHAT TO DO:
If person uses much larger amount than prescribed or if accidentally swallowed, call poison control center 1-800-222-1222 for instructions or dial 911 (emergency) for help.

- Chewable tablet—Follow instructions on prescription.
- Liquid and oral suspension—Use measuring spoon. Mix according to package instructions.

When to take:
- At same times each day, 1 hour before or 2 hours after eating.
- Take until gone or as directed.

If you forget a dose:
Take as soon as you remember. If it is almost time for the next dose, wait for that dose (don't double this dose) and resume regular schedule.

What drug does:
Kills susceptible bacteria.

Time lapse before drug works:
May require several days to affect infection.

Don't take with:
Any other medicine or any dietary supplement without consulting your doctor or pharmacist.

POSSIBLE ADVERSE REACTIONS OR SIDE EFFECTS

SYMPTOMS	WHAT TO DO
Life-threatening:	
Rare, severe allergic reaction (anaphylaxis)—difficulty breathing, hives, itching, swelling (throat, face), vomiting, dizziness.	Seek emergency treatment immediately.
Common:	
Mild diarrhea, nausea, vomiting, sore mouth or tongue, mild stomach cramps (all less common with some cephalosporins).	Continue. Call doctor when convenient.
Infrequent:	
None expected.	
Rare:	
• Severe stomach cramps, severe diarrhea with mucus or blood in stool, fever, unusual weakness or tiredness, weight loss, bleeding or bruising, increased thirst, decreased urine, dizziness, joint pain, appetite loss, skin symptoms (rash, itching, redness, swelling), yellow skin or eyes.	Discontinue. Call doctor right away.
• Genital itching or vaginal discharge.	Continue. Call doctor when convenient.

WARNINGS & PRECAUTIONS

Don't take if:
You are allergic to any cephalosporin antibiotic.

Before you start, consult your doctor if:
- You are allergic to any penicillin antibiotic.
- You have a kidney disorder.
- You have colitis or enteritis.

Over age 60:
No special problems expected.

Pregnancy:
Decide with your doctor if drug benefits justify risk to unborn baby. Risk category B (see page xviii).

Breastfeeding; Lactation; Nursing Mothers:
Drugs in this group are generally acceptable with breastfeeding. Discuss risks and benefits with your doctor.

Infants & children up to age 18:
Follow instructions provided by your child's doctor.

Prolonged use:
- Kills beneficial bacteria that protect body against other germs. Unchecked germs may cause secondary infections.
- Talk to your doctor about the need for follow-up medical examinations or laboratory studies to check prothrombin time.

Skin & sunlight:
No problems expected.

Driving, piloting or hazardous work:
No problems expected.

Discontinuing:
Don't discontinue without doctor's advice until you complete prescribed dose, even though symptoms diminish or disappear.

Others:
- Don't use drug for other medical problems without doctor's approval.
- Advise any doctor, dentist or pharmacist whom you consult that you take this medicine.
- If diarrhea occurs, consult doctor.

POSSIBLE INTERACTION WITH OTHER DRUGS

GENERIC NAME OR DRUG CLASS	COMBINED EFFECT
Anticoagulants*	Increased anticoagulant effect.
Anti-inflammatory drugs, nonsteroidal (NSAIDs)*	Increased risk of peptic ulcer.
Chloramphenicol	Decreased antibiotic effect of cephalosporin.
Probenecid	Increased cephalosporin effect.
Tetracyclines*	Decreased antibiotic effect of cephalosporin.

POSSIBLE INTERACTION WITH OTHER SUBSTANCES

INTERACTS WITH	COMBINED EFFECT
Alcohol:	Increased kidney toxicity, likelihood of disulfiram-like* effect.
Beverages:	None expected.
Cocaine:	Unknown. Cocaine may slow body's recovery. Avoid.
Foods:	Slow absorption. Take with liquid 1 hour before or 2 hours after eating.
Marijuana:	Consult doctor.
Tobacco:	None expected.

***See Glossary**

CHARCOAL, ACTIVATED

BRAND NAMES

Acta-Char
Acta-Char Liquid
Actidose with Sorbitol
Actidose-Aqua
Aqueous Charcodote
Charac-50
Charac-tol 50
Charcoaid
Charcocaps
Charcodote
Charcodote TFS
Insta-Char
Liqui-Char
Pediatric Aqueous Charcodote
Pediatric Charcodote
SuperChar

BASIC INFORMATION

Habit forming? No
Prescription needed? No
Available as generic? Yes
Drug class: Antidote (adsorbent)

USES

- Treatment of poisonings from medication.
- Treatment (infrequent) for diarrhea or excessive gaseousness.

DOSAGE & USAGE INFORMATION

How to take:
- Tablet or capsule—Swallow with liquid. If you can't swallow whole, crumble tablet or open capsule and take with liquid or food.
- Liquid—Take as directed on label. Don't mix with chocolate syrup, ice cream or sherbet.

When to take:
- For poisoning—Take immediately after poisoning. If your doctor or emergency poison control center has also recommended syrup of ipecac, don't take charcoal for 30 minutes or until vomiting from ipecac stops.
- For diarrhea or gas—Take at same times each day.
- Take 2 or more hours after taking other medicines.

Continued next column

OVERDOSE

SYMPTOMS:
Unknown.
WHAT TO DO:
If person uses much larger amount than prescribed or if accidentally swallowed, call poison control center 1-800-222-1222 for instructions or dial 911 (emergency) for help.

If you forget a dose:
- For poisonings—Not applicable.
- For diarrhea or gas—Take as soon as you remember. If it is almost time for the next dose, wait for that dose (don't double this dose) and resume regular schedule.

What drug does:
- Helps prevent poison from being absorbed from stomach and intestines.
- Helps absorb gas in intestinal tract.

Time lapse before drug works:
Begins immediately.

Don't take with:
Ice cream or sherbet.

POSSIBLE ADVERSE REACTIONS OR SIDE EFFECTS

SYMPTOMS	WHAT TO DO
Life-threatening:	
None expected.	
Always:	
Black bowel movements.	No action necessary.
Infrequent:	
None expected.	
Rare:	
Unless taken with cathartic, can cause constipation when taken for overdose of other medicine.	Take a laxative after crisis is over.

WARNINGS & PRECAUTIONS

Don't take if:
The poison was lye or other strong alkali, strong acids (such as sulfuric acid), cyanide, iron, ethyl alcohol or methyl alcohol. Charcoal will not prevent these poisons from causing ill effects.

Before you start, consult your doctor if:
You are taking it as an antidote for poison.

Over age 60:
No problems expected.

Pregnancy:
Decide with your doctor if drug benefits justify risk to unborn baby. Risk category not designated (see page xviii).

Breastfeeding; Lactation; Nursing Mothers:
This drug is generally acceptable with breastfeeding. Discuss risks and benefits with your doctor.

Infants & children up to age 18:
Follow instructions provided by your child's doctor.

Prolonged use:
No problems expected.

Skin & sunlight:
No problems expected.

Driving, piloting or hazardous work:
No problems expected.

Discontinuing:
No problems expected.

Others:
No problems expected.

POSSIBLE INTERACTION WITH OTHER DRUGS

GENERIC NAME OR DRUG CLASS	COMBINED EFFECT
Any medicine taken at the same time	May decrease absorption of medicine. Take drugs 2 hours apart.

POSSIBLE INTERACTION WITH OTHER SUBSTANCES

INTERACTS WITH	COMBINED EFFECT
Alcohol:	None expected.
Beverages:	None expected.
Cocaine:	None expected.
Foods: Chocolate syrup, ice cream or sherbet.	Decreased charcoal effect.
Marijuana:	None expected.
Tobacco:	None expected.

***See Glossary**

CHLORAMBUCIL

BRAND NAMES

Leukeran

BASIC INFORMATION

Habit forming? No
Prescription needed? Yes
Available as generic? No
Drug class: Antineoplastic, immuno-suppressant

USES

- Treatment for certain types of cancers.
- Suppresses immune response after transplant and in immune disorders.

DOSAGE & USAGE INFORMATION

How to take:
Tablet—Swallow with liquid after light meal. Don't drink fluids with meals. Drink extra fluids between meals.

When to take:
At the same time each day.

If you forget a dose:
Take as soon as you remember. If it is almost time for the next dose, wait for that dose (don't double this dose) and resume regular schedule.

What drug does:
Inhibits abnormal cell reproduction. May suppress immune system.

Time lapse before drug works:
Up to 6 weeks for full effect.

Don't take with:
Any other medicine or any dietary supplement without consulting your doctor or pharmacist.

OVERDOSE

SYMPTOMS:
Lack of coordination or loss of balance, agitation, vomiting, seizures (convulsions).
WHAT TO DO:
- **Dial 911 for all medical emergencies or call poison control center 1-800-222-1222 for instructions.**
- **See emergency information on last 3 pages of this book.**

POSSIBLE ADVERSE REACTIONS OR SIDE EFFECTS

SYMPTOMS	WHAT TO DO
Life-threatening: None expected.	
Common:	
• Unusual bleeding or bruising, mouth or lip sores, sore throat, chills and fever, black stools, menstrual changes, back pain.	Discontinue. Call doctor right away.
• Hair loss, joint pain, nausea, vomiting, diarrhea, tiredness, weakness.	Continue. Call doctor when convenient.
Infrequent:	
• Mental confusion, shortness of breath, rash.	Discontinue. Call doctor right away.
• Cough, foot swelling.	Continue. Call doctor when convenient.
Rare:	
• Convulsions.	Seek emergency care.
• Hallucinations, muscle twitching, yellow eyes and skin.	Discontinue. Call doctor right away.

WARNINGS & PRECAUTIONS

Don't take if:
- You are allergic to chlorambucil. Allergic reactions are rare, but may occur.
- Your doctor has not explained serious nature of your medical problem and risks of taking this medicine.

Before you start, consult your doctor if:
- You have gout.
- You have had kidney stones.
- You have active infection.
- You have impaired kidney or liver function.
- You have taken other antineoplastic drugs or had radiation treatment in last 3 weeks.

Over age 60:
Adverse reactions and side effects may be more frequent and severe than in younger persons.

Pregnancy:
Consult doctor. Use of the drug during pregnancy is not recommended. It could cause harm to the unborn baby. Risk category D (see page xviii).

Breastfeeding; Lactation; Nursing Mothers:
This drug is generally not recommended with breastfeeding. Discuss risks and benefits with your doctor.

Infants & children up to age 18:
Follow instructions provided by your child's doctor.

Prolonged use:
- Adverse reactions more likely the longer drug is required.
- Talk to your doctor about the need for follow-up medical examinations or laboratory studies to check for adverse effects and effectiveness of drug.

Skin & sunlight:
No problems expected.

Driving, piloting or hazardous work:
No problems expected.

Discontinuing:
Don't discontinue without doctor's advice until you complete prescribed dose, even though symptoms diminish or disappear. Some side effects may follow discontinuing. Report to doctor blurred vision, convulsions, confusion, persistent headache.

Others:
- Drug can cause decrease in red blood cells in bone marrow. Follow-up lab tests are needed.
- Drug can increase risk of developing other types of cancer. It may cause changes in menstrual cycle and may stop sperm production in men. It may cause problem in getting pregnant or cause permanent infertility. Consult doctor about your risks.
- Advise any doctor, dentist or pharmacist whom you consult that you take this medicine.
- Consult your doctor before you or a household member gets any immunization/vaccine.

POSSIBLE INTERACTION WITH OTHER DRUGS

GENERIC NAME OR DRUG CLASS	COMBINED EFFECT
Antigout drugs*	Decreased antigout effect.
Antineoplastic drugs, other*	Increased effect of all drugs (may be beneficial).
Chloramphenicol	Increased likelihood of toxic effects of both drugs.
Clozapine	Toxic effect on bone marrow.
Cyclosporine	May increase risk of infection.
Immuno-suppressants*	Increased chance of infection.
Lovastatin	Increased heart and kidney damage.
Tiopronin	Increased risk of toxicity to bone marrow.

POSSIBLE INTERACTION WITH OTHER SUBSTANCES

INTERACTS WITH	COMBINED EFFECT
Alcohol:	May increase chance of intestinal bleeding.
Beverages:	None expected.
Cocaine:	Increased risk of side effects. Avoid
Foods:	None expected.
Marijuana:	Consult doctor.
Tobacco:	None expected.

*See Glossary

CHLORAMPHENICOL

BRAND NAMES

Chloromycetin   Novochlorocap

BASIC INFORMATION

Habit forming? No
Prescription needed? Yes
Available as generic? Yes
Drug class: Antibacterial

USES

Treatment of infections susceptible to chloramphenicol. Will not treat viral infections such as cold or flu.

DOSAGE & USAGE INFORMATION

How to take:
Suspension or capsule—Take with a full glass of water.

When to take:
Capsule or suspension—1 hour before or 2 hours after eating.

If you forget a dose:
Take as soon as you remember. If it is almost time for the next dose, wait for that dose (don't double this dose) and resume regular schedule.

What drug does:
Prevents bacteria from growing and reproducing. Will not kill viruses.

Time lapse before drug works:
2 to 5 days, depending on type and severity of infection.

Don't take with:
Any other medicine or any dietary supplement without consulting your doctor or pharmacist.

OVERDOSE

SYMPTOMS:
Nausea, vomiting, diarrhea.
WHAT TO DO:
If person uses much larger amount than prescribed or if accidentally swallowed, call poison control center 1-800-222-1222 for instructions or dial 911 (emergency) for help.

POSSIBLE ADVERSE REACTIONS OR SIDE EFFECTS

SYMPTOMS	WHAT TO DO
Life-threatening:	
Rare, severe allergic reaction (anaphylaxis)—difficulty breathing, hives, itching, swelling (throat, face), vomiting, dizziness.	Seek emergency treatment immediately.
Common:	
None expected.	
Infrequent:	
• Swollen face or extremities, diarrhea, nausea, vomiting, numbness or tingling or burning pain or weakness in hands and feet, pale skin, unusual bleeding or bruising.	Discontinue. Call doctor right away.
• Headache, confusion.	Continue. Call doctor when convenient.
Rare:	
• Pain, blurred vision, possible vision loss, delirium, rash, sore throat, fever, jaundice, anemia.	Discontinue. Call doctor right away.
• In babies: Bloated stomach, uneven breathing, drowsiness, low temperature, gray skin.	Discontinue. Call doctor right away.

WARNINGS & PRECAUTIONS

Don't take if:
- You are allergic to chloramphenicol.
- It is prescribed for a minor disorder such as flu, cold or mild sore throat.

Before you start, consult your doctor if:
- You have had a blood disorder or bone-marrow disease.
- You have had kidney or liver disease.
- You have diabetes.

Over age 60:
Adverse reactions and side effects may be more frequent and severe than in younger persons, particularly skin irritation around rectum.

Pregnancy:
Decide with your doctor if drug benefits justify risk to unborn baby. Risk category C (see page xviii).

Breastfeeding; Lactation; Nursing Mothers:
This drug is generally not recommended with breastfeeding. Discuss risks and benefits with your doctor.

Infants & children up to age 18:
Follow instructions provided by your child's doctor.

Prolonged use:
- You may become more susceptible to infections caused by germs not responsive to chloramphenicol.
- Talk to your doctor about the need for follow-up medical examinations or laboratory studies to check complete blood counts (white blood cell count, platelet count, red blood cell count, hemoglobin, hematocrit), chloramphenicol serum levels.

Skin & sunlight:
No problems expected.

Driving, piloting or hazardous work:
Don't drive or pilot aircraft until you learn how medicine affects you. Don't work around dangerous machinery. Don't climb ladders or work in high places. Danger increases if you drink alcohol or take medicine affecting alertness and reflexes.

Discontinuing:
Don't discontinue without doctor's advice until you complete prescribed dose, even though symptoms diminish or disappear.

Others:
- Chloramphenicol can cause serious anemia. Frequent laboratory blood studies, liver and kidney tests recommended.
- Advise any doctor, dentist or pharmacist whom you consult that you take this medicine.
- Second medical opinion recommended before starting.

POSSIBLE INTERACTION WITH OTHER DRUGS

GENERIC NAME OR DRUG CLASS	COMBINED EFFECT
Anticoagulants*	Increased anticoagulant effect.
Antidiabetics, oral*	Increased antidiabetic effect.
Anticonvulsants*	Increased chance of toxicity to bone marrow.
Antivirals, HIV/AIDS*	Increased risk of peripheral neuropathy.
Cefixime	Decreased antibiotic effect of cefixime.
Cephalosporins*	Decreased chloramphenicol effect.
Clindamycin	Decreased clindamycin effect.
Clozapine	Toxic effect on bone marrow.
Cyclophosphamide	Increased cyclophosphamide effect.
Erythromycins	Decreased erythromycin effect.
Flecainide	Possible decreased blood cell production in bone marrow.
Levamisole	Increased risk of bone marrow depression.
Lincomycin	Decreased lincomycin effect.
Lisinopril	Possible blood disorders.
Penicillins*	Decreased penicillin effect.
Phenobarbital	Increased phenobarbital effect.
Phenytoin	Increased phenytoin effect.
Rifampin	Decreased chloramphenicol effect.
Thioguanine	More likelihood of toxicity of both drugs.
Tiopronin	Increased risk of toxicity to bone marrow.
Tocainide	Possible decreased blood cell production in bone marrow.

POSSIBLE INTERACTION WITH OTHER SUBSTANCES

INTERACTS WITH	COMBINED EFFECT
Alcohol:	Possible liver problems. Possible disulfiram reaction.*
Beverages:	None expected.
Cocaine:	Unknown. Avoid.
Foods:	None expected.
Marijuana:	Consult doctor.
Tobacco:	None expected.

*See Glossary

CHLORHEXIDINE

BRAND NAMES

Peridex
Periochip
Periogard

BASIC INFORMATION

Habit forming? No
Prescription needed? Yes
Available as generic? Yes
Drug class: Antibacterial (dental)

USES

Treatment for gingivitis (inflammation of the gums), periodontal disease and other infections of the mouth.

DOSAGE & USAGE INFORMATION

How to use:

- Oral rinse—Swish in mouth for 30 seconds, then spit out. Do not swallow the solution, and do not rinse mouth with water after using. Use product at full strength; do not dilute.
- Implants—Inserted by dentist.

When to use:
Twice a day after brushing and flossing teeth.

If you forget a dose:
Use as soon as you remember. If it is almost time for the next dose, wait for that dose (don't double this dose) and resume regular schedule.

What drug does:
Kills or prevents growth of susceptible bacteria.

Time lapse before drug works:
Antibacterial action begins within an hour, but full benefit may take several weeks.

Don't use with:
Other mouthwashes without consulting your dentist or pharmacist.

OVERDOSE

SYMPTOMS:
May have slurred speech, staggering or stumbling walk, sleepiness.
WHAT TO DO:
If person uses much larger amount than prescribed or if accidentally swallowed, call poison control center 1-800-222-1222 for instructions or dial 911 (emergency) for help.

POSSIBLE ADVERSE REACTIONS OR SIDE EFFECTS

SYMPTOMS	WHAT TO DO
Life-threatening: None expected.	
Common: Staining of teeth and other oral surfaces, increased tartar, taste changes, minor mouth irritation.	Continue. Call dentist when convenient.
Infrequent: None expected.	
Rare: Allergic reaction (stuffy nose, shortness of breath, skin rash, hives, itching, face swelling), swollen glands on side of face or neck.	Discontinue. Call doctor right away.

WARNINGS & PRECAUTIONS

Don't take if:
You are allergic to chlorhexidine or skin cleaners that contain chlorhexidine.

Before you start, consult your dentist if:
- You have front tooth fillings (may become discolored).
- You have periodontitis.

Over age 60:
No special problems expected.

Pregnancy:
Decide with your doctor if drug benefits justify risk to unborn baby. Risk category B (see page xviii).

Breastfeeding; Lactation; Nursing Mothers:
This drug is generally acceptable with breastfeeding. Discuss risks and benefits with your doctor

Infants & children up to age 18:
Safety and effectiveness in this age group have not been established. Consult doctor.

Prolonged use:
See your dentist every 6 months.

Skin & sunlight:
No special problems expected.

Driving, piloting or hazardous work:
No special problems expected.

Discontinuing:
No special problems expected.

Others:
Brush teeth with a tartar-control toothpaste, and floss daily to help reduce tartar buildup.

POSSIBLE INTERACTION WITH OTHER DRUGS

GENERIC NAME OR DRUG CLASS	COMBINED EFFECT
None expected.	

POSSIBLE INTERACTION WITH OTHER SUBSTANCES

INTERACTS WITH	COMBINED EFFECT
Alcohol:	None expected.
Beverages:	Avoid drinking any fluids right after using mouthwash.
Cocaine:	None expected.
Foods:	Avoid eating any foods right after using mouthwash.
Marijuana:	None expected.
Tobacco:	None expected.

*See Glossary

CHLOROQUINE

BRAND NAMES

Aralen

BASIC INFORMATION

Habit forming? No
Prescription needed? Yes
Available as generic? Yes
Drug class: Antiprotozoal, antirheumatic

USES

- Treatment for protozoal infections, such as malaria and amebiasis.
- Treatment for some forms of arthritis and lupus.

DOSAGE & USAGE INFORMATION

How to take:
Tablet—Swallow with food or milk to lessen stomach irritation.

When to take:
- Depends on condition. Is adjusted during treatment.
- Malaria prevention—Begin taking medicine 2 weeks before traveling to areas where malaria is present and until 8 weeks after return.

If you forget a dose:
- 1 or more doses a day—Take as soon as you remember. If it is almost time for the next dose, wait for the next scheduled dose (don't double this dose).
- 1 dose weekly—Take as soon as possible, then return to regular dosing schedule.

What drug does:
- Inhibits parasite multiplication.
- Decreases inflammatory response in diseased joint.

Continued next column

OVERDOSE

SYMPTOMS:
Severe breathing difficulty, drowsiness, faintness, headache, seizures.
WHAT TO DO:
- **Dial 911 for all medical emergencies or call poison control center 1-800-222-1222 for instructions.**
- **See emergency information on last 3 pages of this book.**

Time lapse before drug works:
1 to 2 hours. For treatment of arthritis symptoms, may take up to 6 months for maximum effectiveness.

Don't take with:
Any other medicine or any dietary supplement without consulting your doctor or pharmacist.

POSSIBLE ADVERSE REACTIONS OR SIDE EFFECTS

SYMPTOMS	WHAT TO DO
Life-threatening:	
Rare, severe allergic reaction (anaphylaxis)—difficulty breathing, hives, itching, swelling (throat, face), vomiting, dizziness.	Seek emergency treatment immediately.
Common:	
Headache, appetite loss, abdominal pain.	Continue. Call doctor when convenient.
Infrequent:	
• Blurred or changed vision.	Discontinue. Call doctor right away.
• Rash or itch, diarrhea, nausea, vomiting, decreased blood pressure, hair loss, blue-black skin or mouth, dizziness, nervousness.	Continue. Call doctor when convenient.
Rare:	
• Mood or mental changes, seizures, sore throat, fever, unusual bleeding or bruising, muscle weakness.	Discontinue. Call doctor right away.
• Ringing or buzzing in ears, hearing loss.	Continue. Call doctor when convenient.

WARNINGS & PRECAUTIONS

Don't take if:
You are allergic to chloroquine or hydroxychloroquine.

Before you start, consult your doctor if:
- You plan to become pregnant within the medication period.
- You have blood disease.
- You have eye or vision problems.
- You have a G6PD deficiency.
- You have liver disease.
- You have nerve or brain disease (including seizure disorders).
- You have porphyria.
- You have psoriasis.
- You have stomach or intestinal disease.
- You drink more than 3 oz. of alcohol daily.

Over age 60:
Adverse reactions and side effects may be more frequent and severe than in younger persons.

Pregnancy:
Decide with your doctor if drug benefits justify risk to unborn baby. Risk category C (see page xviii).

Breastfeeding; Lactation; Nursing Mothers:
This drug is generally acceptable with breastfeeding. Discuss risks and benefits with your doctor

Infants & children up to age 18:
Follow instructions provided by your child's doctor.

Prolonged use:
- May increase risk for certain eye or hearing problems. Consult doctor about your risks.
- Talk to your doctor about the need for follow-up medical examinations or laboratory tests to check for adverse effects.

Skin & sunlight:
May cause rash or intensify sunburn in areas exposed to sun or ultraviolet light (photosensitivity reaction). Avoid overexposure. Notify doctor if reaction occurs.

Driving, piloting or hazardous work:
Don't drive or pilot aircraft until you learn how medicine affects you. Don't work around dangerous machinery. Don't climb ladders or work in high places. Danger increases if you drink alcohol or take medicine affecting alertness and reflexes.

Discontinuing:
Don't discontinue without doctor's advice until you complete prescribed dose, even though symptoms diminish or disappear.

Others:
- Periodic physical and blood examinations recommended.
- Advise any doctor, dentist or pharmacist whom you consult that you take this medicine.
- If you are in a malaria area for a long time, you may need to change to another preventive drug every 2 years.

POSSIBLE INTERACTION WITH OTHER DRUGS

GENERIC NAME OR DRUG CLASS	COMBINED EFFECT
Penicillamine	Possible blood or kidney toxicity.

POSSIBLE INTERACTION WITH OTHER SUBSTANCES

INTERACTS WITH	COMBINED EFFECT
Alcohol:	Possible liver toxicity. Avoid.
Beverages:	None expected.
Cocaine:	Unknown. Avoid.
Foods:	None expected.
Marijuana:	Consult doctor.
Tobacco:	None expected.

CHOLESTYRAMINE

BRAND NAMES

Cholybar
Questran
Questran Light

BASIC INFORMATION

Habit forming? No
Prescription needed? Yes
Available as generic? Yes
Drug class: Antihyperlipidemic, antipruritic

USES

- Removes excess bile acids that occur with some liver problems. Reduces persistent itch caused by bile acids.
- Lowers cholesterol level.
- Treatment for one form of colitis (rare).

DOSAGE & USAGE INFORMATION

How to take:
Powder, granules—Sprinkle into 8 oz. liquid. Let stand for 2 minutes, then mix with liquid before swallowing. Or mix with cereal, soup or pulpy fruit. Don't swallow dry.

When to take:
- 3 or 4 times a day on an empty stomach, 1 hour before or 2 hours after eating.
- If taking other medicines, take 1 hour before or 4 to 6 hours after taking cholestyramine.

If you forget a dose:
Take as soon as you remember. If it is almost time for the next dose, wait for that dose (don't double this dose) and resume regular schedule.

What drug does:
Binds with bile acids to prevent their absorption.

Time lapse before drug works:
- Cholesterol reduction—1 day.
- Bile-acid reduction—3 to 4 weeks.

Continued next column

OVERDOSE

SYMPTOMS:
Increased side effects and adverse reactions.
WHAT TO DO:
If person uses much larger amount than prescribed or if accidentally swallowed, call poison control center 1-800-222-1222 for instructions or dial 911 (emergency) for help.

Don't take with:
- Another medicine at the same time. Space doses 2 hours apart.
- Any other medicine or any dietary supplement without consulting your doctor or pharmacist.

POSSIBLE ADVERSE REACTIONS OR SIDE EFFECTS

SYMPTOMS	WHAT TO DO
Life-threatening: None expected.	
Common: Constipation.	Continue. Call doctor when convenient.
Infrequent:	
• Belching, bloating, diarrhea, mild nausea, vomiting, stomach pain, rapid weight gain.	Discontinue. Call doctor when convenient.
• Heartburn (mild).	Continue. Call doctor when convenient.
Rare:	
• Severe stomach pain, black or tarry stool, rash, hives.	Discontinue. Call doctor right away.
• Sore tongue.	Continue. Call doctor when convenient.

WARNINGS & PRECAUTIONS

Don't take if:
You are allergic to cholestyramine. Allergic reactions are rare, but may occur.

Before you start, consult your doctor if:
- You plan to become pregnant within medication period.
- You have angina or heart or blood-vessel disease.
- You have stomach problems (including ulcer).
- You have tartrazine sensitivity.
- You have constipation or hemorrhoids.
- You have kidney disease.

Over age 60:
Adverse reactions and side effects may be more frequent and severe than in younger persons.

Pregnancy:
Decide with your doctor if drug benefits justify risk to unborn baby. Risk category C (see page xviii).

Breastfeeding; Lactation; Nursing Mothers:
This drug is generally acceptable with breastfeeding. Discuss risks and benefits with your doctor.

Infants & children up to age 18:
Follow instructions provided by your child's doctor.

Prolonged use:
- May decrease absorption of folic acid.
- Talk to your doctor about the need for follow-up medical examinations or laboratory studies to check serum cholesterol and triglycerides.

Skin & sunlight:
No problems expected.

Driving, piloting or hazardous work:
No problems expected.

Discontinuing:
Don't discontinue without doctor's advice until you complete prescribed dose, even though symptoms diminish or disappear.

Others:
Advise any doctor, dentist or pharmacist whom you consult that you take this medicine.

POSSIBLE INTERACTION WITH OTHER DRUGS

GENERIC NAME OR DRUG CLASS	COMBINED EFFECT
Acetaminophen	Decreased effect of acetaminophen.
Adrenocorticoids, systemic	Decreased adrenocorticoid effect.
Anticoagulants, oral*	Increased anticoagulant effect.
Beta carotene	Decreased absorption of beta carotene.
Dexfenfluramine	May require dosage change as weight loss occurs.
Dextrothyroxine	Decreased dextrothyroxine effect.
Digitalis preparations*	Decreased digitalis effect.
Indapamide	Decreased indapamide effect.
Penicillins*	May decrease penicillin effect.
Raloxifene	Decreased effect of raloxifene.
Thiazides*	Decreased absorption of cholestyramine.
Thyroid hormones*	Decreased thyroid effect.
Trimethoprim	Decreased absorption of cholestyramine.
Ursodiol	Decreased absorption of ursodiol.
Vancomycin	Increased chance of hearing loss or kidney damage. Decreased therapeutic effect of vancomycin.
Vitamins	Decreased absorption of fat-soluble vitamins (A,D,E,K).
All other medicines	Decreased absorption, so dosages or dosage intervals may require adjustment.

POSSIBLE INTERACTION WITH OTHER SUBSTANCES

INTERACTS WITH	COMBINED EFFECT
Alcohol:	None expected.
Beverages:	None expected.
Cocaine:	Unknown. Avoid.
Foods:	Absorption of vitamins in foods may be decreased.
Marijuana:	Consult doctor.
Tobacco:	None expected.

*See Glossary

CHOLINESTERASE INHIBITORS

GENERIC AND BRAND NAMES

DONEPEZIL	**GALANTAMINE**
Aricept	**Razadyne**
Aricept ODT	**Razadyne ER**
Namzaric	**RIVASTIGMINE**
	Exelon Patch

BASIC INFORMATION

Habit forming? No
Prescription needed? Yes
Available as generic? Yes, for some
Drug class: Cholinesterase inhibitor

USES

Treats mild, moderate or severe symptoms of dementia (such as problems with memory, judgment, reasoning and other cognitive functions) in patients with Alzheimer's disease or Parkinson's disease.

DOSAGE & USAGE INFORMATION

How to take:

- Capsule or tablet—Swallow with liquid. If unable to swallow whole, open capsule or crumble tablet and take with liquid or food.
- Extended-release capsule (brand Namzaric)—Swallow whole with liquid. Capsule can be opened and sprinkled in or on food.
- Patch—Always follow label instructions.
- Disintegrating tablet—Let dissolve on tongue.
- Oral solution—Take as directed on label.

When to take:

- Donepezil—Once a day in evening before bedtime.
- Rivastigmine or galantamine—At the same times each day.

Continued next column

OVERDOSE

SYMPTOMS:
Severe nausea and vomiting, excessive saliva, sweating, blood pressure decrease, slow heartbeat, fainting, convulsions, muscle weakness.
WHAT TO DO:

- **Dial 911 for all medical emergencies or call poison control center 1-800-222-1222 for instructions.**
- **See emergency information on last 3 pages of this book.**

What drug does:
Slows breakdown of a brain chemical (acetylcholine) that gradually disappears from the brains of people with Alzheimer's disease.

If you forget a dose:
Take as soon as you remember. If it is almost time for the next dose, wait for that dose (don't double this dose) and resume regular schedule.

Time lapse before drug works:
May take several weeks or months before beneficial results are observed. Dosage is normally increased over a period of time to help prevent adverse reactions.

Don't take with:
Any other medicine or any dietary supplement without consulting your doctor or pharmacist.

POSSIBLE ADVERSE REACTIONS OR SIDE EFFECTS

SYMPTOMS	WHAT TO DO
Life-threatening: None expected.	
Common: Nausea, vomiting, diarrhea, lack of coordination.	Continue. Call doctor when convenient.
Infrequent: Rash, indigestion, headache, muscle aches, loss of appetite, stomach pain, nervousness, chills, dizziness, drowsiness, dry or itching eyes, increased sweating, joint pain, runny nose, sore throat, swelling of feet or legs, insomnia, weight loss, unusual tiredness or weakness, flushing of face.	Continue. Call doctor when convenient.
Rare: Changes in liver function (yellow skin or eyes, black or dark or light stool color), lack of coordination, convulsions, speech problems, irregular heartbeat, vision changes, increased libido, changes in blood pressure, hot flashes, breathing difficulty.	Disontinue. Call doctor right away.

WARNINGS & PRECAUTIONS

Don't take if:
You are allergic to cholinesterase inhibitors. Allergic reactions are rare, but may occur.

Before you start, consult your doctor if:
- You have heart rhythm problems.
- You have a history of ulcer disease or are at risk of developing ulcers.
- You have a history of liver disease.
- You have a history of urinary tract problems.
- You have epilepsy or seizure disorder.
- You have a history of asthma.
- You have had a head injury with loss of consciousness.
- You have had previous treatment with any cholinesterase inhibitor that caused jaundice (yellow skin and eyes) or elevated bilirubin.

Over age 60:
No special problems expected.

Pregnancy:
These drugs are usually not prescribed for women of childbearing age. Consult doctor. Risk category B and C (see page xviii).

Breastfeeding; Lactation; Nursing Mothers:
These drugs are usually not prescribed for women of childbearing age.

Infants & children up to age 18:
Not used in this age group.

Prolonged use:
- Drug may lose its effectiveness.
- Talk to your doctor about the need for follow-up medical examinations, laboratory blood studies and liver function tests.

Skin & sunlight:
No problems expected.

Driving, piloting or hazardous work:
Don't drive or pilot aircraft until you learn how medicine affects you. Don't work around dangerous machinery. Don't climb ladders or work in high places. Danger increases if you drink alcohol or take other medicines affecting alertness and reflexes such as antihistamines, tranquilizers, sedatives, pain medicine, narcotics and mind-altering drugs.

Discontinuing:
Do not discontinue drug unless advised by doctor. Abrupt decreases in dosage may cause a cognitive decline.

Others:
- Advise any doctor, dentist or pharmacist whom you consult that you take this medicine.
- May affect the results in some medical tests.
- Do not increase dosage without doctor's approval.
- Treatment may need to be discontinued or the dosage lowered if weekly blood tests indicate a sensitivity to the drug or liver toxicity develops.

POSSIBLE INTERACTION WITH OTHER DRUGS

GENERIC NAME OR DRUG CLASS	COMBINED EFFECT
Anticholinergics*	Decreased anticholinergic effect.
Anti-inflammatories, nonsteroidal (NSAIDs)	May increase gastric acid secretions.
Enzyme inducers*	May decrease effect of cholinesterase inhibitor.
Ketoconazole	May interact, but effect unknown.
Quinidine	May interact, but effect unknown.
Theophylline	Increased theophylline effect or toxicity.

POSSIBLE INTERACTION WITH OTHER SUBSTANCES

INTERACTS WITH	COMBINED EFFECT
Alcohol:	None expected.
Beverages:	None expected.
Cocaine:	Unknown. Avoid.
Foods:	None expected.
Marijuana:	Consult doctor.
Tobacco:	None expected.

***See Glossary**

CITRATES

GENERIC AND BRAND NAMES

POTASSIUM CITRATE
Citra Forte
Urocit-K
POTASSIUM CITRATE & CITRIC ACID
Polycitra-K
POTASSIUM CITRATE & SODIUM CITRATE
Citrolith
SODIUM CITRATE & CITRIC ACID
Albright's Solution
Bicitra
Modified Shohl's Solution
Oracit
TRICITRATES
Polycitra
Polycitra LC

BASIC INFORMATION

Habit forming? No
Prescription needed? Yes
Available as generic? No
Drug class: Urinary alkalizer, antiurolithic

USES

- To make urine more alkaline (less acid).
- To treat or prevent recurrence of some types of kidney stones.

DOSAGE & USAGE INFORMATION

How to take:

- Tablet—Take right after a meal or with a bedtime snack. Swallow tablet whole (do not crush or chew). Take with a full glass of water.
- Solution—Dilute with 6 ounces of water or juice and drink all of the mixture.
- Crystals—Stir into 6 ounces of water or juice and drink all of the mixture.

When to take:
On full stomach, usually after meals or with food.

Continued next column

OVERDOSE

SYMPTOMS:
Listlessness, weakness, confusion, tingling in arms or legs, irregular heartbeat, chest pain, seizures.
WHAT TO DO:

- **Dial 911 for all medical emergencies or call poison control center 1-800-222-1222 for instructions.**
- **See emergency information on last 3 pages of this book.**

If you forget a dose:
Take as soon as you remember. If it is almost time for the next dose, wait for that dose (don't double this dose) and resume regular schedule.

What drug does:
Increases urinary alkalinity by excretion of bicarbonate ions.

Time lapse before drug works:
1 hour.

Don't take with:
Any other medicine or any dietary supplement without consulting your doctor or pharmacist.

POSSIBLE ADVERSE REACTIONS OR SIDE EFFECTS

SYMPTOMS	WHAT TO DO
Life-threatening:	
None expected.	
Common:	
Nausea or mild vomiting, diarrhea, mild stomach cramps.	Continue. Call doctor when convenient.
Infrequent:	
None expected.	
Rare:	
• Black or tarry stools, vomiting (may be bloody), severe stomach cramps.	Discontinue. Seek emergency. treatment.
• Confusion, dizziness, swollen feet and ankles, irritability, depression, muscle pain, nervousness, numbness or tingling in hands or feet, unpleasant taste, unusual weakness or tiredness.	Discontinue. Call doctor right away.

WARNINGS & PRECAUTIONS

Don't take if:
You are allergic to any citrate. Allergic reactions are rare, but may occur.

Before you start, consult your doctor if:
You have any disease involving the adrenal glands, diabetes, chronic diarrhea, heart problems, hypertension, kidney disease, stomach ulcer or gastritis, urinary tract infection, intestinal or esophageal blockage or toxemia of pregnancy.

Over age 60:
Adverse reactions and side effects may be more frequent and severe than in younger persons.

Pregnancy:
Decide with your doctor if drug benefits justify risk to unborn baby. Risk category C for most; others are not designated (see page xviii).

Breastfeeding; Lactation; Nursing Mothers:
It is unknown if drugs in this group are acceptable with breastfeeding. Discuss risks and benefits with your doctor.

Infants & children up to age 18:
Follow instructions provided by your child's doctor.

Prolonged use:
- Adverse reactions more likely.
- Talk to your doctor about the need for follow-up medical examinations or laboratory studies.

Skin & sunlight:
No problems expected.

Driving, piloting or hazardous work:
Don't drive or pilot aircraft until you learn how medicine affects you. Don't work around dangerous machinery. Don't climb ladders or work in high places. Danger increases if you drink alcohol or take medicine affecting alertness and reflexes, such as antihistamines, tranquilizers, sedatives, pain medicine, narcotics and mind-altering drugs.

Discontinuing:
Don't discontinue without consulting doctor. Dose may require gradual reduction if you have taken drug for a long time. Doses of other drugs may also require adjustment.

Others:
- Liquid may be chilled (don't freeze) to improve taste.
- Advise any doctor, dentist or pharmacist whom you consult that you take this medicine.
- Monitor potassium in blood with frequent laboratory studies.

POSSIBLE INTERACTION WITH OTHER DRUGS

GENERIC NAME OR DRUG CLASS	COMBINED EFFECT
Antacids*	Increased risk of side effects.
Angiotensin-converting enzyme (ACE) inhibitors*	Increased risk of side effects.
Anti-inflammatories, nonsteroidal (NSAIDs)	Increased risk of side effects.
Digitalis preparations*	Increased risk of too much potassium in blood.
Diuretics, potassium sparing*	Increased risk of too much potassium in blood.
Methenamine	Decreased effects of methenamine.
Potassium supplements	Increased risk of too much potassium in blood.
Pseudoephedrine	Urinary retention. Increased effect of pseudoephedrine.
Quinidine	Prolonged quinidine effect.

POSSIBLE INTERACTION WITH OTHER SUBSTANCES

INTERACTS WITH	COMBINED EFFECT
Alcohol:	Decreased alertness.
Beverages: Salt-free milk.	Increased risk of side effects.
Cocaine:	Unknown, Avoid.
Foods: Salt substitutes, low salt foods.	Increased risk of side effects.
Marijuana:	Consult doctor.
Tobacco:	Increased risk of stomach irritation.

*See Glossary

CLIDINIUM

BRAND NAMES

Apo-Chlorax
Clindex
Clinoxide
Corium
Librax
Lidox
Lodoxide
Quarzan
Zebrax

BASIC INFORMATION

Habit forming? No
Prescription needed?
Low strength: No
High strength: Yes
Available as generic? No
Drug class: Antispasmodic, anticholinergic

USES

Reduces spasms of digestive system, bladder and urethra.

DOSAGE & USAGE INFORMATION

How to take:
Capsule—Swallow with liquid or food to lessen stomach irritation.

When to take:
30 minutes before meals (unless directed otherwise by doctor).

If you forget a dose:
Take as soon as you remember. If it is almost time for the next dose, wait for that dose (don't double this dose) and resume regular schedule.

What drug does:
Blocks nerve impulses at parasympathetic nerve endings, preventing muscle contractions and gland secretions of organs involved.

Time lapse before drug works:
15 to 30 minutes.

Continued next column

OVERDOSE

SYMPTOMS:
Dry mouth, drowsiness, slow breathing, irregular heartbeat, blurred vision, confusion, lightheadedness, weak pulse, nausea, vomiting, seizures (convulsions).

WHAT TO DO:
- **Dial 911 for all medical emergencies or call poison control center 1-800-222-1222 for instructions.**
- **See emergency information on last 3 pages of this book.**

Don't take with:
Any other medicine or any dietary supplement without consulting your doctor or pharmacist.

POSSIBLE ADVERSE REACTIONS OR SIDE EFFECTS

SYMPTOMS	WHAT TO DO
Life-threatening:	
None expected.	
Common:	
• Confusion, delirium, rapid heartbeat.	Discontinue. Call doctor right away.
• Nausea, vomiting, decreased sweating, constipation.	Continue. Call doctor when convenient.
• Dryness in ears, nose, throat, mouth.	No action necessary.
Infrequent:	
• Lightheadedness.	Discontinue. Call doctor right away.
• Nasal congestion, altered taste, difficult urination, headache, impotence.	Continue. Call doctor when convenient.
Rare:	
Rash or hives, eye pain, blurred vision.	Discontinue. Call doctor right away.

WARNINGS & PRECAUTIONS

Don't take if:
- You are allergic to any anticholinergic. Allergic reactions are rare, but may occur.
- You have trouble with stomach bloating.
- You have difficulty emptying your bladder completely (enlarged prostate).
- You have narrow-angle glaucoma.
- You have severe ulcerative colitis.

Before you start, consult your doctor if:
- You have open-angle glaucoma.
- You have angina, chronic bronchitis or asthma, kidney or thyroid disease, hiatal hernia, liver disease, enlarged prostate, myasthenia gravis, peptic ulcer.
- You will have surgery within 2 months, including dental surgery, requiring general or spinal anesthesia.

Over age 60:
Adverse reactions and side effects may be more frequent and severe than in younger persons.

Pregnancy:
Decide with your doctor if drug benefits justify risk to unborn baby. See page xviii for risk categories.

Breastfeeding; Lactation; Nursing Mothers:
This drug is generally acceptable with breastfeeding. Discuss risks and benefits with your doctor.

Infants & children up to age 18:
Follow instructions provided by your child's doctor.

Prolonged use:
Chronic constipation, possible fecal impaction. Consult doctor immediately.

Skin & sunlight:
No problems expected.

Driving, piloting or hazardous work:
Don't drive or pilot aircraft until you learn how medicine affects you. Don't work around dangerous machinery. Don't climb ladders or work in high places. Danger increases if you drink alcohol or take medicine affecting alertness and reflexes, such as antihistamines, tranquilizers, sedatives, pain medicine, narcotics, or mind-altering drugs.

Discontinuing:
May be unnecessary to finish medicine. Follow doctor's instructions.

Others:
Advise any doctor, dentist or pharmacist whom you consult that you take this medicine.

POSSIBLE INTERACTION WITH OTHER DRUGS

GENERIC NAME OR DRUG CLASS	COMBINED EFFECT
Amantadine	Increased clidinium effect.
Antacids*	Decreased clidinium effect.
Anticholinergics, other*	Increased clidinium effect.
Antidepressants, tricyclic*	Increased clidinium effect. Increased sedation.
Antidiarrheals*	Increased clidinium effect.
Antihistamines*	Increased clidinium effect.
Attapulgite	Decreased clidinium effect.
Haloperidol	Increased internal eye pressure.
Ketoconazole	Decreased ketoconazole effect.
Meperidine	Increased clidinium effect.
Methylphenidate	Increased clidinium effect.
Molindone	Increased anticholinergic effect.
Monoamine oxidase (MAO) inhibitors*	Increased clidinium effect.
Nitrates*	Increased internal eye pressure.
Nizatidine	Increased nizatidine effect.
Orphenadrine	Increased clidinium effect.
Pilocarpine	Loss of pilocarpine effect in glaucoma treatment.
Potassium supplements*	Possible intestinal ulcers with oral potassium tablets.
Tranquilizers*	Increased clidinium effect.
Vitamin C	Decreased clidinium effect. Avoid large doses of vitamin C.

POSSIBLE INTERACTION WITH OTHER SUBSTANCES

INTERACTS WITH	COMBINED EFFECT
Alcohol:	None expected.
Beverages:	None expected.
Cocaine:	Unknown. Avoid.
Foods:	None expected.
Marijuana:	Consult doctor.
Tobacco:	None expected.

*See Glossary

CLINDAMYCIN

BRAND NAMES

Cleocin
Cleocin Pediatric
Cleocin (Vaginal)
Clindesse Vaginal Cream
Dalacin C
Dalacin C Palmitate
Dalacin C Phosphate

BASIC INFORMATION

Habit forming? No
Prescription needed? Yes
Available as generic? Yes
Drug class: Antibacterial

USES

Treatment of bacterial infections that are susceptible to clindamycin.

DOSAGE & USAGE INFORMATION

How to take:

- Capsule or liquid—Swallow with liquid. Take with a full glass of water or a meal to avoid gastric irritation.
- Vaginal cream—Use applicator supplied with product to insert cream into vagina. Wash hands immediately after using.

When to take:

- Oral—At the same times each day.
- Vaginal cream—Apply at bedtime unless directed differently by your doctor.

If you forget a dose:
Take as soon as you remember. If it is almost time for the next dose, wait for that dose (don't double this dose) and resume regular schedule.

What drug does:
Destroys susceptible bacteria. Does not kill viruses.

Time lapse before drug works:
3 to 5 days.

Continued next column

OVERDOSE

SYMPTOMS:
Severe nausea, vomiting, diarrhea.
WHAT TO DO:
If person uses much larger amount than prescribed or if accidentally swallowed, call poison control center 1-800-222-1222 for instructions or dial 911 (emergency) for help.

Don't take with:
Any other medicine or any dietary supplement without consulting your doctor or pharmacist.

POSSIBLE ADVERSE REACTIONS OR SIDE EFFECTS

SYMPTOMS	WHAT TO DO
Life-threatening:	
Rare, severe allergic reaction (anaphylaxis)—difficulty breathing, hives, itching, swelling (throat, face), vomiting, dizziness.	Seek emergency treatment immediately.
Common:	
Diarrhea, nausea, vomiting, stomach cramps.	Continue. Call doctor when convenient.
Infrequent:	
• Unusual thirst, vomiting, severe and watery diarrhea with blood or mucus, painful and swollen joints, fever, yellow skin or eyes, tiredness, weakness.	Discontinue. Call doctor right away.
• White patches in mouth, rash or itch around groin or rectum or armpits, vaginal discharge, dizziness, pain during intercourse.	Continue. Call doctor when convenient.
Rare:	
None expected.	

WARNINGS & PRECAUTIONS

Don't take if:
- You are allergic to lincomycins, clindamycin or doxorubicin.
- You have had ulcerative colitis.

Before you start, consult your doctor if:
- You have had yeast infections of mouth, skin or vagina.
- You will have surgery within 2 months, including dental surgery, requiring general or spinal anesthesia.
- You have kidney or liver disease.
- You have allergies of any kind.
- You have a history of gastrointestinal disorders.

Over age 60:
No special problems expected.

Pregnancy:
Decide with your doctor if drug benefits justify risk to unborn baby. Risk category B (see page xviii).

Breastfeeding; Lactation; Nursing Mothers:
This drug is generally acceptable with breastfeeding. Discuss risks and benefits with your doctor.

Infants & children up to age 18:
Don't give to infants younger than 1 month. Follow instructions provided by your child's doctor.

Prolonged use:
- Severe colitis with diarrhea and bleeding.
- You may become more susceptible to infections caused by germs not responsive to clindamycin.
- Talk to your doctor about the need for follow-up medical examinations or laboratory studies to check stools and perform proctosigmoidoscopy.

Skin & sunlight:
No problems expected.

Driving, piloting or hazardous work:
No problems expected.

Discontinuing:
- Don't discontinue without doctor's advice until you complete prescribed dose, even though symptoms diminish or disappear.
- If vaginal discharge, itching or pain occurs after discontinuing medicine, consult doctor.

Others:
- Advise any doctor, dentist or pharmacist whom you consult that you take this medicine.
- Vaginal cream product may decrease the effectiveness of condoms, cervical caps or diaphragms. Wait 72 hours after treatment to use any of these devices.

POSSIBLE INTERACTION WITH OTHER DRUGS

GENERIC NAME OR DRUG CLASS	COMBINED EFFECT
Antidiarrheal preparations*	Decreased clindamycin effect.
Antimyasthenics*	Decreased antimyasthenic effect.
Chloramphenicol	Decreased clindamycin effect.
Erythromycins*	Decreased clindamycin effect.
Muscle relaxants*	Increased actions of muscle blockers to unsafe degree. Avoid.
Narcotics*	Increased risk of respiratory problems.

POSSIBLE INTERACTION WITH OTHER SUBSTANCES

INTERACTS WITH	COMBINED EFFECT
Alcohol:	None expected.
Beverages:	None expected.
Cocaine:	Unknown. Avoid.
Foods:	None expected.
Marijuana:	Consult doctor.
Tobacco:	None expected.

CLOMIPHENE

BRAND NAMES

Clomid
Milophene
Serophene

BASIC INFORMATION

Habit forming? No
Prescription needed? Yes
Available as generic? Yes
Drug class: Gonad stimulant

USES

- Treatment for men with low sperm counts.
- Treatment for ovulatory failure in women who wish to become pregnant.

DOSAGE & USAGE INFORMATION

How to take:
Tablet—Swallow with liquid.

When to take:
- Men—Take at the same time each day.
- Women—Follow your doctor's instructions.

If you forget a dose:
Take as soon as you remember. If it is almost time for the next dose, wait for that dose (don't double this dose) and resume regular schedule. If you miss 2 or more doses, consult doctor.

What drug does:
Antiestrogen effect stimulates ovulation and sperm production.

Time lapse before drug works:
Usually 3 to 6 months. Ovulation may occur 6 to 10 days after last day of treatment in any cycle.

Don't take with:
Any other medicine or any dietary supplement without consulting your doctor or pharmacist.

OVERDOSE

SYMPTOMS:
Increased severity of adverse reactions and side effects.
WHAT TO DO:
If person uses much larger amount than prescribed or if accidentally swallowed, call poison control center 1-800-222-1222 for instructions or dial 911 (emergency) for help.

POSSIBLE ADVERSE REACTIONS OR SIDE EFFECTS

SYMPTOMS	WHAT TO DO
Life-threatening: None expected.	
Common: Hot flashes.	Continue. Call doctor when convenient.
Infrequent:	
• Rash, itch, vomiting, yellow skin or eyes, bloating, abdominal pain, pelvic pain.	Discontinue. Call doctor right away.
• Constipation, diarrhea, increased appetite, heavy menstrual flow, frequent urination, breast discomfort, weight change, hair loss, nausea, eyes sensitive to light.	Continue. Call doctor when convenient.
Rare:	
• Vision changes.	Discontinue. Call doctor right away.
• Dizziness, headache, tiredness, depression, nervousness, trouble with sleeping.	Continue. Call doctor when convenient.

WARNINGS & PRECAUTIONS

Don't take if:
You are allergic to clomiphene. Allergic reactions are rare, but may occur.

Before you start, consult your doctor if:
- You have an ovarian cyst, fibroid uterine tumors or unusual vaginal bleeding.
- You have inflamed veins caused by blood clots.
- You have liver disease.
- You are depressed.

Over age 60:
Not recommended.

Pregnancy:
Consult doctor. Drug should not be used during pregnancy. Can cause harm to unborn baby. Risk category X (see page xviii).

Breastfeeding; Lactation; Nursing Mothers:
This drug is generally not recommended with breastfeeding. Discuss risks and benefits with your doctor.

Infants & children up to age 18:
Safety and effectiveness in this age group have not been established. Consult doctor.

Prolonged use:
- Not recommended.
- Talk to your doctor about the need for follow-up medical examinations or laboratory studies to check basal body temperature, endometrial biopsy, kidney function, eyes.

Skin & sunlight:
No special problems expected.

Driving, piloting or hazardous work:
- Avoid if you feel dizzy.
- May cause blurred vision.

Discontinuing:
May be unnecessary to finish medicine. Follow doctor's instructions.

Others:
- Have a complete pelvic examination before treatment.
- Advise any doctor, dentist or pharmacist whom you consult that you take this medicine.
- If you become pregnant, twins or triplets are possible.

POSSIBLE INTERACTION WITH OTHER DRUGS

GENERIC NAME OR DRUG CLASS	COMBINED EFFECT
Thyroglobulin	May increase serum thyroglobulin.
Thyroxine (T-4)	May increase serum thyroxine.

POSSIBLE INTERACTION WITH OTHER SUBSTANCES

INTERACTS WITH	COMBINED EFFECT
Alcohol:	None expected.
Beverages:	None expected.
Cocaine:	Unknown. Avoid.
Foods:	None expected.
Marijuana:	Consult doctor.
Tobacco:	None expected.

CLOTRIMAZOLE (Oral-Local)

BRAND NAMES

Mycelex Troches

BASIC INFORMATION

Habit forming? No
Prescription needed? Yes
Available as generic? Yes
Drug class: Antifungal

USES

- Treats thrush, white mouth (candidiasis).
- Used primarily in immunosuppressed patients to treat and prevent mouth infection.

DOSAGE & USAGE INFORMATION

How to take:
Lozenge—Dissolve slowly and completely in the mouth, 5 times a day (usually for 14 days or longer). Swallow saliva during this time. Don't swallow lozenge whole and don't chew.

When to take:
At the same times each day.

If you forget a dose:
Take as soon as you remember. If it is almost time for the next dose, wait for that dose (don't double this dose) and resume regular schedule.

What drug does:
Kills fungus by interfering with cell wall membrane and its permeability.

Time lapse before drug works:
1 to 3 hours.

Don't take with:
Any other medicine for your mouth without consulting your doctor or pharmacist.

OVERDOSE

SYMPTOMS:
Unknown, but if large dose has been taken, follow instructions below.
WHAT TO DO:
If person uses much larger amount than prescribed or if accidentally swallowed, call poison control center 1-800-222-1222 for instructions or dial 911 (emergency) for help.

POSSIBLE ADVERSE REACTIONS OR SIDE EFFECTS

SYMPTOMS	WHAT TO DO
Life-threatening: None expected.	
Common: None expected.	
Infrequent: Abdominal pain, diarrhea, nausea, vomiting.	Discontinue. Call doctor right away.
Rare: None expected.	

WARNINGS & PRECAUTIONS

Don't take if:
You have severe liver disease.

Before you start, consult your doctor if:
You have had a recent organ transplant.

Over age 60:
No special problems expected.

Pregnancy:
Decide with your doctor if drug benefits justify risk to unborn baby. Risk category B (see page xviii).

Breastfeeding; Lactation; Nursing Mothers:
This drug is generally acceptable with breastfeeding. Discuss risks and benefits with your doctor.

Infants & children up to age 18:
Follow instructions provided by your child's doctor.

Prolonged use:
No problems expected.

Skin & sunlight:
No problems expected.

Driving, piloting or hazardous work:
No problems expected.

Discontinuing:
Don't discontinue without consulting doctor. Dose may require gradual reduction if you have taken drug for a long time. Doses of other drugs may also require adjustment.

Others:
- Continue for full term of treatment. May require several months.
- Check with physician if not improved in 1 week.

POSSIBLE INTERACTION WITH OTHER DRUGS

GENERIC NAME OR DRUG CLASS	COMBINED EFFECT
None expected.	

POSSIBLE INTERACTION WITH OTHER SUBSTANCES

INTERACTS WITH	COMBINED EFFECT
Alcohol:	Decreased effects of clotrimazole.
Beverages:	None expected.
Cocaine:	Unknown. Avoid.
Foods:	None expected.
Marijuana:	Consult doctor.
Tobacco:	Decreased effects of clotrimazole.

*See Glossary

CLOZAPINE

BRAND NAMES

Clozaril
FazaClo
Leponex
Versacloz

BASIC INFORMATION

Habit forming? No
Prescription needed? Yes, prescribed only through a special program.
Available as generic? Yes
Drug class: Antipsychotic

USES

- Treats severe schizophrenia in patients not helped by other medicines.
- Reduces the risk of recurrent suicide behavior in patients with schizophrenia or schizoaffective disorder.

DOSAGE & USAGE INFORMATION

How to take:
- Tablet—Swallow with liquid. If you can't swallow whole, crumble tablet and take with liquid or food.
- Orally disintegrating tablet—Let tablet dissolve in your mouth. Don't chew or swallow it whole.
- Oral suspension—Carefully follow the instructions provided with the prescription.

When to take:
Once or twice daily as directed.

If you forget a dose:
Take as soon as you remember. If it is almost time for the next dose, wait for that dose (don't double this dose) and resume regular schedule.

What drug does:
Interferes with binding of dopamine. May produce significant improvement, but may at times also make schizophrenia worse.

Continued next column

OVERDOSE

SYMPTOMS:
Fast heartbeat, drowsiness, delirium, excess saliva, low blood pressure, breathing difficulty, coma.
WHAT TO DO:
- **Dial 911 for all medical emergencies or call poison control center 1-800-222-1222 for instructions.**
- **See emergency information on last 3 pages of this book.**

Time lapse before drug works:
Weeks to months before improvement is evident. Your doctor may increase the dosage to obtain optimal effectiveness.

Don't take with:
Any other medicine or any dietary supplement without consulting your doctor or pharmacist.

POSSIBLE ADVERSE REACTIONS OR SIDE EFFECTS

SYMPTOMS	WHAT TO DO
Life-threatening:	
In case of overdose, see previous column.	
Common:	
Dry mouth, blurred vision, constipation, difficulty urinating, sedation, low blood pressure, dizziness.	Continue. Call doctor when convenient.
Infrequent:	
None expected.	
Rare:	
• Jerky or involuntary movements (especially of the face, lips, jaw, tongue), slow-frequency tremor of head or limbs, (especially while moving), muscle rigidity, lack of facial expression and slow inflexible movements, seizures, high blood sugar (thirstiness, frequent urination, increased hunger, weakness).	Discontinue. Call doctor right away.
• Restlessness or pacing (akathisia), intermittent spasms (of muscles of face, eyes, tongue, jaw, neck, body or limbs).	Continue. Call doctor when convenient.

WARNINGS & PRECAUTIONS

Don't take if:
You are allergic to clozapine. Allergic reactions are rare, but may occur.

Before you start, consult your doctor if:
- You are significantly mentally depressed.
- You have bone marrow depression.
- You have a family history of, or have diabetes.
- You have glaucoma or an enlarged prostate.
- You have ever had seizures from any cause.
- You have liver, heart or gastrointestinal disease or any type of blood disorder.

Over age 60:
- May be more at risk of weakness or dizziness upon standing after sitting or lying down and of excitement, confusion or urination difficulty.
- Use of antipsychotic drugs in elderly patients with dementia-related psychosis may increase risk of death. Consult doctor.

Pregnancy:
Decide with your doctor if drug benefits justify risk to unborn baby. Risk category B (see page xviii).

Breastfeeding; Lactation; Nursing Mothers:
This drug is generally not recommended with breastfeeding. Discuss risks and benefits with your doctor.

Infants & children up to age 18:
Safety and effectiveness in this age group have not been established. Consult doctor.

Prolonged use:
Effects unknown.

Skin & sunlight:
No problems expected.

Driving, piloting or hazardous work:
Don't drive or pilot aircraft until you learn how medicine affects you. Don't work around dangerous machinery. Don't climb ladders or work in high places. Danger increases if you drink alcohol or take medicine affecting alertness and reflexes.

Discontinuing:
Don't discontinue without consulting doctor. Dose may require gradual reduction if you have taken drug for a long time. Doses of other drugs may also require adjustment.

Others:
- This medicine is available only through a special management program for monitoring and distributing this drug.
- You will need laboratory studies each week for white blood cell and differential counts.

POSSIBLE INTERACTION WITH OTHER DRUGS

GENERIC NAME OR DRUG CLASS	COMBINED EFFECT
Antihypertensives*	Risk of low blood pressure.
Bone marrow depressants*	Toxic bone marrow depression.
Bupropion	Increased risk of seizures.
Carbamazepine	Decreased clozapine effect.
Central nervous system (CNS) depressants*	Toxic effects on the central nervous system.
Fluoxetine	Increased risk of adverse reactions.
Fluvoxamine	Increased risk of adverse reactions.
Haloperidol	Increased risk of seizures.
Lithium	Increased risk of seizures.
Phenytoin	Decreased clozapine effect.
Risperidone	Increased risperidone effect.

POSSIBLE INTERACTION WITH OTHER SUBSTANCES

INTERACTS WITH	COMBINED EFFECT
Alcohol:	Avoid. Increases toxic effect on the central nervous system.
Beverages: Caffeine drinks.	Excess (more than 3 cups of coffee or equivalent) increases risk of heartbeat irregularities.
Cocaine:	Heartbeat irregularities. Avoid.
Foods:	None expected.
Marijuana:	Heartbeat irregularities. Avoid.
Tobacco:	Decreased effect of clozapine. Avoid.

*See Glossary

COAL TAR (Topical)

BRAND NAMES

See full list of brand names in the *Generic and Brand Name Directory*, page 856.

BASIC INFORMATION

Habit forming? No
Prescription needed? No (for most)
Available as generic? Yes
Drug class: Antiseborrheic, antipsoriatic, keratolytic

USES

Applied to the skin to treat dandruff, seborrhea, dermatitis, eczema and other skin diseases.

DOSAGE & USAGE INFORMATION

How to use:

- Follow package instructions.
- Don't apply to blistered, oozing, infected or raw skin.
- Keep away from eyes.
- Protect treated area from sunshine for 72 hours.

When to use:
According to package instructions.

If you forget a dose:
Use as soon as you remember. If it is almost time for the next dose, wait for that dose (don't double this dose) and resume regular schedule.

What drug does:

- Kills bacteria and fungus organisms on contact.
- Suppresses overproduction of skin cells.

Time lapse before drug works:
None. Works immediately. May take several days before maximum effect.

Don't use with:
Any other topical medicine without consulting your doctor or pharmacist.

OVERDOSE

SYMPTOMS:
Unknown.
WHAT TO DO:
If person uses much larger amount than prescribed or if accidentally swallowed, call poison control center 1-800-222-1222 for instructions or dial 911 (emergency) for help.

POSSIBLE ADVERSE REACTIONS OR SIDE EFFECTS

SYMPTOMS	WHAT TO DO
Life-threatening: None expected.	
Common: Skin stinging.	Continue. Call doctor if symptoms persist.
Infrequent: Skin more irritated.	Continue. Call doctor if symptoms persist.
Rare: Pus forms in lesions on skin.	Continue. Call doctor when convenient.

WARNINGS & PRECAUTIONS

Don't use if:
You have intolerance to coal tar.

Before you start, consult your doctor if:
You have infected skin or open wounds.

Over age 60:
No special problems expected.

Pregnancy:
Decide with your doctor if drug benefits justify risk to unborn baby. Risk category C (see page xviii).

Breastfeeding; Lactation; Nursing Mothers:
This drug is generally acceptable with breastfeeding. Discuss risks and benefits with your doctor.

Infants & children up to age 18:
Not to be used on infants. For other children, follow doctor's instructions.

Prolonged use:
No special problems expected.

Skin & sunlight:
One or more drugs in this group may cause rash or intensify sunburn in areas exposed to sun or ultraviolet light (photosensitivity reaction). Avoid overexposure. Notify doctor if reaction occurs.

Driving, piloting or hazardous work:
No problems expected.

Discontinuing:
No special problems expected.

Others:
- May affect results in some medical tests.
- Protect treated area from direct sunlight for 72 hours.

POSSIBLE INTERACTION WITH OTHER DRUGS

GENERIC NAME OR DRUG CLASS	COMBINED EFFECT
Psoralens (methoxsalen, trioxsalen)	Excess sensitivity to sun.

POSSIBLE INTERACTION WITH OTHER SUBSTANCES

INTERACTS WITH	COMBINED EFFECT
Alcohol:	None expected.
Beverages:	None expected.
Cocaine:	None expected.
Foods:	None expected.
Marijuana:	None expected.
Tobacco:	None expected.

COLCHICINE

BRAND NAMES

Colcrys
Col-Probenecid
Mitigare

BASIC INFORMATION

Habit forming? No
Prescription needed? Yes
Available as generic? Yes
Drug class: Antigout

USES

- Used to prevent or treat gout flares.
- Treats familial Mediterranean fever (FMF).
- May be used for dermatitis herpetiformis, calcium pyrophosphate deposition disease, amyloidosis, Paget's disease of bone, recurrent pericarditis, Behcet's syndrome, others.

DOSAGE & USAGE INFORMATION

How to take:
Tablet—Swallow with liquid or food to lessen stomach irritation.

When to take:
- For gout prevention, it is usually taken once or twice a day. For gout flares, it is usually taken at first sign of flare and then one hour later. Follow your doctor's advice.
- For other disorders, follow instructions on your prescription label.

If you forget a dose:
Take as soon as you remember. If it is almost time for the next dose, wait for that dose (don't double this dose) and resume regular schedule.

Continued next column

OVERDOSE

SYMPTOMS:
Bloody urine; severe or bloody diarrhea; burning feeling in the throat, skin or stomach; severe nausea or vomiting; muscle weakness; fever; shortness of breath; stupor; convulsions; coma.

WHAT TO DO:
- **Dial 911 for all medical emergencies or call poison control center 1-800-222-1222 for instructions.**
- **See emergency information on last 3 pages of this book.**

What drug does:
The exact way it works is unknown. It reduces the inflammatory response and relieves pain caused by gout or familial Mediterranean fever.

Time lapse before drug works:
12 to 24 hours.

Don't take with:
Any other medicine or any dietary supplement without consulting your doctor or pharmacist.

POSSIBLE ADVERSE REACTIONS OR SIDE EFFECTS

SYMPTOMS	WHAT TO DO
Life-threatening:	
In case of overdose, see previous column.	
Common:	
Diarrhea, nausea, vomiting, abdominal pain.	Discontinue. Call doctor right away.
Infrequent:	
Hair loss with long-term use, appetite loss.	Continue. Call doctor when convenient.
Rare:	
Black or tarry stool, blood in urine, fever, chills, skin symptoms (hives, rash, burning, tingling, peeling, pinpoint red spots, redness), sore throat, muscle weakness or pain, mouth sores, unusual bruising or bleeding, unusual tiredness or weakness, numbness or tingly feeling in fingers or toes, increased infections, pale or gray color of lips or tongue or skin, yellow eyes or skin.	Discontinue. Call doctor right away.

WARNINGS & PRECAUTIONS

Don't take if:
You are allergic to colchicine. Allergic reactions are rare, but may occur.

Before you start, consult your doctor if:
- You have stomach problems (e.g., ulcers) or bowel problems (e.g., ulcerative colitis).
- You have heart, liver or kidney disorder.
- You have any blood disorder.
- You have muscle or nerve problems.
- You drink large amounts of alcohol.

Over age 60:
No special problems expected.

Pregnancy:
Decide with your doctor if drug benefits justify risk to unborn baby. Risk category C (see page xviii).

Breastfeeding; Lactation; Nursing Mothers:
This drug is generally not recommended with breastfeeding. Discuss risks and benefits with your doctor.

Infants & children up to age 18:
Used to treat familial Mediterranean fever in children. Follow doctor's instructions.

Prolonged use:
- May increase risk of side effects.
- Talk to your doctor about the need for follow-up medical exams or lab studies.

Skin & sunlight:
No problems expected.

Driving, piloting or hazardous work:
No problems expected.

Discontinuing:
- Follow doctor's instructions.
- Stop taking drug and consult doctor if severe abdominal pain, diarrhea or vomiting occurs.

Others:
- Don't increase dose without medical advice.
- May affect sperm production in males. Consult doctor.
- Advise any doctor, dentist or pharmacist whom you consult that you take this medicine.
- May interfere with the accuracy of some medical tests.

POSSIBLE INTERACTION WITH OTHER DRUGS

GENERIC NAME OR DRUG CLASS	COMBINED EFFECT
Digoxin	Risk of serious muscle disorders—rhabdomyolysis (which can be fatal) or myopathy.
Enzyme inhibitors*	Some enzyme inhibitors increase risk of toxic effect of colchicine (can be fatal). Risk continues 14 days after enzyme inhibitor stopped. Consult doctor or pharmacist.
Fibrates	Risk of serious muscle disorders—rhabdomyolysis (which can be fatal) or myopathy.
Gemfibrozil	Risk of serious muscle disorders—rhabdomyolysis (which can be fatal) or myopathy.
HMG-CoA reductase inhibitors	Risk of serious muscle disorders—rhabdomyolysis (which can be fatal) or myopathy.
P-glycoprotein inhibitors*	Risk of toxic effect of colchicine (can be fatal). Risk continues 14 days after stopping p-glycoprotein inhibitor.
Vitamin B-12	Decreased absorption of vitamin B-12.

POSSIBLE INTERACTION WITH OTHER SUBSTANCES

INTERACTS WITH	COMBINED EFFECT
Alcohol:	Increased risk of gastrointestinal problems. Avoid
Beverages: Grapefruit juice.	Increased colchicine effect. Avoid.
Cocaine:	Unknown. Avoid.
Foods: Grapefruit.	Increased colchicine effect. Avoid.
Marijuana:	Consult doctor.
Tobacco:	None expected.

*See Glossary

COLESEVELAM

BRAND NAMES

Welchol
Welchol Oral Suspension

BASIC INFORMATION

Habit forming? No
Prescription needed? Yes
Available as generic? No
Drug class: Antihyperlipidemic

USES

- Reduces low density lipoprotein (LDL) cholesterol. Should be used in addition to diet and exercise.
- Helps lower blood sugar levels in adults with type 2 diabetes.

DOSAGE & USAGE INFORMATION

How to take:
- Tablet—Take with food and a full glass of water. If you can't swallow whole, crumble tablet and take with liquid and food.
- Oral suspension—Mix well with 4 to 8 ounces of water. Don't take it in the dry form.

When to take:
As directed by your doctor. Usually once or twice daily with meals.

If you forget a dose:
Take as soon as you remember. If it is almost time for the next dose, wait for that dose (don't double this dose) and resume regular schedule.

What drug does:
Attaches to cholesterol and bile fluid in the intestine and passes out of the body without being absorbed.

Time lapse before drug works:
2 to 4 weeks.

Don't take with:
Any other medicine or any dietary supplement without consulting your doctor or pharmacist.

OVERDOSE

SYMPTOMS:
May possibly cause severe constipation.
WHAT TO DO:
If person uses much larger amount than prescribed or if accidentally swallowed, call poison control center 1-800-222-1222 for instructions or dial 911 (emergency) for help.

POSSIBLE ADVERSE REACTIONS OR SIDE EFFECTS

SYMPTOMS	WHAT TO DO
Life-threatening: None expected.	
Common: Acid or sour stomach, belching, constipation, indigestion, stomach upset or pain.	Continue. Call doctor if symptoms persist.
Infrequent: Congestion, cough, dry or sore throat, hoarseness, muscle aches or pain, trouble swallowing.	Continue. Call doctor when convenient.
Rare: None expected.	

WARNINGS & PRECAUTIONS

Don't take if:
You are allergic to colesevelam. Allergic reactions are rare, but may occur.

Before you start, consult your doctor if:
- You have been diagnosed with a bowel obstruction.
- You have had recent gastrointestinal surgery.
- You have had gastrointestinal motility disorders.
- You have a vitamin deficiency.
- You have difficulty swallowing.

Over age 60:
No special problems expected.

Pregnancy:
Decide with your doctor if drug benefits justify risk to unborn baby. Risk category B (see page xviii).

Breastfeeding; Lactation; Nursing Mothers:
This drug is generally acceptable with breastfeeding. Discuss risks and benefits with your doctor.

Infants & children up to age 18:
Approved for boys and girls (who have had a menstrual period) ages 10 to 18 years to treat high cholesterol. Follow instructions provided by your child's doctor.

Prolonged use:
Talk to your doctor about the need for follow-up medical examinations to determine the effect of colesevelam on your body.

Skin & sunlight:
No problems expected.

Driving, piloting or hazardous work:
No problems expected.

Discontinuing:
Don't discontinue without consulting your doctor.

Others:
- Advise any doctor, dentist or pharmacist whom you consult that you take this medicine.
- May affect the results in some medical tests.

POSSIBLE INTERACTION WITH OTHER DRUGS

GENERIC NAME OR DRUG CLASS	COMBINED EFFECT
Cyclosporine	May affect cyclosporine level. Take 4 hours before colesevelam or as directed.
Glyburide	May affect glyburide level. Take 4 hours before colesevelam or as directed.
Oral contraceptives*	May affect oral contraceptive level. Take 4 hours before colesevelam or as directed.
Phenytoin	May affect phenytoin level. Take 4 hours before colesevelam or as directed.
Thyroid hormones*	May affect thyroid hormone level. Take 4 hours before colesevelam or as directed.
Warfarin	May affect warfarin level. Take 4 hours before colesevelam or as directed.

POSSIBLE INTERACTION WITH OTHER SUBSTANCES

INTERACTS WITH	COMBINED EFFECT
Alcohol:	Unknown effect Avoid.
Beverages:	None expected.
Cocaine:	Unknown. Avoid.
Foods:	None expected.
Marijuana:	Consult doctor.
Tobacco:	None expected.

***See Glossary**

COLESTIPOL

BRAND NAMES

Colestid

BASIC INFORMATION

Habit forming? No
Prescription needed? Yes
Available as generic? Yes
Drug class: Antihyperlipidemic

USES

- Reduces cholesterol level in blood in patients with type IIa hyperlipidemia.
- Treats overdose of digitalis.
- Reduces skin itching associated with some forms of liver disease.
- Treats diarrhea after some surgical operations.
- Treatment of one form of colitis (rare).

DOSAGE & USAGE INFORMATION

How to take:
Oral suspension—Mix well with 6 ounces or more or water or liquid, or in soups, pulpy fruits, with milk or in cereals. Will not dissolve.

When to take:
- Before meals.
- If taking other medicine, take it 1 hour before or 4 to 6 hours after taking colestipol.

If you forget a dose:
Take as soon as you remember. If it is almost time for the next dose, wait for that dose (don't double this dose) and resume regular schedule.

What drug does:
Binds with bile acids in intestines, preventing reabsorption.

Time lapse before drug works:
3 to 12 months.

Don't take with:
Any other medicine or any dietary supplement without consulting your doctor or pharmacist.

OVERDOSE

SYMPTOMS:
Stomach pain, constipation.
WHAT TO DO:
If person uses much larger amount than prescribed or if accidentally swallowed, call poison control center 1-800-222-1222 for instructions or dial 911 (emergency) for help.

POSSIBLE ADVERSE REACTIONS OR SIDE EFFECTS

SYMPTOMS	WHAT TO DO
Life-threatening:	
None expected.	
Common:	
None expected.	
Infrequent:	
• Black, tarry stools from gastrointestinal bleeding.	Discontinue. Seek emergency treatment.
• Severe abdominal pain.	Discontinue. Call doctor right away.
• Constipation, belching, diarrhea, nausea, unexpected weight loss.	Continue. Call doctor when convenient.
Rare:	
Hives, skin rash, hiccups.	Discontinue. Call doctor right away.

WARNINGS & PRECAUTIONS

Don't take if:
You are allergic to colestipol. Allergic reactions are rare, but may occur.

Before you start, consult your doctor if:
- You have liver disease such as cirrhosis.
- You are jaundiced.
- You will have surgery within 2 months, including dental surgery, requiring general or spinal anesthesia.
- You are constipated.
- You have peptic ulcer.
- You have coronary artery disease.

Over age 60:
Constipation more likely. Other adverse effects more likely.

Pregnancy:
Decide with your doctor if drug benefits justify risk to unborn baby. Risk category not designated (see page xviii).

Breastfeeding; Lactation; Nursing Mothers:
This drug is generally acceptable with breastfeeding. Discuss risks and benefits with your doctor.

Infants & children up to age 18:
Approved for children over age 2. Follow instructions provided by your child's doctor.

Prolonged use:
- Request lab studies to determine serum cholesterol and serum triglycerides.
- May decrease absorption of folic acid.

Skin & sunlight:
No problems expected.

Driving, piloting or hazardous work:
No problems expected.

Discontinuing:
Don't discontinue without consulting doctor. Dose may require gradual reduction if you have taken drug for a long time. Doses of other drugs may also require adjustment, particularly digitalis.

Others:
- This medicine does not cure disorders, but helps to control them.
- May interfere with the accuracy of some medical tests.
- Advise any doctor, dentist or pharmacist whom you consult that you take this drug.

POSSIBLE INTERACTION WITH OTHER DRUGS

GENERIC NAME OR DRUG CLASS	COMBINED EFFECT
Anticoagulants, oral*	Decreased anticoagulant effect.
Beta carotene	Decreased absorption of beta carotene.
Dexfenfluramine	May require dosage change as weight loss occurs.
Dextrothyroxine	Decreased dextrothyroxine effect.
Digitalis preparations*	Decreased absorption of digitalis preparations.
Diuretics, thiazide*	Decreased absorption of thiazide diuretics.
Penicillins*	Decreased absorption of penicillins.
Tetracyclines*	Decreased absorption of tetracyclines.
Thiazides*	Decreased absorption of colestipol.
Thyroid hormones*	Decreased thyroid effect.
Trimethoprim	Decreased absorption of colestipol.
Ursodiol	Decreased absorption of ursodiol.
Vancomycin	Increased chance of hearing loss or kidney damage. Decreased therapeutic effect of vancomycin.
Vitamins	Decreased absorption of fat-soluble vitamins (A,D,E,K).
Other medicines	May delay or reduce absorption.

POSSIBLE INTERACTION WITH OTHER SUBSTANCES

INTERACTS WITH	COMBINED EFFECT
Alcohol:	None expected.
Beverages:	None expected.
Cocaine:	Unknown. Avoid.
Foods:	Interferes with absorption of vitamins. Take supplements.
Marijuana:	Consult doctor.
Tobacco:	None expected.

***See Glossary**

COMT INHIBITORS

GENERIC AND BRAND NAMES

ENTACAPONE
- Comtan
- Stalevo

TOLCAPONE
- Tasmar

BASIC INFORMATION

Habit forming? No
Prescription needed? Yes
Available as generic? Yes (entacapone)
Drug class: Antiparkinsonism

USES

Used in combination with levodopa and carbidopa for the treatment of the symptoms of Parkinson's disease.

DOSAGE & USAGE INFORMATION

How to take:
Tablet—Swallow whole with water. Do not crush or chew tablet before swallowing. May be taken with or without food.

When to take:
At the same times each day as directed by your doctor.

If you forget a dose:
Take as soon as you remember. If it is almost time for the next dose, wait for that dose (don't double this dose) and resume regular schedule.

What drug does:
It inhibits the action of an enzyme called catechol O-methyltransferase (COMT). This action helps prolong the effect of levodopa thereby increasing its effectiveness. The drug has no effect on Parkinson's without taking levodopa.

Time lapse before drug works:
Up to two hours. May take several weeks to get desired effect

Don't take with:
Any other medicine or any dietary supplement without consulting your doctor or pharmacist.

OVERDOSE

SYMPTOMS:
Nausea, vomiting, loose stools and dizziness.
WHAT TO DO:
If person uses much larger amount than prescribed or if accidentally swallowed, call poison control center 1-800-222-1222 for instructions or dial 911 (emergency) for help.

POSSIBLE ADVERSE REACTIONS OR SIDE EFFECTS

SYMPTOMS	WHAT TO DO
Life-threatening: None expected.	
Common:	
• Stomach pain, body movements that can't be controlled (new or worsening), nausea or vomiting, twitching, fainting, hyperactivity, headache, diarrhea, hallucinations, sleep problems, feeling dizzy or lightheaded (e.g., when rising after sitting or lying down), drowsiness, infection (cough, fever, runny nose, congestion, sore throat, sneezing).	Continue, but call doctor right away.
• Constipation, fatigue, vivid dreams, excess sweating, dry mouth, muscle cramps.	Continue. Call doctor when convenient.
• Change in urine color to bright yellow or brownish-orange.	No action necessary.
Infrequent:	
• Chest pain, falling, loss of balance control, difficulty breathing, blood in urine, confusion, high fever, lower back or side pain painful or difficult urination, unusual weakness.	Continue, but call doctor right away.
• Heartburn, gas, emotional changes.	Continue. Call doctor when convenient.
Rare:	
• Liver problem (may have yellow skin or eyes, fatigue, light-color stools, appetite loss, dark urine, right-side stomach pain).	Discontinue. Call doctor right away.
• Low blood pressure, muscle stiffness or pain, skin changes, unusual urges (e.g., sexual or gambling), other new symptoms.	Continue, but call doctor right away.

WARNINGS & PRECAUTIONS

Don't take if:
You are allergic to entacapone or tolcapone. Allergic reactions are rare, but may occur.

Before you start, consult your doctor if:
- You have any liver problem, alcoholism or have a history of abnormal liver function tests.
- You have kidney problems.
- You have low blood pressure or orthostatic hypotension (lightheadedness or dizziness when rising quickly after sitting or lying down).
- You have uncontrolled muscle movements.
- You suffer from hallucinations.

Over age 60:
May be more at risk for certain side effects such as hallucinations, confusion or drowsiness.

Pregnancy:
Decide with your doctor if drug benefits justify risk to unborn baby. Risk category C (see page xviii).

Breastfeeding; Lactation; Nursing Mothers:
Drugs in this group are generally not recommended with breastfeeding. Discuss risks and benefits with your doctor.

Infants & children up to age 18:
Drug is not used in this age group.

Prolonged use:
See your doctor on a regular basis while you take this drug. You may need frequent liver-function blood tests and skin exams.

Skin & sunlight:
No problems expected.

Driving, piloting or hazardous work:
Don't drive or pilot aircraft until you learn how medicine affects you. Don't work around dangerous machinery. Don't climb ladders or work in high places. Danger increases if you drink alcohol or take medicine affecting alertness and reflexes.

Discontinuing:
Don't stop taking drug suddenly. Your doctor may reduce (taper) the dose gradually to avoid withdrawal reaction (e.g., worsening of Parkinson's or fever and confusion). Other drugs you take may require a dosage change.

Others:
- Tolcapone is usually withdrawn if it shows no benefits after 3 weeks of use. Use of this drug may cause severe and fatal liver problems. Ask your doctor about your risks. While taking this drug, consult your doctor if you develop symptoms of liver problems listed under Possible Adverse Reactions or Side Effects.
- Advise any doctor, dentist or pharmacist whom you consult that you take this drug.

POSSIBLE INTERACTION WITH OTHER DRUGS

GENERIC NAME OR DRUG CLASS	COMBINED EFFECT
Apomorphine	Increased risk of side effects.
CNS depressants*	Increased risk of side effects.
Methyldopa	Increased risk of side effects.
Monoamine oxidase (MAO) inhibitors*	Serious adverse effects with some MAO inhibitors. Consult doctor.

POSSIBLE INTERACTION WITH OTHER SUBSTANCES

INTERACTS WITH	COMBINED EFFECT
Alcohol:	Increased risk of side effects. Avoid.
Beverages:	None expected.
Cocaine:	Increased risk of side effects. Avoid.
Foods:	None expected.
Marijuana:	Increased risk of side effects. Avoid.
Tobacco:	None expected.

CONDYLOMA ACUMINATUM AGENTS

GENERIC AND BRAND NAMES

IMIQUIMOD
Aldara
Zyclara

PODOFILOX
Condylox

BASIC INFORMATION

Habit forming? No
Prescription needed? Yes
Available as generic? Yes, for some
Drug class: Cytotoxic (topical)

USES

- Treatment for condylomata acuminata (external genital and perianal warts).
- Treatment of superficial basal cell carcinoma.
- Treatment for actinic keratoses.
- May be used for other skin disorders as prescribed by your doctor.

DOSAGE & USAGE INFORMATION

How to use:
For all treatments—Always follow instructions provided with prescription. Wash hands before and after applying medicine. Let the solution dry before allowing other skin surfaces to touch the treated area.
- Imiquimod—Apply a thin film of cream to the wart and rub in well. Leave cream on the treated skin 6 to 10 hours, then wash with soap and water.

Continued next column

- Podofilox—Apply drug to warts with cotton applicator (supplied with drug). Allow drug to remain on warts for 1 to 6 hours, then remove with soap and water.

When to use:
- Imiquimod cream—Once every other day 3 times a week. Apply at bedtime for up to 16 weeks.
- Podofilox topical solution—Apply twice a day (12 hours apart) for 3 consecutive days, then discontinue use for 4 consecutive days. May repeat this cycle of treatment 4 times (4 weeks).

If you forget a dose:
Use as soon as you remember. If it is almost time for the next dose, wait for that dose (don't double this dose) and resume regular schedule.

What drug does:
Kills cells and erodes tissue.

Time lapse before drug works:
Several weeks.

Don't use with:
Any other topical medicine without consulting your doctor or pharmacist.

OVERDOSE

SYMPTOMS:
- **Imiquimod: Overdose unlikely. It may increase adverse skin reactions.**
- **Podofilox: The following symptoms may occur when the body absorbs too much—painful urination, breathing difficulty, dizziness, severe nausea or vomiting, fever, heartbeat irregularity, numbness and tingling of hands and feet, abdominal pain, excitement, irritability, seizures.**

WHAT TO DO:
- **Dial 911 for all medical emergencies or call poison control center 1-800-222-1222 for instructions.**
- **See emergency information on last 3 pages of this book.**

POSSIBLE ADVERSE REACTIONS OR SIDE EFFECTS

SYMPTOMS	WHAT TO DO
Life-threatening: None expected.	
Common: Mild skin reactions (slight stinging, mild redness, tenderness or slight swelling in treated area).	No action necessary. If they continue, call doctor.
Infrequent: Skin rash, burning, red skin, severe skin reaction. If too much of drug absorbed into body: hallucinations, diarrhea, nausea and vomiting, unusual bleeding, fever, muscle pain, flu-like symptoms.	Discontinue. Call doctor right away.
Rare: Lightening of normal skin at application site.	Continue. Call doctor when convenient.

WARNINGS & PRECAUTIONS

Don't use if:
- Warts are crumbled and bleeding.
- If warts have just been biopsied or had other surgery performed on them.
- If you are allergic to condyloma acuminatum agents. Allergic reactions are rare, but may occur.

Before you start, consult your doctor if:
You have used one of these drugs previously and have a new outbreak of warts.

Over age 60:
No special problems expected.

Pregnancy:
Decide with your doctor if drug benefits justify risk to unborn baby. Risk category C (see page xviii).

Breastfeeding; Lactation; Nursing Mothers:
Drugs in this group are generally acceptable with breastfeeding. Discuss risks and benefits with your doctor.

Infants & children up to age 18:
Follow instructions provided by your child's doctor.

Prolonged use:
Increased risk of adverse reactions.

Skin & sunlight:
No problems expected.

Driving, piloting or hazardous work:
No problems expected.

Discontinuing:
No problems expected. Discontinue medicine once warts are healed.

Others:
- Advise any doctor whom you consult that you are using this medicine.
- Keep medicine away from unaffected skin, eyes, nose and mouth. Wash hands before and after using.
- Don't bandage or cover the treated warts with material that is occlusive. If covering is needed, use cotton gauze or cotton underwear.
- Don't use medicine on moles or birthmarks.
- Don't use near heat or open flame.
- Do not apply more of the medicine than prescribed. It will increase risk of side effects.
- Avoid sexual contact while medicine is on the warts.
- Imiquimod may weaken contraceptive devices such as cervical caps, condoms and diaphragms and reduce their contraceptive effect.
- Recurrence of genital warts is common after treatment.

POSSIBLE INTERACTION WITH OTHER DRUGS

GENERIC NAME OR DRUG CLASS	COMBINED EFFECT
Other topical medicines used in same skin area.	Increases risk of side effects. Avoid.

POSSIBLE INTERACTION WITH OTHER SUBSTANCES

INTERACTS WITH	COMBINED EFFECT
Alcohol:	None expected.
Beverages:	None expected.
Cocaine:	None expected.
Foods:	None expected.
Marijuana:	None expected.
Tobacco:	None expected.

*See Glossary

CONTRACEPTIVES, ORAL & SKIN

GENERIC AND BRAND NAMES

See full list of generic and brand names in the *Generic and Brand Name Directory*, page 856.

BASIC INFORMATION

Habit forming? No
Prescription needed? Yes
Available as generic? Yes, for some
Drug class: Female sex hormone, contraceptive (oral & skin)

USES

- Prevents pregnancy.
- Helps regulate menstrual periods.
- Treats premenstrual dysphoric disorder.
- May be used to treat acne vulgaris in females.

DOSAGE & USAGE INFORMATION

How to take:
- Tablet—Swallow with liquid or food.
- Chewable tablet—May be swallowed whole or chewed and swallowed. If pill is chewed, drink a full glass of liquid right afterwards.
- Extended-cycle tablet—Take as directed.
- Skin patch—Follow instructions on package.

When to take:
- Tablet—At same time each day, usually for 21 days of 28-day cycle.
- Other forms—Take or use as directed.

If you forget a dose:
Follow instructions provided with your product. May want to call doctor for advice about other protection against pregnancy.

What drug does:
Blocks release of eggs from the ovaries. Alters cervix mucus to resist sperm entry. Alters uterus lining to resist implantation of fertilized egg.

Time lapse before drug works:
10 days or more to provide contraception.

Don't take with:
Any other medicine or any dietary supplement without consulting your doctor or pharmacist.

OVERDOSE

SYMPTOMS:
Drowsiness, nausea, vomiting, bleeding.
WHAT TO DO:
If person uses much larger amount than prescribed or if accidentally swallowed, call poison control center 1-800-222-1222 for instructions or dial 911 (emergency) for help.

POSSIBLE ADVERSE REACTIONS OR SIDE EFFECTS

SYMPTOMS	WHAT TO DO
Life-threatening:	
Blood clot (sudden and severe pain in leg or chest or stomach or groin, severe headache, shortness of breath, coughing blood, sudden weakness or numbness or vision changes or slurring speech).	Seek emergency treatment immediately.
Common:	
• Menstrual bleeding changes (during first 3 months of drug use).	Continue, but call doctor right away.
• Cramping or bloating, breast tenderness or swelling, dizziness, fluid retention, acne, nausea or vomiting, feeling tired or weak.	Continue. Call doctor when convenient.
Infrequent:	
• Headache or migraine (increase in number), vaginal discharge or itching or irritation, high blood pressure, higher blood sugar (nausea, sweating, pale skin, faintness).	Continue, but call doctor right away.
• Brown blotches on skin, gain or loss of body hair, skin sensitive to sun, weight gain or loss, increase or decrease in sexual desire, dizziness.	Continue. Call doctor when convenient.
Rare:	
• Breast lumps, pain in stomach or side, yellow eyes or skin, pain or swelling in upper abdomen, depression.	Continue, but call doctor right away.
• Irritated skin from patch, swelling or bleeding gums, increased or decreased appetite, problem wearing contact lenses, other unexplained symptoms.	Continue. Call doctor when convenient.

WARNINGS & PRECAUTIONS

Don't take if:

- You are allergic to any female hormone. Allergic reactions are rare, but may occur.
- You have cancer of breast, uterus or ovaries.
- You have or have had a stroke, a history of (or risk of) blood clots, sickle-cell disease, heart attack or heart disease, blood circulation problems, jaundice, liver disease, adrenal problems (Yasmin brand), or unexplained vaginal bleeding.

Before you start, consult your doctor if:

- You have or have had benign breast problems or family history of breast cancer, medical problems in pregnancy, epilepsy, asthma, migraines or other headaches, kidney or gall-bladder disease, high cholesterol, high blood pressure, diabetes or other medical disorders.
- You will have surgery in the near future.
- You are over age 35 or smoke cigarettes.

Over age 60:
Not used in this age group.

Pregnancy:
Stop drug at first sign of pregnancy. Consult doctor. Drug should not be used during pregnancy. Risk category X (see page xviii).

Breastfeeding; Lactation; Nursing Mothers:
Drugs in this group are generally acceptable with breastfeeding. Discuss risks and benefits with your doctor.

Infants & children up to age 18:
May be used for birth control in teenage females.

Prolonged use:

- Possibly cause gallstones or gradual blood pressure rise and possible difficulty becoming pregnant after discontinuing.
- Talk to your doctor about the need for follow-up medical exams or lab studies.

Skin & sunlight:
One or more drugs in this group may cause rash or intensify sunburn in areas exposed to sun or ultraviolet light (photosensitivity reaction). Avoid overexposure. Notify doctor if reaction occurs.

Driving, piloting or hazardous work:
No problems expected.

Discontinuing:

- Use another form of birth control if you want to avoid unintended pregnancy.
- Fertility returns rapidly after discontinuing, but there may be a delay in getting pregnant.

Others:

- Failure to take drug for 1 day may reduce birth control effect. Use backup type of birth control.
- Drospirenone (a progestin)-containing pills have higher blood clot risk than other pills with progestins. Consult doctor.
- Advise any doctor or pharmacist you consult that you take this drug.
- Risk of severe adverse effects is higher in smokers and in women using the skin patch.
- Use of the skin patch exposes you to about 60% more estrogen than typical birth control pills and puts you at higher risk of blood clots and other serious side effects. Consult doctor.

POSSIBLE INTERACTION WITH OTHER DRUGS

GENERIC NAME OR DRUG CLASS	COMBINED EFFECT
Ampicillin	Decreased contraceptive effect.
Antibacterials*	Decreased contraceptive effect.
Anticoagulants*	Decreased anticoagulant effect.
Anticonvulsants, hydantoin*	Decreased contraceptive effect.
Antidepressants, tricyclic*	Increased toxicity of antidepressants.
Antidiabetics* oral	Decreased antidiabetic effect.
Antifibrinolytic agents*	Increased possibility of blood clotting.
Anti-inflammatory drugs nonsteroidal (NSAIDs)*	Decreased contraceptive effect.
Antihistamines*	Decreased contraceptive effect.
Barbiturates*	Decreased contraceptive effect.
Chloramphenicol	Decreased contraceptive effect.

Continued on page 897

POSSIBLE INTERACTION WITH OTHER SUBSTANCES

INTERACTS WITH	COMBINED EFFECT
Alcohol:	Stomach ulcer risk.
Beverages:	None expected.
Cocaine:	Unknown. Avoid.
Foods:	None expected.
Marijuana:	Consult doctor.
Tobacco:	Risk of heart attack, blood clots and stroke. *Don't smoke.*

***See Glossary**

CONTRACEPTIVES, VAGINAL

GENERIC AND BRAND NAMES

See full list of generic and brand names in the *Generic and Brand Name Directory,* page 858.

BASIC INFORMATION

Habit forming? No
Prescription needed? Not for most
Available as generic? Many are available
Drug class: Contraceptive (vaginal)

USES

- Provides a degree of protection against pregnancy.
- Spermicides used alone are not recommended for prevention of sexually transmitted diseases including human immunodeficiency virus (HIV).

DOSAGE & USAGE INFORMATION

How to take:
Read package insert carefully. Some cautions to remember:
- Do not douche for 6 to 8 hours after intercourse.
- Do not remove sponge, cervical cap or diaphragm for 6 to 8 hours after intercourse.
- Follow product label instructions for storage.
- Plastic ring device—Follow special patient brochure instructions.

When to take:
For barrier forms, use consistently with every sexual exposure. The plastic ring device is replaced every month.

If you forget to use:
Follow instructions provided with the product.

Continued next column

OVERDOSE

SYMPTOMS:
Unknown.
WHAT TO DO:
If person uses much larger amount than prescribed or if accidentally swallowed, call poison control center 1-800-222-1222 for instructions or dial 911 (emergency) for help.

What drug does:
- Spermicides form a chemical barrier between sperm in semen and the mucous membranes in the vagina. The chemical acts to inactivate viable sperm and also kills some bacteria, viruses, yeast and fungus.
- The plastic ring device releases hormones that go into the bloodstream and provide the same protection as birth control pills.

Time lapse before vaginal contraceptive works:
- Immediate for foam, gels, jellies and sponges.
- 5 to 15 minutes for film and suppositories.
- Plastic ring device takes 7 days to be effective. Use another form of birth control during that time.

Don't use with:
Other vaginal products without consulting your doctor or pharmacist.

POSSIBLE ADVERSE REACTIONS OR SIDE EFFECTS

SYMPTOMS	WHAT TO DO
Life-threatening:	
Toxic shock syndrome —chills, fever, skin rash, muscle aches, extreme weakness, confusion, redness (of vagina, inside of mouth, nose, throat or eyes). (Very rare.)	Seek emergency treatment immediately.
Common:	
None expected.	
Infrequent:	
None expected.	
Rare:	
Vaginal discharge or irritation or rash, painful urination, cloudy or bloody urine.	Discontinue. Call doctor right away.

WARNINGS & PRECAUTIONS

Don't take if:
You are allergic to any form of octoxynol, nonoxynol or benzalkonium chloride or female hormones. Allergic reactions are rare, but may occur.

Before you start, consult your doctor if:
You desire complete protection against pregnancy. A combination of methods gives better protection than vaginal contraceptives alone.

Over age 60:
Not used in this age group.

Pregnancy:
Risk factor not designated. See categories on page xviii and consult doctor.

Breastfeeding; Lactation; Nursing Mothers:
Drugs in this group are generally acceptable with breastfeeding. Discuss risks and benefits with your doctor.

Infants & children up to age 18:
Consult doctor about use in teenage females.

Prolonged use:
Allergic reactions and irritation more likely.

Skin & sunlight:
No problems expected.

Driving, piloting or hazardous work:
No special problems expected.

Discontinuing:
No special problems expected.

Others:
- Failure rate when used alone is relatively high. Therefore a vaginal cream, sponge, suppository, foam, gel, jelly or other product should be used with a mechanical barrier, such as a condom, cervical cap, vaginal diaphragm or other form of pregnancy protection.
- Spermicidal products containing the chemical ingredient nonoxynol 9 (N9) do not provide protection against infection from HIV or other sexually transmitted diseases.
- Nonoxynol 9 (N9) in stand-alone vaginal contraceptives and spermicides can irritate the vagina and rectum, which may increase the risk of contracting HIV/AIDS from an infected partner.
- Don't use a cervical cap, sponge or diaphragm during menstruation. Consider using a condom instead if additional protection is desired.

POSSIBLE INTERACTION WITH OTHER DRUGS

GENERIC NAME OR DRUG CLASS	COMBINED EFFECT
Topical vaginal medications that include any of the following: sulfa drugs, soaps or disinfectants, nitrates,* permanganates, lanolin, hydrogen peroxide, iodides, cotton dressings, aluminum citrates, salicylates*	Spermicidal activity may be reduced or negated. Avoid combinations.
Vaginal douche products	May prevent spermicidal effect. Avoid until 8 hours following intercourse.

POSSIBLE INTERACTION WITH OTHER SUBSTANCES

INTERACTS WITH	COMBINED EFFECT
Alcohol:	None expected.
Beverages:	None expected.
Cocaine:	None expected.
Foods:	None expected.
Marijuana:	None expected.
Tobacco:	None expected.

***See Glossary**

CROMOLYN

BRAND NAMES

Crolom
Gastrocrom
Novo-Cromolyn
Opticrom
PMS-Sodium Cromoglycate

BASIC INFORMATION

Habit forming? No
Prescription needed? Yes, for some
Available as generic? Yes, for some
Drug class: Nasal decongestant, anti-inflammatory (nonsteroidal)

USES

- Powdered form and nebulizer solution prevent asthma attacks. Will not stop an active asthma attack.
- Eye drops treat inflammation of covering to eye and cornea.
- Nasal spray reduces nasal allergic symptoms.

DOSAGE & USAGE INFORMATION

How to take:
Inhaler
Follow instructions enclosed with inhaler. Don't swallow cartridges for inhaler. Gargle and rinse mouth after inhalations.
Eye drops
- Wash hands.
- Apply pressure to inside corner of eye with middle finger.
- Continue pressure for 1 minute after placing medicine in eye.
- Tilt head backward. Pull lower lid away from eye with index finger of the same hand.
- Drop eye drops into pouch and close eye. Don't blink.
- Keep eyes closed for 1 to 2 minutes.
- Don't touch applicator tip to any surface (including the eye). If you accidentally touch tip, clean with warm water and soap.

Continued next column

- Keep container tightly closed.
- Keep cool, but don't freeze.
- Wash hands immediately after using.

Nasal solution
Follow prescription instructions.

When to take:
At the same times each day. If you also use a bronchodilator inhaler, use the bronchodilator before the cromolyn.

If you forget a dose:
Take as soon as you remember. If it is almost time for the next dose, wait for that dose (don't double this dose) and resume regular schedule.

What drug does:
Blocks histamine release from mast cells.

Time lapse before drug works:
- For inhaler forms: 4 weeks for prevention of asthma attacks. However, if taken 10-15 minutes before exercise or exposure to known allergens, may prevent wheezing.
- 1 to 2 weeks for nasal symptoms; only a few days for eye symptoms.

Don't take with:
Any other medicine or any dietary supplement without consulting your doctor or pharmacist.

OVERDOSE

SYMPTOMS:
Increased side effects and adverse reactions listed.
WHAT TO DO:
If person uses much larger amount than prescribed or if accidentally swallowed, call poison control center 1-800-222-1222 for instructions or dial 911 (emergency) for help.

POSSIBLE ADVERSE REACTIONS OR SIDE EFFECTS

SYMPTOMS	WHAT TO DO
Life-threatening:	
Rare, severe allergic reaction (anaphylaxis)—difficulty breathing, hives, itching, swelling (throat, face), vomiting, dizziness.	Seek emergency treatment immediately.
Common:	
Inhaler—cough, stuffy nose, dry mouth or throat; nasal—burning or stinging inside nose, increased sneezing; oral—diarrhea, headache.	Continue. Call doctor when convenient.
Infrequent:	
Inhaler—hoarseness, watery eyes; nasal—headache, bad taste, postnasal drip; oral—stomach pain, nausea, insomnia, rash.	Continue. Call doctor when convenient.

Rare:

Inhaler—rash, hives, swallowing difficulty, increased wheezing, joint pain or swelling, weakness, muscle pain, difficult or painful urination, difficulty breathing; nasal—difficulty swallowing, hives, itching, skin rash, facial swelling, wheezing, nosebleed; oral—cough, difficulty swallowing, facial swelling, wheezing, breathing difficulty.	Discontinue. Call doctor right away.

WARNINGS & PRECAUTIONS

Don't take if:
You are allergic to cromolyn.

Before you start, consult your doctor if:
You have kidney or liver disease.

Over age 60:
No special problems expected.

Pregnancy:
Decide with your doctor if drug benefits justify risk to unborn baby. Risk category B (see page xviii).

Breastfeeding; Lactation; Nursing Mothers:
This drug is generally acceptable with breastfeeding. Discuss risks and benefits with your doctor.

Infants & children up to age 18:
Approved for use in children over age 2, Follow instructions provided by your child's doctor.

Prolonged use:
Consult doctor on a regular basis while using this drug.

Skin & sunlight:
No problems expected.

Driving, piloting or hazardous work:
No problems expected.

Discontinuing:
Don't discontinue without doctor's approval if drug is used to prevent asthma symptoms. Dosages of other drugs may need to be adjusted.

Others:
- Inhaler must be cleaned and work well for drug to be effective.
- Treatment with inhalation cromolyn does not stop an acute asthma attack and may aggravate it.
- Advise any doctor, dentist or pharmacist whom you consult about the use of this drug.
- Be sure you and the doctor discuss benefits and risks of this drug before starting.
- Call doctor if symptoms worsen or new symptoms develop with use of this medicine.
- Wear a medical identification that indicates the use of this medicine.
- Use medicine only as directed. Don't increase or decrease dosage without doctor's approval.

POSSIBLE INTERACTION WITH OTHER DRUGS

GENERIC NAME OR DRUG CLASS	COMBINED EFFECT
None significant.	

POSSIBLE INTERACTION WITH OTHER SUBSTANCES

INTERACTS WITH	COMBINED EFFECT
Alcohol:	None expected.
Beverages:	None expected.
Cocaine:	Unknown. Avoid.
Foods:	None expected.
Marijuana:	Consult doctor.
Tobacco:	None expected, but tobacco smoke aggravates asthma and eye irritation. Avoid.

*See Glossary

CYCLOBENZAPRINE

BRAND NAMES

Amrix
Cycoflex
Flexeril

BASIC INFORMATION

Habit forming? No
Prescription needed? Yes
Available as generic? Yes
Drug class: Muscle relaxant

USES

Treatment for pain and limited motion caused by spasms in voluntary muscles

DOSAGE & USAGE INFORMATION

How to take:

- Tablet—Swallow with liquid. Take with food if stomach upset occurs.
- Extended-release tablet—Swallow whole with liquid.

When to take:
Take as directed on label.

If you forget a dose:
Take as soon as you remember. If it is almost time for the next dose, wait for that dose (don't double this dose) and resume regular schedule.

What drug does:
Blocks body's pain messages to brain.

Time lapse before drug works:
30 to 60 minutes.

Don't take with:
Any other medicine or any dietary supplement without consulting your doctor or pharmacist.

OVERDOSE

SYMPTOMS:
Drowsiness, confusion, shaking or trembling, visual problems, vomiting, blood pressure drop, low body temperature, fast heartbeat, convulsions, coma.
WHAT TO DO:

- **Dial 911 for all medical emergencies or call poison control center 1-800-222-1222 for instructions.**
- **See emergency information on last 3 pages of this book.**

POSSIBLE ADVERSE REACTIONS OR SIDE EFFECTS

SYMPTOMS	WHAT TO DO
Life-threatening:	
Rare, severe allergic reaction (anaphylaxis)—difficulty breathing, hives, itching, swelling (throat, face), vomiting, dizziness.	Seek emergency treatment immediately.
Common:	
Drowsiness, dizziness, dry mouth.	Continue. Call doctor when convenient.
Infrequent:	
• Blurred vision, fast heartbeat.	Discontinue. Call doctor right away.
• Insomnia, numbness in extremities, bad taste in mouth, fatigue, nausea, sweating.	Continue. Call doctor when convenient.
Rare:	
• Unsteadiness, confusion, rash, itch, depression, hallucinations, swelling, breathing difficulty.	Discontinue. Call doctor right away.
• Difficult urination.	Continue. Call doctor when convenient.

WARNINGS & PRECAUTIONS

Don't take if:

- You are allergic to any skeletal muscle relaxant.*
- You have taken a monoamine oxidase (MAO) inhibitor* in last 2 weeks.
- You have had a heart attack within 6 weeks, or suffer from congestive heart failure.
- You have an overactive thyroid.

Before you start, consult your doctor if:

- You have a heart problem.
- You have reacted to tricyclic antidepressants.
- You have glaucoma.
- You have a prostate condition and urination difficulty.
- You intend to pilot aircraft.

Over age 60:
Adverse reactions and side effects may be more frequent and severe than in younger persons.

Pregnancy:
Decide with your doctor if drug benefits justify risk to unborn baby. Risk category B (see page xviii).

Breastfeeding; Lactation; Nursing Mothers:
This drug is generally acceptable with breastfeeding. Discuss risks and benefits with your doctor.

Infants & children up to age 18:
Don't use for children younger than 15. Follow instructions provided by your child's doctor.

Prolonged use:
Do not take for longer than 2 to 3 weeks.

Skin & sunlight:
No special problems expected.

Driving, piloting or hazardous work:
Don't drive or pilot aircraft until you learn how medicine affects you. Don't work around dangerous machinery. Don't climb ladders or work in high places. Danger increases if you drink alcohol or take medicine affecting alertness and reflexes.

Discontinuing:
May be unnecessary to finish medicine. Follow doctor's instructions.

Others:
Advise any doctor, dentist or pharmacist whom you consult that you take this medicine.

POSSIBLE INTERACTION WITH OTHER DRUGS

GENERIC NAME OR DRUG CLASS	COMBINED EFFECT
Anticholinergics*	Increased anticholinergic effect.
Antidepressants*	Increased sedation.
Antihistamines*	Increased antihistamine effect.
Barbiturates*	Increased sedation.
Central nervous system (CNS) depressants*	Increased sedation.
Cimetidine	Possible increased cyclobenzaprine effect.
Cisapride	Decreased cyclobenzaprine effect.
Clonidine	Decreased clonidine effect.
Dronabinol	Increased effect of dronabinol on central nervous system. Avoid combination.
Guanethidine	Decreased guanethidine effect.
Methyldopa	Decreased methyldopa effect.
Mind-altering drugs*	Increased mind-altering effect.
Monoamine oxidase (MAO) inhibitors*	High fever, convulsions, possible death.
Narcotics*	Increased sedation.
Pain relievers*	Increased pain reliever effect.
Procainamide	Possible increased conduction disturbance.
Quinidine	Possible increased conduction disturbance.
Rauwolfia alkaloids*	Decreased effect of rauwolfia alkaloids.
Sedatives*	Increased sedative effect.
Sleep inducers*	Increased sedation.
Tranquilizers*	Increased tranquilizer effect.

POSSIBLE INTERACTION WITH OTHER SUBSTANCES

INTERACTS WITH	COMBINED EFFECT
Alcohol:	Depressed brain function. Avoid.
Beverages:	None expected.
Cocaine:	Increased risk of side effects. Avoid.
Foods:	None expected.
Marijuana:	Increased risk of side effects. Avoid.
Tobacco:	None expected.

CYCLOPENTOLATE (Ophthalmic)

BRAND NAMES

Ak-Pentolate
Cyclogyl
I-Pentolate
Minims Cyclopentolate
Ocu-Pentolate
Pentolair
Spectro-Pentolate

BASIC INFORMATION

Habit forming? No
Prescription needed? Yes
Available as generic? Yes
Drug class: Cycloplegic, mydriatic

USES

- Enlarges (dilates) pupil.
- Temporarily paralyzes the normal pupil accommodation to light before eye examinations and to treat some eye conditions.

DOSAGE & USAGE INFORMATION

How to use:
Eye drops
- Wash hands.
- Apply pressure to inside corner of eye with middle finger.
- Continue pressure for 1 minute after placing medicine in eye.
- Tilt head backward. Pull lower lid away from eye with index finger of the same hand.
- Drop eye drops into pouch and close eye. Don't blink.
- Keep eyes closed for 1 to 2 minutes.
- Don't touch applicator tip to any surface (including the eye). If you accidentally touch tip, clean with warm water and soap.
- Keep container tightly closed.
- Keep cool, but don't freeze.
- Wash hands immediately after using.

When to use:
As directed on bottle.

Continued next column

OVERDOSE

SYMPTOMS:
Unknown.
WHAT TO DO:
If person uses much larger amount than prescribed or if accidentally swallowed, call poison control center 1-800-222-1222 for instructions or dial 911 (emergency) for help.

If you forget a dose:
Use as soon as you remember. If it is almost time for the next dose, wait for that dose (don't double this dose) and resume regular schedule.

What drug does:
Blocks sphincter muscle of the iris and ciliary body.

Time lapse before drug works:
Within 30 to 60 minutes. Effects usually disappear in 24 hours.

Don't use with:
Other eye medicines such as carbachol, demecarium, echothiophate, isoflurophate, physostigmine, pilocarpine without doctor's approval.

POSSIBLE ADVERSE REACTIONS OR SIDE EFFECTS

SYMPTOMS	WHAT TO DO
Life-threatening:	
None expected.	
Common:	
Increased sensitivity to light, burning eyes.	Continue. Call doctor when convenient.
Infrequent:	
None expected.	
Rare:	
Extremely rare symptoms of excess drug absorbed by the body (confusion, clumsiness, fever, rash, flushed face, hallucinations, fast heartbeat, slurred speech, swollen stomach in children, drowsiness).	Discontinue. Call doctor right away.

WARNINGS & PRECAUTIONS

Don't use if:
You are allergic to cyclopentolate.

Before you start, consult your doctor if:
- Medicine is for a brain-damaged child or child with Down syndrome or child with spastic paralysis.
- You have glaucoma.

Over age 60:
May have increased risk of side effects.

Pregnancy:
Decide with your doctor if drug benefits justify risk to unborn baby. Risk category C (see page xviii).

Breastfeeding; Lactation; Nursing Mothers:
This drug is generally acceptable with breastfeeding. Discuss risks and benefits with your doctor.

Infants & children up to age 18:
- Follow instructions provided by your child's doctor.
- Children with blond hair and blue eyes may be more at risk for side effects.

Prolonged use:
Consult doctor. May increase absorption into body.

Skin & sunlight:
No special problems expected.

Driving, piloting or hazardous work:
Don't drive or pilot aircraft until you learn how medicine affects you. Don't work around dangerous machinery. Don't climb ladders or work in high places. Danger increases if you drink alcohol or take medicine affecting alertness and reflexes, such as antihistamines, tranquilizers, sedatives, pain medicine, narcotics and mind-altering drugs.

Discontinuing:
If effects last longer than 36 hours after last drops, consult doctor.

Others:
- Wear sunglasses to protect eyes from sunlight and bright light.
- Advise any doctor, dentist or pharmacist whom you consult that you use this medicine.

POSSIBLE INTERACTION WITH OTHER DRUGS

GENERIC NAME OR DRUG CLASS	COMBINED EFFECT
Antiglaucoma agents*	Decreased antiglaucoma effect.

POSSIBLE INTERACTION WITH OTHER SUBSTANCES

INTERACTS WITH	COMBINED EFFECT
Alcohol:	None expected.
Beverages:	None expected.
Cocaine:	None expected.
Foods:	None expected.
Marijuana:	None expected.
Tobacco:	None expected.

***See Glossary**

CYCLOPHOSPHAMIDE

GENERIC NAMES

CYCLOPHOSPHAMIDE

BASIC INFORMATION

Habit forming? No
Prescription needed? Yes
Available as generic? Yes
Drug class: Immunosuppressant, antineoplastic

USES

- Treatment for cancer.
- Treatment for severe rheumatoid arthritis, blood-vessel disease and for skin disease.

DOSAGE & USAGE INFORMATION

How to take:
Tablet—Swallow with liquid. If you can't swallow tablet whole, crumble tablet and take with liquid or food.

When to take:
Works best if taken first thing in morning. Should be taken on an empty stomach. However, may take with food to lessen stomach irritation. Don't take at bedtime.

If you forget a dose:
Take as soon as you remember. If it is almost time for the next dose, wait for that dose (don't double this dose) and resume regular schedule.

What drug does:
- Kills cancer cells.
- Suppresses spread of cancer cells.
- Suppresses immune system.

Time lapse before drug works:
7 to 10 days continual use.

Don't take with:
Any other medicine or any dietary supplement without consulting your doctor or pharmacist.

OVERDOSE

SYMPTOMS:
May have symptoms of an infection or various heart symptoms.
WHAT TO DO:
- **Dial 911 for all medical emergencies or call poison control center 1-800-222-1222 for instructions.**
- **See emergency information on last 3 pages of this book.**

POSSIBLE ADVERSE REACTIONS OR SIDE EFFECTS

SYMPTOMS	WHAT TO DO
Life-threatening:	
Rare, severe allergic reaction (anaphylaxis)—difficulty breathing, hives, itching, swelling (throat, face), vomiting, dizziness.	Seek emergency treatment immediately.
Common:	
• Sore throat, fever.	Continue, but call doctor right away.
• Dark skin or nails, nausea, appetite loss, vomiting, missed menstrual period.	Continue. Call doctor when convenient.
Infrequent:	
• Rash, hives, itch, shortness of breath, rapid heartbeat, foot or ankle swelling, blood in urine, painful urination, increased sweating, easy bleeding or bruising, hoarseness, unusual side pain.	Continue, but call doctor right away.
• Confusion, agitation, headache, flushed face, dizziness, stomach pain, joint pain, fatigue, weakness, diarrhea.	Continue. Call doctor when convenient.
Rare:	
• Mouth or lip sores, black stool, unusual thirst, yellow skin or eyes, blurred vision,	Continue, but call doctor right away.
• Increased urination, hair loss.	Continue. Call doctor when convenient.

WARNINGS & PRECAUTIONS

Don't take if:
You are allergic to cyclophosphamide.

Before you start, consult your doctor if:
- You have an infection or bloody urine.
- You will have surgery within 2 months, including dental surgery, requiring general or spinal anesthesia.
- You have impaired liver or kidney function.
- You have impaired bone marrow or blood cell production.
- You have had chemotherapy or x-ray therapy.
- You have taken cortisone drugs in past year.
- You plan to become pregnant.

Over age 60:
No special problems expected.

Pregnancy:
Consult doctor. Use of the drug during pregnancy is not recommended. It could cause harm to the unborn baby. Risk category D (see page xviii).

Breastfeeding; Lactation; Nursing Mothers:
This drug is generally not recommended with breastfeeding. Discuss risks and benefits with your doctor.

Infants & children up to age 18:
Follow instructions provided by your child's doctor.

Prolonged use:
- May increase risk for other cancers or fertility problems in men and women.
- Talk to your doctor about the need for follow-up medical examinations or laboratory studies.

Skin & sunlight:
No problems expected.

Driving, piloting or hazardous work:
Avoid if you feel dizzy or have blurred vision. Otherwise, no problems expected.

Discontinuing:
Don't discontinue without consulting doctor. Dose may require gradual reduction if you have taken drug for a long time. Doses of other drugs may also require adjustment.

Others:
- Frequently causes hair loss. After treatment ends, hair should grow back.
- Advise any doctor, dentist or pharmacist whom you consult about the use of this drug.
- Avoid vaccinations. Consult doctor if needed.
- You will need to drink extra fluids so you will pass more urine. Follow doctor's instructions.

POSSIBLE INTERACTION WITH OTHER DRUGS

GENERIC NAME OR DRUG CLASS	COMBINED EFFECT
Allopurinol or other medicines to treat gout	Possible anemia; decreased antigout effect.
Antidiabetics, oral*	Increased antidiabetic effect.
Bone marrow depressants,* other	Increased bone marrow depressant effect.
Clozapine	Toxic effect on bone marrow.
Cyclosporine	May increase risk of infection.
Digoxin	Possible decreased digoxin absorption.
Immuno-suppressants,* other	Increased risk of infection.
Insulin	Increased insulin effect.
Levamisole	Increased risk of bone marrow depression.
Lovastatin	Increased heart and kidney damage.
Phenobarbital	Increased cyclophosphamide effect.
Probenecid	Increased blood uric acid.
Sulfinpyrazone	Increased blood uric acid.
Tiopronin	Increased risk of toxicity to bone marrow.

POSSIBLE INTERACTION WITH OTHER SUBSTANCES

INTERACTS WITH	COMBINED EFFECT
Alcohol:	None expected.
Beverages:	None expected. Drink plenty of fluids every day.
Cocaine:	Increased risk of side effects. Avoid.
Foods:	None expected.
Marijuana:	Consult doctor.
Tobacco:	None expected.

*See Glossary

CYCLOPLEGIC, MYDRIATIC (Ophthalmic)

GENERIC AND BRAND NAMES

ATROPINE (ophthalmic)
Atropair
Atropine Care Eye Drops & Ointment
Atropine Sulfate S.O.P.
Atropisol
Atrosulf
Isopto Atropine
I-Tropine
Minims Atropine
Ocu-Tropine

HOMATROPINE (ophthalmic)
AK Homatropine
I-Homatrine
I-Homatropine
Minims Homatropine
Spectro Homatropine

BASIC INFORMATION

Habit forming? No
Prescription needed? Yes
Available as generic? Yes, for some
Drug class: Cycloplegic, mydriatic

USES

- Dilates pupil of the eye.
- Used before some eye examinations, before and after some eye surgical procedures and, rarely, to treat some eye problems such as glaucoma.

DOSAGE & USAGE INFORMATION

How to use:
Eye drops
- Wash hands.
- Apply pressure to inside corner of eye with middle finger.
- Continue pressure for 1 minute after placing medicine in eye.
- Tilt head backward. Pull lower lid away from eye with index finger of the same hand.
- Drop eye drops into pouch and close eye. Don't blink.
- Keep eyes closed for 1 to 2 minutes.

Continued next column

Eye ointment
- Wash hands.
- Pull lower lid down from eye to form a pouch.
- Squeeze tube to apply thin strip of ointment into pouch.
- Close eye for 1 to 2 minutes.
- Don't touch applicator tip to any surface (including the eye). If you accidentally touch tip, clean with warm water and soap.
- Keep container tightly closed.
- Keep cool, but don't freeze.
- Wash hands immediately after using.

When to use:
As directed on label.

If you forget a dose:
Use as soon as you remember. If it is almost time for the next dose, wait for that dose (don't double this dose) and resume regular schedule.

What drug does:
Blocks normal response to sphincter muscle of the iris of the eye and the accommodative muscle of the ciliary body.

Time lapse before drug works:
Begins within 1 minute. Residual effects may last up to 14 days.

Don't use with:
Other eye medicines such as carbachol, demecarium, echothiophate, isoflurophate, physostigmine, pilocarpine.

OVERDOSE

SYMPTOMS:
Unknown.
WHAT TO DO:
If person uses much larger amount than prescribed or if accidentally swallowed, call poison control center 1-800-222-1222 for instructions or dial 911 (emergency) for help.

POSSIBLE ADVERSE REACTIONS OR SIDE EFFECTS

SYMPTOMS	WHAT TO DO
Life-threatening:	
None expected.	
Common:	
Increased sensitivity to light, burning eyes.	Continue. Call doctor when convenient.
Infrequent:	
None expected.	
Rare:	
Extremely rare symptoms of excess drug absorbed by the body (confusion, clumsiness, fever, rash, flushed face, hallucinations, fast heartbeat, slurred speech, swollen stomach in children, drowsiness.	Discontinue. Call doctor right away.

CYCLOPLEGIC, MYDRIATIC (Ophthalmic)

WARNINGS & PRECAUTIONS

Don't use if:
You are allergic to any mydriatic cycloplegic drug.

Before you start, consult your doctor if:
- Medicine is for a brain-damaged child or child with Down syndrome or child with spastic paralysis.
- You have glaucoma.

Over age 60:
No problems expected.

Pregnancy:
Decide with your doctor if drug benefits justify risk to unborn baby. Risk category C (see page xviii).

Breastfeeding; Lactation; Nursing Mothers:
Drugs in this group are generally acceptable with breastfeeding. Discuss risks and benefits with your doctor.

Infants & children up to age 18:
Follow instructions provided by your child's doctor. Children may be more at risk for side effects.

Prolonged use:
Avoid. May increase absorption into body.

Skin & sunlight:
No special problems expected.

Driving, piloting or hazardous work:
Don't drive or pilot aircraft until you learn how medicine affects you. Don't work around dangerous machinery. Don't climb ladders or work in high places. Danger increases if you drink alcohol or take medicine affecting alertness and reflexes, such as antihistamines, tranquilizers, sedatives, pain medicine, narcotics and mind-altering drugs.

Discontinuing:
Effects may last up to 14 days later.

Others:
- Wear sunglasses to protect eyes from sunlight and bright light.
- Advise any doctor, dentist or pharmacist whom you consult that you take this drug.

POSSIBLE INTERACTION WITH OTHER DRUGS

GENERIC NAME OR DRUG CLASS	COMBINED EFFECT

Clinically significant interactions with oral or injected medicines unlikely.

POSSIBLE INTERACTION WITH OTHER SUBSTANCES

INTERACTS WITH	COMBINED EFFECT
Alcohol:	None expected.
Beverages:	None expected.
Cocaine:	None expected.
Foods:	None expected.
Marijuana:	None expected.
Tobacco:	None expected.

CYCLOSERINE

BRAND NAMES

Seromycin

BASIC INFORMATION

Habit forming? No
Prescription needed? Yes
Available as generic? No
Drug class: Antibacterial

USES

- Treats urinary tract infections.
- Treats tuberculosis.

DOSAGE & USAGE INFORMATION

How to take:
Capsule—Swallow with liquid or food to lessen stomach irritation. If you can't swallow whole, open capsule and take with liquid or food.

When to take:
- Once or twice daily as prescribed.
- At the same times each day after meals to prevent stomach irritation.

If you forget a dose:
Take as soon as you remember. If it is almost time for the next dose, wait for that dose (don't double this dose) and resume regular schedule.

What drug does:
Interferes with bacterial wall synthesis and keeps germs from multiplying.

Time lapse before drug works:
3 to 4 hours.

Don't take with:
Any other medicine or any dietary supplement without consulting your doctor or pharmacist.

OVERDOSE

SYMPTOMS:
Numbness and tingling of skin, agitation, hallucinations, confusion, seizures.
WHAT TO DO:
- **Dial 911 for all medical emergencies or call poison control center 1-800-222-1222 for instructions.**
- **See emergency information on last 3 pages of this book.**

POSSIBLE ADVERSE REACTIONS OR SIDE EFFECTS

SYMPTOMS	WHAT TO DO
Life-threatening:	
Seizures, muscle twitching or trembling.	Seek emergency treatment immediately.
Common:	
Gum inflammation, pale skin, depression, confusion, dizziness, restlessness, anxiety, nightmares, severe headache, drowsiness.	Continue, but call doctor right away.
Infrequent:	
Visual changes, skin rash, numbness or tingling or burning in hands and feet, yellow skin or eyes, eye pain.	Continue, but call doctor right away.
Rare:	
Seizures, thoughts of suicide.	Discontinue. Seek emergency treatment.

WARNINGS & PRECAUTIONS

Don't take if:
- You are allergic to cycloserine. Allergic reactions are rare, but may occur.
- You are a frequent user of alcohol.
- You have a convulsive disorder.

Before you start, consult your doctor if:
- You are depressed.
- You have kidney disease.
- You have severe anxiety.

Over age 60:
No special problems expected.

Pregnancy:
Decide with your doctor if drug benefits justify risk to unborn baby. Risk category C (see page xviii).

Breastfeeding; Lactation; Nursing Mothers:
This drug is generally acceptable with breastfeeding. Discuss risks and benefits with your doctor.

Infants & children up to age 18:
Follow instructions provided by your child's doctor.

Prolonged use:
- May cause liver or kidney damage.
- May cause anemia.

Skin & sunlight:
No special problems expected.

Driving, piloting or hazardous work:
Don't drive or pilot aircraft until you learn how medicine affects you. Don't work around dangerous machinery. Don't climb ladders or work in high places. Danger increases if you drink alcohol or take medicine affecting alertness and reflexes.

Discontinuing:
Don't discontinue without consulting doctor. Dose may require gradual reduction if you have taken drug for a long time. Doses of other drugs may also require adjustment.

Others:
- May have to take anticonvulsants, sedatives and/or pyridoxine to prevent or minimize toxic effects on the brain.
- If you must take more than 500 mg per day, toxicity is much more likely to occur.
- Talk to your doctor about taking pyridoxine as a supplement.
- Advise any doctor, dentist or pharmacist whom you consult that you take this drug.

POSSIBLE INTERACTION WITH OTHER DRUGS

GENERIC NAME OR DRUG CLASS	COMBINED EFFECT
Ethionamide	Increased risk of seizures.
Isoniazid	Increased risk of central nervous system effects.
Pyridoxine	Reduces effects of pyridoxine. Since pyridoxine is a vital vitamin, patients on cycloserine require pyridoxine supplements to prevent anemia or peripheral neuritis.

POSSIBLE INTERACTION WITH OTHER SUBSTANCES

INTERACTS WITH	COMBINED EFFECT
Alcohol:	Toxic. May increase risk of seizures. Avoid.
Beverages:	None expected.
Cocaine:	Unknown. Avoid.
Foods:	None expected.
Marijuana:	Consult doctor.
Tobacco:	None expected.

CYCLOSPORINE

BRAND NAMES

Gengraf
Neoral
Sandimmune

BASIC INFORMATION

Habit forming? No
Prescription needed? Yes
Available as generic? Yes
Drug class: Immunosuppressant

USES

- Suppresses the immune response in patients who have transplants of the heart, lung, kidney, liver, pancreas. Cyclosporine treats rejection as well as helps prevent it.
- Treats severe psoriasis and rheumatoid arthritis when regular treatment is ineffective.

DOSAGE & USAGE INFORMATION

How to take:
- Oral solution—Take after meals with liquid to decrease stomach irritation. May mix with orange or apple juice. Don't mix in Styrofoam cups. Use special dropper for exact dosage.
- Capsule—Take with water or other fluid. Don't break capsule open.

When to take:
At the same time each day, according to instructions on prescription label.

If you forget a dose:
Take as soon as you remember. If it is almost time for the next dose, wait for that dose (don't double this dose) and resume regular schedule.

What drug does:
Exact mechanism is unknown, but believed to inhibit interleuken II to affect T-lymphocytes.

Time lapse before drug works:
3 to 3-1/2 hours.

Continued next column

OVERDOSE

SYMPTOMS:
Nausea, vomiting, yellow skin and eyes, stomach pain, less urine, loss of appetite.
WHAT TO DO:
- **Dial 911 for all medical emergencies or call poison control center 1-800-222-1222 for instructions.**
- **See emergency information on last 3 pages of this book.**

Don't take with:
Any other medicine or any dietary supplement without consulting your doctor or pharmacist.

POSSIBLE ADVERSE REACTIONS OR SIDE EFFECTS

SYMPTOMS	WHAT TO DO
Life-threatening:	
None expected.	
Common:	
Gum inflammation, blood in urine, yellow skin or eyes, tremors.	Continue, but call doctor right away.
Infrequent:	
• Fever, chills, sore throat, shortness of breath.	Continue, but call doctor right away.
• Frequent urination, headache, leg cramps. increased hair growth.	Continue. Call doctor when convenient.
Rare:	
• Confusion, irregular heartbeat, numbness of hands and feet, nervousness, face flushing, severe abdominal pain, weakness.	Continue, but call doctor right away.
• Acne, headache.	Continue. Call doctor when convenient.

WARNINGS & PRECAUTIONS

Don't take if:
You are allergic to cyclosporine. Allergic reactions are rare, but may occur.

Before you start, consult your doctor if:
- You have kidney or liver disease, high blood pressure or psoriasis.
- You have or have had cancer.

Over age 60:
No special problems expected.

Pregnancy:
Decide with your doctor if drug benefits justify risk to unborn child. Risk category C (see page xviii).

Breastfeeding; Lactation; Nursing Mothers:
This drug is generally acceptable with breastfeeding. Discuss risks and benefits with your doctor.

Infants & children up to age 18:
Follow instructions provided by your child's doctor.

Prolonged use:
- Can cause reduced function of kidney.
- Talk to your doctor about the need for follow-up medical exams or lab tests.

Skin & sunlight:
No problems expected.

Driving, piloting or hazardous work:
Don't drive or pilot aircraft until you learn how medicine affects you. Don't work around dangerous machinery. Don't climb ladders or work in high places. Danger increases if you drink alcohol or take drugs affecting alertness.

Discontinuing:
Don't discontinue without consulting doctor. You probably will require this drug for life.

Others:
- Drug is available in two forms non-modified (Sandimmune, generic) and modified (Gengraf, Neoral, generic) which is better absorbed in the body. Be sure your refills are same form. Forms are not interchangeable.
- Use of the drug may increase the risk of developing an infection or cancer (e.g., lymphoma or skin cancer). Call doctor right away if new symptoms occur.
- Check blood pressure regularly. Cyclosporine sometimes causes hypertension.
- Don't store solution in the refrigerator.
- Avoid any immunizations except those specifically recommended by your doctor.
- Maintain good dental hygiene. Cyclosporine can cause gum problems.
- Kidney toxicity occurs commonly after 12 months of taking cyclosporine.
- Immunosuppressed patients are at increased risk for opportunistic infections, such as activation of latent viral infections, including BK virus-associated nephropathy.
- Don't mix drug in Styrofoam cups.

POSSIBLE INTERACTION WITH OTHER DRUGS

GENERIC NAME OR DRUG CLASS	COMBINED EFFECT
Anti-inflammatory drugs, nonsteroidal (NSAIDs)*	Increased effect of cyclosporine.
Androgens	Increased effect of cyclosporine.
Anticonvulsants*	Decreased effect of cyclosporine.
Danazol	Increased effect of cyclosporine.
Diltiazem	Increased effect of cyclosporine.
Diuretics, potassium-sparing	Increased effect of cyclosporine.
Erythromycin	Increased effect of cyclosporine.
Estrogens	Increased effect of cyclosporine.
Fluconazole	Increased effect of cyclosporine. Cyclosporine dosage must be adjusted.
HMG-CoA reductase inhibitors	Increased risk of muscle and kidney problems.
Imatinib	Increased effect of cyclosporine.
Immunosuppressants,* other	Increased risk of adverse effects.
Itraconazole	Increased cyclosporine toxicity.
Leukotriene modifiers	Increased effect of cyclosporine.
Losartan	Increased potassium levels.
Lovastatin	Increased heart and kidney damage.
Nephrotoxic drugs*	Increased risk of toxicity to kidneys.

Continued on page 897

POSSIBLE INTERACTION WITH OTHER SUBSTANCES

INTERACTS WITH	COMBINED EFFECT
Alcohol:	May increase possibility of toxic effects. Avoid.
Beverages: Grapefruit juice.	Increased cyclosporine effect.
Cocaine:	May increase possibility of toxic effects. Avoid.
Foods:	None expected.
Marijuana:	May increase possibility of toxic effects. Avoid.
Tobacco:	May increase possibility of toxic effects. Avoid.

***See Glossary**

DABIGATRAN

BRAND NAMES

Pradaxa

BASIC INFORMATION

Habit forming? No
Prescription needed? Yes
Available as generic? No
Drug class: Anticoagulant

USES

- Used for the prevention of blood clots and stroke.
- Used to treat and prevent deep vein thrombosis (DVT) and pulmonary embolism (PE).
- Other uses as recommended by your doctor.

DOSAGE & USAGE INFORMATION

How to take:
Capsule—Swallow whole with liquid. May be taken with or without food. Do not crush, chew, break open or empty the pellets from the capsule.

When to take:
Twice a day at the same times each day or as directed by your doctor.

If you forget a dose:
Take as soon as you remember. If it is almost time for the next dose, wait for that dose (don't double this dose) and resume regular schedule.

What drug does:
It blocks thrombin (a substance in the blood) that is involved in the process of blood clotting thereby reducing the risk of blood clots.

Time lapse before drug works:
One to two hours.

Continued next column

OVERDOSE

SYMPTOMS:
Bleeding problems that may be severe (e.g., vomiting blood, nosebleed, bright-red blood in stool).
WHAT TO DO:
- **Dial 911 for all medical emergencies or call poison control center 1-800-222-1222 for instructions.**
- **See emergency information on last 3 pages of this book.**

Don't take with:
Any other medicine or any dietary supplement without consulting your doctor or pharmacist.

POSSIBLE ADVERSE REACTIONS OR SIDE EFFECTS

SYMPTOMS	WHAT TO DO
Life-threatening:	
Rare, severe allergic reaction (anaphylaxis)—difficulty breathing, hives, itching, swelling (throat, face), vomiting, dizziness.	Seek emergency treatment immediately.
Common:	
• Bleeding from gums, frequent nosebleeds, vaginal bleeding, heavy menstrual period, pink or brown urine, red or black or tarry stools, coughing up blood, vomiting blood or coffee ground-like material, unusual pain or swelling, joint pain, unexpected bruising, headaches, feeling dizzy or weak.	Continue, but call doctor right away.
• Nausea, upset stomach, diarrhea, indigestion, stomach pain or burning, cuts or scrapes that take a long time to stop bleeding.	Continue. Call doctor when convenient.
Infrequent:	
None expected.	
Rare:	
Other unexpected symptoms.	Continue. Call doctor when convenient.

WARNINGS & PRECAUTIONS

Don't take if:
- You are allergic to dabigatran.
- You have mechanical heart valves.

Before you start, consult your doctor if:
- You have any kidney (renal) problems.
- You have or have had stomach ulcers.
- You have or have had any bleeding problems.
- You have plans for surgery or any invasive procedure in the next few months.
- You are allergic to any food, medicine or other substance.

Over age 60:
Risk of stroke and bleeding increases with age. Consult doctor about your risks.

Pregnancy:
Decide with your doctor if drug benefits justify risk to unborn baby. Risk category C (see page xviii).

Breastfeeding; Lactation; Nursing Mothers:
This drug is generally acceptable with breastfeeding. Discuss risks and benefits with your doctor.

Infants & children up to age 18:
Safety and effectiveness in this age group have not been established. Consult doctor.

Prolonged use:
Consult with your doctor on a regular basis while taking this drug to monitor your progress, check for side effects, check your renal (kidney) function, and for recommended lab tests.

Skin & sunlight:
No problems expected.

Driving, piloting or hazardous work:
No problems expected.

Discontinuing:
Do not stop taking drug without first talking with your doctor. Stopping drug may increase risk of stroke.

Others:
- Store capsules in original container and keep it tightly closed. Do not put capsules in a pill box or pill organizer.
- Safely throw away any unused capsules after 4 months (or as directed by doctor) and start using a new container.
- Advise any doctor, dentist or pharmacist whom you consult that you take this drug.
- Your doctor may have you stop taking drug for a short time if you plan to have surgery or a medical or a dental procedure. You will be advised when to stop and when to resume taking the drug.
- Drug lessens the ability of your blood to clot. You may bruise more easily and it may take longer for any bleeding to stop.

POSSIBLE INTERACTION WITH OTHER DRUGS

GENERIC NAME OR DRUG CLASS	COMBINED EFFECT
Anticoagulants,* other	Increased anti-coagulant effect (bleeding or bruising).
Anti-inflammatory drugs, nonsteroidal (NSAIDs)*	Increased risk of bleeding.
Mifepristone	Increased risk of vaginal bleeding.
P-glycoprotein inducers*	May decrease dabigatran effect.
P-glycoprotein inhibitors*	May increase dabigatran effect.
Platelet inhibitors	Increased risk of bleeding.

POSSIBLE INTERACTION WITH OTHER SUBSTANCES

INTERACTS WITH	COMBINED EFFECT
Alcohol:	None expected.
Beverages:	None expected.
Cocaine:	Unknown. Avoid.
Foods:	None expected.
Marijuana:	Consult doctor.
Tobacco:	No interaction, but smoking is a risk factor for blood clots. Avoid.

*See Glossary

DANAZOL

BRAND NAMES

Cyclomen
Danocrine

BASIC INFORMATION

Habit forming? No
Prescription needed? Yes
Available as generic? Yes
Drug class: Gonadotropin inhibitor

USES

Treatment of endometriosis, fibrocystic breast disease, angioneurotic edema except in pregnant women, gynecomastia, infertility, excessive menstruation, precocious puberty.

DOSAGE & USAGE INFORMATION

How to take:
Capsule—Swallow with liquid or food to lessen stomach irritation. If you can't swallow whole, open capsule and take with liquid or food.

When to take:
At the same times each day.

If you forget a dose:
Take as soon as you remember. If it is almost time for the next dose, wait for that dose (don't double this dose) and resume regular schedule.

What drug does:
Partially prevents output of pituitary follicle-stimulating hormone and leuteinizing hormone reducing estrogen production.

Time lapse before drug works:
- 2 to 3 months to treat endometriosis.
- 1 to 2 months to treat other disorders.

Don't take with:
- Birth control pills.
- Any other medicine or any dietary supplement without consulting your doctor or pharmacist.

OVERDOSE

SYMPTOMS:
Unknown.
WHAT TO DO:
If person uses much larger amount than prescribed or if accidentally swallowed, call poison control center 1-800-222-1222 for instructions or dial 911 (emergency) for help.

POSSIBLE ADVERSE REACTIONS OR SIDE EFFECTS

SYMPTOMS	WHAT TO DO
Life-threatening: None expected.	
Common: Irregular menstrual periods, decreased breast size, weight gain.	Continue. Call doctor when convenient.
Infrequent:	
• Muscle spasms, acne, dark urine, swelling of feet or legs, oily skin or hair.	Continue, but call doctor right away.
• Flushed or red skin, nervousness, mood or mental changes, vaginal burning or itching or bleeding.	Continue. Call doctor when convenient.
Rare:	
• Enlarged clitoris, decreased testicle size, men have semen changes, bleeding gums, abdomen is tender or painful, changes in vision, tingling in fingers, cough or coughing up blood, chest pain, chills, hives or skin rash, eye pain, diarrhea, trouble speaking, pale stools, loss of coordination, nosebleeds, sores in mouth or nose, shortness of breath, sore throat, weakness in legs or arms, unusual tiredness or weakness or bruising or bleeding.	Continue, but call doctor right away.
• Skin sensitive to sunlight.	Continue. Call doctor when convenient.

WARNINGS & PRECAUTIONS

Don't take if:

- You are allergic to danazol. Allergic reactions are rare, but may occur.
- You become pregnant.
- You have breast cancer.

Before you start, consult your doctor if:

- You take birth control pills.
- You have diabetes.
- You have heart disease.
- You have epilepsy.
- You have kidney disease.
- You have liver disease.
- You have migraine headaches.

Over age 60:
No special problems expected.

Pregnancy:
Consult doctor. Drug should not be used during pregnancy. Can cause harm to unborn baby. Risk category X (see page xviii).

Breastfeeding; Lactation; Nursing Mothers:
This drug is generally not recommended with breastfeeding. Discuss risks and benefits with your doctor.

Infants & children up to age 18:
Safety and effectiveness in this age group have not been established. Consult doctor.

Prolonged use:

- Required for full effect. Don't discontinue without consulting doctor.
- Talk to your doctor about the need for follow-up medical examinations or laboratory studies.

Skin & sunlight:
No problems expected.

Driving, piloting or hazardous work:
No problems expected.

Discontinuing:
Don't discontinue without consulting doctor. Menstrual periods may be absent for 2 to 3 months after discontinuation.

Others:

- May alter blood sugar levels in persons with diabetes.
- Advise any doctor, dentist or pharmacist whom you consult that you take this medicine.
- May interfere with the accuracy of some medical tests.

POSSIBLE INTERACTION WITH OTHER DRUGS

GENERIC NAME OR DRUG CLASS	COMBINED EFFECT
Anticoagulants, oral*	Increased anti-coagulant effect.
Antidiabetic agents, oral*	Decreased anti-diabetic effect.
Cyclosporine	Increased risk of kidney damage.
Insulin	Decreased insulin effect.

POSSIBLE INTERACTION WITH OTHER SUBSTANCES

INTERACTS WITH	COMBINED EFFECT
Alcohol:	Excessive nervous system depression. Avoid.
Beverages:	
Caffeine.	Rapid, irregular heartbeat. Avoid.
Cocaine:	May interfere with expected action of danazol. Avoid.
Foods:	None expected.
Marijuana:	May interfere with expected action of danazol. Avoid.
Tobacco:	Rapid, irregular heartbeat. Increased leg cramps. Avoid.

*See Glossary

DANTROLENE

BRAND NAMES

Dantrium

BASIC INFORMATION

Habit forming? No
Prescription needed? Yes
Available as generic? Yes
Drug class: Muscle relaxant, antispastic

USES

- Relieves muscle spasticity caused by diseases such as multiple sclerosis, cerebral palsy, stroke.
- Relieves muscle spasticity caused by injury to spinal cord.
- Relieves or prevents excess body temperature brought on by some surgical procedures.

DOSAGE & USAGE INFORMATION

How to take:
Capsule—Swallow with liquid.

When to take:
One to 4 times a day as prescribed for muscle spasticity. For excess body temperature, follow label instructions.

If you forget a dose:
Take as soon as you remember. If it is almost time for the next dose, wait for that dose (don't double this dose) and resume regular schedule.

What drug does:
Acts directly on muscles to prevent excess contractions.

Time lapse before drug works:
1 or more weeks.

Don't take with:
Any other medicine or any dietary supplement without consulting your doctor or pharmacist.

OVERDOSE

SYMPTOMS:
Vomiting, diarrhea, tiredness, muscle weakness, loss of consciousness, coma.
WHAT TO DO:

- **Dial 911 for all medical emergencies or call poison control center 1-800-222-1222 for instructions.**
- **See emergency information on last 3 pages of this book.**

POSSIBLE ADVERSE REACTIONS OR SIDE EFFECTS

SYMPTOMS	WHAT TO DO
Life-threatening:	
Rare, severe allergic reaction (anaphylaxis)—difficulty breathing, hives, itching, swelling (throat, face), vomiting, dizziness.	Seek emergency treatment immediately.
Common:	
Drowsiness, dizziness, muscle weakness, tiredness, nausea, vomiting, general ill feeling.	Continue. Call doctor when convenient.
Infrequent:	
• Rash, hives, black or bloody stools, chest pain, blood in urine, confusion, severe constipation, depression, yellow eyes or skin, difficult urination.	Discontinue. Call doctor right away.
• Stomach cramps, changes in vision, chills and fever, mild constipation, frequent urination, headache, loss of appetite, speaking problems, insomnia, newvousness.	Continue. Call doctor when convenient.
Rare:	
Seizures, blood clot symptoms (swelling, pain and redness in leg, shortness of breath).	Seek emergency care.

WARNINGS & PRECAUTIONS

Don't take if:
You are allergic to dantrolene or any muscle relaxant or antispastic medication.

Before you start, consult your doctor if:
- You have liver disease.
- You have heart disease.
- You have lung disease (especially emphysema).
- You are over age 35.
- You will have surgery within 2 months, including dental surgery, requiring general or spinal anesthesia.

Over age 60:
Adverse reactions and side effects may be more frequent and severe than in younger persons.

Pregnancy:
Decide with your doctor if drug benefits justify risk to unborn baby. Risk category C (see page xviii).

Breastfeeding; Lactation; Nursing Mothers:
This drug is generally not recommended with breastfeeding. Discuss risks and benefits with your doctor.

Infants & children up to age 18:
Used in children age 5 and over. Follow instructions provided by your child's doctor.

Prolonged use:
Recommended periodically during prolonged use—Blood counts, G6PD* tests, liver function studies.

Skin & sunlight:
No special problems expected.

Driving, piloting or hazardous work:
Don't drive or pilot aircraft until you learn how medicine affects you. Don't work around dangerous machinery. Don't climb ladders or work in high places. Danger increases if you drink alcohol or take medicine affecting alertness and reflexes, such as antihistamines, tranquilizers, sedatives, pain medicine, narcotics and mind-altering drugs.

Discontinuing:
Don't discontinue without consulting doctor. Dose may require gradual reduction if you have taken drug for a long time. Doses of other drugs may also require adjustment.

Others:
- No problems expected.
- Advise any doctor, dentist or pharmacist whom you consult that you take this medicine.

POSSIBLE INTERACTION WITH OTHER DRUGS

GENERIC NAME OR DRUG CLASS	COMBINED EFFECT
Central nervous system (CNS) depressants *	Increased risk of side effects.
Estrogens	Increased dantrolene effect.

POSSIBLE INTERACTION WITH OTHER SUBSTANCES

INTERACTS WITH	COMBINED EFFECT
Alcohol:	Increased sedation, low blood pressure. Avoid.
Beverages:	None expected.
Cocaine:	Increased risk of side effects. Avoid.
Foods:	None expected.
Marijuana:	Consult doctor.
Tobacco:	May interfere with absorption of drug.

*See Glossary

DAPSONE

BRAND NAMES

Aczone
DDS
Avlosulfon

BASIC INFORMATION

Habit forming? No
Prescription needed? Yes
Available as generic? Yes
Drug class: Antibacterial (antileprosy), sulfone

USES

- Treatment of dermatitis herpetiformis.
- Treatment of leprosy.
- Prevention and treatment of *Pneumocystis carinii* pneumonia.
- Gel form of the drug treats acne vulgaris.
- Other uses include granuloma annulare, pemphigoid, pyoderma gangrenosum, polychondritis, eye ulcerations, systemic lupus erythematosus.

DOSAGE & USAGE INFORMATION

How to take:
- Tablet—Swallow with liquid or food to lessen stomach irritation.
- Topical gel—Follow directions provided with prescription.

When to take:
Once a day at same time.

If you forget a dose:
Take as soon as you remember. If it is almost time for the next dose, wait for that dose (don't double this dose) and resume regular schedule.

Continued next column

OVERDOSE

SYMPTOMS:
Bleeding, vomiting, seizures, blue skin color, coma.
WHAT TO DO:
- **Dial 911 for all medical emergencies or call poison control center 1-800-222-1222 for instructions.**
- **See emergency information on last 3 pages of this book.**

What drug does:
It helps destroy or inhibit the growth of bacteria that causes leprosy. It is unknown how it works to help dermatitis herpetiformis.

Time lapse before drug works:
- 3 years for leprosy.
- 1 to 2 weeks for dermatitis herpetiformis.

Don't take with:
Any other medicine or any dietary supplement without consulting your doctor or pharmacist.

POSSIBLE ADVERSE REACTIONS OR SIDE EFFECTS

SYMPTOMS	WHAT TO DO
Life-threatening: In case of overdose, see previous column.	
Common:	
Pale skin, unusual tiredness or weakness, pain in back or leg or stomach, bluish color (nails, lips or skin), difficult breathing.	Discontinue. Call doctor right away.
Infrequent:	
Headache, loss of appetite, nausea or vomiting, insomnia.	Continue. Call doctor when convenient.
Rare:	
Skin symptoms (oily, peeling, dry or redness), hair loss, mood or mental changes, tingling or burning feeling in hands or feet, sore throat, easy bleeding or bruising, yellow skin or eyes.	Discontinue. Call doctor right away.

WARNINGS & PRECAUTIONS

Don't take if:
- You have G6PD* deficiency.
- You are allergic to dapsone, furosemide, thiazide diuretics, sulfonylureas, carbonic anhydrase inhibitors, sulfonamides. Allergic reactions are rare, but may occur.

Before you start, consult your doctor if:
- You take any other medicine.
- You are anemic.
- You have liver or kidney disease.
- You are of Mediterranean heritage.
- You will have surgery within 2 months, including dental surgery, requiring general or spinal anesthesia.

Over age 60:
Adverse reactions and side effects may be more frequent and severe than in younger persons.

Pregnancy:
Decide with your doctor if drug benefits justify risk to unborn baby. Risk category C (see page xviii).

Breastfeeding; Lactation; Nursing Mothers:
This drug is generally acceptable with breastfeeding. Discuss risks and benefits with your doctor.

Infants & children up to age 18:
Follow instructions provided by your child's doctor.

Prolonged use:
- Request liver function studies.
- Talk to your doctor about the need for follow-up medical examinations or laboratory studies to check complete blood counts (white blood cell count, platelet count, red blood cell count, hemoglobin, hematocrit).

Skin & sunlight:
May cause rash or intensify sunburn in areas exposed to sun or ultraviolet light (photosensitivity reaction). Avoid overexposure. Notify doctor if reaction occurs.

Driving, piloting or hazardous work:
Don't drive or pilot aircraft until you learn how medicine affects you. Don't work around dangerous machinery. Don't climb ladders or work in high places. Danger increases if you drink alcohol or take medicine affecting alertness and reflexes, such as antihistamines, tranquilizers, sedatives, pain medicine, narcotics and mind-altering drugs.

Discontinuing:
Don't discontinue without consulting doctor. Dose may require gradual reduction if you have taken drug for a long time. Doses of other drugs may also require adjustment.

Others:
- This drug has been associated with serious, and sometimes fatal blood or liver problems.
- Contact your doctor right away if you develop a rash while using the gel. In rare cases it has been associated with serious, and sometimes fatal, skin reactions.
- For full effect you may need to take dapsone for many months or years.

POSSIBLE INTERACTION WITH OTHER DRUGS

GENERIC NAME OR DRUG CLASS	COMBINED EFFECT
Aminobenzoic acid (PABA)	Decreased dapsone effect. Avoid.
Antivirals, HIV/AIDS*	Increased risk of peripheral neuropathy. Reduced absorption of both drugs.
Dideoxyinosine (ddI)	Decreased dapsone effect.
Hemolytics*	May increase adverse effects on blood cells.
Methotrexate	May increase blood toxicity.
Probenecid	Increased toxicity of dapsone.
Pyrimethamine	May increase blood toxicity.
Rifampin	Decreased effect of dapsone.
Trimethoprim	May increase blood toxicity.

POSSIBLE INTERACTION WITH OTHER SUBSTANCES

INTERACTS WITH	COMBINED EFFECT
Alcohol:	Increased chance of toxicity to liver.
Beverages:	None expected.
Cocaine:	Increased risk of side effects. Avoid.
Foods:	None expected.
Marijuana:	Increased risk of side effects. Avoid.
Tobacco:	May interfere with absorption of medicine.

***See Glossary**

DECONGESTANTS (Ophthalmic)

GENERIC AND BRAND NAMES

ANTAZOLINE
Vasocon-A
NAPHAZOLINE
Ak-Con
Albalon
Albalon Liquifilm
Allerest
Clear Eyes
Comfort Eye Drops
Degest 2
Estivin II
I-Naphline
Murine Plus
Muro's Opcon
Nafazair
Naphcon
Naphcon A
Naphcon Forte
Ocu-Zoline
Vasoclear
Vasoclear A
Vasocon
Vasocon Regular
OXYMETAZOLINE
OcuClear
Visine L.R.
TETRAHYDROZOLINE
Visine

BASIC INFORMATION

Habit forming? No
Prescription needed? Yes, for some
Available as generic? Yes
Drug class: Decongestant (ophthalmic)

USES

Treats eye redness, itching, burning or other irritation due to dust, colds, allergies, rubbing eyes, wearing contact lenses, swimming or eye strain from close work, watching TV, reading.

OVERDOSE

SYMPTOMS:
Lowered body temperature, drowsiness, slow heartbeat, excessive weakness.
WHAT TO DO:
If person uses much larger amount than prescribed or if accidentally swallowed, call poison control center 1-800-222-1222 for instructions or dial 911 (emergency) for help.

DOSAGE & USAGE INFORMATION

How to use:
Eye drops

- Wash hands.
- Apply pressure to inside corner of eye with middle finger.
- Tilt head backward. Pull lower lid away from eye with index finger of the same hand.
- Drop eye drops into pouch and close eye. Don't blink.
- Keep eyes closed for 1 to 2 minutes.
- Continue pressure for 1 minute after placing medicine in eye.
- Don't touch applicator tip to any surface (including the eye). If you accidentally touch tip, clean with warm water and soap.
- Keep container tightly closed.
- Keep cool, but don't freeze.
- Wash hands immediately after using.

When to use:
As directed. Usually every 3 or 4 hours.

If you forget a dose:
Use as soon as you remember. If it is almost time for the next dose, wait for that dose (don't double this dose) and resume regular schedule.

What drug does:
Acts on small blood vessels to make them constrict or become smaller.

Time lapse before drug works:
2 to 10 minutes.

Don't use with:
Other eye drops without consulting your doctor.

POSSIBLE ADVERSE REACTIONS OR SIDE EFFECTS

SYMPTOMS	WHAT TO DO
Life-threatening: None expected.	
Common: None expected.	
Infrequent: None expected.	
Rare:	
• If too much of drug absorbed by body (headache, nausea, nervousness, excess sweating, dizziness).	Discontinue. Call doctor right away.
• Increased eye irritation, blurred vision, enlarged pupils.	Discontinue. Call doctor if symptoms persist.

WARNINGS & PRECAUTIONS

Don't use if:
You are allergic to any decongestant eye drops.

Before you start, consult your doctor if:
- You take antidepressants or maprotiline.
- You have glaucoma, eye disease, infection or injury.
- You have heart disease, high blood pressure, thyroid disease.

Over age 60:
No problems expected.

Pregnancy:
Decide with your doctor if drug benefits justify risk to unborn baby. Risk category C (see page xviii).

Breastfeeding; Lactation; Nursing Mothers:
Drugs in this group are generally acceptable with breastfeeding. Discuss risks and benefits with your doctor.

Infants & children up to age 18:
Use in children is not recommended.

Prolonged use:
Don't use for more than 3 or 4 days.

Skin & sunlight:
No problems expected.

Driving, piloting or hazardous work:
No problems expected.

Discontinuing:
May not need all the medicine in container. If symptoms disappear, stop using.

Others:
- Check with your doctor if eye irritation continues or becomes worse.
- Store product out of the reach of children.

POSSIBLE INTERACTION WITH OTHER DRUGS

GENERIC NAME OR DRUG CLASS	COMBINED EFFECT
Clinically significant interactions with oral or injected medicines unlikely.	

POSSIBLE INTERACTION WITH OTHER SUBSTANCES

INTERACTS WITH	COMBINED EFFECT
Alcohol:	None expected.
Beverages:	None expected.
Cocaine:	None expected.
Foods:	None expected.
Marijuana:	None expected.
Tobacco:	Smoke may increase eye irritation. Avoid.

DEHYDROEPIANDROSTERONE (DHEA)

BRAND NAMES

Numerous brand names are available

BASIC INFORMATION

Habit forming? No
Prescription needed? No
Available as generic? Yes
Drug class: Adrenal steroid

USES

DHEA is a steroid produced in the human body by the adrenal glands (which sit on top of each kidney). DHEA concentration peaks at about age 20 and then decreases progressively with age. Supplements are sold as an antiaging remedy claimed by some to improve energy, strength, and immunity. DHEA is also said to increase muscle and decrease fat. Studies to date do not provide a clear picture of the risks and benefits of DHEA.

DOSAGE & USAGE INFORMATION

How to take:
For tablet or capsule—Follow instructions on the label or consult your doctor or pharmacist. Different brands supply different doses. DHEA, as a product, is marketed as a dietary supplement and is not reviewed by the U.S. Food & Drug Administration (FDA) for effectiveness and safety. The best dosage amounts are unknown. Use with caution.

When to take:
At the same times each day according to label directions.

If you forget a dose:
Follow label instructions for your particular brand of DHEA. Usually, take it as soon as you remember. If it is almost time for the next dose, wait for the next scheduled dose (don't double this dose).

Continued next column

OVERDOSE

SYMPTOMS:
Unknown.
WHAT TO DO:
If person uses much larger amount than prescribed or if accidentally swallowed, call poison control center 1-800-222-1222 for instructions or dial 911 (emergency) for help.

What drug does:

- Although it is not known whether DHEA itself causes hormonal effects, the body breaks DHEA down into two hormones—estrogen and testosterone. Some people's bodies make large amounts of estrogen and testosterone from DHEA, while others make smaller amounts.
- Hormone supplements may not have the same effects on the body as naturally produced hormones have, because the body processes them differently. Higher doses of supplements may result in higher amounts of hormones in the blood than are healthy.

Time lapse before drug works:
Effectiveness will vary from person to person and will also depend on the reason for taking DHEA, such as for a chronic health problem.

Don't take with:
Any other medicine or any dietary supplement without consulting your doctor or pharmacist.

POSSIBLE ADVERSE REACTIONS OR SIDE EFFECTS

SYMPTOMS	WHAT TO DO
Life-threatening: None expected.	
Common: Unknown.	
Infrequent: In women: acne, hair loss, facial hair growth (hirsutism), deepening of voice (the last two may be irreversible).	Discontinue. Call doctor when convenient.
Rare: Unknown. If symptoms occur that you are concerned about, talk to your doctor or pharmacist. Further research may uncover other side effects.	

WARNINGS & PRECAUTIONS

Don't use if:
You are allergic to DHEA. Allergic reactions are rare, but may occur.

Before you start, consult your doctor if:
- You have any chronic health problem.
- You have a family history of cancer.
- You are allergic to any medication, food or other substance.
- You have or have had prostate, ovarian, breast, cervical or uterine cancer.

Over age 60:
A lower starting dosage may be recommended until a response is determined.

Pregnancy:
Decide with your doctor if any possible benefits of DHEA justify risk to unborn child. Risk category is not designated (see page xviii).

Breastfeeding; Lactation; Nursing Mothers:
This drug is generally not recommended with breastfeeding. Discuss risks and benefits with your doctor.

Infants & children up to age 18:
Not recommended for children.

Prolonged use:
Effects are unknown. More research is needed o determine long-term effects of DHEA use.

Skin & sunlight:
No problems expected.

Driving, piloting or hazardous work:
No problems expected.

Discontinuing:
No problems expected, but effects after long-term use are unknown.

Others:
- Advise any doctor, dentist or pharmacist whom you consult that you take DHEA.
- DHEA is not researched carefully as yet for use in humans. Most research has been performed on animals. Studies are ongoing to find more definite answers about its effect on aging, muscles, and the immune system. Studies in men and women have shown an improvement in the feeling of well being. Studies in AIDS patients and those with multiple sclerosis also have shown improvement in well being, but without an outcome change.
- Researchers are concerned that DHEA supplements may cause high levels of estrogen or testosterone in some people. The body's own testosterone plays a role in prostate cancer and high levels of naturally produced estrogen are suspected of increasing breast cancer risk. The effect of DHEA is unknown.

POSSIBLE INTERACTION WITH OTHER DRUGS

GENERIC NAME OR DRUG CLASS	COMBINED EFFECT
All medications	Effects are unknown. Talk to your doctor or pharmacist.

POSSIBLE INTERACTION WITH OTHER SUBSTANCES

INTERACTS WITH	COMBINED EFFECT
Alcohol:	Unknown.
Beverages:	None expected.
Cocaine:	Unknown. Avoid.
Foods:	None expected.
Marijuana:	Consult doctor.
Tobacco:	Unknown.

*See Glossary

DESMOPRESSIN

BRAND NAMES

DDAVP Stimate

BASIC INFORMATION

Habit forming? No
Prescription needed? Yes
Available as generic? Yes
Drug class: Antidiuretic, antihemorrhagic

USES

- Prevents and controls symptoms associated with central diabetes insipidus.
- The tablet form (not the nasal spray) of the drug is used to treat primary nocturnal enuresis (bedwetting during sleep).

DOSAGE & USAGE INFORMATION

How to take:

- Tablet—Swallow with liquid.
- Nasal spray—Fill the rhinyle (a flexible, calibrated catheter) with a measured dose of the nasal spray. Blow on the other end of the catheter to deposit the solution deep in the nasal cavity.

When to take:
At the same time each day, according to instructions on prescription label. Be sure to restrict fluid intake from 1 hour before to 8 hours after taking drug.

If you forget a dose:
Take as soon as you remember. If it is almost time for the next dose, wait for that dose (don't double this dose) and resume regular schedule.

What drug does:
Increases water reabsorption in the kidney and decreases urine output.

Continued next column

OVERDOSE

SYMPTOMS:
Confusion, headache, drowsiness, sudden weight gain, problem with urinating.
WHAT TO DO:
If person uses much larger amount than prescribed or if accidentally swallowed, call poison control center 1-800-222-1222 for instructions or dial 911 (emergency) for help.

Time lapse before drug works:
Within 1 hour. Effect may last from 6 to 24 hours.

Don't take with:
Any other medicine or any dietary supplement without consulting your doctor or pharmacist.

POSSIBLE ADVERSE REACTIONS OR SIDE EFFECTS

SYMPTOMS	WHAT TO DO
Life-threatening:	
Rare, severe allergic reaction (anaphylaxis)—difficulty breathing, hives, itching, swelling (throat, face), vomiting, dizziness.	Seek emergency treatment immediately.
Common:	
None expected.	
Infrequent:	
Flushing or redness of skin, headache, nausea, nasal congestion.	Continue. Call doctor when convenient.
Rare:	
• Very rare water intoxication symptoms (confusion, drowsiness, headache, seizures, rapid weight gain, less urination).	Discontinue. Seek emergency care.
• Low sodium level in the body (symptoms include nausea, vomiting, muscle cramps, fatigue, weakness).	Discontinue. Call doctor right away.

WARNINGS & PRECAUTIONS

Don't take if:
Person is allergic to desmopressin.

Before you start, consult your doctor if:
- You have allergic rhinitis.
- You have nasal congestion.
- You have a cold or other upper respiratory infection or are dehydrated.
- You have heart disease or high blood pressure.
- You have a history of hyponatremia, which is low sodium (salt) levels in the body.
- You have polydipsia (excessive or abnormal thirstiness).
- You have cystic fibrosis.

Over age 60:
No special problems expected.

Pregnancy:
Decide with your doctor if drug benefits justify risk to unborn baby. Risk category B (see page xviii).

Breastfeeding; Lactation; Nursing Mothers:
This drug is generally acceptable with breastfeeding. Discuss risks and benefits with your doctor.

Infants & children up to age 18:
Treats bedwetting in children age 6 and over. Follow instructions provided by your child's doctor.

Prolonged use:
May require increasing dosage for same effect.

Skin & sunlight:
No special problems expected.

Driving, piloting or hazardous work:
Don't drive or pilot aircraft until you learn how medicine affects you. Don't work around dangerous machinery. Don't climb ladders or work in high places. Danger increases if you drink alcohol or take medicine affecting alertness and reflexes.

Discontinuing:
No special problems expected.

Others:
- Advise any doctor, dentist or pharmacist whom you consult that you take this medicine.
- Be sure to carefully follow all instructions for use of intranasal desmopressin.
- Consult doctor if the amount of water the patient is drinking changes.
- Treatment with desmopressin tablets for primary nocturnal enuresis should be stopped during illnesses with symptoms of fever, recurrent vomiting, or diarrhea that may lead to fluid and/or electrolyte imbalance.
- If sodium levels fall too much causing hyponatremia, the patient could have seizures and in extreme cases, may be life-threatening.
- The drug should be used cautiously in patients who may be drinking a lot of fluids (such as during hot weather or when doing vigorous exercising) due to a higher risk of hyponatremia.

POSSIBLE INTERACTION WITH OTHER DRUGS

GENERIC NAME OR DRUG CLASS	COMBINED EFFECT
Antidepressants, tricyclic (TCA)*	Increased risk of thirstiness which may lead to drinking excess fluids.
Carbamazepine	May increase desmopressin effect.
Chlorpropamide	May increase desmopressin effect.
Clofibrate	May increase desmopressin effect.
Demeclocycline	May decrease desmopressin effect.
Lithium	May decrease desmopressin effect.
Norepinephrine	May decrease desmopressin effect.
Selective serotonin reuptake inhibitors (SSRIs)	Increased risk of thirstiness which may lead to drinking excess fluids.

POSSIBLE INTERACTION WITH OTHER SUBSTANCES

INTERACTS WITH	COMBINED EFFECT
Alcohol:	May decrease desmopressin effect.
Beverages: Caffeine drinks.	May decrease desmopressin effect.
Cocaine:	Unknown. Avoid.
Foods:	None expected.
Marijuana:	Consult doctor.
Tobacco:	May decrease desmopressin effect.

***See Glossary**

DEXTROMETHORPHAN

BRAND NAMES

See full list of brand names in the *Generic and Brand Name Directory*, page 858.

BASIC INFORMATION

Habit forming? No
Prescription needed? No
Available as generic? Yes
Drug class: Cough suppressant, antitussive

USES

Suppresses cough associated with allergies or infections such as colds, bronchitis, flu and lung disorders. Used in many cough, cold and allergy combination medicines.

DOSAGE & USAGE INFORMATION

How to take:

- Chewable tablet—Chew well, then swallow.
- Oral suspension, lozenge or syrup—Take as directed on label.
- Thin strip—Allow strips to dissolve on the tongue. Water or other fluid is not needed.
- Capsule—Swallow with liquid.

When to take:
As needed, no more often than every 4 hours or as directed on label.

If you forget a dose:
Take as soon as you remember. If it is almost time for the next dose, wait for that dose (don't double this dose) and resume regular schedule.

What drug does:
Reduces sensitivity of brain's cough control center, suppressing urge to cough.

Continued next column

OVERDOSE

SYMPTOMS:
Euphoria, overactivity, nausea, vomiting, sense of intoxication, hallucinations, lack of coordination, stagger, stupor, shallow breathing, rapid heartbeat, muscle twitching, convulsions, coma.
WHAT TO DO:

- **Dial 911 for all medical emergencies or call poison control center 1-800-222-1222 for instructions.**
- **See emergency information on last 3 pages of this book.**

Time lapse before drug works:
15 to 30 minutes.

Don't take with:
Any other medicine or any dietary supplement without consulting your doctor or pharmacist.

POSSIBLE ADVERSE REACTIONS OR SIDE EFFECTS

SYMPTOMS	WHAT TO DO
Life-threatening: None expected.	
Common: None expected.	
Infrequent: Mild dizziness or drowsiness, nausea or vomiting, stomach cramps, constipation, headache.	Discontinue. Call doctor if symptoms persist.
Rare: None expected.	

WARNINGS & PRECAUTIONS

Don't take if:
You are allergic to dextromethorphan. Allergic reactions are rare, but may occur.

Before you start, consult your doctor if:
- You have asthma attacks.
- You have impaired liver function.

Over age 60:
Adverse reactions and side effects may be more frequent and severe than in younger persons.

Pregnancy:
Decide with your doctor if drug benefits justify risk to unborn baby. Risk category C (see page xviii).

Breastfeeding; Lactation; Nursing Mothers:
This drug is generally acceptable with breastfeeding. Discuss risks and benefits with your doctor.

Infants & children up to age 18:
Read nonprescription product's label to be sure it is approved for your child's age. Follow label instructions on dosage. Consult doctor or pharmacist if unsure.

Prolonged use:
No problems expected.

Skin & sunlight:
No problems expected.

Driving, piloting or hazardous work:
Don't drive or pilot aircraft until you learn how medicine affects you. Don't work around dangerous machinery. Don't climb ladders or work in high places. Danger increases if you drink alcohol or take medicine affecting alertness and reflexes, such as antihistamines, tranquilizers, sedatives, pain medicine, narcotics and mind-altering drugs.

Discontinuing:
May be unnecessary to finish medicine. Follow doctor's instructions.

Others:
- If cough persists or if you cough blood or brown-yellow, thick mucus, call your doctor.
- Excessive use may lead to functional dependence.*

POSSIBLE INTERACTION WITH OTHER DRUGS

GENERIC NAME OR DRUG CLASS	COMBINED EFFECT
Doxepin (topical)	Increased risk of toxicity of both drugs.
Memantine	May lead to adverse effects of either drug.
Monoamine oxidase (MAO) inhibitors*	Disorientation, high fever, drop in blood pressure and loss of consciousness.
Sedatives* and other central nervous system (CNS) depressants*	Increased sedative effect of both drugs.

POSSIBLE INTERACTION WITH OTHER SUBSTANCES

INTERACTS WITH	COMBINED EFFECT
Alcohol:	None expected.
Beverages: Grapefruit juice.	Risk of toxicity. Avoid.
Cocaine:	Unknown. Avoid.
Foods:	None expected.
Marijuana:	Consult doctor.
Tobacco:	Smoking increases mucus in lungs. Avoid.

*See Glossary

DICLOFENAC (Topical)

BRAND NAMES

Flector Patch
Pennsaid
Solaraze Gel
Voltaren Gel

BASIC INFORMATION

Habit forming? No
Prescription needed? Yes
Available as generic? Yes
Drug class: Nonsteroidal anti-inflammatory drug (NSAID), topical

USES

- Used for joint pain in the hands, wrists, elbows, knees, ankles or feet caused by osteoarthritis.
- Patch is used to treat pain due to minor strains, sprains and bruises.
- May be used for treatment of other disorders as determined by your doctor.
- A brand named Solaraze is used to treat actinic keratosis. It is not covered in this topic.

DOSAGE & USAGE INFORMATION

How to use:

- Gel—Follow instructions provided with the prescription. The gel comes with dosing cards that show you how much to use for a 2-gram- or a 4-gram dose. Do not use more than 4 times a day on a single joint. Do not wear gloves for at least 10 minutes after applying gel to the hands. Wait at least 10 minutes before dressing. Do not bathe or shower for at least 1 hour after application. Don't use drug on wounded or sore skin. Don't cover treated area with bandage.
- Patch—Apply to most painful area twice daily. Wash hands after applying patch. Do not wear patch in bath or shower.
- Solution—Apply as directed on label.

When to take:
Follow your doctor's or label's instructions.

Continued next column

If you forget a dose:
Apply as soon as you remember. If it is almost time for the next dose, wait for that dose (don't double this dose) and resume regular schedule.

What drug does:
It is applied to and absorbed by the affected joint. It works by reducing certain hormones that cause inflammation and pain in the body.

Time lapse before drug works:
It may take several weeks or more of treatment to determine full benefit of the drug.

Don't take with:

- Any other medicine or diet supplement without consulting your doctor or pharmacist.
- Other skin products (cosmetics, sunscreen, lotions, insect repellant or other topical drugs) not prescribed by your doctor.

OVERDOSE

SYMPTOMS:
May have lethargy, drowsiness, nausea, vomiting, stomach pain.
WHAT TO DO:
If person uses much larger amount than prescribed or if accidentally swallowed, call poison control center 1-800-222-1222 for instructions or dial 911 (emergency) for help.

POSSIBLE ADVERSE REACTIONS OR SIDE EFFECTS

SYMPTOMS	WHAT TO DO
Life-threatening:	
Rare, severe allergic reaction (anaphylaxis)—difficulty breathing, hives, itching, swelling (throat, face), vomiting, dizziness.	Seek emergency treatment immediately.
Common:	
Skin reaction where product is applied.	Continue. Call doctor when convenient.
Infrequent:	
None expected.	
Rare:	
These side effects may occur if drug is absorbed into blood-stream: chest pain, shortness of breath, weakness, slurred speech, vision changes, loss of balance, stool changes (black, bloody, tarry or clay-color), vomiting blood or what looks like coffee grounds, swelling, rapid weight gain, change in urination, dark urine, nausea, stomach pain, appetite loss, yellow skin or eyes, sore throat and headache, severe skin rash (blistering, red, peeling), bruising, severe tingling, numbness, fever, muscle weakness, any skin rash.	Discontinue. Call doctor right away.

WARNINGS & PRECAUTIONS

Don't take if:
You are allergic to diclofenac or have had an allergic reaction after taking aspirin or any nonsteroidal anti-inflammatory drug (NSAID).

Before you start, consult your doctor if:
- You have had or will have heart bypass surgery.
- You have a history of asthma.
- You have a history of stomach ulcer or gastrointestinal bleeding.
- You have kidney or liver disease.
- You have high blood pressure, congestive heart failure or other heart or blood vessel disorder.

Over age 60:
Adverse reactions and side effects may be more frequent and severe than in younger persons.

Pregnancy:
Decide with your doctor if drug benefits justify risk to unborn baby. Risk category C (see page xviii).

Breastfeeding; Lactation; Nursing Mothers:
This drug is generally acceptable with breastfeeding. Discuss risks and benefits with your doctor.

Infants & children up to age 18:
Not approved for children under age 18.

Prolonged use:
- Talk to your doctor about the need for follow-up medical exams and/or blood tests and liver and kidney function studies.
- Long-term continuous use may increase the risk of heart attack or stroke. Consult doctor.

Skin & sunlight:
Avoid exposure of treated skin areas to sunlight or artificial UV rays (sunlamps or tanning beds).

Driving, piloting or hazardous work:
No special problems expected.

Discontinuing:
No special problems expected.

Others:
- Though rare, the drug may be absorbed into your bloodstream and increase the risk of life-threatening heart or circulation problems, including heart attack or stroke. There is also increased risk of serious gastrointestinal problems (e.g., bleeding, ulcers and perforation of the stomach or intestines) which may be fatal.
- Rarely, serious liver problems (necrosis, jaundice, hepatitis, failure) may occur with drug use. May lead to need for liver transplant or even death. Consult doctor about the risks.
- Advise any doctor, dentist or pharmacist whom you consult that you use this medicine.
- Avoid getting the drug in your mouth or eyes. If it does get into these areas, rinse with water. Do not apply to wounded or broken skin. Follow product instructions for proper disposal of gel dose cards and patches.
- Do not expose treated areas to heat from a hot tub, heating pad, sauna, or heated water bed. Heat can increase the amount of drug absorbed through the skin and may cause adverse effects.

POSSIBLE INTERACTION WITH OTHER DRUGS

GENERIC NAME OR DRUG CLASS	COMBINED EFFECT
Angiotensin-converting enzyme (ACE) inhibitors*	May decrease ACE inhibitor effect.
Anticoagulants*	May increase risk of internal bleeding.
Anti-inflammatory drugs, nonsteroidal (NSAIDs)* oral	May increase risk of side effects.
Aspirin	May increase risk of side effects.
Cyclosporine	May increase effect of cyclosporine.
Diuretics*	May decrease effect of diuretic.
Hepatotoxics*	Increased risk of liver problems.
Lithium	May increase effect of lithium.
Methotrexate	May increase effect of methotrexate.
Topical skin products	Unknown effect. Consult doctor

POSSIBLE INTERACTION WITH OTHER SUBSTANCES

INTERACTS WITH	COMBINED EFFECT
Alcohol:	May increase risk of side effects.
Beverages:	None expected.
Cocaine:	Unknown. Avoid.
Foods:	None expected.
Marijuana:	Consult doctor.
Tobacco:	May increase risk of side effects.

***See Glossary**

DICYCLOMINE

BRAND NAMES

See full list of brand names in the *Generic and Brand Name Directory*, page 861.

BASIC INFORMATION

Habit forming? No
Prescription needed? Yes
Available as generic? Yes
Drug class: Antispasmodic, anticholinergic

USES

- Reduces spasms of digestive system, bladder and urethra.
- Treats irritable bowel syndrome.

DOSAGE & USAGE INFORMATION

How to take:
Tablet, syrup or capsule—Swallow with liquid or food to lessen stomach irritation.

When to take:
30 minutes before meals (unless directed otherwise by doctor).

If you forget a dose:
Take as soon as you remember. If it is almost time for the next dose, wait for that dose (don't double this dose) and resume regular schedule.

What drug does:
Blocks nerve impulses at parasympathetic nerve endings, preventing muscle contractions and gland secretions of organs involved.

Time lapse before drug works:
15 to 30 minutes.

Don't take with:
Any other medicine or any dietary supplement without consulting your doctor or pharmacist.

OVERDOSE

SYMPTOMS:
Dilated pupils, blurred vision, rapid pulse and breathing, dizziness, fever, hallucinations, confusion, slurred speech, agitation, flushed face, convulsions, coma.
WHAT TO DO:
- **Dial 911 for all medical emergencies or call poison control center 1-800-222-1222 for instructions.**
- **See emergency information on last 3 pages of this book.**

POSSIBLE ADVERSE REACTIONS OR SIDE EFFECTS

SYMPTOMS	WHAT TO DO
Life-threatening: Rare, severe allergic reaction (anaphylaxis)—difficulty breathing, hives, itching, swelling (throat, face), vomiting, dizziness.	Seek emergency treatment immediately.
Common: Constipation, loss of taste, decreased sweating, dryness (ears, nose, throat, mouth, skin).	Continue. Call doctor when convenient.
Infrequent: Headache, difficult urination, blurred vision, swallowing difficulty, eyes sensitive to light, nausea or vomiting, memory problems, feeling tired or weak.	Continue. Call doctor when convenient.
Rare:	
• Confusion, eye pain, rash or hives, fast or irregular heartbeat, joint pain, swelling around face, shortness of breath, fever, chest tightness, wheezing or cough.	Discontinue. Call doctor right away.
• Loss of sexual ability or interest, loss of taste.	Continue. Call doctor when convenient.

WARNINGS & PRECAUTIONS

Don't take if:
You are allergic to dicyclomine.

Before you start, consult your doctor if:
- You have trouble with stomach bloating.
- You have difficulty emptying your bladder completely.
- You have glaucoma.
- You have severe ulcerative colitis.
- You have angina, chronic bronchitis or asthma.
- You have hiatal hernia, liver disease, kidney or thyroid disease, enlarged prostate, myasthenia gravis, peptic ulcer.
- You will have surgery within 2 months, including dental surgery, requiring general or spinal anesthesia.

Over age 60:
Adverse reactions and side effects may be more frequent and severe than in younger persons.

Pregnancy:
Decide with your doctor if drug benefits justify risk to unborn baby. Risk category B (see page xviii).

Breastfeeding; Lactation; Nursing Mothers:
This drug is generally not recommended with breastfeeding. Discuss risks and benefits with your doctor.

Infants & children up to age 18:
Safety and effectiveness in this age group have not been established. Consult doctor.

Prolonged use:
Chronic constipation, possible fecal impaction. Consult doctor immediately.

Skin & sunlight:
No problems expected.

Driving, piloting or hazardous work:
Use disqualifies you for piloting aircraft. Otherwise, no problems expected.

Discontinuing:
May be unnecessary to finish medicine. Follow doctor's instructions.

Others:
Advise any doctor, dentist or pharmacist whom you consult that you take this medicine.

POSSIBLE INTERACTION WITH OTHER DRUGS

GENERIC NAME OR DRUG CLASS	COMBINED EFFECT
Adrenocorticoids, systemic	Possible glaucoma.
Amantadine	Increased dicyclomine effect.
Antacids*	Decreased dicyclomine effect.
Anticholinergics, other*	Increased dicyclomine effect.
Antidepressants, tricyclic*	Increased dicyclomine effect. Increased sedation.
Antidiarrheals*	Decreased dicyclomine effect.
Antihistamines*	Increased dicyclomine effect.
Attapulgite	Decreased dicyclomine effect.
Buclizine	Increased dicyclomine effect.
Digitalis	Possible decreased absorption of digitalis.
Haloperidol	Increased internal eye pressure.
Ketoconazole	Decreased ketoconazole effect.
Meperidine	Increased dicyclomine effect.
Methylphenidate	Increased dicyclomine effect.
Monoamine oxidase (MAO) inhibitors*	Increased dicyclomine effect.
Nitrates*	Increased internal eye pressure.
Nizatidine	Increased nizatidine effect.
Orphenadrine	Increased dicyclomine effect.
Phenothiazines*	Increased dicyclomine effect.
Pilocarpine	Loss of pilocarpine effect in glaucoma treatment.
Potassium supplements*	Possible intestinal ulcers with oral potassium tablets.
Quinidine	Increased dicyclomine effect.
Sedatives* or central nervous system (CNS) depressants*	Increased sedative effect of both drugs.
Vitamin C	Decreased dicyclomine effect. Avoid large doses of vitamin C.

POSSIBLE INTERACTION WITH OTHER SUBSTANCES

INTERACTS WITH	COMBINED EFFECT
Alcohol:	None expected.
Beverages:	None expected.
Cocaine:	Unknown. Avoid.
Foods:	None expected.
Marijuana:	Consult doctor.
Tobacco:	None expected.

*See Glossary

DIFENOXIN & ATROPINE

BRAND NAMES

Motofen

BASIC INFORMATION

Habit forming? Yes
Prescription needed? Yes
Available as generic? No
Drug class: Antidiarrheal

USES

- Reduces spasms of digestive system.
- Treats severe diarrhea.

DOSAGE & USAGE INFORMATION

How to take:
Tablet—Swallow with liquid or food to lessen stomach irritation.

When to take:
After each loose stool or every 3 to 4 hours. No more than 5 tablets in 12 hours.

If you forget a dose:
Take as soon as you remember. If it is almost time for the next dose, wait for that dose (don't double this dose) and resume regular schedule.

What drug does:
- Blocks nerve impulses at parasympathetic nerve endings, preventing muscle contractions and gland secretions of organs involved.
- Acts on brain to decrease spasm of smooth muscle.

Time lapse before drug works:
40 to 60 minutes.

Don't take with:
Any other medicine or diet supplement without consulting your doctor or pharmacist.

OVERDOSE

SYMPTOMS:
Pinpoint pupils, rapid pulse. trouble breathing, dizziness, fever, hallucinations, confusion, slurred speech, agitation, flushed face, convulsions, coma.
WHAT TO DO:
- **Dial 911 for all medical emergencies or call poison control center 1-800-222-1222 for instructions.**
- **See emergency information on last 3 pages of this book.**

POSSIBLE ADVERSE REACTIONS OR SIDE EFFECTS

SYMPTOMS	WHAT TO DO
Life-threatening:	
Rare, severe allergic reaction (anaphylaxis)—difficulty breathing, hives, itching, swelling (throat, face), vomiting, dizziness.	Seek emergency treatment immediately.
Common:	
Dizziness, drowsiness.	Continue. Call doctor when convenient.
Infrequent:	
• Bloating, appetite loss, constipation, abdominal pain, blurred vision, warm or flushed skin, fast heartbeat, dry mouth.	Discontinue. Call doctor right away.
• Frequent urination, lightheadedness, dry skin, insomnia, headache.	Continue. Call doctor when convenient.
Rare:	
Weakness, confusion, fever.	Discontinue. Call doctor right away.

WARNINGS & PRECAUTIONS

Don't take if:
You are allergic to difenoxin or atropine.

Before you start, consult your doctor if:
- You have a history of alcohol or drug abuse.
- You have severe colitis.
- Patient has Down syndrome.
- You have dysentery.
- You have asthma, emphysema, bronchitis or other chronic lung disease.
- You have enlarged prostate, urinary tract blockage, difficult urination, incontinence or an intestinal blockage.
- You have liver or kidney disease or gallbladder disease or gallstones.
- You have glaucoma.
- You have heart disease, hiatal hernia, high blood pressure (hypertension).
- You have myasthenia gravis.
- You have overactive or underactive thyroid.

Over age 60:
Adverse reactions and side effects may be more frequent and severe than in younger persons.

Pregnancy:
Decide with your doctor if drug benefits justify risk to unborn baby. Risk category C (see page xviii).

Breastfeeding; Lactation; Nursing Mothers:
This drug is generally acceptable with breastfeeding. Discuss risks and benefits with your doctor.

Infants & children up to age 18:
Safety and effectiveness in children under age 12 have not been established. Consult doctor.

Prolonged use:
- Chronic constipation, possible fecal impaction. Consult doctor immediately.
- Talk to your doctor about the need for follow-up medical examinations or laboratory studies to check liver function.

Skin & sunlight:
No problems expected.

Driving, piloting or hazardous work:
Don't drive until you learn how medicine affects you. Don't work around dangerous machinery. Don't climb ladders or work in high places. Danger increases if you drink alcohol or take medicine affecting alertness and reflexes, such as antihistamines, tranquilizers, sedatives, pain medicine, narcotics and mind-altering drugs.

Discontinuing:
May be unnecessary to finish medicine. Follow doctor's instructions.

Others:
- Advise any doctor, dentist or pharmacist whom you consult that you take this drug.
- Atropine included in drug is at doses below therapeutic level to prevent abuse.

POSSIBLE INTERACTION WITH OTHER DRUGS

GENERIC NAME OR DRUG CLASS	COMBINED EFFECT
Addictive substances (narcotics,* others)	Increased chance of abuse.
Amantadine	Increased atropine effect.
Anticholinergics, other*	Increased atropine effect.
Antidepressants, tricyclic (TCA)*	Increased atropine effect. Increased sedation.
Antihistamines*	Increased atropine effect.
Antihypertensives*	Increased sedation.
Clozapine	Toxic effect on the central nervous system.
Cortisone drugs*	Increased internal eye pressure.
Ethinamate	Dangerous increased effects of ethinamate. Avoid combining.
Fluoxetine	Increased depressant effects of both drugs.
Guanfacine	May increase depressant effects of either drug.
Haloperidol	Increased internal eye pressure.
Leucovorin	High alcohol content of leucovorin may cause adverse effects.
Meperidine	Increased atropine effect.
Methylphenidate	Increased atropine effect.
Methyprylon	Increased sedative effect, perhaps to dangerous level. Avoid.
Monoamine oxidase (MAO) inhibitors*	Increased atropine effect.
Nabilone	Greater depression of central nervous system.
Naltrexone	Triggers withdrawal symptoms.
Narcotics*	Increased sedation. Avoid.

Continued on page 898

POSSIBLE INTERACTION WITH OTHER SUBSTANCES

INTERACTS WITH	COMBINED EFFECT
Alcohol:	Increased sedation. Avoid.
Beverages:	None expected.
Cocaine:	None expected.
Foods:	None expected.
Marijuana:	None expected.
Tobacco:	None expected.

***See Glossary**

DIGITALIS PREPARATIONS (Digitalis Glycosides)

GENERIC AND BRAND NAMES

DIGITOXIN
Crystodigin

DIGOXIN
Lanoxicaps
Lanoxin
Novodigoxin

BASIC INFORMATION

Habit forming? No
Prescription needed? Yes
Available as generic? Yes
Drug class: Digitalis preparation

USES

- Strengthens weak heart muscle contractions to prevent congestive heart failure.
- Corrects irregular heartbeat.

DOSAGE & USAGE INFORMATION

How to take:
- Tablet or capsule—Swallow with liquid. If you can't swallow whole, crumble tablet or open capsule and take with liquid or food.
- Liquid—Dilute dose in beverage and swallow.

When to take:
At the same time each day.

If you forget a dose:
Take as soon as you remember. If it is almost time for the next dose, wait for that dose (don't double this dose) and resume regular schedule.

What drug does:
- Strengthens heart muscle contraction.
- Delays nerve impulses to heart.

Time lapse before drug works:
May require regular use for a week or more.

Continued next column

OVERDOSE

SYMPTOMS:
Nausea, vomiting, diarrhea, vision disturbances, halos around lights, fatigue, irregular heartbeat, confusion, hallucinations, convulsions.
WHAT TO DO:
- **Dial 911 for all medical emergencies or call poison control center 1-800-222-1222 for instructions.**
- **See emergency information on last 3 pages of this book.**

Don't take with:
Any other medicine or any dietary supplement without consulting your doctor or pharmacist.

POSSIBLE ADVERSE REACTIONS OR SIDE EFFECTS

SYMPTOMS	WHAT TO DO
Life-threatening: In case of overdose, see previous column.	
Common: Heartbeat changes (fast, slow, irregular), fainting.	Discontinue. Call doctor right away.
Infrequent: Nausea, vomiting, dizziness, anxious, headache, weakness, depression, diarrhea, loss of appetite.	Continue. Call doctor when convenient.
Rare: • Black or bloody or tarry stools, blurred or yellow vision, hallucinations, confusion, unusual behaviors, bloody vomit or urine, pain in stomach.	Discontinue. Call doctor right away.
• Enlarged breasts in males, mild skin rash.	Continue. Call doctor when convenient.

WARNINGS & PRECAUTIONS

Don't take if:
- You are allergic to any digitalis preparation. Allergic reactions are rare, but may occur.
- Your heartbeat is slower than 50 beats per minute.

Before you start, consult your doctor if:
- You have taken another digitalis preparation in past 2 weeks.
- You have taken a diuretic within 2 weeks.
- You have liver or kidney disease.
- You have a thyroid disorder.
- You will have surgery within 2 months, including dental surgery, requiring general or spinal anesthesia.

Over age 60:
Adverse reactions and side effects may be more frequent and severe than in younger persons.

Pregnancy:
Decide with your doctor if drug benefits justify risk to unborn baby. Risk category C (see page xviii).

DIGITALIS PREPARATIONS (Digitalis Glycosides)

Breastfeeding; Lactation; Nursing Mothers:
Drugs in this group are generally acceptable with breastfeeding. Discuss risks and benefits with your doctor.

Infants & children up to age 18:
May be used to treat heart failure. Follow instructions provided by your child's doctor.

Prolonged use:
Talk to your doctor about the need for follow-up medical exam or lab studies.

Skin & sunlight:
No problems expected.

Driving, piloting or hazardous work:
Possible vision disturbances. Otherwise, no problems expected.

Discontinuing:
Don't stop without doctor's advice.

Others:
- Advise any doctor, dentist or pharmacist whom you consult that you use this medicine.
- Some digitalis products contain tartrazine dye. Avoid, especially if you are allergic to aspirin.

POSSIBLE INTERACTION WITH OTHER DRUGS

GENERIC NAME OR DRUG CLASS	COMBINED EFFECT
Adrenocorticoids, systemic	Dangerous potassium depletion. Possible digitalis toxicity.
Amiodarone	Increased digitalis effect.
Amphotericin B	Decreased potassium. Increased toxicity of amphotericin B.
Antacids*	Decreased digitalis effect.
Anticonvulsants, hydantoin*	Increased digitalis effect at first, then decreased.
Anticholinergics*	Possible increased digitalis effect.
Anti-inflammatory drugs, nonsteroidal (NSAIDs)*	Increased effect of digoxin.
Attapulgite	May decrease effect of digitalis.
Beta-adrenergic blocking agents*	Increased digitalis effect.
Beta-agonists*	Increased risk of heartbeat irregularity.
Calcium supplements*	Decreased digitalis effects.
Carteolol	Can increase or decrease heart rate.
Cholestyramine	Decreased digitalis effect.
Colestipol	Decreased digitalis effect.
Dextrothyroxine	Decreased digitalis effect.
Disopyramide	Possible decreased digitalis effect.
Diuretics*	Excessive potassium loss that may cause irregular heartbeat.
DPP-4 inhibitors	Digoxin dosage may need to be adjusted.
Ephedrine	Disturbed heart rhythm. Avoid.
Epinephrine	Disturbed heart rhythm. Avoid.
Erythromycins*	May increase digitalis absorption.
Flecainide	May increase digitalis effect.
Fluoxetine	May increase digitalis effect.
Hydroxychloroquine	Possible increased digitalis toxicity.
Itraconazole	Possible toxic levels of digitalis.

Continued on page 898

POSSIBLE INTERACTION WITH OTHER SUBSTANCES

INTERACTS WITH	COMBINED EFFECT
Alcohol:	None expected.
Beverages:	None expected.
Cocaine:	Irregular heartbeat. Avoid.
Food: High in fiber.	May decrease digitalis effect.
Marijuana:	Consult doctor.
Tobacco:	Irregular heartbeat. Avoid.

***See Glossary**

DIPHENOXYLATE & ATROPINE

BRAND NAMES

Lomotil
Lonox

BASIC INFORMATION

Habit forming? Yes
Prescription needed? Yes
Available as generic? Yes
Drug class: Antidiarrheal

USES

Relieves diarrhea and intestinal cramps.

DOSAGE & USAGE INFORMATION

How to take:

- Tablet—Swallow with liquid or food to lessen stomach irritation.
- Drops or liquid—Follow label instructions and use marked dropper.

When to take:
No more often than directed on label.

If you forget a dose:
Take as soon as you remember. If it is almost time for the next dose, wait for that dose (don't double this dose) and resume regular schedule.

What drug does:
Blocks digestive tract's nerve supply, which reduces propelling movements.

Time lapse before drug works:
May require 12 to 24 hours of regular doses to control diarrhea.

Don't take with:
Any other medicine or any dietary supplement without consulting your doctor or pharmacist.

OVERDOSE

SYMPTOMS:
Pinpoint pupils, rapid pulse. trouble breathing, dizziness, fever, hallucinations, confusion, slurred speech, agitation, flushed face, convulsions, coma.
WHAT TO DO:

- **Dial 911 for all medical emergencies or call poison control center 1-800-222-1222 for instructions.**
- **See emergency information on last 3 pages of this book.**

POSSIBLE ADVERSE REACTIONS OR SIDE EFFECTS

SYMPTOMS	WHAT TO DO
Life-threatening:	
Rare, severe allergic reaction (anaphylaxis)—difficulty breathing, hives, itching, swelling (throat, face), vomiting, dizziness.	Seek emergency treatment immediately.
Common:	
None expected.	
Infrequent:	
Dry mouth or skin, numbness of hands or feet, dizziness, depression, rash or itch, blurred vision, decreased urination, drowsiness, headache, swollen gums (these symptoms usually mean too much of the drug has been taken).	Discontinue. Call doctor right away.
Rare:	
Severe stomach pain, nausea, vomiting, constipation, bloating, loss of appetite.	Discontinue. Call doctor right away.

WARNINGS & PRECAUTIONS

Don't take if:
You are allergic to diphenoxylate or atropine.

Before you start, consult your doctor if:

- You have a history of alcohol or drug abuse.
- You have severe colitis.
- Patient has Down syndrome.
- You have dysentery.
- You have asthma, emphysema, bronchitis or other chronic lung disease.
- You have enlarged prostate, urinary tract blockage, difficult urination, incontinence or an intestinal blockage.
- You have liver or kidney disease or gallbladder disease or gallstones.
- You have glaucoma.
- You have heart disease, hiatal hernia, high blood pressure (hypertension).
- You have myasthenia gravis.
- You have overactive or underactive thyroid.

Over age 60:
Adverse reactions and side effects may be more frequent and severe than in younger persons.

Pregnancy:
Decide with your doctor if drug benefits justify risk to unborn baby. Risk category C (see page xviii).

Breastfeeding; Lactation; Nursing Mothers:
This drug is generally acceptable with breastfeeding. Discuss risks and benefits with your doctor.

Infants & children up to age 18:
Don't give to children under 2 years of age. Follow instructions provided by your child's doctor.

Prolonged use:
- May be habit forming if larger doses than recommended are taken for a long period of time.
- Talk to your doctor about the need for follow-up medical examinations or laboratory studies to check liver function.

Skin & sunlight:
No problems expected.

Driving, piloting or hazardous work:
Don't drive or pilot aircraft until you learn how medicine affects you. Don't work around dangerous machinery. Don't climb ladders or work in high places. Danger increases if you drink alcohol or take medicine affecting alertness and reflexes.

Discontinuing:
- May be unnecessary to finish medicine. Follow doctor's instructions.
- After discontinuing, consult doctor if you experience muscle cramps, nausea, vomiting, trembling, stomach cramps or unusual sweating.

Others:
If diarrhea lasts longer than a few days, discontinue drug and call doctor.

POSSIBLE INTERACTION WITH OTHER DRUGS

GENERIC NAME OR DRUG CLASS	COMBINED EFFECT
Barbiturates*	Increased effect of both drugs.
Clozapine	Toxic effect on the central nervous system.
Ethinamate	Dangerous increased effects of ethinamate. Avoid combining.
Fluoxetine	Increased depressant effects of both drugs.
Guanfacine	May increase depressant effects of either drug.
Leucovorin	High alcohol content of leucovorin may cause adverse effects.
Methyprylon	Increased sedative effect, perhaps to dangerous level. Avoid.
Monoamine oxidase (MAO) inhibitors*	May increase blood pressure excessively.
Naltrexone	Triggers withdrawal symptoms.
Narcotics*	Increased sedation. Avoid.
Sedatives*	Increased effect of both drugs.
Sertraline	Increased depressive effects of both drugs.
Tranquilizers*	Increased effect of both drugs.

POSSIBLE INTERACTION WITH OTHER SUBSTANCES

INTERACTS WITH	COMBINED EFFECT
Alcohol:	None expected.
Beverages:	None expected.
Cocaine:	Unknown. Avoid.
Foods:	None expected.
Marijuana:	Consult doctor.
Tobacco:	None expected.

DIPYRIDAMOLE

BRAND NAMES

Aggrenox
Apo-Dipyridamole
Dipimol
Dipridacot
Novodipiradol
Persantine
Pyridamole

BASIC INFORMATION

Habit forming? No
Prescription needed?
U.S.: Yes
Canada: No
Available as generic? Yes
Drug class: Platelet aggregation inhibitor

USES

- May reduce frequency and intensity of angina attacks.
- May reduce the risk of blood clots after heart surgery.

DOSAGE & USAGE INFORMATION

How to take:
Tablet—Swallow with a full glass of water. If you can't swallow whole, crumble tablet and take with liquid.

When to take:
1 hour before or 2 hours after meals.

If you forget a dose:
Take as soon as you remember. If it is almost time for the next dose, wait for that dose (don't double this dose) and resume regular schedule.

Continued next column

OVERDOSE

SYMPTOMS:
Flushing, sweating, warm feeling, restlessness, weakness, dizziness.
WHAT TO DO:

- **Dial 911 for all medical emergencies or call poison control center 1-800-222-1222 for instructions.**
- **See emergency information on last 3 pages of this book.**

What drug does:

- Probably dilates blood vessels to increase oxygen to heart.
- May reduce platelet clumping, which causes blood clots.

Time lapse before drug works:
3 months of continual use.

Don't take with:
Any other medicine or any dietary supplement without consulting your doctor or pharmacist.

POSSIBLE ADVERSE REACTIONS OR SIDE EFFECTS

SYMPTOMS	WHAT TO DO
Life-threatening:	
In case of overdose, see previous column.	
Common:	
Dizziness.	Continue. Call doctor when convenient.
Infrequent:	
• Fainting, headache.	Discontinue. Call doctor right away.
• Red flush, rash, nausea, vomiting, cramps, weakness.	Continue. Call doctor when convenient.
Rare:	
Chest pain.	Discontinue. Call doctor right away.

WARNINGS & PRECAUTIONS

Don't take if:
- You are allergic to dipyridamole. Allergic reactions are rare, but may occur.
- You are recovering from a heart attack.

Before you start, consult your doctor if:
- You have low blood pressure.
- You have liver disease.

Over age 60:
No special problems expected.

Pregnancy:
Decide with your doctor if drug benefits justify risk to unborn baby. Risk category B (see page xviii).

Breastfeeding; Lactation; Nursing Mothers:
This drug is generally acceptable with breastfeeding. Discuss risks and benefits with your doctor.

Infants & children up to age 18:
Approved for use in children age 12 and over. Follow instructions provided by your child's doctor.

Prolonged use:
Talk to your doctor about the need for follow-up medical examinations or laboratory studies.

Skin & sunlight:
No problems expected.

Driving, piloting or hazardous work:
Avoid if you feel dizzy. Otherwise, no problems expected.

Discontinuing:
Don't discontinue without doctor's advice until you complete prescribed dose, even though symptoms diminish or disappear.

Others:
- Drug increases your ability to be active without angina pain. Avoid excessive physical exertion that might injure heart.
- Advise any doctor, dentist or pharmacist whom you consult that you take this medicine.

WARNINGS & PRECAUTIONS

GENERIC NAME OR DRUG CLASS	COMBINED EFFECT
Anticoagulants, oral*	Increased anticoagulant effect. Bleeding tendency.
Aspirin and combination drugs containing aspirin	Increased dipyridamole effect. Dose may need adjustment.

POSSIBLE INTERACTION WITH OTHER SUBSTANCES

INTERACTS WITH	COMBINED EFFECT
Alcohol:	May lower blood pressure excessively.
Beverages:	None expected.
Cocaine:	Unknown. Avoid.
Foods:	Decreased dipyridamole absorption unless taken 1 hour before eating.
Marijuana:	Consult doctor.
Tobacco: Nicotine.	May decrease dipyridamole effect.

*See Glossary

DISOPYRAMIDE

BRAND NAMES

Norpace
Norpace CR
Rythmodan
Rythmodan-LA

BASIC INFORMATION

Habit forming? No
Prescription needed? Yes
Available as generic? Yes
Drug class: Antiarrhythmic

USES

Corrects heart rhythm disorders.

DOSAGE & USAGE INFORMATION

How to take:
- Extended-release tablet or capsule—Swallow with liquid. Do not crush tablet or open capsule.
- Capsule—Swallow with liquid. If you can't swallow whole, ask your pharmacist to prepare a liquid suspension for your use.

When to take:
At the same times each day.

If you forget a dose:
Take as soon as you remember. If it is almost time for the next dose, wait for that dose (don't double this dose) and resume regular schedule.

What drug does:
Delays nerve impulses to heart to regulate heartbeat.

Time lapse before drug works:
Begins in 30 to 60 minutes. Must use for 5 to 7 days to determine effectiveness.

Don't take with:
Any other medicine or any dietary supplement without consulting your doctor or pharmacist.

OVERDOSE

SYMPTOMS:
Dizziness, dry mouth and eyes, blurred vision, constipation, trouble urinating, irregular heartbeat, loss of consciousness.
WHAT TO DO:
- **Dial 911 for all medical emergencies or call poison control center 1-800-222-1222 for instructions.**
- **See emergency information on last 3 pages of this book.**

POSSIBLE ADVERSE REACTIONS OR SIDE EFFECTS

SYMPTOMS	WHAT TO DO
Life-threatening: In case of overdose, see previous column.	
Common:	
• Dizziness, fainting, shortness of breath, nervousness, slow or fast heartbeat, unusual tiredness.	Discontinue. Call doctor right away.
• Dryness (mouth, nose, eyes, throat), changes in urination, constipation.	Continue. Call doctor when convenient.
Infrequent:	
Chest pain, swelling of feet or legs, rapid weight gain, rash or itching, hypoglycemia/low blood sugar (cold sweats, fatigue, extreme hunger, fast heart rate, confusion, cool and pale skin).	Discontinue. Call doctor right away.
Rare:	
• Yellow eyes or skin, fever, sore throat, nosebleeds, bleeding gums, depression.	Discontinue. Call doctor right away.
• Bloating, diarrhea, loss of appetite, impotence, nausea, headache, muscle weakness, insomnia, nervousness.	Continue. Call doctor when convenient.

WARNINGS & PRECAUTIONS

Don't take if:
You are allergic to disopyramide. Allergic reactions are rare, but may occur.

Before you start, consult your doctor if:
- You have heart failure or heart disease.
- You have low blood pressure.
- You have kidney or liver disease.
- You have glaucoma.
- You have enlarged prostate or difficulty with urination.
- You have myasthenia gravis.
- You have diabetes.
- You have an electrolyte disorder or suffer from malnutrition.

Over age 60:
Adverse reactions and side effects may be more frequent and severe than in younger persons.

Pregnancy:
Decide with your doctor if drug benefits justify risk to unborn child. Risk category C (see page xviii).

Breastfeeding; Lactation; Nursing Mothers:
This drug is generally acceptable with breastfeeding. Discuss risks and benefits with your doctor.

Infants & children up to age 18:
Follow instructions provided by your child's doctor.

Prolonged use:
Talk to your doctor about the need for follow-up medical examinations or laboratory studies to check liver function, kidney function, ECG,* blood pressure, serum potassium.

Skin & sunlight:
May cause rash or intensify sunburn in areas exposed to sun or ultraviolet light (photosensitivity reaction). Avoid overexposure. Notify doctor if reaction occurs.

Driving, piloting or hazardous work:
Don't drive or pilot aircraft until you learn how medicine affects you. Don't work around danger-ous machinery. Don't climb ladders or work in high places. Danger increases if you drink alcohol or take medicine affecting alertness and reflexes, such as antihistamines, tranquilizers, sedatives, pain medicine, narcotics, or mind-altering drugs.

Discontinuing:
Don't discontinue without doctor's advice until you complete prescribed dose, even though symptoms diminish or disappear.

Others:
Advise any doctor, dentist or pharmacist whom you consult that you take this medicine.

WARNINGS & PRECAUTIONS

GENERIC NAME OR DRUG CLASS	COMBINED EFFECT
Enzyme inducers*	May decrease effect of disopyramide.
Enzyme inhibitors*	Increased effect of disopyramide that could be life-threatening. Consult doctor.

POSSIBLE INTERACTION WITH OTHER SUBSTANCES

INTERACTS WITH	COMBINED EFFECT
Alcohol:	Use with caution.
Beverages:	None expected.
Cocaine:	Unknown. Avoid.
Foods:	None expected.
Marijuana:	Consult doctor.
Tobacco:	None expected.

DISULFIRAM

BRAND NAMES

Antabuse

BASIC INFORMATION

Habit forming? No
Prescription needed? Yes
Available as generic? Yes
Drug class: Antialcoholic agent

USES

Treatment for alcoholism. Will not cure alcoholism, but is a powerful deterrent to drinking.

DOSAGE & USAGE INFORMATION

How to take:
Tablet—Swallow with liquid.

When to take:
Morning or bedtime. Avoid if you have used *any* alcohol, tonics, cough syrups, fermented vinegar, after-shave lotion or backrub solutions within 12 hours.

If you forget a dose:
Take as soon as you remember. If it is almost time for the next dose, wait for that dose (don't double this dose) and resume regular schedule.

What drug does:
In combination with alcohol, produces a metabolic change that causes severe, temporary toxicity in the body.

Time lapse before drug works:
3 to 12 hours.

Continued next column

OVERDOSE

SYMPTOMS:
Nausea, vomiting, dizziness, numbness and tingling, lack of coordination, seizures.
WHAT TO DO:

- **Dial 911 for all medical emergencies or call poison control center 1-800-222-1222 for instructions.**
- **See emergency information on last 3 pages of this book.**

Don't take with:

- Nonprescription drugs that contain *any* alcohol.
- Any other medicine or any dietary supplement without consulting your doctor or pharmacist.
- Any other central nervous system (CNS) depressant drugs.*

POSSIBLE ADVERSE REACTIONS OR SIDE EFFECTS

SYMPTOMS	WHAT TO DO
Life-threatening:	
In case of overdose, see previous column.	
Common:	
Drowsiness.	Continue. Call doctor when convenient.
Infrequent:	
• Eye pain, vision changes, abdominal discomfort, throbbing headache, numbness in hands and feet, mood changes, decreased sexual ability in men, tiredness.	Continue. Call doctor when convenient.
• Bad taste in mouth (metal or garlic).	No action necessary.
Rare:	
Rash, yellow skin or eyes.	Discontinue. Call doctor right away.

WARNINGS & PRECAUTIONS

Don't take if:

- You are allergic to disulfiram. Allergic reactions are rare, but may occur. Alcohol-disulfiram combination is not an allergic reaction.
- Patient is in a state of alcohol intoxication.

Before you start, consult your doctor if:

- No one has explained to you how disulfiram reacts with alcohol.
- Patient has any type of brain damage.
- You plan to become pregnant within medication period.
- You think you cannot avoid drinking.
- You have low thyroid problem, epilepsy, diabetes, heart, liver or kidney disease.
- You take other drugs.

Over age 60:
Adverse reactions and side effects may be more frequent and severe than in younger persons.

Pregnancy:
Decide with your doctor if drug benefits justify risk to unborn baby. Risk category C (see page xviii).

Breastfeeding; Lactation; Nursing Mothers:
This drug is generally not recommended with breastfeeding. Discuss risks and benefits with your doctor.

Infants & children up to age 18:
Safety and effectiveness in this age group have not been established. Consult doctor.

Prolonged use:
Periodic blood cell counts and liver function tests recommended if you take this drug a long time.

Skin & sunlight:
No problems expected.

Driving, piloting or hazardous work:
Avoid if you feel drowsy or have vision side effects. Otherwise, no restrictions.

Discontinuing:
Don't discontinue without consulting doctor. Dose may require gradual reduction if you have taken drug for a long time. Doses of other drugs may also require adjustment. Avoid alcohol at least 14 days following last dose.

Others:
- Check all liquids that you take or rub on for presence of alcohol.
- Advise any doctor, dentist or pharmacist whom you consult that you take this medicine.

POSSIBLE INTERACTION WITH OTHER DRUGS

GENERIC NAME OR DRUG CLASS	COMBINED EFFECT
Anticoagulants*	Possible unexplained bleeding.
Anticonvulsants*	Excessive sedation.
Barbiturates*	Excessive sedation.
Central nervous system (CNS) depressants*	Increased depressive effect.
Clozapine	Toxic effect on the central nervous system.
Guanfacine	May increase depressant effects of either drug.
Isoniazid	Unsteady walk and disturbed behavior.
Leucovorin	High alcohol content of leucovorin may cause disulfiram reaction.*
Methyprylon	Increased sedative effect, perhaps to dangerous level. Avoid.
Metronidazole	Disulfiram reaction.*
Nabilone	Greater depression of central nervous system.
Sedatives*	Excessive sedation.
Theophylline	Increased theophylline effect; possibly toxic levels.

POSSIBLE INTERACTION WITH OTHER SUBSTANCES

INTERACTS WITH	COMBINED EFFECT
Alcohol: *Any* form or amount.	Possible life-threatening toxicity. See disulfiram reaction.*
Beverages: Punch or fruit drink that may contain alcohol.	Disulfiram reaction.*
Cocaine:	Increased risk of side effects. Avoid.
Foods: Sauces, fermented vinegar, marinades, desserts or other foods prepared with *any* alcohol.	Disulfiram reaction.*
Marijuana:	Consult doctor.
Tobacco:	None expected.

*See Glossary

DIURETICS, LOOP

GENERIC AND BRAND NAMES

BUMETANIDE
Bumex
ETHACRYNIC ACID
Edecrin
FUROSEMIDE
Apo-Furosemide
Furoside
Lasix
Lasix Special
Myrosemide
Novosemide
Uritol
TORSEMIDE
Demadex

BASIC INFORMATION

Habit forming? No
Prescription needed? Yes
Available as generic? Yes
Drug class: Diuretic (loop), antihypertensive

USES

- Lowers high blood pressure.
- Decreases fluid retention.

DOSAGE & USAGE INFORMATION

How to take:
Tablet or liquid—Swallow with liquid. If you can't swallow tablet whole, crumble tablet and take with liquid or food.

When to take:
- 1 dose a day—Take after breakfast.
- More than 1 dose a day—Take last dose no later than 6 p.m. unless otherwise directed.

If you forget a dose:
Take as soon as you remember. If it is almost time for the next dose, wait for that dose (don't double this dose) and resume regular schedule.

Continued next column

What drug does:
Increases elimination of sodium, potassium and water from body. Decreased body fluid reduces blood pressure.

Time lapse before drug works:
1 hour to increase water loss. Requires 2 to 3 weeks to lower blood pressure.

Don't take with:
- Nonprescription drugs with aspirin.
- Any other medicine or any dietary supplement without consulting your doctor or pharmacist.

OVERDOSE

SYMPTOMS:
Weakness, dehydration, dizziness, stomach cramps, confusion, loss of appetite, fainting, vomiting, ringing in the ears.
WHAT TO DO:
- **Dial 911 for all medical emergencies or call poison control center 1-800-222-1222 for instructions.**
- **See emergency information on last 3 pages of this book.**

POSSIBLE ADVERSE REACTIONS OR SIDE EFFECTS

SYMPTOMS	WHAT TO DO
Life-threatening:	
Rare, severe allergic reaction (anaphylaxis)—difficulty breathing, hives, itching, swelling (face, throat), wheezing, vomiting, dizziness.	Seek emergency treatment immediately.
Common:	
Dizziness.	Continue. Call doctor when convenient.
Infrequent:	
Mood change, fatigue, appetite loss, diarrhea, irregular heartbeat, muscle cramps, low blood pressure, pain in abdomen, weakness.	Discontinue. Call doctor right away.
Rare:	
Rash or hives, yellow vision, ringing in ears, hearing loss, sore throat, fever, dry mouth, thirst, side or stomach pain, nausea, vomiting, unusual bleeding or bruising, joint pain, yellow skin or eyes, numbness or tingling in hands or feet.	Discontinue. Call doctor right away.

WARNINGS & PRECAUTIONS

Don't take if:
You are allergic to loop diuretics.

Before you start, consult your doctor if:
- You are taking any other prescription or nonprescription medicine.
- You are allergic to any sulfa drug.
- You have liver or kidney disease.
- You have gout, diabetes or impaired hearing.
- You will have surgery within 2 months, including dental surgery.

Over age 60:
Adverse reactions and side effects may be more frequent and severe than in younger persons.

Pregnancy:
Decide with your doctor if drug benefits justify risk to unborn baby. Risk category C for bumetanide, furosemide; category B for ethacrynic acid, torsemide (see page xviii).

Breastfeeding; Lactation; Nursing Mothers:
Drugs in this group are generally acceptable with breastfeeding. Discuss risks and benefits with your doctor.

Infants & children up to age 18:
Follow instructions provided by your child's doctor.

Prolonged use:
Impaired balance of water and salt, with low potassium level in blood and body tissues.

Skin & sunlight:
One or more drugs in this group may cause rash or intensify sunburn in areas exposed to sun or ultraviolet light (photosensitivity reaction). Avoid overexposure. Notify doctor if reaction occurs.

Driving, piloting or hazardous work:
Avoid if you feel dizzy; otherwise no problems expected.

Discontinuing:
Don't discontinue without doctor's advice until you complete prescribed dose, even though symptoms diminish or disappear.

Others:
Frequent laboratory studies to monitor potassium level in blood recommended. Consult doctor about potassium in diet or supplements.

POSSIBLE INTERACTION WITH OTHER DRUGS

GENERIC NAME OR DRUG CLASS	COMBINED EFFECT
Adrenocorticoids, systemic	Potassium depletion.
Allopurinol	Decreased allopurinol effect.
Amiodarone	Increased risk of heartbeat irregularity.
Angiotensin-converting enzyme (ACE) inhibitors*	Possible excessive potassium in blood.
Anticoagulants*	Abnormal clotting.
Antidepressants, tricyclic*	Excessive blood pressure drop.
Antidiabetics, oral*	Decreased antidiabetic effect.
Antihypertensives*	Increased anti-hypertensive effect. Dosages may require adjustment.
Anti-inflammatory drugs, nonsteroidal (NSAIDs)*	Decreased diuretic effect.
Antivirals, HIV/AIDS*	Increased risk of pancreatitis with furosemide.
Barbiturates*	Low blood pressure.
Beta-adrenergic blocking agents*	Increased anti-hypertensive effect. Dosages may require adjustment.
Corticosteroids*	Decreased potassium.
Digitalis preparations*	Excessive potassium loss could lead to serious heart rhythm disorders.
Diuretics, other*	Increased diuretic effect.
Hypokalemia-causing medicines*	Increased risk of excess potassium loss.
Insulin	Decreased insulin effect.
Lithium	Increased lithium toxicity.
Meloxicam	Decreased effect of diuretic.
Metformin	Increased metformin effect with furosemide.

Continued on page 899

POSSIBLE INTERACTION WITH OTHER SUBSTANCES

INTERACTS WITH	COMBINED EFFECT
Alcohol:	Blood pressure drop. Avoid.
Beverages:	None expected.
Cocaine:	Unknown. Avoid.
Foods:	None expected.
Marijuana:	Consult doctor.
Tobacco:	None expected.

***See Glossary**

DIURETICS, POTASSIUM-SPARING

GENERIC AND BRAND NAMES

AMILORIDE
Midamor
SPIRONOLACTONE
Aldactone
Novospiroton
TRIAMTERENE
Dyrenium

BASIC INFORMATION

Habit forming? No
Prescription needed? Yes
Available as generic? Yes
Drug class: Diuretic, antihypertensive, antihypokalemic

USES

- Treatment for hypertension (high blood pressure) and congestive heart failure.
- Treats low potassium, polycystic ovary syndrome and hirsutism in women.

DOSAGE & USAGE INFORMATION

How to take:
Capsule or tablet—Swallow with liquid. If you can't swallow whole, open capsule or crush tablet and take with liquid or food. May take with meal to lessen stomach irritation.

When to take:
At the same times each day. May interfere with sleep if taken after 6 p.m.

If you forget a dose:
Take as soon as you remember. If it is almost time for the next dose, wait for that dose (don't double this dose) and resume regular schedule.

Continued next column

OVERDOSE

SYMPTOMS:
Drowsiness, nausea, vomiting, confusion, rash, dizziness, dehydration.
WHAT TO DO:
- **Dial 911 for all medical emergencies or call poison control center 1-800-222-1222 for instructions.**
- **See emergency information on last 3 pages of this book.**

What drug does:
- Decreases fluid retention and prevents potassium loss.
- In polycystic ovary syndrome and hirsutism, blocks androgen hormones.

Time lapse before drug works:
2 to 4 hours.

Don't take with:
Any other medicine or any dietary supplement without consulting your doctor or pharmacist.

POSSIBLE ADVERSE REACTIONS OR SIDE EFFECTS

SYMPTOMS	WHAT TO DO
Life-threatening:	
Rare, severe allergic reaction (anaphylaxis)—difficulty breathing, hives, itching, swelling (throat, face), vomiting, dizziness.	Seek emergency treatment immediately.
Common:	
Headache, nausea, appetite loss, mild diarrhea, vomiting.	Continue. Call doctor when convenient.
Infrequent:	
Dizziness, muscle cramps, dry mouth, decreased sexual drive, constipation.	Continue. Call doctor when convenient.
Rare:	
Shortness of breath, skin rash or itch (with amiloride), cough or hoarseness, painful urination, back or side pain (with triamterene and spironolactone), potassium changes (confusion, dry mouth, breathing difficulty, irregular heartbeat, unusual tiredness or weakness, mood or mental changes, muscle cramps, tingling in body), red or burning or inflamed tongue (with triamterene).	Discontinue. Call doctor right away.

DOSAGE & USAGE INFORMATION

Don't take if:
- You are allergic to potassium-sparing diuretics.
- Your serum potassium level is high.

Before you start, consult your doctor if:
- You have diabetes.
- You have heart disease, kidney or liver disease or gout.

Over age 60:
Adverse reactions and side effects may be more frequent and severe than in younger persons.

Pregnancy:
Decide with your doctor if drug benefits justify risk to unborn baby. Risk category B for amiloride; category C for others (see page xviii).

Breastfeeding; Lactation; Nursing Mothers:
Drugs in this group are generally acceptable with breastfeeding. Discuss risks and benefits with your doctor.

Infants & children up to age 18:
Safety and effectiveness in this age group have not been established. Consult doctor.

Prolonged use:
Talk to your doctor about the need for follow-up medical examinations or laboratory studies to check blood pressure, kidney function, ECG* and serum electrolytes.

Skin & sunlight:
One or more of these drugs may cause increased sensitivity to sunlight (photosensitivity reaction). Avoid overexposure. If reaction occurs, notify doctor.

Driving, piloting or hazardous work:
Don't drive or pilot aircraft until you learn how medicine affects you. Don't work around dangerous machinery. Don't climb ladders or work in high places. Danger increases if you drink alcohol or take medicine affecting alertness and reflexes.

Discontinuing:
Don't discontinue without doctor's advice until you complete prescribed dose, even though symptoms diminish or disappear.

Others:
- Periodic physical checkups and potassium-level tests recommended.
- Advise any doctor, dentist or pharmacist whom you consult that you take this medicine.
- If you experience an illness with severe vomiting and diarrhea, consult doctor.
- A special diet may be recommended in addition to this medicine. Follow doctor's advice.

POSSIBLE INTERACTION WITH OTHER DRUGS

GENERIC NAME OR DRUG CLASS	COMBINED EFFECT
Amantadine	Increased effect of amantadine (with triamterene).
Angiotensin-converting enzyme (ACE) inhibitors*	Possible excessive potassium in blood.
Anticoagulants,* oral	Decreased anticoagulant effect.
Antigout drugs*	Decreased antigout effect.
Antihypertensives*	Increased effect of both drugs.
Anti-inflammatory drugs, nonsteroidal (NSAIDs)*	Increased potassium levels.
Cyclosporine	Increased potassium levels.
Digoxin	Increased digoxin effect (with spironolactone).
Diuretics,* other	Increased effect of both drugs.
Dofetilide	Increased risk of heart problems.
Folic acid	Decreased effect of folic acid.
Lithium	Possible lithium toxicity.
Memantine	Increased effect of memantine or triamterene.
Metformin	Increased metformin effect.
Potassium supplements*	Increased potassium levels.

POSSIBLE INTERACTION WITH OTHER SUBSTANCES

INTERACTS WITH	COMBINED EFFECT
Alcohol:	Increased blood pressure drop. Avoid.
Beverages:	None expected.
Cocaine:	Blood pressure rise. Avoid.
Foods: Salt substitutes.	Possible excess potassium levels.
Marijuana:	Consult doctor.
Tobacco:	None expected.

*See Glossary

DIURETICS, POTASSIUM-SPARING & HYDROCHLOROTHIAZIDE

GENERIC AND BRAND NAMES

AMILORIDE & HYDROCHLOROTHIAZIDE
Moduret
Moduretic
SPIRONOLACTONE & HYDROCHLOROTHIAZIDE
Aldactazide
Spirozide
TRIAMTERENE & HYDROCHLOROTHIAZIDE
Apo-Triazide
Diazide
Maxzide
Novo-Triamzide

BASIC INFORMATION

Habit forming? No
Prescription needed? Yes
Available as generic? Yes
Drug class: Diuretic, antihypertensive, antihypokalemic

USES

- Treats high blood pressure (hypertension) and congestive heart failure. Decreases fluid retention and prevents potassium loss.
- Treatment for hypokalemia (low potassium).

DOSAGE & USAGE INFORMATION

How to take:
Capsule or tablet—Swallow with liquid. Take with meals or milk if stomach irritation occurs.

When to take:
At the same time or times each day. May interfere with sleep if taken after 6 p.m.

If you forget a dose:
Take as soon as you remember. If it is almost time for the next dose, wait for that dose (don't double this dose) and resume regular schedule.

Continued next column

OVERDOSE

SYMPTOMS:
Drowsiness, nausea, vomiting, confusion, rash, dizziness, dehydration, weakness, excess urination, fever, flushing.
WHAT TO DO:
- **Dial 911 for all medical emergencies or call poison control center 1-800-222-1222 for instructions.**
- **See emergency information on last 3 pages of this book.**

What drug does:
This is a combination of 2 diuretics that blocks exchange of certain chemicals in the kidneys so sodium and water are excreted. Conserves potassium.

Time lapse before drug works:
Starts in 2 to 4 hours; several days for full effect.

Don't take with:
Any other medicine or any dietary supplement without consulting your doctor or pharmacist.

POSSIBLE ADVERSE REACTIONS OR SIDE EFFECTS

SYMPTOMS	WHAT TO DO
Life-threatening:	
Rare, severe allergic reaction (anaphylaxis)—difficulty breathing, hives, itching, swelling (throat, face), vomiting, dizziness.	Seek emergency treatment immediately.
Common:	
Nausea or mild vomiting, appetite loss, stomach cramps, mild diarrhea, constipation (with amiloride).	Continue. Call doctor when convenient.
Infrequent:	
Dizziness, muscle cramps, headache, skin sensitive to sun, dry mouth, decreased interest in sex. With spironolactone—tender breasts, deepening of voice, menstrual changes and increased hair growth in females, breast enlargement in males, increased sweating in both sexes.	Continue. Call doctor when convenient.
Rare:	
Black or tarry or bloody stools, blood in urine, pain or difficulty in urinating, fever or chills, pain in back or side or joints, rash or hives on skin, severe stomach pain, unusual bleeding or bruising, yellow skin or eyes, potassium changes (confusion, dry mouth, breathing difficulty, irregular heartbeat, unusual tiredness or weakness, mood or mental changes, muscle cramps,	Discontinue. Call doctor right away.

tingling in body), red or burning feeling of tongue (with triamterene).

WARNINGS & PRECAUTIONS

Don't take if:

- You are allergic to potassium-sparing or thiazide diuretics* or sulfa drugs.*
- Your serum potassium level is high.

Before you start, consult your doctor if:

- You have diabetes or lupus erythematosus.
- You have menstrual problems (in females) or enlarged breast (in males).
- You have heart or blood vessel disease, kidney or liver disease, pancreatitis or gout.

Over age 60:
Adverse reactions and side effects may be more frequent and severe than in younger persons.

Pregnancy:
Decide with your doctor if drug benefits justify risk to unborn baby. Risk category B or C (see page xviii).

Breastfeeding; Lactation; Nursing Mothers:
Drugs in this group are generally acceptable with breastfeeding. Discuss risks and benefits with your doctor.

Infants & children up to age 18:
Safety and effectiveness in this age group have not been established. Consult doctor.

Prolonged use:
Talk to your doctor about the need for follow-up medical examinations or laboratory studies to check blood pressure, kidney function, ECG* and serum electrolytes.

Skin & sunlight:
One or more of these drugs may cause increased sensitivity to sunlight (photosensitivity reaction). Avoid overexposure. If reaction occurs, notify doctor.

Driving, piloting or hazardous work:
Don't drive or pilot aircraft until you learn how medicine affects you. Don't work around dangerous machinery. Don't climb ladders or work in high places. Danger increases if you drink alcohol or take medicine affecting alertness and reflexes, such as antihistamines, tranquilizers, sedatives, pain medicine, narcotics and mind-altering drugs.

Discontinuing:
Don't discontinue without doctor's approval, even though symptoms diminish or disappear.

Others:

- Your doctor may prescribe a special diet.
- Advise any doctor, dentist or pharmacist whom you consult that you take this medicine.
- If you experience an illness with severe vomiting or diarrhea, consult doctor.

POSSIBLE INTERACTION WITH OTHER DRUGS

GENERIC NAME OR DRUG CLASS	COMBINED EFFECT
Amantadine	Increased effect of amantadine (with triamterene).
Angiotensin-converting enzyme (ACE) inhibitors*	Possible excessive potassium in blood.
Anticoagulants,* oral	Decreased anticoagulant effect.
Antigout drugs*	Decreased antigout effect.
Antihypertensives*	Increased effect of both drugs.
Anti-inflammatory drugs, nonsteroidal (NSAIDs)*	Increased potassium levels.

Continued on page 899

POSSIBLE INTERACTION WITH OTHER SUBSTANCES

INTERACTS WITH	COMBINED EFFECT
Alcohol:	Increased blood pressure drop. Avoid.
Beverages:	None expected.
Cocaine:	Blood pressure rise. Avoid.
Foods: Salt substitutes.	Possible excess potassium levels.
Marijuana:	Consult doctor.
Tobacco:	None expected.

***See Glossary**

DIURETICS, THIAZIDE

GENERIC AND BRAND NAMES

See full list of generic and brand names in the *Generic and Brand Name Directory*, page 861.

BASIC INFORMATION

Habit forming? No
Prescription needed? Yes
Available as generic? Yes, for some.
Drug class: Antihypertensive, diuretic (thiazide)

USES

- Controls, but doesn't cure, high blood pressure.
- Reduces fluid retention (edema) caused by conditions such as heart disorders and liver disease.

DOSAGE & USAGE INFORMATION

How to take:
Tablet, capsule or liquid—Swallow with liquid. If you can't swallow whole, crumble tablet or open capsule and take with liquid or food. Don't exceed dose.

When to take:
At the same time each day.

If you forget a dose:
Take as soon as you remember. If it is almost time for the next dose, wait for that dose (don't double this dose) and resume regular schedule.

What drug does:
- Forces sodium and water excretion, reducing body fluid.
- Relaxes muscle cells of small arteries.
- Reduced body fluid and relaxed arteries lower blood pressure.

Continued next column

OVERDOSE

SYMPTOMS:
Muscle pain or cramps, weakness, dry mouth, drowsiness, thirst, stomach pain, dehydration, excess urination, flushing.
WHAT TO DO:
- **Dial 911 for all medical emergencies or call poison control center 1-800-222-1222 for instructions.**
- **See emergency information on last 3 pages of this book.**

Time lapse before drug works:
4 to 6 hours. May require several weeks to lower blood pressure.

Don't take with:
Any other medicine or any dietary supplement without consulting your doctor or pharmacist.

POSSIBLE ADVERSE REACTIONS OR SIDE EFFECTS

SYMPTOMS	WHAT TO DO
Life-threatening:	
Rare, severe allergic reaction (anaphylaxis)—difficulty breathing, hives, itching, swelling (throat, face), vomiting, dizziness.	Seek emergency treatment immediately.
Common:	
None expected.	
Infrequent:	
• Confusion, seizures, mental/mood changes, muscle cramps, unusual tiredness or weakness, dry mouth, nausea or vomiting, thirstiness, irregular heartbeat.	Discontinue. Call doctor right away.
• Dizziness, loss of appetite, diarrhea, decreased sex drive, skin sensitive to sunlight.	Continue. Call doctor when convenient.
Rare:	
Fever, chills, cough, low back or side pain, difficult urination, severe stomach pain, yellow eyes or skin, bloody stools or urine, joint pain, red spots on skin.	Discontinue. Call doctor right away.

WARNINGS & PRECAUTIONS

Don't take if:
You are allergic to any thiazide diuretic drug.

Before you start, consult your doctor if:
- You are allergic to any sulfa drug or tartrazine dye.
- You have systemic lupus erythematosus.
- You have gout, diabetes or a liver, pancreas or kidney disorder.

Over age 60:
Adverse reactions and side effects may be more frequent and severe than in younger persons.

Pregnancy:
Risk factors vary for drugs in this group. See category list on page xviii and consult doctor.

Breastfeeding; Lactation; Nursing Mothers:
Drugs in this group are generally acceptable with breastfeeding. Discuss risks and benefits with your doctor.

Infants & children up to age 18:
Safety and effectiveness in this age group have not been established. Consult doctor.

Prolonged use:
- You may need medicine to treat high blood pressure for the rest of your life.
- Talk to your doctor about the need for follow-up medical exams or lab studies.

Skin & sunlight:
One or more drugs in this group may cause rash or intensify sunburn in areas exposed to sun or ultraviolet light (photosensitivity reaction). Avoid overexposure. Notify doctor if reaction occurs.

Driving, piloting or hazardous work:
Don't drive or pilot aircraft until you learn how medicine affects you. Don't work around dangerous machinery. Don't climb ladders or work in high places. Danger increases if you drink alcohol or take medicine affecting alertness and reflexes.

Discontinuing:
Don't discontinue without medical advice.

Others:
- Hot weather and fever may cause dehydration and drop in blood pressure. Dose may require temporary adjustment. Weigh daily and report any unexpected weight loss to your doctor.
- May cause rise in uric acid, leading to gout.
- May cause blood-sugar rise in diabetics.
- May affect results in some medical tests.
- Advise any doctor, dentist or pharmacist whom you consult that you take this medicine.

POSSIBLE INTERACTION WITH OTHER DRUGS

GENERIC NAME OR DRUG CLASS	COMBINED EFFECT
Adrenocorticoids, systemic	Potassium depletion.
Allopurinol	Decreased allopurinol effect.
Amiodarone	Increased risk of heartbeat irregularity.
Amphotericin B	Increased potassium.
Angiotensin-converting enzyme (ACE) inhibitors*	Decreased blood pressure.
Antidepressants, tricyclic*	Dangerous drop in blood pressure. Avoid combination unless under medical supervision.
Antidiabetic agents, oral*	Increased blood sugar.
Antihypertensives*	Increased hypertensive effect.
Antivirals, HIV/AIDS*	Increased risk of pancreatitis.
Barbiturates*	Increased anti-hypertensive effect.
Beta-adrenergic blocking agents*	Increased anti-hypertensive effect. Dosages of both drugs may require adjustments.
Calcium supplements*	Increased calcium in blood.
Carteolol	Increased anti-hypertensive effect.
Cholestyramine	Decreased anti-hypertensive effect.
Colestipol	Decreased anti-hypertensive effect.
Digitalis preparations*	Excessive potassium loss that causes dangerous heart rhythms.
Diuretics, thiazide,* other	Increased effect of other thiazide diuretics.

Continued on page 899

POSSIBLE INTERACTION WITH OTHER SUBSTANCES

INTERACTS WITH	COMBINED EFFECT
Alcohol:	Dangerous blood pressure drop.
Beverages:	None expected.
Cocaine	Increased risk of high blood pressure or heart block. Avoid.
Foods: Licorice.	Excessive potassium loss that causes dangerous heart rhythms.
Marijuana:	May increase blood pressure. Avoid.
Tobacco:	None expected.

*See Glossary

DIVALPROEX

BRAND NAMES

Depakote
Depakote ER
Depakote Sprinkle
Epival

BASIC INFORMATION

Habit forming? No
Prescription needed? Yes
Available as generic? Yes
Drug class: Anticonvulsant

USES

- Treatment of epilepsy.
- Treatment of bipolar disorder.
- Treats migraine headaches.

DOSAGE & USAGE INFORMATION

How to take:
Delayed-release tablet/capsule or extended-release tablet—Swallow whole with liquid or food to lessen stomach irritation. Do not crush or chew capsule or tablets. You may open capsule and sprinkle contents on food (such as a teaspoon of applesauce), then swallow right away.

When to take:
Delayed form 2-3 times each day; extended form one time a day. The two different forms (delayed or extended) have different actions in the body.

If you forget a dose:
Take as soon as you remember. If it is almost time for the next dose, wait for that dose (don't double this dose) and resume regular schedule.

What drug does:
It helps stabilize electrical activity in the brain.

Time lapse before drug works:
1 to 4 hours, but full effect may take weeks.

Don't take with:
Any other medicine or any dietary supplement without consulting your doctor or pharmacist.

OVERDOSE

SYMPTOMS:
Extreme drowsiness, heart problems, loss of consciousness.
WHAT TO DO:
- **Dial 911 for all medical emergencies or call poison control center 1-800-222-1222 for instructions.**
- **See emergency information on last 3 pages of this book.**

POSSIBLE ADVERSE REACTIONS OR SIDE EFFECTS

SYMPTOMS	WHAT TO DO
Life-threatening:	
Rare, severe allergic reaction (anaphylaxis)—difficulty breathing, hives, itching, swelling (throat, face), vomiting, dizziness.	Seek emergency treatment immediately.
Common:	
Loss of appetite, indigestion, nausea, vomiting, stomach cramps, diarrhea, tremor, headache, weight gain or loss, menstrual changes.	Continue. Call doctor when convenient.
Infrequent:	
Clumsiness or unsteadiness, constipation, mild skin rash, dizziness, drowsiness, irritable or excited, hair loss.	Continue. Call doctor when convenient.
Rare:	
Mood or behavior changes, continued nausea and vomiting and appetite loss, increase in number of seizures, swelling (of face, feet or legs) yellow skin or eyes, tiredness or weakness, back-and-forth eye movements (nystagmus), seeing spots/seeing double, unusual bleeding or bruising, dark urine, light color stools, fever, severe stomach cramps, confusion.	Continue, but call doctor right away.

WARNINGS & PRECAUTIONS

Don't take if:
You are allergic to divalproex or valproic acid.

Before you start, consult your doctor if:
- You have liver, kidney, blood or brain disorder, pancreatitis or urea cycle disorder.
- Drug is to be used for a young child.
- You have a history of depression or suicide thoughts or suicidal behavior.
- You are a woman of childbearing age.

Over age 60:
Adverse reactions and side effects may be more frequent and severe than in younger persons.

Pregnancy:
Consult doctor. Use of the drug during pregnancy is not recommended. It could cause harm to the unborn baby. Risk category D (see page xviii).

Breastfeeding; Lactation; Nursing Mothers:
This drug is generally acceptable with breastfeeding. Discuss risks and benefits with your doctor.

Infants & children up to age 18:
Treats epilepsy in children over age 10. Follow instructions provided by your child's doctor.

Prolonged use:
Request periodic blood tests, liver and kidney function tests. These tests are necessary for safe and effective use.

Skin & sunlight:
No problems expected.

Driving, piloting or hazardous work:
Don't drive or pilot aircraft until you learn how medicine affects you. Don't work around dangerous machinery. Don't climb ladders or work in high places. Danger increases if you drink alcohol or take drugs affecting alertness.

Discontinuing:
Don't discontinue without consulting doctor. Dose may require gradual reduction if you have taken drug for a long time. Doses of other drugs may also require adjustment.

Others:
- Advise any doctor, dentist or pharmacist whom you consult that you take this medicine.
- Be sure you and your doctor discuss benefits and risks of this drug before starting.
- Wear or carry medical ID that indicates your disorder and the use of this medicine.
- Read and follow prescription instructions. Do not change dosage without doctor's approval.
- In rare cases, the drug can cause life-threatening liver failure (especially in children under age 2) or life-threatening pancreatitis (inflammation of the pancreas).
- Rarely, antiepileptic (anticonvulsant) drugs may lead to suicidal thoughts and behaviors. Call doctor right away if suicidal symptoms or unusual behaviors occur.

POSSIBLE INTERACTION WITH OTHER DRUGS

GENERIC NAME OR DRUG CLASS	COMBINED EFFECT
Anticoagulants,* oral	Increased risk of bleeding problems.
Anticonvulsants,* other	Each drug may need dosage adjusted.
Anti-inflammatory drugs, nonsteroidal (NSAIDs)*	Increased risk of bleeding problems.
Aspirin	Increased effect of divalproex.
Carbamazepine	Decreased effect of divalproex.
Central nervous system (CNS) depressants*	Increased sedative effect.
Clonazepam	May prolong seizure.
Diazepam	Increased effect of diazepam.
Enzyme inducers*	May Increase effect of some enzyme inducers and decrease effect of divalproex.
Felbamate	Increased effect of divalproex.
Hepatotoxics*	Increased risk of liver problems.
Lamotrigine	Increased effect of lamotrigine and risk of life-threatening rash.
Phenobarbital	Increased effect of phenobarbital and decreased effect of divalproex.
Phenytoin	Increased phenytoin effect; decreased divalproex effect.
Primidone	Increased effect of primidone.
Rifampin	Decreased effect of divalproex.
Salicylates*	Increased effect of divalproex.
Zidovudine	Increased effect of zidovudine.

POSSIBLE INTERACTION WITH OTHER SUBSTANCES

INTERACTS WITH	COMBINED EFFECT
Alcohol:	Sedation. Avoid.
Beverages:	None expected.
Cocaine:	Unknown. Avoid.
Foods:	None expected.
Marijuana:	Consult doctor.
Tobacco:	None expected.

***See Glossary**

DOFETILIDE

BRAND NAMES

Tikosyn

BASIC INFORMATION

Habit forming? No
Prescription needed? Yes
Available as generic? No
Drug class: Antiarrhythmic

USES

Corrects irregular heartbeats to a normal rhythm.

DOSAGE & USAGE INFORMATION

How to take:
Capsule—This drug is first used in a hospital or other setting where the patient can be monitored for any heart problems. Read the information provided with the prescription.

When to take:
At the same times each day.

If you forget a dose:
Take as soon as you remember. If it is almost time for the next dose, wait for that dose (don't double this dose) and resume regular schedule.

What drug does:
Slows the nerve impulses in the heart.

Time lapse before drug works:
2 to 3 hours.

Don't take with:
Any other medicine or any dietary supplement without consulting your doctor or pharmacist.

OVERDOSE

SYMPTOMS:
Irregular heartbeat, fainting, shortness of breath, unusual tiredness or weakness.
WHAT TO DO:

- **Dial 911 for all medical emergencies or call poison control center 1-800-222-1222 for instructions.**
- **See emergency information on last 3 pages of this book.**

POSSIBLE ADVERSE REACTIONS OR SIDE EFFECTS

SYMPTOMS	WHAT TO DO
Life-threatening: In case of overdose, see previous column.	
Common: Dizziness, fainting, fast or irregular heartbeat.	Discontinue. Call doctor right away.
Infrequent: • Unusual swelling of the extremities, chest pain, slow heartbeat, sudden numbness or tingling (hands, feet or face), paralysis, confusion, weakness, slurred speech, shortness of breath, yellow eyes or skin.	Discontinue. Call doctor right away.
• Abdominal pain, back pain, diarrhea, chills, cough, fever, general feeling of illness, joint pain, headache, nausea, runny nose, sore throat, vomiting, sleeplessness, rash.	Continue. Call doctor if symptoms persist.
Rare: None expected.	

WARNINGS & PRECAUTIONS

Don't take if:
You are allergic to dofetilide.

Before you start, consult your doctor if:
- You are using any other medication.
- You have electrolyte disorder such as low levels of potassium or magnesium.
- You have been diagnosed with kidney or liver disease.

Over age 60:
No special problems expected.

Pregnancy:
Decide with your doctor if drug benefits justify risk to unborn baby. Risk category C (see page xviii).

Breastfeeding; Lactation; Nursing Mothers:
This drug is generally not recommended with breastfeeding. Discuss risks and benefits with your doctor.

Infants & children up to age 18:
Safety and effectiveness in this age group have not been established. Consult doctor.

Prolonged use:
Talk to your doctor about the need for follow up laboratory studies to determine the effect of the medicine on your body.

Skin & sunlight:
None expected.

Driving, piloting or hazardous work:
Don't drive or pilot aircraft until you learn how medicine affects you. Don't work around dangerous machinery. Don't climb ladders or work in high places. Danger increases if you drink alcohol or take medicine affecting alertness and reflexes.

Discontinuing:
Don't discontinue without consulting doctor.

Others:
Advise any doctor, dentist or pharmacist whom you consult that you take this medicine.

POSSIBLE INTERACTION WITH OTHER DRUGS

GENERIC NAME OR DRUG CLASS	COMBINED EFFECT
Antiarrhythmics, other*	Increased dofetilide effect.
Antidepressants, tricyclic*	Increased risk of heart problems.
Calcium channel blockers*	Increased risk of heart problems.
Cimetidine	Increased risk of heart problems.
Diuretics*	Increased risk of heart problems.
Enzyme inhibitors*	Increased dofetilide effect.
Ketoconazole	Increased risk of heart problems.
Macrolides, oral*	Increased risk of heart problems.
Megestrol	Increased dofetilide effect.
Metformin	Increased dofetilide effect.
Norfloxacin	Increased dofetilide effect.
Phenothiazines*	Increased risk of heart problems.
Progestins*	Increased risk of heart problems.
Selective serotonin reuptake inhibitors*	Increased dofetilide effect.
Trimethoprim	Increased dofetilide effect.
Zafirlukast	Increased dofetilide effect.

POSSIBLE INTERACTION WITH OTHER SUBSTANCES

INTERACTS WITH	COMBINED EFFECT
Alcohol:	Increases the chance of liver problems.
Beverages: Grapefruit juice.	May increase effect of dofetilide.
Cocaine:	Unknown. Avoid.
Foods: Grapefruit.	May increase effect of dofetilide.
Marijuana:	Consult doctor.
Tobacco:	None expected.

*See Glossary

DOPAMINE AGONISTS, NONERGOT

GENERIC AND BRAND NAMES

PRAMIPEXOLE
- **Mirapex**
- **Mirapex ER**

ROPINIROLE
- **Repreve**
- **Requip**
- **Requip XL**

ROTIGOTINE
- **Neupro**

BASIC INFORMATION

Habit forming? No
Prescription needed? Yes
Available as generic? Yes, for some
Drug class: Dopamine agonist

USES

- Treats symptoms of Parkinson's disease.
- Treats restless legs syndrome (RLS); also known as Willis-Ekbom disease.

DOSAGE & USAGE INFORMATION

How to take:
- Tablet—Swallow with liquid. May take with or without food (food may lessen stomach upset).
- Extended-release tablet—Swallow whole with liquid. Do not break, crush or chew tablet. Take with or without food (food stops upset stomach).
- Skin (transdermal) patch—Carefully follow instructions provided with the product.

When to take:
- Parkinson's—tablet is taken 3 times a day at same times each day; extended-release tablet is taken once daily at same time each day; skin patch is changed every 24 hours.
- RLS—take a tablet 1 to 3 hours before bedtime.
- Skin patch is changed every 24 hours.

If you forget a dose:
Take as soon as you remember. If it is almost time for the next dose, wait for that dose (don't double this dose) and resume regular schedule.

Continued next column

OVERDOSE

SYMPTOMS:
Agitation, weakness, confusion, cough, chest pain, increased body movements, sweating, hallucinations, low blood pressure.
WHAT TO DO:
If person uses much larger amount than prescribed or if accidentally swallowed, call poison control center 1-800-222-1222 for instructions or dial 911 (emergency) for help.

What drug does:
Mimics the action of dopamine in the brain to improve the symptoms of Parkinson's and RLS. Dopamine helps control body movements.

Time lapse before drug works:
May take several weeks for full benefit. Drug dosage starts low and is gradually increased.

Don't take with:
Any other medicine or any dietary supplement without consulting your doctor or pharmacist.

POSSIBLE ADVERSE REACTIONS OR SIDE EFFECTS

SYMPTOMS	WHAT TO DO
Life-threatening: None expected.	
Common:	
• Confusion, dizziness or lightheadedness, faintness, falling, hallucinations, skin reaction from patch, drowsiness, swelling (ankles, feet, legs, hands), unusual body movements, twitching, nausea, unusual tiredness or weakness, loss of appetite, vomiting, trouble sleeping.	Continue, but call doctor right away.
• Constipation, dry mouth, indigestion.	Continue. Call doctor when convenient.
Infrequent:	
• Abdominal pain or bloating, blurred or changed vision, cold or flu symptoms, falling asleep without warning, urination is painful or difficult or frequent, blood in urine, muscle or joint pain, muscle weakness, mental changes, difficulty swallowing, headache, high or low blood pressure, slow or fast heartbeat, chest pain, nervousness, memory problems, depression, noises in the ears, rapid weight gain, restlessness, shortness of breath.	Continue, but call doctor right away.
• Decreased libido, increased sweating, abnormal dreams, skin rash or itching, weight loss.	Continue. Call doctor when convenient.

Rare:

Abnormal thinking, anxiety, unusual urges (e.g., sexual or gambling).	Continue, but call doctor right away.

WARNINGS & PRECAUTIONS

Don't take if:
You are allergic to pramipexole, ropinirole or rotigotine. Allergic reactions are rare, but may occur.

Before you start, consult your doctor if:
- You have a sleep disorder other than restless legs syndrome (e.g., sleep apnea, insomnia or narcolepsy).
- You have a kidney or liver disorder.
- You have heart problem, high or low blood pressure or orthostatic hypotension (feeling lightheaded or dizzy when rising quickly after sitting or lying down).
- You have uncontrolled muscle movements.
- You have lung disorder such as asthma.
- You have sulfite sensitivity (for drug patch).
- You have a mental or mood disorder.
- You have a problem with alcohol abuse.
- You suffer from hallucinations.

Over age 60:
May be more at risk for certain side effects such as hallucinations.

Pregnancy:
Decide with your doctor if drug benefits justify risk to unborn baby. Risk category C (see page xviii).

Breastfeeding; Lactation; Nursing Mothers:
Drugs in this group are generally not recommended with breastfeeding. Discuss risks and benefits with your doctor.

Infants & children up to age 18:
Safety and effectiveness in this age group have not been established. Consult doctor.

Prolonged use:
See your doctor on a regular basis while you take this drug to monitor effectiveness, dosage, any unwanted effects, and for skin check.

Skin & sunlight:
For skin patch, avoid exposing patch to sunlight.

Driving, piloting or hazardous work:
Don't drive or pilot aircraft until you learn how medicine affects you. Don't work around dangerous machinery. Don't climb ladders or work in high places. Danger increases if you drink alcohol or take medicine affecting alertness and reflexes.

Discontinuing:
Don't stop taking or using drug suddenly. Your doctor may reduce (taper) the dose gradually to avoid withdrawal symptoms. Other drugs you take may require a dosage change.

Others:
- Advise any doctor, dentist or pharmacist whom you consult that you take/use this drug.
- Use of this drug may cause a person to fall asleep without warning during activities of daily living (e.g., working, talking, eating or driving [which has resulted in accidents]). Use caution where needed.
- Get up slowly from a lying or sitting position to avoid dizziness, lightheadedness or fainting.

POSSIBLE INTERACTION WITH OTHER DRUGS

GENERIC NAME OR DRUG CLASS	COMBINED EFFECT
Antipsychotic drugs*	May decrease effect of both drugs.
Central nervous system (CNS) depressants*	Increased risk of side effects of both drugs.
Dopamine antagonists*	May decrease effect of nonergot dopamine agonist.
Enzyme inhibitors*	May increase effect of ropinirole.
Estrogens*	May increase effect of ropinirole.
Levodopa	Increased risk of side effects of levodopa.

POSSIBLE INTERACTION WITH OTHER SUBSTANCES

INTERACTS WITH	COMBINED EFFECT
Alcohol:	Increased risk of side effects. Avoid.
Beverages:	None expected.
Cocaine:	Increased risk of side effects. Avoid.
Foods:	None expected.
Marijuana:	Increased risk of side effects. Avoid.
Tobacco:	May affect drug dosage of ropinirole. Consult doctor.

DOXEPIN (Topical)

BRAND NAMES

Prudoxin	Zonalon

BASIC INFORMATION

Habit forming? No
Prescription needed? Yes
Available as generic? No
Drug class: Antipruritic (topical)

USES

- Treats itching of the skin caused by certain types of eczema (an inflammation of the skin).
- Treatment of moderate itching of atopic dermatitis and lichen simplex chronicus in adult patients.

DOSAGE & USAGE INFORMATION

How to use:
Cream—Apply a thin layer to the affected area of skin and gently rub it in. Do not cover the treated area with a bandage or other dressing.

When to use:
Up to 4 times a day. Allow 3 to 4 hours between applications. Not to be used longer than 8 days.

If you forget a dose:
Use as soon as you remember. If it is almost time for the next dose, wait for that dose (don't double this dose) and resume regular schedule.

Continued next column

OVERDOSE

SYMPTOMS:
An overdose of topical medicine is unlikely to occur, but if too much is applied, it can be absorbed into the system.

- **Mild effects include blurred vision, drowsiness, very dry mouth, decreased awareness or responsiveness.**
- **More severe effects include irregular or fast heartbeat, enlarged pupils, jerking movements, dizziness, fainting, abdominal pain or swelling, weak or feeble pulse, high fever or low temperature, vomiting, incurable constipation, seizures, breathing difficulty, unconsciousness.**

WHAT TO DO:

- **Dial 911 for all medical emergencies or call poison control center 1-800-222-1222 for instructions.**
- **See emergency information on last 3 pages of this book.**

What drug does:
The exact mechanism is unknown. It appears to block histamine* reactions, which can cause the itching. The drug also has a sedating effect on some people, which can help to relieve the itching symptoms. Variable and sometimes significant amounts of the drug are absorbed through the skin.

Time lapse before drug works:
May take up to 8 days for maximum benefit.

Don't take with:
Any other oral or topical medication without consulting your doctor or pharmacist. This includes nonprescription drugs such as cold or allergy remedies that may contain alcohol or antihistamines. They increase the risk of drowsiness.

POSSIBLE ADVERSE REACTIONS OR SIDE EFFECTS

SYMPTOMS	WHAT TO DO
Life-threatening:	
In case of overdose, see previous column.	
Common:	
• Swelling of the skin where drug applied, worsening of the itching or burning or tingling or crawling feeling in the skin.	Discontinue. Call doctor right away.
• Stinging of skin where drug is applied, dryness or tightness of skin, taste changes, dizziness, dry mouth or lips, drowsiness, thirst, emotional changes, fatigue, headache.	Continue. Call doctor when convenient.
Infrequent:	
Scaling or cracking of the skin, nausea, anxiety, irritation.	Continue. Call doctor when convenient.
Rare:	
Fever.	Discontinue. Call doctor right away.

Note: Though the drug is applied topically, it is absorbed into the bloodstream and can cause systemic reactions. Adverse reactions are more likely in patients who use the drug on more than 10% of their body surface.

WARNINGS & PRECAUTIONS

Don't use if:
You are allergic to doxepin. Allergic reactions are rare, but may occur.

Before you start, consult your doctor if:
- You have narrow-angle glaucoma.
- You are allergic to any other medications.
- You have a problem with urinary retention.

Over age 60:
No special problems expected.

Pregnancy:
Decide with your doctor if drug benefits justify risk to unborn baby. Risk category B (see page xviii).

Breastfeeding; Lactation; Nursing Mothers:
This drug is generally acceptable with breastfeeding. Discuss risks and benefits with your doctor.

Infants & children up to age 18:
Safety and effectiveness in this age group have not been established. Consult doctor.

Prolonged use:
Don't use for more than 8 days unless directed by doctor. Longer use can increase the risk of side effects or adverse reactions.

Skin & sunlight:
No special problems expected.

Driving, piloting or hazardous work:
Don't drive or pilot aircraft until you learn how medicine affects you. Don't work around dangerous machinery. Don't climb ladders or work in high places. Danger increases if you drink alcohol or take medicine affecting alertness and reflexes.

Discontinuing:
To be sure of maximum benefit, don't discontinue this medicine before the treatment has been completed unless advised to do so by your doctor.

Others:
- Use medicine only on affected area. It is not to be used in the mouth, the eyes or the vagina.
- Advise any doctor, dentist or pharmacist whom you consult that you are using this medicine.
- If your skin condition doesn't improve within 8 days, consult doctor.
- Use medicine only as directed. Do not increase or reduce dosage without doctor's approval.

POSSIBLE INTERACTION WITH OTHER DRUGS

GENERIC NAME OR DRUG CLASS	COMBINED EFFECT
Antidepressants*	Increased risk of toxicity of both drugs.
Carbamazepine	Increased risk of toxicity of both drugs.
Central nervous system (CNS) depressants*	Increased sedation. May need dosage adjustment.
Cimetidine	Increased risk of doxepin toxicity.
Dextromethorphan	Increased risk of toxicity of both drugs.
Flecainide	Increased risk of toxicity of both drugs.
Monoamine oxidase (MAO) inhibitors*	Potentially life-threatening. Allow 14 days between use of the 2 drugs.
Phenothiazines*	Increased risk of toxicity of both drugs.
Propafenone	Increased risk of toxicity of both drugs.
Quinidine	Increased risk of toxicity of both drugs

POSSIBLE INTERACTION WITH OTHER SUBSTANCES

INTERACTS WITH	COMBINED EFFECT
Alcohol:	Increased sedative effect. Avoid.
Beverages:	None expected.
Cocaine:	Unknown. Avoid.
Foods:	None expected.
Marijuana:	Consult doctor.
Tobacco:	None expected.

***See Glossary**

DPP-4 INHIBITORS

GENERIC AND BRAND NAMES

ALOGLIPTIN
- Kazano
- Nesina
- Oseni

LINAGLIPTIN
- Glyxambi
- Jentadueto
- Tradjenta

SAXAGLIPTIN
- Kombiglyze XR
- Onglyza

SITAGLIPTIN
- Janumet
- Janumet XR
- Januvia

BASIC INFORMATION

Habit forming? No
Prescription needed? Yes
Available as generic? No
Drug class: Antidiabetic; incretin enhancer

USES

Used in addition to diet and exercise to improve blood sugar levels in patients with type 2 diabetes. It may be prescribed alone or along with other oral diabetes drugs. (Note: The drug is not used for treating type 1 diabetes or diabetic ketoacidosis.)

DOSAGE & USAGE INFORMATION

How to take:
- Tablet—Swallow with liquid. May be taken with or without food or as directed by your doctor.
- Extended-release tablet—Swallow whole with liquid. Do not cut, chew or crush tablet.

Continued next column

OVERDOSE

SYMPTOMS:
Rarely may cause hypoglycemia (low blood sugar)—See list of symptoms in next column under Rare.
WHAT TO DO:
- **Eat some type of sugar immediately, such as a glucose product, orange juice (add some sugar), nondiet sodas, candy (such as 5 Lifesavers), honey.**
- **If patient loses consciousness, give glucagon if you have it and know how to use it.**
- **Dial 911 (emergency) for medical help or call poison control center 1-800-222-1222 for instructions.**
- **See emergency information on last 3 pages of this book.**

When to take:
Usually once a day at the same time each day.

If you forget a dose:
Take as soon as you remember. If it is almost time for the next dose, wait for that dose (don't double this dose) and resume regular schedule.

What drug does:
It works by increasing incretins. Incretins increase insulin release when blood sugar levels are high, especially after meals. They also decrease the amount of sugar made by the liver.

Time lapse before drug works:
One to four hours.

Don't take with:
Any other medicine or diet supplement without consulting your doctor or pharmacist.

POSSIBLE ADVERSE REACTIONS OR SIDE EFFECTS

SYMPTOMS	WHAT TO DO
Life-threatening:	
Rare, severe allergic reaction (anaphylaxis)—difficulty breathing, hives, itching, swelling (throat, face), vomiting, dizziness.	Seek emergency treatment immediately.
Common:	
Sore throat, headache, runny or stuffy nose, upper respiratory infection (e.g., cold).	Continue. Call doctor if symptoms persist.
Infrequent:	
Nausea, mild stomach pain, diarrhea, joint pain, urinary tract infection, swelling of hands or feet.	Continue. Call doctor if symptoms persist.
Rare:	
• Symptoms of low blood sugar—nervousness, hunger (excessive), cold sweats, rapid pulse, anxiety, cold skin, chills, confusion, drowsiness, loss of concentration, headache, nausea, weakness, shakiness, vision changes.	Seek treatment (eat some form of quick-acting sugar—glucose tablets, sugar, fruit juice, corn syrup, honey).

• Symptoms of high blood sugar—increased urination, unusual thirst, dry mouth, drowsiness, flushed or dry skin, fruit-like breath odor, appetite loss, stomach pain or vomiting, tiredness, trouble breathing, increased blood sugar level.	Check your blood sugar immediately. Call doctor right away.
• Other symptoms that cause concern.	Continue. Call doctor when convenient.

WARNINGS & PRECAUTIONS

Don't take if:
You are allergic to DPP-4 inhibitors.

Before you start, consult your doctor if:
- You have any kidney problems.
- You have HIV or a long-term infection.
- You are pregnant or plan to become pregnant.
- You are an alcoholic.

Over age 60:
A reduced drug dosage may be recommended for patients with decreased kidney function.

Pregnancy:
Decide with your doctor if drug benefits justify risk to unborn baby. Risk category B (see page xviii).

Breastfeeding; Lactation; Nursing Mothers:
Drugs in this group are generally acceptable with breastfeeding. Discuss risks and benefits with your doctor.

Infants & children up to age 18:
Safety and effectiveness in this age group have not been established. Consult doctor.

Prolonged use:
Talk to your doctor about the need for follow-up medical examinations and/or laboratory studies to determine continued effectiveness of drug.

Skin & sunlight:
No problems expected.

Driving, piloting or hazardous work:
No problems expected. You do need to be cautious for symptoms of hypoglycemia.

Discontinuing:
Don't discontinue without doctor's advice, even though symptoms diminish or disappear.

Others:
- Along with taking drugs for diabetes, be sure to follow your doctor's instructions for lifestyle changes such as a proper diet, weight control measures and a regular exercise program.
- Notify your doctor if you have a fever, infection, diarrhea, or experience vomiting.
- Advise any doctor, dentist or pharmacist whom you consult that you take this medicine.
- Wear or carry medical identification that indicates you have type 2 diabetes and the drugs you take.
- See your diabetes doctor regularly to review your treatment and check for complications.
- You and your family should educate yourselves about diabetes; learn to recognize the symptoms of hypoglycemia and how to treat it. Hypoglycemia may occur in the treatment of diabetes as a result of skipped meals, excessive exercise, or alcohol consumption. Carry non-dietetic candy or glucose tablets to treat episodes of low blood sugar.
- Inflammation of the pancreas (acute pancreatitis) may occur with use of sitagliptin. Call doctor right away if symptoms develop (severe abdominal pain, nausea, vomiting).

POSSIBLE INTERACTION WITH OTHER DRUGS

GENERIC NAME OR DRUG CLASS	COMBINED EFFECT
Digoxin	Digoxin dosage may need to be adjusted.
Enzyme inhibitors*	Increased effect of saxagliptin.
Hypoglycemia-causing medications*	May increase risk of low blood sugar or side effects.

POSSIBLE INTERACTION WITH OTHER SUBSTANCES

INTERACTS WITH	COMBINED EFFECT
Alcohol:	May cause severe low blood sugar. Avoid.
Beverages:	None expected.
Cocaine:	Unknown. Avoid.
Foods:	None expected.
Marijuana:	Consult doctor.
Tobacco:	No drug interaction. Tobacco use does raise risk of diabetes complications. Avoid smoking.

*See Glossary

DRONABINOL (THC, Marijuana)

BRAND NAMES

Marinol

BASIC INFORMATION

Habit forming? Yes
Prescription needed? Yes
Available as generic? Yes
Drug class: Antiemetic

USES

- Prevents nausea and vomiting that may accompany taking anticancer medication (cancer chemotherapy). Should not be used unless other antinausea medicines fail.
- Appetite stimulant. Used to treat appetite loss in AIDS patients.

DOSAGE & USAGE INFORMATION

How to take:
Capsule—Swallow with liquid.

When to take:
Under supervision, a total of no more than 4 to 6 doses per day, every 2 to 4 hours after cancer chemotherapy for prescribed number of days.

If you forget a dose:
Take as soon as you remember. If it is almost time for the next dose, wait for that dose (don't double this dose) and resume regular schedule.

What drug does:
Affects nausea and vomiting center in brain to make it less irritable following cancer chemotherapy. Exact mechanism is unknown.

Time lapse before drug works:
2 to 4 hours.

Don't take with:
Any other medicine or any dietary supplement without consulting your doctor or pharmacist.

OVERDOSE

SYMPTOMS:
Pounding, rapid heart rate, high or low blood pressure, confusion, hallucinations, drastic mood changes, nervousness, anxiety.
WHAT TO DO:
- **Dial 911 for all medical emergencies or call poison control center 1-800-222-1222 for instructions.**
- **See emergency information on last 3 pages of this book.**

POSSIBLE ADVERSE REACTIONS OR SIDE EFFECTS

SYMPTOMS	WHAT TO DO
Life-threatening:	
In case of overdose, see previous column.	
Common:	
• Rapid, pounding heartbeat.	Discontinue. Call doctor right away.
• Dizziness, irritability, drowsiness, euphoria, decreased coordination.	Continue. Call doctor when convenient.
• Red eyes, dry mouth.	No action necessary.
Infrequent:	
Depression, anxiety, nervousness, headache, hallucinations, dramatic mood changes, blurred or changed vision.	Discontinue. Call doctor right away.
Rare:	
• Rapid heartbeat, fainting, frequent or difficult urination, convulsions, shortness of breath, paranoia.	Discontinue. Call doctor right away.
• Nausea, loss of appetite, dizziness when standing after sitting or lying down, diarrhea.	Continue. Call doctor when convenient.

WARNINGS & PRECAUTIONS

Don't take if:
- Your nausea and vomiting is caused by anything other than cancer chemotherapy.
- You are sensitive or allergic to any form of marijuana or sesame oil. Allergic reactions are rare, but may occur.
- Your cycle of chemotherapy is longer than 7 consecutive days. Harmful side effects may occur.

Before you start, consult your doctor if:
- You have heart disease or high blood pressure.
- You are an alcoholic or drug addict.
- You are pregnant or intend to become pregnant.
- You are nursing an infant.
- You have schizophrenia or a manic-depressive disorder.

Over age 60:
Adverse reactions and side effects may be more frequent and severe than in younger persons.

Pregnancy:
Decide with your doctor if drug benefits justify risk to unborn baby. Risk category C (see page xviii).

Breastfeeding; Lactation; Nursing Mothers:
It is unknown if this drug is generally acceptable with breastfeeding. Discuss risks and benefits with your doctor.

Infants & children up to age 18:
Not recommended.

Prolonged use:
- Avoid. Habit forming.
- Talk to your doctor about the need for follow-up medical examinations or laboratory studies to check heart function.

Skin & sunlight:
No problems expected.

Driving, piloting or hazardous work:
Don't drive or pilot aircraft until you learn how medicine affects you. Don't work around dangerous machinery. Don't climb ladders or work in high places. Danger increases if you drink alcohol or take medicine affecting alertness and reflexes, such as antihistamines, tranquilizers, sedatives, pain medicine, narcotics and mind-altering drugs.

Discontinuing:
Withdrawal effects such as irritability, insomnia, restlessness, sweating, diarrhea, hiccups, loss of appetite and hot flashes may follow abrupt withdrawal within 12 hours. Should they occur, these symptoms will probably subside within 96 hours.

Others:
- Store in refrigerator.
- Advise any doctor, dentist or pharmacist whom you consult that you take this medicine.

POSSIBLE INTERACTION WITH OTHER DRUGS

GENERIC NAME OR DRUG CLASS	COMBINED EFFECT
Anesthetics*	Oversedation.
Anticonvulsants*	Oversedation.
Antidepressants, tricyclic*	Oversedation.
Antihistamines*	Oversedation.
Barbiturates*	Oversedation.
Clozapine	Toxic effect on the central nervous system.
Ethinamate	Dangerous increased effects of ethinamate. Avoid combining.
Fluoxetine	Increased depressant effects of both drugs.
Guanfacine	May increase depressant effects of either drug.
Leucovorin	High alcohol content of leucovorin may cause adverse effects.
Methyprylon	Increased sedative effect, perhaps to dangerous level. Avoid.
Molindone	Increased effects of both drugs. Avoid.
Muscle relaxants*	Oversedation.
Nabilone	Greater depression of central nervous system.
Narcotics*	Oversedation.
Sedatives*	Oversedation.
Sertraline	Increased depressive effects of both drugs.
Tranquilizers*	Oversedation.

POSSIBLE INTERACTION WITH OTHER SUBSTANCES

INTERACTS WITH	COMBINED EFFECT
Alcohol:	Oversedation.
Beverages:	None expected.
Cocaine:	Unknown. Avoid.
Foods:	None expected.
Marijuana:	Consult doctor.
Tobacco:	None expected.

*See Glossary

EFLORNITHINE (Topical)

BRAND NAMES

Vaniqa

BASIC INFORMATION

Habit forming? No
Prescription needed? Yes
Available as generic? No
Drug class: Enzyme inhibitor (topical)

USES

Treatment for unwanted facial hair (hirsutism is the medical term) on women.

DOSAGE & USAGE INFORMATION

How to take:
Cream—Follow directions on package label. Usually requires twice-daily application. Limit application to facial area and avoid getting medication in eyes, nose or mouth. Apply at least five minutes after hair removal technique (shaving). Do not apply cosmetics until the medication dries. Do not wash face for at least four hours after applying medication.

When to take:
At the same times each day at least 8 hours apart.

If you forget a dose:
Use as soon as you remember. If it is almost time for the next dose, wait for that dose (don't double this dose) and resume regular schedule.

What drug does:
Inhibits an enzyme that encourages hair growth.

Time lapse before drug works:
4-8 weeks for improvement to be seen.

Don't take with:
Any other medicine or any dietary supplement without consulting your doctor or pharmacist.

OVERDOSE

SYMPTOMS:
Unknown.
WHAT TO DO:
If person uses much larger amount than prescribed or if accidentally swallowed, call poison control center 1-800-222-1222 for instructions or dial 911 (emergency) for help.

POSSIBLE ADVERSE REACTIONS OR SIDE EFFECTS

SYMPTOMS	WHAT TO DO
Life-threatening: None expected.	
Common: Stinging skin, acne breakout.	Continue. Call doctor if condition persists.
Infrequent: Skin symptoms (tingling, redness, chapped, swollen, burning, bleeding, rash, hair bumps, continued acne).	Continue. Call doctor when convenient.
Rare: None expected.	

WARNINGS & PRECAUTIONS

Don't take if:
You are allergic to eflornithine. Allergic reactions are rare, but may occur.

Before you start, consult your doctor if:
You have facial abrasions, cuts or scrapes.

Over age 60:
No special problems expected.

Pregnancy:
Decide with your doctor if drug benefits justify risk to unborn baby. Risk category C (see page xviii).

Breastfeeding; Lactation; Nursing Mothers:
This drug is generally acceptable with breastfeeding. Discuss risks and benefits with your doctor.

Infants & children up to age 18:
Safety and efficacy has not been established in children under 12 years of age. Follow instructions provided by your child's doctor.

Prolonged use:
If no improvement is seen after six months of using this medicine, consult doctor.

Skin & sunlight:
No problems expected. However, if skin irritation occurs after prolonged exposure to sun, consult doctor.

Driving, piloting or hazardous work:
No problems expected.

Discontinuing:
- Consult doctor before discontinuing.
- In about 8 weeks, your hair growth will probably return to the same as it was before you started the medication.

Others:
Advise any doctor, dentist or pharmacist whom you consult that you use this drug.

POSSIBLE INTERACTION WITH OTHER DRUGS

GENERIC NAME OR DRUG CLASS	COMBINED EFFECT
None significant.	

POSSIBLE INTERACTION WITH OTHER SUBSTANCES

INTERACTS WITH	COMBINED EFFECT
Alcohol:	None expected.
Beverages:	None expected.
Cocaine:	None expected.
Foods:	None expected.
Marijuana:	None expected.
Tobacco:	None expected.

*See Glossary

ENDOTHELIN RECEPTOR ANTAGONIST

GENERIC AND BRAND NAMES

AMBRISENTAN
Letairis

BOSENTAN
Tracleer

BASIC INFORMATION

Habit forming? No
Prescription needed? Yes
Available as generic? No
Drug class: Antihypertensive (pulmonary)

USES

Treats the symptoms of pulmonary arterial hypertension (PAH), which is high blood pressure in the lungs. It improves the breathing and exercise capacity of patients with PAH. The drug does not cure the disorder. These drugs are available only through restricted programs which will be explained by your doctor.

DOSAGE & USAGE INFORMATION

How to take:
Tablet—Swallow with liquid. Take with or without food. Do not break, crush or chew the tablet.

When to take:
Usually once or twice daily at the same times each day. Follow your doctor's instructions.

If you forget a dose:
Take as soon as you remember. If it is almost time for the next dose, wait for that dose (don't double this dose) and resume regular schedule.

What drug does:
It works by blocking the effect of endothelin (a substance made by the body). In patients with PAH, the blood vessels become narrowed (constricted) due to an excess production of endothelin.

Continued next column

OVERDOSE

SYMPTOMS:
Headache, dizziness, flushing, stuffy nose, nausea.
WHAT TO DO:
If person uses much larger amount than prescribed or if accidentally swallowed, call poison control center 1-800-222-1222 for instructions or dial 911 (emergency) for help.

Time lapse before drug works:
It may take 1 to 2 months or longer to notice effects. The dosage may be increased after 4 weeks if there are no problems in taking the drug.

Don't take with:
Any other medicine or any dietary supplement without consulting your doctor or pharmacist.

POSSIBLE ADVERSE REACTIONS OR SIDE EFFECTS

SYMPTOMS	WHAT TO DO
Life-threatening:	
None expected.	
Common:	
Feeling of warmth, flushing, stuffy or sore nose, sore throat, stomach upset, swelling of legs and ankles, mild dizziness.	Continue. Call doctor when convenient.
Infrequent:	
• Difficult breathing, irregular heartbeat, unusual tiredness or weakness, ongoing dizziness or lightheadedness, wheezing or unusual coughing.	Discontinue. Call doctor right away.
• Stomach pain, heartburn, mild heart palpitations, constipation, hoarseness, headache, low blood pressure.	Continue. Call doctor when convenient.
Rare:	
Liver problems (symptoms include loss of appetite, nausea, vomiting, light colored stools, fever, extreme tiredness, right upper stomach pain, yellow skin or eyes, dark urine, itching).	Discontinue. Call doctor right away.

WARNINGS & PRECAUTIONS

Don't take if:
- You are allergic to endothelin receptor antagonists. Allergic reactions are rare, but may occur.
- You have moderate or severe liver (hepatic) impairment.
- You are pregnant or plan to become pregnant.
- You are of childbearing age and have not had a negative pregnancy test or are not willing to use two reliable methods of birth control.

Before you start, consult your doctor if:
- You have mild liver impairment.
- You have not had a liver function test.
- You have anemia or edema (swelling in hands, lower legs or feet).

Over age 60:
Adverse reactions and side effects may be more frequent and severe than in younger persons.

Pregnancy:
- Consult doctor. Drug should not be used during pregnancy. Can cause harm to unborn baby. Risk category X (see page xviii).
- If this drug is used during pregnancy, or if you become pregnant while taking this drug, you need to talk to your doctor about the potential harm to the fetus.

Breastfeeding; Lactation; Nursing Mothers:
Drugs in this group are generally not recommended with breastfeeding. Discuss risks and benefits with your doctor.

Infants & children up to age 18:
Not approved for ages under 18.

Prolonged use:
- Your doctor will check your progress at regular visits to make sure this drug is working properly and to check for unwanted effects.
- Your doctor will advise you about any liver function tests needed before you start the drug and while taking the drug.

Skin & sunlight:
No problems expected.

Driving, piloting or hazardous work:
Use caution if you experience dizziness.

Discontinuing:
Consult doctor before discontinuing. The dosage may need to be slowly reduced before stopping.

Others:
- These drugs may decrease hemoglobin* and hematocrit.* Follow doctor's advice about routine blood testing.
- Advise any doctor, dentist or pharmacist whom you consult that you take this drug.
- The drug may decrease the amount of sperm men make and affect their ability to have children. Consult doctor.
- Females of childbearing age will be required to have a pregnancy test every month during treatment. If you miss a menstrual period while using this drug, call your doctor right away. Use two forms of effective birth control to keep from getting pregnant while you are using this drug (even if the drug is temporarily stopped), and for at least one month after you stop taking the drug.
- Use of bosentan may lead to liver problems. Follow your doctor's advice about periodic liver function studies. Call doctor right away if symptoms of liver problems develop (see list in Possible Adverse Reactions or Side Effects).

POSSIBLE INTERACTION WITH OTHER DRUGS

GENERIC NAME OR DRUG CLASS	COMBINED EFFECT
Cyclosporine A	Increased endothelin receptor antagonist effect; decreased cyclosporine effect. Avoid use with bosentan.
Enzyme inducers*	Decreased endothelin receptor antagonist effect.
Enzyme inhibitors*	Increased endothelin receptor antagonist effect.
Glyburide	Decreased effect of both glyburide and endothelin receptor antagonist. May increase risk of liver problems. Avoid with bosentan.
Contraceptives, hormonal*	Decreased birth control effect.

POSSIBLE INTERACTION WITH OTHER SUBSTANCES

INTERACTS WITH	COMBINED EFFECT
Alcohol:	No interaction, but avoid excess alcohol.
Beverages:	None expected.
Cocaine:	Unknown. Avoid.
Foods:	None expected.
Marijuana:	Consult doctor.
Tobacco:	No interaction, but should not smoke.

***See Glossary**

EPHEDRINE

BRAND NAMES

Broncholate
Rynatuss
Rynatuss Pediatric

BASIC INFORMATION

Habit forming? No
Prescription needed? Yes
Available as generic? Yes, for some
Drug class: Sympathomimetic

USES

- Relieves bronchial asthma.
- Decreases congestion of breathing passages.
- Suppresses allergic reactions.

DOSAGE & USAGE INFORMATION

How to take:
- Tablet or capsule—Swallow with liquid. You may chew or crush tablet.
- Extended-release tablet or capsule—Swallow each dose whole.
- Syrup—Take as directed on bottle.
- Drops—Dilute dose in beverage.

When to take:
As needed, no more often than every 4 hours. To prevent insomnia, take last dose at least 2 hours before bedtime.

If you forget a dose:
Take as soon as you remember. If it is almost time for the next dose, wait for that dose (don't double this dose) and resume regular schedule.

What drug does:
- Prevents cells from releasing allergy-causing chemicals (histamines).
- Relaxes muscles of bronchial tubes.
- Decreases blood vessel size and blood flow, thus causing decongestion.

Continued next column

OVERDOSE

SYMPTOMS:
Severe anxiety, confusion, delirium, muscle tremors, rapid and irregular pulse.
WHAT TO DO:
- **Dial 911 for all medical emergencies or call poison control center 1-800-222-1222 for instructions.**
- **See emergency information on last 3 pages of this book.**

Time lapse before drug works:
30 to 60 minutes.

Don't take with:
- Nonprescription drugs with ephedrine, pseudoephedrine or epinephrine.
- Any other medicine or any dietary supplement without consulting your doctor or pharmacist.

POSSIBLE ADVERSE REACTIONS OR SIDE EFFECTS

SYMPTOMS	WHAT TO DO
Life-threatening: None expected.	
Common: Nervousness, Insomnia.	Continue. Call doctor when convenient.
Infrequent: Dizziness, appetite loss, nausea, vomiting, painful or difficult urination, sweating, high blood pressure, trembling, headache, paleness, being thirsty, dry throat, feeling weak.	Continue. Call doctor when convenient.
Rare: • Irregular or fast heartbeat, seizures, chest pain, mental changes.	Discontinue. Call doctor right away.
• Redness of neck and face, feeling anxious, stomach pain.	Continue. Call doctor when convenient.

WARNINGS & PRECAUTIONS

Don't take if:
You are allergic to ephedrine or any sympathomimetic* drug. Allergic reactions are rare, but may occur.

Before you start, consult your doctor if:
- You have high blood pressure.
- You have diabetes.
- You have overactive thyroid gland.
- You have difficulty urinating.
- You have taken any MAO inhibitor in past 2 weeks.
- You have taken digitalis preparations in the last 7 days.
- You will have surgery within 2 months, including dental surgery, requiring general or spinal anesthesia.

Over age 60:
More likely to develop high blood pressure, heart-rhythm disturbances, angina and to feel drug's stimulant effects.

Pregnancy:
Decide with your doctor if drug benefits justify risk to unborn baby. Risk category C (see page xviii).

Breastfeeding; Lactation; Nursing Mothers:
This drug is generally not recommended with breastfeeding. Discuss risks and benefits with your doctor.

Infants & children up to age 18:
Read nonprescription product's label to be sure it is approved for your child's age. Follow label instructions on dosage. Consult doctor or pharmacist if unsure.

Prolonged use:
- Excessive doses—Rare toxic psychosis.
- Men with enlarged prostate gland may have more urination difficulty.

Skin & sunlight:
No problems expected.

Driving, piloting or hazardous work:
Avoid if you feel dizzy. Otherwise, no problems expected.

Discontinuing:
May be unnecessary to finish medicine. Follow doctor's instructions.

Others:
Advise any doctor, dentist or pharmacist whom you consult that you take this drug.

POSSIBLE INTERACTION WITH OTHER DRUGS

GENERIC NAME OR DRUG CLASS	COMBINED EFFECT
Adrenocorticoids, systemic	Decreased adrenocorticoid effect.
Antidepressants, tricyclic*	Increased effect of ephedrine. Excessive stimulation of heart and blood pressure.
Antihypertensives*	Decreased anti-hypertensive effect.
Beta-adrenergic blocking agents*	Decreased effects of both drugs.
Dextrothyroxine	Increased ephedrine effect.
Digitalis preparations*	Serious heart rhythm disturbances.
Epinephrine	Increased epinephrine effect.
Ergot preparations*	Serious blood pressure rise.
Furazolidone	Sudden, severe increase in blood pressure.
Guanadrel	Decreased effect of both drugs.
Guanethidine	Decreased effect of both drugs.
Methyldopa	Possible increased blood pressure.
Monoamine oxidase (MAO) inhibitors*	Increased ephedrine effect. Dangerous blood pressure rise.
Nitrates*	Possible decreased effects of both drugs.
Phenothiazines*	Possible increased ephedrine toxicity. Possible decreased ephedrine effect.
Pseudoephedrine	Increased pseudoephedrine effect.
Rauwolfia	Decreased rauwolfia effect.
Sympathomimetics*	Increased ephedrine effect.
Terazosin	Decreased effectiveness of terazosin.
Theophylline	Increased gastro-intestinal intolerance.

POSSIBLE INTERACTION WITH OTHER SUBSTANCES

INTERACTS WITH	COMBINED EFFECT
Alcohol:	None expected.
Beverages: Caffeine drinks.	Nervousness or insomnia.
Cocaine:	Risk of heartbeat irregularities and high blood pressure. Avoid.
Foods:	None expected.
Marijuana:	Rapid heartbeat, possible heart rhythm disturbance. Avoid.
Tobacco:	None expected.

*See Glossary

EPLERENONE

BRAND NAMES

Inspra

BASIC INFORMATION

Habit forming? No
Prescription needed? Yes
Available as generic? Yes
Drug class: Antihypertensive; selective aldosterone blocker

USES

- Treatment for hypertension (high blood pressure). May be used alone or along with other antihypertensive medications.
- Used to treat further complications after a myocardial infarction.
- May be used for treatment of other disorders as determined by your doctor.

DOSAGE & USAGE INFORMATION

How to take:
Tablet—Swallow with liquid. May be taken with or without food.

When to take:
Once or twice daily as directed by your doctor.

If you forget a dose:
Take as soon as you remember. If it is almost time for the next dose, wait for that dose (don't double this dose) and resume regular schedule.

What drug does:
It blocks aldosterone, a hormone in the body that increases blood pressure by causing fluid and salt retention. The drug causes the kidneys to remove the excess water and salt from the body.

Time lapse before drug works:
May take several weeks for full effectiveness.

Don't take with:
Any other medicine or any dietary supplement without consulting your doctor or pharmacist.

OVERDOSE

SYMPTOMS:
Unknown. Could cause very low blood pressure (hypotension).
WHAT TO DO:
If person uses much larger amount than prescribed or if accidentally swallowed, call poison control center 1-800-222-1222 for instructions or dial 911 (emergency) for help.

POSSIBLE ADVERSE REACTIONS OR SIDE EFFECTS

SYMPTOMS	WHAT TO DO
Life-threatening: None expected.	
Common: Dizziness.	Continue. Call doctor when convenient.
Infrequent: Diarrhea, flu-like symptoms, cough, fatigue, headache, stomach pain.	Continue. Call doctor when convenient.
Rare:	
• Symptoms of too much potassium in the body (confusion, feeling very weak and tired, shortness of breath, nervousness, low blood pressure, numbness and tingling, irregular heartbeat).	Discontinue. Call doctor right away or seek emergency help if symptoms are severe.
• Abnormal vaginal bleeding in women, enlarged breasts or breast pain in men.	Continue. Call doctor when convenient.

WARNINGS & PRECAUTIONS

Don't take if:
You are allergic to eplerenone. Allergic reactions are rare, but may occur.

Before you start, consult your doctor if:
- You have diabetes, kidney or liver disease.
- You have high blood potassium levels or low blood sodium levels.
- You are on any special diet using low salt or salt substitutes.
- You have high cholesterol or heart disease.
- You are allergic to any medication, food or other substance.

Over age 60:
No special problems expected.

Pregnancy:
Decide with your doctor if drug benefits justify risk to unborn baby. Risk category B (see page xviii).

Breastfeeding; Lactation; Nursing Mothers:
This drug is generally acceptable with breastfeeding. Discuss risks and benefits with your doctor.

Infants & children up to age 18:
Not approved for children under age 18.

Prolonged use:
- No special problems expected. Hypertension usually requires life-long treatment.
- Schedule regular doctor visits to determine if drug is continuing to be effective in controlling the hypertension and to check your potassium levels and kidney function.

Skin & sunlight:
No special problems expected.

Driving, piloting or hazardous work:
Use caution if you feel dizzy or are experiencing other side effects.

Discontinuing:
Don't discontinue without consulting your doctor, even if you feel well. You can have hypertension without feeling any symptoms. Untreated high blood pressure can cause serious problems.

Others:
- Advise any doctor, dentist or pharmacist whom you consult that you take this medicine.
- See your doctor regularly, especially when you first start taking this drug.
- Follow any diet or exercise plan your doctor prescribes. It can help control hypertension.
- Get up slowly from a sitting or lying position to avoid any dizziness, faintness or lightheadedness.
- Can elevate triglycerides and cholesterol; these may need to be monitored regularly.
- Consult doctor if you become ill with vomiting or diarrhea.

POSSIBLE INTERACTION WITH OTHER DRUGS

GENERIC NAME OR DRUG CLASS	COMBINED EFFECT
Angiotensin-converting enzyme (ACE) inhibitors*	Decreased anti-hypertensive effect.
Angiotensin II receptor antagonists	Decreased anti-hypertensive effect.
Diuretics, potassium-sparing*	Excess potassium levels in the body.
Enzyme inhibitors*	Increased effect of eplerenone.
Potassium supplements*	Excess potassium levels in the body.
St. Johns Wort	Decreased antihypertensive effect.

POSSIBLE INTERACTION WITH OTHER SUBSTANCES

INTERACTS WITH	COMBINED EFFECT
Alcohol:	Increased risk of side effects. Avoid.
Beverages: Grapefruit juice.	Increased effect of eplerenone.
Cocaine:	Unknown. Avoid.
Foods: Salt substitutes containing potassium or foods high in potassium (e.g., bananas).	Excess potassium in the body. Avoid.
Marijuana:	Consult doctor.
Tobacco:	None expected. Best to avoid.

*See Glossary

ERECTILE DYSFUNCTION AGENTS

GENERIC AND BRAND NAMES

AVANAFIL
Stendra
SILDENAFIL
Revatio
Viagra
TADALAFIL
Cialis
VARDENAFIL
Levitra
Staxyn

BASIC INFORMATION

Habit forming? No
Prescription needed? Yes
Available as generic? Yes, for some
Drug class: Impotence therapy

USES

- Treats male sexual function (erection) problems.
- Tadalafil is used for treatment of signs and symptoms of benign prostatic hyperplasia.
- Brand name Revatio treats adult pulmonary arterial hypertension (rare fatal lung disorder).

DOSAGE & USAGE INFORMATION

How to take:
- Tablet—Swallow with water. If you can't swallow whole, crumble tablet and take with liquid or food.
- Orally disintegrating tablet—Place on your tongue. It will dissolve rapidly. Do not swallow, crush or split tablet. Do not take it with liquid.

When to take:
- As directed; usually 30 minutes to 1 hour before sexual activity.
- Daily dose tadalafil, take as directed on label.

If you forget a dose:
- Does not apply for the tablet taken before sexual activity.
- For once daily tadalafil, take as soon as you remember. If it is almost time for the next dose, wait for that dose (don't double this dose) and resume regular schedule.

Continued next column

OVERDOSE

SYMPTOMS:
May include nausea, irregular heartbeat, chest pain, faintness, lightheadedness.
WHAT TO DO:
If person uses much larger amount than prescribed or if accidentally swallowed, call poison control center 1-800-222-1222 for instructions or dial 911 (emergency) for help.

What drug does:
- Increases blood flow to the penis that may help men develop and sustain an erection. Sexual stimulation is still needed for these drugs to be effective.
- In pulmonary arterial hypertension, the drug works by dilating (widening) blood vessels, thereby lowering blood pressure in the lungs.

Time lapse before drug works:
- Avanafill, 15-30 minutes, lasts up to 6 hours.
- Sildenafil, 30-60 minutes, lasts up to 4 hours.
- Tadalafil, 16-60 minutes, lasts up to 36 hours.
- Vardenafil, 25-30 minutes, lasts up to 4-5 hours.

Don't take with:
Any other medicine or any dietary supplement without consulting your doctor or pharmacist.

POSSIBLE ADVERSE REACTIONS OR SIDE EFFECTS

SYMPTOMS	WHAT TO DO
Life-threatening:	
Fatalities have been reported when used with nitrate medications.	
Common:	
Headache, flushing, stomach upset, stuffy or runny nose, back pain, muscle aches.	Continue. Call doctor if symptoms persist.
Infrequent:	
• Urination problems, blurred vision, changes in color perception, light sensitivity, skin rash, dizziness, prolonged erection (lasting more than 4 hours).	Discontinue. Call doctor right away.
• Diarrhea.	Continue. Call doctor if symptoms persist.
Rare:	
• Chest pain, fainting, foot or ankle swelling, allergic reaction, (shortness of breath, skin rash, hives, itching, face swelling), changes in hearing, ringing or buzzing in the ears.	Discontinue. Call doctor right away.
• Unexpected symptoms occur while having sexual intercourse.	Discontinue. Call doctor when convenient.

WARNINGS & PRECAUTIONS

Don't take if:
You are allergic to sildenafil, vardenafil, or tadalafil. Allergic reactions are rare, but may occur.

Before you start, consult your doctor if:
- You have any other medical problem.
- You have or have had heart, blood pressure or blood cell problems or have had a stroke.
- You have a stomach ulcer.
- You have vision problems.
- You have retinitis pigmentosa.
- You have a deformed shape of the penis or have had an erection last more than 4 hours.
- You use drugs like amyl nitrate or butyl nitrate recreationally.
- You have liver or kidney disease.

Over age 60:
No special problems expected.

Pregnancy:
Not indicated for use in females.

Breastfeeding; Lactation; Nursing Mothers:
Not indicated for use in females.

Infants & children up to age 18:
Not recommended or indicated.

Prolonged use:
Talk to your doctor about the need for follow-up medical examinations or laboratory studies to determine drug's effectiveness.

Skin & sunlight:
No problems expected.

Driving, piloting or hazardous work:
No problems expected.

Discontinuing:
No problems expected.

Others:
- Advise any doctor, dentist or pharmacist whom you consult that you take this medicine.
- Do not increase dose without doctor's approval.
- Consult doctor if there are any significant changes in your vision. A small number of men taking one or more of these drugs have developed NAION (non-arteritic ischemic optic neuropathy), a loss of vision that is frequently irreversible. Ask your doctor about your risks.
- Do not combine the drug with any other impotence therapy unless approved by your doctor.
- There are many causes of impotence. Your doctor should perform a complete exam before prescribing this medication.
- These drugs do not protect against sexually transmitted diseases.
- Call doctor right away in the event of sudden decrease or loss of hearing. You may be also experience tinnitus (ringing/buzzing in the ears) and dizziness.

POSSIBLE INTERACTION WITH OTHER DRUGS

GENERIC NAME OR DRUG CLASS	COMBINED EFFECT
Alpha adrenergic receptor blockers	Sudden drop in blood pressure. Do not use together.
Enzyme inhibitors*	Increased effect of erectile dysfunction agent.
Protease inhibitors	Increased effect of erectile dysfunction agent.
Nitrates	Sudden drop in blood pressure. Do not use together.
Rifampin	Decreased effect of erectile dysfunction agent.

POSSIBLE INTERACTION WITH OTHER SUBSTANCES

INTERACTS WITH	COMBINED EFFECT
Alcohol:	May decrease effect of erectile dysfunction agents.
Beverages: Grapefruit juice	May increase blood levels of erectile dysfunction agents.
Cocaine:	Unknown. Avoid.
Foods:	None expected.
Marijuana:	Consult doctor.
Tobacco:	None expected.

*See Glossary

ERGOLOID MESYLATES

BRAND NAMES

Gerimal
Hydergine
Hydergine LC
Niloric

BASIC INFORMATION

Habit forming? No
Prescription needed? Yes
Available as generic? Yes
Drug class: Ergot preparation

USES

Treatment for reduced alertness, poor memory, confusion, depression or lack of motivation in the elderly.

DOSAGE & USAGE INFORMATION

How to take:

- Tablet or capsule—Swallow with liquid. If you can't swallow whole, crumble tablet or open capsule and take with liquid or food.
- Liquid—Take as directed on label.
- Sublingual tablet—Dissolve tablet under tongue.

When to take:
At the same times each day.

If you forget a dose:
Take as soon as you remember. If it is almost time for the next dose, wait for that dose (don't double this dose) and resume regular schedule.

What drug does:
Stimulates brain-cell metabolism to increase use of oxygen and nutrients.

Time lapse before drug works:
Gradual improvements over 3 to 4 months.

Continued next column

OVERDOSE

SYMPTOMS:
Dizziness, blurred vision, fainting, flushing, loss of appetite, stomach cramps, nausea or vomiting, headache, stuffy nose.
WHAT TO DO:
If person uses much larger amount than prescribed or if accidentally swallowed, call poison control center 1-800-222-1222 for instructions or dial 911 (emergency) for help.

Don't take with:

- Nonprescription drugs containing alcohol without consulting doctor.
- Any other medicine or any dietary supplement without consulting your doctor or pharmacist.

POSSIBLE ADVERSE REACTIONS OR SIDE EFFECTS

SYMPTOMS	WHAT TO DO
Life-threatening: None expected,	
Common: None expected.	
Infrequent: Soreness under tongue (with sublingual form).	Continue. Call doctor when convenient.
Rare: Rash, dizziness when getting up from sitting or lying, drowsiness and slow heartbeat.	Discontinue. Call doctor right away.

WARNINGS & PRECAUTIONS

Don't use if:
- If you are allergic to any ergot preparation. Allergic reactions are rare, but may occur.
- Your heartbeat is less than 60 beats per minute.
- Your systolic blood pressure is consistently below 100.

Before you start, consult your doctor if:
- You have had low blood pressure.
- You have liver disease.
- You have severe mental illness.

Over age 60:
Primarily used in persons older than 60. Results unpredictable, but many patients show improved brain function.

Pregnancy:
Consult doctor. Pregnancy safety unknown. Generally not used in this age group.

Breastfeeding; Lactation; Nursing Mothers:
This drug is generally not recommended with breastfeeding. Discuss risks and benefits with your doctor.

Infants & children up to age 18:
Safety and effectiveness in this age group have not been established. Consult doctor.

Prolonged use:
Talk to your doctor about the need for follow-up medical examinations or laboratory studies.

Skin & sunlight:
No problems expected.

Driving, piloting or hazardous work:
Avoid if you feel dizzy, faint or have blurred vision. Otherwise, no problems expected.

Discontinuing:
No problems expected.

Others:
- May lessen your body's ability to adjust to cold temperatures.
- Advise any doctor, dentist or pharmacist whom you consult that you use this medicine.

POSSIBLE INTERACTION WITH OTHER DRUGS

GENERIC NAME OR DRUG CLASS	COMBINED EFFECT
Ergot preparations,* other	May cause serious side effects. Avoid.
Sympathomimetics*	May cause decreased circulation to arms, legs, feet and hands. Avoid.

POSSIBLE INTERACTION WITH OTHER SUBSTANCES

INTERACTS WITH	COMBINED EFFECT
Alcohol:	Use caution. May drop blood pressure excessively.
Beverages:	None expected.
Cocaine:	Overstimulation. Avoid.
Foods:	None expected.
Marijuana:	Consult doctor.
Tobacco:	Decreased ergoloid effect. Don't smoke.

ERGOT ALKALOIDS

GENERIC AND BRAND NAMES

ERGONOVINE
- Ergometrine
- Ergotrate
- Ergotrate Maleate

METHYL-ERGONOVINE
- Methylergometrine

BASIC INFORMATION

Habit forming? No
Prescription needed? Yes
Available as generic? Yes
Drug class: Ergot preparation (uterine stimulant)

USES

Retards excessive post-delivery bleeding.

DOSAGE & USAGE INFORMATION

How to take:
Tablet—Swallow with liquid or food to lessen stomach irritation.

When to take:
At the same times each day.

If you forget a dose:
Take as soon as you remember. If it is almost time for the next dose, wait for that dose (don't double this dose) and resume regular schedule.

What drug does:
Causes smooth muscle cells of uterine wall to contract and surround bleeding blood vessels of relaxed uterus.

Time lapse before drug works:
20 to 30 minutes.

Don't take with:
Any other medicine or any dietary supplement without consulting your doctor or pharmacist.

OVERDOSE

SYMPTOMS:
Nausea, vomiting, stomach pain, numbness and tingling of hands and feet. Severe cases may have difficult breathing, convulsions, coma.

WHAT TO DO:
- **Dial 911 for all medical emergencies or call poison control center 1-800-222-1222 for instructions.**
- **See emergency information on last 3 pages of this book.**

POSSIBLE ADVERSE REACTIONS OR SIDE EFFECTS

SYMPTOMS	WHAT TO DO
Life-threatening: In case of overdose, see previous column.	
Common: Nausea, vomiting, severe lower abdominal menstrual-like cramps.	Discontinue. Call doctor right away.
Infrequent:	
• Confusion, ringing in ears, diarrhea, muscle cramps.	Discontinue. Call doctor right away.
• Unusual sweating.	Continue. Call doctor when convenient.
Rare: Sudden and severe headache, shortness of breath, chest pain, hands and feet are numb and cold.	Discontinue. Seek emergency treatment.

WARNINGS & PRECAUTIONS

Don't take if:
You are allergic to any ergot alkaloid. Allergic reactions are rare, but may occur.

Before you start, consult your doctor if:
- You have coronary artery or blood vessel disease.
- You have liver or kidney disease.
- You have high blood pressure.
- You have postpartum infection.

Over age 60:
Not used in this age group.

Pregnancy:
Consult doctor. Ergotamine should not be used during pregnancy. Can cause harm to unborn baby. Risk category X. Methylergonovine is risk category C (decide with doctor). See page xviii.

Breastfeeding; Lactation; Nursing Mothers:
Drugs in this group are generally not recommended with breastfeeding. Discuss risks and benefits with your doctor.

Infants & children up to age 18:
Safety and effectiveness in this age group have not been established. Consult doctor.

Prolonged use:
Talk to your doctor about the need for follow-up medical exams or lab studies.

Skin & sunlight:
No problems expected.

Driving, piloting or hazardous work:
No problems expected.

Discontinuing:
May be unnecessary to finish medicine. Follow doctor's instructions.

Others:
Drug should be used for short time only following childbirth or miscarriage.

POSSIBLE INTERACTION WITH OTHER DRUGS

GENERIC NAME OR DRUG CLASS	COMBINED EFFECT
Antifungals, azoles	Can cause serious or life-threatening problems with blood circulation. Avoid.
Beta-adrenergic blocking agents*	Possible vasospasm (peripheral and cardiac).
Ergot preparations,* other	May cause serious side effects. Avoid.
Macrolide antibiotics	Can cause serious or life-threatening problems with blood circulation. Avoid.
Protease Inhibitors	Can cause serious or life-threatening problems with blood circulation. Avoid.
Sympathomimetics*	May cause decreased circulation to arms, legs, feet and hands. Avoid.
Triptans	May cause serious side effects if taken within 24 hours of ergot alkaloid.

POSSIBLE INTERACTION WITH OTHER SUBSTANCES

INTERACTS WITH	COMBINED EFFECT
Alcohol:	None expected.
Beverages:	None expected.
Cocaine:	Unknown. Avoid.
Foods:	None expected.
Marijuana:	Consult doctor.
Tobacco:	Decreased ergot alkaloid effect. Don't smoke.

*See Glossary

ERGOT DERIVATIVES

GENERIC AND BRAND NAMES

ERGOTAMINE
- Cafergot
- Ergomar
- Migergot
- Wigraine

DIHYDRO-ERGOTAMINE
- Levadex
- Migranal

BASIC INFORMATION

Habit forming? Unlikely
Prescription needed? Yes
Available as generic? Yes, for some
Drug class: Antimigraine

USES

- Treatment for migraine (with or without aura) and cluster headaches.
- Treats other disorders as determined by doctor.

DOSAGE & USAGE INFORMATION

How to take:
- Tablet—Swallow with liquid.
- Sublingual tablet—Don't swallow whole. Let dissolve under tongue. Don't eat or drink while tablet is dissolving.
- Suppository—Remove wrapper and use finger to gently push into rectum.
- Nasal spray or oral inhaler—Use only as directed on prescription label.
- Injection (dihydroergotamine)—Normally given by a healthcare provider.

When to take:
Take or use drug at the first warning sign or symptom of the headache. Lie down in a quiet, dark room. If headache persists, follow your doctor's instructions as to when you should take or use additional dosages.

Continued next column

OVERDOSE

SYMPTOMS:
Nausea, vomiting, numbness of fingers and toes, tingling, confusion, drowsiness, headache, shock, convulsions, coma.
WHAT TO DO:
- **Dial 911 for all medical emergencies or call poison control center 1-800-222-1222 for instructions.**
- **See emergency information on last 3 pages of this book.**

If you forget a dose:
The drug is taken or used as needed and not on a regular schedule.

What drug does:
It constricts (narrows) blood vessels in the head that have become dilated (widened). This action helps abort (stop) an impending headache or stops a headache in progress.

Time lapse before drug works:
15 to 60 minutes (depends on dosage form).

Don't take with:
Any other medicine or any dietary supplement without consulting your doctor or pharmacist.

POSSIBLE ADVERSE REACTIONS OR SIDE EFFECTS

SYMPTOMS	WHAT TO DO
Life-threatening:	
In case of overdose, see previous column.	
Common:	
• Feet and ankle swelling.	Discontinue. Call doctor right away.
• Dizziness, nausea, diarrhea, vomiting, drowsiness, dry mouth.	Continue. Call doctor when convenient.
Infrequent:	
• Itchy or swollen skin, or pale hands or fingers or feet, pain or weakness (in arms, legs, back).	Discontinue. Call cold doctor right away.
• Headaches increase or are more severe, nose irritation or taste changes (with nasal spray), drug aftertaste (with oral inhaler).	Continue. Call doctor when convenient.
Rare:	
Anxiety, confusion, changes in vision, stomach pain, fast or slow heartbeat, chest pain, numbness or tingling (in face, fingers, toes), high or low blood pressure, shortness of breath, problems with speech or balance.	Discontinue. Call doctor right away.

Note: Side effects vary depending on dosage form, or possible overuse, of the drug. Some symptoms may result from the headache itself (e.g., nausea).

WARNINGS & PRECAUTIONS

Don't take if:
You are allergic to any ergot derivative. Allergic reactions are rare, but may occur.

Before you start, consult your doctor if:
- You have angina, heart problems, high blood pressure, coronary artery disease, peripheral artery disease, blood circulation problem, high cholesterol, or have had a heart attack or stroke.
- You have diabetes, kidney or liver disease.
- You smoke cigarettes.
- You plan to become pregnant.
- You have a serious infection.
- You are allergic to other spray inhalants.

Over age 60:
Adverse reactions and side effects may be more frequent and severe than in younger persons.

Pregnancy:
Consult doctor. Drug should not be used during pregnancy. Can cause harm to unborn baby. Risk category X (see page xviii).

Breastfeeding; Lactation; Nursing Mothers:
Drugs in this group are generally not recommended with breastfeeding. Discuss risks and benefits with your doctor.

Infants & children up to age 18:
Safety and effectiveness in this age group have not been established. Consult doctor.

Prolonged use:
To avoid risk of serious adverse effects, do not exceed prescribed dosage or overuse the drug. It is not intended for chronic daily use. There have been reports of psychological dependence and drug abuse.

Skin & sunlight:
No problems expected.

Driving, piloting or hazardous work:
Don't drive or pilot aircraft until you learn how drug affects you. Don't work around dangerous machinery. Don't climb ladders or work in high places. Danger increases if you drink alcohol or take drugs affecting alertness and reflexes.

Discontinuing:
If drug is used for long period of time or is overused, rebound headaches (also called medication overuse headaches) may occur when drug is stopped. Consult doctor for advice.

Others:
- Advise any doctor, dentist or pharmacist whom you consult that you take this drug.
- These drugs constrict blood vessels which can lead to serious adverse effects (possibly fatal) due to decrease in blood flow. The drugs can cause fibrosis (scarring) in lungs, heart or kidneys. Consult doctor about your risks.
- It may take several episodes of using the drug to determine full effectiveness.
- Do not exceed the prescribed maximum dose.
- The drug product may also contain caffeine. Read the Caffeine Drug Chart for complete information about the drug.

POSSIBLE INTERACTION WITH OTHER DRUGS

GENERIC NAME OR DRUG CLASS	COMBINED EFFECT
Enzyme inhibitors*	Dangerous risk of blood circulation problem. Must avoid.
Ergot preparations,* other	Increased risk of side effects. Take drugs 24 hours apart.
Sympathomimetics*	Risk of dangerous high blood pressure.
Triptans	Increased risk of side effects. Take drugs 24 hours apart.

POSSIBLE INTERACTION WITH OTHER SUBSTANCES

INTERACTS WITH	COMBINED EFFECT
Alcohol:	Increased risk of side effects. Avoid.
Beverages: Grapefruit juice.	Increased risk of side effects. Consult doctor.
Cocaine:	Increased risk of side effects. Avoid.
Foods: Grapefruit.	Increased risk of side effects. Consult doctor.
Marijuana:	Increased risk of side effects. Avoid.
Tobacco:	Increased risk of blood circulation problems. Avoid.

***See Glossary**

ERYTHROMYCINS

GENERIC AND BRAND NAMES

See full list of generic and brand names in the *Generic and Brand Name Directory*, page 862.

BASIC INFORMATION

Habit forming? No
Prescription needed? Yes
Available as generic? Yes
Drug class: Antibacterial; antiacne agent

USES

- Treatment of a variety of bacterial infections.
- Treatment for acne.
- May be used to treat other disorders as determined by your doctor.

DOSAGE & USAGE INFORMATION

How to take:

- Tablet, capsule, extended-release tablet or extended-release capsule—Swallow each dose whole with liquid. Do not crush or chew.
- Oral suspension or pediatric drop—Shake well before using. Use dropper supplied with prescription or use a special measuring device to measure dose.
- Chewable tablet—Crush or chew completely and then swallow.

When to take:
At the same times each day, 1 hour before or 2 hours after eating. May be taken with food if stomach upset occurs. Enteric-coated tablets may be taken with or without food.

If you forget a dose:
Take as soon as you remember. If it is almost time for the next dose, wait for that dose (don't double this dose) and resume regular schedule.

Continued next column

OVERDOSE

SYMPTOMS:
Nausea, vomiting, abdominal discomfort, diarrhea.
WHAT TO DO:
If person uses much larger amount than prescribed or if accidentally swallowed, call poison control center 1-800-222-1222 for instructions or dial 911 (emergency) for help.

What drug does:
Prevents growth and reproduction of susceptible bacteria.

Time lapse before drug works:
Depends on the type of infection or acne symptoms; may take 7 to 21 days or longer.

Don't take with:
Any other medicine or any dietary supplement without consulting your doctor or pharmacist.

POSSIBLE ADVERSE REACTIONS OR SIDE EFFECTS

SYMPTOMS	WHAT TO DO
Life-threatening:	
Rare, severe allergic reaction (anaphylaxis)—difficulty breathing, hives, itching, swelling (face, throat), wheezing, vomiting, dizziness.	Seek emergency treatment immediately.
Common:	
Mild nausea.	Continue. Call doctor when convenient.
Infrequent:	
• Fever with severe nausea and vomiting, severe stomach pain, unusual tiredness or weakness, yellow eyes or skin, pale stools, skin rash (redness and itching).	Discontinue. Call doctor right away.
• Mild diarrhea, stomach cramps or vomiting, sore mouth or tongue or have white patches.	Continue. Call doctor when convenient.
Rare:	
Irregular or slow heartbeat, any hearing loss, fainting.	Discontinue. Call doctor right away.

WARNINGS & PRECAUTIONS

Don't take if:
You are allergic to any erythromycin or macrolide antibiotics.

Before you start, consult your doctor if:
- You have had liver disease or impaired liver function or stomach problems.
- You have a hearing loss.
- You have a history of heart rhythm problems, a history of long QT syndrome or myasthenia gravis or an electrolyte imbalance.

Over age 60:
Adverse reactions and side effects may be more frequent and severe than in younger persons.

Pregnancy:
Decide with your doctor if drug benefits justify risk to unborn baby. Risk category B (see page xviii).

Breastfeeding; Lactation; Nursing Mothers:
Drugs in this group are generally acceptable with breastfeeding. Discuss risks and benefits with your doctor.

Infants & children up to age 18:
Follow instructions provided by your child's doctor.

Prolonged use:
- You may become more susceptible to infections caused by germs not responsive to erythromycin.
- Talk to your doctor about the need for follow-up medical examinations or laboratory studies to check liver function.

Skin & sunlight:
No problems expected.

Driving, piloting or hazardous work:
No problems expected.

Discontinuing:
- Don't discontinue without doctor's advice until you complete prescribed dose, even though symptoms diminish or disappear.
- Diarrhea may occur 2 months or more after stopping drug. Consult doctor if it occurs.

Others:
- Advise any doctor, dentist or pharmacist whom you consult that you take this medicine.
- If infection symptoms don't start to improve in a few days or they worsen, call your doctor.

POSSIBLE INTERACTION WITH OTHER DRUGS

GENERIC NAME OR DRUG CLASS	COMBINED EFFECT
Benzodiazepines*	Increased effect of benzodiazepine.
Carbamazepine	Increased effect of carbamazepine.
Chloramphenicol	Decreased effect of chloramphenicol.
Cyclosporine	May increase cyclosporin toxicity.
Digoxin	Increased effect of digoxin.
Enzyme inhibitors*	Increased effect of erythromycin. Avoid.
Hepatotoxics*	Increased risk of liver problems.
HMG-CoA reductase inhibitors	Increased risk of muscle and kidney problems (with lovastatin).
Leukotriene modifiers	Decreased effect of zafirlukast.
Lincomycins*	Decreased lincomycin effect.
Penicillins*	Decreased penicillin effect.
Sibutramine	Increased effect of sibutramine.
Sildenafil	Increased effect of sildenafil.
Xanthines*	Increased effect of xanthine.
Warfarin	Increased risk of bleeding.

POSSIBLE INTERACTION WITH OTHER SUBSTANCES

INTERACTS WITH	COMBINED EFFECT
Alcohol:	Possible liver damage. Avoid.
Beverages: Grapefruit juice.	May increase effect of erythromycin.
Cocaine:	Unknown. Avoid.
Foods: Grapefruit.	May increase effect of erythromycin.
Marijuana:	Consult doctor.
Tobacco:	None expected.

*See Glossary

ESTRAMUSTINE

BRAND NAMES

Emcyt

BASIC INFORMATION

Habit forming? No
Prescription needed? Yes
Available as generic? No
Drug class: Antineoplastic

USES

Treats prostate cancer.

DOSAGE & USAGE INFORMATION

How to take:
Capsule—Swallow with liquid. If you can't swallow whole, open capsule and take with liquid or food. Instructions to take on empty stomach mean 1 hour before or 2 hours after eating.

When to take:
According to doctor's instructions. Try to take 1 hour before or 2 hours after eating or drinking any milk products.

If you forget a dose:
Take as soon as you remember. If it is almost time for the next dose, wait for that dose (don't double this dose) and resume regular schedule.

What drug does:
Suppresses growth of cancer cells.

Time lapse before drug works:
Within 20 hours.

Don't take with:
Any other medicines (including over-the-counter drugs such as cough and cold medicines, laxatives, antacids, diet pills, caffeine, nose drops or vitamins) without consulting your doctor or pharmacist.

OVERDOSE

SYMPTOMS:
Unknown. May be more severe adverse reactions.
WHAT TO DO:

- **Dial 911 for all medical emergencies or call poison control center 1-800-222-1222 for instructions.**
- **See emergency information on last 3 pages of this book.**

POSSIBLE ADVERSE REACTIONS OR SIDE EFFECTS

SYMPTOMS	WHAT TO DO
Life-threatening:	
Sudden headaches, chest pain, shortness of breath, leg pain, vision changes, slurred speech.	Seek emergency treatment immediately.
Common:	
• Skin rash, itching.	Discontinue. Call doctor right away.
• Diarrhea, dizziness, nausea, vomiting, headache, swelling in hands or feet.	Continue. Call doctor when convenient.
Infrequent:	
• Joint or muscle pain, difficulty swallowing, sore throat and fever, peeling skin, yellow eyes or skin.	Discontinue. Call doctor right away.
• Mouth or tongue irritation, breast tenderness or enlargement, decreased interest in sex.	Continue. Call doctor when convenient.
Rare:	
Bloody urine, hearing loss, back pain, abdominal pain.	Discontinue. Call doctor right away.

WARNINGS & PRECAUTIONS

Don't take if:
- You are allergic to estramustine, estrogens or mechlorethamine. (Estramustine is a combination of an estrogen and mechlorethamine.) Allergic reactions are rare, but may occur.
- You have active thromboembolic disorder.

Before you start, consult your doctor if:
- You have jaundice or hepatitis.
- You have history of thrombophlebitis or blood clots.
- You have an ulcer.
- You have asthma.
- You have epilepsy.
- You have bone disease or depressed bone marrow.
- You have kidney or gallbladder disease.
- You have mental depression.
- You suffer from migraine headaches.
- You have or recently had chicken pox.
- You have shingles (herpes zoster).
- You have had a recent heart attack or stroke or have heart or blood vessel disease.

Over age 60:
Adverse reactions and side effects may be more frequent and severe than in younger persons.

Pregnancy:
Used in men only. If a male is taking this drug at the time of conception, there may be a risk of birth defects. Consult doctor.

Breastfeeding; Lactation; Nursing Mothers:
Used in men only.

Infants & children up to age 18:
Not used in this age group.

Prolonged use:
- Talk to your doctor about the need for follow-up medical examinations or laboratory studies to check complete blood counts (white blood cell count, platelet count, red blood cell count, hemoglobin, hematocrit), blood pressure, liver function, serum acid, serum calcium and alkaline phosphatase.
- The drug may cause permanent sterility after it has been taken for a while. Discuss with your doctor any future plans you have for having children.

Skin & sunlight:
No problems expected.

Driving, piloting or hazardous work:
No problems expected.

Discontinuing:
No problems expected.

Others:
- Advise any doctor, dentist or pharmacist whom you consult that you take this medicine.
- May affect results in some medical tests.
- May decrease sperm count in males.
- Do not get any immunizations (vaccinations) while taking this drug unless approved by your doctor.
- Persons in your household should not take an oral polio vaccine while you are taking this drug. They could pass the polio virus on to you. Avoid any persons who have recently taken oral polio vaccine.

POSSIBLE INTERACTION WITH OTHER DRUGS

GENERIC NAME OR DRUG CLASS	COMBINED EFFECT
Adrenocorticoids, systemic	Increased effect of adrenocorticoid.
Calcium supplements*	Decreased absorption of estramustine.
Hepatotoxic medications*	Increased risk of liver toxicity.
Vaccines (killed virus)	Decreased vaccine effect.

POSSIBLE INTERACTION WITH OTHER SUBSTANCES

INTERACTS WITH	COMBINED EFFECT
Alcohol:	Increased "hangover effect" and other gastrointestinal symptoms.
Beverages: Milk and milk products.	Decreased absorption of estramustine. Take drug 1 hour before or 2 hours after milk product.
Cocaine:	Unknown. Avoid.
Foods: Foods high in calcium.	Decreased absorption of estramustine. Take drug 1 hour before or 2 hours after milk product.
Marijuana:	Consult doctor.
Tobacco:	Increased risk of heart attack and blood clots.

*See Glossary

ESTROGENS

GENERIC AND BRAND NAMES

See full list of generic and brand names in *Generic and Brand Name Directory*, page 862.

BASIC INFORMATION

Habit forming? No
Prescription needed? Yes
Available as generic? Yes
Drug class: Female sex hormone (estrogen)

USES

- Treatment for estrogen deficiency.
- Treatment for symptoms of menopause and menstrual cycle irregularity.
- Treatment for estrogen-deficiency osteoporosis (bone softening from calcium loss).
- Treatment for vulvar or vaginal atrophy.
- Treatment for certain breast or prostate cancers.

DOSAGE & USAGE INFORMATION

How to take:
- Tablet or capsule—Take with or after food to reduce nausea. Swallow with liquid. If you can't swallow whole, crumble tablet or open capsule and take with liquid or food.
- Vaginal cream or suppository—Use as directed on label.
- Gel or emulsion—Apply to skin as directed on product's label.
- Injection—Given by medical professional.
- Transdermal patch or transdermal spray—Follow label instructions.
- Vaginal insert—Follow label instructions.

When to take:
- Oral estrogen—Take at the same time each day.
- Other forms—Follow label instructions for correct dosage schedule.

Continued next column

OVERDOSE

SYMPTOMS:
Nausea, vomiting, fluid retention, breast enlargement and discomfort, abnormal vaginal bleeding, headache, drowsiness.
WHAT TO DO:
If person uses much larger amount than prescribed or if accidentally swallowed, call poison control center 1-800-222-1222 for instructions or dial 911 (emergency) for help.

If you forget a dose:
For oral dosage, take as soon as you remember. If it is almost time for the next dose, wait for the next scheduled dose (don't double this dose). For other forms, follow label instructions.

What drug does:
- Increases estrogen levels in the body.
- Combined with progestins for contraception.

Time lapse before drug works:
10 to 20 days.

Don't take with:
Any other medicine or any dietary supplement without consulting your doctor or pharmacist.

POSSIBLE ADVERSE REACTIONS OR SIDE EFFECTS

SYMPTOMS	WHAT TO DO
Life-threatening:	
Blood clots (severe or sudden headache, severe pain in calf or chest or other body parts, shortness of breath, slurred speech, weakness or numbness, sudden vision changes).	Seek emergency help.
Common:	
• Painful or swollen breasts, swollen feet or ankles, rapid weight gain.	Discontinue. Call doctor right away.
• Appetite loss, nausea, stomach, cramps or bloating, skin irritation in patch or spray users.	Continue. Call doctor when convenient.
Infrequent:	
• Breast lumps or discharge, changes in vaginal bleeding (more, less, spotting, prolonged), migraine headache.	Continue, but call doctor right away.
• Dizziness, contact lens intolerance, vomiting, mild diarrhea, headache, increased or decreased sexual desire, slow weight gain, other unusual symptoms.	Continue. Call doctor when convenient.
Rare:	
Stomach or side pain, joint or muscle pain, jaundice (yellow skin or eyes).	Discontinue. Call doctor right away.

WARNINGS & PRECAUTIONS

Don't take if:
You are allergic to any estrogen-containing drugs. Allergic reactions are rare, but may occur.

Before you start, consult your doctor if:
- You have or have had cancer of the breast or reproductive organs, fibrocystic breast disease, fibroid tumors of the uterus or endometriosis, unexplained vaginal bleeding.
- You have migraine headaches, epilepsy or porphyria.
- You have had blood clots, stroke, high blood pressure, congestive heart failure or heart attack.
- You have diabetes, asthma, kidney, liver or gallbladder disease.
- You plan to become pregnant within 3 months.

Over age 60:
Adverse reactions and side effects may be more frequent and severe than in younger persons.

Pregnancy:
Consult doctor. Drug should not be used during pregnancy. Can cause harm to unborn baby. Risk category X (see page xviii).

Breastfeeding; Lactation; Nursing Mothers:
Drugs in this group have some precautions and concerns with breastfeeding. Discuss risks and benefits with your doctor.

Infants & children up to age 18:
Not recommended in this age group.

Prolonged use:
Talk to your doctor about the need for follow-up medical examinations or laboratory studies to check for drug's effectiveness.

Skin & sunlight:
One or more drugs in this group may cause rash or intensify sunburn in areas exposed to sun or ultraviolet light (photosensitivity reaction). Avoid overexposure. Notify doctor if reaction occurs.

Driving, piloting or hazardous work:
No problems expected.

Discontinuing:
Consult your doctor before discontinuing.

Others:
- Hormone therapy (HRT) is not recommended for the prevention of chronic disease, but may be considered for the short-term management of menopausal symptoms for younger women. Before deciding on hormone replacement therapy, discuss your specific risks and benefits (e.g., age, lifestyle and risk for disease) with your doctor. Get regular checkups to make sure that you should continue with HRT.
- May interfere with the accuracy of some medical tests.
- Carefully read the paper called "Information for the Patient" that was given to you with your first prescription or a refill. If you lose it, ask your pharmacist for a copy.
- Advise any doctor, dentist or pharmacist whom you consult that you take this medicine.

POSSIBLE INTERACTION WITH OTHER DRUGS

GENERIC NAME OR DRUG CLASS	COMBINED EFFECT
Adrenocorticoids	Increased effect of adrenocorticoid.
Anticoagulants, oral*	Decreased anti-coagulant effect.
Anticonvulsants, hydantoin*	Decreased estrogen effect.
Antidepressants, tricyclic*	Increased toxicity of antidepressants.
Antidiabetics, oral*	Unpredictable increase or decrease in blood sugar.
Antivirals, HIV/AIDS*	Increased risk of pancreatitis.
Bromocriptine	May need to adjust bromocriptine dose.
Cyclosporine	Increased effect of cyclosporine.
Hepatotoxics*	Increased risk of liver problems.
Tamoxifen	Decreased effect of tamoxifen.

POSSIBLE INTERACTION WITH OTHER SUBSTANCES

INTERACTS WITH	COMBINED EFFECT
Alcohol:	None expected.
Beverages: Grapefruit juice.	Possible increased estrogen effect.
Cocaine:	Unknown. Avoid.
Foods:	None expected.
Marijuana:	Possible menstrual irregularities; bleed-ing between periods.
Tobacco:	Increased risk of blood clots leading to stroke or heart attack.

*See Glossary

ESZOPICLONE

BRAND NAMES

Lunesta

BASIC INFORMATION

Habit forming? Yes
Prescription needed? Yes
Available as generic? Yes
Drug class: Sedative-hypnotic agent

USES

Treatment for insomnia symptoms such as trouble falling asleep, waking up too often during the night and waking up too early in the morning.

DOSAGE & USAGE INFORMATION

How to take:
Tablet—Swallow whole with liquid.

When to take:
Take immediately before bedtime. For best results do not take with or right after a heavy meal.

If you forget a dose:
Take as soon as you remember. Take drug only when you are able to get 7 hours of sleep before your daily activity begins. Do not exceed prescribed dosage.

What drug does:
The exact mechanism is not known. Acts as a central nervous system depressant which helps decrease sleep problems.

Time lapse before drug works:
Usually within 1 hour. The sleep-inducing affect should last for 6 to 8 hours.

Don't take with:
Any other medicine or any dietary supplement without consulting your doctor or pharmacist.

OVERDOSE

SYMPTOMS:
Drowsiness, weakness, stupor (not able to respond), coma.
WHAT TO DO:
- **Dial 911 for all medical emergencies or call poison control center 1-800-222-1222 for instructions.**
- **See emergency information on last 3 pages of this book.**

POSSIBLE ADVERSE REACTIONS OR SIDE EFFECTS

SYMPTOMS	WHAT TO DO
Life-threatening:	
Rare, severe allergic reaction (anaphylaxis)—difficulty breathing, hives, itching, swelling (throat, face), vomiting, dizziness.	Seek emergency treatment immediately.
Common:	
Headache, unpleasant taste, drowsiness in the daytime, dizziness.	Continue. Call doctor when convenient.
Infrequent:	
Indigestion, nausea, nervousness, dry mouth, diarrhea, depression, cold-like symptoms or infection, problems with coordination, lightheadedness, anxiety.	Continue. Call doctor when convenient.
Rare:	
• Behavioral changes, (agitation, confusion, aggressiveness, suicidal thoughts, other bizarre behaviors), hallucinations, chest pain, swelling of arms or legs, enlarged breasts in men, depression worsens, painful menstruation, sleep-related behaviors.*	Discontinue. Call doctor right away.
• Memory problems (may occur if you wake before the effect of drug is gone), any other symptoms occur that cause concern (they may be due to the drug, an underlying disorder or the effects of lack of sleep).	Continue. Call doctor when convenient.

WARNINGS & PRECAUTIONS

Don't take if:
You are allergic to eszopiclone or zopiclone.

Before you start, consult your doctor if:
- You have respiratory problems.
- You have liver disease.
- You suffer from depression or psychiatric disorder.
- You are an active or recovering alcoholic or drug or substance abuser.

Over age 60:
Adverse reactions and side effects may be more frequent and severe than in younger persons.

Pregnancy:
Decide with your doctor if drug benefits justify risk to unborn baby. Risk category C (see page xviii).

Breastfeeding; Lactation; Nursing Mothers:
This drug is generally acceptable with breastfeeding. Discuss risks and benefits with your doctor.

Infants & children up to age 18:
Safety and effectiveness in this age group have not been established. Consult doctor.

Prolonged use:
You and your doctor will decide if there is a need to take the drug for prolonged period for chronic insomnia (insomnia at least three nights a week for a period of one month or longer).

Skin & sunlight:
No special problems expected.

Driving, piloting or hazardous work:
Don't drive or pilot aircraft until you learn how medicine affects you. Don't work around dangerous machinery. Don't climb ladders or work in high places. Danger increases if you drink alcohol or take other medicines affecting alertness and reflexes.

Discontinuing:
- Don't discontinue without consulting doctor. Dose may require gradual reduction if you have taken drug for a long time.
- You may have sleeping problems for 1 or 2 nights after stopping drug.
- Withdrawal symptoms may occur after you stop the drug. Consult your doctor if any emotional or physical symptoms do occur.

Others:
- Advise any doctor, dentist or pharmacist whom you consult that you take this medicine.
- Don't take drug if you are traveling on an overnight airplane trip of less than 7 or 8 hours. A temporary memory loss may occur (called traveler's amnesia).

POSSIBLE INTERACTION WITH OTHER DRUGS

GENERIC NAME OR DRUG CLASS	COMBINED EFFECT
Central nervous system (CNS) depressants*	Increased sedative effect.
Enzyme inducers*	Decreased effect of eszopiclone.
Enzyme inhibitors*	Increased effect of eszopiclone.
Ketoconazole	Increased effect of eszopiclone.
Olanzapine	Decreased alertness.

POSSIBLE INTERACTION WITH OTHER SUBSTANCES

INTERACTS WITH	COMBINED EFFECT
Alcohol:	Increased sedation. Avoid.
Beverages:	None expected.
Cocaine:	Increased risk of side effects. Avoid.
Foods:	Decreased drug effect if taken with a heavy meal.
Marijuana:	Increased risk of side effects. Avoid.
Tobacco:	None expected.

ETHIONAMIDE

BRAND NAMES

Trecator

BASIC INFORMATION

Habit forming? No
Prescription needed? Yes
Available as generic? No
Drug class: Antimycobacterial (antituberculosis)

USES

Treats tuberculosis. Used in combination with other antituberculosis drugs such as isoniazid, streptomycin, rifampin, ethambutol.

DOSAGE & USAGE INFORMATION

How to take:
Tablet—Swallow with liquid or food to lessen stomach irritation. If you can't swallow whole, crumble tablet and take with liquid or food.

When to take:
Usually every 8 to 12 hours, with or after meals.

If you forget a dose:
Take as soon as you remember. If it is almost time for the next dose, wait for that dose (don't double this dose) and resume regular schedule.

What drug does:
Kills germs that cause tuberculosis.

Time lapse before drug works:
Within 3 hours.

Don't take with:
Any other medicine or any dietary supplement without consulting your doctor or pharmacist.

OVERDOSE

SYMPTOMS:
Unknown.
WHAT TO DO:
If person uses much larger amount than prescribed or if accidentally swallowed, call poison control center 1-800-222-1222 for instructions or dial 911 (emergency) for help.

POSSIBLE ADVERSE REACTIONS OR SIDE EFFECTS

SYMPTOMS	WHAT TO DO
Life-threatening: None expected.	
Common: Dizziness, sore mouth, nausea or vomiting, metallic taste.	Continue. Call doctor when convenient.
Infrequent: Jaundice (yellow eyes and skin), numbness and tingling or pain in hands or feet, depression, confusion, unsteadiness, mood or mental changes.	Discontinue. Call doctor right away.
Rare:	
• Shakiness, rapid heartbeat, blurred vision or other eye changes, skin rash, increased hunger, changes in menstrual periods, decreased sexual ability, puffy skin, unusual weight gain.	Discontinue. Call doctor right away.
• Gradual swelling in the neck (thyroid gland), enlargement of breasts (male).	Continue. Call doctor when convenient.

WARNINGS & PRECAUTIONS

Don't take if:
You are allergic to ethionamide. Allergic reactions are rare, but may occur.

Before you start, consult your doctor if:
- You have diabetes.
- You have liver disease.

Over age 60:
Adverse reactions and side effects may be more frequent and severe than in younger persons.

Pregnancy:
Decide with your doctor if drug benefits justify risk to unborn baby. Risk category C (see page xviii).

Breastfeeding; Lactation; Nursing Mothers:
This drug is generally acceptable with breastfeeding. Discuss risks and benefits with your doctor.

Infants & children up to age 18:
Follow instructions provided by your child's doctor.

Prolonged use:
No special problems expected.

Skin & sunlight:
No special problems expected.

Driving, piloting or hazardous work:
Don't drive or pilot aircraft until you learn how medicine affects you. Don't work around dangerous machinery. Don't climb ladders or work in high places. Danger increases if you drink alcohol or take medicine affecting alertness and reflexes.

Discontinuing:
Don't discontinue without consulting doctor.

Others:
- Advise any doctor, dentist or pharmacist whom you consult that you take this medicine.
- Request occasional laboratory studies for liver function.
- Request occasional eye examinations.
- Treatment may take months or years.
- You should take pyridoxine (vitamin B-6) supplements while taking ethionamide.

POSSIBLE INTERACTION WITH OTHER DRUGS

GENERIC NAME OR DRUG CLASS	COMBINED EFFECT
Antivirals, HIV/AIDS*	Increased risk of peripheral neuropathy.
Cycloserine	Increased risk of seizures.
Pyridoxine	Increased excretion by kidney. (Should take pyridoxine supplements while on ethionamide to prevent development of neuritis in feet and hands).

POSSIBLE INTERACTION WITH OTHER SUBSTANCES

INTERACTS WITH	COMBINED EFFECT
Alcohol:	Increased incidence of liver diseases.
Beverages: Any alcoholic beverage.	Increased incidence of liver diseases.
Cocaine:	Unknown. Avoid.
Foods:	None expected.
Marijuana:	Consult doctor.
Tobacco:	No interaction expected, but may slow body's recovery.

***See Glossary**

ETOPOSIDE

BRAND NAMES

VePesid VP-16

BASIC INFORMATION

Habit forming? No
Prescription needed? Yes
Available as generic? No
Drug class: Antineoplastic

USES

- Treats testicular, lung and bladder cancer.
- Treats Hodgkin's disease and some other forms of cancer.

DOSAGE & USAGE INFORMATION

How to take:

- Capsule—Swallow with liquid. If you can't swallow whole, open capsule and take with liquid or food. Instructions to take on empty stomach mean 1 hour before or 2 hours after eating.
- Injection—Given under doctor's supervision.

When to take:
According to your doctor's instructions.

If you forget a dose:
Take as soon as you remember. If it is almost time for the next dose, wait for that dose (don't double this dose) and resume regular schedule.

What drug does:
Inhibits DNA in cancer cells.

Time lapse before drug works:
Unpredictable.

Don't take with:
Any other medicines (including over-the-counter drugs such as cough and cold medicines, laxatives, antacids, diet pills, caffeine, nose drops or vitamins) without consulting your doctor or pharmacist.

OVERDOSE

SYMPTOMS:
Unknown.
WHAT TO DO:

- **Dial 911 for all medical emergencies or call poison control center 1-800-222-1222 for instructions.**
- **See emergency information on last 3 pages of this book.**

POSSIBLE ADVERSE REACTIONS OR SIDE EFFECTS

SYMPTOMS	WHAT TO DO
Life-threatening:	
Rare, severe allergic reaction (anaphylaxis)—difficulty breathing, hives, itching, swelling (throat, face), vomiting, dizziness.	Seek emergency treatment immediately.
Common:	
Appetite loss, nausea, vomiting, loss of hair.	Continue. Call doctor when convenient.
Infrequent:	
Symptoms of low white blood cell count and low platelet count (black or tarry stools, bloody urine, cough, chills, fever, low back pain, bruising).	Discontinue. Call doctor right away.
Rare:	
Mouth sores.	Continue. Call doctor when convenient.

WARNINGS & PRECAUTIONS

Don't take if:
You are allergic to etoposide.

Before you start, consult your doctor if:
You have liver or kidney disease.

Over age 60:
Adverse reactions and side effects may be more frequent and severe than in younger persons.

Pregnancy:
Consult doctor. Use of the drug during pregnancy is not recommended. It could cause harm to the unborn baby. Risk category D (see page xviii).

Breastfeeding; Lactation; Nursing Mothers:
This drug is generally not recommended with breastfeeding. Discuss risks and benefits with your doctor.

Infants & children up to age 18:
Follow instructions provided by your child's doctor.

Prolonged use:
- Talk to your doctor about the need for follow-up medical examinations or laboratory studies to check complete blood counts (white blood cell count, platelet count, red blood cell count, hemoglobin, hematocrit).
- Check mouth frequently for ulcers.

Skin & sunlight:
No problems expected.

Driving, piloting or hazardous work:
Avoid if you feel confused, drowsy or dizzy.

Discontinuing:
May still experience symptoms of bone marrow depression, such as blood in stools, fever or chills, blood spots under the skin, back pain, hoarseness, bloody urine. If any of these occur, call your doctor right away.

Others:
- Advise any doctor, dentist or pharmacist whom you consult that you take this medicine.
- May affect results in some medical tests.
- Etoposide may be used in combinations with other antineoplastic treatments. The incidence and severity of side effects may be different when used in combinations. For further information, consult your doctor.

POSSIBLE INTERACTION WITH OTHER DRUGS

GENERIC NAME OR DRUG CLASS	COMBINED EFFECT
Angiotensin-converting enzyme (ACE) inhibitors*	May increase bone marrow depression or make kidney damage more likely.
Antineoplastic (cancer-treating) drugs*	May increase bone marrow depression or make kidney damage more likely.
Clozapine	Toxic effect on bone marrow.
Tiopronin	Increased risk of toxicity to bone marrow.
Vaccines, live or killed virus	Increased likelihood of toxicity or reduced effectiveness of vaccine. Wait 3 months to 1 year after etoposide treatment before getting vaccine.

POSSIBLE INTERACTION WITH OTHER SUBSTANCES

INTERACTS WITH	COMBINED EFFECT
Alcohol:	Increased likelihood of adverse reactions. Avoid.
Beverages: Grapefruit juice.	Decreased effect of etoposide. Avoid.
Cocaine:	Increased risk of adverse reactions. Avoid.
Foods:	None expected.
Marijuana:	Increased risk of adverse reactions. Avoid.
Tobacco:	None expected.

***See Glossary**

EZETIMIBE

BRAND NAMES

Liptruzet
Vytorin
Zetia

BASIC INFORMATION

Habit forming? No
Prescription needed? Yes
Available as generic? No
Drug class: Antihyperlipidemic

USES

Lowers blood cholesterol levels caused by low-density lipoproteins (LDL) in persons who haven't improved by exercising, dieting or using other measures. May be used alone or with other cholesterol lowering drugs (e.g., HMG-CoA reductase inhibitors).

DOSAGE & USAGE INFORMATION

How to take:
Tablet—Swallow with liquid. May be taken on an empty stomach or with food.

When to take:
Once a day at the same time each day.

If you forget a dose:
Take as soon as you remember. If it is almost time for the next dose, wait for that dose (don't double this dose) and resume regular schedule.

What drug does:
It lowers the amount of cholesterol your body absorbs from your diet by reducing absorption in the intestines. Other cholesterol lowering drugs alter the production and metabolism of cholesterol in the liver.

Time lapse before drug works:
2 to 4 weeks.

Don't take with:
- A high-fat diet.
- Any other medicine or any dietary supplement without consulting your doctor or pharmacist.

OVERDOSE

SYMPTOMS:
Unknown.
WHAT TO DO:
If person uses much larger amount than prescribed or if accidentally swallowed, call poison control center 1-800-222-1222 for instructions or dial 911 (emergency) for help.

POSSIBLE ADVERSE REACTIONS OR SIDE EFFECTS

SYMPTOMS	WHAT TO DO
Life-threatening:	
Rare, severe allergic reaction (anaphylaxis)—difficulty breathing, hives, itching, swelling (throat, face), vomiting, dizziness.	Seek emergency treatment immediately.
Common:	
None expected.	
Infrequent:	
Headache, cold or other upper respiratory infection, muscle aches, chest pains, fatigue, dizziness, diarrhea, stomach pain, nausea, tingling sensation.	Continue. Call doctor when convenient.
Rare:	
None expected.	

WARNINGS & PRECAUTIONS

Don't take if:
You are allergic to ezetimibe or its components.

Before you start, consult your doctor if:
- You have liver disease.
- You are allergic to any medication, food or other substance.

Over age 60:
No special problems expected.

Pregnancy:
Decide with your doctor if drug benefits justify risk to unborn baby. Risk category C (see page xviii).

Breastfeeding; Lactation; Nursing Mothers:
This drug is generally not recommended with breastfeeding. Discuss risks and benefits with your doctor.

Infants & children up to age 18:
Not recommended for under age 10. Follow instructions provided by your child's doctor.

Prolonged use:
Talk to your doctor about the need for follow-up medical examinations or laboratory studies to check liver function and serum cholesterol.

Skin & sunlight:
No problems expected.

Driving, piloting or hazardous work:
No special problems expected.

Discontinuing:
No special problems expected.

Others:
- Advise any doctor, dentist or pharmacist whom you consult that you take this medicine.
- Continue to follow your doctor's instructions about dietary intake, reduced intake of saturated fats, increased fiber intake, weight reduction, and increased physical activity.
- Cholesterol lowering drugs may cause muscle problems or rhabdomyolysis (muscle injury). If you develop persistent muscle aches, pain or weakness or urine turns dark, call the doctor right away.

POSSIBLE INTERACTION WITH OTHER DRUGS

GENERIC NAME OR DRUG CLASS	COMBINED EFFECT
Cholestyramine	Take ezetimibe 2 hours before or 4 hours after.
Colestipol	Take ezetimibe 2 hours before or 4 hours after.
Colesevelam	Take ezetimibe 2 hours before or 4 hours after.
Cyclosporine	Increased effect of ezetimibe.
Fibrates	May require dosage adjustment of either drug.
Gemfibrozil	Increased effect of ezetimibe.

POSSIBLE INTERACTION WITH OTHER SUBSTANCES

INTERACTS WITH	COMBINED EFFECT
Alcohol:	None expected.
Beverages:	None expected.
Cocaine:	Unknown. Avoid.
Foods:	None expected.
Marijuana:	Consult doctor.
Tobacco:	None expected.

*See Glossary

FACTOR Xa INHIBITORS

GENERIC AND BRAND NAMES

APIXABAN
Eliquis
EDOXABAN
Savaysa
RIVAROXABAN
Xarelto

BASIC INFORMATION

Habit forming? No
Prescription needed? Yes
Available as generic? No
Drug class: Anticoagulant

USES

- Used to treat and prevent blood clots (e.g., deep venous thrombosis [DVT]) that can form in blood vessels. Blood clots can travel to the brain and cause a stroke or to the lungs causing pulmonary embolism.
- Used for patients with atrial fibrillation (heart rhythm disorder) to reduce risk of stroke.

DOSAGE & USAGE INFORMATION

How to take:
Tablet—Swallow whole with water. May be taken with or without food. If it is prescribed for atrial fibrillation, take it with your evening meal.

When to take:
Once a day at the same time each day.

If you forget a dose:
Take as soon as you remember. If it is almost time for the next dose, wait for that dose (don't double this dose) and resume regular schedule.

What drug does:
It works by decreasing the clotting ability of the blood and helps prevent harmful clots from forming in the blood vessels or heart.

Time lapse before drug works:
It starts working within 2 to 4 hours.

Continued next column

OVERDOSE

SYMPTOMS:
Increased risk of dangerous bleeding (may be internal bleeding).
WHAT TO DO:
If person uses much larger amount than prescribed or if accidentally swallowed, call poison control center 1-800-222-1222 for instructions or dial 911 (emergency) for help.

Don't take with:
Any other medicine or dietary supplement without consulting your doctor or pharmacist.

POSSIBLE ADVERSE REACTIONS OR SIDE EFFECTS

SYMPTOMS	WHAT TO DO
Life-threatening: In case of overdose, see previous column.	
Common: Bleeding following surgery, surgical wound oozing or bleeding, anemia (pale skin, shortness of breath, weakness).	Continue, but call doctor right away.
Infrequent:	
• Unusual bleeding from the gums or vagina or rectum or eyes, black or tarry stools, red or brown urine, coughing up blood or coffee-ground material, unusual bruising, purple or red spots under the skin, fast or irregular heartbeat, chest pain.	Continue, but call doctor right away.
• Dizziness, faintness, headache, diarrhea, upset stomach, fever, unusual tiredness or weakness, pain (in back, legs, arms), dry mouth, muscle spasms, swelling in arms or legs, rash or itchy skin, low blood pressure.	Continue. Call doctor when convenient.
Rare:	
• Bleeding in the brain (may have symptoms of sudden numbness or weakness on one side of the body, sudden and severe headache, confusion, balance or vision problems).	Seek emergency help.
• Liver problem (yellow skin or eyes, nausea, vomiting, weight loss), skin peels or blisters or loosens.	Continue, but call doctor right away.

WARNINGS & PRECAUTIONS

Don't take if:
You are allergic to factor Xa inhibitors. Allergic reactions are rare, but may occur.

Before you start, consult your doctor if:
- You have a history of intracranial hemorrhage (bleeding in the brain), spinal deformity or spinal surgery.
- You have a liver or kidney disorder.
- You have or have had conditions that have an increased risk of bleeding (e.g., ulcers, recent surgery, retinopathy or stroke).
- You have rare inherited disorders of lactose or galactose intolerance.
- You have a spinal or epidural catheter.
- You plan on having surgery (including dental surgery) in the near future.

Over age 60:
Adverse reactions and side effects may be more frequent and severe than in younger persons.

Pregnancy:
Decide with your doctor if drug benefits justify risk to unborn baby. Apixaban is risk category B; rivaroxaban is category C (see page xviii).

Breastfeeding; Lactation; Nursing Mothers:
Drugs in this group are generally not recommended with breastfeeding. Discuss risks and benefits with your doctor.

Infants & children up to age 18:
Safety and effectiveness in this age group have not been established. Consult doctor.

Prolonged use:
Talk to your doctor about the need for follow-up medical exams or laboratory studies to check the effectiveness of the treatment.

Skin & sunlight:
No problems expected.

Driving, piloting or hazardous work:
Avoid if you experience dizziness or faintness, otherwise no problems expected.

Discontinuing:
Don't discontinue without doctor's approval. Stopping the drug can increase the risk of blood clots and stroke. If drug used for atrial fibrillation, be sure to refill prescription before you run out.

Others:
- Advise any doctor, dentist or pharmacist whom you consult that you take this drug.
- While taking this drug you will likely bruise and bleed more easily than usual or bleed for longer than usual. Avoid rough sports or other situations where you could be bruised, cut or injured. Be extra careful when using sharp objects. Bleeding complications can be serious or fatal. Consult doctor if you have any concerns or questions about side effects.
- Use of this drug in patients who undergo certain spinal procedures (e.g., lumbar puncture or epidural injections for pain) increases risk for hematoma (trapped blood). Consult doctor about your risk and possible complications.
- Drug use may increase risk of serious bleeding during a surgery, other medical procedures or some types of dental work. You may be advised to stop using this drug at least 24 hours before a surgery, medical procedure or dental work.

POSSIBLE INTERACTION WITH OTHER DRUGS

GENERIC NAME OR DRUG CLASS	COMBINED EFFECT
Anticoagulants,* other	Increased risk of bleeding.
Anti-inflammatory drugs, nonsteroidal (NSAIDs)*	Increased risk of bleeding.
Antiplatelet drugs*	Increased risk of bleeding.
Enzyme inducers*	Decreased effect of rivaroxaban.
Enzyme inhibitors*	Increased effect of rivaroxaban.

POSSIBLE INTERACTION WITH OTHER SUBSTANCES

INTERACTS WITH	COMBINED EFFECT
Alcohol:	None expected.
Beverages: Grapefruit juice.	May increase effect of rivaroxaban.
Cocaine:	Unknown. Avoid.
Foods: Grapefruit.	May increase effect of rivaroxaban.
Marijuana:	Consult doctor.
Tobacco:	None expected, but people with heart disorders should not smoke.

***See Glossary**

FELBAMATE

BRAND NAMES

FBM   Felbatol

BASIC INFORMATION

Habit forming? No
Prescription needed? Yes
Available as generic? Yes
Drug class: Anticonvulsant

USES

- Treatment for partial epileptic seizures.
- Treatment for Lennox-Gastaut syndrome (a severe form of epilepsy in children).

DOSAGE & USAGE INFORMATION

How to take:

- Tablet—Swallow with liquid. May be taken with food to lessen stomach upset unless the doctor has directed taking on an empty stomach.
- Oral suspension—Shake bottle well before measuring. Use specially marked measuring device to measure each dose accurately. Don't measure with a regular household teaspoon.

When to take:
At the same times each day. Your doctor will determine the best schedule. Dosages will gradually be increased over the first 3 weeks.

If you forget a dose:
Take as soon as you remember. If it is almost time for the next dose, wait for that dose (don't double this dose) and resume regular schedule.

What drug does:

- Decreases the frequency of partial seizures that start in a localized part of the brain, including those that progress into more generalized grand mal seizures.
- Decreases seizure activity and improves quality of life in children with Lennox-Gastaut syndrome.

Continued next column

OVERDOSE

SYMPTOMS:
Gastric distress, increased heart rate.
WHAT TO DO:
If person uses much larger amount than prescribed or if accidentally swallowed, call poison control center 1-800-222-1222 for instructions or dial 911 (emergency) for help.

Time lapse before drug works:
May take several weeks for maximum effectiveness.

Don't take with:
Any other medicine or any dietary supplement without consulting your doctor or pharmacist.

POSSIBLE ADVERSE REACTIONS OR SIDE EFFECTS

SYMPTOMS	WHAT TO DO
Life-threatening:	
Rare, severe allergic reaction (anaphylaxis)—difficulty breathing, hives, itching, swelling (throat, face), vomiting, dizziness.	Seek emergency treatment immediately.
Common:	
• Fever, red or purple spots on skin, walking in unusual manner.	Continue, but call doctor right away.
• Abdominal pain, taste changes, constipation, sleeping difficulty, dizziness, headache, nausea or vomiting, indigestion, appetite loss.	Continue. Call doctor when convenient.
Infrequent:	
• Mood or mental changes, clumsiness, skin rash, tremor.	Continue, but call doctor right away.
• Vision changes, diarrhea, drowsiness, coughing or sneezing, ear pain or fullness, runny nose, weight loss.	Continue. Call doctor when convenient.
Rare:	
Black or tarry stools, bloody or dark-colored urine, unusual bruising or bleeding, breathing difficulty, wheezing, pain or tightness in chest, sore throat, mouth or lip sores, swollen face, swollen or painful glands or lymph nodes, yellow skin or eyes, chills, general tired feeling, continuing headache or abdominal pain, hives, itching, muscle cramps, stuffy nose, skin reaction to sunlight.	Continue, but call doctor right away.

WARNINGS & PRECAUTIONS

Don't take if:
You are allergic to felbamate.

Before you start, consult your doctor if:
- You have a sensitivity to other carbamate drugs,* other medications or other substances.
- You have any blood disorder.
- You have a history of bone marrow depression.
- You have or have had any liver disease.

Over age 60:
Adverse reactions and side effects may be more frequent and severe than in younger persons.

Pregnancy:
Decide with your doctor if drug benefits justify risk to unborn baby. Risk category C (see page xviii).

Breastfeeding; Lactation; Nursing Mothers:
This drug is generally not recommended with breastfeeding. Discuss risks and benefits with your doctor.

Infants & children up to age 18:
Follow instructions provided by your child's doctor.

Prolonged use:
Talk to your doctor about the need for follow-up medical examinations or laboratory studies to check complete blood counts (white blood cell count, platelet count, red blood cell count, hemoglobin, hematocrit), iron concentrations and liver function studies.

Skin & sunlight:
No problems expected.

Driving, piloting or hazardous work:
Don't drive or pilot aircraft until you learn how medicine affects you. Don't work around dangerous machinery. Don't climb ladders or work in high places. Danger increases if you drink alcohol or take other medicines affecting alertness and reflexes such as antihistamines, tranquilizers, sedatives, pain medicine, narcotics and mind-altering drugs.

Discontinuing:
Don't discontinue without doctor's approval due to risk of increased seizure activity.

Others:
- Felbamate may cause serious side effects including blood problems and liver problems (rarely fatal). Decide with your doctor if drug benefits justify risks.
- Advise any doctor, dentist or pharmacist whom you consult that you take this medicine.
- Felbamate may be used alone or combined with other antiepileptic drugs. The dosages of other antiepileptic drugs you currently use will be reduced to minimize side effects and adverse reactions due to interactions.
- Rarely, antiepileptic drugs may lead to suicidal thoughts and behaviors. Call doctor right away if suicidal symptoms or unusual behaviors occur.
- Wear medical identification that indicates the use of this medicine.

POSSIBLE INTERACTION WITH OTHER DRUGS

GENERIC NAME OR DRUG CLASS	COMBINED EFFECT
Carbamazepine	Increased side effects and adverse reactions.
Phenytoin	Increased side effects and adverse reactions.
Valproic acid*	Increased effect of valproic acid.

POSSIBLE INTERACTION WITH OTHER SUBSTANCES

INTERACTS WITH	COMBINED EFFECT
Alcohol:	None expected.
Beverages:	None expected.
Cocaine:	Unknown. Avoid.
Foods:	None expected.
Marijuana:	Consult doctor.
Tobacco:	None expected.

***See Glossary**

FIBRATES

GENERIC AND BRAND NAMES

FENOFIBRATE
- **Antara**
- **Fenoglide**
- **Lipidil**
- **Lipofen**
- **Lofibra**
- **Tricor**
- **Triglide**

FENOFIBRIC ACID
- **Fibricor**
- **TriLipix**

BASIC INFORMATION

Habit forming? No
Prescription needed? Yes
Available as generic? Yes
Drug class: Antihyperlipidemic

USES

Used in addition to diet changes to help control levels of blood fats (e.g., lipid disorders such as elevated levels of cholesterol and triglycerides).

DOSAGE & USAGE INFORMATION

How to take:
- Capsule or tablet—Swallow with liquid. Follow the instructions provided with your prescription about taking the drug with or without food.
- Delayed-release capsule—Swallow with liquid. Do not crush or chew capsule. Take with or without food.

When to take:
At the same time(s) each day. Follow directions on prescription as to when to take each day.

If you forget a dose:
Take as soon as you remember. If it is almost time for the next dose, wait for that dose (don't double this dose) and resume regular schedule.

Continued next column

OVERDOSE

SYMPTOMS:
Unknown.
WHAT TO DO:
If person uses much larger amount than prescribed or if accidentally swallowed, call poison control center 1-800-222-1222 for instructions or dial 911 (emergency) for help.

What drug does:
Helps break down the fats in the blood.

Time lapse before drug works:
3 months or more as measured by laboratory testing.

Don't take with:
Any other medicine or any dietary supplement without consulting your doctor or pharmacist.

POSSIBLE ADVERSE REACTIONS OR SIDE EFFECTS

SYMPTOMS	WHAT TO DO
Life-threatening: None expected.	
Common: None expected.	
Infrequent:	
• Chest pain, shortness of breath, irregular heartbeat.	Discontinue. Seek emergency treatment.
• Severe symptoms that include nausea, flu-like illness and vomiting.	Discontinue. Call doctor right away.
• Diarrhea, stomach pain, belching, bloating, constipation, muscle aches and pains.	Continue. Call doctor when convenient.
Rare:	
• Rash, itch, mouth or lip sores, sore throat, swollen feet and legs, blood in urine, painful urination, fever, chills, paleness, yellow eyes or skin.	Discontinue. Call doctor right away.
• Dizziness, weakness, drowsiness, hair loss, headache, diminished sex drive, dry mouth, loss of appetite, unusual bleeding or bruising, unusual tiredness, mild vomiting.	Continue. Call doctor when convenient.

WARNINGS & PRECAUTIONS

Don't take if:
- You are allergic to any fibrates. Allergic reactions are rare, but may occur.
- You have had serious liver disease.

Before you start, consult your doctor if:
- You have had liver or kidney disease.
- You have had peptic ulcer disease.
- You have had gallbladder disease or gallstones.
- You have diabetes.

Over age 60:
No special problems expected.

Pregnancy:
Decide with your doctor if drug benefits justify risk to unborn baby. Risk category C (see page xviii).

Breastfeeding; Lactation; Nursing Mothers:
Drugs in this group are generally not recommended with breastfeeding. Discuss risks and benefits with your doctor.

Infants & children up to age 18:
Safety and effectiveness in this age group have not been established. Consult doctor.

Prolonged use:
- May cause gallbladder infection.
- Possible cause of stomach cancer.
- Talk to your doctor about the need for follow-up medical examinations or laboratory studies to check cholesterol and triglyceride levels.

Skin & sunlight:
May cause rash or intensify sunburn in areas exposed to sun or ultraviolet light (photosensitivity reaction). Avoid overexposure. Notify doctor if reaction occurs.

Driving, piloting or hazardous work:
Avoid if you feel drowsy or dizzy. Otherwise, no problems expected.

Discontinuing:
Don't discontinue without doctor's advice until you complete prescribed dose, even though symptoms diminish or disappear.

Others:
- Advise any doctor, dentist or pharmacist whom you consult that you take this medicine.
- Periodic blood cell counts and liver-function studies recommended if you take clofibrate for a long time.
- Some medical studies question effectiveness. A number of medical studies warn of toxicity.

POSSIBLE INTERACTION WITH OTHER DRUGS

GENERIC NAME OR DRUG CLASS	COMBINED EFFECT
Anticoagulants, oral*	Increased effect of anticoagulant (dose should be reduced).
Antidiabetics, oral*	Increased antidiabetic effect.
Contraceptives, oral*	Decreased fibrate effect.
Cyclosporine	May cause or worsen kidney problems.
Desmopressin	May decrease desmopressin effect.
Dexfenfluramine	Dose may need to be adjusted.
Estrogens*	Decreased fibrate effect.
Furosemide	Possible toxicity of both drugs.
HMG-CoA reductase inhibitors	May cause muscle or kidney problems or make them worse.
Insulin	Increased insulin effect.
Insulin lispro	May need decreased dosage of insulin.
Probenecid	Increased effect and toxicity of fibrate.
Thyroid hormones*	Increased fibrate effect.
Ursodiol	Decreased effect of ursodiol.

POSSIBLE INTERACTION WITH OTHER SUBSTANCES

INTERACTS WITH	COMBINED EFFECT
Alcohol:	None expected.
Beverages:	None expected.
Cocaine:	Unknown. Avoid.
Foods: Fatty foods.	Decreased fibrate effect.
Marijuana:	Consult doctor.
Tobacco:	None expected.

***See Glossary**

FLECAINIDE ACETATE

BRAND NAMES

Tambocor

BASIC INFORMATION

Habit forming? No
Prescription needed? Yes
Available as generic? Yes
Drug class: Antiarrhythmic

USES

Stabilizes irregular heartbeat.

DOSAGE & USAGE INFORMATION

How to take:
Tablet—Swallow with liquid. If you can't swallow whole, crumble tablet and take with liquid or food.

When to take:
At the same time each day, according to instructions on prescription label. Take tablets approximately 12 hours apart.

If you forget a dose:
Take as soon as you remember. If it is almost time for the next dose, wait for that dose (don't double this dose) and resume regular schedule.

What drug does:
Decreases conduction of abnormal electrical activity in the heart muscle or its regulating systems.

Time lapse before drug works:
1 to 6 hours. May take 3 to 5 days for maximum effect.

Don't take with:
Any other medicine or any dietary supplement without consulting your doctor or pharmacist.

OVERDOSE

SYMPTOMS:
Severe nausea and vomiting and dizziness, fainting, slow heartbeat, seizures.
WHAT TO DO:
- **Dial 911 for all medical emergencies or call poison control center 1-800-222-1222 for instructions.**
- **See emergency information on last 3 pages of this book.**

POSSIBLE ADVERSE REACTIONS OR SIDE EFFECTS

SYMPTOMS	WHAT TO DO
Life-threatening:	
In case of overdose, see previous column.	
Common:	
Blurred vision, dizziness.	Continue. Call doctor when convenient.
Infrequent:	
• Shakiness, rash, nausea, vomiting, irregular heartbeat, chest pain.	Continue, but call doctor right away.
• Anxiety, depression, loss of appetite or taste, weakness (muscles or joints), swollen feet or legs or ankles, headache, numbness or tingling in hands or feet, stomach pain, constipation.	Continue. Call doctor when convenient.
Rare:	
Yellow skin and eyes.	Continue, but call doctor right away.

WARNINGS & PRECAUTIONS

Don't take if:
You are allergic to flecainide or a local anesthetic such as novocaine, lidocaine or other drug whose generic name ends with "caine." Allergic reactions are rare, but may occur.

Before you start, consult your doctor if:
- You have kidney disease.
- You have liver disease.
- You have had a heart attack.
- You have a heart rhythm irregularity.
- You have a pacemaker.

Over age 60:
Adverse reactions and side effects may be more frequent and severe than in younger persons.

Pregnancy:
Decide with your doctor if drug benefits justify risk to unborn baby. Risk category C (see page xviii).

Breastfeeding; Lactation; Nursing Mothers:
This drug is generally acceptable with breastfeeding. Discuss risks and benefits with your doctor.

Infants & children up to age 18:
Safety and effectiveness in this age group have not been established. Consult doctor.

Prolonged use:
Talk to your doctor about the need for follow-up medical exams or lab studies.

Skin & sunlight:
No problems expected.

Driving, piloting or hazardous work:
Don't drive or pilot aircraft until you learn how medicine affects you. Don't work around dangerous machinery. Don't climb ladders or work in high places. Danger increases if you drink alcohol or take medicine affecting alertness and reflexes, such as antihistamines, tranquilizers, sedatives, pain medicine, narcotics and mind-altering drugs.

Discontinuing:
Don't discontinue without consulting doctor. Dose may require gradual reduction if you have taken drug for a long time. Doses of other drugs may also require adjustment.

Others:
- Wear identification bracelet or carry an identification card with inscription of medicine you take.
- Use of the drug in people who have had a heart attack may increase risk of another heart attack. Discuss your risks with doctor.
- Advise any doctor, dentist or pharmacist whom you consult that you take this medicine.

POSSIBLE INTERACTION WITH OTHER DRUGS

GENERIC NAME OR DRUG CLASS	COMBINED EFFECT
Antacids* (high dose)	Possible increased flecainide acetate effect.
Antiarrhythmics, other*	Possible irregular heartbeat.
Beta-adrenergic blocking agents*	Possible decreased efficiency of heart muscle contraction, leading to congestive heart failure.
Bone marrow depressants*	Possible decreased production of blood cells in bone marrow.
Carbonic anhydrase inhibitors*	Possible increased flecainide acetate effect.
Cimetidine	Increased effect of flecainide.
Digitalis preparations*	Possible increased digitalis effect. Possible irregular heartbeat.
Disopyramide	Possible decreased efficiency of heart muscle contraction, leading to congestive heart failure.
Doxepin (topical)	Increased risk of toxicity of both drugs.
Enzyme inhibitors*	Increased effect of flecainide acetate.
Nicardipine	Possible increased effect and toxicity of each drug.
Paroxetine	Increased effect of both drugs.
Propafenone	Increased effect of both drugs and increased risk of toxicity.
Sodium bicarbonate	Possible increased flecainide acetate effect.
Verapamil	Possible decreased efficiency of heart muscle contraction, leading to congestive heart failure.

POSSIBLE INTERACTION WITH OTHER SUBSTANCES

INTERACTS WITH	COMBINED EFFECT
Alcohol:	May further depress normal heart function.
Beverages: Caffeine-containing beverages.	Possible decreased flecainide effect.
Cocaine:	Unknown. Avoid.
Foods:	None expected.
Marijuana:	Consult doctor.
Tobacco:	Possible decreased flecainide effect.

*See Glossary

FLUOROQUINOLONES

GENERIC AND BRAND NAMES

CIPROFLOXACIN
- **Cipro**
- **Cipro XR**
- **Proquin XR**

ENOXACIN
- **Penetrex**

GEMIFLOXACIN
- **Factive**

LEVOFLOXACIN
- **Levaquin**

LOMEFLOXACIN
- **Maxaquin**

MOXIFLOXACIN
- **Avelox**

NORFLOXACIN
- **Noroxin**

OFLOXACIN
- **Floxin**

SPARFLOXACIN
- **Zagam**

TROVAFLOXACIN
- **Trovan**

BASIC INFORMATION

Habit forming? No
Prescription needed? Yes
Available as generic? Yes, for some
Drug class: Antibacterial

USES

- Treats a wide range of bacteria that may cause diarrhea, pneumonia, skin and soft tissue infections, urinary tract infections, conjunctivitis and bone infections.
- Treatment for specific agents that could be used in biologic warfare.

DOSAGE & USAGE INFORMATION

How to take:
- Tablet—Take with full glass of water. Take enoxacin and ofloxacin on an empty stomach. Others may be taken with or without meals.
- Extended release tablet—Swallow whole. Do not crush, break or chew tablet.
- Oral and otic suspension—Take as directed on label.

Continued next column

When to take:
As directed by your doctor.

If you forget a dose:
Take as soon as you remember. If it is almost time for the next dose, wait for that dose (don't double this dose) and resume regular schedule.

What drug does:
Destroys bacteria that is infecting the body.

Time lapse before drug works:
1 to 2 weeks for most infections, but some infections may take 6 weeks or more for cure.

Don't take with:
Any other medicine or any dietary supplement without consulting your doctor or pharmacist.

OVERDOSE

SYMPTOMS:
Confusion and hallucinations, headache, abdominal pain, convulsions.
WHAT TO DO:
- **Dial 911 for all medical emergencies or call poison control center 1-800-222-1222 for instructions.**
- **See emergency information on last 3 pages of this book.**

POSSIBLE ADVERSE REACTIONS OR SIDE EFFECTS

SYMPTOMS	WHAT TO DO
Life-threatening:	
Rare, severe allergic reaction (anaphylaxis)—difficulty breathing, hives, itching, swelling (throat, face), vomiting, dizziness.	Seek emergency treatment immediately.
Common:	
• Sparfloxacin may cause fainting, slow, irregular heart rate.	Discontinue. Call doctor right away.
• Mild stomach discomfort, nausea, dizziness, drowsiness, nervousness, insomnia, headache, vaginal discharge or pain, lightheadedness, mild diarrhea.	Continue. Call doctor when convenient.
Infrequent:	
Skin (itching, red, blisters, burning, rash, swelling, peeling).	Discontinue. Call doctor right away.
Rare:	
• Abdominal pain or cramps or tenderness, shortness of breath, sweating, swelling (neck, face, calves or legs), pale stools, bloody or dark or cloudy urine, joint or calf pain, diarrhea, tired or weak feeling, tendon problems (inflammation, pain, rupture), yellow eyes or skin, seizures, fast or irregular heartbeat, vomiting, fever, agitation, confusion,	Discontinue. Call doctor right away.

hallucinations, tremors, peripheral neuropathy (hands or feet have pain, tingling, burning, numbness, change in sensation).	
• Appetite loss, dreams abnormal, muscle or back pain, skin sensitive to sun, sore mouth or tongue, vision problems, vaginal infection, taste changes, flushing.	Continue. Call doctor when convenient.

WARNINGS & PRECAUTIONS

Don't take if:
You are allergic to fluoroquinolones or quinolone derivatives.

Before you start, consult your doctor if:
- You have any disorder of the central nervous system such as epilepsy or stroke.
- You have had sun sensitivity or tendinitis.
- You have diabetes or liver, kidney or heart disease.

Over age 60:
Risk of tendinitis or tendon rupture (see Others).

Pregnancy:
Decide with your doctor if drug benefits justify risk to unborn baby. Risk category C (see page xviii).

Breastfeeding; Lactation; Nursing Mothers:
Drugs in this group are generally acceptable with breastfeeding. Discuss risks and benefits with your doctor.

Infants & children up to age 18:
Follow instructions provided by your child's doctor.

Prolonged use:
Usually not prescribed for long-term use.

Skin & sunlight:
One or more drugs in this group may cause rash or intensify sunburn in areas exposed to sun or ultraviolet light (photosensitivity reaction). Avoid overexposure. Notify doctor if reaction occurs.

Driving, piloting or hazardous work:
Don't drive or pilot aircraft until you learn how medicine affects you. Don't work around dangerous machinery. Don't climb ladders or work in high places. Risk increases if you drink alcohol or take medicine affecting alertness.

Discontinuing:
Don't discontinue without consulting doctor or completing prescribed dosage. Call doctor if symptoms occur after you stop drug (such as stomach cramps, fever, swelling, calf pain, diarrhea that is watery or bloody).

Others:
- Advise any doctor, dentist or pharmacist whom you consult that you take this medicine.
- May affect accuracy of some medical tests.
- Drink plenty of fluids while taking drug.
- These drugs increase the risk of tendinitis or tendon rupture in some patients. If tendon pain or inflammation occurs, stop exercising, discontinue drug and call doctor.
- Serious liver problems associated with use of trovafloxacin can lead to liver transplantation and/or death. Consult your doctor about risks.

POSSIBLE INTERACTION WITH OTHER DRUGS

GENERIC NAME OR DRUG CLASS	COMBINED EFFECT
Aminophylline	Increased effect of aminophylline.
Antacids*	Decreased fluoroquinolone effect.
Antidiabetic agents*	Adverse diabetic reactions.
Anti-inflammatory drugs, nonsteroidal (NSAIDs)*	Increased risk of central nervous system problems.
Caffeine	Increased risk of central nervous system problems.
Calcium supplements	Decreased fluoroquinolone effect.
Citrates	Decreased trovafloxacin effect.

Continued on page 900

POSSIBLE INTERACTION WITH OTHER SUBSTANCES

INTERACTS WITH	COMBINED EFFECT
Alcohol:	Increased possibility of central nervous system side effects.
Beverages: Caffeine drinks.	Increased effect of caffeine. Don't use with enoxacin.
Cocaine:	Increased risk of side effects. Avoid
Foods: Dairy foods.	Decreased effect of fluoroquinolone. Take 2 hours apart.
Marijuana:	Increased risk of side effects. Avoid.
Tobacco:	None expected.

***See Glossary**

FLUOROURACIL (Topical)

BRAND NAMES

5-FU
Efudex
Fluoroplex

BASIC INFORMATION

Habit forming? No
Prescription needed? Yes
Available as generic? No
Drug class: Antineoplastic (topical)

USES

- Treats precancerous actinic keratoses on skin.
- Treats superficial basal cell carcinomas (skin cancers that don't spread to distant organs and, therefore, do not threaten life).

DOSAGE & USAGE INFORMATION

How to use:
- Apply with cotton-tipped applicator.
- Cream, lotion, ointment—Bathe and dry area before use. Apply small amount and rub gently.
- Wash hands (if fingertips are used to apply) after applying medicine to other parts of body.

When to use:
Once or twice a day or as directed by doctor.

If you forget a dose:
Use as soon as you remember. If it is almost time for the next dose, wait for that dose (don't double this dose) and resume regular schedule.

What drug does:
It stops precancerous and cancerous cells from multiplying.

Time lapse before drug works:
2 to 3 days.

Don't use with:
Other topical medications unless prescribed by your doctor.

OVERDOSE

SYMPTOMS:
Unknown.
WHAT TO DO:
If person uses much larger amount than prescribed or if accidentally swallowed, call poison control center 1-800-222-1222 for instructions or dial 911 (emergency) for help.

POSSIBLE ADVERSE REACTIONS OR SIDE EFFECTS

SYMPTOMS	WHAT TO DO
Life-threatening	
None expected.	
Common	
• Skin redness or swelling.	Discontinue. Call doctor right away.
• After 1 or 2 weeks of use, skin itching or oozing, rash, tenderness, soreness.	Continue. Call doctor when convenient.
Infrequent	
Skin darkening or scaling.	Continue. Call doctor when convenient.
Rare	
Watery eyes.	Discontinue. Call doctor right away.

WARNINGS & PRECAUTIONS

Don't use if:
You are allergic to fluorouracil. Allergic reactions are rare, but may occur.

Before you start, consult your doctor if:
- You have chloasma or acne rosacea.
- You have any other skin problems.

Over age 60:
No problems expected.

Pregnancy:
Consult doctor. Drug should not be used during pregnancy. Can cause harm to unborn baby. Risk category X (see page xviii).

Breastfeeding; Lactation; Nursing Mothers:
This drug is generally not recommended with breastfeeding. Discuss risks and benefits with your doctor.

Infants & children up to age 18:
Follow instructions provided by your child's doctor.

Prolonged use:
No problems expected, but check with doctor.

Skin & sunlight:
May cause rash or intensify sunburn in areas exposed to sun or ultraviolet light (photosensitivity reaction). Avoid overexposure. Notify doctor if reaction occurs.

Driving, piloting or hazardous work:
No problems expected, but check with doctor.

Discontinuing:
Pink, smooth area remains after treatment (usually fades in 1 to 2 months).

Others:
- Skin lesions may need biopsy before treatment.
- Keep medicine out of eyes or mouth.
- Heat and moisture in bathroom medicine cabinet can cause breakdown of medicine. Store someplace else.

POSSIBLE INTERACTION WITH OTHER DRUGS

GENERIC NAME OR DRUG CLASS	COMBINED EFFECT
None significant.	

POSSIBLE INTERACTION WITH OTHER SUBSTANCES

INTERACTS WITH	COMBINED EFFECT
Alcohol:	None expected.
Beverages:	None expected.
Cocaine:	None expected.
Foods:	None expected.
Marijuana:	None expected.
Tobacco:	None expected.

FOLIC ACID (Vitamin B-9)

BRAND NAMES

Apo-Folic
Beyaz
Folvite
Novo-Folacid
Safyral

Numerous brands of single vitamin and multivitamin combinations may be available.

BASIC INFORMATION

Habit forming? No
Prescription needed?
High strength: Yes
Vitamin mixtures: No
Available as generic? Yes
Drug class: Vitamin supplement

USES

- Dietary supplement to promote normal growth, development and good health.
- Dietary supplement during pregnancy to prevent spinal defects.
- Treatment for anemias due to folic acid deficiency occurring from alcoholism, liver disease, hemolytic anemia, sprue, infants on artificial formula, pregnancy, breast feeding and use of oral contraceptives.
- Some studies have found that folic acid supplementation alone or in combination with other vitamins taken before conception and during early pregnancy may reduce the incidence of neural tube defects in infants.

OVERDOSE

SYMPTOMS:
Mouth or tongue pain, numbness or tingling, confusion, weakness, tiredness, trouble with concentration.
WHAT TO DO:
If person uses much larger amount than prescribed or if accidentally swallowed, call poison control center 1-800-222-1222 for instructions or dial 911 (emergency) for help.

DOSAGE & USAGE INFORMATION

How to take:
Tablet—Swallow with liquid or food to lessen stomach irritation. If you can't swallow whole, crumble tablet and take with liquid or food.

When to take:
At the same time each day.

If you forget a dose:
Take as soon as you remember. If it is almost time for the next dose, wait for that dose (don't double this dose) and resume regular schedule.

What drug does:
Essential to normal red blood cell formation.

Time lapse before drug works:
Not determined.

Don't take with:
Any other medicine or any dietary supplement without consulting your doctor or pharmacist.

POSSIBLE ADVERSE REACTIONS OR SIDE EFFECTS

SYMPTOMS	WHAT TO DO
Life-threatening:	
None expected.	
Common:	
Large dose may produce yellow urine.	Continue. Call doctor when convenient.
Infrequent:	
None expected.	
Rare:	
Rash, itching, bronchospasm (trouble breathing, wheezing).	Discontinue. Call doctor right away.

WARNINGS & PRECAUTIONS

Don't take if:
You are allergic to any folic acid or B vitamins. Allergic reactions are rare, but may occur.

Before you start, consult your doctor if:
- You have liver disease.
- You have pernicious anemia. (Folic acid corrects anemia, but nerve damage of pernicious anemia continues.)

Over age 60:
No problems expected.

Pregnancy:
Consult doctor. No problems expected. Risk category A (see page xviii).

Breastfeeding; Lactation; Nursing Mothers:
This drug is generally acceptable with breastfeeding. Discuss risks and benefits with your doctor.

Infants & children up to age 18:
Read nonprescription product's label to be sure it is approved for your child's age. Follow label instructions on dosage. Consult doctor or pharmacist if unsure.

Prolonged use:
No problems expected.

Skin & sunlight:
No problems expected.

Driving, piloting or hazardous work:
No problems expected.

Discontinuing:
Don't discontinue without doctor's advice until you complete prescribed dose, even though symptoms diminish or disappear.

Others:
- Folic acid removed by kidney dialysis. Dialysis patients should increase intake to 300% of RDA or take as directed.
- A balanced diet should provide all the folic acid a healthy person needs and make supplements unnecessary. Best sources are green, leafy vegetables; fruits; liver and kidney.

POSSIBLE INTERACTION WITH OTHER DRUGS

GENERIC NAME OR DRUG CLASS	COMBINED EFFECT
Analgesics*	Decreased effect of folic acid.
Anticonvulsants, hydantoin*	Decreased effect of folic acid. Possible increased seizure frequency.
Chloramphenicol	Possible decreased folic acid effect.
Contraceptives, oral*	Decreased effect of folic acid.
Cortisone drugs*	Decreased effect of folic acid.
Methotrexate	Decreased effect of folic acid.
Para-aminosalicylic acid (PAS)	Decreased effect of folic acid.
Pyrimethamine	Decreased effect of folic acid.
Sulfasalazine	Decreased dietary absorption of folic acid.
Triamterene	Decreased effect of folic acid.
Trimethoprim	Decreased effect of folic acid.
Zinc supplements	Increased need for zinc.

POSSIBLE INTERACTION WITH OTHER SUBSTANCES

INTERACTS WITH	COMBINED EFFECT
Alcohol:	None expected.
Beverages:	None expected.
Cocaine:	Unknown. Avoid.
Foods:	None expected.
Marijuana:	Consult doctor.
Tobacco:	None expected.

***See Glossary**

FUSION INHIBITOR

GENERIC AND BRAND NAMES

ENFUVIRTIDE
Fuzeon

BASIC INFORMATION

Habit forming? No
Prescription needed? Yes
Available as generic? No
Drug class: Fusion inhibitor

USES

For treatment of HIV and AIDS patients. Used in combination with one or more of the other HIV and AIDS drugs. It is used when other anti-HIV drugs are not effective.

DOSAGE & USAGE INFORMATION

How to take:
Injection—The drug is injected under the skin using small hypodermic needles. Follow all instructions carefully on the prescription label. Dispose of all needles and syringes as instructed by your doctor.

When to take:
Two shots a day; one taken in the morning and the other 12 hours later at night.

If you forget a dose:
Take as soon as you remember. If it is almost time for the next dose, wait for that dose (don't double this dose) and resume regular schedule.

What drug does:
It stops HIV from entering healthy cells. It helps slow the progress of HIV disease, but does not cure it.

Time lapse before drug works:
May require several weeks or months before full benefits are apparent.

Continued next column

OVERDOSE

SYMPTOMS:
Unknown.
WHAT TO DO:
If person uses much larger amount than prescribed or if accidentally swallowed, call poison control center 1-800-222-1222 for instructions or dial 911 (emergency) for help.

Don't take with:
Any other medicine or any dietary supplement without consulting your doctor or pharmacist. This is very important with antiviral drugs.

POSSIBLE ADVERSE REACTIONS OR SIDE EFFECTS

SYMPTOMS	WHAT TO DO
Life-threatening: None expected.	
Common:	
• Peripheral neuropathy (burning, tingling, numbness, pain or weakness in arms, hands, feet or legs), pain around cheeks or eyes, fever, chills, runny or stuffy nose, cough, wheezing, tightness in chest.	Continue, but call doctor right away.
• Anxiety, weakness, depression, sores on skin, cold sores, trouble sleeping, weight loss, skin symptoms at site of drug injections, muscle pain, itching.	Continue. Call doctor when convenient.
Infrequent:	
• Eye inflammation (swelling, red and painful), stomach or back pain, dark urine, constipation, appetite loss, nausea or vomiting, indigestion, yellow skin or eyes, skin lump or growth.	Continue, but call doctor right away.
• General ill feeling, diarrhea, headache, joint pain, changes in taste, swollen or painful lymph glands.	Continue. Call doctor when convenient.
Rare:	
Pneumonia symptoms (rapid breathing, shortness of breath, cough with fever). Any other unusual symptoms that occur (may be due to the illness, this drug or other drugs being taken).	Continue, but call doctor right away.

WARNINGS & PRECAUTIONS

Don't take if:
You are allergic to fusion inhibitors. Allergic reactions are rare, but may occur.

Before you start, consult your doctor if:
- You have or have had lung disease.
- You have high viral load or low CD4 cell count.
- You use intravenous (IV) drugs.
- You smoke.

Over age 60:
The drug has not been studied in this age group.

Pregnancy:
Decide with your doctor if drug benefits justify risk to unborn baby. Risk category B (see page xviii).

Breastfeeding; Lactation; Nursing Mothers:
It is not recommended that HIV-infected mothers breastfeed. Consult your doctor.

Infants & children up to age 18:
Follow instructions provided by your child's doctor. Approved for use in ages 6 and over.

Prolonged use:
- Long-term effects have not been established.
- Talk to your doctor about frequent blood counts and liver function studies.

Skin & sunlight:
No problems expected.

Driving, piloting or hazardous work:
Don't drive or pilot aircraft until you learn how medicine affects you. Don't work around dangerous machinery. Don't climb ladders or work in high places. Danger increases if you drink alcohol or take medicine affecting alertness and reflexes.

Discontinuing:
Don't discontinue without doctor's advice. Doses of other drugs may require adjustment.

Others:
- Advise any doctor, dentist or pharmacist whom you consult that you take this medicine.
- Avoid sexual intercourse or use condoms to help prevent the transmission of HIV. Don't share needles or equipment for injections with other persons.
- This drug is combined with others for the best treatment. This increases the risk of side effects or adverse reactions.
- May interfere with results of some blood tests.
- Numerous medical studies are ongoing concerning the use of these and other anti-HIV drugs. Full safety and effectiveness are still being determined.

POSSIBLE INTERACTION WITH OTHER DRUGS

GENERIC NAME OR DRUG CLASS	COMBINED EFFECT
None expected.	

POSSIBLE INTERACTION WITH OTHER SUBSTANCES

INTERACTS WITH	COMBINED EFFECT
Alcohol:	None expected.
Beverages:	None expected.
Cocaine:	Unknown. Avoid.
Foods:	None expected.
Marijuana:	Consult doctor.
Tobacco:	None expected.

*See Glossary

GABAPENTIN

BRAND NAMES

Gralise
Horizant
Neurontin

BASIC INFORMATION

Habit forming? No
Prescription needed? Yes
Available as generic? Yes
Drug class: Anticonvulsant, antiepileptic

USES

- Treatment for partial (focal) epileptic seizures. Used in combination with other antiepileptic drugs.
- Treats restless legs syndrome (Willis-Ekbom disease).
- Treats postherpetic neuralgia.

DOSAGE & USAGE INFORMATION

How to take:
- Capsule or tablet—Swallow with liquid. May be taken with or without food.
- Oral solution—Follow directions on your prescription label.
- Extended-release tablet—Swallow whole with liquid. Do not split, chew or crush tablet. Take with food.

When to take:
- For seizure treatment, your doctor will determine the best schedule. Dosages will be increased rapidly over the first 3 days of use. Further increases may be necessary to achieve maximum benefits.
- For restless legs syndrome, take once a day with food at about 5:00 p.m.

If you forget a dose:
Take as soon as you remember. If it is almost time for the next dose, wait for that dose (don't double this dose) and resume regular schedule.

Continued next column

OVERDOSE

SYMPTOMS:
Drowsiness, double vision, slurred speech, tiredness, diarrhea.
WHAT TO DO:
If person uses much larger amount than prescribed or if accidentally swallowed, call poison control center 1-800-222-1222 for instructions or dial 911 (emergency) for help.

What drug does:
- The exact mechanism in treating seizures is unknown. The anticonvulsant action may result from an altered transport of brain amino acids. Amino acids play an important part in chemical reactions within the cells.
- It is unknown just how the drug works to treat restless legs syndrome.

Time lapse before drug works:
May take several weeks for effectiveness.

Don't take with:
Any other medicine or any dietary supplement without consulting your doctor or pharmacist.

POSSIBLE ADVERSE REACTIONS OR SIDE EFFECTS

SYMPTOMS	WHAT TO DO
Life-threatening: None expected.	
Common: Sleepiness, dizziness, fatigue, clumsiness, lack of coordination.	Continue. Call doctor when convenient.
Infrequent: Rapid eye movement (nystagmus), double or blurred vision.	Continue. Call doctor when convenient.
Rare: Rash, nervousness, depression, twitching or swelling (in hands, feet or legs), runny nose, dry or sore throat, nausea or vomiting, coughing, constipation, muscle or back ache, weight gain, forgetfulness, impotence, increased appetite, indigestion.	Continue. Call doctor when convenient.

WARNINGS & PRECAUTIONS

Don't take if:
You are allergic to gabapentin. Allergic reactions are rare, but may occur.

Before you start, consult your doctor if:
You have kidney disease.

Over age 60:
Adverse reactions and side effects may be more frequent and severe than in younger persons.

Pregnancy:
Decide with your doctor if drug benefits justify risk to unborn baby. Risk category C (see page xviii).

Breastfeeding; Lactation; Nursing Mothers:
This drug is generally acceptable with breastfeeding. Discuss risks and benefits with your doctor.

Infants & children up to age 18:
Approved for use for age 12 and over. Follow instructions provided by your child's doctor.

Prolonged use:
No special problems expected. Follow-up laboratory blood studies may be recommended by your doctor.

Skin & sunlight:
No problems expected.

Driving, piloting or hazardous work:
Don't drive or pilot aircraft until you learn how medicine affects you. Don't work around dangerous machinery. Don't climb ladders or work in high places. Danger increases if you drink alcohol or take other medicines affecting alertness and reflexes such as antihistamines, tranquilizers, sedatives, pain medicine, narcotics and mind-altering drugs.

Discontinuing:
Don't discontinue without doctor's approval due to risk of increased seizure activity.

Others:
- Advise any doctor, dentist or pharmacist whom you consult that you take this medicine.
- The brand name Horizant is not interchangeable with other gabapentin brand names. Consult your doctor or pharmacist if you have any concerns.
- Side effects of gabapentin are usually mild to moderate. Because it is normally used with other anticonvulsant drugs, additional side effects may also occur. If they do, discuss them with your doctor.
- Rarely, antiepileptic drugs may lead to suicidal thoughts and behaviors. Call doctor right away if suicidal symptoms or unusual behaviors occur.

POSSIBLE INTERACTION WITH OTHER DRUGS

GENERIC NAME OR DRUG CLASS	COMBINED EFFECT
Antacids*	Allow at least 2 hours between the 2 drugs.
Central nervous system (CNS) depressants*	Increased sedation.

POSSIBLE INTERACTION WITH OTHER SUBSTANCES

INTERACTS WITH	COMBINED EFFECT
Alcohol:	Increased sedation.
Beverages:	None expected.
Cocaine:	Unknown. Avoid.
Foods:	None expected.
Marijuana:	Consult doctor.
Tobacco:	None expected.

***See Glossary**

GEMFIBROZIL

BRAND NAMES

Lopid

BASIC INFORMATION

Habit forming? No
Prescription needed? Yes
Available as generic? Yes
Drug class: Antihyperlipidemic

USES

Reduces fatty substances in the blood (triglycerides) and raises high-density lipoprotein (HDL) cholesterol levels.

DOSAGE & USAGE INFORMATION

How to take:
Tablet or capsule—Swallow with liquid or food to lessen stomach irritation.

When to take:
Take 30 minutes before morning and evening meals.

If you forget a dose:
Take as soon as you remember. If it is almost time for the next dose, wait for that dose (don't double this dose) and resume regular schedule.

What drug does:
Inhibits formation of fatty substances.

Time lapse before drug works:
3 months or more.

Don't take with:
Any other medicine or any dietary supplement without consulting your doctor or pharmacist.

OVERDOSE

SYMPTOMS:
Diarrhea, stomach cramps, nausea, vomiting, muscle and joint pain.
WHAT TO DO:
If person uses much larger amount than prescribed or if accidentally swallowed, call poison control center 1-800-222-1222 for instructions or dial 911 (emergency) for help.

POSSIBLE ADVERSE REACTIONS OR SIDE EFFECTS

SYMPTOMS	WHAT TO DO
Life-threatening: None expected.	
Common: Indigestion.	Continue. Call doctor when convenient.
Infrequent: Shortness of breath, chest pain, irregular heartbeat, nausea, vomiting, diarrhea, stomach pain.	Discontinue. Call doctor right away.
Rare: • Rash, itch, sores in mouth or lips, sore throat, swollen feet or legs, fever, chills, blood in urine, painful urination.	Discontinue. Call doctor right away.
• Dizziness, headache, drowsiness, muscle cramps, dry skin, backache, unusual tiredness, decreased sex drive.	Continue. Call doctor when convenient.

WARNINGS & PRECAUTIONS

Don't take if:
You are allergic to gemfibrozil. Allergic reactions are rare, but may occur.

Before you start, consult your doctor if:
- You have had liver or kidney disease.
- You have gallstones or gallbladder disease.
- You have had peptic-ulcer disease.
- You have diabetes.

Over age 60:
Adverse reactions and side effects may be more frequent and severe than in younger persons.

Pregnancy:
Decide with your doctor if drug benefits justify risk to unborn baby. Risk category C (see page xviii).

Breastfeeding; Lactation; Nursing Mothers:
This drug is generally acceptable with breastfeeding. Discuss risks and benefits with your doctor.

Infants & children up to age 18:
Safety and effectiveness in this age group have not been established. Consult doctor.

Prolonged use:
Periodic blood cell counts and liver function studies recommended if you take gemfibrozil for a long time.

Skin & sunlight:
No problems expected.

Driving, piloting or hazardous work:
Avoid if you feel drowsy or dizzy. Otherwise, no problems expected.

Discontinuing:
Don't discontinue without doctor's advice until you complete prescribed dose.

Others:
- May affect results in some medical tests.
- Advise any doctor, dentist or pharmacist whom you consult that you take this medicine.

POSSIBLE INTERACTION WITH OTHER DRUGS

GENERIC NAME OR DRUG CLASS	COMBINED EFFECT
Anticoagulants, oral*	Increased anticoagulant effect. Dose reduction of anticoagulant necessary.
Antidiabetics, oral*	Increased antidiabetic effect.
Contraceptives, oral*	Decreased gemfibrozil effect.
Dexfenfluramine	May require dosage change as weight loss occurs.
Estrogens*	Decreased gemfibrozil effect.
Furosemide	Possible toxicity of both drugs.
HMG-CoA reductase inhibitors	Increased risk of muscle inflammation and kidney failure.
Insulin	Increased insulin effect.
Lovastatin	Increased risk of kidney problems.
Thyroid hormones*	Increased gemfibrozil effect.

POSSIBLE INTERACTION WITH OTHER SUBSTANCES

INTERACTS WITH	COMBINED EFFECT
Alcohol:	None expected.
Beverages:	None expected.
Cocaine:	Decreased effect of gemfibrozil. Avoid.
Foods: Fatty foods.	Decreased gemfibrozil effect.
Marijuana:	Consult doctor.
Tobacco:	Decreased gemfibrozil absorption. Avoid.

*See Glossary

GLP-1 RECEPTOR AGONISTS

GENERIC AND BRAND NAMES

ALBIGLUTIDE
Tanzeum
DULAGLUTIDE
Trulicity
EXENATIDE
Bydureon
Byetta
LIRAGLUTIDE
Saxenda
Victoza

BASIC INFORMATION

Habit forming? No
Prescription needed? Yes
Available as generic? No
Drug class: Antidiabetic; incretin mimetic

USES

- Used alone or in combination with other antidiabetic agents to control blood sugar levels in adults who have type 2 diabetes.
- Used for weight loss in certain adult patients.

DOSAGE & USAGE INFORMATION

How to take:
Self injection—Injected under the skin (subcutaneous) of the upper leg (thigh), stomach area (abdomen), or upper arm. The drug comes in a prefilled pen that is used with a small needle. Read and follow the instructions provided with the prescription.

When to take:
- Inject Byetta (exenatide) twice a day, at any time within the 60 minutes before morning and evening meals. Do not inject it after eating the meal.
- Inject Bydureon (exenatide), albiglutide or dulaglutide once a week at any time of day with or without food.
- Inject liraglutide once daily at any time of day, without regard to meals.

Continued next column

If you forget a dose:
- If you forget to inject Byetta (exenatide) before a meal, skip that dose and inject next dose as scheduled.
- Inject a once a week dose as soon as you remember. If next dose is due in less than 3 days, skip missed dose and resume schedule. Don't double any doses.
- Inject liraglutide as soon as you remember. If more than 12 hours late, skip the missed dose and resume schedule. Don't double next dose.

What drug does:
- Helps the pancreas produce insulin in response to rising blood sugar levels; inhibits the liver's production of sugar; reduces the rate at which sugar enters the bloodstream by slowing the release of food from the stomach.
- It appears to decrease appetite, which may lead to weight loss.

Time lapse before drug works:
The effects begin right after it is injected.

Don't take with:
Any other medicine or any dietary supplement without consulting your doctor or pharmacist.

OVERDOSE

SYMPTOMS:
Nausea, vomiting, low blood sugar (hypoglycemia) that may be severe—See list of symptoms in next column under Infrequent.
WHAT TO DO:
- **Dial 911 for all medical emergencies or call poison control center 1-800-222-1222 for instructions.**
- **See emergency information on last 3 pages of this book.**

POSSIBLE ADVERSE REACTIONS OR SIDE EFFECTS

SYMPTOMS	WHAT TO DO
Life-threatening: None expected.	
Common: Nausea, vomiting.	Continue. Call doctor if symptoms persist.
Infrequent:	
• Stomach symptoms (acid, sour, upset, ache, belching), diarrhea, feeling jittery, loss of strength, headache, heartburn, appetite decreased, sweating, dizziness, constipation, injection site soreness, upper respiratory infection.	Continue. Call doctor when convenient.
• Symptoms of low blood sugar–hunger, (excessive), cold sweats and skin, rapid pulse, anxiety, nervousness, chills, confusion, drowsiness, loss of concentration, headache, nausea, weakness, shakiness, vision changes.	Seek treatment (eat some form of quick-acting sugar—glucose tablets, sugar, fruit juice, corn syrup, honey).

• Symptoms of high blood sugar—increased urination, unusual thirst, dry mouth, drowsiness, flushed or dry skin, fruit-like breath odor, appetite loss, stomach pain or vomiting, tiredness, trouble breathing, increased blood sugar level.	Check your blood sugar immediately. Call doctor right away.

Rare:

• Abdominal pain (persistent, severe, may have vomiting), neck swelling or lump, hoarseness, difficulty breathing or swallowing.	Discontinue. Call doctor right away.
• Other symptoms that cause concern.	Continue. Call doctor when convenient.

WARNINGS & PRECAUTIONS

Don't take if:
You are allergic to exenatide or liraglutide.

Before you start, consult your doctor if:
- You have type 1 diabetes.
- You require insulin for diabetes treatment.
- You have any kidney or liver problems.
- You have a history of pancreatitis.
- You have a personal or family history of multiple endocrine neoplasia syndrome type 2 or medullary thyroid carcinoma (for liraglutide).
- You have severe gastrointestinal disease.

Over age 60:
No special problems expected.

Pregnancy:
Decide with your doctor if drug benefits justify risk to unborn baby. Risk category C (see page xviii).

Breastfeeding; Lactation; Nursing Mothers:
Drugs in this group are generally acceptable with breastfeeding. Discuss risks and benefits with your doctor.

Infants & children up to age 18:
Safety and effectiveness in this age group have not been established. Consult doctor.

Prolonged use:
Talk to your doctor about the need for follow-up medical examinations and/or laboratory studies to determine continued effectiveness of drug.

Skin & sunlight:
No problems expected.

Driving, piloting or hazardous work:
No problems expected.

Discontinuing:
Don't discontinue without doctor's advice, even though symptoms diminish or disappear.

Others:
- Notify your doctor if you have a fever, infection, diarrhea or experience vomiting.
- Advise any doctor, dentist or pharmacist whom you consult that you take this medicine.
- You and your family should educate yourselves about diabetes; learn to recognize the symptoms of hypoglycemia and how to treat it.
- Wear or carry medical identification that says you have type 2 diabetes.
- Follow your diabetic diet, medication, and exercise routines closely. Changing any of these things can affect blood sugar levels.
- Weight loss patients will be evaluated after 16 weeks to see if this drug treatment is working.
- Use of these drugs has risk for certain thyroid tumors or cancer. Consult doctor about risks.

POSSIBLE INTERACTION WITH OTHER DRUGS

GENERIC NAME OR DRUG CLASS	COMBINED EFFECT
Digoxin	May decrease effect of digoxin.
Drugs taken by mouth that need to pass quickly through the stomach (such as oral contraceptives or antibiotics).	May need to take them 1 hour before injecting GLP-1 receptor agonist. Consult doctor or pharmacist.
Hypoglycemia-causing medications*	Risk of low blood sugar.
Hypoglycemics*	Risk of low blood sugar.

Continued on page 900

POSSIBLE INTERACTION WITH OTHER SUBSTANCES

INTERACTS WITH	COMBINED EFFECT
Alcohol:	May cause low blood sugar. Avoid.
Beverages:	None expected.
Cocaine:	Unknown. Avoid.
Foods:	None expected. Follow your diabetic diet instructions.
Marijuana:	Possible increase in blood sugar. Avoid.
Tobacco:	None expected.

*See Glossary

GLUCAGON

BRAND NAMES

Glucagon for Injection

BASIC INFORMATION

Habit forming? No
Prescription needed? Yes
Available as generic? Yes
Drug class: Antihypoglycemic, diagnostic aid

USES

- Treats low blood sugar (hypoglycemia) in diabetics.
- Used as antidote for overdose of beta-adrenergic blockers, quinidine and tricyclic antidepressants.

DOSAGE & USAGE INFORMATION

How to take:
Injection—As directed by your doctor.

When to take:
When there are signs of low blood sugar (anxiety, chills, cool and pale skin, hunger, nausea, tremors, sweating, weakness, stomach pain, confusion, drowsiness, fast heartbeat, continuing headache, unsteady walk, unusual tiredness, vision changes, unconsciousness) in diabetics who don't respond to eating some form of sugar.

If you forget a dose:
Single dose only.

What drug does:
Forces liver to make more sugar and release it into the bloodstream.

Time lapse before drug works:
- For hypoglycemic condition—5 to 20 minutes.
- For muscle relaxant—1 to 10 minutes.

Continued next column

OVERDOSE

SYMPTOMS:
Nausea, vomiting, severe weakness, irregular heartbeat, hoarseness, cramps.
WHAT TO DO:
- **Dial 911 for all medical emergencies or call poison control center 1-800-222-1222 for instructions.**
- **See emergency information on last 3 pages of this book.**

Don't take with:
Any other medicines (including over-the-counter drugs such as cough and cold medicines, laxatives, antacids, diet pills, caffeine, nose drops or vitamins) without consulting your doctor or pharmacist.

POSSIBLE ADVERSE REACTIONS OR SIDE EFFECTS

SYMPTOMS	WHAT TO DO
Life-threatening:	
Rare, severe allergic reaction (anaphylaxis)—difficulty breathing, hives, itching, swelling (throat, face), vomiting, dizziness.	Seek emergency treatment immediately.
Common:	
Nausea.	Continue. Call doctor if symptoms persist.
Infrequent:	
Lightheadedness, breathing difficulty, skin rash.	Discontinue. Call doctor right away.
Rare:	
None expected.	

WARNINGS & PRECAUTIONS

Don't take if:
You can't tolerate glucagon.

Before you start, consult your doctor if:
- You are allergic to beef or pork.
- You have pheochromocytoma.*

Over age 60:
No special problems expected.

Pregnancy:
Decide with your doctor if drug benefits justify risk to unborn baby. Risk category B (see page xviii).

Breastfeeding; Lactation; Nursing Mothers:
This drug is generally acceptable with breastfeeding. Discuss risks and benefits with your doctor.

Infants & children up to age 18:
Follow instructions provided by your child's doctor.

Prolonged use:
To be used intermittently and not for prolonged periods.

Skin & sunlight:
No problems expected.

Driving, piloting or hazardous work:
Don't drive or pilot aircraft until you learn how medicine affects you. Don't work around dangerous machinery. Don't climb ladders or work in high places. Danger increases if you drink alcohol or take medicine affecting alertness and reflexes.

Discontinuing:
No special problems expected.

Others:
- May affect results in some medical tests.
- Explain to other family members how to inject glucagon.
- Before injecting, try to eat some form of sugar, such as glucose tablets, corn syrup, honey, orange juice, hard candy or sugar cubes.
- Store unmixed glucagon at room temperature. Store mixed glucagon in refrigerator, but don't freeze. Mixed solution is only good for 48 hours.
- Check expiration date regularly and replace drug before it expires.

POSSIBLE INTERACTION WITH OTHER DRUGS

GENERIC NAME OR DRUG CLASS	COMBINED EFFECT
Anticoagulants*	Increased anticoagulant effect.

POSSIBLE INTERACTION WITH OTHER SUBSTANCES

INTERACTS WITH	COMBINED EFFECT
Alcohol:	Decreased glucagon effect.
Beverages:	None expected.
Cocaine:	Increased adverse reactions. Avoid
Foods: Sugar, fruit juice, candy.	Enhances glucagon effect.
Marijuana:	Consult doctor.
Tobacco:	None expected.

*See Glossary

GLYCOPYRROLATE

BRAND NAMES

Robinul
Robinul Forte

BASIC INFORMATION

Habit forming? No
Prescription needed? Yes
Available as generic? Yes
Drug class: Antispasmodic, anticholinergic

USES

- Reduces spasms of digestive system.
- Reduces production of saliva during dental procedures.
- Treats peptic ulcer by reducing gastric acid.

DOSAGE & USAGE INFORMATION

How to take:
Tablet—Swallow with liquid. If you can't swallow whole, crumble tablet and take with small amount of liquid or food.

When to take:
30 minutes before meals (unless directed otherwise by doctor).

If you forget a dose:
Take as soon as you remember. If it is almost time for the next dose, wait for that dose (don't double this dose) and resume regular schedule.

What drug does:
Blocks nerve impulses at parasympathetic nerve endings, preventing smooth (involuntary) muscle contractions and gland secretions of organs involved.

Time lapse before drug works:
15 to 30 minutes.

Don't take with:
Any other medicine or any dietary supplement without consulting your doctor or pharmacist.

OVERDOSE

SYMPTOMS:
Dry mouth, blurred vision, confusion, anxiety, low blood pressure, slowed breathing, irregular heartbeat, hot and dry skin, drowsiness, seizures.
WHAT TO DO:
- **Dial 911 for all medical emergencies or call poison control center 1-800-222-1222 for instructions.**
- **See emergency information on last 3 pages of this book.**

POSSIBLE ADVERSE REACTIONS OR SIDE EFFECTS

SYMPTOMS	WHAT TO DO
Life-threatening:	
Rare, severe allergic reaction (anaphylaxis)—difficulty breathing, hives, itching, swelling (throat, face), vomiting, dizziness.	Seek emergency treatment immediately.
Common:	
Dry mouth, loss of taste, constipation, difficult urination.	Continue. Call doctor when convenient.
Infrequent:	
• Confusion, dizziness, drowsiness, eye pain, headache, rash, sleep disturbance (such as nightmares or frequent waking), nausea, vomiting, rapid heartbeat, lightheadedness.	Discontinue. Call doctor right away.
• Insomnia, blurred vision, diminished sex drive, decreased sweating, nasal congestion, change in taste.	Continue. Call doctor when convenient.
Rare:	
Rash, hives.	Discontinue. Call doctor right away.

WARNINGS & PRECAUTIONS

Don't take if:
You are allergic to glycopyrrolate.

Before you start, consult your doctor if:
- You have glaucoma.
- You have angina, chronic bronchitis or asthma, liver disease, hiatal hernia, enlarged prostate, myasthenia gravis, peptic ulcer, kidney or thyroid disease.
- You have trouble with stomach bloating.
- You have difficulty emptying your bladder completely.
- You have ulcerative colitis.
- You will have surgery within 2 months, including dental surgery, requiring general or spinal anesthesia.

Over age 60:
Adverse reactions and side effects may be more frequent and severe than in younger persons.

Pregnancy:
Decide with your doctor if drug benefits justify risk to unborn baby. Risk category B (see page xviii).

Breastfeeding; Lactation; Nursing Mothers:
This drug is generally acceptable with breastfeeding. Discuss risks and benefits with your doctor.

Infants & children up to age 18:
Follow instructions provided by your child's doctor.

Prolonged use:
Chronic constipation, possible fecal impaction. Consult doctor immediately.

Skin & sunlight:
No problems expected.

Driving, piloting or hazardous work:
No problems expected.

Discontinuing:
May be unnecessary to finish medicine. Follow doctor's instructions.

Others:
- Heatstroke more likely if you become overheated during exertion.
- Advise any doctor, dentist or pharmacist whom you consult that you take this medicine.

POSSIBLE INTERACTION WITH OTHER DRUGS

GENERIC NAME OR DRUG CLASS	COMBINED EFFECT
Adrenocorticoids, systemic	Possible glaucoma.
Antacids*	Decreased glycopyrrolate absorption effect.
Amantadine	Increased glycopyrrolate effect.
Anticholinergics, other*	Increased glycopyrrolate effect.
Antidepressants, tricyclic*	Increased glycopyrrolate effect.
Antidiarrheals*	Decreased glycopyrrolate absorption effect.
Attapulgite	Decreased effect of anticholinergic.
Buclizine	Increased glycopyrrolate effect.
Digitalis	Possible decreased absorption of digitalis.
Haloperidol	Increased internal eye pressure.
Ketoconazole	Decreased ketoconazole effect.
Meperidine	Increased glycopyrrolate effect.
Methylphenidate	Increased anticholinergic effect.
Molindone	Increased nizatidine effect.
Monoamine oxidase (MAO) inhibitors*	Increased glycopyrrolate effect.
Orphenadrine	Increased glycopyrrolate effect.
Phenothiazines	Increased glycopyrrolate effect.
Pilocarpine	Increased glycopyrrolate effect. Loss of pilocarpine effect in glaucoma treatment.
Potassium chloride tabs	Increased side effects of potassium tablets.
Quinidine	Increased glycopyrrolate effect.
Sedatives* or central nervous system (CNS) depressants*	Increased sedative effect of both drugs.
Vitamin C	Increased glycopyrrolate effect. Avoid large vitamin C doses.

POSSIBLE INTERACTION WITH OTHER SUBSTANCES

INTERACTS WITH	COMBINED EFFECT
Alcohol:	None expected.
Beverages:	None expected.
Cocaine:	Excessively rapid heartbeat. Avoid.
Foods:	None expected.
Marijuana:	Drowsiness and dry mouth. Avoid.
Tobacco:	None expected.

*See Glossary

GOLD COMPOUNDS

GENERIC AND BRAND NAMES

AURANOFIN
Ridaura-Oral

GOLD SODIUM THIOMALATE
Myocrisin

BASIC INFORMATION

Habit forming? No
Prescription needed? Yes
Available as generic? No
Drug class: Gold compounds

USES

Treatment for rheumatoid arthritis and juvenile idiopathic arthritis.

DOSAGE & USAGE INFORMATION

How to take:
- Capsule—Swallow with full glass of fluid. Follow prescription directions. Taking too much can cause serious adverse reactions.
- Injection—Under medical supervision.

When to take:
Once or twice daily, morning and night.

If you forget a dose:
Take as soon as you remember. If it is almost time for the next dose, wait for that dose (don't double this dose) and resume regular schedule.

What drug does:
Modifies disease activity of rheumatoid arthritis by mechanisms not yet understood.

Time lapse before drug works:
3 to 6 months.

Don't take with:
Any other medicine or any dietary supplement without consulting your doctor or pharmacist.

OVERDOSE

SYMPTOMS:
Rash, diarrhea, stomach upset, unusual bleeding or bruising, mouth sores, bloody urine, fainting.
WHAT TO DO:
- **Dial 911 for all medical emergencies or call poison control center 1-800-222-1222 for instructions.**
- **See emergency information on last 3 pages of this book.**

POSSIBLE ADVERSE REACTIONS OR SIDE EFFECTS

SYMPTOMS	WHAT TO DO
Life-threatening: None expected.	
Common:	
• Sores or white spots in mouth or throat, metallic taste, skin rash or itching, sore or red gums, sore tongue.	Discontinue. Call doctor right away.
• Indigestion, stomach cramps, appetite loss, diarrhea, mild nausea or vomiting.	Continue. Call doctor when convenient.
Infrequent:	
• Urine is bloody or cloudy, hives.	Discontinue. Call doctor right away.
• Constipation, joint pain, taste changes.	Continue. Call doctor when convenient.
Rare:	
• Bloody or tarry or pale stools, wheezing or severe coughing, seizures, urine is dark or decreased, painful urination, numbness or tingling or swelling in hands or feet, eye irritation, changes in vision, fever, cold symptoms, yellow eyes and skin, bloody vomit, vaginal irritation, lack of coordination.	Discontinue. Call doctor right away or seek emergency care.
• Hair loss, feeling tired or weak.	Continue. Call doctor when convenient.

WARNINGS & PRECAUTIONS

Don't take if:
You are allergic to gold or other metals. Allergic reactions are rare, but may occur.

Before you start, consult your doctor if:
- You have any blood disorder.
- You have kidney disease.
- You are pregnant or may become pregnant.
- You have lupus erythematosus.
- You have Sjögren's syndrome.
- You have chronic skin disease.
- You are debilitated.
- You have blood dyscrasias.

Over age 60:
Adverse reactions and side effects may be more frequent and severe than in younger persons.

Pregnancy:
Decide with your doctor if drug benefits justify risk to unborn baby. Risk category C (see page xviii).

Breastfeeding; Lactation; Nursing Mothers:
Drugs in this group are generally not recommended with breastfeeding. Discuss risks and benefits with your doctor.

Infants & children up to age 18:
Follow instructions provided by your child's doctor.

Prolonged use:
Request periodic laboratory studies of blood counts, urine and liver function. These should be done before use and at least once a month during treatment.

Skin & sunlight:
- One or more drugs in this group may cause rash or intensify sunburn in areas exposed to sun or ultraviolet light (photosensitivity reaction). Avoid overexposure. Notify doctor if reaction occurs.
- Blue-gray pigmentation in skin exposed to sunlight.

Driving, piloting or hazardous work:
Avoid if you have serious adverse reactions or side effects. Otherwise, no problems expected.

Discontinuing:
- Don't discontinue without doctor's advice until you complete prescribed dose.
- Side effects and adverse reactions may appear during treatment or for many months after discontinuing.

Others:
- Gold has been shown to cause kidney tumors and cancer in animals given excessive doses.
- Advise any doctor, dentist or pharmacist whom you consult that you take this drug.
- May interfere with the accuracy of some medical tests.

POSSIBLE INTERACTION WITH OTHER DRUGS

GENERIC NAME OR DRUG CLASS	COMBINED EFFECT
Bone marrow depressants*	Increased risk of toxicity of both drugs.
Hepatotoxics*	Increased risk of toxicity of both drugs.
Nephrotoxics*	Increased risk of toxicity of both drugs.
Penicillamine	Increased likelihood of kidney damage.
Phenytoin	Increased phenytoin blood levels. Phenytoin dosage may require adjustment.

POSSIBLE INTERACTION WITH OTHER SUBSTANCES

INTERACTS WITH	COMBINED EFFECT
Alcohol:	None expected.
Beverages:	None expected.
Cocaine:	Unknown. Avoid.
Foods:	None expected.
Marijuana:	Consult doctor.
Tobacco:	None expected.

***See Glossary**

GRISEOFULVIN

BRAND NAMES

Fulvicin P/G
Fulvicin U/F
Grifulvin V
Grisactin
Grisactin Ultra
Grisovin-FP
Gris-PEG

BASIC INFORMATION

Habit forming? No
Prescription needed? Yes
Available as generic? Yes
Drug class: Antifungal

USES

Treatment for fungal infections susceptible to griseofulvin.

DOSAGE & USAGE INFORMATION

How to take:

- Tablet or capsule—Swallow with liquid or food to lessen stomach irritation. If you can't swallow whole, crumble tablet or open capsule and take with liquid or food.
- Liquid—Follow label instructions.

When to take:
With or immediately after meals.

If you forget a dose:
Take as soon as you remember. If it is almost time for the next dose, wait for that dose (don't double this dose) and resume regular schedule.

What drug does:
Prevents fungi from growing and reproducing.

Time lapse before drug works:
2 to 10 days for skin infections. 2 to 4 weeks for infections of fingernails or toenails. Complete cure of either may require several months.

Don't take with:
Any other medicine or any dietary supplement without consulting your doctor or pharmacist.

OVERDOSE

SYMPTOMS:
Nausea, vomiting, diarrhea, numbness and tingling, confusion.
WHAT TO DO:
If person uses much larger amount than prescribed or if accidentally swallowed, call poison control center 1-800-222-1222 for instructions or dial 911 (emergency) for help.

POSSIBLE ADVERSE REACTIONS OR SIDE EFFECTS

SYMPTOMS	WHAT TO DO
Life-threatening:	
Rare, severe allergic reaction (anaphylaxis)—difficulty breathing, hives, itching, swelling (throat, face), vomiting, dizziness.	Seek emergency treatment immediately.
Common:	
Headache.	Continue. Call doctor when convenient.
Infrequent:	
• Confusion, rash or hives, mouth or tongue sores.	Discontinue. Call doctor right away.
• Insomnia, tiredness, nausea, vomiting, diarrhea, stomach discomfort, dizziness.	Continue. Call doctor when convenient.
Rare:	
Sore throat, fever, numbness or tingling in hands or feet, cloudy urine, yellow skin or eyes, sensitivity of skin to sunlight (these symptoms are more likely to occur with high doses taken for long periods).	Discontinue. Call doctor right away.

WARNINGS & PRECAUTIONS

Don't take if:
You are allergic to griseofulvin.

Before you start, consult your doctor if:
- You plan to become pregnant within medication period.
- You are allergic to penicillin.
- You have liver disease.
- You have porphyria.
- You have lupus.

Over age 60:
Adverse reactions and side effects may be more frequent and severe than in younger persons.

Pregnancy:
Consult doctor. Drug should not be used during pregnancy. Can cause harm to unborn baby. Risk category X (see page xviii).

Breastfeeding; Lactation; Nursing Mothers:
This drug is generally acceptable with breastfeeding. Discuss risks and benefits with your doctor.

Infants & children up to age 18:
Follow instructions provided by your child's doctor.

Prolonged use:
- You may become susceptible to infections caused by germs not responsive to griseofulvin.
- Talk to your doctor about the need for follow-up medical examinations or laboratory studies to check complete blood counts (white blood cell count, platelet count, red blood cell count, hemoglobin, hematocrit), liver function, kidney function.

Skin & sunlight:
May cause rash or intensify sunburn in areas exposed to sun or ultraviolet light (photosensitivity reaction). Avoid overexposure. Notify doctor if reaction occurs.

Driving, piloting or hazardous work:
Don't drive or pilot aircraft until you learn how medicine affects you. Don't work around dangerous machinery. Don't climb ladders or work in high places. Danger increases if you drink alcohol or take medicine affecting alertness and reflexes.

Discontinuing:
Don't discontinue without doctor's advice until you complete prescribed dose, even though symptoms diminish or disappear.

Others:
- Periodic laboratory blood studies and liver and kidney function tests recommended.
- Advise any doctor, dentist or pharmacist whom you consult that you take this medicine.

POSSIBLE INTERACTION WITH OTHER DRUGS

GENERIC NAME OR DRUG CLASS	COMBINED EFFECT
Anticoagulants, oral*	Decreased anticoagulant effect.
Barbiturates*	Decreased griseofulvin effect.
Contraceptives, oral*	Decreased contraceptive effect.
Photosensitizing medications*	Increased sun hazard.

POSSIBLE INTERACTION WITH OTHER SUBSTANCES

INTERACTS WITH	COMBINED EFFECT
Alcohol:	Increased intoxication. Possible disulfiram reaction.*
Beverages:	None expected.
Cocaine:	Unknown. Avoid.
Foods:	None expected, but foods high in fat will improve drug absorption.
Marijuana:	Consult doctor.
Tobacco:	None expected.

GUAIFENESIN

BRAND NAMES

See full list of brand names in the *Generic and Brand Name Directory*, page 863.

BASIC INFORMATION

Habit forming? No
Prescription needed? No
Available as generic? Yes
Drug class: Expectorant

USES

Loosens mucus in respiratory passages from allergies and infections (hay fever, cough or cold). Guaifenesin may be a single ingredient or it can be combined with other drugs, such as in a cough and cold product.

DOSAGE & USAGE INFORMATION

How to take:

- Tablet or capsule—Swallow with liquid. If you can't swallow whole, crumble tablet or open capsule and take with liquid or food.
- Extended-release tablet or extended-release capsule—Swallow whole with liquid.
- Syrup, oral solution or lozenge—Take as directed on label. Follow with 8 oz. water.
- Soft chew—Take as directed on label.

When to take:
As needed, no more often than every 4 hours for regular forms. The extended-release forms are usually taken every 12 hours.

If you forget a dose:
Take as soon as you remember. If it is almost time for the next dose, wait for that dose (don't double this dose) and resume regular schedule.

Continued next column

OVERDOSE

SYMPTOMS:
Drowsiness, mild weakness, nausea, vomiting.
WHAT TO DO:
If person uses much larger amount than prescribed or if accidentally swallowed, call poison control center 1-800-222-1222 for instructions or dial 911 (emergency) for help.

What drug does:
Increases production of watery fluids to thin mucus so it can be coughed out or absorbed.

Time lapse before drug works:
15 to 30 minutes. Regular use for 5 to 7 days may be necessary for maximum benefit.

Don't take with:
Any other medicine or any dietary supplement without consulting your doctor or pharmacist.

POSSIBLE ADVERSE REACTIONS OR SIDE EFFECTS

SYMPTOMS	WHAT TO DO
Life-threatening:	
None expected.	
Common:	
None expected.	
Infrequent:	
Drowsiness, rash, stomach pain, diarrhea, nausea, vomiting, dizziness, headache, hives.	Discontinue. Call doctor if symptoms persist.
Rare:	
None expected.	

WARNINGS & PRECAUTIONS

Don't take if:
You are allergic to any cough or cold product containing guaifenesin. Allergic reactions are rare, but may occur.

Before you start, consult your doctor if:
You are allergic to any medicine, food or other substance.

Over age 60:
No special problems expected.

Pregnancy:
Decide with your doctor if drug benefits justify risk to unborn baby. Risk category C (see page xviii).

Breastfeeding; Lactation; Nursing Mothers:
This drug is generally acceptable with breastfeeding. Discuss risks and benefits with your doctor.

Infants & children up to age 18:
Read nonprescription product's label to be sure it is approved for your child's age. Follow label instructions on dosage. Consult doctor or pharmacist if unsure.

Prolonged use:
No problems expected.

Skin & sunlight:
No problems expected.

Driving, piloting or hazardous work:
Avoid if you feel drowsy. Otherwise, no problems expected.

Discontinuing:
May be unnecessary to finish medicine. Discontinue when symptoms disappear. If symptoms persist more than 1 week, consult doctor.

Others:
- Some guaifenesin syrup products contain alcohol. Read labels for alcohol content if you want to avoid these products.
- Advise any doctor, dentist or pharmacist whom you consult that you take this medicine.

POSSIBLE INTERACTION WITH OTHER DRUGS

GENERIC NAME OR DRUG CLASS	COMBINED EFFECT
Anticoagulants*	Possible risk of bleeding.

POSSIBLE INTERACTION WITH OTHER SUBSTANCES

INTERACTS WITH	COMBINED EFFECT
Alcohol:	None expected.
Beverages:	None expected.
Cocaine:	Unknown. Avoid.
Foods:	None expected.
Marijuana:	Consult doctor.
Tobacco:	None expected.

***See Glossary**

HALOPERIDOL

BRAND NAMES

Apo-Haloperidol
Haldol
Haldol Decanoate
Haldol LA
Halperon
Novo-Peridol
Peridol
PMS Haloperidol

BASIC INFORMATION

Habit forming? No
Prescription needed? Yes
Available as generic? Yes
Drug class: Antipsychotic

USES

- Reduces severe anxiety, agitation and psychotic behavior.
- Treatment for Tourette's syndrome.
- Treatment for infantile autism.
- Treatment for Huntington's chorea.

DOSAGE & USAGE INFORMATION

How to take:
- Tablet—Swallow with liquid. If you can't swallow whole, crumble tablet and take with liquid or food.
- Drops—Dilute dose in beverage before swallowing.

When to take:
At the same times each day.

If you forget a dose:
Take as soon as you remember. If it is almost time for the next dose, wait for that dose (don't double this dose) and resume regular schedule.

What drug does:
Corrects an imbalance in nerve impulses from brain; blocks effect of dopamine.

Continued next column

OVERDOSE

SYMPTOMS:
Weak, rapid pulse; shallow, slow breathing; tremor or muscle weakness; very low blood pressure; convulsions; deep sleep, coma.
WHAT TO DO:
- **Dial 911 for all medical emergencies or call poison control center 1-800-222-1222 for instructions.**
- **See emergency information on last 3 pages of this book.**

Time lapse before drug works:
Up to 4 weeks.

Don't take with:
Any other medicine or any dietary supplement without consulting your doctor or pharmacist.

POSSIBLE ADVERSE REACTIONS OR SIDE EFFECTS

SYMPTOMS	WHAT TO DO
Life-threatening:	
High fever, rapid pulse, profuse sweating, muscle rigidity, confusion and irritability, seizures.	Discontinue. Seek emergency treatment.
Common:	
• Jerky or involuntary movements (of the face, lips, tongue, jaw), slow-frequency tremor of head or limbs, slow inflexible movements, lack of facial expression, trembling and shaking, weak arms and legs, twisting moves of body.	Discontinue. Call doctor right away.
• Blurred vision, dry mouth, weight gain, breast pain in females, constipation.	Discontinue. Call doctor right away.
Infrequent:	
• Less thirst, difficulty urinating, loss of balance control, lip smacking, cheek puffing.	Continue. Call doctor when convenient.
• Drowsiness, nausea, vomiting, skin has redness or rash or itching. libido problem.	Continue. Call doctor when convenient.
Rare:	
• Severe muscle stiff-ness, loss of bladder control, fast heartbeat, unusual tiredness or weakness, seizures, difficult breathing, excess sweating, high or low blood pressure.	Discontinue. Call doctor right away or seek emergency care.
• Confusion, skin is hot and dry, eyelid spasms, sore throat and fever, yellow eyes or skin, easy bruising or bleeding.	Discontinue. Call doctor right away.
• Unusual symptoms not present before starting drug.	Continue. Call doctor when convenient.

WARNINGS & PRECAUTIONS

Don't take if:
You are allergic to haloperidol. Allergic reactions are rare, but may occur.

Before you start, consult your doctor if:
- You have a history of mental depression or have kidney or liver problems, diabetes, epilepsy, glaucoma, asthma, prostate trouble or Parkinson's disease.
- You have high blood pressure, heart disease or cardiac abnormalities.
- You have QT-prolonging conditions, including electrolyte imbalance.
- You have hyper- or hypothyroidism.
- You drink alcoholic beverages frequently.

Over age 60:
- Adverse reactions and side effects may be more frequent and severe than in younger persons.
- Use of antipsychotic drugs in elderly patients with dementia-related psychosis may increase risk of death. Consult doctor.

Pregnancy:
Decide with your doctor if drug benefits justify risk to unborn baby. Risk category C (see page xviii).

Breastfeeding; Lactation; Nursing Mothers:
This drug is generally acceptable with breastfeeding. Discuss risks and benefits with your doctor.

Infants & children up to age 18:
Safety and effectiveness in this age group have not been established. Consult doctor.

Prolonged use:
- May develop tardive dyskinesia (involuntary movements of jaws, lips and tongue).
- Talk to your doctor about the need for follow-up medical examinations or laboratory studies to check blood pressure, liver function.

Skin & sunlight:
- May cause rash or intensify sunburn in areas exposed to sun or ultraviolet light (photosensitivity). Avoid overexposure and use sunscreen. Consult doctor if reaction occurs.
- Avoid getting overheated. The drug affects body temperature and sweating.

Driving, piloting or hazardous work:
Don't drive or pilot aircraft until you learn how medicine affects you. Don't work around dangerous machinery. Don't climb ladders or work in high places. Danger increases if you drink alcohol or take medicine affecting alertness.

Discontinuing:
Don't discontinue without consulting doctor. Dose may require gradual reduction if you have taken drug for a long time. Doses of other drugs may also require adjustment.

Others:
- For dry mouth, suck on sugarless hard candy or chew sugarless gum. If dry mouth persists, consult your dentist.
- Higher doses of this drug may increase risk of serious heart rhythm problems (including cases of sudden death). Call doctor right away if heart symptoms occur.
- Advise any doctor, dentist or pharmacist whom you consult that you take this medicine.

POSSIBLE INTERACTION WITH OTHER DRUGS

GENERIC NAME OR DRUG CLASS	COMBINED EFFECT
Anticholinergics*	Increased anti-chollnergic effect. May cause elevated eye pressure.
Anticonvulsants*	Changed seizure pattern.
Antidepressants*	Excessive sedation.
Antihistamines*	Excessive sedation.
Antihypertensives*	May cause severe blood pressure drop.
Barbiturates*	Excessive sedation.
Bupropion	Increased risk of seizures.
Central nervous system (CNS) depressants*	Increased CNS depression; blood pressure drop.
Clozapine	Toxic effect on the nervous system.

Continued on page 901

POSSIBLE INTERACTION WITH OTHER SUBSTANCES

INTERACTS WITH	COMBINED EFFECT
Alcohol:	Excessive sedation and depressed brain function. Avoid.
Beverages:	None expected.
Cocaine:	Increased risk of side effects. Avoid.
Foods:	None expected.
Marijuana:	Increased risk of side effects. Avoid.
Tobacco:	None expected.

***See Glossary**

HISTAMINE H_2 RECEPTOR ANTAGONISTS

GENERIC AND BRAND NAMES

CIMETIDINE
- Apo-Cimetidine
- Liquid Tagamet
- Novocimetine
- Peptol
- Tagamet
- Tagamet HB

FAMOTIDINE
- Duexis
- Fluxid
- Mylanta-AR
- Pepcid
- Pepcid AC
- Pepcid Complete
- Pepcid RPD
- Tums Dual Action
- Ulcidine

NIZATIDINE
- Axid

RANITIDINE
- Apo-Ranitidine
- Zantac
- Zantac 75
- Zantac-C
- Zantac Efferdose
- Zantac Geldose

BASIC INFORMATION

Habit forming? No
Prescription needed? Yes, for some
Available as generic? Yes, for some
Drug class: Histamine H_2 antagonist

USES

- Treatment for duodenal, gastric and peptic ulcers and other conditions in which stomach produces excess acid.
- Treatment for and prevention of heartburn.
- Maintenance of healing of erosive esophagitis.

DOSAGE & USAGE INFORMATION

How to take:
- Tablet, capsule or liquid—Swallow with liquid.
- Chewable tablet—Chew thoroughly and swallow with water.
- Disintegrating tablet—Let dissolve on tongue.
- Oral suspension or effervescent tablets for oral solution—Follow instructions on prescription.

Continued next column

OVERDOSE

SYMPTOMS:
Confusion, slurred speech, breathing difficulty, rapid heartbeat, delirium.
WHAT TO DO:
- **Dial 911 for all medical emergencies or call poison control center 1-800-222-1222 for instructions.**
- **See emergency information on last 3 pages of this book.**

When to take:
- 1 dose per day—Take at bedtime.
- 2 or more doses per day—Take at the same times each day.

If you forget a dose:
Take as soon as you remember. If it is almost time for the next dose, wait for that dose (don't double this dose) and resume regular schedule.

What drug does:
Decreases amount of acid produced in stomach.

Time lapse before drug works:
- Begins in 30 minutes. May require several days to relieve pain.
- Lower dosages in nonprescription medicines may take 45 minutes to relieve heartburn.

Don't take with:
Any other medicine or any dietary supplement without consulting your doctor or pharmacist.

POSSIBLE ADVERSE REACTIONS OR SIDE EFFECTS

SYMPTOMS	WHAT TO DO
Life-threatening:	
Rare, severe allergic reaction (anaphylaxis)—difficulty breathing, hives, itching, swelling (throat, face), vomiting, dizziness.	Seek emergency treatment immediately.
Common:	
None expected.	
Infrequent:	
Dizziness, diarrhea, headache, loss of sex drive, unusual milk flow in females, hair loss.	Continue. Call doctor when convenient.
Rare:	
• Confusion, rash or hives, sore throat, fever, slow or fast or irregular heartbeat, unusual bleeding or bruising, muscle cramps or pain, fatigue, weakness.	Discontinue. Call doctor right away.
• Constipation.	Continue. Call doctor when convenient.

WARNINGS & PRECAUTIONS

Don't take if:
You are allergic to any histamine H_2 antagonist.

Before you start, consult your doctor if:
- You plan to become pregnant while on drug.
- You take aspirin. Aspirin may irritate stomach.

Over age 60:
Adverse reactions and side effects may be more frequent and severe than in younger persons.

Pregnancy:
Decide with your doctor if drug benefits justify risk to unborn baby. Risk category B (see page xviii).

Breastfeeding; Lactation; Nursing Mothers:
Drugs in this group are generally acceptable with breastfeeding. Discuss risks and benefits with your doctor.

Infants & children up to age 18:
Follow instructions provided by your child's doctor.

Prolonged use:
- Possible liver damage.
- Talk to your doctor about the need for follow-up medical examinations or laboratory studies.

Skin & sunlight:
No problems expected.

Driving, piloting or hazardous work:
Don't drive or pilot aircraft until you learn how medicine affects you. Don't work around dangerous machinery. Don't climb ladders or work in high places. Danger increases if you drink alcohol or take medicine affecting alertness and reflexes.

Discontinuing:
Don't discontinue without consulting a doctor. Dose may require gradual reduction if you have taken drug for a long time. Doses of other drugs may also require adjustment.

Others:
- Patients on kidney dialysis—Take at end of dialysis treatment.
- May interfere with the accuracy of some medical tests.
- Advise any doctor, dentist or pharmacist whom you consult that you take this medicine.

POSSIBLE INTERACTION WITH OTHER DRUGS

GENERIC NAME OR DRUG CLASS	COMBINED EFFECT
Alprazolam	Increased effect and toxicity of alprazolam.
Antacids*	Decreased absorption of histamine H_2 receptor antagonist.
Anticoagulants, oral*	Increased anticoagulant effect.
Anticholinergics*	Increased histamine H_2 receptor antagonist effect.
Antivirals, HIV/AIDS*	Increased antiviral effect with cimetidine.
Azelastine	Increased azelastine effect.
Bupropion	Increased bupropion effect.
Carbamazepine	Increased effect and toxicity of carbamazepine.
Carmustine (BCNU)	Severe impairment of red blood cell production; some interference with white blood cell formation.
Chlordiazepoxide	Increased effect and toxicity of chlordiazepoxide.
Cisapride	Decreased histamine H_2 receptor effect.
Citalopram	Increased effect of citalopram.

Continued on page 901

POSSIBLE INTERACTION WITH OTHER SUBSTANCES

INTERACTS WITH	COMBINED EFFECT
Alcohol:	May increase level of alcohol in blood.
Beverages:	
Milk.	Enhanced effectiveness. Small amounts useful for taking medication.
Caffeine drinks.	May increase acid secretion and delay healing.
Cocaine:	Decreased effect of histamine H_2 receptor antagonist. Avoid.
Foods:	Enhanced effectiveness. Protein-rich foods should be eaten in moderation to minimize secretion of stomach acid.
Marijuana:	Consult doctor.
Tobacco:	Reversed effect of histamine H2 receptor antagonist. Tobacco may slow body's recovery. Avoid.

***See Glossary**

HMG-CoA REDUCTASE INHIBITORS

GENERIC AND BRAND NAMES

ATORVASTATIN
- Caduet
- Lipitor
- Liptruzet

FLUVASTATIN
- Lescol
- Lescol XL

LOVASTATIN
- Advicor
- Altocor
- Mevacor
- Mevinolin

PITAVASTATIN
- Livalo

PRAVASTATIN
- Eptastatin
- Pravachol
- Pravigard PAC

ROSUVASTATIN
- Crestor

SIMVASTATIN
- Epistatin
- Simcor
- Synvinolin
- Vytorin
- Zocor

BASIC INFORMATION

Habit forming? No
Prescription needed? Yes
Available as generic? Yes, for some
Drug class: Antihyperlipidemic

USES

- Lowers blood cholesterol levels caused by low-density lipoproteins (LDL) in persons who haven't improved by exercising, dieting or using other measures. It raises high density lipoproteins (HDL).
- Used along with a low-fat diet to slow the progression of atherosclerosis in patients with coronary heart disease and high cholesterol.
- Reduces risk of heart attacks and strokes in certain patients.
- Lowers triglyceride levels.

DOSAGE & USAGE INFORMATION

How to take:
- Tablet or capsule—Swallow with liquid. If you can't swallow whole, crumble tablet and take with liquid or food.
- Extended release tablet—Swallow with liquid. Do not crumble or chew tablet.

Continued next column

When to take:
According to directions on prescription.

If you forget a dose:
Take as soon as you remember. If it is almost time for the next dose, wait for that dose (don't double this dose) and resume regular schedule.

What drug does:
Inhibits an enzyme in the liver.

Time lapse before drug works:
2 to 4 weeks.

Don't take with:
- A high-fat diet.
- Any other medicine or any dietary supplement without consulting your doctor or pharmacist.

OVERDOSE

SYMPTOMS:
Unknown.
WHAT TO DO:
If person uses much larger amount than prescribed or if accidentally swallowed, call poison control center 1-800-222-1222 for instructions or dial 911 (emergency) for help.

POSSIBLE ADVERSE REACTIONS OR SIDE EFFECTS

SYMPTOMS	WHAT TO DO
Life-threatening:	
Rare, severe allergic reaction (anaphylaxis)—difficulty breathing, hives, itching, swelling (throat, face), vomiting, dizziness.	Seek emergency treatment immediately.
Common:	
None expected.	
Infrequent:	
• Aching muscles, fever, blurred vision.	Discontinue. Call doctor right away.
• Constipation, nausea, dizziness, skin rash, headache, diarrhea, heartburn.	Continue. Call doctor when convenient.
Rare:	
• Muscle or stomach pain, unusual tiredness or weakness.	Discontinue. Call doctor right away.
• Impotence, insomnia, memory loss, forgetfulness, confusion.	Continue. Call doctor when convenient.

WARNINGS & PRECAUTIONS

Don't take if:
You are allergic to HMG-CoA reductase inhibitor.

Before you start, consult your doctor if:
- You take immunosuppressive drugs, particularly following an organ transplant.
- You have low blood pressure, hormone abnormalities, active infection, active liver disease or a seizure disorder.
- You have a history of alcohol abuse.
- You have had recent surgery.

Over age 60:
May be more sensitive to drug's side effects.

Pregnancy:
Consult doctor. Drug should not be used during pregnancy. Can cause harm to unborn baby. Risk category X (see page xviii).

Breastfeeding; Lactation; Nursing Mothers:
Drugs in this group are generally not recommended with breastfeeding. Discuss risks and benefits with your doctor.

Infants & children up to age 18:
Follow instructions provided by your child's doctor.

Prolonged use:
Talk to your doctor about the need for follow-up medical examinations or laboratory studies.

Skin & sunlight:
No problems expected.

Driving, piloting or hazardous work:
No special problems expected.

Discontinuing:
Don't discontinue without consulting doctor. Dose may require gradual reduction if you have taken drug for a long time. Doses of other drugs may also require adjustment.

Others:
- Advise any doctor, dentist or pharmacist whom you consult that you take this medicine.
- Drugs to lower cholesterol may cause muscle problems or rhabdomyolysis (muscle injury). If you have persistent muscle aches, weakness, pain, or dark urine, call doctor right away.
- Taking these drugs may slightly increase the risk of raised blood sugar and diabetes type 2.
- Use an effective form of birth control while taking this drug. Contact doctor right away if you become pregnant during drug use.

POSSIBLE INTERACTION WITH OTHER DRUGS

GENERIC NAME OR DRUG CLASS	COMBINED EFFECT
Amiodarone	Risk of muscle injury and kidney failure.
Anticoagulants*	May increase bleeding risk.
Antifungals, azole*	Increased effect of HMG-CoA reductase inhibitor.
Cholestyramine	Decreased HMG-CoA reductase inhibitor effect if taken at same time.
Colestipol	Decreased effect of HMG-CoA reductase inhibitor if taken at same time.
Cyclosporine	Increased risk of muscle and kidney problems.
Digoxin	Increased digoxin effect.
Enzyme inhibitors*	Increased effect of HMG-CoA reductase inhibitor.
Erythromycins*	Risk of muscle injury and kidney failure.
Fibrates	Risk of muscle injury and kidney failure.
Gemfibrozil	Risk of muscle injury and kidney failure.
Macrolides*	Risk of muscle injury and kidney failure.
Immunosuppressants*	Risk of muscle injury and kidney failure.
Niacin	Risk of muscle injury and kidney failure.
Contraceptives, oral*	May increase levels of oral contraceptive with atorvastatin.
Orlistat	Increased effect of HMG-CoA reductase inhibitor.
Protease inhibitors	Risk of muscle injury and kidney failure.
Ranolazine	Increased effect of simvastatin.
Telithromycin	Increased effect of HMG-CoA reductase inhibitor.

POSSIBLE INTERACTION WITH OTHER SUBSTANCES

INTERACTS WITH	COMBINED EFFECT
Alcohol:	Avoid alcohol while taking this drug.
Beverages: Grapefruit juice.	May increase effect of HMG-CoA reductase inhibitor.
Cocaine:	Unknown. Avoid.
Foods: Grapefruit.	May increase effect of HMG-CoA reductase inhibitor.
Marijuana:	Consult doctor.
Tobacco:	None expected.

*See Glossary

HYDRALAZINE

GENERIC AND BRAND NAMES

BiDil	HYDRALAZINE
Hydra-Zide	

BASIC INFORMATION

Habit forming? No
Prescription needed? Yes
Available as generic? Yes
Drug class: Antihypertensive

USES

- Treatment for high blood pressure and congestive heart failure.
- The brand name BiDil is approved to treat heart failure specifically in African Americans.

DOSAGE & USAGE INFORMATION

How to take:
Tablet—Swallow with liquid. If you can't swallow whole, crumble tablet and take with liquid or food.

When to take:
At the same time each day. Should be taken with food.

If you forget a dose:
Take as soon as you remember. If it is almost time for the next dose, wait for that dose (don't double this dose) and resume regular schedule.

What drug does:
Relaxes and expands blood vessel walls, lowering blood pressure.

Time lapse before drug works:
Regular use for several weeks may be necessary to determine drug's effectiveness.

Continued next column

OVERDOSE

SYMPTOMS:
Rapid and weak heartbeat, fainting, headache, weakness, cold and sweaty skin, flushing.
WHAT TO DO:
- **Dial 911 for all medical emergencies or call poison control center 1-800-222-1222 for instructions.**
- **See emergency information on last 3 pages of this book.**

Don't take with:
- Nonprescription drugs containing alcohol without consulting doctor.
- Any other medicine or any dietary supplement without consulting your doctor or pharmacist.

POSSIBLE ADVERSE REACTIONS OR SIDE EFFECTS

SYMPTOMS	WHAT TO DO
Life-threatening:	
In case of overdose, see previous column.	
Common:	
Nausea or vomiting, rapid or irregular heartbeat, headache, diarrhea, appetite loss, painful or difficult urination.	Discontinue. Call doctor right away.
Infrequent:	
• Hives or rash, flushed face, sore throat, fever, chest pain, swelling of lymph glands, skin blisters, swelling in feet or legs.	Discontinue. Call doctor right away.
• Confusion, dizziness, anxiety, depression, joint pain, general discomfort or weakness, muscle pain, watery and irritated eyes, constipation.	Continue. Call doctor when convenient.
Rare:	
• Weakness and faintness when arising from bed or chair, yellow skin or eyes.	Discontinue. Call doctor right away.
• Numbness or tingling in hands or feet, nasal congestion, impotence.	Continue. Call doctor when convenient.

WARNINGS & PRECAUTIONS

Don't take if:
You are allergic to hydralazine or tartrazine dye. Allergic reactions are rare, but may occur.

Before you start, consult your doctor if:
- You feel pain in chest, neck or arms on physical exertion.
- You have had lupus.
- You have had a stroke.
- You have had kidney disease or impaired kidney function.

- You have a history of coronary artery disease or rheumatic heart disease.
- You will have surgery within 2 months, including dental surgery, requiring general or spinal anesthesia.

Over age 60:
Adverse reactions and side effects may be more frequent and severe than in younger persons.

Pregnancy:
Decide with your doctor if drug benefits justify risk to unborn baby. Risk category C (see page xviii).

Breastfeeding; Lactation; Nursing Mothers:
This drug is generally acceptable with breastfeeding. Discuss risks and benefits with your doctor.

Infants & children up to age 18:
Follow instructions provided by your child's doctor.

Prolonged use:
- May cause lupus (arthritis-like illness).
- Possible psychosis.
- May cause numbness, tingling in hands or feet.
- Talk to your doctor about the need for follow-up medical examinations or laboratory studies to check blood pressure, complete blood counts (white blood cell count, platelet count, red blood cell count, hemoglobin, hematocrit), ANA titers.*

Skin & sunlight:
No problems expected.

Driving, piloting or hazardous work:
Don't drive or pilot aircraft until you learn how medicine affects you. Don't work around dangerous machinery. Don't climb ladders or work in high places. Danger increases if you drink alcohol or take medicine affecting alertness and reflexes, such as antihistamines, tranquilizers, sedatives, pain medicine, narcotics and mind-altering drugs.

Discontinuing:
Don't discontinue without doctor's advice until you complete prescribed dose, even though symptoms diminish or disappear.

Others:
- Vitamin B-6 diet supplement may be advisable. Consult doctor.
- Some products contain tartrazine dye. Avoid, especially if you are allergic to aspirin.
- May interfere with the accuracy of some medical tests.
- Advise any doctor, dentist or pharmacist whom you consult that you take this medicine.

POSSIBLE INTERACTION WITH OTHER DRUGS

GENERIC NAME OR DRUG CLASS	COMBINED EFFECT
Amphetamines*	Decreased hydralazine effect.
Antihypertensives, other*	Increased anti-hypertensive effect.
Anti-inflammatory drugs, nonsteroidal (NSAIDs)*	Decreased effect of hydralazine.
Antivirals, HIV/AIDS*	Increased risk of peripheral neuropathy.
Carteolol	Increased anti-hypertensive effect.
Diazoxide & other anti-hypertensive drugs	Increased anti-hypertensive effect.
Diuretics, oral*	Increased effects of both drugs. When monitored carefully, combination may be beneficial in controlling hypertension.
Guanfacine	Increased effects of both drugs.

Continued on page 902

POSSIBLE INTERACTION WITH OTHER SUBSTANCES

INTERACTS WITH	COMBINED EFFECT
Alcohol:	May lower blood pressure excessively. Use extreme caution.
Beverages:	None expected.
Cocaine:	Increased risk of heart block and high blood pressure. Avoid.
Foods:	Increased hydralazine absorption.
Marijuana:	Consult doctor.
Tobacco:	Possible angina attacks.

***See Glossary**

HYDROXYCHLOROQUINE

BRAND NAMES

Plaquenil

BASIC INFORMATION

Habit forming? No
Prescription needed? Yes
Available as generic? No
Drug class: Antiprotozoal, antirheumatic

USES

- Treatment for protozoal infections, such as malaria and amebiasis.
- Treatment for some forms of arthritis and lupus.

DOSAGE & USAGE INFORMATION

How to take:
Tablet—Swallow with food or milk to lessen stomach irritation.

When to take:
- Depends on condition. Is adjusted during treatment.
- Malaria prevention—Begin taking medicine 2 weeks before entering areas where malaria is present and until 8 weeks after return.

If you forget a dose:
- 1 or more doses a day—Take as soon as you remember. If it is almost time for the next dose, wait for that dose (don't double this dose) and resume regular schedule.
- 1 dose weekly—Take as soon as possible, then return to regular dosing schedule.

What drug does:
- Inhibits parasite multiplication.
- Decreases inflammatory response in diseased joint.

Time lapse before drug works:
1 to 2 hours.

Don't take with:
Any other medicine or any dietary supplement without consulting your doctor or pharmacist.

OVERDOSE

SYMPTOMS:
Severe breathing difficulty, drowsiness, faintness, visual changes, headache, seizures.
WHAT TO DO:
- **Dial 911 for all medical emergencies or call poison control center 1-800-222-1222 for instructions.**
- **See emergency information on last 3 pages of this book.**

POSSIBLE ADVERSE REACTIONS OR SIDE EFFECTS

SYMPTOMS	WHAT TO DO
Life-threatening: In case of overdose, see previous column.	
Common:	
Headache, appetite loss, abdominal pain.	Continue. Call doctor when convenient.
Infrequent:	
• Blurred vision, changes in vision.	Discontinue. Call doctor right away.
• Rash or itch, diarrhea, nausea, vomiting, hair loss, blue-black skin or mouth, dizziness, nervousness.	Continue. Call doctor when convenient.
Rare:	
• Mood or mental changes, seizures, sore throat, fever, unusual bleeding or bruising, muscle weakness, convulsions.	Discontinue. Call doctor right away.
• Ringing or buzzing in ears, hearing loss.	Continue. Call doctor when convenient.

WARNINGS & PRECAUTIONS

Don't take if:
You are allergic to chloroquine or hydroxychloroquine. Allergic reactions are rare, but may occur.

Before you start, consult your doctor if:
- You plan to become pregnant within the medication period.
- You have blood disease.
- You have eye or vision problems.
- You have a G6PD* deficiency.
- You have liver disease.
- You have nerve or brain disease (including seizure disorders).
- You have porphyria.
- You have psoriasis.
- You have stomach or intestinal disease.
- You drink more than 3 oz. of alcohol daily.

Over age 60:
Adverse reactions and side effects may be more frequent and severe than in younger persons.

Pregnancy:
Decide with your doctor if drug benefits justify risk to unborn baby. Risk category C (see page xviii).

Breastfeeding; Lactation; Nursing Mothers:
This drug is generally acceptable with breastfeeding. Discuss risks and benefits with your doctor.

Infants & children up to age 18:
Not recommended. Children have increased risk of overdose and other adverse effects.

Prolonged use:
- Permanent damage to the retina (back part of the eye) or nerve deafness.
- Talk to your doctor about the need for follow-up medical examinations or laboratory studies to check complete blood counts (white blood cell count, platelet count, red blood cell count, hemoglobin, hematocrit), eyes.

Skin & sunlight:
No special problems expected.

Driving, piloting or hazardous work:
Don't drive or pilot aircraft until you learn how medicine affects you. Don't work around dangerous machinery. Don't climb ladders or work in high places. Danger increases if you drink alcohol or take medicine affecting alertness and reflexes.

Discontinuing:
Don't discontinue without doctor's advice until you complete prescribed dose, even though symptoms diminish or disappear.

Others:
- Periodic physical and blood examinations recommended.
- If you are in a malaria area for a long time, you may need to change to another preventive drug every 2 years.

POSSIBLE INTERACTION WITH OTHER DRUGS

GENERIC NAME OR DRUG CLASS	COMBINED EFFECT
Estrogens*	Possible liver toxicity.
Gold compounds*	Risk of severe rash and itch.
Kaolin	Decreased absorption of hydroxychloroquine.
Magnesium trisilicate	Decreased absorption of hydroxychloroquine.
Penicillamine	Possible blood or kidney toxicity.

POSSIBLE INTERACTION WITH OTHER SUBSTANCES

INTERACTS WITH	COMBINED EFFECT
Alcohol:	Possible liver toxicity. Avoid.
Beverages:	None expected.
Cocaine:	Unknown. Avoid.
Foods:	None expected.
Marijuana:	Consult doctor.
Tobacco:	None expected.

***See Glossary**

HYDROXYUREA

BRAND NAMES

Droxia
Hydrea

BASIC INFORMATION

Habit forming? No
Prescription needed? Yes
Available as generic? Yes
Drug class: Antineoplastic

USES

- Treats head, neck, ovarian and cervical cancer.
- Treats leukemia, melanoma and polycythemia vera.
- Treats sickle cell disease.

DOSAGE & USAGE INFORMATION

How to take:
Capsule—Swallow with liquid. If you can't swallow whole, open capsule and take with liquid or food. Instructions to take on empty stomach mean 1 hour before or 2 hours after eating.

When to take:
According to doctor's instructions.

If you forget a dose:
Take as soon as you remember. If it is almost time for the next dose, wait for that dose (don't double this dose) and resume regular schedule.

What drug does:
Probably interferes with synthesis of DNA.

Time lapse before drug works:
2 hours.

Don't take with:
Any other medicines (including over-the-counter drugs such as cough and cold medicines, laxatives, antacids, diet pills, caffeine, nose drops or vitamins) without consulting your doctor or pharmacist.

OVERDOSE

SYMPTOMS:
Mouth sores; drowsiness; hands and feet may have pain, swelling and purple color.
WHAT TO DO:
- **Dial 911 for all medical emergencies or call poison control center 1-800-222-1222 for instructions.**
- **See emergency information on last 3 pages of this book.**

POSSIBLE ADVERSE REACTIONS OR SIDE EFFECTS

SYMPTOMS	WHAT TO DO
Life-threatening:	
None expected.	
Common:	
• Skin rash, fever, chills, cough, back pain, painful urination.	Discontinue. Call doctor right away.
• Diarrhea, drowsiness, nausea, vomiting.	Continue. Call doctor when convenient.
Infrequent:	
Mouth sores, red spots on skin, unusual bleeding or bruising, black or tarry stools, blood in urine, nails start to blacken.	Discontinue. Call doctor right away.
Rare:	
Confusion, hallucinations, difficulty in urinating, seizures, dizziness, headache, joint pain, feet or leg swelling, numbness or tingling or pain in fingers or toes, unusual tiredness or weakness, itching.	Discontinue. Call doctor right away.

WARNINGS & PRECAUTIONS

Don't take if:
You are allergic to hydroxyurea. Allergic reactions are rare, but may occur.

Before you start, consult your doctor if:
- You have chicken pox.
- You have shingles (herpes zoster).
- You have anemia or blood disorder.
- You have gout.
- You have an infection.
- You have kidney disease or kidney stones.
- You have taken interferon in the past.

Over age 60:
Adverse reactions and side effects may be more frequent and severe than in younger persons.

Pregnancy:
Consult doctor. Use of the drug during pregnancy is not recommended. It could cause harm to the unborn baby. Risk category D (see page xviii).

Breastfeeding; Lactation; Nursing Mothers:
This drug is generally acceptable with breastfeeding. Discuss risks and benefits with your doctor.

Infants & children up to age 18:
Safety and effectiveness in this age group have not been established. Consult doctor.

Prolonged use:
Talk to your doctor about the need for follow-up medical examinations or laboratory studies to check kidney function, complete blood counts (white blood cell count, platelet count, red blood cell count, hemoglobin, hematocrit) and serum uric acid.

Skin & sunlight:
No problems expected.

Driving, piloting or hazardous work:
Avoid if you feel confused, drowsy or dizzy.

Discontinuing:
May still experience symptoms of bone marrow depression, such as: blood in stools, fever or chills, blood spots under the skin, back pain, hoarseness, bloody urine. If any of these occur, call your doctor right away.

Others:
- Advise any doctor, dentist or pharmacist whom you consult that you take this medicine.
- May affect results in some medical tests.

POSSIBLE INTERACTION WITH OTHER DRUGS

GENERIC NAME OR DRUG CLASS	COMBINED EFFECT
Bone marrow depressants, other*	Dangerous suppression of bone marrow activity.
Clozapine	Toxic effect on bone marrow.
Levamisole	Increased risk of bone marrow depression.
Probenecid	May require increased dosage to treat gout.
Sulfinpyrazone	May require increased dosage to treat gout.
Tiopronin	Increased risk of toxicity to bone marrow.
Vaccines, live or killed virus	Increased risk of side effects.

POSSIBLE INTERACTION WITH OTHER SUBSTANCES

INTERACTS WITH	COMBINED EFFECT
Alcohol:	None expected.
Beverages:	None expected.
Cocaine:	Unknown. Avoid.
Foods:	None expected.
Marijuana:	Consult doctor.
Tobacco:	None expected.

***See Glossary**

HYDROXYZINE

BRAND NAMES

Anxanil
Apo-Hydroxyzine
Atarax
Multipax
Novo-Hydroxyzin
Vistaril
Vistaril

BASIC INFORMATION

Habit forming? No
Prescription needed? Yes
Available as generic? Yes
Drug class: Tranquilizer, antihistamine

USES

- Treatment for anxiety, tension and agitation.
- Relieves itching from allergic reactions.

DOSAGE & USAGE INFORMATION

How to take:
- Tablet, syrup or capsule—Swallow with liquid. If you can't swallow whole, crumble tablet or open capsule and take with liquid or food.
- Liquid—If desired, dilute dose in beverage before swallowing.

When to take:
At the same times each day.

If you forget a dose:
Take as soon as you remember. If it is almost time for the next dose, wait for that dose (don't double this dose) and resume regular schedule.

What drug does:
Blocks action of histamine after an allergic response triggers histamine release. Histamines cause itching, sneezing, runny nose and eyes and other symptoms.

Time lapse before drug works:
15 to 30 minutes.

Continued next column

OVERDOSE

SYMPTOMS:
Severe drowsiness, unsteadiness, nausea and vomiting, stupor, convulsions.
WHAT TO DO:
- **Dial 911 for all medical emergencies or call poison control center 1-800-222-1222 for instructions.**
- **See emergency information on last 3 pages of this book.**

Don't take with:
Any other medicine or any dietary supplement without consulting your doctor or pharmacist.

POSSIBLE ADVERSE REACTIONS OR SIDE EFFECTS

SYMPTOMS	WHAT TO DO
Life-threatening: In case of overdose, see previous column.	
Common:	
Drowsiness, dizziness, dryness of mouth or nose or throat, nausea.	Continue. Call doctor when convenient.
Infrequent:	
• Change in vision, clumsiness, rash.	Discontinue. Call doctor right away.
• Less tolerance for contact lenses, painful or difficult urination, appetite loss.	Continue. Call doctor when convenient.
Rare:	
Nightmares, agitation, irritability, sore throat, fever, rapid heartbeat, unusual bleeding or bruising, fatigue, weakness, confusion, fainting, seizures.	Discontinue. Call doctor right away.

WARNINGS & PRECAUTIONS

Don't take if:
You are allergic to any antihistamine. Allergic reactions are rare, but may occur.

Before you start, consult your doctor if:
- You have asthma or kidney disease.
- You will have surgery within 2 months, including dental surgery, requiring general or spinal anesthesia.

Over age 60:
Adverse reactions and side effects may be more frequent and severe.

Pregnancy:
Decide with your doctor if drug benefits justify risk to unborn baby. Risk category C (see page xviii).

Breastfeeding; Lactation; Nursing Mothers:
This drug is generally acceptable with breastfeeding. Discuss risks and benefits with your doctor.

Infants & children up to age 18:
Follow instructions provided by your child's doctor.

Prolonged use:
Tolerance* may develop and reduce effectiveness.

Skin & sunlight:
No problems expected.

Driving, piloting or hazardous work:
Don't drive or pilot aircraft until you learn how medicine affects you. Don't work around dangerous machinery. Don't climb ladders or work in high places. Danger increases if you drink alcohol or take medicine affecting alertness and reflexes, such as antihistamines, tranquilizers, sedatives, pain medicine, narcotics and mind-altering drugs.

Discontinuing:
Don't discontinue without consulting doctor. Dose may require gradual reduction if you have taken drug for a long time. Doses of other drugs may also require adjustment.

Others:
Advise any doctor, dentist or pharmacist whom you consult that you take this medicine.

POSSIBLE INTERACTION WITH OTHER DRUGS

GENERIC NAME OR DRUG CLASS	COMBINED EFFECT
Antidepressants, tricyclic*	Increased effects of both drugs.
Antihistamines*	Increased hydroxyzine effect.
Attapulgite	Decreased hydroxyzine effect.
Carteolol	Decreased antihistamine effect.
Central nervous system (CNS) depressants*	Greater depression of central nervous system.
Clozapine	Toxic effect on the central nervous system.
Fluoxetine	Increased depressant effects of both drugs.
Guanfacine	May increase depressant effects of either drug.
Leucovorin	High alcohol content of leucovorin may cause adverse effects.
Narcotics*	Increased effects of both drugs.
Pain relievers*	Increased effects of both drugs.
Sertraline	Increased depressive effects of both drugs.
Sotalol	Increased antihistamine effect.

POSSIBLE INTERACTION WITH OTHER SUBSTANCES

INTERACTS WITH	COMBINED EFFECT
Alcohol:	Increased sedation and intoxication. Use with caution.
Beverages: Caffeine drinks.	Decreased tranquilizer effect of hydroxyzine.
Cocaine:	Increased risk of side effects. Avoid.
Foods:	None expected.
Marijuana:	Increased risk of side effects. Avoid.
Tobacco:	None expected.

***See Glossary**

HYOSCYAMINE

BRAND NAMES

Anaspaz
Anaspaz PB
Cystospaz
Cystospaz-M
Donnatal
Donnatal Elixir
Donnatal Extentabs
Gastrosed
Levbid
Levsin S/L
Levsinex
Levsinex Timecaps
Neoquesss
Nulev

BASIC INFORMATION

Habit forming? No
Prescription needed?
Low strength: No
High strength: Yes
Available as generic? Yes
Drug class: Antispasmodic, anticholinergic

USES

Reduces spasms of digestive system, bladder and urethra.

DOSAGE & USAGE INFORMATION

How to take:
- Tablet or liquid—Swallow with liquid or food to lessen stomach irritation. You may chew or crush tablets.
- Extended-release capsule or tablet—Swallow each dose whole.
- Drops—Dilute dose in beverage before swallowing.

When to take:
30 minutes before meals (unless directed otherwise by doctor).

Continued next column

OVERDOSE

SYMPTOMS:
Dilated pupils, rapid pulse and breathing, dizziness, fever, hallucinations, confusion, slurred speech, agitation, flushed face, convulsions, coma.
WHAT TO DO:
- **Dial 911 for all medical emergencies or call poison control center 1-800-222-1222 for instructions.**
- **See emergency information on last 3 pages of this book.**

If you forget a dose:
Take as soon as you remember. If it is almost time for the next dose, wait for that dose (don't double this dose) and resume regular schedule.

What drug does:
Blocks nerve impulses at parasympathetic nerve endings, preventing muscle contractions and gland secretions of organs involved.

Time lapse before drug works:
15 to 30 minutes.

Don't take with:
- Antacids* or antidiarrheals* at the same time.
- Any other medicine or any dietary supplement without consulting your doctor or pharmacist.

POSSIBLE ADVERSE REACTIONS OR SIDE EFFECTS

SYMPTOMS	WHAT TO DO
Life-threatening:	
Rare, severe allergic reaction (anaphylaxis)—difficulty breathing, hives, itching, swelling (throat, face), vomiting, dizziness.	Seek emergency treatment immediately.
Common:	
• Confusion, delirium, rapid heartbeat.	Discontinue. Call doctor right away.
• Nausea, vomiting, decreased sweating, constipation, dryness (in ears, nose, throat, mouth).	Continue. Call doctor when convenient.
Infrequent:	
Headache, painful or difficult urination, nasal congestion, altered taste.	Continue. Call doctor when convenient.
Rare:	
Rash or hives, eye pain, blurred vision, lightheadedness.	Discontinue. Call doctor right away.

WARNINGS & PRECAUTIONS

Don't take if:
- You are allergic to hyoscyamine.
- You have trouble with stomach bloating.
- You have difficulty emptying your bladder completely.
- You have narrow-angle glaucoma.
- You have severe ulcerative colitis.

Before you start, consult your doctor if:

- You have open-angle glaucoma.
- You have angina.
- You have chronic bronchitis or asthma.
- You have hiatal hernia.
- You have liver, kidney or thyroid disease.
- You have enlarged prostate.
- You have myasthenia gravis.
- You have peptic ulcer.
- You will have surgery within 2 months, including dental surgery, requiring general or spinal anesthesia.

Over age 60:
Adverse reactions and side effects may be more frequent and severe than in younger persons.

Pregnancy:
Decide with your doctor if drug benefits justify risk to unborn baby. Risk category C (see page xviii).

Breastfeeding; Lactation; Nursing Mothers:
This drug is generally acceptable with breastfeeding. Discuss risks and benefits with your doctor.

Infants & children up to age 18:
Follow instructions provided by your child's doctor.

Prolonged use:
Chronic constipation, possible fecal impaction. Consult doctor immediately.

Skin & sunlight:
No problems expected.

Driving, piloting or hazardous work:
Use disqualifies you for piloting aircraft. Otherwise, no problems expected.

Discontinuing:
May be unnecessary to finish medicine. Follow doctor's instructions.

Others:
Advise any doctor, dentist or pharmacist whom you consult that you take this medicine.

POSSIBLE INTERACTION WITH OTHER DRUGS

GENERIC NAME OR DRUG CLASS	COMBINED EFFECT
Amantadine	Increased hyoscyamine effect.
Anticholinergics, other*	Increased hyoscyamine effect.
Antidepressants, tricyclic*	Increased hyoscyamine effect.
Antihistamines*	Increased hyoscyamine effect.
Cortisone drugs*	Increased internal eye pressure.
Haloperidol	Increased internal eye pressure.
Ketoconazole	Decreased ketoconazole effect.
Meperidine	Increased hyoscyamine effect.
Methylphenidate	Increased hyoscyamine effect.
Molindone	Increased anticholinergic effect.
Monoamine oxidase (MAO) inhibitors*	Increased hyoscyamine effect.
Nizatidine	Increased nizatidine effect.
Orphenadrine	Increased hyoscyamine effect.
Phenothiazines*	Increased hyoscyamine effect.
Pilocarpine	Loss of pilocarpine effect in glaucoma treatment.
Sedatives* or central nervous system (CNS) depressants*	Increased sedative effect of both drugs.
Vitamin C	Decreased hyoscyamine effect. Avoid large doses of vitamin C.

POSSIBLE INTERACTION WITH OTHER SUBSTANCES

INTERACTS WITH	COMBINED EFFECT
Alcohol:	None expected.
Beverages:	None expected.
Cocaine:	Increased risk of side effects. Avoid.
Foods:	None expected.
Marijuana:	Increased risk of side effects. Avoid.
Tobacco:	None expected.

*See Glossary

IMATINIB

BRAND NAMES

Gleevec

BASIC INFORMATION

Habit forming? No
Prescription needed? Yes
Available as generic? No
Drug class: Antineoplastic

USES

- Treatment for some types of leukemia (cancer of white blood cells).
- Treatment for gastrointestinal stomal tumors.
- Treatment for certain rare cancers and blood diseases.

DOSAGE & USAGE INFORMATION

How to take:
Capsule—Swallow with large glass of water to minimize the risk of stomach and gastrointestinal irritation.

When to take:
Usually once a day at mealtime or according to doctor's instructions.

If you forget a dose:
Take as soon as you remember. If it is almost time for the next dose, wait for that dose (don't double this dose) and resume regular schedule.

What drug does:
Reduces substantially the level of cancerous cells in the bone marrow and blood of treated patients.

Time lapse before drug works:
Starts working in 2-4 hours, but effectiveness may take 1-3 months.

Don't take with:
Any other medicines or dietary supplements without consulting your doctor or pharmacist.

OVERDOSE

SYMPTOMS:
Nausea, vomiting, diarrhea, rash, swelling, tiredness, muscle spasms, stomach pain, headache, weakness.

WHAT TO DO:
- **Dial 911 for all medical emergencies or call poison control center 1-800-222-1222 for instructions.**
- **See emergency information on last 3 pages of this book.**

POSSIBLE ADVERSE REACTIONS OR SIDE EFFECTS

SYMPTOMS	WHAT TO DO
Life-threatening:	
Some adverse effects from advanced cancer and/or the medicine can be serious or life threatening.	Seek emergency treatment for any symptoms that appear severe or critical.
Common:	
• Chest pain, shortness of breath, swelling (face, hands, legs. feet), black or tarry stools, nausea or vomiting, muscle pain or cramps, blood in urine, decreased or painful urination, fever, chills, pale skin, quick weight gain, stomach cramps, sore throat, sores on body, ulcers or white spots on lips or in mouth, swollen glands, unusual bleeding or tiredness or weakness.	Continue. Call doctor right away.
• Joint or bone pain, diarrhea, skin rash, fatigue.	Continue. Call doctor when convenient.
Infrequent:	
• Convulsions, irregular heartbeat, wheezing, tightness in chest, numbness or tingling (in hands, feet, or lips), pinpoint red spots on skin.	Continue. Call doctor right away.
• Bloody nose, mood changes, slow weight gain, weight loss, sneezing, constipation, loss of appetite, headache, increased thirst, dry mouth.	Continue. Call doctor when convenient.
Rare:	
Acid indigestion or upset stomach, stuffy nose, itchy skin.	Continue. Call doctor when convenient.

Note: Side effects are often unavoidable with drugs used to treat cancer. Discuss any concerns or questions or other symptoms with your doctor.

WARNINGS & PRECAUTIONS

Don't take if:
You are allergic to imatinib. Allergic reactions are rare, but may occur.

Before you start, consult your doctor if:
- You have any infection.
- You have anemia, leukopenia, neutropenia or thrombocytopenia.
- You have bone marrow depression.
- You have recent chickenpox or herpes zoster.
- You have liver problems.

Over age 60:
May be more at risk for edema (swelling caused by excess fluid in the body).

Pregnancy:
Consult doctor. Use of the drug during pregnancy is not recommended. It could cause harm to the unborn baby. Risk category D (see page xviii).

Breastfeeding; Lactation; Nursing Mothers:
This drug is generally not recommended with breastfeeding. Discuss risks and benefits with your doctor.

Infants & children up to age 18:
Follow instructions provided by your child's doctor.

Prolonged use:
- Talk to your doctor about the need for follow-up medical examinations or laboratory studies to check your response to the drug, blood studies, weight gain and liver function.
- The long term effects of the drug are not yet known.

Skin & sunlight:
No problems expected.

Driving, piloting or hazardous work:
No problems expected.

Discontinuing:
No special problems expected. Don't discontinue drug without doctor's approval.

Others:
- Advise any doctor, dentist or pharmacist whom you consult that you take this medicine. Consult doctor before having dental work.
- Do not have any immunizations without your doctor's approval. Imatinib may lower your resistance to the infection that you are getting the immunization for.
- Avoid persons who have recently taken the oral polio virus vaccine; they may pass the virus on to you.
- Avoid people with infections, because you have an increased chance of getting the infection. Advise your doctor of any unusual symptoms, infections or injuries.
- Be careful when brushing or flossing teeth, or using a razor or other sharp object to avoid being injured.
- May affect results in some medical tests.
- Avoid contact sports or activities that could cause, injury, bruising or bleeding.

POSSIBLE INTERACTION WITH OTHER DRUGS

GENERIC NAME OR DRUG CLASS	COMBINED EFFECT
Anticoagulants*	Blood clotting problems.
Blood dyscrasia-causing medications*	Increased risk of adverse effects of imatinib.
Bone marrow depressants, other*	Increased bone marrow suppression.
Cyclosporine	Increased effect of cyclosporine.
Enzyme inducers*	Decreased effect of imatinib.
Enzyme inhibitors*	Increased effect of imatinib.
HMG-CoA reductase inhibitors	Increased effect of HMG-CoA reductase inhibitor.
Pimozide	Increased effect of pimozide.
Vaccines (killed)	Decreased effectiveness of vaccine.
Vaccines (live)	Increased risk of getting the disease the vaccine prevents.

POSSIBLE INTERACTION WITH OTHER SUBSTANCES

INTERACTS WITH	COMBINED EFFECT
Alcohol:	None expected.
Beverages:	None expected.
Cocaine:	Unknown. Avoid.
Foods:	None expected.
Marijuana:	Consult doctor.
Tobacco:	None expected.

*See Glossary

IMMUNOSUPPRESSIVE AGENTS

GENERIC AND BRAND NAMES

MYCOPHENOLATE
- CellCept
- Myfortic
- Myfortic Delayed Release

TACROLIMUS
- Astagraf XL
- Prograf

BASIC INFORMATION

Habit forming? No
Prescription needed? Yes
Available as generic? Yes
Drug class: Immunosuppressant

USES

Helps to suppress the immune system and prevent rejection in patients who have undergone organ transplants. Carefully follow doctor's instructions about use of these drugs.

DOSAGE & USAGE INFORMATION

How to take:
- Capsule or tablet—Swallow with liquid. Do not open capsule or crush tablet.
- Delayed-release tablet—Swallow whole with liquid. Do not crush or chew tablet.
- Extended-release capsule—Swallow whole with liquid. Do not crush or chew capsule.
- Oral suspension—Take as directed on label.

When to take:
At the same time(s) each day, according to prescription label. To take on an empty stomach means 1 hour before or 2 hours after a meal. Some may be taken with or without food.

If you forget a dose:
Take as soon as you remember (within 14 hours for Astagraf XL). If it is almost time for the next dose, wait for next scheduled dose (don't double this dose). Consult doctor if you are unsure about dosing schedule.

Continued next column

OVERDOSE

SYMPTOMS:
Increased severity of adverse reactions.
WHAT TO DO:
- **Dial 911 for all medical emergencies or call poison control center 1-800-222-1222 for instructions.**
- **See emergency information on last 3 pages of this book.**

What drug does:
Tacrolimus and mycophenolate are 2 different types of immunosuppressive drugs. They suppress immune reactions (to transplanted organs) in certain cells by inhibiting their growth.

Time lapse before drug works:
3 to 3-1/2 hours. May take several weeks to evaluate effectiveness against organ rejection.

Don't take with:
Any other medicine or any dietary supplement without consulting your doctor or pharmacist.

POSSIBLE ADVERSE REACTIONS OR SIDE EFFECTS

SYMPTOMS	WHAT TO DO
Life-threatening:	
Rare, severe allergic reaction (anaphylaxis)—difficulty breathing, hives, itching, swelling (throat, face), vomiting, dizziness.	Seek emergency treatment immediately.
Common:	
• Infections (fever, chills, hoarseness, cough, trouble urinating, back or side pain), headache, tingling or numbness (hands or feet), chest pain, blood in urine, trouble sleeping, trembling of hands, shortness of breath.	Continue, but call doctor right away.
• Mild back pain, constipation or diarrhea, stomach pain, heartburn.	Continue. Call doctor when convenient.
Infrequent:	
• Bloody vomit, anxiety, nervousness, white patches (on mouth, tongue or throat), weakness, seizures.	Continue, but call doctor right away.
• Hair loss or excess hair growth, nausea, mild vomiting, skin rash or itch, muscle or joint pain, dizziness.	Continue. Call doctor when convenient.
Rare:	
• Bloody or black or tarry stools, small red spots on skin, unusual bleeding or bruising, irregular heartbeat.	Continue, but call doctor right away.
• Mood or mental changes.	Continue. Call doctor when convenient.

WARNINGS & PRECAUTIONS

Don't take if:
You are allergic to mycophenolate or tacrolimus.

Before you start, consult your doctor if:
- You have liver or kidney disease.
- You have a digestive system disease.
- You have an infection.
- You have chickenpox (or have recently been exposed) or herpes zoster (shingles).
- You are female and able to bear children.

Over age 60:
Adverse reactions and side effects may be more frequent and severe than in younger persons.

Pregnancy:
Consult doctor. Use of mycophenolate during pregnancy is not recommended. It could cause harm to the unborn baby. Risk category D. Decide with your doctor if tacrolimus benefits justify risk to unborn baby. Risk category C (see page xviii).

Breastfeeding; Lactation; Nursing Mothers:
Drugs in this group are generally acceptable with breastfeeding. Discuss risks and benefits with your doctor.

Infants & children up to age 18:
Follow instructions provided by your child's doctor.

Prolonged use:
- Can cause reduced function of kidneys.
- Can increase risk of developing lymphoma (cancer of lymph glands).
- Talk to your doctor about the need for follow-up medical exams or lab studies.

Skin & sunlight:
No special problems expected.

Driving, piloting or hazardous work:
Don't drive or pilot aircraft until you learn how medicine affects you. Don't work around dangerous machinery. Don't climb ladders or work in high places. Danger increases if you drink alcohol or take drugs affecting alertness and reflexes.

Discontinuing:
Don't discontinue without consulting doctor. You will usually require this drug for your lifetime.

Others:
- Advise any doctor, dentist or pharmacist whom you consult that you take this drug.
- Wear medical identification stating that you have had a transplant and take this drug.
- Check blood pressure routinely at home. Drug may sometimes cause hypertension.
- Immunosuppressed patients are at increased risk for opportunistic infections and malignancies. Discuss your risks with doctor.
- Avoid any immunizations except those specifically recommended by your doctor.
- Maintain good dental hygiene. Immunosuppression can cause gum problems.
- Talk to your doctor about forms of birth control before starting treatment with mycophenolate.

POSSIBLE INTERACTION WITH OTHER DRUGS

GENERIC NAME OR DRUG CLASS	COMBINED EFFECT
Acyclovir	Increased effect of mycophenolate.
Antacids*	Decreased effect of mycophenolate.
Cholestyramine	Decreased effect of mycophenolate.
Diuretics, potassium-sparing*	Increased risk of potassium toxicity.
Enzyme inducers*	Decreased effect of tacrolimus.
Enzyme inhibitors*	Increased effect of tacrolimus.
Ganciclovir	Increased effect of mycophenolate.
Immunosuppressants,* other	Increased risk of adverse effects.
Nephrotoxics*	Increased risk of kidney problems.
Nitroimidazoles	Increased tacrolimus effect.
Potassium supplements	Increased risk of potassium toxicity.

Continued on page 902

POSSIBLE INTERACTION WITH OTHER SUBSTANCES

INTERACTS WITH	COMBINED EFFECT
Alcohol:	Increased risk of toxic effects. Avoid.
Beverages: Grapefruit juice.	Increased effect of drug. Avoid.
Cocaine:	Unknown. Avoid.
Foods: Grapefruit.	Increased effect of drug. Avoid.
Marijuana:	Consult doctor.
Tobacco:	None expected.

*See Glossary

INDAPAMIDE

BRAND NAMES

Lozide

BASIC INFORMATION

Habit forming? No
Prescription needed? Yes
Available as generic? Yes
Drug class: Antihypertensive, diuretic

USES

- Controls, but doesn't cure, high blood pressure.
- Reduces fluid retention (edema) caused by conditions such as heart disorders.

DOSAGE & USAGE INFORMATION

How to take:
Tablet—Swallow with liquid or food to lessen stomach irritation.

When to take:
At the same times each day, usually at bedtime.

If you forget a dose:
Bedtime dose—If you forget your once-a-day bedtime dose, don't take it more than 3 hours late. Never double dose.

What drug does:
Forces kidney to excrete more sodium and causes excess salt and fluid to be excreted.

Time lapse before drug works:
2 hours for effect to begin. May require 1 to 4 weeks for full effects.

Don't take with:
Any other medicine or any dietary supplement without consulting your doctor or pharmacist.

OVERDOSE

SYMPTOMS:
Nausea, vomiting, diarrhea, very dry mouth, thirst, weakness, fatigue, slowed breathing, low blood pressure.
WHAT TO DO:
- **Dial 911 for all medical emergencies or call poison control center 1-800-222-1222 for instructions.**
- **See emergency information on last 3 pages of this book.**

POSSIBLE ADVERSE REACTIONS OR SIDE EFFECTS

SYMPTOMS	WHAT TO DO
Life-threatening:	
Rare, severe allergic reaction (anaphylaxis)—difficulty breathing, hives, itching, swelling (throat, face), vomiting, dizziness.	Seek emergency treatment immediately.
Common:	
Excessive tiredness or weakness, muscle cramps.	Discontinue. Call doctor right away.
Infrequent:	
Insomnia, mood change, dizziness on changing position, headache, excessive thirst, diarrhea, appetite loss, nausea, dry mouth, decreased sex drive, frequent urination.	Continue. Call doctor when convenient.
Rare:	
Itching, rash or hives, irregular heartbeat, weak pulse.	Discontinue. Call doctor right away.

WARNINGS & PRECAUTIONS

Don't take if:
You are allergic to indapamide or to any sulfa drug or thiazide diuretic.*

Before you start, consult your doctor if:
- You have severe kidney disease.
- You have diabetes, gout or liver disease.
- You will have surgery within 2 months, including dental surgery, requiring general or spinal anesthesia.
- You have lupus erythematosus.
- You are pregnant or plan to become pregnant.

Over age 60:
Adverse reactions and side effects may be more frequent and severe than in younger persons.

Pregnancy:
Decide with your doctor if drug benefits justify risk to unborn baby. Risk category B (see page xviii).

Breastfeeding; Lactation; Nursing Mothers:
This drug is generally acceptable with breastfeeding. Discuss risks and benefits with your doctor.

Infants & children up to age 18:
Safety and effectiveness in this age group have not been established. Consult doctor.

Prolonged use:
Consult your doctor on a regular basis.

Skin & sunlight:
May cause rash or intensify sunburn in areas exposed to sun or ultraviolet light (photosensitivity reaction). Avoid overexposure. Notify doctor if reaction occurs.

Driving, piloting or hazardous work:
Don't drive or pilot aircraft until you learn how medicine affects you. Don't work around dangerous machinery. Don't climb ladders or work in high places. Danger increases if you drink alcohol or take medicine affecting alertness and reflexes, such as antihistamines, tranquilizers, sedatives, pain medicine, narcotics and mind-altering drugs.

Discontinuing:
Don't discontinue without consulting doctor. Dose may require gradual reduction if you have taken drug for a long time. Doses of other drugs may also require adjustment.

Others:
Advise any doctor, dentist or pharmacist whom you consult that you take this medicine.

POSSIBLE INTERACTION WITH OTHER DRUGS

GENERIC NAME OR DRUG CLASS	COMBINED EFFECT
Adrenocorticoids, systemic	Possible excessive potassium loss.
Allopurinol	Decreased allopurinol effect.
Amiodarone	Increased risk of irregular heartbeat due to low potassium.
Amphotericin B	Increased potassium.
Angiotensin-converting enzyme (ACE) inhibitors*	Decreased blood pressure. Possible excess potassium.
Antidepressants, tricyclic*	Dangerous drop in blood pressure.
Antidiabetic agents, oral*	Increased blood sugar.
Antihypertensives, other*	Increased anti-hypertensive effect.
Barbiturates*	Increased indapamide effect.
Beta-adrenergic blocking agents*	Increased effect of indapamide.
Calcium supplements*	Increased calcium in blood.
Carteolol	Increased anti-hypertensive effect.
Cholestyramine	Decreased indapamide effect.
Colestipol	Decreased indapamide effect.
Digitalis preparations*	Excessive potassium loss that may cause dangerous heart rhythms.
Diuretics, thiazide*	Increased effect of thiazide diuretics.
Indomethacin	Decreased indapamide effect.
Lithium	High risk of lithium toxicity.
Monoamine oxidase (MAO) inhibitors*	Increased indapamide effect.
Nicardipine	Dangerous blood pressure drop. Dosages may require adjustment.
Nimodipine	Dangerous blood pressure drop.
Opiates*	Weakness and faintness when arising from bed or chair.

Continued on page 903

POSSIBLE INTERACTION WITH OTHER SUBSTANCES

INTERACTS WITH	COMBINED EFFECT
Alcohol:	Dangerous blood pressure drop. Avoid.
Beverages:	No problems expected.
Cocaine:	Increased risk of side effects. Avoid.
Foods:	
Licorice.	Excessive potassium loss that may cause dangerous heart rhythms.
Marijuana:	Increased risk of side effects. Avoid.
Tobacco:	Reduced effect of indapamide. Avoid.

*See Glossary

INSULIN

BRAND NAMES

See full list of brand names in the *Generic and Brand Name Directory*, page 865.

BASIC INFORMATION

Habit forming? No
Prescription needed? No
Available as generic? Yes
Drug class: Antidiabetic

USES

Treats diabetes, a metabolic disorder, in which patients have high levels of sugar in their blood.

DOSAGE & USAGE INFORMATION

How to take:

- Injection—Injected under the skin. Use sterile, disposable needles. Rotate injection sites.
- Inhaled powder—Follow instructions provided.

When to take:
At the same times each day.

If you forget a dose:
Take as soon as you remember. If it is almost time for the next dose, wait for that dose (don't double this dose) and resume regular schedule.

What drug does:
Facilitates passage of blood sugar through cell membranes so sugar is usable.

Time lapse before drug works:
30 minutes to 8 hours, depending on type of insulin used.

Continued next column

OVERDOSE

SYMPTOMS:
See next column under Infrequent for symptoms of low blood sugar.
WHAT TO DO:

- **Eat some type of sugar immediately, such as glucose product, orange juice (add some sugar), nondiet sodas, candy (such as 5 Lifesavers), honey.**
- **If patient loses consciousness, give glucagon if you have it and know how to use it.**
- **Dial 911 (emergency) for medical help or call poison control center 1-800-222-1222 for instructions.**
- **See emergency information on last 3 pages of this book.**

Don't take with:
Any other medicine or any dietary supplement without consulting your doctor or pharmacist.

POSSIBLE ADVERSE REACTIONS OR SIDE EFFECTS

SYMPTOMS	WHAT TO DO
Life-threatening:	
Rare, severe allergic reaction (anaphylaxis)—difficulty breathing, hives, itching, swelling (throat, face), vomiting, dizziness.	Seek emergency treatment immediately.
Common:	
Throat pain with inhaled insulin.	Continue. Call doctor when convenient.
Infrequent:	
• Symptoms of low blood sugar—nervousness, hunger (excessive), cold sweats, rapid pulse, anxiety, cold skin, chills, confusion, concentration loss, drowsiness, headache, nausea, weakness, shakiness, vision changes.	Seek treatment (eat some form of quick-acting sugar—glucose tablets, sugar, fruit juice, corn syrup, honey).
• Symptoms of high blood sugar—increased urination, unusual thirst, dry mouth, drowsiness, flushed or dry skin, fruit-like breath odor, appetite loss, stomach pain or vomiting, tiredness, trouble breathing, increased blood sugar level.	Seek emergency treatment immediately.
• Swelling, redness, itch or warmth at injection site.	Continue. Call doctor when convenient.
Rare:	
None expected.	

WARNINGS & PRECAUTIONS

Don't take if:

- You are allergic to insulin.
- It is the inhalation form and you have chronic lung disease (e.g., asthma or COPD).

Before you start, consult your doctor if:

- You have liver or kidney problems or lung cancer.
- You are or have been a tobacco smoker.

Over age 60:
Older adults may need close monitoring for drug's effectiveness and for adverse effects.

Pregnancy:
Decide with your doctor if drug benefits justify risk to unborn baby. Risk category B (see page xviii).

Breastfeeding; Lactation; Nursing Mothers:
This drug is generally acceptable with breastfeeding. Discuss risks and benefits with your doctor.

Infants & children up to age 18:
Follow instructions provided by your child's doctor.

Prolonged use:
Talk to your doctor about the need for follow-up medical exams and lab studies to check blood sugar, serum potassium, urine and drug's effectiveness.

Skin & sunlight:
No problems expected.

Driving, piloting or hazardous work:
No problems expected after dose is established.

Discontinuing:
Don't discontinue without doctor's advice until you complete prescribed dose, even though symptoms diminish or disappear.

Others:
- Diet and exercise affect how much insulin you need. Work with your doctor to determine accurate dose.
- Notify your doctor if you skip a dose, overeat, have fever or infection.
- Notify doctor if you develop symptoms of high blood sugar: drowsiness, dry skin, orange fruit-like odor to breath, increased urination, appetite loss, unusual thirst.
- Never freeze insulin.
- Inhaled insulin may cause bronchospasm (sudden tightening of the chest).
- May interfere with the accuracy of some medical tests.

POSSIBLE INTERACTION WITH OTHER DRUGS

GENERIC NAME OR DRUG CLASS	COMBINED EFFECT
Adrenocorticoids, systemic	Decreased insulin effect.
Anticonvulsants, hydantoin*	Decreased insulin effect.
Antidiabetics, oral*	Increased antidiabetic effect.
Beta-adrenergic blocking agents*	Possible increased difficulty in regulating blood sugar levels.
Bismuth subsalicylate	Increased insulin effect. May require dosage adjustment.
Carteolol	Hypoglycemic effects may be prolonged.
Contraceptives, oral*	Decreased insulin effect.
Dexfenfluramine	May require dosage change as weight loss occurs.
Diuretics, thiazide*	Decreased insulin effect.
Furosemide	Decreased insulin effect.
Insulin analogs	May require dosage adjustment.
Monoamine oxidase (MAO) inhibitors*	Increased insulin effect.
Nicotine	Increased insulin effect.
Oxyphenbutazone	Increased insulin effect.
Phenylbutazone	Increased insulin effect.
Salicylates*	Increased insulin effect.
Smoking deterrents	May require insulin dosage adjustment.
Sulfa drugs*	Increased insulin effect.
Tetracyclines*	Increased insulin effect.
Thyroid hormones*	Decreased insulin effect.

POSSIBLE INTERACTION WITH OTHER SUBSTANCES

INTERACTS WITH	COMBINED EFFECT
Alcohol:	Increased insulin effect. Blood sugar problems. Avoid.
Beverages:	None expected.
Cocaine:	Unknown. Avoid.
Foods:	None expected. Follow your diabetic diet instructions.
Marijuana:	Consult doctor.
Tobacco:	Decreased insulin absorption. Avoid.

*See Glossary

INSULIN ANALOGS

GENERIC AND BRAND NAMES

INSULIN ASPART
- **Novolog**
- **Novolog FlexPen**
- **Novolog FlexTouch**
- **Novolog Mix 50/50**

INSULIN DETEMIR
- **Levemir**
- **Levemir FlexTouch**

INSULIN GLARGINE
- **Lantus**
- **Lantus OptiClik**
- **Lantus Solostar Pen**
- **Toujeo**

INSULIN GLULISINE
- **Apidra**
- **Apidra SoloStar**

INSULIN LISPRO
- **Humalog**
- **Humalog Mix 50/50**
- **Humalog Mix 75/25**

BASIC INFORMATION

Habit forming? No
Prescription needed? Yes
Available as generic? No
Drug class: Antidiabetic

USES

Treats diabetes, a metabolic disorder, in which patients have high levels of sugar in their blood. The drug keeps blood sugar levels from going too high after eating.

DOSAGE & USAGE INFORMATION

How to take:
Injection—Inject under the skin. Use disposable, sterile needles or disposable pen. Rotate injection sites. Carefully follow instructions provided with the drug.

Continued next column

OVERDOSE

SYMPTOMS:
Low blood sugar (hypoglycemia)—See symptoms next column under Infrequent.
WHAT TO DO:
- **Eat some type of sugar immediately, such as glucose tablets.**
- **If person is unconscious, give glucagon if you have it and know how to use it.**
- **Dial 911 (emergency) for medical help or call poison control center 1-800-222-1222 for instructions.**
- **See emergency information on last 3 pages of this book.**

When to take:
At the same time(s) each day. If taken at mealtime, use it within a 15 minute period prior to the meal. May need to be taken in combination with a long-acting insulin to prevent hyperglycemia.

If you forget a dose:
Follow your doctor's instructions. If unsure, consult your doctor or pharmacist.

What drug does:
These drugs are rapid- or fast-acting and work quickly in the body after injection. They are variations (analogs) of human insulin and more closely mimic the time action of natural insulin that comes from the pancreas.

Time lapse before drug works:
30 minutes to 1 hour (which is faster than regular insulin).

Don't take with:
Any other medicine or any dietary supplement without consulting your doctor or pharmacist.

POSSIBLE ADVERSE REACTIONS OR SIDE EFFECTS

SYMPTOMS	WHAT TO DO
Life-threatening:	
Rare, severe allergic reaction (anaphylaxis)—difficulty breathing, hives, itching, swelling (throat, face), vomiting, dizziness.	Seek emergency treatment immediately.
Common:	
None expected.	
Infrequent:	
• Symptoms of low blood sugar—nervousness, hunger (excessive), cold sweats, rapid pulse, anxiety, cold skin, chills, confusion, concentration loss, drowsiness, headache, nausea, weakness, shakiness, vision changes.	Seek treatment (eat some form of quick-acting sugar—glucose tablets, sugar, fruit juice, corn syrup, honey).
• Symptoms of high blood sugar—increased urination, unusual thirst, dry mouth, drowsiness, flushed or dry skin, fruit-like breath odor, appetite loss, stomach pain or vomiting, tiredness, trouble breathing, increased blood sugar level.	Seek emergency treatment immediately.

• Swelling or redness, itch or warmth at injection site, other skin changes at injection site (e.g., thinning or thickened skin).	Continue. Call doctor when convenient.

Rare:

Dry mouth, excessive thirst, weak or fast pulse, heartbeat irregularities, mental or mood changes, nausea or vomiting, unusual tiredness or weakness, muscle cramps (may be symptoms of hypokalemia).	Continue, but call doctor right away.

WARNINGS & PRECAUTIONS

Don't take if:
If you are allergic to insulin.

Before you start, consult your doctor if:
- Your diagnosis and dose schedule are not established or you don't know how to deal with overdose emergencies.
- You take MAO inhibitors.*
- You have hypoglycemia, liver or kidney disease or low thyroid function.

Over age 60:
Insulin requirements may change. The family should notify the doctor if abnormal behavior or confusion occurs in an older person.

Pregnancy:
Decide with your doctor if drug benefits justify risk to unborn baby. Risk category B for aspart, lispro, detemir and category C for glargine, glulisine (see page xviii).

Breastfeeding; Lactation; Nursing Mothers:
Drugs in this group are generally acceptable with breastfeeding. Discuss risks and benefits with your doctor.

Infants & children up to age 18:
Follow instructions provided by your child's doctor.

Prolonged use:
Talk to your doctor about the need for follow-up medical exams or lab tests.

Skin & sunlight:
No problems expected.

Driving, piloting or hazardous work:
No problems expected after dose is established. Need to be cautious for signs of hypoglycemia.

Discontinuing:
Don't discontinue without doctor's advice, even though symptoms diminish or disappear.

Others:
- Diet and exercise affect how much insulin you need. Work with your doctor to determine accurate dose. Monitor your glucose levels as directed.
- Notify your doctor if you have a fever, infection, diarrhea, or experience vomiting.
- Advise any doctor, dentist or pharmacist whom you consult that you take this medicine.
- Never freeze insulin.
- Wear or carry medical ID that indicates you have diabetes and take insulin.
- You and your family should educate yourselves about diabetes and learn to recognize hypoglycemia and treat it with sugar or glucagon.
- May interfere with the accuracy of some medical tests.

POSSIBLE INTERACTION WITH OTHER DRUGS

GENERIC NAME OR DRUG CLASS	COMBINED EFFECT
Antidiabetics, oral*	Increased antidiabetic effect.
Hyperglycemia-causing agents*	May need increased dosage of insulin.
Hypoglycemia-causing agents*	May need decreased dosage of insulin.
Insulin	May need dosage adjustment.
Sympatholytics*	Increased insulin effect.
Smoking deterrents	May require insulin dosage adjustment.

POSSIBLE INTERACTION WITH OTHER SUBSTANCES

INTERACTS WITH	COMBINED EFFECT
Alcohol:	Increased insulin effect. Blood sugar problems. Avoid.
Beverages:	None expected.
Cocaine:	Unknown. Avoid.
Foods:	None expected. Follow your diabetic diet instructions.
Marijuana:	Consult doctor.
Tobacco:	Decreased insulin absorption. Avoid.

***See Glossary**

INTEGRASE INHIBITORS

GENERIC AND BRAND NAMES

DOLUTEGRAVIR
- **Tivicay**
- **Triumeq**

ELVITEGRAVIR
- **Stribild**
- **Vitekta**

RALTEGRAVIR
- **Dutrebis**
- **Isentress**

BASIC INFORMATION

Habit forming? No
Prescription needed? Yes
Available as generic? No
Drug class: Antiviral agent

USES

Treatment of human immunodeficiency virus (HIV). HIV is the virus that causes acquired immunodeficiency syndrome (AIDS).

DOSAGE & USAGE INFORMATION

How to take:
- Tablet—Swallow whole with liquid. Dolutegravir and raltegravir may be taken with or without food. Take elvitegravir with food.
- Chewable tablet—Chew or swallow tablet whole.
- Suspension—Follow instructions provided with prescription.

When to take:
Once or twice a day (as per prescription) at the same time(s) each day.

If you forget a dose:
Take as soon as you remember. If it is almost time for the next dose, wait for that dose (don't double this dose) and resume regular schedule.

Continued next column

OVERDOSE

SYMPTOMS:
Unknown
WHAT TO DO:
If person uses much larger amount than prescribed or if accidentally swallowed, call poison control center 1-800-222-1222 for instructions or dial 911 (emergency) for help.

What drug does:
It helps control HIV infection by inhibiting an enzyme that is required for HIV replication. This may reduce the amount of HIV in the blood and may increase immune system cells. The drug does not cure HIV or AIDS.

Time lapse before drug works:
May require several weeks or months before full benefits are apparent.

Don't take with:
Any other medicine or any dietary supplement without consulting your doctor or pharmacist. This is very important with HIV drugs.

POSSIBLE ADVERSE REACTIONS OR SIDE EFFECTS

SYMPTOMS	WHAT TO DO
Life-threatening:	
None expected.	
Common:	
Diarrhea, nausea, headache, fever.	Continue. Call doctor when convenient.
Infrequent:	
Lightheadedness or dizziness, mild stomach pain, muscle or joint pain, tiredness, vomiting, dark urine, increased hunger or thirst.	Continue. Call doctor when convenient.
Rare:	
• Chest pain, fast heartbeat, shortness of breath.	Discontinue. Call doctor right away.
• Bruising or bleeding, signs of infection, confusion, mood changes, decreased urination, yellow skin or eyes.	Continue, but call doctor right away.
• Body aches, loss of appetite, a feeling of discomfort, body fat increases or fat moves to different areas of body, bloating, weight gain, excessive sweating, swelling, drowsiness, constipation, other unexplained symptoms.	Continue. Call doctor when convenient.

WARNINGS & PRECAUTIONS

Don't take if:
You are allergic to integrase inhibitors. Allergic reactions are rare, but may occur.

Before you start, consult your doctor if:
- You are allergic to any medicine, food or other substance, or have a family history of allergies.
- You have a muscle disorder.
- You have kidney or liver disease.

Over age 60:
Adverse reactions and side effects may be more frequent and severe than in younger persons.

Pregnancy:
Decide with your doctor if drug benefits justify risk to unborn baby. Risk category B for dolutegravir and elvitegravir; category C for raltegravir (see page xviii).

Breastfeeding; Lactation; Nursing Mothers:
It is not recommended that HIV-infected mothers breastfeed. Consult your doctor.

Infants & children up to age 18:
- Dolutegravir is approved for ages over 12 (with certain restrictions). Consult doctor.
- Elvitegravir not approved for ages under 18.
- Raltegravir is approved for children over 4 weeks of age depending on child's weight and form of drug. Consult doctor.

Prolonged use:
- Long-term effects of using this drug have not been established.
- Talk to your doctor about the need for follow-up blood tests and liver function studies.

Skin & sunlight:
No problems expected.

Driving, piloting or hazardous work:
No problems expected.

Discontinuing:
Don't discontinue without doctor's advice.

Others:
- Advise any doctor, dentist or pharmacist whom you consult that you take this medicine.
- Taking this drug does not prevent you from passing HIV to another person through sexual contact or sharing needles. Avoid sexual contact or practice safe sex (e.g., using condoms) to help prevent the transmission of HIV. Never share or re-use needles. If you have questions, ask your doctor for advice.
- Consult your doctor right away if you develop a new infection (e.g., fever, chills, sore throat or other symptoms).
- Consult your doctor right away if you develop muscle pain, especially if you take HMG-CoA reductase inhibitors or fibrates.
- Brand name Stribild contains four drugs. One is cobicistat* which enhances drug's potency.
- Take drug daily as prescribed. Do not increase or decrease dosage of drug without doctor's approval.

POSSIBLE INTERACTION WITH OTHER DRUGS

GENERIC NAME OR DRUG CLASS	COMBINED EFFECT
Antacids*	Decreased effect of integrase inhibitor.
Other drugs	Increased or decreased effect of an integrase inhibitor or other drug. Consult doctor.
Rifampin	Decreased effect of integrase inhibitor.

POSSIBLE INTERACTION WITH OTHER SUBSTANCES

INTERACTS WITH	COMBINED EFFECT
Alcohol:	None expected.
Beverages:	None expected.
Cocaine:	Unknown. Avoid.
Foods:	None expected.
Marijuana:	Consult doctor.
Tobacco:	None expected.

***See Glossary**

INTERMITTENT CLAUDICATION AGENTS

GENERIC AND BRAND NAMES

CILOSTAZOL
Pletal

BASIC INFORMATION

Habit forming? No
Prescription needed? Yes
Available as generic? No
Drug class: Vasodilator

USES

- Treats intermittent claudication (leg pain caused by poor circulation).
- Improves blood vessel function.
- Treatment for other disorders as determined by your doctor.

DOSAGE & USAGE INFORMATION

How to take:
Tablet—Swallow with liquid. If you can't swallow tablet whole, ask your pharmacist for advice.

When to take:
Twice a day. Take one half hour before or 2 hours after breakfast and dinner.

If you forget a dose:
Take as soon as you remember. If it is almost time for the next dose, wait for that dose (don't double this dose) and resume regular schedule.

What drug does:
Exact way it works is unknown. It increases blood flow by relaxing and expanding blood vessel walls. It keeps blood from clotting.

Time lapse before drug works:
3-4 weeks, but takes 12 weeks for full benefit.

Don't take with:
Any other medicine, herbal remedy or dietary supplements without consulting your doctor or pharmacist.

OVERDOSE

SYMPTOMS:
Severe headache, severe diarrhea, dizziness, change in heartbeat, nausea or vomiting, flushing, hot face.
WHAT TO DO:
- **Dial 911 for all medical emergencies or call poison control center 1-800-222-1222 for instructions.**
- **See emergency information on last 3 pages of this book.**

POSSIBLE ADVERSE REACTIONS OR SIDE EFFECTS

SYMPTOMS	WHAT TO DO
Life-threatening: None expected.	
Common:	
• Fever, rapid or irregular heartbeat.	Discontinue. Call doctor right away.
• Back pain, gas, dizziness, cough, diarrhea, headache, muscle stiffness or pain, runny or stuffy nose, sore throat.	Continue. Call doctor when convenient.
Infrequent:	
Fainting, stools are bloody or black or tarry, nausea or indigestion or heartburn (severe or ongoing), tongue swelling, stiff neck, nosebleeds, severe stomach pain or cramping, vomiting blood or material like coffee grounds, unusual bleeding or bruising.	Discontinue. Call doctor right away.
Rare:	
Burning feeling in throat or chest, bone pain, difficulty in swallowing, ringing or buzzing in the ears, hives, joint pain or stiffness, swelling (face, arms, fingers or lower legs).	Continue. Call doctor when convenient.

INTERMITTENT CLAUDICATION AGENTS

WARNINGS & PRECAUTIONS

Don't take if:
- You are allergic to cilostazol. Allergic reactions are rare, but may occur.
- You have congestive heart failure.

Before you start, consult your doctor if:
- You have heart disease, heart rhythm disorder or a bleeding disorder.
- You have liver or kidney disease.
- You suffer from migraines.
- You are a smoker.

Over age 60:
Adverse reactions and side effects may be more frequent and severe than in younger persons.

Pregnancy:
Decide with your doctor if drug benefits justify risk to unborn baby. Risk category C (see page xviii).

Breastfeeding; Lactation; Nursing Mothers:
This drug is generally not recommended with breastfeeding. Discuss risks and benefits with your doctor.

Infants & children up to age 18:
Safety and effectiveness in this age group have not been established. Consult doctor.

Prolonged use:
No problems expected.

Skin & sunlight:
No problems expected.

Driving, piloting or hazardous work:
Avoid if you feel dizzy or weak. Otherwise, no problems expected.

Discontinuing:
Don't discontinue without doctor's advice until you complete prescribed dose, even though symptoms diminish or disappear.

Others:
- Response to drug varies. If your symptoms don't improve after 3 weeks of use, consult doctor.
- Avoid smoking while taking this medication, as it may worsen your condition.
- Advise any doctor, dentist or pharmacist whom you consult that you take this medicine.

POSSIBLE INTERACTION WITH OTHER DRUGS

GENERIC NAME OR DRUG CLASS	COMBINED EFFECT
Enzyme inhibitors*	Increased effect of intermittent claudication agent.
Sertraline	Increased effect of intermittent claudication agent

POSSIBLE INTERACTION WITH OTHER SUBSTANCES

INTERACTS WITH	COMBINED EFFECT
Alcohol:	None expected.
Beverages: Grapefruit juice.	Increased effect of intermittent claudication agent.
Cocaine:	Unknown. Avoid.
Foods: Grapefruit.	Increased effect of intermittent claudication agent.
Marijuana:	Consult doctor.
Tobacco:	Decreased effect of intermittent claudication agent. Also, nicotine narrows your blood vessels. Avoid.

*See Glossary

IODOQUINOL

BRAND NAMES

Diiodohydroxyquin
Diodoquin
Diquinol
Yodoquinol
Yodoxin

BASIC INFORMATION

Habit forming? No
Prescription needed? Yes
Available as generic? Yes
Drug class: Antiprotozoal, antiparasitic

USES

Treatment for intestinal amebiasis and balantidiasis.

DOSAGE & USAGE INFORMATION

How to take:
Tablet—Mix with applesauce or chocolate syrup if unable to swallow tablets.

When to take:
Three times daily after meals for 20 days. Treatment may be repeated after 2 to 3 weeks.

If you forget a dose:
Take as soon as you remember. If it is almost time for the next dose, wait for that dose (don't double this dose) and resume regular schedule.

What drug does:
Kills amoeba (microscopic parasites) in intestinal tract.

Time lapse before drug works:
May require full course of treatment (20 days) to cure.

Don't take with:
Any other medicine or any dietary supplement without consulting your doctor or pharmacist.

OVERDOSE

SYMPTOMS:
Unknown.
WHAT TO DO:
If person uses much larger amount than prescribed or if accidentally swallowed, call poison control center 1-800-222-1222 for instructions or dial 911 (emergency) for help.

POSSIBLE ADVERSE REACTIONS OR SIDE EFFECTS

SYMPTOMS	WHAT TO DO
Life-threatening:	
None expected.	
Common:	
Diarrhea, nausea, vomiting, abdominal pain.	Continue. Call doctor when convenient.
Infrequent:	
• Clumsiness, rash, hives, itching, blurred vision, muscle pain, numbness or tingling in hands or feet, chills, fever, weakness.	Discontinue. Call doctor right away.
• Swelling of neck (thyroid gland).	Continue. Call doctor when convenient.
Rare:	
Dizziness, headache, rectal itching.	Continue. Call doctor when convenient.

WARNINGS & PRECAUTIONS

Don't take if:
You are allergic to iodoquinol. Allergic reactions are rare, but may occur.

Before you start, consult your doctor if:
- You have optic atrophy or thyroid disease.
- You have kidney or liver disease.

Over age 60:
Adverse reactions and side effects may be more frequent and severe than in younger persons.

Pregnancy:
Decide with your doctor if drug benefits justify risk to unborn baby. Risk category not designated (see page xviii).

Breastfeeding; Lactation; Nursing Mothers:
This drug is generally not recommended with breastfeeding. Discuss risks and benefits with your doctor.

Infants & children up to age 18:
Follow instructions provided by your child's doctor.

Prolonged use:
Not recommended.

Skin & sunlight:
No problems expected.

Driving, piloting or hazardous work:
No problems expected.

Discontinuing:
Don't discontinue without consulting doctor.

Others:
- Thyroid tests may be inaccurate for as long as 6 months after discontinuing iodoquinol treatment.
- May interfere with the accuracy of some medical tests.

POSSIBLE INTERACTION WITH OTHER DRUGS

GENERIC NAME OR DRUG CLASS	COMBINED EFFECT
None expected.	

POSSIBLE INTERACTION WITH OTHER SUBSTANCES

INTERACTS WITH	COMBINED EFFECT
Alcohol:	None expected.
Beverages:	None expected.
Cocaine:	Unknown. Avoid.
Foods:	Taking with food may decrease gastrointestinal side effects.
Marijuana:	Consult doctor.
Tobacco:	None expected.

IPRATROPIUM

BRAND NAMES

Apo-Ipravent
Atrovent
Atrovent Inhalation Aerosol
Combivent Respimat
Duoneb
Kendral-Ipratropium

BASIC INFORMATION

Habit forming? No
Prescription needed? Yes
Available as generic? Yes
Drug class: Bronchodilator, anticholinergic

USES

- Treats asthma, bronchitis and emphysema.
- Should not be used alone for acute asthma attacks. May be used with inhalation forms of albuterol or fenoterol.
- May be used to treat rhinorrhea (runny nose).

DOSAGE & USAGE INFORMATION

How to use:
- Inhalation aerosol or solution—Carefully follow the printed instructions provided with the inhaler. Avoid contact with eyes.
- Nasal spray—Prime the nasal spray pump as directed, then spray in each nostril. Avoid contact with eyes.

When to take:
Inhalation form may be started at 2 inhalations, 4 times a day and increased by your doctor as needed. Usual dose of nasal form is 2 sprays in each nostril, 2 to 3 times a day.

If you forget a dose:
Use as soon as you remember. If it is almost time for the next dose, wait for that dose (don't double this dose) and resume regular schedule.

What drug does:
Dilates (opens or widens) bronchial tubes or nasal passages by direct effect on them.

Continued next column

OVERDOSE

SYMPTOMS:
Unknown.
WHAT TO DO:
If person uses much larger amount than prescribed or if accidentally swallowed, call poison control center 1-800-222-1222 for instructions or dial 911 (emergency) for help.

Time lapse before drug works:
Inhalation form may take 2 to 3 days for full effectiveness. Effect of nasal form may begin right away or take a few days.

Don't take with:
Any other medicines (including over-the-counter drugs such as cough and cold medicines, laxatives, antacids, diet pills, caffeine, nose drops or vitamins) without consulting your doctor or pharmacist.

POSSIBLE ADVERSE REACTIONS OR SIDE EFFECTS

SYMPTOMS	WHAT TO DO
Life-threatening:	
Rare, severe allergic reaction (anaphylaxis)—difficulty breathing, hives, itching, swelling (throat, face), vomiting, dizziness.	Seek emergency treatment immediately.
Common:	
Cough, dry mouth, unpleasant taste.	Continue. Call doctor when convenient.
Infrequent:	
Blurred vision or other vision changes, difficult urination, stuffy nose, sweating, tremors, weakness, dizziness, nervousness, headache.	Continue. Call doctor when convenient.
Rare:	
Skin rash or hives, ongoing constipation, increased wheezing, chest tightness or difficulty in breathing, pounding heartbeat, stomach pain or bloating, severe eye pain.	Discontinue. Call doctor right away.

WARNINGS & PRECAUTIONS

Don't take if:
You are sensitive to ipratropium, belladonna, atropine or soybeans, soy lecithin or peanuts.

Before you start, consult your doctor if:
- You have prostate trouble.
- You have glaucoma.
- You have difficulty in urinating.

Over age 60:
Adverse reactions and side effects may be more frequent and severe than in younger persons.

Pregnancy:
Decide with your doctor if drug benefits justify risk to unborn baby. Risk category B (see page xviii).

Breastfeeding; Lactation; Nursing Mothers:
This drug is generally acceptable with breastfeeding. Discuss risks and benefits with your doctor.

Infants & children up to age 18:
Inhalation form may be used in children over age 12. Nasal spray may be used in children over age 6. Follow instructions provided by your child's doctor.

Prolonged use:
- For inhalation form, see your doctor to verify drug's effectiveness and for dose adjustment if needed and for eye pressure exams.
- Nasal form should not be used for more than 4 days.

Skin & sunlight:
No problems expected.

Driving, piloting or hazardous work:
Avoid if you feel confused, drowsy or dizzy.

Discontinuing:
No special problems expected.

Others:
- Advise any doctor, dentist or pharmacist whom you consult that you take this medicine.
- Do not increase the drug dosage without your doctor's approval.
- Allow 5-minute intervals between ipratropium inhalations and inhalations of cromolyn, cortisone or other inhalant medicines.

POSSIBLE INTERACTION WITH OTHER DRUGS

GENERIC NAME OR DRUG CLASS	COMBINED EFFECT
Anticholinergics*	Increased anticholinergic effect.
Cromolyn (inhalation form)	Wait 5 minutes after using ipratropium before using cromolyn.

POSSIBLE INTERACTION WITH OTHER SUBSTANCES

INTERACTS WITH	COMBINED EFFECT
Alcohol:	None expected.
Beverages:	None expected.
Cocaine:	Unknown. Avoid.
Foods:	None expected.
Marijuana:	Consult doctor.
Tobacco:	None expected, but smoking should be avoided.

***See Glossary**

IRON SUPPLEMENTS

GENERIC AND BRAND NAMES

See full list of generic and brand names in the *Generic and Brand Name Directory*, page 865.

BASIC INFORMATION

Habit forming? No
Prescription needed? No
Available as generic? Yes
Drug class: Mineral supplement (iron)

USES

Treatment for dietary iron deficiency or iron-deficiency anemia from other causes.

DOSAGE & USAGE INFORMATION

How to take:

- Tablet, capsule or syrup—Swallow with liquid or food to lessen stomach irritation. If you can't swallow whole, crumble tablet or open capsule and take with liquid or food. Place medicine far back on tongue to avoid staining teeth.
- Extended-release capsule—Swallow whole with liquid. Do not crush.
- Chewable tablet—Chew well before swallowing.
- Liquid—Dilute dose in beverage before swallowing and drink through a straw.

When to take:
1 hour before or 2 hours after eating.

If you forget a dose:
Take as soon as you remember. If it is almost time for the next dose, wait for that dose (don't double this dose) and resume regular schedule.

Continued next column

OVERDOSE

SYMPTOMS:

- **Moderate overdose—Stomach pain, vomiting, diarrhea, black stools, lethargy.**
- **Serious overdose—Weakness and collapse, pallor, weak and rapid heartbeat, shallow breathing, convulsions, coma.**

WHAT TO DO:

- **Dial 911 for all medical emergencies or call poison control center 1-800-222-1222 for instructions.**
- **See emergency information on last 3 pages of this book.**

What drug does:
Stimulates bone marrow's production of hemoglobin (red blood cell pigment that carries oxygen to body cells).

Time lapse before drug works:
3 to 7 days. May require 3 weeks for maximum benefit.

Don't take with:
Any other medicine or any dietary supplement without consulting your doctor or pharmacist.

POSSIBLE ADVERSE REACTIONS OR SIDE EFFECTS

SYMPTOMS	WHAT TO DO
Life-threatening:	
In case of overdose, see previous column.	
Common:	
• Stomach pain that is continuing.	Discontinue. Call doctor right away.
• Dark green or black stool, teeth stained with liquid iron, constipation, diarrhea, mild nausea or vomiting.	Continue. Call doctor when convenient.
Infrequent:	
None expected.	
Rare:	
• Throat pain on swallowing, chest pain, cramps, blood in stool or black stool that has sticky consistency.	Discontinue. Call doctor right away.
• Darkened urine, heartburn.	Continue. Call doctor when convenient.

WARNINGS & PRECAUTIONS

Don't take if:

- You are allergic to any iron supplement or tartrazine dye. Allergic reactions are rare, but may occur.
- You take iron injections.
- You have acute hepatitis, hemosiderosis or hemochromatosis (conditions involving excess iron in body).
- You have hemolytic anemia.

Before you start, consult your doctor if:

- You plan to become pregnant while on medication.
- You have had stomach surgery.
- You have had peptic ulcer, enteritis or colitis.

Over age 60:
No special problems expected when taking recommended normal daily amounts.

Pregnancy:
Decide with your doctor if drug benefits justify risk to unborn baby. Risk category not designated (see page xviii).

Breastfeeding; Lactation; Nursing Mothers:
Drugs in this group are generally acceptable with breastfeeding. Discuss risks and benefits with your doctor.

Infants & children up to age 18:
Follow instructions provided by your child's doctor. Overdose common and dangerous. Keep out of children's reach.

Prolonged use:
- May cause hemochromatosis (iron storage disease) with bronze skin, liver damage, diabetes, heart problems and impotence.
- Talk to your doctor about the need for follow-up medical examinations or laboratory studies to check complete blood counts (white blood cell count, platelet count, red blood cell count, hemoglobin, hematocrit), serum iron, total iron-binding capacity.

Skin & sunlight:
No problems expected.

Driving, piloting or hazardous work:
No problems expected.

Discontinuing:
May be unnecessary to finish medicine. Follow doctor's instructions.

Others:
- Liquid form stains teeth. Mix with water or juice to lessen the effect. Brush with baking soda or hydrogen peroxide to help remove stain.
- Some products contain tartrazine dye. Avoid, especially if you are allergic to aspirin.
- May interfere with the accuracy of some medical tests.
- If using extended-release form or coated tablet and your stools don't turn black, consult doctor. The tablet may not be breaking down, and an underdose may result.

POSSIBLE INTERACTION WITH OTHER DRUGS

GENERIC NAME OR DRUG CLASS	COMBINED EFFECT
Acetohydroxamic acid	Decreased effects of both drugs.
Antacids*	Poor iron absorption.
Chloramphenicol	Decreased effect of iron. Interferes with formation of red blood cells and hemoglobin.
Cholestyramine	Decreased iron effect.
Etidronate	Decreased etidronate effect. Take at least 2 hours after iron supplement.
H_2 antagonists*	Decreased iron effect.
Iron supplements, other*	Possible excess iron storage in liver.
Proton pump inhibitors	May decrease effect of iron supplement.
Tetracyclines*	Decreased tetracycline effect. Take iron 3 hours before or 2 hours after taking tetracycline.
Vitamin E	Decreased iron and vitamin E effect.
Zinc supplements	Increased need for zinc.

POSSIBLE INTERACTION WITH OTHER SUBSTANCES

INTERACTS WITH	COMBINED EFFECT
Alcohol:	Increased iron absorption. May cause organ damage. Avoid or use in moderation.
Beverages: Milk, tea.	Decreased iron effect.
Cocaine:	Unknown. Avoid.
Foods: Dairy foods, eggs, whole-grain bread and cereal.	Decreased iron effect.
Marijuana:	Consult doctor.
Tobacco:	None expected.

*See Glossary

ISONIAZID

BRAND NAMES

INH
Isotamine
Laniazid
Nydrazid
PMS Isoniazid
Rifamate
Tubizid

BASIC INFORMATION

Habit forming? No
Prescription needed? Yes
Available as generic? Yes
Drug class: Antitubercular

USES

Kills tuberculosis germs.

DOSAGE & USAGE INFORMATION

How to take:
- Tablet—Swallow with liquid to lessen stomach irritation.
- Syrup—Follow label directions.

When to take:
At the same time each day.

If you forget a dose:
Take as soon as you remember. If it is almost time for the next dose, wait for that dose (don't double this dose) and resume regular schedule.

What drug does:
Interferes with TB germ metabolism. Eventually destroys the germ.

Time lapse before drug works:
3 to 6 months. You may need to take drug as long as 2 years.

Don't take with:
Any other medicine or any dietary supplement without consulting your doctor or pharmacist.

OVERDOSE

SYMPTOMS:
Nausea, vomiting, dizziness, slurred speech, hallucinations, blurred vision, difficult breathing, convulsions, stupor, coma.
WHAT TO DO:
- **Dial 911 for all medical emergencies or call poison control center 1-800-222-1222 for instructions.**
- **See emergency information on last 3 pages of this book.**

POSSIBLE ADVERSE REACTIONS OR SIDE EFFECTS

SYMPTOMS	WHAT TO DO
Life-threatening: In case of overdose, see previous column.	
Common: Unsteadiness, tingling or numbness or pain in arms and legs, dark urine, appetite loss, unusual tiredness or weakness, yellow skin or eyes, nausea, vomiting.	Discontinue. Call doctor right away.
Infrequent: Changes in vision, eye pain, fever, sore throat, seizures, mood or mental changes, depression, skin rash easy bruising or bleeding.	Discontinue. Call doctor right away.
Rare: Diarrhea, stomach pain.	Continue. Call doctor when convenient.

WARNINGS & PRECAUTIONS

Don't take if:
You are allergic to isoniazid. Allergic reactions are rare, but may occur.

Before you start, consult your doctor if:
- You plan to become pregnant within medication period.
- You are allergic to ethionamide, pyrazinamide or nicotinic acid.
- You drink alcohol.
- You have liver or kidney disease.
- You have epilepsy, diabetes or lupus.

Over age 60:
Adverse reactions and side effects may be more frequent and severe than in younger persons.

Pregnancy:
Decide with your doctor if drug benefits justify risk to unborn baby. Risk category C (see page xviii).

Breastfeeding; Lactation; Nursing Mothers:
This drug is generally acceptable with breastfeeding. Discuss risks and benefits with your doctor.

Infants & children up to age 18:
Follow instructions provided by your child's doctor.

Prolonged use:
- Numbness and tingling of hands and feet.
- Talk to your doctor about the need for follow-up medical examinations or laboratory studies to check liver function, eyes.

Skin & sunlight:
No problems expected.

Driving, piloting or hazardous work:
Avoid if you feel dizzy. Otherwise, no problems expected.

Discontinuing:
Don't discontinue without doctor's advice until you complete prescribed dose, even though symptoms diminish or disappear.

Others:
- Diabetic patients may have false blood sugar tests.
- Periodic liver function tests and laboratory blood studies recommended.
- Prescription for vitamin B-6 (pyridoxine) recommended to prevent nerve damage.
- Advise any doctor, dentist or pharmacist whom you consult that you take this medicine.

POSSIBLE INTERACTION WITH OTHER DRUGS

GENERIC NAME OR DRUG CLASS	COMBINED EFFECT
Acetaminophen	Increased risk of liver damage.
Adrenocorticoids, systemic	Decreased isoniazid effect.
Alfentanil	Prolonged duration of alfentanil effect (undesirable).
Antacids* (aluminum-containing)	Decreased absorption of isoniazid.
Anticholinergics*	May increase pressure within eyeball.
Anticoagulants, oral*	Increased anticoagulant effect.
Antidiabetics*	Increased antidiabetic effect.
Antihypertensives*	Increased antihypertensive effect.
Antivirals, HIV/AIDS*	Increased risk of peripheral neuropathy.
Carbamazepine	Increased risk of liver damage.
Cycloserine	Increased risk of central nervous system effects.
Disulfiram	Increased effect of disulfiram.
Laxatives*	Decreased absorption and effect of isoniazid.
Hepatotoxics*	Increased risk of liver damage.
Ketoconazole	Increased risk of liver damage.
Narcotics*	Increased narcotic effect.
Phenytoin	Increased phenytoin effect.
Pyridoxine (Vitamin B-6)	Decreases risk of nerve damage in extremities.
Rifampin	Increased isoniazid toxicity to liver.
Sedatives*	Increased sedative effect.
Stimulants*	Increased stimulant effect.

POSSIBLE INTERACTION WITH OTHER SUBSTANCES

INTERACTS WITH	COMBINED EFFECT
Alcohol:	Increased incidence of liver disease and seizures.
Beverages:	None expected.
Cocaine:	Unknown. Avoid.
Foods: Swiss or Cheshire cheese, fish.	Red or itching skin, fast heartbeat. Seek emergency treatment.
Marijuana:	Consult doctor.
Tobacco:	No interactions expected, but tobacco may slow body's recovery.

***See Glossary**

ISOTRETINOIN

BRAND NAMES

Absorica
Amnesteem
Claravis
Myorisan
Sotret
Zenatane

BASIC INFORMATION

Habit forming? No
Prescription needed? Yes
Available as generic? Yes
Drug classification: Antiacne (systemic)

USES

- Decreases cystic acne formation in severe cases.
- Treats certain other skin disorders involving an overabundance of outer skin layer.

DOSAGE & USAGE INFORMATION

How to take:
Capsule—Swallow whole with a glass of water or liquid (this helps decrease risk of esophagus irritation). Do not chew, crush or open capsule.

When to take:
Twice a day. Follow prescription directions.

If you forget a dose:
Take as soon as you remember. If it is almost time for the next dose, wait for that dose (don't double this dose) and resume regular schedule.

What drug does:
Reduces sebaceous gland activity and size.

Time lapse before drug works:
May require 15 to 20 weeks to experience full benefit.

Don't take with:
- Vitamin A or supplements containing Vitamin A.
- Any other medicine or any dietary supplement without consulting your doctor or pharmacist.

OVERDOSE

SYMPTOMS:
Dizziness, vomiting, headache, stomach pain, tingling or warmth of the skin, lip swelling, loss of balance or coordination.
WHAT TO DO:
- **Dial 911 for all medical emergencies or call poison control center 1-800-222-1222 for instructions.**
- **See emergency information on last 3 pages of this book.**

POSSIBLE ADVERSE REACTIONS OR SIDE EFFECTS

SYMPTOMS	WHAT TO DO
Life-threatening:	
Rare, severe allergic reaction (anaphylaxis)—difficulty breathing, hives, itching, swelling (throat, face), vomiting, dizziness.	Seek emergency treatment immediately.
Common:	
Burning or red or itching eyes, skin or lips have burning or pain or redness, nosebleeds, difficulty in moving, joint pain.	Discontinue. Call doctor right away.
Infrequent:	
Itchy skin, dry mouth or eyes or nose, hair thinning.	Continue. Call doctor when convenient.
Rare:	
• Severe stomach pain, bleeding gums, blurred vision, severe diarrhea, continuing headache, vomiting, eye pain, rectal bleeding, yellow skin or eyes, serious depression, mental changes, thoughts of suicide.	Discontinue. Call doctor right away.
• Mild headache, increased sensitivity to light, stomach upset, peeling of skin on palms or soles of feet.	Continue. Call doctor when convenient.

WARNINGS & PRECAUTIONS

Don't take if:
- You are allergic to isotretinoin, etretinate, tretinoin or vitamin A derivatives.
- You are pregnant or plan pregnancy.
- *You are even able to bear children. Read, understand and follow the patient information enclosed with your prescription.*

Before you start, consult your doctor if:
- You have diabetes.
- You or any member of family have high triglyceride levels in blood.
- You or family members have a history of severe depression.

Over age 60:
Adverse reactions and side effects may be more frequent and severe than in younger persons.

Pregnancy:
Consult doctor. Drug should not be used during pregnancy. Can cause harm to unborn baby. Risk category X (see page xviii).

Breastfeeding; Lactation; Nursing Mothers:
This drug is generally not recommended with breastfeeding. Discuss risks and benefits with your doctor.

Infants & children up to age 18:
Approved for ages over 12. Follow instructions provided by your child's doctor.

Prolonged use:
- Possible damage to cornea of the eye.
- Talk to your doctor about the need for follow-up medical examinations or laboratory studies to check complete blood counts (white blood cell count, platelet count, red blood cell count, hemoglobin, hematocrit), liver function, blood lipids, blood sugar.

Skin & sunlight:
May cause rash or intensify sunburn in areas exposed to sun or ultraviolet light (photosensitivity reaction). Avoid overexposure. Notify doctor if reaction occurs.

Driving, piloting or hazardous work:
Use caution if there is a decrease in your night vision or you are unable to see well. Consult doctor.

Discontinuing:
Single course of treatment is usually all that's needed. If second course required, wait 8 weeks after completing first course.

Others:
- Use only for severe cases of cystic acne that have not responded to less hazardous forms of acne treatment.
- May interfere with the accuracy of some medical tests.
- May cause bone problems (osteoporosis, fractures, delayed healing).
- Don't donate blood for at least 30 days after discontinuing medicine.
- Acne may worsen at the start of treatment.
- Contact lens wearers may experience discomfort during treatment with this drug.
- Contact doctor right away if a person taking this drug develops symptoms of depression, psychosis (severe mental problems) or has any suicide thoughts or suicide behaviors.
- Advise any doctor, dentist or pharmacist whom you consult that you take this medicine.
- If you are planning pregnancy or are at risk of pregnancy, don't take this drug.

POSSIBLE INTERACTION WITH OTHER DRUGS

GENERIC NAME OR DRUG CLASS	COMBINED EFFECT
Antiacne topical preparations* (other), cosmetics (medicated), skin preparations with alcohol, soaps or cleansers (abrasive)	Severe skin irritation.
Etretinate	Increased chance of toxicity of each drug.
Tetracyclines*	Increased risk of developing pseudotumor cerebri.*
Topical drugs or cosmetics	May interact with isotretinoin.
Tretinoin	Increased chance of toxicity.
Vitamin A	Additive toxic effect of each. Avoid.

POSSIBLE INTERACTION WITH OTHER SUBSTANCES

INTERACTS WITH	COMBINED EFFECT
Alcohol:	Increase in triglycerides in blood. Avoid.
Beverages:	None expected.
Cocaine:	Unknown. Avoid.
Foods:	None expected.
Marijuana:	Consult doctor.
Tobacco:	May decrease absorption of drug. Avoid tobacco during treatment.

***See Glossary**

KAOLIN & PECTIN

BRAND NAMES

Donnagel-MB
Kao-Con
Kaotin
Kapectolin
Kapectolin with Paregoric
K-C
K-P
K-Pek
Parepectolin

BASIC INFORMATION

Habit forming? No
Prescription needed? No
Available as generic? Yes
Drug class: Antidiarrheal

USES

Treats mild to moderate diarrhea. Used in conjunction with fluids, appropriate diet and rest. Treats symptoms only. Does not cure any disorder that causes diarrhea.

DOSAGE & USAGE INFORMATION

How to take:
Liquid—Swallow prescribed dosage (without diluting) after each loose bowel movement.

When to take:
After each loose bowel movement.

If you forget a dose:
Usually not taken on a routine schedule.

What drug does:
Makes loose stools less watery, but may not prevent loss of fluids.

Time lapse before drug works:
15 to 30 minutes.

Don't take with:
Any other medicine or any dietary supplement without consulting your doctor or pharmacist.

OVERDOSE

SYMPTOMS:
Unknown.
WHAT TO DO:
If person uses much larger amount than prescribed or if accidentally swallowed, call poison control center 1-800-222-1222 for instructions or dial 911 (emergency) for help.

POSSIBLE ADVERSE REACTIONS OR SIDE EFFECTS

SYMPTOMS	WHAT TO DO
Life-threatening: None expected.	
Common: None expected.	
Infrequent: None expected.	
Rare: Constipation (mild).	Continue. Call doctor when convenient.

WARNINGS & PRECAUTIONS

Don't take if:
You are allergic to kaolin or pectin. Allergic reactions are rare, but may occur.

Before you start, consult your doctor if:
- Patient is child or infant.
- You have any chronic medical problem with heart disease, peptic ulcer, asthma or others.
- You have fever over 101°F.

Over age 60:
No special problems expected.

Pregnancy:
Decide with your doctor if drug benefits justify risk to unborn baby. Risk category B (see page xviii).

Breastfeeding; Lactation; Nursing Mothers:
This drug is generally acceptable with breastfeeding. Discuss risks and benefits with your doctor.

Infants & children up to age 18:
Consult doctor before giving any medicine for diarrhea. Fluid loss caused by diarrhea in infants and children can cause serious dehydration.

Prolonged use:
Not recommended.

Skin & sunlight:
No problems expected.

Driving, piloting or hazardous work:
No problems expected.

Discontinuing:
May be unnecessary to finish medicine. Follow doctor's instructions.

Others:
Consult doctor about fluids, diet and rest.

POSSIBLE INTERACTION WITH OTHER DRUGS

GENERIC NAME OR DRUG CLASS	COMBINED EFFECT
Digoxin	Decreases absorption of digoxin. Separate doses by at least 2 hours.
Lincomycins*	Decreases absorption of lincomycin. Separate doses by at least 2 hours.
All other oral medicines	May decrease absorption of other medicines. Separate doses by at least 2 hours.

POSSIBLE INTERACTION WITH OTHER SUBSTANCES

INTERACTS WITH	COMBINED EFFECT
Alcohol:	Increased diarrhea. Prevents action of kaolin and pectin.
Beverages:	None expected.
Cocaine:	Unknown. Avoid.
Foods:	None expected.
Marijuana:	Consult doctor.
Tobacco:	None expected.

***See Glossary**

KERATOLYTICS

GENERIC AND BRAND NAMES

See full list of generic and brand names in the *Generic and Brand Name Directory*, page 866.

BASIC INFORMATION

Habit forming? No
Prescription needed? Yes, on some.
Available as generic? Yes
Drug class: Keratolytic, antiacne (topical), antiseborrheic

USES

Treatment for skin disorders such as acne, psoriasis, ichthyosis, keratosis, folliculitis, flat warts, eczema, urticaria, calluses, corns, seborrheic dermatitis, dandruff and others.

DOSAGE & USAGE INFORMATION

How to use:
Cream, gel, lotion, ointment, pads, plaster, shampoo, soap, topical solution, suspension—Always follow instructions on the label or use as directed by your doctor.

When to use:
At the same time each day or as needed.

If you forget an application:
Use as soon as you remember. If it is almost time for the next dose, wait for that dose (don't double this dose) and resume regular schedule.

What drug does:
Keratolytics are drugs that soften, loosen and remove keratin (the tough outer layer of the skin).

Time lapse before drug works:
2 to 3 weeks. May require 6 weeks for maximum improvement.

Don't use with:
- Benzoyl peroxide. Apply 12 hours apart.
- Any other topical medicine without consulting your doctor or pharmacist.

OVERDOSE

SYMPTOMS:
Unknown.
WHAT TO DO:
If person uses much larger amount than prescribed or if accidentally swallowed, call poison control center 1-800-222-1222 for instructions or dial 911 (emergency) for help.

POSSIBLE ADVERSE REACTIONS OR SIDE EFFECTS

SYMPTOMS	WHAT TO DO
Life-threatening:	
Rare, severe allergic reaction (anaphylaxis)—difficulty breathing, hives, itching, swelling (throat, face), vomiting, dizziness (with use of salicylic acid).	Seek emergency treatment immediately.
Common:	
Pigment change in treated area, warmth or stinging or peeling of skin, sensitivity to wind or cold.	Continue. call doctor when convenient.
Infrequent:	
Blistering or crusting or burning or swelling skin, skin irritation that begins after treatment.	Discontinue. Call or doctor right away.
Rare:	
Symptoms of drug absorbed by body (diarrhea, nausea, dizziness, headache, breathing difficulty, tiredness, weakness).	Discontinue. Call doctor right away.

WARNINGS & PRECAUTIONS

Don't take if:
- You are allergic to resorcinol or salicylic acid. Allergic reactions are rare, but may occur.
- You are sunburned or windburned or have an open skin wound, skin irritation or infection.

Before you start, consult your doctor if:
- You have eczema.
- You have diabetes.
- You have peripheral vascular disease (blood vessel disease).

Over age 60:
No problems expected.

Pregnancy:
Risk factors vary for drugs in this group. Always consult doctor. See category list on page xviii.

Breastfeeding; Lactation; Nursing Mothers:
Drugs in this group are generally acceptable with breastfeeding. Discuss risks and benefits with your doctor.

Infants & children up to age 18:
- Young children are more at risk for adverse effects.
- If prescribed, follow instructions provided by your child's doctor.
- Read nonprescription product's label to be sure it is approved for your child's age. Follow label instructions on dosage. Consult doctor or pharmacist if unsure.

Prolonged use:
No problems expected.

Skin & sunlight:
No special problems expected.

Driving, piloting or hazardous work:
No problems expected.

Discontinuing:
Follow your doctor's instructions or the directions on the label.

Others:
- Acne may get worse before improvement starts in 2 or 3 weeks. Don't wash face more than 2 or 3 times daily.
- Keep medicine away from mouth or eyes. If it accidentally gets into the eyes, flush immediately with clear water.
- Keep medicine away from heat or flame.

POSSIBLE INTERACTION WITH OTHER DRUGS

GENERIC NAME OR DRUG CLASS	COMBINED EFFECT
Antiacne topical preparations (other)	Severe skin irritation.
Cosmetics (medicated)	Severe skin irritation.
Skin preparations with alcohol	Severe skin irritation.
Soaps or cleansers (abrasive)	Severe skin irritation.

POSSIBLE INTERACTION WITH OTHER SUBSTANCES

INTERACTS WITH	COMBINED EFFECT
Alcohol:	None expected.
Beverages:	None expected.
Cocaine:	None expected.
Foods:	None expected.
Marijuana:	None expected.
Tobacco:	None expected.

LACOSAMIDE

BRAND NAMES

Vimpat

BASIC INFORMATION

Habit forming? Possibly
Prescription needed? Yes
Available as generic? No
Drug class: Anticonvulsant; antiepileptic

USES

- Treatment of partial-onset seizures in patients with epilepsy. It is used in combination with other antiepileptic drugs.
- Other uses as recommended by your doctor.

DOSAGE & USAGE INFORMATION

How to take:
- Tablet—Swallow the tablet with a liquid. May be taken with or without food, and on a full or empty stomach.
- Oral solution—Follow instructions on label.
- Injection—Given by medical professional.

When to take:
Tablet is usually taken twice a day at the same times each day. Your doctor will determine the best schedule. Dosages may be increased weekly to achieve maximum benefits.

If you forget a dose:
Take as soon as you remember. If it is almost time for the next dose, wait for that dose (don't double this dose) and resume regular schedule.

What drug does:
It helps stabilize electrical activity in the brain, but the exact way it controls seizures is unknown.

Time lapse before drug works:
May take several weeks for full effectiveness.

Don't take with:
Any other medicine or any dietary supplement without consulting your doctor or pharmacist.

OVERDOSE

SYMPTOMS:
Unknown (may be similar to side effects).
WHAT TO DO:
If person uses much larger amount than prescribed or if accidentally swallowed, call poison control center 1-800-222-1222 for instructions or dial 911 (emergency) for help.

POSSIBLE ADVERSE REACTIONS OR SIDE EFFECTS

SYMPTOMS	WHAT TO DO
Life-threatening: None expected.	
Common:	
• Dizziness, unsteady walk, shakiness or trembling, unusual drowsiness, lack of coordination.	Continue, but call doctor right away.
• Blurred or double vision, headache, nausea or vomiting.	Continue. Call doctor when convenient.
Infrequent:	
• Mood or mental changes, feeling sad or irritable, forgetfulness, itchy skin, tiredness, trouble sleeping or concentrating, unusual eye movements, depression.	Continue, but call doctor right away.
• Diarrhea, weakness, spinning sensation.	Continue. Call doctor when convenient.
Rare:	
• Tingling or prickling feelings, noises in ears, chills, fever, changes in heartbeat (fast, slow, irregular, pounding), unusual bleeding or bruising, yellow skin or eyes, new or worsening seizures, shortness of breath, fainting, behavior changes, over-excited.	Continue, but call doctor right away.
• Indigestion, heart-burn, dry mouth, constipation, muscle spasms.	Continue. Call doctor when convenient.

WARNINGS & PRECAUTIONS

Don't take if:
You are allergic to lacosamide. Allergic reactions are rare, but may occur.

Before you start, consult your doctor if:
- You have diabetes or kidney or liver problems.
- You have a history of any heart disorder or blood vessel problem.
- You have a history of mental or mood problems (such as depression) or suicidal thoughts or attempts.
- You have a condition that requires you to limit or avoid use of aspartame (oral solution form of drug contains aspartame).
- You are allergic to any medication, food or other substance.

Over age 60:
No special problems expected.

Pregnancy:
Decide with your doctor if drug benefits justify risk to unborn baby. Risk category C (see page xviii).

Breastfeeding; Lactation; Nursing Mothers:
This drug is generally acceptable with breastfeeding. Discuss risks and benefits with your doctor.

Infants & children up to age 18:
Safety and effectiveness in this age group have not been established. Consult doctor.

Prolonged use:
No special problems expected. Follow-up with your doctor on a regular basis to monitor your condition and check for drug side effects.

Skin & sunlight:
No problems expected.

Driving, piloting or hazardous work:
This drug may cause dizziness, coordination problems or blurred or double vision. Don't drive or pilot aircraft until you learn how medicine affects you. Don't work around dangerous machinery. Don't climb ladders or work in high places. The risk of dizziness increases if you drink alcohol.

Discontinuing:
Don't discontinue without doctor's approval due to risk of increased seizure activity. The dosage may need to be gradually decreased before stopping the drug completely.

Others:
- This drug cannot cure epilepsy and will only work to control seizures for as long as you continue to take it.
- Lacosamide is used with other anticonvulsant drugs and additional side effects may occur. If they do, consult your doctor.
- Advise any doctor, dentist or pharmacist whom you consult that you take this medicine.
- Rarely, antiepileptic drugs may lead to suicidal thoughts and behaviors. Call doctor right away if suicidal symptoms or unusual behaviors occur.
- Carry or wear medical identification that lists your seizure disorder and drugs you take.

POSSIBLE INTERACTION WITH OTHER DRUGS

GENERIC NAME OR DRUG CLASS	COMBINED EFFECT
QT interval prolongation-causing drugs*	Increased risk of cardiac (heart) side effects.

POSSIBLE INTERACTION WITH OTHER SUBSTANCES

INTERACTS WITH	COMBINED EFFECT
Alcohol:	Increased risk of dizziness. Avoid.
Beverages:	None expected.
Cocaine:	Unknown. Avoid.
Foods:	None expected.
Marijuana:	Consult doctor.
Tobacco:	None expected.

***See Glossary**

LAMOTRIGINE

BRAND NAMES

Lamictal
Lamictal Chewable Dispersible Tablet
Lamictal ODT
Lamictal XR

BASIC INFORMATION

Habit forming? No
Prescription needed? Yes
Available as generic? Yes
Drug class: Anticonvulsant, antiepileptic

USES

- Treatment for partial (focal) epileptic seizures. May be used in combination with other antiepileptic drugs.
- Treatment of primary generalized tonic-clonic (PGTC) seizures, also known as "grand mal" seizures.
- Maintenance therapy for bipolar disorder.

DOSAGE & USAGE INFORMATION

How to take:

- Tablet—Swallow whole with liquid. May be taken with or without food. Do not crush or chew tablet as it can have a bitter taste.
- Chewable dispersible tablet—Swallow whole or chew. If you chew the tablets, drink a small amount of water or diluted fruit juice to aid in swallowing. To mix it in a liquid, follow directions provided with prescription.
- Orally disintegrating tablet—Place tablet on tongue and let it dissolve. Can be taken with or without food or a liquid.
- Extended-release tablet—Swallow whole with liquid. Do not cut, chew or crush tablet.

When to take:
Your doctor will determine the best schedule. Dosages may be increased gradually over the first few weeks of use to achieve maximum benefits.

Continued next column

OVERDOSE

SYMPTOMS:
Severe drowsiness, severe headache, severe dizziness, increased seizures, coma.
WHAT TO DO:

- **Dial 911 for all medical emergencies or call poison control center 1-800-222-1222 for instructions.**
- **See emergency information on last 3 pages of this book.**

If you forget a dose:
Take as soon as you remember. If it is almost time for the next dose, wait for that dose (don't double this dose) and resume regular schedule.

What drug does:
The exact mechanism is unknown. The anticonvulsant action may result from a decrease in the release of stimulatory neurotransmitters (substances that stimulate nerve cells).

Time lapse before drug works:
May take several weeks for effectiveness.

Don't take with:
Any other medicine or any dietary supplement without consulting your doctor or pharmacist.

POSSIBLE ADVERSE REACTIONS OR SIDE EFFECTS

SYMPTOMS	WHAT TO DO
Life-threatening: In case of overdose, see previous column.	
Common:	
• Skin rash, double vision or blurred vision, clumsiness.	Continue, but call doctor right away.
• Dizziness, nausea or vomiting, headache, drowsiness.	Continue. Call doctor when convenient.
Infrequent:	
Anxiety, depression, confusion, irritability, other mood or mental changes, increase in seizure activity, back-and-forth eye movements (nystagmus).	Continue, but call doctor right away.
Rare:	
• Swelling (hands, face, mouth, feet), breathing difficulty, tiredness or weakness, fever, chills, sore throat, unusual bruising or bleeding, skin peeling or blistering or loosening, muscle cramps or pain, sores on mouth or lips, small red or purple dots on skin, slurred speech, symptoms of aseptic meningitis (headache, fever, chills, nausea, vomiting, stiff neck, and sensitivity to light).	Continue, but call doctor right away.

• Indigestion, runny nose, trembling, trouble sleeping, weakness.	Continue. Call doctor when convenient.

WARNINGS & PRECAUTIONS

Don't take if:
You are allergic to lamotrigine. Allergic reactions are rare, but may occur.

Before you start, consult your doctor if:
- You have kidney or liver disease.
- You have any heart disorder.
- You are allergic to any medication, food, or other substance.

Over age 60:
No special problems expected.

Pregnancy:
Decide with your doctor if drug benefits justify risk to unborn baby. Risk category C (see page xviii). Use of this drug during the first 3 months of pregnancy may increase risk of baby being born with cleft lip or cleft palate.

Breastfeeding; Lactation; Nursing Mothers:
This drug is generally acceptable with breastfeeding. Discuss risks and benefits with your doctor.

Infants & children up to age 18:
May be used for children age 2 and older. Follow instructions provided by your child's doctor.

Prolonged use:
Schedule regular visits to your doctor to determine if drug is continuing to be effective in controlling seizures. Follow-up laboratory blood studies may be recommended by your doctor.

Skin & sunlight:
No special problems expected.

Driving, piloting or hazardous work:
Don't drive or pilot aircraft until you learn how medicine affects you. Don't work around dangerous machinery. Don't climb ladders or work in high places. Danger increases if you drink alcohol or take other medicines affecting alertness and reflexes, such as antihistamines, tranquilizers, sedatives, pain medicine, narcotics and mind-altering drugs.

Discontinuing:
Don't discontinue without doctor's approval due to risk of increased seizure activity. Dosage may need to be gradually reduced.

Others:
- Advise any doctor, dentist or pharmacist whom you consult that you take this medicine.
- Use as directed. Don't increase or decrease dosage without doctor's approval.
- A skin rash may indicate a serious, and potentially life-threatening, medical problem. If a skin rash develops, it is usually during the first 4 to 6 weeks after treatment with the drug is started. Call your doctor promptly if you develop any skin rash.
- Rarely, antiepileptic drugs may lead to suicidal thoughts and behaviors. Call doctor right away if suicidal symptoms or unusual behaviors occur.
- Wear or carry medical identification stating your seizure disorder and drugs you take.

POSSIBLE INTERACTION WITH OTHER DRUGS

GENERIC NAME OR DRUG CLASS	COMBINED EFFECT
Carbamazepine	Decreased effect of lamotrigine. Increase in risk of side effects of carbamazepine.
Central nervous system (CNS) depressants*	Increased sedation.
Folate antagonists,* other	Folic acid deficiency.
Phenobarbital	Decreased effect of lamotrigine.
Phenytoin	Decreased effect of lamotrigine.
Primidone	Decreased effect of lamotrigine.
Valproic acid*	Increased effect of lamotrigine.

POSSIBLE INTERACTION WITH OTHER SUBSTANCES

INTERACTS WITH	COMBINED EFFECT
Alcohol:	Increased sedation. Avoid.
Beverages:	None expected.
Cocaine:	Unknown. Avoid.
Foods:	None expected.
Marijuana:	Consult doctor.
Tobacco:	None expected.

***See Glossary**

LAXATIVES, BULK-FORMING

GENERIC AND BRAND NAMES

See full list of generic and brand names in the *Generic and Brand Name Directory*, page 868.

BASIC INFORMATION

Habit forming? No
Prescription needed? No
Available as generic? Yes
Drug class: Laxative, bulk-forming

USES

For short-term relief of simple constipation (bowel movements that are abnormally difficult or infrequent). Normal frequency of bowel movements may vary from 2 to 3 times a day to 2 to 3 times a week. Laxatives treat the symptoms of constipation, not the cause.

DOSAGE & USAGE INFORMATION

How to take:
Powder, oral solution, tablet, capsule, granules, chewable tablet, caramel, effervescent powder, wafer—Follow package instructions. Swallow with full glass of water or fruit juice. Drink 6 to 8 glasses of water each day in addition to one taken with each dose. Mix all powders thoroughly to avoid any risk of unmixed powder causing intestinal blockage.

When to take:
As directed on the label or according to doctor's instructions.

If you forget a dose:
Take as soon as you remember. If it is almost time for the next dose, wait for that dose (don't double this dose) and resume regular schedule.

What drug does:
Adds dietary fiber that is not digested. Once in the intestine, it helps to increase fecal bulk, lubricate and soften the intestinal contents and facilitate the passage of stools.

Continued next column

OVERDOSE

SYMPTOMS:
Weakness, increased sweating, confusion, irregular heartbeat, muscle cramps.
WHAT TO DO:
If person uses much larger amount than prescribed or if accidentally swallowed, call poison control center 1-800-222-1222 for instructions or dial 911 (emergency) for help.

Time lapse before drug works:
May work in 12 to 24 hours. Sometimes does not work for 2 to 3 days.

Don't take with:
- Any other medicine or any dietary supplement without consulting your doctor or pharmacist.
- Don't take within 2 hours of taking another medicine. Laxative interferes with absorption of medicine.

POSSIBLE ADVERSE REACTIONS OR SIDE EFFECTS

SYMPTOMS	WHAT TO DO
Life-threatening: None expected.	
Common: None expected.	
Infrequent: Mild stomach cramps, throat irritation with liquid form.	Continue. Call doctor when convenient.
Rare: Allergic skin rash or itching, trouble breathing, swallowing difficulty.	Discontinue. Call doctor right away.

WARNINGS & PRECAUTIONS

Don't take if:
- You have symptoms of appendicitis (abdominal pain, cramping, soreness, bloating, nausea, vomiting). Consult doctor.
- You have dysphagia (swallowing difficulty).
- You are allergic to bulk-forming laxatives. Allergic reactions are rare, but may occur.
- You have missed a bowel movement for just 1 or 2 days.

Before you start, consult your doctor if:
- You are allergic to any medicine, food, or other substance or have a family history of allergies.
- You have diabetes or heart or kidney disease.
- You have hypertension (high blood pressure) and the laxative contains sodium.
- You have an intestinal obstruction or undiagnosed rectal bleeding.
- You are taking other laxatives.

Over age 60:
No special problems expected.

Pregnancy:
Most bulk-forming laxatives contain sodium or sugars, which may cause fluid retention. Risk factors vary or may not be designated for these laxatives. Read categories on page xviii and consult doctor.

Breastfeeding; Lactation; Nursing Mothers:
Drugs in this group are generally acceptable with breastfeeding. Discuss risks and benefits with your doctor.

Infants & children up to age 18:
- Don't give to children under age 6 without doctor's approval. Young children are not able to describe their symptoms accurately, and a proper diagnosis needs to be made before starting any treatment.
- Don't give to a child who refuses to have a bowel movement (toileting refusal). May force a painful bowel movement and cause the child to hold back even more. Consult doctor.
- For children over age 6, follow package instructions or doctor's directions for correct dosage amount.

Prolonged use:
Don't take for more than 1 week unless under doctor's supervision. Bulk-form laxatives are sometimes used for long-term therapy.

Skin & sunlight:
No special problems expected.

Driving, piloting or hazardous work:
No special problems expected.

Discontinuing:
May be unnecessary to finish medicine. Follow doctor's instructions or instructions on label.

Others:
- Don't give to "flush out" the system or as a "tonic."
- Use as directed. Don't increase or decrease dosage without doctor's approval.
- Excessive use of laxatives in a teenager may indicate an eating disorder such as anorexia nervosa or bulimia nervosa. Consult doctor.
- If there is a sudden change in bowel habits or bowel function that lasts longer than 2 weeks, consult doctor.

POSSIBLE INTERACTION WITH OTHER DRUGS

GENERIC NAME OR DRUG CLASS	COMBINED EFFECT
Antacids*	Irritation of stomach or small intestine.
Anticoagulants*	Decreased anti-coagulant effect. Take 2 hours apart.
Digitalis preparations*	Decreased digitalis effect. Take 2 hours apart.
Diuretics, potassium-sparing*	Decreased potassium effect.
Potassium supplements*	Decreased potassium effect.
Salicylates*	Decreased salicylate effect. Take 2 hours apart.
Tetracyclines*	Decreased tetracycline effect. Take 2 hours apart.

POSSIBLE INTERACTION WITH OTHER SUBSTANCES

INTERACTS WITH	COMBINED EFFECT
Alcohol:	None expected.
Beverages:	None expected.
Cocaine:	Unknown. Avoid.
Foods:	None expected.
Marijuana:	Consult doctor.
Tobacco:	None expected.

***See Glossary**

LAXATIVES, OSMOTIC

GENERIC AND BRAND NAMES

See full list of generic and brand names in the *Generic and Brand Name Directory*, page 868.

BASIC INFORMATION

Habit forming? No
Prescription needed? No
Available as generic? Yes
Drug class: Laxative, hyperosmotic

USES

For short-term relief of simple constipation (bowel movements that are abnormally difficult or infrequent). Laxatives treat the symptoms of constipation, not the cause.

DOSAGE & USAGE INFORMATION

How to take:

- Oral solution, tablet, crystals, effervescent powder, milk of magnesia—Follow package instructions. Swallow with full glass of water or fruit juice. A second glass of liquid is often recommended for best effect. Drink 6 to 8 glasses of water each day, in addition to one taken with each dose.
- Enema or suppository—Read and follow package instructions.

When to take:
Since drug produces stool within 30 minutes to 3 hours following a dose, take it at a time that will not interfere with sleep or scheduled activities. Don't take late in the day on an empty stomach.

If you forget a dose:
Take as soon as you remember. If it is almost time for the next dose, wait for that dose (don't double this dose) and resume regular schedule.

Continued next column

OVERDOSE

SYMPTOMS:
Irregular heartbeat, sweating, confusion, muscle cramps, weakness. Sodium phosphate product can cause serious and possible fatal effects if person exceeds prescribed dosage.
WHAT TO DO:

- **Dial 911 for all medical emergencies or call poison control center 1-800-222-1222 for instructions.**
- **See emergency information on last 3 pages of this book.**

What drug does:
Draws water into the bowel from surrounding tissue to help loosen and soften the stool and increases bowel action.

Time lapse before drug works:

- Oral forms—30 minutes to 3 hours. May take longer if taken with a meal.
- Rectal forms—2 to 15 minutes.

Don't take with:

- Any other medicine or any dietary supplement without consulting your doctor or pharmacist.
- Don't take within 2 hours of taking another medicine. Laxative interferes with absorption of medicine.

POSSIBLE ADVERSE REACTIONS OR SIDE EFFECTS

SYMPTOMS	WHAT TO DO
Life-threatening:	
None expected.	
Common:	
None expected.	
Infrequent:	
• Rectal bleeding or burning or itching or pain (with rectal forms).	Discontinue. Call doctor right away.
• Belching, cramps, nausea, diarrhea, increased thirst.	Continue. Call doctor when convenient.
Rare:	
When used too often or dose is too high (confusion, irregular heartbeat, muscle cramps, unusual tiredness or weakness, dehydration).	Discontinue. Call doctor right away.

WARNINGS & PRECAUTIONS

Don't take if:

- You are having symptoms of appendicitis (abdominal pain, cramping, soreness, bloating, nausea and vomiting). Consult doctor.
- You are allergic to osmotic laxatives. Allergic reactions are rare, but may occur.
- You have missed a bowel movement for just 1 or 2 days.

Before you start, consult your doctor if:

- You are allergic to any medicine, food or other substance or have a family history of allergies.
- You have hypertension (high blood pressure) and the laxative contains sodium.
- You have an intestinal obstruction, rectal bleeding or a colostomy or ileostomy.
- You have diabetes or heart or kidney disease.
- You are taking other laxatives.

Over age 60:

- Rectal solutions could cause excess fluid in the body. Consult doctor before using.
- No special problems expected with laxatives taken by mouth.

Pregnancy:
Risk factors vary or may not be designated for these laxatives. Read categories on page xviii and consult doctor.

Breastfeeding; Lactation; Nursing Mothers:
Drugs in this group are generally acceptable with breastfeeding. Discuss risks and benefits with your doctor.

Infants & children up to age 18:

- Don't give to children under age 6 without doctor's approval. Young children are not able to describe their symptoms accurately, and a proper diagnosis needs to be made before starting any treatment.
- Don't give to a child who refuses to have a bowel movement (toileting refusal). May force a painful bowel movement and cause the child to hold back even more. Consult doctor.
- For children over age 6, follow package instructions or doctor's directions for correct dosage amount.

Prolonged use:
Don't take for more than 1 week unless under doctor's supervision. May cause laxative dependence in which normal bowel function depends on the laxative to produce a bowel movement.

Skin & sunlight:
No special problems expected.

Driving, piloting or hazardous work:
No special problems expected.

Discontinuing:
May be unnecessary to finish medicine. Follow doctor's instructions or instructions on label.

Others:

- Don't use to "flush out" the body or as a "tonic."
- Use as directed on label or by doctor. Don't use more than prescribed dosage.
- Taking too much sodium phosphate can cause serious and life threatening effects.
- Excessive use of laxatives in a teenager may indicate an eating disorder such as anorexia nervosa or bulimia nervosa. Consult doctor.
- Consult doctor for changes in bowel habits or function that last longer than 2 weeks.

POSSIBLE INTERACTION WITH OTHER DRUGS

GENERIC NAME OR DRUG CLASS	COMBINED EFFECT
Antacids*	Irritation of stomach or small intestine.
Anticoagulants*	Decreased anti-coagulant effect with aluminum- or magnesium-containing laxatives. Avoid.
Ciprofloxacin	Decreased ciprofloxacin effect with magnesium-containing laxatives. Avoid.
Digitalis preparations*	Decreased digitalis effect with aluminum- or magnesium-containing laxatives. Avoid.
Diuretics, potassium-sparing*	Decreased potassium effect.
Etidronate	Decreased etidronate effect if taken with magnesium-containing laxatives. Take 2 hours apart.
Phenothiazines*	Decreased phenothiazine effect with aluminum- or magnesium-containing laxatives. Avoid.
Potassium supplements*	Decreased potassium effect.
Sodium polystyrene	Fluid imbalance in body with magnesium-containing laxatives. Avoid.
Tetracyclines*	Decreased tetracycline effect. Take 2 hours apart.

POSSIBLE INTERACTION WITH OTHER SUBSTANCES

INTERACTS WITH	COMBINED EFFECT
Alcohol:	None expected.
Beverages:	None expected.
Cocaine:	Unknown. Avoid.
Foods:	None expected.
Marijuana:	Consult doctor.
Tobacco:	None expected.

***See Glossary**

LAXATIVES, SOFTENER/LUBRICANT

GENERIC AND BRAND NAMES

See full list of generic and brand names in the *Generic and Brand Name Directory,* page 869.

BASIC INFORMATION

Habit forming? No
Prescription needed? No
Available as generic? Yes
Drug class: Laxative (stool softener-emollient), lubricant

USES

For short-term relief of simple constipation (bowel movements that are abnormally difficult or infrequent). Normal frequency of bowel movements may vary from 2 to 3 times a day to 2 to 3 times a week. Laxatives treat the symptoms of constipation, not the cause.

DOSAGE & USAGE INFORMATION

How to take:

- Tablet, capsule, syrup, chewable tablet, oral solution—Follow package instructions. Swallow with full glass of water, fruit juice or milk. Drink 6 to 8 glasses of water each day in addition to one taken with each dose.
- Enema or suppository—Read and follow package instructions.

When to take:
Produces stool within 30 minutes to 3 hours following a dose. Take drug at a time that will not interfere with sleep or scheduled activities. Don't take late in the day on an empty stomach.

If you forget a dose:
Take as soon as you remember. If it is almost time for the next dose, wait for that dose (don't double this dose) and resume regular schedule.

Continued next column

OVERDOSE

SYMPTOMS:
Weakness, increased sweating, confusion, irregular heartbeat, muscle cramps.
WHAT TO DO:
If person uses much larger amount than prescribed or if accidentally swallowed, call poison control center 1-800-222-1222 for instructions or dial 911 (emergency) for help.

What drug does:
Softener laxatives help liquids mix into the stool to help prevent hard stool masses. Lubricant laxatives coat the stool surface with a thin film that helps ease the passage of the stool through the intestines.

Time lapse before drug works:

- When taken by mouth, usually works within 1 to 2 days after first dose, but may take 3 to 5 days for full effectiveness.
- Rectal dosage forms work in 2 to 15 minutes.

Don't take with:

- Any other medicine or any dietary supplement without consulting your doctor or pharmacist.
- Other stool softener laxatives or mineral oil.
- Don't take within 2 hours of taking another medicine. Laxative interferes with absorption of medicine.

POSSIBLE ADVERSE REACTIONS OR SIDE EFFECTS

SYMPTOMS	WHAT TO DO
Life-threatening:	
None expected.	
Common:	
None expected.	
Infrequent:	
• Rectal bleeding or burning or itching or pain (with rectal forms).	Discontinue. Call doctor right away.
• Mild stomach cramps, throat irritation with liquid forms, diarrhea.	Continue. Call doctor when convenient.
Rare:	
Skin rash.	Discontinue. Call doctor right away.

WARNINGS & PRECAUTIONS

Don't take if:

- You have symptoms of appendicitis (abdominal pain, cramping, soreness, bloating, nausea and vomiting). Consult doctor.
- You are allergic to a softener-emollient or lubricant laxative. Allergic reactions are rare, but may occur.
- You have missed a bowel movement for just 1 or 2 days.

Before you start, consult your doctor if:
- You are allergic to any medicine, food or other substance or have a family history of allergies.
- You have hypertension (high blood pressure) and the laxative contains sodium.
- You have an intestinal obstruction, undiagnosed rectal bleeding or a colostomy or ileostomy.
- You are taking other laxatives.
- You have diabetes or heart or kidney disease.
- You have dysphagia (swallowing difficulty) and want to take mineral oil.

Over age 60:
Oral mineral oil is not recommended for bedridden elderly patients; otherwise, no special problems expected.

Pregnancy:
Risk factors vary or may not be designated for these laxatives. Read categories on page xviii and consult doctor.

Breastfeeding; Lactation; Nursing Mothers:
Drugs in this group are generally acceptable with breastfeeding. Discuss risks and benefits with your doctor.

Infants & children up to age 18:
- Don't give to children under age 6 without doctor's approval. Young children are not able to describe their symptoms accurately, and a proper diagnosis needs to be made before starting any treatment.
- Don't give to a child who refuses to have a bowel movement (toileting refusal). May force a painful bowel movement and cause the child to hold back even more. Consult doctor.
- For children over age 6, follow package instructions or doctor's directions for correct dosage amount.

Prolonged use:
Don't take for more than 1 week unless under doctor's supervision.

Skin & sunlight:
No special problems expected.

Driving, piloting or hazardous work:
No special problems expected.

Discontinuing:
May be unnecessary to finish medicine. Follow doctor's instructions or instructions on label.

Others:
- Don't give to "flush out" the system or as a "tonic."
- Use as directed. Don't increase or decrease dosage without doctor's approval.
- Excessive use of laxatives in a teenager may indicate an eating disorder such as anorexia nervosa or bulimia nervosa. Consult doctor.
- If there is a sudden change in bowel habits or bowel function that lasts longer than 2 weeks, consult doctor.

POSSIBLE INTERACTION WITH OTHER DRUGS

GENERIC NAME OR DRUG CLASS	COMBINED EFFECT
Antacids*	Irritation of stomach or small intestine.
Anticoagulants*	Decreased anticoagulant effect with mineral oil.
Contraceptives, oral*	Decreased contraceptive effect with mineral oil.
Danthron	Increased danthron effect.
Digitalis preparations*	Decreased digitalis effect with mineral oil.
Diuretics, potassium-sparing*	Decreased potassium effect.
Phenolphthalein	Increased phenolphthalein effect.
Potassium supplements*	Decreased potassium effect.
Vitamins A, D, E, K	Decreased vitamin effect with mineral oil.

POSSIBLE INTERACTION WITH OTHER SUBSTANCES

INTERACTS WITH	COMBINED EFFECT
Alcohol:	None expected.
Beverages:	None expected.
Cocaine:	Unknown. Avoid.
Foods:	None expected.
Marijuana:	Consult doctor.
Tobacco:	None expected.

***See Glossary**

LAXATIVES, STIMULANT

GENERIC AND BRAND NAMES

See full list of generic and brand names in the *Generic and Brand Name Directory*, page 870.

BASIC INFORMATION

Habit forming? Potentially
Prescription needed? No
Available as generic? Yes
Drug class: Laxative (stimulant)

USES

For short-term relief of simple constipation (bowel movements that are abnormally difficult or infrequent). Normal frequency of bowel movements may vary from 2 to 3 times a day to 2 to 3 times a week. Laxatives treat the symptoms of constipation, not the cause.

DOSAGE & USAGE INFORMATION

How to take:

- Tablet, chewable tablet, syrup, chewing gum, oral solution, granules, fluidextract, emulsion, wafer—Follow package instructions. Swallow with full glass of water, fruit juice or milk. Give child 6 to 8 glasses of fluid each day in addition to the one taken with each dose to keep stool soft. Give on an empty stomach. Results may be delayed if given with food.
- Enema—Lubricate rectal area with petroleum jelly before inserting enema applicator. Insert carefully to avoid damage to rectal wall. To mix powder for rectal solution, follow instructions on package.
- Suppository—Remove wrapper and moisten suppository with water. Gently insert tapered end into rectum. Push well into rectum with finger. Retain in rectum 20 to 30 minutes.

Continued next column

OVERDOSE

SYMPTOMS:
Weakness, increased sweating, confusion, irregular heartbeat, muscle cramps.
WHAT TO DO:
If person uses much larger amount than prescribed or if accidentally swallowed, call poison control center 1-800-222-1222 for instructions or dial 911 (emergency) for help.

When to take:
Usually at bedtime on an empty stomach, unless directed otherwise. Castor oil is usually taken late in the day, as it works within 2 to 6 hours.

If you forget a dose:
Take as soon as you remember. If it is almost time for the next dose, wait for that dose (don't double this dose) and resume regular schedule.

What drug does:
Acts on smooth muscles of intestinal wall to cause vigorous bowel movement.

Time lapse before drug works:
Oral form within 6 to 10 hours (castor oil 2 to 6 hours). Rectal form within 15 minutes to 1 hour.

Don't take with:

- Any other medicine or any dietary supplement without consulting your doctor or pharmacist.
- Don't take within 2 hours of another medicine. Laxative interferes with absorption of medicine.

POSSIBLE ADVERSE REACTIONS OR SIDE EFFECTS

SYMPTOMS	WHAT TO DO
Life-threatening:	
None expected.	
Common:	
None expected.	
Infrequent:	
• Rectal bleeding or burning or itching or pain (with rectal forms).	Discontinue. Call doctor right away.
• Belching, cramps, nausea, diarrhea, throat irritation.	Continue. Call doctor when convenient.
Rare:	
Confusion, irregular heartbeat, muscle cramps, unusual tiredness or weakness, pink to red color of urine and stools (with phenolphthalein), pink, red or violet to brown urine color (with cascara, danthron or senna), yellow to brown color of urine (with phenolphthalein, cascara,or senna), skin rash (allergy).	Discontinue. Call doctor right away.

WARNINGS & PRECAUTIONS

Don't take if:
- You have symptoms of appendicitis (abdominal pain, cramping, soreness, bloating, nausea and vomiting). Consult doctor.
- You are allergic to a stimulant laxative. Allergic reactions are rare, but may occur.
- You have missed a bowel movement for just 1 or 2 days.

Before you start, consult your doctor if:
- You are a allergic to any medicine, food or other substance or have a family history of allergies.
- You have hypertension (high blood pressure) and the laxative contains sodium.
- You have an intestinal obstruction or undiagnosed rectal bleeding.
- You have diabetes or heart or kidney disease.
- You are taking other laxatives.

Over age 60:
Excessive use of stimulant laxatives may cause excess loss of body fluid, resulting in weakness and lack of coordination.

Pregnancy:
Risk factors vary or may not be designated for these laxatives. Read categories on page xviii and consult doctor.

Breastfeeding; Lactation; Nursing Mothers:
Drugs in this group are generally acceptable with breastfeeding. Discuss risks and benefits with your doctor.

Infants & children up to age 18:
- Don't give to children under age 6 without doctor's approval. Young children are not able to describe their symptoms accurately, and a proper diagnosis needs to be made before starting any treatment.
- Don't give to a child who refuses to have a bowel movement (toileting refusal). May force a painful bowel movement and cause the child to hold back even more. Consult doctor.
- For children over age 6, follow package instructions or doctor's directions for correct dosage amount.

Prolonged use:
Don't take for more than 1 week unless under doctor's supervision.

Skin & sunlight:
No special problems expected.

Driving, piloting or hazardous work:
No special problems expected.

Discontinuing:
May be unnecessary to finish medicine. Follow doctor's instructions or instructions on label.

Others:
- Don't give to "flush out" the system or as a "tonic."
- Use as directed. Don't increase or decrease dosage without doctor's approval.
- Excessive use of laxatives in a teenager may indicate an eating disorder such as anorexia nervosa or bulimia nervosa. Consult doctor.
- If there is a sudden change in bowel habits or bowel function that lasts longer than 2 weeks, consult doctor.

POSSIBLE INTERACTION WITH OTHER DRUGS

GENERIC NAME OR DRUG CLASS	COMBINED EFFECT
Antacids*	Irritation of stomach or small intestine.
Diuretics, potassium-sparing*	Decreased potassium effect.
Histamine H_2 receptor antagonists*	Stomach irritation with bisacodyl. Take 1 hour apart.
Potassium supplements*	Decreased potassium effect.

POSSIBLE INTERACTION WITH OTHER SUBSTANCES

INTERACTS WITH	COMBINED EFFECT
Alcohol:	None expected.
Beverages:	
Milk.	Stomach irritation with bisacodyl. Take 1 hour apart.
Cocaine:	Unknown. Avoid.
Foods:	None expected.
Marijuana:	Consult doctor.
Tobacco:	None expected.

*See Glossary

LEFLUNOMIDE

BRAND NAMES

Arava

BASIC INFORMATION

Habit forming? No
Prescription needed? Yes
Available as generic? Yes
Drug class: Antirheumatic

USES

Treats symptoms caused by rheumatoid arthritis, such as inflammation, swelling, stiffness and joint pain. Slows deterioration of joint.

DOSAGE & USAGE INFORMATION

How to take:
Tablet—Take with full glass of water. If you can't swallow whole, crumble tablet and take with liquid or food.

When to take:
At the same time each day.

If you forget a dose:
Take as soon as you remember. If it is almost time for the next dose, wait for that dose (don't double this dose) and resume regular schedule.

What drug does:
Stops the body from producing too many of the immune cells that are responsible for the swelling and inflammation (immunosuppressive and antiinflammatory).

Time lapse before drug works:
6 to 12 hours.

Don't take with:
Any other medicine or any dietary supplement without consulting your doctor or pharmacist.

OVERDOSE

SYMPTOMS:
Easy bruising or bleeding, diarrhea, stomach pain, jaundice (yellow skin or eyes), dark urine.

WHAT TO DO:
- **Dial 911 for all medical emergencies or call poison control center 1-800-222-1222 for instructions.**
- **See emergency information on last 3 pages of this book.**

POSSIBLE ADVERSE REACTIONS OR SIDE EFFECTS

SYMPTOMS	WHAT TO DO
Life-threatening:	
Rare, severe allergic reaction (anaphylaxis)—difficulty breathing, hives, itching, swelling (throat, face), vomiting, dizziness.	Seek emergency treatment immediately.
Common:	
• Chest congestion, cough, difficulty in breathing, loss of appetite, nausea, vomiting, yellow eyes or skin, dizziness, fever, sneezing, sore throat, pain or burning while urinating, frequent urge to urinate.	Continue. Call doctor right away.
• Abdominal pain, hair loss, back pain, diarrhea, heartburn, rash, unexplained weight loss.	Continue. Call doctor if symptoms persist.
Infrequent:	
• Unusual tiredness or weakness, pounding or rapid heartbeat, burning or tingling sensation in fingers and toes, joint or muscle pain.	Continue. Call doctor right away.
• Acne, loss of appetite, anxiety, red or irritated eyes, constipation, dry mouth, gas, mouth sores, pain or burning in throat, itching, runny nose.	Continue. Call doctor when convenient.
Rare:	
None expected.	

WARNINGS & PRECAUTIONS

Don't take if:
You are allergic to leflunomide.

Before you start, consult your doctor if:
- You have immune system problems.
- You have severe or uncontrolled infections.
- You have a liver disease or you have elevated liver enzymes (per medical tests).
- You have kidney disease.

Over age 60:
No special problems expected.

Pregnancy:
Consult doctor. Drug should not be used during pregnancy. Can cause harm to unborn baby. Risk category X (see page xviii).

Breastfeeding; Lactation; Nursing Mothers:
This drug is generally not recommended with breastfeeding. Discuss risks and benefits with your doctor.

Infants & children up to age 18:
Safety and effectiveness in this age group have not been established. Consult doctor.

Prolonged use:
Follow-up with your doctor on a regular basis to monitor your condition and check for drug side effects.

Skin & sunlight:
None expected.

Driving, piloting or hazardous work:
None expected.

Discontinuing:
Don't discontinue without consulting doctor or before completing prescribed dosage.

Others:
- May affect accuracy of some laboratory tests.
- Women of childbearing age are advised to use reliable contraception before receiving leflunomide. If you become pregnant while taking this drug, notify your doctor immediately.
- Use of leflunomide by men during time of conception may cause birth defects in their children. Therefore, men taking leflunomide should use condoms as a form of birth control.
- Severe liver injury may occur while taking this drug. Blood tests to check liver enzymes should be done at least monthly for 3 months after starting the drug and every 3 months thereafter.
- Don't have any immunizations during or after treatment with this drug without doctor's approval.
- Advise any doctor, dentist or pharmacist whom you consult that you take this medicine.

POSSIBLE INTERACTION WITH OTHER DRUGS

GENERIC NAME OR DRUG CLASS	COMBINED EFFECT
Charcoal, activated	Decreased leflunomide effect.
Cholestyramine	Decreased leflunomide effect.
Hepatotoxics*	Increased risk of liver injury.
Methotrexate	Increased risk of side effects.
Rifampin	May increase risk of leflunomide toxicity.

POSSIBLE INTERACTION WITH OTHER SUBSTANCES

INTERACTS WITH	COMBINED EFFECT
Alcohol:	Increased risk of liver problems. Avoid.
Beverages:	None expected.
Cocaine:	Unknown. Avoid.
Foods:	None expected.
Marijuana:	Consult doctor.
Tobacco:	None expected.

***See Glossary**

LEUCOVORIN

BRAND NAMES

Citrocovorin Calcium
Citrovorum Factor
Folinic Acid
Wellcovorin

BASIC INFORMATION

Habit forming? No
Prescription needed? Yes
Available as generic? Yes
Drug class: Antianemic

USES

- Used to treat harmful effects of methotrexate (which is used in cancer treatment).
- Used to prevent and treat anemia.

DOSAGE & USAGE INFORMATION

How to take:
Tablet—Swallow with liquid or food to lessen stomach irritation. If you can't swallow whole, crumble tablet and take with liquid or food.

When to take:
At the same time each day, according to instructions on prescription label.

If you forget a dose:
Take as soon as you remember. If it is almost time for the next dose, wait for that dose (don't double this dose) and resume regular schedule.

What drug does:
Favors development of DNA, RNA and protein synthesis.

Time lapse before drug works:
20 to 30 minutes.

Don't take with:
Any other medicine or any dietary supplement without consulting your doctor or pharmacist.

OVERDOSE

SYMPTOMS:
Symptoms are unlikely to occur.
WHAT TO DO:
If person uses much larger amount than prescribed or if accidentally swallowed, call poison control center 1-800-222-1222 for instructions or dial 911 (emergency) for help.

POSSIBLE ADVERSE REACTIONS OR SIDE EFFECTS

SYMPTOMS	WHAT TO DO
Life-threatening:	
Rare, severe allergic reaction (anaphylaxis)—difficulty breathing, hives, itching, swelling (throat, face), vomiting, dizziness.	Seek emergency treatment immediately.
Common:	
None expected.	
Infrequent:	
None expected.	
Rare:	
• Skin rash, hives, seizures.	Discontinue. Call doctor right away.
• Nausea.	Continue. Call doctor when convenient.

WARNINGS & PRECAUTIONS

Don't take if:
You are allergic to leucovorin.

Before you start, consult your doctor if:
- You have acid urine, ascites, dehydration.
- You have kidney function impairment.
- You have pernicious anemia.
- You have vitamin B-12 deficiency.

Over age 60:
Adverse reactions and side effects may be more frequent and severe than in younger persons.

Pregnancy:
Decide with your doctor if drug benefits justify risk to unborn baby. Risk category C (see page xviii).

Breastfeeding; Lactation; Nursing Mothers:
This drug is generally acceptable with breastfeeding. Discuss risks and benefits with your doctor.

Infants & children up to age 18:
Follow instructions provided by your child's doctor. May increase number of seizures in children with seizure disorder.

Prolonged use:
No problems expected.

Skin & sunlight:
No problems expected.

Driving, piloting or hazardous work:
Don't drive or pilot aircraft until you learn how medicine affects you. Don't work around dangerous machinery. Don't climb ladders or work in high places. Danger increases if you drink alcohol or take medicine affecting alertness and reflexes.

Discontinuing:
Don't discontinue without consulting doctor. Dose may require gradual reduction if you have taken drug for a long time. Doses of other drugs may also require adjustment.

Others:
Advise any doctor, dentist or pharmacist whom you consult that you take this medicine.

POSSIBLE INTERACTION WITH OTHER DRUGS

GENERIC NAME OR DRUG CLASS	COMBINED EFFECT
Anticonvulsants, barbiturate and hydantoin*	Large doses of leucovorin may counteract the effects of these medicines.
Central nervous system (CNS) depressants*	High alcohol content of leucovorin may cause adverse effects.
Fluorouracil	Increased levels of fluorouracil.
Primidone	Large doses of leucovorin may counteract the effects of both drugs.

POSSIBLE INTERACTION WITH OTHER SUBSTANCES

INTERACTS WITH	COMBINED EFFECT
Alcohol:	Increased adverse reactions of both.
Beverages:	None expected.
Cocaine:	Increased adverse reactions of both drugs.
Foods:	None expected.
Marijuana:	Increased adverse reactions of both drugs.
Tobacco:	Increased adverse reactions of both.

***See Glossary**

LEUKOTRIENE MODIFIERS

GENERIC AND BRAND NAMES

MONTELUKAST
Singulair
ZAFIRLUKAST
Accolate
ZILEUTON
Zyflo CR

BASIC INFORMATION

Habit forming? No
Prescription needed? Yes
Available as generic? Yes, for some
Drug class: Antiasthmatic

USES

- Treatment of mild to moderate asthma. Not used to treat an active asthma attack. May be used with other asthma medications as directed by your doctor. Montelukast is used for prophylaxis (preventive) and chronic treatment of asthma.
- Prevention of exercise-induced broncho-constriction (EIB) in patients age 15 and older.
- Montelukast is used for treatment of seasonal allergies (hay fever) and perennial allergic rhinitis (PAR), also known as indoor allergies.

DOSAGE & USAGE INFORMATION

How to take:
- Tablet—Swallow with water, with or without food, except for zafirlukast which must be taken on an empty stomach 1 hour before or 2 hours after a meal.
- Chewable tablet—Chew tablet before you swallow it.
- Extended-release tablet—Swallow whole with liquid. Do not cut, chew or crush tablet.
- Oral granules—Take granules directly in the mouth or mix with a spoonful of cold or room temperature soft food.

Continued next column

OVERDOSE

SYMPTOMS:
Headache, upset stomach, vomiting, feeling restless or agitated, thirst, sleepiness.
WHAT TO DO:
If person uses much larger amount than prescribed or if accidentally swallowed, call poison control center 1-800-222-1222 for instructions or dial 911 (emergency) for help.

When to take:
- Montelukast—At the same time each day.
- Zafirlukast—Twice a day at the same times.
- Zileuton—Four times a day at the same times. Take extended-release zileuton within one hour after morning and evening meals.

If you forget a dose:
Take as soon as you remember. If it is almost time for the next dose, wait for that dose (don't double this dose) and resume regular schedule.

What drug does:
Inhibits inflammatory cells associated with asthma. Inhibits reflex reactions to irritants, exercise and cold.

Time lapse before drug works:
30 minutes to 4 hours.

Don't take with:
Any other medicine or any dietary supplement without consulting your doctor or pharmacist.

POSSIBLE ADVERSE REACTIONS OR SIDE EFFECTS

SYMPTOMS	WHAT TO DO
Life-threatening:	
Rare, severe allergic reaction (anaphylaxis)—difficulty breathing, hives, itching, swelling (face, throat), wheezing, vomiting, dizziness.	Seek emergency treatment immediately.
Common:	
Headache, stomach upset, nausea.	Continue. Call doctor when convenient.
Infrequent:	
Weak feeling, pain in abdomen, headache, unusual tiredness, cough, dental pain, dizziness, heartburn, fever, stuffy nose, skin rash.	Continue. Call doctor when convenient.
Rare:	
Liver problems (yellow eyes or skin, fatigue, symptoms of flu, itching, pain in upper right abdominal area), agitation, aggression, abnormal dreams, being anxious, hallucinations, irritable, depression, insomnia, restlessness, tremor, suicidal thinking and behavior (including suicide).	Discontinue. Call doctor right away.

WARNINGS & PRECAUTIONS

Don't take if:
You are allergic to leukotriene modifiers.

Before you start, consult your doctor if:
- You have any other medical problem.
- You have a liver or kidney disease.
- You have a history of alcoholism.

Over age 60:
In some cases older patients taking zafirlukast experienced more infections; otherwise, no problems expected.

Pregnancy:
Decide with your doctor if drug benefits justify risk to unborn baby. Montelukast and zafirlukast are risk category B; zileuton is risk category C (see page xviii).

Breastfeeding; Lactation; Nursing Mothers:
Drugs in this group are generally acceptable with breastfeeding. Discuss risks and benefits with your doctor.

Infants & children up to age 18:
Follow instructions provided by your child's doctor.

Prolonged use:
Schedule regular visits with your doctor to determine if the drug continues its effectiveness in controlling asthma symptoms.

Skin & sunlight:
No problems expected.

Driving, piloting or hazardous work:
No problems expected.

Discontinuing:
Don't discontinue without consulting doctor.

Others:
- In a few rare instances, patients taking zafirlukast while having their oral steroid dosage reduced developed Churg-Strauss syndrome (a rare and sometimes fatal condition). It is not known if the problem is caused by zafirlukast. Symptoms of Churg-Strauss syndrome are similar to those caused by flu. Before starting zafirlukast and reducing oral steroids, discuss the benefits and risk factors with your doctor.
- Advise any doctor or dentist you consult that you take this medicine.
- Talk to your doctor if your asthma attacks are not being controlled by the usual dosage of your fast-acting bronchodilator.
- This drug may affect results in some medical tests.
- This drug should be taken every day, even if you are not having asthma symptoms.

POSSIBLE INTERACTION WITH OTHER DRUGS

GENERIC NAME OR DRUG CLASS	COMBINED EFFECT
Beta adrenergic blocking agents	Increased beta blocker effect.
Calcium channel blockers	Increased effect of calcium channel blocker.
Carbamazepine	Increased effect of carbamazepine.
Cisapride	Increased effect of cisapride.
Cyclosporine	Increased effect of cyclosporine.
Dofetilide	Increased dofetilide effect.
Erythromycin	Decreased effect of zafirlukast.
Phenobarbital	Decreased effect of montelukast.
Phenytoin	Increased effect of phenytoin.
Tolbutamide	Increased effect of tolbutamide.
Warfarin	Increased effect of warfarin.

POSSIBLE INTERACTION WITH OTHER SUBSTANCES

INTERACTS WITH	COMBINED EFFECT
Alcohol:	None expected.
Beverages:	None expected.
Cocaine:	Unknown. Avoid.
Foods:	None expected.
Marijuana:	Consult doctor.
Tobacco:	None expected.

LEVETIRACETAM

BRAND NAMES

Elepsia XR
Keppra
Keppra XR

BASIC INFORMATION

Habit forming? No
Prescription needed? Yes
Available as generic? Yes
Drug class: Anticonvulsant; antiepileptic

USES

Used to help control some types of seizures in the treatment of epilepsy. This medicine cannot cure epilepsy and will only work to control seizures for as long as you continue to take it.

DOSAGE & USAGE INFORMATION

How to take:

- Tablet or extended-release tablet—Swallow whole with a liquid. Do not crush, chew or break tablet. May be taken with or without food, and on a full or empty stomach.
- Oral solution—Follow instructions on label.

When to take:
At the same time(s) each day. Your doctor will determine the schedule. Dose may be increased every two weeks to achieve maximum benefits.

If you forget a dose:
Take as soon as you remember. If it is almost time for the next dose, wait for that dose (don't double this dose) and resume regular schedule.

What drug does:
The exact mechanism of the anticonvulsant activity of levetiracetam is unknown.

Time lapse before drug works:
May take several weeks for effectiveness.

Don't take with:
Any other medicine or any dietary supplement without consulting your doctor or pharmacist.

OVERDOSE

SYMPTOMS:
Drowsiness, agitation, aggression, slow breathing, stupor, coma.
WHAT TO DO:

- **Dial 911 for all medical emergencies or call poison control center 1-800-222-1222 for instructions.**
- **See emergency information on last 3 pages of this book.**

POSSIBLE ADVERSE REACTIONS OR SIDE EFFECTS

SYMPTOMS	WHAT TO DO
Life-threatening:	
In case of overdose, see previous column.	
Common:	
Cough, dizziness, dry or sore throat, hoarseness, loss of strength or energy, muscle pain or weakness, runny nose, sleepiness, tender and swollen glands in neck, trouble swallowing, unusual tiredness or weakness, voice changes.	Continue. Call doctor when convenient.
Infrequent:	
• Clumsiness or unsteadiness, crying, depression, double vision, fever or chills, headache, loss of memory or problems with memory, lower back or side pain, mood or mental changes, suicidal thoughts or feelings, nervousness, angry outbursts, pain or tenderness around eyes and cheekbones, painful or difficult urination, paranoia, muscle control problems, overreacting, shortness of breath or trouble breathing, chest tightness, wheezing.	Continue, but call doctor right away. Seek emergency care for severe symptoms.
• Burning or crawling or itching or numbness or prickling or tingling feelings, feeling of constant movement of self or surroundings, loss of appetite, spinning sensation, weight loss.	Continue. Call doctor when convenient.
Rare:	
Other symptoms.	Continue. Call doctor when convenient.

WARNINGS & PRECAUTIONS

Don't take if:
You are allergic to levetiracetam. Allergic reactions are rare, but may occur.

Before you start, consult your doctor if:
- You have kidney disease.
- You are allergic to any medication, food or other substance.
- You have any other medical problems.

Over age 60:
No special problems expected.

Pregnancy:
Decide with your doctor if drug benefits justify risk to unborn baby. Risk category C (see page xviii).

Breastfeeding; Lactation; Nursing Mothers:
This drug is generally acceptable with breastfeeding. Discuss risks and benefits with your doctor.

Infants & children up to age 18:
Approved for children one month of age and older. Follow instructions provided by your child's doctor.

Prolonged use:
No special problems expected. Follow-up laboratory blood studies may be recommended by your doctor.

Skin & sunlight:
No problems expected.

Driving, piloting or hazardous work:
Don't drive or pilot aircraft until you learn how medicine affects you. Don't work around dangerous machinery. Don't climb ladders or work in high places. Danger increases if you drink alcohol or take other medicines affecting alertness and reflexes such as antihistamines, tranquilizers, sedatives, pain medicine, narcotics and mind-altering drugs.

Discontinuing:
Don't discontinue without doctor's approval due to risk of increased seizure activity. The dosage may need to be gradually decreased before stopping the drug completely.

Others:
- Advise any doctor, dentist or pharmacist whom you consult that you take this medicine.
- Levetiracetam may be used with other anticonvulsant drugs and additional side effects may also occur. If they do, discuss them with your doctor.
- Rarely, antiepileptic drugs may lead to suicidal thoughts and behaviors. Call doctor right away if suicidal symptoms or unusual behaviors occur.
- Wear or carry medical identification to show your seizure disorder and the drugs you take.

POSSIBLE INTERACTION WITH OTHER DRUGS

GENERIC NAME OR DRUG CLASS	COMBINED EFFECT
Anticonvulsants,* other	May decrease or increase effect of both drugs.
CNS Depressants*	Increased sedative effect.

POSSIBLE INTERACTION WITH OTHER SUBSTANCES

INTERACTS WITH	COMBINED EFFECT
Alcohol:	Increased sedative effect. Avoid.
Beverages:	None expected.
Cocaine:	Unknown. Avoid.
Foods:	None expected.
Marijuana:	Consult doctor.
Tobacco:	None expected.

***See Glossary**

LEVOCARNITINE

BRAND NAMES

Carnitor
Carnitor Sugar-free Oral Solution
L-Carnitine
VitaCarn

BASIC INFORMATION

Habit forming? No
Prescription needed? Yes
Available as generic? Yes
Drug class: Nutritional supplement

USES

Treats carnitine deficiency, a genetic impairment preventing normal utilization from diet.

DOSAGE & USAGE INFORMATION

How to take:

- Oral solution—Take after meals with liquid to decrease stomach irritation.
- Tablet—Swallow with liquid or food to lessen stomach irritation. If you can't swallow whole, crumble tablet and take with liquid or food.
- Injection—Given by medical professional.

When to take:
Immediately following or during meals to reduce stomach irritation.

If you forget a dose:
Take as soon as you remember. If it is almost time for the next dose, wait for that dose (don't double this dose) and resume regular schedule.

What drug does:
Facilitates normal use of fat to produce energy. Dietary source is meat and milk.

Time lapse before drug works:
Immediate action.

Don't take with:
Any other medicine or any dietary supplement without consulting your doctor or pharmacist.

OVERDOSE

SYMPTOMS:
Unknown; may cause diarrhea.
WHAT TO DO:
If person uses much larger amount than prescribed or if accidentally swallowed, call poison control center 1-800-222-1222 for instructions or dial 911 (emergency) for help.

POSSIBLE ADVERSE REACTIONS OR SIDE EFFECTS

SYMPTOMS	WHAT TO DO
Life-threatening:	
None expected.	
Common:	
Changed body odor.	Continue. Call doctor when convenient.
Infrequent:	
Diarrhea, abdominal pain, nausea, vomiting.	Continue, but call doctor right away.
Rare:	
None expected.	

WARNINGS & PRECAUTIONS

Don't take if:
You are allergic to levocarnitine. Allergic reactions are rare, but may occur.

Before you start, consult your doctor if:
- You have a seizure disorder.
- You have severe kidney disorder.

Over age 60:
No problems expected.

Pregnancy:
Decide with your doctor if drug benefits justify risk to unborn baby. Risk category B (see page xviii).

Breastfeeding; Lactation; Nursing Mothers:
This drug is generally acceptable with breastfeeding. Discuss risks and benefits with your doctor.

Infants & children up to age 18:
Follow instructions provided by your child's doctor.

Prolonged use:
Talk to your doctor about the need for follow-up medical examinations or laboratory studies to check triglycerides.

Skin & sunlight:
No problems expected.

Driving, piloting or hazardous work:
No problems expected.

Discontinuing:
Don't discontinue without consulting doctor. Dose may require gradual reduction if you have taken drug for a long time. Doses of other drugs may also require adjustment.

Others:
- Advise any doctor, dentist or pharmacist whom you consult that you take this medicine.
- Do not use a product named "vitamin B-T" sold in health food stores. It contains dextro- and levo-carnitine. This product completely negates the effectiveness of levocarnitine (L-carnitine). Only the L-carnitine form is effective in treating carnitine deficiency.

POSSIBLE INTERACTION WITH OTHER DRUGS

GENERIC NAME OR DRUG CLASS	COMBINED EFFECT
None expected.	

POSSIBLE INTERACTION WITH OTHER SUBSTANCES

INTERACTS WITH	COMBINED EFFECT
Alcohol:	None expected.
Beverages:	None expected.
Cocaine:	Unknown. Avoid.
Foods:	None expected.
Marijuana:	Consult doctor.
Tobacco:	None expected.

LEVODOPA

BRAND NAMES

Dopar
Larodopa

BASIC INFORMATION

Habit forming? No
Prescription needed? Yes
Available as generic? Yes
Drug class: Antiparkinsonism

USES

Controls Parkinson's disease symptoms such as rigidity, tremor and unsteady gait.

DOSAGE & USAGE INFORMATION

How to take:
Tablet or capsule—Swallow with liquid or food to lessen stomach irritation. If you can't swallow whole, crumble tablet or open capsule and take with liquid or food.

When to take:
At the same times each day.

If you forget a dose:
Take as soon as you remember. If it is almost time for the next dose, wait for that dose (don't double this dose) and resume regular schedule.

What drug does:
Restores chemical balance necessary for normal nerve impulses.

Time lapse before drug works:
2 to 3 weeks to improve; 6 weeks or longer for maximum benefit.

Don't take with:
Any other medicine or any dietary supplement without consulting your doctor or pharmacist.

OVERDOSE

SYMPTOMS:
Muscle twitch, spastic eyelid closure, nausea, vomiting, diarrhea, irregular and rapid pulse, weakness, fainting, confusion, agitation, hallucination, coma.
WHAT TO DO:
- **Dial 911 for all medical emergencies or call poison control center 1-800-222-1222 for instructions.**
- **See emergency information on last 3 pages of this book.**

POSSIBLE ADVERSE REACTIONS OR SIDE EFFECTS

SYMPTOMS	WHAT TO DO
Life-threatening: In case of overdose, see previous column.	
Common:	
• Uncontrollable body movements.	Discontinue. Call doctor right away.
• Mood change, diarrhea, depression, anxiety.	Continue. Call doctor when convenient.
• Dry mouth, body odor.	No action necessary.
Infrequent:	
• Fainting, severe dizziness, headache, insomnia, nightmares, itchy skin, rash, nausea, vomiting, irregular heartbeat, eyelid spasm.	Discontinue. Call doctor right away.
• Flushed face, muscle twitching, discolored or dark urine, difficult urination, blurred vision, appetite loss, constipation, tiredness.	Continue. Call doctor when convenient.
Rare:	
• High blood pressure.	Discontinue. Call doctor right away.
• Upper abdominal pain, increased sex drive.	Continue. Call doctor when convenient.

WARNINGS & PRECAUTIONS

Don't take if:
- You are allergic to levodopa or carbidopa. Allergic reactions are rare, but may occur.
- You have taken a monoamine oxidase (MAO) inhibitor* in past 2 weeks.
- You have glaucoma (narrow-angle type).

Before you start, consult your doctor if:
- You have diabetes or epilepsy.
- You have had high blood pressure, heart or lung disease.
- You have had liver or kidney disease.
- You have a peptic ulcer.
- You have malignant melanoma.
- You will have surgery within 2 months, including dental surgery, requiring general or spinal anesthesia.

Over age 60:
No special problems expected.

Pregnancy:
Decide with your doctor if drug benefits justify risk to unborn baby. Risk category C (see page xviii).

Breastfeeding; Lactation; Nursing Mothers:
This drug is generally not recommended with breastfeeding. Discuss risks and benefits with your doctor.

Infants & children up to age 18:
Safety and effectiveness in this age group have not been established. Consult doctor.

Prolonged use:
- May lead to uncontrolled movements of head, face, mouth, tongue, arms or legs.
- Talk to your doctor about the need for follow-up medical examinations or laboratory studies.

Skin & sunlight:
No problems expected.

Driving, piloting or hazardous work:
Don't drive or pilot aircraft until you learn how medicine affects you. Don't work around dangerous machinery. Don't climb ladders or work in high places. Danger increases if you drink alcohol or take medicine affecting alertness and reflexes, such as antihistamines, tranquilizers, sedatives, pain medicine, narcotics and mind-altering drugs.

Discontinuing:
Don't discontinue without doctor's advice until you complete prescribed dose, even though symptoms diminish or disappear.

Others:
- Expect to start with small dose and increase gradually to lessen frequency and severity of adverse reactions.
- Advise any doctor, dentist or pharmacist whom you consult that you take this medicine.

POSSIBLE INTERACTION WITH OTHER DRUGS

GENERIC NAME OR DRUG CLASS	COMBINED EFFECT
Antidepressants, tricyclic (TCA)*	Decreased blood pressure. Weakness and faintness when arising from bed or chair.
Antiparkinsonism drugs, other*	Increased levodopa effect.
Bupropion	Increased levodopa effect.
Haloperidol	Decreased levodopa effect.
Loxapine	Decreased levodopa effect.
MAO inhibitors*	Dangerous rise in blood pressure.
Methyldopa	Decreased levodopa effect.
Molindone	Decreased levodopa effect.
Olanzapine	May decrease levodopa effect.
Papaverine	Decreased levodopa effect.
Phenothiazines*	Decreased levodopa effect.
Phenytoin	Decreased levodopa effect.
Pyridoxine (Vitamin B-6)	Decreased levodopa effect.
Quetiapine	Decreased levodopa effect.
Rauwolfia alkaloids*	Decreased levodopa effect.
Selegiline	May require reduced dosage of levodopa.
Thioxanthenes*	Decreased levodopa effect.
Ziprasidone	Decreased levodopa effect.

POSSIBLE INTERACTION WITH OTHER SUBSTANCES

INTERACTS WITH	COMBINED EFFECT
Alcohol:	None expected.
Beverages:	None expected.
Cocaine:	Increased risk of heartbeat irregularity. Avoid.
Foods: High-protein diet.	Decreased levodopa effect.
Marijuana:	Increased fatigue, lethargy, fainting. Avoid.
Tobacco:	None expected.

LINACLOTIDE

BRAND NAMES

Linzess

BASIC INFORMATION

Habit forming? No
Prescription needed? Yes
Available as generic? No
Drug class: Guanylate cyclase-C agonist

USES

- Relieves symptoms of irritable bowel syndrome (IBS) with constipation.
- Treats chronic constipation of unknown cause (idiopathic constipation).

DOSAGE & USAGE INFORMATION

How to take:
Capsule—Swallow whole with a glass of water. Do not break, chew or crush capsule.

When to take:
Drug is taken once a day (at the same time each day) on an empty stomach and at least 30 minutes before your first meal of the day.

If you forget a dose:
Take as soon as you remember. If it is almost time for the next dose, wait for that dose (don't double this dose) and resume regular schedule.

What drug does:
It works in the intestine (it is not absorbed in the body) by increasing intestinal fluid secretion which eases the passage of stool and relieves the symptoms associated with constipation.

Time lapse before drug works:
A few days to a week.

Don't take with:
Any other medicine or any dietary supplement without consulting your doctor or pharmacist.

OVERDOSE

SYMPTOMS:
Diarrhea and possibly other symptoms.
WHAT TO DO:
If person uses much larger amount than prescribed or if accidentally swallowed, call poison control center 1-800-222-1222 for instructions or dial 911 (emergency) for help.

POSSIBLE ADVERSE REACTIONS OR SIDE EFFECTS

SYMPTOMS	WHAT TO DO
Life-threatening: None expected.	
Common:	
• Diarrhea (may be severe), pain in abdomen or stomach.	Discontinue. Call doctor right away.
• Passing gas, feeling bloated.	Continue. Call doctor when convenient.
Infrequent: Upset stomach, vomiting, heartburn, unable to control bowel movements, gastrointestinal infections, headache.	Continue. Call doctor when convenient.
Rare:	
• Bleeding from the rectum, black or tarry stools.	Discontinue. Call doctor right away.
• Unusual symptoms that cause concern.	Continue. Call doctor when convenient.

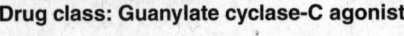

WARNINGS & PRECAUTIONS

Don't take if:
- You are allergic to linaclotide. Allergic reactions are rare, but may occur.
- Patient is under age 6.

Before you start, consult your doctor if:
- You have or have had any problem with bowel or stomach blockage (may be referred to as mechanical gastrointestinal obstruction).
- You have severe diarrhea.

Over age 60:
No special problems expected.

Pregnancy:
Decide with your doctor if drug benefits justify risk to unborn baby. Risk category C (see page xviii).

Breastfeeding; Lactation; Nursing Mothers:
This drug is generally acceptable with breastfeeding. Discuss risks and benefits with your doctor.

Infants & children up to age 18:
- Do not give to children under age 6.
- Safety and efficacy for children ages 6 to 17 has not been established. Consult doctor.

Prolonged use:
See your doctor for regular visits to make sure the drug is working properly and to check for unwanted effects.

Skin & sunlight:
No problems expected.

Driving, piloting or hazardous work:
No problems expected.

Discontinuing:
No problems expected, but consult your doctor before discontinuing.

Others:
- Advise any doctor, dentist or pharmacist whom you consult that you take this drug.
- Keep drug in its original container. Do not remove the drying agent in the container.

POSSIBLE INTERACTION WITH OTHER DRUGS

GENERIC NAME OR DRUG CLASS	COMBINED EFFECT
None expected.	

POSSIBLE INTERACTION WITH OTHER SUBSTANCES

INTERACTS WITH	COMBINED EFFECT
Alcohol:	None expected.
Beverages:	None expected.
Cocaine:	Unknown. Avoid.
Foods:	None expected.
Marijuana:	Consult doctor.
Tobacco:	None expected.

LINCOMYCIN

BRAND NAMES

Lincocin

BASIC INFORMATION

Habit forming? No
Prescription needed? Yes
Available as generic? Yes
Drug class: Antibacterial

USES

Treatment of bacterial infections that are susceptible to lincomycin.

DOSAGE & USAGE INFORMATION

How to take:
Capsule—Swallow with liquid 1 hour before or 2 hours after eating. Drink 8 ounces of water with each dose.

When to take:
At the same times each day.

If you forget a dose:
Take as soon as you remember. If it is almost time for the next dose, wait for that dose (don't double this dose) and resume regular schedule.

What drug does:
Destroys susceptible bacteria. Does not kill viruses.

Time lapse before drug works:
3 to 5 days.

Don't take with:
Any other medicine or any dietary supplement without consulting your doctor or pharmacist.

OVERDOSE

SYMPTOMS:
Severe nausea, vomiting, diarrhea.
WHAT TO DO:
If person uses much larger amount than prescribed or if accidentally swallowed, call poison control center 1-800-222-1222 for instructions or dial 911 (emergency) for help.

POSSIBLE ADVERSE REACTIONS OR SIDE EFFECTS

SYMPTOMS	WHAT TO DO
Life-threatening:	
Rare, severe allergic reaction (anaphylaxis)—difficulty breathing, hives, itching, swelling (throat, face), vomiting, dizziness.	Seek emergency treatment immediately.
Common:	
Mild diarrhea, mild stomach cramps, nausea.	Continue. Call doctor when convenient.
Infrequent:	
• Unusual thirst, vomiting, stomach cramps, severe and watery diarrhea with blood or mucus, painful and swollen joints, yellow skin or eyes, fever, weight loss, tiredness, weakness.	Discontinue. Call doctor right away.
• Itch around groin or rectum or armpits, white patches in mouth, vaginal discharge or itching.	Continue. Call doctor when convenient.
Rare:	
Skin rash.	Discontinue. Call doctor right away.

WARNINGS & PRECAUTIONS

Don't take if:
- You are allergic to lincomycins.
- You have had ulcerative colitis.
- Prescribed for infant under 1 month old.

Before you start, consult your doctor if:
- You have had yeast infections of mouth, skin or vagina.
- You will have surgery within 2 months, including dental surgery, requiring general or spinal anesthesia.
- You have kidney or liver disease.
- You have allergies of any kind.

Over age 60:
Adverse reactions and side effects may be more frequent and severe than in younger persons.

Pregnancy:
Decide with your doctor if drug benefits justify risk to unborn baby. Risk category C (see page xviii).

Breastfeeding; Lactation; Nursing Mothers:
This drug is generally acceptable with breastfeeding. Discuss risks and benefits with your doctor.

Infants & children up to age 18:
Follow instructions provided by your child's doctor.

Prolonged use:
- Severe colitis with diarrhea and bleeding.
- You may become more susceptible to infections caused by germs not responsive to lincomycin.
- Talk to your doctor about the need for follow-up medical examinations or laboratory studies or proctosigmoidoscopy.

Skin & sunlight:
No problems expected.

Driving, piloting or hazardous work:
No problems expected.

Discontinuing:
Don't discontinue without doctor's advice until you complete prescribed dose, even though symptoms diminish or disappear.

Others:
- May interfere with the accuracy of some medical tests.
- Advise any doctor, dentist or pharmacist whom you consult that you take use medicine.

POSSIBLE INTERACTION WITH OTHER DRUGS

GENERIC NAME OR DRUG CLASS	COMBINED EFFECT
Antidiarrheal preparations*	Decreased lincomycin effect.
Attapulgite	May decrease effectiveness of lincomycin.
Chloramphenicol	Decreased lincomycin effect.
Erythromycins*	Decreased lincomycin effect.
Narcotics*	Increased risk of respiratory problems.

POSSIBLE INTERACTION WITH OTHER SUBSTANCES

INTERACTS WITH	COMBINED EFFECT
Alcohol:	None expected.
Beverages:	None expected.
Cocaine:	Unknown. Avoid.
Foods:	None expected.
Marijuana:	Consult doctor.
Tobacco:	None expected.

***See Glossary**

LINEZOLID

BRAND NAMES

Zyvox

BASIC INFORMATION

Habit forming? No
Prescription needed? Yes
Available as generic? No
Drug class: Antibacterial, antibiotic (oxazolidinone)

USES

Treats bacterial infections of the blood, lungs and skin. It may also be used for other conditions as determined by your doctor.

DOSAGE & USAGE INFORMATION

How to take:

- Tablet—Take with full glass of water. If you can't swallow whole, crumble tablet and take with liquid or food.
- Oral suspension—Take as directed on label. The medicine should be gently mixed by inverting the bottle 3 to 5 times before each dose. Do not shake the bottle.

When to take:
As directed by your doctor. Usually every 12 hours.

If you forget a dose:
Take as soon as you remember. If it is almost time for the next dose, wait for that dose (don't double this dose) and resume regular schedule.

What drug does:
Destroys bacteria in the body, probably by blocking protein production inside bacteria.

Time lapse before drug works:
10 to 14 days for most infections, but some infections may take longer. Continue taking this medicine for the full time of treatment even if you begin to feel better after a few days.

Continued next column

OVERDOSE

SYMPTOMS:
Unknown.
WHAT TO DO:

- **Dial 911 for all medical emergencies or call poison control center 1-800-222-1222 for instructions.**
- **See emergency information on last 3 pages of this book.**

Don't take with:
Any other medicine or any dietary supplement without consulting your doctor or pharmacist.

POSSIBLE ADVERSE REACTIONS OR SIDE EFFECTS

SYMPTOMS	WHAT TO DO
Life-threatening:	
Rare, severe allergic reaction (anaphylaxis)—difficulty breathing, hives, itching, swelling (throat, face), vomiting, dizziness.	Seek emergency treatment immediately.
Common:	
• Diarrhea, headache.	Continue. Call doctor when convenient.
• Nausea.	Continue. Call doctor if symptoms persist.
Infrequent:	
• Fever, sore mouth or tongue, rash, black or tarry stools, chest pain, chills, cough, painful or difficult urination, unusual bleeding or bruising, unusual tiredness or weakness, vomiting.	Discontinue. Call doctor right away.
• Constipation, change in taste, sleeplessness, vaginal yeast infection.	Continue. Call doctor when convenient.
Rare:	
None expected.	

WARNINGS & PRECAUTIONS

Don't take if:
You are allergic to linezolid.

Before you start, consult your doctor if:
- You are using any other medication.
- You have a history of bleeding problems, diarrhea, high blood pressure or any other medical problems.

Over age 60:
No special problems expected.

Pregnancy:
Decide with your doctor if drug benefits justify risk to unborn baby. Risk category C (see page xviii).

Breastfeeding; Lactation; Nursing Mothers:
This drug is generally acceptable with breastfeeding. Discuss risks and benefits with your doctor.

Infants & children up to age 18:
Follow instructions provided by your child's doctor.

Prolonged use:
Usually not prescribed for long-term use.

Skin & sunlight:
None expected.

Driving, piloting or hazardous work:
None expected.

Discontinuing:
Don't discontinue without consulting doctor or completing prescribed dosage.

Others:
- May affect accuracy of some laboratory test values.
- Do not store in the bathroom, near the kitchen sink or in other damp places. Heat or moisture may cause the medicine to break down.
- Advise any doctor, dentist or pharmacist whom you consult that you take this medicine.

POSSIBLE INTERACTION WITH OTHER DRUGS

GENERIC NAME OR DRUG CLASS	COMBINED EFFECT
Pseudoephedrine	May increase blood pressure.
Serotonergics*	Serotonin syndrome.*

POSSIBLE INTERACTION WITH OTHER SUBSTANCES

INTERACTS WITH	COMBINED EFFECT
Alcohol:	None expected.
Beverages:	None expected.
Cocaine:	Unknown. Avoid.
Foods:	None expected.
Marijuana:	Consult doctor.
Tobacco:	None expected.

***See Glossary**

LITHIUM

BRAND NAMES

Carbolith
Cibalith-S
Duralith
Eskalith
Eskalith CR
Lithane
Lithizine
Lithobid
Lithonate
Lithotabs

BASIC INFORMATION

Habit forming? No
Prescription needed? Yes
Available as generic? Yes
Drug class: Mood stabilizer

USES

- Normalizes mood and behavior in bipolar (manic-depressive) disorder.
- Treats alcohol toxicity and addiction.
- Treats schizoid personality disorders.

DOSAGE & USAGE INFORMATION

How to take:

- Tablet or capsule—Swallow with liquid or food to lessen stomach irritation. If you can't swallow whole, crumble tablet or open capsule and take with liquid or food. Drink plenty of liquids each day, especially in hot weather.
- Extended-release tablet—Swallow each dose whole. Do not crush.
- Syrup—Take at mealtime. Follow with 8 oz. water.

When to take:
At the same times each day, preferably at mealtime.

If you forget a dose:
Take as soon as you remember. If it is almost time for the next dose, wait for that dose (don't double this dose) and resume regular schedule.

Continued next column

OVERDOSE

SYMPTOMS:
Moderate overdose increases some side effects and may cause diarrhea, nausea. Large overdose may cause vomiting, muscle weakness, convulsions, stupor and coma.
WHAT TO DO:

- **Dial 911 for all medical emergencies or call poison control center 1-800-222-1222 for instructions.**
- **See emergency information on last 3 pages of this book.**

What drug does:
May correct chemical imbalance in brain's transmission of nerve impulses that influence mood and behavior.

Time lapse before drug works:
1 to 3 weeks. May require 3 months before depressive phase of illness improves.

Don't take with:
Any other medicine or any dietary supplement without consulting your doctor or pharmacist.

POSSIBLE ADVERSE REACTIONS OR SIDE EFFECTS

SYMPTOMS	WHAT TO DO
Life-threatening:	
In case of overdose, see previous column.	
Common:	
Dizziness, diarrhea, nausea, vomiting, shakiness, tremor, dry mouth, thirst, decreased sexual ability, increased urination, loss of appetite.	Continue. Call doctor when convenient.
Infrequent:	
• Rash, stomach pain, fainting, heartbeat irregularities, shortness of breath, ear noises, swollen hands or feet, slurred speech.	Discontinue. Call doctor right away.
• Thyroid impairment (coldness, dry or puffy skin), muscle aches, headache, weight gain, fatigue, menstrual irregularities, acnelike breakouts, drowsiness, confusion, weakness.	Continue. Call doctor when convenient.
Rare:	
• Blurred vision, eye pain.	Discontinue. Call doctor right away.
• Jerking of arms and legs, worsening of psoriasis, hair loss.	Continue. Call doctor when convenient.

WARNINGS & PRECAUTIONS

Don't take if:

- You are allergic to lithium or tartrazine dye. Allergic reactions are rare, but may occur.
- You have kidney or heart disease.
- Patient is younger than 12.

Before you start, consult your doctor if:

- You plan to become pregnant within medication period.

- You have diabetes, thyroid disorder, epilepsy, brain disease, schizophrenia, difficult urination, heart disease, kidney disorder, Parkinson's or history of leukemia.
- You are on a low-salt diet or drink more than 4 cups of coffee per day.
- You plan surgery within 2 months.

Over age 60:
Adverse reactions and side effects may be more frequent and severe than in younger persons.

Pregnancy:
Consult doctor. Use of the drug during pregnancy is not recommended. It could cause harm to the unborn baby. Risk category D (see page xviii).

Breastfeeding; Lactation; Nursing Mothers:
This drug is generally acceptable with breastfeeding. Discuss risks and benefits with your doctor.

Infants & children up to age 18:
Approved for children over age 12. Follow instructions provided by your child's doctor.

Prolonged use:
- Increased risk of thyroid disorder.
- Talk to your doctor about the need for follow-up medical exams and lab studies to check progress of drug use.

Skin & sunlight:
No problems expected.

Driving, piloting or hazardous work:
Don't drive or pilot aircraft until you learn how medicine affects you. Don't work around dangerous machinery. Don't climb ladders or work in high places. Danger increases if you drink alcohol or take medicine affecting alertness.

Discontinuing:
Don't discontinue without consulting doctor. Dose may require gradual reduction if you have taken drug for a long time. Doses of other drugs may also require adjustment. Quitting this medication when feeling well creates risk of relapse which may not respond to restarting the medication.

Others:
- Regular checkups, periodic blood tests, and tests of lithium levels and thyroid function recommended.
- Avoid exercise in hot weather and other activities that cause heavy sweating. This contributes to lithium poisoning. It is essential to take adequate fluids during hot weather to avoid toxicity.
- Call your doctor if you have an illness that causes heavy sweating, vomiting, or diarrhea. The loss of too much salt and water from your body could cause lithium toxicity.
- Advise any doctor, dentist or pharmacist whom you consult that you take this medicine.
- Some products contain tartrazine dye. Avoid, especially if allergic to aspirin.

POSSIBLE INTERACTION WITH OTHER DRUGS

GENERIC NAME OR DRUG CLASS	COMBINED EFFECT
Acetazolamide	Decreased lithium effect.
Antihistamines*	Possible excessive sedation.
Anti-inflammatory drugs, nonsteroidal (NSAIDs)*	Increased effect of lithium.
Bupropion	Increased risk of seizures.
Carbamazepine	Increased lithium effect.
Desmopressin	Possible decreased desmopressin effect.
Diazepam	Possible hypothermia.
Diclofenac	Possible increase in effect and toxicity.
Didanosine	Increased risk of peripheral neuropathy.

Continued on page 903

POSSIBLE INTERACTION WITH OTHER SUBSTANCES

INTERACTS WITH	COMBINED EFFECT
Alcohol:	Possible lithium poisoning.
Beverages: Caffeine drinks.	Decreased lithium effect.
Cocaine:	Increased risk of side effects. Avoid.
Foods: Salt.	High intake could decrease lithium effect. Low intake could increase lithium effect. *Don't* restrict intake.
Marijuana:	Increased risk of side effects. Avoid.
Tobacco:	None expected.

***See Glossary**

LOMUSTINE

BRAND NAMES

CCNU
CeeNU

BASIC INFORMATION

Habit forming? No
Prescription needed? Yes
Available as generic? No
Drug class: Antineoplastic

USES

- Treats brain cancer and Hodgkin's lymphoma.
- Sometimes used to treat breast, lung, skin and gastrointestinal cancer.

DOSAGE & USAGE INFORMATION

How to take:
Capsule—Swallow with liquid. If you can't swallow whole, open capsule and take with liquid or food. Instructions to take on empty stomach mean 1 hour before or 2 hours after eating. Note: There may be two or more different types of capsules in the container. This is not an error.

When to take:
According to doctor's instructions. Usual course of treatment requires single dosage repeated every 6 weeks.

If you forget a dose:
Take as soon as you remember. If it is almost time for the next dose, wait for that dose (don't double this dose) and resume regular schedule.

What drug does:
Interferes with growth of cancer cells.

Time lapse before drug works:
None. Works immediately.

Don't take with:
Any other medicine or any dietary supplement without consulting your doctor or pharmacist.

OVERDOSE

SYMPTOMS:
Fever and chills (infection), diarrhea, cough, stomach pain, dizziness, shortness of breath, tiredness.
WHAT TO DO:
- **Dial 911 for all medical emergencies or call poison control center 1-800-222-1222 for instructions.**
- **See emergency information on last 3 pages of this book.**

POSSIBLE ADVERSE REACTIONS OR SIDE EFFECTS

SYMPTOMS	WHAT TO DO
Life-threatening:	
In case of overdose, see previous column.	
Common:	
• Fever, chills, difficult urination, unusual bleeding.	Continue. but call doctor right away.
• Appetite loss, nausea, hair loss.	Continue. Call doctor when convenient.
Infrequent:	
• Paleness, confusion, slurred speech, mouth sores, skin rash.	Continue. but call doctor right away.
• Darkened skin.	Continue. Call doctor when convenient.
Rare:	
Shortness of breath, yellow skin and eyes, cough.	Continue. but call doctor right away.

WARNINGS & PRECAUTIONS

Don't take if:
- You are allergic to lomustine. Allergic reactions are rare, but may occur.
- You have chicken pox.
- You have shingles (herpes zoster).

Before you start, consult your doctor if:
- You have an infection.
- You have kidney or lung disease.
- You have had previous cancer chemotherapy or radiation treatment.

Over age 60:
Adverse reactions and side effects may be more frequent and severe than in younger persons.

Pregnancy:
Consult doctor. Use of the drug during pregnancy is not recommended. It could cause harm to the unborn baby. Risk category D (see page xviii).

Breastfeeding; Lactation; Nursing Mothers:
This drug is generally not recommended with breastfeeding. Discuss risks and benefits with your doctor.

Infants & children up to age 18:
Follow instructions provided by your child's doctor.

Prolonged use:
Talk to your doctor about the need for follow-up medical examinations or laboratory studies to check kidney function, liver function and complete blood counts (white blood cell count, platelet count, red blood cell count, hemoglobin, hematocrit).

Skin & sunlight:
No problems expected.

Driving, piloting or hazardous work:
Don't drive or pilot aircraft until you learn how medicine affects you. Don't work around dangerous machinery. Don't climb ladders or work in high places. Danger increases if you drink alcohol or take medicine affecting alertness and reflexes.

Discontinuing:
Call doctor if any of these occur after discontinuing: black or tarry stools, bloody urine, hoarseness, bleeding or bruising, fever or chills.

Others:
- Advise any doctor, dentist or pharmacist whom you consult that you take this medicine.
- May affect results in some medical tests.
- Avoid immunizations, if possible.
- Avoid persons with infections.
- Check with doctor about instructions for brushing or flossing teeth.
- Avoid contact sports.

POSSIBLE INTERACTION WITH OTHER DRUGS

GENERIC NAME OR DRUG CLASS	COMBINED EFFECT
Antineoplastic drugs, other*	Increased chance of drug toxicity.
Blood dyscrasia-causing medicines*	Adverse effect on bone marrow, causing decreased white cells and platelets.
Bone marrow depressants,* other	Increased risk of bone marrow depression.
Clozapine	Toxic effect on bone marrow.
Levamisole	Increased risk of bone marrow depression.
Tiopronin	Increased risk of toxicity to bone marrow.
Vaccines, live or killed virus	Increased chance of toxicity or reduced effectiveness of vaccine. Wait 3 to 12 months after lomustine treatment before getting vaccination.

POSSIBLE INTERACTION WITH OTHER SUBSTANCES

INTERACTS WITH	COMBINED EFFECT
Alcohol:	Increased chance of liver damage.
Beverages:	None expected.
Cocaine:	Unknown. Avoid.
Foods:	None expected.
Marijuana:	Consult doctor.
Tobacco:	None expected.

***See Glossary**

LOPERAMIDE

BRAND NAMES

Apo-Loperamide Caplets
Imodium
Imodium A-D
Imodium Advanced
Imodium Multi Symptom Relief
Kaopectate II Caplets
Pepto Diarrhea Control

BASIC INFORMATION

Habit forming? No, unless taken in high doses for long periods.
Prescription needed? Yes, for some
Available as generic? Yes
Drug class: Antidiarrheal

USES

- Treats mild to moderate diarrhea. Used in conjunction with fluids, appropriate diet and rest. Treats symptoms only. Does not cure any disorder that causes diarrhea.
- Treats chronic diarrhea associated with inflammatory bowel disease.

DOSAGE & USAGE INFORMATION

How to take:

- Tablet or capsule—Swallow with food to lessen stomach irritation.
- Liquid—Follow label instructions and use marked dropper.

When to take:
No more often than directed on label.

If you forget a dose:
Take as soon as you remember. If it is almost time for the next dose, wait for that dose (don't double this dose) and resume regular schedule.

Continued next column

OVERDOSE

SYMPTOMS:
Constipation, lethargy, drowsiness or unconsciousness.
WHAT TO DO:

- **Dial 911 for all medical emergencies or call poison control center 1-800-222-1222 for instructions.**
- **See emergency information on last 3 pages of this book.**

What drug does:
Blocks digestive tract's nerve supply, which reduces irritability and contractions in intestinal tract.

Time lapse before drug works:
1 to 2 hours.

Don't take with:
Any other medicine or any dietary supplement without consulting your doctor or pharmacist.

POSSIBLE ADVERSE REACTIONS OR SIDE EFFECTS

SYMPTOMS	WHAT TO DO
Life-threatening:	
Rare, severe allergic reaction (anaphylaxis)—difficulty breathing, hives, itching, swelling (throat, face), vomiting, dizziness.	Seek emergency treatment immediately.
Common:	
None expected.	
Infrequent:	
None expected.	
Rare:	
• Nausea, vomiting, bloating, constipation, appetite loss, rash, abdominal pain.	Discontinue. Call doctor right away.
• Drowsiness, dizziness, dry mouth.	Continue. Call doctor when convenient.

WARNINGS & PRECAUTIONS

Don't take if:
- You have severe colitis.
- You have colitis resulting from antibiotic treatment or infection.
- You are allergic to loperamide.

Before you start, consult your doctor if:
- You are dehydrated from fluid loss caused by diarrhea.
- You have liver disease.

Over age 60:
Adverse reactions and side effects may be more frequent and severe than in younger persons.

Pregnancy:
Decide with your doctor if drug benefits justify risk to unborn baby. Risk category B (see page xviii).

Breastfeeding; Lactation; Nursing Mothers:
This drug is generally acceptable with breastfeeding. Discuss risks and benefits with your doctor.

Infants & children up to age 18:
- Don't give to children under age 6 without doctor's approval. Young children are not able to describe their symptoms accurately, and a proper diagnosis needs to be made before starting any treatment.
- For children over age 6, follow package instructions or doctor's directions for correct dosage amount.

Prolonged use:
Habit forming at high dose.

Skin & sunlight:
No problems expected.

Driving, piloting or hazardous work:
Don't drive or pilot aircraft until you learn how medicine affects you. Don't work around dangerous machinery. Don't climb ladders or work in high places. Danger increases if you drink alcohol or take medicine affecting alertness and reflexes.

Discontinuing:
- May be unnecessary to finish medicine. Follow doctor's instructions.
- After discontinuing, consult doctor if you experience muscle cramps, nausea, vomiting, trembling, stomach cramps or unusual sweating.

Others:
- Advise any doctor, dentist or pharmacist whom you consult that you take this drug.
- Call doctor if acute diarrhea lasts longer than 48 hours after starting drug.
- In chronic diarrhea, loperamide is unlikely to be effective if diarrhea doesn't improve in 10 days.

POSSIBLE INTERACTION WITH OTHER DRUGS

GENERIC NAME OR DRUG CLASS	COMBINED EFFECT
Antibiotics*	Increased risk of diarrhea.
Narcotic analgesics	Increased risk of severe constipation.

POSSIBLE INTERACTION WITH OTHER SUBSTANCES

INTERACTS WITH	COMBINED EFFECT
Alcohol:	Depressed brain function. Avoid.
Beverages:	None expected.
Cocaine:	Unknown. Avoid.
Foods:	None expected.
Marijuana:	Consult doctor.
Tobacco:	None expected.

***See Glossary**

LORCASERIN

BRAND NAMES

Belviq

BASIC INFORMATION

Habit forming? May be habit forming
Prescription needed? Yes
Available as generic? No
Drug class: Central nervous systemic stimulant; anorexiant

USES

It is used along with a reduced-calorie diet and exercise program to treat obesity in adults who have a BMI (body mass index) over 30. It is also used in adults with a BMI over 27 who have high blood pressure, high cholesterol or diabetes.

DOSAGE & USAGE INFORMATION

How to take:
Tablet—Swallow whole with water. It may be taken with or without food.

When to take:
Twice a day at the same times each day (in the morning and evening).

If you forget a dose:
Take as soon as you remember. If it is almost time for the next dose, wait for that dose (don't double this dose) and resume regular schedule.

What drug does:
It works by activating an area of the brain called the serotonin 2C receptor. This action helps a person feel full faster so they eat less than usual.

Time lapse before drug works:
It starts working within hours, but its weight loss effect is known after taking the drug for 12 weeks

Don't take with:
Any other medicine or any dietary supplement without consulting your doctor or pharmacist.

OVERDOSE

SYMPTOMS:
Specific symptoms are unclear; may have more severe side effects (e.g., nausea, headache, dizziness, euphoria, hallucinations).
WHAT TO DO:
- **Dial 911 for all medical emergencies or call poison control center 1-800-222-1222 for instructions.**
- **See emergency information on last 3 pages of this book.**

POSSIBLE ADVERSE REACTIONS OR SIDE EFFECTS

SYMPTOMS	WHAT TO DO
Life-threatening: None expected.	
Common: Headache, dizziness, fatigue, nausea, cold symptoms, back pain, dry mouth, cough, constipation, low blood sugar in diabetic patients.	Continue, but call doctor right away.
Infrequent: Diarrhea, urinary tract infection, tooth or throat pain, upset stomach, sinus infection, patients with diabetes may have high blood pressure or swelling of feet and ankles or worsening of diabetes.	Continue, but call doctor right away.
Rare:	
• Confusion, difficulty with concentration or memory, mood changes, depression, hallucinations, feeling euphoric, thoughts of suicide, an erection lasts over 4 hours, enlarged breasts in males, breast milk without childbirth, slow heart rate, serotonin syndrome or neuroleptic malignant syndrome (see Glossary for list of symptoms of each), heart valve problem (shortness of breath, swelling of feet or ankles, chest pain).	Discontinue. Call doctor right away or seek emergency help.
• Feeling anxious or stressed, insomnia, rash, dry eyes or blurred vision or other eye symptoms, muscle pain or spasms.	Continue, but call doctor right away.

WARNINGS & PRECAUTIONS

Don't take if:
- You are allergic to lorcaserin. Allergic reactions are rare, but may occur.
- You are pregnant or plan to become pregnant.

Before you start, consult your doctor if:
- You have liver or kidney problems.
- You have or have had slow heartbeat, heart block, congestive heart failure or heart valvular disease.
- You suffer from depression or mental illness.
- You have pulmonary hypertension.
- You have diabetes or low blood sugar.
- You have cancer or a blood disorder.
- You have a physical abnormality of the penis.

Over age 60:
No problems expected, but dosage may need to be adjusted depending on health status.

Pregnancy:
Consult doctor. Drug should not be used during pregnancy. Can cause harm to unborn baby. Risk category X (see page xviii).

Breastfeeding; Lactation; Nursing Mothers:
This drug is generally not recommended with breastfeeding. Discuss risks and benefits with your doctor.

Infants & children up to age 18:
Safety and effectiveness in this age group have not been established. Consult doctor.

Prolonged use:
Effects of long term use are unknown. Talk to your doctor about the need for follow-up medical exams or laboratory studies to check the effectiveness of the treatment.

Skin & sunlight:
No problems expected.

Driving, piloting or hazardous work:
Don't drive or pilot aircraft until you learn how medicine affects you. Don't work around dangerous machinery. Don't climb ladders or work in high places. Danger increases if you drink alcohol or take medicine affecting alertness and reflexes.

Discontinuing:
No problems expected, but consult your doctor before stopping the drug.

Others:
- Advise any doctor, dentist or pharmacist whom you consult that you take this drug.
- In addition to taking the drug, be sure to follow the diet and exercise plan your doctor has recommended for you.
- If you have not lost 5% of your body weight by 12 weeks of use, your doctor will likely have you stop the drug. Further use of this drug is unlikely to be of benefit in weight loss.

POSSIBLE INTERACTION WITH OTHER DRUGS

GENERIC NAME OR DRUG CLASS	COMBINED EFFECT
Antidiabetic agents*	Increased risk of low blood sugar.
Erectile dysfunction agents*	Increased risk of side effects.
Serotonergics*	Increased risk of serotonin syndrome* or neuroleptic malignant syndrome.*
Thioridazine	Increased effect of thioridazine. Avoid.
Weight loss products, other	Unknown effect. Consult doctor.

POSSIBLE INTERACTION WITH OTHER SUBSTANCES

INTERACTS WITH	COMBINED EFFECT
Alcohol:	None expected.
Beverages:	None expected.
Cocaine:	Unknown. Avoid.
Foods:	None expected.
Marijuana:	Consult doctor.
Tobacco:	None expected.

LOXAPINE

BRAND NAMES

Adasuve
Loxapac
Loxitane
Loxitane C

BASIC INFORMATION

Habit forming? No
Prescription needed? Yes
Available as generic? Yes
Drug class: Tranquilizer, antidepressant

USES

- Treats serious mental illness.
- Treats anxiety and depression.

DOSAGE & USAGE INFORMATION

How to take:

- Oral solution—Take after meals with liquid to decrease stomach irritation.
- Tablet—Swallow with liquid or food to lessen stomach irritation. If you can't swallow whole, crumble tablet and take with liquid or food.
- Capsule—Swallow with liquid or food to lessen stomach irritation. If you can't swallow whole, open capsule and take with liquid or food.
- Adasuve brand—Given by health professional.

When to take:
At the same times each day, according to instructions on prescription label.

If you forget a dose:
Take as soon as you remember. If it is almost time for the next dose, wait for that dose (don't double this dose) and resume regular schedule.

What drug does:
Blocks the effects of dopamine* in the brain.

Time lapse before drug works:
1/2 to 3 hours.

Don't take with:
Any other medicine or any dietary supplement without consulting your doctor or pharmacist.

OVERDOSE

SYMPTOMS:
Dizziness, drowsiness, severe shortness of breath, muscle spasms, coma.
WHAT TO DO:

- **Dial 911 for all medical emergencies or call poison control center 1-800-222-1222 for instructions.**
- **See emergency information on last 3 pages of this book.**

POSSIBLE ADVERSE REACTIONS OR SIDE EFFECTS

SYMPTOMS	WHAT TO DO
Life-threatening:	
In case of overdose, see previous column.	
Common:	
• Swallowing difficulty, expressionless face, stiff arms and legs, dizziness.	Discontinue. Call doctor right away.
• Increased dental problems because of dry mouth and less saliva.	Consult your dentist about a prevention program.
Infrequent:	
• Chewing movements with lip smacking, loss of balance, shuffling walk, tremor of fingers and hands, uncontrolled tongue movements.	Discontinue. Call doctor right away.
• Constipation, difficult urination, blurred vision, confusion, loss of sex drive, headache, insomnia, menstrual irregularities, weight gain, light sensitivity, nausea, inhalent may cause taste changes or irritated throat.	Continue. Call doctor when convenient.
Rare:	
Rapid heartbeat, fever, sore throat, yellow eyes or skin, unusual bleeding.	Discontinue. Call doctor right away.

WARNINGS & PRECAUTIONS

Don't take if:
You are allergic to loxapine. Allergic reactions are rare, but may occur.

Before you start, consult your doctor if:
You have a seizure disorder, an enlarged prostate, glaucoma, Parkinson's disease, heart disease, asthma, emphysema, bronchitis, urinary tract problem, blood pressure problem, blood or blood vessel disorder, kidney or liver disorder or abuse alcohol.

Over age 60:

- Adverse reactions and side effects may be more frequent and severe than in younger persons.
- Use of antipsychotic drugs in elderly patients with dementia-related psychosis may increase risk of death. Consult doctor.

Pregnancy:
Decide with your doctor if drug benefits justify risk to unborn baby. Risk category not designated (see page xviii).

Breastfeeding; Lactation; Nursing Mothers:
This drug is generally not recommended with breastfeeding. Discuss risks and benefits with your doctor.

Infants & children up to age 18:
Safety and effectiveness in this age group have not been established. Consult doctor.

Prolonged use:
Talk to your doctor about the need for follow-up medical examinations or laboratory studies.

Skin & sunlight:
May cause rash or intensify sunburn in areas exposed to sun or ultraviolet light (photosensitivity reaction). Avoid overexposure and use sunscreen. Notify doctor if reaction occurs.

Driving, piloting or hazardous work:
Don't drive or pilot aircraft until you learn how medicine affects you. Don't work around dangerous machinery. Don't climb ladders or work in high places. Danger increases if you drink alcohol or take medicine affecting alertness and reflexes.

Discontinuing:
- Don't discontinue without consulting doctor. Dose may require gradual reduction if you have taken drug for a long time. Doses of other drugs may also require adjustment.
- These symptoms may occur after medicine has been discontinued: dizziness; nausea; abdominal pain; uncontrolled movements of mouth, tongue and jaw.

Others:
- Use careful oral hygiene.
- Advise any doctor, dentist or pharmacist whom you consult that you take this medicine.

POSSIBLE INTERACTION WITH OTHER DRUGS

GENERIC NAME OR DRUG CLASS	COMBINED EFFECT
Anticonvulsants*	Decreased effect of anticonvulsant.
Antidepressants, tricyclic*	May increase toxic effects of both drugs.
Bupropion	Increased risk of seizures.
Central nervous system (CNS) depressants*	Increased sedative effects of both drugs.
Epinephrine	Rapid heart rate; drop in blood pressure.
Extrapyramidal reaction*-causing drugs	Increased risk of side effects.
Fluoxetine	Increased depressant effects of both drugs.
Guanadrel	Decreased effect of guanadrel.
Guanethidine	Decreased effect of guanethidine.
Guanfacine	Increased effects of both drugs.
Haloperidol	May increase toxic effects of both drugs.
Leucovorin	High alcohol content of leucovorin may cause adverse effects.
Methyldopa	May increase toxic effects of both drugs.
Metoclopramide	May increase toxic effects of both drugs.
Molindone	May increase toxic effects of both drugs.
Pemoline	Increased central nervous stimulation.
Pergolide	Decreased pergolide effect.
Phenothiazines*	May increase toxic effects of both drugs.
Pimozide	May increase toxic effects of both drugs.

Continued on page 903

POSSIBLE INTERACTION WITH OTHER SUBSTANCES

INTERACTS WITH	COMBINED EFFECT
Alcohol:	May decrease effect of loxapine. Avoid.
Beverages:	None expected.
Cocaine:	May increase toxicity of both drugs. Avoid.
Foods:	None expected.
Marijuana:	May increase toxicity of both drugs. Avoid.
Tobacco:	May increase toxicity.

***See Glossary**

LUBIPROSTONE

BRAND NAMES

Amitiza

BASIC INFORMATION

Habit forming? No
Prescription needed? Yes
Available as generic? No
Drug class: Laxative

USES

- Treatment for chronic idiopathic constipation and its associated symptoms. Idiopathic means the cause of the constipation is unknown or unapparent. The constipation is not caused by a disease or medications.
- Treatment for irritable bowel syndrome with constipation in adult women.
- May be used for treatment of other disorders as determined by your doctor.

DOSAGE & USAGE INFORMATION

How to take:
Capsule—Swallow with liquid and take with food to help reduce any nausea symptoms.

When to take:
At the same time each day, usually twice a day with a meal or snack.

If you forget a dose:
Take as soon as you remember. If it is almost time for the next dose, wait for that dose (don't double this dose) and resume regular schedule.

What drug does:
It works in the gastrointestinal system to increase fluid secretion. The increased fluid softens the stool and stimulates bowel activity which helps produce bowel movements. The constipation symptoms of bloating, straining and abdominal discomfort are reduced.

Continued next column

OVERDOSE

SYMPTOMS:
Nausea, vomiting, diarrhea, dizziness, loose or watery stools, headache, retching, hot flush or flushing, abdominal pain and possibly other symptoms.
WHAT TO DO:
If person uses much larger amount than prescribed or if accidentally swallowed, call poison control center 1-800-222-1222 for instructions or dial 911 (emergency) for help.

Time lapse before drug works:
Spontaneous bowel movements may occur within 24 hours of taking the first dose. Several spontaneous bowel movements should occur within the first week.

Don't take with:
Any other medicine or any dietary supplement without consulting your doctor or pharmacist.

POSSIBLE ADVERSE REACTIONS OR SIDE EFFECTS

SYMPTOMS	WHAT TO DO
Life-threatening:	
None expected.	
Common:	
Nausea, headache, diarrhea, abdominal pain and swelling, flatulence.	Continue. Call doctor when convenient.
Infrequent:	
Severe diarrhea.	Discontinue. Call doctor right away.
Rare:	
Other symptoms that cause concern.	Continue. Call doctor when convenient.

WARNINGS & PRECAUTIONS

Don't take if:
You are allergic to lubiprostone. Allergic reactions are rare, but may occur.

Before you start, consult your doctor if:
You have a mechanical gastrointestinal obstruction (this can include problems such as adhesions, carcinomas [cancers] or hernias).

Over age 60:
No special problems expected.

Pregnancy:
Decide with your doctor if drug benefits justify risk to unborn baby. Risk category C (see page xviii).

Breastfeeding; Lactation; Nursing Mothers:
This drug is generally acceptable with breastfeeding. Discuss risks and benefits with your doctor.

Infants & children up to age 18:
Safety and effectiveness in this age group have not been established. Consult doctor.

Prolonged use:
Talk to your doctor periodically to determine if continued use of the drug is necessary.

Skin & sunlight:
No problems expected.

Driving, piloting or hazardous work:
No problems expected.

Discontinuing:
No problems expected.

Others:
- Advise any doctor, dentist or pharmacist whom you consult that you take this medicine.
- Consult your doctor if severe diarrhea occurs.

POSSIBLE INTERACTION WITH OTHER DRUGS

GENERIC NAME OR DRUG CLASS	COMBINED EFFECT
None expected.	

POSSIBLE INTERACTION WITH OTHER SUBSTANCES

INTERACTS WITH	COMBINED EFFECT
Alcohol:	None expected.
Beverages:	None expected.
Cocaine:	Unknown. Avoid.
Foods:	None expected.
Marijuana:	Consult doctor.
Tobacco:	None expected.

LURASIDONE

BRAND NAMES

Latuda

BASIC INFORMATION

Habit forming? No
Prescription needed? Yes
Available as generic? No
Drug class: Antipsychotic

USES

- Treatment for schizophrenia.
- Treatment for adults with bipolar depression.
- Other uses as recommended by your doctor.

DOSAGE & USAGE INFORMATION

How to take:
Tablet—Swallow with liquid. It should be taken with a meal to help your body absorb the drug. If you can't swallow tablet whole, ask your doctor or pharmacist for advice.

When to take:
Usually once a day at the same time each day.

If you forget a dose:
Take as soon as you remember. If it is almost time for the next dose, wait for that dose (don't double this dose) and resume regular schedule.

What drug does:
The exact way it works is unknown. It appears to suppress levels of certain brain chemicals (e.g., dopamine and serotonin) that may be elevated in people with schizophrenia.

Time lapse before drug works:
Starts working within hours, but can take up to several weeks for full effect.

Don't take with:
Any other medicine or any dietary supplement without consulting your doctor or pharmacist.

OVERDOSE

SYMPTOMS:
May include fast heart rate, drowsiness, lightheadedness, faintness, or uncontrolled muscle movements.
WHAT TO DO:
- **Dial 911 for all medical emergencies or call poison control center 1-800-222-1222 for instructions.**
- **See emergency information on last 3 pages of this book.**

POSSIBLE ADVERSE REACTIONS OR SIDE EFFECTS

SYMPTOMS	WHAT TO DO
Life-threatening: In case of overdose, see previous column.	
Common: Drowsiness, jittery feeling, restlessness, nausea or vomiting, insomnia, indigestion, heartburn, tremor, slow movement, muscle stiffness, fast heartbeat.	Continue. Call doctor when convenient.
Infrequent: Fatigue, excess saliva, back pain, agitation, dizziness, anxiety, rash, blurred vision, abdominal pain, diarrhea, decreased appetite, itching, drooling.	Continue. Call doctor when convenient.
Rare:	
• Trouble swallowing, neuroleptic malignant syndrome (high fever, stiff muscles, confusion, blood pressure changes, sweating, muscle pain, weakness), tardive dyskinesia (uncontrolled movements of the face, tongue and other body parts), seizures, thoughts of suicide.	Discontinue. Call doctor right away.
• Weight gain, high blood sugar (frequent urination, feeling unusually thirsty or weak or hungry), lack of menstrual periods, breasts are leaking or enlarged, impotence, symptoms of infection or other unexplained symptoms.	Continue. Call doctor when convenient.

WARNINGS & PRECAUTIONS

Don't take if:
You are allergic to lurasidone. Allergic reactions are rare, but may occur.

Before you start, consult your doctor if:
- You have or have had liver or kidney disease.
- You have heart disease, low blood pressure, congestive heart failure, other heart problems.
- You take certain drugs classified as enzyme inducers* or enzyme inhibitors.*
- You have cerebrovascular disease or history of stroke.
- You have a history of breast cancer.
- You have seizures or epilepsy.
- You have Parkinson's disease.
- You have high cholesterol or triglycerides.
- You have or have had low white blood cell counts.
- You have trouble swallowing.
- You have or have had suicidal thoughts or attempts or alcohol abuse or dependence.
- Patient has Alzheimer's or dementia.
- You have a family history of, or have diabetes.
- You have tardive dyskinesia.
- You have had neuroleptic malignant syndrome (serious or fatal problems may occur).

Over age 60:
- Adverse reactions and side effects may be more severe than in younger persons.
- Use of antipsychotic drugs in elderly patients with dementia-related psychosis may increase risk of death. Consult doctor.

Pregnancy:
Decide with your doctor if drug benefits justify risk to unborn baby. Risk category B (see page xviii).

Breastfeeding; Lactation; Nursing Mothers:
This drug is generally acceptable with breastfeeding. Discuss risks and benefits with your doctor.

Infants & children up to age 18:
Safety and effectiveness in this age group have not been established. Consult doctor.

Prolonged use:
Consult with your doctor on a regular basis while taking this drug to monitor your progress, check for side effects and for recommended lab tests.

Skin & sunlight:
No problems expected.

Driving, piloting or hazardous work:
Don't drive or pilot aircraft until you learn how medicine affects you. Don't work around dangerous machinery. Don't climb ladders or work in high places. Danger increases if you drink alcohol or take medicine affecting alertness and reflexes.

Discontinuing:
Don't discontinue this drug without consulting doctor. Dosage may require a gradual reduction before stopping.

Others:
- Get up slowly from a sitting or lying position to avoid dizziness, faintness or lightheadedness.
- Advise any doctor, dentist or pharmacist whom you consult that you take this medicine.
- Drug can affect body's ability to maintain normal temperature. You may be more sensitive to temperature extremes such as very hot or cold conditions. Avoid getting too cold or becoming overheated or dehydrated. Drink plenty of fluids.
- Drug is not approved for the treatment of patients with dementia-related psychosis.
- Take medicine only as directed. Do not change the dosage without doctor's approval.

POSSIBLE INTERACTION WITH OTHER DRUGS

GENERIC NAME OR DRUG CLASS	COMBINED EFFECT
Antihypertensives*	Increased risk of low blood pressure.
Central nervous system (CNS) depressants*	Increased sedative effect.
Dopamine agonists*	Decreased dopamine agonist effect.
Enzyme inhibitors*	May increase lurasidone effect.
Enzyme inducers*	May decrease lurasidone effect.

POSSIBLE INTERACTION WITH OTHER SUBSTANCES

INTERACTS WITH	COMBINED EFFECT
Alcohol:	Increased sedative affect. Avoid.
Beverages: Grapefruit juice.	May increase effect of drug.
Cocaine:	Unknown. Avoid.
Foods: Grapefruit.	May increase effect of drug.
Marijuana:	Sedation. Avoid.
Tobacco:	None expected.

***See Glossary**

MACROLIDE ANTIBIOTICS

GENERIC AND BRAND NAMES

AZITHROMYCIN
- **Zithromax**
- **Zmax**

CLARITHROMYCIN
- **Biaxin**
- **Omeclamox-Pak**

DIRITHROMYCIN
- **Dynabac**

FIDAXOMICIN
- **Dificid**

BASIC INFORMATION

Habit forming? No
Prescription needed? Yes
Available as generic? Yes, for some
Drug class: Antibiotic (macrolide)

USES

Treatment for mild to moderate bacterial infections responsive to macrolide antibiotics. These include bronchitis, tonsillitis, some pneumonias, ear infections, skin infections (e.g., acne), sinusitis, streptococcal sore throat, urethritis, *Clostridium difficile*-associated diarrhea and others. (Note: Erythromycin is also a macrolide antibiotic and has its own chart.)

DOSAGE & USAGE INFORMATION

How to take:
- Enteric-coated tablet or delayed-release tablet—Swallow with liquid. Do not crush or chew tablet.
- Tablet or capsule—Swallow with liquid. If you can't swallow whole, ask your doctor or pharmacist for advice.
- Oral suspension or extended-release oral suspension—Follow directions on your prescription label.

When to take:
At the same time each day. Follow directions on your prescription label about taking the drug with food or on an empty stomach or if it makes no difference.

Continued next column

OVERDOSE

SYMPTOMS:
Possibly diarrhea, nausea, vomiting, abdominal pain.
WHAT TO DO:
If person uses much larger amount than prescribed or if accidentally swallowed, call poison control center 1-800-222-1222 for instructions or dial 911 (emergency) for help.

If you forget a dose:
Take as soon as you remember. If it is almost time for the next dose, wait for that dose (don't double this dose) and resume regular schedule.

What drug does:
Prevents growth and reproduction of susceptible bacteria.

Time lapse before drug works:
Usually 2 to 5 days. Some infections may take 10 days or longer to resolve.

Don't take with:
Any other medicine or any dietary supplement without consulting your doctor or pharmacist.

POSSIBLE ADVERSE REACTIONS OR SIDE EFFECTS

SYMPTOMS	WHAT TO DO
Life-threatening:	
Rare, severe allergic reaction (anaphylaxis)—difficulty breathing, hives, itching, swelling (throat, face), vomiting, dizziness.	Seek emergency treatment immediately.
Common:	
None expected.	
Infrequent:	
Nausea, vomiting, abdominal discomfort, diarrhea.	Continue. Call doctor when convenient.
Rare:	
• Allergic reaction (skin rash, itching), liver damage (yellow skin or eyes, nausea, abdominal pain, unusual bleeding or bruising, severe fatigue).	Discontinue. Call doctor right away.
• Headache, dizziness.	Continue. Call doctor when convenient.

WARNINGS & PRECAUTIONS

Don't take if:
You are allergic to macrolide antibiotics.

Before you start, consult your doctor if:
- You have any liver or kidney disorder.
- You have had jaundice.
- You have QT prolongation (a heart disorder) or cardiac arrhythmia.
- You have any muscle disorder, immune system problem or blood disorder.
- You are allergic to any medication, food or other substance.

Over age 60:
Adverse reactions and side effects may be more frequent and severe than in younger persons.

Pregnancy:
Decide with your doctor if drug benefits justify risk to unborn baby. Risk category B for azithromycin and fidaxomicin; category C for clarithromycin and dirithromycin (see page xviii).

Breastfeeding; Lactation; Nursing Mothers:
Drugs in this group are generally acceptable with breastfeeding. Discuss risks and benefits with your doctor.

Infants & children up to age 18:
Follow instructions provided by your child's doctor. Young children may not complain or recognize adverse effects of drug. Observe child closely for any reactions.

Prolonged use:
Not recommended. The drug is discontinued once the infection is cured.

Skin & sunlight:
No special problems expected.

Driving, piloting or hazardous work:
Avoid if you experience dizziness. Otherwise, no problems expected.

Discontinuing:
Don't discontinue without doctor's advice until you complete prescribed dose, even though symptoms diminish or disappear.

Others:
- Advise any doctor, dentist or pharmacist whom you consult that you take this medicine.
- Some macrolide antibiotics may cause QT interval prolongation and heart arrhythmia (very rarely can be fatal). Consult doctor about your risks.
- May affect the results of some medical tests.

POSSIBLE INTERACTION WITH OTHER DRUGS

GENERIC NAME OR DRUG CLASS	COMBINED EFFECT
Antacids,* aluminum- or magnesium-containing	Take one hour apart.
Bromocriptine	Increased effect of bromocriptine.
Carbamazepine	Increased effect of carbamazepine.
Cisapride	May increase toxic effects. Avoid.
Cyclosporine	Increased effect of cyclosporine.
Digoxin	Increased effect of digoxin
Disopyramide	Unknown effect. Use with caution.
Enzyme inhibitors*	Increased effect of enzyme inhibitor (with clarithromycin).
Ergot preparations*	Can cause serious or life-threatening problems with blood circulation. Avoid.
HMG-CoA reductase inhibitors	Increased effect of HMG-CoA reductase inhibitor and risk of serious muscle injury. Avoid.
Iron supplements	Decreased effect of dirithromycin. Take 1 hour apart.
Phenytoin	Decreased effect of phenytoin.
Pimozide	May increase toxic effects. Avoid.
Rifabutin	Decreased effect of antibiotic.
Rifampin	Decreased effect of antibiotic.
Tacrolimus	Increased effect of tacrolimus.
Theophylline	Increased effect of theophylline.
Triazolam	Increased risk of triazolam.
Warfarin	Increased risk of bleeding.
Zidovudine	Decreased effect of zidovudine.

POSSIBLE INTERACTION WITH OTHER SUBSTANCES

INTERACTS WITH	COMBINED EFFECT
Alcohol:	None expected.
Beverages:	None expected.
Cocaine:	Unknown. Avoid.
Foods:	None expected.
Marijuana:	Consult doctor.
Tobacco:	None expected.

*See Glossary

MAPROTILINE

BRAND NAMES

Ludiomil

BASIC INFORMATION

Habit forming? No
Prescription needed? Yes
Available as generic? Yes
Drug class: Antidepressant

USES

Treatment for depression or anxiety associated with depression.

DOSAGE & USAGE INFORMATION

How to take:
Tablet—Swallow with liquid.

When to take:
At the same time each day, usually bedtime.

If you forget a dose:
Bedtime dose—If you forget your once-a-day bedtime dose, don't take it more than 3 hours late. If more than 3 hours, wait for next scheduled dose. Don't double this dose.

What drug does:
Probably affects part of brain that controls messages between nerve cells.

Time lapse before drug works:
Begins in 1 to 2 weeks. May require 4 to 6 weeks for maximum benefit.

Don't take with:
Any other medicine or any dietary supplement without consulting your doctor or pharmacist.

OVERDOSE

SYMPTOMS:
Respiratory failure, fever, heart arrhythmia, muscle stiffness, drowsiness, hallucinations, convulsions, coma.
WHAT TO DO:
- **Dial 911 for all medical emergencies or call poison control center 1-800-222-1222 for instructions.**
- **See emergency information on last 3 pages of this book.**

POSSIBLE ADVERSE REACTIONS OR SIDE EFFECTS

SYMPTOMS	WHAT TO DO
Life-threatening:	
In case of overdose, see previous column.	
Common:	
• Tremor.	Discontinue. Call doctor right away.
• Headache, dry mouth or unpleasant taste, constipation, diarrhea, indigestion, fatigue, weakness, nausea, drowsiness, nervousness, anxiety, excessive sweating, insomnia, craving sweets.	Continue. Call doctor when convenient.
Infrequent:	
• Hallucinations, shakiness, dizziness, fainting, blurred vision, eye pain, vomiting, irregular heartbeat or slow pulse, inflamed tongue, abdominal pain, yellow skin or eyes, hair loss, rash, chills, joint pain, vision changes, hiccups, palpitations.	Discontinue. Call doctor right away.
• Painful or difficult urination, abnormal dreams, decreased sex drive, stuffy nose, back pain, muscle aches, frequent urination, painful or absent or irregular menstruation.	Continue. Call doctor when convenient.
Rare:	
Itchy skin, sore throat, fever, involuntary movements (of jaw, lips or tongue), nightmares, confusion, swollen breasts in men.	Discontinue. Call doctor right away.

WARNINGS & PRECAUTIONS

Don't take if:
- You are allergic to maprotiline. Allergic reactions are rare, but may occur.
- You have had a heart attack within 6 weeks.
- You have taken a monoamine oxidase (MAO) inhibitor* within 2 weeks.

Before you start, consult your doctor if:
- You will have surgery within 2 months, including dental surgery, requiring anesthesia.
- You have a history of drug or alcohol abuse.
- You have an enlarged prostate, heart disease or high blood pressure, stomach or intestinal problems, overactive thyroid, asthma, liver disease, schizophrenia, urinary retention, glaucoma, respiratory disorder, seizure disorders, diabetes, or kidney disease.

Over age 60:
Adverse reactions and side effects may be more frequent and severe than in younger persons.

Pregnancy:
Decide with your doctor if drug benefits justify risk to unborn baby. Risk category B (see page xviii).

Breastfeeding; Lactation; Nursing Mothers:
This drug is generally acceptable with breastfeeding. Discuss risks and benefits with your doctor.

Infants & children up to age 18:
Not approved in ages under 18. If prescribed, carefully read information provided with prescription. Contact doctor right away if symptoms get worse or any there is any talk of suicide or suicide behaviors. Read information under Others.

Prolonged use:
Request blood cell counts, liver function studies; monitor blood pressure closely.

Skin & sunlight:
May cause rash or intensify sunburn in areas exposed to sun or ultraviolet light (photosensitivity reaction). Avoid overexposure and use sunscreen. Notify doctor if reaction occurs.

Driving, piloting or hazardous work:
Don't drive or pilot aircraft until you learn how medicine affects you. Don't work around dangerous machinery. Don't climb ladders or work in high places. Danger increases if you drink alcohol or take medicine affecting alertness and reflexes.

Discontinuing:
Don't discontinue without consulting doctor. Dose may require gradual reduction if you have taken drug for a long time. Doses of other drugs may also require adjustment.

Others:
- Advise any doctor, dentist or pharmacist whom you consult that you take this drug.
- Adults and children taking antidepressants may experience a worsening of the depression symptoms and may have increased suicidal thoughts or behaviors. Call doctor right away if these symptoms or behaviors occur.
- For dry mouth, suck sugarless hard candy or chew sugarless gum. If dry mouth persists, consult your dentist.

POSSIBLE INTERACTION WITH OTHER DRUGS

GENERIC NAME OR DRUG CLASS	COMBINED EFFECT
Anticholinergics*	Increased sedation.
Antiglaucoma agents	Heart rhythm problems, high blood pressure.
Antihistamines*	Increased antihistamine effect.
Barbiturates*	Decreased anti-depressant effect.
Benzodiazepines*	Increased sedation.
Bupropion	Increased risk of seizures.
Central nervous system (CNS) depressants*	Increased sedation.
Cimetidine	Possible increased antidepressant effect and toxicity.

Continued on page 904

POSSIBLE INTERACTION WITH OTHER SUBSTANCES

INTERACTS WITH	COMBINED EFFECT
Alcohol: Beverages or medicines with alcohol.	Excessive intoxication. Avoid.
Beverages:	None expected.
Cocaine:	Increased risk of side effects. Avoid.
Foods:	None expected.
Marijuana:	Excessive drowsiness. Avoid.
Tobacco:	May decrease absorption of maprotiline. Avoid.

***See Glossary**

MARAVIROC

BRAND NAMES

Selzentry

BASIC INFORMATION

Habit forming? No
Prescription needed? Yes
Available as generic? No
Drug class: Antiviral agent

USES

Treatment of adults infected with human immunodeficiency virus (HIV) who have failed other antiviral drug therapy. This drug specifically treats patients with CCR5-tropic HIV-1. The drug is used in combination with other antiviral drugs. HIV is the virus that causes acquired immunodeficiency syndrome (AIDS).

DOSAGE & USAGE INFORMATION

How to take:
Tablet—Swallow with liquid. May be taken with or without food.

When to take:
Take twice a day at the same times each day.

If you forget a dose:
Take as soon as you remember. If it is almost time for the next dose, wait for that dose (don't double this dose) and resume regular schedule.

What drug does:
It helps control HIV infection by blocking the virus from entering your immune system cells. This helps your immune system stay healthy so it can fight the infection. The drug does not cure HIV or AIDS.

Time lapse before drug works:
May require several weeks or months before full benefits are apparent.

Don't take with:
Any other medicine or any dietary supplement without consulting your doctor or pharmacist. This is very important with HIV drugs.

OVERDOSE

SYMPTOMS:
Feeling faint or lightheaded, cold sweats.
WHAT TO DO:
If person uses much larger amount than prescribed or if accidentally swallowed, call poison control center 1-800-222-1222 for instructions or dial 911 (emergency) for help.

POSSIBLE ADVERSE REACTIONS OR SIDE EFFECTS

SYMPTOMS	WHAT TO DO
Life-threatening: None expected.	
Common: Diarrhea, dizziness, fatigue, nausea, cough, fever, other mild cold symptoms, abdominal pain.	Continue. Call doctor when convenient.
Infrequent: Swelling, stomach pain, muscle or joint pain, white patches or sores inside mouth or on lips, constipation, urination problems, changes in body fat, sleeping problems, dizziness when getting up from a sitting or lying down position, cold sores or sores on genital or anal area, itching.	Continue. Call doctor when convenient.
Rare:	
• Skin rash, liver problem (symptoms may include nausea, stomach pain, loss of appetite, skin rash, dark urine, clay-colored stools, yellow skin or eyes), chest pain, new infection (fever, chills, cough, flu-like symptoms), light-headedness or near fainting, breathing difficulty.	Discontinue. Call doctor right away.
• Other unexplained symptoms.	Continue. Call doctor when convenient.

WARNINGS & PRECAUTIONS

Don't take if:
You are allergic to maraviroc. Allergic reactions are rare, but may occur.

Before you start, consult your doctor if:
- You are allergic to any medicine, food or other substance, or have a family history of allergies.
- You have kidney or liver disease (especially hepatitis B or C), or diabetes.
- You have heart disease, low blood pressure, circulation problems, or a history of heart attack or stroke.

Over age 60:
Adverse reactions and side effects may be more frequent and severe than in younger persons.

Pregnancy:
Decide with your doctor if drug benefits justify risk to unborn baby. Risk category B (see page xviii). HIV can be passed to the baby if the mother is not properly treated during pregnancy.

Breastfeeding; Lactation; Nursing Mothers:
It is not recommended that HIV-infected mothers breast-feed. Consult your doctor.

Infants & children up to age 18:
Safety and efficacy have not been established for children under age 16. Consult doctor.

Prolonged use:
- Long-term effects of using this drug have not been established.
- Talk to your doctor about the need for follow up blood tests and liver function studies.

Skin & sunlight:
No problems expected.

Driving, piloting or hazardous work:
Don't drive or pilot aircraft until you learn how medicine affects you. Don't work around dangerous machinery. Don't climb ladders or work in high places. Danger increases if you drink alcohol or take medicine affecting alertness and reflexes, such as antihistamines, tranquilizers, sedatives, pain medicine, narcotics and mind-altering drugs.

Discontinuing:
Don't discontinue (even for a short time) without doctor's advice.

Others:
- Advise any doctor, dentist or pharmacist whom you consult that you take this medicine.
- Taking this drug does not prevent you from passing HIV to another person through sexual contact or sharing needles. Avoid sexual contact or practice safe sex (e.g., using condoms) to help prevent the transmission of HIV. Never share or re-use needles. If you have questions, ask your doctor for advice.
- Liver damage may develop with use of this drug. See symptoms in Possible Adverse Reactions or Side Effects—Rare.
- Consult your doctor right away if you develop symptoms of a new infection.
- Take drug daily as prescribed. Do not increase or decrease dosage of drug without doctor's approval.

POSSIBLE INTERACTION WITH OTHER DRUGS

GENERIC NAME OR DRUG CLASS	COMBINED EFFECT
Antihypertensives*	Increased risk of dizziness. Use with caution.
Enzyme inducers*	Decreased effect of maraviroc.
Enzyme inhibitors*	Increased effect of maraviroc.
St. John's wort	Decreased effect of maraviroc. Avoid.

POSSIBLE INTERACTION WITH OTHER SUBSTANCES

INTERACTS WITH	COMBINED EFFECT
Alcohol:	None expected.
Beverages:	None expected.
Cocaine:	Unknown. Avoid.
Foods:	None expected.
Marijuana:	Consult doctor.
Tobacco:	None expected.

*See Glossary

MECHLORETHAMINE (Topical)

BRAND NAMES

Mustargen **Valchlor**

BASIC INFORMATION

Habit forming? No
Prescription needed? Yes
Available as generic? No
Drug class: Antineoplastic (topical)

USES

- Treats mycosis fungoides.
- Treats other malignancies (by injection).

DOSAGE & USAGE INFORMATION

How to use:
- Solution—It is mixed according to doctor's instructions. Don't inhale vapors or powder. Shower and rinse before treatment. Use rubber gloves to apply over entire body. Avoid contact with eyes, nose and mouth.
- Ointment—Use according to doctor's instructions.
- Gel—Detailed instructions for storage and use are provided with the prescription. Follow them carefully.

When to use:
Usually once a day.

If you forget a dose:
Apply as soon as you remember. If it is almost time for the next dose, wait for that dose (don't double this dose) and resume regular schedule.

What drug does:
Destroys cells that produce mycosis fungoides.

Time lapse before drug works:
Starts to work right away, but response to treatment may take 3 to 6 months.

Continued next column

OVERDOSE

SYMPTOMS:
Infection (fever, chills, cough), unusual bleeding or bruising, bloody stools or vomit.
WHAT TO DO:
- **Dial 911 for all medical emergencies or call poison control center 1-800-222-1222 for instructions.**
- **See emergency information on last 3 pages of this book.**

Don't use with:
Any other topical or oral medicines (including over-the-counter drugs such as cough and cold medicines, laxatives, antacids, diet pills, caffeine, nose drops or vitamins) without consulting your doctor or pharmacist.

POSSIBLE ADVERSE REACTIONS OR SIDE EFFECTS

SYMPTOMS	WHAT TO DO
Life-threatening:	
Rare, severe allergic reaction (anaphylaxis)—difficulty breathing, hives, itching, swelling (throat, face), vomiting, dizziness.	Seek emergency treatment immediately.
Common:	
Skin symptoms (infection, itching, reaction, sores, blisters, darkening).	Continue. Call doctor when convenient.
Infrequent:	
None expected.	
Rare:	
None expected.	

MECHLORETHAMINE (Topical)

WARNINGS & PRECAUTIONS

Don't use if:
You are allergic to mechlorethamine.

Before you start, consult your doctor if:
- You have chicken pox.
- You have shingles (herpes zoster).
- You have skin infection.

Over age 60:
No special problems expected.

Pregnancy:
Consult doctor. Use of the drug during pregnancy is not recommended. It could cause harm to the unborn baby. Risk category D (see page xviii).

Breastfeeding; Lactation; Nursing Mothers:
This drug is generally not recommended with breastfeeding. Discuss risks and benefits with your doctor.

Infants & children up to age 18:
Safety and effectiveness in this age group have not been established. Consult doctor.

Prolonged use:
- Allergic or hypersensitive reactions more likely. The ointment form is less likely to cause a reaction.
- Talk to your doctor about the need for follow-up medical examinations or laboratory studies to check liver function, complete blood counts (white blood cell count, platelet count, red blood cell count, hemoglobin, hematocrit) and hearing tests.

Skin & sunlight:
No problems expected.

Driving, piloting or hazardous work:
Don't drive or pilot aircraft until you learn how medicine affects you. Don't work around dangerous machinery. Don't climb ladders or work in high places. Danger increases if you drink alcohol or take medicine affecting alertness and reflexes.

Discontinuing:
No special problems expected.

Others:
- Advise any doctor, dentist or pharmacist whom you consult that you take this medicine.
- May affect results in some medical tests.
- Don't use if solution is discolored.

POSSIBLE INTERACTION WITH OTHER DRUGS

GENERIC NAME OR DRUG CLASS	COMBINED EFFECT
None significant.	

POSSIBLE INTERACTION WITH OTHER SUBSTANCES

INTERACTS WITH	COMBINED EFFECT
Alcohol:	None expected.
Beverages:	None expected.
Cocaine:	None expected.
Foods:	None expected.
Marijuana:	None expected.
Tobacco:	None expected.

*See Glossary

MEGLITINIDES

GENERIC AND BRAND NAMES

NATEGLINIDE
Starlix

REPAGLINIDE
PrandiMet
Prandin

BASIC INFORMATION

Habit forming? No
Prescription needed? Yes
Available as generic? No
Drug class: Antidiabetic

USES

Helps control, but does not cure type 2 (non-insulin dependent) diabetes. Used alone or in combination with other antidiabetic drugs, along with diet and exercise.

DOSAGE & USAGE INFORMATION

How to take:
Tablet—Take with water between 15-30 minutes before a meal. If you skip the meal, also skip the dose of medicine. If you have an extra meal, take an extra dose.

When to use:
15-30 minutes before each meal or as directed by doctor.

If you forget a dose:
Take as soon as you remember. If it is almost time for the next dose, wait for that dose (don't double this dose) and resume regular schedule.

What drug does:
Increases amount of insulin secreted from the pancreas, which helps to control blood sugar.

Continued next column

OVERDOSE

SYMPTOMS:
Mild to severe symptoms of low blood sugar—cold sweats, confusion, cool pale skin, difficulty in concentrating, drowsiness, excessive hunger, rapid heartbeat, nausea, nervousness, nightmares, restless sleep, seizures, shakiness, slurred speech, unusual tiredness or weakness, coma.
WHAT TO DO:
- **Dial 911 for all medical emergencies or call poison control center 1-800-222-1222 for instructions.**
- **See emergency information on last 3 pages of this book.**

Time lapse before drug works:
10-30 minutes; peaks in 1 hour.

Don't take with:
Any other medicine or any dietary supplement without consulting your doctor or pharmacist.

POSSIBLE ADVERSE REACTIONS OR SIDE EFFECTS

SYMPTOMS	WHAT TO DO
Life-threatening: In case of overdose or low blood sugar, see previous column.	
Common: Symptoms of a cold (sore throat, runny or stuffy nose, cough), back pain, diarrhea, joint pain.	Continue. Call doctor when convenient.
Frequent: Low blood sugar symptoms: anxiety, cold sweats, shakiness, rapid heartbeat, blurred vision, pale skin, behavior changes similar to being drunk, confusion or difficulty thinking, drowsiness, excessive hunger, headache, nausea, nightmares, restless sleep, unusual tiredness or weakness.	Treat the low blood sugar. If symptoms are severe, seek emergency treatment.
Infrequent Bloody or cloudy urine, urination problems (burning, painful, difficult, frequent, urge to urinate), wheezing, chills, skin rash or itching or hives, eyes tearing, vomiting.	Discontinue. Call doctor right away.
Rare: • Unusual bleeding or bruising, red spots on skin, black or tarry stools, hoarseness, lower back or side pain.	Continue, but call doctor right away.
• Indigestion, feeling of warmth or heat or burning, stomach pain, constipation, dizziness.	Continue. Call doctor when convenient.

WARNINGS & PRECAUTIONS

Don't use if:
You are allergic to meglitinides. Allergic reactions are rare, but may occur.

Before you start, consult your doctor if:
- You have type 1 diabetes or diabetic ketoacidosis (ketones in the blood).
- You have an infection, fever, an injury or trauma, high stress levels or are planning surgery.
- You have a nervous system disorder.
- You have kidney or liver disease or underactive adrenal or pituitary gland.
- You are weak or undernourished.

Over age 60:
No special problems expected.

Pregnancy:
Decide with your doctor if drug benefits justify risk to unborn baby. Risk category C (see page xviii).

Breastfeeding; Lactation; Nursing Mothers:
Drugs in this group are generally not recommended with breastfeeding. Discuss risks and benefits with your doctor.

Infants & children up to age 18:
Safety and effectiveness in this age group have not been established. Consult doctor.

Prolonged use:
Talk to your doctor about the need for follow up medical examinations or laboratory studies to check blood glucose levels and glycosylated hemoglobin (HbA1c) values.

Skin & sunlight:
No problems expected.

Driving, piloting or hazardous work:
Don't drive or pilot aircraft until you learn how medicine affects you. Don't work around dangerous machinery. Don't climb ladders or work in high places. Danger increases if you drink alcohol or take medicine affecting alertness and reflexes.

Discontinuing:
Don't discontinue without consulting your doctor even if you feel well. You can have diabetes without feeling any symptoms. Untreated diabetes can cause serious problems.

Others:
- Advise any doctor, dentist or pharmacist whom you consult that you take this medicine. It may interfere with the accuracy of some medical tests.
- Follow any special diet your doctor may prescribe. It can help control diabetes.
- Consult doctor if you become ill with vomiting or diarrhea while taking this drug.
- Use caution when exercising. Ask your doctor about an appropriate exercise program.
- Wear or carry medical identification stating that you have diabetes and take this drug.
- Learn to recognize the symptoms of low and high blood sugar. You and your family need to know what to do if these symptoms occur and when to call the doctor for help.
- Have a glucagon kit and syringe in the event severe low blood sugar occurs.

POSSIBLE INTERACTION WITH OTHER DRUGS

GENERIC NAME OR DRUG CLASS	COMBINED EFFECT
Anti-inflammatory drugs nonsteroidal (NSAIDs)	Increased risk of low blood sugar.
Barbiturates	Blood sugar problems.
Beta adrenergic blocking agents	Blood sugar control problems.
Carbamazepine	Blood sugar problems.
Corticosteroids*	Decreased effect of meglitinide.
Diuretics, thiazide	Decreased effect of repaglinide.
Gemfibrozil	Increased effect of meglitinide.
Hyperglycemia-causing medications*	Increased risk of loss of glycemic control.
Monoamine oxidase (MAO) inhibitors*	Increased risk of low blood sugar.
Salicylates	Increased risk of low blood sugar.
Sympathomimetics*	Decreased effect of meglitinide.
Thyroid hormones	Decreased effect of meglitinide

POSSIBLE INTERACTION WITH OTHER SUBSTANCES

INTERACTS WITH	COMBINED EFFECT
Alcohol:	Low blood sugar. Avoid.
Beverages:	None expected.
Cocaine:	Unknown. Avoid.
Foods:	None expected.
Marijuana:	Consult doctor.
Tobacco:	None expected.

*See Glossary

MELATONIN

BRAND NAMES

Numerous brand names are available.

BASIC INFORMATION

Habit forming? No
Prescription needed? No
Available as generic? Yes
Drug class: Hormone

USES

- Melatonin is a hormone produced in the human body by the pineal gland and secreted at night. In most people, the melatonin levels are highest during the normal hours of sleep. The levels increase rapidly in the late evening, peaking after midnight and decreasing toward morning.
- Jet lag: Some research studies have shown that taking melatonin before a flight and continuing for a few days after arrival at the destination helped control jet lag symptoms of fatigue and sleep disturbances. Appears to work best after plane trips that crossed more than six time zones. Timing of doses very important for effectiveness.
- Insomnia and restless leg syndrome: Some research studies have shown that taking melatonin about 2 hours before bedtime decreased the time needed to fall asleep and improved quality of sleep (less wakefulness).
- Other claims that it can slow aging, fight disease, and enhance one's sex life have been less studied and more difficult to prove.

OVERDOSE

SYMPTOMS:
It is unknown what symptoms may occur.
WHAT TO DO:
If person uses much larger amount than prescribed or if accidentally swallowed, call poison control center 1-800-222-1222 for instructions or dial 911 (emergency) for help.

DOSAGE & USAGE INFORMATION

How to take:
For tablet or capsule—Follow instructions on the label or consult your doctor or pharmacist. Different brands supply different doses. Melatonin, as a product, is marketed as a dietary supplement and is not reviewed by the U.S. Food & Drug Administration (FDA) for effectiveness and safety. The melatonin products being sold are made from animal pineal glands or synthesized. The best dosage amounts are unknown. Use with caution.

When to take:
At the same time each day according to label directions. It is recommended that melatonin be taken at night before bedtime.

If you forget a dose:
Follow label instructions for your particular brand of melatonin. Usually you can take as soon as you remember. If it is almost time for the next dose, wait for that dose (don't double this dose) and resume regular schedule.

What drug does:
- Glands in the body make chemicals called hormones and release them into the bloodstream. Hormones taken as supplements also end up in the bloodstream. In either case, the blood then carries hormones to different parts of the body. There, hormones influence the way organs and tissues work.
- Hormone supplements may not have the same effects on the body as naturally produced hormones have, because the body processes them differently. Higher doses of supplements may result in higher amounts of hormones in the blood than are healthy.

Time lapse before drug works:
Effectiveness will vary from person to person and will also depend on the reason for taking melatonin.

Don't take with:
Any other medicine or any dietary supplement without consulting your doctor or pharmacist.

MELATONIN

POSSIBLE ADVERSE REACTIONS OR SIDE EFFECTS

SYMPTOMS	WHAT TO DO
Life-threatening: None expected.	
Common: Unknown.	
Infrequent: Drowsiness, confusion, headache or grogginess may occur the following morning.	Reduce dosage or discontinue taking.
Rare: Unknown. If symptoms occur that you are concerned about, talk to your doctor or pharmacist. Further research may uncover other side effects.	

WARNINGS & PRECAUTIONS

Don't use if:
You are allergic to melatonin. Allergic reactions are rare, but may occur.

Before you start, consult your doctor if:
- You have any chronic health problem.
- You have high blood pressure (hypertension) or cardiovascular disease. Some studies in animals suggest that melatonin may constrict blood vessels (a problem that could be dangerous for people with these conditions).
- You are allergic to any medication, food or other substance.

Over age 60:
A lower starting dosage is often recommended until a response is determined.

Pregnancy:
Decide with your doctor if drug benefits justify risk to unborn child. Risk category is not designated (see page xviii).

Breastfeeding; Lactation; Nursing Mothers:
This drug is generally acceptable with breastfeeding. Discuss risks and benefits with your doctor.

Infants & children up to age 18:
Follow instructions provided by your child's doctor.

Prolonged use:
Effects are unknown. More research is needed to determine long-term effects of melatonin use.

Skin & sunlight:
No problems expected.

Driving, piloting or hazardous work:
Since it causes drowsiness, don't drive or pilot aircraft until you learn how medicine affects you. Don't work around dangerous machinery. Don't climb ladders or work in high places. Danger increases if you drink alcohol or take medicine affecting alertness and reflexes.

Discontinuing:
No problems expected.

Others:
- Advise any doctor, dentist or pharmacist whom you consult that you take melatonin.
- Melatonin is not researched carefully as yet, but there does not appear to be any particular problem. Some studies are promising as to its effect on health.
- Before starting melatonin, talk to your doctor about your sleep problems or try other things that can help sleep, such as avoidance of caffeine, chocolate, and especially alcohol in any amount.

POSSIBLE INTERACTION WITH OTHER DRUGS

GENERIC NAME OR DRUG CLASS	COMBINED EFFECT
All medications	Effects are unknown. Talk to your doctor or pharmacist.

POSSIBLE INTERACTION WITH OTHER SUBSTANCES

INTERACTS WITH	COMBINED EFFECT
Alcohol:	Disrupts the nighttime melatonin effect. Avoid.
Beverages:	None expected.
Cocaine:	Unknown. Avoid.
Foods:	None expected.
Marijuana:	Consult doctor.
Tobacco:	Smoking can disrupt your normal melatonin cycle. Avoid.

MELOXICAM

BRAND NAMES

Mobic

BASIC INFORMATION

Habit forming? No
Prescription needed? Yes
Available as generic? Yes
Drug class: Nonsteroidal anti-inflammatory, antirheumatic

USES

Treatment for joint pain, stiffness, inflammation and swelling of rheumatoid arthritis, osteoarthritis and gout.

DOSAGE & USAGE INFORMATION

How to take:
Tablet—Swallow whole with liquid. May be taken with or without food.

When to take:
At the same time each day.

If you forget a dose:
Take as soon as you remember. If it is almost time for the next dose, wait for that dose (don't double this dose) and resume regular schedule.

What drug does:
Reduces tissue concentration of prostaglandins (hormones which produce inflammation and pain).

Continued next column

OVERDOSE

SYMPTOMS:
Tiredness, drowsiness, nausea, vomiting, stomach pain, high blood pressure, breathing difficulty, seizures, coma.
WHAT TO DO:
- **Dial 911 for all medical emergencies or call poison control center 1-800-222-1222 for instructions.**
- **See emergency information on last 3 pages of this book.**

Time lapse before drug works:
Begins in 2 to 3 hours. May require 3 weeks of regular use for maximum benefit.

Don't take with:
Any other medicine or any dietary supplement without consulting your doctor or pharmacist.

POSSIBLE ADVERSE REACTIONS OR SIDE EFFECTS

SYMPTOMS	WHAT TO DO
Life-threatening:	
Rare, severe allergic reaction (anaphylaxis)—difficulty breathing, hives, itching, swelling (throat, face), vomiting, dizziness.	Seek emergency treatment immediately.
Common:	
Diarrhea, heartburn, indigestion, gas.	Continue. Call doctor when convenient.
Infrequent:	
Abdominal pain, anxiety, confusion, constipation, nausea, nervousness, sleepiness.	Continue. Call doctor when convenient.
Rare:	
Difficulty swallowing, swelling around the face, shortness of breath, pain (chest, left arm, jaw), unusual tiredness, bloody or tarry stools, vomiting blood (looks like coffee-grounds), severe stomach pain.	Discontinue. Call doctor right away or seek emergency care.

WARNINGS & PRECAUTIONS

Don't take if:
- You are allergic to aspirin or any other nonsteroidal anti-inflammatory drug (NSAIDs).
- You have nasal polyps.

Before you start, consult your doctor if:
- You have a history of alcohol abuse.
- You have bleeding problems or ulcers.
- You have any condition that causes fluid retention (heart problems or high blood pressure) or dehydration.
- You have used tobacco recently.
- You have impaired kidney or liver function.
- You have asthma.

Over age 60:
No special problems expected.

Pregnancy:
Decide with your doctor if drug benefits justify risk to unborn baby. Risk category C (see page xviii).

Breastfeeding; Lactation; Nursing Mothers:
This drug is generally acceptable with breastfeeding. Discuss risks and benefits with your doctor.

Infants & children up to age 18:
Follow instructions provided by your child's doctor.

Prolonged use:
Talk to your doctor about the need for follow-up medical exams or laboratory studies to check complete blood counts, liver function, stools for blood, and eyes.

Skin & sunlight:
No problems expected.

Driving, piloting or hazardous work:
Don't drive or pilot aircraft until you learn how medicine affects you. Don't work around dangerous machinery. Don't climb ladders or work in high places. Danger increases if you drink alcohol or take medicine affecting alertness and reflexes, such as antihistamines, tranquilizers, sedatives, pain medicine, narcotics and mind-altering drugs.

Discontinuing:
No problems expected. If drug has been taken for a long time, consult doctor before discontinuing.

Others:
- May affect results in some medical tests.
- Advise any doctor, dentist or pharmacist whom you consult that you take this medicine.
- Do not refrigerate drug and do not store in the bathroom, near the kitchen sink or in other damp places.

POSSIBLE INTERACTION WITH OTHER DRUGS

GENERIC NAME OR DRUG CLASS	COMBINED EFFECT
Angiotensin-converting enzyme (ACE) inhibitors*	May decrease ACE inhibitor effect.
Anti-inflammatory drugs, nonsteroidal (NSAIDs)* other	Increased risk of side effects.
Aspirin	Increased risk of stomach ulcer.
Furosemide	Decreased effect of furosemide.
Lithium	Increased lithium effect.
Warfarin	Increased risk of bleeding problems.

POSSIBLE INTERACTION WITH OTHER SUBSTANCES

INTERACTS WITH	COMBINED EFFECT
Alcohol:	Possible stomach ulcer or bleeding. Avoid.
Beverages:	None expected.
Cocaine:	Unknown. Avoid.
Foods:	None expected.
Marijuana:	Consult doctor.
Tobacco:	None expected.

*See Glossary

MELPHALAN

BRAND NAMES

Alkeran
L-PAM
Phenylalanine Mustard

BASIC INFORMATION

Habit forming? No
Prescription needed? Yes
Available as generic? No
Drug class: Antineoplastic

USES

Treatment for certain types of cancer.

DOSAGE & USAGE INFORMATION

How to take:

- Tablet—Swallow with a full glass of water to help prevent vomiting. Take on an empty stomach, 1 hour before or 2 hours after a meal. If vomiting occurs shortly after taking a dose, consult your doctor for advice.
- IV form—Is given by medical professional.

When to take:
As directed by your doctor. The dosage regimen is individualized. The drug is usually taken for several weeks to start and then followed by a drug rest period for up to 4 weeks.

If you forget a dose:
Take as soon as you remember. If it is almost time for the next dose, wait for that dose (don't double this dose) and resume regular schedule.

What drug does:
It interferes with the growth of cancer cells and causes them to die.

Time lapse before drug works:
May be gradual over weeks or months before your response to the drug can be determined.

Don't take with:
Any other medicine or any dietary supplement without consulting your doctor or pharmacist.

OVERDOSE

SYMPTOMS:
Nausea, vomiting, diarrhea, mouth sores, intestinal bleeding.
WHAT TO DO:

- **Dial 911 for all medical emergencies or call poison control center 1-800-222-1222 for instructions.**
- **See emergency information on last 3 pages of this book.**

POSSIBLE ADVERSE REACTIONS OR SIDE EFFECTS

SYMPTOMS	WHAT TO DO
Life-threatening:	
Rare, severe allergic reaction (anaphylaxis)—difficulty breathing, hives, itching, swelling (throat, face), vomiting, dizziness.	Seek emergency treatment immediately.
Common:	
Nausea and vomiting.	Continue. Call doctor when convenient.
Infrequent:	
Fever or chills occur with other symptoms (e.g., back or side pain, painful or difficult urination, cough or hoarseness), stools are bloody or tarry or black, bloody urine, fast or irregular heartbeat, small-red spots on skin, shortness of breath, sudden skin rash or itching, trouble breathing, unusual bleeding or bruising, ongoing nausea or vomiting.	Continue, but call doctor right away. Seek emergency care if symptoms are more severe.
Rare:	
• Mouth or lip sores, difficulty swallowing, joint pain, diarrhea, arms or legs become sore or red, feet or lower legs are swollen, unusual lumps or masses, yellow skin or eyes.	Continue, but call doctor right away.
• Menstrual periods stop, weight loss, other symptoms that cause concern.	Continue. Call doctor when convenient.

WARNINGS & PRECAUTIONS

Don't take if:
- You are allergic to melphalan or prior use of the drug has been ineffective.
- Your doctor has not explained the benefits and risks of taking this medicine.

Before you start, consult your doctor if:
- You have gout.
- You have had kidney stones.
- You have impaired kidney or liver function.
- You have taken other anticancer drugs or had radiation treatment in last 3 weeks.
- You have herpes zoster (shingles) or chicken pox (or been exposed).
- You have heart disease, congestive heart failure or other forms of cancer.
- You have bone marrow depression.
- You have an active infection.

Over age 60:
Adverse reactions and side effects may be more frequent and severe than in younger persons.

Pregnancy:
Consult doctor. Use of the drug during pregnancy is not recommended. It could cause harm to the unborn baby. Risk category D (see page xviii).

Breastfeeding; Lactation; Nursing Mothers:
This drug is generally not recommended with breastfeeding. Discuss risks and benefits with your doctor.

Infants & children up to age 18:
Follow instructions provided by your child's doctor.

Prolonged use:
- Adverse reactions may be more likely the longer drug is required.
- Talk to your doctor about the need for follow-up medical examinations or laboratory studies to check complete blood counts, kidney function and drug's effectiveness.

Skin & sunlight:
No problems expected.

Driving, piloting or hazardous work:
Use caution until you determine how drug affects you.

Discontinuing:
- Don't discontinue without doctor's advice until you complete prescribed dose.
- Some side effects may follow discontinuing. Report to doctor blurred vision, convulsions, confusion, persistent headache, fever or chills, blood in urine, unusual bleeding or other unexpected symptoms.

Others:
- Can cause sterility which could be permanent.
- Your doctor may advise you to drink extra fluids so that you pass more urine (helps prevent kidney problems).
- The drug increases risk for infections and other malignancies. Consult doctor if you develop new or unexpected symptoms.
- Do not have any immunizations (vaccinations) without doctor's approval. Avoid persons who have taken oral polio vaccine within the last several months.
- Advise any doctor, dentist or pharmacist whom you consult that you take this medicine.

POSSIBLE INTERACTION WITH OTHER DRUGS

GENERIC NAME OR DRUG CLASS	COMBINED EFFECT
None expected	With use of tablet (oral) form of melphalan, but do consult your doctor about any other drugs you take.

POSSIBLE INTERACTION WITH OTHER SUBSTANCES

INTERACTS WITH	COMBINED EFFECT
Alcohol:	None expected.
Beverages:	None expected.
Cocaine:	Unknown. Avoid.
Foods:	None expected.
Marijuana:	Consult doctor.
Tobacco:	None expected.

MEMANTINE

BRAND NAMES

Namenda
Namenda Oral Solution
Namenda XR
Namzaric

BASIC INFORMATION

Habit forming? No
Prescription needed? Yes
Available as generic? Yes
Drug class: N-methyl-D-aspartate (NMDA) receptor antagonist

USES

Treats the symptoms of moderate to severe Alzheimer's disease. Alzheimer's is a progressive disease of the brain. Memantine may be used alone or in combination with other drugs for Alzheimer's.

DOSAGE & USAGE INFORMATION

How to take:

- Tablet—Swallow with liquid. It may be taken with or without food.
- Oral solution—Follow the detailed directions provided with the prescription.
- Extended-release capsule—Swallow with liquid. It may be taken with or without food. Do not crush, chew or divide capsule. The capsule may be opened and sprinkled on or in food (e.g., applesauce) and then swallowed.

When to take:
Once or twice daily at the same times each day. A caregiver should monitor usage.

Continued next column

OVERDOSE

SYMPTOMS:
May include agitation, confusion, slow movements, loss of consciousness, restlessness, unsteady gait, vomiting, sleepiness, dizziness, visual hallucinations, weakness, stupor, coma.
WHAT TO DO:

- **Dial 911 for all medical emergencies or call poison control center 1-800-222-1222 for instructions.**
- **See emergency information on last 3 pages of this book.**

If you forget a dose:
Take as soon as you remember. If it is almost time for the next dose, wait for that dose (don't double this dose) and resume regular schedule.

What drug does:
Blocks excess amounts of a brain chemical called glutamate that can damage or kill nerve cells. It does not cure Alzheimer's disease or treat the underlying cause.

Time lapse before drug works:
Improvement may be seen in weeks, but may take months for maximum benefits. Dosage of the drug may be increased in one-week time periods for the first few weeks.

Don't take with:
Any other medicine or any dietary supplement without consulting your doctor or pharmacist.

POSSIBLE ADVERSE REACTIONS OR SIDE EFFECTS

SYMPTOMS	WHAT TO DO
Life-threatening: In case of overdose, see previous column.	
Common: None expected.	
Infrequent: Dizziness, headache, constipation, confusion.	Continue. Call doctor when convenient.
Rare: Fatigue, back pain, vomiting, sleepiness, high blood pressure. Other symptoms may occur. They may be due to progression of the disease or may be due to the drug.	Continue. Call doctor when convenient.

WARNINGS & PRECAUTIONS

Don't take if:
You are allergic to memantine or its components. Allergic reactions are rare, but may occur.

Before you start, consult your doctor if:
- You have kidney, liver or heart problems.
- You have a seizure disorder.
- You are allergic to any medication, food or other substance.

Over age 60:
No special problems expected.

Pregnancy:
This drug is usually not prescribed for reproductive-age women.

Breastfeeding; Lactation; Nursing Mothers:
This drug is usually not prescribed for reproductive-age women.

Infants & children up to age 18:
Not used for this age group.

Prolonged use:
Visit the doctor regularly to determine if the drug is continuing to be effective.

Skin & sunlight:
No special problems expected.

Driving, piloting or hazardous work:
Avoid if you feel dizzy, drowsy or confused.

Discontinuing:
No problems expected. Consult doctor.

Others:
- Advise any doctor, dentist or pharmacist whom the patient consults about the use of this medicine.
- The effects of the drug will differ for different patients. Some patients will improve, others may stay the same, and others continue to deteriorate (get worse).

POSSIBLE INTERACTION WITH OTHER DRUGS

GENERIC NAME OR DRUG CLASS	COMBINED EFFECT
Amantadine	May lead to adverse effects of either drug.
Carbonic anhydrase inhibitors	Increased effect of memantine.
Dextromethorphan	May lead to adverse effects of either drug.
Histamine H_2 receptor antagonists	Increased effect of either drug.
Hydrochlorothiazide	Increased effect of both drugs.
Sodium bicarbonate	Increased effect of memantine.
Triamterene	Increased effect of either drug.

POSSIBLE INTERACTION WITH OTHER SUBSTANCES

INTERACTS WITH	COMBINED EFFECT
Alcohol:	None expected. Best to avoid.
Beverages:	None expected.
Cocaine:	Unknown. Avoid.
Foods:	None expected.
Marijuana:	Consult doctor.
Tobacco:	None expected.

MEPROBAMATE

BRAND NAMES

See full list of brand names in the *Generic and Brand Name Directory*, page 871.

BASIC INFORMATION

Habit forming? Yes
Prescription needed? Yes
Available as generic? Yes
Drug class: Tranquilizer, antianxiety agent

USES

Reduces mild anxiety, tension and insomnia.

DOSAGE & USAGE INFORMATION

How to take:

- Tablet—Swallow with liquid.
- Extended-release capsule—Swallow each dose whole.

When to take:
At the same times each day.

If you forget a dose:
Take as soon as you remember. If it is almost time for the next dose, wait for that dose (don't double this dose) and resume regular schedule.

What drug does:
Sedates brain centers that control behavior and emotions.

Time lapse before drug works:
1 to 2 hours.

Don't take with:

- Nonprescription drugs containing alcohol or caffeine without consulting doctor.
- Any other medicine or any dietary supplement without consulting your doctor or pharmacist.

OVERDOSE

SYMPTOMS:
Dizziness, slurred speech, stagger, confusion, depressed breathing and heart function, stupor, coma.
WHAT TO DO:

- **Dial 911 for all medical emergencies or call poison control center 1-800-222-1222 for instructions.**
- **See emergency information on last 3 pages of this book.**

POSSIBLE ADVERSE REACTIONS OR SIDE EFFECTS

SYMPTOMS	WHAT TO DO
Life-threatening:	
Rare, severe allergic reaction (anaphylaxis)—difficulty breathing, hives, itching, swelling (throat, face), vomiting, dizziness.	Seek emergency treatment immediately.
Common:	
Dizziness, confusion, agitation, drowsiness, unsteadiness, fatigue, weakness.	Continue. Call doctor when convenient.
Infrequent:	
• Rash, hives, itchy skin, change in vision, diarrhea, nausea or vomiting, slurred speech, blurred vision.	Discontinue. Call doctor right away.
• False sense of well-being, headache.	Continue. Call doctor when convenient.
Rare:	
Sore throat, fever, heartbeat is rapid and pounding or very slow or irregular, difficult breathing, unusual bleeding or bruising.	Discontinue. Call doctor right away.

WARNINGS & PRECAUTIONS

Don't take if:
You are allergic to meprobamate, tybamate, carbromal or carisoprodol.

Before you start, consult your doctor if:
- You have epilepsy.
- You have impaired liver or kidney function.
- You have tartrazine dye allergy.
- You suffer from drug abuse or alcoholism, active or in remission.
- You have porphyria.

Over age 60:
Adverse reactions and side effects may be more frequent and severe than in younger persons.

Pregnancy:
Consult doctor. Use of the drug during pregnancy is not recommended. It could cause harm to the unborn baby. Risk category D (see page xviii).

Breastfeeding; Lactation; Nursing Mothers:
This drug is generally acceptable with breastfeeding. Discuss risks and benefits with your doctor.

Infants & children up to age 18:
Safety and effectiveness in this age group have not been established. Consult doctor.

Prolonged use:
- Habit forming.
- May impair blood cell production.

Skin & sunlight:
No problems expected.

Driving, piloting or hazardous work:
Don't drive or pilot aircraft until you learn how medicine affects you. Don't work around dangerous machinery. Don't climb ladders or work in high places. Danger increases if you drink alcohol or take medicine affecting alertness and reflexes, such as antihistamines, tranquilizers, sedatives, pain medicine, narcotics and mind-altering drugs.

Discontinuing:
Don't discontinue without consulting doctor. Dose may require gradual reduction if you have taken drug for a long time. Doses of other drugs may also require adjustment. Report to your doctor any unusual symptom that begins in the first week you discontinue this medicine. These symptoms may include convulsions, confusion, nightmares, insomnia.

Others:
- Advise any doctor, dentist or pharmacist whom you consult that you take this drug.
- For dry mouth, suck sugarless hard candy or chew sugarless gum. If dry mouth persists, consult your dentist.

POSSIBLE INTERACTION WITH OTHER DRUGS

GENERIC NAME OR DRUG CLASS	COMBINED EFFECT
Addictive drugs*	Increased risk of addictive effect.
Antidepressants, tricyclic*	Increased anti-depressant effect.
Antihistamines*	Possible excessive sedation.
Central nervous system (CNS) depressants*	Increased depressive effects of both drugs.
Monoamine oxidase (MAO) inhibitors*	Increased meprobamate effect.
Narcotics*	Increased narcotic effect.
Sertraline	Increased depressive effects of both drugs.

POSSIBLE INTERACTION WITH OTHER SUBSTANCES

INTERACTS WITH	COMBINED EFFECT
Alcohol:	Dangerous increased effect of meprobamate.
Beverages: Caffeine drinks.	Decreased calming effect of meprobamate.
Cocaine:	Increased risk of side effects. Avoid
Foods:	None expected.
Marijuana:	Increased risk of side effects. Avoid.
Tobacco:	None expected.

*See Glossary

MERCAPTOPURINE

BRAND NAMES

Purinethol
Purixan

BASIC INFORMATION

Habit forming? No
Prescription needed? Yes
Available as generic? Yes, for tablet
Drug class: Antineoplastic, immunosuppressant

USES

- Treatment for some kinds of cancer.
- Treatment for regional enteritis and ulcerative colitis and other immune disorders.

DOSAGE & USAGE INFORMATION

How to take:
- Tablet—Swallow with liquid.
- Oral solution—Carefully follow instructions provided with prescription. If unsure, consult doctor or pharmacist.

When to take:
At the same time each day.

If you forget a dose:
Take as soon as you remember. If it is almost time for the next dose, wait for that dose (don't double this dose) and resume regular schedule.

What drug does:
Inhibits abnormal cell reproduction.

Time lapse before drug works:
May require 6 weeks for maximum effect.

Don't take with:
Any other medicine or any dietary supplement without consulting your doctor or pharmacist.

OVERDOSE

SYMPTOMS:
Nausea, vomiting, diarrhea, severe stomach pain, yellow eyes or skin.
WHAT TO DO:
- **Dial 911 for all medical emergencies or call poison control center 1-800-222-1222 for instructions.**
- **See emergency information on last 3 pages of this book.**

POSSIBLE ADVERSE REACTIONS OR SIDE EFFECTS

SYMPTOMS	WHAT TO DO
Life-threatening:	
In case of overdose, see previous column.	
Common:	
• Unusual bleeding or bruising, unusual tiredness or weakness, yellow skin or eyes, infection (fever, chills, headaches, muscle aches).	Discontinue. Call doctor right away.
• Stomach discomfort, nausea, vomiting.	Continue. Call doctor when convenient.
Infrequent:	
• Diarrhea, headache, confusion, blurred vision, shortness of breath, joint pain, blood in urine, back pain, appetite loss, feet and leg swelling, itching, rash.	Discontinue. Call doctor right away.
• Cough, acne, hair loss.	Continue. Call doctor when convenient.
Rare:	
Mouth sores, red or swollen mouth, pain in upper chest (pancreas problem).	Discontinue. Call doctor right away.

WARNINGS & PRECAUTIONS

Don't take if:
You are allergic to mercaptopurine. Allergic reactions are rare, but may occur.

Before you start, consult your doctor if:
- You are an alcoholic.
- You have blood, liver or kidney disease.
- You have colitis or peptic ulcer.
- You have gout.
- You have an infection.
- You plan to become pregnant within 3 months.

Over age 60:
Adverse reactions and side effects may be more frequent and severe than in younger persons.

Pregnancy:
Consult doctor. Use of the drug during pregnancy is not recommended. It could cause harm to the unborn baby. Risk category D (see page xviii).

Breastfeeding; Lactation; Nursing Mothers:
This drug is generally acceptable with breastfeeding. Discuss risks and benefits with your doctor.

Infants & children up to age 18:
Safety and effectiveness in this age group have not been established. Consult doctor.

Prolonged use:
- Adverse reactions more likely the longer drug is required.
- Talk to your doctor about the need for follow-up medical exams or lab tests.

Skin & sunlight:
No problems expected.

Driving, piloting or hazardous work:
Avoid if you feel dizzy, drowsy or confused. Otherwise, no problems expected.

Discontinuing:
Don't discontinue without doctor's advice until you complete prescribed dose, even though symptoms diminish or disappear. Some side effects may follow discontinuing. Report to doctor blurred vision, convulsions, confusion, persistent headache, chills or fever, bloody urine or stools, back pain, yellow eyes or skin.

Others:
- Drink more fluid than usual so you will have more frequent urination.
- Don't give this medicine to anyone else for any purpose. It is a strong drug that requires close medical supervision.
- Advise any doctor, dentist or pharmacist whom you consult that you take this medicine.

POSSIBLE INTERACTION WITH OTHER DRUGS

GENERIC NAME OR DRUG CLASS	COMBINED EFFECT
Acetaminophen	Increased likelihood of liver toxicity.
Anticoagulants,* oral	May increase or decrease anticoagulant effect.
Antineoplastic drugs,* other	Increased effect of both drugs (may be desirable) or increased toxicity of each.
Chloramphenicol	Increased toxicity of each.
Clozapine	Toxic effect on bone marrow.
Cyclosporine	May increase risk of infection.
Hepatotoxic drugs*	Increased risk of liver toxicity.
Immunosuppressants,* other	Increased risk of infections and neoplasms.*
Isoniazid	Increased risk of liver damage.
Levamisole	Increased risk of bone marrow depression.
Lovastatin	Increased heart and kidney damage.
Probenecid	Increased toxic effect of mercaptopurine.
Sulfinpyrazone	Increased toxic effect of mercaptopurine.
Tiopronin	Increased risk of toxicity to bone marrow.
Vaccines, live or killed	Increased risk of toxicity or reduced effectiveness of vaccine.

POSSIBLE INTERACTION WITH OTHER SUBSTANCES

INTERACTS WITH	COMBINED EFFECT
Alcohol:	May increase chance of intestinal bleeding.
Beverages:	None expected.
Cocaine:	Increased risk of side effects. Avoid.
Foods:	Reduced irritation in stomach.
Marijuana:	Increased risk of side effects. Avoid.
Tobacco:	None expected.

***See Glossary**

MESALAMINE

BRAND NAMES

5-ASA
Apriso
Asacol
Asacol HD
Canasa
Delzicol
Lialda
Mesalazine
Pentasa
Rowasa
Salofalk
sfRowasa

BASIC INFORMATION

Habit forming? No
Prescription needed? Yes
Available as generic? Yes
Drug class: Anti-inflammatory (nonsteroidal)

USES

- Treats ulcerative colitis.
- Reduces inflammatory conditions of the lower colon and rectum.

DOSAGE & USAGE INFORMATION

How to use:

- Rectal—Use as an enema. Insert the tip of the pre-packaged medicine container into the rectum. Squeeze container to empty contents. Retain in rectum all night or as long as possible.
- Rectal suppository—Follow instructions on package.
- Delayed-release tablet or capsule; extended-release capsule—Swallow with liquid. Do not crush or chew tablet or open capsule. Take Lialda brand with food.

Continued next column

OVERDOSE

SYMPTOMS:
Confusion, severe diarrhea, lightheadedness or dizziness, severe sleepiness, severe headache, ringing or buzzing in ears, hearing loss, nausea and vomiting, sweating, fast or deep breathing.

WHAT TO DO:

- **Dial 911 for all medical emergencies or call poison control center 1-800-222-1222 for instructions.**
- **See emergency information on last 3 pages of this book.**

When to use:

- Rectal—Each night, preferably after a bowel movement. Continue for 3 to 6 weeks according to your doctor's instructions.
- Rectal suppository—Use 1-3 times a day according to doctor's instructions.
- Oral forms—Follow instructions on label. It may be once-a-day dose or 3 to 4 times a day as directed by doctor.

If you forget a dose:
Use as soon as you remember. If it is almost time for the next dose, wait for that dose (don't double this dose) and resume regular schedule.

What drug does:
It works in the intestinal tract to reduce inflammation and also helps keep the disorder in remission and prevent relapses.

Time lapse before drug works:
3 to 21 days.

Don't take with:

- Oral sulfasalazine concurrently. To do so may increase chances of kidney damage.
- Any other medicine or any dietary supplement without consulting your doctor or pharmacist.

POSSIBLE ADVERSE REACTIONS OR SIDE EFFECTS

SYMPTOMS	WHAT TO DO
Life-threatening: None expected.	
Common: None expected.	
Infrequent: None expected.	
Rare:	
• Abdominal pain, bloody diarrhea, fever, skin rash, anal irritation, chest pain, shortness of breath.	Discontinue. Call doctor right away.
• Gaseousness, nausea, headache, mild hair loss, diarrhea.	Continue. Call doctor when convenient.

WARNINGS & PRECAUTIONS

Don't take if:
You are allergic to mesalamine. Allergic reactions are rare, but may occur.

Before you start, consult your doctor if:
- You have had chronic kidney disease.
- You have had allergic response to other drugs, supplements or foods.
- You have pancreatitis.
- You have heart inflammation (pericarditis).

Over age 60:
Adverse reactions and side effects may be more frequent and severe than in younger persons.

Pregnancy:
Decide with your doctor if drug benefits justify risk to unborn baby. Risk category B (see page xviii).

Breastfeeding; Lactation; Nursing Mothers:
This drug is generally acceptable with breastfeeding. Discuss risks and benefits with your doctor.

Infants & children up to age 18:
Follow instructions provided by your child's doctor.

Prolonged use:
Talk to your doctor about the need for follow-up medical examinations or laboratory studies to check urine.

Skin & sunlight:
No problems expected.

Driving, piloting or hazardous work:
Don't drive or pilot aircraft until you learn how medicine affects you. Don't work around dangerous machinery. Don't climb ladders or work in high places. Danger increases if you drink alcohol or take medicine affecting alertness and reflexes.

Discontinuing:
Don't discontinue without consulting doctor. Dose may require gradual reduction if you have taken drug for a long time. Doses of other drugs may also require adjustment.

Others:
- Internal eye pressure should be measured regularly.
- Advise any doctor, dentist or pharmacist whom you consult that you use/take this drug.
- Canasa may stain things it touches.

POSSIBLE INTERACTION WITH OTHER DRUGS

GENERIC NAME OR DRUG CLASS	COMBINED EFFECT
Antacids*	Affects the dosage of brand name Apriso. Avoid.

POSSIBLE INTERACTION WITH OTHER SUBSTANCES

INTERACTS WITH	COMBINED EFFECT
Alcohol:	None expected.
Beverages:	None expected.
Cocaine:	None expected.
Foods:	None expected.
Marijuana:	None expected.
Tobacco:	None expected.

***See Glossary**

METFORMIN

BRAND NAMES

See full list of brand names in the *Generic and Brand Name Directory*, page 871.

BASIC INFORMATION

Habit forming? No
Prescription needed? Yes
Available as generic? Yes
Drug class: Antihyperglycemic, antidiabetic

USES

Treatment for hyperglycemia (excess sugar in the blood) that cannot be controlled by diet alone in patients with diabetes type 2.

DOSAGE & USAGE INFORMATION

How to take:
- Tablet—Swallow with liquid. Take with food.
- Extended release tablet—Swallow whole. Take with food. Do not break, crush or chew before swallowing.
- Chewable tablet—Follow product instructions.
- Oral solution—Follow product instructions.

Continued next column

When to take:
Usually 1 to 3 times a day as directed by doctor. Take with meals to lessen stomach irritation. Dosage may be increased on a weekly basis until maximum benefits are achieved.

If you forget a dose:
Take as soon as you remember. If it is almost time for the next dose, wait for that dose (don't double this dose) and resume regular schedule.

What drug does:
Helps to lower blood sugar when it is too high. Treats symptoms of diabetes, but is not a cure.

Time lapse before drug works:
May take several weeks for full effectiveness.

Don't take with:
Any other medicine or any dietary supplement without consulting your doctor or pharmacist.

OVERDOSE

SYMPTOMS:
- **Symptoms of lactic acidosis (acid in the blood)—chills, diarrhea, fatigue, muscle pain, sleepiness, slow heartbeat, breathing difficulty, unusual weakness.**
- **Hypoglycemia symptoms (low blood sugar) —stomach pain, nervousness, shakiness anxious feeling, confusion, cold sweats, chills, convulsions, cool pale skin, excess hunger, unsteady walk, nausea or vomiting, rapid heartbeat, unusual weakness or tiredness, vision changes, unconsciousness.**

WHAT TO DO:
- **For mild low blood sugar symptoms, drink or eat something containing sugar right away.**
- **Dial 911 for all medical emergencies or call poison control center 1-800-222-1222 for instructions.**
- **See emergency information on last 3 pages of this book.**

POSSIBLE ADVERSE REACTIONS OR SIDE EFFECTS

SYMPTOMS	WHAT TO DO
Life-threatening:	
In case of overdose or low blood sugar, see previous column.	
Common:	
• Stomach pain, diarrhea, vomiting.	Continue, but call doctor right away.
• Decreased appetite, changes in taste, gas, headache, weight loss, nausea, feeling of fullness or stomach discomfort.	Continue. Call doctor when convenient.
Infrequent:	
None expected.	
Rare:	
Lactic acidosis or severe low blood sugar (see symptoms under Overdose).	Discontinue. Call doctor right away or seek emergency care.

WARNINGS & PRECAUTIONS

Don't take if:
You are allergic to metformin. Allergic reactions are rare, but may occur.

Before you start, consult your doctor if:
- You have any kidney or liver disease or any heart or blood vessel disorder.
- You have any chronic health problem.
- You have an infection, illness or any condition that can cause low blood sugar.
- You have a history of acid in the blood (metabolic acidosis or ketoacidosis).
- You are allergic to any medication, food or other substance.

Over age 60:
No special problems expected. A lower starting dosage may be recommended by your doctor.

Pregnancy:
Decide with your doctor if drug benefits justify risk to unborn baby. Risk category B (see page xviii).

Breastfeeding; Lactation; Nursing Mothers:
This drug is generally acceptable with breastfeeding. Discuss risks and benefits with your doctor.

Infants & children up to age 18:
Most forms of metformin are approved for children age 10 and over. Follow instructions provided by your child's doctor.

Prolonged use:
- Schedule regular doctor visits to determine if the drug is continuing to be effective in controlling the diabetes and to check for any problems in kidney function.
- You will most likely require an antidiabetic medicine for the rest of your life.
- You will need to test your blood glucose levels several times a day or, for some, once to several times a week.

Skin & sunlight:
No special problems expected.

Driving, piloting or hazardous work:
No special problems expected.

Discontinuing:
Don't discontinue without consulting your doctor, even if you feel well. You can have diabetes without feeling any symptoms. Untreated diabetes can cause serious problems.

Others:
- Advise any doctor, dentist or pharmacist whom you consult that you take this medicine. Drug may interfere with the accuracy of some medical tests.
- Follow any special diet your doctor may prescribe. It can help control diabetes.
- Consult doctor if you become ill with vomiting or diarrhea.
- Use caution when exercising. Ask your doctor about an appropriate exercise program.
- Wear medical identification stating that you have diabetes and take this medication.
- Learn to recognize the symptoms of low blood sugar. You and your family need to know what to do if these symptoms occur.
- Have a glucagon kit and syringe in the event severe low blood sugar occurs. Carry a quick-acting sugar to treat symptoms of mild low blood sugar.
- High blood sugar (hyperglycemia) may occur with diabetes. Ask your doctor about symptoms to watch for and treatment steps to take.
- This drug may be discontinued temporarily prior to x-ray studies or some surgeries.
- Educate yourself about diabetes.

POSSIBLE INTERACTION WITH OTHER DRUGS

GENERIC NAME OR DRUG CLASS	COMBINED EFFECT
Amiloride	Increased metformin effect.
Calcium channel blockers*	Increased metformin effect.
Cimetidine	Increased metformin effect.
Dexfenfluramine	May require dosage change as weight loss occurs.
Digoxin	Increased metformin effect.
Dofetilide	Increased dofetilide effect.
Furosemide	Increased metformin effect.
Hyperglycemia-causing medications*	Increased risk of hyperglycemia.
Hypoglycemia-causing medications*	Increased risk of hypoglycemia.
Morphine	Increased metformin effect.
Procainamide	Increased metformin effect.
Quinidine	Increased metformin effect.
Quinine	Increased metformin effect.

Continued on page 904

POSSIBLE INTERACTION WITH OTHER SUBSTANCES

INTERACTS WITH	COMBINED EFFECT
Alcohol:	Increased risk of hypoglycemia. Avoid excessive amounts.
Beverages:	None expected.
Cocaine:	Unknown. Avoid.
Foods:	None expected.
Marijuana:	Consult doctor.
Tobacco:	None expected.

***See Glossary**

METHENAMINE

BRAND NAMES

Hiprex
Mandelamine
Urex

BASIC INFORMATION

Habit forming? No
Prescription needed? Yes
Available as generic? Yes
Drug class: Anti-infective (urinary)

USES

Suppresses chronic urinary tract infections.

DOSAGE & USAGE INFORMATION

How to take:
- Tablet—Swallow with liquid or food to lessen stomach irritation. If you can't swallow whole, crumble tablet and take with liquid or food. If enteric-coated tablet, swallow whole.
- Liquid form—Use a measuring spoon to ensure correct dose.
- Granules—Dissolve dose in 4 oz. of water. Drink all the liquid.

When to take:
At the same times each day.

If you forget a dose:
Take as soon as you remember. If it is almost time for the next dose, wait for that dose (don't double this dose) and resume regular schedule.

What drug does:
A chemical reaction in the urine changes methenamine into formaldehyde, which destroys certain bacteria.

Time lapse before drug works:
Continual use for 3 to 6 months.

Don't take with:
Any other medicine or any dietary supplement without consulting your doctor or pharmacist.

OVERDOSE

SYMPTOMS:
Unknown.
WHAT TO DO:
- **Dial 911 for all medical emergencies or call poison control center 1-800-222-1222 for instructions.**
- **See emergency information on last 3 pages of this book.**

POSSIBLE ADVERSE REACTIONS OR SIDE EFFECTS

SYMPTOMS	WHAT TO DO
Life-threatening:	
In case of overdose, see previous column.	
Common:	
• Rash.	Discontinue. Call doctor right away.
• Nausea, difficult urination.	Continue. Call doctor when convenient.
Infrequent:	
• Blood in urine.	Discontinue. Call doctor right away.
• Burning on urination, lower back pain.	Continue. Call doctor when convenient.
Rare:	
None expected.	

WARNINGS & PRECAUTIONS

Don't take if:
- You are allergic to methenamine. Allergic reactions are rare, but may occur.
- You have a severe impairment of kidney or liver function.
- Your urine cannot or should not be acidified (check with your doctor).

Before you start, consult your doctor if:
- You have had kidney or liver disease.
- You plan to become pregnant within medication period.
- You have had gout.

Over age 60:
No special problems expected.

Pregnancy:
Decide with your doctor if drug benefits justify risk to unborn baby. Risk category C (see page xviii).

Breastfeeding; Lactation; Nursing Mothers:
This drug is generally acceptable with breastfeeding. Discuss risks and benefits with your doctor.

Infants & children up to age 18:
Follow instructions provided by your child's doctor.

Prolonged use:
No problems expected.

Skin & sunlight:
No problems expected.

Driving, piloting or hazardous work:
No problems expected.

Discontinuing:
Don't discontinue without doctor's advice until you complete prescribed dose, even though symptoms diminish or disappear.

Others:
- Requires an acid urine to be effective. Eat more protein foods, cranberries, cranberry juice with vitamin C, plums, prunes.
- Advise any doctor, dentist or pharmacist whom you consult that you take this medicine.

POSSIBLE INTERACTION WITH OTHER DRUGS

GENERIC NAME OR DRUG CLASS	COMBINED EFFECT
Antacids*	Decreased methenamine effect.
Carbonic anhydrase inhibitors*	Decreased methenamine effect.
Citrates*	Decreases effects of methenamine.
Diuretics, thiazide*	Decreased urine acidity.
Sodium bicarbonate	Decreased methenamine effect.
Sulfadoxine and pyrimethamine	Increased risk of kidney toxicity.
Sulfa drugs*	Possible kidney damage.

POSSIBLE INTERACTION WITH OTHER SUBSTANCES

INTERACTS WITH	COMBINED EFFECT
Alcohol:	Possible brain depression. Avoid or use with caution.
Beverages: Milk and other dairy products.	Decreased methenamine effect.
Cocaine:	Unknown. Avoid.
Foods: Citrus, cranberries, plums, prunes.	Increased methenamine effect.
Marijuana:	Drowsiness, muscle weakness or blood pressure drop. Avoid.
Tobacco:	None expected.

*See Glossary

METHOTREXATE

BRAND NAMES

Amethopterin
Folex
Folex PFS
Mexate
Mexate AQ
Otrexup
Rasuvo
Rheumatrex
Trexall

BASIC INFORMATION

Habit forming? No
Prescription needed? Yes
Available as generic? Yes
Drug class: Antimetabolite, antipsoriatic

USES

- Treatment for certain types of cancer.
- Treatment for psoriasis in patients with severe cases.
- Treatment for severe rheumatoid arthritis.

DOSAGE & USAGE INFORMATION

How to take:
- Tablet—Swallow with liquid.
- Subcutaneous injection—Self-injected weekly with provided injector. Follow detailed instructions provided with prescription.
- Injectable form—Is sometimes self-injected or used as an oral dose. Consult doctor.

When to take:
Follow instructions with your prescription.

If you forget a dose:
Take as soon as you remember. If it is almost time for the next dose, wait for that dose (don't double this dose) and resume regular schedule.

What drug does:
Inhibits abnormal cell reproduction.

Time lapse before drug works:
May require 6 weeks for maximum effect.

Don't take with:
Any other medicine or any dietary supplement without consulting your doctor or pharmacist.

OVERDOSE

SYMPTOMS:
Symptoms are similar to adverse reactions.
WHAT TO DO:
- **Dial 911 for all medical emergencies or call poison control center 1-800-222-1222 for instructions.**
- **See emergency information on last 3 pages of this book.**

POSSIBLE ADVERSE REACTIONS OR SIDE EFFECTS

SYMPTOMS	WHAT TO DO
Life-threatening:	
Rare, severe allergic reaction (anaphylaxis)—difficulty breathing, hives, itching, swelling (throat, face), vomiting, dizziness.	Seek emergency treatment immediately.
Common:	
Sore throat, fever, mouth sores, chills, black or tarry stools, unusual bleeding or bruising, abdominal pain, nausea, vomiting (may be bloody).	Discontinue. Call doctor right away.
Infrequent:	
• Seizures.	Discontinue. Seek emergency care.
• Dizziness when standing after sitting or lying, drowsiness, joint pain, headache, confusion, blurred vision, shortness of breath, bloody urine, yellow skin or eyes, diarrhea, reddish skin, back pain.	Discontinue. Call doctor right away.
• Cough, rash, sexual difficulties in males, acne, boils, hair loss, itchy skin.	Continue. Call doctor when convenient.
Rare:	
Painful urination.	Discontinue. Call doctor right away.

WARNINGS & PRECAUTIONS

Don't take if:
You are allergic to methotrexate.

Before you start, consult your doctor if:
- You are an alcoholic.
- You have blood, liver or kidney disease.
- You have colitis or peptic ulcer.
- You have gout.
- You have an infection.
- You plan to become pregnant within 3 months.

Over age 60:
Adverse reactions and side effects may be more frequent and severe than in younger persons.

Pregnancy:
Consult doctor. Drug should not be used during pregnancy. Can cause harm to unborn baby. Risk category X (see page xviii).

Breastfeeding; Lactation; Nursing Mothers:
This drug is generally not recommended with breastfeeding. Discuss risks and benefits with your doctor.

Infants & children up to age 18:
Follow instructions provided by your child's doctor.

Prolonged use:
- Adverse reactions more likely the longer drug is required.
- Talk to your doctor about the need for follow-up medical examinations or laboratory studies to check liver function, kidney function, complete blood counts (white blood cell count, platelet count, red blood cell count, hemoglobin, hematocrit).

Skin & sunlight:
May cause rash or intensify sunburn in areas exposed to sun or ultraviolet light (photosensitivity reaction). Avoid overexposure. Notify doctor if reaction occurs.

Driving, piloting or hazardous work:
Avoid if you feel dizzy, drowsy or confused. Otherwise, no problems expected.

Discontinuing:
Don't discontinue without doctor's advice until you complete prescribed dose, even though symptoms diminish or disappear. Some side effects may follow discontinuing. Consult doctor if blurred vision, convulsions, confusion, or persistent headache occur.

Others:
- This drug can be toxic to body's organs. Its use requires close medical supervision.
- Drug use can rarely cause severe skin reaction that can be fatal. Consult doctor.
- Advise any doctor, dentist or pharmacist whom you consult that you take this medicine.

POSSIBLE INTERACTION WITH OTHER DRUGS

GENERIC NAME OR DRUG CLASS	COMBINED EFFECT
Anticoagulants,* oral	Increased anticoagulant effect.
Anticonvulsants,* hydantoin	Possible methotrexate toxicity.
Antigout drugs*	Increased effect of methotrexate.
Anti-inflammatory drugs, nonsteroidal (NSAIDs)*	Possible increased methotrexate toxicity.
Asparaginase	Decreased methotrexate effect.
Bone marrow depressants,* other	Increased risk of bone marrow depression.
Clozapine	Toxic effect on bone marrow.
Diclofenac	May increase toxicity.
Etretinate	Increased chance of toxicity to liver.
Fluorouracil	Decreased methotrexate effect.
Folic acid	Possible decreased methotrexate effect.
Isoniazid	Increased risk of liver damage.
Leflunomide	Increased risk of side effects.
Leucovorin calcium	Decreased methotrexate toxicity.
Levamisole	Increased risk of bone marrow depression.
Oxyphenbutazone	Possible methotrexate toxicity.
Penicillins*	Increased risk of methotrexate toxicity.
Phenylbutazone	Possible methotrexate toxicity.

Continued on page 904

POSSIBLE INTERACTION WITH OTHER SUBSTANCES

INTERACTS WITH	COMBINED EFFECT
Alcohol:	Likely liver damage. Avoid.
Beverages:	Extra fluid intake decreases chance of methotrexate toxicity.
Cocaine:	Increased chance of methotrexate adverse reactions. Avoid.
Foods:	None expected.
Marijuana:	Consult doctor.
Tobacco:	None expected.

*See Glossary

METOCLOPRAMIDE

BRAND NAMES

Apo-Metoclop
Clopra
Emex
Maxeran
Metozolv ODT
Octamide
Octamide PFS
Reclomide
Reglan

BASIC INFORMATION

Habit forming? No
Prescription needed? Yes
Available as generic? Yes
Drug class: Antiemetic; dopaminergic blocker

USES

Treatment for gastroesophageal reflux disease (GERD) and for diabetic gastroparesis.

DOSAGE & USAGE INFORMATION

How to take:

- Tablet or syrup—Swallow with liquid or food to lessen stomach irritation.
- Oral disintegrating tablet—Let tablet dissolve on tongue. Don't swallow with water.

When to take:
30 minutes before each meal and at bedtime or take before symptoms expected, up to 4 times a day.

If you forget a dose:
Take as soon as you remember. If it is almost time for the next dose, wait for that dose (don't double this dose) and resume regular schedule.

What drug does:
Prevents smooth muscle in stomach from relaxing, thereby helping to empty stomach more quickly.

Continued next column

OVERDOSE

SYMPTOMS:
Severe drowsiness, muscle spasms, mental confusion, trembling, seizure.
WHAT TO DO:

- **Dial 911 for all medical emergencies or call poison control center 1-800-222-1222 for instructions.**
- **See emergency information on last 3 pages of this book.**

Time lapse before drug works:
30 to 60 minutes.

Don't take with:
Any other medicine or any dietary supplement without consulting your doctor or pharmacist.

POSSIBLE ADVERSE REACTIONS OR SIDE EFFECTS

SYMPTOMS	WHAT TO DO
Life-threatening: None expected.	
Common: Drowsiness, restlessness, rash.	Continue. Call doctor when convenient.
Infrequent:	
• Wheezing, shortness of breath.	Discontinue. Call doctor right away.
• Dizziness, headache, insomnia, tender and swollen breasts, increased milk flow, menstrual changes, decreased sex drive.	Continue. Call doctor when convenient.
Rare:	
• Abnormal, involuntary movements (of jaw, lips and tongue), depression, Parkinson syndrome.*	Discontinue. Call doctor right away.
• Constipation, nausea, diarrhea, dry mouth.	Continue. Call doctor when convenient.

WARNINGS & PRECAUTIONS

Don't take if:
You are allergic to metoclopramide. Allergic reactions are rare, but may occur.

Before you start, consult your doctor if:

- You have Parkinson's disease.
- You have liver or kidney disease or epilepsy.
- You have bleeding from gastrointestinal tract or intestinal obstruction.
- You will have surgery within 2 months, including dental surgery, requiring general or spinal anesthesia.

Over age 60:
Adverse reactions and side effects may be more frequent and severe than in younger persons.

Pregnancy:
Decide with your doctor if drug benefits justify risk to unborn baby. Risk category B (see page xviii).

Breastfeeding; Lactation; Nursing Mothers:
This drug is generally acceptable with breastfeeding. Discuss risks and benefits with your doctor.

Infants & children up to age 18:
Safety and effectiveness in this age group have not been established. Consult doctor.

Prolonged use:
Adverse reactions including muscle spasms and trembling hands more likely to occur.

Skin & sunlight:
No problems expected.

Driving, piloting or hazardous work:
Don't drive or pilot aircraft until you learn how medicine affects you. Don't work around dangerous machinery. Don't climb ladders or work in high places. Danger increases if you drink alcohol or take medicine affecting alertness and reflexes, such as antihistamines, tranquilizers, sedatives, pain medicine, narcotics and mind-altering drugs.

Discontinuing:
May be unnecessary to finish medicine. Follow doctor's instructions.

Others:
Advise any doctor, dentist or pharmacist whom you consult that you take this medicine.

POSSIBLE INTERACTION WITH OTHER DRUGS

GENERIC NAME OR DRUG CLASS	COMBINED EFFECT
Acetaminophen	Increased absorption of acetaminophen.
Anticholinergics*	Decreased metoclopramide effect.
Aspirin	Increased absorption of aspirin.
Bromocriptine	Decreased bromocriptine effect.
Butyrophenone	Increased chance of muscle spasm and trembling.
Central nervous system (CNS) depressants*	Excess sedation.
Clozapine	Toxic effect on the central nervous system.
Digitalis preparations*	Decreased absorption of digitalis.
Ethinamate	Dangerous increased effects of ethinamate. Avoid combining.
Fluoxetine	Increased depressant effects of both drugs.
Guanfacine	May increase depressant effects of either drug.
Insulin	Unpredictable changes in blood glucose. Dosages may require adjustment.
Leucovorin	High alcohol content of leucovorin may cause adverse effects.
Levodopa	Increased absorption of levodopa.
Lithium	Increased absorption of lithium.
Loxapine	May increase toxic effects of both drugs.
Methyprylon	Increased sedative effect, perhaps to dangerous level. Avoid.
Nabilone	Greater depression of central nervous system.
Narcotics*	Decreased metoclopramide effect.
Nizatidine	Decreased nizatidine absorption.
Pergolide	Decreased pergolide effect.

Continued on page 905

POSSIBLE INTERACTION WITH OTHER SUBSTANCES

INTERACTS WITH	COMBINED EFFECT
Alcohol:	Excess sedation. Avoid.
Beverages:	
Coffee.	Decreased metoclopramide effect.
Cocaine:	Unknown. Avoid.
Foods:	None expected.
Marijuana:	Consult doctor.
Tobacco:	None expected.

*See Glossary

METYRAPONE

BRAND NAMES

Metopirone

BASIC INFORMATION

Habit forming? No
Prescription needed? Yes
Available as generic? No
Drug class: Antiadrenal

USES

- To diagnose the function of the pituitary gland.
- Treats Cushing's syndrome, a disorder characterized by higher than normal concentrations of cortisol (one of the hormones secreted by the adrenal glands) in the blood.

DOSAGE & USAGE INFORMATION

How to take:
- For medical testing purposes—Take the prescribed number of tablets with milk or food on the day before the scheduled test. On the day of the test, blood and urine studies will show the amount of hormones in your blood. Results of the test will help establish your diagnosis.
- For treatment of Cushing's syndrome—Swallow tablet with liquid. If you can't swallow whole, crumble tablet and take with liquid or food.

When to take:
- For medical testing—Take the prescribed number of tablets on the day before the scheduled test.
- For treatment of Cushing's syndrome—Take total daily amount in divided doses. Follow prescription directions carefully.

Continued next column

OVERDOSE

SYMPTOMS:
Severe nausea or vomiting or diarrhea, abdominal pain, dehydration, sudden weakness, confusion, irregular heartbeat.
WHAT TO DO:
- **Dial 911 for all medical emergencies or call poison control center 1-800-222-1222 for instructions.**
- **See emergency information on last 3 pages of this book.**

If you forget a dose:
Take as soon as you remember. If it is almost time for the next dose, wait for that dose (don't double this dose) and resume regular schedule.

What drug does:
Prevents one of the chemical reactions in the production of cortisol by the adrenal glands.

Time lapse before drug works:
Approximately 1 hour.

Don't take with:
- Cortisone*-like medicines for 48 hours prior to testing.
- Any other medicine or any dietary supplement without consulting your doctor or pharmacist.

POSSIBLE ADVERSE REACTIONS OR SIDE EFFECTS

SYMPTOMS	WHAT TO DO
Life-threatening: In case of overdose, see previous column.	
Common: Dizziness, headache, nausea.	Continue. Call doctor when convenient.
Infrequent: Drowsiness.	Continue. Call doctor when convenient.
Rare: Hair loss or excess hair growth, loss of appetite, confusion, acne (may begin or may worsen if already present).	Continue. Call doctor when convenient.

WARNINGS & PRECAUTIONS

Don't take if:
You are allergic to metyrapone. Allergic reactions are rare, but may occur.

Before you start, consult your doctor if:
- You have porphyria.
- You have adrenal insufficiency (Addison's disease), thyroid problems, heart disease, type 2 diabetes or liver disease.
- You have decreased pituitary function.

Over age 60:
No special problems expected.

Pregnancy:
Decide with your doctor if drug benefits justify risk to unborn baby. Risk category C (see page xviii).

Breastfeeding; Lactation; Nursing Mothers:
This drug is generally acceptable with breastfeeding. Discuss risks and benefits with your doctor.

Infants & children up to age 18:
Follow instructions provided by your child's doctor.

Prolonged use:
No special problems expected.

Skin & sunlight:
No special problems expected.

Driving, piloting or hazardous work:
Don't drive or pilot aircraft until you learn how medicine affects you. Don't work around dangerous machinery. Don't climb ladders or work in high places. Danger increases if you drink alcohol or take medicine affecting alertness and reflexes.

Discontinuing:
No special problems expected.

Others:
Advise any doctor, dentist or pharmacist whom you consult that you take this medicine.

POSSIBLE INTERACTION WITH OTHER DRUGS

GENERIC NAME OR DRUG CLASS	COMBINED EFFECT
Acetaminophen	Increased risk of liver problems.
Antidiabetics, oral*	Increased risk of adverse reactions.
Contraceptives, oral*	May decrease effect of metyrapone.
Corticosteroids*	May decrease effect of metyrapone.
Estrogens*	May decrease effect of metyrapone.
Insulin	Increased risk of adverse reactions.
Phenytoin	May decrease effect of metyrapone.

POSSIBLE INTERACTION WITH OTHER SUBSTANCES

INTERACTS WITH	COMBINED EFFECT
Alcohol:	None expected.
Beverages:	None expected.
Cocaine:	Unknown. Avoid.
Foods:	Increased appetite and absorption of nutrients, causing difficulty with weight control.
Marijuana:	Consult doctor.
Tobacco:	None expected.

***See Glossary**

MEXILETINE

GENERIC NAME

MEXILETINE

BASIC INFORMATION

Habit forming? No
Prescription needed? Yes
Available as generic? Yes
Drug class: Antiarrhythmic

USES

Stabilizes irregular heartbeat.

DOSAGE & USAGE INFORMATION

How to take:
Capsule—Swallow whole with food, milk or antacid to lessen stomach irritation.

When to take:
At the same times each day as directed by your doctor.

If you forget a dose:
Take as soon as you remember. If it is almost time for the next dose, wait for that dose (don't double this dose) and resume regular schedule.

What drug does:
Blocks the fast sodium channel in heart tissue.

Time lapse before drug works:
30 minutes to 2 hours.

Don't take with:
Any other medicine or any dietary supplement without consulting your doctor or pharmacist.

OVERDOSE

SYMPTOMS:
Nausea, vomiting, drowsiness, confusion, slow heartbeat, seizures, cardiac arrest.
WHAT TO DO:
- **Dial 911 for all medical emergencies or call poison control center 1-800-222-1222 for instructions.**
- **See emergency information on last 3 pages of this book.**

POSSIBLE ADVERSE REACTIONS OR SIDE EFFECTS

SYMPTOMS	WHAT TO DO
Life-threatening:	
Chest pain, shortness of breath, irregular or fast heartbeat.	Discontinue. Seek emergency treatment.
Common:	
Dizziness, anxiety, shakiness, nausea, vomiting, heartburn, unsteadiness when walking.	Discontinue. Call doctor right away.
Infrequent:	
• Sore throat, fever, mouth sores, blurred vision, confusion, constipation, diarrhea, headache, numbness or tingling in hands or feet, ringing in ears, unexplained bleeding or bruising, rash, slurred speech, insomnia, weakness, difficult swallowing.	Discontinue. Call doctor right away.
• Loss of taste.	Continue. Call doctor when convenient.
Rare:	
• Seizures.	Discontinue. Seek emergency care.
• Hallucinations, mental changes, memory loss, difficult breathing, swollen feet and ankles, hiccups, yellow eyes or skin.	Discontinue. Call doctor right away.
• Hair loss, impotence.	Continue. Call doctor when convenient.

WARNINGS & PRECAUTIONS

Don't take if:
You are allergic to mexiletine, lidocaine or tocainide. Allergic reactions are rare, but may occur.

Before you start, consult your doctor if:
- You have had liver or kidney disease or impaired kidney function.
- You have had lupus.
- You have a history of seizures.
- You will have surgery within 2 months, including dental surgery, requiring general or spinal anesthesia.
- You have heart disease or low blood pressure.

Over age 60:
Adverse reactions and side effects may be more frequent and severe than in younger persons.

Pregnancy:
Decide with your doctor if drug benefits justify risk to unborn baby. Risk category C (see page xviii).

Breastfeeding; Lactation; Nursing Mothers:
This drug is generally acceptable with breastfeeding. Discuss risks and benefits with your doctor.

Infants & children up to age 18:
Safety and effectiveness in this age group have not been established. Consult doctor.

Prolonged use:
- May cause lupus*-like illness.
- Talk to your doctor about the need for follow-up medical examinations or laboratory studies to check ECG,* liver function.

Skin & sunlight:
No problems expected.

Driving, piloting or hazardous work:
Use caution if you feel dizzy or weak. Otherwise, no problems expected.

Discontinuing:
Don't discontinue without consulting doctor. Dose may require gradual reduction if you have taken drug for a long time. Doses of other drugs may also require adjustment.

Others:
Advise any doctor, dentist or pharmacist whom you consult that you take this medicine.

POSSIBLE INTERACTION WITH OTHER DRUGS

GENERIC NAME OR DRUG CLASS	COMBINED EFFECT
Cimetidine	Increased mexiletine effect and toxicity.
Encainide	Increased effect of toxicity on the heart muscle.
Nicardipine	Possible increased effect and toxicity of each drug.
Phenobarbital	Decreased mexiletine effect.
Phenytoin	Decreased mexiletine effect.
Propafenone	Increased effect of both drugs and increased risk of toxicity.
Rifampin	Decreased mexiletine effect.
Urinary acidifiers*	May decrease effectiveness of medicine.
Urinary alkalizers*	May slow elimination of mexiletine and cause need to adjust dosage.

POSSIBLE INTERACTION WITH OTHER SUBSTANCES

INTERACTS WITH	COMBINED EFFECT
Alcohol:	Causes irregular effectiveness of mexiletine. Avoid.
Beverages: Caffeine drinks, iced drinks.	Irregular heartbeat.
Cocaine:	Unknown. Avoid.
Foods:	None expected.
Marijuana:	Consult doctor.
Tobacco:	Dangerous combination. May lead to liver problems and reduce excretion of mexiletine.

*See Glossary

MIFEPRISTONE (RU-486)

BRAND NAMES

Mifeprex

BASIC INFORMATION

Habit forming? No
Prescription needed? Yes
Available as generic? No
Drug class: Abortifacient

USES

- Terminates pregnancy in the early stages (up to 7 weeks or 49 days since the beginning of the last menstrual period). Mifepristone is not approved for ending later pregnancies. Requires three trips to your doctor's office: day one for administration, day three for a second medication (if you are still pregnant) and day fourteen for follow-up and determination of the status of your pregnancy.
- Note: The brand name drug Korlym is used to treat Cushing's syndrome and is not covered in this drug chart. Consult your doctor or pharmacist if you have questions.

DOSAGE & USAGE INFORMATION

How to take:
Tablet—Taken on day one under strict compliance in the presence of your doctor. On day three misoprostol is taken in your doctor's office (if you are still pregnant).

When to take:
At your doctor's office.

What drug does:
Mifepristone blocks a hormone (progesterone) needed for your pregnancy to continue.

If you forget a dose:
Medication is taken at a medical office.

Continued next column

OVERDOSE

SYMPTOMS:
Symptoms are unknown. Drug is taken in medical office so overdose is not likely to occur.
WHAT TO DO:
If person uses much larger amount than prescribed or if accidentally swallowed, call poison control center 1-800-222-1222 for instructions or dial 911 (emergency) for help.

Time lapse before drug works:
A few days to two weeks. If you are still pregnant after two weeks, your doctor will discuss other options you have including a surgical alternative to terminate your pregnancy.

Don't take with:
Any other medicine or any dietary supplement without consulting your doctor or pharmacist.

POSSIBLE ADVERSE REACTIONS OR SIDE EFFECTS

SYMPTOMS	WHAT TO DO
Life-threatening:	
None expected.	
Common:	
Nausea or vomiting, diarrhea, abdominal pain, back pain, dizziness, unusual tiredness or weakness, headache.	Call doctor if symptoms persist.
Infrequent:	
• Excessive and heavy vaginal bleeding.	Call doctor right away.
• Pale skin, troubled breathing, unusual bleeding or bruising, anxiety, upset stomach, acid indigestion, fever, insomnia, leg pain, increased clear or white vaginal discharge, shaking, stuffy nose, cough, fainting, genital pain or itching, chills, cold or flu-like symptoms.	Call doctor if symptoms persist.
Rare:	
None expected.	

WARNINGS & PRECAUTIONS

Don't take if:
You are allergic to mifepristone or misoprostol. Allergic reactions are rare, but may occur.

Before you start, consult your doctor if:
- You have a history of adrenal failure.
- You have a history of hemorrhagic (bleeding) disorders.
- You have an ectopic pregnancy (a pregnancy outside the uterus).
- You have any other medical problems.
- You have anemia.
- You have an in-place intra-uterine device (IUD).
- You cannot easily get emergency medical help during the two weeks after you take the drug.
- You have a family history of porphyria.

Over age 60:
Not used in this age group.

Pregnancy:
Mifepristone is used to terminate pregnancy.

Breastfeeding; Lactation; Nursing Mothers:
This drug is generally acceptable with breastfeeding. Discuss risks and benefits with your doctor.

Infants & children up to age 18:
Safety and effectiveness in this age group have not been established. Consult doctor.

Prolonged use:
Not intended for prolonged use.

Skin & sunlight:
No problems expected.

Driving, piloting or hazardous work:
No problems expected.

Discontinuing:
Drug is administered in the presence of a licensed medical professional who has registered with the manufacturer.

Others:
- Prior to using this medication, you will be required to sign a statement that you have decided to end your pregnancy.
- Advise any doctor, dentist or pharmacist whom you consult that you take this medicine.
- The follow-up doctor visits are very important. Don't miss them.
- May affect the results in some medical tests.
- An ultrasonographic scan may be scheduled 14 days after mifepristone administration to confirm termination of pregnancy and assess bleeding.
- If you do not want to become pregnant again, start using a birth control method as soon as your pregnancy ends.
- Serious bacterial infection and sepsis (blood infection) may occur without the usual signs of infection, such as fever and pelvic tenderness.
- If you are still pregnant after the two weeks, there may be birth defects if the pregnancy continues. Your doctor will discuss other options to end your pregnancy.

POSSIBLE INTERACTION WITH OTHER DRUGS

GENERIC NAME OR DRUG CLASS	COMBINED EFFECT
Anticoagulants*	Excessive bleeding.
Corticosteroids* (long term use)	Effects unknown. Avoid.
Dabigatran	Increased risk of vaginal bleeding.
Enzyme inducers*	May decrease effect of mifepristone.
Enzyme inhibitors*	May increase effect of mifepristone.

POSSIBLE INTERACTION WITH OTHER SUBSTANCES

INTERACTS WITH	COMBINED EFFECT
Alcohol:	None expected. Best to avoid.
Beverages: Grapefruit juice.	May increase effect of mifepristone.
Cocaine:	Unknown. Avoid.
Foods:	None expected.
Marijuana:	Consult doctor.
Tobacco:	None expected.

*See Glossary

MIGLITOL

BRAND NAMES

Glyset

BASIC INFORMATION

Habit forming? No
Prescription needed? Yes
Available as generic? No
Drug class: Antidiabetic

USES

Treatment for hyperglycemia (excess sugar in the blood) that cannot be controlled by diet alone in patients with type 2 diabetes. The drug may be used alone or in combination with other antidiabetic drugs.

DOSAGE & USAGE INFORMATION

How to take:
Tablet—Swallow with liquid. Take at the very beginning of a meal.

When to take:
Usually 3 times a day or as directed by doctor. Dosage may be increased at 4 to 8 week intervals until maximum benefits are achieved.

If you forget a dose:
If your meal is finished, then skip the missed dose and wait for your next meal and next scheduled dose (don't double this dose).

What drug does:
Impedes the digestion and absorption of carbohydrates and their subsequent conversion into glucose. This improves control of blood glucose and may reduce the complications of diabetes. However, miglitol does not cure diabetes.

Continued next column

OVERDOSE

SYMPTOMS:
An overdose may cause flatulence, diarrhea and abdominal pain. It is unlikely to produce serious side effects. The drug itself will not produce low blood sugar (hypoglycemia).
WHAT TO DO:
If person uses much larger amount than prescribed or if accidentally swallowed, call poison control center 1-800-222-1222 for instructions or dial 911 (emergency) for help.

Time lapse before drug works:
May take several weeks for full effectiveness.

Don't take with:
Any other medicine or any dietary supplement without consulting your doctor or pharmacist.

POSSIBLE ADVERSE REACTIONS OR SIDE EFFECTS

SYMPTOMS	WHAT TO DO
Life-threatening:	
None expected from the drug by itself.	
Common:	
Diarrhea, stomach cramps, gas, bloating feeling.	Continue. Call doctor when convenient.
Infrequent:	
Skin rash.	Continue. Call doctor when convenient.
Rare:	
Low blood sugar (hypoglycemia) may occur if you are taking other drugs for diabetes and you do not consume enough calories. Symptoms may include stomach pain, anxious feeling, cold sweats, chills, confusion, convulsions, cool pale skin, excessive hunger, nausea or vomiting, rapid heartbeat, nervousness, shakiness, unsteady walk, unusual weakness or tiredness, vision changes.	For low blood sugar, take glucose or eat honey or drink orange juice. For more severe symptoms, call doctor right away or seek emergency help.

WARNINGS & PRECAUTIONS

Don't take if:
You are allergic to miglitol. Allergic reactions are rare, but may occur.

Before you start, consult your doctor if:
- You have any kidney or liver disease or any heart or blood vessel disorder.
- You have any chronic health problem.
- You have an infection, illness or any condition that can cause low blood sugar.
- You have a history of acid in the blood (metabolic acidosis or ketoacidosis).
- You have inflammatory bowel disease or any other intestinal disorder.
- You are allergic to any medication, food or other substance.

Over age 60:
No special problems expected.

Pregnancy:
Decide with your doctor if drug benefits justify risk to unborn baby. Risk category B (see page xviii).

Breastfeeding; Lactation; Nursing Mothers:
This drug is generally acceptable with breastfeeding. Discuss risks and benefits with your doctor.

Infants & children up to age 18:
Safety and effectiveness in this age group have not been established. Consult doctor.

Prolonged use:
- Schedule regular doctor visits to check if the drug is effective in controlling the diabetes and to test for any problems in kidney function.
- You will most likely require an antidiabetic medicine for the rest of your life.
- You will need to test your blood glucose levels several times a day, or for some, once to several times a week.

Skin & sunlight:
No special problems expected.

Driving, piloting or hazardous work:
No special problems expected.

Discontinuing:
Don't discontinue without consulting your doctor even if you feel well. You can have diabetes without feeling any symptoms. Untreated diabetes can cause serious problems.

Others:
- Advise any doctor, dentist or pharmacist whom you consult that you take this medicine.
- It may interfere with the accuracy of some medical tests.
- Follow any special diet your doctor may prescribe. It can help control diabetes.
- Consult doctor if you become ill with vomiting or diarrhea while taking this drug.
- Use caution when exercising. Ask your doctor about an appropriate exercise program.
- Wear medical identification stating that you have diabetes and take this medication.
- Learn to recognize the symptoms of low blood sugar. You and your family need to know what to do if these symptoms occur.
- Have a glucagon kit and syringe in the event severe low blood sugar occurs.
- High blood sugar (hyperglycemia) may occur with diabetes. Ask your doctor about symptoms to watch for and treatment steps to take.
- Educate yourself about diabetes.

POSSIBLE INTERACTION WITH OTHER DRUGS

GENERIC NAME OR DRUG CLASS	COMBINED EFFECT
Amylase (Pancreatic enzyme)	Decreased miglitol effect.
Antidiabetic agents, sulfonylurea	May cause hypoglycemia.
Charcoal, activated	Decreased miglitol effect.
Pancreatin (Pancreatic enzyme)	Decreased miglitol effect.
Pramlintide	Decreased absorption of nutrients.
Propranolol	Decreased effect of propranolol.
Ranitidine	Decreased effect of ranitidine.

POSSIBLE INTERACTION WITH OTHER SUBSTANCES

INTERACTS WITH	COMBINED EFFECT
Alcohol:	May increase effect of miglitol. Avoid excessive amounts.
Beverages:	None expected.
Cocaine:	Unknown. Avoid.
Foods:	None expected.
Marijuana:	Consult doctor.
Tobacco:	None expected. People with diabetes should not smoke.

MINOXIDIL

BRAND NAMES

Loniten

BASIC INFORMATION

Habit forming? No
Prescription needed? Yes
Available as generic? Yes
Drug class: Antihypertensive

USES

- Treatment for high blood pressure in conjunction with other drugs, such as beta-adrenergic blockers and diuretics.
- Treatment for congestive heart failure.
- Can stimulate hair growth.

DOSAGE & USAGE INFORMATION

How to take:
Tablet—Swallow with liquid. If you can't swallow whole, crumble tablet and take with liquid or food.

When to take:
At the same time each day, according to instructions on prescription label.

If you forget a dose:
Take as soon as you remember. If it is almost time for the next dose, wait for that dose (don't double this dose) and resume regular schedule.

What drug does:
Relaxes small blood vessels (arterioles) so blood can pass through more easily.

Time lapse before drug works:
2 to 3 hours for effect to begin; 3 to 7 days of continuous use may be necessary for maximum blood pressure response.

Don't take with:
Any other medicine or any dietary supplement without consulting your doctor or pharmacist.

OVERDOSE

SYMPTOMS:
Low blood pressure, dizziness, fainting, fast heartbeat, shortness of breath.
WHAT TO DO:
- **Dial 911 for all medical emergencies or call poison control center 1-800-222-1222 for instructions.**
- **See emergency information on last 3 pages of this book.**

POSSIBLE ADVERSE REACTIONS OR SIDE EFFECTS

SYMPTOMS	WHAT TO DO
Life-threatening: In case of overdose, see previous column.	
Common:	
• Excessive hair growth, flushing or redness of skin.	Continue. Call doctor when convenient.
• Bloating.	Discontinue. Call doctor right away.
Infrequent:	
• Chest pain, slow or irregular heartbeat, shortness of breath, swollen feet or legs, rapid weight gain.	Discontinue. Call doctor right away.
• Numbness of hands, feet or face, tender breasts, headache, darkening of skin.	Continue. Call doctor when convenient.
Rare:	
Rash.	Discontinue. Call doctor right away.

WARNINGS & PRECAUTIONS

Don't take if:
You are allergic to minoxidil. Allergic reactions are rare, but may occur.

Before you start, consult your doctor if:
- You have had a recent stroke or heart attack or angina pectoris in past 3 weeks.
- You have impaired kidney function.
- You have pheochromocytoma.*

Over age 60:
Adverse reactions and side effects may be more frequent and severe than in younger persons.

Pregnancy:
Decide with your doctor if drug benefits justify risk to unborn baby. Risk category C (see page xviii).

Breastfeeding; Lactation; Nursing Mothers:
This drug is generally acceptable with breastfeeding. Discuss risks and benefits with your doctor.

Infants & children up to age 18:
Follow instructions provided by your child's doctor.

Prolonged use:
Request periodic blood lab tests that include potassium levels.

Skin & sunlight:
May cause rash or intensify sunburn in areas exposed to sun or ultraviolet light (photosensitivity reaction). Avoid overexposure. Notify doctor if reaction occurs.

Driving, piloting or hazardous work:
Avoid if you become dizzy or faint. Otherwise, no problems expected.

Discontinuing:
Don't discontinue without consulting doctor. Dose may require gradual reduction if you have taken drug for a long time. Doses of other drugs may also require adjustment.

Others:
- Advise any doctor, dentist or pharmacist whom you consult that you take this medicine.
- Check blood pressure frequently.

POSSIBLE INTERACTION WITH OTHER DRUGS

GENERIC NAME OR DRUG CLASS	COMBINED EFFECT
Anesthesia	Drastic blood pressure drop.
Antihypertensives,* other	May need dosage change.
Carteolol	Increased anti-hypertensive effect.
Diuretics*	Dosage adjustments may be necessary to keep blood pressure at desired level.
Estrogens*	May increase blood pressure.
Guanadrel	Weakness and faintness when arising from bed or chair.
Guanethidine	Weakness and faintness when arising from bed or chair.
Lisinopril	Increased anti-hypertensive effect. Dosage of each may require adjustment.
Nicardipine	Blood pressure drop. Dosages may require adjustment.
Nimodipine	Dangerous blood pressure drop.
Nitrates*	Drastic blood pressure drop.
Sotalol	Increased anti-hypertensive effect.
Sympathomimetics*	Possible decreased minoxidil effect.
Terazosin	Decreased effectiveness of terazosin.

POSSIBLE INTERACTION WITH OTHER SUBSTANCES

INTERACTS WITH	COMBINED EFFECT
Alcohol:	Possible excessive blood pressure drop.
Beverages:	None expected.
Cocaine:	Increased risk of side effects. Avoid.
Foods: Salt substitutes.	Possible excessive potassium in blood.
Marijuana:	Increased risk of side effects. Avoid.
Tobacco:	May decrease minoxidil effect. Avoid.

*See Glossary

MINOXIDIL (Topical)

BRAND NAMES

Dermal
Men's Rogaine Topical Foam
Rogaine
Rogaine Extra Strength for Men
Rogaine for Men
Rogaine for Women

BASIC INFORMATION

Habit forming? No
Prescription needed? No
Available as generic? Yes
Drug class: Hair growth stimulant

USES

Treats hair loss on scalp from male and female pattern baldness (alopecia androgenetica).

DOSAGE & USAGE INFORMATION

How to use:
Topical solution

- Apply only to dry hair and scalp. With the provided applicator, apply the amount prescribed to the scalp area being treated. Begin in center of the treated area.
- Wash hands immediately after use.
- Don't use a blow dryer.
- If you are using at bedtime, wait 30 minutes after applying before retiring.

When to use:
Twice a day or as directed.

If you forget a dose:
Use as soon as you remember. If it is almost time for the next dose, wait for that dose (don't double this dose) and resume regular schedule.

What drug does:
Stimulates hair growth by possibly dilating small blood capillaries, thereby providing more blood to hair follicles.

Continued next column

OVERDOSE

SYMPTOMS:
Unknown.
WHAT TO DO:
If person uses much larger amount than prescribed or if accidentally swallowed, call poison control center 1-800-222-1222 for instructions or dial 911 (emergency) for help.

Time lapse before drug works:
Varies with individuals.

Don't use with:
Other hair-growth products without consulting your doctor or pharmacist.

POSSIBLE ADVERSE REACTIONS OR SIDE EFFECTS

SYMPTOMS	WHAT TO DO
Life-threatening:	
None expected.	
Common:	
None expected.	
Infrequent:	
Itching scalp or flaking or reddened skin.	Continue. Call doctor if symptoms persist.
Rare:	
• Very rarely too much of drug absorbed into body and these symptoms may occur (heartbeat changes, vision changes, dizziness, fainting, flushing, headache, swelling [face, hands, feet, legs], hands and feet numb or tingling, rapid weight gain).	Discontinue. Call doctor right away.
• Burning scalp, skin rash, more hair loss.	Discontinue. Call doctor when convenient.

WARNINGS & PRECAUTIONS

Don't use if:
You are allergic to minoxidil. Allergic reactions are rare, but may occur.

Before you start, consult your doctor if:
- You are allergic to anything.
- You have heart disease or high blood pressure.
- You have skin irritation or abrasion or severe sunburn (systemic absorption may be increased).

Over age 60:
No problems expected.

Pregnancy:
Decide with your doctor if drug benefits justify risk to unborn baby. Risk category C (see page xviii).

Breastfeeding; Lactation; Nursing Mothers:
This drug is generally acceptable with breastfeeding. Discuss risks and benefits with your doctor.

Infants & children up to age 18:
Safety and effectiveness in this age group have not been established. Consult doctor.

Prolonged use:
No problems expected.

Skin & sunlight:
No problems expected.

Driving, piloting or hazardous work:
No problems expected.

Discontinuing:
No problems expected.

Others:
- Keep away from eyes, nose and mouth. Flush with plain water if accident occurs.
- New hair will drop out when you stop using minoxidil.
- Keep solution cool, but don't freeze.

POSSIBLE INTERACTION WITH OTHER DRUGS

GENERIC NAME OR DRUG CLASS	COMBINED EFFECT
Adrenocorticoids,* topical	May cause undesirable absorption of minoxidil.
Minoxidil, oral	Increased risk of toxicity.
Petrolatum, topical	May cause undesirable absorption of minoxidil.
Retinoids,* topical	May cause undesirable absorption of minoxidil.

POSSIBLE INTERACTION WITH OTHER SUBSTANCES

INTERACTS WITH	COMBINED EFFECT
Alcohol:	None expected.
Beverages:	None expected.
Cocaine:	None expected.
Foods:	None expected.
Marijuana:	None expected.
Tobacco:	None expected.

*See Glossary

MIRABEGRON

BRAND NAMES

Myrbetriq

BASIC INFORMATION

Habit forming? No
Prescription needed? Yes
Available as generic? No
Drug class: Antispasmodic (urinary tract)

USES

It is used to treat overactive bladder (symptoms include an urgent need to urinate, a need to urinate often or leakage of urine).

DOSAGE & USAGE INFORMATION

How to take:
Extended-release tablet—Swallow whole with liquid. Do not break, crush or chew tablet. It may be taken with or without food.

When to take:
The drug is taken once a day at the same time each day.

If you forget a dose:
Take as soon as you remember. If it is almost time for the next dose, wait for that dose (don't double this dose) and resume regular schedule.

What drug does:
It works by relaxing the detrusor (bladder muscle) which helps the bladder hold more urine. This helps reduce the symptoms of an overactive bladder.

Time lapse before drug works:
It can take up to 8 weeks for symptoms to improve. The starting dosage may be increased after 8 weeks to achieve full effectiveness.

Don't take with:
Any other medicine or any dietary supplement without consulting your doctor or pharmacist.

OVERDOSE

SYMPTOMS:
Heart palpitations, fast pulse rate.
WHAT TO DO:
If person uses much larger amount than prescribed or if accidentally swallowed, call poison control center 1-800-222-1222 for instructions or dial 911 (emergency) for help.

POSSIBLE ADVERSE REACTIONS OR SIDE EFFECTS

SYMPTOMS	WHAT TO DO
Life-threatening: None expected.	
Common:	
• Urinary tract infection, high blood pressure.	Continue, but call doctor right away.
• Cold symptoms, mild headache, nausea.	Continue. Call doctor when convenient.
Infrequent:	
• Heartbeat is fast or slow or irregular, unable to completely empty bladder (urinary retention), weak urine stream.	Continue, but call doctor right away.
• Constipation or diarrhea, back pain, muscle aches, symptoms of flu or sinus infection, blurred vision, dry mouth, dizziness,	Continue. Call doctor when convenient.
Rare:	
• New, severe and unusual symptoms.	Discontinue. Call doctor right away.
• Abdominal pain, fatigue, rash or itching.	Continue. Call doctor when convenient.

WARNINGS & PRECAUTIONS

Don't take if:
You are allergic to mirabegron. Allergic reactions are rare, but may occur.

Before you start, consult your doctor if:
- You have bladder problems (such as blockage) or an enlarged prostate.
- You have high blood pressure.
- You have liver or kidney problems.

Over age 60:
No special problems expected.

Pregnancy:
Decide with your doctor if drug benefits justify risk to unborn baby. Risk category C (see page xviii).

Breastfeeding; Lactation; Nursing Mothers:
This drug is generally not recommended with breastfeeding. Discuss risks and benefits with your doctor.

Infants & children up to age 18:
Safety and effectiveness in this age group have not been established. Consult doctor.

Prolonged use:
See your doctor for regular visits to make sure the drug is working properly and to check for unwanted effects (such as high blood pressure).

Skin & sunlight:
No problems expected.

Driving, piloting or hazardous work:
Don't drive or pilot aircraft until you learn how medicine affects you. Don't work around dangerous machinery. Don't climb ladders or work in high places. Danger increases if you drink alcohol or take medicine affecting alertness and reflexes.

Discontinuing:
No problems expected, but consult your doctor before stopping the drug.

Others:
- Advise any doctor, dentist or pharmacist whom you consult that you take this drug.
- Follow your doctor's recommendations for other treatment steps, such as diet changes, bladder training and/or pelvic floor exercises.

POSSIBLE INTERACTION WITH OTHER DRUGS

GENERIC NAME OR DRUG CLASS	COMBINED EFFECT
Digoxin	Increased effect of digoxin.
Desipramine	Increased effect of desipramine.
Flecainide	Increased effect of flecainide.
Metoprolol	Increased effect of metoprolol.
Muscarinic receptor antagonists	Increased risk of urinary retention. Use caution.
Propafenone	Increased effect of propafenone.
Thioridazine	Increased effect of thioridazine. Avoid.
Warfarin	May increase warfarin effect.

POSSIBLE INTERACTION WITH OTHER SUBSTANCES

INTERACTS WITH	COMBINED EFFECT
Alcohol:	None expected.
Beverages:	None expected.
Cocaine:	Unknown. Avoid.
Foods:	None expected.
Marijuana:	Consult doctor.
Tobacco:	None expected.

MIRTAZAPINE

BRAND NAMES

Remeron
Remeron SolTab

BASIC INFORMATION

Habit forming? Not expected
Prescription needed? Yes
Available as generic? Yes
Drug class: Antidepressant

USES

Treats symptoms of mental depression.

DOSAGE & USAGE INFORMATION

How to take:

- Tablet—Swallow with liquid. May be taken with or without food.
- Oral disintegrating tablet—Let tablet dissolve in your mouth.

When to take:
At the same time each day, usually in the evening before bedtime.

If you forget a dose:
Take as soon as you remember. If it is almost time for the next dose, wait for that dose (don't double this dose) and resume regular schedule.

What drug does:
The exact mechanism is unknown. It appears to block certain chemicals in the brain, which in turn helps production of other brain chemicals that play a role in helping to relieve symptoms of depression.

Time lapse before drug works:
Will take up to several weeks to show improvement of the depression symptoms.

Don't take with:
Any other medicine or any dietary supplement without consulting your doctor or pharmacist.

OVERDOSE

SYMPTOMS:
Drowsiness, disorientation, memory impairment, rapid heartbeat.
WHAT TO DO:

- **Dial 911 for all medical emergencies or call poison control center 1-800-222-1222 for instructions.**
- **See emergency information on last 3 pages of this book.**

POSSIBLE ADVERSE REACTIONS OR SIDE EFFECTS

SYMPTOMS	WHAT TO DO
Life-threatening: None expected.	
Common:	
• Abdominal pain, vomiting, joint pain, increased cough, rash, itching, apathy, agitation, anxiety, twitching.	Discontinue. Call doctor right away.
• Sleepiness, increased appetite, weight gain, dizziness, dry mouth, constipation, tiredness, increased thirst.	Continue. Call doctor when convenient.
Infrequent:	
• Slow heartbeat, migraine, dehydration, weight loss, unusual weakness, pain in any part of the body, changes in menstrual periods, changes in vision or hearing, mouth sores, mood or mental changes, breathing difficulty, lack of coordination.	Discontinue. Call doctor right away.
• Forgetfulness, lightheadedness.	Continue. Call doctor when convenient.
Rare:	
• Swelling of hands or feet or legs, infection (fever, chills, aches or pains, sore throat), swollen or discolored tongue, changes in urinary function, hives.	Discontinue. Call doctor right away.
• Muscle aches, strange dreams, headache.	Continue. Call doctor when convenient.

WARNINGS & PRECAUTIONS

Don't take if:
You are allergic to mirtazapine. Allergic reactions are rare, but may occur.

Before you start, consult your doctor if:
- You have seizure disorder, heart disease, blood circulation problem or had a stroke.
- You are dehydrated.
- You have a history of drug dependence or drug abuse.
- You have kidney or liver disease.
- You have a history of mood disorders, such as mania, or thoughts of suicide.
- You are allergic to any other medication, food or other substances.

Over age 60:
A lower starting dosage is usually recommended until a response is determined.

Pregnancy:
Decide with your doctor if drug benefits justify risk to unborn baby. Risk category C (see page xviii).

Breastfeeding; Lactation; Nursing Mothers:
This drug is generally acceptable with breastfeeding. Discuss risks and benefits with your doctor.

Infants & children up to age 18:
Safety and effectiveness in this age group have not been established. Consult doctor.

Prolonged use:
Consult with your doctor on a regular basis while taking this drug to check your progress, to discuss any increase or changes in side effects and the need for continued treatment.

Skin & sunlight:
May cause a rash or intensify sunburn in areas exposed to sun or ultraviolet light (photosensitivity reaction). Avoid excessive sun exposure. Consult doctor if reaction occurs.

Driving, piloting or hazardous work:
Don't drive or pilot aircraft until you learn how medicine affects you. Don't work around dangerous machinery. Don't climb ladders or work in high places. Danger increases if you drink alcohol or take medicine affecting alertness and reflexes.

Discontinuing:
Consult doctor before discontinuing this drug.

Others:
- Get up slowly from a sitting or lying position to avoid dizziness, faintness or lightheadedness.
- Advise any doctor, dentist or pharmacist whom you consult that you take this medicine.
- Adults and children taking antidepressants may experience a worsening of the depression symptoms and may have increased suicidal thoughts or behaviors. Call doctor right away if these symptoms or behaviors occur.
- Do not increase or reduce dosage without doctor's approval.

POSSIBLE INTERACTION WITH OTHER DRUGS

GENERIC NAME OR DRUG CLASS	COMBINED EFFECT
Central nervous system (CNS) depressants*	Excess sedation.
Clonidine	May decrease effect of clonidine.
Enzyme inducers*	May decrease effect of mirtazapine.
Enzyme inhibitors*	May increase effect of mirtazapine.
Monoamine oxidase (MAO) inhibitors*	Potentially life-threatening. Allow 14 days between use of 2 drugs.
Serotonergics*	May increase risk of serotonin syndrome.*
Warfarin	May increase effect of warfarin.

POSSIBLE INTERACTION WITH OTHER SUBSTANCES

INTERACTS WITH	COMBINED EFFECT
Alcohol:	Increased sedative affect. Avoid.
Beverages:	None expected.
Cocaine:	Unknown. Avoid.
Foods:	None expected.
Marijuana:	Consult doctor.
Tobacco:	None expected.

***See Glossary**

MISOPROSTOL

BRAND NAMES

Arthrotec
Cytotec

BASIC INFORMATION

Habit forming? No
Prescription needed? Yes
Available as generic? Yes
Drug class: Antiulcer agent

USES

Prevents development of stomach ulcers in persons taking nonsteroidal anti-inflammatory drugs (NSAIDs), including aspirin.

DOSAGE & USAGE INFORMATION

How to take:
Tablet—Swallow with liquid. Swallow Arthrotec brand whole. If you can't swallow Cytotec brand whole, crumble tablet and take with liquid or food. Instructions to take on empty stomach mean 1 hour before or 2 hours after eating.

When to take:
Usually 4 times a day while awake, with or after meals and at bedtime.

If you forget a dose:
Take as soon as you remember. If it is almost time for the next dose, wait for that dose (don't double this dose) and resume regular schedule.

What drug does:

- Improves defense against peptic ulcers by strengthening natural defenses of the stomach lining.
- Decreases stomach acid production.

Time lapse before drug works:
10 to 15 minutes.

Continued next column

OVERDOSE

SYMPTOMS:
May have tremor, sleepiness, breathing difficulty, stomach pain, diarrhea, fever, heart palpitations, low blood pressure, slow heartbeat.
WHAT TO DO:

- **Dial 911 for all medical emergencies or call poison control center 1-800-222-1222 for instructions.**
- **See emergency information on last 3 pages of this book.**

Don't take with:
Any other medicines (including over-the-counter drugs such as cough and cold medicines, laxatives, antacids, diet pills, caffeine, nose drops or vitamins) without consulting your doctor or pharmacist.

POSSIBLE ADVERSE REACTIONS OR SIDE EFFECTS

SYMPTOMS	WHAT TO DO
Life-threatening:	
Rare, severe allergic reaction (anaphylaxis)—difficulty breathing, hives, itching, swelling (throat, face), vomiting, dizziness.	Seek emergency treatment immediately.
Common:	
Abdominal pain, diarrhea.	Continue. Call doctor when convenient.
Infrequent:	
Nausea or vomiting, constipation, headache, gaseousness.	Continue. Call doctor when convenient.
Rare:	
Vaginal bleeding, stools are bloody or black or tarry, bloody vomit.	Discontinue. Call doctor right away.

WARNINGS & PRECAUTIONS

Don't take if:
- You are allergic to any prostaglandin.*
- You are pregnant or of child-bearing age.

Before you start, consult your doctor if:
- You have epilepsy.
- You have heart disease.
- You have blood vessel disease of any kind.

Over age 60:
No special problems expected.

Pregnancy:
Consult doctor. Drug should not be used during pregnancy. Can cause harm to unborn baby. Risk category X (see page xviii).

Breastfeeding; Lactation; Nursing Mothers:
This drug is generally acceptable with breastfeeding. Discuss risks and benefits with your doctor.

Infants & children up to age 18:
Safety and effectiveness in this age group have not been established. Consult doctor.

Prolonged use:
Talk to your doctor about the need for follow-up medical examinations or laboratory studies to check gastric analysis.

Skin & sunlight:
No problems expected.

Driving, piloting or hazardous work:
Avoid if you feel confused, drowsy or dizzy.

Discontinuing:
No special problems expected.

Others:
Advise any doctor, dentist or pharmacist whom you consult that you take this medicine.

POSSIBLE INTERACTION WITH OTHER DRUGS

GENERIC NAME OR DRUG CLASS	COMBINED EFFECT
Antacids,* magnesium-containing	Severe diarrhea.

POSSIBLE INTERACTION WITH OTHER SUBSTANCES

INTERACTS WITH	COMBINED EFFECT
Alcohol:	Decreases misoprostol effect. Avoid.
Beverages:	
Caffeine-containing.	Decreases misoprostol effect. Avoid.
Cocaine:	Unknown. Avoid.
Foods:	None expected.
Marijuana:	Consult doctor.
Tobacco:	Decreases misoprostol effect. Avoid.

*See Glossary

MONOAMINE OXIDASE (MAO) INHIBITORS

GENERIC AND BRAND NAMES

ISOCARBOXAZID
Marplan
PHENELZINE
Nardil
TRANYLCYPROMINE
Parnate

BASIC INFORMATION

Habit forming? No
Prescription needed? Yes
Available as generic? No
Drug class: MAO (monoamine oxidase) inhibitor, antidepressant

USES

- Treatment for depression and panic disorder.
- Prevention of vascular or tension headaches.

DOSAGE & USAGE INFORMATION

How to take:
Tablet—Swallow with liquid. If you can't swallow whole, crumble tablet and take with liquid or food.

When to take:
At the same times each day.

If you forget a dose:
Take as soon as you remember. If it is almost time for the next dose, wait for that dose (don't double this dose) and resume regular schedule.

What drug does:
Inhibits nerve transmissions in brain that may cause depression.

Time lapse before drug works:
4 to 6 weeks for maximum effect.

Don't take with:
- Foods containing tyramine.* Life-threatening elevation of blood pressure may result.
- Any other medicine or any dietary supplement without consulting your doctor or pharmacist.

OVERDOSE

SYMPTOMS:
Drowsiness, dizziness, hyperactivity, severe headache, agitation, heartbeat irregularities, hallucinations, fever, sweating, breathing difficulties, convulsions, coma.
WHAT TO DO:
- **Dial 911 for all medical emergencies or call poison control center 1-800-222-1222 for instructions.**
- **See emergency information on last 3 pages of this book.**

POSSIBLE ADVERSE REACTIONS OR SIDE EFFECTS

SYMPTOMS	WHAT TO DO
Life-threatening: In case of overdose, see previous column. Also read information about dangers associated with tyramine.	
Common:	
Dizziness when changing position, restlessness, tremors, dry mouth, fatigue, weakness, blurred vision, constipation, difficult urination.	Continue. Call doctor when convenient.
Infrequent:	
• Fainting, enlarged pupils, severe headache, chest pain, rapid or pounding heartbeat.	Discontinue. Call doctor right away or seek emergency care.
• Insomnia, nightmares, diarrhea, swollen feet or legs, joint pain.	Continue. Call doctor when convenient.
Rare:	
Nausea, vomiting, stiff neck, yellow, skin or eyes, fever, increased sweating, dark urine, slurred speech, staggering gait, rash, hallucinations.	Discontinue. Call doctor right away.

WARNINGS & PRECAUTIONS

Don't take if:
You are allergic to any MAO inhibitor. Allergic reactions are rare, but may occur.

Before you start, consult your doctor if:
- You have heart disease, congestive heart failure, heart rhythm irregularities or high blood pressure or have had a stroke.
- You have liver or kidney disease.
- You are an alcoholic.
- You have diabetes, epilepsy, asthma, overactive thyroid, schizophrenia, Parkinson's disease, adrenal gland tumor.
- You will have surgery within 2 months, including dental surgery, requiring anesthesia.

Over age 60:
Adverse reactions and side effects may be more frequent and severe than in younger persons.

MONOAMINE OXIDASE (MAO) INHIBITORS

Pregnancy:
Decide with your doctor if drug benefits justify risk to unborn baby. Risk category C (see page xviii).

Breastfeeding; Lactation; Nursing Mothers:
Drugs in this group are generally not recommended with breastfeeding. Discuss risks and benefits with your doctor.

Infants & children up to age 18:
Not approved for ages under 16. If prescribed, carefully read information provided with prescription. Contact doctor right away if depression symptoms get worse or there is any talk of suicide or suicide behaviors. Also, read information under Others.

Prolonged use:
- May be toxic to liver.
- Talk to your doctor about the need for follow-up medical examinations or laboratory studies to check blood pressure, liver function.

Skin & sunlight:
No special problems expected.

Driving, piloting or hazardous work:
Don't drive or pilot aircraft until you learn how medicine affects you. Don't work around dangerous machinery. Don't climb ladders or work in high places. Danger increases if you drink alcohol or take medicine affecting alertness and reflexes.

Discontinuing:
- Don't discontinue without doctor's advice until you complete prescribed dose, even though symptoms diminish or disappear.
- Follow precautions regarding foods, drinks and other medicines for 2 weeks after discontinuing.
- Adverse symptoms caused by this medicine may occur even after discontinuation. If you develop any of the symptoms listed under Overdose, notify your doctor immediately.

Others:
- May affect blood sugar level in diabetics.
- Advise any doctor, dentist or pharmacist whom you consult about the use of this drug.
- Adults and children taking antidepressants may experience a worsening of the depression symptoms and may have increased suicidal thoughts or behaviors. Call doctor right away if these symptoms or behaviors occur.
- Fever may indicate that MAO inhibitor dose requires adjustment.

POSSIBLE INTERACTION WITH OTHER DRUGS

GENERIC NAME OR DRUG CLASS	COMBINED EFFECT
Amphetamines*	Blood pressure rise to high levels.
Anticholinergics*	Increased anticholinergic effect.
Anticonvulsants*	Changed seizure pattern.
Antidepressants, tricyclic*	Blood pressure rise to life-threatening level. Possible fever, convulsions, delirium.
Antidiabetic agents, oral* and insulin	Excessively low blood sugar.
Antihypertensives*	Excessively low blood pressure.
Beta-adrenergic blocking agents*	Possible blood pressure rise if MAO inhibitor is discontinued after simultaneous use with acebutolol.
Bupropion	Increased risk of side effects.
Buspirone	Very high blood pressure.
Caffeine	Irregular heartbeat or high blood pressure.
Carbamazepine	Fever, seizures. Avoid.

Continued on page 905

POSSIBLE INTERACTION WITH OTHER SUBSTANCES

INTERACTS WITH	COMBINED EFFECT
Alcohol:	Increased sedation to dangerous level.
Beverages:	
Caffeine drinks.	Irregular heartbeat or high blood pressure.
Drinks containing tyramine.*	Blood pressure rise to life-threatening level.
Cocaine:	Increased risk of side effects. Avoid.
Foods:	
Foods containing tyramine.*	Blood pressure rise to life-threatening level.
Marijuana:	Increased risk of side effects. Avoid.
Tobacco:	None expected.

***See Glossary**

MONOAMINE OXIDASE TYPE B (MAO-B) INHIBITORS

GENERIC AND BRAND NAMES

RASAGILINE
Azilect
SELEGILINE
Apo-Selegiline
Atapryl
Carbex
Eldepryl
Emsam
SELEGILINE (con't)
Gen-Selegiline
Movergan
Novo-Selegiline
Nu-Selegiline
Selpak
Zelepar

BASIC INFORMATION

Habit forming? No
Prescription needed? Yes
Available as generic? Yes, for some
Drug class: Antiparkinsonism; antidyskinetic

USES

- Treats symptoms of Parkinson's disease.
- Treats major depressive disorder (MDD).
- Treats other disorders per doctor's advice.

DOSAGE & USAGE INFORMATION

How to take:
- Tablet or capsule—Swallow whole with liquid.
- Oral disintegrating tablet—Let it dissolve on tongue. Do not swallow tablet whole.
- Skin patch—Follow prescription instructions.

When to take:
Take as directed. The oral form is usually taken at breakfast and lunch to help avoid interfering with sleep. Change patch daily at same time.

If you forget a dose:
Take as soon as you remember. If it is almost time for the next dose, wait for that dose (don't double this dose) and resume regular schedule.

Continued next column

OVERDOSE

SYMPTOMS:
May cause sweating, dizziness, insomnia, severe headache, hallucinations, excitement, nervousness, irritability, weakness, seizures, high or low blood pressure, muscle spasm, breathing problems, cold/clammy skin, coma.
WHAT TO DO:
- **Dial 911 for all medical emergencies or call poison control center 1-800-222-1222 for instructions.**
- **See emergency information on last 3 pages of this book.**

What drug does:
It helps prevent breakdown of dopamine levels in the brain. Dopamine is a brain chemical having to do with control of movement and coordination.

Time lapse before drug works:
Starts working in 1-2 hours, but may take 4-6 weeks to determine drug's full effectiveness.

Don't take with:
- Foods containing tyramine.* Rarely may cause severe high blood pressure (hypertension).
- Any other medicine or diet supplement without consulting your doctor or pharmacist.

POSSIBLE ADVERSE REACTIONS OR SIDE EFFECTS

SYMPTOMS	WHAT TO DO
Life-threatening:	
Rare hypertensive crisis (chest pain, large pupils, heartbeat fast or slow, severe headache, light sensitivity, fever, cold skin, stiff or sore neck, severe nausea/vomiting).	Discontinue. Seek emergency care.
Common:	
• Mood or mental changes, increase in uncontrolled body movements.	Discontinue. Call doctor right away.
• Stomach pain, dry mouth, dizziness, insomnia, feeling faint, mild nausea or vomiting, skin patch causes mild skin reaction.	Continue. Call doctor when convenient.
Infrequent:	
Bloody or tarry stools, urination difficult or frequent, trouble with breathing or speaking, dizziness or lightheaded when getting up from sitting or lying down, lip smacking, hallucinations, lack of balance, unusual tongue or chewing movements, stomach pain, restlessness, swollen feet or legs, new or odd movements of arms or legs or other body parts, vomiting blood or coffee-ground-like material, wheezing.	Discontinue. Call doctor right away.

Rare:	
Constipation, anxiety, tiredness, eyelid spasm, changes in taste, blurred or double vision, leg pain, ringing in ears, diarrhea, burning of lips or mouth or throat, drowsiness, sensitivity to light, sweating, loss of appetite, memory problems, muscle cramps, nervousness, numbness in fingers or toes, red or raised or itchy skin, unusual weight loss, heartburn, jaw clenching or teeth gnashing, excess feeling of well-being.	Continue. Call doctor when convenient.

WARNINGS & PRECAUTIONS

Don't take if:
You are allergic to rasagiline or selegiline. Allergic reactions are rare, but may occur.

Before you start, consult your doctor if:
- You have a history of ulcers.
- You have tardive dyskinesia or a tremor.
- Patient has profound dementia or severe psychosis.

Over age 60:
No special problems expected.

Pregnancy:
Decide with your doctor if drug benefits justify risk to unborn baby. Risk category C (see page xviii).

Breastfeeding; Lactation; Nursing Mothers:
Drugs in this group are generally not recommended with breastfeeding. Discuss risks and benefits with your doctor.

Infants & children up to age 18:
Safety and effectiveness in this age group have not been established. Consult doctor.

Prolonged use:
Follow up with your doctor on a regular basis to verify the continued effectiveness of the drug.

Skin & sunlight:
May cause rash or intensify sunburn in areas exposed to sun or ultraviolet light (photosensitivity reaction). Avoid overexposure. Notify doctor if reaction occurs.

Driving, piloting or hazardous work:
Don't drive or pilot aircraft until you learn how medicine affects you. Don't work around dangerous machinery. Don't climb ladders or work in high places. Danger increases if you drink alcohol or take drugs affecting alertness.

Discontinuing:
Don't discontinue without your doctor's approval as symptoms may worsen. A gradual reduction in dosage may be required.

Others:
- Avoid sudden rises from lying-down or sitting positions.
- Advise any doctor, dentist or pharmacist whom you consult that you take this medicine.

POSSIBLE INTERACTION WITH OTHER DRUGS

GENERIC NAME OR DRUG CLASS	COMBINED EFFECT
Antidepressants, tricyclic	Serious reactions. Avoid or take at least 14 days apart.
Enzyme inducers*	May decrease effect of rasagiline.
Enzyme inhibitors*	May increase effect of rasagiline.
Levodopa	Increased risk of adverse reactions.
Meperidine	Life-threatening. reactions. Avoid.
Narcotics*	Serious reactions. Avoid.
Serotonergic agents,* other	May cause a serotonin syndrome* type reaction. Avoid.

POSSIBLE INTERACTION WITH OTHER SUBSTANCES

INTERACTS WITH	COMBINED EFFECT
Alcohol:	Increased sedation. Avoid.
Beverages: Caffeine	Increased effect of caffeine. Limit use.
Cocaine:	Increased risk of side effects. Avoid.
Foods: Tyramine-containing*	Severe hypertension. Avoid.
Marijuana:	Excess drowsiness. Avoid.
Tobacco:	None expected.

***See Glossary**

MUSCARINIC RECEPTOR ANTAGONISTS

GENERIC AND BRAND NAMES

DARIFENACIN
Enablex
FESOTERODINE
Toviaz
OXYBUTYNIN
Anturol
Ditropan
Ditropan XL
Gelnique
Oxytrol
Oxytrol for Women
SOLIFENACIN
Vesicare
TOLTERODINE
Detrol
Detrol LA
TROSPIUM
Sanctura
Sanctura XR

BASIC INFORMATION

Habit forming? No
Prescription needed? Yes, for most
Available as generic? Yes, for some
Drug class: Antispasmodic; anticholinergic

USES

Used to treat an overactive bladder in men and women.

DOSAGE & USAGE INFORMATION

How to take:

- Tablet—Swallow whole with liquid. May be taken with or without food. Take trospium 1 hour before eating or on an empty stomach.
- Extended-release capsule or tablet—Swallow whole with liquid. Take with or without food.
- Syrup, skin patch, or gel (oxybutynin)—Follow instructions on prescription.

When to take:
Take one or more times (as directed) a day at the same time each day. The patch should be applied twice a week. The gel is applied daily.

Continued next column

OVERDOSE

SYMPTOMS:
Unsteadiness, confusion, dizziness, severe drowsiness, irregular heartbeat, fever, red face, hallucinations, lack of urine, difficult breathing, vomiting.
WHAT TO DO:

- **Dial 911 for all medical emergencies or call poison control center 1-800-222-1222 for instructions.**
- **See emergency information on last 3 pages of this book.**

If you forget a dose:
Take as soon as you remember. If it is almost time for the next dose, wait for that dose (don't double this dose) and resume regular schedule.

What drug does:
The drugs increase the amount of urine the bladder can hold and also decrease the pressure involved with the urge to urinate.

Time lapse before drug works:
Symptoms should start to improve in about a week. It may take 4 to 6 weeks for full benefits.

Don't take with:
Any other medicine or any dietary supplement without consulting your doctor or pharmacist.

POSSIBLE ADVERSE REACTIONS OR SIDE EFFECTS

SYMPTOMS	WHAT TO DO
Life-threatening:	
Rare, severe allergic reaction (anaphylaxis)—difficulty breathing, hives, itching, swelling (throat, face), vomiting, dizziness.	Seek emergency treatment immediately.
Common:	
Constipation, dry mouth, dry eyes or throat, blurred vision, decreased sweating, urinary retention, drowsiness, skin patch causes itching at application site.	Continue. Call doctor if symptoms persist.
Infrequent:	
Flu-like symptoms, decreased sexual ability, difficulty in urinating, headache, sensitivity to light, nausea, vomiting, abdominal pain, insomnia, dizziness, fatigue, depression, diarrhea, gas.	Continue. Call doctor when convenient.
Rare:	
• Eye pain, fast heart rate, urinary tract infection, swelling of arms or legs, changes in mental status, confusion, fainting, skin changes (blistering, peeling or loosening), dark color urine, chest pain, high blood pressure.	Discontinue. Call doctor right away.
• Back or muscle pain, dry skin, skin rash.	Continue. Call doctor when convenient.

MUSCARINIC RECEPTOR ANTAGONISTS

WARNINGS & PRECAUTIONS

Don't take if:
You are allergic to muscarinic receptor antagonists or anticholinergic drugs.

Before you start, consult your doctor if:
- You have heart disease, bleeding disorder or high blood pressure.
- You have hiatal hernia; liver, kidney or thyroid disease; enlarged prostate or myasthenia gravis.
- You have any gastrointestinal disorder or intestinal or urinary tract blockage.
- You have a bladder emptying problem.
- You have glaucoma.
- You have severe ulcerative colitis.
- Patient has Alzheimer's or dementia.

Over age 60:
Adverse reactions and side effects may be more frequent and severe than in younger persons.

Pregnancy:
Decide with your doctor if drug benefits justify risk to unborn baby. Risk category B for oxybutynin, others are category C (see page xviii).

Breastfeeding; Lactation; Nursing Mothers:
Drugs in this group are generally acceptable with breastfeeding. Discuss risks and benefits with your doctor.

Infants & children up to age 18:
Oxybutynin is approved for use in children over age 5. Safety and effectiveness of other drugs in this group for use in children under age 18 has not been established. Consult doctor.

Prolonged use:
No special problems expected.

Skin & sunlight:
No special problems expected.

Driving, piloting or hazardous work:
Don't drive or pilot aircraft until you learn how drug affects you. Don't work around dangerous machinery. Don't climb ladders or work in high places. Danger increases if you drink alcohol or take drugs affecting alertness and reflexes.

Discontinuing:
No special problems expected. Follow doctor's instructions.

Others:
- Advise any doctor, dentist or pharmacist whom you consult that you take this drug.
- Be careful in hot weather, hot tubs or saunas. The drug can increase your risk of heat stroke (due to decreased sweating).
- Chew sugarless gum or suck on ice chips to relieve a dry mouth. Call your doctor or dentist if the dry mouth lasts longer than 2 weeks.

POSSIBLE INTERACTION WITH OTHER DRUGS

GENERIC NAME OR DRUG CLASS	COMBINED EFFECT
Anticholinergics,* other	Increased risk of side effects from either drug.
Antidepressants, tricyclic*	Increased effect of antidepressant.
Cationic drugs*	May increase effect of trospium.
Central nervous system (CNS) depressants*	Increased sedative effect.
Enzyme inducers*	Decreased effect of darifenacin or solifenacin.
Enzyme inhibitors*	Increased effect of darifenacin, solifenacin or tolterodine.
Flecainide	Increased effect of flecainide. Use caution.
Ketoconazole	Increased effect of solifenacin.
QT prolongation-causing drugs*	Increased risk of heart problems.
Thioridazine	Increased effect of thioridazine.

POSSIBLE INTERACTION WITH OTHER SUBSTANCES

INTERACTS WITH	COMBINED EFFECT
Alcohol:	Increased sedative effect. Avoid.
Beverages: Grapefruit juice.	May increase effect of darifenacin, solifenacin or tolterodine.
Cocaine:	Unknown. Avoid.
Foods:	None expected.
Marijuana:	Consult doctor.
Tobacco:	None expected.

*See Glossary

MUSCLE RELAXANTS, SKELETAL

GENERIC AND BRAND NAMES

CARISOPRODOL
- **Rela**
- **Sodol**
- **Soma**
- **Soma Compound with Codeine**
- **Sopridol**
- **Soridol**

CHLORPHENESIN
- **Maolate**

CHLORZOXAZONE
- **Paraflex**
- **Parafon Forte**

METAXALONE
- **Skelaxin**

METHOCARBAMOL
- **Carbacot**
- **Delaxin**
- **Marbaxin**
- **Robamol**
- **Robaxin**
- **Robaxisal**
- **Robomol**
- **Skelex**

BASIC INFORMATION

Habit forming? Possibly
Prescription needed? Yes
Available as generic? Yes, for some.
Drug class: Muscle relaxant

USES

Adjunctive treatment to rest, analgesics and physical therapy for muscle spasms.

DOSAGE & USAGE INFORMATION

How to take:
- Tablet—Swallow with liquid. If you can't swallow whole, crumble tablet and take with liquid or food. Instructions to take on empty stomach mean 1 hour before or 2 hours after eating.
- Extended-release tablet—Swallow whole with liquid. Don't crumble tablet.

When to take:
As needed, no more often than every 4 hours.

Continued next column

If you forget a dose:
Take as soon as you remember. If it is almost time for the next dose, wait for that dose (don't double this dose) and resume regular schedule.

What drug does:
Blocks body's pain messages to brain. Also causes sedation.

Time lapse before drug works:
30 to 60 minutes.

Don't take with:
Any other medicine or any dietary supplement without consulting your doctor or pharmacist.

OVERDOSE

SYMPTOMS:
Blurred vision, uncontrolled eye movements, large pupils, euphoria, hallucinations, delirium, headache, weakness, rigid or uncontrolled muscles, seizures, coma.
WHAT TO DO:
- **Dial 911 for all medical emergencies or call poison control center 1-800-222-1222 for instructions.**
- **See emergency information on last 3 pages of this book.**

POSSIBLE ADVERSE REACTIONS OR SIDE EFFECTS

SYMPTOMS	WHAT TO DO
Life-threatening:	
Rare, severe allergic reaction (anaphylaxis)—difficulty breathing, hives, itching, swelling (throat, face), vomiting, dizziness.	Seek emergency treatment immediately.
Common:	
• Drowsiness, dizziness.	Continue. Call doctor when convenient.
• Orange or red-purple urine.	No action necessary.
Infrequent:	
• Agitation, constipation or diarrhea, nausea, cramps, vomiting, wheezing, shortness of breath, headache, depression, muscle weakness, trembling, insomnia, uncontrolled eye movements, fainting.	Discontinue. Call doctor right away.
• Blurred vision.	Continue. Call Doctor when convenient.
Rare:	
• Seizures.	Discontinue. Seek emergency care.
• Rash, hives, itching, fever, yellow skin or eyes, sore throat, tiredness, black or tarry or bloody stools, hiccups.	Discontinue. Call doctor right away.

MUSCLE RELAXANTS, SKELETAL

WARNINGS & PRECAUTIONS

Don't take if:
- You are allergic to any skeletal muscle relaxant.
- You have porphyria.

Before you start, consult your doctor if:
- You have had liver or kidney disease.
- You plan pregnancy within medication period.
- You are allergic to tartrazine dye.
- You suffer from depression.

Over age 60:
Adverse reactions and side effects may be more frequent and severe than in younger persons.

Pregnancy:
Risk factors vary for drugs in this group. Always consult doctor. See category list on page xviii.

Breastfeeding; Lactation; Nursing Mothers:
It is unknown if all drugs in this group are generally acceptable with breastfeeding. Discuss risks and benefits with your doctor.

Infants & children up to age 18:
Safety and effectiveness in ages below 16 have not been established. Consult doctor.

Prolonged use:
Talk to your doctor about the need for follow-up medical examinations or laboratory studies to check liver function, kidney function, complete blood counts (white blood cell count, platelet count, red blood cell count, hemoglobin, hematocrit).

Skin & sunlight:
No problems expected.

Driving, piloting or hazardous work:
Don't drive or pilot aircraft until you learn how medicine affects you. Don't work around dangerous machinery. Don't climb ladders or work in high places. Danger increases if you drink alcohol or take medicine affecting alertness and reflexes, such as antihistamines, tranquilizers, sedatives, pain medicine, narcotics and mind-altering drugs.

Discontinuing:
Don't discontinue without doctor's advice until you complete prescribed dose, even though symptoms diminish or disappear.

Others:
- May affect results in some medical tests.
- Advise any doctor, dentist or pharmacist whom you consult that you take this medicine.

POSSIBLE INTERACTION WITH OTHER DRUGS

GENERIC NAME OR DRUG CLASS	COMBINED EFFECT
Antidepressants*	Increased sedation.
Antihistamines*	Increased sedation.
Central nervous system (CNS) depressants*	Increased depressive effects of both drugs.
Clozapine	Toxic effect on the central nervous system.
Dronabinol	Increased effect of dronabinol on central nervous system. Avoid combination.
Mind-altering drugs*	Increased sedation.
Muscle relaxants,* others	Increased sedation.
Narcotics*	Increased sedation.
Sedatives*	Increased sedation.
Sertraline	Increased depressive effects of both drugs.
Sleep inducers*	Increased sedation.
Tranquilizers*	Increased sedation.

POSSIBLE INTERACTION WITH OTHER SUBSTANCES

INTERACTS WITH	COMBINED EFFECT
Alcohol:	Increased sedation.
Beverages:	None expected.
Cocaine:	Increased risk of side effects. Avoid.
Foods:	None expected.
Marijuana:	Increased risk of side effects. Avoid.
Tobacco:	None expected.

NABILONE

BRAND NAMES

Cesamet

BASIC INFORMATION

Habit forming? No
Prescription needed? Yes
Available as generic? No
Drug class: Antiemetic

USES

- Treats nausea and vomiting.
- Prevents nausea and vomiting in patients receiving cancer chemotherapy.

DOSAGE & USAGE INFORMATION

How to take:
Capsule—Swallow with liquid. If you can't swallow whole, open capsule and take with liquid or food.

When to take:
At the same times each day, according to instructions on prescription label.

If you forget a dose:
Take as soon as you remember. If it is almost time for the next dose, wait for that dose (don't double this dose) and resume regular schedule.

What drug does:
Chemically related to marijuana, it probably regulates the vomiting control center in the brain.

Time lapse before drug works:
2 hours.

Don't take with:
- Alcohol or any drug that depresses the central nervous system. See Central Nervous System (CNS) Depressants in the Glossary.
- Any other medicine or any dietary supplement without consulting your doctor or pharmacist.

OVERDOSE

SYMPTOMS:
Severe mental changes, confusion, fearfulness, hallucinations, fainting, disorientation, trouble breathing, coma.
WHAT TO DO:
- **Dial 911 for all medical emergencies or call poison control center 1-800-222-1222 for instructions.**
- **See emergency information on last 3 pages of this book.**

POSSIBLE ADVERSE REACTIONS OR SIDE EFFECTS

SYMPTOMS	WHAT TO DO
Life-threatening: None expected.	
Common: Dry mouth, drowsiness, clumsiness, false sense of well-being, headache.	Continue. Call doctor when convenient.
Infrequent: Blurred vision, dizziness on standing, appetite loss, muscle pain.	Continue. Call doctor when convenient.
Rare: Mood changes, fast or pounding heartbeat, delusions, seizures, hallucinations, unusual weakness or tiredness, depression, faintness, anxiety, nervousness.	Discontinue. Call doctor right away or seek emergency care.

WARNINGS & PRECAUTIONS

Don't take if:
- You are allergic to nabilone. Allergic reactions are rare, but may occur.
- You have schizophrenic, manic or depressive states.

Before you start, consult your doctor if:
- You have abused drugs or are dependent on them, including alcohol.
- You have had high blood pressure or heart disease.
- You have had impaired liver function.

Over age 60:
Adverse reactions and side effects may be more frequent and severe than in younger persons.

Pregnancy:
Decide with your doctor if drug benefits justify risk to unborn baby. Risk category C (see page xviii).

Breastfeeding; Lactation; Nursing Mothers:
This drug is generally not recommended with breastfeeding. Discuss risks and benefits with your doctor.

Infants & children up to age 18:
Safety and effectiveness in this age group have not been established. Consult doctor.

Prolonged use:
- Avoid prolonged use. This medicine is intended to be used only during a cycle of cancer chemotherapy.
- Talk to your doctor about the need for follow-up medical examinations or laboratory studies to check blood pressure, heart function.

Skin & sunlight:
No problems expected.

Driving, piloting or hazardous work:
Don't drive or pilot aircraft until you learn how medicine affects you. Don't work around dangerous machinery. Don't climb ladders or work in high places. Danger increases if you drink alcohol or take medicine affecting alertness and reflexes.

Discontinuing:
After stopping the drug, adverse effects can last for 48 to 72 hours. They may include dizziness, drowsiness, euphoria, disorientation, problems with coordination, anxiety, depression, hallucinations or psychosis. Consult doctor right away if symptoms occur.

Others:
- Blood pressure should be measured regularly.
- Advise any doctor, dentist or pharmacist whom you consult that you take this medicine.
- Get up from bed or chair slowly to avoid fainting.

POSSIBLE INTERACTION WITH OTHER DRUGS

GENERIC NAME OR DRUG CLASS	COMBINED EFFECT
Apomorphine	Decreased effect of apomorphine.
Central nervous system (CNS) depressants,* other	Greater depression of the central nervous system.

POSSIBLE INTERACTION WITH OTHER SUBSTANCES

INTERACTS WITH	COMBINED EFFECT
Alcohol:	Dangerous depression of the central nervous system. Avoid.
Beverages:	None expected.
Cocaine:	Unknown. Avoid.
Foods:	None expected.
Marijuana:	Consult doctor.
Tobacco:	None expected.

*See Glossary

NAFARELIN

BRAND NAMES

Synarel

BASIC INFORMATION

Habit forming? No
Prescription needed? Yes
Available as generic? No
Drug class: Gonadotropin inhibitor

USES

- Treatment for endometriosis to relieve pain and reduce scattered implants of endometrial tissue.
- Treatment for central precocious puberty.

DOSAGE & USAGE INFORMATION

How to take:
Nasal spray—Follow instructions on package insert provided with your medicine.

When to take:
As directed by your doctor. Usually two times a day.

If you forget a dose:
Take as soon as you remember. If it is almost time for the next dose, wait for that dose (don't double this dose) and resume regular schedule.

What drug does:
Reduces estrogen production by ovaries.

Time lapse before drug works:
May require 6 months for full effect.

Don't take with:
- Birth control pills.
- Nasal sprays to decongest the membranes in the nose.
- Any other medicine or any dietary supplement without consulting your doctor or pharmacist.

OVERDOSE

SYMPTOMS:
Unknown.
WHAT TO DO:
If person uses much larger amount than prescribed or if accidentally swallowed, call poison control center 1-800-222-1222 for instructions or dial 911 (emergency) for help.

POSSIBLE ADVERSE REACTIONS OR SIDE EFFECTS

SYMPTOMS	WHAT TO DO
Life-threatening: None expected.	
Common: Hot flashes.	No action necessary.
Infrequent: Decreased sexual desire, vaginal dryness, headache, acne, swelling of hands and feet, reduction of breast size, weight gain, itchy and flaking scalp, muscle aches, nasal irritation.	Continue. Call doctor when convenient.
Rare: Insomnia, depression, weight loss.	Continue. Call doctor when convenient.

WARNINGS & PRECAUTIONS

Don't take if:
- You are allergic to nafarelin. Allergic reactions are rare, but may occur.
- You become pregnant.
- You have breast cancer.
- You have undiagnosed abnormal vaginal bleeding.

Before you start, consult your doctor if:
- You take birth control pills.
- You have diabetes.
- You have heart disease.
- You have epilepsy.
- You have kidney disease.
- You have liver disease.
- You have migraine headaches.
- You need to use topical nasal decongestants.

Over age 60:
Not used in this age group.

Pregnancy:
Consult doctor. Drug should not be used during pregnancy. Can cause harm to unborn baby. Risk category X (see page xviii).

Breastfeeding; Lactation; Nursing Mothers:
This drug is generally not recommended with breastfeeding. Discuss risks and benefits with your doctor.

Infants & children up to age 18:
Follow instructions provided by your child's doctor.

Prolonged use:
- Full effect requires prolonged use. Don't discontinue without consulting doctor.
- Talk to your doctor about the need for mammogram, follow-up medical examinations or laboratory studies to check liver function.

Skin & sunlight:
No special problems expected.

Driving, piloting or hazardous work:
No special problems expected.

Discontinuing:
Don't discontinue without consulting doctor. Menstrual periods may be absent for 2 to 3 months after discontinuation.

Others:
- May alter blood sugar levels in diabetic persons.
- Interferes with accuracy of laboratory tests to study pituitary gonadotropic and gonadal functions.
- Bone density decreases during treatment phase, but recovers following treatment.

POSSIBLE INTERACTION WITH OTHER DRUGS

GENERIC NAME OR DRUG CLASS	COMBINED EFFECT
Decongestant nasal sprays*	Decreased absorption of nafarelin.

POSSIBLE INTERACTION WITH OTHER SUBSTANCES

INTERACTS WITH	COMBINED EFFECT
Alcohol:	Excessive nervous system depression. Avoid.
Beverages:	
Caffeine drinks.	Rapid, irregular heartbeat. Avoid.
Cocaine:	Increased risk of side effects. Avoid.
Foods:	None expected.
Marijuana:	Increased risk of side effects. Avoid.
Tobacco:	Rapid, irregular heartbeat; increased leg cramps. Avoid.

NALOXONE

BRAND NAMES

Bunavail
Evzio
Suboxone
Targiniq ER
Zubsolv

BASIC INFORMATION

Habit forming? No
Prescription needed? Yes
Available as generic? No
Drug class: Opiate antagonist

USES

- Used as an emergency medical treatment to reverse the life-threatening effects of a narcotic (opioid) overdose. An overdose can occur as a result of drug abuse or accidentally taking too much of a drug.
- Naloxone is also combined with certain narcotic analgesic (painkiller) drugs into single dosage forms. This is done to deter abuse and misuse of the narcotic drug.
- Other uses as determined by your doctor.

DOSAGE & USAGE INFORMATION

How to take:

- Auto-injection or nasal spray—Follow the instructions provided with the prescription. Ask your doctor or pharmacist for advice if you are unsure of how to use.
- Buccal film, sublingual tablet, sublingual film, extended-release tablet—Follow instructions on the prescription. Do not cut, break, chew, crush, dissolve, snort or inject the drug.

When to take:

- For an overdose, use auto-injector or nasal spray as instructed. Always seek emergency medical help.
- Other forms are taken on prescribed schedule.

Continued next column

OVERDOSE

SYMPTOMS:
None expected from the use of naloxone. May have symptoms from original overdose.
WHAT TO DO:

- **Dial 911 for all medical emergencies or call poison control center 1-800-222-1222 for instructions.**
- **See emergency information on last 3 pages of this book.**

If you forget a dose:

- Emergency dosage is used when needed.
- Take other dosage forms of the drug as soon as you remember. If it is almost time for the next dose, wait for the next scheduled dose (don't double this dose).

What drug does:

- As an emergency treatment for overdose, it reverses the effects (including difficult breathing and excessive sleepiness) of opioids (narcotic drugs) in the body. Opioids bind to certain receptors in the body to provide pain relief and can also provide a euphoric response. The euphoria ("high" feeling) is what leads people to abuse or misuse the drugs.
- When combined with narcotic drugs in different dosage forms, naloxone has no action or effect if the medicine is taken as prescribed. If the dosage form is injected or snorted, naloxone is activated and helps block the narcotic effect in the body, especially the euphoria. It deters narcotic abuse and misuse.

Time lapse before drug works:
A few minutes in emergency use. For other dosage forms, it depends on the specific drug.

Don't take with:
Any other medicine or any dietary supplement without consulting your doctor or pharmacist.

POSSIBLE ADVERSE REACTIONS OR SIDE EFFECTS

SYMPTOMS	WHAT TO DO
Life-threatening:	
Rare, severe allergic reaction (anaphylaxis)—difficulty breathing, hives, itching, swelling (throat, face), vomiting, dizziness.	Seek emergency treatment immediately.
Common:	
None expected when used as prescribed.	No action necessary.
Infrequent:	
None expected when used as prescribed.	No action necessary.
Rare:	
With emergency use—If given to people who are addicted to opioids may induce withdrawal symptoms (e.g., sweating, being nervous or restless or irritable, trembling, body aches, nausea or vomiting, feeling tired and weak, runny nose, cramps).	Call doctor right away.

WARNINGS & PRECAUTIONS

Don't take if:
You are allergic to naloxone.

Before you start, consult your doctor if:
You have heart or liver problems.

Over age 60:
No problems expected.

Pregnancy:
Decide with your doctor if drug benefits justify risk to unborn baby. Risk category B (see page xviii).

Breastfeeding; Lactation; Nursing Mothers:
This drug is generally acceptable with breastfeeding (except when combined with narcotics). Discuss risks and benefits with your doctor.

Infants & children up to age 18:
- Emergency dosage approved for use in ages under 18. It is vital to seek medical care after it is used. Different children can have different responses to naloxone.
- For combination drug products, follow instructions provided by your child's doctor.

Prolonged use:
- Emergency dosage not used long-term.
- For other dosage forms, it depends on the specific drug prescribed.

Skin & sunlight:
No problems expected.

Driving, piloting or hazardous work:
No problems expected from use of naloxone alone. If naloxone is combined with a narcotic drug, read the prescription information for any restrictions.

Discontinuing:
Emergency form used only once. For other dosage forms, consult your doctor before stopping the drug.

Others:
- Overdose victims will usually be unable to use emergency naloxone to treat themselves. Be sure that family members, caregivers or others who spend time with you know the symptoms of an overdose, how to use naloxone (repeat doses may be needed). They need to call for emergency care and know what to do until medical help arrives. Consult doctor or pharmacist how to safely dispose of used emergency treatment devices.
- Advise any doctor, dentist or pharmacist whom you consult that you take this drug (if you take it on a regular basis).
- If you take a drug that combines naloxone with a narcotic analgesic, be sure to read the instructions provided with the prescription. In addition, review the Narcotic Analgesics drug chart in this book.

POSSIBLE INTERACTION WITH OTHER DRUGS

GENERIC NAME OR DRUG CLASS	COMBINED EFFECT
None expected for emergency naloxone.	

POSSIBLE INTERACTION WITH OTHER SUBSTANCES

INTERACTS WITH	COMBINED EFFECT
Alcohol:	None expected.
Beverages:	None expected.
Cocaine:	Unknown. Avoid.
Foods:	None expected.
Marijuana:	Consult doctor.
Tobacco:	None expected.

***See Glossary**

NALTREXONE

BRAND NAMES

Barr
Contrave
Embeda
ReVia
Vivitrol

BASIC INFORMATION

Habit forming? No
Prescription needed? Yes
Available as generic? Yes
Drug class: Narcotic antagonist

USES

- Treats detoxified former opioid (narcotics) addicts (along with a counseling program). It helps you maintain a drug-free life.
- Used in a combination drug for weight loss in obese and overweight adults.
- May be used to treat alcoholism (along with a counseling program).
- The brand name Embeda treats moderate to severe chronic pain (when treatment is needed for an extended period of time).

DOSAGE & USAGE INFORMATION

How to take:
- For former addicts—*Don't take drug until detoxification has been accomplished.*
- Tablet—Swallow with liquid. If you can't swallow whole, ask doctor or pharmacist for advice.
- Extended-release injectable—Given by a health care provider.
- Extended-release capsule (Embeda)—Swallow whole or open capsule and sprinkle contents on applesauce. Do not crush, chew or dissolve contents (pellets) as this can lead to potentially fatal dose.

When to take:
- Varies depending on dosage and use. Follow instructions on the prescription.
- Extended-release injectable is given monthly.

Continued next column

OVERDOSE

SYMPTOMS:
Unknown.
WHAT TO DO:
- **Dial 911 for all medical emergencies or call poison control center 1-800-222-1222 for instructions.**
- **See emergency information on last 3 pages of this book.**

If you forget a dose:
Take as soon as you remember. If it is almost time for the next dose, wait for that dose (don't double this dose) and resume regular schedule.

What drug does:
- It works in the brain to block the pleasurable effects or high feeling you get when you use narcotics and decreases craving for alcohol.
- It is unknown how it works for pain relief.

Time lapse before drug works:
1 hour.

Don't take with:
Narcotics, other drugs or any dietary supplement without consulting your doctor or pharmacist.

POSSIBLE ADVERSE REACTIONS OR SIDE EFFECTS

SYMPTOMS	WHAT TO DO
Life-threatening: None expected.	
Common: Stomach cramps or mild pain, anxiety, nervousness, joint or muscle pain, restlessness, headache, trouble sleeping, unusual tiredness, nausea or vomiting, depression.	Continue. Call doctor when convenient.
Infrequent:	
• Skin rash.	Continue, but call doctor right away.
• Constipation or diarrhea, dizziness, increased thirst, cough, hoarseness, runny or stuffy nose, chills, sneezing, sore throat, irritability, appetite loss, male sexual problems, injection site reaction (pain, tender, itching or hard lump).	Continue. Call doctor when convenient.
Rare:	
• Severe stomach pain, eye symptoms (blurred vision, aching, burning, swollen), chest pain, shortness of breath, frequent or painful urination, mood or mental changes, swelling (face, feet or lower legs), fever, confusion, itching, hallucinations, ringing in ears.	Continue, but call doctor right away.

• Other symptoms that cause concern.	Continue. Call doctor when convenient.

WARNINGS & PRECAUTIONS

Don't take if:
- You are allergic to naltrexone. Allergic reactions are rare, but may occur.
- You are still using opioids or are in opioid withdrawal or have failed a naloxone challenge test.
- You are actively drinking alcohol.
- You have liver disease.

Before you start, consult your doctor if:
- You have milder liver disease, kidney disease or bleeding disorder such as hemophilia.
- You have a history of depression or suicidal thoughts or attempts.

Over age 60:
Effects unknown. Consult doctor.

Pregnancy:
Decide with your doctor if drug benefits justify risk to unborn baby. Risk category C (see page xviii).

Breastfeeding; Lactation; Nursing Mothers:
This drug is generally acceptable with breastfeeding. Discuss risks and benefits with your doctor.

Infants & children up to age 18:
Safety and effectiveness in this age group have not been established. Consult doctor.

Prolonged use:
Consult your doctor about long-term use.

Skin & sunlight:
No problems expected.

Driving, piloting or hazardous work:
Don't drive or pilot aircraft until you learn how medicine affects you. Don't work around dangerous machinery. Don't climb ladders or work in high places. Danger increases if you drink alcohol or take medicine affecting alertness and reflexes.

Discontinuing:
Don't discontinue without consulting doctor. Dose may require gradual reduction if you have taken drug for a long time. Doses of other drugs may also require adjustment.

Others:
- Must be given under close supervision by people experienced in using naltrexone to treat addicts.
- See your doctor on a regular basis to check treatment progress and for recommended medical exams and lab tests.
- Excessive doses of this drug can cause liver injury. Consult doctor about your risks.
- Attempting to use narcotics to overcome effects of naltrexone may lead to life-threatening reactions.
- The drug does not treat narcotic or alcohol withdrawal symptoms.
- Advise any doctor, dentist or pharmacist whom you consult that you take this drug.
- Wear or carry medical ID that indicates you are taking this drug.
- Consult doctor about stopping this drug several days prior to an expected surgery.

POSSIBLE INTERACTION WITH OTHER DRUGS

GENERIC NAME OR DRUG CLASS	COMBINED EFFECT
Hepatotoxics,* other	Increased risk of liver damage.
Narcotic medicines*	1. Can cause withdrawal symptoms. May lead to cardiac arrest, coma and death (if naltrexone taken while person is dependent on these drugs). 2. If these drugs are taken while person is taking naltrexone, opioid effect (pain relief) will be blocked.

POSSIBLE INTERACTION WITH OTHER SUBSTANCES

INTERACTS WITH	COMBINED EFFECT
Alcohol:	Severe side effects. Must avoid.
Beverages:	None expected.
Cocaine:	Severe side effects. Must avoid.
Foods:	None expected.
Marijuana:	Unpredictable effects. Avoid.
Tobacco:	None expected.

***See Glossary**

NARCOTIC ANALGESICS

GENERIC AND BRAND NAMES

See full list of generic and brand names in the *Generic and Brand Name Directory*, page 871.

BASIC INFORMATION

Habit forming? Yes
Prescription needed? Yes
Available as generic? Yes, for most
Drug class: Narcotic

USES

- Treats pain symptoms from many causes. Used in drug combination-products to relieve cough. Treats other conditions as determined by your doctor.
- Buprenorphine is used to treat opioid addition.

DOSAGE & USAGE INFORMATION

How to take:

- Immediate-release tablet or capsule—Swallow whole with liquid.
- Extended-release, controlled release, sustained-release (tablets, capsules, pellets)—Swallow whole with liquid. Do not crush, chew, crumble or open. If unable to swallow whole, ask doctor or pharmacist for advice.
- Liquid—Be sure dose is accurately measured and taken as prescribed.
- Dispersible tablets—Stir into water or fruit juice just before taking each dose.
- Suppositories—Remove wrapper and moisten suppository with water. Lie on your side and gently insert into rectum.
- Nasal, transmucosal, transdermal, lozenge form—Follow instructions on prescription.
- Buccal tablet—Let tablet dissolve in mouth per instructions. Do not chew or swallow whole.
- Injection—Patient may be trained to self-inject.

Continued next column

When to take:
Depends on dosage form and dosage strength. May take with food to lessen stomach irritation.

If you forget a dose:
Take as soon as you remember. If it is almost time for the next dose, wait for that dose (don't double this dose) and resume regular schedule.

What drug does:

- Blocks pain messages in the brain and nervous system.
- Decreases activity of the brain's cough center.

Time lapse before drug works:
Symptoms may improve within several minutes.

Don't take with:
Any other medicine or any dietary supplement without consulting your doctor or pharmacist.

OVERDOSE

SYMPTOMS:
May have slow or troubled breathing, severe weakness and drowsiness, small pupils, cold and clammy skin, seizures, unconsciousness.
WHAT TO DO:

- **Dial 911 for all medical emergencies or call poison control center 1-800-222-1222 for instructions.**
- **See emergency information on last 3 pages of this book.**

POSSIBLE ADVERSE REACTIONS OR SIDE EFFECTS

SYMPTOMS	WHAT TO DO
Life-threatening:	
Rare, severe allergic reaction (anaphylaxis)—difficulty breathing, hives, itching, swelling (throat, face), vomiting, dizziness.	Seek emergency treatment immediately.
Common:	
Dizziness, drowsiness, lightheadedness, nausea or vomiting.	Continue. Call doctor when convenient.
Infrequent:	
• Breathing difficulty, confusion, fast or slow or pounding heartbeat, flushing, increased sweating.	Discontinue. Call doctor right away.
• Stomach cramps, painful or frequent or decreased urination, general ill feeling, constipation, dry mouth, feeling euphoric or nervous or restless, loss of appetite, headache, tiredness or weakness, excess sedation.	Continue. Call doctor when convenient.
Rare:	
• Seizures.	Discontinue. Seek emergency help.
• Ringing or buzzing in ears, overexcited, hallucinations, dark urine, yellow skin or eyes, pale stools, skin rash or hives or itching, changes in vision.	Discontinue. Call doctor right away.

• Trouble sleeping, mood changes, trembling, heartburn, nightmares, depression.	Continue. Call doctor when convenient.

WARNINGS & PRECAUTIONS

Don't take if:

- You are allergic to the narcotic prescribed.
- You have severe respiratory or diarrhea disorder.

Before you start, consult your doctor if:
You have asthma or any lung disorder, liver or kidney problems, heart disorder, seizure disorder, drug abuse or dependence, emotional instability, suicidal thoughts or attempts, gall bladder problems, recent renal or gastrointestinal surgery, head injury or other brain disorder, low thyroid, severe inflammatory bowel disease, high or low blood pressure, prostate problems.

Over age 60:
Increased risk of drug's adverse effects.

Pregnancy:
Risk factors vary for drugs in this group. See category list on page xviii and consult doctor.

Breastfeeding; Lactation; Nursing Mothers:
Drugs in this group are generally not recommended with breastfeeding. Discuss risks and benefits with your doctor.

Infants & children up to age 18:
Follow instructions provided by your child's doctor. Children are more at risk of side effects.

Prolonged use:

- High doses and long-term use can be habit forming. Consult doctor on a regular basis.
- May cause chronic constipation.

Skin & sunlight:
No special problems expected.

Driving, piloting or hazardous work:
Don't drive or pilot aircraft until you learn how medicine affects you. Don't work around dangerous machinery. Don't climb ladders or work in high places. Danger increases if you drink alcohol or take drugs affecting alertness.

Discontinuing:

- If used for several weeks or more, consult doctor before stopping. Dose may need to be slowly reduced.
- Advise doctor if any symptoms develop after stopping (e.g., agitation, anxiety, goose bumps, sweating, large eye pupils, irritability, insomnia, yawning, weakness).

Others:

- Lying down after the first few doses may decrease unwanted effects of nausea, vomiting, lightheadedness or dizziness.
- Follow directions exactly for using the fentanyl skin patch to avoid an overdose that could lead to severe side effects (including death). Follow all other safety precautions carefully.
- Advise any doctor, dentist or pharmacist whom you consult that you take this medicine.
- Get up slowly from sitting or lying position.

POSSIBLE INTERACTION WITH OTHER DRUGS

GENERIC NAME OR DRUG CLASS	COMBINED EFFECT
Analgesics,* other	Increased analgesic effect.
Anticholinergics*	Increased risk of side effects.
Antidepressants*	Increased sedative effect.
Antidiarrheals*	Increased risk of side effects.
Antihistamines*	Increased sedative effect.
Antihypertensives*	Increased effect of antihypertensive.
Central nervous system (CNS) depressants*	Increased sedative effect.
Enzyme inducers*	May decrease effect of some narcotic analgesics.
Enzyme inhibitors*	May increase effect of some narcotic analgesics.

Continued on page 906

POSSIBLE INTERACTION WITH OTHER SUBSTANCES

INTERACTS WITH	COMBINED EFFECT
Alcohol:	Increased risk of side effects. Avoid drinks and drugs containing alcohol.
Beverages:	None expected.
Cocaine:	Increased adverse effects. Avoid.
Foods:	None expected.
Marijuana:	Increased adverse effects. Avoid.
Tobacco:	None expected.

***See Glossary**

NEFAZODONE

BRAND NAMES

Serzone

BASIC INFORMATION

Habit forming? No
Prescription needed? Yes
Available as generic? Yes
Drug class: Antidepressant (phenylpiperazine)

USES

Treats symptoms of mental depression.

DOSAGE & USAGE INFORMATION

How to take:
Tablet—Swallow with liquid. May be taken with or without food.

When to take:
At the same times each day. The prescribed dosage may be increased weekly until maximum benefits are achieved.

If you forget a dose:
Take as soon as you remember. If it is almost time for the next dose, wait for that dose (don't double this dose) and resume regular schedule.

What drug does:
The exact mechanism is unknown. It appears to block reuptake of serotonin and norepinephrine (stimulating chemicals in the brain that play a role in emotions and psychological disturbances).

Time lapse before drug works:
Will take up to several weeks to relieve the depression.

Don't take with:
Any other medicine or any dietary supplement without consulting your doctor or pharmacist.

OVERDOSE

SYMPTOMS:
Drowsiness, nausea, vomiting, low blood pressure (faintness, weakness, dizziness, lightheadedness) or increased severity of adverse reactions.

WHAT TO DO:
- **Dial 911 for all medical emergencies or call poison control center 1-800-222-1222 for instructions.**
- **See emergency information on last 3 pages of this book.**

POSSIBLE ADVERSE REACTIONS OR SIDE EFFECTS

SYMPTOMS	WHAT TO DO
Life-threatening:	
Rare, severe allergic reaction (anaphylaxis)—difficulty breathing, hives, itching, swelling (throat, face), vomiting, dizziness.	Seek emergency treatment immediately.
Common:	
• Clumsiness or unsteadiness, blurred vision or other vision changes, fainting, lightheadedness, ringing in the ears, skin rash or itching.	Discontinue. Call doctor right away.
• Strange dreams, constipation or diarrhea, dry mouth, heartburn, fever, chills, flushing or feeling warm, headache, increased appetite, insomnia, coughing, tingling or prickly sensations, sore throat, trembling, drowsiness, confusion or agitation, memory lapses.	Continue. Call doctor when convenient.
Infrequent:	
• Tightness in chest, trouble breathing, wheezing, eye pain, stomach problem (nausea, vomiting, diarrhea and stomach pain).	Discontinue. Call doctor right away.
• Joint pain, breast pain, increased thirst.	Continue. Call doctor when convenient.
Rare:	
• Face swelling, hives, muscle pain or stiffness, chest pain, fast heartbeat, mood or mental changes, difficulty speaking, hallucinations, uncontrolled excited behavior, twitching, ear pain, increased hearing sensitivity, bleeding or bruising, irritated red eyes, eyes sensitive to light, pain in back or side, swollen glands, problems with urination.	Discontinue. Call doctor right away.

• Unusual tiredness or weakness, false sense of well-being, menstrual changes, change in sexual desire or function.	Continue. Call doctor when convenient.

WARNINGS & PRECAUTIONS

Don't use if:
You are allergic to nefazodone or trazodone (phenylpiperazine antidepressants).

Before you start, consult your doctor if:
- You have or have had liver disease.
- You have a seizure disorder, heart disease or blood circulation problem or have had a stroke.
- You are dehydrated.
- You have a history of drug abuse.
- You have a history of mood disorders (such as mania) or thoughts of suicide.
- You are allergic to any medication.

Over age 60:
A lower starting dosage is usually recommended until a response is determined.

Pregnancy:
Decide with your doctor if drug benefits justify risk to unborn baby. Risk category C (see page xviii).

Breastfeeding; Lactation; Nursing Mothers:
This drug is generally not recommended with breastfeeding. Discuss risks and benefits with your doctor.

Infants & children up to age 18:
Not approved for ages under 18. If prescribed, carefully read information provided with prescription. Contact doctor right away if depression symptoms get worse or there is any talk of suicide or suicide behaviors. Also, read information under Others.

Prolonged use:
Consult with your doctor on a regular basis while taking this drug to check your progress and to discuss any increase or changes in side effects and the need for continued treatment.

Skin & sunlight:
May cause a rash or intensify sunburn in areas exposed to sun or ultraviolet light (photosensitivity reaction). Avoid excess sun exposure and use sunscreen. Call doctor if reaction occurs.

Driving, piloting or hazardous work:
Don't drive or pilot aircraft until you learn how medicine affects you. Don't work around dangerous machinery. Don't climb ladders or work in high places. Danger increases if you drink alcohol or take medicine affecting alertness and reflexes.

Discontinuing:
Don't discontinue this drug without consulting doctor. Dosage may require a gradual reduction before stopping.

Others:
- Get up slowly from a sitting or lying position to avoid dizziness, faintness or lightheadedness.
- Advise any doctor, dentist or pharmacist whom you consult that you take this medicine.
- Take drug only as directed. Do not increase or reduce dosage without doctor's approval.
- Adults and children taking antidepressants may experience a worsening of the depression symptoms and may have increased suicidal thoughts or behaviors. Call doctor right away if these symptoms or behaviors occur.
- Use of this drug may result in hepatic (liver) problems that can be life-threatening. Call doctor right away if yellow skin or eyes, stomach problems, appetite loss or fatigue occur.

POSSIBLE INTERACTION WITH OTHER DRUGS

GENERIC NAME OR DRUG CLASS	COMBINED EFFECT
Alprazolam	Increased effect of alprazolam.
Antihypertensives*	Possible too-low blood pressure.
Carbamazepine	Decreased effect of nefazodone. Avoid.
Central nervous system (CNS) depressants*	Increased sedation.
Cisapride	Increased effect of cisapride. Avoid.
Digoxin	Increased digoxin effect.

Continued on page 906

POSSIBLE INTERACTION WITH OTHER SUBSTANCES

INTERACTS WITH	COMBINED EFFECT
Alcohol:	Increased sedative affect. Avoid.
Beverages:	None expected.
Cocaine:	Unknown. Avoid.
Foods:	None expected.
Marijuana:	Consult doctor.
Tobacco:	None expected.

***See Glossary**

NEOMYCIN (Oral)

BRAND NAMES

Mycifradin

BASIC INFORMATION

Habit forming? No
Prescription needed? Yes
Available as generic? Yes
Drug class: Antibacterial

USES

- Clears intestinal tract of germs prior to surgery.
- Treats some causes of diarrhea.
- Lowers blood cholesterol.
- Lessens symptoms of hepatic coma.

DOSAGE & USAGE INFORMATION

How to take:
Tablet—Swallow with liquid or food to lessen stomach irritation. If you can't swallow whole, crumble tablet and take with liquid or food.

When to take:
According to directions on prescription.

If you forget a dose:
Take as soon as you remember. If it is almost time for the next dose, wait for that dose (don't double this dose) and resume regular schedule.

What drug does:
Kills germs susceptible to neomycin.

Time lapse before drug works:
2 to 3 days.

Don't take with:
Any other medicine or any dietary supplement without consulting your doctor or pharmacist.

OVERDOSE

SYMPTOMS:
Loss of hearing, numbness or tingling, decreased urination, muscle twitching, seizures.
WHAT TO DO:
- **Dial 911 for all medical emergencies or call poison control center 1-800-222-1222 for instructions.**
- **See emergency information on last 3 pages of this book.**

POSSIBLE ADVERSE REACTIONS OR SIDE EFFECTS

SYMPTOMS	WHAT TO DO
Life-threatening: None expected.	
Common: Sore mouth or rectum, nausea, vomiting.	Continue. Call doctor when convenient.
Infrequent: None expected.	
Rare: Clumsiness, dizziness, rash, hearing loss, ringing or noises in ear, frothy stools, gaseousness, decreased frequency of urination, diarrhea.	Discontinue. Call doctor right away.

WARNINGS & PRECAUTIONS

Don't take if:
You are allergic to neomycin or any aminoglycoside.* Allergic reactions are rare, but may occur.

Before you start, consult your doctor if:
- You will have surgery within 2 months, including dental surgery, requiring general or spinal anesthesia.
- You have hearing loss or loss of balance secondary to 8th cranial nerve disease.
- You have intestinal obstruction.
- You have myasthenia gravis, Parkinson's disease, kidney disease, inflammatory bowel disease, ulcers in the bowel or electrolyte disorder.

Over age 60:
Adverse reactions and side effects may be more frequent and severe than in younger persons.

Pregnancy:
Consult doctor. Use of the drug during pregnancy is not recommended. It could cause harm to the unborn baby. Risk category D (see page xviii).

Breastfeeding; Lactation; Nursing Mothers:
This drug is generally acceptable with breastfeeding. Discuss risks and benefits with your doctor.

Infants & children up to age 18:
Follow instructions provided by your child's doctor.

Prolonged use:
- Adverse effects more likely.
- Talk to your doctor about the need for follow-up medical examinations or laboratory studies to check hearing, kidney function.

Skin & sunlight:
No problems expected.

Driving, piloting or hazardous work:
No problems expected.

Discontinuing:
May be unnecessary to finish medicine. Follow doctor's instructions.

Others:
- Advise any doctor, dentist or pharmacist whom you consult that you take this medicine.
- Taking this drug increases risk of permanent hearing loss, nerve damage, kidney damage, severe muscle relaxation (can lead to paralysis) and breathing problems. Contact doctor right away if new and unusual symptoms develop while taking this drug,

POSSIBLE INTERACTION WITH OTHER DRUGS

GENERIC NAME OR DRUG CLASS	COMBINED EFFECT
Aminoglycosides*	Increased chance of toxic effect on hearing, kidneys, muscles.
Beta carotene	Decreased absorption of beta carotene.
Capreomycin	Increased chance of toxic effects on hearing, kidneys.
Cephalothin	Increased chance of toxic effect on kidneys.
Cisplatin	Increased chance of toxic effects on hearing, kidneys.
Ethacrynic acid	Increased chance of toxic effects on hearing, kidneys.
Furosemide	Increased chance of toxic effects on hearing, kidneys.
Mercaptomerin	Increased chance of toxic effects on hearing, kidneys.
Penicillins*	Decreased antibiotic effect.
Tiopronin	Increased risk of toxicity to kidneys.
Vancomycin	Increased chance of toxic effects on hearing, kidneys.

POSSIBLE INTERACTION WITH OTHER SUBSTANCES

INTERACTS WITH	COMBINED EFFECT
Alcohol:	Increased chance of toxicity. Avoid.
Beverages:	None expected.
Cocaine:	Unknown. Avoid.
Foods:	None expected.
Marijuana:	Consult doctor.
Tobacco:	None expected.

***See Glossary**

NIACIN
(Vitamin B-3, Nicotinic Acid, Nicotinamide)

BRAND NAMES

Advicor
Endur-Acin
Nia-Bid
Niac
Niacels
Niacin
Niacor
Niaspan
Nico-400
Nicobid
Nicolar
Nicotinex
Nicotinyl alcohol
Papulex
Roniacol
Ronigen
Rycotin
Simcor
Slo-Niacin
Span-Niacin
Tega-Span
Tri-B3
Numerous brands of single vitamin and multivitamin combinations are available.

BASIC INFORMATION

Habit forming? No
Prescription needed? Yes, for some
Available as generic? Yes
Drug class: Vitamin supplement, vasodilator, antihyperlipidemic

USES

- Replacement for niacin lost due to inadequate diet.
- Treatment for vertigo (dizziness) and ringing in ears.
- Prevention of premenstrual headache.
- Reduction of blood levels of cholesterol and triglycerides.
- Treatment for pellagra.

OVERDOSE

SYMPTOMS:
Body flush, nausea, vomiting, abdominal cramps, diarrhea, weakness, sweating, lightheadedness, fainting, possibly irregular heartbeat.
WHAT TO DO:
- **Dial 911 for all medical emergencies or call poison control center 1-800-222-1222 for instructions.**
- **See emergency information on last 3 pages of this book.**

DOSAGE & USAGE INFORMATION

How to take:
- Tablet, capsule or liquid—Swallow with liquid or food to lessen stomach irritation.
- Extended-release tablets or capsules—Swallow each dose whole.

When to take:
At the same times each day.

If you forget a dose:
Take as soon as you remember. If it is almost time for the next dose, wait for that dose (don't double this dose) and resume regular schedule.

What drug does:
- Corrects niacin deficiency.
- Dilates blood vessels.
- In large doses, decreases cholesterol production.

Time lapse before drug works:
15 to 20 minutes.

Don't take with:
Any other medicine or any dietary supplement without consulting your doctor or pharmacist.

POSSIBLE ADVERSE REACTIONS OR SIDE EFFECTS

SYMPTOMS	WHAT TO DO
Life-threatening:	
Rare, severe allergic reaction (anaphylaxis)—difficulty breathing, hives, itching, swelling (throat, face), vomiting, dizziness.	Seek emergency treatment immediately.
Common:	
Dry skin.	Continue. Call doctor when convenient.
Infrequent:	
• Upper abdominal pain, diarrhea.	Discontinue. Call doctor right away.
• Headache, dizziness, faintness, temporary numbness and tingling in hands and feet.	Continue. Call doctor when convenient.
• "Hot" feeling, flush.	No action necessary.
Rare:	
Rash, itching, double vision, weakness and faintness when arising from bed or chair, yellow skin or eyes.	Discontinue. Call doctor right away.

NIACIN (Vitamin B-3, Nicotinic Acid, Nicotinamide)

WARNINGS & PRECAUTIONS

Don't take if:
You are allergic to niacin or any niacin-containing vitamin mixtures.

Before you start, consult your doctor if:
- You have sensitivity to tartrazine dye.
- You have diabetes.
- You have gout.
- You have gallbladder or liver disease.
- You have impaired liver function.
- You have active peptic ulcer.

Over age 60:
Response to drug cannot be predicted. Dose must be individualized.

Pregnancy:
Decide with your doctor if drug benefits justify risk to unborn baby. Risk category C (see page xviii).

Breastfeeding; Lactation; Nursing Mothers:
This drug is generally acceptable with breastfeeding. Discuss risks and benefits with your doctor.

Infants & children up to age 18:
- If prescribed, follow instructions provided by your child's doctor.
- Read nonprescription product's label to be sure it is approved for your child's age. Follow label instructions on dosage. Consult doctor or pharmacist if unsure.

Prolonged use:
- May cause impaired liver function.
- Talk to your doctor about the need for follow-up medical examinations or laboratory studies to check liver function, blood sugar.

Skin & sunlight:
No problems expected.

Driving, piloting or hazardous work:
Avoid if you feel dizzy or faint. Otherwise, no problems expected.

Discontinuing:
May be unnecessary to finish medicine. Follow doctor's instructions.

Others:
- A balanced diet should provide all the niacin a healthy person needs and make supplements unnecessary. Best sources are meat, eggs and dairy products.
- Store in original container in cool, dry, dark place.
- Obesity reduces effectiveness.
- Some nicotinic acid products contain tartrazine dye. Read labels carefully if sensitive to tartrazine.

POSSIBLE INTERACTION WITH OTHER DRUGS

GENERIC NAME OR DRUG CLASS	COMBINED EFFECT
Antidiabetics*	Decreased antidiabetic effect.
Beta-adrenergic blocking agents*	Excessively low blood pressure.
Dexfenfluramine	May require dosage change as weight loss occurs.
HMG-CoA reductase inhibitors*	Increased risk of muscle or kidney problems.
Mecamylamine	Excessively low blood pressure.
Methyldopa	Excessively low blood pressure.
Probenecid	Decreased effect of probenecid.
Sulfinpyrazone	Decreased effect of sulfinpyrazone.

POSSIBLE INTERACTION WITH OTHER SUBSTANCES

INTERACTS WITH	COMBINED EFFECT
Alcohol:	Excessively low blood pressure. Use caution.
Beverages:	None expected.
Cocaine:	Unknown. Avoid.
Foods:	None expected.
Marijuana:	Consult doctor.
Tobacco:	None expected.

***See Glossary**

NICOTINE

BRAND NAMES

Commit
Habitrol
Nicoderm
Nicoderm CQ
Nicoderm CQ Thin Flex patch
Nicorette
Nicorette DS
Nicorette Fresh Mint
Nicorette Lozenge
Nicotrol
Nicotrol NS
Prostep
Thrive Gum
Thrive Lozenge

BASIC INFORMATION

Habit Forming? Yes
Prescription needed? No
Available as generic? Yes, for some
Drug class: Antismoking agent

USES

Treatment aid to giving up smoking. Nicotine replacement is to be used in conjunction with a medically supervised behavioral modification program for smoking cessation.

DOSAGE & USAGE INFORMATION

How to use:

- Skin patch—Apply to clean, nonhairy site on the trunk or upper outer arm. Fold old patch in half (sticky sides together) and dispose of where children and pets cannot get to it.
- Chewing gum—Chew gum pieces slowly and intermittently (chew several times, then place between cheek and gum) for best effect.
- Nasal spray—Use 2-3 sprays in each nostril.
- Lozenge—Use as directed on package.

Continued next column

When to use:

- Skin patch—Daily. Remove old patch and apply new patch to new location on the skin.
- Chewing gum—When there is an urge to smoke, chew the gum for about 30 minutes. For the brand name Thrive Gum, follow instructions on label.
- Nasal spray—Hourly, or as directed. Dosage adjustments should be made as needed.
- Lozenge—Follow label instructions.
- For all—Always follow product's directions.

If you forget a dose:

- Skin patch—Remove old patch and apply new patch as soon as you remember, then return to regular schedule.
- Chewing gum or nasal spray—Use as soon as you remember (don't double dosages).

What drug does:
Delivers a supply of nicotine to the body for relief of smoking withdrawal symptoms (irritability, headache, nervousness, drowsiness, fatigue). Reduces craving for cigarettes.

Time lapse before drug works:
Minutes to hours depending on type of product.

Don't take with:
Any other medicine or any dietary supplement without consulting your doctor or pharmacist.

OVERDOSE

SYMPTOMS:
Nausea, vomiting, severe diarrhea, increased mouth watering, abdominal pain, diarrhea, confusion, vision and hearing changes, Irregular or fast pulse, fainting, breathing difficulty, convulsions.
WHAT TO DO:

- **Dial 911 for all medical emergencies or call poison control center 1-800-222-1222 for instructions.**
- **See emergency information on last 3 pages of this book.**

POSSIBLE ADVERSE REACTIONS OR SIDE EFFECTS

SYMPTOMS	WHAT TO DO
Life-threatening: None expected.	
Common:	
• Skin patch—itching, redness, burning or skin rash at site of patch.	Continue. Call doctor when convenient.
• Chewing gum—dental problems, sore mouth or throat, belching, mouth watering.	Continue. Call doctor when convenient.
• Nasal spray—runny nose, watering eyes, throat irritation, sneezing and cough.	Continue. Call doctor when convenient.
Infrequent: Diarrhea, dizziness, indigestion, strange dreams, nervousness, muscle aches, nausea, constipation, increased cough, tiredness, irritability, changes in menstruation, insomnia, headache, increase in sweating.	Continue. Call doctor when convenient.

Rare:

• Allergic reaction (swelling, hives, rash, itching, vomiting, irregular or fast heartbeat). Symptoms of overdose occur (high doses of nicotine can cause toxic effects, even in people who are nicotine tolerant).	Discontinue. Call doctor right away.
• Hiccups or hoarseness with chewing gum.	Continue. Call doctor when convenient.

WARNINGS & PRECAUTIONS

Don't take if:
You are allergic to nicotine or any of the components in the skin patch. Allergic reactions are rare, but may occur.

Before you start, consult your doctor if:
- You are pregnant.
- You have a skin disorder; mouth, throat, dental or TMJ disorder; long-term nasal disorder (allergy, hay fever, sinusitis, polyps) or asthma.
- You have cardiovascular or peripheral vascular disease or high blood pressure.
- You have liver or kidney disease.
- You have hyperthyroidism, insulin-dependent diabetes, pheochromocytoma, peptic ulcer disease or endocrine disorder.

Over age 60:
Adverse reactions and side effects may be more frequent and severe than in younger persons.

Pregnancy:
Tobacco smoke and nicotine are harmful to the fetus. The specific effects of nicotine from these drugs are unknown. Discuss the risks of both with your doctor. Risk category varies for drugs in this group (see page xviii).

Breastfeeding; Lactation; Nursing Mothers:
This drug is generally acceptable with breastfeeding. Discuss risks and benefits with your doctor.

Infants & children up to age 18:
Safety and effectiveness in this age group have not been established.

Prolonged use:
Treatment may take several months. It is not intended for long-term use.

Skin & sunlight:
No problems expected.

Driving, piloting or hazardous work:
Avoid if you feel dizzy or lightheaded. Otherwise, no problems expected.

Discontinuing:
Adverse reactions and side effects related to nicotine withdrawal may continue for some time after discontinuing.

Others:
- Advise any doctor, dentist or pharmacist whom you consult that you take this medicine.
- May affect results of some medical tests.
- Keep both the used and unused skin patches out of the reach of children and pets. Dispose of old patches according to directions.
- For full benefit from this treatment and to decrease risk of side effects, stop cigarette smoking as soon as you begin treatment.

POSSIBLE INTERACTION WITH OTHER DRUGS

GENERIC NAME OR DRUG CLASS	COMBINED EFFECT
Acetaminophen	Increased effect of acetaminophen.
Beta-adrenergic blocking agents*	Increased effect of beta blocker.
Bronchodilators, xanthine* (except dyphylline)	Increased bronchodilator effect.
Imipramine	Increased effect of imipramine.
Insulin & insulin lispro	May require insulin dosage adjustment.
Isoproterenol	Decreased effect of isoproterenol.
Oxazepam	Increased effect of oxazepam.

Continued on page 907

POSSIBLE INTERACTION WITH OTHER SUBSTANCES

INTERACTS WITH	COMBINED EFFECT
Alcohol:	Increased cardiac irritability. Avoid.
Beverages:	None expected.
Cocaine:	Increased cardiac irritability. Avoid.
Foods:	None expected.
Marijuana:	Increased risk of side effects. Avoid.
Tobacco:	Increased adverse effects of nicotine. Must avoid.

***See Glossary**

NITAZOXANIDE

BRAND NAMES

Alinia

BASIC INFORMATION

Habit forming? No
Prescription needed? Yes
Available as generic? No
Drug class: Antiprotozoal; antidiarrheal

USES

- Treatment of diarrhea and other symptoms caused by two parasitic infections (Cryptosporidium parvum and Giardia lamblia). These two types of parasites are common causes of persistent diarrhea in children and adults. Outbreaks have been associated with day care centers, swimming pools, water parks and public water supplies.
- May be used for treatment of other disorders as determined by your doctor.

DOSAGE & USAGE INFORMATION

How to take:
- Oral suspension—Follow label instructions and take with food.
- Tablet—Swallow with liquid and take with food.

When to take:
Usual dose is twice a day for 3 days.

If you forget a dose:
Take as soon as you remember. If it is almost time for the next dose, wait for that dose (don't double this dose) and resume regular schedule.

What drug does:
Prevents enzyme reactions that are needed for the parasites to survive. Thousands of enzymes are present in the body with a range of functions including food digestion and toxin elimination.

Continued next column

OVERDOSE

SYMPTOMS:
Unknown.
WHAT TO DO:
If person uses much larger amount than prescribed or if accidentally swallowed, call poison control center 1-800-222-1222 for instructions or dial 911 (emergency) for help.

Time lapse before drug works:
Starts working in a few hours, but takes 3 days for effective treatment.

Don't take with:
Any other medicine or any dietary supplement without consulting your doctor or pharmacist.

POSSIBLE ADVERSE REACTIONS OR SIDE EFFECTS

SYMPTOMS	WHAT TO DO
Life-threatening: None expected.	
Common: None expected.	
Infrequent: None expected.	
Rare:	
Abdominal pain, diarrhea, vomiting, headache.	Continue. Call doctor if symptoms persist or are severe.

WARNINGS & PRECAUTIONS

Don't take if:
You are allergic to nitazoxanide. Allergic reactions are rare, but may occur.

Before you start, consult your doctor if:
- You have liver or kidney disease.
- You have bile or gallbladder problems.
- You have diabetes (the drug contains sucrose).
- You are immunosuppressed due to illness or drugs.

Over age 60:
No special problems expected.

Pregnancy:
Decide with your doctor if drug benefits justify risk to unborn baby. Risk category B (see page xviii).

Breastfeeding; Lactation; Nursing Mothers:
This drug is generally acceptable with breastfeeding. Discuss risks and benefits with your doctor.

Infants & children up to age 18:
Approved for children over age one. Follow instructions provided by your child's doctor.

Prolonged use:
Not used for more than 3 days.

Skin & sunlight:
No problems expected.

Driving, piloting or hazardous work:
No problems expected.

Discontinuing:
No problems expected.

Others:
- Safety and effectiveness of nitazoxanide for HIV positive patients and immunodeficient patients have not been established.
- Advise any doctor, dentist or pharmacist whom you consult that you take this drug.

POSSIBLE INTERACTION WITH OTHER DRUGS

GENERIC NAME OR DRUG CLASS	COMBINED EFFECT
Protein bound drugs*	May increase effect of nitazoxanide.

POSSIBLE INTERACTION WITH OTHER SUBSTANCES

INTERACTS WITH	COMBINED EFFECT
Alcohol:	None expected.
Beverages:	None expected.
Cocaine:	Unknown. Avoid.
Foods:	None expected.
Marijuana:	Consult doctor.
Tobacco:	None expected.

*See Glossary

NITRATES

GENERIC AND BRAND NAMES

See full list of generic and brand names in the *Generic and Brand Name Directory*, page 872.

BASIC INFORMATION

Habit forming? No
Prescription needed? Yes
Available as generic? Yes
Drug class: Antianginal (nitrate)

USES

- Reduces frequency and severity of angina attacks.
- Treats congestive heart failure.
- The brand name BiDil is approved to treat heart failure specifically in African Americans.

DOSAGE & USAGE INFORMATION

How to take:
- Extended-release tablet or capsule—Swallow each dose whole with liquid.
- Chewable tablet—Chew tablet at earliest sign of angina, and hold in mouth for 2 minutes.
- Regular tablet or capsule—Swallow whole with liquid. Don't crush, chew or open.
- Buccal tablet—Allow to dissolve inside of mouth.
- Lingual spray—Spray under tongue according to instructions enclosed with prescription.
- Ointment—Apply as directed.
- Patch—Apply to skin according to package instructions.
- Sublingual tablet—Place under tongue every 3 to 5 minutes at earliest sign of angina. If you don't have complete relief with 3 or 4 tablets, call doctor.

Continued next column

OVERDOSE

SYMPTOMS:
Dizziness, nausea, vomiting, throbbing headache, feeling of pressure in head, fever, confusion, fainting, shortness of breath, heartbeat changes, skin may be flushed or cold and clammy, convulsions, coma.
WHAT TO DO:
- **Dial 911 for all medical emergencies or call poison control center 1-800-222-1222 for instructions.**
- **See emergency information on last 3 pages of this book.**

When to take:
- Swallowed tablet—Take at the same times each day, 1 or 2 hours after meals.
- Sublingual tablet or spray—At onset of angina.
- Ointment—Follow prescription directions.
- Patch—According to physician's instructions.

If you forget a dose:
Take as soon as you remember. If it is almost time for the next dose, wait for that dose (don't double this dose) and resume regular schedule.

What drug does:
Relaxes blood vessels, increasing blood flow to heart muscle.

Time lapse before drug works:
- Sublingual tablets and spray—1 to 3 minutes.
- Other forms—15 to 30 minutes. Will not stop an attack, but may prevent attacks.

Don't take with:
Any other medicine or any dietary supplement without consulting your doctor or pharmacist.

POSSIBLE ADVERSE REACTIONS OR SIDE EFFECTS

SYMPTOMS	WHAT TO DO
Life-threatening:	
In case of overdose, see previous column.	
Common:	
Flushed face and neck, mild headache, nausea, vomiting, lightheadedness or dizziness, restlessness.	Continue. Call doctor when convenient.
Infrequent:	
None expected.	
Rare:	
• Skin rash, blurred vision, dry mouth, severe and ongoing headache, fast heartbeat.	Discontinue. Call doctor right away.
• Sore and red skin (with topical product).	Continue. Call doctor if symptoms persist.

WARNINGS & PRECAUTIONS

Don't take if:
You are allergic to nitrates, including nitroglycerin. Allergic reactions are rare, but may occur.

Before you start, consult your doctor if:
- You are taking nonprescription drugs.
- You plan to become pregnant within medication period.
- You have glaucoma.
- You drink alcoholic beverages or smoke marijuana.

Over age 60:
Adverse reactions and side effects may be more frequent and severe than in younger persons.

Pregnancy:
Risk factors vary for drugs in this group. Always consult doctor. See category list on page xviii.

Breastfeeding; Lactation; Nursing Mothers:
Drugs in this group are generally acceptable with breastfeeding. Discuss risks and benefits with your doctor.

Infants & children up to age 18:
Safety and effectiveness in this age group have not been established. Consult doctor.

Prolonged use:
- Drug may become less effective and require higher doses.
- Talk to your doctor about the need for follow-up medical examinations or laboratory studies to check blood pressure, heart rate.

Skin & sunlight:
No problems expected.

Driving, piloting or hazardous work:
Don't drive or pilot aircraft until you learn how medicine affects you. Don't work around dangerous machinery. Don't climb ladders or work in high places. Danger increases if you drink alcohol or take medicine affecting alertness and reflexes.

Discontinuing:
Except for sublingual tablets, don't discontinue without doctor's advice until you complete prescribed dose, even though symptoms diminish or disappear.

Others:
- If discomfort is not caused by angina, nitrate medication will not bring relief. Call doctor if discomfort persists.
- Periodic urine and lab blood studies of white cell counts recommended if you take nitrates.
- Keep sublingual tablets in original container. Always carry them with you, but keep from body heat if possible.
- Sublingual tablets produce a burning, stinging sensation when placed under the tongue. Replace supply if no burning or stinging is noted.
- To avoid development of tolerance, drug-free intervals of 10 hours are sufficient.

POSSIBLE INTERACTION WITH OTHER DRUGS

GENERIC NAME OR DRUG CLASS	COMBINED EFFECT
Anticholinergics*	Increased internal eye pressure.
Antihypertensives*	Excessive blood pressure drop.
Beta-adrenergic blocking agents*	Excessive blood pressure drop.
Calcium channel blockers*	Decreased blood pressure.
Carteolol	Possible excessive blood pressure drop.
Guanfacine	Increased effects of both drugs.
Narcotics*	Excessive blood pressure drop.
Phenothiazines*	May decrease blood pressure.
Sildenafil	Increased effect of nitrates.
Sympathomimetics*	Possible reduced effects of both medicines.

POSSIBLE INTERACTION WITH OTHER SUBSTANCES

INTERACTS WITH	COMBINED EFFECT
Alcohol:	Excessive blood pressure drop.
Beverages:	None expected.
Cocaine:	Unknown. Avoid.
Foods:	None expected.
Marijuana:	Consult doctor.
Tobacco:	None expected.

***See Glossary**

NITROFURANTOIN

BRAND NAMES

Apo-Nitrofurantoin
Cyantin
Furadantin
Furalan
Furaloid
Furan
Furanite
Furantoin
Furatine
Furaton
Macrobid
Macrodantin
Nephronex
Nifuran
Nitrex
Nitrofan
Nitrofor
Nitrofuracot
Novofuran
Ro-Antoin
Sarodant
Trantoin
Urotoin

BASIC INFORMATION

Habit forming? No
Prescription needed? Yes
Available as generic? Yes
Drug class: Antimicrobial, antibacterial (antibiotic)

USES

Treatment for urinary tract infections.

DOSAGE & USAGE INFORMATION

How to take:

- Tablet or capsule—Swallow with food or milk to lessen stomach irritation. If you can't swallow whole, crumble tablet or open capsule and take with liquid or food.
- Extended-release capsule—Swallow with liquid. Do not open capsule.
- Liquid—Shake well and take with food. Use a measuring spoon to ensure accuracy.

When to take:
At the same times each day.

If you forget a dose:
Take as soon as you remember. If it is almost time for the next dose, wait for that dose (don't double this dose) and resume regular schedule.

Continued next column

OVERDOSE

SYMPTOMS:
Nausea, vomiting.
WHAT TO DO:
If person uses much larger amount than prescribed or if accidentally swallowed, call poison control center 1-800-222-1222 for instructions or dial 911 (emergency) for help.

What drug does:
Prevents susceptible bacteria in the urinary tract from growing and multiplying.

Time lapse before drug works:
1 to 2 weeks.

Don't take with:
Any other medicine or any dietary supplement without consulting your doctor or pharmacist.

POSSIBLE ADVERSE REACTIONS OR SIDE EFFECTS

SYMPTOMS	WHAT TO DO
Life-threatening:	
Rare, severe allergic reaction (anaphylaxis)—difficulty breathing, hives, itching, swelling (throat, face), vomiting, dizziness.	Seek emergency treatment immediately.
Common:	
• Diarrhea, appetite loss, nausea, vomiting, chest pain, cough, difficult breathing, chills or unexplained fever, abdominal pain.	Discontinue. Call doctor right away.
• Rusty-colored or brown urine.	No action necessary.
Infrequent:	
• Rash, itchy skin, numbness or tingling or burning of face or mouth, fatigue, weakness.	Discontinue. Call doctor right away.
• Dizziness, headache, drowsiness, paleness (in children), discolored teeth (from liquid form).	Continue. Call doctor when convenient.
Rare:	
Yellow skin or eyes.	Discontinue. Call doctor right away.

WARNINGS & PRECAUTIONS

Don't take if:
- You are allergic to nitrofurantoin.
- You have impaired kidney function.
- You drink alcohol.

Before you start, consult your doctor if:
- You are prone to allergic reactions.
- You are pregnant and within 2 weeks of delivery.
- You have had kidney disease, lung disease, anemia, nerve damage, or G6PD* deficiency (a metabolic deficiency).
- You have diabetes. Drug may affect urine sugar tests.

Over age 60:
Adverse reactions and side effects may be more frequent and severe than in younger persons.

Pregnancy:
Decide with your doctor if drug benefits justify risk to unborn baby. Risk category B (see page xviii).

Breastfeeding; Lactation; Nursing Mothers:
This drug is generally acceptable with breastfeeding. Discuss risks and benefits with your doctor.

Infants & children up to age 18:
Don't give to infants younger than 1 month. Follow instructions provided by your child's doctor.

Prolonged use:
- Chest pain, cough, shortness of breath.
- Talk to your doctor about the need for follow-up medical examinations or laboratory studies to check liver function, lung function.

Skin & sunlight:
No problems expected.

Driving, piloting or hazardous work:
Avoid if you feel dizzy or drowsy. Otherwise, no problems expected.

Discontinuing:
Don't discontinue without consulting doctor. Dose may require gradual reduction if you have taken drug for a long time. Doses of other drugs may also require adjustment.

Others:
- Periodic blood counts, liver function tests, and chest x-rays recommended.
- Advise any doctor, dentist or pharmacist whom you consult that you take this drug.

POSSIBLE INTERACTION WITH OTHER DRUGS

GENERIC NAME OR DRUG CLASS	COMBINED EFFECT
Antivirals, HIV/AIDS*	Increased risk of pancreatitis and peripheral neuropathy.
Hemolytics,* other	Increased risk of toxicity.
Nalidixic acid	Decreased nitrofurantoin effect.
Neurotoxic medicines*	Increased risk of damage to nerve cells.
Probenecid	Increased nitrofurantoin effect.
Sulfinpyrazone	Possible nitrofurantoin toxicity.

POSSIBLE INTERACTION WITH OTHER SUBSTANCES

INTERACTS WITH	COMBINED EFFECT
Alcohol:	Possible disulfiram reaction.* Avoid.
Beverages:	None expected.
Cocaine:	Unknown. Avoid.
Foods:	None expected.
Marijuana:	Consult doctor.
Tobacco:	None expected.

NITROIMIDAZOLES

GENERIC AND BRAND NAMES

METRONIDAZOLE
Apo-Metronidazole
Flagyl
Helidac
Metizol
Metric 21
Metro Cream
MetroGel
MetroGel-Vaginal
Neo-Metric
Noritate

METRONIDAZOLE (con't)
Novonidazol
PMS Metronidazole
Protostat
Pylera
Satric
Trikacide
Vandazole

TINIDAZOLE
Tindamax

BASIC INFORMATION

Habit forming? No
Prescription needed? Yes
Available as generic? Yes, for some
Drug class: Antiprotozoal; antibacterial

USES

- Treatment for parasitic infections (such as amebiasis, trichomoniasis or giardiasis).
- Treatment for certain bacterial infections.
- Treatment (combined with other drugs) for ulcer caused by *Helicobacter pylori* infection.
- Topical form treats acne rosacea.
- Vaginal forms treats vaginal infections.
- Other uses as determined by your doctor.

DOSAGE & USAGE INFORMATION

How to take:

- Tablet or capsule—Swallow with liquid. Take with food to lessen stomach irritation. The metronidazole tablet may be crushed and taken with food. Ask your pharmacist about crushing tinidazole tablet.
- Extended-release tablet—Swallow with liquid. Take with food to lessen stomach irritation. Do not crush or crumble tablet.
- Cream—Apply to affected area.
- Vaginal form—Follow instructions provided.

Continued next column

When to take:

- Oral form—At the same times each day.
- Topical cream—Apply twice a day.
- Vaginal cream, gel or tablet—Use as directed.

If you forget a dose:
Take as soon as you remember. If it is almost time for the next dose, wait for that dose (don't double this dose) and resume regular schedule.

What drug does:
Kills or stops the growth of protozoa or bacteria.

Time lapse before drug works:
Depends on the infection. May take one day or up to 14 days or longer for a complete cure.

Don't take with:

- Drugs (such as some cough remedies) or supplements containing alcohol.
- Any other medicine or any dietary supplement without consulting your doctor or pharmacist.

OVERDOSE

SYMPTOMS:
Nausea, vomiting, problem with coordination.
WHAT TO DO:
If person uses much larger amount than prescribed or if accidentally swallowed, call poison control center 1-800-222-1222 for instructions or dial 911 (emergency) for help.

POSSIBLE ADVERSE REACTIONS OR SIDE EFFECTS

SYMPTOMS	WHAT TO DO
Life-threatening:	
Rare, severe allergic reaction (anaphylaxis)—difficulty breathing, hives, itching, swelling (throat, face), vomiting, dizziness.	Seek emergency treatment immediately.
Common:	
Appetite loss, nausea, stomach pain or cramps, diarrhea, vomiting, lightheaded, mild dizziness.	Continue. Call doctor when convenient.
Infrequent:	
• Numbness, tingling, weakness or pain in hands or feet (peripheral neuropathy).	Discontinue. Call doctor right away.
• Metallic or bitter taste, dry mouth, headache.	Continue. Call doctor when convenient.
• Dark urine (will go away after treatment).	No action necessary.
Rare:	
Unsteadiness or clumsiness, mood or mental changes, skin symptoms (rash, hives, redness, itching), sore throat and fever, seizures, severe stomach or back pain, unusual bleeding or bruising, vaginal irritation or dryness or discharge, changes in urination (painful, frequent, unable to control or decreased).	Discontinue. Call doctor right away.

WARNINGS & PRECAUTIONS

Don't take if:
You are allergic to nitroimidazoles or in the first trimester of pregnancy.

Before you start, consult your doctor if:
- You have seizure, nervous system, or brain disorder.
- You have a stomach or intestinal disorder.
- You have a blood disorder.
- You have liver, kidney or heart disease.
- You have a history of alcoholism.

Over age 60:
No special problems expected.

Pregnancy:
Decide with your doctor if drug benefits justify any possible risk to unborn child. The drugs are not recommended during the first trimester of pregnancy. Metronidazole is risk category B and tinidazole is category C (see page xviii).

Breastfeeding; Lactation; Nursing Mothers:
Drugs in this group are generally acceptable with breastfeeding. Discuss risks and benefits with your doctor.

Infants & children up to age 18:
Follow instructions provided by your child's doctor.

Prolonged use:
These drugs are not intended for long-term use.

Skin & sunlight:
No problems expected.

Driving, piloting or hazardous work:
Don't drive or pilot aircraft until you learn how medicine affects you. Don't work around dangerous machinery. Don't climb ladders or work in high places. Danger increases if you drink alcohol or take other medicines affecting alertness and reflexes.

Discontinuing:
Don't discontinue without doctor's advice until you complete prescribed dose, even though symptoms diminish or disappear.

Others:
- Advise any doctor, dentist or pharmacist whom you consult that you take this medicine. May affect results in some medical tests.
- Contact your doctor if symptoms do not improve in 24 to 48 hours.
- If you are being treated for trichomoniasis, your doctor may want your sexual partner treated at the same time (even if the person has no symptoms).
- These drugs may cause a yeast infection (oral or vaginal) to worsen. Contact your doctor if this occurs.

POSSIBLE INTERACTION WITH OTHER DRUGS

GENERIC NAME OR DRUG CLASS	COMBINED EFFECT
Anticoagulants,* oral	Increased anti-coagulant effect. (bleeding or bruising).
Antivirals, HIV/AIDS*	Increased risk of peripheral neuropathy.*
Cholestyramine	Decreased effect of nitroimidazole. Take several hours apart.
Cyclosporine	Increased cyclosporine effect.
Disulfiram	Disulfiram reaction.* Take 2 weeks apart.
Enzyme inducers*	Decreased effect of nitroimidazole.
Enzyme inhibitors*	Increased effect of nitroimidazole.
Fluorouracil	Increased risk of side effects.
Lithium	Increased lithium effect.
Neurotoxic medications*	Increased risk of side effects.
Oxytetracycline	Decreased metronidazole effect.
Phenytoin	Increased phenytoin effect. Decreased nitroimidazole effect.
Tacrolimus	Increased tacrolimus effect.

POSSIBLE INTERACTION WITH OTHER SUBSTANCES

INTERACTS WITH	COMBINED EFFECT
Alcohol:	Possible disulfiram reaction.* Avoid alcohol while taking drug and for 3 days after finishing.
Beverages:	None expected.
Cocaine:	Unknown. Avoid.
Foods:	None expected.
Marijuana:	Consult doctor.
Tobacco:	None expected.

*See Glossary

NON-NUCLEOSIDE REVERSE TRANSCRIPTASE INHIBITORS

GENERIC AND BRAND NAMES

DELAVIRDINE
- **Rescriptor**

EFAVIRENZ
- **Sustiva**
- **Atripla**

ETRAVIRINE
- **Intelence**

NEVIRAPINE
- **Viramune**
- **Viramune XR**

RILPIVIRINE
- **Complera**
- **Edurant**

BASIC INFORMATION

Habit forming? No
Prescription needed? Yes
Available as generic? Yes, for some
Drug class: Antiviral, HIV and AIDS

USES

For treatment of HIV and AIDS patients. Used in combination with one or more of the other AIDS drugs. May be used to prevent HIV transmission.

DOSAGE & USAGE INFORMATION

How to take:
- Oral suspension—Swallow with liquid.
- Tablet—Swallow with liquid. May be taken with or without food. Efavirenz should not be taken with a high fat meal.
- Extended-release tablet—Swallow whole with liquid. Do not crush, chew or divide tablet. May be taken with or without food.

When to take:
At the same time each day. At the start of treatment, one tablet is taken daily for 2 weeks and then increased to 2 tablets daily. This helps to decrease risk of side effects.

Continued next column

If you forget a dose:
Take as soon as you remember. If it is almost time for the next dose, wait for that dose (don't double this dose) and resume regular schedule.

What drug does:
Interferes with HIV replication. It helps slow the progress of HIV disease, but does not cure it.

Time lapse before drug works:
May require several weeks or months before full benefits are apparent.

Don't take with:
Any other medicine or any dietary supplement without consulting your doctor or pharmacist. This is very important with these antiviral drugs.

OVERDOSE

SYMPTOMS:
Unknown for some drugs in this group. May have swelling, tiredness, nausea, vomiting, fever, insomnia, rash, dizziness, weight loss.
WHAT TO DO:
If person uses much larger amount than prescribed or if accidentally swallowed, call poison control center 1-800-222-1222 for instructions or dial 911 (emergency) for help.

POSSIBLE ADVERSE REACTIONS OR SIDE EFFECTS

SYMPTOMS	WHAT TO DO
Life-threatening:	
Rare, severe allergic reaction (anaphylaxis)—difficulty breathing, hives, itching, swelling (throat, face), vomiting, dizziness.	Seek emergency treatment immediately.
Common:	
Mild to moderate skin rash, chills, fever, sore throat.	Discontinue. Call doctor right away.
Infrequent:	
• Fever, blistering skin, mouth sores, aching joints or muscles, eye inflammation, unusual tiredness.	Continue, but call doctor right away.
• Headache, nausea, diarrhea, burning or tingling feeling, numbness, sleepiness, loss of appetite, constipation, mood or mental changes, intense dreams, anxiety, difficulty concentrating.	Continue. Call doctor when convenient.
Rare:	
• Yellow skin or eyes, dark urine, heart palpitations, thoughts of suicide.	Discontinue. Call doctor right away.
• Any other unusual symptoms that occur (may be due to the illness, this drug or other drugs taken).	Continue, but call doctor right away.

NON-NUCLEOSIDE REVERSE TRANSCRIPTASE INHIBITORS

WARNINGS & PRECAUTIONS

Don't take if:
You are allergic to non-nucleoside reverse transcriptase inhibitors.

Before you start, consult your doctor if:
- You are allergic to any medicine, food or other substance, or have a family history of allergies.
- You have kidney or liver disease.

Over age 60:
Adverse reactions and side effects may be more frequent and severe than in younger persons.

Pregnancy:
Risk factors vary for drugs in this group. Always consult doctor. See category list on page xviii.

Breastfeeding; Lactation; Nursing Mothers:
It is not recommended that HIV-infected mothers breastfeed. Consult your doctor.

Infants & children up to age 18:
Follow instructions provided by your child's doctor.

Prolonged use:
Talk to your doctor about frequent blood counts and liver function studies.

Skin & sunlight:
No problems expected.

Driving, piloting or hazardous work:
Don't drive or pilot aircraft until you learn how medicine affects you. Don't work around dangerous machinery. Don't climb ladders or work in high places. Danger increases if you drink alcohol or take medicine affecting alertness and reflexes.

Discontinuing:
Don't discontinue without doctor's advice until you complete prescribed dose, even though symptoms diminish or disappear.

Others:
- Advise any doctor, dentist or pharmacist whom you consult that you take this medicine.
- Taking this drug does not prevent you from passing HIV to another person through sexual contact or sharing needles. Avoid sexual contact or practice safe sex (e.g., using condoms) to help prevent the transmission of HIV. Never share or re-use needles. If you have questions, ask your doctor for advice.
- Consult your doctor right away if you develop a new infection (e.g., fever, chills, sore throat or other symptoms).
- Do not increase or decrease dosage of drug without doctor's approval.

POSSIBLE INTERACTION WITH OTHER DRUGS

GENERIC NAME OR DRUG CLASS	COMBINED EFFECT
Amphetamines	May require dosage adjustment of amphetamine.
Antacids	Take 1-2 hours apart.
Benzodiazepines	May require dosage adjustment of benzodiazepine.
Calcium channel blockers	May require dosage adjustment of calcium channel blocker.
Carbamazepine	Decreased antiviral drug effect.
Clarithromycin	Interaction effects vary.
Contraceptives, oral*	Decreased contraceptive effect. Use alternative birth control method.
Didanosine	Take at least 1 hour apart.
Ergot preparations*	May require dosage adjustment of ergot drug.
Fluoxetine	Increased antiviral effect.
Histamine H_2 receptor antagonists	May require dosage adjustment of antiviral drug.

POSSIBLE INTERACTION WITH OTHER SUBSTANCES

INTERACTS WITH	COMBINED EFFECT
Alcohol:	None expected.
Beverages:	None expected.
Cocaine:	Unknown. Avoid.
Foods:	None expected.
Marijuana:	Consult doctor.
Tobacco:	None expected.

*See Glossary

NUCLEOSIDE REVERSE TRANSCRIPTASE INHIBITORS

GENERIC AND BRAND NAMES

ABACAVIR
- **Epzicom**
- **Triumeq**
- **Trizivir**
- **Ziagen**

DIDANOSINE
- **Videx**
- **Videx EC**

EMTRICITABINE
- **Atripla**
- **Complera**
- **Emtriva**
- **Stribild**
- **Truvada**

LAMIVUDINE
- **Dutrebis**
- **Epivir**
- **Epzicom**
- **Triumeq**
- **Trizivir**

STAVUDINE
- **Zerit**

ZIDOVUDINE
- **Apo-Zidovudine**
- **AZT**
- **Combivir**
- **Novo-AZT**
- **Retrovir**
- **Trizivir**
- **Combivir**

BASIC INFORMATION

Habit forming? No
Prescription needed? Yes
Available as generic? Yes, for some
Drug class: Antiviral

USES

Treats human immunodeficiency virus (HIV) and acquired immunodeficiency syndrome (AIDS).

DOSAGE & USAGE INFORMATION

How to take:
- Tablet or capsule—Swallow with water. Take with or without food as directed.
- Didanosine tablet—Chew or manually crumble tablet. If you crumble tablet, mix with at least 1 oz. of water. Swallow right away. Take 1 hour before or 2 hours after meal.
- Syrup—Use marked measuring device.

Continued next column

- Buffered didanosine for oral solution—Follow instructions on label. Swallow immediately.

When to take:
At the same time each day, according to instructions on prescription label. Follow directions on label for taking with or without food.

If you forget a dose:
Take as soon as you remember. If it is almost time for the next dose, wait for that dose (don't double this dose) and resume regular schedule.

What drug does:
Suppresses replication of human immunodeficiency virus.

Time lapse before drug works:
Depends on the progress of the disease.

Don't take with:
Any other medicine or any dietary supplement without consulting your doctor or pharmacist.

OVERDOSE

SYMPTOMS:
Specific symptoms for drugs in this group have not been identified.
WHAT TO DO:
- **Dial 911 for all medical emergencies or call poison control center 1-800-222-1222 for instructions.**
- **See emergency information on last 3 pages of this book.**

POSSIBLE ADVERSE REACTIONS OR SIDE EFFECTS

SYMPTOMS	WHAT TO DO
Life-threatening:	
Rare, severe allergic reaction (anaphylaxis)—difficulty breathing, hives, itching, swelling (throat, face), vomiting, dizziness.	Seek emergency treatment immediately.
Common:	
• Tingling or numbness and burning in the feet and ankles.	Discontinue. Call doctor right away.
• Headache, anxiety, restlessness, digestive problems, diarrhea.	Continue. Call doctor when convenient.
Infrequent:	
• Unusual tiredness and weakness, fever, chills, sore throat, unusual bleeding or bruising, yellow skin and eyes, skin rash, pale skin, muscle or joint pain, mouth or throat sores, stomach pain, nausea and vomiting.	Discontinue. Call doctor right away.
• Lack of strength or energy, insomnia, discolored nails.	Continue. Call doctor when convenient.
Rare:	
Seizures, mood or mental changes, confusion.	Discontinue. Call doctor right away.

NUCLEOSIDE REVERSE TRANSCRIPTASE INHIBITORS

WARNINGS & PRECAUTIONS

Don't take if:
You are allergic to antivirals for HIV and AIDS.

Before you start, consult your doctor if:
- You have a history of alcoholism.
- You have hypertriglyceridemia, anemia, liver or kidney disease, gout, phenylketonuria, peripheral neuropathy or pancreatitis.
- You have a condition that limits sodium intake.
- A gene test is needed before starting abacavir.

Over age 60:
No special problems expected.

Pregnancy:
Risk factors vary for drugs in this group. Always consult doctor. See category list on page xviii.

Breastfeeding; Lactation; Nursing Mothers:
It is not recommended that HIV-infected mothers breastfeed. Consult your doctor.

Infants & children up to age 18:
Follow instructions provided by your child's doctor.

Prolonged use:
Talk with your doctor about the need for follow-up medical examination or laboratory studies to check blood serum and uric acid levels.

Skin & sunlight:
No special problems expected.

Driving, piloting or hazardous work:
Don't drive or pilot aircraft until you learn how medicine affects you. Don't work around danger-ous machinery. Don't climb ladders or work in high places. Danger increases if you drink alcohol or take medicine affecting alertness and reflexes.

Discontinuing:
Consult doctor before stopping. Dose may need gradual reduction if drug taken for a long time. Doses of other drugs may need adjustment.

Others:
- Advise any doctor, dentist or pharmacist whom you consult that you take this medicine.
- Taking this drug does not prevent you from passing HIV to another person through sexual contact or sharing needles. Avoid sexual contact or practice safe sex (e.g., using condoms) to help prevent the transmission of HIV. Never share or re-use needles. If you have questions, ask your doctor for advice.
- Consult your doctor right away if you develop a new infection (e.g., fever, chills, sore throat or other symptoms).
- Severe liver complications (non-cirrhotic portal hypertension) may occur with didanosine use. Consult doctor about your risks.
- Cobicistat* (Tybost), a drug enhancer, is combined in some of the brand name drugs.
- Do not increase or decrease dosage of drug without doctor's approval.

POSSIBLE INTERACTION WITH OTHER SUBSTANCES

GENERIC NAME OR DRUG CLASS	COMBINED EFFECT
Bone marrow depressants,* other	Drugs may need dosage changes.
Cimetidine	Increased effect of zidovudine.
Clarithromycin	Decreased effect of zidovudine.
Dapsone	Risk of peripheral neuropathy. Decreased effect of both drugs.
Fluoroquinolones	Decreased effect of fluoroquinolone.
Ganciclovir	Increased toxicity of both drugs. Use with caution.
Itraconazole	Decreased absorption of itraconazole.
Ketoconazole	Decreased effect of ketoconazole.
Pancreatitis-associated drugs*	Increased risk of pancreatitis with didanosine.
Peripheral neuropathy-associated drugs*	Increased risk of peripheral neuropathy.

Continued on page 907

POSSIBLE INTERACTION WITH OTHER SUBSTANCES

INTERACTS WITH	COMBINED EFFECT
Alcohol:	Risk of adverse effects. Avoid.
Beverages:	None expected.
Cocaine:	Unknown. Avoid.
Foods:	None expected.
Marijuana:	Consult doctor.
Tobacco:	None expected.

***See Glossary**

NUCLEOTIDE REVERSE TRANSCRIPTASE INHIBITORS

GENERIC AND BRAND NAMES

TENOFOVIR

Atripla
Complera
Stribild
Truvada
Viread

BASIC INFORMATION

Habit forming? No
Prescription needed? Yes
Available as generic? No
Drug class: Antiviral

USES

- Treats human immunodeficiency virus (HIV) and acquired immunodeficiency syndrome (AIDS). Used in combination with other antiretroviral agents. Does not cure or prevent HIV or AIDS.
- Treatment of chronic hepatitis B.

DOSAGE & USAGE INFORMATION

How to take:
- Tablets—Swallow with water. Take with or without food as directed.
- Oral powder—Follow instructions for product.

When to take:
At the same time each day, according to instructions on prescription label.

If you forget a dose:
Take as soon as you remember. If it is almost time for the next dose, wait for that dose (don't double this dose) and resume regular schedule.

What drug does:
Suppresses replication of human immunodeficiency virus.

Continued next column

OVERDOSE

SYMPTOMS:
May have more severe forms of adverse reaction symptoms.
WHAT TO DO:
- **Dial 911 for all medical emergencies or call poison control center 1-800-222-1222 for instructions.**
- **See emergency information on last 3 pages of this book.**

Time lapse before drug works:
Depends on the progress of the disease.

Don't take with:
Any other medicine or any dietary supplement without consulting your doctor or pharmacist.

POSSIBLE ADVERSE REACTIONS OR SIDE EFFECTS

SYMPTOMS	WHAT TO DO
Life-threatening:	
In case of overdose, see previous column.	
Common:	
Vomiting, lack or loss of strength.	Continue. Call doctor when convenient.
Infrequent:	
Gaseousness, weight loss, diarrhea.	Continue. Call doctor when convenient.
Rare:	
Breathing is fast or shallow, shortness of breath, unusual tiredness, sleepiness, stomach discomfort, loss of appetite, muscle cramping or pain, overall feeling of discomfort, liver problems (nausea, stomach pain, loss of appetite, low fever, yellow skin and eyes, pale stools, dark urine), any unusual symptoms.	Discontinue. Call doctor right away or seek emergency care.

WARNINGS & PRECAUTIONS

Don't take if:
You are allergic to tenofovir. Allergic reactions are rare, but may occur.

Before you start, consult your doctor if:
You have liver or kidney disease.

Over age 60:
Adverse reactions and side effects may be more frequent and severe than in younger persons.

Pregnancy:
Decide with your doctor if drug benefits justify risk to unborn baby. Risk category B (see page xviii).

Breastfeeding; Lactation; Nursing Mothers:
It is not recommended that HIV-infected mothers breastfeed. Consult your doctor.

Infants & children up to age 18:
Follow instructions provided by your child's doctor.

Prolonged use:
- Talk with your doctor about the need for follow-up medical examination or laboratory studies to check drug's effectiveness.
- Long-term effects of this drug are unknown. Studies are ongoing.

Skin & sunlight:
No special problems expected.

Driving, piloting or hazardous work:
No problems expected.

Discontinuing:
Don't discontinue without consulting doctor.

Others:
- Advise any doctor, dentist or pharmacist whom you consult that you take this medicine.
- Taking this drug does not prevent you from passing HIV to another person through sexual contact or sharing needles. Avoid sexual contact or practice safe sex (e.g., using condoms) to help prevent the transmission of HIV. Never share or re-use needles. If you have questions, ask your doctor for advice.
- Consult your doctor right away if you develop a new infection (e.g., fever, chills, sore throat or other symptoms).
- Cobicistat* (Tybost), a drug enhancer, is combined in some of the brand name drugs.
- Do not increase or decrease dosage of drug without doctor's approval.

POSSIBLE INTERACTION WITH OTHER DRUGS

GENERIC NAME OR DRUG CLASS	COMBINED EFFECT
Antivirals for herpes virus	Increased effect of tenofovir.
Didanosine	Increased effect of didanosine. Take tenofovir 2 hours before or 1 hour after didanosine.

POSSIBLE INTERACTION WITH OTHER SUBSTANCES

INTERACTS WITH	COMBINED EFFECT
Alcohol:	None expected.
Beverages:	None expected.
Cocaine:	Unknown. Avoid.
Foods:	None expected.
Marijuana:	Consult doctor.
Tobacco:	None expected.

*See Glossary

NYSTATIN

BRAND NAMES

Dermacomb
Myco II
Mycobiotic II
Mycogen II
Mycolog II
Mycostatin
Myco-Triacet II
Mykacet
Mykacet II
Mytrex
Nadostine
Nilstat
Nystaform
Nystex
Tristatin II

BASIC INFORMATION

Habit forming? No
Prescription needed? Yes
Available as generic? Yes
Drug class: Antifungal

USES

Treatment of fungus infections of the mouth or vagina that are susceptible to nystatin.

DOSAGE & USAGE INFORMATION

How to take:

- Tablet—Swallow with liquid. May take with or without food.
- Ointment, cream, lotion or powder—Use as directed by doctor and label.
- Liquid or powder for oral suspension—Take as directed. Instruction varies by preparation.
- Lozenge—Take as directed on label.

When to take:
At the same time each day.

If you forget a dose:
Take as soon as you remember. If it is almost time for the next dose, wait for that dose (don't double this dose) and resume regular schedule.

Continued next column

OVERDOSE

SYMPTOMS:
May cause nausea, vomiting, diarrhea.
WHAT TO DO:
If person uses much larger amount than prescribed or if accidentally swallowed, call poison control center 1-800-222-1222 for instructions or dial 911 (emergency) for help.

What drug does:
Prevents growth and reproduction of fungus.

Time lapse before drug works:
Begins immediately. May require 3 weeks for maximum benefit, depending on location and severity of infection.

Don't take with:
Any other medicine or any dietary supplement without consulting your doctor or pharmacist.

POSSIBLE ADVERSE REACTIONS OR SIDE EFFECTS

SYMPTOMS	WHAT TO DO
Life-threatening:	
None expected.	
Common:	
(at high doses) Nausea, stomach pain, vomiting, diarrhea.	Discontinue. Call doctor right away.
Infrequent:	
Mild irritation or itch at application site.	Continue. Call doctor if symptoms persist.
Rare:	
None expected.	

WARNINGS & PRECAUTIONS

Don't take if:
You are allergic to nystatin. Allergic reactions are rare, but may occur.

Before you start, consult your doctor if:
You have any chronic illness.

Over age 60:
No special problems expected.

Pregnancy:
Decide with your doctor if drug benefits justify risk to unborn baby. Risk category C (see page xviii).

Breastfeeding; Lactation; Nursing Mothers:
This drug is generally acceptable with breastfeeding. Discuss risks and benefits with your doctor.

Infants & children up to age 18:
Follow instructions provided by your child's doctor.

Prolonged use:
No problems expected.

Skin & sunlight:
No problems expected.

Driving, piloting or hazardous work:
No problems expected.

Discontinuing:
Don't discontinue without doctor's advice until you complete prescribed dose, even though symptoms diminish or disappear.

Others:
No problems expected.

POSSIBLE INTERACTION WITH OTHER DRUGS

GENERIC NAME OR DRUG CLASS	COMBINED EFFECT
None reported.	

POSSIBLE INTERACTION WITH OTHER SUBSTANCES

INTERACTS WITH	COMBINED EFFECT
Alcohol:	None expected.
Beverages:	None expected.
Cocaine:	Unknown. Avoid.
Foods:	None expected.
Marijuana:	Consult doctor.
Tobacco:	None expected.

***See Glossary**

OLANZAPINE

BRAND NAMES

Symbyax
Zyprexa
Zyprexa Relprevv
Zyprexa Zydis

BASIC INFORMATION

Habit forming? No
Prescription needed? Yes
Available as generic? Yes
Drug class: Antipsychotic

USES

- Treatment for symptoms of schizophrenia, acute mania and other psychotic disorders.
- Treatment for bipolar disorder.

DOSAGE & USAGE INFORMATION

How to take:
- Tablet or capsule—Swallow with liquid. May be taken with or without food.
- Oral disintegrating tablet—Dissolve in mouth.
- Injection—Given by health care professional.

When to take:
- Oral—once a day at the same time each day.
- Injection—Given every 2 to 4 weeks. Must wait in medical office 3 hours after each dose.

If you forget a dose:
Take as soon as you remember. If it is almost time for the next dose, wait for that dose (don't double this dose) and resume regular schedule.

What drug does:
The exact mechanism is unknown. It appears to relieve symptoms by blocking certain nerve impulses between nerve cells.

Time lapse before drug works:
One to 7 days. Further increases in dosage may be needed to relieve symptoms in some patients.

Don't take with:
Any other medicine or any dietary supplement without consulting your doctor or pharmacist.

OVERDOSE

SYMPTOMS:
Drowsiness, slurred speech, rapid heartbeat, agitation, severe sweating, rigid muscles, breathing difficulty, stupor, seizures, coma.
WHAT TO DO:
- **Dial 911 for all medical emergencies or call poison control center 1-800-222-1222 for instructions.**
- **See emergency information on last 3 pages of this book.**

POSSIBLE ADVERSE REACTIONS OR SIDE EFFECTS

SYMPTOMS	WHAT TO DO
Life-threatening:	
Rare—neuroleptic malignant syndrome (high fever, sweating, muscle cramps and rigidity, mental changes, seizures, fast heartbeat, coma).	Discontinue. Seek emergency treatment.
Common:	
• Dizziness, difficulty in speaking or swallowing, shaking hands and fingers, trembling, vision problems, weakness, lightheadedness when arising from a sitting or lying position.	Continue, but call doctor right away.
• Drowsiness, constipation, weight gain, agitation, insomnia, headache, nervousness, runny nose, anxiety, dry mouth, arm or leg stiffness, injection site redness or swelling.	Continue. Call doctor when convenient.
Infrequent:	
• Jerky or involuntary movements (in face, lips, jaw, tongue), fast heartbeat, chest pain.	Continue, but call doctor right away.
• Fever, flu-like symptoms, twitching, mood or mental changes, speech unclear, swollen feet or ankles, appetite increased, cough, saliva increased, muscle tightness, muscle spasms (face, neck, back), joint pain, nausea, vomiting, sore throat, incontinence, abdominal pain.	Continue. Call doctor when convenient.
Rare:	
• New symptoms occur after drug injection.	Call doctor right away.
• Breathing difficulty, high blood sugar (thirstiness, frequent urination, increased hunger, weakness), swollen face, rash, confusion.	Discontinue. Call doctor right away.
• Decreased sex drive, menstrual changes, sluggishness.	Continue. Call doctor when convenient.

WARNINGS & PRECAUTIONS

Don't take if:
You are allergic to olanzapine. Allergic reactions are rare, but may occur.

Before you start, consult your doctor if:
- You have liver disease, heart disease, a blood vessel disorder or history of seizures.
- You have a history of breast cancer.
- You have intestinal blockage.
- You are subject to dehydration or low body temperature.
- You have a history of drug abuse/ dependence.
- You have a family history of, are at risk for, or have diabetes.
- You have glaucoma or prostate problems.
- Patient has Alzheimer's.
- You are allergic to any medication, food or other substance.

Over age 60:
- Adverse reactions and side effects may be more severe. A lower starting dose is usually recommended until a response is determined.
- Use of antipsychotic drugs in elderly patients with dementia-related psychosis may increase risk of death. Consult doctor.

Pregnancy:
Decide with your doctor if drug benefits justify risk to unborn baby. Risk category C (see page xviii).

Breastfeeding; Lactation; Nursing Mothers:
This drug is generally acceptable with breastfeeding. Discuss risks and benefits with your doctor.

Infants & children up to age 18:
Not approved for ages under 13. If prescribed, carefully read information provided with prescription. Contact doctor right away if depression symptoms get worse or there is any talk of suicide or suicide behaviors. Also, read information under Others.

Prolonged use:
Consult with your doctor on a regular basis while taking this drug to check your progress or to discuss any increase or changes in side effects and the need for continued treatment.

Skin & sunlight:
May cause rash or intensify sunburn in areas exposed to sun or ultraviolet light (photosensitivity reaction). Use sunscreen and avoid over-exposure. Notify doctor if reaction occurs.

Driving, piloting or hazardous work:
Don't drive or pilot aircraft until you learn how medicine affects you. Don't work around dangerous machinery. Don't climb ladders or work in high places. Danger increases if you drink alcohol or take drug affecting alertness.

Discontinuing:
Don't discontinue this drug without consulting doctor. Dosage may require a gradual reduction before stopping.

Others:
- Get up slowly from a sitting or lying position to avoid any dizziness or lightheadedness.
- Hot temperatures, exercise, and hot baths can increase risk of heatstroke. Drug may affect body's ability to maintain normal temperature.
- Advise any doctor, dentist or pharmacist whom you consult that you take this medicine.
- Take drug only as directed. Do not increase or reduce dosage without doctor's approval.
- Adults and children taking antidepressants may experience a worsening of the depression symptoms and may have increased suicidal thoughts or behaviors. Call doctor right away if these symptoms or behaviors occur.
- Injected form has risk of PDSS (post-injection delirium/sedation syndrome). Get medical help if severe drowsiness (may be unconscious or in a coma), confusion or disorientation occurs.

POSSIBLE INTERACTION WITH OTHER DRUGS

GENERIC NAME OR DRUG CLASS	COMBINED EFFECT
Anticholinergics,* other	Increased risk of side effects.
Antihypertensives*	Increased effect of antihypertensive.
Carbamazepine	Decreased effect of olanzapine.
Enzyme inducers*	May decrease olanzapine effect.
Enzyme inhibitors*	May increase olanzapine effect.

Continued on page 907

POSSIBLE INTERACTION WITH OTHER SUBSTANCES

INTERACTS WITH	COMBINED EFFECT
Alcohol:	Increased sedation and dizziness. Avoid.
Beverages:	None expected.
Cocaine:	Depression. Avoid.
Foods:	None expected.
Marijuana:	Depression. Avoid.
Tobacco:	Decreased effect of olanzapine. Avoid.

*See Glossary

OLSALAZINE

BRAND NAMES

Dipentum

BASIC INFORMATION

Habit forming? No
Prescription needed? Yes
Available as generic? No
Drug class: Inflammatory bowel disease suppressant

USES

- Treatment for ulcerative colitis.
- Treatment for other disorders as determined by your doctor.

DOSAGE & USAGE INFORMATION

How to take:
Capsule—Swallow with liquid or food to lessen stomach irritation. If you can't swallow whole, open capsule and take with liquid or food.

When to take:
Usually twice a day, or as directed on prescription label.

If you forget a dose:
Take as soon as you remember. If it is almost time for the next dose, wait for that dose (don't double this dose) and resume regular schedule.

What drug does:
It reduces inflammation inside the bowel which helps control symptoms.

Time lapse before drug works:
1 hour.

Don't take with:
Any other medicine or any dietary supplement without consulting your doctor or pharmacist.

OVERDOSE

SYMPTOMS:
Unknown.
WHAT TO DO:
If person uses much larger amount than prescribed or if accidentally swallowed, call poison control center 1-800-222-1222 for instructions or dial 911 (emergency) for help.

POSSIBLE ADVERSE REACTIONS OR SIDE EFFECTS

SYMPTOMS	WHAT TO DO
Life-threatening:	
Fever, sore throat, paleness, unusual bleeding (representing effect on blood, a rare complication).	Seek emergency treatment.
Common:	
Diarrhea, appetite loss, nausea, vomiting.	Continue. Call doctor when convenient.
Infrequent:	
Mood changes, sleeplessness, headache.	Continue. Call doctor when convenient.
Rare:	
Skin eruption (acne-like), muscle aches.	Continue. Call doctor when convenient.

WARNINGS & PRECAUTIONS

Don't take if:
You are allergic to olsalazine, mesalamine, aspirin or any other salicylate. Allergic reactions are rare, but may occur.

Before you start, consult your doctor if:
- You have kidney disease.
- You are taking any other prescription or nonprescription medicine.

Over age 60:
No special problems expected.

Pregnancy:
Decide with your doctor if drug benefits justify risk to unborn baby. Risk category C (see page xviii).

Breastfeeding; Lactation; Nursing Mothers:
This drug is generally acceptable with breastfeeding. Discuss risks and benefits with your doctor.

Infants & children up to age 18:
Safety and effectiveness in this age group have not been established. Consult doctor.

Prolonged use:
Follow through with full prescribed course of treatment.

Skin & sunlight:
No special problems expected.

Driving, piloting or hazardous work:
Don't drive or pilot aircraft until you learn how medicine affects you. Don't work around dangerous machinery. Don't climb ladders or work in high places. Danger increases if you drink alcohol or take medicine affecting alertness and reflexes.

Discontinuing:
Don't discontinue without consulting doctor. Dose may require gradual reduction if you have taken drug for a long time. Doses of other drugs may also require adjustment.

Others:
Request your doctor to check blood counts and kidney function on a regular basis.

POSSIBLE INTERACTION WITH OTHER DRUGS

GENERIC NAME OR DRUG CLASS	COMBINED EFFECT
None expected since olsalazine is not appreciably absorbed from the gastrointestinal tract into the bloodstream.	

POSSIBLE INTERACTION WITH OTHER SUBSTANCES

INTERACTS WITH	COMBINED EFFECT
Alcohol:	Increased risk of gastrointestinal upset and/or bleeding.
Beverages: Highly spiced beverages.	Will irritate underlying condition that olsalazine treats.
Cocaine:	Unknown. Avoid.
Foods: Highly spiced foods.	Will irritate underlying condition that olsalazine treats.
Marijuana:	Consult doctor.
Tobacco:	Will irritate underlying condition that olsalazine treats. Avoid.

OMEGA-3 FATTY ACIDS

GENERIC AND BRAND NAMES

ICOSAPENT ETHYL
Vascepa
OMEGA-3-ACID ETHYL ESTERS
Lovaza
Omtryg
OMEGA-3-CARBOXYLIC ACIDS
Epanova

BASIC INFORMATION

Habit forming? No
Prescription needed? Yes
Available as generic? Yes, for some
Drug class: Antihyperlipidemic

USES

- Helps to lower high triglyceride (fat-like substance) levels in the blood. Its use is recommended in combination with an appropriate heart-healthy diet and exercise program.
- May be used for treatment of other disorders as determined by your doctor.
- The information in this chart pertains only to the prescription form of these products. Numerous nonprescription products are available, but brand quality may vary.

DOSAGE & USAGE INFORMATION

How to take:
Capsule—Swallow whole with liquid. Do not break open, chew or crush capsule. It is advised that it be taken with food.

When to take:
It may be taken 1 to 2 times a day. Follow the instructions on your prescription.

If you forget a dose:
Take as soon as you remember. If it is almost time for the next dose, wait for that dose (don't double this dose) and resume regular schedule.

Continued next column

OVERDOSE

SYMPTOMS:
Unknown.
WHAT TO DO:
If person uses much larger amount than prescribed or if accidentally swallowed, call poison control center 1-800-222-1222 for instructions or dial 911 (emergency) for help.

What drug does:
Omega-3 fatty acids are derived from fish oil. They help reduce the production of fatty substances (triglycerides) in the liver. The exact way they work is not completely understood. Some may also increase the level of low-density lipoprotein (LDL) cholesterol. LDL is considered the "bad" cholesterol.

Time lapse before drug works:
About two months for measurable results as determined by laboratory testing.

Don't take with:
Any other medicine or any dietary supplement without consulting your doctor or pharmacist.

POSSIBLE ADVERSE REACTIONS OR SIDE EFFECTS

SYMPTOMS	WHAT TO DO
Life-threatening:	
Rare, severe allergic reaction (anaphylaxis)—difficulty breathing, hives, itching, swelling (throat, face), vomiting, dizziness.	Seek emergency treatment immediately.
Common:	
Burping, infection, flu-like symptoms, upset stomach, taste changes, back pain.	Continue. Call doctor when convenient
Infrequent:	
Chest pain.	Discontinue. Call doctor right away.
Rare:	
None expected.	

WARNINGS & PRECAUTIONS

Don't take if:
You are allergic to any omega-3 fatty acids, fish oil or fish.

Before you start, consult your doctor if:
- You have medical problems that may contribute to high triglycerides (e.g., diabetes or hypothyroidism).
- You take drugs such as beta blockers, estrogens or thiazide diuretics. They can be a risk factor for high triglycerides.
- You have liver problems.
- You have a blood clotting disorder.
- You are obese.

Over age 60:
No special problems expected.

Pregnancy:
Decide with your doctor if drug benefits justify risk to unborn baby. Risk category C (see page xviii).

Breastfeeding; Lactation; Nursing Mothers:
This drug is generally acceptable with breastfeeding. Discuss risks and benefits with your doctor.

Infants & children up to age 18:
Safety and effectiveness in this age group have not been established. Consult doctor.

Prolonged use:
No problems expected. Ask your doctor about routine laboratory studies to verify continued effectiveness of the drug therapy and to check your triglyceride levels.

Skin & sunlight:
No problems expected.

Driving, piloting or hazardous work:
No problems expected.

Discontinuing:
No problems expected.

Others:
- Follow your doctor's advice carefully regarding proper diet and an exercise program. Drugs such as this one are usually prescribed only when more help is needed.
- Consult your doctor about routine monitoring of LDL cholesterol levels to be sure they are not increasing excessively.
- Diabetic patients may experience worsening of blood sugar control.
- Advise any doctor, dentist or pharmacist whom you consult that you take this medicine.

POSSIBLE INTERACTION WITH OTHER DRUGS

GENERIC NAME OR DRUG CLASS	COMBINED EFFECT
Anticoagulants, oral*	May increase anticoagulant effect. Consult doctor.

POSSIBLE INTERACTION WITH OTHER SUBSTANCES

INTERACTS WITH	COMBINED EFFECT
Alcohol:	None expected. Decreased alcohol intake is often recommended to help reduce triglyceride levels.
Beverages:	None expected.
Cocaine:	Unknown. Avoid.
Foods:	None expected.
Marijuana:	Consult doctor.
Tobacco:	None expected. Smoking can contribute to high triglycerides.

*See Glossary

OREXIN RECEPTOR ANTAGONISTS

GENERIC AND BRAND NAMES

SUVOREXANT
Belsomra

BASIC INFORMATION

Habit forming? Yes, maybe
Prescription needed? Yes
Available as generic? No
Drug class: Anti-insomnia agent

USES

Treatment of insomnia (difficulty in falling and staying asleep).

DOSAGE & USAGE INFORMATION

How to take:
Tablet—Swallow whole with a small amount of water. Take with or without food (however, drug will work faster on an empty stomach).

When to take:
Take once a night within 30 minutes of going to bed. Allow at least 7 hours before you plan to wake up. Ask your doctor if you should take drug only as needed or on a regular schedule.

If you forget a dose:
Take as soon as you remember. If there is not enough time to get a full night's sleep of 7 hours, skip the missed dose.

What drug does:
It blocks the action of orexins (brain chemicals) that help keep the body awake. Blocking orexins helps a person to fall asleep faster and stay asleep longer.

Time lapse before drug works:
About 30 minutes to one hour.

Don't take with:
Any other medicine or any dietary supplement without consulting your doctor or pharmacist.

OVERDOSE

SYMPTOMS:
Increased sleepiness and possibly other symptoms depending on amount taken.
WHAT TO DO:
- **Dial 911 for all medical emergencies or call poison control center 1-800-222-1222 for instructions.**
- **See emergency information on last 3 pages of this book.**

POSSIBLE ADVERSE REACTIONS OR SIDE EFFECTS

SYMPTOMS	WHAT TO DO
Life-threatening:	
Rare, severe allergic reaction (anaphylaxis)—difficulty breathing, hives, itching, swelling (throat, face), vomiting, dizziness.	Seek emergency treatment immediately.
Common:	
Daytime drowsiness (may interfere with daily activities such as driving), headache.	Discontinue. Call doctor when convenient.
Infrequent:	
Abnormal dreams, aches or pain, cough, dry mouth, upper respiratory infection, tiredness, diarrhea.	Discontinue. Call doctor when convenient.
Rare:	
Changes in behavior, confusion, feeling sad or fearful or nervous or irritable, loss of appetite, trouble concentrating, may be unable to move or talk for up to several minutes while going to sleep or waking up, temporary weakness in legs, worsening of depression, thoughts of suicide, problems with memory, hallucinations, sleep-related behaviors.*	Discontinue. Call doctor right away.

OREXIN RECEPTOR ANTAGONISTS

WARNINGS & PRECAUTIONS

Don't take if:
- You are allergic to orexin receptor antagonists.
- You have narcolepsy (excessive sleepiness or sudden attacks of daytime sleep).

Before you start, consult your doctor if:
- You have breathing or lung problems (e.g., sleep apnea or COPD) or have liver problems.
- You have cataplexy (sudden onset of muscle weakness).
- You suffer from depression or have a history of mental illness or thoughts of suicide.
- You are overweight or obese (especially women).
- You have a history of alcohol or drug abuse or addiction.

Over age 60:
No problems expected.

Pregnancy:
Decide with your doctor if drug benefits justify risk to unborn baby. Risk category C (see page xviii).

Breastfeeding; Lactation; Nursing Mothers:
This drug is generally not recommended with breastfeeding. Discuss risks and benefits with your doctor.

Infants & children up to age 18:
Safety and effectiveness in this age group have not been established. Consult doctor.

Prolonged use:
- Don't take for longer than 1 to 2 weeks unless under doctor's supervision. Schedule regular visits to make sure the drug is working properly and to check for unwanted effects.
- Drug may become habit-forming (mental or physical dependence) with long-term use.

Skin & sunlight:
No problems expected.

Driving, piloting or hazardous work:
Don't drive or pilot aircraft until you learn how medicine affects you (including the day after you take the drug). Don't work around dangerous machinery. Don't climb ladders or work in high places. Danger increases if you drink alcohol or take other medicines affecting alertness and reflexes.

Discontinuing:
You may experience certain side effects such as daytime sleepiness for several days after stopping the drug.

Others:
- Advise any doctor, dentist or pharmacist whom you consult that you take this drug.
- Risk of side effects is increased if you are female, get less than a full night's sleep after taking the drug or you take a dose higher than prescribed.
- Ask your doctor about avoiding next-day activities such as driving or performing tasks that require full mental alertness.
- Consult doctor if insomnia continues after taking the drug for 7 to 10 days.

POSSIBLE INTERACTION WITH OTHER DRUGS

GENERIC NAME OR DRUG CLASS	COMBINED EFFECT
Central nervous system (CNS) depressants*	Increased risk of side effects. Avoid.
Digoxin	May increase digoxin effect.
Enzyme inducers*	Decreased effect of orexin receptor antagonist.
Enzyme inhibitors*	Increased effect of orexin receptor antagonist. Avoid orexin receptor antagonist or take reduced dose per doctor's instruction.

POSSIBLE INTERACTION WITH OTHER SUBSTANCES

INTERACTS WITH	COMBINED EFFECT
Alcohol:	Increased risk of side effects. Do not take drug if you drank alcohol anytime that same evening.
Beverages: Grapefruit juice.	May increase effect of orexin receptor antagonist.
Cocaine:	Unknown. Avoid.
Foods: Grapefruit.	May increase effect of orexin receptor antagonist.
Marijuana:	Consult doctor.
Tobacco:	None expected.

*See Glossary

ORLISTAT

BRAND NAMES

Alli
Xenical

BASIC INFORMATION

Habit forming? No
Prescription needed? Yes, for Xenical
Available as generic? No
Drug class: Antiobesity; lipase inhibitor

USES

- Treatment for obesity and weight loss. To be used in conjunction with a reduced-calorie diet. Treatment with this drug is not recommended for cosmetic weight loss.
- Used to delay onset of type 2 diabetes.

DOSAGE & USAGE INFORMATION

How to take:
Capsule—Swallow with liquid. If you can't swallow whole, open capsule and take with liquid.

When to take:
With or shortly following meals (containing fats) up to three times daily.

If you forget a dose:
Take as soon as you remember. If it is almost time for the next dose, wait for that dose (don't double this dose) and resume regular schedule.

What drug does:
Blocks some of the normal absorption of fats from the intestines, causing them to be excreted in the feces.

Time lapse before drug works:
Several months or longer for maximum benefit.

Don't take with:
Any other medicine or any dietary supplement without consulting your doctor or pharmacist.

OVERDOSE

SYMPTOMS:
Unknown.
WHAT TO DO:
If person uses much larger amount than prescribed or if accidentally swallowed, call poison control center 1-800-222-1222 for instructions or dial 911 (emergency) for help.

POSSIBLE ADVERSE REACTIONS OR SIDE EFFECTS

SYMPTOMS	WHAT TO DO
Life-threatening:	
Rare, severe allergic reaction (anaphylaxis)—difficulty breathing, hives, itching, swelling (throat, face), vomiting, dizziness.	Seek emergency treatment immediately.
Common:	
• Flu-like symptoms, runny nose, congestion, sneezing, sore throat, cough, fever.	Continue. Call doctor when convenient.
• Abdominal pain, oily bowel movements, inability to hold bowel movements, immediate need to have bowel movements, gas with leaky bowel movements, oily spotting of underwear, headaches.	Continue. Call doctor if symptoms persist.
Infrequent:	
• Troubled breathing, tightness in chest, wheezing.	Discontinue. Call doctor right away.
• Anxiety, back pain, menstrual changes, rectal discomfort or pain, tooth or gum problems.	Continue. Call doctor if symptoms persist.
Rare:	
• Diarrhea, hearing changes, pain in ear, bloody or cloudy urine, difficult or painful urination, frequent urge to urinate, possible liver damage (itching, yellow skin or eyes, dark urine, appetite loss, light-colored stools).	Discontinue. Call doctor right away.
• Joint pain, dizziness, dry skin, fatigue, insomnia, muscle pain, nausea, skin rash, vomiting.	Continue. Call doctor if symptoms persist.

WARNINGS & PRECAUTIONS

Don't take if:

- You are allergic to orlistat.
- You have been diagnosed with malabsorption.
- You have cholestasis (blocked bile flow).

Before you start, consult your doctor if:

- You are allergic to any foods or dyes.
- You are taking any other medications or dietary supplements for weight loss.
- You have a history of anorexia or bulimia.
- You have any liver disorder, kidney stones or gallbladder problems.
- You are pregnant or are planning to become pregnant.

Over age 60:
Adverse reactions and side effects may be more frequent and severe than in younger persons.

Pregnancy:
Consult doctor. Drug should not be used during pregnancy. Risk category X (see page xviii).

Breastfeeding; Lactation; Nursing Mothers:
This drug is generally acceptable with breastfeeding. Discuss risks and benefits with your doctor.

Infants & children up to age 18:
Follow instructions provided by your child's doctor.

Prolonged use:
Talk to your doctor about the need for follow-up medical examinations or laboratory studies.

Skin & sunlight:
No problems expected.

Driving, piloting or hazardous work:
No special problems expected.

Discontinuing:
Don't discontinue without doctor's advice.

Others:

- Expect to start with small doses and increase gradually to lessen frequency and severity of adverse reactions.
- Orlistat may interfere with your body's absorption of certain vitamins; therefore, you should take a multivitamin supplement daily, two hours before or after taking orlistat.
- During treatment, you should be on a nutritionally balanced reduced-calorie diet that contains no more than 30 percent of calories from fat.
- Severe liver injury has been reported rarely with the use of this drug. Discontinue drug and call doctor if you develop signs or symptoms of liver injury (see Rare under Possible Adverse Reactions or Side Effects).
- Patients with diabetes may require a reduced dosage of oral hypoglycemic medicine or insulin due to weight loss.
- Advise any doctor, dentist or pharmacist whom you consult that you take this medicine.

POSSIBLE INTERACTION WITH OTHER DRUGS

GENERIC NAME OR DRUG CLASS	COMBINED EFFECT
Cyclosporine	Unknown effect. Monitor closely.
Hepatotoxics*	Increased risk of liver injury.
Pravastatin	Increased pravastatin effect.
Vitamins A, D, E, K	Decreased vitamin effect.
Warfarin	Increased warfarin effect.

POSSIBLE INTERACTION WITH OTHER SUBSTANCES

INTERACTS WITH	COMBINED EFFECT
Alcohol:	None expected.
Beverages:	None expected.
Cocaine:	Unknown. Avoid.
Foods:	None expected.
Marijuana:	Consult doctor.
Tobacco:	None expected.

***See Glossary**

OSPEMIFENE

BRAND NAMES

Osphena

BASIC INFORMATION

Habit forming? No
Prescription needed? Yes
Available as generic? No
Drug class: Estrogen agonist/antagonist

USES

- Treatment for painful sexual intercourse (dyspareunia) in menopausal women. The pain is due to vulvar and vaginal atrophy (dryness, loss of flexibility and thinning) that results from decreased estrogen levels.
- Other uses as determined by your doctor.

DOSAGE & USAGE INFORMATION

How to take:
Tablet—Swallow whole with liquid and take with food.

When to take:
Once a day at the same time each day (preferably at a meal time).

If you forget a dose:
Take as soon as you remember. If it is almost time for the next dose, wait for that dose (don't double this dose) and resume regular schedule.

What drug does:
The drug is not an estrogen, but acts like estrogen on vulvar and vaginal tissues. It can increase tissue thickness, improve elasticity and add lubrication—all of which help ease the pain during intercourse.

Time lapse before drug works:
May take up to 12 weeks for full benefit depending on severity of symptoms.

Don't take with:
Any other medicine or any dietary supplement without consulting your doctor or pharmacist.

OVERDOSE

SYMPTOMS:
Unknown.
WHAT TO DO:
If person takes much larger amount than prescribed or if accidentally swallowed, call doctor or poison control center 1-800-222-1222 for help.

POSSIBLE ADVERSE REACTIONS OR SIDE EFFECTS

SYMPTOMS	WHAT TO DO
Life-threatening: None expected.	
Common: Increased sweating, vaginal discharge, hot flushes or flashes, muscle spasms.	Continue. Call doctor when convenient.
Infrequent: None expected.	
Rare: Changes in vision or speech, sudden new and severe headache, severe pains in chest or legs (with or without shortness of breath), weakness or fatigue, acute vaginal bleeding, symptoms of deep vein thrombosis (pain, warmth, swelling or redness in one or both legs).	Discontinue. Call doctor right away or seek emergency treatment.

WARNINGS & PRECAUTIONS

Don't take if:
You are allergic to ospemifene. Allergic reactions are rare, but may occur.

Before you start, consult your doctor if:
- You have or have had a history of blood clots (deep vein thrombosis).
- You have a history of stroke, heart attack or heart disease.
- You use tobacco.
- You have unusual vaginal bleeding.
- You have or have had cancer.
- You have an intact uterus.
- You have liver disease, lupus or diabetes.

Over age 60:
No special problems expected.

Pregnancy:
Consult doctor. Drug should not be used during pregnancy. Can cause harm to unborn baby. Risk category X (see page xviii).

Breastfeeding; Lactation; Nursing Mothers:
This drug is generally not recommended with breastfeeding. Discuss risks and benefits with your doctor.

Infants & children up to age 18:
Safety and effectiveness in this age group have not been established.

Prolonged use:
The drug is usually prescribed for the shortest time needed to relieve painful symptoms. Visit your doctor on a regular basis to verify the drug is working properly and if you should continue to take it.

Skin & sunlight:
No problems expected.

Driving, piloting or hazardous work:
No problems expected.

Discontinuing:
No problems expected.

Others:
- Advise any doctor, dentist or pharmacist whom you consult that you take this drug.
- Call your doctor right away if you have any unusual vaginal bleeding.
- Drug may increase the risk of blood clots in the veins (deep vein thrombosis) or having a stroke. Consult doctor about your risks.
- Drug may increase the risk of cancer of the lining of the uterus (endometrial cancer). Consult doctor about your risks.
- The drug is not intended for treatment of decreased libido (sexual drive) in women.

POSSIBLE INTERACTION WITH OTHER DRUGS

GENERIC NAME OR DRUG CLASS	COMBINED EFFECT
Enzyme inducers*	Reduced effect of ospemifene.
Enzyme inhibitors*	Increased effect of ospemifene.
Estrogens*	Unknown effect. Avoid.
Estrogen agonists/ antagonists,* other	Unknown effect. Avoid.

POSSIBLE INTERACTION WITH OTHER SUBSTANCES

INTERACTS WITH	COMBINED EFFECT
Alcohol:	None expected.
Beverages:	None expected.
Cocaine:	Unknown effect. Avoid.
Foods:	None expected.
Marijuana:	Unknown effect. Avoid.
Tobacco:	No drug interaction. Tobacco use increases risk of blood clot. Avoid.

OXCARBAZEPINE

BRAND NAMES

Oxtellar XR **Trileptal**

BASIC INFORMATION

Habit forming? No
Prescription needed? Yes
Available as generic? Yes
Drug class: Anticonvulsant, antiepileptic

USES

- Treatment for partial (focal) epileptic seizures. May be used alone or in combination with other antiepileptic drugs.
- Other uses as determined by your doctor.

DOSAGE & USAGE INFORMATION

How to take:
- Tablet—Swallow with liquid. May be taken with or without food.
- Oral suspension—Follow product instructions.
- Extended-release tablet—Swallow whole with liquid. Take on an empty stomach. Do not chew, crush or split tablet.

When to take:
Your doctor will determine the best schedule. Dosages will be increased rapidly over the first few days of use. Further increases may be necessary to achieve maximum benefits.

If you forget a dose:
Take as soon as you remember. If it is almost time for the next dose, wait for that dose (don't double this dose) and resume regular schedule.

What drug does:
The exact mechanism is unknown, but it helps decrease abnormal electrical activity in the brain.

Continued next column

OVERDOSE

SYMPTOMS:
Unknown. May be similar to adverse effects.
WHAT TO DO:
- **Dial 911 for all medical emergencies or call poison control center 1-800-222-1222 for instructions.**
- **See emergency information on last 3 pages of this book.**

Time lapse before drug works:
May take several weeks for effectiveness.

Don't take with:
Any other medicine or any dietary supplement without consulting your doctor or pharmacist.

POSSIBLE ADVERSE REACTIONS OR SIDE EFFECTS

SYMPTOMS	WHAT TO DO
Life-threatening:	
Rare, severe allergic reaction (anaphylaxis)—difficulty breathing, hives, itching, swelling (throat, face), vomiting, dizziness.	Seek emergency treatment immediately.
Common:	
• Cough with fever and sneezing and sore throat, clumsiness, vision changes, dizziness, crying, depression, false sense of well-being, spinning sensation, uncontrolled eye movement, feeling of constant movement of self or surroundings.	Continue, but call doctor right away.
• Runny or stuffy nose, nausea or vomiting, sleepiness.	Continue. Call doctor when convenient.
Infrequent:	
• Blurred vision, cloudy or bloody urine, urination changes (decreased or increased, painful, urgent), confusion, falling, bruising, ill feeling, thirstiness, hoarseness, vaginal itching, heartbeat irregularities, facial pain, memory loss, coordination problems, trembling, shortness of breath, unusual tiredness or weakness.	Continue, but call doctor right away.
• Upset stomach, acne, changes in taste, dry mouth, constipation or diarrhea, heartburn, sweating, feeling of warmth in face or chest, back pain, belching, bloody nose.	Continue. Call doctor when convenient.

Rare:

Sore or bleeding lips, chills, chest pain, irritability, hives or itching, muscle or joint pain, nervousness, rectal bleeding, peeling or blistering skin, sores in mouth, swollen legs, purple spots on skin, burning feeling in chest or stomach.	Continue, but call doctor right away.

WARNINGS & PRECAUTIONS

Don't take if:
You are allergic to oxcarbazepine or carbamazepine.

Before you start, consult your doctor if:
- You have a history of kidney or liver disease.
- You have hyponatremia (too little sodium in the body).
- You are allergic to any medication, food or other substance.
- You have any other medical problems.

Over age 60:
No special problems expected.

Pregnancy:
Decide with your doctor if drug benefits justify risk to unborn baby. Risk category C (see page xviii).

Breastfeeding; Lactation; Nursing Mothers:
This drug is generally acceptable with breastfeeding. Discuss risks and benefits with your doctor.

Infants & children up to age 18:
Follow instructions provided by your child's doctor.

Prolonged use:
No special problems expected. Follow-up laboratory blood studies may be recommended by your doctor.

Skin & sunlight:
No problems expected.

Driving, piloting or hazardous work:
Don't drive or pilot aircraft until you learn how medicine affects you. Don't work around dangerous machinery. Don't climb ladders or work in high places. Danger increases if you drink alcohol or take other medicines affecting alertness and reflexes.

Discontinuing:
Don't discontinue without doctor's approval due to risk of increased seizure activity. The dosage may need to be gradually decreased before stopping the drug completely.

Others:
- Advise any doctor, dentist or pharmacist whom you consult that you take this medicine.
- The effectiveness of oral contraceptives that contain estrogen may be reduced. Talk to your doctor about other forms of birth control.
- Oxcarbazepine may be used with other anticonvulsant drugs and additional side effects may occur. If they do, discuss them with your doctor.
- There is a small risk of life threatening skin reactions while taking this drug. If you develop an allergic reaction or a skin rash, contact your doctor right away.
- This medicine alone may cause drowsiness, dizziness, or lightheadedness, especially when getting up from a sitting or lying position.
- Rarely, antiepileptic drugs may lead to suicidal thoughts and behaviors. Call doctor right away if suicidal symptoms or unusual behaviors occur.
- Wear or carry medical identification to show your seizure disorder and the drugs you take.

POSSIBLE INTERACTION WITH OTHER DRUGS

GENERIC NAME OR DRUG CLASS	COMBINED EFFECT
Anticonvulsants,* other	Increased risk of side effects.
CNS Depressants*	Increased sedative effect.
Contraceptives, oral*	Decreased effect of contraceptive.
Felodipine	Decreased effect of felodipine.
Verapamil	Decreased effect of oxcarbazepine.

POSSIBLE INTERACTION WITH OTHER SUBSTANCES

INTERACTS WITH	COMBINED EFFECT
Alcohol:	Increased sedative effect. Avoid.
Beverages:	None expected.
Cocaine:	Unknown. Avoid.
Foods:	None expected.
Marijuana:	Consult doctor.
Tobacco:	None expected.

*See Glossary

OXYMETAZOLINE (Nasal)

BRAND NAMES

See full list of brand names in the *Generic and Brand Name Directory*, page 872.

BASIC INFORMATION

Habit forming? No
Prescription needed? No
Available as generic? Yes
Drug class: Sympathomimetic

USES

Relieves congestion of nose, sinuses and throat from allergies and infections.

DOSAGE & USAGE INFORMATION

How to take:
Nasal solution, nasal spray—Use as directed on label. Avoid contamination. Don't use same container for more than 1 person.

When to take:
When needed, no more often than every 4 hours.

If you forget a dose:
Take as soon as you remember. If it is almost time for the next dose, wait for that dose (don't double this dose) and resume regular schedule.

What drug does:
Constricts walls of small arteries in nose, sinuses and eustachian tubes.

Time lapse before drug works:
5 to 30 minutes. May last 8 to 12 hours.

Don't take with:
- Nonprescription drugs for allergy, cough or cold without consulting doctor.
- Any other medicine or any dietary supplement without consulting your doctor or pharmacist.

OVERDOSE

SYMPTOMS:
If too much is absorbed by the body or if accidentally swallowed: drowsiness, dizziness, slow heartbeat, fainting.
WHAT TO DO:
- **Dial 911 for all medical emergencies or call poison control center 1-800-222-1222 for instructions.**
- **See emergency information on last 3 pages of this book.**

POSSIBLE ADVERSE REACTIONS OR SIDE EFFECTS

SYMPTOMS	WHAT TO DO
Life-threatening:	
In case of overdose, see previous column.	
Common:	
None expected.	
Infrequent:	
Burning or stinging or dryness of nasal passages, sneezing.	Continue. Call doctor if symptoms persist.
Rare:	
• Headache, insomnia, nervousness (all may occur if body absorbs too much of drug).	Discontinue. Call doctor if symptoms persist.
• Rebound congestion (increased runny or stuffy nose).	Discontinue.

WARNINGS & PRECAUTIONS

Don't take if:
You are allergic to oxymetazoline or other nasal spray.

Before you start, consult your doctor if:
- You have heart disease or high blood pressure.
- You have diabetes.
- You have overactive thyroid.
- You have taken MAO inhibitors in past 2 weeks.
- You have glaucoma.

Over age 60:
No special problems expected.

Pregnancy:
Decide with your doctor if drug benefits justify risk to unborn baby. Risk category C (see page xviii).

Breastfeeding; Lactation; Nursing Mothers:
This drug is generally acceptable with breastfeeding. Discuss risks and benefits with your doctor.

Infants & children up to age 18:
Read nonprescription product's label to be sure it is approved for your child's age. Follow label instructions on dosage. Consult doctor or pharmacist if unsure.

Prolonged use:
Drug may lose effectiveness, cause increased congestion (rebound effect*) and irritate nasal membranes.

Skin & sunlight:
No problems expected.

Driving, piloting or hazardous work:
No problems expected.

Discontinuing:
May be unnecessary to finish medicine. Follow doctor's instructions.

Others:
- Don't use for more than 3 days in a row.
- Keep product out of the reach of children.

POSSIBLE INTERACTION WITH OTHER DRUGS

GENERIC NAME OR DRUG CLASS	COMBINED EFFECT
Antidepressants, tricyclic*	Possible rise in blood pressure.
Butorphanol	Delays start of butorphanol effect.
Maprotiline	Possible increased blood pressure.

POSSIBLE INTERACTION WITH OTHER SUBSTANCES

INTERACTS WITH	COMBINED EFFECT
Alcohol:	None expected.
Beverages: Caffeine drinks.	Nervousness or insomnia.
Cocaine:	Risk of heartbeat irregularities and high blood pressure. Avoid.
Foods:	None expected.
Marijuana:	Overstimulation. Avoid.
Tobacco:	None expected.

*See Glossary

PACLITAXEL

BRAND NAMES

Abraxane Taxol

BASIC INFORMATION

Habit forming? No
Prescription needed? Yes
Available as generic? Yes
Drug class: Antineoplastic

USES

Treats ovarian cancer, breast cancer and some lung cancers.

DOSAGE & USAGE INFORMATION

How to take:
Injection—Administered only by a doctor or under the supervision of a doctor.

When to take:
Your doctor will determine the schedule. Usually the drug is infused over a 24-hour period at 21-day intervals. Other drugs may be given prior to paclitaxel injection to help prevent adverse effects.

If you forget a dose:
Not a concern since drug is administered by a health care provider.

What drug does:
Interferes with the growth of cancer cells, which are eventually destroyed.

Time lapse before drug works:
Results may not show for several weeks or months.

Don't take with:
Any other medicine or any dietary supplement without consulting your doctor or pharmacist.

OVERDOSE

SYMPTOMS:
Unknown.
WHAT TO DO:
Overdose is unlikely. You will be monitored by medical personnel during the time of the infusion.

POSSIBLE ADVERSE REACTIONS OR SIDE EFFECTS

SYMPTOMS	WHAT TO DO
Life-threatening:	
Anaphylactic reaction soon after an injection (hives, rash, intense itching, faintness, breathing difficulty).	Emergency care will be provided.
Common:	
• Paleness, tiredness, flushing of face, skin rash or itching, shortness of breath, fever, chills, cough or hoarseness, back or side pain, difficult or painful urination, unusual bleeding or bruising, black or tarry stools, blood in urine, pinpoint red spots on skin, bleeding gums, delayed wound healing.	Call doctor right away.
• Pain in joints or muscles, diarrhea, nausea and vomiting, numbness or burning or tingling in hands or feet.	Call doctor when convenient.
• Loss of hair (should regrow after therapy completed).	No action necessary.
Infrequent:	
Slow or irregular heartbeat, chest pain.	Call doctor right away.
Rare:	
Pain or redness at injection site, mouth or lip sores.	Call doctor right away.

WARNINGS & PRECAUTIONS

Don't take if:
You are allergic to paclitaxel.

Before you start, consult your doctor if:
- You have an infection or any other medical problem.
- You have or recently had chickenpox or herpes zoster (shingles).
- You have heart problems.
- You are pregnant or if you plan to become pregnant.
- You have had radiation therapy or previously taken anticancer drugs.

Over age 60:
No special problems expected.

Pregnancy:
Consult doctor. Use of the drug during pregnancy is not recommended. It could cause harm to the unborn baby. Risk category D (see page xviii).

Breastfeeding; Lactation; Nursing Mothers:
This drug is generally not recommended with breastfeeding. Discuss risks and benefits with your doctor.

Infants & children up to age 18:
Safety and effectiveness in this age group have not been established. Consult doctor.

Prolonged use:
Not recommended for long-term use.

Skin & sunlight:
No problems expected.

Driving, piloting or hazardous work:
Avoid if you feel side effects such as nausea and vomiting.

Discontinuing:
Your doctor will determine the schedule.

Others:
- Advise any doctor, dentist or pharmacist whom you consult that you take this medicine.
- May affect the results in some medical tests.
- Do not have any immunizations (vaccinations) without doctor's approval. Other household members should not take oral polio vaccine. It could pass the polio virus on to you. Avoid any contact with persons who have taken oral polio vaccine.
- Possible delayed effects (including some types of cancers) may occur months to years after use. Your doctor should discuss with you all risks involving this drug.
- You will have increased risk of infections. Take extra precautions (handwashing), and avoid people with infections. Avoid crowds if possible. Contact your doctor immediately if you develop signs or symptoms of infection.
- Use care in the use of toothbrushes, dental floss and toothpicks. Talk to your medical doctor before you have dental work done.
- Do not touch your eyes, mouth or the inside of your nose without first carefully washing your hands.
- Avoid activities (e.g., contact sports) that could cause bruising or injury.
- Avoid cutting yourself when using a safety razor, fingernail or toenail clippers.

POSSIBLE INTERACTION WITH OTHER DRUGS

GENERIC NAME OR DRUG CLASS	COMBINED EFFECT
Blood dyscrasia-causing medicines*	Increased risk of paclitaxel toxicity.
Bone marrow depressants,* (other)	Increased risk of paclitaxel toxicity.

POSSIBLE INTERACTION WITH OTHER SUBSTANCES

INTERACTS WITH	COMBINED EFFECT
Alcohol:	None expected.
Beverages:	None expected.
Cocaine:	Unknown. Avoid.
Foods:	None expected.
Marijuana:	Consult doctor.
Tobacco:	None expected.

PANCRELIPASE

BRAND NAMES

Creon
Pancreaze
Pertyze
Ultresa
Viokace
Zenpep

BASIC INFORMATION

Habit forming? No
Prescription needed? Yes
Available as generic? Yes
Drug class: Enzyme (pancreatic)

USES

- Replaces pancreatic enzymes lost due to surgery or disease.
- Treats fatty stools (steatorrhea).

DOSAGE & USAGE INFORMATION

How to take:
- Tablet, capsule or delayed-release capsule—Swallow whole. Do not take with milk or milk products.
- Powder—Sprinkle on liquid or soft food.

When to take:
Before meals.

If you forget a dose:
Take as soon as you remember. If it is almost time for the next dose, wait for that dose (don't double this dose) and resume regular schedule.

What drug does:
Enhances digestion of proteins, carbohydrates and fats.

Time lapse before drug works:
30 minutes.

Don't take with:
Any other medicine or any dietary supplement without consulting your doctor or pharmacist.

OVERDOSE

SYMPTOMS:
Upset stomach, diarrhea.
WHAT TO DO:
If person uses much larger amount than prescribed or if accidentally swallowed, call poison control center 1-800-222-1222 for instructions or dial 911 (emergency) for help.

POSSIBLE ADVERSE REACTIONS OR SIDE EFFECTS

SYMPTOMS	WHAT TO DO
Life-threatening: None expected.	
Common: None expected.	
Infrequent:	
Diarrhea, stomach cramps or pain, nausea (all may occur with high doses).	Continue. Call doctor when convenient.
Rare:	
• Rash, hives, with very high doses (blood in urine, swollen feet or legs, joint pain).	Discontinue. Call doctor right away.
• Breathing in powder form may cause shortness of breath, stuffy nose, tightness in chest.	Continue. Call doctor when convenient.

WARNINGS & PRECAUTIONS

Don't take if:
You are allergic to pancreatin, pancrelipase, or pork. Allergic reactions are rare, but may occur.

Before you start, consult your doctor if:
You take any other medicines.

Over age 60:
Adverse reactions and side effects may be more frequent and severe than in younger persons.

Pregnancy:
Decide with your doctor if drug benefits justify risk to unborn baby. Risk category C (see page xviii).

Breastfeeding; Lactation; Nursing Mothers:
This drug is generally acceptable with breastfeeding. Discuss risks and benefits with your doctor.

Infants & children up to age 18:
Follow instructions provided by your child's doctor.

Prolonged use:
No additional problems expected.

Skin & sunlight:
No problems expected.

Driving, piloting or hazardous work:
No problems expected.

Discontinuing:
Don't discontinue without consulting doctor. Dose may require gradual reduction if you have taken drug for a long time. Doses of other drugs may also require adjustment.

Others:
- If you take powder form, avoid inhaling.
- Advise any doctor, dentist or pharmacist whom you consult that you take this medicine.

POSSIBLE INTERACTION WITH OTHER DRUGS

GENERIC NAME OR DRUG CLASS	COMBINED EFFECT
Calcium carbonate antacids*	Decreased effect of pancrelipase.
Iron supplements	Decreased iron absorption.
Magnesium hydroxide antacids*	Decreased effect of pancrelipase.

POSSIBLE INTERACTION WITH OTHER SUBSTANCES

INTERACTS WITH	COMBINED EFFECT
Alcohol:	Unknown. Consult doctor.
Beverages: Milk.	Decreased effect of pancrelipase.
Cocaine:	Unknown. Avoid.
Foods: Ice cream, milk products.	Decreased effect of pancrelipase.
Marijuana:	Consult doctor.
Tobacco:	Decreased absorption of pancrelipase.

*See Glossary

PANTOTHENIC ACID (Vitamin B-5)

GENERIC AND BRAND NAMES

CALCIUM PANTOTHENATE **Dexol T.D.**

Numerous brands of single vitamin and multivitamin combinations may be available.

BASIC INFORMATION

Habit forming? No
Prescription needed? No
Available as generic? Yes
Drug class: Vitamin supplement

USES

Prevents and treats vitamin B-5 deficiency.

DOSAGE & USAGE INFORMATION

How to take:
Tablet—Swallow with liquid.

When to take:
At the same times each day.

If you forget a dose:
Take as soon as you remember. If it is almost time for the next dose, wait for that dose (don't double this dose) and resume regular schedule.

What drug does:
Acts as co-enzyme in carbohydrate, protein and fat metabolism.

Time lapse before drug works:
15 to 20 minutes.

Don't take with:
- Levodopa—Small amounts of pantothenic acid will nullify levodopa effect. Carbidopa-levodopa combination not affected by this interaction.
- Any other medicine or any dietary supplement without consulting your doctor or pharmacist.

OVERDOSE

SYMPTOMS:
May have heartburn, diarrhea, nausea.
WHAT TO DO:
If person uses much larger amount than prescribed or if accidentally swallowed, call poison control center 1-800-222-1222 for instructions or dial 911 (emergency) for help.

POSSIBLE ADVERSE REACTIONS OR SIDE EFFECTS

SYMPTOMS	WHAT TO DO
Life-threatening: None expected.	
Common: None expected.	
Infrequent: None expected.	
Rare: None expected.	

PANTOTHENIC ACID (Vitamin B-5)

WARNINGS & PRECAUTIONS

Don't take if:
You are allergic to pantothenic acid. Allergic reactions are rare, but may occur.

Before you start, consult your doctor if:
You have hemophilia.

Over age 60:
No problems expected.

Pregnancy:
Decide with your doctor. Risk category A (see page xviii).

Breastfeeding; Lactation; Nursing Mothers:
This drug is generally acceptable with breastfeeding. Discuss risks and benefits with your doctor.

Infants & children up to age 18:
Read nonprescription product's label to be sure it is approved for your child's age. Follow label instructions on dosage. Consult doctor or pharmacist if unsure.

Prolonged use:
Large doses for more than 1 month may cause toxicity.

Skin & sunlight:
No problems expected.

Driving, piloting or hazardous work:
No problems expected.

Discontinuing:
No problems expected.

Others:
Regular pantothenic acid supplements are recommended if you take chloramphenicol, cycloserine, ethionamide, hydralazine, immunosuppressants,* isoniazid or penicillamine. These decrease pantothenic acid absorption and can cause anemia or tingling and numbness in hands and feet.

POSSIBLE INTERACTION WITH OTHER DRUGS

GENERIC NAME OR DRUG CLASS	COMBINED EFFECT
None significant.	

POSSIBLE INTERACTION WITH OTHER SUBSTANCES

INTERACTS WITH	COMBINED EFFECT
Alcohol:	None expected.
Beverages:	None expected.
Cocaine:	Unknown. Avoid.
Foods:	None expected.
Marijuana:	Consult doctor.
Tobacco:	None expected.

*See Glossary

PAPAVERINE

BRAND NAMES

Cerespan
Genabid
Pavabid
Pavabid Plateau Caps
Pavacot
Pavagen
Pavarine
Pavased
Pavatine
Pavatym
Paverolan

BASIC INFORMATION

Habit forming? No
Prescription needed? Yes
Available as generic? Yes
Drug class: Vasodilator

USES

- Used to improve blood circulation in the extremities, heart or brain.
- Injected into penis to produce erections.

DOSAGE & USAGE INFORMATION

How to take:

- Tablet—Swallow with liquid or food to lessen stomach irritation. If you can't swallow whole, crumble tablet and take with liquid or food.
- Extended-release capsule—Swallow whole with liquid.
- Injections to penis—Follow doctor's instructions.

When to take:
At the same times each day.

If you forget a dose:
Take as soon as you remember. If it is almost time for the next dose, wait for that dose (don't double this dose) and resume regular schedule.

What drug does:
Relaxes and expands blood vessel walls, allowing better distribution of oxygen and nutrients.

Continued next column

OVERDOSE

SYMPTOMS:
Nausea, vomiting, weakness, fainting, flush, sweating, dizziness, vision changes, fast heartbeat.
WHAT TO DO:

- **Dial 911 for all medical emergencies or call poison control center 1-800-222-1222 for instructions.**
- **See emergency information on last 3 pages of this book.**

Time lapse before drug works:
30 to 60 minutes.

Don't take with:
Any other medicine or any dietary supplement without consulting your doctor or pharmacist.

POSSIBLE ADVERSE REACTIONS OR SIDE EFFECTS

SYMPTOMS	WHAT TO DO
Life-threatening: None expected.	
Common: Drowsiness, dry mouth and throat, dizziness, flushed face, headache, stomach upset or indigestion, nausea, mild constipation.	Continue. Call doctor when convenient.
Infrequent: Rash, itchy skin, blurred or double vision, weakness, fast heartbeat.	Discontinue. Call doctor right away.
Rare: Jaundice (yellow skin and eyes).	Discontinue. Call doctor right away.

WARNINGS & PRECAUTIONS

Don't take if:
You are allergic to papaverine. Allergic reactions are rare, but may occur.

Before you start, consult your doctor if:
- You plan to become pregnant within medication period.
- You have had a heart attack, heart disease, angina or stroke.
- You have Parkinson's disease.

Over age 60:
Adverse reactions and side effects may be more frequent and severe than in younger persons.

Pregnancy:
Decide with your doctor if drug benefits justify risk to unborn baby. Risk category C (see page xviii).

Breastfeeding; Lactation; Nursing Mothers:
This drug (tablet) is generally acceptable with breastfeeding. Discuss risks and benefits with your doctor.

Infants & children up to age 18:
Follow instructions provided by your child's doctor.

Prolonged use:
No problems expected.

Skin & sunlight:
No problems expected.

Driving, piloting or hazardous work:
Don't drive or pilot aircraft until you learn how medicine affects you. Don't work around dangerous machinery. Don't climb ladders or work in high places. Danger increases if you drink alcohol or take medicine affecting alertness and reflexes, such as antihistamines, tranquilizers, sedatives, pain medicine, narcotics and mind-altering drugs.

Discontinuing:
May be unnecessary to finish medicine. If drug does not help in 1 to 2 weeks, consult doctor about discontinuing.

Others:
- Periodic liver function tests recommended.
- Internal eye pressure measurements recommended if you have glaucoma.
- Advise any doctor, dentist or pharmacist whom you consult that you take this medicine.

POSSIBLE INTERACTION WITH OTHER DRUGS

GENERIC NAME OR DRUG CLASS	COMBINED EFFECT
Levodopa	Decreased levodopa effect.
Narcotics*	Increased sedation.
Pain relievers*	Increased sedation.
Pergolide	Decreased pergolide effect.
Sedatives*	Increased sedation.
Sympathomimetics*	Reversal of the effect of papaverine.
Tranquilizers*	Increased sedation.

POSSIBLE INTERACTION WITH OTHER SUBSTANCES

INTERACTS WITH	COMBINED EFFECT
Alcohol:	None expected.
Beverages:	None expected.
Cocaine:	Decreased papaverine effect. Avoid.
Foods:	None expected.
Marijuana:	Consult doctor.
Tobacco:	Decrease in papaverine's dilation of blood vessels.

***See Glossary**

PAREGORIC

BRAND NAMES

Brown Mixture
Camphorated Opium Tincture
Kapectolin with Paregoric
Parepectolin

BASIC INFORMATION

Habit forming? Yes
Prescription needed? Yes
Available as generic? Yes
Drug class: Narcotic, antidiarrheal

USES

Reduces intestinal cramps and diarrhea.

DOSAGE & USAGE INFORMATION

How to take:
Drops or liquid—Dilute dose in beverage before swallowing.

When to take:
As needed for diarrhea, no more often than every 4 hours.

If you forget a dose:
Take as soon as you remember. If it is almost time for the next dose, wait for that dose (don't double this dose) and resume regular schedule.

What drug does:
Anesthetizes surface membranes of intestines and blocks nerve impulses.

Time lapse before drug works:
2 to 6 hours.

Don't take with:
Any other medicine or any dietary supplement without consulting your doctor or pharmacist.

OVERDOSE

SYMPTOMS:
Deep sleep, slow and difficult breathing, cold and clammy skin, weak muscles, slow heartbeat, low blood pressure, stupor, coma.
WHAT TO DO:
- **Dial 911 for all medical emergencies or call poison control center 1-800-222-1222 for instructions.**
- **See emergency information on last 3 pages of this book.**

POSSIBLE ADVERSE REACTIONS OR SIDE EFFECTS

SYMPTOMS	WHAT TO DO
Life-threatening:	
Rare, severe allergic reaction (anaphylaxis)—difficulty breathing, hives, itching, swelling (throat, face), vomiting, dizziness.	Seek emergency treatment immediately.
Common:	
Dizziness, flushed face, unusual tiredness, difficult urination.	Continue. Call doctor when convenient.
Infrequent:	
Severe constipation, abdominal pain, vomiting.	Discontinue. Call doctor right away.
Rare:	
• Hives, rash, itchy skin, slow heartbeat, irregular breathing.	Discontinue. Call doctor right away.
• Depression.	Continue. Call doctor when convenient

WARNINGS & PRECAUTIONS

Don't take if:
You are allergic to any narcotic.*

Before you start, consult your doctor if:
You have impaired liver or kidney function.

Over age 60:
Unknown effect.

Pregnancy:
Decide with your doctor if drug benefits justify risk to unborn baby. Risk category C (see page xviii).

Breastfeeding; Lactation; Nursing Mothers:
This drug is generally not recommended with breastfeeding. Discuss risks and benefits with your doctor.

Infants & children up to age 18:
Follow instructions provided by your child's doctor.

Prolonged use:
Causes psychological and physical dependence.

Skin & sunlight:
No problems expected.

Driving, piloting or hazardous work:
Don't drive or pilot aircraft until you learn how medicine affects you. Don't work around dangerous machinery. Don't climb ladders or work in high places. Danger increases if you drink alcohol or take medicine affecting alertness and reflexes, such as antihistamines, tranquilizers, sedatives, pain medicine, narcotics and mind-altering drugs.

Discontinuing:
May be unnecessary to finish medicine. Follow doctor's instructions.

Others:
Advise any doctor, dentist or pharmacist whom you consult that you take this medicine.

POSSIBLE INTERACTION WITH OTHER DRUGS

GENERIC NAME OR DRUG CLASS	COMBINED EFFECT
Analgesics*	Increased analgesic effect.
Anticholinergics*	Increased risk of constipation.
Antidepressants*	Increased sedation.
Antidiarrheal preparations*	Increased sedative effect. Avoid.
Antihistamines*	Increased sedation.
Central nervous system (CNS) depressants*	Increased central nerve system depression.
Naloxone	Decreased paregoric effect.
Naltrexone	Decreased paregoric effect.
Narcotics,* other	Increased narcotic effect.

POSSIBLE INTERACTION WITH OTHER SUBSTANCES

INTERACTS WITH	COMBINED EFFECT
Alcohol:	Increases alcohol's intoxicating effect. Avoid.
Beverages:	None expected.
Cocaine:	Increased risk of side effects. Avoid.
Foods:	None expected.
Marijuana:	Increased risk of side effects. Avoid.
Tobacco:	None expected.

*See Glossary

PEDICULICIDES (Topical)

GENERIC AND BRAND NAMES

BENZOYL ALCOHOL
Ulesfia
IVERMECTIN (topical)
Sklice Lotion
Soolantra
LINDANE
GBH
G-Well
Kwellada
Kwildane
PMS Lindane
MALATHION
Derbac
Ovide
PERMETHRIN
Acticin
Elimite Cream
Nix Cream Rinse
PYRETHRINS & PIPERONYL BUTOXIDE
A-200 Gel
A-200 Shampoo
Barc
Blue
Lice-Enz Foam and Comb Lice Killing Shampoo Kit
Pyrinyl
R&C
TISIT
TISIT Blue
Triple X
SPINOSAD
Natroba Topical Suspension

BASIC INFORMATION

Habit forming? No
Prescription needed? No, for most
Available as generic? Yes, for some
Drug class: Pediculicide; scabicide; anti-inflammatory

USES

- Treats scabies and lice infections of skin or scalp.
- Cream and lotion treats scabies.
- Shampoo treats lice infections.
- Brand name Soolantra treats rosacea.

OVERDOSE

SYMPTOMS:
Rarely toxic effects from too much absorbed through skin or if swallowed—Vomiting, muscle cramps, dizziness, seizure, rapid heartbeat.
WHAT TO DO:
- **Dial 911 for all medical emergencies or call poison control center 1-800-222-1222 for instructions.**
- **See emergency information on last 3 pages of this book.**

DOSAGE & USAGE INFORMATION

How to use:
- For lice, all household members should be examined for infestation and treated if infested. Read directions on product for proper application technique and length of time to leave on the body. Wear plastic gloves when you apply. Use care not to apply more than directed. Avoid contact with eyes, nose and mouth. Flush eyes with water if product gets in the eyes. Bathe before applying. Wash hands after applying. Use in well-ventilated room.
- For rosacea, for each area to be treated, use a pea-size amount of cream and spread on in a thin layer. Avoid eyes and lips.

When to use:
As directed on package for lice treatment. Once daily for rosacea.

If you forget a dose:
Use as soon as you remember, then continue regular schedule. If it is almost time for the next dose, wait for that dose (don't double that dose).

What drug does:
- Most types are absorbed into bodies of lice and scabies organisms, killing them. Benzoyl alcohol kills lice by suffocation.
- For rosacea, the drug's action may be as an anti-inflammatory which helps heal the lesions, bumps and acne.

Time lapse before drug works:
- For lice, cream or lotion requires 8 to 12 hours contact with skin.
- For rosacea, results may be seen in some people within 2 to 3 weeks.

Don't use with:
Other topical medicines without consulting your doctor or pharmacist.

POSSIBLE ADVERSE REACTIONS OR SIDE EFFECTS

SYMPTOMS	WHAT TO DO
Life-threatening: None expected.	
Common: None expected.	
Infrequent: None expected.	
Rare:	
• Skin irritation or rash or burning.	Discontinue. Call doctor right away.
• Skin itch that continues 1 week to several weeks after treatment.	Call doctor when convenient.

WARNINGS & PRECAUTIONS

Don't use if:
You are allergic to any pediculicide. Allergic reactions are rare, but may occur.

Before you start, consult your doctor if:
- You are allergic to anything that touches your skin.
- You are using any other medicines, creams, lotions or oils.

Over age 60:
No problems expected.

Pregnancy:
Decide with your doctor if drug benefits justify risk to unborn baby. Risk category B or C (see page xviii).

Breastfeeding; Lactation; Nursing Mothers:
Drugs in this group are generally not recommended with breastfeeding. Discuss risks and benefits with your doctor.

Infants & children up to age 18:
- Read nonprescription product's label to be sure it is approved for your child's age. Follow label instructions on dosage. Consult doctor or pharmacist if unsure.
- Lindane is not recommended for use in infants. For other children, follow package instructions. Never use more of the product than instructed.

Prolonged use:
- Not recommended with lice treatment.
- For rosacea, the drug can be used up to a year. Consult doctor for longer term use.

Skin & sunlight:
No problems expected, but check with doctor.

Driving, piloting or hazardous work:
No problems expected, but check with doctor.

Discontinuing:
No problems expected, but check with doctor.

Others:
- Don't use on open sores or wounds.
- Even after successful lice treatment, itching can continue due to remaining inflammation in the skin. This should not be confused with a reinfestation. Consult doctor if you are unsure.

POSSIBLE INTERACTION WITH OTHER DRUGS

GENERIC NAME OR DRUG CLASS	COMBINED EFFECT
Antimyasthenics*	Excessive absorption and chance of toxicity (with malathion only).
Cholinesterase inhibitors*	Excessive absorption and chance of toxicity (with malathion only).

POSSIBLE INTERACTION WITH OTHER SUBSTANCES

INTERACTS WITH	COMBINED EFFECT
Alcohol:	None expected.
Beverages:	None expected.
Cocaine:	None expected.
Foods:	None expected.
Marijuana:	None expected.
Tobacco:	None expected.

*See Glossary

PENICILLAMINE

BRAND NAMES

Cuprimine
Depen

BASIC INFORMATION

Habit forming? No
Prescription needed? Yes
Available as generic? No
Drug class: Chelating agent, antirheumatic, antidote (heavy metal)

USES

- Treatment for rheumatoid arthritis.
- Prevention of kidney stones.
- Treatment for heavy metal poisoning.

DOSAGE & USAGE INFORMATION

How to take:
Tablet or capsule—Swallow with liquid on an empty stomach 1 hour before or 2 hours after eating.

When to take:
At the same times each day.

If you forget a dose:
Take as soon as you remember. If it is almost time for the next dose, wait for that dose (don't double this dose) and resume regular schedule.

What drug does:
- Combines with heavy metals so kidney can excrete them.
- Combines with cysteine (amino acid found in many foods) to prevent cysteine kidney stones.
- May improve protective function of some white blood cells against rheumatoid arthritis.

Continued next column

OVERDOSE

SYMPTOMS:
Unknown.
WHAT TO DO:
- **Dial 911 for all medical emergencies or call poison control center 1-800-222-1222 for instructions.**
- **See emergency information on last 3 pages of this book.**

Time lapse before drug works:
2 to 3 months.

Don't take with:
Any other medicine or any dietary supplement without consulting your doctor or pharmacist.

POSSIBLE ADVERSE REACTIONS OR SIDE EFFECTS

SYMPTOMS	WHAT TO DO
Life-threatening:	
In case of overdose, see previous column.	
Common:	
• Rash, itchy skin, swollen lymph glands, sores on body, sores or white spots in mouth, fever, joint pain.	Discontinue. Call doctor right away.
• Appetite loss, nausea, diarrhea, vomiting, decreased taste.	Continue. Call doctor when convenient.
Infrequent:	
Sore throat, fatigue, unusual bruising, swollen feet or legs, bloody or cloudy urine, weight gain, weakness.	Discontinue. Call doctor right away.
Rare:	
Eye pain, changes in vision, ringing in ears, difficult breathing, coughing up blood, yellow skin or eyes, abdominal pain, skin blisters, peeling skin, black or tarry stools, dark urine, lower back or side pain.	Discontinue. Call doctor right away.

WARNINGS & PRECAUTIONS

Don't take if:
- You are allergic to penicillamine. Allergic reactions are rare, but may occur.
- You have severe anemia.

Before you start, consult your doctor if:
- You have kidney disease.
- You are allergic to any penicillin antibiotic.

Over age 60:
Adverse reactions and side effects may be more frequent and severe than in younger persons.

Pregnancy:
Consult doctor. Use of the drug during pregnancy is not recommended. It could cause harm to the unborn baby. Risk category D (see page xviii).

Breastfeeding; Lactation; Nursing Mothers:
This drug is generally acceptable with breastfeeding. Discuss risks and benefits with your doctor.

Infants & children up to age 18:
Follow instructions provided by your child's doctor.

Prolonged use:
- May damage blood cells, kidney, liver.
- Talk to your doctor about the need for follow-up medical examinations or laboratory studies.

Skin & sunlight:
No problems expected.

Driving, piloting or hazardous work:
No problems expected.

Discontinuing:
No problems expected.

Others:
- Drug use carries risk of toxicity. It can be life threatening in people with certain disorders (e.g., aplastic anemia, agranulocytosis, thrombocytopenia and others). You should be closely supervised by a medical professional.
- Request laboratory studies on blood and urine every 2 weeks. Kidney and liver function studies recommended every 6 months.
- Advise any doctor or dentist you consult that you use this medicine.

POSSIBLE INTERACTION WITH OTHER DRUGS

GENERIC NAME OR DRUG CLASS	COMBINED EFFECT
Gold compounds*	Damage to blood cells and kidney.
Immuno-suppressants*	Damage to blood cells and kidney.
Iron supplements*	Decreased effect of penicillamine. Wait 2 hours between doses.
Pyridoxine (vitamin B-6)	Increased need for pyridoxine.

POSSIBLE INTERACTION WITH OTHER SUBSTANCES

INTERACTS WITH	COMBINED EFFECT
Alcohol:	Increased side effects of penicillamine.
Beverages:	None expected.
Cocaine:	Increased side effects of penicillamine. Avoid.
Foods:	Possible decreased penicillamine effect due to decreased absorption.
Marijuana:	Increased side effects of penicillamine. Avoid.
Tobacco:	None expected.

*See Glossary

PENICILLINS

GENERIC AND BRAND NAMES

See full list of generic and brand names in the *Generic and Brand Name Directory*, page 873.

BASIC INFORMATION

Habit forming? No
Prescription needed? Yes
Available as generic? Yes
Drug class: Antibacterial

USES

Treatment of bacterial infections that are susceptible to penicillin, including lower respiratory tract infections, otitis media, sinusitis, skin and skin structure infections, urinary tract infections, gastrointestinal disorders, ulcers, endocarditis, pharyngitis. Different penicillins treat different kinds of infections.

DOSAGE & USAGE INFORMATION

How to take:

- Tablet or capsule—Swallow with liquid on an empty stomach 1 hour before or 2 hours after eating. You may take amoxicillin, penicillin V, pivampicillin or pivmecillinam on a full stomach.
- Chewable tablet—Chew or crush before swallowing.
- Oral suspension—Measure each dose with an accurate measuring device (not a household teaspoon). Store according to instructions.
- Tablets for oral suspension—Mix one tablet in 2 teaspoonfuls of water. Drink right away.

When to take:
Follow instructions on prescription label, or take as directed by doctor. The number of doses, the time between doses and the length of treatment will depend on the problem being treated.

Continued next column

OVERDOSE

SYMPTOMS:
Nausea, vomiting, diarrhea, skin rash, changes in behavior, confusion, less urine.
WHAT TO DO:

- **Dial 911 for all medical emergencies or call poison control center 1-800-222-1222 for instructions.**
- **See emergency information on last 3 pages of this book.**

If you forget a dose:
Take as soon as you remember. If it is almost time for the next dose, wait for that dose (don't double this dose) and resume regular schedule.

What drug does:
Destroys susceptible bacteria. Does not kill viruses (e.g., colds or influenza), fungi or parasites.

Time lapse before drug works:
May be several days before medicine affects infection.

Don't take with:
Any other medicine or any dietary supplement without consulting your doctor or pharmacist.

POSSIBLE ADVERSE REACTIONS OR SIDE EFFECTS

SYMPTOMS	WHAT TO DO
Life-threatening:	
Rare, severe allergic reaction (anaphylaxis)—difficulty breathing, hives, itching, swelling (throat, face), vomiting, dizziness.	Seek emergency treatment immediately.
Common:	
Nausea, vomiting or diarrhea (all mild), sore mouth or tongue, white patches in mouth or on tongue, vaginal itching or discharge, stomach upset, black tongue.	Continue. Call doctor when convenient.
Infrequent:	
None expected.	
Rare:	
New symptoms that cause concern (may be from illness or possibly from drug).	Continue. Call doctor when convenient.

WARNINGS & PRECAUTIONS

Don't take if:
You are allergic to penicillins* or cephalosporins.* A life-threatening reaction may occur.

Before you start, consult your doctor if:
- You are allergic to any substance or drug.
- You have mononucleosis.
- You have congestive heart failure.
- You have high blood pressure or any bleeding disorder.
- You have cystic fibrosis.
- You have kidney disease or a stomach or intestinal disorder.

Over age 60:
No special problems expected.

Pregnancy:
Decide with your doctor if drug benefits justify risk to unborn baby. Risk category B (see page xviii).

Breastfeeding; Lactation; Nursing Mothers:
Drugs in this group are generally acceptable with breastfeeding. Discuss risks and benefits with your doctor.

Infants & children up to age 18:
Follow instructions provided by your child's doctor.

Prolonged use:
- You may become more susceptible to infections caused by germs not responsive to penicillins.
- Talk to your doctor about the need for follow-up medical examinations or laboratory studies.

Skin & sunlight:
No problems expected.

Driving, piloting or hazardous work:
Usually not dangerous. Most hazardous reactions likely to occur a few minutes after taking penicillin.

Discontinuing:
Don't discontinue without doctor's advice until you complete prescribed dose, even though symptoms diminish or disappear.

Others:
- Urine sugar test for diabetes may show false positive result.
- If your symptoms don't improve within a few days (or if they worsen), call your doctor.
- Don't take medicines for diarrhea without your doctor's approval.
- Birth control pills may not be effective. Use additional birth control methods.

POSSIBLE INTERACTION WITH OTHER DRUGS

GENERIC NAME OR DRUG CLASS	COMBINED EFFECT
Chloramphenicol	Decreased effect of both drugs.
Cholestyramine	May decrease penicillin effect.
Colestipol	May decrease penicillin effect.
Contraceptives, oral*	Impaired contraceptive efficiency.
Erythromycins*	Decreased effect of both drugs.
Methotrexate	Increased risk of methotrexate toxicity.
Probenecid	Increased effect of all penicillins.
Sodium benzoate & sodium phenylacetate	May reduce effect of sodium benzoate & sodium phenylacetate.
Sulfonamides*	Decreased penicillin effect.
Tetracyclines*	Decreased effect of both drugs.

POSSIBLE INTERACTION WITH OTHER SUBSTANCES

INTERACTS WITH	COMBINED EFFECT
Alcohol:	Occasional stomach irritation.
Beverages:	None expected.
Cocaine:	Unknown. Avoid.
Foods: Acidic fruits or juices, aged cheese, wines, syrups (if taken with penicillin G).	Decreased antibiotic effect.
Marijuana:	Consult doctor.
Tobacco:	None expected.

*See Glossary

PENICILLINS & BETA-LACTAMASE INHIBITORS

GENERIC AND BRAND NAMES

AMOXICILLIN & CLAVULANATE
- **Augmentin**
- **Augmentin ES-600**
- **Augmentin SR**
- **Clavulin**

BASIC INFORMATION

Habit forming? No
Prescription needed? Yes
Available as generic? Yes
Drug class: Antibacterial

USES

Treatment of bacterial infections that are susceptible to penicillin and beta-lactamase inhibitors, including lower respiratory tract infections, otitis media, sinusitis, skin and skin structure infections, and urinary tract infections.

DOSAGE & USAGE INFORMATION

How to take:
- Tablet—Swallow with liquid on a full or empty stomach. Taking with food may lessen any stomach irritation.
- Chewable tablet—Chew or crush before swallowing.
- Oral suspension—Measure each dose with an accurate device (not a household teaspoon).

When to take:
Follow instructions on prescription label, or take as directed by doctor. Normally the drug is taken every 8 hours for 7 to 10 days.

Continued next column

OVERDOSE

SYMPTOMS:
Nausea, vomiting, diarrhea, skin rash, changes in behavior, confusion, less urine.
WHAT TO DO:
- **Dial 911 for all medical emergencies or call poison control center 1-800-222-1222 for instructions.**
- **See emergency information on last 3 pages of this book.**

If you forget a dose:
Take as soon as you remember. If it is almost time for the next dose, wait for that dose (don't double this dose) and resume regular schedule.

What drug does:
Destroys susceptible bacteria. Does not kill viruses, fungi or parasites. Beta-lactamase inhibitors increase penicillin's effectiveness by inactivating beta-lactamase (a substance in some bacteria which destroys the penicillin).

Time lapse before drug works:
May be several days before medicine affects infection.

Don't take with:
Any other medicine or any dietary supplement without consulting your doctor or pharmacist.

POSSIBLE ADVERSE REACTIONS OR SIDE EFFECTS

SYMPTOMS	WHAT TO DO
Life-threatening:	
Rare, severe allergic reaction (anaphylaxis)—difficulty breathing, hives, itching, swelling (throat, face), vomiting, dizziness.	Seek emergency treatment immediately.
Common:	
Nausea, vomiting or diarrhea (all mild), sore mouth or tongue, white patches in mouth or on tongue, vaginal itching or discharge, stomach upset, black tongue.	Continue. Call doctor when convenient.
Infrequent:	
None expected.	
Rare:	
Unexplained bleeding or bruising, tiredness, sore throat, fever, severe abdominal cramps, yellow eyes or skin, dark urine, severe vomiting.	Discontinue. Call doctor right away.

PENICILLINS & BETA-LACTAMASE INHIBITORS

WARNINGS & PRECAUTIONS

Don't take if:
You are allergic to penicillins or cephalosporins. Life-threatening reaction may occur.

Before you start, consult your doctor if:
- You are allergic to any substance or drug.
- You have mononucleosis.
- You have congestive heart failure.
- You have high blood pressure or any bleeding disorder.
- You have cystic fibrosis.
- You have kidney disease or a stomach or intestinal disorder.

Over age 60:
No special problems expected.

Pregnancy:
Decide with your doctor if drug benefits justify risk to unborn baby. Risk category B (see page xviii).

Breastfeeding; Lactation; Nursing Mothers:
Drugs in this group are generally acceptable with breastfeeding. Discuss risks and benefits with your doctor.

Infants & children up to age 18:
Follow instructions provided by your child's doctor.

Prolonged use:
- You may become more susceptible to infections caused by germs not responsive to penicillins.
- Talk to your doctor about the need for follow-up medical examinations or laboratory studies.

Skin & sunlight:
No problems expected.

Driving, piloting or hazardous work:
Usually not dangerous. Most hazardous reactions likely to occur a few minutes after taking.

Discontinuing:
Don't discontinue without doctor's advice until you complete prescribed dose, even though symptoms diminish or disappear.

Others:
- Urine sugar test for diabetes may show false positive result.
- If your symptoms don't improve within a few days (or if they worsen), call your doctor.
- Don't take for diarrhea without your doctor's approval.
- Birth control pills may not be effective. Use additional birth control methods.

POSSIBLE INTERACTION WITH OTHER DRUGS

GENERIC NAME OR DRUG CLASS	COMBINED EFFECT
Chloramphenicol	Decreased effect of both drugs.
Cholestyramine	May decrease penicillin effect.
Colestipol	May decrease penicillin effect.
Contraceptives, oral*	Impaired contraceptive efficiency.
Erythromycins*	Decreased effect of both drugs.
Methotrexate	Increased risk of methotrexate toxicity.
Probenecid	Increased effect of all penicillins.
Sodium benzoate & sodium phenylacetate	May reduce effect of sodium benzoate & sodium phenylacetate.
Tetracyclines*	Decreased effect of both drugs.

POSSIBLE INTERACTION WITH OTHER SUBSTANCES

INTERACTS WITH	COMBINED EFFECT
Alcohol:	Occasional stomach irritation.
Beverages:	None expected.
Cocaine:	Unknown. Avoid.
Foods:	None expected.
Marijuana:	Consult doctor.
Tobacco:	None expected.

*See Glossary

PENTAMIDINE

BRAND NAMES

NebuPent
Pentacarinat
Pneumopent

BASIC INFORMATION

Habit forming? No
Prescription needed? Yes
Available as generic? No
Drug class: Antiprotozoal

USES

- Treats pneumocystis pneumonia caused by *Pneumocystis jirovecii*.
- Treats some tropical diseases such as leishmaniasis, African sleeping sickness and others.

DOSAGE & USAGE INFORMATION

How to take:
Inhalation—Follow package instructions.

When to take:
According to doctor's instructions.

If you forget a dose:
Take as soon as you remember. If it is almost time for the next dose, wait for that dose (don't double this dose) and resume regular schedule.

What drug does:
Interferes with RNA and DNA of infecting organisms.

Time lapse before drug works:
30 minutes to 1 hour.

Don't take with:
Any other medicines (including over-the-counter drugs such as cough and cold medicines, laxatives, antacids, diet pills, caffeine, nose drops or vitamins) without consulting your doctor or pharmacist.

OVERDOSE

SYMPTOMS:
Unknown.
WHAT TO DO:
If person uses much larger amount than prescribed or if accidentally swallowed, call poison control center 1-800-222-1222 for instructions or dial 911 (emergency) for help.

POSSIBLE ADVERSE REACTIONS OR SIDE EFFECTS

SYMPTOMS	WHAT TO DO
Life-threatening:	
Rare, severe allergic reaction (anaphylaxis)—difficulty breathing, hives, itching, swelling (throat, face), vomiting, dizziness.	Seek emergency treatment immediately.
Common:	
Chest pain or congestion, wheezing, coughing, difficulty in breathing, skin rash, throat pain or dryness or sensation of lump.	Discontinue. Call doctor right away.
Infrequent:	
• Abdomen or back pain, nausea, vomiting, anxiety, cold sweats, chills, headache, appetite changes, decreased urination, unusual tiredness.	Discontinue. Call doctor right away.
• Bitter or metallic taste.	No action necessary.
Rare:	
None expected.	

WARNINGS & PRECAUTIONS

Don't take if:
You are allergic to pentamidine.

Before you start, consult your doctor if:
You have asthma.

Over age 60:
Adverse reactions and side effects may be more frequent and severe than in younger persons.

Pregnancy:
Decide with your doctor if drug benefits justify risk to unborn baby. Risk category C (see page xviii).

Breastfeeding; Lactation; Nursing Mothers:
This drug is generally not recommended with breastfeeding. Discuss risks and benefits with your doctor.

Infants & children up to age 18:
Follow instructions provided by your child's doctor.

Prolonged use:
Talk to your doctor about the need for follow-up medical examinations or laboratory studies.

Skin & sunlight:
No problems expected.

Driving, piloting or hazardous work:
Avoid if you feel confused, drowsy or dizzy.

Discontinuing:
No special problems expected.

Others:
- Advise any doctor, dentist or pharmacist whom you consult that you take this medicine.
- May affect results in some medical tests.
- Avoid exposure to people who have infectious diseases.
- To help decrease bitter taste in mouth, suck on a hard candy after taking medicine.
- Consult your doctor for additional information on the injectable form of this drug.

POSSIBLE INTERACTION WITH OTHER DRUGS

GENERIC NAME OR DRUG CLASS	COMBINED EFFECT
None reported with inhalation form of drug.	

POSSIBLE INTERACTION WITH OTHER SUBSTANCES

INTERACTS WITH	COMBINED EFFECT
Alcohol:	Increased risk of adverse reactions. Avoid.
Beverages:	None expected.
Cocaine:	Increased risk of adverse reactions. Avoid.
Foods:	None expected.
Marijuana:	Increased risk of adverse reactions. Avoid.
Tobacco:	Increased risk of adverse reactions. Avoid.

*See Glossary

PENTOXIFYLLINE

BRAND NAMES

Trental

BASIC INFORMATION

Habit forming? No
Prescription needed? Yes
Available as generic? Yes
Drug class: Hemorheologic agent

USES

- Reduces pain in legs caused by poor blood circulation (usually due to intermittent claudication).
- May be used for other disorders as determined by your doctor.

DOSAGE & USAGE INFORMATION

How to take:
Extended-release tablet—Swallow whole with liquid. Do not crush or crumble tablet.

When to take:
At mealtimes. Taking with food decreases the likelihood of irritating the stomach to cause nausea. May take with antacids to help prevent stomach irritation.

If you forget a dose:
Take as soon as you remember. If it is almost time for the next dose, wait for that dose (don't double this dose) and resume regular schedule.

What drug does:
- Reduces "stickiness" of red blood cells and improves flexibility of the red cells.
- Improves blood flow through blood vessels.

Time lapse before drug works:
Several weeks for full effect on circulation.

Don't take with:
Any other medicine or any dietary supplement without consulting your doctor or pharmacist.

OVERDOSE

SYMPTOMS:
Drowsiness, flushed face, fainting, unusual agitation, loss of consciousness, convulsions.
WHAT TO DO:
- **Dial 911 for all medical emergencies or call poison control center 1-800-222-1222 for instructions.**
- **See emergency information on last 3 pages of this book.**

POSSIBLE ADVERSE REACTIONS OR SIDE EFFECTS

SYMPTOMS	WHAT TO DO
Life-threatening: None expected.	
Common: None expected.	
Infrequent: Drowsiness, headache, nausea, vomiting, stomach upset.	Continue. Call doctor when convenient.
Rare: Chest pain, irregular heartbeat.	Discontinue. Call doctor right away.

WARNINGS & PRECAUTIONS

Don't take if:
You are allergic to pentoxifylline or other xanthines (caffeine, theophylline, theobromine, aminophylline, dyphylline, or oxtriphylline). Allergic reactions are rare, but may occur.

Before you start, consult your doctor if:
- You have coronary artery disease.
- You have active bleeding or any condition where there is a risk of bleeding (e.g., stroke).
- You have cerebrovascular (blood vessels in the brain) disease.
- You have liver or kidney disease.

Over age 60:
Adverse reactions and side effects may be more frequent and severe than in younger persons.

Pregnancy:
Decide with your doctor if drug benefits justify risk to unborn baby. Risk category C (see page xviii).

Breastfeeding; Lactation; Nursing Mothers:
This drug is generally acceptable with breastfeeding. Discuss risks and benefits with your doctor.

Infants & children up to age 18:
Safety and effectiveness in this age group have not been established. Consult doctor.

Prolonged use:
No problems expected.

Skin & sunlight:
No problems expected.

Driving, piloting or hazardous work:
Wait to see if drug causes drowsiness. If not, no problems expected.

Discontinuing:
Don't discontinue without doctor's approval.

Others:
- Don't smoke. Nicotine constricts blood vessels and worsens your condition.
- Advise any doctor or dentist you consult that you use this medicine.

POSSIBLE INTERACTION WITH OTHER DRUGS

GENERIC NAME OR DRUG CLASS	COMBINED EFFECT
Anticoagulants,* oral	Possible decreased effect of anticoagulant.
Antihypertensives*	Possible increased effect of hypertensive drug.
Cimetidine	Increased risk of side effects.
Xanthines*	Increased nervous system stimulation.

POSSIBLE INTERACTION WITH OTHER SUBSTANCES

INTERACTS WITH	COMBINED EFFECT
Alcohol:	Unknown. Best to avoid.
Beverages: Coffee, tea or other caffeine-containing beverages.	Increased nervous system stimulation.
Cocaine:	Unknown Avoid.
Foods:	None expected.
Marijuana:	Consult doctor.
Tobacco:	Decreased effect of pentoxifylline. Avoid.

*See Glossary

PHENAZOPYRIDINE

BRAND NAMES

Azo-Cheragan
Azo-Gantrisin
Azo-Standard
Baridium
Eridium
Geridium
Phen-Azo
Phenazodine
Pyrazodine
Pyridiate
Pyridium
Pyronium
Urodine
Urogesic
Viridium

BASIC INFORMATION

Habit forming? No
Prescription needed? Yes
Available as generic? Yes
Drug class: Analgesic (urinary)

USES

Relieves pain of lower urinary tract irritation, as in cystitis, urethritis or prostatitis. Relieves symptoms only. Phenazopyridine alone does not cure infections.

DOSAGE & USAGE INFORMATION

How to take:
Tablet—Swallow with liquid or food to lessen stomach irritation.

When to take:
At the same times each day.

If you forget a dose:
Take as soon as you remember. If it is almost time for the next dose, wait for that dose (don't double this dose) and resume regular schedule.

What drug does:
Anesthetizes lower urinary tract. Relieves pain, burning, pressure and urgency to urinate.

Continued next column

OVERDOSE

SYMPTOMS:
Changes in skin color, tiredness, shortness of breath, easy bruising or bleeding, fast heartbeat, yellow skin or eyes, seizures.
WHAT TO DO:

- **Dial 911 for all medical emergencies or call poison control center 1-800-222-1222 for instructions.**
- **See emergency information on last 3 pages of this book.**

Time lapse before drug works:
1 to 2 hours.

Don't take with:
Any other medicine or any dietary supplement without consulting your doctor or pharmacist.

POSSIBLE ADVERSE REACTIONS OR SIDE EFFECTS

SYMPTOMS	WHAT TO DO
Life-threatening:	
Rare, severe allergic reaction (anaphylaxis)—difficulty breathing, hives, itching, swelling (throat, face), vomiting, dizziness.	Seek emergency treatment immediately.
Common:	
Red-orange color of urine or stool.	No action necessary.
Infrequent:	
Indigestion, dizziness, headache, itching skin, stomach cramps.	Continue. Call doctor when convenient.
Rare:	
Rash, yellow skin or eyes, bluish skin color, change in amount of urine, weight gain, fever, confusion, swelling (face, feet, legs, fingers), shortness of breath, skin rash, unusual tiredness or weakness.	Discontinue. Call doctor right away.

WARNINGS & PRECAUTIONS

Don't take if:

- You have hepatitis.
- You are allergic to any urinary analgesic.

Before you start, consult your doctor if:
You have kidney or liver disease.

Over age 60:
Adverse reactions and side effects may be more frequent and severe than in younger persons.

Pregnancy:
Decide with your doctor if drug benefits justify risk to unborn baby. Risk category B (see page xviii).

Breastfeeding; Lactation; Nursing Mothers:
This drug is generally not recommended with breastfeeding. Discuss risks and benefits with your doctor.

Infants & children up to age 18:
Follow instructions provided by your child's doctor.

Prolonged use:
Rarely, it may lead to yellowish color of the skin or sclera of the eye. Consult doctor on regular basis to determine effectiveness of drug and for any adverse effects.

Skin & sunlight:
No problems expected.

Driving, piloting or hazardous work:
No problems expected.

Discontinuing:
Follow doctor's instructions.

Others:

- Advise any doctor, dentist or pharmacist whom you consult that you take this medicine.
- Drug may stain clothing.
- Drug may discolor soft contact lenses.

POSSIBLE INTERACTION WITH OTHER DRUGS

GENERIC NAME OR DRUG CLASS	COMBINED EFFECT
None significant.	

POSSIBLE INTERACTION WITH OTHER SUBSTANCES

INTERACTS WITH	COMBINED EFFECT
Alcohol:	None expected.
Beverages:	None expected.
Cocaine:	Unknown. Avoid.
Foods:	None expected.
Marijuana:	Consult doctor.
Tobacco:	None expected.

PHENOTHIAZINES

GENERIC AND BRAND NAMES

See full list of generic and brand names in the *Generic and Brand Name Directory*, page 874.

BASIC INFORMATION

Habit forming? No
Prescription needed? Yes
Available as generic? Yes, for some.
Drug class: Antipsychotic, antiemetic (phenothiazine)

USES

- Treatment for mental and emotional disorders.
- Treats nausea, vomiting, hiccups.
- May be used for other conditions as determined by your doctor.

DOSAGE & USAGE INFORMATION

How to take:

- Tablet, sustained release capsule, extended-release capsule—Swallow with liquid or food to lessen stomach irritation.
- Suppository—Remove wrapper and moisten suppository with water. Gently insert into rectum, pointed end first.
- Drops or liquid—Dilute dose in beverage.

When to take:
Times will vary. Take at the same times each day as directed by your doctor.

If you forget a dose:
Take as soon as you remember. If it is almost time for the next dose, wait for that dose (don't double this dose) and resume regular schedule.

What drug does:

- Suppresses brain centers that control abnormal emotions and behavior.
- Suppresses brain's vomiting center.

Continued next column

OVERDOSE

SYMPTOMS:
Agitation, restlessness, dry mouth, fever, irregular heartbeat, stupor, convulsions, coma.

WHAT TO DO:

- **Dial 911 for all medical emergencies or call poison control center 1-800-222-1222 for instructions.**
- **See emergency information on last 3 pages of this book.**

Time lapse before drug works:
Some benefit seen within a week; takes 4 to 6 weeks for full effect.

Don't take with:

- Antacid or medicine for diarrhea at same time.
- Any other medicine or any dietary supplement without consulting your doctor or pharmacist.

POSSIBLE ADVERSE REACTIONS OR SIDE EFFECTS

SYMPTOMS	WHAT TO DO
Life-threatening:	
High fever, rapid pulse, profuse sweating, muscle rigidity, confusion and irritability, seizures.	Discontinue. Seek emergency treatment.
Common:	
Dry mouth, blurred vision, constipation, difficulty urinating, sedation, dizziness, low blood pressure.	Continue. Call doctor when convenient.
Infrequent:	
Continuous jerky or involuntary movements (especially of the face, lips, jaw, tongue), slow-frequency tremor of head or limbs, (especially while moving), muscle rigidity, lack of facial expression, slow and inflexible movements, pacing or restlessness, spasms of muscles (of face, eyes, tongue, jaw, neck, body or limbs), yellow skin or eyes.	Discontinue. Call doctor right away.
Rare:	
Other symptoms not listed above.	Continue. Call doctor when convenient.

WARNINGS & PRECAUTIONS

Don't take if:

- You are allergic to any phenothiazine. Allergic reactions are rare, but may occur.
- You have a blood or bone marrow disease.

Before you start, consult your doctor if:
- You will have surgery within 2 months, including dental surgery, requiring general or spinal anesthesia.
- You have asthma or emphysema or other lung disorder, glaucoma, or prostate trouble.
- You take nonprescription ulcer medicine, asthma medicine or amphetamines.

Over age 60:
- Adverse reactions and side effects may be more frequent and severe than in younger persons. More likely to develop involuntary movement of jaws, lips, tongue. Report this to your doctor immediately. Early treatment can help.
- Use of antipsychotic drugs in elderly patients with dementia-related psychosis may increase risk of death. Consult doctor.

Pregnancy:
Risk factors vary for drugs in this group. Always consult doctor. See category list on page xviii.

Breastfeeding; Lactation; Nursing Mothers:
Some of the drugs in this group are generally acceptable with breastfeeding. Discuss risks and benefits with your doctor.

Infants & children up to age 18:
Follow instructions provided by your child's doctor.Children more likely than adults to have adverse reactions from these drugs.

Prolonged use:
May lead to tardive dyskinesia (involuntary movement of jaws, lips, tongue).

Skin & sunlight:
- One or more drugs in this group may cause rash or intensify sunburn in areas exposed to sun or ultraviolet light (photosensitivity reaction). Use sunscreen and avoid over-exposure. Notify doctor if reaction occurs. Sensitivity may remain for 3 months after discontinuing drug.
- Avoid getting overheated or chilled. These drugs affect body temperature and sweating.

Driving, piloting or hazardous work:
Don't drive or pilot aircraft until you learn how medicine affects you. Don't work around dangerous machinery. Don't climb ladders or work in high places. Danger increases if you drink alcohol or take medicine affecting alertness and reflexes.

Discontinuing:
- Nervous and mental disorders—Don't discontinue without doctor's advice until you complete prescribed dose, even though symptoms diminish or disappear.
- Other disorders—Follow doctor's instructions about discontinuing.
- Adverse reactions may occur after drug is discontinued. Consult doctor if new symptoms develop, such as dizziness, nausea, stomach pain, trembling or tardive dyskinesia.*

Others:
- To relieve mouth dryness, chew or suck sugarless gum, candy, or ice.
- Avoid getting the liquid form of the drug on the skin. It may cause a skin rash or irritation.
- Advise any doctor, dentist or pharmacist whom you consult that you take this medicine.

POSSIBLE INTERACTION WITH OTHER DRUGS

GENERIC NAME OR DRUG CLASS	COMBINED EFFECT
Anticholinergics*	Increased phenothiazine effect.
Anticonvulsants*	Increased risk of seizures. May need to increase dosage of anticonvulsant.
Antidepressants, tricyclic*	Increased anti-depressant effect.
Antihistamines*	Increased antihistamine effect.
Antihypertensives*	Severe low blood pressure.
Appetite suppressants*	Decreased appetite suppressant effect.
Bupropion	Increased risk of seizures.
Clozapine	Toxic effect on the central nervous system.
Dofetilide	Increased risk of heart problems.

Continued on page 907

POSSIBLE INTERACTION WITH OTHER SUBSTANCES

INTERACTS WITH	COMBINED EFFECT
Alcohol:	Dangerous oversedation. Avoid.
Beverages:	None expected.
Cocaine:	Increased risk of side effects. Avoid.
Foods:	None expected.
Marijuana:	Drowsiness. May increase antinausea effect.
Tobacco:	None expected.

***See Glossary**

PHENYLEPHRINE

BRAND NAMES

See full list of brand names in the *Generic and Brand Name Directory*, page 875.

BASIC INFORMATION

Habit forming? No
Prescription needed? No
Available as generic? Yes
Drug class: Sympathomimetic, decongestant

USES

- Temporary relief of congestion of nose and sinuses caused by allergies, colds, hay fever or sinus infection.
- Treats congestion of eustachian tubes caused by middle ear infections.

DOSAGE & USAGE INFORMATION

How to take:
- Nasal drops or spray—Wash hands before use. Blow nose gently, and then use the drops or spray according to instructions on the label.
- Combination drug products—Use according to label directions. Phenylephrine is the decongestant ingredient in many combination cough, cold and hay fever remedies.

When to take:
Follow directions on product's label.

If you forget a dose:
Take as soon as you remember. If it is almost time for the next dose, wait for that dose (don't double this dose) and resume regular schedule.

What drug does:
Narrows the blood vessels in the nasal passages or ears which helps relieve the stuffy feeling caused by congestion.

Continued next column

OVERDOSE

SYMPTOMS:
Dizziness, fainting, hallucinations, slow and shallow breathing, irregular or fast heartbeat, vomiting, seizures.
WHAT TO DO:
- **Dial 911 for all medical emergencies or call poison control center 1-800-222-1222 for instructions.**
- **See emergency information on last 3 pages of this book.**

Time lapse before drug works:
5 to 30 minutes.

Don't take with:
Nonprescription drugs for asthma, cough, cold, allergy, appetite suppressants, sleeping pills or drugs containing caffeine without consulting your doctor or pharmacist.

POSSIBLE ADVERSE REACTIONS OR SIDE EFFECTS

SYMPTOMS	WHAT TO DO
Life-threatening:	
None expected.	
Common:	
Dizziness, insomnia, nervousness, lightheadedness.	Discontinue. Call doctor if symptoms persist.
Infrequent:	
Nasal product may cause burning, dryness, stinging inside nose.	Discontinue. Call doctor if symptoms persist.
Rare:	
• Unusual behavior, fast or pounding heartbeat, seizure, severe shaking.	Discontinue. Call doctor right away.
• Headache, nausea, vomiting, anxiety, mild shaking, sweating.	Discontinue. Call doctor if symptoms persist.

WARNINGS & PRECAUTIONS

Don't take if:
You are allergic to phenylephrine or any sympathomimetic.* Allergic reactions are rare, but may occur.

Before you start, consult your doctor if:
- You have high blood pressure.
- You have heart disease.
- You have diabetes.
- You have overactive thyroid.
- You have prostate problems.
- You have urination problems.
- You have glaucoma.
- You have taken MAO inhibitors in past 2 weeks.

Over age 60:
Adverse reactions and side effects may be more frequent and severe than in younger persons.

Pregnancy:
Decide with your doctor if drug benefits justify risk to unborn baby. Risk category C (see page xviii).

Breastfeeding; Lactation; Nursing Mothers:
This drug is generally acceptable with breastfeeding. Discuss risks and benefits with your doctor.

Infants & children up to age 18:
Read nonprescription product's label to be sure it is approved for your child's age. Follow label instructions on dosage. Consult doctor or pharmacist if unsure.

Prolonged use:
Do not use product for longer than advised on label or by doctor. Nasal spray may cause rebound congestion* if used longer than recommended on label.

Skin & sunlight:
No problems expected.

Driving, piloting or hazardous work:
Drug may cause dizziness or lightheadedness. Don't drive or pilot aircraft until you learn how medicine affects you. Don't work around dangerous machinery. Don't climb ladders or work in high places.

Discontinuing:
May be unnecessary to finish medicine. Follow label or doctor's instructions.

Others:
- Call the doctor if symptoms worsen or new symptoms develop with use of this medicine.
- Heed all warnings on the product label.
- Advise any doctor, dentist or pharmacist whom you consult that you take this medicine.

POSSIBLE INTERACTION WITH OTHER DRUGS

GENERIC NAME OR DRUG CLASS	COMBINED EFFECT
Antidepressants, tricyclic*	Increased phenylephrine effect.
Antihypertensives*	Increased risk of high blood pressure.
Monoamine oxidase (MAO) inhibitors*	Serious interaction (possibly fatal). Don't use within 14 days.

POSSIBLE INTERACTION WITH OTHER SUBSTANCES

INTERACTS WITH	COMBINED EFFECT
Alcohol:	Increased dizziness. Avoid.
Beverages:	None expected.
Cocaine:	Unknown. Avoid.
Foods:	None expected.
Marijuana:	Consult doctor.
Tobacco:	None expected.

***See Glossary**

PHENYLEPHRINE (Ophthalmic)

BRAND NAMES

Ak-Dilate
Ak-Nefrin
Dilatair
Dionephrine
I-Phrine
Isopto Frin
Minims Phenylephrine
Mydfrin
Neofrin
Ocugestrin
Ocu-Phrin
Phenoptic
Prefrin Liquifilm
Relief Eye Drops for Red Eyes
Spersaphrine

BASIC INFORMATION

Habit forming? No
Prescription needed? Yes, some strengths
Available as generic? No
Drug class: Mydriatic, decongestant (ophthalmic)

USES

- High-concentration drops—Dilates pupils.
- Low-concentration drops (available without prescription)—Relieves minor eye irritations caused by colds, hay fever, dust, wind, swimming, sun, smog, hard contact lenses, eye strain, smoke.

DOSAGE & USAGE INFORMATION

How to use:
Eye drops
- Wash hands.
- Apply pressure to inside corner of eye with middle finger.
- Continue pressure for 1 minute after placing medicine in eye.
- Tilt head backward. Pull lower lid away from eye with index finger of the same hand.
- Drop eye drops into pouch and close eye. Don't blink.
- Keep eyes closed for 1 to 2 minutes.

Continued next column

OVERDOSE

SYMPTOMS:
None expected unless too much is absorbed or drops are accidentally swallowed.
WHAT TO DO:
If person uses much larger amount than prescribed or if accidentally swallowed, call poison control center 1-800-222-1222 for instructions or dial 911 (emergency) for help.

- Don't touch applicator tip to any surface (including the eye). If you accidentally touch tip, clean with warm water and soap.
- Keep container tightly closed.
- Keep cool, but don't freeze.
- Wash hands immediately after using.

When to use:
As directed on label.

If you forget a dose:
Use as soon as you remember. If it is almost time for the next dose, wait for that dose (don't double this dose) and resume regular schedule.

What drug does:
Acts on small blood vessels to make them constrict.

Time lapse before drug works:
15 to 90 minutes.

Don't use with:
- Other eye drops or ointment without consulting your eye doctor.
- Antidepressants,* guanadrel, guanethidine, maprotiline, pargyline, any monoamine oxidase (MAO) inhibitor.*

POSSIBLE ADVERSE REACTIONS OR SIDE EFFECTS

SYMPTOMS	WHAT TO DO
Life-threatening:	
None expected, unless you use much more than directed.	Discontinue. Call doctor right away.
Common:	
None expected.	
Infrequent:	
Burning or stinging eyes, headache, eyes more sensitive to light, watery eyes, eye irritation not present before.	Continue. Call doctor when convenient.
Rare:	
None expected, unless too much gets absorbed. If so, symptoms may be paleness, dizziness, tremor, increased sweating, irregular or fast heartbeat.	Discontinue. Call doctor right away.

WARNINGS & PRECAUTIONS

Don't use if:

- You are allergic to phenylephrine. Allergic reactions are rare, but may occur.
- You have glaucoma.

Before you start, consult your doctor if:

- You have heart disease with irregular heartbeat, high blood pressure, diabetes.
- You take antidepressants,* guanadrel, guanethidine, maprotiline, pargyline, any monoamine oxidase (MAO) inhibitor.*

Over age 60:
Adverse reactions and side effects may be more frequent and severe than in younger persons.

Pregnancy:
Decide with your doctor if drug benefits justify risk to unborn baby. Risk category C (see page xviii).

Breastfeeding; Lactation; Nursing Mothers:
This drug is generally acceptable with breastfeeding. Discuss risks and benefits with your doctor.

Infants & children up to age 18:
Follow instructions provided by your child's doctor.

Prolonged use:
Avoid if possible.

Skin & sunlight:
No special problems expected.

Driving, piloting or hazardous work:
No problems expected.

Discontinuing:
No problems expected.

Others:
Consult doctor if condition doesn't improve in 3 to 4 days.

POSSIBLE INTERACTION WITH OTHER DRUGS

GENERIC NAME OR DRUG CLASS	COMBINED EFFECT

Clinically significant interactions with oral or injected medicines unlikely.

POSSIBLE INTERACTION WITH OTHER SUBSTANCES

INTERACTS WITH	COMBINED EFFECT
Alcohol:	None expected.
Beverages:	None expected.
Cocaine:	None expected.
Foods:	None expected.
Marijuana:	None expected.
Tobacco:	None expected.

PILOCARPINE (Oral)

BRAND NAMES

Salagen

BASIC INFORMATION

Habit forming? No
Prescription needed? Yes
Available as generic? Yes
Drug class: Cholinergic

USES

Treatment for dry mouth caused by radiation treatment of patients with cancer (of the head or neck) or patients with Sjogren's syndrome.

DOSAGE & USAGE INFORMATION

How to take:
Tablet—Follow doctor's directions. May require dosing several times a day.

When to take:
At the same times each day.

If you forget a dose:
Take as soon as you remember. If it is almost time for the next dose, wait for that dose (don't double this dose) and resume regular schedule.

What drug does:
Stimulates the salivary glands to increase their secretions.

Time lapse before drug works:
20-60 minutes.

Don't take with:
Any other medicine or any dietary supplement without consulting your doctor or pharmacist.

OVERDOSE

SYMPTOMS:
Stomach cramps or pain, diarrhea, severe nausea or vomiting, rapid heartbeat, chest pain, confusion, fainting, severe headache, shortness of breath, unusual trembling or shaking, visual problems.
WHAT TO DO:
- **Dial 911 for all medical emergencies or call poison control center 1-800-222-1222 for instructions.**
- **See emergency information on last 3 pages of this book.**

POSSIBLE ADVERSE REACTIONS OR SIDE EFFECTS

SYMPTOMS	WHAT TO DO
Life-threatening:	
In case of overdose, see previous column.	
Common:	
Runny nose, cough, fever, chills, diarrhea, indigestion, nausea, tiredness or weakness, warm feeling, sweating, skin flushing or red, urinary frequency, joint pain or muscle aches.	Continue. Call doctor when convenient.
Infrequent:	
• Swelling (ankles, feet, face or fingers), rapid heartbeat, visual problems, bloody nose, vomiting.	Discontinue. Call doctor right away.
• Trembling or shaking, trouble with swallowing, voice change, headache.	Continue. Call doctor when convenient.
Rare:	
None expected.	

WARNINGS & PRECAUTIONS

Don't take if:
- You are allergic to pilocarpine (ophthalmic or oral) Allergic reactions are rare, but may occur.
- You have uncontrolled asthma.

Before you start, consult your doctor if:
- You have gallbladder problems.
- You have iritis or glaucoma.
- You have had heart or blood vessel disease.
- You have asthma.
- You have any cognitive or psychiatric problem.
- You have or have had kidney disease.
- You have been told you have a tendency for retinal detachment or have retinal disease.
- You have peptic ulcer disease.

Over age 60:
No special problems expected.

Pregnancy:
Decide with your doctor if drug benefits justify risk to unborn baby. Risk category C (see page xviii).

Breastfeeding; Lactation; Nursing Mothers:
This drug is generally not recommended with breastfeeding. Discuss risks and benefits with your doctor.

Infants & children up to age 18:
Safety and effectiveness in this age group have not been established. Consult doctor.

Prolonged use:
If no improvement is seen after twelve weeks of using this medicine, consult doctor.

Skin & sunlight:
No problems expected.

Driving, piloting or hazardous work:
May cause visual disturbances, especially at night. Don't drive or pilot aircraft until you learn how medicine affects you. Don't work around dangerous machinery. Don't climb ladders or work in high places. Danger increases if you drink alcohol or take medicine affecting alertness and reflexes.

Discontinuing:
Consult doctor before discontinuing.

Others:
- Advise any doctor, dentist or pharmacist whom you consult that you take this medicine.
- Since this drug causes sweating, be sure to drink plenty of fluids.

POSSIBLE INTERACTION WITH OTHER DRUGS

GENERIC NAME OR DRUG CLASS	COMBINED EFFECT
Anticholinergics*	Decreased effect of both drugs.
Antiglaucoma agents*	Increased antiglaucoma effect.
Antiglaucoma, beta blockers	Increased risk of side effects.
Beta adrenergic blocking agents	Increased risk of side effects.
Bethanechol	Increased risk of side effects.
Cholinergics,* other	Increased effect of both drugs.

POSSIBLE INTERACTION WITH OTHER SUBSTANCES

INTERACTS WITH	COMBINED EFFECT
Alcohol:	None expected.
Beverages:	None expected.
Cocaine:	Unknown. Avoid.
Foods:	None expected.
Marijuana:	Consult doctor.
Tobacco:	None expected.

*See Glossary

PLATELET INHIBITORS

GENERIC AND BRAND NAMES

CLOPIDOGREL
Plavix
PRASUGREL
Effient
TICLOPIDINE
Ticlid

BASIC INFORMATION

Habit forming? No
Prescription needed? Yes
Available as generic? Yes, for some
Drug class: Antithrombotic; platelet aggregation inhibitor

USES

Used for high risk patients to help prevent blood clots and reduce the risk of stroke, heart attack, or other serious problems with the heart or blood vessels. May be used in combination with aspirin therapy.

DOSAGE & USAGE INFORMATION

How to take:
Tablet—Swallow with liquid. If you can't swallow tablet whole, ask pharmacist for advice. Take ticlopidine with food. Clopidogrel and prasugrel may be taken with or without food.

When to take:
As directed. Usually twice a day for ticlopidine and once a day for clopidogrel and prasugrel.

If you forget a dose:
Take as soon as you remember. If it is almost time for the next dose, wait for that dose (don't double this dose) and resume regular schedule.

What drug does:
Prevents certain blood cells from clumping together which reduces the risk of blood clots.

Time lapse before drug works:
1-2 hours. It will take several days for full benefit.

Don't take with:
Any other medicine or any dietary supplement without consulting your doctor or pharmacist.

OVERDOSE

SYMPTOMS:
Bloody vomit; bleeding from gums, nose or rectum; difficult breathing; seizures.
WHAT TO DO:
Dial 911 (emergency) for medical help or call poison control center 1-800-222-1222 for instructions.

POSSIBLE ADVERSE REACTIONS OR SIDE EFFECTS

SYMPTOMS	WHAT TO DO
Life-threatening:	
Rare, severe allergic reaction (anaphylaxis)—difficulty breathing, hives, itching, swelling (throat, face), vomiting, dizziness.	Seek emergency treatment immediately.
Common:	
• Chest pain, general body pain, red or purple spots on skin, dizziness, blurred vision.	Continue, but call doctor right away.
• Diarrhea, stomach upset, back pain, heartburn, muscle aches, symptoms of a cold.	Continue. Call doctor when convenient.
Infrequent:	
• Unusual bleeding or bruising, slow or fast or irregular heartbeat, shortness of breath, black or tarry stools, bloating or swelling, difficult or painful or decreased urination, swollen glands, chest discomfort, chills, cough, fever, fainting, tingling of hands or feet, unusual weight gain or loss, unusual tiredness or weakness, sore throat, sores on the lips or in mouth, lightheadedness.	Continue, but call doctor right away.
• Mild headache, mild weakness, rash, pain in arms or legs.	Continue. Call doctor when convenient.
Rare:	
• Sudden severe headache or weakness, severe stomach pain, peeling or flaking or blistering skin, being uncoordinated, symptoms of thrombotic thrombocytopenic purpura (mental changes, dark or bloody urine, difficult speaking, pale skin, seizures, yellow eyes or skin, fever, weakness).	Continue, but call doctor right away.

• Loss of appetite, gaseousness, nausea, mild tiredness, constipation, trouble sleeping, anxious or depressed.	Continue. Call doctor when convenient.

WARNINGS & PRECAUTIONS

Don't take if:

- You are allergic to any platelet inhibitor.
- You have active bleeding (e.g., head, bowel or stomach).

Before you start, consult your doctor if:

- You have liver disease, kidney disease, diabetes, stomach ulcers or diverticulitis.
- You have recurrent bleeding (e.g., head, bowel or stomach).
- You have history of stroke, transient ischemic attack (TIA), mini-stroke or heart disorder.
- You weigh under 132 pounds (60 kilograms).
- You have had a recent injury (trauma).
- You have had thrombotic thrombocytopenic purpura.
- You are planning or have had recent surgery or other medical procedure.

Over age 60:
No special problems expected. Prasugrel is not recommended for patients 75 years of age and older due to drug's toxicity.

Pregnancy:
Decide with your doctor if drug benefits justify risk to unborn baby. Risk category B (see page xviii).

Breastfeeding; Lactation; Nursing Mothers:
Drugs in this group are generally not recommended with breastfeeding. Discuss risks and benefits with your doctor.

Infants & children up to age 18:
Safety and effectiveness in this age group have not been established. Consult doctor.

Prolonged use:
Regular follow up visits to your doctor are important to monitor the effects of the drug.

Skin & sunlight:
No special problems expected.

Driving, piloting or hazardous work:
Don't drive or pilot aircraft until you learn how the medicine affects you. Don't work around dangerous machinery. Don't climb ladders or work in high places. Danger increases if you drink alcohol or take medicine affecting alertness and reflexes.

Discontinuing:
Don't discontinue without medical advice. It can increase your risk for heart attack or stroke.

Others:

- Advise any doctor, dentist or pharmacist whom you consult that you take this medicine. You may be advised to discontinue the drug 1-2 weeks before elective surgery (including dental surgery).
- Wear a medical identification bracelet or tag that indicates you are taking this drug.
- When possible, avoid situations or activities that can increase risk of bleeding or bruising.
- Report any unusual bleeding to your doctor.

POSSIBLE INTERACTION WITH OTHER DRUGS

GENERIC NAME OR DRUG CLASS	COMBINED EFFECT
Antacids*	Decreased ticlopidine effect.
Anticoagulants	Increased risk of bleeding.
Anti-inflammatory drugs, nonsteroidal (NSAIDs)*	Increased risk of bleeding.
Aspirin	Increased risk of bleeding.
Enzyme inhibitors*	Decreased effect of clopidogrel.
Phenytoin	Increased phenytoin effect with ticlopidine.
Proton pump inhibitors	Decreased effect of clopidogrel.
Thrombolytic agents*	Increased risk of bleeding.
Xanthines*	Increased xanthine effect with ticlopidine.

POSSIBLE INTERACTION WITH OTHER SUBSTANCES

INTERACTS WITH	COMBINED EFFECT
Alcohol:	Increased risk of side effects. Avoid.
Beverages:	None expected.
Cocaine:	Unknown. Avoid.
Foods:	None expected.
Marijuana:	Consult doctor.
Tobacco:	None expected.

POLLEN ALLERGEN EXTRACTS

GENERIC AND BRAND NAMES

GRASS POLLEN ALLERGEN EXTRACT
Oralair
RAGWEED POLLEN ALLERGEN EXTRACT
Ragwitek
TIMOTHY GRASS POLLEN ALLERGEN EXTRACT
Grastek

BASIC INFORMATION

Habit forming? No
Prescription needed? Yes
Available as generic? No
Drug class: Anti-allergen agent

USES

Treatment for allergic reactions (often called hay fever) caused by certain types of pollen (an allergen). Allergy symptoms include runny or stuffy nose, sneezing, and watery or itchy eyes.

DOSAGE & USAGE INFORMATION

How to take:
Sublingual tablet—Place the tablet under the tongue and allow it to completely dissolve. Do not swallow for at least 1 minute. Then wait 5 more minutes before eating any food or drinking any beverage. Wash hands after handling tablet.

When to take:
- The first tablet will be given in the doctor's office. You will be observed for at least 30 minutes to check for serious allergic reaction.
- Take tablet once a day at the same time each day. Begin taking 3 to 4 months prior to start of the pollen season and take until it ends.

Continued next column

OVERDOSE

SYMPTOMS:
May have milder allergic symptoms in mouth or throat or more severe symptoms as listed in next column under Life-threatening.
WHAT TO DO:
- **Dial 911 for all medical emergencies or call poison control center 1-800-222-1222 for instructions.**
- **See emergency information on last 3 pages of this book.**

If you forget a dose:
Take as soon as you remember. If it is almost time for the next dose, wait for next scheduled dose (don't double this dose). If you miss more than two doses, consult your doctor for advice.

What drug does:
The action is called immunotherapy. Over time, it changes the way the body's immune system reacts to allergens such as pollen and helps to reduce symptoms of an allergic reaction. It does not treat the immediate symptoms.

Time lapse before drug works:
It takes many weeks to months for maximum benefit.

Don't take with:
Any other medicine or any dietary supplement without consulting your doctor or pharmacist.

POSSIBLE ADVERSE REACTIONS OR SIDE EFFECTS

SYMPTOMS	WHAT TO DO
Life-threatening:	
Rare, severe allergic reaction (anaphylaxis)—difficulty breathing, hives, itching, swelling (throat, face), vomiting, dizziness.	Seek emergency treatment immediately.
Common:	
• Mouth swelling or numbness, cough, throat pain.	Discontinue. Call doctor right away.
• Itching (mouth, lips, tongue, ear), throat irritation.	Continue. Call doctor when convenient.
Infrequent:	
• Throat or lip or tongue swelling.	Discontinue. Call doctor right away.
• Mild symptoms of itching skin or headache or heartburn.	Continue. Call doctor when convenient.
Rare:	
• Throat tightness, swallowing is painful or difficult, trouble speaking, severe diarrhea or stomach cramps, severe skin itching or flushing, severe heartburn, fainting, rapid or weak heartbeat, chest pain.	Discontinue. Call doctor right away.
• Stuffy nose, dry throat, sneezing, fatigue, nausea.	Continue. Call doctor when convenient.

WARNINGS & PRECAUTIONS

Don't take if:
- You are allergic to any pollen allergen extract.
- You have severe, unstable or uncontrolled asthma.
- You have a history of severe allergic reactions.
- You have a history of eosinophilic esophagitis.

Before you start, consult your doctor if:
- You have not been tested for a pollen allergy.
- You have oral inflammation (e.g., mouth ulcers, oral lichen planus or thrush) or oral wounds that occur after oral surgery or dental extraction.
- You have heart or lung problems.
- You have acute asthma exacerbation.
- You have a history of allergic reactions to other drugs, substances or food.

Over age 60:
Pollen allergen extracts are approved for adults up to age 65. If over age 65, consult your doctor.

Pregnancy:
Decide with your doctor if drug benefits justify risk to unborn baby. Risk category B for Grastek and Oralair; category C for Ragwitek (see page xviii).

Breastfeeding; Lactation; Nursing Mothers:
It is unknown if drugs in this group are acceptable with breastfeeding. Discuss risks and benefits with your doctor.

Infants & children up to age 18:
- Grastek approved for ages 5 and over; Oralair approved for ages 10 and over; Ragwitek not approved for ages under 18.
- An adult should always watch the child as he or she takes this drug.

Prolonged use:
It is usually taken prior to and throughout a specific pollen season. Consult your doctor about possible continued daily use of the drug.

Skin & sunlight:
No problems expected.

Driving, piloting or hazardous work:
No problems expected.

Discontinuing:
No problems expected. Stopping the drug will decrease the anti-allergy effect.

Others:
- Advise any doctor, dentist or pharmacist whom you consult that you take this drug.
- Your doctor may recommend that you have injectable epinephrine available and are trained in its use for yourself or your child in case of a serious allergic reaction. Be sure your doctor knows about all other drugs you are taking as certain drugs may worsen an allergic reaction, and also, epinephrine may not work as well when you are taking certain drugs. If you have an allergic reaction, get emergency medical help. Do not resume taking this drug without consulting your doctor.

POSSIBLE INTERACTION WITH OTHER DRUGS

GENERIC NAME OR DRUG CLASS	COMBINED EFFECT
None expected.	

POSSIBLE INTERACTION WITH OTHER SUBSTANCES

INTERACTS WITH	COMBINED EFFECT
Alcohol:	None expected.
Beverages:	None expected.
Cocaine:	Unknown. Avoid.
Foods:	None expected.
Marijuana:	Consult doctor.
Tobacco:	None expected.

POTASSIUM SUPPLEMENTS

GENERIC AND BRAND NAMES

See full list of generic and brand names in the *Generic and Brand Name Directory*, page 876.

BASIC INFORMATION

Habit forming? No
Prescription needed? Yes
Available as generic? Yes
Drug class: Mineral supplement (potassium), electrolyte replenisher, antihyperthyroid.

USES

- Treatment for potassium deficiency due to diuretics, cortisone or digitalis medicines.
- Treatment for hypercalcemia due to cancer.
- Treats overactive thyroid disease.
- Treats iodine deficiency.
- Treatment for low potassium associated with some illnesses.

DOSAGE & USAGE INFORMATION

How to take:
- Tablet or capsule—Take as directed on label.
- Effervescent tablets, granules, powder or liquid—Dilute dose in water.

When to take:
At the same time each day, preferably with food or immediately after meals.

If you forget a dose:
Take as soon as you remember. If it is almost time for the next dose, wait for that dose (don't double this dose) and resume regular schedule.

What drug does:
Preserves or restores normal function of nerve cells, thyroid, heart and skeletal muscle cells and kidneys, as well as stomach juice secretions.

Continued next column

OVERDOSE

SYMPTOMS:
Tingling of arms and legs, irregular heartbeat, blood pressure drop, convulsions, coma, cardiac arrest.
WHAT TO DO:
- **Dial 911 for all medical emergencies or call poison control center 1-800-222-1222 for instructions.**
- **See emergency information on last 3 pages of this book.**

Time lapse before drug works:
30 minutes to 2 hours. Full benefit may require 12 to 24 hours.

Don't take with:
Any other medicine or any dietary supplement without consulting your doctor or pharmacist.

POSSIBLE ADVERSE REACTIONS OR SIDE EFFECTS

SYMPTOMS	WHAT TO DO
Life-threatening: In case of overdose, see previous column.	
Common: Diarrhea, nausea, vomiting, stomach upset.	Continue. Call doctor if symptoms persist.
Infrequent: Numbness or tingling, in hands or feet, slow or irregular heartbeat, confusion, anxiety, unusual tiredness or weakness, legs feel week or heavy, short of breath.	Discontinue. Call doctor right away.
Rare: Ongoing stomach pain, throat or chest pain (especially when swallowing), black or tarry stools.	Discontinue. Call doctor right away.

WARNINGS & PRECAUTIONS

Don't take if:
- You are allergic to any potassium supplement. Allergic reactions are rare, but may occur.
- You have acute or chronic kidney disease.

Before you start, consult your doctor if:
- You have Addison's disease or familial periodic paralysis.
- You have heart disease.
- You have intestinal blockage.
- You have a stomach ulcer.
- You have high blood pressure.
- You have kidney disease.
- You have pancreatitis.
- You use laxatives or have chronic diarrhea.
- You use salt substitutes or low-salt milk.

Over age 60:
No special problems expected.

Pregnancy:
Decide with your doctor if drug benefits justify risk to unborn baby. Risk category C (see page xviii).

Breastfeeding; Lactation; Nursing Mothers:
Drugs in this group are generally acceptable with breastfeeding. Discuss risks and benefits with your doctor.

Infants & children up to age 18:
Follow instructions provided by your child's doctor.

Prolonged use:
- Talk to your doctor about the need for follow-up medical examinations or laboratory studies to check serum potassium levels.
- If burning mouth, headache or salivation occur, call doctor.

Skin & sunlight:
No problems expected.

Driving, piloting or hazardous work:
Don't drive or pilot aircraft until you learn how medicine affects you. Don't work around dangerous machinery. Don't climb ladders or work in high places. Danger increases if you drink alcohol or take medicine affecting alertness and reflexes.

Discontinuing:
Don't discontinue without consulting doctor. Dose may require gradual reduction if you have taken drug for a long time. Doses of other drugs may also require adjustment.

Others:
- Overdose or underdose can have serious effect. Frequent EKGs and lab blood studies are recommended to measure serum electrolytes and kidney function.
- Prolonged diarrhea may call for increased dosage of potassium.
- Advise any doctor or dentist you consult that you take this medicine.
- Serious injury may necessitate temporary decrease in potassium.
- Some products contain tartrazine dye. Avoid, especially if you are allergic to aspirin.

POSSIBLE INTERACTION WITH OTHER DRUGS

GENERIC NAME OR DRUG CLASS	COMBINED EFFECT
Adrenocorticoids, systemic	Decreased potassium effect.
Amiloride	Dangerous rise in blood potassium.
Angiotensin-converting enzyme (ACE) inhibitors*	Possible increased potassium effect.
Antacids*	May decrease potassium absorption.
Anticholinergics, other*	Increased possibility of intestinal ulcers, which sometimes occur with oral potassium tablets.
Anti-inflammatory drugs, nonsteroidal (NSAIDs)*	Increased risk of stomach irritation.
Antithyroid drugs*	Excessive effect of antithyroid drugs.
Beta-adrenergic blocking agents*	Increased potassium levels.
Calcium	Decreased potassium effect.
Cortisone drugs*	Increased fluid retention.
Digitalis preparations*	Possible irregular heartbeat.
Diuretics, thiazide or loop*	Decreased potassium effect.
Laxatives*	Possible decreased potassium effect.
Lithium	Increased chance of producing a thyroid goiter.
Losartan	Increased potassium levels.

Continued on page 908

POSSIBLE INTERACTION WITH OTHER SUBSTANCES

INTERACTS WITH	COMBINED EFFECT
Alcohol:	None expected.
Beverages: Salty drinks.	Increased fluid retention.
Cocaine:	May cause irregular heartbeat. Avoid.
Foods: Salty foods.	Increased fluid retention.
Marijuana:	May cause irregular heartbeat. Avoid.
Tobacco:	None expected.

*See Glossary

PRAMLINTIDE

BRAND NAMES

Symlin SymlinPen

BASIC INFORMATION

Habit forming? No
Prescription needed? Yes
Available as generic? No
Drug class: Antidiabetic; amylinomimetic

USES

Treatment for certain patients with type 1 or type 2 diabetes who already use insulin, but still need better blood sugar control.

DOSAGE & USAGE INFORMATION

How to take:
Self-injection—Injected under the skin (subcutaneous) of the upper leg (thigh) or stomach area (abdomen). Inject pramlintide at a site that is more than 2 inches away from your insulin injection. Follow the instructions provided. Your doctor will advise you of any changes in insulin dosages. Never mix insulin and pramlintide. Use different syringes. Rotate injection sites.

When to take:
Inject it just before major meals. A major meal must have at least 250 calories or 30 grams of carbohydrate. Adjust your pre-meal insulin dose and check blood sugar before and after every meal and at bedtime (or as advised by doctor).

If you forget a dose:
If you forget to inject the dose before you start eating a meal, skip that dose and then inject the next dose as scheduled. Don't double that dose.

Continued next column

OVERDOSE

SYMPTOMS:
Severe nausea, vomiting, diarrhea, dizziness.
WHAT TO DO:

- **If person uses much larger amount than prescribed or if accidentally swallowed, call poison control center 1-800-222-1222 for instructions or dial 911 (emergency) for help.**
- **Be alert to hypoglycemia (low blood sugar) symptoms listed in next column under Infrequent. Pramlintide alone dose not cause hypoglycemia, but it is used with insulin and insulin can induce hypoglycemia.**

What drug does:

- Slows down movement of food through the stomach. This affects how fast sugar enters the blood after eating. It reduces blood sugar output by the liver.
- It reduces appetite by creating a feeling of fulness resulting in potential weight loss.

Time lapse before drug works:
Within the 3 hours after a meal.

Don't take with:
Any other medicine or any dietary supplement without consulting your doctor or pharmacist.

POSSIBLE ADVERSE REACTIONS OR SIDE EFFECTS

SYMPTOMS	WHAT TO DO
Life-threatening:	
None expected.	
Common:	
Nausea.	Continue. Call doctor if symptom persists.
Infrequent:	
• Indigestion, loss of appetite, vomiting, stomach pain, tiredness, dizziness, injection site reaction (redness, bruising, pain), joint pain, cough, sore throat, headache.	Continue. Call doctor when convenient.
• Symptoms of low blood sugar—nervousness, hunger (excessive), cold sweats, rapid pulse, anxiety, cold skin, chills, confusion, loss of concentration, drowsiness, headache, nausea, weakness, shakiness, vision changes.	Seek treatment (eat some form of quick-acting sugar—glucose tablets, sugar, fruit juice, corn syrup, honey).
• Symptoms of high blood sugar—increased urination, unusual thirst, dry mouth, drowsiness, flushed or dry skin, fruit-like breath odor, appetite loss, stomach pain or vomiting, tiredness, trouble breathing, increased blood sugar level.	Check your blood sugar immediately. Call doctor right away.
Rare:	
Other symptoms that cause concern.	Continue. Call doctor when convenient.

WARNINGS & PRECAUTIONS

Don't take if:
You are allergic to pramlintide. Allergic reactions are rare, but may occur.

Before you start, consult your doctor if:
- You suffer from gastroparesis (a condition in which the stomach does not empty properly).
- You have difficulty recognizing symptoms of hypoglycemia (low blood sugar).
- You have poor control of your diabetes.
- You have difficulty with your insulin regimen or monitoring your blood sugar levels or have recurrent episodes of hypoglycemia.

Over age 60:
No special problems expected.

Pregnancy:
Decide with your doctor if drug benefits justify risk to unborn baby. Risk category C (see page xviii).

Breastfeeding; Lactation; Nursing Mothers:
This drug is generally acceptable with breastfeeding. Discuss risks and benefits with your doctor.

Infants & children up to age 18:
Safety and effectiveness in this age group have not been established. Consult doctor.

Prolonged use:
Talk to your doctor about the need for follow-up medical examinations and/or laboratory studies to check effectiveness of the drug.

Skin & sunlight:
No problems expected.

Driving, piloting or hazardous work:
No problems expected. You do need to be cautious for symptoms of hypoglycemia (especially at the start of drug treatment).

Discontinuing:
Don't discontinue without doctor's advice.

Others:
- Notify your doctor if you have a fever, infection, diarrhea, or experience vomiting.
- Advise any doctor, dentist or pharmacist whom you consult that you take this medicine.
- Wear or carry medical identification that indicates you have type 1 or type 2 diabetes and the drugs you take.
- You and your family should educate yourselves about diabetes; learn hypoglycemia symptoms and how to treat it. Hypoglycemia may occur in the treatment of diabetes as a result of skipped meals, excess exercise or alcohol use. Carry non-dietetic candy or glucose tablets to treat episodes of low blood sugar.
- Follow your prescribed diet, drug regimen and exercise routines closely. Changing any of these things can affect blood sugar levels.

POSSIBLE INTERACTION WITH OTHER DRUGS

GENERIC NAME OR DRUG CLASS	COMBINED EFFECT
Acarbose	Decreased absorption of nutrients.
Anticholinergics*	Decreased stomach emptying.
Antidiabetics, oral*	Increased risk of hypoglycemia.
Drugs taken by mouth that need to pass quickly through the stomach (such as oral contraceptives or antibiotics)	May need to take them 1 hour before or 2 hours after injecting pramlintide.
Drugs that may increase blood sugar lowering or increase risk of hypoglycemia	Increased risk of hypoglycemia.
Insulin	Possible severe hypoglycemia can occur within 3 hours of pramlintide dose.
Miglitol	Decreased absorption of nutrients.

POSSIBLE INTERACTION WITH OTHER SUBSTANCES

INTERACTS WITH	COMBINED EFFECT
Alcohol:	May cause severe low blood sugar. Avoid.
Beverages:	None expected.
Cocaine:	Unknown. Avoid.
Foods:	None expected.
Marijuana:	Consult doctor.
Tobacco:	None expected.

*See Glossary

PREGABALIN

BRAND NAMES

Lyrica

BASIC INFORMATION

Habit forming? No
Prescription needed? Yes
Available as generic? Yes
Drug class: Antiepileptic

USES

- Treatment for neuropathic (nerve) pain that is associated with disorders such as diabetic peripheral neuropathy, postherpetic neuralgia and spinal cord injuries.
- Treatment for fibromyalgia (a disorder that causes pain, fatigue and sleep problems).
- Used along with other drugs to treat epilepsy.
- Treats other disorders as advised by doctor.

DOSAGE & USAGE INFORMATION

How to take:

- Capsule—Swallow with liquid. Take with or without food.
- Solution—Follow directions on label.

When to take:
2 to 3 times a day at the same times each day. Dosage may be increased by your doctor after 3 to 7 days depending on your response.

If you forget a dose:
Take as soon as you remember. If it is almost time for the next dose, wait for that dose (don't double this dose) and resume regular schedule.

What drug does:
The exact mechanism is unclear. It has an affect on certain nerve transmissions in the brain and spinal cord which results in the analgesic (pain relief), anticonvulsant and anti-anxiety activity.

Time lapse before drug works:
Starts working the first day, but may take a week or more to determine full effectiveness.

Continued next column

OVERDOSE

SYMPTOMS:
Unknown (may be similar to side effects).
WHAT TO DO:
If person uses much larger amount than prescribed or if accidentally swallowed, call poison control center 1-800-222-1222 for instructions or dial 911 (emergency) for help.

Don't take with:
Any other medicine or any dietary supplement without consulting your doctor or pharmacist.

POSSIBLE ADVERSE REACTIONS OR SIDE EFFECTS

SYMPTOMS	WHAT TO DO
Life-threatening:	
Rare, severe allergic reaction (anaphylaxis)—difficulty breathing, hives, itching, swelling (throat, face), vomiting, dizziness.	Seek emergency treatment immediately.
Common:	
Dizziness, dry mouth, sleepiness, swelling (hands, feet, ankles), blurred or double vision, weight gain, headache, problems with concentration or attention, appetite increased, vomiting, constipation, erectile dysfunction.	Continue. Call doctor when convenient.
Infrequent:	
Emotional changes, diarrhea, unsteady movements (ataxia), euphoric mood, other mood changes, fatigue, sexual function change, hallucinations, unusual dreams, flushing or burning feeling, body symptoms (aches, pain, stiffness, weakness, twitching, tightness), urination changes, dry nose, thirstiness, stomach upset or swollen, eye symptoms (pain, dry, tearing), heartbeat is faster, mild breathing difficulty, rash or hives, insomnia worsens, decreased appetite.	Continue. Call doctor when convenient.
Rare:	
Cold-like symptoms, heartbeat is slower or change in rhythm, changes in blood pressure, menstrual changes, breast pain or discharge, hands or feet feel cold, other new or unexplained symptoms.	Continue. Call doctor when convenient.

WARNINGS & PRECAUTIONS

Don't take if:
- You are allergic to pregabalin.
- You have certain rare hereditary problems (galactose intolerance, the Lapp lactase deficiency or glucose-galactose malabsorption).

Before you start, consult your doctor if:
- You have diabetes.
- You have any kidney disorder.

Over age 60:
If you experience dizziness or sleepiness, take precautions to prevent accidents such as falls.

Pregnancy:
Decide with your doctor if drug benefits justify risk to unborn baby. Risk category C (see page xviii).

Breastfeeding; Lactation; Nursing Mothers:
This drug is generally acceptable with breastfeeding. Discuss risks and benefits with your doctor.

Infants & children up to age 18:
Safety and effectiveness in this age group have not been established. Consult doctor.

Prolonged use:
Talk to your doctor about the need for follow-up examinations to determine continued effectiveness of the drug in treating your disorder.

Skin & sunlight:
No special problems expected.

Driving, piloting or hazardous work:
Don't drive or pilot aircraft until you learn how medicine affects you. Don't work around dangerous machinery. Don't climb ladders or work in high places. Danger increases if you drink alcohol or take other medicines affecting alertness and reflexes.

Discontinuing:
Don't discontinue drug without doctor's advice. Dosage may need to be gradually reduced.

Others:
- Don't increase or decrease drug dosage without doctor's approval.
- Advise any doctor, dentist or pharmacist whom you consult that you take this medicine.
- Rarely, antiepileptic drugs may lead to suicidal thoughts and behaviors. Call doctor right away if suicidal symptoms or unusual behaviors occur.
- Wear or carry medical identification to show your seizure disorder and the drugs you take.

POSSIBLE INTERACTION WITH OTHER DRUGS

GENERIC NAME OR DRUG CLASS	COMBINED EFFECT
Central nervous system (CNS) depressants*	May add to any sedative effect.
Lorazepam	Increased effect of lorazepam.
Oxycodone	Increased risk of side effects.

POSSIBLE INTERACTION WITH OTHER SUBSTANCES

INTERACTS WITH	COMBINED EFFECT
Alcohol:	Increased risk of side effects, such as sedation. Avoid.
Beverages:	None expected.
Cocaine:	Unknown. Avoid.
Foods:	None expected.
Marijuana:	Consult doctor.
Tobacco:	None expected.

*See Glossary

PRIMAQUINE

GENERIC NAMES

PRIMAQUINE

BASIC INFORMATION

Habit forming? No
Prescription needed? Yes
Available as generic? Yes
Drug class: Antiprotozoal (antimalarial)

USES

- Treats some forms of malaria.
- Prevents relapses of some forms of malaria.
- Treats *Pneumocystis pneumonia* (used in combination with clindamycin).

DOSAGE & USAGE INFORMATION

How to take:
Tablet—Take with meals or antacids to minimize stomach irritation.

When to take:
At the same time each day, according to instructions on prescription label.

If you forget a dose:
Take as soon as you remember. If it is almost time for the next dose, wait for that dose (don't double this dose) and resume regular schedule.

What drug does:
Alters the properties of DNA in malaria organisms to prevent them from multiplying.

Time lapse before drug works:
2 to 3 hours.

Don't take with:
Any other medicine or any dietary supplement without consulting your doctor or pharmacist.

OVERDOSE

SYMPTOMS:
Nausea, vomiting, upset or cramping stomach, drowsiness, fast or irregular heartbeat, seizures.
WHAT TO DO:

- **Dial 911 for all medical emergencies or call poison control center 1-800-222-1222 for instructions.**
- **See emergency information on last 3 pages of this book.**

POSSIBLE ADVERSE REACTIONS OR SIDE EFFECTS

SYMPTOMS	WHAT TO DO
Life-threatening:	
None expected.	
Common:	
None expected.	
Infrequent:	
Dark urine, back or leg or stomach pain, appetite loss, pale skin, fever.	Discontinue. Call doctor right away.
Rare:	
Blue fingernails or lips or skin, dizziness, difficult breathing, unusual tiredness, sore throat.	Discontinue. Call doctor right away.

WARNINGS & PRECAUTIONS

Don't take if:
- You have G6PD* deficiency.
- You are allergic to primaquine. Allergic reactions are rare, but may occur.

Before you start, consult your doctor if:
You are Black, Oriental, Asian or of Mediterranean origin.

Over age 60:
No special problems expected.

Pregnancy:
Decide with your doctor if drug benefits justify risk to unborn baby. Risk category C (see page xviii).

Breastfeeding; Lactation; Nursing Mothers:
This drug is generally acceptable with breastfeeding. Discuss risks and benefits with your doctor.

Infants & children up to age 18:
Follow instructions provided by your child's doctor.

Prolonged use:
No special problems expected.

Skin & sunlight:
No special problems expected.

Driving, piloting or hazardous work:
No special problems expected.

Discontinuing:
No special problems expected.

Others:
- If you are Black, Asian, Oriental or of Mediterranean origin, insist on a test for G6PD* deficiency before taking this medicine.
- Advise any doctor, dentist or pharmacist whom you consult that you take this medicine.

POSSIBLE INTERACTION WITH OTHER DRUGS

GENERIC NAME OR DRUG CLASS	COMBINED EFFECT
Hemolytics,* other	Increased risk of serious side effects affecting the blood.
Quinacrine	Increased toxic effects of primaquine.

POSSIBLE INTERACTION WITH OTHER SUBSTANCES

INTERACTS WITH	COMBINED EFFECT
Alcohol:	Possible liver toxicity. Avoid.
Beverages:	None expected.
Cocaine:	Unknown. Avoid.
Foods:	None expected.
Marijuana:	Consult doctor.
Tobacco:	None expected.

***See Glossary**

PRIMIDONE

BRAND NAMES

Apo-Primidone
Myidone
Mysoline
PMS Primidone
Sertan

BASIC INFORMATION

Habit forming? No
Prescription needed? Yes
Available as generic? Yes
Drug class: Anticonvulsant

USES

Prevents some forms of epileptic seizures.

DOSAGE & USAGE INFORMATION

How to take:
- Tablet—Swallow with liquid. If you can't swallow whole, crumble tablet and take with liquid or food.
- Liquid—If desired, dilute dose in beverage before swallowing.

When to take:
Daily in regularly spaced doses, according to doctor's prescription.

If you forget a dose:
Take as soon as you remember. If it is almost time for the next dose, wait for that dose (don't double this dose) and resume regular schedule.

What drug does:
The exact way it works is unknown. It appears to reduce seizures by controlling certain electrical impulses in the brain.

Time lapse before drug works:
2 to 3 weeks for full effectiveness.

Don't take with:
Any other medicine or any dietary supplement without consulting your doctor or pharmacist.

OVERDOSE

SYMPTOMS:
May have severe tiredness and/or dizziness, unable to wake up, slowed breathing.
WHAT TO DO:
- **Dial 911 for all medical emergencies or call poison control center 1-800-222-1222 for instructions.**
- **See emergency information on last 3 pages of this book.**

POSSIBLE ADVERSE REACTIONS OR SIDE EFFECTS

SYMPTOMS	WHAT TO DO
Life-threatening: In case of overdose, see previous column.	
Common:	
• Unsteadiness, shaking.	Discontinue. Call doctor right away.
• Clumsiness, feeling of constant movement, dizziness, spinning sensation.	Continue. Call doctor when convenient.
Infrequent:	
• Unusual excitement, restlessness.	Discontinue. Call doctor right away.
• Decreased sexual ability, appetite loss, drowsiness, mood or mental changes, nausea or vomiting.	Continue. Call doctor when convenient.
Rare:	
Rash, sores on mouth or lips, fever, chills, fainting, pale skin, cough, hoarseness, lower back or side pain, irregular heartbeat, shortness of breath, difficult or painful urination, unusual tiredness or weakness, unusual bleeding or bruising, worsening seizures.	Discontinue. Call doctor right away.

WARNINGS & PRECAUTIONS

Don't take if:
- You are allergic to primidone. Allergic reactions are rare, but may occur.
- You have had porphyria.

Before you start, consult your doctor if:
You have had liver, kidney or lung disease, asthma or lupus.

Over age 60:
Adverse reactions and side effects may be more frequent and severe than in younger persons.

Pregnancy:
Consult doctor. Use of the drug during pregnancy is not recommended. It could cause harm to the unborn baby. Risk category D (see page xviii).

Breastfeeding; Lactation; Nursing Mothers:
This drug is generally not recommended with breastfeeding. Discuss risks and benefits with your doctor.

Infants & children up to age 18:
Follow instructions provided by your child's doctor.

Prolonged use:
- May increase risk for anemia or problems with bones. Consult doctor about your risks.
- Talk to your doctor about the need for follow-up medical examinations or laboratory studies.

Skin & sunlight:
No special problems expected.

Driving, piloting or hazardous work:
Don't drive or pilot aircraft until you learn how medicine affects you. Don't work around dangerous machinery. Don't climb ladders or work in high places. Danger increases if you drink alcohol or take medicine affecting alertness and reflexes.

Discontinuing:
Don't discontinue abruptly or without doctor's advice until you complete prescribed dose, even though symptoms diminish or disappear.

Others:
- Tell doctor if you become ill or injured and must interrupt dose schedule.
- Periodic laboratory blood tests of drug level recommended.
- Rarely, antiepileptic drugs may lead to suicidal thoughts and behaviors. Call doctor right away if suicidal symptoms or unusual behaviors occur.
- Wear or carry medical identification to show your seizure disorder and the drugs you take.

POSSIBLE INTERACTION WITH OTHER DRUGS

GENERIC NAME OR DRUG CLASS	COMBINED EFFECT
Adrenocorticoids, systemic	Decreased adrenocorticoid effect.
Anticoagulants,* oral	Decreased primidone effect.
Anticonvulsants,* other	Changed seizure pattern.
Antidepressants*	Increased effect of antidepressant.
Antihistamines*	Increased sedation effect of primidone.
Aspirin	Decreased aspirin effect.
Carbamazepine	May increase or decrease of primidone effect.
Carbonic anhydrase inhibitors*	Possible decreased primidone effect.
Central nervous system (CNS) depressants*	Increased CNS depressant effects.
Contraceptives, oral*	Decreased contraceptive effect.
Cyclosporine	Decreased cyclosporine effect.
Digitalis preparations*	Decreased digitalis effect.
Disulfiram	Possible increased primidone effect.
Estrogens*	Decreased estrogen effect.
Griseofulvin	Possible decreased griseofulvin effect.
Isoniazid	Decreased primidone effect.
Lamotrigine	Decreased lamotrigine effect.
Leucovorin (large dose)	May counteract anticonvulsant effect of primidone.
Loxapine	Decreased anticonvulsant effect of primidone.
Metronidazole	Possible decreased metronidazole effect.
Mind-altering drugs*	Increased effect of mind-altering drug.
Monoamine oxidase (MAO) inhibitors*	Increased sedation effect of primidone.
Nabilone	Greater depression of central nervous system.

Continued on page 908

POSSIBLE INTERACTION WITH OTHER SUBSTANCES

INTERACTS WITH	COMBINED EFFECT
Alcohol:	Dangerous sedative effect. Avoid.
Beverages:	None expected.
Cocaine:	Unknown. Avoid.
Foods:	Possible need for more vitamin D.
Marijuana:	Consult doctor.
Tobacco:	None expected.

***See Glossary**

PROBENECID

BRAND NAMES

Benemid
Benuryl
Col-Probenecid
Probalan

BASIC INFORMATION

Habit forming? No
Prescription needed? Yes
Available as generic? Yes
Drug class: Antigout

USES

- Treats chronic gout.
- Increases blood levels of penicillins and cephalosporins.

DOSAGE & USAGE INFORMATION

How to take:
Tablet—Swallow with liquid or food to lessen stomach irritation. If you can't swallow whole, crumble tablet and take with liquid or food.

When to take:
At the same time each day.

If you forget a dose:
Take as soon as you remember. If it is almost time for the next dose, wait for that dose (don't double this dose) and resume regular schedule.

What drug does:
- Forces kidneys to excrete uric acid.
- Reduces amount of penicillin excreted in urine.

Time lapse before drug works:
May require several months of regular use to prevent acute gout.

Don't take with:
- Nonprescription drugs containing aspirin or caffeine.
- Any other medicine or any dietary supplement without consulting your doctor or pharmacist.

OVERDOSE

SYMPTOMS:
Unknown.
WHAT TO DO:
- **Dial 911 for all medical emergencies or call poison control center 1-800-222-1222 for instructions.**
- **See emergency information on last 3 pages of this book.**

POSSIBLE ADVERSE REACTIONS OR SIDE EFFECTS

SYMPTOMS	WHAT TO DO
Life-threatening:	
Rare, severe allergic reaction (anaphylaxis)—difficulty breathing, hives, itching, swelling (throat, face), vomiting, dizziness.	Seek emergency treatment immediately.
Common:	
Headache, appetite loss, nausea, vomiting.	Continue. Call doctor when convenient.
Infrequent:	
• Blood in urine, low back pain, worsening gout.	Discontinue. Call doctor right away.
• Dizziness, flushed face, itchy skin, painful or frequent urination, sore gums.	Continue. Call doctor when convenient.
Rare:	
Sore throat, fever and chills, difficult breathing, unusual bleeding or bruising, painful joint, yellow skin or eyes, foot or leg or face swelling.	Discontinue. Call doctor right away.

WARNINGS & PRECAUTIONS

Don't take if:
- You are allergic to any uricosuric.*
- You have acute gout.
- Patient is younger than 2.

Before you start, consult your doctor if:
- You have had kidney stones or kidney disease or peptic ulcer.
- You have bone marrow or blood cell disease.
- You are undergoing chemotherapy for cancer.

Over age 60:
Adverse reactions and side effects may be more frequent and severe than in younger persons.

Pregnancy:
Decide with your doctor if drug benefits justify risk to unborn baby. Risk category B (see page xviii).

Breastfeeding; Lactation; Nursing Mothers:
This drug is generally acceptable with breastfeeding. Discuss risks and benefits with your doctor.

Infants & children up to age 18:
Follow instructions provided by your child's doctor.

Prolonged use:
- Possible kidney damage.
- Talk to your doctor about the need for follow-up medical exams and lab studies.

Skin & sunlight:
No problems expected.

Driving, piloting or hazardous work:
Avoid if you feel dizzy. Otherwise, no problems expected.

Discontinuing:
Don't discontinue without consulting doctor. Dose may require gradual reduction if you have taken drug for a long time. Doses of other drugs may also require adjustment.

Others:
- If signs of gout attack develop while taking medicine, consult doctor.
- Advise any doctor, dentist or pharmacist whom you consult that you take this medicine.

POSSIBLE INTERACTION WITH OTHER DRUGS

GENERIC NAME OR DRUG CLASS	COMBINED EFFECT
Allopurinol	Increased effect of each drug.
Anticoagulants,* oral	Increased anticoagulant effect.
Anti-inflammatory drugs, nonsteroidal (NSAIDs)*	Increased toxic risk.
Aspirin	Decreased probenecid effect.
Bismuth subsalicylate	Decreased probenecid effect.
Cephalosporins*	Increased cephalosporin effect.
Ciprofloxacin	May cause kidney dysfunction.
Dapsone	Increased dapsone effect. Increased toxicity.
Diclofenac	Increased diclofenac effect.
Diuretics, thiazide*	Decreased probenecid effect.
Hypoglycemics, oral*	Increased hypoglycemic effect.
Indomethacin	Increased adverse effects of indomethacin.
Ketoprofen	Increased risk of ketoprofen toxicity.
Loracarbef	Increased loracarbef effect.
Methotrexate	Increased methotrexate toxicity.
Nitrofurantoin	Increased effect of nitrofurantoin.
Para-aminosalicylic acid	Increased effect of para-aminosalicylic acid.
Penicillins*	Enhanced penicillin effect.
Pyrazinamide	Decreased probenecid effect.
Salicylates*	Decreased probenecid effect.
Sodium benzoate & sodium phenylacetate	May reduce effect of sodium benzoate & sodium phenylacetate.
Sulfa drugs*	Slows elimination. May cause harmful accumulation of sulfa.
Thioguanine	More likelihood of toxicity of both drugs.
Valacyclovir	Increased valacyclovir effect.
Zidovudine	Increased zidovudine toxicity risk.

POSSIBLE INTERACTION WITH OTHER SUBSTANCES

INTERACTS WITH	COMBINED EFFECT
Alcohol:	Decreased probenecid effect.
Beverages:	None expected.
Cocaine:	Unknown. Avoid.
Foods:	None expected.
Marijuana:	Consult doctor.
Tobacco:	None expected.

***See Glossary**

PROCARBAZINE

BRAND NAMES

Matulane
Natulan

BASIC INFORMATION

Habit forming? No
Prescription needed? Yes
Available as generic? No
Drug class: Antineoplastic

USES

Treatment for certain types of cancer.

DOSAGE & USAGE INFORMATION

How to take:
Capsule—Swallow with liquid after light meal. Don't drink fluids with meals. Drink extra fluids between meals. Avoid sweet or fatty foods.

When to take:
At the same time each day.

If you forget a dose:
Take as soon as you remember. If it is almost time for the next dose, wait for that dose (don't double this dose) and resume regular schedule.

What drug does:
Inhibits abnormal cell reproduction. Procarbazine is an alkylating agent* and an MAO inhibitor.

Time lapse before drug works:
Up to 6 weeks for full effect.

Don't take with:
Any other medicine or any dietary supplement without consulting your doctor or pharmacist.

OVERDOSE

SYMPTOMS:
Nausea, vomiting, diarrhea, low blood pressure, trembling, convulsions, coma.
WHAT TO DO:
- **Dial 911 for all medical emergencies or call poison control center 1-800-222-1222 for instructions.**
- **See emergency information on last 3 pages of this book.**

POSSIBLE ADVERSE REACTIONS OR SIDE EFFECTS

SYMPTOMS	WHAT TO DO
Life-threatening:	
In case of overdose, see previous column.	
Common:	
• Confusion, seizures, hallucinations, shortness of breath, severe tiredness or weakness, coughing up mucus, missing menstrual periods.	Discontinue. Call doctor right away.
• Drowsiness, nausea, vomiting, joint or muscle pain, fatigue, nightmares, trouble sleeping, muscle twitching, being nervous.	Continue. Call doctor when convenient.
Infrequent:	
• Black or tarry stools, bloody urine or vomit, cough or hoarseness, fever and chills, urination is painful or difficult, unusual bleeding or bruising, diarrhea, mouth or lip sores, tingling or numbness in fingers or toes, yellow eyes or skin, being unsteady.	Discontinue. Call doctor right away.
• Constipation, warmth or redness of face, dizziness, dry mouth, difficulty swallowing, headache, loss of appetite, depression.	Continue. Call doctor when convenient.
Rare:	
• Severe high blood pressure.	Seek emergency treatment.
• Fainting, skin rash or hives or itching, wheezing.	Discontinue. Call doctor right away.

WARNINGS & PRECAUTIONS

Don't take if:
You are allergic to any MAO inhibitor. Allergic reactions are rare, but may occur.

Before you start, consult your doctor if:
- You are an alcoholic.
- You have asthma, heart disease, congestive heart failure, heart rhythm irregularities, high blood pressure, liver or kidney disease.
- You have had a stroke.
- You have diabetes or epilepsy.

- You have overactive thyroid.
- You have schizophrenia.
- You have Parkinson's disease.
- You have adrenal gland tumor.
- You will have surgery within 2 months, including dental surgery.

Over age 60:
Adverse reactions and side effects may be more frequent and severe than in younger persons.

Pregnancy:
Consult doctor. Use of the drug during pregnancy is not recommended. It could cause harm to the unborn baby. Risk category D (see page xviii).

Breastfeeding; Lactation; Nursing Mothers:
This drug is generally not recommended with breastfeeding. Discuss risks and benefits with your doctor.

Infants & children up to age 18:
Follow instructions provided by your child's doctor.

Prolonged use:
- May be toxic to liver.
- Talk to your doctor about the need for follow-up medical examinations or laboratory studies to check complete blood counts (white blood cell count, platelet count, red blood cell count, hemoglobin, hematocrit), bone marrow, kidney function.

Skin & sunlight:
No special problems expected.

Driving, piloting or hazardous work:
Don't drive or pilot aircraft until you learn how medicine affects you. Don't work around dangerous machinery. Don't climb ladders or work in high places. Danger increases if you drink alcohol or take medicine affecting alertness and reflexes.

Discontinuing:
- Don't discontinue without doctor's advice until you complete prescribed dose, even though symptoms diminish or disappear.
- Follow precautions regarding foods, drinks and other medicines for 2 weeks after discontinuing.

Others:
- May affect blood sugar levels in patients with diabetes.
- Advise any doctor, dentist or pharmacist whom you consult that you take this drug.

POSSIBLE INTERACTION WITH OTHER DRUGS

GENERIC NAME OR DRUG CLASS	COMBINED EFFECT
Amphetamines*	Blood pressure rise to life-threatening level.
Anticonvulsants,* oral	Changed seizure pattern.
Antidepressants, tricyclic*	Blood pressure rise to life-threatening level.
Antidiabetics,* oral and insulin	Excessively low blood sugar.
Antihistamines*	Increased sedation.
Barbiturates*	Increased sedation.
Bone marrow depressants*	Increased toxicity to bone marrow.
Buspirone	Elevated blood pressure.
Caffeine	Irregular heartbeat or high blood pressure.
Carbamazepine	Fever, seizures. Avoid.
Central nervous system (CNS) depressants*	Increased CNS depression.
Clozapine	Toxic effect on bone marrow and central nervous system.
Cyclobenzaprine	Fever, seizures. Avoid.
Dextromethorphan	Fever, hypertension.

Continued on page 908

POSSIBLE INTERACTION WITH OTHER SUBSTANCES

INTERACTS WITH	COMBINED EFFECT
Alcohol:	Increased sedation to dangerous level. Disulfiram-like reaction.*
Beverages:	
Caffeine drinks.	Irregular heartbeat or high blood pressure.
Drinks containing tyramine.*	Blood pressure rise to life-threatening level.
Cocaine:	Increased risk of side effects. Avoid.
Foods:	
Foods containing tyramine.*	Blood pressure rise to life-threatening level.
Marijuana:	Increased risk of side effects. Avoid.
Tobacco:	None expected.

***See Glossary**

PROGESTINS

GENERIC AND BRAND NAMES

See full list of generic and brand names in *Generic and Brand Name Directory*, page 877.

BASIC INFORMATION

Habit forming? No
Prescription needed? Yes, for most
Available as generic? Yes, for some.
Drug class: Female sex hormone (progestin)

USES

- Treatment for menstrual or uterine disorders.
- Contraceptive (used alone or with estrogen). May be used for emergency contraception.
- Treatment for symptoms of menopause.
- Treatment for several types of cancer.
- Treatment for female hormone imbalance.
- Megestrol is used for treatment of weight loss in AIDS and cancer patients.
- Treatment for female infertility caused by progesterone deficiency.
- Treatment for endometrial hyperplasia.

DOSAGE & USAGE INFORMATION

How to take:

- Capsule—Swallow with liquid. Do not crush, chew or break capsule.
- Tablet—Swallow with liquid or food to lessen stomach irritation. You may crumble tablet.
- Injection—Given by medical provider.
- Transdermal patch—Follow label instructions.
- Implant—Inserted by a health care provider.
- Oral suspension—Follow label instructions.

When to take:
Daily dose at the same time each day. For other forms, follow instructions on label.

Continued next column

If you forget a dose:

- Contraceptive—Consult your doctor or label instructions. You may need to use another birth control method until your next period.
- If used for menstrual disorders—Take as soon as you remember. If it is almost time for the next dose, wait for that dose (don't double this dose) and resume regular schedule.

What drug does:

- Progesterone is a female hormone produced in the body. Progestins are synthetic hormones that have progesterone-like actions. They can have multiple effects on the female reproductive system.
- The mechanism that produces weight gain or helps in cancer treatment is unknown.

Time lapse before drug works:

- Menstrual disorders—24 to 48 hours.
- Contraception—3 weeks.
- Cancer—May require 2 to 3 months of regular use for maximum benefit.

Don't take with:
Any other medicine or any dietary supplement without consulting your doctor or pharmacist.

OVERDOSE

SYMPTOMS:
Nausea, vomiting, fluid retention, breast discomfort or enlargement, vaginal bleeding.
WHAT TO DO:

- **Dial 911 for all medical emergencies or call poison control center 1-800-222-1222 for instructions.**
- **See emergency information on last 3 pages of this book.**

POSSIBLE ADVERSE REACTIONS OR SIDE EFFECTS

SYMPTOMS	WHAT TO DO
Life-threatening:	
Rare, severe allergic reaction (anaphylaxis)—difficulty breathing, hives, itching, swelling (throat, face), vomiting, dizziness.	Seek emergency treatment immediately.
Common:	
• Vaginal bleeding changes (heavy, irregular, spotting, stopped).	Continue, but call doctor right away.
• Abdominal cramping, bloating, swollen feet or ankles, tiredness or weakness, mild headache, nausea, mood changes, skin irritation with injection, skin pain with implant.	Continue. Call doctor when convenient.
Infrequent:	
Depression, acne, tender breasts, changes in facial or body hair, brown spots on skin, loss of sexual desire, insomnia.	Continue. Call doctor when convenient.

Rare:

• Blood clot (with high doses): sudden headache, pain in calf, vision changes, breathing or speech problems.	Discontinue. Seek emergency help.
• Rash, changes in breast milk.	Continue, but call doctor right away.

WARNINGS & PRECAUTIONS

Don't take if:
You are allergic to any progestin hormone.

Before you start, consult your doctor if:
- You have diabetes, heart or kidney disease.
- You have liver or gallbladder disease.
- You have had thrombophlebitis, embolism or stroke, bleeding disorder or high cholesterol.
- You have unexplained vaginal bleeding.
- You have had breast or uterine cancer.
- You have varicose veins.
- You have a seizure disorder.
- You suffer from migraines or depression.
- You have breast disease (lumps, cysts).

Over age 60:
No special problems expected.

Pregnancy:
Risk factors vary for drugs in this group. Always consult doctor. See category list on page xviii.

Breastfeeding; Lactation; Nursing Mothers:
Drugs in this group are generally acceptable with breastfeeding. Discuss risks and benefits with your doctor.

Infants & children up to age 18:
Follow instructions provided by your child's doctor.

Prolonged use:
No problems expected.

Skin & sunlight:
No problems expected.

Driving, piloting or hazardous work:
No problems expected.

Discontinuing:
Consult doctor. This medicine stays in the body and may cause fetal abnormalities. Wait at least 3 months before becoming pregnant. Side effects may also occur (dizziness, nausea, unusual menstrual bleeding).

Others:
- Hormone therapy (HRT) is not recommended for the prevention of chronic disease, but may be considered for the short-term management of menopausal symptoms for younger women. Before deciding on hormone replacement therapy, discuss your specific risks and benefits (e.g., age, lifestyle and risk for disease) with your doctor. Get regular checkups to make sure that you should continue with HRT.
- Injection form (Depo-Provera) may result in the loss of bone density. The risk increases, the longer the drug is used. The bone loss may not be reversible.
- Patients with diabetes must be monitored closely. Consult doctor if changes in blood glucose levels occur.
- May affect results in some medical tests.
- Advise any doctor, dentist or pharmacist whom you consult that you take this medicine.
- Carefully read the paper called "Information for the Patient" that was given to you with your first prescription or a refill. If you lose it, ask your pharmacist for a copy.

POSSIBLE INTERACTION WITH OTHER DRUGS

GENERIC NAME OR DRUG CLASS	COMBINED EFFECT
Aminoglutethimide	Decreased progestin effect.
Enzyme inducers*	Decreased progestin effect.

POSSIBLE INTERACTION WITH OTHER SUBSTANCES

INTERACTS WITH	COMBINED EFFECT
Alcohol:	None expected.
Beverages:	None expected.
Cocaine:	Unknown. Avoid.
Foods:	None expected.
Marijuana:	Consult doctor.
Tobacco:	
All forms.	Possible blood clots in lung, brain, legs (with high drug doses). Avoid.

*See Glossary

PROGUANIL

BRAND NAMES

Malarone
Paludrine

BASIC INFORMATION

Habit forming? No
Prescription needed? Yes
Available as generic? No
Drug class: Antimalarial

USES

Prevents and treats malaria.

DOSAGE & USAGE INFORMATION

How to take:
Tablet—Swallow whole with liquid after meals.

When to take:
At the same time each day.

If you forget a dose:
Take as soon as you remember. If it is almost time for the next dose, wait for that dose (don't double this dose) and resume regular schedule.

What drug does:
Exact mechanism unknown.

Time lapse before drug works:
1 to 2 weeks.

Don't take with:
Any other medicine or any dietary supplement without consulting your doctor or pharmacist.

OVERDOSE

SYMPTOMS:
Abdominal pain, vomiting, hair loss, peeling skin on hands and feet, sores in mouth, easy bruising.
WHAT TO DO:
If person uses much larger amount than prescribed or if accidentally swallowed, call poison control center 1-800-222-1222 for instructions or dial 911 (emergency) for help.

POSSIBLE ADVERSE REACTIONS OR SIDE EFFECTS

SYMPTOMS	WHAT TO DO
Life-threatening:	
Rare, severe allergic reaction (anaphylaxis)—difficulty breathing, hives, itching, swelling (throat, face), vomiting, dizziness.	Seek emergency treatment immediately.
Common:	
Abdominal pain, back pain, coughing, diarrhea, fever, headache, loss of strength, nausea, muscle pain, sore throat, sneezing, vomiting.	Continue. Call doctor if symptoms persist.
Infrequent:	
Acid or sour stomach, belching, dizziness, flu-like symptoms, heartburn, indigestion, loss of appetite, weight loss, temporary hair loss.	Continue. Call doctor if symptoms persist.
Rare:	
Skin rash or itching.	Continue. Call doctor when convenient.

WARNINGS & PRECAUTIONS

Don't take if:
You are allergic to proguanil.

Before you start, consult your doctor if:
- You are pregnant or breast-feeding.
- You have kidney problems.

Over age 60:
No special problems expected.

Pregnancy:
Decide with your doctor if drug benefits justify risk to unborn baby. Risk category C (see page xviii).

Breastfeeding; Lactation; Nursing Mothers:
This drug is generally acceptable with breastfeeding. Discuss risks and benefits with your doctor.

Infants & children up to age 18:
Follow instructions provided by your child's doctor.

Prolonged use:
Not intended for long term use.

Skin & sunlight:
No problems expected.

Driving, piloting or hazardous work:
Don't drive or pilot aircraft until you learn how medicine affects you. Don't work around dangerous machinery. Don't climb ladders or work in high places. Danger increases if you drink alcohol or take medicine affecting alertness and reflexes, such as antihistamines, tranquilizers, sedatives, pain medicine, narcotics and mind-altering drugs.

Discontinuing:
Don't discontinue without doctor's advice until you complete the prescribed dosage.

Others:
- Advise any doctor, dentist or pharmacist whom you consult that you take this medicine.
- Persons of Asian or African descent metabolize this drug rapidly, therefore the drug may not reach effective blood levels for protection against malaria.

POSSIBLE INTERACTION WITH OTHER DRUGS

GENERIC NAME OR DRUG CLASS	COMBINED EFFECT
None expected.	

POSSIBLE INTERACTION WITH OTHER SUBSTANCES

INTERACTS WITH	COMBINED EFFECT
Alcohol:	None expected.
Beverages:	None expected.
Cocaine:	Unknown. Avoid.
Foods:	None expected.
Marijuana:	Consult doctor.
Tobacco:	None expected.

PROPAFENONE

BRAND NAMES

Rhythmol

BASIC INFORMATION

Habit forming? No
Prescription needed? Yes
Available as generic? Yes
Drug class: Antiarrhythmic

USES

Treats severe heartbeat irregularities (life-threatening ventricular rhythm disturbances).

DOSAGE & USAGE INFORMATION

How to take:
Tablet—Swallow with liquid or food to lessen stomach irritation. If you can't swallow whole, crumble tablet and take with liquid or food.

When to take:
At the same time each day, according to instructions on prescription label.

If you forget a dose:
Take as soon as you remember. If it is almost time for the next dose, wait for that dose (don't double this dose) and resume regular schedule.

What drug does:
Slows electrical activity in the heart to decrease the excitability of the heart muscle.

Time lapse before drug works:
3-1/2 hours to 1 week for full effect. Begins working almost immediately.

Don't take with:
Any other medicine (including nonprescription drugs such as cough and cold medicines, nose drops, vitamins, laxatives, antacids, diet pills, or caffeine) without consulting your doctor or pharmacist.

OVERDOSE

SYMPTOMS:
Sleepiness, low blood pressure, slow heartbeat that may be irregular, rarely seizures.
WHAT TO DO:
- **Dial 911 for all medical emergencies or call poison control center 1-800-222-1222 for instructions.**
- **See emergency information on last 3 pages of this book.**

POSSIBLE ADVERSE REACTIONS OR SIDE EFFECTS

SYMPTOMS	WHAT TO DO
Life-threatening:	
Severe chest pain, severe shortness of breath.	Seek emergency treatment.
Common:	
• Faster or more irregular heartbeat.	Discontinue. Call doctor right away.
• Taste change, dizziness.	Continue. Call doctor when convenient.
Infrequent:	
• Blurred vision, skin rash.	Discontinue. Call doctor right away.
• Constipation, diarrhea. dry mouth, nausea.	Continue. Call doctor when convenient.
Rare:	
Fever, chills, trembling, joint pain, slow heartbeat.	Discontinue. Call doctor right away.

WARNINGS & PRECAUTIONS

Don't take if:
You are allergic to propafenone. Allergic reactions are rare, but may occur.

Before you start, consult your doctor if:
- You have asthma or bronchospasm.
- You have congestive heart failure.
- You have liver disease or kidney disease.
- You have a recent history of heart attack.
- You have a pacemaker.

Over age 60:
No special problems expected.

Pregnancy:
Decide with your doctor if drug benefits justify risk to unborn baby. Risk category C (see page xviii).

Breastfeeding; Lactation; Nursing Mothers:
This drug is generally acceptable with breastfeeding. Discuss risks and benefits with your doctor.

Infants & children up to age 18:
Safety and effectiveness in this age group have not been established. Consult doctor.

Prolonged use:
Don't discontinue without consulting doctor. Dose may require gradual reduction if you have taken drug for a long time. Dosages of other drugs may also require adjustment.

Skin & sunlight:
No special problems expected.

Driving, piloting or hazardous work:
Don't drive or pilot aircraft until you learn how medicine affects you. Don't work around dangerous machinery. Don't climb ladders or work in high places. Danger increases if you drink alcohol or take medicine affecting alertness and reflexes.

Discontinuing:
Don't discontinue without consulting doctor. Dose may require gradual reduction if you have taken drug for a long time. Doses of other drugs may also require adjustment.

Others:
- Advise any doctor, dentist or pharmacist whom you consult that you take this medicine, especially if you are to be anesthetized.
- Report changes in symptoms to your doctor and return for periodic visits to check progress.
- Carry or wear a medical I.D. card or bracelet that indicate your disorder and the drugs you take.

POSSIBLE INTERACTION WITH OTHER DRUGS

GENERIC NAME OR DRUG CLASS	COMBINED EFFECT
Anesthetics, local (e.g., prior to dental procedures)	May increase risk of side effects.
Antiarrhythmics,* other	Increased risk of adverse reactions.
Beta-adrenergic blocking agents*	Increased beta blocker effect.
Digitalis preparations*	Increased digitalis absorption. May require decreased dosage of digitalis preparation.
Doxepin (topical)	Increased risk of toxicity of both drugs.
Enzyme inhibitors*	Increased effect of propafenone.
Warfarin	Increased warfarin effect.

POSSIBLE INTERACTION WITH OTHER SUBSTANCES

INTERACTS WITH	COMBINED EFFECT
Alcohol:	Unpredictable effect on heartbeat. Avoid.
Beverages: Caffeine drinks.	Increased heartbeat irregularity. Avoid.
Cocaine:	Increased heartbeat irregularity. Avoid.
Foods:	None expected.
Marijuana:	Increased heartbeat irregularity. Avoid.
Tobacco:	Increased heartbeat irregularity. Avoid.

*See Glossary

PROPANTHELINE

BRAND NAMES

Pro-Banthine
Propanthel

BASIC INFORMATION

Habit forming? No
Prescription needed?
High strength: Yes
Low strength: No
Available as generic? Yes
Drug class: Antispasmodic, anticholinergic

USES

Reduces spasms of digestive system, bladder and urethra.

DOSAGE & USAGE INFORMATION

How to take:
Tablet—Swallow with liquid or food to lessen stomach irritation.

When to take:
30 minutes before meals (unless directed otherwise by doctor).

If you forget a dose:
Take as soon as you remember. If it is almost time for the next dose, wait for that dose (don't double this dose) and resume regular schedule.

What drug does:
Blocks nerve impulses at parasympathetic nerve endings, preventing muscle contractions and gland secretions of organs involved.

Time lapse before drug works:
15 to 30 minutes.

Don't take with:
Any other medicine or any dietary supplement without consulting your doctor or pharmacist.

OVERDOSE

SYMPTOMS:
Dilated pupils, blurred vision, rapid pulse and breathing, dizziness, fever, hallucinations, confusion, slurred speech, agitation, flushed face, convulsions, coma.
WHAT TO DO:
- **Dial 911 for all medical emergencies or call poison control center 1-800-222-1222 for instructions.**
- **See emergency information on last 3 pages of this book.**

POSSIBLE ADVERSE REACTIONS OR SIDE EFFECTS

SYMPTOMS	WHAT TO DO
Life-threatening:	
Rare, severe allergic reaction (anaphylaxis)—difficulty breathing, hives, itching, swelling (throat, face), vomiting, dizziness.	Seek emergency treatment immediately.
Common:	
• Confusion, delirium, rapid heartbeat.	Discontinue. Call doctor right away.
• Nausea, vomiting, decreased sweating, constipation, loss of taste, dryness (ears, nose, throat, mouth).	Continue. Call doctor when convenient.
Infrequent:	
• Lightheadedness.	Discontinue. Call doctor right away.
• Headache, difficult urination, nasal congestion, altered taste, impotence.	Continue. Call doctor when convenient.
Rare:	
Rash or hives, eye pain, blurred vision.	Discontinue. Call doctor right away.

WARNINGS & PRECAUTIONS

Don't take if:
- You are allergic to any anticholinergic.
- You have trouble with stomach bloating.
- You have difficulty emptying your bladder completely.
- You have narrow-angle glaucoma.
- You have severe ulcerative colitis.

Before you start, consult your doctor if:
- You have open-angle glaucoma.
- You have angina.
- You have chronic bronchitis or asthma.
- You have hiatal hernia.
- You have liver, kidney or thyroid disease.
- You have enlarged prostate.
- You have myasthenia gravis.
- You have peptic ulcer.
- You will have surgery within 2 months, including dental surgery, requiring general or spinal anesthesia.

Over age 60:
Adverse reactions and side effects may be more frequent and severe than in younger persons.

Pregnancy:
Decide with your doctor if drug benefits justify risk to unborn baby. Risk category C (see page xviii).

Breastfeeding; Lactation; Nursing Mothers:
This drug is generally not recommended with breastfeeding. Discuss risks and benefits with your doctor.

Infants & children up to age 18:
Safety and effectiveness in this age group have not been established. Consult doctor.

Prolonged use:
Chronic constipation, possible fecal impaction. Consult doctor immediately.

Skin & sunlight:
No problems expected.

Driving, piloting or hazardous work:
No problems expected.

Discontinuing:
May be unnecessary to finish medicine. Follow doctor's instructions.

Others:
Advise any doctor, dentist or pharmacist whom you consult that you take this medicine.

POSSIBLE INTERACTION WITH OTHER DRUGS

GENERIC NAME OR DRUG CLASS	COMBINED EFFECT
Adrenocorticoids, systemic	Possible glaucoma.
Amantadine	Increased propantheline effect.
Antacids*	Decreased propantheline effect.
Anticholinergics,* other	Increased propantheline effect.
Antidepressants, tricyclic*	Increased propantheline effect. Increased sedation.
Antidiarrhea preparations*	Reduced propantheline effect.
Antihistamines*	Increased propantheline effect.
Attapulgite	Decreased propantheline effect.
Buclizine	Increased propantheline effect.
Digitalis preparations*	Possible decreased absorption of digitalis.
Haloperidol	Increased internal eye pressure.
Ketoconazole	Decreased ketoconazole effect.
Meperidine	Increased propantheline effect.
Methylphenidate	Increased propantheline effect.
Molindone	Increased anti-cholinergic effect.
Monoamine oxidase (MAO) inhibitors*	Increased propantheline effect.
Nitrates*	Increased internal eye pressure.
Nizatidine	Increased nizatidine effect.
Orphenadrine	Increased propantheline effect.
Phenothiazines*	Increased propantheline effect.
Pilocarpine	Loss of pilocarpine effect in glaucoma treatment.
Potassium supplements*	Increased possibility of intestinal ulcers with oral potassium tablets.
Quinidine	Increased propantheline effect.
Sedatives* or central nervous system (CNS) depressants*	Increased sedative effect of both drugs.
Vitamin C	Decreased propantheline effect. Avoid large doses of vitamin C.

POSSIBLE INTERACTION WITH OTHER SUBSTANCES

INTERACTS WITH	COMBINED EFFECT
Alcohol:	None expected.
Beverages:	None expected.
Cocaine:	Excessively rapid heartbeat. Avoid.
Foods:	None expected.
Marijuana:	Drowsiness and dry mouth. Avoid.
Tobacco:	None expected.

*See Glossary

PROTEASE INHIBITORS

GENERIC AND BRAND NAMES

See full list of generic and brand names in the *Generic and Brand Name Directory*, page 878.

BASIC INFORMATION

Habit forming? No
Prescription needed? Yes
Available as generic? Yes, for some
Drug class: Protease inhibitor

USES

Used in combination with other drugs as a treatment for HIV infection.

DOSAGE & USAGE INFORMATION

How to take:

- Tablet or capsule—Swallow with liquid. Take with food or meal to help drug's absorption. Take tipranavir or darunavir at the same time as ritonavir. Take indinavir 1 hour before or 2 hours after eating. Take fosamprenavir tablet with or without food. Your doctor may advise other methods to help the body absorb drugs.
- Film-coated tablet—Swallow with liquid. Can be taken with or without food.
- Liquid ritonavir—Swallow with chocolate milk or liquid nutritional supplement to disguise unpleasant taste.
- Oral solution—Take as directed on label.

When to take:
At the same times each day, according to instructions on prescription label.

If you forget a dose:
Take as soon as you remember. If it is almost time for the next dose, wait for that dose (don't double this dose) and resume regular schedule.

Continued next column

OVERDOSE

SYMPTOMS:
Unknown.
WHAT TO DO:

- **Dial 911 for all medical emergencies or call poison control center 1-800-222-1222 for instructions.**
- **See emergency information on last 3 pages of this book.**

What drug does:
Blocks an enzyme called protease that is vital to the final stages of HIV replication (reproduction). Blocking protease causes HIV to make copies of itself that can't infect new cells.

Time lapse before drug works:
It will take weeks to months of treatment with the drug to determine the benefits of this therapy.

Don't take with:
Any other medicine or any dietary supplement without consulting your doctor or pharmacist.

POSSIBLE ADVERSE REACTIONS OR SIDE EFFECTS

SYMPTOMS	WHAT TO DO
Life-threatening:	
Rare, severe allergic reaction (anaphylaxis)—difficulty breathing, hives, itching, swelling (throat, face), vomiting, dizziness.	Seek emergency treatment immediately.
Common:	
Diarrhea, abdominal discomfort, nausea, sores in mouth, dizziness, dry mouth, tiredness, appetite loss.	Continue. Call doctor when convenient.
Infrequent:	
Rash, muscle or joint pain, headache, abdominal pain, weakness, back pain, numbness or tingling in hands or feet, tingling around mouth.	Continue. Call doctor when convenient.
Rare:	
• Confusion, yellow skin or eyes, severe skin reaction, lack of coordination, seizures, liver problems (fatigue, loss of appetite, nausea, dark urine, yellow skin or eyes, stomach pain). Watch for warning signs of hyperglycemia or diabetes (increased thirst and hunger, unexplained weight loss, increased urination, fatigue and dry itchy skin).	Continue, but call doctor right away.
• Other symptoms not listed (may be due to drug or infection).	Continue. Call doctor when convenient.

WARNINGS & PRECAUTIONS

Don't take if:
You are allergic to protease inhibitors.

Before you start, consult your doctor if:
You have liver or kidney disease, diabetes, hypertension or peripheral neuropathy.*

Over age 60:
No special problems expected.

Pregnancy:
Decide with your doctor if drug benefits justify risk to unborn baby. Risk category B or C (see page xviii).

Breastfeeding; Lactation; Nursing Mothers:
Breastfeeding is not recommended in HIV-infected women. Consult doctor.

Infants & children up to age 18:
Some of these drugs are approved for use in infants and children. Consult your child's doctor.

Prolonged use:
Medical studies have shown that HIV can become resistant to the effects of these drugs.

Skin & sunlight:
No special problems expected.

Driving, piloting or hazardous work:
Don't drive or pilot aircraft until you learn how medicine affects you. Don't work around dangerous machinery. Don't climb ladders or work in high places. Danger increases if you drink alcohol or take medicine affecting alertness.

Discontinuing:
Don't discontinue without consulting doctor.

Others:
- Advise any doctor, dentist or pharmacist whom you consult that you take this medicine.
- These drugs do not reduce the risk of HIV transmission to others via sexual contact. Avoid sexual contact or use condoms to help prevent HIV infection. Don't share needles or equipment for injections with other persons.
- Cobicistat* (Tybost), a drug enhancer, is combined in some of the brand name drugs.
- These drugs may cause or aggravate diabetes or hypertension.

POSSIBLE INTERACTION WITH OTHER DRUGS

GENERIC NAME OR DRUG CLASS	COMBINED EFFECT
Alfuzosin	Increased alfuzosin effect.
Antacids*	Take 2 hours apart of protease inhibitor.
Anticonvulsants*	Decreased effect of protease inhibitor.
Colchicine	Increased effect of colchicine.
Contraceptives, oral	Less contraceptive effect.
Dexamethasone	Decreased effect of protease inhibitor.
Digoxin	Increased digoxin effect.
Enzyme inducers*	Decreased effect of protease inhibitor.
Enzyme inhibitors*	Increased effect of enzyme inhibitor.
Ergot preparations*	Serious or life-threatening problems. Avoid.
Fluticasone	Increased effect of fluticasone.
Histamine H_2 receptor antagonists	May need dosage adjustment of protease inhibitor.
HMG-CoA reductase inhibitors	Increased risk of muscle damage and kidney failure. Consult doctor.
Methadone	Decreased effect of methadone.
Non-nucleoside reverse transcriptase inhibitors	May need dosage adjustment of protease inhibitor.

Continued on page 909

POSSIBLE INTERACTION WITH OTHER SUBSTANCES

INTERACTS WITH	COMBINED EFFECT
Alcohol:	None expected.
Beverages: Grapefruit juice.	May increase effect of protease inhibitor.
Cocaine:	Unknown. Avoid.
Foods: Grapefruit.	May increase effect of protease inhibitor.
Marijuana:	Consult doctor.
Tobacco:	Decreased effect of ritonavir.

***See Glossary**

PROTECTANT (Ophthalmic)

GENERIC AND BRAND NAMES

HYDROXYPROPYL CELLULOSE
Lacrisert

HYDROXYPROPYL METHYL-CELLULOSE
Artificial Tears
Bion Tears
Eye Lube
Gonak
Goniosoft
Goniosol
Isopto Alkaline
Isopto Plain
Isopto Tears
Just Tears
Lacril
Methocel
Moisture Drops
Nature's Tears
Ocutears
Tearisol
Tears Naturale
Tears Naturale Free
Tears Naturale II
Tears Renewed
Ultra Tears

BASIC INFORMATION

Habit forming? No
Prescription needed? Yes
Available as generic? Yes, for some
Drug class: Protectant (ophthalmic), artificial tears

USES

- Relieves eye dryness and irritation caused by inadequate flow of tears.
- Moistens contact lenses and artificial eyes.

OVERDOSE

SYMPTOMS:
Unknown.
WHAT TO DO:
If person uses much larger amount than prescribed or if accidentally swallowed, call poison control center 1-800-222-1222 for instructions or dial 911 (emergency) for help.

DOSAGE & USAGE INFORMATION

How to use:
Eye drops
- Wash hands.
- Apply pressure to inside corner of eye with middle finger.
- Continue pressure for 1 minute after placing medicine in eye.
- Tilt head backward. Pull lower lid away from eye with index finger of the same hand.
- Drop eye drops into pouch and close eye. Don't blink.
- Keep eyes closed for 1 to 2 minutes.
- Don't touch applicator tip to any surface (including the eye). If you accidentally touch tip, clean with warm water and soap.
- Keep container tightly closed.
- Keep cool, but don't freeze.
- Wash hands immediately after using.

When to use:
As directed. Usually every 3 or 4 hours.

If you forget a dose:
Use as soon as you remember. If it is almost time for the next dose, wait for that dose (don't double this dose) and resume regular schedule.

What drug does:
- Stabilizes and thickens tear film.
- Lubricates and protects eye.

Time lapse before drug works:
2 to 10 minutes.

Don't use with:
Other eye drops without consulting your doctor or pharmacist.

POSSIBLE ADVERSE REACTIONS OR SIDE EFFECTS

SYMPTOMS	WHAT TO DO
Life-threatening: None expected.	
Common: None expected.	
Infrequent: Eye irritation not present before using artificial tears.	Discontinue. Call doctor right away.
Rare: None expected.	

WARNINGS & PRECAUTIONS

Don't use if:
You are allergic to any artificial tears. Allergic reactions are rare, but may occur.

Before you start, consult your doctor if:
You use any other eye drops.

Over age 60:
No problems expected.

Pregnancy:
Consult doctor. Risk factor not designated. See category list on page xviii.

Breastfeeding; Lactation; Nursing Mothers:
Drugs in this group are generally acceptable with breastfeeding. Discuss risks and benefits with your doctor.

Infants & children up to age 18:
Follow instructions provided by your child's doctor.

Prolonged use:
Don't use for more than 3 or 4 days.

Skin & sunlight:
No problems expected.

Driving, piloting or hazardous work:
No problems expected.

Discontinuing:
May not need all the medicine in container. If symptoms disappear, stop using.

Others:
Check with your doctor if eye irritation continues or becomes worse.

POSSIBLE INTERACTION WITH OTHER DRUGS

GENERIC NAME OR DRUG CLASS	COMBINED EFFECT

Clinically significant interactions with oral or injected medicines unlikely.

POSSIBLE INTERACTION WITH OTHER SUBSTANCES

INTERACTS WITH	COMBINED EFFECT
Alcohol:	None expected.
Beverages:	None expected.
Cocaine:	None expected.
Foods:	None expected.
Marijuana:	None expected.
Tobacco:	None expected.

PROTON PUMP INHIBITORS

GENERIC AND BRAND NAMES

DEXLANSOPRAZOLE
Dexilant
ESOMEPRAZOLE
Nexium Delayed-Release Capsules
Nexium for Delayed-Release Oral Suspension
Nexium 24HR
Vimovo
LANSOPRAZOLE
Prevacid
Prevacid Solutab
Prevacid 24 Hour
OMEPRAZOLE
Losec
Omeclamox-Pak
Prilosec
OMEPRAZOLE (con't)
Prilosec OTC
Rapinex Powder for Oral Suspension
Zegerid Capsules
Zegerid Chewable Tablets
Zegerid OTC
Zegerid Powder
PANTOPRAZOLE
Pantoloc
Protonix
Protonix Delayed Release Oral Suspension
RABEPRAZOLE
Aciphex
Aciphex Sprinkle

BASIC INFORMATION

Habit forming? No
Prescription needed? Yes, for some
Available as generic? Yes, for some
Drug class: Antiulcer agent; proton pump inhibitor

USES

- Treats gastroesophageal reflux disease (GERD).
- Treats ulcers in the stomach and duodenum.
- Treats disorders with excess stomach acid (such as Zollinger-Ellison syndrome).

DOSAGE & USAGE INFORMATION

How to take:
- Delayed-release and extended-release capsule—Swallow whole with liquid. Do not crush, chew or open (unless allowed on label).

Continued next column

- Delayed-release oral suspension—Follow instructions on prescription.
- Immediate-release capsule—Swallow whole with water (not other liquids). Do not crush, chew or break open.
- Powder—Follow instructions on prescription.
- Tablet (enteric coated)—Swallow whole with liquid. Do not crush, crumble or chew tablet.

When to take:
Once daily, right before a meal (preferably breakfast), unless otherwise directed by your doctor. Dexlansoprazole can be taken without regard to food. With once-a-day dosing, it is important to take the medicine on schedule.

If you forget a dose:
Take as soon as you remember. If it is almost time for the next dose, wait for that dose (don't double this dose) and resume regular schedule.

What drug does:
Stops the production of stomach acid.

Time lapse before drug works:
Thirty minutes to 3 hours.

Don't take with:
Any other medicine or any dietary supplement without consulting your doctor or pharmacist.

OVERDOSE

SYMPTOMS:
Drowsiness, nausea, sweating, headache, dry mouth, fast heartbeat, vision changes, flush.
WHAT TO DO:
- **Dial 911 for all medical emergencies or call poison control center 1-800-222-1222 for instructions.**
- **See emergency information on last 3 pages of this book.**

POSSIBLE ADVERSE REACTIONS OR SIDE EFFECTS

SYMPTOMS	WHAT TO DO
Life-threatening:	
Rare, severe allergic reaction (anaphylaxis)—difficulty breathing, hives, itching, swelling (throat, face), vomiting, dizziness.	Seek emergency treatment immediately.
Common:	
Diarrhea, stomach pain.	Continue. Call doctor when convenient.
Infrequent:	
Nausea, loss of appetite, headache, heartburn, muscle pain, skin rash, drowsiness.	Continue. Call doctor when convenient.
Rare:	
Weakness or unusual tiredness, sore throat and fever, sores in mouth, unusual bleeding or bruising, cloudy or bloody urine, urination changes (difficult, frequent or painful).	Discontinue. Call doctor right away.

WARNINGS & PRECAUTIONS

Don't take if:
You are allergic to proton pump inhibitors (PPIs).

Before you start, consult your doctor if:
- You are allergic to any medicines, foods or other substances.
- You have or have had liver disease.
- You have a stomach infection.

Over age 60:
No special problems expected.

Pregnancy:
Decide with your doctor if drug benefits justify risk to unborn baby. Risk category B (C for omeprazole) (see page xviii).

Breastfeeding; Lactation; Nursing Mothers:
Drugs in this group are generally acceptable with breastfeeding. Discuss risks and benefits with your doctor.

Infants & children up to age 18:
Follow instructions provided by your child's doctor. Some drugs in this group are used for short-term treatment of gastroesophageal reflux disease and erosive esophagitis in age over 1.

Prolonged use:
- The length of treatment can run from 4 to 8 weeks or may be indefinite. Symptoms may improve in 1 to 2 weeks, but your doctor will determine when healing is complete.
- Risk of low magnesium blood levels (hypomagnesemia). Symptoms may include muscle spasms, heart rhythm problems or seizures. Consult doctor about your risks.

Skin & sunlight:
No special problems expected.

Driving, piloting or hazardous work:
No special problems expected.

Discontinuing:
Don't discontinue without consulting doctor until you complete prescribed dose, even though symptoms diminish or disappear.

Others:
- Advise any doctor, dentist or pharmacist whom you consult that you take this medicine.
- If directed by your doctor, it is permissible and sometimes helpful to take with antacids* to relieve upper abdominal pain. Antacids may be used more than once daily if needed.
- Brand name Zegerid contains sodium bicarbonate (a form of salt). If you are on a diet that restricts salt or sodium, consult your doctor before using this drug.
- May increase risk of fractures of the hip, wrist, and spine with high doses or long-term use.
- May affect results of some medical tests.
- Drug increases risk of *Clostridium difficile* diarrhea. Consult doctor if diarrhea persists.

POSSIBLE INTERACTION WITH OTHER DRUGS

GENERIC NAME OR DRUG CLASS	COMBINED EFFECT
Antifungals, azole	Decreased effect of azole antifungal.
Atazanavir	Decreased effect of atazanavir. Avoid.
Clopidogrel	Decreased effect of clopidogrel with some proton pump inhibitors.
Diazepam	Increased effect of diazepam.
Digoxin	Increased effect of digoxin.
Hypomagnesemia-causing drugs,* other	Increased risk of low magnesium.
Iron supplements	Decreased effect of iron supplement.
Phenytoin	Increased effect of phenytoin with omeprazole.
Sucralfate	Decreased effect of some proton pump inhibitors. Take 30 minutes before sucralfate.
Tacrolimus	Increased effect of tacrolimus.
Theophylline	May require dosage adjustment of theophylline with lansoprazole.
Warfarin	May cause abnormal bleeding.

POSSIBLE INTERACTION WITH OTHER SUBSTANCES

INTERACTS WITH	COMBINED EFFECT
Alcohol:	None expected.
Beverages:	None expected.
Cocaine:	Unknown. Avoid.
Foods:	None expected.
Marijuana:	Consult doctor.
Tobacco:	None expected.

*See Glossary

PSEUDOEPHEDRINE

BRAND NAMES

See full list of brand names in the *Generic and Brand Name Directory*, page 878.

BASIC INFORMATION

Habit forming? No
Prescription needed? Yes, for high strength
Available as generic? Yes
Drug class: Sympathomimetic, decongestant

USES

Reduces congestion of nose, sinuses and ears (eustachian tubes) from infections and allergies.

DOSAGE & USAGE INFORMATION

How to take:

- Tablet or capsule—Swallow with liquid. You may chew or crush tablet or open capsule.
- Extended-release tablet or capsule—Swallow whole. Do not crush or chew tablet. The capsule may be opened and the contents mixed with jam or jelly and taken with no chewing.
- Syrup—Take as directed on label.
- Drops—Place directly on tongue and swallow.
- Oral solution—Take as directed on label.
- Combination products—Follow instructions on label.

When to take:

- At the same times each day.
- To prevent insomnia, take last dose of day a few hours before bedtime.

If you forget a dose:
Take as soon as you remember. If it is almost time for the next dose, wait for that dose (don't double this dose) and resume regular schedule.

Continued next column

OVERDOSE

SYMPTOMS:
Nervousness, restlessness, other symptoms that may be due to additional drug ingredients in the product.
WHAT TO DO:
If person uses much larger amount than prescribed or if accidentally swallowed, call poison control center 1-800-222-1222 for instructions or dial 911 (emergency) for help.

What drug does:
Decreases blood volume in nasal tissues, shrinking tissues and enlarging airways.

Time lapse before drug works:
15 to 20 minutes.

Don't take with:
Any other medicine or any dietary supplement without consulting your doctor or pharmacist.

POSSIBLE ADVERSE REACTIONS OR SIDE EFFECTS

SYMPTOMS	WHAT TO DO
Life-threatening:	
Rare, severe allergic reaction (anaphylaxis)—difficulty breathing, hives, itching, swelling (throat, face), vomiting, dizziness.	Seek emergency treatment immediately.
Common:	
None expected.	
Infrequent:	
Nervousness, restlessness, trouble sleeping.	Discontinue. Call doctor if symptoms persist.
Rare:	
• Hallucinations, seizures, slow or irregular heartbeat, difficult breathing or shortness of breath.	Discontinue. Seek emergency treatment.
• Dizziness or lightheadedness, headache, nausea, vomiting, excess sweating, painful or difficult urination, paleness, weakness, trembling.	Discontinue. Call doctor if symptoms persist.

WARNINGS & PRECAUTIONS

Don't take if:
You are allergic to pseudoephedrine or any sympathomimetic* drug.

Before you start, consult your doctor if:
- You have diabetes or overactive thyroid.
- You have taken any monoamine oxidase (MAO) inhibitor* in past 2 weeks.
- You have high blood pressure or heart or blood vessel disease.
- You have glaucoma.
- You have prostate problems.

Over age 60:
Adverse reactions and side effects may be more frequent and severe than in younger persons.

Pregnancy:
Decide with your doctor if drug benefits justify risk to unborn baby. Risk category C (see page xviii).

Breastfeeding; Lactation; Nursing Mothers:
This drug is generally acceptable with breastfeeding. Discuss risks and benefits with your doctor.

Infants & children up to age 18:
Read nonprescription product's label to be sure it is approved for your child's age. Follow label instructions on dosage. Consult doctor or pharmacist if unsure.

Prolonged use:
Not intended for long-term use. Consult doctor.

Skin & sunlight:
No problems expected.

Driving, piloting or hazardous work:
Avoid if you feel dizzy. Otherwise, no problems expected.

Discontinuing:
May be unnecessary to finish medicine. Follow label directions or doctor's instructions.

Others:
- Call the doctor if symptoms worsen or new symptoms develop with use of this medicine.
- Advise any doctor, dentist or pharmacist whom you consult that you take this medicine.
- Heed all warnings on the product label.
- Most pseudoephedrine-containing products are available without a prescription, but there are restrictions on their sales. This is because pseudoephedrine is a substance often used in the illegal manufacture of methamphetamine or "speed." You will need to ask a pharmacist for the product, show identification, sign a logbook and be limited in the amount you can purchase.

POSSIBLE INTERACTION WITH OTHER DRUGS

GENERIC NAME OR DRUG CLASS	COMBINED EFFECT
Antihypertensives*	Decreased anti-hypertensive effect.
Beta-adrenergic blocking agents*	Decreased effect of beta-blocker.
Citrates	Urinary retention. Increased effect of pseudoephedrine.
Digitalis preparations*	Irregular heartbeat.
Methyldopa	Possible increased blood pressure.
Monoamine oxidase (MAO) inhibitors*	Serious reactions (potentially fatal). Take at least 2 weeks apart.
Nitrates*	Possible decreased nitrate effect.
Rauwolfia alkaloids	Decreased effect of pseudoephedrine.
Sympathomimetics,* other	Increased risk of side effects.
Thyroid hormones	Increased effect of either drug.

POSSIBLE INTERACTION WITH OTHER SUBSTANCES

INTERACTS WITH	COMBINED EFFECT
Alcohol:	None expected.
Beverages: Caffeine drinks.	Nervousness or insomnia.
Cocaine:	High risk of heartbeat irregularities and high blood pressure. Avoid
Foods:	None expected.
Marijuana:	Rapid heartbeat. Avoid.
Tobacco:	None expected.

*See Glossary

PSORALENS

GENERIC AND BRAND NAMES

METHOXSALEN
Oxsoralen
Oxsoralen Topical
Oxsoralen Ultra
UltraMOP

TRIOXSALEN
Trisoralen

BASIC INFORMATION

Habit forming? No
Prescription needed? Yes
Available as generic? No
Drug class: Repigmenting agent (psoralen)

USES

- Repigmenting skin affected with vitiligo (absence of skin pigment).
- Treatment for psoriasis, when other treatments haven't helped.
- Treatment for mycosis fungoides.

DOSAGE & USAGE INFORMATION

How to take or apply:
- Tablet or capsule—Swallow with liquid or food to lessen stomach irritation.
- Topical—As directed by doctor.

When to take or apply:
2 to 4 hours before exposure to sunlight or sunlamp.

If you forget a dose:
Take as soon as you remember. If it is almost time for the next dose, wait for that dose (don't double this dose) and resume regular schedule.

What drug does:
Helps pigment cells when used in conjunction with ultraviolet light.

Time lapse before drug works:
- For vitiligo, 6 to 9 months.
- For psoriasis, 10 weeks or longer.

Don't take with:
Any other medicine that causes skin sensitivity to sun. Ask your pharmacist if you have questions.

OVERDOSE

SYMPTOMS:
Skin blisters, peeling, burning, redness.
WHAT TO DO:
If person uses much larger amount than prescribed or if accidentally swallowed, call poison control center 1-800-222-1222 for instructions or dial 911 (emergency) for help.

POSSIBLE ADVERSE REACTIONS OR SIDE EFFECTS

SYMPTOMS	WHAT TO DO
Life-threatening: None expected.	
Common:	
• Increased skin sensitivity to sun.	Always protect from overexposure.
• Increased eye sensitivity to sunlight.	Always protect with wrap-around sunglasses.
• Nausea.	Continue. Call doctor when convenient.
Infrequent:	
• Skin red and sore.	Discontinue. Call doctor right away.
• Dizziness, headache, depression, leg cramps, insomnia.	Continue. Call doctor when convenient.
Rare:	
Hepatitis with jaundice, blistering and peeling.	Discontinue. Call doctor right away.

WARNINGS & PRECAUTIONS

Don't take if:
- You are allergic to any psoralen. Allergic reactions are rare, but may occur.
- You are unwilling or unable to remain under close medical supervision.

Before you start, consult your doctor if:
- You have heart or liver disease.
- You have allergy to sunlight.
- You have cataracts.
- You have lupus erythematosus, porphyria, chronic infection, skin cancer or peptic ulcer.
- You will have surgery within 2 months, including dental surgery, requiring general or spinal anesthesia.
- You have skin cancer.

Over age 60:
Adverse reactions and side effects may be more frequent and severe than in younger persons.

Pregnancy:
Risk factors vary for drugs in this group. Always consult doctor. See category list on page xviii.

Breastfeeding; Lactation; Nursing Mothers:
Drugs in this group are generally not recommended with breastfeeding. Discuss risks and benefits with your doctor.

Infants & children up to age 18:
Follow instructions provided by your child's doctor.

Prolonged use:
- Increased chance of toxic effects.
- Talk to your doctor about the need for follow-up medical examinations or laboratory studies to check ANA titers,* complete blood counts (white blood cell count, platelet count, red blood cell count, hemoglobin, hematocrit), liver function, kidney function, eyes.

Skin & sunlight:
- One or more drugs in this group may cause rash or intensify sunburn in areas exposed to sun or ultraviolet light (photosensitivity reaction). Avoid overexposure. Notify doctor if reaction occurs.
- Too much can burn skin. Cover skin for 24 hours before and 8 hours following treatments.

Driving, piloting or hazardous work:
No problems expected. Protect eyes and skin from bright light.

Discontinuing:
Skin may remain sensitive for some time after treatment stops. Use extra protection from sun.

Others:
- Advise any doctor, dentist or pharmacist whom you consult that you take this medicine.
- Use sunblock on lips.
- Don't use just to make skin tan.
- Don't use hard gelatin capsules interchangeably with soft gelatin capsules.

POSSIBLE INTERACTION WITH OTHER DRUGS

GENERIC NAME OR DRUG CLASS	COMBINED EFFECT
Photosensitizing medications*	Greatly increased likelihood of extreme sensitivity to sunlight.

POSSIBLE INTERACTION WITH OTHER SUBSTANCES

INTERACTS WITH	COMBINED EFFECT
Alcohol:	May increase chance of liver toxicity.
Beverages:	None expected.
Cocaine:	Unknown. Avoid.
Foods: Those containing furocoumarin (limes, parsley, figs, parsnips, carrots, celery, mustard).	May increase risk of photosensitivity.
Marijuana:	Consult doctor.
Tobacco:	May cause uneven absorption of medicine. Avoid.

*See Glossary

PYRIDOXINE (Vitamin B-6)

BRAND NAMES

Beesix
Diclegis
Hexa-Betalin
Pyroxine
Rodex
Vitabec 6

Numerous brands of single vitamin and multivitamin combinations may be available.

BASIC INFORMATION

Habit forming? No
Prescription needed?
High strength: Yes
Low strength: No
Available as generic? Yes
Drug class: Vitamin supplement

USES

- Prevention and treatment of pyridoxine deficiency.
- Treatment of some forms of anemia.
- Treatment of nausea and vomiting in pregnancy.
- Treatment of INH (isonicotinic acid hydrazide), cycloserine poisoning.

DOSAGE & USAGE INFORMATION

How to take:
- Tablet—Swallow with liquid.
- Extended-release capsule or tablet—Swallow each dose whole with liquid.

When to take:
At the same times each day.

If you forget a dose:
Take as soon as you remember. If it is almost time for the next dose, wait for that dose (don't double this dose) and resume regular schedule.

Continued next column

OVERDOSE

SYMPTOMS:
Overdose symptoms may occur after several months of excess use. Numbness and tingling in hands or feet, loss of balance, lack of coordination, difficulty walking, seizures.
WHAT TO DO:
- **Dial 911 for all medical emergencies or call poison control center 1-800-222-1222 for instructions.**
- **See emergency information on last 3 pages of this book.**

What drug does:
It is an essential nutrient needed by the body. It acts as co-enzyme in carbohydrate, protein and fat metabolism.

Time lapse before drug works:
15 to 20 minutes.

Don't take with:
Any other medicine or any dietary supplement without consulting your doctor or pharmacist.

POSSIBLE ADVERSE REACTIONS OR SIDE EFFECTS

SYMPTOMS	WHAT TO DO
Life-threatening: None expected.	
Common: None expected.	
Infrequent: None expected.	
Rare: Large doses may lead to numbness or tingling in hands or feet, clumsiness.	Discontinue. Call doctor if symptoms persist.

WARNINGS & PRECAUTIONS

Don't take if:
You are allergic to pyridoxine. Allergic reactions are rare, but may occur.

Before you start, consult your doctor if:
You are pregnant or breastfeeding.

Over age 60:
No problems expected.

Pregnancy:
Consult doctor. Use is considered safe. Risk category A (see page xviii).

Breastfeeding; Lactation; Nursing Mothers:
This drug is generally acceptable with breastfeeding. Discuss risks and benefits with your doctor.

Infants & children up to age 18:
Read nonprescription product's label to be sure it is approved for your child's age. Follow label instructions on dosage. Consult doctor or pharmacist if unsure.

Prolonged use:
Large doses for more than 1 month may cause toxicity.

Skin & sunlight:
No problems expected.

Driving, piloting or hazardous work:
No problems expected.

Discontinuing:
No problems expected.

Others:
- Advise any doctor, dentist or pharmacist whom you consult that you take this medicine.
- Regular pyridoxine supplements recommended if you take chloramphenicol, cycloserine, ethionamide, hydralazine, immunosuppressants, isoniazid or penicillamine. These decrease pyridoxine absorption and can cause anemia or tingling and numbness in hands and feet.

POSSIBLE INTERACTION WITH OTHER DRUGS

GENERIC NAME OR DRUG CLASS	COMBINED EFFECT
Contraceptives, oral*	Decreased pyridoxine effect.
Cycloserine	Decreased pyridoxine effect.
Estrogens*	Decreased pyridoxine effect.
Ethionamide	Decreased pyridoxine effect.
Hydralazine	Decreased pyridoxine effect.
Hypnotics, barbiturates*	Decreased hypnotic effect.
Immuno-suppressants*	Decreased pyridoxine effect.
Isoniazid	Decreased pyridoxine effect.
Levodopa	Decreased levodopa effect.
Penicillamine	Decreased pyridoxine effect.
Phenobarbital	Possible decreased phenobarbital effect.
Phenytoin	Decreased phenytoin effect.

POSSIBLE INTERACTION WITH OTHER SUBSTANCES

INTERACTS WITH	COMBINED EFFECT
Alcohol:	None expected.
Beverages:	None expected.
Cocaine:	Unknown. Avoid.
Foods:	None expected.
Marijuana:	Consult doctor.
Tobacco:	None expected.

*See Glossary

QUETIAPINE

BRAND NAMES

Seroquel
Seroquel XR

BASIC INFORMATION

Habit forming? No
Prescription needed? Yes
Available as generic? Yes
Drug class: Antipsychotic

USES

- Treatment for symptoms of schizophrenia.
- Treatment for bipolar disorder and major depressive disorder.

DOSAGE & USAGE INFORMATION

How to take:

- Tablet—Swallow with liquid. May be taken with or without food.
- Extended-release tablet—Swallow each dose whole. Don't crush or chew. Take without food or with a light meal or as advised by doctor.

When to take:
As directed by your doctor—2 to 3 times a day at the same times each day. The dosage may be increased over the first few days of use.

If you forget a dose:
Take as soon as you remember. If it is almost time for the next dose, wait for that dose (don't double this dose) and resume regular schedule.

What drug does:
The exact mechanism is unknown. It appears to alleviate symptoms of schizophrenia by blocking certain nerve impulses between nerve cells.

Time lapse before drug works:
One to 7 days. Further increases in the dosage may be needed to relieve symptoms.

Don't take with:
Any other medicine or any dietary supplement without consulting your doctor or pharmacist.

OVERDOSE

SYMPTOMS:
Drowsiness, fast heartbeat, low blood pressure, sedation.
WHAT TO DO:

- **Dial 911 for all medical emergencies or call poison control center 1-800-222-1222 for instructions.**
- **See emergency information on last 3 pages of this book.**

POSSIBLE ADVERSE REACTIONS OR SIDE EFFECTS

SYMPTOMS	WHAT TO DO
Life-threatening:	
High fever, rapid pulse, profuse sweating, muscle rigidity, confusion and irritability, seizures (rare neuroleptic malignant syndrome).	Discontinue. Seek emergency treatment.
Common:	
• Dizziness, difficulty in speaking or swallowing, shaking hands and fingers, trembling, vision problems, weakness, lightheadedness when arising from a sitting or lying position.	Continue, but call doctor right away.
• Drowsiness, constipation, weight gain, agitation, insomnia, headache, nervousness, runny nose, anxiety, dry mouth, arm or leg stiffness.	Continue. Call doctor when convenient.
Infrequent:	
• Jerky or involuntary movements (in face, lips, jaw, tongue), chest pain, fast heartbeat.	Continue, but call doctor right away.
• Fever, flu-like symptoms, twitching, mood or mental changes, speech unclear, swollen feet or ankles, appetite increased, cough, saliva increased, muscle tightness, muscle spasms (face, neck, back), joint pain, nausea, vomiting, sore throat, incontinence, abdominal pain.	Continue. Call doctor when convenient.
Rare:	
• Breathing difficulty, high blood sugar (thirstiness, frequent urination, increased hunger, weakness), swollen face, rash, confusion.	Discontinue. Call doctor right away.
• Decreased sex drive, menstrual changes, sluggishness.	Continue. Call doctor when convenient.

WARNINGS & PRECAUTIONS

Don't take if:
You are allergic to quetiapine. Allergic reactions are rare, but may occur.

Before you start, consult your doctor if:
- You have a heart problem or disease, high or low blood pressure or blood vessel problem.
- Patient has Alzheimer's or dementia.
- You have a history of breast cancer.
- You are subject to dehydration or low body temperature.
- You have had suicidal thoughts or behaviors.
- You have a history of alcohol or drug abuse.
- You have diabetes or high blood sugar, thyroid problems, liver or kidney disease or high levels of blood fats.
- You have glaucoma or cataracts.
- You have prostate problems.
- You are allergic to any medication, food or other substance.
- You have a history of seizures.

Over age 60:
- Adverse reactions and side effects may be more severe than in younger persons. A lower starting dosage is usually recommended until a response is determined.
- Use of antipsychotic drugs in elderly patients with dementia-related psychosis may increase risk of death. Consult doctor.

Pregnancy:
Decide with your doctor if drug benefits justify risk to unborn baby. Risk category C (see page xviii).

Breastfeeding; Lactation; Nursing Mothers:
This drug is generally acceptable with breastfeeding. Discuss risks and benefits with your doctor.

Infants & children up to age 18:
- Follow instructions provided by your child's doctor.
- Contact doctor right away if depression symptoms get worse or there is any talk of suicide or suicide behaviors.

Prolonged use:
- Consult with your doctor on a regular basis to check your progress or to discuss any increase or changes in side effects and the need for continued treatment.
- Get eyes examined every 6 months.

Skin & sunlight:
May cause rash or intensify sunburn in areas exposed to sun or ultraviolet light (photo-sensitivity reaction). Use sunscreen and avoid overexposure. Notify doctor if reaction occurs.

Driving, piloting or hazardous work:
Don't drive or pilot aircraft until you learn how medicine affects you. Don't work around dangerous machinery. Don't climb ladders or work in high places. Danger increases if you drink alcohol or take drug affecting alertness.

Discontinuing:
Don't discontinue drug without doctor's advice. Dosage may need to be slowly reduced first.

Others:
- Get up slowly from a sitting or lying position to avoid any dizziness or lightheadedness.
- Advise any doctor, dentist or pharmacist whom you consult that you take this medicine.
- Hot temperatures, exercise, and hot baths can increase risk of heatstroke. Drug may affect body's ability to maintain normal temperature.
- Adults and children taking antidepressants may experience a worsening of the depression symptoms and may have increased suicidal thoughts or behaviors. Call doctor right away if these symptoms or behaviors occur.
- Take drug only as directed. Do not increase or reduce dosage without doctor's approval.

POSSIBLE INTERACTION WITH OTHER DRUGS

GENERIC NAME OR DRUG CLASS	COMBINED EFFECT
Antihypertensives*	Increased risk of low blood pressure.
Central nervous system (CNS) depressants,* other	Increased sedative effect.
Enzyme inducers*	May decrease quetiapine effect.
Enzyme inhibitors*	May increase quetiapine effect.

Continued on page 909

POSSIBLE INTERACTION WITH OTHER SUBSTANCES

INTERACTS WITH	COMBINED EFFECT
Alcohol:	Increased sedation and dizziness. Avoid.
Beverages: Grapefruit juice.	Increased effect of drug. Avoid.
Cocaine:	Unknown. Avoid.
Foods:	None expected.
Marijuana:	Consult doctor.
Tobacco:	None expected.

***See Glossary**

QUINIDINE

BRAND NAMES

Apo-Quinidine
Cardioquin
Cin-Quin
Duraquin
Novoquinidin
Quinaglute Dura-Tabs
Quinalan
Quinate
Quinidex Extentabs
Quinora

BASIC INFORMATION

Habit forming? No
Prescription needed?
U.S.: Yes
Canada: No
Available as generic? Yes
Drug class: Antiarrhythmic

USES

- Corrects heart rhythm disorders.
- May be used in treatment of malaria.

DOSAGE & USAGE INFORMATION

How to take:

- Tablet or capsule—Swallow with liquid or food to lessen stomach irritation. If you can't swallow whole, crush tablet or open capsule and take with small amount of food.
- Extended-release tablet—Swallow each dose whole. Don't crush them.

When to take:
At the same times each day.

If you forget a dose:
Take as soon as you remember. If it is almost time for the next dose, wait for that dose (don't double this dose) and resume regular schedule.

Continued next column

OVERDOSE

SYMPTOMS:
Irregular heartbeat, severe blood pressure drop, vomiting, diarrhea, ringing in ears, hearing loss, dizziness, vision changes, confusion, headache, delirium.
WHAT TO DO:

- **Dial 911 for all medical emergencies or call poison control center 1-800-222-1222 for instructions.**
- **See emergency information on last 3 pages of this book.**

What drug does:
Delays nerve impulses to the heart to regulate heartbeat.

Time lapse before drug works:
2 to 4 hours.

Don't take with:
Any other medicine or any dietary supplement without consulting your doctor or pharmacist.

POSSIBLE ADVERSE REACTIONS OR SIDE EFFECTS

SYMPTOMS	WHAT TO DO
Life-threatening:	
Rare, severe allergic reaction (anaphylaxis)—difficulty breathing, hives, itching, swelling (throat, face), vomiting, dizziness.	Seek emergency treatment immediately.
Common:	
Diarrhea, nausea, vomiting, appetite loss, muscle weakness.	Continue. Call doctor when convenient.
Infrequent:	
Yellow skin or eyes, abdominal pain, dizziness, fainting, fever, changes in vision, eyes sensitive to light, confusion, delirium, ringing in ears.	Discontinue. Call doctor right away.
Rare:	
Joint pain or swelling, chest pain, muscle pain, headache, pale skin, unusual tiredness or weakness.	Discontinue. Call doctor right away.

WARNINGS & PRECAUTIONS

Don't take if:

- You are allergic to quinidine.
- You have an active infection.

Before you start, consult your doctor if:

- You have an electrolyte disorder.
- You have heart disease or myasthenia gravis.
- You have kidney or liver disease.

Over age 60:
Adverse reactions and side effects may be more frequent and severe than in younger persons.

Pregnancy:
Decide with your doctor if drug benefits justify risk to unborn baby. Risk category C (see page xviii).

Breastfeeding; Lactation; Nursing Mothers:
This drug is generally acceptable with breastfeeding. Discuss risks and benefits with your doctor.

Infants & children up to age 18:
Follow instructions provided by your child's doctor.

Prolonged use:
Talk to your doctor about the need for follow-up medical exams and lab studies.

Skin & sunlight:
May cause rash or intensify sunburn in areas exposed to sun or ultraviolet light (photosensitivity reaction). Avoid overexposure. Notify doctor if reaction occurs.

Driving, piloting or hazardous work:
Don't drive or pilot aircraft until you learn how medicine affects you. Don't work around dangerous machinery. Don't climb ladders or work in high places. Danger increases if you drink alcohol or take medicine affecting alertness and reflexes.

Discontinuing:
Don't discontinue without doctor's advice until you complete prescribed dose, even though symptoms diminish or disappear.

Others:
Advise any doctor, dentist or pharmacist whom you consult that you take this medicine.

POSSIBLE INTERACTION WITH OTHER DRUGS

GENERIC NAME OR DRUG CLASS	COMBINED EFFECT
Alkalizers, urinary*	Slows quinidine elimination, increasing its effect and toxicity.
Amiodarone	Increased effect of quinidine. Risk of heart rhythm problems.
Antacids	Take at least 2 hours apart.
Anticholinergics	Increased anti-cholinergic effect.
Anticoagulants,* oral	Possible increased anticoagulant effect.
Antidepressants, tricyclic	Increased risk of heart rhythm problems.
Cimetidine	Increased quinidine effect.
Digitalis preparations	May slow heartbeat. Dose adjustments may be needed.
Diltiazem	Increased quinidine effect.
Enzyme inhibitors*	Increased effect of quinidine.
Erythromycin	Possible irregular heartbeat.
Haloperidol	Possible irregular heartbeat.
Mefloquine	Possible irregular heartbeat. Avoid.
Memantine	Increased effect of either drug.
Metformin	Increased metformin effect.
Nifedipine	Possible decreased quinidine effect.
Phenobarbital	Decreased quinidine effect.
Phenothiazines*	Possible increased quinidine effect.
QT interval prolongation-causing drugs*	Heartbeat irregularities.
Verapamil	Increased quinidine effect.

POSSIBLE INTERACTION WITH OTHER SUBSTANCES

INTERACTS WITH	COMBINED EFFECT
Alcohol:	None expected.
Beverages:	
Caffeine drinks.	Causes rapid heart-beat. Use with care.
Grapefruit juice.	Toxicity risk. Avoid.
Cocaine:	Unknown. Avoid.
Foods:	None expected.
Marijuana:	Consult doctor.
Tobacco:	None expected.

*See Glossary

QUININE

BRAND NAMES

Qualaquin

BASIC INFORMATION

Habit forming? No
Prescription needed? Yes
Available as generic? Yes
Drug class: Antiprotozoal

USES

Treatment or prevention of malaria.

DOSAGE & USAGE INFORMATION

How to take:
Tablet or capsule—Swallow with liquid or food to lessen stomach irritation.

When to take:
- Prevention—At the same time each day, usually at bedtime.
- Treatment—At the same times each day in evenly spaced doses.

If you forget a dose:
Take as soon as you remember. If it is almost time for the next dose, wait for that dose (don't double this dose) and resume regular schedule.

What drug does:
- Reduces contractions of skeletal muscles.
- Increases blood flow.
- Interferes with genes in malaria micro-organisms.

Continued next column

OVERDOSE

SYMPTOMS:
Severe impairment of vision and/or hearing, severe nausea, vomiting, diarrhea, headache, stomach pain, dizziness, sweating, flushing, irregular heartbeat, low blood pressure, drowsiness, balance problems, breathing difficulty, coma.
WHAT TO DO:
- **Dial 911 for all medical emergencies or call poison control center 1-800-222-1222 for instructions.**
- **See emergency information on last 3 pages of this book.**

Time lapse before drug works:
May require several days or weeks for maximum effect.

Don't take with:
Any other medicine or any dietary supplement without consulting your doctor or pharmacist.

POSSIBLE ADVERSE REACTIONS OR SIDE EFFECTS

SYMPTOMS	WHAT TO DO
Life-threatening:	
Rare, severe allergic reaction (anaphylaxis)—difficulty breathing, hives, itching, swelling (throat, face), vomiting, dizziness.	Seek emergency treatment immediately.
Common:	
• Blurred vision or change in vision, eyes sensitive to light.	Discontinue. Call doctor right away.
• Dizziness, headache, abdominal discomfort, mild nausea, vomiting, diarrhea, ringing or buzzing in ears, impaired hearing.	Continue. Call doctor when convenient.
Infrequent:	
Rash, hives, itchy skin, difficult breathing.	Discontinue. Call doctor right away.
Rare:	
Sore throat, fever, unusual bleeding or bruising, unusual tiredness or weakness, angina.	Discontinue. Call doctor right away.

WARNINGS & PRECAUTIONS

Don't take if:
You are allergic to quinine or quinidine.

Before you start, consult your doctor if:
- You plan to become pregnant within medication period.
- You have asthma.
- You have eye disease, hearing problems or ringing in the ears.
- You have heart disease.
- You have myasthenia gravis.

Over age 60:
Adverse reactions and side effects may be more frequent and severe than in younger persons.

Pregnancy:
Decide with your doctor if drug benefits justify risk to unborn baby. Risk category C (see page xviii).

Breastfeeding; Lactation; Nursing Mothers:
This drug is generally acceptable with breastfeeding. Discuss risks and benefits with your doctor.

Infants & children up to age 18:
Follow instructions provided by your child's doctor.

Prolonged use:
May develop headache, blurred vision, nausea, temporary hearing loss, but seldom need to discontinue because of these symptoms.

Skin & sunlight:
May cause rash or intensity sunburn in areas exposed to sun or ultraviolet light (photosensitivity reaction). Avoid overexposure. Notify doctor if reaction occurs.

Driving, piloting or hazardous work:
Avoid if you feel dizzy or have blurred vision. Otherwise, no problems expected.

Discontinuing:
Don't discontinue without doctor's advice until you complete prescribed dose, even though symptoms diminish or disappear.

Others:
- Advise any doctor, dentist or pharmacist whom you consult that you take this medicine.
- Quinine is only approved for treating malaria. It is sometimes used as an unapproved treatment for restless leg syndrome or restless leg cramps. This use can lead to serious and life-threatening side effects. Consult your doctor.
- Don't confuse with quinidine, a medicine for heart rhythm problems.

POSSIBLE INTERACTION WITH OTHER DRUGS

GENERIC NAME OR DRUG CLASS	COMBINED EFFECT
Alkalizers,* urinary	Possible toxic effects of quinine.
Antacids* (with aluminum hydroxide)	Decreased quinine effect.
Anticoagulants,* oral	Increased anticoagulant effect.
Dapsone	Increased risk of adverse effect on blood cells.
Digitalis	Possible increased digitalis effect.
Digoxin	Possible increased digoxin effect.
Mefloquine	Increased risk of heartbeat irregularities.
Metformin	Increased metformin effect.
Quinidine	Possible toxic effects of quinine.

POSSIBLE INTERACTION WITH OTHER SUBSTANCES

INTERACTS WITH	COMBINED EFFECT
Alcohol:	None expected.
Beverages:	None expected.
Cocaine:	Unknown. Avoid.
Foods:	None expected.
Marijuana:	Consult doctor.
Tobacco:	None expected.

***See Glossary**

RALOXIFENE

BRAND NAMES

Evista

BASIC INFORMATION

Habit forming? No
Prescription needed? Yes
Available as generic? Yes
Drug class: Osteoporosis therapy, prophylactic

USES

- Prevents and treats osteoporosis in postmenopausal women. Does not treat hot flashes of menopause.
- Lowers low-density lipoprotein (LDL) cholesterol blood levels.
- Used for reduction of breast cancer risk in postmenopausal women.
- Other uses as determined by your doctor.

DOSAGE & USAGE INFORMATION

How to take:
Tablet—Swallow with water, with or without food. If you can't swallow whole, crumble tablet and take with liquid or food.

When to take:
At the same time each day.

If you forget a dose:
Take as soon as you remember. If it is almost time for the next dose, wait for that dose (don't double this dose) and resume regular schedule.

What drug does:
It is a selective estrogen receptor modulator or SERM. These drugs can act like the hormone estrogen (such as in the bones) or block estrogen (such as in the breast tissue).

Time lapse before drug works:
Up to twelve months.

Don't take with:
Any other medicine or any dietary supplement without consulting your doctor or pharmacist.

OVERDOSE

SYMPTOMS:
May have dizziness, leg cramps and others.
WHAT TO DO:
If person uses much larger amount than prescribed or if accidentally swallowed, call poison control center 1-800-222-1222 for instructions or dial 911 (emergency) for help.

POSSIBLE ADVERSE REACTIONS OR SIDE EFFECTS

SYMPTOMS	WHAT TO DO
Life-threatening:	
Blood clot formation. (see Rare below for symptoms).	Discontinue. Seek emergency treatment
Common:	
• Chest pain, bloody or cloudy urine, burning or painful urination, frequent urge to urinate, infection, cold- or flu-like symptoms, leg cramping, skin rash, swelling of hands or ankles or feet, vaginal itching.	Discontinue. Call doctor right away.
• Joint or muscle pain, swollen joints, gas, upset stomach, vomiting, hot flashes, insomnia, white vaginal discharge, depression, sweating, unexplained weight gain.	Continue. Call doctor when convenient.
Infrequent:	
Abdominal pain, diarrhea, loss of appetite, nausea, weakness, migraine headache, difficulty breathing, fever, congestion.	Discontinue. Call doctor right away.
Rare:	
Blood clot formation (symptoms include swelling in legs, sharp pain in legs, sudden chest pain, coughing up blood, changes in vision).	Discontinue. Seek emergency care.

WARNINGS & PRECAUTIONS

Don't take if:
- You are allergic to raloxifene. Allergic reactions are rare, but may occur.
- You are scheduled for surgery within 72 hours.

Before you start, consult your doctor if:
- You plan to become pregnant within the medication period.
- You have or have had a history of blood clot (deep vein thrombosis) formation.
- You have a history of stroke, transient ischemic attacks (TIA) or heart disease.
- You have high triglycerides (a blood fat).
- You have or have had cancer or tumors.
- You have liver disease.

Over age 60:
No special problems expected.

Pregnancy:
Consult doctor. Drug should not be used during pregnancy. Can cause harm to unborn baby. Risk category X (see page xviii).

Breastfeeding; Lactation; Nursing Mothers:
This drug is generally acceptable with breastfeeding. Discuss risks and benefits with your doctor.

Infants & children up to age 18:
Safety and effectiveness in this age group have not been established. Consult doctor.

Prolonged use:
No problems expected. Your doctor should periodically evaluate your response to the drug and adjust the dose if necessary.

Skin & sunlight:
No problems expected.

Driving, piloting or hazardous work:
No problems expected.

Discontinuing:
Don't discontinue without consulting doctor.

Others:
- Advise any doctor, dentist or pharmacist whom you consult that you take this medicine.
- This drug is associated with an increased risk of developing blood clots in the veins (deep vein thrombosis) or stroke. Consult doctor.
- In addition to taking the drug, weight-bearing exercise and adequate intake of calcium and vitamin D are essential in preventing bone loss. Periods of prolonged inactivity may worsen your condition. Daily dietary supplements of elemental calcium and vitamin D may be recommended by your doctor.
- May help lower low-density lipoprotein (LDL) cholesterol levels.

POSSIBLE INTERACTION WITH OTHER DRUGS

GENERIC NAME OR DRUG CLASS	COMBINED EFFECT
Cholestyramine	Lessens the effect of raloxifene.
Estrogens	Not recommended for use with raloxifene.
Protein bound drugs*	Caution is recommended. Consult doctor before taking any of these in conjunction with raloxifene.
Warfarin	May lessen the effect of warfarin.

POSSIBLE INTERACTION WITH OTHER SUBSTANCES

INTERACTS WITH	COMBINED EFFECT
Alcohol:	Increased risk of adverse effects. Avoid.
Beverages:	None expected.
Cocaine:	Unknown. Avoid.
Foods:	None expected.
Marijuana:	Consult doctor.
Tobacco:	Increases risk of blood clot. Avoid.

***See Glossary**

RAMELTEON

BRAND NAMES

Rozerem

BASIC INFORMATION

Habit forming? No
Prescription needed? Yes
Available as generic? No
Drug class: Melatonin receptor agonist; nonbenzodiazepine hypnotic

USES

It is used to help you fall asleep faster when you have trouble falling asleep and experience insomnia.

DOSAGE & USAGE INFORMATION

How to take:
Tablet—Swallow whole with liquid. Do not crush or chew the tablet. Take it with or without food. Do not take it with, or immediately after, a high fat meal. Doing so can reduce the effectiveness of the drug.

When to take:
Take within 30 minutes of bedtime. Take the drug only when you know that you will get 8 full hours (or more) of sleep after the dose.

If you forget a dose:
Skip the missed dose. Resume dosage schedule the next night if needed to help you sleep. Do not take more than one dose in a 24 hour period.

What drug does:
It stimulates melatonin receptors (chemicals) in the brain that are responsible for the regulation of the body's 24 hour sleep-wake cycle.

Time lapse before drug works:
Usually within 30 to 90 minutes.

Continued next column

OVERDOSE

SYMPTOMS:
Unknown.
WHAT TO DO:
- **Dial 911 for all medical emergencies or call poison control center 1-800-222-1222 for instructions.**
- **See emergency information on last 3 pages of this book.**

Don't take with:
The drug fluvoxamine or any other medicine or any dietary supplement without consulting your doctor or pharmacist.

POSSIBLE ADVERSE REACTIONS OR SIDE EFFECTS

SYMPTOMS	WHAT TO DO
Life-threatening:	
Rare, severe allergic reaction (anaphylaxis)—difficulty breathing, hives, itching, swelling (throat, face), vomiting, dizziness.	Seek emergency treatment immediately.
Common:	
Dizziness, drowsiness, nausea, headache, fatigue, worsening insomnia.	Discontinue. Call doctor when convenient.
Infrequent:	
Decreased sex drive, menstrual changes, milky discharge from breasts, cold- or flu-like symptoms, muscle or joint or body aches or pain, taste changes, stomach upset.	Discontinue. Call doctor when convenient.
Rare:	
The class of drugs used to treat insomnia are called hypnotics. They can cause a variety of adverse reactions in one's emotions, behavior, cognition (thinking) and mood. Symptoms can include worsening depression, memory problems, confusion, bizarre behaviors, suicidal thoughts, hallucinations, unusual excitement, irritability, aggressiveness, nervousness, sleep-related behaviors,* and possibly others.	Discontinue. Call doctor right away.

WARNINGS & PRECAUTIONS

Don't take if:
- You are allergic to ramelteon.
- You are taking the drug fluvoxamine.

Before you start, consult your doctor if:
- You have or have had liver disease.
- You suffer from depression or a psychiatric disorder or abuse alcohol or drugs.
- You suffer from sleep apnea or emphysema, asthma, bronchitis or other chronic lung disease.

Over age 60:
No special problems expected.

Pregnancy:
Decide with your doctor if drug benefits justify risk to unborn baby. Risk category C (see page xviii).

Breastfeeding; Lactation; Nursing Mothers:
This drug is generally acceptable with breastfeeding. Discuss risks and benefits with your doctor.

Infants & children up to age 18:
Safety and effectiveness in this age group have not been established. Consult doctor.

Prolonged use:
No special problems expected. You and your doctor will decide if there is a need to take the drug for a prolonged period for chronic insomnia (insomnia at least three nights a week for a period of one month or longer).

Skin & sunlight:
No special problems expected.

Driving, piloting or hazardous work:
Don't drive or pilot aircraft until you learn how medicine affects you. Don't work around dangerous machinery. Don't climb ladders or work in high places. Danger increases if you drink alcohol or take other medicines affecting alertness and reflexes.

Discontinuing:
No problems expected. Ramelteon does not appear to produce withdrawal symptoms or lead to physical dependence.

Others:
- Advise any doctor, dentist or pharmacist whom you consult that you take this medicine.
- Consult your doctor if your insomnia symptoms do not improve after taking the drug a few nights.

POSSIBLE INTERACTION WITH OTHER DRUGS

GENERIC NAME OR DRUG CLASS	COMBINED EFFECT
Enzyme inducers*	Decreased effect of ramelteon.
Enzyme inhibitors*	Increased effect of ramelteon. Consult your doctor or pharmacist before using.
Fluvoxamine	Increased effect of ramelteon. Avoid.

POSSIBLE INTERACTION WITH OTHER SUBSTANCES

INTERACTS WITH	COMBINED EFFECT
Alcohol:	Increased sedation. Avoid.
Beverages:	
Grapefruit juice.	Unknown effect. Consult doctor.
Cocaine:	Unknown. Avoid.
Foods:	
Grapefruit.	Unknown effect. Consult doctor.
High fat meal	Drug effect is decreased if taken with, or right after, the meal.
Marijuana:	Consult doctor.
Tobacco:	None expected.

***See Glossary**

RANOLAZINE

BRAND NAMES

Ranexa

BASIC INFORMATION

Habit forming? No
Prescription needed? Yes
Available as generic? Yes
Drug class: Antianginal

USES

Treatment of chronic angina (chest pain caused by an insufficient supply of oxygen to the heart). It may be used in combination with other drug treatments, such as amlodipine, beta blockers or nitrates.

DOSAGE & USAGE INFORMATION

How to take:
Extended-release tablet—Swallow whole with liquid. It may be taken with or without food. Do not crush, crumble or chew tablet.

When to take:
Usually twice a day at the same times each day. Follow the instructions on your prescription.

If you forget a dose:
Take as soon as you remember. If it is almost time for the next dose, wait for that dose (don't double this dose) and resume regular schedule.

What drug does:
Exact method of action is unknown. It appears to relax contracted heart muscle and return blood flow to normal levels which relieves the pain.

Time lapse before drug works:
Within a few hours. Dosage may be increased by your doctor as needed for pain relief.

Don't take with:
Any other medicine or any dietary supplement without consulting your doctor or pharmacist.

OVERDOSE

SYMPTOMS:
May include dizziness, nausea, vomiting, tremor, being unsteady, hallucinations, speaking problems.
WHAT TO DO:
If person uses much larger amount than prescribed or if accidentally swallowed, call poison control center 1-800-222-1222 for instructions or dial 911 (emergency) for help.

POSSIBLE ADVERSE REACTIONS OR SIDE EFFECTS

SYMPTOMS	WHAT TO DO
Life-threatening: None expected.	
Common: Dizziness, headache, constipation, nausea, weakness.	Continue. Call doctor when convenient.
Infrequent: Abdominal pain, vomiting, dry mouth, tremor, dizziness or lightheadedness, ringing in ear.	Continue. Call doctor when convenient.
Rare: Palpitations, slow heartbeat, blood in urine, blurred vision, fainting.	Discontinue. Call doctor right away.

WARNINGS & PRECAUTIONS

Don't take if:
You are allergic to ranolazine. Allergic reactions are rare, but may occur.

Before you start, consult your doctor if:
- You have pre-existing QT interval prolongation (an abnormality of the heart's electrical system as diagnosed by medical testing).
- You have kidney or liver disease.
- You have uncorrected hypokalemia (low potassium level).
- You have a history of ventricular tachycardia.

Over age 60:
Adverse reactions and side effects may be more frequent and severe than in younger persons.

Pregnancy:
Decide with your doctor if drug benefits justify risk to unborn baby. Risk category C (see page xviii).

Breastfeeding; Lactation; Nursing Mothers:
This drug is generally not recommended with breastfeeding. Discuss risks and benefits with your doctor.

Infants & children up to age 18:
Safety and effectiveness in this age group have not been established. Consult doctor.

Prolonged use:
No problems expected.

Skin & sunlight:
No problems expected.

Driving, piloting or hazardous work:
Don't drive motor vehicles or pilot aircraft until you learn how medicine affects you. Don't work around dangerous machinery. Don't climb ladders or work in high places.

Discontinuing:
No problems expected. Consult your doctor before stopping the drug.

Others:
- Other antianginal drugs should be tried first due to the risk of QT interval prolongation (abnormality of the heart's electrical system).
- An ECG* should be performed prior to starting the drug and periodically during drug therapy.
- It is important to follow your doctor's advice on diet, exercise, smoking and weight control.
- Advise any doctor, dentist or pharmacist whom you consult that you take this medicine.

POSSIBLE INTERACTION WITH OTHER DRUGS

GENERIC NAME OR DRUG CLASS	COMBINED EFFECT
Antidepressants, tricyclics	Increased effect of tricyclic antidepressant.
Antipsychotics*	Increased effect of antipsychotic.
Digoxin	Increased effect and blood levels of digoxin.
Enzyme inhibitors*	Risk of abnormal heart rhythm. Avoid.
QT interval prolongation-causing drugs,* other	Risk of abnormal heart rhythm. Avoid.
Simvastatin	Increased effect of simvastatin.

POSSIBLE INTERACTION WITH OTHER SUBSTANCES

INTERACTS WITH	COMBINED EFFECT
Alcohol:	None expected.
Beverages: Grapefruit juice.	Risk of abnormal heart rhythm. Avoid.
Cocaine:	Unknown. Avoid.
Foods: Grapefruit.	Risk of abnormal heart rhythm. Avoid.
Marijuana:	Consult doctor.
Tobacco:	None expected. Persons with angina should not smoke.

*See Glossary

RENIN INHIBITORS

GENERIC AND BRAND NAMES

ALISKIREN
- **Amturnide**
- **Tekamlo**
- **Tekturna**
- **Tekturna HCT**
- **Valturna**

BASIC INFORMATION

Habit forming? No
Prescription needed? Yes
Available as generic? No
Drug class: Antihypertensive; renin inhibitor

USES

Treatment for hypertension (high blood pressure). May be used alone or along with other antihypertensive medications.

DOSAGE & USAGE INFORMATION

How to take:
Tablet—Swallow with liquid. May be taken with or without food. Do not remove the special drying agent from the bottle.

When to take:
Once daily at the same time each day, or as directed by your doctor.

If you forget a dose:
Take as soon as you remember. If it is almost time for the next dose, wait for that dose (don't double this dose) and resume regular schedule.

What drug does:
It inhibits renin, a kidney enzyme associated with the regulation of blood pressure. This helps the blood vessels to relax and widen so blood pressure is lowered.

Continued next column

OVERDOSE

SYMPTOMS:
Unknown. Could cause very low blood pressure (hypotension).
WHAT TO DO:
If person uses much larger amount than prescribed or if accidentally swallowed, call poison control center 1-800-222-1222 for instructions or dial 911 (emergency) for help.

Time lapse before drug works:
It starts working within a few hours, but may take several weeks for full effectiveness.

Don't take with:
Any other medicine or any dietary supplement without consulting your doctor or pharmacist.

POSSIBLE ADVERSE REACTIONS OR SIDE EFFECTS

SYMPTOMS	WHAT TO DO
Life-threatening:	
Rare, severe allergic reaction (anaphylaxis)—difficulty breathing, hives, itching, swelling (throat, face), vomiting, dizziness.	Seek emergency treatment immediately.
Common:	
Diarrhea, headache.	Continue. Call doctor when convenient.
Infrequent:	
Cough, rash.	Continue. Call doctor when convenient.
Rare:	
• Low blood pressure (feeling faint, dizzy or lightheaded), seizure.	Discontinue. Call doctor right away.
• Abdominal pain, bloating, heartburn, nausea, burping, reflux, fatigue, upper respiratory tract infection, back pain, runny or stuffy nose.	Continue. Call doctor when convenient.

WARNINGS & PRECAUTIONS

Don't take if:
- You are allergic to aliskiren.
- You are pregnant.

Before you start, consult your doctor if:
- You have kidney (renal) disease or disorder.
- You have a history of dialysis, nephrotic syndrome, or renovascular hypertension (high blood pressure caused by narrowing of the arteries that carry blood to the kidneys).
- You have hyperkalemia (high level of potassium in the blood).
- You plan to become pregnant.
- You are allergic to any medication, food or other substance.

Over age 60:
No special problems expected.

Pregnancy:
Consult doctor. Use of the drug during pregnancy is not recommended. It could cause harm to the unborn baby. Risk category D for second and third trimester and category C for first trimester (see page xviii).

Breastfeeding; Lactation; Nursing Mothers:
Drugs in this group are generally not recommended with breastfeeding. Discuss risks and benefits with your doctor.

Infants & children up to age 18:
Safety and effectiveness in this age group have not been established. Consult doctor.

Prolonged use:
- No special problems expected. Hypertension usually requires life-long treatment.
- Schedule regular doctor visits to determine if drug is continuing to be effective in controlling the hypertension.

Skin & sunlight:
No special problems expected.

Driving, piloting or hazardous work:
Use caution if you feel dizzy or are experiencing other side effects.

Discontinuing:
Don't discontinue without consulting your doctor, even if you feel well. You can have hypertension without feeling any symptoms. Untreated high blood pressure can cause serious problems.

Others:
- Advise any doctor, dentist or pharmacist whom you consult that you take this medicine. May interfere with the accuracy of some medical tests.
- Follow any diet or exercise plan your doctor prescribes. It can help control hypertension.
- Get up slowly from a sitting or lying position to avoid any dizziness, faintness or lightheadedness.

POSSIBLE INTERACTION WITH OTHER DRUGS

GENERIC NAME OR DRUG CLASS	COMBINED EFFECT
Antihypertensives,* other	Increased antihypertensive effect.
Furosemide	May decrease furosemide effect.

POSSIBLE INTERACTION WITH OTHER SUBSTANCES

INTERACTS WITH	COMBINED EFFECT
Alcohol:	None expected.
Beverages:	None expected.
Cocaine:	Unknown. Avoid.
Foods:	None expected.
Marijuana:	Consult doctor.
Tobacco:	None expected.

***See Glossary**

RETINOIDS (Oral)

GENERIC AND BRAND NAMES

ACITRETIN
Soriatane

BASIC INFORMATION

Habit forming? No
Prescription needed? Yes
Available as generic? Yes
Drug class: Antipsoriatic

USES

- Treats psoriasis in patients who don't respond well to standard or usual treatment. It may be combined with phototherapy or other antipsoriatic drugs.
- It may be used to improve arthritis symptoms that accompany psoriasis.
- Treatment for certain other skin disorders.

DOSAGE & USAGE INFORMATION

How to take:
Capsule—Swallow with liquid and take with food to lessen stomach irritation.

When to take:
At the same time each day, usually once a day.

If you forget a dose:
Take as soon as you remember. If it is almost time for the next dose, wait for that dose (don't double this dose) and resume regular schedule.

What drug does:
The exact mechanism of action is unknown. The drug appears to help growth of normal skin cells. It also has anti-inflammatory action.

Time lapse before drug works:
The skin may show improvement in 2 weeks, but full benefit can take 2 to 3 months.

Don't take with:
Any other medicine or any dietary supplement without consulting your doctor or pharmacist.

OVERDOSE

SYMPTOMS:
May include nausea, vomiting, severe headache, drowsiness, loss of balance.
WHAT TO DO:
If person uses much larger amount than prescribed or if accidentally swallowed, call poison control center 1-800-222-1222 for instructions or dial 911 (emergency) for help.

POSSIBLE ADVERSE REACTIONS OR SIDE EFFECTS

SYMPTOMS	WHAT TO DO
Life-threatening: None expected.	
Common:	
• Severe headache, severe nausea or vomiting.	Discontinue. Call doctor right away.
• Stiffness or pain in bones or joints or muscles, mild headaches, lip symptoms (chapped, redness, sore, cracking, swollen), dryness of nose or eyes, hair loss, skin sensitivity to sunlight, sore mouth, nosebleeds, peeling or scaling (eyelids, palms, fingertips or soles of feet), runny nose, sore or swollen gums, thirstiness.	Continue. Call doctor when convenient.
Infrequent:	
• Blurred vision, eye pain.	Discontinue. Call doctor right away.
• Fingernails loose or skin around them is sore or red, loss of eyebrows or eye lashes, redness of eye or inside eyelid, eyes watery or sensitive to light, swollen eyelids, problem with contact lens use, psoriasis gets worse in early drug use.	Continue. Call doctor when convenient.
Rare:	
• Double vision, decreased night vision, yellow skin or eyes, stomach pain, dark urine, unusual bruising.	Discontinue. Call doctor right away.
• Skin symptoms (spots, infection, sores, stinging, odd odor, burning, rash), ear pain or itching, flu-like symptoms, stye, cough or hoarseness, trouble in speaking, vaginal discharge or itching or irritation.	Continue. Call doctor when convenient.

WARNINGS & PRECAUTIONS

Don't take if:

- You are female and pregnant or may get pregnant in the next 3 years. Severe birth defects have occurred while using and after discontinuing this drug. Don't start drug until you have 2 negative pregnancy tests. Read all the prescribing information carefully.
- If you are allergic to retinoids or parabens (used as preservative in gelatin capsule). Allergic reactions are rare, but may occur.

Before you start, consult your doctor if:

- You have high cholesterol or triglycerides.
- You have diabetes or are an alcoholic.
- You have had problems with too much vitamin A in the body (hypervitaminosis).
- You are a female of reproductive age.
- You have kidney, liver or pancreas disease.

Over age 60:
Adverse reactions and side effects may be more frequent and severe than in younger persons.

Pregnancy:
Consult doctor. Drug should not be used during pregnancy. Can cause harm to unborn baby. Risk category X (see page xviii).

Breastfeeding; Lactation; Nursing Mothers:
This drug is generally not recommended with breastfeeding. Discuss risks and benefits with your doctor.

Infants & children up to age 18:
Safety and effectiveness in this age group have not been established. Drug may prevent normal bone growth in children. Consult doctor.

Prolonged use:
Talk to your doctor about follow-up medical exams or laboratory studies to check blood lipids, liver function, eyes or pregnancy tests.

Skin & sunlight:
May cause rash or intensify sunburn in areas exposed to sun or ultraviolet light (photosensitivity reaction). Use sunscreen with SPF of 15 or higher. Avoid overexposure. Notify doctor if reaction occurs.

Driving, piloting or hazardous work:
Don't drive motor vehicles or pilot aircraft until you learn how medicine affects you. Don't work around dangerous machinery. Don't climb ladders or work in high places.

Discontinuing:
Drug use is discontinued once skin has healed sufficiently. Your psoriasis may recur after stopping the drug. Consult doctor for advice. Don't use leftover drug without doctor's approval.

Others:

- Certain methods of birth control may fail while using this drug, including tubal ligation and progestin (minipill) preparations. Other birth control methods may be affected also. Use two different methods of birth control.
- Don't donate blood during drug treatment or for 3 years thereafter.
- Advise any doctor, dentist or pharmacist whom you consult that you take this medicine.
- May cause liver damage or problems in controlling blood sugar.
- Laboratory blood studies for cholesterol and triglyceride levels should be obtained prior to and during treatment.
- Avoid skin products that cause skin dryness or sensitivity. Consult your doctor for advice.

POSSIBLE INTERACTION WITH OTHER DRUGS

GENERIC NAME OR DRUG CLASS	COMBINED EFFECT
Acne preparations*	Excessive drying effect on skin.
Contraceptives, oral* (progestin-only type)	Decreased effect of oral contraceptive.
Isotretinoin	Increased risk of adverse reactions.
Methotrexate	Risk of hepatitis.
Phenytoin	Increased effect of phenytoin.
Tetracyclines*	Risk of pseudotumor cerebri (pressure in the brain).
Tretinoin	Increased risk of adverse reactions.
Vitamin A	Increased risk of adverse reactions.

POSSIBLE INTERACTION WITH OTHER SUBSTANCES

INTERACTS WITH	COMBINED EFFECT
Alcohol:	Risk of severe side effects during drug use and after stopping. Avoid alcohol.
Beverages:	None expected.
Cocaine:	Unknown. Avoid.
Foods:	None expected.
Marijuana:	Consult doctor.
Tobacco:	None expected.

*See Glossary

RETINOIDS (Topical)

GENERIC AND BRAND NAMES

ADAPALENE
- **Differin**
- **Epiduo**

BEXAROTENE
- **Targretin**

TAZAROTENE
- **Avage**
- **Fabior**
- **Tazorac**

TRETINOIN
- **Atralin Gel**
- **Avita**
- **Renova**
- **Retin-A Cream**
- **Retin-A Cream Regimen Kit**
- **Retin-A Gel**
- **Retin-A Gel Regimen Kit**

TRETINOIN (Con't)
- **Retin-A Solution**
- **Retinoic Acid**
- **Solage**
- **Stieva-A Cream**
- **Stieva-A Cream Forte**
- **Stieva-A Gel**
- **Stieva-A Solution**
- **Tretin-X**
- **Tri-Luma**
- **Veltin Gel**
- **Vitamin A Acid Cream**
- **Vitamin A Acid Gel**
- **Ziana Gel**

BASIC INFORMATION

Habit forming? No
Prescription needed? Yes
Available as generic? Yes, for some
Drug class: Antiacne (topical), antipsoriatic

USES

- Treatment for acne, psoriasis, ichthyosis, keratosis, folliculitis, flat warts.
- Treatment for sun-damaged skin (wrinkles), mottled skin, rough skin and pigmented skin.
- Bexarotene treats skin cancer (cutaneous T-cell lymphoma).
- Tri-Luma is a combination drug—tretinoin and fluocinolone and hydroquinone (not covered in this book).
- Solage is a combination drug—tretinoin plus mequinol (which is not covered in this book).

OVERDOSE

SYMPTOMS:
Unknown.
WHAT TO DO:
If person uses much larger amount than prescribed or if accidentally swallowed, call poison control center 1-800-222-1222 for instructions or dial 911 (emergency) for help.

DOSAGE & USAGE INFORMATION

How to use:
Wash skin with nonmedicated soap, pat dry, wait 20 minutes before applying or as directed.
- Cream, gel or foam—Apply to affected areas with fingertips and rub in gently.
- Solution—Apply to affected areas with gauze pad or cotton swab.
- Avoid getting product into eyes or mouth, onto lips, inside nose or on vagina.

When to use:
Apply once daily, usually in the evening before going to bed.

If you forget a dose:
Use as soon as you remember. If it is almost time for the next dose, wait for that dose (don't double this dose) and resume regular schedule.

What drug does:
Helps control acne inflammation and prevent new acne outbreaks. Increases skin cell turnover so skin layer peels off more easily.

Time lapse before drug works:
May require 2 to 6 weeks for minimum benefit and 3 to 12 months for full benefit.

Don't use with:
Any other topical medicine without consulting your doctor or pharmacist.

POSSIBLE ADVERSE REACTIONS OR SIDE EFFECTS

SYMPTOMS	WHAT TO DO
Life-threatening:	
None expected.	
Common:	
• Mild redness, itching, chapping, dryness of skin during first few weeks of use.	Depending on severity, may reduce frequency of use.
• Worsening of acne or psoriasis during first few weeks of use (due to action of the drug on previous unseen breakouts on skin).	Continue. Call doctor if symptoms persist.
Infrequent:	
Painful skin irritation, darkening or lightening of skin where treated.	Discontinue. Call doctor when convenient.
Rare:	
None expected.	

WARNINGS & PRECAUTIONS

Don't take if:
You are allergic to topical retinoids or any of the components of the gel product. Allergic reactions are rare, but may occur.

Before you start, consult your doctor if:
- You are using any other prescription or nonprescription medicine for the skin.
- You are using abrasive skin cleansers or medicated cosmetics.

Over age 60:
No problems expected.

Pregnancy:
- Adapalene and tretinoin—discuss with your doctor if benefits outweigh risk to unborn baby. Risk category C (see page xviii).
- Bexarotene and tazarotene—consult doctor. Should not be used during pregnancy. Can cause harm to unborn baby. Risk category X (see page xviii).

Breastfeeding; Lactation; Nursing Mothers:
Drugs in this group are generally acceptable with breastfeeding. Discuss risks and benefits with your doctor.

Infants & children up to age 18:
- Follow instructions provided by your child's doctor.
- Safety and effectiveness under age 12 have not been established.

Prolonged use:
No problems expected.

Skin & sunlight:
- May cause rash or intensify sunburn in areas exposed to sun or ultraviolet light. Avoid overexposure. Notify doctor if reaction occurs.
- If you are normally exposed to considerable sunlight, use extra caution. Use broad spectrum of sunscreen on treated areas and wear protective clothing (e.g., hat).

Driving, piloting or hazardous work:
No problems expected.

Discontinuing:
- May be unnecessary to finish medicine. Discontinue when acne improves. For some patients, this medicine may be used indefinitely to control acne.
- If skin problem doesn't improve after first few weeks of use, consult doctor.

Others:
- Cold or windy weather may further irritate the skin. Avoid if possible.
- Products with a drying effect on the skin may cause irritation when used with retinoids. These include cosmetics, abrasive soaps and cleansers, astringents, and topical products that contain alcohol, spices or lime.
- Advise any doctor, dentist or pharmacist whom you consult that you use this medicine.
- Brand name Solage also contains the drug mequinol.
- Don't apply drug to skin area that has cuts, abrasions, a rash or is sunburned.

POSSIBLE INTERACTION WITH OTHER DRUGS

GENERIC NAME OR DRUG CLASS	COMBINED EFFECT
Antiacne topical preparations, other	Excessive skin irritation.
Cosmetics (medicated)	Severe skin irritation.
Insect repellents containing DEET	Repellent can be absorbed into skin.
Skin-peeling agents (salicylic acid, sulfur, resorcinol)	Excessive skin irritation.
Skin preparations with alcohol	Severe skin irritation.
Soaps or cleansers (abrasive)	Severe skin irritation.

POSSIBLE INTERACTION WITH OTHER SUBSTANCES

INTERACTS WITH	COMBINED EFFECT
Alcohol:	None expected.
Beverages:	None expected.
Cocaine:	None expected.
Foods:	None expected.
Marijuana:	None expected.
Tobacco:	None expected.

RIBAVIRIN

BRAND NAMES

Tribavirin
Virazid
Virazole

BASIC INFORMATION

Habit forming? No
Prescription needed? Yes
Available as generic? Yes
Drug class: Antiviral

USES

- Treats severe viral pneumonia.
- Treats respiratory syncytial virus (RSV) infections in hospitalized infants or children.
- Treats influenza A and B with some success.
- Does not treat other viruses such as the common cold.
- Some brand names of ribavirin are used to treat chronic hepatitis C when combined with another drug. The information in this chart does not include any facts about these brands. Your doctor will provide you the information and instructions and will follow your therapy carefully.

DOSAGE & USAGE INFORMATION

How to take:
By inhalation of a fine mist through mouth. Requires a special sprayer attached to oxygen mask; face mask for infants or hood.

When to take:
As ordered by your doctor.

If you forget a dose:
Not appropriate.

What drug does:
Kills virus or prevents its growth.

Time lapse before drug works:
Begins working in 1 hour. May require treatment for 12 to 18 hours per day for 3 to 7 days.

Don't take with:
Any other medicine or any dietary supplement without consulting your doctor or pharmacist.

OVERDOSE

SYMPTOMS:
None expected.
WHAT TO DO:
Drug is given in a medical facility.

POSSIBLE ADVERSE REACTIONS OR SIDE EFFECTS

SYMPTOMS	WHAT TO DO
Life-threatening:	
None expected.	
Common:	
None expected.	
Infrequent:	
• Unusual tiredness or weakness.	Discontinue. Call doctor right away.
• Headache, insomnia, appetite loss, nausea.	Continue. Call doctor when convenient.
Rare:	
Skin irritation or rash.	Continue. Call doctor when convenient.

WARNINGS & PRECAUTIONS

Don't take if:
You are allergic to ribavirin. Allergic reactions are rare, but may occur.

Before you start, consult your doctor if:
- You are now on low-salt, low-sugar or other special diet.
- You have severe anemia.

Over age 60:
Adverse reactions and side effects may be more frequent and severe than in younger persons.

Pregnancy:
Consult doctor. Drug should not be used during pregnancy. Can cause harm to unborn baby. Risk category X (see page xviii).

Breastfeeding; Lactation; Nursing Mothers:
This drug is generally not recommended with breastfeeding. Discuss risks and benefits with your doctor.

Infants & children up to age 18:
Follow instructions provided by your child's doctor.

Prolonged use:
No problems expected.

Skin & sunlight:
No special problems expected.

Driving, piloting or hazardous work:
Don't drive or pilot aircraft until you learn how medicine affects you. Don't work around dangerous machinery. Don't climb ladders or work in high places. Danger increases if you drink alcohol or take medicine affecting alertness and reflexes, such as antihistamines, tranquilizers, sedatives, pain medicine, narcotics and mind-altering drugs.

Discontinuing:
Don't discontinue without consulting doctor. Dose may require gradual reduction if you have taken drug for a long time. Doses of other drugs may also require adjustment.

Others:
- Health care workers exposed to ribavirin may experience headache, eye itching, redness or swelling.
- Female health care workers who are pregnant or may become pregnant should avoid exposure to drug.
- The information in this chart does not cover or include the brand names Rebetol or Ribasphere. Your doctor must provide that information for you.

POSSIBLE INTERACTION WITH OTHER DRUGS

GENERIC NAME OR DRUG CLASS	COMBINED EFFECT
Zidovudine	Decreased effect of ribavirin and zidovudine.

POSSIBLE INTERACTION WITH OTHER SUBSTANCES

INTERACTS WITH	COMBINED EFFECT
Alcohol:	None expected.
Beverages:	None expected.
Cocaine:	Unknown. Avoid.
Foods:	None expected.
Marijuana:	Consult doctor.
Tobacco:	None expected.

RIBOFLAVIN (Vitamin B-2)

BRAND NAMES

Numerous brands of single vitamin and multivitamin combinations are available.

BASIC INFORMATION

Habit forming? No
Prescription needed? No
Available as generic? Yes
Drug class: Vitamin supplement

USES

- Dietary supplement to ensure normal growth and health.
- Dietary supplement to treat symptoms caused by deficiency of B-2: sores in mouth, eyes sensitive to light, itching and peeling skin.

DOSAGE & USAGE INFORMATION

How to take:
Tablet—Swallow with liquid or food to lessen stomach irritation. If you can't swallow whole, crumble tablet and take with liquid or food.

When to take:
At the same times each day.

If you forget a dose:
Take as soon as you remember. If it is almost time for the next dose, wait for that dose (don't double this dose) and resume regular schedule.

What drug does:
Promotes normal growth and health.

Time lapse before drug works:
Requires continual intake.

Don't take with:
Any other medicine or any dietary supplement without consulting your doctor or pharmacist.

OVERDOSE

SYMPTOMS:
Specific symptoms are unknown. May have itching, burning, prickling sensations; eyes sensitive to light.
WHAT TO DO:
If person uses much larger amount than prescribed or if accidentally swallowed, call poison control center 1-800-222-1222 for instructions or dial 911 (emergency) for help.

POSSIBLE ADVERSE REACTIONS OR SIDE EFFECTS

SYMPTOMS	WHAT TO DO
Life-threatening: None expected.	
Common: None expected.	
Infrequent: None expected.	
Rare: Urine yellow in color.	No action necessary.

WARNINGS & PRECAUTIONS

Don't take if:
- You are allergic to any B vitamin. Allergic reactions are rare, but may occur.
- You have chronic kidney failure.

Before you start, consult your doctor if:
You are pregnant or plan pregnancy.

Over age 60:
No problems expected.

Pregnancy:
Consult doctor. Use is generally considered safe. Risk category A (see page xviii).

Breastfeeding; Lactation; Nursing Mothers:
This drug is generally acceptable with breastfeeding. Discuss risks and benefits with your doctor.

Infants & children up to age 18:
Read nonprescription product's label to be sure it is approved for your child's age. Follow label instructions on dosage. Consult doctor or pharmacist if unsure.

Prolonged use:
No problems expected.

Skin & sunlight:
No problems expected.

Driving, piloting or hazardous work:
No problems expected.

Discontinuing:
No problems expected.

Others:
- Advise any doctor, dentist or pharmacist whom you consult that you take this medicine.
- A balanced diet should provide all the vitamin B-2 a healthy person needs and makes supplements unnecessary during periods of good health. Best sources are milk, meats and green leafy vegetables.

POSSIBLE INTERACTION WITH OTHER DRUGS

GENERIC NAME OR DRUG CLASS	COMBINED EFFECT
Anticholinergics*	Possible increased riboflavin absorption.
Antidepressants, tricyclic*	Decreased riboflavin effect.
Phenothiazines*	Decreased riboflavin effect.
Probenecid	Decreased riboflavin effect.

POSSIBLE INTERACTION WITH OTHER SUBSTANCES

INTERACTS WITH	COMBINED EFFECT
Alcohol:	Prevents uptake and absorption of vitamin B-2.
Beverages:	None expected.
Cocaine:	Unknown. Avoid.
Foods:	None expected.
Marijuana:	Consult doctor.
Tobacco:	None expected.

***See Glossary**

RIFAMYCINS

GENERIC AND BRAND NAMES

RIFAMPIN
Rifadin
Rifamate
Rifampicin
Rimactane

RIFAPENTINE
Priftin

BASIC INFORMATION

Habit forming? No
Prescription needed? Yes
Available as generic? Yes
Drug class: Antibacterial, antitubercular

USES

Treatment for tuberculosis and other infections. Used in combination with other antitubercular medications.

DOSAGE & USAGE INFORMATION

How to take:
Capsule or tablet—Swallow with liquid or food. If you can't swallow whole, open capsule or crumble tablet and take with liquid or small amount of food. For child, mix with small amount of applesauce or jelly.

When to take:
1 hour before or 2 hours after a meal.

If you forget a dose:
Take as soon as you remember. If it is almost time for the next dose, wait for that dose (don't double this dose) and resume regular schedule.

What drug does:
Prevents multiplication of tuberculosis germs.

Time lapse before drug works:
Usually 2 weeks. May require 1 to 2 years without missed doses for maximum benefit.

Continued next column

OVERDOSE

SYMPTOMS:
Nausea, vomiting, heartburn, mental changes, swelling of face, itching all over, red-orange color of skin or whites of eyes or mucous membranes.
WHAT TO DO:
- **Dial 911 for all medical emergencies or call poison control center 1-800-222-1222 for instructions.**
- **See emergency information on last 3 pages of this book.**

Don't take with:
Any other medicine or any dietary supplement without consulting your doctor or pharmacist.

POSSIBLE ADVERSE REACTIONS OR SIDE EFFECTS

SYMPTOMS	WHAT TO DO
Life-threatening:	
Rare, severe allergic reaction (anaphylaxis)—difficulty breathing, hives, itching, swelling (throat, face), vomiting, dizziness.	Seek emergency treatment immediately.
Common:	
• Blood in urine, joint pain, back or side pain, swelling of feet or legs.	Continue, but call doctor right away.
• Diarrhea, reddish urine or stool or tears or sweat or saliva.	Continue. Call doctor when convenient.
Infrequent:	
• Skin of face and scalp (rash, flushed, itchy) blurred vision, difficulty breathing, nausea, vomiting, abdominal cramps, tiredness, bleeding or bruising.	Continue, but call doctor right away.
• Dizziness, unsteady gait, confusion, muscle or bone pain, heart-burn, flatulence, chills, headache, fever, mood or behavior changes.	Continue. Call doctor when convenient.
Rare:	
• Sore throat or mouth or tongue, yellow skin or eyes.	Continue, but call doctor right away.
• Appetite loss, less urination.	Continue. Call doctor when convenient.

WARNINGS & PRECAUTIONS

Don't take if:
You are allergic to any of the rifamycins.

Before you start, consult your doctor if:
You have an alcohol problem, diabetes, liver disease or porphyria.

Over age 60:
Adverse reactions and side effects may be more frequent and severe than in younger persons.

Pregnancy:
Decide with your doctor if drug benefits justify risk to unborn baby. Risk category C (see page xviii).

Breastfeeding; Lactation; Nursing Mothers:
Drugs in this group are generally acceptable with breastfeeding. Discuss risks and benefits with your doctor.

Infants & children up to age 18:
Follow instructions provided by your child's doctor. Safety and effectiveness under age 12 have not been established.

Prolonged use:
- May be more susceptible to infections caused by germs not responsive to rifamycins.
- Talk to your doctor about the need for follow-up medical examinations or laboratory studies to check liver function.

Skin & sunlight:
No problems expected.

Driving, piloting or hazardous work:
Don't drive or pilot aircraft until you learn how medicine affects you. Don't work around dangerous machinery. Don't climb ladders or work in high places. Danger increases if you drink alcohol or take medicine affecting alertness.

Discontinuing:
Don't discontinue without doctor's advice until you complete prescribed dose, even though symptoms diminish or disappear.

Others:
- Reddish tears may discolor soft contact lenses.
- Advise any doctor or dentist you consult that you are using this medication.

POSSIBLE INTERACTION WITH OTHER DRUGS

GENERIC NAME OR DRUG CLASS	COMBINED EFFECT
Adrenocorticoids, systemic	Decreased adreno-corticoid effect.
Anticoagulants,* oral	Decreased anticoagulant effect.
Antidepressants,* tricyclic	Decreased anti-depressant effect.
Antidiabetics,* oral	Decreased antidiabetic effect.
Antifungals,* azole	Decreased antifungal effect.
Barbiturates*	Decreased barbiturate effect.
Calcium channel* blockers	Decreased channel blocker effect.
Chloramphenicol	Decreased effect of both drugs.
Clarithromycin	Decreased antibiotic effect.
Clofibrate	Decreased clofibrate effect.
Clozapine	Toxic effect on bone marrow.
Contraceptives, oral*	Decreased contraceptive effect.
Cyclosporine	Decreased effect of both drugs.
Dapsone	Decreased dapsone effect.
Diazepam	Decreased diazepam effect.
Digitalis preparations*	Decreased digitoxin effect.
Disopyramide	Decreased disopyramide effect.
Doxycycline	Decreased antibiotic effect.
Estrogens*	Decreased effect of both drugs.
Haloperidol	Decreased haloperidol effect.
Hepatotoxics*	Increased risk of liver toxicity.
Leflunomide	Increased risk of leflunomide toxicity.
Levothyroxine	Decreased levothyroxine effect.
Methadone	Decreased methadone effect.
Mexiletine	Decreased mexiletine effect.
Non-nucleoside reverse transcriptase inhibitors	May require dosage change of rifampin.

Continued on page 909

POSSIBLE INTERACTION WITH OTHER SUBSTANCES

INTERACTS WITH	COMBINED EFFECT
Alcohol:	Possible liver problems.
Beverages:	None expected.
Cocaine:	Unknown. Avoid.
Foods:	None expected.
Marijuana:	Consult doctor.
Tobacco:	None expected.

***See Glossary**

RIFAXIMIN

BRAND NAMES

Xifaxan

BASIC INFORMATION

Habit forming? No
Prescription needed? Yes
Available as generic? No
Drug class: Antidiarrheal; antibacterial

USES

- Treatment for diarrhea (often called travelers' diarrhea) caused by drinking fluids or eating food contaminated by a bacteria (most often *Escherichia coli*). It may not be effective for diarrhea caused by other bacteria or viruses.
- Reduction in risk of overt hepatic encephalopathy recurrence in patients age 18 and older.

DOSAGE & USAGE INFORMATION

How to take:
Tablet—Swallow with liquid. Take with or without food.

When to take:
Three times a day at the same times each day. It is normally taken for three days.

If you forget a dose:
Take as soon as you remember. If it is almost time for the next dose, wait for that dose (don't double this dose) and resume regular schedule.

What drug does:
It works in the gastrointestinal tract to kill the bacteria causing the diarrhea. It is not absorbed into the bloodstream like most antibiotic drugs.

Continued next column

OVERDOSE

SYMPTOMS:
Unknown, but may include: nausea, vomiting, diarrhea and abdominal discomfort.
WHAT TO DO:
If person uses much larger amount than prescribed or if accidentally swallowed, call poison control center 1-800-222-1222 for instructions or dial 911 (emergency) for help.

Time lapse before drug works:
Starts working in a few hours, but takes three days to cure the infection being treated.

Don't take with:
Any other medicine or any dietary supplement without consulting your doctor or pharmacist.

POSSIBLE ADVERSE REACTIONS OR SIDE EFFECTS

SYMPTOMS	WHAT TO DO
Life-threatening: None expected.	
Common: Swelling of legs or or ankles, nausea, dizziness, fatigue, stomach swelling.	Continue, but call feet doctor right away.
Infrequent: Diarrhea, headache, gas, stomach pain, urgent need to have bowel movements, rectal spasms.	Continue. Call doctor when convenient.
Rare: None expected.	

WARNINGS & PRECAUTIONS

Don't take if:
You are allergic to rifaximin or other antibiotics called rifamycins (including rifampin or rifabutin). Allergic reactions are rare, but may occur.

Before you start, consult your doctor if:
- You have a fever.
- You have blood in your stool.
- You have dysentery (a severe form of diarrhea).
- You have pseudomembranous colitis.

Over age 60:
No special problems expected.

Pregnancy:
Decide with your doctor if drug benefits justify risk to unborn baby. Risk category C (see page xviii).

Breastfeeding; Lactation; Nursing Mothers:
This drug is generally acceptable with breastfeeding. Discuss risks and benefits with your doctor.

Infants & children up to age 18:
Safety and effectiveness for age under 12 has not been established. Consult doctor.

Prolonged use:
Not recommended. Medicine is discontinued once the infection is cured.

Skin & sunlight:
No special problems expected.

Driving, piloting or hazardous work:
No problems expected.

Discontinuing:
- Don't discontinue without doctor's advice until you complete prescribed dose, even though symptoms diminish or disappear.
- Do discontinue and call your doctor if diarrhea gets worse, you have bloody diarrhea or develop a fever.

Others:
- Be sure to drink plenty of fluids while you have diarrhea symptoms to prevent dehydration.
- Contact your doctor if symptoms do not improve in 24 to 48 hours.
- Advise any doctor, dentist or pharmacist whom you consult that you take this medicine.

POSSIBLE INTERACTION WITH OTHER DRUGS

GENERIC NAME OR DRUG CLASS	COMBINED EFFECT
None expected.	

POSSIBLE INTERACTION WITH OTHER SUBSTANCES

INTERACTS WITH	COMBINED EFFECT
Alcohol:	None expected.
Beverages:	None expected.
Cocaine:	Unknown. Avoid.
Foods:	None expected.
Marijuana:	Consult doctor.
Tobacco:	None expected.

RILUZOLE

BRAND NAMES

Rilutek

BASIC INFORMATION

Habit forming? No
Prescription needed? Yes
Available as generic? Yes
Drug class: Amyotrophic lateral sclerosis therapy agent

USES

Treatment for amyotrophic lateral sclerosis (ALS or Lou Gehrig's disease), a motor neuron disorder that causes weakness and atrophy of the muscles. Riluzole may extend a patient's survival time and delay the need for surgery for complications. It does not cure the disorder.

DOSAGE & USAGE INFORMATION

How to take:
Tablet—Swallow with liquid. Instructions to take on an empty stomach mean 1 hour before or 2 hours after eating.

When to take:
Usually every 12 hours.

If you forget a dose:
Take as soon as you remember. If it is almost time for the next dose, wait for that dose (don't double this dose) and resume regular schedule.

What drug does:
The cause of ALS is unknown and the exact mechanism of how the drug works is also unknown. It appears to decrease production of certain chemicals in the body and may affect some cellular activity that leads to the progressive weakness caused by the disorder.

Continued next column

OVERDOSE

SYMPTOMS:
Unknown.
WHAT TO DO:
- **Dial 911 for all medical emergencies or call poison control center 1-800-222-1222 for instructions.**
- **See emergency information on last 3 pages of this book.**

Time lapse before drug works:
6 to 18 months for maximum benefits.

Don't take with:
Any other medicine or any dietary supplement without consulting your doctor or pharmacist.

POSSIBLE ADVERSE REACTIONS OR SIDE EFFECTS

SYMPTOMS	WHAT TO DO
Life-threatening:	
Rare, severe allergic reaction (anaphylaxis)—difficulty breathing, hives, itching, swelling (throat, face), vomiting, dizziness.	Seek emergency treatment immediately.
Common:	
• Difficulty breathing, abdominal pain, increased coughing, blood pressure increased.	Discontinue. Call doctor right away.
• Mouth has a burning, or tingling feeling, constipation, nausea, vomiting, dizziness, mild weakness, headache, diarrhea.	Continue. Call doctor when convenient.
Infrequent:	
Skin symptoms (redness, itching, scaling, bruising, oozing or thickening), fever, sore feet or legs, sores in mouth or on lips, fast heartbeat.	Discontinue. Call doctor right away.
Rare:	
Changes in vision, tightness in chest, wheezing, memory loss, mental or mood changes, severe drowsiness, swelling (face, hands, fingers, feet or legs), nosebleed, hallucinations, blood in urine, unusual bleeding, unusual tiredness or weakness, severe headache, pain in various parts of the body, eyes red or irritated, continued or painful penile erection, noisy breathing, loss of bladder control, hives.	Discontinue. Call doctor right away.

WARNINGS & PRECAUTIONS

Don't take if:
You are sensitive to riluzole.

Before you start, consult your doctor if:
- You are allergic to any medicine, food or other substance, or have a family history of allergies.
- You have liver or kidney disease.

Over age 60:
No special problems expected.

Pregnancy:
Decide with your doctor if drug benefits justify risk to unborn baby. Risk category C (see page xviii).

Breastfeeding; Lactation; Nursing Mothers:
This drug is generally not recommended with breastfeeding. Discuss risks and benefits with your doctor.

Infants & children up to age 18:
Safety and effectiveness in this age group have not been established. Consult doctor.

Prolonged use:
Talk to your doctor about the need for follow-up medical examinations or laboratory studies to check liver function.

Skin & sunlight:
No problems expected.

Driving, piloting or hazardous work:
Don't drive or pilot aircraft until you learn how medicine affects you. Don't work around dangerous machinery. Don't climb ladders or work in high places. Danger increases if you drink alcohol or take other medicines affecting alertness and reflexes.

Discontinuing:
No problems expected.

Others:
- May affect the results of some medical tests.
- Use as directed. Don't increase or decrease dosage without doctor's approval.
- Advise any doctor, dentist or pharmacist whom you consult that you take this medicine.
- Call your doctor if you develop a fever.

POSSIBLE INTERACTION WITH OTHER DRUGS

GENERIC NAME OR DRUG CLASS	COMBINED EFFECT
Hepatotoxics*	Increased risk of adverse effects.
Other medications	Complete studies have not been done to evaluate interactions with other drugs, but the potential exists for a variety of possible interactions. Consult doctor.

POSSIBLE INTERACTION WITH OTHER SUBSTANCES

INTERACTS WITH	COMBINED EFFECT
Alcohol:	Increased risk of adverse effects.
Beverages:	None expected.
Cocaine:	Unknown. Avoid.
Foods:	None expected.
Marijuana:	Consult doctor.
Tobacco:	None expected.

ROFLUMILAST

BRAND NAMES

Dalirespp

BASIC INFORMATION

Habit forming? No
Prescription needed? Yes
Available as generic? No
Drug class: Phosphodiesterase-4 (PDE4) inhibitor

USES

Reduces frequency of flare-ups (exacerbations) in patients with severe chronic obstructive pulmonary disease (COPD) that is linked to chronic bronchitis. This drug is an add-on to bronchodilator treatment. It is not to be taken for the relief of acute bronchospasms.

DOSAGE & USAGE INFORMATION

How to take:
Tablet—Swallow whole with liquid. May be taken with or without food.

When to take:
Once a day at the same time each day.

If you forget a dose:
Take as soon as you remember. If it is almost time for the next dose, wait for that dose (don't double this dose) and resume regular schedule.

What drug does:
It helps block the inflammatory process in COPD thereby reducing inflammation in the lungs that typically leads to symptoms (e.g., coughing and excess mucus).

Continued next column

OVERDOSE

SYMPTOMS:
May have headache, dizziness, palpitations, gastrointestinal problem, lightheadedness, clamminess and low blood pressure.
WHAT TO DO:
If person uses much larger amount than prescribed or if accidentally swallowed, call poison control center 1-800-222-1222 for instructions or dial 911 (emergency) for help.

Time lapse before drug works:
Starts working within hours, but noticeable improvements in lung function may take 4 to 8 weeks.

Don't take with:
Any other medicine or any dietary supplement without consulting your doctor or pharmacist.

POSSIBLE ADVERSE REACTIONS OR SIDE EFFECTS

SYMPTOMS	WHAT TO DO
Life-threatening:	
None expected.	
Common:	
Weight loss, nausea, headache, diarrhea, back pain.	Continue. Call doctor when convenient.
Infrequent:	
Dizziness or vertigo, insomnia, appetite loss, flu or cold-like symptoms.	Continue. Call doctor when convenient.
Rare:	
• Unusual changes in behavior or mood, depression or anxiety symptoms worsen, talk or thoughts of suicide.	Discontinue. Call doctor right away.
• Feeling nervous, stomach pain, indigestion, vomiting, muscle spasms, tremors.	Continue. Call doctor when convenient.

WARNINGS & PRECAUTIONS

Don't take if:
You are allergic to roflumilast. Allergic reactions are rare, but may occur.

Before you start, consult your doctor if:
- You have any liver disorder.
- You have a history of anxiety, depression or psychiatric disorders.
- You have a history of suicidal thoughts or behavior.

Over age 60:
No special problems expected.

Pregnancy:
Decide with your doctor if drug benefits justify risk to unborn baby. Risk category C (see page xviii).

Breastfeeding; Lactation; Nursing Mothers:
This drug is generally not recommended with breastfeeding. Discuss risks and benefits with your doctor.

Infants & children up to age 18:
Safety and effectiveness in this age group have not been established. COPD does not usually occur in children.

Prolonged use:
- Consult with your doctor on a regular basis while taking this drug to monitor your progress, to check for side effects and to get recommended lab tests.
- The drug may cause weight loss in some patients. If excess weight loss occurs, consult your doctor.

Skin & sunlight:
No problems expected.

Driving, piloting or hazardous work:
No problems expected.

Discontinuing:
No problems expected, but do consult your doctor before stopping the drug.

Others:
- Advise any doctor, dentist or pharmacist whom you consult that you take this medicine.
- Caregivers and families should be alert to unusual changes in the patient's mood or behavior. If new symptoms occur or milder symptoms worsen, consult doctor.

POSSIBLE INTERACTION WITH OTHER DRUGS

GENERIC NAME OR DRUG CLASS	COMBINED EFFECT
Enzyme inhibitors*	Increased effect of roflumilast.
Enzyme inducers*	Decreased effect of roflumilast.
Oral contraceptives* containing ethinyl estradiol or gestodyne	Increased effect of roflumilast.

POSSIBLE INTERACTION WITH OTHER SUBSTANCES

INTERACTS WITH	COMBINED EFFECT
Alcohol:	None expected.
Beverages: Grapefruit juice.	May increase effect of roflumilast.
Cocaine:	Unknown. Avoid.
Foods: Grapefruit.	May increase effect of roflumilast.
Marijuana:	Consult doctor.
Tobacco:	None expected. People with COPD should not smoke.

*See Glossary

SALICYLATES

GENERIC AND BRAND NAMES

See full list of generic and brand names in the *Generic and Brand Name Directory*, page 880.

BASIC INFORMATION

Habit forming? No
Prescription needed? For some
Available as generic? Yes
Drug class: Analgesic, anti-inflammatory (nonsteroidal)

USES

- Reduces pain, fever, inflammation.
- Relieves swelling, stiffness, joint pain of arthritis or rheumatism.
- Decreases risk of myocardial infarction (aspirin only).

DOSAGE & USAGE INFORMATION

How to take:
- Tablet or capsule—Swallow with liquid.
- Extended-release tablet—Swallow each dose whole.
- Suppository—Remove wrapper and moisten suppository with water. Gently insert into rectum, large end first.

When to take:
Pain, fever, inflammation—As needed, no more often than every 4 hours.

If you forget a dose:
Take as soon as you remember. If it is almost time for the next dose, wait for that dose (don't double this dose) and resume regular schedule.

Continued next column

OVERDOSE

SYMPTOMS:
Ringing in ears, nausea, vomiting, dizziness, fever, deep and rapid breathing, hallucinations, convulsions, coma.
WHAT TO DO:
- **Dial 911 for all medical emergencies or call poison control center 1-800-222-1222 for instructions.**
- **See emergency information on last 3 pages of this book.**

What drug does:
- Affects hypothalamus, the part of the brain that regulates temperature by dilating small blood vessels in skin.
- Prevents clumping of platelets (small blood cells) so blood vessels remain open.
- Decreases prostaglandin effect.
- Suppresses body's pain messages.

Time lapse before drug works:
30 minutes for pain, fever, arthritis.

Don't take with:
- Tetracyclines. Space doses 1 hour apart.
- Any other medicine or any dietary supplement without consulting your doctor or pharmacist.

POSSIBLE ADVERSE REACTIONS OR SIDE EFFECTS

SYMPTOMS	WHAT TO DO
Life-threatening:	
Rare, severe allergic reaction (anaphylaxis)—difficulty breathing, hives, itching, swelling (throat, face), vomiting, dizziness.	Seek emergency treatment immediately.
Common:	
Heartburn, indigestion, mild nausea or vomiting or headache, drowsiness.	Continue. Call doctor when convenient.
Infrequent:	
Rectal irritation (with suppository).	Continue. Call doctor when convenient.
Rare:	
Severe headache, convulsions, extreme drowsiness, flushing or change in skin color, any loss of hearing, severe vomiting, swelling of face, vision problems, bloody or black stools, ringing in ears, severe or ongoing stomach cramps or pain (all symptoms are more likely with repeated doses for long periods).	Discontinue. Call doctor right away or seek emergency treatment for severe. symptoms

WARNINGS & PRECAUTIONS

Don't take if:
- You are allergic to salicylates or aspirin.
- You need to restrict sodium in your diet. Buffered effervescent tablets and sodium salicylate are high in sodium.

Before you start, consult your doctor if:
You have stomach or duodenal ulcers, gout, asthma, nasal polyps, a bleeding or blood clotting disorder, overactive thyroid, anemia, heart disease, high blood pressure, kidney or liver disease, hemophilia, glucose-6-phosphate dehydrogenase (G6PD) deficiency or are sensitive to aspirin (non-immune drug reaction).

Over age 60:
Adverse reactions and side effects may be more frequent and severe than in younger persons.

Pregnancy:
Risk factors vary for drugs in this group. Always consult doctor. See category list on page xviii.

Breastfeeding; Lactation; Nursing Mothers:
Most drugs in this group are generally acceptable with breastfeeding. Discuss risks and benefits with your doctor.

Infants & children up to age 18:
- Safety and effectiveness in this age group have not been established. Consult doctor.
- Do not give to persons under age 18 who have fever and discomfort of viral illness, especially chicken pox and influenza. Probably increases risk of Reye's syndrome.

Prolonged use:
May increase risk of kidney or liver problems, ulcers or bleeding problems. Ask doctor about having follow-up medical exams or lab tests.

Skin & sunlight:
No special problems expected.

Driving, piloting or hazardous work:
No restrictions unless you feel drowsy.

Discontinuing:
For prescribed dosage, don't discontinue without doctor's advice.

Others:
- Advise any doctor, dentist or pharmacist whom you consult that you take this medicine.
- Salicylates including aspirin are available in numerous combination drug products . Be sure to read drug charts on other ingredients.

POSSIBLE INTERACTION WITH OTHER DRUGS

GENERIC NAME OR DRUG CLASS	COMBINED EFFECT
Acetaminophen	Increased risk of kidney damage (with high, long-term doses).
Adrenocorticoids, systemic	Decreased salicylate effect.
Allopurinol	Decreased allopurinol effect.
Angiotensin-converting enzyme (ACE) inhibitors*	Decreased ACE inhibitor effect.
Antacids*	Decreased salicylate effect.
Anticoagulants,* oral	Increased effect of anticoagulant and abnormal bleeding.
Antidiabetics,* oral	Low blood sugar.
Anti-inflammatory drugs, nonsteroidal (NSAIDs)*	Risk of stomach bleeding and ulcers.
Aspirin, other	Likely salicylate toxicity.
Beta-adrenergic blocking agents*	Decreased anti-hypertensive effect.
Bismuth subsalicylate	Increased risk of salicylate toxicity.
Bumetanide	Decreased diuretic effect.
Calcium supplements*	Increased salicylate effect.
Carteolol	Decreased effect of carteolol.
Ethacrynic acid	Decreased diuretic effect.
Furosemide	Possible salicylate toxicity.
Gold compounds*	Increased likelihood of kidney damage.
Indomethacin	Risk of stomach bleeding and ulcers.
Insulin	Decreased blood sugar.

Continued on page 910

POSSIBLE INTERACTION WITH OTHER SUBSTANCES

INTERACTS WITH	COMBINED EFFECT
Alcohol:	Possible stomach irritation and bleeding. Avoid.
Beverages:	None expected.
Cocaine:	Unknown. Avoid.
Foods:	None expected.
Marijuana:	Consult doctor.
Tobacco:	None expected.

SCOPOLAMINE (Hyoscine)

BRAND NAMES

See full list of brand names in the *Generic and Brand Name Directory*, page 880.

BASIC INFORMATION

Habit forming? No
Prescription needed?
High strength: Yes
Low strength: No
Available as generic? Yes
Drug class: Antispasmodic, anticholinergic

USES

- Reduces spasms of digestive system, bladder and urethra.
- Relieves painful menstruation.
- Prevents motion sickness.

DOSAGE & USAGE INFORMATION

How to take:

- Tablet or capsule—Swallow with liquid or food to lessen stomach irritation.
- Drops—Dilute dose in beverage.
- Skin disc—Clean application site. Change application sites with each dose.

When to take:

- Motion sickness—Apply disc 30 minutes before departure.
- Other uses—Take 30 minutes before meals (unless directed otherwise by doctor).

If you forget a dose:
Take as soon as you remember. If it is almost time for the next dose, wait for that dose (don't double this dose) and resume regular schedule.

Continued next column

OVERDOSE

SYMPTOMS:
Tiredness, drowsiness, confusion, agitation, hallucinations, vision changes, dry and flushed skin, dry mouth, fast and/or irregular heartbeat, seizures, coma.
WHAT TO DO:

- **Dial 911 for all medical emergencies or call poison control center 1-800-222-1222 for instructions.**
- **See emergency information on last 3 pages of this book.**

What drug does:
Blocks nerve impulses at parasympathetic nerve endings, preventing muscle contractions and gland secretions of organs involved.

Time lapse before drug works:
15 to 30 minutes.

Don't take with:
Any other medicine or any dietary supplement without consulting your doctor or pharmacist.

POSSIBLE ADVERSE REACTIONS OR SIDE EFFECTS

SYMPTOMS	WHAT TO DO
Life-threatening:	
Rare, severe allergic reaction (anaphylaxis)—difficulty breathing, hives, itching, swelling (throat, face), vomiting, dizziness.	Seek emergency treatment immediately.
Common:	
• Confusion, delirium, rapid heartbeat.	Discontinue. Call doctor right away.
• Nausea, vomiting, decreased sweating, constipation, changes in taste, dryness (in ears, nose, throat, mouth).	Continue. Call doctor when convenient.
Infrequent:	
Headache, difficult urination, stuffy nose, feeling lightheaded.	Continue. Call doctor when convenient.
Rare:	
Rash or hives, eye pain, blurred vision.	Discontinue. Call doctor right away.

WARNINGS & PRECAUTIONS

Don't take if:

- You are allergic to any anticholinergic.
- You have trouble with stomach bloating.
- You have difficulty emptying your bladder completely.
- You have narrow-angle glaucoma.
- You have severe ulcerative colitis.

Before you start, consult your doctor if:

- You have open-angle glaucoma, angina, chronic bronchitis or asthma, hiatal hernia, liver disease, enlarged prostate, myasthenia gravis, peptic ulcer, kidney or thyroid disease.
- You will have surgery within 2 months, including dental surgery, requiring general or spinal anesthesia.

Over age 60:
Adverse reactions and side effects may be more frequent and severe than in younger persons.

Pregnancy:
Decide with your doctor if drug benefits justify risk to unborn baby. Risk category C (see page xviii).

Breastfeeding; Lactation; Nursing Mothers:
This drug is generally acceptable with breastfeeding. Discuss risks and benefits with your doctor.

Infants & children up to age 18:
Safety and effectiveness in this age group have not been established. Consult doctor.

Prolonged use:
Chronic constipation, possible fecal impaction. Consult doctor immediately.

Skin & sunlight:
No problems expected.

Driving, piloting or hazardous work:
Use disqualifies you for piloting aircraft. Don't drive until you learn how medicine affects you. Don't work around dangerous machinery. Don't climb ladders or work in high places.

Discontinuing:
Follow doctor's instructions.

Others:
Advise any doctor, dentist or pharmacist whom you consult that you take this medicine.

POSSIBLE INTERACTION WITH OTHER DRUGS

GENERIC NAME OR DRUG CLASS	COMBINED EFFECT
Adrenocorticoids, systemic	Possible glaucoma.
Amantadine	Increased scopolamine effect.
Antacids*	Decreased scopolamine effect.
Anticholinergics,* other	Increased scopolamine effect.
Antidepressants, tricyclic*	Increased scopolamine effect. Increased sedation.
Antidiarrheals*	Decreased scopolamine effect.
Antihistamines*	Increased scopolamine effect.
Attapulgite	Decreased scopolamine effect.
Buclizine	Increased scopolamine effect.
Clozapine	Toxic effect on the central nervous system.
Digitalis preparations*	Possible decreased absorption of scopolamine.
Encainide	Increased effect of toxicity on heart muscle.
Ethinamate	Dangerous increased effects of ethinamate. Avoid combining.
Fluoxetine	Increased depressant effects of both drugs.
Guanfacine	May increase depressant effects of either medicine.
Haloperidol	Increased internal eye pressure.
Ketoconazole	Decreased ketoconazole effect.
Leucovorin	High alcohol content of leucovorin may cause adverse effects.
Meperidine	Increased scopolamine effect.
Methylphenidate	Increased scopolamine effect.
Methyprylon	May increase sedative effect to dangerous level. Avoid.
Molindone	Increased anticholinergic effect.
Monoamine oxidase (MAO) inhibitors*	Increased scopolamine effect.

Continued on page 910

POSSIBLE INTERACTION WITH OTHER SUBSTANCES

INTERACTS WITH	COMBINED EFFECT
Alcohol:	None expected.
Beverages:	None expected.
Cocaine:	Excessively rapid heartbeat. Avoid.
Foods:	None expected.
Marijuana:	Drowsiness, dry mouth. Avoid.
Tobacco:	None expected.

***See Glossary**

SELECTIVE PROGESTERONE RECEPTOR MODULATORS

GENERIC AND BRAND NAMES

ULIPRISTAL
Ella

BASIC INFORMATION

Habit forming? No
Prescription needed? Yes
Available as generic? No
Drug class: Contraceptive

USES

Emergency contraceptive that is used to prevent pregnancy after unprotected sex or after failure of another birth control method.

DOSAGE & USAGE INFORMATION

How to take:
Tablet—Swallow whole with liquid. May be taken with or without food.

When to use:
As soon as possible within 5 days (120 hours) after unprotected sex or after failure of another birth control method. It may be used any time during the menstrual cycle. Repeated use within the same menstrual cycle is not recommended.

If you forget a dose:
It is a one time dose.

What drug does:
It works primarily by stopping or delaying the release of an egg from the ovary. It may also work by preventing egg fertilization or preventing the implantation of a fertilized egg in the uterus.

Time lapse before drug works:
About 1 to 3 hours.

Don't take with:
Any other medicine or any dietary supplement without consulting your doctor or pharmacist.

OVERDOSE

SYMPTOMS:
Unknown. An overdose is unlikely as each package contains one tablet.
WHAT TO DO:
If person uses much larger amount than prescribed or if accidentally swallowed, call poison control center 1-800-222-1222 for instructions or dial 911 (emergency) for help.

POSSIBLE ADVERSE REACTIONS OR SIDE EFFECTS

SYMPTOMS	WHAT TO DO
Life-threatening:	
None expected.	
Common:	
• Heavy bleeding, pain.	Call doctor right away.
• Headache, nausea, unusual tiredness or weakness, cramps.	Call doctor if symptoms continue.
Infrequent:	
• Next menstrual period is early or less than a week late.	No action necessary.
• Next menstrual period is over a week late.	Call doctor right away.
• Dizziness, acne or skin breakouts, spotting between menstrual periods.	Call doctor if symptoms continue.
Rare:	
Abdominal or stomach pain 3 to 5 weeks after taking drug.	Call doctor right away.

SELECTIVE PROGESTERONE RECEPTOR MODULATORS

WARNINGS & PRECAUTIONS

Don't use if:
- You are allergic to ulipristal. Allergic reactions are rare, but may occur.
- You are pregnant.

Before you start, consult your doctor if:
- You suspect you are pregnant.
- You are overweight (may be less effective).

Over age 60:
Not used in this age group.

Pregnancy:
Consult doctor. Drug should not be used during pregnancy. Can cause harm to unborn baby. Risk category X (see page xviii).

Breastfeeding; Lactation; Nursing Mothers:
This drug is generally acceptable with breastfeeding. Discuss risks and benefits with your doctor.

Infants & children up to age 18:
May be used by females who have started their menstrual periods. Follow instructions provided by your child's doctor.

Prolonged use:
Not used long term.

Skin & sunlight:
No problems expected.

Driving, piloting or hazardous work:
No problems expected.

Discontinuing:
It is a one time dose.

Others:
- Call doctor right away if you have vomiting or diarrhea within three hours of taking this drug. You may need to take another dose.
- Drug will not protect you from getting HIV/AIDS or other sexually transmitted disease.
- The drug is not intended for routine use as a method of birth control.
- A pregnancy test is recommended if your menstrual period is more than a week late.
- Fertility is likely to return to normal rather quickly after taking this drug. Other methods of birth control should be used as soon as possible after taking ulipristal, as it will not prevent future pregnancies. If you are using hormonal contraceptives,* your doctor may recommend that you use a barrier method of birth control until your next menstrual period.
- Call doctor right away if you have severe lower abdominal or stomach pain 3 to 5 weeks after taking drug. You may have an ectopic pregnancy (outside of the uterus). This can be serious and life-threatening. It may lead to problems that make it harder for you to become pregnant in the future.

POSSIBLE INTERACTION WITH OTHER DRUGS

GENERIC NAME OR DRUG CLASS	COMBINED EFFECT
Contraceptives, hormonal*	Decreased contraceptive effect.
Enzyme inducers*	May decrease contraceptive effect of ulipristal.
Enzyme inhibitors*	May increase risk of adverse effects of ulipristal.

POSSIBLE INTERACTION WITH OTHER SUBSTANCES

INTERACTS WITH	COMBINED EFFECT
Alcohol:	None expected.
Beverages: Grapefruit juice.	May increase risk of side effects.
Cocaine:	Unknown. Avoid.
Food: Grapefruit.	May increase risk of side effects.
Marijuana:	Consult doctor.
Tobacco:	None expected.

SELECTIVE SEROTONIN REUPTAKE INHIBITORS (SSRIs)

GENERIC AND BRAND NAMES

See full list of brand names in the *Generic and Brand Name Directory*, page 880.

BASIC INFORMATION

Habit forming? No
Prescription needed? Yes
Available as generic? Yes, for some
Drug class: Antidepressant, antiobsessional agent, antianxiety agent.

USES

Treats depression, obsessive compulsive disorder, generalized anxiety disorder, panic disorder, post traumatic stress disorder, social phobia, premenstrual dysphoric disorder, seasonal affective disorder, depressive episode of bipolar disorder, menopause symptoms.

DOSAGE & USAGE INFORMATION

How to take:

- Capsule or tablet—Swallow with water. Take with or without food. If you can't swallow whole, ask doctor or pharmacist for advice.
- Oral disintegrating tablet—Dissolve in mouth.
- Oral solution, extended-release capsule or controlled-release tablet—Follow instructions.

When to take:
At the same time each day or weekly, usually in the a.m. Some dosages may be twice daily.

If you forget a dose:
Take as soon as you remember. If it is almost time for the next dose, wait for that dose (don't double this dose) and resume regular schedule.

Continued next column

OVERDOSE

SYMPTOMS:
Dizziness, drowsiness, sweating, nausea, vomiting, tremor, heart rhythm changes, amnesia, delirium, convulsions, coma.

WHAT TO DO:

- **Dial 911 for all medical emergencies or call poison control center 1-800-222-1222 for instructions.**
- **See emergency information on last 3 pages of this book.**

What drug does:
Affects serotonin, one of the chemicals in the brain called neurotransmitters, that plays a role in emotions and psychological disturbances.

Time lapse before drug works:
1 to 4 weeks.

Don't take with:
Any other medicine or any dietary supplement without consulting your doctor or pharmacist.

POSSIBLE ADVERSE REACTIONS OR SIDE EFFECTS

SYMPTOMS	WHAT TO DO
Life-threatening:	
Rare, severe allergic reaction (anaphylaxis)—difficulty breathing, hives, itching, swelling (throat, face), vomiting, dizziness.	Seek emergency treatment immediately.
Common:	
Drowsiness, nausea, cough or hoarseness, lower back or side pain, sores on lips or mouth, constipation or diarrhea, headache, anxiety, changes in sexual desire or function, insomnia, dry mouth, unusual weakness or tiredness.	Continue. Call doctor when convenient.
Infrequent:	
• Vision changes, confusion, apathy (lack of emotion), breathing difficulty, chills, black or tarry stools, fever, enlarged lymph glands, heart rhythm changes, vomiting, skin rash or itching.	Discontinue. Call doctor right away.
• Abdominal pain, loss of appetite, yawning, tingling, skin burning or prickly feeling, stuffy nose, change in sense of taste, tooth grinding, trembling, increased saliva, gas, heartburn, sweating, urinary changes, hair loss, muscle or joint pain, menstrual changes, weight changes.	Continue. Call doctor when convenient.
Rare:	
• Seizures (convulsions).	Discontinue. Seek emergency help.

• Abnormal bleeding, breast tenderness or enlargement, red or peeling skin, red or irritated eyes, sore throat, sudden body or facial spasms, low blood sugar (anxiety, chills, nervousness, difficulty concentrating), dizziness, clumsiness.	Discontinue. Call doctor right away.

WARNINGS & PRECAUTIONS

Don't take if:

- You are allergic to any SSRIs.
- You currently take (or took in the last two weeks) a monoamine oxidase (MAO) inhibitor.

Before you start, consult your doctor if:

- You have had kidney or liver problems.
- You have a history of seizure disorders.
- You have a history of drug or alcohol abuse.
- You have a history of mood disorders, mania, or thoughts of suicide.

Over age 60:
Adverse reactions and side effects may be more severe and frequent than in younger patients.

Pregnancy:
Decide with your doctor if drug benefits justify risk to unborn child. Risk category B and C for drugs in this group except paroxetine is risk category D (use of the drug during pregnancy is not recommended. It could cause harm to the unborn baby). See page xviii.

Breastfeeding; Lactation; Nursing Mothers:
Drugs in this group are generally acceptable with breastfeeding. Discuss risks and benefits with your doctor.

Infants & children up to age 18:
For children under age 18, follow instructions provided by your child's doctor. Carefully read information provided with prescription. Contact doctor right away if depression symptoms get worse or there is any talk of suicide or suicide behaviors. Read also the information under Others on this page.

Prolonged use:
No problems expected. Your doctor should periodically evaluate your response to the drug and adjust the dose if necessary.

Skin & sunlight:
One or more drugs in this group may cause rash or intensify sunburn in areas exposed to sun or ultraviolet light (photosensitivity reaction). Avoid overexposure. Notify doctor if reaction occurs.

Driving, piloting or hazardous work:
Don't drive or pilot aircraft until you learn how medicine affects you. Don't work around dangerous machinery. Don't climb ladders or work in high places. Danger increases if you drink alcohol or take medicine affecting alertness and reflexes.

Discontinuing:

- Don't discontinue without consulting doctor. You may need to reduce the dose gradually to avoid side effects.
- After discontinuing the drug, call your doctor right away if any new or unusual symptoms develop (emotional or physical).

Others:

- Advise any doctor, dentist or pharmacist whom you consult that you take this medicine.
- Take drug only as directed. Do not increase or reduce dosage without doctor's approval.
- Adults and children taking antidepressants may experience a worsening of the depression symptoms and may have increased suicidal thoughts or behaviors. Call doctor right away if these symptoms or behaviors occur.

POSSIBLE INTERACTION WITH OTHER DRUGS

GENERIC NAME OR DRUG CLASS	COMBINED EFFECT
Anticoagulants, oral*	Increased risk of side effects of both drugs.
Antidepressants, tricyclic*	Increased risk of side effects.
Anti-inflammatory drugs, nonsteroidal (NSAIDs)*	Risk of stomach bleeding and ulcers.

Continued on page 911

POSSIBLE INTERACTION WITH OTHER SUBSTANCES

INTERACTS WITH	COMBINED EFFECT
Alcohol:	Contributes to depression. Avoid.
Beverages: Grapefruit juice.	Toxicity risk. Avoid.
Cocaine:	Unknown. Avoid.
Foods: Grapefruit.	Toxicity risk. Avoid.
Marijuana:	Consult doctor.
Tobacco:	None expected.

*See Glossary

SEROTONIN-DOPAMINE ANTAGONISTS

GENERIC AND BRAND NAMES

ILOPERIDONE
Fanapt
PALIPERIDONE
Invega
Invega Sustenna
RISPERIDONE
Risperdal
Risperdal Consta
Risperdal M-TAB

BASIC INFORMATION

Habit forming? No
Prescription needed? Yes
Available as generic? Yes, for some
Drug class: Antipsychotic

USES

- Treats schizophrenia, schizoaffective disorder and bipolar disorder.
- Treatment of irritability in children with autism.
- Treats other disorders per doctor's advice.

DOSAGE & USAGE INFORMATION

How to take:
- Tablet—Swallow with liquid. May be taken with or without food.
- Injection—Patients will go to a medical office as scheduled for injection.
- Orally disintegrating tablet—Dissolves in the mouth in seconds. Do not chew.
- Oral solution—Dilute in 3 to 4 ounces of water, orange juice, or low fat milk (no cola or tea).
- Extended-release tablet—Swallow whole with liquid.

When to take:
At the same times each day. Take once-a-day dose in morning. Prescribed dosage may be increased gradually over first several days.

If you forget a dose:
Take as soon as you remember. If it is almost time for the next dose, wait for that dose (don't double this dose) and resume regular schedule.

Continued next column

OVERDOSE

SYMPTOMS:
Drowsiness, dizziness, rapid heartbeat, low blood pressure, convulsions, muscle spasms and uncontrolled body movements.
WHAT TO DO:
- **Dial 911 for all medical emergencies or call poison control center 1-800-222-1222 for instructions.**
- **See emergency information on last 3 pages of this book.**

What drug does:
It appears to act on neurotransmitters (serotonin and dopamine) in the brain to help restore more normal thinking and more normal mood.

Time lapse before drug works:
One to 7 days. A gradual increase in the dosage amount may be necessary to relieve symptoms.

Don't take with:
Any other medicine or diet supplement without consulting your doctor or pharmacist.

POSSIBLE ADVERSE REACTIONS OR SIDE EFFECTS

SYMPTOMS	WHAT TO DO
Life-threatening:	
Rare, severe allergic reaction (anaphylaxis)—difficulty breathing, hives, itching, swelling (throat, face), vomiting, dizziness.	Seek emergency treatment immediately.
Common:	
• Difficulty speaking or swallowing, loss of balance, vision changes, mask-like face, shuffling walk, arms or legs are stiff or weak, trembling or twitching, muscle spasms (face, neck, back), unable to move eyes, body twisting.	Discontinue. Call doctor right away.
• Constipation or diarrhea, drowsiness, dry mouth, headache, heartburn, cough, dreaming more, sore throat, nausea, stuffy or runny nose, unusual tiredness or weakness, weight gain, anxiety or nervousness, mood or mental changes, sexual dysfunction, urination problems, restlessness, insomnia, sweating.	Continue. Call doctor when convenient.
Infrequent:	
• Fast or irregular heartbeat, dizziness, lightheadedness, chest pain, trouble breathing.	Discontinue. Call doctor right away.
• Menstrual changes, skin is dry or darker or oily or has rash, excess saliva, joint or back or stomach pain, vomiting, appetite loss, weight loss, unexpected breast milk.	Continue. Call doctor when convenient.

Rare:

• High blood sugar (thirstiness, frequent urination, increased hunger, weakness), lip smacking, uncontrolled movements (arms, legs, tongue and chewing), cheek puffing, increased blinking, eyelid spasms, unusual facial or body positions, manic behavior, high or low body temperature, unusual bleeding or bruising.	Discontinue. Call doctor right away.
• Other symptoms that cause concern.	Continue. Call doctor when convenient.

WARNINGS & PRECAUTIONS

Don't take if:
You are allergic to serotonin-dopamine antagonists.

Before you start, consult your doctor if:
- You have or have had liver, kidney, heart or blood vessel disease; stroke; diabetes or pre-diabetes; high or low blood pressure; seizures; Parkinson's disease; breast cancer; electrolyte disorder; Alzheimer's; neuroleptic malignant syndrome; phenylketonuria; suicide thoughts; tardive dyskinesia or trouble swallowing.
- Patient is elderly and has dementia.

Over age 60:
- Adverse reactions and side effects may be more severe than in younger persons.
- Use of antipsychotic drugs in elderly patients with dementia-related psychosis may increase risk of death. Consult doctor.

Pregnancy:
Decide with your doctor if drug benefits justify risk to unborn baby. Risk category C (see page xviii).

Breastfeeding; Lactation; Nursing Mothers:
Drugs in this group are generally not recommended with breastfeeding. Discuss risks and benefits with your doctor.

Infants & children up to age 18:
Follow instructions provided by your child's doctor.

Prolonged use:
See your doctor on a regular basis to monitor drug's effectiveness and any side effects.

Skin & sunlight:
- May cause rash or intensify sunburn in areas exposed to sun or ultraviolet light (photosensitivity reaction). Use sunscreen and avoid overexposure. Notify doctor if reaction occurs.
- Hot temperatures and exercise, hot baths can increase risk of heatstroke. Drug may affect body's ability to maintain normal temperature.

Driving, piloting or hazardous work:
Don't drive or pilot aircraft until you learn how drug affects you. Don't work around dangerous machinery. Don't climb ladders or work in high places. Danger increases if you drink alcohol or take drugs affecting alertness and reflexes.

Discontinuing:
- Don't discontinue this drug without doctor's approval. Dosage may require a gradual reduction before stopping.
- Withdrawal effects may occur after stopping drug. Consult doctor if new symptoms develop that cause you concern.

Others:
- Get up slowly from a sitting or lying position to avoid feeling dizzy, faint or lightheaded.
- Advise any doctor, dentist or pharmacist whom you consult that you take this medicine.
- Take drug only as directed. Do not increase or reduce dosage without doctor's approval.

POSSIBLE INTERACTION WITH OTHER DRUGS

GENERIC NAME OR DRUG CLASS	COMBINED EFFECT
Antihypertensives*	Increased risk of low blood pressure.
Carbamazepine	Decreased effect of risperidone.
Central nervous system (CNS) depressants,* other	Increased sedative effect.
Clozapine	Increased effect of risperidone.
Dopamine agonists*	Decreased effect of dopamine agonist.
Enzyme Inhibitors*	Increased effect of iloperidone.
QT interval prolongation-causing drugs*	Heartbeat irregularities with iloperidone.

POSSIBLE INTERACTION WITH OTHER SUBSTANCES

INTERACTS WITH	COMBINED EFFECT
Alcohol:	Serious risks. Avoid.
Beverages:	None expected.
Cocaine:	Unknown. Avoid.
Foods:	None expected.
Marijuana:	Consult doctor.
Tobacco:	None expected.

***See Glossary**

SEROTONIN & NOREPINEPHRINE REUPTAKE INHIBITORS (SNRIs)

GENERIC AND BRAND NAMES

DESVENLAFAXINE
- Khedezla
- Pristiq

DULOXETINE
- Cymbalta

LEVOMILNACIPRAN
- Fetzima

MILNACIPRAN
- Savella

VENLAFAXINE
- Effexor XR

BASIC INFORMATION

Habit forming? No
Prescription needed? Yes
Available as generic? Yes, for some
Drug class: Antidepressant

USES

- Treatment for major depressive disorder.
- Treatment for diabetic peripheral neuropathy.
- Treatment for generalized anxiety disorder.
- May be used for chronic pain syndrome, social anxiety disorder, fibromyalgia, hot flashes, stress incontinence or other disorders.

DOSAGE & USAGE INFORMATION

How to take:
Tablet, extended-release capsule, extended-release tablet or delayed-release capsule—Swallow whole with liquid. Take with food if stomach upset occurs. Do not open, crush or chew medicine. If can't swallow whole, ask doctor or pharmacist for advice. (You may open Effexor XR capsule and sprinkle on spoonful of applesauce; mix it, swallow and drink glass of water.)

When to take:
At the same times each day.

Continued next column

If you forget a dose:
Take as soon as you remember. If it is almost time for the next dose, wait for that dose (don't double this dose) and resume regular schedule.

What drug does:
Increases level of two brain chemicals (serotonin and norepinephrine) that affect behavior and mood and play a role in depression.

Time lapse before drug works:
Begins in 1 to 3 weeks, but may take 4 to 6 weeks for maximum benefit.

Don't take with:
Any other medicine or any dietary supplement without consulting your doctor or pharmacist.

OVERDOSE

SYMPTOMS:
Extreme drowsiness or tiredness or weakness, fainting, seizure, fast heartbeat, tingling or burning sensation, tremor, nausea, vomiting, agitation, hyperactive, enlarged pupils. In some cases, may be no symptoms.
WHAT TO DO:
- **Dial 911 for all medical emergencies or call poison control center 1-800-222-1222 for instructions.**
- **See emergency information on last 3 pages of this book.**

POSSIBLE ADVERSE REACTIONS OR SIDE EFFECTS

SYMPTOMS	WHAT TO DO
Life-threatening:	
Rare, severe allergic reaction (anaphylaxis)—difficulty breathing, hives, itching, swelling (throat, face), vomiting, dizziness.	Seek emergency treatment immediately.
Common:	
Nausea, dry mouth, increased sweating, appetite loss, insomnia or drowsiness, fatigue, headache, constipation, diarrhea, rash or itching.	Continue. Call doctor when convenient.
Infrequent:	
Mood or behavior or mental changes, dizziness, impotence, less interest in sex or changes in orgasm, skin flushing, stomach upset or pain, vomiting, weight loss, muscle aches or pain, joint pain or swelling, trembling or shaking, vision changes, abnormal dreams, sore throat, stuffy or runny nose, fever, cough, frequent or hesitant urination, nervousness, weakness, lightheadedness.	Continue. Call doctor when convenient.
Rare:	
Seizures, fainting, irregular heartbeat, abnormal behaviors, severe symptoms such as suicide thoughts or behaviors.	Discontinue. Call doctor right away or seek emergency care.

SEROTONIN & NOREPINEPHRINE REUPTAKE INHIBITORS (SNRIs)

WARNINGS & PRECAUTIONS

Don't take if:
You are allergic to serotonin and norepinephrine reuptake inhibitors or take MAO inhibitors.*

Before you start, consult your doctor if:
- You have diabetes; heart, liver, or kidney disease; glaucoma; a blood clotting or bleeding problem or high cholesterol.
- You have high or low blood pressure.
- You have thoughts about suicide.
- You have a brain disorder or brain damage or mental retardation.
- You are losing weight.
- You have bipolar disorder or mania.
- You have a history of seizures or epilepsy.
- You drink excess amounts of alcohol.

Over age 60:
No special problems expected.

Pregnancy:
Decide with your doctor if drug benefits justify risk to unborn baby. Risk category C (see page xviii).

Breastfeeding; Lactation; Nursing Mothers:
Drugs in this group are generally acceptable with breastfeeding. Discuss risks and benefits with your doctor.

Infants & children up to age 18:
Follow instructions provided by your child's doctor.Carefully read information provided with prescription. Contact doctor right away if depression symptoms get worse or there is any talk of suicide or suicide behaviors. Also, read information under Others.

Prolonged use:
Consult with your doctor on a regular basis while taking this drug to check blood pressure and to determine the need for continued treatment.

Skin & sunlight:
No special problems expected.

Driving, piloting or hazardous work:
Don't drive or pilot aircraft until you learn how medicine affects you. Don't work around dangerous machinery. Don't climb ladders or work in high places. Danger increases if you drink alcohol or take medicine affecting alertness and reflexes.

Discontinuing:
- Don't discontinue without consulting doctor. You may need to reduce the dose gradually.
- If any new or unusual symptoms develop (emotional or physical) after you discontinue the drug, call your doctor right away.

Others:
- Rise slowly from a sitting or lying position to avoid dizziness, faintness or lightheadedness.
- Diabetic patients should consult doctor if blood sugar levels are affected by taking this drug.
- Advise any doctor, dentist or pharmacist whom you consult that you take this drug.
- Adults and children taking antidepressants may experience a worsening of the depression symptoms and may have increased suicidal thoughts or behaviors. Call doctor right away if these symptoms or behaviors occur.
- Take drug as directed. Do not increase or reduce dosage without doctor's approval.

POSSIBLE INTERACTION WITH OTHER DRUGS

GENERIC NAME OR DRUG CLASS	COMBINED EFFECT
Antiarrhythmics*	Increased side effect risk (with duloxetine).
Antidepressants, other	Increased sedative effect.
Antidepressants, tricyclic	Increased side effect risk (with duloxetine).
Central nervous system (CNS) depressants,* other	Increased sedative effect.
Cimetidine	Increased effect of venlafaxine.
Enzyme inhibitors*	Increased effect of either drug.
Monoamine oxidase (MAO) inhibitors*	Severe adverse reactions. Allow 14 days between use.
Phenothiazines	Increased side effect risk (with duloxetine).

Continued on page 912

POSSIBLE INTERACTION WITH OTHER SUBSTANCES

INTERACTS WITH	COMBINED EFFECT
Alcohol:	Possible severe liver damage. Avoid.
Beverages:	None expected.
Cocaine:	Unknown. Avoid.
Foods:	None expected.
Marijuana:	Consult doctor.
Tobacco:	None expected.

***See Glossary**

SGLT2 INHIBITORS

GENERIC AND BRAND NAMES

CANAGLIFLOZIN
- **Invokamet**
- **Invokana**

DAPAGLIFOZIN
- **Farxiga**
- **Xigduo XR**

EMPAGLIFLOZIN
- **Glyxambi**
- **Jardiance**

BASIC INFORMATION

Habit forming? No
Prescription needed? Yes
Available as generic? No
Drug class: Antidiabetic

USES

Used in addition to diet and exercise to improve blood sugar (glucose) levels in patients with type 2 diabetes. It may be prescribed alone or with other antidiabetic drugs.

DOSAGE & USAGE INFORMATION

How to take:
Tablet—Swallow whole with liquid. If you can't swallow whole, ask your doctor or pharmacist for advice. May be taken with or without food.

When to take:
Once a day in the morning. Take canagliflozin before the first meal of the day.

If you forget a dose:
Take as soon as you remember. If it is almost time for the next dose, wait for that dose (don't double this dose) and resume regular schedule.

What drug does:
It inhibits the action of a protein in the kidneys called sodium-glucose co-transporter 2 or SGLT2. This stops the kidneys from reabsorbing glucose, increases glucose excretion in the urine and reduces blood glucose levels.

Time lapse before drug works:
One to two hours; several weeks for full benefit.

Continued next column

OVERDOSE

SYMPTOMS:
Unknown.
WHAT TO DO:
If person takes much larger amount than prescribed or if accidentally swallowed, call doctor or poison control center 1-800-222-1222 for help.

Don't take with:
Any other medicine or dietary supplement without consulting your doctor or pharmacist.

POSSIBLE ADVERSE REACTIONS OR SIDE EFFECTS

SYMPTOMS	WHAT TO DO
Life-threatening:	
None expected.	
Common:	
• Yeast infection of the vagina or penis (itching, discharge, pain in men, odor in women), urinary tract infection (burning sensation), changes in urination (urgency, frequency, larger amount, a need to urinate at night), stuffy nose, sore throat.	Continue. Call doctor when convenient.
• Symptoms of low blood sugar—shakiness, hunger, sweating, anxiety, dizziness, loss of concentration, nervousness, confusion, cold and clammy skin, drowsiness, headache, weakness, vision changes, fast heartbeat.	Seek treatment (eat some form of quick-acting sugar—glucose tablets, sugar, fruit juice, corn syrup, honey).
Infrequent:	
• Feeling lightheaded or faint or dizzy (especially when you stand up), low blood pressure.	Discontinue. Call doctor right away.
• Nausea, fatigue, constipation, dry mouth, thirstiness, stomach discomfort, dehydration, feeling weak.	Continue. Call doctor when convenient.
Rare:	
• Bone fracture.	Discontinue. Seek emergency care.
• Rash or hives or itching, blood in urine, symptoms of pancreatitis (pain in abdomen that travels to back, tenderness, nausea, vomiting).	Discontinue. Call doctor right away.

WARNINGS & PRECAUTIONS

Don't take if:

- You are allergic to SGLT2 inhibitors. Allergic reactions are rare, but may occur.
- You have severe renal impairment, are on dialysis or have end stage renal disease.

Before you start, consult your doctor if:

- You have any kidney (renal) problems.
- You have a liver disorder.
- You have low blood pressure.
- You have high cholesterol.
- You have or have had bladder cancer.
- You have a history of genital yeast infections.
- You have any electrolyte imbalance (e.g., high potassium levels or sodium level problems).

Over age 60:
Adverse reactions and side effects may be more frequent and severe than in younger persons.

Pregnancy:
Decide with your doctor if drug benefits justify risk to unborn baby. Risk category C (see page xviii).

Breastfeeding; Lactation; Nursing Mothers:
Drugs in this group are generally not recommended with breastfeeding. Discuss risks and benefits with your doctor.

Infants & children up to age 18:
Safety and effectiveness in this age group have not been established. Consult doctor.

Prolonged use:
Consult with your doctor on a regular basis while taking this drug to check for adverse effects and determine continued effectiveness of drug.

Skin & sunlight:
Canagliflozin may cause rash or intensify sunburn in areas exposed to sun or ultraviolet light (photosensitivity reaction). Avoid over-exposure. Notify doctor if reaction occurs.

Driving, piloting or hazardous work:
Don't drive or pilot aircraft until you learn how medicine affects you. Don't work around dangerous machinery. Don't climb ladders or work in high places. Danger increases if you drink alcohol or take drugs affecting alertness and reflexes.

Discontinuing:
Don't discontinue without doctor's advice, even though symptoms diminish or disappear.

Others:

- Along with taking drugs for diabetes, be sure to follow your doctor's instructions for lifestyle changes such as a proper diet, weight control measures and a regular exercise program.
- Notify your doctor if you have a fever, infection, diarrhea, experience vomiting or experience trauma (such as automobile accident).
- Advise any doctor, dentist or pharmacist whom you consult that you take this medicine.
- Wear or carry medical identification that indicates you have type 2 diabetes and the drugs you take.
- May rarely lead to ketoacidosis (a serious complication of diabetes). Seek medical care if symptoms occur (breathing difficulty, nausea, vomiting, stomach pain, fatigue, confusion).
- Use of the drug will cause your urine to test positive for glucose.
- Drink plenty of fluids daily to avoid dehydration.
- The drug is not used for treating type 1 diabetes or diabetic ketoacidosis.
- You and your family should educate yourselves about diabetes; learn to recognize the symptoms of hypoglycemia and how to treat it. Hypoglycemia may occur in the treatment of diabetes as a result of skipped meals, excessive exercise or alcohol use. Carry non-dietetic candy or glucose tablets to treat episodes of low blood sugar.

POSSIBLE INTERACTION WITH OTHER DRUGS

GENERIC NAME OR DRUG CLASS	COMBINED EFFECT
Digoxin	Digoxin dosage may need to be adjusted (with canagliflozin).
Diuretics*	Increased risk of dehydration and electrolyte imbalance.
Enzyme inducers*	Decreased effect of SGLT2 inhibitor.
Hypoglycemia-causing medications*	May increase risk of low blood sugar or side effects.

POSSIBLE INTERACTION WITH OTHER SUBSTANCES

INTERACTS WITH	COMBINED EFFECT
Alcohol:	Increased risk of side effects. Avoid.
Beverages:	None expected.
Cocaine:	Unknown. Avoid.
Foods:	None expected.
Marijuana:	Unknown. Avoid.
Tobacco:	No drug interaction. Tobacco use does raise risk of diabetes complications. Avoid smoking.

SIMETHICONE

BRAND NAMES

Alka-Seltzer Gas Relief
Degas
Di-Gel
Extra Strength Gas-X
Extra Strength Maalox Anti-Gas
Extra Strength Maalox GRF Gas Relief Formula
Flatulex
Gas Aid
Gas Relief
Gas-X
Gas-X Extra Strength
Gas-X Thin Strips
Gas-X with Maalox
Gelusil
Genasyme
Imodium Advanced
Imodium Multi-Symptom Relief
Maalox Anti-Gas
Maalox GRF Gas Relief Formula
Maximum Strength Mylanta Gas Relief
Maximum Strength Phazyme
Mygel
Mylanta Gas
Mylicon
Mylicon-80
Mylicon-125
Ovol
Ovol 40
Ovol-80
PediaCare Infants' Gas Relief
Phazyme
Phazyme 55
Phazyme 95
Riopan Plus

BASIC INFORMATION

Habit forming? No
Prescription needed? No
Available as generic? Yes, for some
Drug class: Antiflatulent

USES

- Treatment for retention of abdominal gas.
- Used prior to x-ray of abdomen.

OVERDOSE

SYMPTOMS:
Unknown.
WHAT TO DO:
If person uses much larger amount than prescribed or if accidentally swallowed, call poison control center 1-800-222-1222 for instructions or dial 911 (emergency) for help.

DOSAGE & USAGE INFORMATION

How to take:
- Tablet or capsule—Swallow with liquid.
- Liquid—Dissolve in water. Drink all of dose.
- Thin strip—Let it dissolve on your tongue.
- Chewable tablet—Chew completely. Don't swallow whole.

When to take:
After meals and at bedtime.

What drug does:
Reduces surface tension of gas bubbles in stomach.

Time lapse before drug works:
10 minutes.

If you forget a dose:
Take when remembered if needed.

Don't take with:
Any other medicine or any dietary supplement without consulting your doctor or pharmacist.

POSSIBLE ADVERSE REACTIONS OR SIDE EFFECTS

SYMPTOMS | **WHAT TO DO**

Life-threatening:
None expected.

Common:
None expected.

Infrequent:
None expected.

Rare:
None expected.

WARNINGS & PRECAUTIONS

Don't take if:
You are allergic to simethicone. Allergic reactions are rare, but may occur.

Before you start, consult your doctor if:
You have allergies to other drugs or substances.

Over age 60:
No problems expected.

Pregnancy:
Decide with your doctor if drug benefits justify risk to unborn baby. Risk category C (see page xviii).

Breastfeeding; Lactation; Nursing Mothers:
This drug is generally acceptable with breastfeeding. Discuss risks and benefits with your doctor.

Infants & children up to age 18:
Read nonprescription product's label to be sure it is approved for your child's age. Follow label instructions on dosage. Consult doctor or pharmacist if unsure.

Prolonged use:
No problems expected.

Skin & sunlight:
No problems expected.

Driving, piloting or hazardous work:
No problems expected.

Discontinuing:
May be unnecessary to finish medicine. Discontinue when symptoms disappear.

Others:
No problems expected.

POSSIBLE INTERACTION WITH OTHER DRUGS

GENERIC NAME OR DRUG CLASS	COMBINED EFFECT
None significant.	

POSSIBLE INTERACTION WITH OTHER SUBSTANCES

INTERACTS WITH	COMBINED EFFECT
Alcohol:	None expected.
Beverages:	None expected.
Cocaine:	None expected.
Foods:	None expected.
Marijuana:	None expected.
Tobacco:	None expected.

SODIUM BICARBONATE

BRAND NAMES

Alka-Seltzer Original
Arm & Hammer Pure Baking Soda
Bell/ans
Bromo-Seltzer
Citrocarbonate Soda Mint
Zegerid Capsules
Zegerid Chewable Tablets
Zegerid OTC
Zegerid Powder

BASIC INFORMATION

Habit forming? No
Prescription needed? No
Available as generic? Yes
Drug class: Alkalizer, antacid

USES

- Treats metabolic acidosis.
- Alkalinizes urine to reduce uric acid kidney stones.
- Treats hyperacidity of the stomach that is present with indigestion, gastroesophageal reflux and peptic ulcer disease.

DOSAGE & USAGE INFORMATION

How to take:
- Tablet—Swallow with liquid. If you can't swallow whole, crumble tablet and take with liquid or food.
- Powder—Mix in a glass of water and drink.
- Effervescent sodium bicarbonate—Mix in a glass of cold water and drink.

When to take:
- For hyperacidity—1 to 3 hours after meals.
- For kidney stones—According to prescription instructions.

Continued next column

OVERDOSE

SYMPTOMS:
Muscle spasms and weakness, vomiting, irritable, diarrhea or constipation, frequent urination, seizures.
WHAT TO DO:
- **Dial 911 for all medical emergencies or call poison control center 1-800-222-1222 for instructions.**
- **See emergency information on last 3 pages of this book.**

If you forget a dose:
Take as soon as you remember. If it is almost time for the next dose, wait for that dose (don't double this dose) and resume regular schedule.

What drug does:
- Buffers acid in the stomach.
- Increases excretion of bicarbonate in the urine to help dissolve uric acid stones.

Time lapse before drug works:
Works immediately, but the duration of effect is short.

Don't take with:
Any other medicine or any dietary supplement without consulting your doctor or pharmacist.

POSSIBLE ADVERSE REACTIONS OR SIDE EFFECTS

SYMPTOMS	WHAT TO DO
Life-threatening:	
None expected.	
Common:	
None expected.	
Infrequent:	
Stomach cramps, thirstiness.	Continue. Call doctor if symptoms persist.
Rare:	
With large doses and long-term use: urge to urinate, ongoing headaches or loss of appetite, slow breathing, mild swelling of feet or lower legs, twitching, nervousness, tiredness, taste changes.	Discontinue. Call doctor right away.

WARNINGS & PRECAUTIONS

Don't take if:
You are allergic to sodium bicarbonate. Allergic reactions are rare, but may occur.

Before you start, consult your doctor if:
You have heart disease, kidney disease or toxemia of pregnancy.

Over age 60:
Adverse reactions and side effects may be more frequent and severe than in younger persons.

Pregnancy:
Decide with your doctor if drug benefits justify risk to unborn baby. Risk category C (see page xviii).

Breastfeeding; Lactation; Nursing Mothers:
This drug is generally acceptable with breastfeeding. Discuss risks and benefits with your doctor.

Infants & children up to age 18:
Read nonprescription product's label to be sure it is approved for your child's age. Follow label instructions on dosage. Consult doctor or pharmacist if unsure.

Prolonged use:
Don't use for longer than prescribed or recommended. May cause sodium overload.

Skin & sunlight:
No special problems expected.

Driving, piloting or hazardous work:
No special problems expected.

Discontinuing:
May be unnecessary to finish medicine. Follow doctor's instructions.

Others:
- Heat and moisture in bathroom medicine cabinet can cause breakdown of medicine. Store someplace else.
- May interfere with the accuracy of some medical tests (especially acidosis and urinalysis tests).

POSSIBLE INTERACTION WITH OTHER DRUGS

GENERIC NAME OR DRUG CLASS	COMBINED EFFECT
Adrenocorticoids*	Sodium overload.
Cortisone	Sodium overload.
Ketoconazole	Decreased absorption of ketoconazole.
Mecamylamine	Increased mecamylamine effect.
Memantine	Increased effect of memantine.
Methenamine	Decreased methenamine effect.
Tetracyclines*	Greatly reduced absorption of tetracyclines.
Any other medicine	Decreased absorption of other medicine if taken within 1 to 2 hours of taking sodium bicarbonate.

POSSIBLE INTERACTION WITH OTHER SUBSTANCES

INTERACTS WITH	COMBINED EFFECT
Alcohol:	Decreased effectiveness of sodium bicarbonate.
Beverages: Milk and milk products (large amounts).	Increased risk of side effects.
Cocaine:	None expected.
Foods:	None expected.
Marijuana:	None expected.
Tobacco:	Decreased effectiveness of sodium bicarbonate.

*See Glossary

SODIUM FLUORIDE

BRAND NAMES

Fluor-A-Day
Fluorident
Fluoritab
Fluorodex
Fluotic
Flura
Karidium
Listermint with Fluoride
Luride
Luride-SF
Pediaflor
Pedi-Dent
Solu-Flur
Numerous other multiple vitamin-mineral supplements. Check labels.

BASIC INFORMATION

Habit forming? No
Prescription needed? Yes, for some
Available as generic? Yes
Drug class: Mineral supplement (fluoride)

USES

- Reduces tooth cavities.
- Treats osteoporosis.

OVERDOSE

SYMPTOMS:
Stomach cramps or pain, nausea, faintness, vomiting, diarrhea, shallow breathing, muscle spasms, weakness, tremors, seizures, irregular heartbeat.
WHAT TO DO:
- **Dial 911 for all medical emergencies or call poison control center 1-800-222-1222 for instructions.**
- **See emergency information on last 3 pages of this book.**

DOSAGE & USAGE INFORMATION

How to take:
- Tablet—Swallow with liquid or crumble tablet and take with liquid (not milk) or food.
- Liquid—Measure with dropper and take directly or with liquid.
- Chewable tablet—Chew slowly and thoroughly before swallowing.

When to take:
Usually at bedtime after teeth are thoroughly brushed.

If you forget a dose:
Take as soon as you remember. If it is almost time for the next dose, wait for that dose (don't double this dose) and resume regular schedule.

What drug does:
Provides supplemental fluoride to combat tooth decay.

Time lapse before drug works:
8 weeks to provide maximum effect.

Don't take with:
Any other medicine or any dietary supplement without consulting your doctor or pharmacist.

POSSIBLE ADVERSE REACTIONS OR SIDE EFFECTS

SYMPTOMS	WHAT TO DO
Life-threatening:	
In case of overdose, see previous column.	
Common:	
None expected.	
Infrequent:	
None expected.	
Rare:	
Sores in mouth or on lips.	Discontinue. Call doctor or dentist if symptoms persist.

WARNINGS & PRECAUTIONS

Don't take if:
- Your water supply contains 0.7 parts fluoride per million. Too much fluoride stains teeth permanently.
- You are allergic to any fluoride-containing product. Allergic reactions are rare, but may occur.
- You have underactive thyroid.

Before you start, consult your doctor if:
- You have kidney disease.
- You have ulcers.
- You have joint pain.

Over age 60:
No problems expected.

Pregnancy:
Decide with your doctor if drug benefits justify risk to unborn baby. Risk category B (see page xviii).

Breastfeeding; Lactation; Nursing Mothers:
This drug is generally acceptable with breastfeeding. Discuss risks and benefits with your doctor.

Infants & children up to age 18:
Read nonprescription product's label to be sure it is approved for your child's age. Follow label instructions on dosage. Consult doctor or pharmacist if unsure.

Prolonged use:
Excess may cause discolored teeth and decreased calcium in blood.

Skin & sunlight:
No problems expected.

Driving, piloting or hazardous work:
No problems expected.

Discontinuing:
No problems expected.

Others:
- Store in original plastic container. Fluoride decomposes glass.
- Some products contain tartrazine dye. Avoid, especially if you are allergic to aspirin.

POSSIBLE INTERACTION WITH OTHER DRUGS

GENERIC NAME OR DRUG CLASS	COMBINED EFFECT
Calcium supplements*	Decreased effect of calcium and fluoride.

POSSIBLE INTERACTION WITH OTHER SUBSTANCES

INTERACTS WITH	COMBINED EFFECT
Alcohol:	None expected.
Beverages:	
Milk.	Prevents absorption of fluoride. Space dose 2 hours before or after milk.
Cocaine:	None expected.
Foods:	None expected.
Marijuana:	None expected.
Tobacco:	None expected.

***See Glossary**

STIMULANT MEDICATIONS

GENERIC AND BRAND NAMES

DEXMETHYLPHENIDATE
Focalin
Focalin XR
METHYLPHENIDATE
Concerta
Daytrana
Metadate CD
Metadate ER
Methylin Chewable
METHYLPHENIDATE (con't)
Methylin ER
Methylin Oral Suspension
PMS Methylphenidate
Quillivant XR
Ritalin
Ritalin LA
Ritalin SR

BASIC INFORMATION

Habit forming? Yes
Prescription needed? Yes
Available as generic? Yes, for some
Drug class: Central nervous system stimulant, sympathomimetic

USES

- Decreases overactivity and lengthens attention span in children and adults with attention-deficit hyperactivity disorder (ADHD). A total treatment plan may also include educational, social and psychological therapies.
- Treatment of depression in adults.
- Treatment for narcolepsy and other disorders.

DOSAGE & USAGE INFORMATION

How to take:

- Tablet (short-acting)—Swallow with liquid. Take as directed, usually 30 to 45 minutes before meals, or with meals if stomach upset occurs. If swallowing is a problem, ask your pharmacist for advice.
- Oral solution, extended-release oral solution or chewable tablet—Follow instructions on label.
- Skin patch—Follow instructions on label.

Continued next column

- Extended- or sustained-release tablet and capsule—Swallow whole with liquid (or as directed). Do not crush tablet. Do not open capsule (unless label states that it may be opened and sprinkled on cool applesauce and swallowed right away).

When to take:
At the same times each day. Regular tablets are often taken at breakfast and lunch (best not to take late in day). Extended-release forms are usually taken in the morning.

If you forget a dose:
Take as soon as you remember. If it is almost time for the next dose, wait for that dose (don't double this dose) and resume regular schedule.

What drug does:
Stimulates brain to improve alertness, concentration and attention span.

Time lapse before drug works:
May take 2 or more weeks to see effectiveness. Dosage may be increased or decreased depending on the response and side effects.

Don't take with:
Any other medicine or any dietary supplement without consulting your doctor or pharmacist.

OVERDOSE

SYMPTOMS:
Rapid heartbeat, fever, confusion, vomiting, agitation, hallucinations, headache, sweating, convulsions, coma.
WHAT TO DO:

- **Dial 911 for all medical emergencies or call poison control center 1-800-222-1222 for instructions.**
- **See emergency information on last 3 pages of this book.**

POSSIBLE ADVERSE REACTIONS OR SIDE EFFECTS

SYMPTOMS	WHAT TO DO
Life-threatening:	
Rare, severe allergic reaction (anaphylaxis)—difficulty breathing, hives, itching, swelling (throat, face), vomiting, dizziness.	Seek emergency treatment immediately.
Common:	
• Fast heartbeat, blood pressure increased.	Discontinue. Call doctor right away.
• Nervousness, appetite loss, trouble sleeping.	Continue. Call doctor when convenient.
Infrequent:	
• Rash or hives, chest or joint pain, unusual bruising or bleeding, unable to control body movements, fever.	Discontinue. Call doctor right away.
• Nausea, dizziness, headache, stomach pain, drowsiness, muscle cramps.	Continue. Call doctor when convenient.
Rare:	
Changed or blurred vision, unusual vocal outbursts, seizures, abnormal or manic behavior, trouble breathing, fainting, hallucinations, long-lasting penis erection.	Discontinue. Call doctor right away.

WARNINGS & PRECAUTIONS

Don't take if:
You are allergic to stimulant medications.

Before you start, consult your doctor if:
- You have epilepsy or have seizures.
- You have high blood pressure, any heart or blood vessel disorder or liver problems.
- You have glaucoma.
- You take MAO inhibitors.*
- You suffer from anxiety, agitation, tension, depressive or psychotic problems or have Tourette's syndrome or motor tics.
- You have a history of drug or alcohol abuse.

Over age 60:
Adverse reactions and side effects may be more frequent and severe than in younger persons.

Pregnancy:
Decide with your doctor if drug benefits justify risk to unborn baby. Risk category C (see page xviii).

Breastfeeding; Lactation; Nursing Mothers:
Drugs in this group are generally acceptable with breastfeeding. Discuss risks and benefits with your doctor.

Infants & children up to age 18:
Follow instructions provided by your child's doctor. Regular doctor visits are important to monitor drug's effectiveness and side effects.

Prolonged use:
- Increased risk of weight loss and abnormal behaviors. Rare risk of physical growth retardation in children.
- Talk to your doctor about the need for follow up medical examinations or laboratory studies to check drug's effectiveness and monitor any adverse effects.

Skin & sunlight:
No problems expected.

Driving, piloting or hazardous work:
Don't drive, ride a bicycle or pilot aircraft until you learn how drug affects you. Don't work around dangerous machinery. Don't climb ladders or work in high places. Danger increases if you drink alcohol or take drugs affecting alertness and reflexes.

Discontinuing:
- Don't discontinue without doctor's advice even if symptoms diminish or disappear.
- Withdrawal symptoms may occur after you discontinue the drug. Report to your doctor any new physical or emotional symptoms.

Others:
- Drug may cause blood vessel problems in fingers or toes—they feel cold or numb and rarely, can develop wounds. Contact doctor right away if symptoms occur.
- Drug may cause serious heart and psychiatric (mental) problems, including sudden death. Read warning information provided with prescription. Call doctor right away if symptoms develop (e.g., chest pain, shortness of breath, fainting, or hallucinations).
- Dose must be carefully adjusted by doctor.
- Advise any doctor, dentist or pharmacist whom you consult about the use of this drug.

POSSIBLE INTERACTION WITH OTHER DRUGS

GENERIC NAME OR DRUG CLASS	COMBINED EFFECT
Anticholinergics*	Increased anticholinergic effect.
Anticoagulants,* oral	Increased anticoagulant effect.
Anticonvulsants*	Increased anticonvulsant effect, or decreased stimulant effect.
Antidepressants, tricyclic*	Increased anti-depressant effect. Decreased stimulant medication effect.
Antihypertensives*	Decreased anti-hypertensive effect.
Central nervous system (CNS) stimulants*	Overstimulation.
Clonidine	Increased risk of adverse effects.
Dextrothyroxine	Increased stimulant medication effect.
Monoamine oxidase (MAO) inhibitors*	Dangerous rise in blood pressure. Take at least 14 days apart.
Pimozide	May mask the cause of tics.

POSSIBLE INTERACTION WITH OTHER SUBSTANCES

INTERACTS WITH	COMBINED EFFECT
Alcohol:	Unknown. Avoid.
Beverages:	None expected.
Cocaine:	Unknown. Avoid.
Foods:	None expected.
Marijuana:	Consult doctor.
Tobacco:	None expected.

***See Glossary**

STIMULANTS, AMPHETAMINE-RELATED

GENERIC AND BRAND NAMES

ARMODAFINIL
Nuvigil

MODAFINIL
Alertec
Provigil
Sparlon

BASIC INFORMATION

Habit forming? Possibly
Prescription needed? Yes
Available as generic? Yes, for some
Drug class: Antinarcoleptic; central nervous system stimulant

USES

- Treatment to help people who have narcolepsy to stay awake during the day. It does not cure narcolepsy.
- Used to improve wakefulness in patients with excessive sleepiness disorders, improve wakefulness for obstructive sleep apnea and shift work sleep disorder.
- Treatment for attention deficit hyperactivity disorder (ADHD).

DOSAGE & USAGE INFORMATION

How to take:
Tablet—Swallow whole with liquid. You may take it with or without food.

When to take:
At the same time each day, usually in the morning. Follow instructions on the label.

If you forget a dose:
Take as soon as you remember, until noon of the same day. If you don't remember until later, skip the missed dose to avoid problems getting to sleep. Return to your regular dosing schedule the next day. Do not double doses.

Continued next column

OVERDOSE

SYMPTOMS:
Agitation, nausea, diarrhea, insomnia, restlessness, confusion, increased blood pressure, hallucinations, fast or slow heart rate, chest pain.

WHAT TO DO:

- **Dial 911 for all medical emergencies or call poison control center 1-800-222-1222 for instructions.**
- **See emergency information on last 3 pages of this book.**

What drug does:
Stimulates the central nervous system. The exact way these drugs work is unknown.

Time lapse before drug works:
2 to 4 hours.

Don't take with:
Any other medicine or any dietary supplement without consulting your doctor or pharmacist.

POSSIBLE ADVERSE REACTIONS OR SIDE EFFECTS

SYMPTOMS	WHAT TO DO
Life-threatening:	
Rare, severe allergic reaction (anaphylaxis)—difficulty breathing, hives, itching, swelling (throat, face), vomiting, dizziness.	Seek emergency treatment immediately.
Common:	
Anxiety, headache, nausea, nervousness, trouble sleeping.	Continue. Call doctor when convenient.
Infrequent:	
Appetite changes, diarrhea, dry mouth, skin symptoms (dryness, flushing or tingling), muscle stiffness, stuffy or runny nose, trembling or shaking, vomiting.	Continue. Call doctor when convenient.
Rare:	
Vision changes, chills or fever, confusion, abnormal heart rate, dizziness, fainting, increased thirst or urination, depression, memory or mood changes, shortness of breath, trouble in urinating, uncontrolled movements (face, mouth or tongue).	Discontinue. Call doctor right away.

WARNINGS & PRECAUTIONS

Don't take if:
You are allergic to armodafinil or modafinil or other central nervous system stimulants.

Before you start, consult your doctor if:

- You have heart disease or have had a heart attack.
- You have or have had high blood pressure.
- You take oral contraceptives.
- You have liver or kidney disease.

- You have a history of psychosis, depression, mania, or other severe mental illness.

Over age 60:
No special problems expected.

Pregnancy:
Decide with your doctor if drug benefits justify risk to unborn baby. Risk category C (see page xviii).

Breastfeeding; Lactation; Nursing Mothers:
Drugs in this group are generally not recommended with breastfeeding. Discuss risks and benefits with your doctor.

Infants & children up to age 18:
Follow instructions provided by your child's doctor. The brand name Sparlon is approved for treatment of attention deficit hyperactivity disorder in children over age 6.

Prolonged use:
May lead to physical or mental dependence. Consult your doctor if any of the following signs of dependence occur:
- A strong desire to continue taking this drug.
- A need to increase the dose to receive the effects of the medicine.
- Withdrawal side effects when you stop taking the medicine.

Skin & sunlight:
No problems expected.

Driving, piloting or hazardous work:
Don't drive or pilot aircraft until you learn how medicine affects you. Don't work around dangerous machinery. Don't climb ladders or work in high places. Danger increases if you drink alcohol or take medicine affecting alertness and reflexes such as antihistamines, tranquilizers or sedatives, pain medicine, narcotics and mind-altering drugs.

Discontinuing:
- Consult your doctor if any new or unusual symptoms occur after discontinuing the drug.
- Dose may require gradual reduction if you have taken drug for a long time.

Others:
- If you are using a birth control method, such as pills or implants, they may not be as effective while taking these drugs and for up to one month after stopping them.
- Stop drug and contact your doctor if you experience any sort of unusual rash or mood changes. Rare cases of serious or life-threatening rash and serious psychiatric adverse experiences (including anxiety, mania, hallucinations and thoughts of suicide) have been reported.
- Advise any doctor, dentist or pharmacist whom you consult that you take this medicine.
- May affect the results in some medical tests.

POSSIBLE INTERACTION WITH OTHER DRUGS

GENERIC NAME OR DRUG CLASS	COMBINED EFFECT
Antidepressants, tricyclic*	Increased effect of antidepressant.
CNS stimulants*	Increased stimulant effect.
Contraceptives*	Decreased contraceptive effect.
Diazepam	Decreased diazepam effect.
Enzyme inducers*	Decreased stimulant effect.
Enzyme inhibitors*	Increased stimulant effect.
MAO inhibitors*	Unknown effect. Avoid.
Mephenytoin	Mephenytoin dose may need adjustment.
Theophylline	Decreased theophylline effect.
Warfarin	Increased warfarin effect.

POSSIBLE INTERACTION WITH OTHER SUBSTANCES

INTERACTS WITH	COMBINED EFFECT
Alcohol:	Effects unknown. Avoid.
Beverages: Grapefruit juice.	Unknown effect. Consult doctor.
Cocaine:	Unknown. Avoid.
Foods: Grapefruit.	Unknown effect. Consult doctor.
Marijuana:	Consult doctor.
Tobacco:	None expected.

*See Glossary

SUCRALFATE

BRAND NAMES

Carafate
Sulcrate
Sulcrate Suspension Plus

BASIC INFORMATION

Habit forming? No
Prescription needed? Yes
Available as generic? Yes
Drug class: Antiulcer agent

USES

- Treatment for duodenal and gastric ulcers.
- Used to relieve side effects of nonsteroidal anti-inflammatory therapy in rheumatoid arthritis.
- Treatment for gastroesophageal reflux disease (GERD).

DOSAGE & USAGE INFORMATION

How to take:

- Tablet—Take as directed on an empty stomach.
- Oral suspension—Follow instructions on package.

When to take:
1 hour before meals and at bedtime. Allow 2 hours to elapse before taking other prescription medicines.

If you forget a dose:
Take as soon as you remember. If it is almost time for the next dose, wait for that dose (don't double this dose) and resume regular schedule.

What drug does:
Covers ulcer site and protects from acid, enzymes and bile salts.

Time lapse before drug works:
Begins in 30 minutes. May require several days to relieve pain.

Don't take with:
Any other medicine or any dietary supplement without consulting your doctor or pharmacist.

OVERDOSE

SYMPTOMS:
May have upset stomach, nausea, vomiting.
WHAT TO DO:
If person uses much larger amount than prescribed or if accidentally swallowed, call poison control center 1-800-222-1222 for instructions or dial 911 (emergency) for help.

POSSIBLE ADVERSE REACTIONS OR SIDE EFFECTS

SYMPTOMS	WHAT TO DO
Life-threatening:	
Rare, severe allergic reaction (anaphylaxis)—difficulty breathing, hives, itching, swelling (throat, face), vomiting, dizziness.	Seek emergency treatment immediately.
Common:	
Constipation.	Continue. Call doctor when convenient.
Infrequent:	
Dizziness, sleepiness, rash or itchy skin, abdominal pain, indigestion, vomiting, nausea, dry mouth, diarrhea.	Continue. Call doctor when convenient.
Rare:	
Back pain.	Continue. Call doctor when convenient.

WARNINGS & PRECAUTIONS

Don't take if:
You are allergic to sucralfate.

Before you start, consult your doctor if:

- You will have surgery within 2 months, including dental surgery, requiring general or spinal anesthesia.
- You have gastrointestinal or kidney disease.

Over age 60:
Adverse reactions and side effects may be more frequent and severe than in younger persons.

Pregnancy:
Decide with your doctor if drug benefits justify risk to unborn baby. Risk category B (see page xviii).

Breastfeeding; Lactation; Nursing Mothers:
This drug is generally acceptable with breastfeeding. Discuss risks and benefits with your doctor.

Infants & children up to age 18:
Follow instructions provided by your child's doctor.

Prolonged use:
Request blood counts if medicine needed longer than 8 weeks.

Skin & sunlight:
No problems expected.

Driving, piloting or hazardous work:
Don't drive or pilot aircraft until you learn how medicine affects you. Don't work around dangerous machinery. Don't climb ladders or work in high places. Danger increases if you drink alcohol or take medicine affecting alertness and reflexes.

Discontinuing:
Don't discontinue without consulting doctor. Dose may require gradual reduction if you have taken drug for a long time. Doses of other drugs may also require adjustment.

Others:
Advise any doctor, dentist or pharmacist whom you consult that you take this medicine.

POSSIBLE INTERACTION WITH OTHER DRUGS

GENERIC NAME OR DRUG CLASS	COMBINED EFFECT
Anagrelide	May interfere with anagrelide absorption.
Antacids*	Take 1/2 hour before or after sucralfate.
Cimetidine	Possible decreased absorption of cimetidine if taken simultaneously.
Ciprofloxacin	Decreased absorption of ciprofloxacin. Take 2 hours before sucralfate.
Digoxin	Decreased absorption of digoxin. Take 2 hours before sucralfate.
Fluoroquinolones	Decreased fluoroquinolone effect.
Ketoconazole	Decreased ketoconazole effect.
Norfloxacin	Decreased absorption of norfloxacin. Take 2 hours before sucralfate.
Ofloxacin	Decreased absorption of ofloxacin. Take 2 hours before sucralfate.
Phenytoin	Possible decreased absorption of phenytoin if taken simultaneously.
Proton pump inhibitors	May decrease effect of some proton pump inhibitors. Take 30 minutes before sucralfate.
Theophylline	Decreased absorption of theophylline. Take 2 hours before sucralfate.
Vitamins A, D, E, K	Decreased vitamin absorption.

POSSIBLE INTERACTION WITH OTHER SUBSTANCES

INTERACTS WITH	COMBINED EFFECT
Alcohol:	None expected.
Beverages:	None expected.
Cocaine:	Unknown. Avoid.
Foods:	None expected.
Marijuana:	Consult doctor.
Tobacco:	None expected.

*See Glossary

SULFASALAZINE

BRAND NAMES

Azaline
Azulfidine
Azulfidine En-Tabs
PMS Sulfasalazine
PMS Sulfasalazine EC
Salazopyrin
Salazosulfapyridine
Salicylazosulfa-pyridine
S.A.S. Enteric-500
S.A.S.-500

BASIC INFORMATION

Habit forming? No
Prescription needed? Yes
Available as generic? Yes
Drug class: Sulfa (sulfonamide)

USES

- Treatment for ulceration and bleeding from ulcerative colitis and Crohn's disease.
- Treatment for rheumatoid arthritis for patients not responding to other treatments.

DOSAGE & USAGE INFORMATION

How to take:
Tablet or enteric-coated tablet—Swallow whole with full glass of water. Do not crush or chew tablet. Take after a meal or with food to lessen stomach irritation.

When to take:
At the same times each day, evenly spaced.

If you forget a dose:
Take as soon as you remember. If it is almost time for the next dose, wait for that dose (don't double this dose) and resume regular schedule.

What drug does:
It has an anti-inflammatory action in the body.

Time lapse before drug works:
2 to 5 days.

Don't take with:
Any other medicine or any dietary supplement without consulting your doctor or pharmacist.

OVERDOSE

SYMPTOMS:
Drowsiness, nausea, vomiting, stomach pain, seizures.
WHAT TO DO:
- **Dial 911 for all medical emergencies or call poison control center 1-800-222-1222 for instructions.**
- **See emergency information on last 3 pages of this book.**

POSSIBLE ADVERSE REACTIONS OR SIDE EFFECTS

SYMPTOMS	WHAT TO DO
Life-threatening:	
Rare, severe allergic reaction (anaphylaxis)—difficulty breathing, hives, itching, swelling (throat, face), vomiting, dizziness.	Seek emergency treatment immediately.
Common:	
• Itchy skin, rash, aching joints, fever, ongoing headaches, vomiting.	Discontinue. Call doctor right away.
• Nausea, upset stomach, diarrhea, appetite loss, skin sensitive to sun.	Continue. Call doctor when convenient.
Infrequent:	
Yellow eyes or skin, bluish color to nails or skin, difficulty breathing, chills, unusual tiredness or weakness, sore throat, general ill feeling, pale skin, unusual bleeding or bruising.	Discontinue. Call doctor right away.
Rare:	
• Pain (back, legs, muscles, stomach, chest), bloody diarrhea, cough, difficulty swallowing, red or peeling or blistering skin.	Discontinue. Call doctor right away.
• Orange color urine.	No action necessary.

WARNINGS & PRECAUTIONS

Don't take if:
You are allergic to any sulfa drug.*

Before you start, consult your doctor if:
- You have asthma or severe allergies.
- You have liver or kidney disease, porphyria, intestinal or urinary blockages, blood problems or G6PD deficiency.

Over age 60:
Adverse reactions and side effects may be more frequent and severe than in younger persons.

Pregnancy:
Decide with your doctor if drug benefits justify risk to unborn baby. Risk category B (see page xviii).

Breastfeeding; Lactation; Nursing Mothers:
This drug is generally acceptable with breastfeeding. Discuss risks and benefits with your doctor.

Infants & children up to age 18:
Follow instructions provided by your child's doctor.

Prolonged use:
- May enlarge thyroid gland.
- You may become more susceptible to infections caused by germs not responsive to this drug.
- Request frequent blood counts, liver and kidney function studies.

Skin & sunlight:
May cause rash or intensify sunburn in areas exposed to sun or ultraviolet light (photosensitivity reaction). Avoid overexposure. Notify doctor if reaction occurs.

Driving, piloting or hazardous work:
Avoid if you feel dizzy. Otherwise, no problems expected.

Discontinuing:
Don't discontinue without doctor's advice until you complete prescribed dose, even though symptoms diminish or disappear.

Others:
- Advise any doctor, dentist or pharmacist whom you consult that you take this medicine.
- Drink plenty of fluids each day to help prevent adverse reactions.
- If you require surgery, tell anesthetist you take sulfa.

POSSIBLE INTERACTION WITH OTHER DRUGS

GENERIC NAME OR DRUG CLASS	COMBINED EFFECT
Aminobenzoates	Possible decreased sulfa effect.
Antibiotics*	Decreased sulfa effect.
Anticoagulants,* oral	Increased anticoagulant effect.
Anticonvulsants, hydantoin*	Toxic effect on brain.
Antidiabetics*	Toxic effect on brain.
Aspirin	Increased sulfa effect.
Calcium supplements*	Decreased sulfa effect.
Clozapine	Toxic effect on the central nervous system.
Digoxin	Decreased digoxin effect.
Hepatotoxic agents*	Increased liver toxicity.
Iron supplements*	Decreased sulfa effect.
Isoniazid	Possible anemia.
Mecamylamine	Decreased antibiotic effect.
Methenamine	Possible kidney blockage.
Methotrexate	Increased methotrexate effect.
Oxyphenbutazone	Increased sulfa effect.
Para-aminosalicylic acid	Decreased sulfa effect.
Penicillins*	Decreased penicillin effect.
Phenylbutazone	Increased sulfa effect.
Probenecid	Increased sulfa effect.
Sulfinpyrazone	Increased sulfa effect.
Sulfonylureas*	May increase hypoglycemic action.
Trimethoprim	Increased sulfa effect.
Vitamin C	Possible kidney damage. Avoid large doses of vitamin C.
Zidovudine	Increased risk of toxic effects of zidovudine.

POSSIBLE INTERACTION WITH OTHER SUBSTANCES

INTERACTS WITH	COMBINED EFFECT
Alcohol:	Increased alcohol effect.
Beverages: Low fluid intake.	May increase risk of kidney damage.
Cocaine:	Unknown. Avoid.
Foods:	None expected.
Marijuana:	Consult doctor.
Tobacco:	None expected.

SULFONAMIDES

GENERIC AND BRAND NAMES

See full list of generic and brand names in the *Generic and Brand Name Directory*, page 880.

BASIC INFORMATION

Habit forming? No
Prescription needed? Yes
Available as generic? Yes, for some
Drug class: Antibacterial (antibiotic), antiprotozoal, sulfa (sulfonamide)

USES

- Treatment of urinary tract and other infections.
- Sulfamethoxazole in combination with trimethoprim may be used to treat bronchitis, certain types of pneumonia, skin infections, middle ear infections, intestinal tract infections and urinary tract infections.

DOSAGE & USAGE INFORMATION

How to take:
- Tablet—Swallow with liquid. Instructions to take on empty stomach mean 1 hour before or 2 hours after eating. Drink an extra amount of water daily so that urine output will be adequate.
- Liquid—Shake carefully before measuring.
- Other forms—Follow label instructions.

When to take:
At the same times each day, evenly spaced.

If you forget a dose:
Take as soon as you remember. If it is almost time for the next dose, wait for that dose (don't double this dose) and resume regular schedule.

What drug does:
Interferes with a nutrient (folic acid) necessary for growth and reproduction of bacteria. Will not attack viruses.

Continued next column

OVERDOSE

SYMPTOMS:
Nausea, vomiting, stomach pain, fever, pain when urinating, lightheadedness, headache, drowsiness, unconsciousness, coma.
WHAT TO DO:
- **Dial 911 for all medical emergencies or call poison control center 1-800-222-1222 for instructions.**
- **See emergency information on last 3 pages of this book.**

Time lapse before drug works:
2 to 5 days to affect infection.

Don't take with:
Any other medicine or any dietary supplement without consulting your doctor or pharmacist.

POSSIBLE ADVERSE REACTIONS OR SIDE EFFECTS

SYMPTOMS	WHAT TO DO
Life-threatening:	
Rare, severe allergic reaction (anaphylaxis)—difficulty breathing, hives, itching, swelling (throat, face), vomiting, dizziness.	Seek emergency treatment immediately.
Common:	
• Itchy skin, rash.	Discontinue. Call doctor right away.
• Headache, nausea, vomiting, diarrhea, appetite loss, skin sensitive to sun, dizziness.	Continue. Call doctor when convenient.
Infrequent:	
• Red or peeling or blistering skin, sore throat, fever, swallowing difficulty, unusual bruising or bleeding, aching joints or muscles, yellow skin or eyes, pale skin.	Discontinue. Call doctor right away.
• Weakness or tiredness.	Continue. Call doctor when convenient.
Rare:	
Painful urination, low back pain, numbness, stomach pain, bloody diarrhea or urine, neck swelling, mood or behavior changes, increased or decreased urine output, thirst.	Discontinue. Call doctor right away.

WARNINGS & PRECAUTIONS

Don't take if:
You are allergic to any sulfa drug.

Before you start, consult your doctor if:
- You are allergic to carbonic anhydrase inhibitors, oral antidiabetics or diuretics (thiazide or loop).
- You are allergic by nature.
- You have liver or kidney disease.
- You have glucose 6-phosphate dehydrogenase (G6PD) disease.
- You have porphyria.
- You have anemia or other blood problems.

Over age 60:
Adverse reactions and side effects may be more frequent and severe than in younger persons.

Pregnancy:
Risk factors vary for drugs in this group. Always consult doctor. See category list on page xviii.

Breastfeeding; Lactation; Nursing Mothers:
Drugs in this group are generally acceptable with breastfeeding. Discuss risks and benefits with your doctor.

Infants & children up to age 18:
Follow instructions provided by your child's doctor.

Prolonged use:
- You may become more susceptible to infections caused by germs not responsive to this drug.
- Drug may enlarge thyroid gland (rare).
- Talk to your doctor about the need for frequent blood counts, liver and kidney function studies.

Skin & sunlight:
May cause rash or intensify sunburn in areas exposed to sun or ultraviolet light (photo-sensitivity reaction). Avoid excess exposure. Notify doctor if reaction occurs.

Driving, piloting or hazardous work:
Avoid if you feel dizzy. Otherwise, no problems expected.

Discontinuing:
Don't discontinue without doctor's advice until you complete prescribed dose, even though symptoms diminish or disappear.

Others:
- Drink plenty of liquid each day to help prevent side effects or adverse reactions.
- Advise any doctor, dentist or pharmacist whom you consult that you take this medicine.
- If you require surgery, tell anesthetist you take sulfa.

POSSIBLE INTERACTION WITH OTHER DRUGS

GENERIC NAME OR DRUG CLASS	COMBINED EFFECT
Aminobenzoate potassium	Possible decreased sulfonamide effect.
Anticoagulants,* oral	Increased anticoagulant effect.
Anticonvulsants,* hydantoin	Increased anticonvulsant effect.
Antidiabetics,* oral	Increased antidiabetic effect.
Bone marrow depressants*	Increased risk of side effects.
Contraceptives,* oral estrogen	Decreased contraceptive effect.
Cyclosporine	Decreased cyclosporine effect.
Hemolytics,* other	Increased risk of side effects.
Hepatotoxic agents*	Increased liver toxicity.
Mecamylamine	Decreased antibiotic effect.
Methenamine	Possible kidney blockage.
Methotrexate	Increased methotrexate effect.
Penicillins*	Decreased penicillin effect.
Phenylbutazone	Increased sulfonamide effect.
Probenecid	Increased sulfonamide effect.
Sulfinpyrazone	Increased sulfonamide effect.

POSSIBLE INTERACTION WITH OTHER SUBSTANCES

INTERACTS WITH	COMBINED EFFECT
Alcohol:	None expected.
Beverages: Inadequate fluid intake.	Increased risk of side effects.
Cocaine:	Unknown. Avoid.
Foods:	None expected.
Marijuana:	Consult doctor.
Tobacco:	None expected.

*See Glossary

SULFONYLUREAS

GENERIC AND BRAND NAMES

See full list of generic and brand names in the *Generic and Brand Name Directory*, page 881.

BASIC INFORMATION

Habit forming? No
Prescription needed? Yes
Available as generic? Yes, for some.
Drug class: Antidiabetic (oral), sulfonylurea

USES

- Treatment for diabetes in adults who can't control blood sugar by diet, weight loss and exercise.
- Treatment for diabetes insipidus (chlorpropamide).

DOSAGE & USAGE INFORMATION

How to take:
- Tablet—Swallow with liquid or food to lessen stomach irritation. If you can't swallow whole, crumble tablet and take with liquid or food.
- Extended-release tablet—Swallow whole with liquid. Do not crush or chew tablet.

When to take:
At the same times each day.

If you forget a dose:
Take as soon as you remember. If it is almost time for the next dose, wait for that dose (don't double this dose) and resume regular schedule.

What drug does:
Stimulates pancreas to produce more insulin. Insulin in blood forces cells to use sugar in blood.

Time lapse before drug works:
3 to 4 hours. May require 2 weeks for maximum benefit.

Don't take with:
Any other medicine or any dietary supplement without consulting your doctor or pharmacist.

OVERDOSE

SYMPTOMS:
Excessive hunger, nausea, anxiety, cool skin, cold sweats, drowsiness, rapid heartbeat, weakness, unconsciousness, seizure, coma.
WHAT TO DO:
- **Dial 911 for all medical emergencies or call poison control center 1-800-222-1222 for instructions.**
- **See emergency information on last 3 pages of this book.**

POSSIBLE ADVERSE REACTIONS OR SIDE EFFECTS

SYMPTOMS	WHAT TO DO
Life-threatening:	
In case of overdose, see previous column.	
Common:	
Diarrhea, appetite loss, nausea, vomiting, heartburn, constipation, changes in taste, upset stomach, dizziness, increase in urine, frequent urination, weight gain, headache.	Continue. Call doctor when convenient.
Infrequent:	
Symptoms of low blood sugar—nervousness, hunger (excessive), cold sweats, rapid pulse, anxiety, cold skin, chills, confusion, drowsiness, loss of concentration, headache, nausea, weakness, shakiness, vision changes.	Seek treatment (eat some form of quick-acting sugar—glucose tablets, sugar, fruit juice, corn syrup, honey).
Rare:	
Fatigue, itchy skin or rash, sore throat, fever, unusual.bleeding or bruising, yellow skin or eyes, swelling of legs or feet, weakness, stomach pain, mental or mood changes, sudden weight gain, seizures.	Discontinue. Call doctor right away.

WARNINGS & PRECAUTIONS

Don't take if:
You are allergic to any sulfonylurea.

Before you start, consult your doctor if:
- You have thyroid disease.
- You take insulin.
- You have kidney, liver or heart disease.

Over age 60:
Adverse reactions and side effects may be more frequent and severe than in younger persons.

Pregnancy:
Risk factors vary for drugs in this group. Always consult doctor. See category list on page xviii.

Breastfeeding; Lactation; Nursing Mothers:
Drugs in this group are generally acceptable with breastfeeding. Discuss risks and benefits with your doctor.

Infants & children up to age 18:
Safety and effectiveness in this age group have not been established. Consult doctor.

Prolonged use:
- Adverse effects more likely.
- Talk to your doctor about the need for follow-up medical examinations or laboratory studies.

Skin and sunlight:
One or more drugs in this group may cause rash or intensify sunburn in areas exposed to sun or ultraviolet light (photosensitivity reaction). Avoid overexposure. Notify doctor if reaction occurs.

Driving, piloting or hazardous work:
No problems expected unless you develop hypoglycemia (low blood sugar). If so, avoid driving or hazardous activity.

Discontinuing:
Don't discontinue without consulting doctor. Dose may require gradual reduction if you have taken drug for a long time. Doses of other drugs may also require adjustment.

Others:
- Don't exceed recommended dose. Hypoglycemia (low blood sugar) may occur, even with proper dose schedule. You must balance medicine, diet and exercise.
- May affect results in some medical tests.
- Advise any doctor, dentist or pharmacist whom you consult that you take this medicine.

POSSIBLE INTERACTION WITH OTHER DRUGS

GENERIC NAME OR DRUG CLASS	COMBINED EFFECT
Adrenocorticoids, systemic	Decreased antidiabetic effect.
Androgens*	Increased blood sugar lowering.
Anticoagulants*	Unpredictable prothrombin times.
Anticonvulsants, hydantoin	Decreased blood sugar lowering.
Antifungals, azoles	Increased blood sugar lowering.
Anti-inflammatory nonsteroidal drugs (NSAIDs)*	Increased blood sugar lowering.
Aspirin	Increased blood sugar lowering.
Beta-adrenergic blocking agents*	Increased blood sugar lowering. Possible increased difficulty in regulating blood sugar levels.
Bismuth subsalicylate	Increased insulin effect. May require dosage adjustment.
Chloramphenicol	Increased blood sugar lowering.
Cimetidine	Increased blood sugar lowering.
Clofibrate	Increased blood sugar lowering.
Contraceptives, oral*	Decreased blood sugar lowering.
Dapsone	Increased risk of adverse effect on blood cells.
Desmopressin	May increase desmopressin effect.
Dexfenfluramine	May require dosage change as weight loss occurs.
Dextrothyroxine	Sulfonylurea may require adjustment.
Digoxin	Possible decreased digoxin effect.
Diuretics* (loop, thiazide)	Decreased blood sugar lowering.
Epinephrine	Increased blood sugar lowering.
Estrogens*	Increased blood sugar lowering.
Guanethidine	Unpredictable blood sugar lowering effect.
Hemolytics*	Increased risk of adverse effect on blood cells.

Continued on page 912

POSSIBLE INTERACTION WITH OTHER SUBSTANCES

INTERACTS WITH	COMBINED EFFECT
Alcohol:	Disulfiram reaction.* Avoid.
Beverages:	None expected.
Cocaine:	Unknown. Avoid.
Foods:	None expected.
Marijuana:	Consult doctor.
Tobacco:	None expected.

TAMOXIFEN

BRAND NAMES

PMS-Tamoxifen　Soltamox

BASIC INFORMATION

Habit forming? No
Prescription needed? Yes
Available as generic? Yes
Drug class: Antineoplastic

USES

- Treats advanced breast cancer.
- Can help prevent breast cancer in those at risk.

DOSAGE & USAGE INFORMATION

How to take:

- Tablet—Swallow with liquid. If you can't swallow whole, crumble tablet and take with liquid or food. Instructions to take on empty stomach mean 1 hour before or 2 hours after eating.
- Oral solution—Follow the directions provided with the prescription.
- Enteric-coated tablet—Swallow whole. Do not crush or crumble tablet.

When to take:
Follow instructions on prescription.

If you forget a dose:
Take as soon as you remember. If it is almost time for the next dose, wait for that dose (don't double this dose) and resume regular schedule.

What drug does:
Blocks uptake of estradiol and inhibits growth of cancer cells.

Time lapse before drug works:

- 4 to 10 weeks.
- With bone metastases—several months.

Continued next column

OVERDOSE

SYMPTOMS:
Unknown.
WHAT TO DO:
If person uses much larger amount than prescribed or if accidentally swallowed, call poison control center 1-800-222-1222 for instructions or dial 911 (emergency) for help.

Don't take with:
Any other medicines (including over-the-counter drugs such as cough and cold medicines, laxatives, antacids, diet pills, caffeine, nose drops or vitamins) without consulting your doctor or pharmacist.

POSSIBLE ADVERSE REACTIONS OR SIDE EFFECTS

SYMPTOMS	WHAT TO DO
Life-threatening:	
Blood clot (leg pain and swelling and redness). Stroke (trouble with moving or speaking or seeing, numbness in face or arm, severe headache, confusion).	Seek emergency treatment immediately.
Common:	
Hot flashes or weight gain in females.	Continue. Call doctor when convenient.
Infrequent:	
Headache, dry skin, nausea, vomiting, tiredness, dizziness, loss of appetite, hair loss or thinning, impotence and loss of sexual desire in males, menstrual period changes, genital itching, bone pain, depression, weight loss, constipation.	Continue. Call doctor when convenient.
Rare:	
Vision changes, yellow skin or eyes, fever, skin symptoms (blisters, peeling, loosens), unusual bruising or bleeding, vaginal discharge or bleeding, menstrual irregularities, pelvic pain or pressure, swelling in legs or other parts of body, confusion, shortness of breath, weakness, sleepiness.	Discontinue. Call doctor right away.

WARNINGS & PRECAUTIONS

Don't take if:
You are allergic to tamoxifen. Allergic reactions are rare, but may occur.

Before you start, consult your doctor if:
- You have cataracts or other eye problem.
- You have blood disorder or have had blood clots or have had a stroke.
- You have endometriosis, irregular menstrual cycles or uterine fibroids.
- You have high cholesterol or high calcium levels or liver disorder.

Over age 60:
Adverse reactions and side effects may be more frequent and severe than in younger persons.

Pregnancy:
Consult doctor. Use of the drug during pregnancy is not recommended. It could cause harm to the unborn baby. Risk category D (see page xviii).

Breastfeeding; Lactation; Nursing Mothers:
This drug is generally not recommended with breastfeeding. Discuss risks and benefits with your doctor.

Infants & children up to age 18:
Safety and effectiveness in this age group have not been established. Consult doctor.

Prolonged use:
Talk to your doctor about the need for follow-up medical examinations (especially pelvic exams) or laboratory studies to check complete blood counts (white blood cell count, platelet count, red blood cell count, hemoglobin, hematocrit) and serum calcium.

Skin & sunlight:
No problems expected.

Driving, piloting or hazardous work:
No problems expected.

Discontinuing:
No problems expected. Consult doctor.

Others:
- Advise any doctor, dentist or pharmacist whom you consult that you take this medicine.
- May affect results in some medical tests.
- Be sure you and your doctor discuss all aspects of using this drug, and read all instructional materials.
- Drug use can increase your risk for other types of cancer. Consult doctor about your risks.
- May make you more fertile. Talk to your doctor about using some type of birth control.

POSSIBLE INTERACTION WITH OTHER DRUGS

GENERIC NAME OR DRUG CLASS	COMBINED EFFECT
Amiodarone	Decreased tamoxifen effect.
Anticoagulants*	Increased effect of anticoagulant.
Antineoplastics, * other	May increase or decrease effect of either drug.
Bromocriptine	Increased tamoxifen effect.
Cimetidine	Decreased tamoxifen effect.
Diphenhydramine	Decreased tamoxifen effect.
Estrogens*	Decreased tamoxifen effect.
Quinidine	Decreased tamoxifen effect.
Selective serotonin reuptake inhibitors (SSRIs)	Decreased effect of tamoxifen.
Thioridazine	Decreased tamoxifen effect.

POSSIBLE INTERACTION WITH OTHER SUBSTANCES

INTERACTS WITH	COMBINED EFFECT
Alcohol:	None expected.
Beverages:	None expected.
Cocaine:	Unknown. Avoid.
Foods:	None expected.
Marijuana:	Consult doctor.
Tobacco:	None expected.

*See Glossary

TAPENTADOL

BRAND NAMES

Nucynta
Nucynta ER

BASIC INFORMATION

Habit forming? Yes, with long term use
Prescription needed? Yes
Available as generic? No
Drug class: Analgesic

USES

Treatment for moderate to severe pain.

DOSAGE & USAGE INFORMATION

How to take:

- Tablet—Swallow whole with liquid. May be taken with or without food. Do not crush, chew or split tablet. Do not mix tablet into a liquid for snorting or injecting. It is for oral use only.
- Extended-release tablet—Swallow whole. If you can't swallow whole, ask doctor or pharmacist for advice. Do not crush, crumble, split or chew tablet (could lead to rapid release of drug and side effects that could be fatal).
- Oral solution—Follow doctor's instructions.

When to take:
Immediate-release form—every 4 to 6 hours as needed for pain; extended-release form—every 12 hours (or as directed for either).

If you forget a dose:
If taken on a dosing schedule, take as soon as you remember. If it is almost time for the next dose, wait for next scheduled dose (don't double this dose).

Continued next column

What drug does:
The exact way it works is unknown. It affects certain chemicals in the brain and central nervous system to help reduce both the perception of pain and the emotional response to pain.

Time lapse before drug works:
Usually within 60 minutes.

Don't take with:
Any other medicine or any dietary supplement without consulting your doctor or pharmacist.

OVERDOSE

SYMPTOMS:
Breathing difficulty, sleepiness, seizures, cold/clammy skin, blurred vision, confusion, vomiting, fainting, weak pulse, slow heartbeat, low blood pressure, stupor, coma. Deaths due to overdose have been reported with abuse and misuse of narcotic drugs, by ingesting, inhaling, or injecting crushed tablets.

WHAT TO DO:

- **Dial 911 for all medical emergencies or call poison control center 1-800-222-1222 for instructions.**
- **See emergency information on last 3 pages of this book.**

POSSIBLE ADVERSE REACTIONS OR SIDE EFFECTS

SYMPTOMS	WHAT TO DO
Life-threatening:	
Rare, severe allergic reaction (anaphylaxis)—difficulty breathing, hives, itching, swelling (throat, face), vomiting, dizziness.	Seek emergency treatment immediately.
Common:	
Constipation, nausea, headache, drowsiness, dizziness, vomiting.	Continue. Call doctor when convenient.
Infrequent:	
• Constant urge to urinate or inability to urinate, blurred vision.	Discontinue. Call doctor right away.
• Loss of appetite, stomach pain, dry mouth, weakness, confusion, sweating, diarrhea, gas, trouble sleeping, changes in mood, nervousness, heartburn, tiredness.	Continue. Call doctor when convenient.
Rare:	
Seizures, problem with balance, memory problems, shortness of breath, difficulty performing tasks, skin symptoms (itching, redness or swelling), fainting, lightheadedness when getting up from a sitting or lying position, hallucinations, fast heartbeat, speech problems, agitation, clumsiness, hot flashes, sensations in hands or feet (burning, pain, tingling, weakness, trembling or shaking).	Discontinue. Call doctor right away.

WARNINGS & PRECAUTIONS

Don't take if:
You are allergic to tapentadol, tramadol or narcotic drugs or you are sensitive to any ingredients in the drugs.

Before you start, consult your doctor if:
- You have kidney or liver disease, seizure disorder, stomach disorder, bowel blockage, thyroid disorder, pancreatitis, gallbladder problems, brain tumor or prior head injury.
- You have a history of depression, mental illness or suicidal thoughts or behaviors.
- You have a history of drug abuse or substance abuse, including alcohol abuse.
- You are severely overweight, have sleep apnea or significant curvature of the spine.
- You have any lung or breathing problem.

Over age 60:
Adverse reactions and side effects may be more frequent and severe than in younger persons.

Pregnancy:
Decide with your doctor if drug benefits justify risk to unborn baby. Risk category C (see page xviii).

Breastfeeding; Lactation; Nursing Mothers:
This drug is generally not recommended with breastfeeding. Discuss risks and benefits with your doctor.

Infants & children up to age 18:
Safety and effectiveness in this age group have not been established. Consult doctor.

Prolonged use:
- Consult with your doctor on a regular basis while using this drug.
- Can cause drug dependence, addiction and withdrawal symptoms.

Skin & sunlight:
No special problems expected.

Driving, piloting or hazardous work:
Don't drive or pilot aircraft until you learn how medicine affects you. Don't work around dangerous machinery. Don't climb ladders or work in high places. Danger increases if you drink alcohol or take medicine affecting alertness and reflexes.

Discontinuing:
If you have taken this drug for a long time, consult with your doctor before discontinuing. Withdrawal symptoms may occur if drug is discontinued abruptly. Symptoms include anxiety, sweating, insomnia, pain, nausea, tremors, diarrhea, breathing problems and hallucinations. Call doctor if symptoms occur.

Others:
- Drug can cause constipation. Consult doctor before using a laxative or stool softener to treat or prevent this side effect.
- Don't increase dosage or frequency of use without your doctor's approval.
- Development of a potentially life-threatening serotonin syndrome* or overdose (which can be fatal) may occur with the use of this drug.
- Do not use within 14 days of taking monoamine oxidase (MAO) inhibitors.*
- Advise any doctor, dentist or pharmacist whom you consult that you take this medicine.

POSSIBLE INTERACTION WITH OTHER DRUGS

GENERIC NAME OR DRUG CLASS	COMBINED EFFECT
Central nervous system (CNS) depressants*	Serious adverse events including breathing difficulty.
Monoamine oxidase (MAO) inhibitors*	Serious adverse events including seizures and serotonin syndrome.*
Narcotics*	Breathing difficulty and increased risk of side effects and seizures.
Serotonergics*	Increased risk of seizures and serotonin syndrome.*

POSSIBLE INTERACTION WITH OTHER SUBSTANCES

INTERACTS WITH	COMBINED EFFECT
Alcohol:	Serious side effects. Avoid.
Beverages:	None expected.
Cocaine:	Unknown. Avoid.
Foods:	None expected.
Marijuana:	Consult doctor.
Tobacco:	None expected.

TEGASEROD

BRAND NAMES

Zelnorm

BASIC INFORMATION

Habit forming? No
Prescription needed? Yes
Available as generic? No
Drug class: Gastrointestinal serotonin receptor agonist

USES

- Treatment for women with constipation-predominant irritable bowel syndrome (IBS or IBS-C). IBS is a common gastrointestinal disorder affecting women more often than men. Symptoms include abdominal pain and discomfort, bloating, constipation or diarrhea.
- Treatment for chronic constipation.
- May be used for treatment of other disorders as determined by your doctor.

Note: This drug is available in the U.S. through a special distribution program. Consult your doctor.

DOSAGE & USAGE INFORMATION

How to take:
Tablet—Take with water on an empty stomach.

When to use:
Twice daily just before a meal (usually morning and evening). Follow prescription instructions.

If you forget a dose:
Take the missed dose just before your next meal (don't double this dose). Then resume your regular schedule.

What drug does:
Helps in the movement of stools through the bowels and prevents constipation by activating certain body nerve cells (5-HT4 receptors) located in the stomach and intestines.

Continued next column

OVERDOSE

SYMPTOMS:
Diarrhea, headache, abdominal pain, nausea, vomiting, gaseousness.
WHAT TO DO:
If person uses much larger amount than prescribed or if accidentally swallowed, call poison control center 1-800-222-1222 for instructions or dial 911 (emergency) for help.

Time lapse before drug works:
Symptoms may improve within a day to a week, but may take up to 4 weeks to determine if the drug is effective in controlling the IBS symptoms.

Don't take with:
Any other medicine or any dietary supplement without consulting your doctor or pharmacist.

POSSIBLE ADVERSE REACTIONS OR SIDE EFFECTS

SYMPTOMS	WHAT TO DO
Life-threatening:	
None expected.	
Common:	
Diarrhea (may stop after one episode), mild abdominal pain, nausea, headache, gaseousness.	Continue. Call doctor when convenient.
Infrequent:	
Dizziness, flu or cold symptoms, back pain, feeling of fullness.	Continue. Call doctor when convenient.
Rare:	
Increased severity or worsening of symptoms of abdominal pain or diarrhea (may be bloody), rectal bleeding, dizziness, lightheadedness, fainting.	Discontinue. Call doctor right away.

WARNINGS & PRECAUTIONS

Don't take if:
You are allergic to tegaserod.

Before you start, consult your doctor if:
- You have liver or kidney disease.
- You often have diarrhea.
- You have bowel disease, bowel obstruction, gallbladder disease with symptoms, sphincter disorder, or abdominal adhesions.

Over age 60:
Not approved for patients over age 55.

Pregnancy:
Decide with your doctor if drug benefits justify risk to unborn baby. Risk category B (see page xviii).

Breastfeeding; Lactation; Nursing Mothers:
This drug is generally acceptable with breastfeeding. Discuss risks and benefits with your doctor.

Infants & children up to age 18:
Safety and effectiveness in this age group have not been established. Consult doctor.

Prolonged use:
The drug is indicated for short term use. Caution is advised for its use longer than 3 months since safety has not been established.

Skin & sunlight:
No problems expected.

Driving, piloting or hazardous work:
No problems expected.

Discontinuing:
No problems expected, but consult your doctor before discontinuing the drug.

Others:
- There has been a small increase in abdominal surgeries in patients taking this drug. It is not known if the drug is involved in the increase. There is some concern that the drug may cause or worsen ovarian cysts. Be sure to discuss any questions with your doctor.
- Advise any doctor, dentist or pharmacist whom you consult that you take this medicine.
- The drug's effectiveness for men with IBS has not been established.
- Follow your doctor's advice and recommendations for any additional treatment steps for IBS. They may involve stress reduction, relaxation techniques, dietary changes and others.
- Take medicine only as directed. Do not increase or reduce dosage without doctor's approval. Have regular medical follow-up while taking this drug.
- Recommended therapy is 4-6 weeks with an additional 4-6 weeks if patient has good response.

POSSIBLE INTERACTION WITH OTHER DRUGS

GENERIC NAME OR DRUG CLASS	COMBINED EFFECT
None expected.	

POSSIBLE INTERACTION WITH OTHER SUBSTANCES

INTERACTS WITH	COMBINED EFFECT
Alcohol:	None expected.
Beverages:	None expected.
Cocaine:	Unknown. Avoid.
Foods:	None expected.
Marijuana:	Consult doctor.
Tobacco:	None expected.

TELITHROMYCIN

BRAND NAMES

Ketek

BASIC INFORMATION

Habit forming? No
Prescription needed? Yes
Available as generic? No
Drug class: Antibacterial; antibiotic (ketolide)

USES

Treatment for pneumonia of mild to moderate severity.

DOSAGE & USAGE INFORMATION

How to take:
Tablet—Swallow with a full glass of water. It may be taken with or without food. Do not crush or chew tablet.

When to take:
Once a day at the same time each day.

If you forget a dose:
Take as soon as you remember. If it is almost time for the next dose, wait for that dose (don't double this dose) and resume regular schedule.

What drug does:
Prevents growth and reproduction of susceptible bacteria.

Time lapse before drug works:
Starts working in a few hours, but takes 5 to 10 days to cure the infection being treated.

Don't take with:
Any other medicine or any dietary supplement without consulting your doctor or pharmacist.

OVERDOSE

SYMPTOMS:
Unknown.
WHAT TO DO:
If person uses much larger amount than prescribed or if accidentally swallowed, call poison control center 1-800-222-1222 for instructions or dial 911 (emergency) for help.

POSSIBLE ADVERSE REACTIONS OR SIDE EFFECTS

SYMPTOMS	WHAT TO DO
Life-threatening:	
Rare, severe allergic reaction (anaphylaxis)—difficulty breathing, hives, itching, swelling (throat, face), vomiting, dizziness.	Seek emergency treatment immediately.
Common:	
Diarrhea, nausea.	Continue. Call doctor when convenient.
Infrequent:	
• Vision changes (blurred or difficulty in focusing or double vision).	Continue, but call doctor right away.
• Abdominal discomfort, belching, heartburn, vomiting, tingling or numbness or prickling sensations, changes in vaginal discharge, headache, dry lips or skin, flushing or red or warm skin, pale urine, urination increased or more frequent, genital itching, appetite loss, changes in sense of smell or taste, facial pain or tenderness or swelling, insomnia, mood changes, confusion, dizziness or feeling lightheaded.	Continue. Call doctor when convenient.
Rare:	
Stomach or chest, pain, breathing or swallowing difficulty, fainting, fever or chills, heartbeat is irregular or fast or slow, urine is dark or decreased, diarrhea is bloody or watery, yellow skin or eyes, unusual weakness or tiredness, unusual bleeding or bruising, joint or muscle pain, sores or patches on mouth or tongue or throat, skin symptoms (rash, peeling, itching blistering).	Discontinue. Call doctor right away.

WARNINGS & PRECAUTIONS

Don't take if:
- You are allergic to ketolide or macrolide antibiotics or take cisapride or pimozide.
- You have myasthenia gravis.

Before you start, consult your doctor if:
- You have kidney or liver problems.
- You have hepatitis or jaundice.
- You have heart rhythm problems or a slow heartbeat (bradycardia).
- You have low potassium or low magnesium levels in your blood.
- You are allergic to any medication, food or other substance.

Over age 60:
No special problems expected.

Pregnancy:
Decide with your doctor if drug benefits justify risk to unborn baby. Risk category C (see page xviii).

Breastfeeding; Lactation; Nursing Mothers:
This drug is generally acceptable with breastfeeding. Discuss risks and benefits with your doctor.

Infants & children up to age 18:
Safety and effectiveness in this age group have not been established. Consult doctor.

Prolonged use:
Drug is discontinued once the infection is cured.

Skin & sunlight:
No special problems expected.

Driving, piloting or hazardous work:
Avoid if you experience vision problems or dizziness. Otherwise, no problems expected.

Discontinuing:
Don't discontinue without doctor's advice until you complete prescribed dose, even though symptoms diminish or disappear.

Others:
- Advise any doctor, dentist or pharmacist whom you consult that you take this medicine.
- May affect the results of some medical tests.

POSSIBLE INTERACTION WITH OTHER DRUGS

GENERIC NAME OR DRUG CLASS	COMBINED EFFECT
Antiarrhythmics	Risk of heart rhythm problems.
Antifungals, azoles	Increased effect of telithromycin.
Benzodiazepines*	Increased effect of benzodiazepine.
Cisapride (this drug has limited availability)	Heart problem risk. Avoid.
Digoxin	Increased effect of digoxin.
Enzyme inducers*	Decreased effect of enzyme inducer or telithromycin.
Enzyme inhibitors*	Increased effect of enzyme inhibitor or telithromycin.
Ergot preparations*	Toxicity risk of ergot preparation. Avoid.
HMG-CoA reductase inhibitors	Increased effect of HMG-CoA reductase inhibitor.
Metoprolol	Increased risk of side effects.
Pimozide	Heart problem risk. Avoid.
QT interval prolongation-causing drugs*	Effects not clear. Use with caution.
Rifampin	Decreased effect of rifampin. Avoid.
Ritonavir	Increased effect of ritonavir.
Sirolimus	Increased effect of sirolimus.
Sotalol	Decreased effect of sotalol.
Tacrolimus	Increased effect of tacrolimus.
Theophylline	Increased risk of side effects. Take one hour apart.
Warfarin	Risk of side effects—bleeding or bruising.

POSSIBLE INTERACTION WITH OTHER SUBSTANCES

INTERACTS WITH	COMBINED EFFECT
Alcohol:	None expected.
Beverages:	None expected.
Cocaine:	Unknown. Avoid.
Foods:	None expected.
Marijuana:	Consult doctor.
Tobacco:	None expected.

***See Glossary**

TERBINAFINE (Oral)

BRAND NAMES

Lamisil Oral Granules
Lamisil Tablets

BASIC INFORMATION

Habit forming? No
Prescription needed? Yes
Available as generic? Yes
Drug class: Antifungal

USES

- Treats fungal infection (called onychomycosis) of the toenails or fingernails.
- May be used for other fungal infections as determined by your doctor.
- Note: Another form of this drug is applied topically to the skin. For information, see the drug chart for Antifungals (Topical).

DOSAGE & USAGE INFORMATION

How to take:
- Tablet—Swallow with liquid. May be taken with or without food. If you can't swallow whole, crush tablet and take with liquid or food.
- Oral granules—Sprinkle on food. Follow instructions on prescription on what foods to use.

When to take:
- It is usually taken once a day at the same time each day for a period of weeks to months.
- It may be taken daily for one week per month for 2, 3 or 4 months (called pulse therapy).

If you forget a dose:
Take as soon as you remember. If it is almost time for the next dose, wait for that dose (don't double this dose) and resume regular schedule.

Continued next column

OVERDOSE

SYMPTOMS:
May include nausea, vomiting, abdominal pain, dizziness, rash, frequent urination and headache.
WHAT TO DO:
- **Dial 911 for all medical emergencies or call poison control center 1-800-222-1222 for instructions.**
- **See emergency information on last 3 pages of this book.**

What drug does:
During the weeks of treatment, the drug will slowly kill the fungus infecting the nail. In many cases, the nail will then grow out normally.

Time lapse before drug works:
Improvement in symptoms may be seen in days or weeks, but fungus infections can be very slow to clear up. The drug is usually prescribed for 6 weeks for fingernail infections and 12 weeks or longer for toenail infections.

Don't take with:
Any other medicine or diet supplement without consulting your doctor or pharmacist.

POSSIBLE ADVERSE REACTIONS OR SIDE EFFECTS

SYMPTOMS	WHAT TO DO
Life-threatening:	
Rare, severe allergic reaction (anaphylaxis)—difficulty breathing, hives, itching, swelling (throat, face), vomiting, dizziness.	Seek emergency treatment immediately.
Common:	
Nausea, vomiting, mild stomach pain, diarrhea, feeling of fullness, mild appetite loss, headache.	Continue. Call doctor when convenient.
Infrequent:	
• Skin rash, itching.	Discontinue. Call doctor right away.
• Taste changes or loss of taste.	Continue. Call doctor when convenient.
Rare:	
• Nausea or vomiting that is persistent, unusual tiredness or weakness, stomach pain or appetite loss that is more severe, yellow eyes or skin, dark urine, pale stools, fever, chills, sore throat, aching muscles, skin symptoms (redness, blistering, peeling, loosening), unusual bleeding or bruising.	Discontinue. Call doctor right away.
• Hair loss.	Continue. Call doctor when convenient.

WARNINGS & PRECAUTIONS

Don't take if:
You are allergic to terbinafine.

Before you start, consult your doctor if:
- You have chronic or active liver disease or kidney disease.
- You have lupus erythematosus.
- You are an alcoholic.

Over age 60:
No special problems expected.

Pregnancy:
Decide with your doctor if drug benefits justify risk to unborn baby. Risk category B (see page xviii).

Breastfeeding; Lactation; Nursing Mothers:
This drug is generally acceptable with breastfeeding. Discuss risks and benefits with your doctor.

Infants & children up to age 18:
Safety and effectiveness in this age group have not been established. Consult doctor.

Prolonged use:
Use of the drug is usually limited to weeks or months of treatment.

Skin & sunlight:
No problems expected.

Driving, piloting or hazardous work:
No problems expected.

Discontinuing:
Don't discontinue the drug without doctor's advice until you complete prescribed dose, even though symptoms diminish or disappear.

Others:
- Periodic laboratory blood studies and liver and kidney function tests may be recommended.
- Be patient and persistent in following the treatment regimen for nail care. Consult your doctor if symptoms do not improve within a few weeks (or months for onychomycosis), or if they become worse.
- Taste changes caused by the drug usually improve within several weeks after stopping the drug, but a few cases may last a year or more.
- Very rarely, severe complications may occur with use of this drug. They include serious (and possibly fatal) liver disease, an abnormal blood disorder and Stevens-Johnson syndrome (a severe allergic skin reaction).
- Advise any doctor, dentist or pharmacist whom you consult that you take this medicine.

POSSIBLE INTERACTION WITH OTHER DRUGS

GENERIC NAME OR DRUG CLASS	COMBINED EFFECT
Caffeine	Increased effect of caffeine.
Enzyme inducers*	May decrease effect of terbinafine or enzyme inducer.
Enzyme inhibitors*	May increase effect of terbinafine or enzyme inhibitor.
Warfarin	May affect blood clotting ability.

POSSIBLE INTERACTION WITH OTHER SUBSTANCES

INTERACTS WITH	COMBINED EFFECT
Alcohol:	Increased risk of side effects. Avoid.
Beverages:	None expected.
Cocaine:	Unknown. Avoid.
Foods:	None expected.
Marijuana:	Consult doctor.
Tobacco:	None expected.

*See Glossary

TETRACYCLINES

GENERIC AND BRAND NAMES

See full list of generic and brand names in the *Generic and Brand Name Directory*, page 881.

BASIC INFORMATION

Habit forming? No
Prescription needed? Yes
Available as generic? Yes
Drug class: Antibacterial, antiacne

USES

- Treatment for bacterial infections susceptible to any tetracycline. Not used for viruses.
- Treatment for acne, ulcers and used as diuretic.
- Treatment for dental bacterial infections.

DOSAGE & USAGE INFORMATION

How to take:

- Tablet or capsule—Take on empty stomach 1 hour before or 2 hours after eating. If you can't swallow whole, crumble tablet or open capsule and take with liquid or food.
- Delayed-release capsule or extended-release tablet or capsule—Swallow whole with liquid. Do not open, crush or chew.
- Dental product—Follow package instructions.
- Liquid—Shake well. Take with measuring spoon.

When to take:
At the same times each day, evenly spaced.

If you forget a dose:
Take as soon as you remember. If it is almost time for the next dose, wait for that dose (don't double this dose) and resume regular schedule.

What drug does:
Prevents bacteria growth and reproduction.

Time lapse before drug works:

- Infections—May require 5 days to affect infection.
- Acne—May require 4 weeks to affect acne.

Continued next column

OVERDOSE

SYMPTOMS:
Severe nausea, vomiting, diarrhea.
WHAT TO DO:
If person uses much larger amount than prescribed or if accidentally swallowed, call poison control center 1-800-222-1222 for instructions or dial 911 (emergency) for help.

Don't take with:
Any other medicine or any dietary supplement without consulting your doctor or pharmacist.

POSSIBLE ADVERSE REACTIONS OR SIDE EFFECTS

SYMPTOMS	WHAT TO DO
Life-threatening:	
Rare, severe allergic reaction (anaphylaxis)—difficulty breathing, hives, itching, swelling (throat, face), vomiting, dizziness.	Seek emergency treatment immediately.
Common:	
• Increased sensitivity to sunlight.	Discontinue. Call doctor right away.
• Mild stomach cramps, diarrhea; with minocycline—being unsteady, dizziness.	Continue. Call doctor when convenient.
Infrequent:	
• With demeclocycline —urination more frequent or increased, more thirst, unusual tiredness or weakness; with minocycline—darker or changed color of skin and mucous membranes.	Discontinue. Call doctor right away.
• Sore mouth or tongue, rectal or genital itch.	Continue. Call doctor when convenient.
• Darkened tongue (will go away when drug is discontinued).	No action necessary.
Rare:	
Changes in vision, yellow skin or eyes, continued vomiting, severe stomach cramps, loss of appetite, ongoing headache, bulging fontanel (soft spot on head of infant).	Discontinue. Call doctor right away.

WARNINGS & PRECAUTIONS

Don't take if:
You are allergic to any tetracycline antibiotic.

Before you start, consult your doctor if:

- You have kidney or liver disease.
- You have lupus or myasthenia gravis.

Over age 60:
No special problems expected.

Pregnancy:
Consult doctor. Use of the drug during pregnancy is not recommended. It could cause harm to the unborn baby. Risk category D (see page xviii).

Breastfeeding; Lactation; Nursing Mothers:
Drugs in this group are generally acceptable with breastfeeding. Discuss risks and benefits with your doctor.

Infants & children up to age 18:
Follow instructions provided by your child's doctor. May cause permanent teeth discoloration and slow bone growth in ages 8 and under.

Prolonged use:
Consult doctor on a regular basis to determine continued effectiveness of drug and check for adverse effects.

Skin & sunlight:
May cause rash or intensify sunburn in areas exposed to sun or ultraviolet light (photosensitivity reaction). Avoid overexposure. Notify doctor if reaction occurs.

Driving, piloting or hazardous work:
No problems expected.

Discontinuing:
Don't discontinue without doctor's advice until you complete prescribed dose, even though symptoms diminish or disappear.

Others:
- May affect results in some medical tests.
- Birth control pill may not be effective. Use additional birth control method.
- Advise any doctor, dentist or pharmacist whom you consult that you take this medicine.

POSSIBLE INTERACTION WITH OTHER DRUGS

GENERIC NAME OR DRUG CLASS	COMBINED EFFECT
Antacids*	Decreased tetracycline effect.
Anticoagulants,* oral	Increased anticoagulant effect.
Antivirals, HIV/AIDS*	Decreased antibiotic effect.
Bismuth subsalicylate	Decreased tetra-cycline absorption.
Calcium supplements*	Decreased tetracycline effect.
Cefixime	Decreased antibiotic effect of cefixime.
Cholestyramine or colestipol	Decreased tetracycline effect.
Contraceptives, oral*	Decreased contraceptive effect.
Desmopressin	Possible decreased desmopressin effect.
Digitalis preparations*	Increased digitalis effect.
Etretinate	Increased chance of adverse reactions of etretinate.
Lithium	Increased lithium effect.
Mineral supplements* (iron, calcium, magnesium, zinc)	Decreased tetracycline absorption. Separate doses by 1 to 2 hours.
Penicillins*	Decreased penicillin effect.
Sodium bicarbonate	Reduced tetracycline absorption.
Tiopronin	Increased risk of toxicity to kidneys (except with doxycycline and minocycline).

Continued on page 913

POSSIBLE INTERACTION WITH OTHER SUBSTANCES

INTERACTS WITH	COMBINED EFFECT
Alcohol:	Possible liver damage. Avoid.
Beverages: Milk.	Decreased tetracycline absorption. Take dose 2 hours after or 1 hour before drinking.
Cocaine:	Unknown. Avoid.
Foods: Dairy products.	Decreased tetracycline absorption. Take dose 2 hours after or 1 hour before eating.
Marijuana:	Consult doctor.
Tobacco:	None expected.

THIAMINE (Vitamin B-1)

BRAND NAMES

Betalin S
Betaxin
Bewon
Biamine

Numerous brands of single vitamin and multivitamin combinations may be available.

BASIC INFORMATION

Habit forming? No
Prescription needed? No
Available as generic? Yes
Drug class: Vitamin supplement

USES

- Dietary supplement to promote normal growth, development and health.
- Treatment for beri-beri (a thiamine-deficiency disease).
- Dietary supplement for alcoholism, cirrhosis, overactive thyroid, infection, breastfeeding, absorption diseases, pregnancy, prolonged diarrhea, burns.

DOSAGE & USAGE INFORMATION

How to take:
Tablet or liquid—Swallow with beverage or food to lessen stomach irritation.

When to take:
At the same time each day.

If you forget a dose:
Take as soon as you remember. If it is almost time for the next dose, wait for that dose (don't double this dose) and resume regular schedule.

What drug does:
- Promotes normal growth and development.
- Combines with an enzyme to metabolize carbohydrates.

Time lapse before drug works:
15 minutes.

Don't take with:
Any other medicine or any dietary supplement without consulting your doctor or pharmacist.

OVERDOSE

SYMPTOMS:
Unknown.
WHAT TO DO:
If person uses much larger amount than prescribed or if accidentally swallowed, call poison control center 1-800-222-1222 for instructions or dial 911 (emergency) for help.

POSSIBLE ADVERSE REACTIONS OR SIDE EFFECTS

SYMPTOMS	WHAT TO DO
Life-threatening:	
Rare, severe allergic reaction (anaphylaxis)—difficulty breathing, hives, itching, swelling (throat, face), vomiting, dizziness.	Seek emergency treatment immediately.
Common:	
None expected with normal doses.	
Infrequent:	
None expected with normal doses.	
Rare:	
None expected with normal doses.	

WARNINGS & PRECAUTIONS

Don't take if:
You are allergic to any B vitamin.

Before you start, consult your doctor if:
You have liver or kidney disease.

Over age 60:
No problems expected.

Pregnancy:
Consult doctor. Use is generally considered safe. Risk category A (see page xviii).

Breastfeeding; Lactation; Nursing Mothers:
This drug is generally acceptable with breastfeeding. Discuss risks and benefits with your doctor.

Infants & children up to age 18:
Read nonprescription product's label to be sure it is approved for your child's age. Follow label instructions on dosage. Consult doctor or pharmacist if unsure.

Prolonged use:
No problems expected.

Skin & sunlight:
No problems expected.

Driving, piloting or hazardous work:
No problems expected.

Discontinuing:
No problems expected.

Others:
A balanced diet should provide enough thiamine for healthy people to make a supplement unnecessary. Best dietary sources of thiamine are whole-grain cereals and meats.

POSSIBLE INTERACTION WITH OTHER DRUGS

GENERIC NAME OR DRUG CLASS	COMBINED EFFECT
Barbiturates*	Decreased thiamine effect.

POSSIBLE INTERACTION WITH OTHER SUBSTANCES

INTERACTS WITH	COMBINED EFFECT
Alcohol:	None expected.
Beverages: Carbonates, citrates (additives listed on many beverage labels).	Decreased thiamine effect.
Cocaine:	Unknown. Avoid.
Foods: Carbonates, citrates (additives listed on many food labels).	Decreased thiamine effect.
Marijuana:	Consult doctor.
Tobacco:	None expected.

***See Glossary**

THIAZOLIDINEDIONES

GENERIC AND BRAND NAMES

PIOGLITAZONE	ROSIGLITAZONE
Actos	Avandamet
ACTOplus Met	Avandaryl
ACTOplus Met XR	Avandia
Duetact	
Oseni	

BASIC INFORMATION

Habit forming? No
Prescription needed? Yes
Available as generic? Yes, for some
Drug class: Antidiabetic

USES

Treatment for type 2 diabetes. Rosiglitazone or pioglitazone may be used alone, with insulin or with other antidiabetic drugs. These drugs do not cure diabetes.

DOSAGE & USAGE INFORMATION

How to take:

- Tablet—Swallow with liquid. Take at mealtime. If you can't swallow whole, crumble tablet and take with liquid or food.
- Extended-release tablet—Swallow whole with liquid. May be taken with or without food. Do not crush, chew or split tablet.

When to take:
Once a day or as directed by doctor. Dosage may be increased after several weeks.

Continued next column

OVERDOSE

SYMPTOMS:
Symptoms of hypoglycemia—stomach pain, anxious feeling, cold sweats, chills, confusion, convulsions, cool pale skin, excessive hunger, nausea or vomiting, rapid heartbeat, nervousness, shakiness, unsteady walk, unusual weakness or tiredness, vision changes, unconsciousness.
WHAT TO DO:

- **For mild hypoglycemia symptoms, drink or eat something with sugar right away.**
- **Dial 911 for all medical emergencies or call poison control center 1-800-222-1222 for instructions.**
- **See emergency information on last 3 pages of this book.**

If you forget a dose:
Wait for your next meal that same day and take dose then. If you forget until the next day, take that day's regular dose on schedule (don't double this dose).

What drug does:
Lowers blood glucose by increasing body's response to insulin.

Time lapse before drug works:
May take several weeks for full effectiveness.

Don't take with:
Any other medicine or any dietary supplement without consulting your doctor or pharmacist.

POSSIBLE ADVERSE REACTIONS OR SIDE EFFECTS

SYMPTOMS	WHAT TO DO
Life-threatening:	
In case of overdose or low blood sugar, see previous column.	
Common:	
• Pain in back or other body part, infection.	Continue. Call doctor right away.
• Headache, dizziness, nausea, unusual tiredness or weakness.	Continue. Call doctor when convenient.
Infrequent:	
Sore throat, runny nose, diarrhea.	Continue. Call doctor when convenient.
Rare:	
• Severe low blood sugar (see symptoms under Overdose).	Discontinue. Call doctor right away or seek emergency care.
• Liver problems including jaundice (yellow skin or eyes) and hepatitis that could lead to liver transplantation or death; symptoms of heart failure (excessive and rapid weight gain, shortness of breath, swelling of legs or feet) after starting drug.	Discontinue. Call doctor right away.

WARNINGS & PRECAUTIONS

Don't take if:
You are allergic to any of the thiazolidinediones. Allergic reactions are rare, but may occur

Before you start, consult your doctor if:

- You have liver disease or any heart disorder.
- You have any chronic health problem.

- You have a history of acid in the blood (metabolic acidosis or ketoacidosis).
- You are allergic to any medication, food or other substance.

Over age 60:
No special problems expected.

Pregnancy:
Decide with your doctor if drug benefits justify risk to unborn baby. Risk category C (see page xviii).

Breastfeeding; Lactation; Nursing Mothers:
Drugs in this group are generally acceptable with breastfeeding. Discuss risks and benefits with your doctor.

Infants & children up to age 18:
Safety and effectiveness in this age group have not been established. Consult doctor.

Prolonged use:
- Schedule regular doctor visits to determine if the drug is continuing to be effective in controlling the diabetes and to check for any liver function problems.
- You will most likely require an antidiabetic medicine for the rest of your life.
- You will need to test your blood glucose levels several times a day, or for some, once to several times a week.

Skin & sunlight:
No special problems expected.

Driving, piloting or hazardous work:
No special problems expected.

Discontinuing:
Don't discontinue without consulting your doctor even if you feel well. You can have diabetes without feeling any symptoms. When untreated, diabetes can cause serious problems.

Others:
- Use of these drugs may lead to liver problems. Currently, it is necessary to get liver function studies prior to starting the drug, then every other month for 6 months and periodically thereafter while on the drug.
- Use of these drugs can cause fluid retention which may lead to or worsen chronic heart failure.
- Advise any doctor or dentist whom you consult that you take this medicine. It may interfere with the accuracy of some medical tests.
- May cause ovulation to resume in some women with ovarian disorders. Discuss the need for nonhormonal contraception with your doctor.
- Follow any special diet your doctor may prescribe. It can help control diabetes.
- Consult doctor if you become ill with vomiting or diarrhea while taking this drug.
- Use caution when exercising. Ask your doctor about an appropriate exercise program.
- Wear or carry medical identification stating that you have diabetes and take this drug.
- Learn to recognize the symptoms of low blood sugar. You and your family need to know what to do if these symptoms occur.
- Have a glucagon kit and syringe in the event severe low blood sugar occurs.
- Use of these drugs may raise both HDL and LDL cholesterol levels.
- High blood sugar (hyperglycemia) may occur with diabetes. Ask your doctor about symptoms to watch for and treatment steps to take.
- Educate yourself about diabetes.

POSSIBLE INTERACTION WITH OTHER DRUGS

GENERIC NAME OR DRUG CLASS	COMBINED EFFECT
Antidiabetic agents, sulfonylurea	May decrease fasting plasma glucose concentrations.
Contraceptives, oral*	Decreased effect of contraceptive.
Cyclosporine	Decreased effect of cyclosporine.
HMG-CoA reductase inhibitors	Decreased effect of HMG-CoA reductase inhibitor.
Tacrolimus	Decreased effect of tacrolimus.

POSSIBLE INTERACTION WITH OTHER SUBSTANCES

INTERACTS WITH	COMBINED EFFECT
Alcohol:	No special problems. Avoid excessive amounts of alcohol.
Beverages:	None expected.
Cocaine:	Unknown. Avoid.
Foods:	None expected.
Marijuana:	Consult doctor.
Tobacco:	None expected. People with diabetes should not smoke.

*See Glossary

THIOGUANINE

GENERIC NAMES

THIOGUANINE

BASIC INFORMATION

Habit forming? No
Prescription needed? Yes
Available as generic? Yes
Drug class: Antineoplastic

USES

- Treats some forms of leukemia.
- Used for other disorders as determined by doctor.

DOSAGE & USAGE INFORMATION

How to take:
Tablet—Swallow with liquid. If you can't swallow whole, crumble tablet and take with liquid or food. Instructions to take on empty stomach mean 1 hour before or 2 hours after eating.

When to take:
According to doctor's instructions.

If you forget a dose:
Take as soon as you remember. If it is almost time for the next dose, wait for that dose (don't double this dose) and resume regular schedule.

What drug does:
Interferes with growth of cancer cells.

Time lapse before drug works:
Varies greatly among patients.

Don't take with:
Any other medicines (including over-the-counter drugs such as cough and cold medicines, laxatives, antacids, diet pills, caffeine, nose drops or vitamins) without consulting your doctor or pharmacist.

OVERDOSE

SYMPTOMS:
Unknown.
WHAT TO DO:
If person uses much larger amount than prescribed or if accidentally swallowed, call doctor or poison control center 1-800-222-1222 for help.

POSSIBLE ADVERSE REACTIONS OR SIDE EFFECTS

SYMPTOMS	WHAT TO DO
Life-threatening:	
Rare, severe allergic reaction (anaphylaxis)—difficulty breathing, hives, itching, swelling (throat, face), vomiting, dizziness.	Seek emergency treatment immediately.
Common:	
• Appetite loss, diarrhea, skin rash.	Continue. Call doctor when convenient.
• Nausea.	Continue. Call doctor if symptom persists.
Infrequent:	
Bloody urine, hoarseness or cough, fever or chills, lower back or side pain, painful or difficult urination, red spots on skin, unusual bleeding or bruising, joint pain, swollen feet and legs, unsteady gait, black or tarry stools.	Discontinue. Call doctor right away.
Rare:	
Mouth and lip sores, jaundice (yellow skin and eyes).	Discontinue. Call doctor right away.

WARNINGS & PRECAUTIONS

Don't take if:
- You are allergic to thioguanine.
- You have chicken pox or shingles.

Before you start, consult your doctor if:
- You have gout.
- You have an infection.
- You have kidney or liver disease.
- You have had radiation or cancer chemotherapy within 6 weeks.

Over age 60:
Adverse reactions and side effects may be more frequent and severe than in younger persons.

Pregnancy:
Consult doctor. Use of the drug during pregnancy is not recommended. It could cause harm to the unborn baby. Risk category D (see page xviii).

Breastfeeding; Lactation; Nursing Mothers:
This drug is generally not recommended with breastfeeding. Discuss risks and benefits with your doctor.

Infants & children up to age 18:
Follow instructions provided by your child's doctor.

Prolonged use:
- Increased likelihood of side effects.
- Talk to your doctor about the need for follow-up medical examinations or laboratory studies to check kidney and liver function, serum uric acid, and complete blood counts (white blood cell count, platelet count, red blood cell count, hemoglobin, hematocrit).

Skin & sunlight:
No problems expected.

Driving, piloting or hazardous work:
No problems expected.

Discontinuing:
Report to your doctor any of these symptoms that occur after discontinuing: black or tarry stools, bloody urine, hoarseness or cough, fever or chills, lower back or side pain, painful or difficult urination, red spots on skin, unusual bleeding or bruising.

Others:
- Advise any doctor, dentist or pharmacist whom you consult that you take this medicine.
- May affect results in some medical tests.
- Use an effective form of birth control. Consult doctor for advice.

POSSIBLE INTERACTION WITH OTHER DRUGS

GENERIC NAME OR DRUG CLASS	COMBINED EFFECT
Antigout drugs*	May need increased antigout dosage.
Bone marrow depressants,* other	Increased risk of bone marrow depression.
Vaccines, live or killed	Increased risk of toxicity or reduced effectiveness of vaccine.
Zidovudine	More likelihood of toxicity of both drugs.

POSSIBLE INTERACTION WITH OTHER SUBSTANCES

INTERACTS WITH	COMBINED EFFECT
Alcohol:	Increased side effects.
Beverages:	None expected.
Cocaine:	Increased risk of side effects.
Foods:	None expected.
Marijuana:	Consult doctor.
Tobacco:	None expected.

THIOTHIXENE

BRAND NAMES

Navane
Thiothixene HCl Intensol

BASIC INFORMATION

Habit forming? No
Prescription needed? Yes
Available as generic? Yes
Drug class: Antipsychotic (thioxanthene)

USES

Reduces anxiety, agitation, psychosis.

DOSAGE & USAGE INFORMATION

How to take:

- Capsule—Swallow with liquid. If you can't swallow whole, open capsule and take with liquid or food.
- Syrup—Dilute dose in beverage before swallowing.

When to take:
At the same times each day.

If you forget a dose:
Take as soon as you remember. If it is almost time for the next dose, wait for that dose (don't double this dose) and resume regular schedule.

What drug does:
Corrects imbalance of nerve impulses.

Time lapse before drug works:
3 weeks.

Don't take with:
Any other medicine or any dietary supplement without consulting your doctor or pharmacist.

OVERDOSE

SYMPTOMS:
Drowsiness, dizziness, weakness, muscle rigidity, twitching, tremors, confusion, dry mouth, blurred vision, rapid pulse, shallow breathing, low blood pressure, walking problems, coma.
WHAT TO DO:

- **Dial 911 for all medical emergencies or call poison control center 1-800-222-1222 for instructions.**
- **See emergency information on last 3 pages of this book.**

POSSIBLE ADVERSE REACTIONS OR SIDE EFFECTS

SYMPTOMS	WHAT TO DO
Life-threatening:	
Rare, severe allergic reaction (anaphylaxis)—difficulty breathing, hives, itching, swelling (throat, face), vomiting, dizziness.	Seek emergency treatment immediately.
Common:	
• Jerky or involuntary movements (especially of the face, lips, jaw, tongue), slow-frequency tremor of head or limbs (especially while moving), muscle rigidity, lack of facial expression and slow, inflexible movements.	Discontinue. Call doctor right away.
• Pacing or restlessness, intermittent spasms of muscles (of face, eyes, tongue, jaw, neck, body or limbs), dry mouth, blurred vision, constipation, difficulty urinating.	Continue. Call doctor when convenient.
Infrequent:	
Sedation, low blood pressure and dizziness.	Continue. Call doctor when convenient.
Rare:	
Other symptoms that cause concern.	Continue. Call doctor when convenient.

WARNINGS & PRECAUTIONS

Don't take if:

- You are allergic to any thioxanthene or phenothiazine tranquilizer.
- You have serious blood disorder.
- You have Parkinson's disease.
- Patient is younger than 12.

Before you start, consult your doctor if:

- You have had liver or kidney disease.
- You have epilepsy, glaucoma, prostate trouble.
- You have high blood pressure or heart disease (especially angina).
- You drink alcohol daily.
- You will have surgery within 2 months, including dental surgery, requiring general or spinal anesthesia.

Over age 60:
- Adverse reactions and side effects may be more frequent and severe.
- Use of antipsychotic drugs in elderly patients with dementia-related psychosis may increase risk of death. Consult doctor.

Pregnancy:
Decide with your doctor if drug benefits justify risk to unborn baby. Risk category C (see page xviii).

Breastfeeding; Lactation; Nursing Mothers:
This drug is generally not recommended with breastfeeding. Discuss risks and benefits with your doctor.

Infants & children up to age 18:
Follow instructions provided by your child's doctor.

Prolonged use:
- Pigment deposits in lens and retina of eye.
- Involuntary movements of jaws, lips, tongue (tardive dyskinesia).
- Talk to your doctor about the need for follow-up medical examinations or laboratory studies to check complete blood counts (white blood cell count, platelet count, red blood cell count, hemoglobin, hematocrit), liver function, eyes.

Skin & sunlight:
- May cause rash or intensify sunburn in areas exposed to sun or ultraviolet light (photosensitivity reaction). Use sunscreen and avoid overexposure. Notify doctor if reaction occurs.
- Hot temperatures and exercise or hot baths can increase risk of heatstroke. Drug may affect body's ability to maintain normal temperature.

Driving, piloting or hazardous work:
Don't drive or pilot aircraft until you learn how medicine affects you. Don't work around dangerous machinery. Don't climb ladders or work in high places. Danger increases if you drink alcohol or take drugs affecting alertness.

Discontinuing:
Don't discontinue without consulting doctor. Dose may require gradual reduction if you have taken drug for a long time. Doses of other drugs may also require adjustment.

Others:
- Advise any doctor, dentist or pharmacist whom you consult that you take this medicine.
- For dry mouth, suck sugarless hard candy or chew sugarless gum. If dry mouth persists, consult your dentist.

POSSIBLE INTERACTION WITH OTHER DRUGS

GENERIC NAME OR DRUG CLASS	COMBINED EFFECT
Anticonvulsants*	Change in seizure pattern.
Antidepressants, tricyclic*	Increased thiothixene effect. Excessive sedation.
Antihistamines*	Increased thiothixene effect. Excessive sedation.
Antihypertensives*	Excessively low blood pressure.
Barbiturates*	Increased thiothixene effect. Excessive sedation.
Bupropion	Increased risk of seizures.
Epinephrine	Excessively low blood pressure.
Guanethidine	Decreased guanethidine effect.
Levodopa	Decreased levodopa effect.
Mind-altering drugs*	Increased thiothixene effect. Excessive sedation.
Monoamine oxidase (MAO) inhibitors*	Excessive sedation.
Narcotics*	Increased thiothixene effect. Excessive sedation.
Pergolide	Decreased pergolide effect.
Quinidine	Increased risk of heartbeat irregularities.

Continued on page 913

POSSIBLE INTERACTION WITH OTHER SUBSTANCES

INTERACTS WITH	COMBINED EFFECT
Alcohol:	Increased risk of side effects. Avoid.
Beverages:	None expected.
Cocaine:	Unknown. Avoid.
Foods:	None expected.
Marijuana:	Consult doctor.
Tobacco:	None expected.

***See Glossary**

THYROID HORMONES

GENERIC AND BRAND NAMES

LEVOTHYROXINE
- Levo-T
- Levoxyl
- Synthroid
- Unithroid

LIOTHYRONINE
- Cytomel
- Triostat

LIOTRIX
- Thyrolar

BASIC INFORMATION

Habit forming? No
Prescription needed? Yes
Available as generic? Yes, for some
Drug class: Thyroid hormone

USES

Replacement for thyroid hormones lost due to deficiency.

DOSAGE & USAGE INFORMATION

How to take:
Tablet—Swallow with liquid. Levothyroxine tablet should be taken with full glass of water.

When to take:
At the same time each day before a meal or on awakening.

If you forget a dose:
Take as soon as you remember. If it is almost time for the next dose, wait for that dose (don't double this dose) and resume regular schedule.

What drug does:
Replaces hormones normally produced by thyroid gland.

Time lapse before drug works:
48 hours.

Don't take with:
Any other medicine or any dietary supplement without consulting your doctor or pharmacist.

OVERDOSE

SYMPTOMS:
Hot feeling, heart palpitations, tremors, nervousness, sweating, insomnia, rapid and irregular pulse, headache, irritability, diarrhea, weight loss, muscle cramps, confusion, seizures, coma.

WHAT TO DO:
- **Dial 911 for all medical emergencies or call poison control center 1-800-222-1222 for instructions.**
- **See emergency information on last 3 pages of this book.**

POSSIBLE ADVERSE REACTIONS OR SIDE EFFECTS

SYMPTOMS	WHAT TO DO
Life-threatening: In case of overdose, see previous column.	
Common: None expected.	
Infrequent: Tremor, headache, irritability, insomnia, appetite change, diarrhea, leg cramps, fever, menstrual irregularities, heat sensitivity, unusual sweating, weight loss, nervousness.	Continue. Call doctor when convenient.
Rare: Hives, rash, vomiting, chest pain, rapid and irregular heartbeat, shortness of breath, yellow eyes or skin, confusion, mood or mental changes.	Discontinue. Call doctor right away.

WARNINGS & PRECAUTIONS

Don't take if:
- You are allergic to thyroid hormones. Allergic reactions are rare, but may occur.
- You have had a heart attack within the past 6 weeks.
- You have no thyroid deficiency, but want to use this to lose weight.

Before you start, consult your doctor if:
- You have heart disease or high blood pressure.
- You have diabetes.
- You have Addison's disease, have had adrenal gland deficiency or use epinephrine, ephedrine or isoproterenol for asthma.

Over age 60:
Adverse reactions and side effects may be more frequent and severe than in younger persons.

Pregnancy:
Consult doctor. Use is generally considered safe. Risk category A (see page xviii).

Breastfeeding; Lactation; Nursing Mothers:
Drugs in this group are generally acceptable with breastfeeding. Discuss risks and benefits with your doctor.

Infants & children up to age 18:
Follow instructions provided by your child's doctor.

Prolonged use:

- No problems expected if dose is correct.
- Talk to your doctor about the need for follow-up medical examinations or laboratory studies to check thyroid, heart.

Skin & sunlight:
No problems expected.

Driving, piloting or hazardous work:
No problems expected.

Discontinuing:
Don't discontinue without consulting doctor. Dose may require gradual reduction if you have taken drug for a long time. Doses of other drugs may also require adjustment.

Others:

- Digestive upsets, tremors, cramps, nervousness, insomnia or diarrhea may indicate need for dose adjustment.
- Different brands can cause different results. Do not change brands without consulting doctor.
- Advise any doctor, dentist or pharmacist whom you consult that you take this medicine.

POSSIBLE INTERACTION WITH OTHER DRUGS

GENERIC NAME OR DRUG CLASS	COMBINED EFFECT
Adrenocorticoids, systemic	May require thyroid hormone dosage change.
Amphetamines*	Increased amphetamine effect.
Anticoagulants,* oral	Increased anticoagulant effect.
Antidepressants, tricyclic*	Increased antidepressant effect. Irregular heartbeat.
Antidiabetics,* oral or insulin	Antidiabetic may require adjustment.
Aspirin (large doses, continuous use)	Increased effect of thyroid hormone.
Barbiturates*	Decreased barbiturate effect.
Beta-adrenergic blocking agents*	Possible decreased effect of beta blocker.
Cholestyramine	Decreased effect of thyroid hormone.
Colestipol	Decreased effect of thyroid hormone.
Contraceptives, oral*	Decreased effect of thyroid hormone.
Digitalis preparations*	Decreased digitalis effect.
Ephedrine	Increased ephedrine effect.
Epinephrine	Increased epinephrine effect.
Estrogens*	Decreased effect of thyroid hormone.
Meglitinides	Increased blood sugar levels.
Methylphenidate	Increased methylphenidate effect.
Phenytoin	Possible decreased effect of thyroid hormone.
Sympathomimetics*	Increased risk of rapid or irregular heartbeat.

POSSIBLE INTERACTION WITH OTHER SUBSTANCES

INTERACTS WITH	COMBINED EFFECT
Alcohol:	None expected.
Beverages:	None expected.
Cocaine:	Unknown. Avoid.
Foods:	None expected.
Marijuana:	Consult doctor.
Tobacco:	None expected.

*See Glossary

TIAGABINE

BRAND NAMES

Gabitril

BASIC INFORMATION

Habit forming? No
Prescription needed? Yes
Available as generic? No
Drug class: Anticonvulsant; antiepileptic

USES

Treatment of partial seizures in patients with epilepsy. It is used in combination with other antiepileptic drugs.

DOSAGE & USAGE INFORMATION

How to take:
Tablet—Swallow whole with a liquid. Take with food or on a full stomach.

When to take:
Usually once a day to start, then taken 2 to 4 times a day at the same times each day. Your doctor will determine the schedule. Dosage may be increased slowly to achieve best results.

If you forget a dose:
Take as soon as you remember. If it is almost time for the next dose, wait for that dose (don't double this dose) and resume regular schedule.

What drug does:
It is not known exactly how the drug works. It increases the amount of a brain chemical (called GABA) that helps prevent seizure activity.

Time lapse before drug works:
May take several weeks for full effectiveness.

Don't take with:
Any other medicine or any dietary supplement without consulting your doctor or pharmacist.

OVERDOSE

SYMPTOMS:
Drowsiness, muscle spasms, hostility, impaired speech, agitation, depression, weakness, impaired consciousness.
WHAT TO DO:
- **Dial 911 for all medical emergencies or call poison control center 1-800-222-1222 for instructions.**
- **See emergency information on last 3 pages of this book.**

POSSIBLE ADVERSE REACTIONS OR SIDE EFFECTS

SYMPTOMS	WHAT TO DO
Life-threatening:	
1. Rare, severe allergic reaction (anaphylaxis)—difficulty breathing, hives, itching, swelling (throat, face), vomiting, dizziness. 2. Stevens-Johnson syndrome (mucous membranes and skin have redness, blisters, peeling, looseness).	Seek emergency treatment immediately.
Common:	
Dizziness, weakness, tremor, drowsiness, concentration problems, infection (chills, fever, headache, sore throat), nervousness.	Continue. Call doctor when convenient.
Infrequent:	
• Numbness or burning or tingling sensations, speech problems, agitation, confusion, hostility, memory problems, vision changes, clumsiness.	Continue, but call doctor right away.
• Stomach pain, increased cough, muscle aches or pain or weakness, nausea, sleeping problems, pain, unusual tiredness, vomiting, diarrhea.	Continue. Call doctor when convenient.
Rare:	
• Status epilepticus (ongoing seizures), severe weakness, suicidal thoughts.	Seek emergency treatment immediately.
• New or increased, seizures, severe rash, depression, unusual behaviors.	Continue, but call doctor right away.
• Itching, flushing, mild rash, appetite increased, mouth sores, emotional upsets.	Continue. Call doctor when convenient.

WARNINGS & PRECAUTIONS

Don't take if:
You are allergic to tiagabine.

Before you start, consult your doctor if:
- You do not have seizures or epilepsy.
- You have any type of liver disorder.
- You have a history of mental or mood problems (such as depression) or suicidal thoughts or attempts.
- You are allergic to any medication, food or other substance.

Over age 60:
No special problems expected.

Pregnancy:
Decide with your doctor if drug benefits justify risk to unborn baby. Risk category C (see page xviii).

Breastfeeding; Lactation; Nursing Mothers:
This drug is generally acceptable with breastfeeding. Discuss risks and benefits with your doctor.

Infants & children up to age 18:
Follow instructions provided by your child's doctor.

Prolonged use:
No special problems expected. Follow-up with your doctor on a regular basis to monitor your condition and check for drug side effects.

Skin & sunlight:
No problems expected.

Driving, piloting or hazardous work:
This drug may cause dizziness and drowsiness. Don't drive or pilot aircraft until you learn how medicine affects you. Don't work around dangerous machinery. Don't climb ladders or work in high places. The risk of dizziness increases if you drink alcohol.

Discontinuing:
Don't discontinue without doctor's approval due to risk of increased seizure activity. The dosage may need to be gradually decreased before stopping the drug completely.

Others:
- This drug cannot cure epilepsy and will only work to control seizures for as long as you continue to take it.
- Tiagabine is used with other anticonvulsant drugs and additional side effects may occur. If they do, consult your doctor.
- Advise any doctor, dentist or pharmacist whom you consult that you take this medicine.
- Rarely, antiepileptic drugs may lead to suicidal thoughts and behaviors. Call doctor right away if suicidal symptoms or unusual behaviors occur.
- Carry or wear medical identification that lists your seizure disorder and drugs you take.
- Use of this drug in patients without epilepsy may lead to new onset seizures and status epilepticus. Nonepileptic patients who develop seizures should discontinue the drug. Consult doctor to see if you have a seizure disorder.

POSSIBLE INTERACTION WITH OTHER DRUGS

GENERIC NAME OR DRUG CLASS	COMBINED EFFECT
Enzyme inducing, antiepileptic drugs*	Decreased effect of tiagabine.
Other drugs	Unknown. Consult doctor or pharmacist.
Protein bound drugs*	Increased effect of both drugs.
Seizure threshold lowering drugs*	Increased risk of seizures.
Valproate	May alter effect of either drug. Consult doctor.

POSSIBLE INTERACTION WITH OTHER SUBSTANCES

INTERACTS WITH	COMBINED EFFECT
Alcohol:	Increased risk of dizziness. Avoid.
Beverages:	Ask your doctor about drinking grapefruit juice.
Cocaine:	Unknown. Avoid.
Foods:	None expected.
Marijuana:	Consult doctor.
Tobacco:	None expected.

*See Glossary

TICAGRELOR

BRAND NAMES

Brilinta

BASIC INFORMATION

Habit forming? No
Prescription needed? Yes
Available as generic? No
Drug class: Platelet aggregation inhibitor

USES

It helps prevent blood clots and reduces the risk of heart attack, stroke or other vascular events. It is used in people who have had a heart attack or have acute coronary syndrome (symptoms include severe chest pain called angina).

DOSAGE & USAGE INFORMATION

How to take:
Tablet—Swallow whole with water. It may be taken with or without food. Take it with food if the tablet upsets your stomach.

When to take:
Twice a day at the same times each day. This drug is intended to be used in combination with aspirin. Your doctor will advise you of how much aspirin to take each day.

If you forget a dose:
Take as soon as you remember. If it is almost time for the next dose, wait for that dose (don't double this dose) and resume regular schedule.

What drug does:
An antiplatelet drug works by preventing platelets (a type of blood cell) from clumping and forming clots that reduce blood flow and may cause a heart attack or stroke.

Time lapse before drug works:
It starts working within 1 to 2 hours.

Continued next column

OVERDOSE

SYMPTOMS:
Nausea, vomiting, irregular heartbeat, and increased risk of bleeding.
WHAT TO DO:
- **Dial 911 for all medical emergencies or call poison control center 1-800-222-1222 for instructions.**
- **See emergency information on last 3 pages of this book.**

Don't take with:
Any other medicine or dietary supplement without consulting your doctor or pharmacist.

POSSIBLE ADVERSE REACTIONS OR SIDE EFFECTS

SYMPTOMS	WHAT TO DO
Life-threatening: Bleeding (see symptoms below).	Seek emergency treatment immediately.
Common: Difficult breathing or shortness of breath, nosebleeds.	Continue, but call doctor right away.
Infrequent:	
• Unusual bleeding from the gums or vagina or rectum or eyes, black or tarry stools, red or brown urine, coughing up blood or coffee-ground material, unusual bruising, purple or red spots under the skin, fast or slow or irregular heartbeat, chest pain.	Continue, but call doctor right away.
• Dizziness, faintness, headache, diarrhea, upset stomach, unusual tiredness or weakness, back pain, cough, nausea.	Continue. Call doctor when convenient.
Rare:	
• Bleeding in the brain (may have symptoms of sudden numbness or weakness on one side of the body, sudden and severe headache, confusion, balance or vision problems).	Seek emergency care immediately.
• Low or high blood pressure.	Continue. Call doctor when convenient.

WARNINGS & PRECAUTIONS

Don't take if:
You are allergic to ticagrelor. Allergic reactions are rare, but may occur.

Before you start, consult your doctor if:
- You have a history of intracranial hemorrhage (bleeding in the brain).
- You have a liver (hepatic) disorder.
- You have bleeding problem (e.g., peptic ulcer).
- You have lung disease or breathing problems (e.g., asthma or COPD).
- You plan on having surgery (including dental surgery) in the near future.

Over age 60:
Adverse reactions and side effects may be more frequent and severe than in younger persons.

Pregnancy:
Decide with your doctor if drug benefits justify risk to unborn baby. Risk category C (see page xviii).

Breastfeeding; Lactation; Nursing Mothers:
This drug is generally not recommended with breastfeeding. Discuss risks and benefits with your doctor.

Infants & children up to age 18:
Safety and effectiveness in this age group have not been established. Consult doctor.

Prolonged use:
Talk to your doctor about the need for follow-up medical exams or laboratory studies to check the effectiveness of the treatment.

Skin & sunlight:
No problems expected.

Driving, piloting or hazardous work:
Avoid if you experience dizziness or faintness, otherwise no problems expected.

Discontinuing:
Don't discontinue without doctor's approval. Stopping the drug can increase the risk of blood clots and their complications.

Others:
- Advise any doctor, dentist or pharmacist whom you consult that you take this drug.
- While taking this drug you will likely bruise and bleed more easily than usual or bleed for longer than usual. Avoid rough sports or other situations where you could be bruised, cut or injured. Be extra careful when using sharp objects. Bleeding complications can be serious or fatal. Consult doctor if you have any concerns or questions about side effects.
- Drug use may increase risk of serious bleeding during a surgery, other medical procedures or some types of dental work. You may be advised to stop using this drug at least 5 days before a surgery, medical procedure or dental work.

POSSIBLE INTERACTION WITH OTHER DRUGS

GENERIC NAME OR DRUG CLASS	COMBINED EFFECT
Anticoagulants*	Increased risk of bleeding.
Anti-inflammatory drugs, nonsteroidal (NSAIDs)*	Increased risk of bleeding.
Antiplatelet drugs,* other	Increased risk of bleeding.
Aspirin	Decreased effect of ticagrelor if you take more than 100 mg of aspirin daily.
Digoxin	Increased effect of digoxin.
Enzyme inducers*	Decreased effect of ticagrelor.
Enzyme inhibitors*	Increased effect of ticagrelor.
HMG-CoA reductase inhibitors	Increased side effects of certain HMG-CoA inhibitors.

POSSIBLE INTERACTION WITH OTHER SUBSTANCES

INTERACTS WITH	COMBINED EFFECT
Alcohol:	None expected.
Beverages: Grapefruit juice.	May increase effect of ticagrelor.
Cocaine:	Unknown. Avoid.
Foods: Grapefruit.	May increase effect of ticagrelor.
Marijuana:	Consult doctor.
Tobacco:	None expected, but people with heart disorders should not smoke.

*See Glossary

TIOPRONIN

BRAND NAMES

Capen
Captimer
Epatiol
Mucolysin
Sutilan
Thiola
Thiosol
Tioglis
Vincol

BASIC INFORMATION

Habit forming? No
Prescription needed? Yes
Available as generic? No
Drug class: Antiurolithic

USES

Prevents the formation of kidney stones when there is too much cystine in the urine.

DOSAGE & USAGE INFORMATION

How to take:
Tablet—Swallow with liquid. If you can't swallow whole, crumble tablet and take with liquid or food. Instructions to take on empty stomach mean 1 hour before or 2 hours after eating.

When to take:
3 times daily (once approximately every 8 hours).

If you forget a dose:
Take as soon as you remember. If it is almost time for the next dose, wait for that dose (don't double this dose) and resume regular schedule.

What drug does:
Removes high levels of cystine from the body.

Time lapse before drug works:
It starts working right away, but full benefit may take weeks to months. Laboratory studies are used to measure the cystine in the urine.

Continued next column

OVERDOSE

SYMPTOMS:
Unknown.
WHAT TO DO:
- **Dial 911 for all medical emergencies or call poison control center 1-800-222-1222 for instructions.**
- **See emergency information on last 3 pages of this book.**

Don't take with:
- Medicines that are known to cause kidney damage or depress bone marrow.
- Any other medicine or any dietary supplement without consulting your doctor or pharmacist.

POSSIBLE ADVERSE REACTIONS OR SIDE EFFECTS

SYMPTOMS	WHAT TO DO
Life-threatening:	
Rare, severe allergic reaction (anaphylaxis)—difficulty breathing, hives, itching, swelling (throat, face), vomiting, dizziness.	Seek emergency treatment immediately.
Common:	
• Skin rash, itching skin, mouth sores.	Discontinue. Call doctor right away.
• Abdominal pain, gaseousness, diarrhea, nausea or vomiting.	Continue. Call doctor when convenient.
Infrequent:	
• Cloudy urine, chills, breathing difficulty, joint pain.	Discontinue. Call doctor right away.
• Impaired smell or taste.	Continue. Call doctor when convenient.
Rare:	
Coughing up blood, fever, unusual tiredness or weakness, double vision, muscle weakness.	Discontinue. Call doctor right away.

WARNINGS & PRECAUTIONS

Don't take if:
You are allergic to tiopronin or penicillamine.

Before you start, consult your doctor if:
You have had any of the following in the past:
- Agranulocytosis.
- Aplastic anemia.
- Thrombocytopenia.
- Impaired kidney function.

Over age 60:
Adverse reactions and side effects may be more frequent and severe than in younger persons.

Pregnancy:
Decide with your doctor if drug benefits justify risk to unborn baby. Risk category C (see page xviii).

Breastfeeding; Lactation; Nursing Mothers:
This drug is generally not recommended with breastfeeding. Discuss risks and benefits with your doctor.

Infants & children up to age 18:
Follow instructions provided by your child's doctor. Safety and effectiveness in age over 9 have not been established.

Prolonged use:
No special problems expected.

Skin & sunlight:
No special problems expected.

Driving, piloting or hazardous work:
Don't drive or pilot aircraft until you learn how medicine affects you. Don't work around dangerous machinery. Don't climb ladders or work in high places. Danger increases if you drink alcohol or take medicine affecting alertness and reflexes.

Discontinuing:
No special problems expected.

Others:
- Advise any doctor, dentist or pharmacist whom you consult that you take this medicine.
- May affect results of some medical tests.

POSSIBLE INTERACTION WITH OTHER DRUGS

GENERIC NAME OR DRUG CLASS	COMBINED EFFECT
Bone marrow depressants*	May increase possibility of toxic effects of tiopronin.
Medications toxic to kidneys (nephrotoxics*)	May increase possibility of toxic effects of tiopronin.

POSSIBLE INTERACTION WITH OTHER SUBSTANCES

INTERACTS WITH	COMBINED EFFECT
Alcohol:	None expected.
Beverages: Water.	Enhances effects of tiopronin. Drink 8 to 10 glasses daily.
Cocaine:	Unknown. Avoid.
Foods:	None expected.
Marijuana:	Consult doctor.
Tobacco:	None expected.

*See Glossary

TIZANIDINE

BRAND NAMES

Zanaflex

BASIC INFORMATION

Habit forming? No
Prescription needed? Yes
Available as generic? Yes
Drug class: Antispastic, muscle relaxant

USES

- Relieves muscle spasticity caused by diseases such as multiple sclerosis. It does not appear to improve muscle weakness.
- Relieves muscle spasticity caused by injury to spinal cord.

DOSAGE & USAGE INFORMATION

How to take:
Tablet or capsule—Swallow with liquid. Follow instructions on your prescription about whether to take with food or not. Food can change the amount of the drug absorbed by your body.

When to take:
Up to three times a day or as directed by your doctor.

If you forget a dose:
Take as soon as you remember. If it is almost time for the next dose, wait for that dose (don't double this dose) and resume regular schedule.

What drug does:
Slows nerve impulses that stimulate skeletal muscles, decreasing cramping.

Time lapse before drug works:
Within hours. Dosage may be increased over a several week period to achieve maximum effectiveness.

Don't take with:
Any other medicine or any dietary supplement without consulting your doctor or pharmacist.

OVERDOSE

SYMPTOMS:
Breathing difficulties, heartbeat irregularities, sleepiness, confusion, coma.
WHAT TO DO:

- **Dial 911 for all medical emergencies or call poison control center 1-800-222-1222 for instructions.**
- **See emergency information on last 3 pages of this book.**

POSSIBLE ADVERSE REACTIONS OR SIDE EFFECTS

SYMPTOMS	WHAT TO DO
Life-threatening: In case of overdose, see previous column.	
Common:	
• Burning or pain with urination, chest pain, fever, chills, nausea, vomiting, unusual tiredness.	Discontinue. Call doctor away.
• Dizziness, weakness, drowsiness, dry mouth, mild fatigue.	Continue. Call doctor when convenient.
Infrequent:	
• Blurred vision, itching skin, hallucinations, irregular heartbeat, shortness of breath, pain in upper abdomen, weight gain.	Discontinue. Call doctor right away
• Constipation, being nervous, sore throat.	Continue. Call doctor when convenient.
Rare:	
General ill feeling, other symptoms that cause concern.	Continue. Call doctor when convenient.

WARNINGS & PRECAUTIONS

Don't take if:
You are allergic to tizanidine. Allergic reactions are rare, but may occur.

Before you start, consult your doctor if:
- You have liver or kidney disease.
- You are allergic to any medication, food or other substance.
- You are taking an alpha-adrenergic blood pressure medicine.

Over age 60:
Adverse reactions and side effects may be more frequent and severe than in younger persons.

Pregnancy:
Decide with your doctor if drug benefits justify risk to unborn baby. Risk category C (see page xviii).

Breastfeeding; Lactation; Nursing Mothers:
This drug is generally not recommended with breastfeeding. Discuss risks and benefits with your doctor.

Infants & children up to age 18:
Safety and effectiveness in this age group have not been established. Consult doctor.

Prolonged use:
Talk to your doctor about the need for liver function studies while using this drug.

Skin & sunlight:
No special problems expected.

Driving, piloting or hazardous work:
Don't drive or pilot aircraft until you learn how medicine affects you. Don't work around dangerous machinery. Don't climb ladders or work in high places. Danger increases if you drink alcohol or take medicine affecting alertness and reflexes, such as antihistamines, tranquilizers, sedatives, pain medicine, narcotics and mind-altering drugs.

Discontinuing:
Don't discontinue without consulting doctor.

Others:
- May interfere with the results in some medical tests.
- Get up slowly from a sitting or lying position to avoid any dizziness, faintness or lightheadedness.
- Advise any doctor, dentist or pharmacist whom you consult that you take this medicine.
- The brand name Zanaflex is available as a tablet and a capsule. The two formulations are not interchangeable. Switching from the capsule to the tablet may increase risk of side effects.
- Take medicine only as directed. Do not increase or reduce dosage without doctor's approval.

POSSIBLE INTERACTION WITH OTHER DRUGS

GENERIC NAME OR DRUG CLASS	COMBINED EFFECT
Central nervous system (CNS) depressants *	Increased sedation.
Ciprofloxacin	Dangerous increased tizanidine effect. Avoid.
Contraceptives, oral	Increased effect of tizanidine.
Enzyme Inhibitors*	Increased effect of tizanidine.
Fluvoxamine	Dangerous increased tizanidine effect. Avoid.
Hypotension-causing drugs,* other	Increased effect of hypotension.

POSSIBLE INTERACTION WITH OTHER SUBSTANCES

INTERACTS WITH	COMBINED EFFECT
Alcohol:	Increased sedation, low blood pressure. Avoid.
Beverages:	None expected.
Cocaine:	Increased risk of side effects. Avoid.
Foods:	Follow prescription instructions about taking with food.
Marijuana:	Increased risk of side effects. Avoid.
Tobacco:	May interfere with absorption of medicine.

***See Glossary**

TOFACITINIB

BRAND NAMES

Xeljanz

BASIC INFORMATION

Habit forming? No
Prescription needed? Yes
Available as generic? No
Drug class: JAK (Janus kinase) inhibitor

USES

- Treatment of moderate to severe rheumatoid arthritis (RA) in adults. It may be used alone or with other medications.
- May be used to treat other conditions as determined by your doctor.

DOSAGE & USAGE INFORMATION

How to take:
Tablet—Swallow whole with liquid. It may be taken with or without food (food may lessen stomach upset).

When to take:
Drug is taken twice a day at the same times each day.

If you forget a dose:
Take as soon as you remember. If it is almost time for the next dose, wait for that dose (don't double this dose) and resume regular schedule.

What drug does:
It works by blocking Janus kinase (JAK) a component of the body's cells that plays a role in activating the inflammation involved with RA.

Time lapse before drug works:
One to 2 days, but it takes 2 weeks or longer for symptoms to start improving and may be months to feel full benefits.

Don't take with:
Any other medicine or any dietary supplement without consulting your doctor or pharmacist.

OVERDOSE

SYMPTOMS:
Unknown.
WHAT TO DO:

- **Dial 911 for all medical emergencies or call poison control center 1-800-222-1222 for instructions.**
- **See emergency information on last 3 pages of this book.**

POSSIBLE ADVERSE REACTIONS OR SIDE EFFECTS

SYMPTOMS	WHAT TO DO
Life-threatening:	
Infections (read information under Others).	Seek emergency treatment immediately.
Common:	
Cold or flu-like symptoms (stuffy nose, fever, chills, cough, sneezing, sore throat), bloody or cloudy urine, difficult or painful or frequent urination, high blood pressure.	Continue, but call doctor right away.
Infrequent:	
Headache, diarrhea, dizziness, heartburn, indigestion, nausea.	Continue. Call doctor when convenient.
Rare:	
Serious infections may occur and symptoms will vary: unusual weakness or tiredness, yellow eyes or skin, change in stool (lighter color or black or tarry), unusual bruising, painful skin rash or blisters, general ill feeling, abdominal pain, chest pain, persistent cough, loss of appetite, unusual weight loss, night sweats, severe headache or dizziness, shortness of breath, numbness or tingling, swelling (feet, ankles, legs, hands or joints), muscle or joint pain, vomiting or coughing up blood, change in bowel habits, swollen lymph nodes, other symptoms that cause concern.	Discontinue. Call doctor right away.

WARNINGS & PRECAUTIONS

Don't take if:
You are allergic to tofacitinib. Allergic reactions are rare, but may occur.

Before you start, consult your doctor if:
- You have a history of anemia or low blood cell counts.
- You have an infection or a history of chronic infections (e.g., bacteria, virus or fungus).
- You have stomach or bowel problems (e.g., ulcers, ulcerative colitis, diverticulitis or perforation).
- You have or have had cancer.
- You have or have had kidney or liver disease (including hepatitis).
- You have weakened immune system due to illness (e.g., diabetes or HIV) or drugs.
- You have hyperlipidemia (high cholesterol or fats in the blood).
- You have recently received or are scheduled to receive a vaccine.
- You have or have had or been exposed to tuberculosis (TB). You will need a skin test for TB before taking drug.

Over age 60:
Adverse reactions and side effects may be more frequent and severe than in younger persons.

Pregnancy:
Decide with your doctor if drug benefits justify risk to unborn baby. Risk category C (see page xviii).

Breastfeeding; Lactation; Nursing Mothers:
This drug is generally not recommended with breastfeeding. Discuss risks and benefits with your doctor.

Infants & children up to age 18:
Safety and effectiveness in this age group have not been established. Consult doctor.

Prolonged use:
- See your doctor for regular visits to make sure the drug is working properly, to obtain recommended blood tests and to check for unwanted effects.
- Long term efficacy of drug is not known.

Skin & sunlight:
No problems expected.

Driving, piloting or hazardous work:
No problems expected.

Discontinuing:
No problems expected, but consult your doctor before discontinuing. Certain lab tests may be required for a period of time after stopping drug.

Others:
- Patients taking this drug are at increased risk for infections and malignancies that can be serious and possibly fatal. These include bacterial, viral and fungal infections; tuberculosis; cancer; lymphoma; herpes zoster (shingles) and others. Also, gastrointestinal perforations have occurred. Consult doctor about your risks.
- It is important to obtain all recommended laboratory blood tests while taking this drug.
- Advise any doctor, dentist or pharmacist whom you consult that you take this drug.

POSSIBLE INTERACTION WITH OTHER DRUGS

GENERIC NAME OR DRUG CLASS	COMBINED EFFECT
Biologic response modifiers*	Increased risk of infection. Avoid.
Enzyme inducers*	May decrease effect of tofacitinib.
Enzyme inhibitors*	May increase effect of tofacitinib.
Immunosuppressants*	Increased risk of infections. Avoid.
Live vaccines	May decrease vaccine effect. Possible infection. Avoid.

POSSIBLE INTERACTION WITH OTHER SUBSTANCES

INTERACTS WITH	COMBINED EFFECT
Alcohol:	None expected.
Beverages:	None expected.
Cocaine:	Unknown. Avoid.
Foods:	None expected.
Marijuana:	Consult doctor.
Tobacco:	None expected.

*See Glossary

TOPIRAMATE

BRAND NAMES

Qsymia
Quadexy XR
Topamax
Topamax Sprinkle Capsules
Trokendi XR

BASIC INFORMATION

Habit forming? No
Prescription needed? Yes
Available as generic? Yes
Drug class: Anticonvulsant, antiepileptic

USES

- Treatment for partial (focal) epileptic seizures. May be used alone or in combination with other antiepileptic drugs.
- Treats overweight and obesity in adults.
- Used to prevent migraine headache.

DOSAGE & USAGE INFORMATION

How to take:

- Tablet—Swallow the tablets whole with a drink of water. Do not crush or chew (the tablet has a bitter taste). May be taken with or without food and on a full or empty stomach.
- Sprinkle capsules—Can be swallowed whole or opened carefully and the contents sprinkled on a small amount of soft food, such as applesauce, pudding, ice cream, oatmeal or yogurt. Swallow this mixture immediately. Do not chew or store for later use.
- Extended-release capsule—Swallow whole with liquid. Do not crush, chew or open capsule.

When to take:
Your doctor will determine the best schedule. Dosages will be increased rapidly over the first weeks of use. Further increases may be necessary to achieve maximum benefits.

Continued next column

OVERDOSE

SYMPTOMS:
Drowsiness, speech problems, confusion, dizziness, depression, vision changes, problem with coordination, tiredness, stomach pain, agitation, seizures, stupor.
WHAT TO DO:

- **Dial 911 for all medical emergencies or call poison control center 1-800-222-1222 for instructions.**
- **See emergency information on last 3 pages of this book.**

If you forget a dose:
Take as soon as you remember. If it is almost time for the next dose, wait for that dose (don't double this dose) and resume regular schedule.

What drug does:
The exact mechanism of the anticonvulsant effect is unknown. It appears to block the spread of seizures rather than raise the seizure threshold like other anticonvulsants. Topiramate's anticonvulsant actions involve several mechanisms.

Time lapse before drug works:
May take several weeks for effectiveness.

Don't take with:
Any other medicine or any dietary supplement without consulting your doctor or pharmacist.

POSSIBLE ADVERSE REACTIONS OR SIDE EFFECTS

SYMPTOMS	WHAT TO DO
Life-threatening: In case of overdose, see previous column.	
Common:	
• Burning or prickling, or tingling sensations, clumsiness or being unsteady, confusion, uncontrolled back-and-forth or rolling eye movements, dizziness, double vision or other vision problems, drowsiness, generalized slowing of mental and physical activity, memory problems, menstrual changes, menstrual pain, nervousness, speech or language problems, trouble in concentrating or paying attention, unusual tiredness or weakness.	Continue, but call doctor right away.
• Breast pain in women, nausea, tremor.	Continue. Call doctor when convenient.
Infrequent:	
• Abdominal pain, fever, chills, sore throat, loss of appetite, mood or mental changes (such as aggression, agitation, apathy, irritability, and depression), red or irritated or bleeding gums, weight loss, chest pain.	Continue, but call doctor right away.

• Back pain, leg pain, constipation, heartburn, hot flushes, increased sweating.	Continue. Call doctor when convenient.
Rare:	
Eye pain, frequent or difficult urination, bloody urine, hearing loss, itching, loss of bladder control, nosebleeds, pale skin, red or irritated eyes, ringing in ears, skin rash, swelling, trouble breathing.	Continue, but call doctor right away.

WARNINGS & PRECAUTIONS

Don't take if:
You are allergic to topiramate. Allergic reactions are rare, but may occur.

Before you start, consult your doctor if:
- You have liver disease.or kidney disease or kidney stones (nephrolithiasis).
- You are allergic to any medication, food or other substance.
- You have any other medical problems.

Over age 60:
No special problems expected.

Pregnancy:
Consult doctor. Use of the drug during pregnancy is not recommended. It could cause harm to the unborn baby. Risk category D (see page xviii).

Breastfeeding; Lactation; Nursing Mothers:
This drug is generally acceptable with breastfeeding. Discuss risks and benefits with your doctor.

Infants & children up to age 18:
- Treatment for seizures in ages 2 and older. Dose is according to body weight.
- Used for migraine prevention in ages over 12.
- Drug is not expected to cause different side effects or problems in children than in adults.

Prolonged use:
No special problems expected. Follow-up laboratory blood studies may be recommended by your doctor.

Skin & sunlight:
No problems expected.

Driving, piloting or hazardous work:
Don't drive or pilot aircraft until you learn how medicine affects you. Don't work around dangerous machinery. Don't climb ladders or work in high places. Danger increases if you drink alcohol or take other medicines affecting alertness and reflexes.

Discontinuing:
Don't discontinue without doctor's approval due to risk of increased seizure activity. The dosage may need to be gradually decreased before stopping the drug completely.

Others:
- Advise any doctor, dentist or pharmacist whom you consult that you take this medicine.
- Drink plenty of fluids while taking topiramate. If you have had kidney stones in the past, this will help to reduce your chances of forming kidney stones.
- Topiramate may be used with other anticonvulsant drugs and additional side effects may also occur. If they do, discuss them with your doctor.
- Rarely, antiepileptic drugs may lead to suicidal thoughts and behaviors. Call doctor right away if suicidal symptoms or unusual behaviors occur.
- Wear or carry medical identification to show your seizure disorder and the drugs you take.

POSSIBLE INTERACTION WITH OTHER DRUGS

GENERIC NAME OR DRUG CLASS	COMBINED EFFECT
Anticonvulsants,* other	May decrease or increase effect of both drugs.
Carbonic anhydrase inhibitors*	Increased risk of kidney stones.
Central nervous system (CNS) depressants*	Increased sedative effect.
Contraceptives, oral*	Decreased effect of contraceptive.
Digoxin	May decrease effect of digoxin.

POSSIBLE INTERACTION WITH OTHER SUBSTANCES

INTERACTS WITH	COMBINED EFFECT
Alcohol:	Increased sedative effect. Avoid.
Beverages:	None expected.
Cocaine:	Unknown. Avoid.
Foods:	None expected.
Marijuana:	Consult doctor.
Tobacco:	None expected.

*See Glossary

TOREMIFENE

BRAND NAMES

Fareston

BASIC INFORMATION

Habit forming? No
Prescription needed? Yes
Available as generic? No
Drug class: Antineoplastic

USES

Used to treat breast cancer in postmenopausal women.

DOSAGE & USAGE INFORMATION

How to take:
Tablet—Swallow with water and take with or without food. If you can't swallow whole, crumble tablet and take with liquid or food.

When to take:
At the same time each day.

If you forget a dose:
Take as soon as you remember. If it is almost time for the next dose, wait for that dose (don't double this dose) and resume regular schedule.

What drug does:
Exact mechanism unknown. Appears to block growth-stimulating effects of estrogen in the tumor.

Time lapse before drug works:
4 to 6 weeks to determine effectiveness.

Don't take with:
Any other medicine or any dietary supplement without consulting your doctor or pharmacist.

OVERDOSE

SYMPTOMS:
Dizziness, headache, nausea and vomiting.
WHAT TO DO:
- **Dial 911 for all medical emergencies or call poison control center 1-800-222-1222 for instructions.**
- **See emergency information on last 3 pages of this book.**

POSSIBLE ADVERSE REACTIONS OR SIDE EFFECTS

SYMPTOMS	WHAT TO DO
Life-threatening:	
Fainting or seizures.	Seek emergency treatment immediately.
Common:	
Hot flashes, nausea.	Continue. Call doctor if symptoms persist.
Infrequent:	
• Change in vaginal discharge, pain or feeling of pressure in pelvis, vaginal bleeding, confusion, increased urination, loss of appetite, unusual tiredness, changes in vision.	Continue, but call doctor right away.
• Dizziness, dry eyes, bone pain, vomiting.	Continue. Call doctor when convenient.
Rare:	
None expected.	

WARNINGS & PRECAUTIONS

Don't take if:
- You are allergic to toremifene. Allergic reactions are rare, but may occur.
- You have a history of blood clots or have been diagnosed with thromboembolic disease.

Before you start, consult your doctor if:
- You have heart failure, irregular heartbeat or liver problem.
- You or a family member has long QT syndrome (an inherited disorder).
- You have had low blood levels of potassium or magnesium.
- You have any blood or bleeding disorder.
- You have ever been diagnosed with endometrial hyperplasia (unusual growth of the lining of the uterus).
- You have a tumor that has spread to your bone.

Over age 60:
No special problems expected.

Pregnancy:
Consult doctor. Use of the drug during pregnancy is not recommended. It could cause harm to the unborn baby. Risk category D (see page xviii).

Breastfeeding; Lactation; Nursing Mothers:
This drug is generally not recommended with breastfeeding. Discuss risks and benefits with your doctor.

Infants & children up to age 18:
There is no identified potential use of toremifene in children.

Prolonged use:
Talk to your doctor about the need for follow-up laboratory studies to check complete blood count, blood calcium concentrations and liver function.

Skin & sunlight:
Avoid prolonged or extended exposure to direct sunlight and/or artificial sunlight while using this medication.

Driving, piloting or hazardous work:
No problems expected.

Discontinuing:
Don't discontinue without consulting doctor.

Others:
- Advise any doctor or dentist you consult that you are taking this medication.
- Drug use may increase risk of heart problem called QT prolongation that can lead to fainting, seizures or sudden death.
- May interfere with the accuracy of some medical tests.

POSSIBLE INTERACTION WITH OTHER DRUGS

GENERIC NAME OR DRUG CLASS	COMBINED EFFECT
Anticoagulants	May increase time it takes blood to clot.
Diuretics, thiazide	Possible increased calcium.
Enzyme inducers*	May lessen the effect of toremifene.
Enzyme inhibitors*	May increase the effect of toremifene.

POSSIBLE INTERACTION WITH OTHER SUBSTANCES

INTERACTS WITH	COMBINED EFFECT
Alcohol:	None expected.
Beverages:	None expected.
Cocaine:	Unknown. Avoid.
Foods:	None expected.
Marijuana:	Consult doctor.
Tobacco:	None expected.

*See Glossary

TRAMADOL

BRAND NAMES

CIP-Tramadol ER
Ryzolt
Ultracet
Ultram
Ultram ER
Ultram ODT

BASIC INFORMATION

Habit forming? Yes
Prescription needed? Yes
Available as generic? Yes
Drug class: Analgesic

USES

Treatment for moderate to moderately severe pain (used more for chronic than acute pain).

DOSAGE & USAGE INFORMATION

How to take:

- Tablet or extended-release tablet—Swallow whole with liquid. May be taken with or without food. Do not crush, chew or split tablet. It is for oral use only. Read prescription instructions.
- Disintegrating tablet—Let tablet dissolve in mouth. Do not crush, chew or split tablet. It is for oral use only. Read prescription instructions.

When to take:
Every 4 to 6 hours as needed for pain. Extended release form is taken once a day.

If you forget a dose:
Take as soon as you remember. If it is almost time for the next dose, wait for that dose (don't double this dose) and resume regular schedule.

What drug does:
Exact mechanism unknown. It appears to block pain messages to the brain and spinal cord.

Continued next column

OVERDOSE

SYMPTOMS:
Breathing difficulty, sleepiness, seizures, cold/clammy skin, slow heartbeat, low blood pressure, stupor, coma. Deaths due to overdose have been reported with abuse and misuse of tramadol, by ingesting, inhaling, or injecting the crushed tablets.

WHAT TO DO:

- **Dial 911 for all medical emergencies or call poison control center 1-800-222-1222 for instructions.**
- **See emergency information on last 3 pages of this book.**

Time lapse before drug works:
Within 60 minutes.

Don't take with:
Any other medicine or any dietary supplement without consulting your doctor or pharmacist.

POSSIBLE ADVERSE REACTIONS OR SIDE EFFECTS

SYMPTOMS	WHAT TO DO
Life-threatening:	
Rare, severe allergic reaction (anaphylaxis)—difficulty breathing, hives, itching, swelling (throat, face), vomiting, dizziness.	Seek emergency treatment immediately.
Common:	
Constipation, nausea, headache, drowsiness, clumsiness, dizziness, itching or flushing or redness of skin, trouble sleeping.	Continue. Call doctor when convenient.
Infrequent:	
• Constant urge to urinate or inability to urinate, blurred vision.	Discontinue. Call doctor right away.
• Loss of appetite, stomach pain, dry mouth, weakness, confusion, sweating, diarrhea, gas, hot flashes, nervousness, heartburn, tiredness.	Continue. Call doctor when convenient.
Rare:	
Seizures, balancing difficulty, memory problems, shortness of breath, difficulty performing tasks, hallucinations, light-headedness when getting up from a sitting or lying position, sensations in hands and feet (burning, tingling, pain, weakness, trembling or shaking), faintness, fast heartbeat.	Discontinue. Call doctor right away.

WARNINGS & PRECAUTIONS

Don't take if:
You are allergic to tramadol or narcotic drugs or you are sensitive to any ingredients in the drug.

Before you start, consult your doctor if:
- You have kidney or liver disease, seizure disorder, stomach disorder or prior head injury.
- You have a history of depression, mental illness or suicidal thoughts or behaviors.
- You have a history of drug abuse or substance abuse, including alcohol abuse.
- You have respiratory problems.

Over age 60:
Adverse reactions and side effects may be more frequent and severe than in younger persons.

Pregnancy:
Decide with your doctor if drug benefits justify risk to unborn baby. Risk category C (see page xviii).

Breastfeeding; Lactation; Nursing Mothers:
This drug is generally acceptable with breastfeeding. Discuss risks and benefits with your doctor.

Infants & children up to age 18:
Safety and effectiveness in this age group have not been established. Consult doctor.

Prolonged use:
- Consult with your doctor on a regular basis while using this drug.
- Can cause drug dependence, addiction and withdrawal symptoms.

Skin & sunlight:
No special problems expected.

Driving, piloting or hazardous work:
Don't drive or pilot aircraft until you learn how medicine affects you. Don't work around dangerous machinery. Don't climb ladders or work in high places. Danger increases if you drink alcohol or take medicine affecting alertness and reflexes.

Discontinuing:
If you have taken this drug for a long time, consult with your doctor before discontinuing. Withdrawal symptoms may occur if drug is discontinued abruptly. Symptoms include: anxiety, sweating, insomnia, pain, nausea, tremors, diarrhea, breathing problems and hallucinations. Call doctor if symptoms occur.

Others:
- Don't increase dosage or frequency of use without your doctor's approval. Development of a potentially life-threatening serotonin syndrome* or overdose (which can be fatal) may occur with the use of tramadol.
- Advise any doctor, dentist or pharmacist whom you consult that you take this medicine.

POSSIBLE INTERACTION WITH OTHER DRUGS

GENERIC NAME OR DRUG CLASS	COMBINED EFFECT
Carbamazepine	Decreased effect of tramadol.
Central nervous system (CNS) depressants*	Serious adverse events including breathing difficulty.
Clozapine	May increase risk of seizures.
Cyclobenzaprine	May increase risk of seizures.
Digoxin	May increase risk of digoxin toxicity.
Enzyme inducers*	May decrease effect of tramadol.
Enzyme inhibitors*	May increase effect of tramadol.
Monoamine oxidase (MAO) inhibitors*	Serious adverse events including seizures and serotonin syndrome.*
Narcotics*	Increased risk of side effects and seizures.
Phenothiazines*	Increased risk of side effects.
Quinidine	Unknown effect. May need to adjust dose.
Serotonergics*	Increased risk of seizures and serotonin syndrome.*
Tranquilizers*	Increased risk of side effects.
Warfarin	May increase risk of bleeding.

POSSIBLE INTERACTION WITH OTHER SUBSTANCES

INTERACTS WITH	COMBINED EFFECT
Alcohol:	Serious side effects. Avoid.
Beverages:	None expected.
Cocaine:	Unknown. Avoid.
Foods:	None expected.
Marijuana:	Consult doctor.
Tobacco:	None expected.

***See Glossary**

TRAZODONE

BRAND NAMES

Oleptro
Trazon

BASIC INFORMATION

Habit forming? No
Prescription needed? Yes
Available as generic? Yes
Drug class: Antidepressant (nontricyclic)

USES

- Treats mental depression.
- Treats anxiety.
- Helps promote sleep.
- Treats some types of chronic pain.

DOSAGE & USAGE INFORMATION

How to take:
- Tablet—Swallow with liquid or food to lessen stomach irritation. If you can't swallow whole, crumble tablet and take with liquid or food.
- Extended release tablet—Swallow whole. Do not crumble or crush tablet.

When to take:
According to prescription directions. Bedtime dose usually higher than other doses.

If you forget a dose:
Take as soon as you remember. If it is almost time for the next dose, wait for that dose (don't double this dose) and resume regular schedule.

What drug does:
Maintains balance of certain brain chemicals.

Time lapse before drug works:
2 to 4 weeks for full effect.

Don't take with:
Any other medicine or any dietary supplement without consulting your doctor or pharmacist.

OVERDOSE

SYMPTOMS:
Drowsiness, vomiting, difficulty breathing, painful and prolonged penis erection, fast heartbeat, seizures.
WHAT TO DO:
- **Dial 911 for all medical emergencies or call poison control center 1-800-222-1222 for instructions.**
- **See emergency information on last 3 pages of this book.**

POSSIBLE ADVERSE REACTIONS OR SIDE EFFECTS

SYMPTOMS	WHAT TO DO
Life-threatening: In case of overdose, see previous column.	
Common:	
Drowsiness, dry mouth, nausea, vomiting, headache, unpleasant taste, dizziness.	Continue. Call doctor when convenient.
Infrequent:	
• Tremors, fainting, confusion.	Discontinue. Call doctor right away.
• Constipation or diarrhea, muscle aches or pain, blurred vision, unusual tiredness or weakness.	Continue. Call doctor when convenient.
Rare:	
• Penile erection is prolonged or painful.	Discontinue. Seek emergency care.
• Unusual excitement, slow heartbeat, skin rash.	Discontinue. Call fast doctor right away.

WARNINGS & PRECAUTIONS

Don't take if:
- You are allergic to trazodone. Allergic reactions are rare, but may occur.
- You are thinking about suicide.

Before you start, consult your doctor if:
- You have heart rhythm problem.
- You have any heart disease.
- You will have surgery within 2 months, including dental surgery, requiring general or spinal anesthesia.
- You have bipolar (manic-depressive) disorder.
- You have liver or kidney disease.

Over age 60:
Adverse reactions and side effects may be more frequent and severe than in younger persons.

Pregnancy:
Decide with your doctor if drug benefits justify risk to unborn baby. Risk category C (see page xviii).

Breastfeeding; Lactation; Nursing Mothers:
This drug is generally acceptable with breastfeeding. Discuss risks and benefits with your doctor.

Infants & children up to age 18:
Safety and effectiveness in this age group have not been established. If prescribed, carefully read information provided with prescription. Contact doctor right away if depression symptoms get worse or there is any talk of suicide or suicide behaviors. Also, read information under Others.

Prolonged use:
See your doctor for occasional blood counts, especially if you have fever and sore throat.

Skin & sunlight:
May cause rash or intensify sunburn in areas exposed to sun or ultraviolet light (photosensitivity reaction). Use sunscreen and avoid overexposure. Notify doctor if reaction occurs.

Driving, piloting or hazardous work:
Don't drive or pilot aircraft until you learn how medicine affects you. Don't work around dangerous machinery. Don't climb ladders or work in high places. Danger increases if you drink alcohol or take medicine affecting alertness and reflexes, such as antihistamines, tranquilizers, sedatives, pain medicine, narcotics and mind-altering drugs.

Discontinuing:
Don't discontinue without consulting doctor. Dose may require gradual reduction if you have taken drug for a long time. Doses of other drugs may also require adjustment.

Others:
- Advise any doctor, dentist or pharmacist whom you consult that you take this medicine.
- Adults and children taking antidepressants may experience a worsening of the depression symptoms and may have increased suicidal thoughts or behaviors. Call doctor right away if these symptoms or behaviors occur.
- For dry mouth, suck on sugarless hard candy or chew sugarless gum.
- Electroconvulsive therapy* should be avoided. Combined effect is unknown.

POSSIBLE INTERACTION WITH OTHER DRUGS

GENERIC NAME OR DRUG CLASS	COMBINED EFFECT
Antidepressants,* other	Excess drowsiness.
Antihistamines*	Excess drowsiness.
Antihypertensives*	Possible too-low blood pressure. Avoid.
Barbiturates*	Too-low blood pressure and drowsiness. Avoid.
Bupropion	Increased risk of seizures.
Central nervous system (CNS) depressants*	Increased sedation.
Digoxin	Possible increased digitalis effect.
Enzyme Inhibitors*	Increased effect of trazodone.
Guanabenz	Increased effects of both medicines.
Monoamine oxidase (MAO) inhibitors*	May add to toxic effect of each.
Narcotics*	Excess drowsiness.
Phenytoin	Possible increased phenytoin effect.
Ritonavir	Possible increased effect of trazodone.

POSSIBLE INTERACTION WITH OTHER SUBSTANCES

INTERACTS WITH	COMBINED EFFECT
Alcohol:	Excess sedation. Avoid.
Beverages: Caffeine.	May add to heartbeat irregularity. Avoid.
Cocaine:	Increased risk of side effects. Avoid.
Foods:	None expected.
Marijuana:	Increase risk of side effects. Avoid.
Tobacco:	May add to heartbeat irregularity. Avoid.

***See Glossary**

TRIAZOLAM

BRAND NAMES

Apo-Triazo
Halcion
Novo-Triolam
Nu-Triazo

BASIC INFORMATION

Habit forming? Yes
Prescription needed? Yes
Available as generic? Yes
Drug class: Sedative-hypnotic agent

USES

- Treatment for insomnia (short term).
- Prevention or treatment of transient insomnia associated with sudden sleep schedule changes, such as travel across several time zones.

DOSAGE & USAGE INFORMATION

How to take:
Tablet—Swallow with liquid. If you can't swallow whole, crumble tablet and take with liquid or food.

When to take:
At the same time each day, according to instructions on prescription label. You should be in bed when you take your dose.

If you forget a dose:
Take as soon as you remember. If it is almost time for the next dose, wait for that dose (don't double this dose) and resume regular schedule.

What drug does:
Affects limbic system of brain, the part that controls emotions.

Time lapse before drug works:
Within 30 minutes.

Don't take with:
Any other medicine or any dietary supplement without consulting your doctor or pharmacist.

OVERDOSE

SYMPTOMS:
Drowsiness, confusion, slurred speech, coordination problems, breathing difficulty, seizures, coma.
WHAT TO DO:
- **Dial 911 for all medical emergencies or call poison control center 1-800-222-1222 for instructions.**
- **See emergency information on last 3 pages of this book.**

POSSIBLE ADVERSE REACTIONS OR SIDE EFFECTS

SYMPTOMS	WHAT TO DO
Life-threatening:	
Rare, severe allergic reaction (anaphylaxis)—difficulty breathing, hives, itching, swelling (throat, face), vomiting, dizziness.	Seek emergency treatment immediately.
Common:	
Drowsiness, dizziness, clumsiness.	Continue. Call doctor when convenient.
Infrequent:	
• Amnesia, confusion, hallucinations, rash, itch, depression, irritability, vision changes, fever, sore throat, chills.	Discontinue. Call doctor right away.
• Constipation or diarrhea, nausea, vomiting, difficult urination, vivid dreams, behavior changes, abdominal pain, headache, dry mouth.	Continue. Call doctor when convenient.
Rare:	
• Slow heartbeat, breathing difficulty, mouth or throat sores, yellow skin or eyes, sleep-related behaviors.*	Discontinue. Call doctor right away.
• Decreased sex drive.	Continue. Call doctor when convenient.

WARNINGS & PRECAUTIONS

Don't take if:
You are allergic to triazolam or benzodiazepines.

Before you start, consult your doctor if:
- You have sleep apnea.
- You are an active or recovering alcoholic or have abused drugs.
- You have liver, kidney or lung disease.
- You have diabetes, epilepsy or porphyria.
- You have a history of depression.

Over age 60:
Adverse reactions and side effects may be more frequent and severe than in younger persons.

Pregnancy:
Consult doctor. Drug should not be used during pregnancy. Can cause harm to unborn baby. Risk category X (see page xviii).

Breastfeeding; Lactation; Nursing Mothers:
This drug is generally acceptable with breastfeeding. Discuss risks and benefits with your doctor.

Infants & children up to age 18:
Safety and effectiveness in this age group have not been established. Consult doctor.

Prolonged use:
May impair liver function.

Skin & sunlight:
No problems expected.

Driving, piloting or hazardous work:
Don't drive or pilot aircraft until you learn how medicine affects you. Don't work around dangerous machinery. Don't climb ladders or work in high places. Danger increases if you drink alcohol or take medicine affecting alertness and reflexes.

Discontinuing:
Don't discontinue without consulting doctor. Dose may require gradual reduction if you have taken drug for a long time. Doses of other drugs may also require adjustment.

Others:
- Hot weather, heavy exercise and profuse sweating may reduce excretion and cause overdose.
- Blood sugar may rise in diabetics, requiring insulin adjustment.
- Don't use for insomnia more than 4-7 days.
- Advise any doctor, dentist or pharmacist whom you consult that you take this medicine.
- Triazolam has a very short duration of action in the body.

POSSIBLE INTERACTION WITH OTHER DRUGS

GENERIC NAME OR DRUG CLASS	COMBINED EFFECT
Anticonvulsants*	Change in seizure frequency or severity.
Antidepressants*	Increased sedative effects of both drugs.
Antihistamines*	Increased sedative effects of both drugs.
Antihypertensives*	Excessively low blood pressure.
Central nervous system (CNS) depressants,* other	Increased central nervous system depression.
Cimetidine	Increased triazolam effect. May be dangerous.
Clozapine	Toxic effect on the central nervous system.
Contraceptives, oral*	Increased triazolam effect and toxicity.
Disulfiram	Increased triazolam effect and toxicity.
Erythromycins*	Increased triazolam effect and toxicity.
Isoniazid	Increased triazolam effect and toxicity.
Ketoconazole	Increased triazolam effect and toxicity.
Levodopa	Possible decreased levodopa effect.
Molindone	Increased tranquilizer effect.
Monoamine oxidase (MAO) inhibitors*	Convulsions, deep sedation, rage.
Narcotics*	Increased sedative effects of both drugs.
Nefazodone	Increased effects of both drugs.
Omeprazole	Delayed excretion of triazolam causing increased amount of triazolam in blood.
Probenecid	Increased triazolam effect.
Zidovudine	Increased toxicity of zidovudine.

POSSIBLE INTERACTION WITH OTHER SUBSTANCES

INTERACTS WITH	COMBINED EFFECT
Alcohol:	Heavy sedation or amnesia. Avoid.
Beverages: Grapefruit juice.	Increased triazolam effect.
Cocaine:	Sedation. Avoid.
Foods:	None expected.
Marijuana:	Sedation. Avoid.
Tobacco:	Decreased triazolam effect.

*See Glossary

TRIMETHOBENZAMIDE

BRAND NAMES

Tigan

BASIC INFORMATION

Habit forming? No
Prescription needed? Yes
Available as generic? Yes
Drug class: Antiemetic

USES

Reduces nausea and vomiting.

DOSAGE & USAGE INFORMATION

How to take:
Capsule—Swallow with liquid. If you can't swallow whole, open capsule and take with liquid or food.

When to take:
When needed, no more often than label directs.

If you forget a dose:
Take as soon as you remember. If it is almost time for the next dose, wait for that dose (don't double this dose) and resume regular schedule.

What drug does:
Exact mechanism unknown. Possibly blocks nerve impulses to brain's vomiting centers.

Time lapse before drug works:
20 to 40 minutes.

Don't take with:
Any other medicine or any dietary supplement without consulting your doctor or pharmacist.

OVERDOSE

SYMPTOMS:
Confusion, drowsiness, muscle spasms, uncontrolled movements, convulsions, blurred vision, trouble breathing, coma.
WHAT TO DO:
- **Dial 911 for all medical emergencies or call poison control center 1-800-222-1222 for instructions.**
- **See emergency information on last 3 pages of this book.**

POSSIBLE ADVERSE REACTIONS OR SIDE EFFECTS

SYMPTOMS	WHAT TO DO
Life-threatening: In case of overdose, see previous column.	
Common: Drowsiness.	Continue. Call doctor when convenient.
Infrequent: Rash, blurred vision, diarrhea, dizziness, headache, muscle cramps, unusual tiredness.	Discontinue. Call doctor right away.
Rare: Seizures, tremor, depression, sore throat, fever, repeated vomiting, back pain, yellow skin or eyes, body spasm (head and heels bent backward and body bowed forward).	Discontinue. Call doctor right away.

WARNINGS & PRECAUTIONS

Don't take if:
You are allergic to trimethobenzamide. Allergic reactions are rare, but may occur.

Before you start, consult your doctor if:
You have reacted badly to antihistamines.

Over age 60:
Adverse reactions and side effects may be more frequent and severe than in younger persons.

Pregnancy:
Decide with your doctor if drug benefits justify risk to unborn baby. Risk category C (see page xviii).

Breastfeeding; Lactation; Nursing Mothers:
This drug is generally not recommended with breastfeeding. Discuss risks and benefits with your doctor.

Infants & children up to age 18:
Safety and effectiveness in this age group have not been established. Consult doctor.

Prolonged use:
- Damages blood cell production of bone marrow.
- Causes Parkinson's-like symptoms of tremors, rigidity.

Skin & sunlight:
No special problems expected.

Driving, piloting or hazardous work:
Don't drive until you learn how medicine affects you. Don't work around dangerous machinery. Don't climb ladders or work in high places. Danger increases if you drink alcohol or take medicine affecting alertness and reflexes, such as antihistamines, tranquilizers, sedatives, pain medicine, narcotics and mind-altering drugs.

Discontinuing:
May be unnecessary to finish medicine. Follow doctor's instructions.

Others:
Advise any doctor, dentist or pharmacist whom you consult that you take this medicine.

POSSIBLE INTERACTION WITH OTHER DRUGS

GENERIC NAME OR DRUG CLASS	COMBINED EFFECT
Antidepressants*	Increased sedative effect.
Antihistamines*	Increased sedative effect.
Barbiturates*	Increased effect of both drugs.
Belladonna	Increased effect of both drugs.
Cholinergics*	Increased effect of both drugs.
Clozapine	Toxic effect on the central nervous system.
Ethinamate	Dangerous increased effects of ethinamate. Avoid combining.
Fluoxetine	Increased depressant effects of both drugs.
Guanfacine	May increase depressant effects of either medicine.
Leucovorin	High alcohol content of leucovorin may cause adverse effects.
Methyprylon	May increase sedative effect to dangerous level. Avoid.
Mind-altering drugs*	Increased effect of mind-altering drug.
Nabilone	Greater depression of central nervous system.
Narcotics*	Increased sedative effect.
Ototoxic medications*	May mask the symptoms of ototoxicity.
Phenothiazines*	Increased effect of both drugs.
Sedatives*	Increased sedative effect.
Sertraline	Increased depressive effects of both drugs.
Sleep inducers*	Increased effect of sleep inducer.
Tranquilizers*	Increased sedative effect.

POSSIBLE INTERACTION WITH OTHER SUBSTANCES

INTERACTS WITH	COMBINED EFFECT
Alcohol:	Oversedation. Avoid.
Beverages:	None expected.
Cocaine:	Unknown. Avoid.
Foods:	None expected.
Marijuana:	Increased antinausea effect.
Tobacco:	None expected.

***See Glossary**

TRIMETHOPRIM

BRAND NAMES

Apo-Sulfatrim
Apo-Sulfatrim DS
Bactrim
Bactrim DS
Bethaprim
Cotrim
Cotrim DS
Co-trimaxizole
Novotrimel
Novotrimel DS
Nu-Cotrimox
Nu-Cotrimox DS
Proloprim
Protrin
Roubac
Septra
Septra DS
SMZ-TMP
Sulfamethoprim
Sulfamethoprim DS
Sulfaprim
Sulfaprim DS
Sulfatrim
Sulfatrim DS
Sulfoxaprim
Sulfoxaprim DS
Sulmeprim
Triazole
Triazole DS
Trimeth-Sulfa
Trimpex
Trisulfam
Uroplus DS
Uroplus SS

BASIC INFORMATION

Habit forming? No
Prescription needed? Yes
Available as generic? Yes
Drug class: Antimicrobial (antibacterial)

USES

- Treats urinary tract infections susceptible to trimethoprim.
- Helps prevent recurrent urinary tract infections if taken once a day.
- Treats *Pneumocystis pneumonia*.

OVERDOSE

SYMPTOMS:
Nausea, vomiting, diarrhea, dizziness, headache, depression, confusion.
WHAT TO DO:
- **Dial 911 for all medical emergencies or call poison control center 1-800-222-1222 for instructions.**
- **See emergency information on last 3 pages of this book.**

DOSAGE & USAGE INFORMATION

How to take:
- Tablet—Swallow with liquid or food to lessen stomach irritation. If can't swallow whole, crush tablet and take with food or liquid.
- Oral suspension (in combination with sulfamethoxazole)—Follow directions on label.

When to take:
Follow directions on label.

If you forget a dose:
Take as soon as you remember. If it is almost time for the next dose, wait for that dose (don't double this dose) and resume regular schedule.

What drug does:
Stops harmful bacterial germs from multiplying. Will not kill viruses.

Time lapse before drug works:
2 to 5 days.

Don't take with:
Any other medicine or any dietary supplement without consulting your doctor or pharmacist.

POSSIBLE ADVERSE REACTIONS OR SIDE EFFECTS

SYMPTOMS	WHAT TO DO
Life-threatening:	
Rare, severe allergic reaction (anaphylaxis)—difficulty breathing, hives, itching, swelling (throat, face), vomiting, dizziness.	Seek emergency treatment immediately.
Common:	
None expected.	
Infrequent:	
• Skin rash or itching.	Discontinue. Call doctor right away.
• Diarrhea, nausea, vomiting, stomach cramps, loss of appetite.	Continue. Call doctor when convenient.
Rare:	
Blue nails or lips or skin, difficult breathing, headache, sore throat, fever, chills, mouth sores, unusual bleeding or bruising, unusual tiredness or weakness, yellow skin or eyes, aching joints or muscles, pale skin.	Discontinue. Call doctor right away.

WARNINGS & PRECAUTIONS

Don't take if:
- You are allergic to trimethoprim or any sulfa drug.*
- You are anemic due to folic acid deficiency.

Before you start, consult your doctor if:
You have had liver or kidney disease.

Over age 60:
Adverse reactions and side effects may be more frequent and severe than in younger persons.

Pregnancy:
Decide with your doctor if drug benefits justify risk to unborn baby. Risk category C (see page xviii).

Breastfeeding; Lactation; Nursing Mothers:
This drug is generally acceptable with breastfeeding. Discuss risks and benefits with your doctor.

Infants & children up to age 18:
Follow instructions provided by your child's doctor.

Prolonged use:
- Anemia.
- Talk to your doctor about the need for follow-up medical examinations or laboratory studies to check complete blood counts (white blood cell count, platelet count, red blood cell count, hemoglobin, hematocrit).

Skin & sunlight:
May cause rash or intensify sunburn in areas exposed to sun or ultraviolet light (photosensitivity reaction). Avoid overexposure. Notify doctor if reaction occurs.

Driving, piloting or hazardous work:
No problems expected.

Discontinuing:
Don't discontinue without doctor's advice until you complete prescribed dose, even though symptoms diminish or disappear.

Others:
Advise any doctor, dentist or pharmacist whom you consult that you take this medicine.

POSSIBLE INTERACTION WITH OTHER DRUGS

GENERIC NAME OR DRUG CLASS	COMBINED EFFECT
Anticonvulsants*	Increased risk of anemia.
Bone marrow depressants*	Increased possibility of bone marrow suppression.
Folate antagonists,* other	Increased risk of anemia.
Metformin	Increased metformin effect.
Phenytoin	Increased phenytoin effect.

POSSIBLE INTERACTION WITH OTHER SUBSTANCES

INTERACTS WITH	COMBINED EFFECT
Alcohol:	Increased alcohol effect with Bactrim or Septra.
Beverages:	None expected.
Cocaine:	Unknown. Avoid.
Foods:	None expected.
Marijuana:	Consult doctor.
Tobacco:	None expected.

*See Glossary

TRIPTANS

GENERIC AND BRAND NAMES

ALMOTRIPTAN
Axert
ELETRIPTAN
Relpax
FROVATRIPTAN
Frova
NARATRIPTAN
Amerge
RIZATRIPTAN
Maxalt
Maxalt-MLT
SUMATRIPTAN
Alsuma
Imitrex
Imitrex Nasal Spray
Sumavel DosePro
Treximet
Zecuity
ZOLMITRIPTAN
Zomig
Zomig-Nasal Spray
Zomig-ZMT

BASIC INFORMATION

Habit forming? No
Prescription needed? Yes
Available as generic? Yes, for some
Drug class: Antimigraine

USES

- Treatment for acute migraine headaches not relieved by other medications (e.g., aspirin or acetaminophen). Does not prevent migraines.
- Treatment for cluster headaches.

DOSAGE & USAGE INFORMATION

How to take:

- Injection or needle-free system—Follow instructions provided by prescription or doctor for injection technique and how to dispose.
- Tablet—Swallow whole with liquid. Do not crush, break or chew tablet.
- Orally disintegrating tablet—Place on tongue, let it dissolve and swallow with saliva.
- Wafers—Place on tongue to dissolve and be swallowed with saliva.
- Skin patch—Follow prescription instructions.
- Nasal spray—One spray into one nostril as a single dose or as instructed by doctor.

Continued next column

When to take:

- At the first sign of a migraine (aura or pain). After using drug, lie down in a quiet, dark room to increase effectiveness of treatment.
- An additional dose may be helpful if the migraine returns. Do not exceed the prescribed quantity or frequency. Do not use additional dose if first dose does not bring substantial relief.
- For treatment of cluster headaches, follow your doctor's instructions.

If you forget a dose:
Triptans are not taken on a routine schedule. They are taken when migraines occur.

What drug does:
Enlarged (dilated) blood vessels in the brain cause migraines. Triptans work by narrowing (constricting) these blood vessels.

Time lapse before drug works:
Relief usually begins within 30 minutes for oral dosage (tablets/wafers), 10 minutes for injection and 15 minutes for nasal spray.

Don't take with:

- Ergotamine-containing drugs. Delay 24 hours.
- Any other medicine or any dietary supplement without consulting your doctor or pharmacist.

OVERDOSE

SYMPTOMS:
Dizziness, sleepiness, other unknown symptoms.
WHAT TO DO:

- **Dial 911 for all medical emergencies or call poison control center 1-800-222-1222 for instructions.**
- **See emergency information on last 3 pages of this book.**

POSSIBLE ADVERSE REACTIONS OR SIDE EFFECTS

SYMPTOMS	WHAT TO DO
Life-threatening:	
Rare, severe allergic reaction (anaphylaxis)—difficulty breathing, hives, itching, swelling (throat, face), vomiting, dizziness.	Seek emergency treatment immediately.
Common:	
Nausea and/or vomiting (vomiting may be from migraine or from drug), drowsiness or dizziness, migraine recurs.	Continue. Call doctor when convenient.
Infrequent:	
Sensations (of burning, warmth, numbness, cold, tingling), light-headedness, flushing, discomfort (of jaw, mouth, throat, nose or sinuses), anxiety, tiredness, vision changes, just feeling ill, muscle weakness or aches or cramps or stiffness, burning pain or redness at injection site.	Continue. Call doctor when convenient.

Rare:

Pain or pressure or tightness in the chest, difficulty swallowing, irregular heartbeat, shortness of breath, severe stomach pain.	Discontinue. Call doctor right away or seek emergency care for severe symptoms.

WARNINGS & PRECAUTIONS

Don't take if:
- You are allergic to any triptan.
- You have angina pectoris, a history of myocardial infarction or myocardial ischemia, Prinzmetal's angina, stroke or uncontrolled hypertension (high blood pressure).

Before you start, consult your doctor if:
- You have heart rhythm problems or coronary artery disease.
- You have liver or kidney disease.

Over age 60:
Adverse reactions and side effects may be more frequent and severe than in younger persons.

Pregnancy:
Decide with your doctor if drug benefits justify risk to unborn baby. Risk category C (see page xviii).

Breastfeeding; Lactation; Nursing Mothers:
Drugs in this group are generally acceptable with breastfeeding. Discuss risks and benefits with your doctor.

Infants & children up to age 18:
Follow instructions provided by your child's doctor.

Prolonged use:
No problems expected.

Skin & sunlight:
No problems expected.

Driving, piloting or hazardous work:
Avoid if you feel drowsy or dizzy. Otherwise no problems expected.

Discontinuing:
No problems expected. Talk to your doctor if you have plans to discontinue use of drug.

Others:
- Advise any doctor, dentist or pharmacist whom you consult that you take this medicine.
- Rarely, the drug may affect results in some medical tests.
- Follow your doctor's recommendations for any additional treatments for prevention of migraines.
- Sensitivity to light is a symptom of migraine and is not drug-related.

POSSIBLE INTERACTION WITH OTHER DRUGS

GENERIC NAME OR DRUG CLASS	COMBINED EFFECT
Dihydroergotamine	Increased vasoconstriction. Delay 24 hours between drugs.
Enzyme inhibitors*	Increased effect of eletriptan.
Ergotamine	Increased vasoconstriction. Delay 24 hours between drugs.
Monoamine oxidase (MAO) inhibitors*	Adverse effects unknown. Avoid.
Propranolol	Increased effect of rizatriptan.
Selective serotonin reuptake inhibitors (SSRIs)	Risk of serotonin syndrome.*
Serotonergics*	Risk of serotonin syndrome.*
Serotonin & norepinephrine reuptake inhibitors (SNRIs)	Risk of serotonin syndrome.*
Triptans, other	Increased vasoconstriction. Delay 24 hours between drugs.

POSSIBLE INTERACTION WITH OTHER SUBSTANCES

INTERACTS WITH	COMBINED EFFECT
Alcohol:	No interaction known, but alcohol aggravates migraines. Avoid.
Beverages: Grapefruit juice.	May increase the effect of eletriptan.
Cocaine:	Unknown. Avoid.
Foods: Grapefruit.	May increase the effect of eletriptan.
Marijuana:	Consult doctor.
Tobacco:	None expected.

*See Glossary

TUMOR NECROSIS FACTOR BLOCKERS

GENERIC AND BRAND NAMES

ADALIMUMAB
Humira
CERTOLIZUMAB
Cimzia
ETANERCEPT
Enbrel
GOLIMUMAB
Simponi
Simponi Aria
INFLIXIMAB
Remicade

BASIC INFORMATION

Habit forming? No
Prescription needed? Yes
Available as generic? No
Drug class: Antirheumatic; biological response modifier

USES

- Treatment of moderately to severely active rheumatoid arthritis. Used for patients who have not responded to one or more disease modifying antirheumatic drugs (DMARDs). May be used alone or in combination with certain other arthritis drugs (e.g., methotrexate).
- Treats Crohn's disease, ulcerative colitis, moderate to severe ankylosing spondylitis, psoriatic arthritis, and moderate to severe plaque psoriasis.
- May be used for other disorders as determined by your doctor.

DOSAGE & USAGE INFORMATION

How to take:
- Self injection—The drug may be self-injected under the skin (subcutaneously). Follow your doctor's instructions and the directions provided with the prescription on how and where to inject. Do not use the drug unless you are sure about the proper method for injection. Store drug in the refrigerator (do not freeze) until you plan to use it. After each use, throw away the syringe and any drug left in it (ask pharmacist about disposal methods). Never reuse the needles or syringes.
- Infusion—The drug is given in a medical office.

Continued next column

OVERDOSE

SYMPTOMS:
Unknown.
WHAT TO DO:
If person uses much larger amount than prescribed or if accidentally swallowed, call poison control center 1-800-222-1222 for instructions or dial 911 (emergency) for help.

When to take:
Once or twice a week for etanercept, every other week for adalimumab and certolizumab, monthly for golimumab and infliximab is given several times a year. Read label for specific instructions.

If you forget a dose:
Inject as soon as you remember. If it is almost time for the next dose, wait for that dose (don't double this dose) and resume regular schedule.

What drug does:
Blocks the damage done to healthy cells by the tumor necrosis factor (TNF), a protein in the body. The drug helps prevent the progressive joint destruction of rheumatoid arthritis.

Time lapse before drug works:
It will take several weeks before full benefits of the drug are noticeable.

Don't use with:
Any other medicine or any dietary supplement without consulting your doctor or pharmacist.

POSSIBLE ADVERSE REACTIONS OR SIDE EFFECTS

SYMPTOMS	WHAT TO DO
Life-threatening:	
Rare, severe allergic reaction (anaphylaxis)—difficulty breathing, hives, itching, swelling (throat, face), vomiting, dizziness.	Seek emergency treatment immediately.
Common:	
Chills, fever, chest pain or tightness, hives or itching, flushed face, breathing difficulty, stuffy or runny nose, headache, sneezing, wheezing, sore throat, unusual tiredness or weakness, vomiting, rash (on face, scalp or stomach), reaction at injection site.	Call doctor right away.
Infrequent:	
Painful or frequent or difficult urination, bloody or colored urine, fast heartbeat, faintness, pain or stiffness in muscles and joints, pain (stomach, back, face, shoulder, rectum), dizziness, diarrhea, sores in mouth or on tongue, vaginal itching or burning, sore fingernails or toenails.	Call doctor right away.

Rare:	
Allergic reaction (itching, rash, hives, swelling of face or lips, wheezing), black or tarry stools, vision problems, constipation, feeling of fullness, yellow skin or eyes, swollen glands, fungal infection (fever, cough, tiredness, weight loss, sweats, short of breath).	Call doctor right away.

WARNINGS & PRECAUTIONS

Don't take if:
You are allergic to tumor necrosis factor blockers or their components.

Before you start, consult your doctor if:
- You have congestive heart failure (CHF).
- You have or have had tuberculosis.
- You have a central nervous system disorder or a blood disorder or diabetes.
- You are scheduled for surgery.
- You are allergic to rubber or latex (it is used in the drug product's needle covering).
- You have an active infection.
- You have a chronic disorder or infection, a malignancy, or are immunosuppressed.

Over age 60:
Use with caution in elderly patients, since infections are more common in this age group.

Pregnancy:
Decide with your doctor if drug benefits justify risk to unborn baby. Risk category B (see page xviii).

Breastfeeding; Lactation; Nursing Mothers:
Drugs in this group are generally acceptable with breastfeeding. Discuss risks and benefits with your doctor.

Infants & children up to age 18:
Follow instructions provided by your child's doctor. Different drugs in this group are approved for use in children with certain medical disorders.

Prolonged use:
- No specific problems expected.
- Talk to your doctor about the need for follow-up medical examinations or laboratory studies to check effectiveness of the drug and to monitor for infections or adverse effects.

Skin & sunlight:
No problems expected.

Driving, piloting or hazardous work:
No problems expected.

Discontinuing:
No problems expected. Consult doctor first.

Others:
- Serious infections (such as bacterial, viral, fungal, and tuberculosis) and death have been reported in patients using these drugs. Consult your doctor if any signs or symptoms of infection occur.
- Infliximab use may lead to liver toxicity and liver failure with symptoms of yellow skin or eyes, vomiting, and abdominal pain. It may also cause blood abnormalities or blood vessel inflammation with symptoms of vision changes, weakness, numbness or tingling, paleness, fever, or easy bruising or bleeding. Call doctor right away if symptoms occur.
- Slight increased risk of cancer and lupus-like syndromes. Ask your doctor about your risks.
- Advise any doctor, dentist or pharmacist whom you consult that you take this medicine.
- See your doctor for regular visits while using this drug.
- Avoid immunizations unless doctor approved.

POSSIBLE INTERACTION WITH OTHER DRUGS

GENERIC NAME OR DRUG CLASS	COMBINED EFFECT
Anakinra	Risk of serious infection. Avoid.
Immunosuppressants*	May increase risk of infection.
Live vaccines	May decrease vaccine effect.

POSSIBLE INTERACTION WITH OTHER SUBSTANCES

INTERACTS WITH	COMBINED EFFECT
Alcohol:	None expected.
Beverages:	None expected.
Cocaine:	Unknown. Avoid.
Foods:	None expected.
Marijuana:	Consult doctor.
Tobacco:	None expected.

***See Glossary**

URSODIOL

BRAND NAMES

Actigall
Urso
Ursofalk

BASIC INFORMATION

Habit forming? No
Prescription needed? Yes
Available as generic? Yes
Drug class: Anticholelithic

USES

- Dissolves cholesterol gallstones in selected patients who either can't tolerate surgery or don't require surgery for other reasons. Not used when surgery is clearly indicated.
- Prevention of gallstone formation during rapid weight loss.
- Treatment for primary biliary cirrhosis.

DOSAGE & USAGE INFORMATION

How to take:
Capsule or tablet—Swallow with a full glass of water. You may take with food to lessen stomach irritation. If you can't swallow whole, open capsule and take with liquid or food. Ask pharmacist for information about crushing tablet form.

When to take:
With meals, 2 or 3 times a day according to your doctor's instructions.

If you forget a dose:
Take as soon as you remember. If it is almost time for the next dose, wait for that dose (don't double this dose) and resume regular schedule.

Continued next column

OVERDOSE

SYMPTOMS:
Severe diarrhea.
WHAT TO DO:
If person uses much larger amount than prescribed or if accidentally swallowed, call poison control center 1-800-222-1222 for instructions or dial 911 (emergency) for help.

What drug does:
Decreases secretion of cholesterol into bile by suppressing production and secretion of cholesterol by the liver. Ursodiol will not help gallstone problems unless the gallstones are made of cholesterol. It works best when the stones are small.

Time lapse before drug works:
Unpredictable. Varies among patients. If taken for gallstones, the drug may need to be taken for a long time.

Don't take with:
Any other medicine or any dietary supplement without consulting your doctor or pharmacist.

POSSIBLE ADVERSE REACTIONS OR SIDE EFFECTS

SYMPTOMS	WHAT TO DO
Life-threatening:	
Rare, severe allergic reaction (anaphylaxis)—difficulty breathing, hives, itching, swelling (throat, face), vomiting, dizziness.	Seek emergency treatment immediately.
Common:	
None expected.	
Infrequent:	
Diarrhea, nausea, upset stomach, headache.	Continue. Call doctor when convenient.
Rare:	
Any unusual symptoms occur.	Continue. Call doctor when convenient.

WARNINGS & PRECAUTIONS

Don't take if:
You are allergic to any bile acids.

Before you start, consult your doctor if:
- You have complications of gallstones, such as infection, cholecystitis or obstruction of the bile ducts.
- You have had pancreatitis.
- You have heart, liver or kidney disease.

Over age 60:
No special problems expected.

Pregnancy:
Decide with your doctor if drug benefits justify risk to unborn baby. Risk category B (see page xviii).

Breastfeeding; Lactation; Nursing Mothers:
This drug is generally acceptable with breastfeeding. Discuss risks and benefits with your doctor.

Infants & children up to age 18:
Safety and effectiveness in this age group have not been established. Consult doctor.

Prolonged use:
- No special problems expected.
- Talk to your doctor about the need for follow-up medical examinations or laboratory studies to check kidney function.

Skin & sunlight:
No problems expected.

Driving, piloting or hazardous work:
Don't pilot aircraft until you learn how medicine affects you. Don't work around dangerous machinery. Don't climb ladders or work in high places. Danger increases if you drink alcohol or take medicine affecting alertness and reflexes, such as antihistamines, tranquilizers, sedatives, pain medicine, narcotics and mind-altering drugs.

Discontinuing:
Don't discontinue without consulting doctor.

Others:
- Plan regular visits to your doctor while you take ursodiol. Have ultrasound and liver function studies done at appropriate intervals. Liver damage is unlikely, but theoretically could happen.
- Advise any doctor, dentist or pharmacist whom you consult that you take this medicine.

POSSIBLE INTERACTION WITH OTHER DRUGS

GENERIC NAME OR DRUG CLASS	COMBINED EFFECT
Antacids* (aluminum-containing)	Decreased absorption of ursodiol.
Cholestyramine	Decreased absorption of ursodiol.
Clofibrate	Decreased effect of ursodiol.
Colestipol	Decreased absorption of ursodiol.
Estrogens*	Decreased effect of ursodiol.
Progestins	Decreased effect of ursodiol.

POSSIBLE INTERACTION WITH OTHER SUBSTANCES

INTERACTS WITH	COMBINED EFFECT
Alcohol:	None expected, unless you have impaired liver function from alcohol abuse.
Beverages:	None expected.
Cocaine:	Unknown. Avoid.
Foods:	None expected.
Marijuana:	Consult doctor.
Tobacco:	None reported. However, tobacco may possibly impair absorption from the intestinal tract. Better to avoid.

*See Glossary

VALPROIC ACID

BRAND NAMES

Depakene
Stavzor

BASIC INFORMATION

Habit forming? No
Prescription needed? Yes
Available as generic? Yes
Drug class: Anticonvulsant

USES

- Treatment of various types of epilepsy.
- Treatment for bipolar (manic-depressive) disorder.
- Prevention of migraine headaches.

DOSAGE & USAGE INFORMATION

How to take:
- Capsule or syrup—Swallow with liquid or food to lessen stomach irritation. Do not crush or chew capsule.
- Delayed-release capsule—Swallow whole with a glass of water. Do not crush or chew.

When to take:
One to three times a day as directed by doctor.

If you forget a dose:
Take as soon as you remember. If it is almost time for the next dose, wait for that dose (don't double this dose) and resume regular schedule.

What drug does:
It helps stabilize electrical and chemical activity in the brain.

Time lapse before drug works:
1 to 4 hours, but full effect may take weeks.

Don't take with:
Any other medicine or any dietary supplement without consulting your doctor or pharmacist.

OVERDOSE

SYMPTOMS:
Extreme drowsiness, heart problems, loss of consciousness.
WHAT TO DO:
- **Dial 911 for all medical emergencies or call poison control center 1-800-222-1222 for instructions.**
- **See emergency information on last 3 pages of this book.**

POSSIBLE ADVERSE REACTIONS OR SIDE EFFECTS

SYMPTOMS	WHAT TO DO
Life-threatening:	
Rare, severe allergic reaction (anaphylaxis)—difficulty breathing, hives, itching, swelling (throat, face), vomiting, dizziness.	Seek emergency treatment immediately.
Common:	
Mild appetite loss, indigestion, nausea, vomiting, abdominal cramps, diarrhea, tremor, weight gain or loss, menstrual changes in girls, headache.	Continue. Call doctor when convenient.
Infrequent:	
Clumsiness or unsteadiness, constipation, skin rash, dizziness, drowsiness, irritable or excited, hair loss.	Continue. Call doctor when convenient.
Rare:	
Mood or behavior changes, continued nausea or vomiting or appetite loss, increase in seizures, swelling (face, feet, legs), clay-color stools, confusion, yellow skin or eyes, tiredness or weakness, back-and-forth eye movements, seeing spots/seeing double, unusual bleeding or bruising, dark urine, fever, severe stomach cramps.	Continue, but call doctor right away or seek emergency care for severe symptoms.

WARNINGS & PRECAUTIONS

Don't take if:
You are allergic to valproic acid.

Before you start, consult your doctor if:
- You have liver, kidney, blood or brain disorder or pancreatitis or urea cycle disorder.
- Drug is to be used for a young child.
- You have a history of depression or suicide thoughts or suicidal behavior.
- You are a woman of childbearing age.

Over age 60:
Adverse reactions and side effects may be more frequent and severe than in younger persons.

Pregnancy:
Consult doctor. Use of the drug during pregnancy is not recommended. It could cause harm to the unborn baby. Risk category D (see page xviii).

Breastfeeding; Lactation; Nursing Mothers:
This drug is generally acceptable with breastfeeding. Discuss risks and benefits with your doctor.

Infants & children up to age 18:
Follow instructions provided by your child's doctor.

Prolonged use:
Request periodic blood tests, liver and kidney function tests. These tests are necessary for safe and effective use.

Skin & sunlight:
No problems expected.

Driving, piloting or hazardous work:
Don't drive or pilot aircraft until you learn how medicine affects you. Don't work around dangerous machinery. Don't climb ladders or work in high places. Danger increases if you drink alcohol or take medicine affecting alertness and reflexes.

Discontinuing:
Don't discontinue without consulting doctor. Dose may require gradual reduction if you have taken drug for a long time. Doses of other drugs may also require adjustment.

Others:
- Read the insert provided with the drug. Follow all instructions and heed all warnings.
- Advise any doctor, dentist or pharmacist whom you consult that you take this drug.
- In rare cases, the drug can cause life-threatening liver failure (especially in children under age 2) or life-threatening pancreatitis (inflammation of the pancreas). Consult your doctor about your risks.
- Rarely, antiepileptic (anticonvulsant) drugs may lead to suicidal thoughts and behaviors. Call doctor right away if suicidal symptoms or unusual behaviors occur.
- Wear or carry medical identification to show your seizure disorder and the drugs you take.

POSSIBLE INTERACTION WITH OTHER DRUGS

GENERIC NAME OR DRUG CLASS	COMBINED EFFECT
Anticoagulants,* oral	Increased risk of bleeding problems.
Anticonvulsants,* other	Each drug may need dosage adjusted.
Anti-inflammatory drugs, nonsteroidal* (NSAIDs)	Increased risk of bleeding problems.
Aspirin	Increased effect of valproic acid.
Carbamazepine	Decreased effect of valproic acid.
Central nervous system (CNS) depressants*	Increased sedative effect.
Clonazepam	May prolong seizure.
Diazepam	Increased effect of diazepam.
Enzyme inducers*	May increase effect of some enzyme inducers and decrease effect of valproic acid.
Felbamate	Increased effect of valproic acid.
Hepatotoxics*	Increased risk of liver problems.
Lamotrigine	Increased effect of lamotrigine and risk of life-threatening rash.
Phenobarbital	Increased effect of phenobarbital and decreased effect of valproic acid.
Phenytoin	Increased phenytoin effect; decreased valproic acid effect.
Primidone	Increased effect of primidone.
Rifampin	Decreased effect of valproic acid.

Continued on page 913

POSSIBLE INTERACTION WITH OTHER SUBSTANCES

INTERACTS WITH	COMBINED EFFECT
Alcohol:	Sedation. Avoid.
Beverages:	None expected.
Cocaine:	Unknown. Avoid.
Foods:	None expected.
Marijuana:	Consult doctor.
Tobacco:	None expected.

***See Glossary**

VANCOMYCIN

BRAND NAMES

Vancocin

BASIC INFORMATION

Habit forming? No
Prescription needed? Yes
Available as generic? Yes
Drug class: Antibacterial

USES

- Treats colitis when caused by *Clostridium* infections.
- Treats some forms of severe diarrhea and other disorders as determined by your doctor.

DOSAGE & USAGE INFORMATION

How to take:

- Capsule—Swallow with liquid. If you can't swallow whole, open capsule and take with liquid or food. Instructions to take on empty stomach mean 1 hour before or 2 hours after eating.
- Oral solution—Use the calibrated measuring device. Swallow with other liquid to prevent nausea.
- There is also an injectable form. This information applies to the oral form only.

When to take:
According to doctor's instructions. Usually every 6 hours.

If you forget a dose:
Take as soon as you remember. If it is almost time for the next dose, wait for that dose (don't double this dose) and resume regular schedule.

What drug does:
Kills bacterial cells.

Time lapse before drug works:
None. Works right away. This medicine is not absorbed to a great extent through the intestinal tract.

Continued next column

OVERDOSE

SYMPTOMS:
Unknown.
WHAT TO DO:
If person uses much larger amount than prescribed or if accidentally swallowed, call poison control center 1-800-222-1222 for instructions or dial 911 (emergency) for help.

Don't take with:
Any other medicines (including over-the-counter drugs such as cough and cold medicines, laxatives, antacids, diet pills, caffeine, nose drops or vitamins) without consulting your doctor or pharmacist.

POSSIBLE ADVERSE REACTIONS OR SIDE EFFECTS

SYMPTOMS	WHAT TO DO
Life-threatening:	
Rare, severe allergic reaction (anaphylaxis)—difficulty breathing, hives, itching, swelling (throat, face), vomiting, dizziness.	Seek emergency treatment immediately.
Common:	
• Swelling (face, arms, hands, legs, feet), bloody or cloudy urine, painful or decreased urination, thirstiness, irregular heartbeat, pain (side, muscle, back, stomach), weight gain, nausea, vomiting, dry mouth, fever, tingling or numbness in feet or hands, appetite loss.	Discontinue. Call doctor right away.
• Bitter taste, headache, gas, sore mouth.	Continue. Call doctor when convenient.
Infrequent:	
Drowsiness, weakness, difficulty breathing.	Discontinue. Call doctor right away
Rare:	
• Skin symptoms (hives, rash, scaling, red, peeling, itching, blisters, loosening), black or tarry stools, unusual bleeding or bruising or sweating, dizziness, confusion, balance problems, ears ringing, cough, irritated eyes, chills, swallowing difficulty, blurred vision, chest tightness, diarrhea, hearing loss.	Discontinue. Call doctor right away.
• Depression, insomnia, constipation.	Continue. Call doctor when convenient.

WARNINGS & PRECAUTIONS

Don't take if:
You are allergic to vancomycin.

Before you start, consult your doctor if:
- You have hearing problems.
- You have severe kidney disease.
- You have intestinal obstruction.

Over age 60:
Adverse reactions and side effects may be more frequent and severe than in younger persons.

Pregnancy:
Decide with your doctor if drug benefits justify risk to unborn baby. Risk category B (see page xviii).

Breastfeeding; Lactation; Nursing Mothers:
This drug is generally acceptable with breastfeeding. Discuss risks and benefits with your doctor.

Infants & children up to age 18:
Safety and effectiveness in this age group have not been established. Consult doctor.

Prolonged use:
Talk to your doctor about the need for follow-up medical examinations or laboratory studies to check hearing acuity, kidney function, vancomycin serum concentration and urinalysis.

Skin & sunlight:
No problems expected.

Driving, piloting or hazardous work:
No problems expected.

Discontinuing:
No special problems expected.

Others:
- Advise any doctor, dentist or pharmacist whom you consult that you take this medicine.
- May affect results in some medical tests.

POSSIBLE INTERACTION WITH OTHER DRUGS

GENERIC NAME OR DRUG CLASS	COMBINED EFFECT
Cholestyramine	Decreased therapeutic effect of vancomycin.
Colestipol	Decreased therapeutic effect of vancomycin.
Metformin	Increased metformin effect.
Nephrotoxics*	Increased risk of kidney problems.

POSSIBLE INTERACTION WITH OTHER SUBSTANCES

INTERACTS WITH	COMBINED EFFECT
Alcohol:	None expected.
Beverages:	None expected.
Cocaine:	Unknown. Avoid.
Foods:	None expected.
Marijuana:	Consult doctor.
Tobacco:	None expected.

***See Glossary**

VARENICLINE

BRAND NAMES

Chantix

BASIC INFORMATION

Habit Forming? No
Prescription needed? Yes
Available as generic? No
Drug class: Antismoking agent

USES

Helps adults quit smoking. It is recommended that patients combine use of this drug with a stop-smoking program (such as counseling, support groups and/or patient education).

DOSAGE & USAGE INFORMATION

How to take:
Tablet—Swallow whole with a full glass (8 ounces) of water.

When to use:
- The tablet is taken twice a day after eating.
- The first step in treatment is to set a date to quit smoking. Then start taking the drug one week before that date. A lower dose is taken at the start of treatment and increased over the first few days. In most cases, treatment time is 12 weeks. It may be continued for another 12 weeks to improve long-term success in quitting smoking.

If you forget a dose:
Take as soon as you remember. If it is almost time for the next dose, wait for that dose (don't double this dose) and resume regular schedule.

Continued next column

OVERDOSE

SYMPTOMS:
Unknown.
WHAT TO DO:
- **Dial 911 for all medical emergencies or call poison control center 1-800-222-1222 for instructions.**
- **See emergency information on last 3 pages of this book.**

What drug does:
It is a nicotine-free drug that acts on the brain to reduce cravings for cigarettes (and other tobacco products). It also blocks the pleasurable effects of smoking. This helps to decrease the desire to smoke and reduces the unpleasant smoking withdrawal symptoms.

Time lapse before drug works:
It may take several days to see the effects. Full benefit may take 12 weeks of treatment.

Don't take with:
Any other medicine or diet supplement without consulting your doctor or pharmacist.

POSSIBLE ADVERSE REACTIONS OR SIDE EFFECTS

SYMPTOMS	WHAT TO DO
Life-threatening:	
Chest pain, calf pain when walking, trouble breathing or speaking, feeling sudden weakness or numbness.	Seek emergency treatment immediately.
Common:	
Nausea (may go on for several months), insomnia, headache.	Continue. Call doctor when convenient.
Infrequent:	
Strange dreams, gas, abdominal pain, upset stomach, taste changes, constipation, weak or tired feeling, vomiting.	Continue. Call doctor when convenient.
Rare:	
Chest pain, fast or slow or irregular heartbeat, memory loss, seizures, severe or persistent nausea, suicidal thoughts, unusual mental or mood changes, vision changes, drowsiness, other unusual symptoms.	Discontinue. Call doctor right away.

WARNINGS & PRECAUTIONS

Don't take if:
You are allergic to varenicline. Allergic reactions are rare, but may occur.

Before you start, consult your doctor if:
- You have kidney disease.
- You have or have had depression or a psychiatric illness.
- You have any chronic medical or health problem.

Over age 60:
Adverse reactions and side effects may be more frequent and severe than in younger persons.

Pregnancy:
Decide with your doctor if drug benefits justify risk to unborn baby. Risk category C (see page xviii). (Tobacco smoke and nicotine are known to be harmful to a fetus.)

Breastfeeding; Lactation; Nursing Mothers:
This drug is generally not recommended with breastfeeding. Discuss risks and benefits with your doctor.

Infants & children up to age 18:
Safety and effectiveness in this age group have not been established. Consult doctor.

Prolonged use:
It is not intended for long-term use. Usually prescribed for one or two 12-week periods of time.

Skin & sunlight:
No problems expected.

Driving, piloting or hazardous work:
Drug may cause drowsiness. Don't drive or pilot aircraft until you learn how medicine affects you. Don't work around dangerous machinery. Don't climb ladders or work in high places.

Discontinuing:
- Don't discontinue abruptly without doctor's advice. It could increase risk of feeling irritable and cause sleep disturbance.
- Nicotine withdrawal symptoms may occur during and after treatment. These include increased appetite, weight gain, tension, irritability, insomnia, headache and others. Consult doctor if withdrawal symptoms persist.

Others:
- Advise any doctor, dentist or pharmacist whom you consult that you take this medicine.
- For full benefit from a stop-smoking program, follow your doctor's advice.
- Even if you smoke after your quit date, continue to try to quit. Advise your doctor if you do continue to smoke after a few weeks of treatment.
- Patients who are attempting to quit smoking with this drug should be observed for serious mood or behavior changes. Symptoms may include depressed mood, agitation, being aggressive, other emotional changes and thoughts about suicide or possible suicide attempts. Consult doctor about your risks or call doctor right away if any symptoms occur.

POSSIBLE INTERACTION WITH OTHER DRUGS

GENERIC NAME OR DRUG CLASS	COMBINED EFFECT
Cimetidine	May increase effect of varenicline.
Insulin	No interaction, but dosage may need adjustment once you quit smoking.
Nicotine replacement	May increase risk of side effects.
Theophylline	No interaction, but dosage may need adjustment once you quit smoking.
Warfarin	No interaction, but dosage may need adjustment once you quit smoking.

POSSIBLE INTERACTION WITH OTHER SUBSTANCES

INTERACTS WITH	COMBINED EFFECT
Alcohol:	None expected.
Beverages:	None expected.
Cocaine:	Unknown. Avoid.
Foods:	None expected.
Marijuana:	Consult doctor.
Tobacco:	As intended, varenicline blocks the pleasurable effect of nicotine.

VILAZODONE

BRAND NAMES

Viibryd

BASIC INFORMATION

Habit forming? No
Prescription needed? Yes
Available as generic? Yes
Drug class: Antidepressant

USES

- Treatment for major depressive disorder.
- Other uses as recommended by your doctor.

DOSAGE & USAGE INFORMATION

How to take:
Tablet—Swallow with liquid and take with food. The drug may not work as well if you take it on an empty stomach. If you can't swallow tablet whole, ask your doctor or pharmacist for advice.

When to take:
Once a day at the same time each day (usually with a meal).

If you forget a dose:
Take as soon as you remember. If it is almost time for the next dose, wait for that dose (don't double this dose) and resume regular schedule.

What drug does:
The exact mechanism is not fully understood. The drug's dual action increases the level and the effects of serotonin (a brain chemical; also called neurotransmitter). Serotonin plays a role in emotions and psychological disturbances.

Time lapse before drug works:
Begins in 1 to 2 weeks. May require 4 to 6 weeks for maximum benefit.

Don't take with:
Any other medicine or any dietary supplement without consulting your doctor or pharmacist.

OVERDOSE

SYMPTOMS:
Restlessness, hallucinations, disorientation, lethargy, diarrhea, fast heartbeat.
WHAT TO DO:

- **Dial 911 for all medical emergencies or call poison control center 1-800-222-1222 for instructions.**
- **See emergency information on last 3 pages of this book.**

POSSIBLE ADVERSE REACTIONS OR SIDE EFFECTS

SYMPTOMS	WHAT TO DO
Life-threatening: In case of overdose, see previous column.	
Common: Diarrhea, nausea, trouble sleeping, dizziness, dry mouth, change in sexual desire or function.	Continue. Call doctor when convenient.
Infrequent: Vomiting, abnormal dreams, blurred vision, dry eyes, migraine, sedation.	Continue. Call doctor when convenient.
Rare: New or sudden changes in (mood, behaviors, actions, thoughts, feelings, speaking, energy), suicidal thinking or behavior, abnormal bleeding, seizures, low sodium levels in your body (weakness, memory problems, headache, mental changes), serotonin syndrome or neuroleptic syndrome (fast heartbeat, muscle twitching or tightness or stiffness, agitation, hallucinations, fever, sweating, confusion, coordination problems).	Discontinue. Call doctor right away.

WARNINGS & PRECAUTIONS

Don't take if:
- You are allergic to vilazodone. Allergic reactions are rare, but may occur.
- You have taken a monoamine oxidase (MAO) inhibitor* within 2 weeks.

Before you start, consult your doctor if:
- You have a history or family history of mania or hypomania or bipolar disorder.
- You have or have had seizures or convulsions.
- You have low sodium levels (per lab tests).
- You drink alcohol.
- You have or have had bleeding problems.
- You have liver or kidney problems.

Over age 60:
No special problems expected.

Pregnancy:
Decide with your doctor if drug benefits justify risk to unborn baby. Risk category C (see page xviii).

Breastfeeding; Lactation; Nursing Mothers:
This drug is generally acceptable with breastfeeding. Discuss risks and benefits with your doctor.

Infants & children up to age 18:
Safety and effectiveness in this age group have not been established. If prescribed, carefully read information provided with prescription. Contact doctor right away if symptoms get worse or any there is any talk of suicide or suicide behaviors. Read information under Others.

Prolonged use:
Consult with your doctor on a regular basis while taking this drug to monitor your progress, check for side effects and for recommended lab tests.

Skin & sunlight:
No problems expected.

Driving, piloting or hazardous work:
Don't drive or pilot aircraft until you learn how drug affects you. Don't work around dangerous machinery. Don't climb ladders or work in high places. Danger increases if you drink alcohol or take drugs affecting alertness and reflexes.

Discontinuing:
- Don't discontinue without consulting doctor. Dose may require gradual reduction. Doses of other drugs may also require adjustment.
- If new or unexplained symptoms occur after stopping the drug, call doctor right away.

Others:
- Adults and children taking antidepressants may experience a worsening of the depression symptoms, unusual behavior changes and may display increased suicidal thoughts or behavior. Call doctor right away if these symptoms or behaviors occur.
- Advise any doctor, dentist or pharmacist whom you consult that you take this drug.

POSSIBLE INTERACTION WITH OTHER DRUGS

GENERIC NAME OR DRUG CLASS	COMBINED EFFECT
Anticoagulants*	Increased risk of abnormal bleeding.
Anti-inflammatory drugs, nonsteroidal* (NSAIDs)	Increased risk of bleeding problems.
Antiplatelet drugs*	Increased risk of abnormal bleeding.
Central nervous system (CNS) depressants*	Increased sedation.
Diuretics*	May increase risk of hyponatremia (low sodium in blood).
Dopamine antagonists*	Risk of neuroleptic malignant syndrome* or serotonin syndrome.*
Enzyme inhibitors*	Increased effect of vilazodone.
Enzyme inducers*	Decrease effect of vilazodone.
Monoamine oxidase (MAO) inhibitors*	Severe adverse reactions. Allow 14 days between use.
Protein bound drugs,* other	Increased effect of protein bound drug.
Serotonergics,* other	Risk of serotonin syndrome.* Avoid.

POSSIBLE INTERACTION WITH OTHER SUBSTANCES

INTERACTS WITH	COMBINED EFFECT
Alcohol:	Sedation. Avoid.
Beverages: Grapefruit juice.	May increase effect of vilazodone.
Cocaine:	Unknown. Avoid.
Food: Grapefruit.	May increase effect of vilazodone.
Marijuana:	Consult doctor.
Tobacco:	None expected.

*See Glossary

VITAMIN A

BRAND NAMES

Acon
Afaxin
Alphalin
Aquasol A
Dispatabs
Sust-A

Numerous brands of single vitamin and multivitamin combinations are available.

BASIC INFORMATION

Habit forming? No
Prescription needed? No
Available as generic? Yes
Drug class: Vitamin supplement

USES

- Dietary supplement to ensure normal growth and health, especially of eyes and skin.
- Beta carotene form decreases severity of sun exposure in patients with porphyria.

OVERDOSE

SYMPTOMS:
Overdose may come from long-term, chronic intake or a massive one time ingestion. Symptoms may include confusion, hair loss, yellow skin or eyes, dizziness, headache, skin irritation, pain in bones or joints, coma.
WHAT TO DO:
- **Dial 911 for all medical emergencies or call poison control center 1-800-222-1222 for instructions.**
- **See emergency information on last 3 pages of this book.**

DOSAGE & USAGE INFORMATION

How to take:
- Drops or capsule—Swallow with liquid. If you can't swallow whole, open capsule and take with liquid or food.
- Oral solution—Swallow with liquid.
- Tablet—Swallow with liquid.

When to take:
At the same time each day.

If you forget a dose:
Take as soon as you remember. If it is almost time for the next dose, wait for that dose (don't double this dose) and resume regular schedule.

What drug does:
Promotes normal growth and health.

Time lapse before drug works:
Requires continual intake.

Don't take with:
Any other medicine or any dietary supplement without consulting your doctor or pharmacist.

POSSIBLE ADVERSE REACTIONS OR SIDE EFFECTS

SYMPTOMS	WHAT TO DO
Life-threatening: In case of overdose, see previous column.	
Common: None expected with normal doses.	
Infrequent: None expected with normal doses.	
Rare: With high doses, symptoms may occur that cause concern.	Discontinue. Call. doctor when convenient.

WARNINGS & PRECAUTIONS

Don't take if:
You are allergic to vitamin A. Allergic reactions are rare, but may occur.

Before you start, consult your doctor if:
You have any kidney disorder.

Over age 60:
No problems expected.

Pregnancy:
Consult doctor. Risk factor determined by length of pregnancy and dosage amount. See category list on page xviii.

Breastfeeding; Lactation; Nursing Mothers:
This drug is generally acceptable with breastfeeding. Discuss risks and benefits with your doctor.

Infants & children up to age 18:
Read nonprescription product's label to be sure it is approved for your child's age. Follow label instructions on dosage. Consult doctor or pharmacist if unsure.

Prolonged use:
No problems expected.

Skin & sunlight:
No special problems expected.

Driving, piloting or hazardous work:
No problems expected.

Discontinuing:
Don't discontinue without doctor's advice until you complete prescribed dose, even though symptoms diminish or disappear.

Others:
- Don't exceed recommended dosage. Too much over a long time may be harmful.
- A balanced diet will help provide vitamin A. Best sources are liver; yellow-orange fruits and vegetables; dark-green, leafy vegetables; milk; butter and margarine.

POSSIBLE INTERACTION WITH OTHER DRUGS

GENERIC NAME OR DRUG CLASS	COMBINED EFFECT
Anticoagulants*	Increased anticoagulant effect with large doses (over 10,000 I.U.) of vitamin A.
Calcium supplements*	Decreased vitamin effect.
Cholestyramine	Decreased vitamin A absorption.
Colestipol	Decreased vitamin absorption.
Contraceptives, oral*	Increased vitamin A levels.
Etretinate	Increased risk of toxic effects.
Isotretinoin	Increased risk of toxic effect of each.
Mineral oil (long-term)	Decreased vitamin A absorption.
Neomycin	Decreased vitamin absorption.
Vitamin A derivatives, other	Increased toxicity risk.
Vitamin E (excess dose)	Vitamin A depletion.

POSSIBLE INTERACTION WITH OTHER SUBSTANCES

INTERACTS WITH	COMBINED EFFECT
Alcohol:	None expected.
Beverages:	None expected.
Cocaine:	Unknown. Avoid.
Foods:	None expected.
Marijuana:	Consult doctor.
Tobacco:	None expected.

*See Glossary

VITAMIN B-12 (Cyanocobalamin)

GENERIC AND BRAND NAMES

CYANOCOBALAMIN
Anocobin
Bedoz
Berubigen
Betalin 12
CaloMist
Cyanabin
Kaybovite
Kaybovite-1000
Nascobal
Redisol
Rubion
Rubramin
Rubramin-PC

HYDROXOCOBALAMIN
Acti-B-12
Alpha Redisol
Alphamin
Codroxomin
Droxomin

Numerous brands of single vitamin and multivitamin combinations may be available.

BASIC INFORMATION

Habit forming? No
Prescription needed? Yes, for some
Available as generic? Yes
Drug class: Vitamin supplement

USES

- Dietary supplement for normal growth, development and health.
- Treatment for nerve damage.
- Treatment for pernicious anemia.
- Treatment and prevention of vitamin B-12 deficiencies in people who have had stomach or intestines surgically removed.
- Prevention of vitamin B-12 deficiency in strict vegetarians and persons with absorption diseases.

OVERDOSE

SYMPTOMS:
Unknown.
WHAT TO DO:
If person uses much larger amount than prescribed or if accidentally swallowed, call poison control center 1-800-222-1222 for instructions or dial 911 (emergency) for help.

DOSAGE & USAGE INFORMATION

How to take:
- Tablet—Swallow with liquid.
- Extended-release tablet—Swallow whole with liquid. Do not crush or chew.
- Injection—Follow doctor's directions.
- Nasal gel or nasal spray—Follow instructions on product. Use nasal spray 1 hour before or 1 hour after eating hot foods or drinking hot liquids.

When to take:
- Oral—At the same time each day.
- Injection—Follow doctor's directions.
- Nasal gel or spray—Use weekly or as directed on prescription.

If you forget a dose:
Take as soon as you remember. If it is almost time for the next dose, wait for that dose (don't double this dose) and resume regular schedule.

What drug does:
Acts as enzyme to promote normal fat and carbohydrate metabolism and protein synthesis.

Time lapse before drug works:
15 minutes.

Don't take with:
Any other medicine or any dietary supplement without consulting your doctor or pharmacist.

POSSIBLE ADVERSE REACTIONS OR SIDE EFFECTS

SYMPTOMS	WHAT TO DO
Life-threatening:	
Rare, severe allergic reaction (anaphylaxis)—difficulty breathing, hives, itching, swelling (throat, face), vomiting, dizziness.	Seek emergency treatment immediately.
Common:	
None expected with normal doses.	
Infrequent:	
None expected with normal doses.	
Rare:	
With high doses, symptoms may occur that cause concern.	Discontinue. Call. doctor when convenient.

VITAMIN B-12 (Cyanocobalamin)

WARNINGS & PRECAUTIONS

Don't take if:
- You are allergic to any B vitamin.
- You have Leber's disease (optic nerve atrophy).

Before you start, consult your doctor if:
- You have gout.
- You have heart disease.

Over age 60:
Don't take more than recommended amount per day unless prescribed by your doctor.

Pregnancy:
Decide with your doctor if drug benefits justify risk to unborn baby. Risk category C (see page xviii).

Breastfeeding; Lactation; Nursing Mothers:
Drugs in this group are generally acceptable with breastfeeding. Discuss risks and benefits with your doctor.

Infants & children up to age 18:
Read nonprescription product's label to be sure it is approved for your child's age. Follow label instructions on dosage. Consult doctor or pharmacist if unsure.

Prolonged use:
No problems expected.

Skin & sunlight:
No problems expected.

Driving, piloting or hazardous work:
No problems expected.

Discontinuing:
Don't discontinue without doctor's advice until you complete prescribed dose, even though symptoms diminish or disappear.

Others:
- A balanced diet should provide all the vitamin B-12 a healthy person needs and make supplements unnecessary. Best sources are meat, fish, egg yolk and cheese.
- Tablets should be used only for diet supplements. All other uses of vitamin B-12 require injections.
- Don't take large doses of vitamin C (1,000 mg or more per day) unless prescribed by your doctor.

POSSIBLE INTERACTION WITH OTHER DRUGS

GENERIC NAME OR DRUG CLASS	COMBINED EFFECT
Anticonvulsants*	Decreased absorption of vitamin B-12.
Chloramphenicol	Decreased vitamin B-12 effect.
Cholestyramine	Decreased absorption of vitamin B-12.
Cimetidine	Decreased absorption of vitamin B-12.
Colchicine	Decreased absorption of vitamin B-12.
Famotidine	Decreased absorption of vitamin B-12.
H_2 antagonists*	Decreased absorption of vitamin B-12.
Neomycin	Decreased absorption of vitamin B-12.
Para-aminosalicylic acid	Decreased effects of para-aminosalicylic acid.
Potassium (extended-release forms)	Decreased absorption of vitamin B-12.
Ranitidine	Decreased absorption of vitamin B-12.
Vitamin C (ascorbic acid)	Destroys vitamin B-12 if taken at same time. Take 2 hours apart.

POSSIBLE INTERACTION WITH OTHER SUBSTANCES

INTERACTS WITH	COMBINED EFFECT
Alcohol:	Decreased absorption of vitamin B-12.
Beverages:	None expected.
Cocaine:	Unknown. Avoid.
Foods:	None expected.
Marijuana:	Consult doctor.
Tobacco:	None expected.

*See Glossary

VITAMIN C (Ascorbic Acid)

BRAND NAMES

Ascorbicap
Cecon
Cemill
Cenolate
Cetane
Cevalin
Cevi-Bid
Ce-Vi-Sol
Cevita
C-Span
Flavorcee
Redoxon
Sunkist

Numerous brands of single vitamin and multivitamin combinations are available.

BASIC INFORMATION

Habit forming? No
Prescription needed? No
Available as generic? Yes
Drug class: Vitamin supplement

USES

- Prevention and treatment of scurvy and other vitamin C deficiencies.
- Treatment of anemia.
- Maintenance of acid urine.

DOSAGE & USAGE INFORMATION

How to take:

- Tablet or capsule—Swallow with liquid.
- Extended-release tablet or extended-release capsule—Swallow whole with liquid. Do not crush or chew.
- Chewable tablet—Chew well, then swallow.
- Effervescent tablet—Follow label instructions. Let tablet dissolve in water and drink entire mixture.
- Suspension, solution or drops—Use a dropper or dose-measuring spoon to measure doses.
- Syrup—Follow label instructions. Use a dose-measuring spoon to measure dosage and then swallow.
- Lozenge—Let dissolve completely in mouth.
- Powder or crystal—Measure and mix as directed. Drink all the mixture right away.

Continued next column

OVERDOSE

SYMPTOMS:
Diarrhea, vomiting, dizziness.
WHAT TO DO:
If person uses much larger amount than prescribed or if accidentally swallowed, call poison control center 1-800-222-1222 for instructions or dial 911 (emergency) for help.

When to take:
1, 2 or 3 times per day, as prescribed on label.

If you forget a dose:
Take as soon as you remember. If it is almost time for the next dose, wait for that dose (don't double this dose) and resume regular schedule.

What drug does:

- May help form collagen.
- Increases iron absorption from intestine.
- Contributes to hemoglobin and red blood cell production in bone marrow.

Time lapse before drug works:
1 week.

Don't take with:
Any other medicine or any dietary supplement without consulting your doctor or pharmacist.

POSSIBLE ADVERSE REACTIONS OR SIDE EFFECTS

SYMPTOMS	WHAT TO DO
Life-threatening: None expected.	
Common: None expected with normal doses.	
Infrequent: None expected with normal doses.	
Rare: With high doses, symptoms may occur that cause concern.	Discontinue. Call doctor when convenient.

WARNINGS & PRECAUTIONS

Don't take if:
You are allergic to vitamin C. Allergic reactions are rare, but may occur.

Before you start, consult your doctor if:
- You have sickle-cell or other anemia.
- You have had kidney stones.
- You have gout.

Over age 60:
Don't take more than recommended amount per day unless prescribed by your doctor.

Pregnancy:
Decide with your doctor if drug benefits justify risk to unborn baby. Risk category C (see page xviii).

Breastfeeding; Lactation; Nursing Mothers:
This drug is generally acceptable with breastfeeding. Discuss risks and benefits with your doctor.

Infants & children up to age 18:
Read nonprescription product's label to be sure it is approved for your child's age. Follow label instructions on dosage. Consult doctor or pharmacist if unsure.

Prolonged use:
Large doses for longer than 2 months may cause kidney stones.

Skin & sunlight:
No problems expected.

Driving, piloting or hazardous work:
No problems expected.

Discontinuing:
No problems expected.

Others:
- May cause inaccurate tests for sugar in urine or blood in stool.
- May cause crisis in patients with sickle-cell anemia.
- A balanced diet should provide all the vitamin C a healthy person needs and make supplements unnecessary. Best sources are citrus, strawberries, cantaloupe and raw peppers.
- Don't take large doses of vitamin C unless prescribed by your doctor.
- Some products contain tartrazine dye. Avoid, if allergic (especially aspirin hypersensitivity).

POSSIBLE INTERACTION WITH OTHER DRUGS

GENERIC NAME OR DRUG CLASS	COMBINED EFFECT
Amphetamines*	Possible decreased amphetamine effect.
Anticholinergics*	Possible decreased anticholinergic effect.
Anticoagulants,* oral	Possible decreased anticoagulant effect.
Antidepressants, tricyclic (TCA)*	Possible decreased antidepressant effect.
Barbiturates*	Decreased vitamin C effect. Increased barbiturate effect.
Cellulose sodium phosphate	Decreased vitamin C effect.
Contraceptives, oral*	Decreased vitamin C effect.
Estrogens*	Increased likelihood of adverse effects from estrogen with 1 g or more of vitamin C per day.
Iron supplements*	Increased iron absorption.
Mexiletine	Possible decreased effectiveness of mexiletine.
Quinidine	Possible decreased quinidine effect.
Salicylates*	Decreased vitamin C effect and salicylate excretion. May lead to salicylate toxicity.
Tranquilizers* (phenothiazine)	May decrease phenothiazine effect if no vitamin C deficiency exists.

POSSIBLE INTERACTION WITH OTHER SUBSTANCES

INTERACTS WITH	COMBINED EFFECT
Alcohol:	None expected.
Beverages:	None expected.
Cocaine:	Unknown. Avoid.
Foods:	None expected.
Marijuana:	Consult doctor.
Tobacco:	None expected.

*See Glossary

VITAMIN D

GENERIC AND BRAND NAMES

ALFACALCIDOL
One-Alpha
CALCIFEDIOL
Calderol
CALCITRIOL
Rocaltrol
CHOLECALCIFEROL
Fosamax Plus D
Fosavance
DIHYDROTACHY-STEROL
DHT
DHT Intensol
Hytakerol
DOXERCALCIFEROL
Hectorol
ERGOCALCIFEROL
Calciferol
Drisdol
Osto Forte
Radiostol
Radiostol Forte

(Many other name brands are available)

BASIC INFORMATION

Habit forming? No
Prescription needed? Only for high strength
Available as generic? Yes
Drug class: Vitamin supplement

USES

- Dietary supplement.
- Prevention of rickets (bone disease).
- Treatment for hypocalcemia (low blood calcium) in kidney disease.
- Supplement in those who use sunscreen daily.

OVERDOSE

SYMPTOMS:
Can cause nonspecific symptoms such as loss of appetite, weight loss, heart arrhythmias. Also can raise blood calcium levels that can lead to damage to heart, blood vessels and kidneys.
WHAT TO DO:
If person uses much larger amount than prescribed or if accidentally swallowed, call poison control center 1-800-222-1222 for instructions or dial 911 (emergency) for help.

DOSAGE & USAGE INFORMATION

How to take:
- Tablet or capsule—Swallow with liquid.
- Extended-release tablet or extended-release capsule—Swallow whole with liquid. Do not crush or chew.
- Chewable tablet—Chew well, then swallow.
- Effervescent tablet—Follow label instructions. Dissolve tablet in water. Drink entire mixture
- Suspension, solution or drops—Use a dropper or dose-measuring spoon to measure doses.
- Syrup—Follow label instructions. Use a dose-measuring spoon to measure dosage and then swallow.
- Lozenge—Let dissolve completely in mouth.
- Powder or crystal—Measure and mix as directed. Drink all the mixture right away.

When to take:
As directed, usually once a day at the same time each day. Some are taken once a week/month.

If you forget a dose:
Take as soon as you remember. If it is almost time for the next dose, wait for that dose (don't double this dose) and resume regular schedule.

What drug does:
- Maintains growth and health.
- Prevents rickets.
- Essential so body can use calcium and phosphate.

Time lapse before drug works:
2 hours. May require 2 to 3 weeks of continual use for maximum effect.

Don't take with:
Any other medicine or any dietary supplement without consulting your doctor or pharmacist.

POSSIBLE ADVERSE REACTIONS OR SIDE EFFECTS

SYMPTOMS	WHAT TO DO
Life-threatening: None expected.	
Common: None expected with normal doses.	
Infrequent: None expected with normal doses.	
Rare: With high doses, symptoms may occur that cause concern.	Discontinue. Call. doctor when convenient.

WARNINGS & PRECAUTIONS

Don't take if:
You are allergic to medicine containing vitamin D. Allergic reactions are rare, but may occur.

Before you start, consult your doctor if:
- You plan to become pregnant.
- You have epilepsy.
- You have heart or blood-vessel disease.
- You have kidney disease.

Over age 60:
Adverse reactions and side effects may be more frequent and severe than in younger persons.

Pregnancy:
Decide with your doctor if drug benefits justify risk to unborn baby. Risk category C (see page xviii).

Breastfeeding; Lactation; Nursing Mothers:
Drugs in this group are generally acceptable with breastfeeding. Discuss risks and benefits with your doctor.

Infants & children up to age 18:
Read nonprescription product's label to be sure it is approved for your child's age. Follow label instructions on dosage. Consult doctor or pharmacist if unsure.

Prolonged use:
- Consult doctor if unexpected symptoms occur.
- Talk to your doctor about the need for follow-up medical exams or lab studies.

Skin & sunlight:
No special problems expected.

Driving, piloting or hazardous work:
No problems expected.

Discontinuing:
If prescribed, don't discontinue without doctor's advice until you complete prescribed dose.

Others:
- Don't exceed dose. Too much over a long time may be harmful.
- Some products contain tartrazine dye. Avoid, if allergic (especially aspirin hypersensitivity).
- Sunscreen prevents the body from manufacturing vitamin D from sunshine. Consider supplementary vitamin D if you use sunscreen daily. Ask doctor for advice.

POSSIBLE INTERACTION WITH OTHER DRUGS

GENERIC NAME OR DRUG CLASS	COMBINED EFFECT
Antacids* (magnesium-containing)	Possible excess magnesium.
Anticonvulsants, hydantoin*	Decreased vitamin D effect.
Calcium (high doses)	Excess calcium in blood.
Calcium channel blockers*	Possible decreased effect of calcium channel blockers.
Calcium supplements*	Excessive absorption of vitamin D.
Cholestyramine	Decreased vitamin D effect.
Colestipol	Decreased vitamin D absorption.
Cortisone	Decreased vitamin D effect.
Digitalis preparations*	Heartbeat irregularities.
Diuretics, thiazide*	Possible increased calcium.
Mineral oil	Decreased vitamin D effect.
Neomycin	Decreased vitamin D absorption.
Nicardipine	Decreased nicardipine effect.
Phenobarbital	Decreased vitamin D effect.
Phosphorus preparations*	Accumulation of excess phosphorus.
Rifampin	Possible decreased vitamin D effect.
Vitamin D, other	Possible toxicity.

POSSIBLE INTERACTION WITH OTHER SUBSTANCES

INTERACTS WITH	COMBINED EFFECT
Alcohol:	None expected.
Beverages:	None expected.
Cocaine:	Unknown. Avoid.
Foods:	None expected.
Marijuana:	Consult doctor.
Tobacco:	None expected.

***See Glossary**

VITAMIN D (Topical)

GENERIC AND BRAND NAMES

CALCIPOTRIENE	CALCITRIOL
Dovonex	(topical)
Sorilux	Vectical
Taclonex	

BASIC INFORMATION

Habit forming? No
Prescription needed? Yes
Available as generic? Yes, for some
Drug class: Antipsoriatic

USES

Treats discoid or "plaque" psoriasis, the most common form of the disorder.

DOSAGE & USAGE INFORMATION

How to use:
Cream, ointment, foam or topical solution—Apply a thin layer to the affected skin or scalp as per instructions. Rub in gently and completely. Avoid your eyes, mouth and vagina. Wash hands after use.

When to use:
Twice a day or as advised by your doctor.

If you forget a dose:
Apply as soon as you remember. If it is almost time for the next dose, wait for that dose (don't double this dose) and resume regular schedule.

What drug does:
It helps slow the growth of abnormal skin cells.

Time lapse before drug works:
2 weeks. May take up to 8 weeks for maximum benefits (e.g., marked improvement in symptoms for most patients or complete clearing for others.)

Continued next column

OVERDOSE

SYMPTOMS:
May be absorbed into the body through excess topical application and increase the levels of calcium and cause nausea, vomiting, loss of appetite, increased thirst and urination, unusual tiredness or weakness.
WHAT TO DO:
If person uses much larger amount than prescribed or if accidentally swallowed, call poison control center 1-800-222-1222 for instructions or dial 911 (emergency) for help.

Don't use with:
Other topical or oral drugs without consulting with your doctor or pharmacist.

POSSIBLE ADVERSE REACTIONS OR SIDE EFFECTS

SYMPTOMS	WHAT TO DO
Life-threatening:	
None expected.	
Common:	
Irritation, burning, itching of the skin.	Discontinue. Call doctor when convenient.
Infrequent:	
Redness, dryness, peeling, rash, worsening of psoriasis.	Discontinue. Call doctor when convenient.
Rare:	
Darkening of treated areas of skin, pus in hair follicles. Very rare symptoms may occur if drug absorbed into body (nausea, vomiting, loss of appetite, increased thirst and urination, unusual tiredness or weakness).	Discontinue. Call doctor when convenient.

WARNINGS & PRECAUTIONS

Don't use if:
- You are allergic to topical vitamin D.
- You have hypercalcemia (excess of calcium in the body).

Before you start, consult your doctor if:
- You have had allergic reaction to other topical drugs, oral drugs, food or other substances.
- You are taking calcium supplements, oral vitamin D or thiazide diuretics.

Over age 60:
Adverse reactions and side effects may be more frequent and severe than in younger persons.

Pregnancy:
Decide with your doctor if drug benefits justify risk to unborn baby. Risk category C (see page xviii).

Breastfeeding; Lactation; Nursing Mothers:
Drugs in this group are generally acceptable with breastfeeding. Discuss risks and benefits with your doctor.

Infants & children up to age 18:
Safety and effectiveness in this age group have not been established. Consult doctor.

Prolonged use:
Talk to your doctor about the need for follow-up laboratory studies to check calcium levels in your blood or urine.

Skin & sunlight:
Your skin may be more sensitive to sunlight when using this drug. Sunlight exposure (even brief periods) may cause a skin rash, itching, redness, other skin discoloration or a severe sunburn. When you begin using this drug, try to stay out of direct sunlight, especially between the hours of 10:00 a.m. and 3:00 p.m. Apply a sunblock product that has an SPF of 30 or higher. Wear protective clothing, including a hat and sunglasses. Do not use a sun lamp or tanning bed or booth. Consult your doctor if you have any concerns.

Driving, piloting or hazardous work:
No special problems expected.

Discontinuing:
No special problems expected.

Others:
- Wash hands after applying the drug.
- Advise any doctor, dentist or pharmacist whom you consult that you use this medicine.
- One or more of these products is flammable. Read label. Do not use near an open flame.

POSSIBLE INTERACTION WITH OTHER DRUGS

GENERIC NAME OR DRUG CLASS	COMBINED EFFECT
Calcium supplements	Increased calcium in the blood.
Diuretics, thiazide	Increased calcium in the blood.
Vitamin D (oral)	Increased calcium in the blood.

POSSIBLE INTERACTION WITH OTHER SUBSTANCES

INTERACTS WITH	COMBINED EFFECT
Alcohol:	None expected.
Beverages:	None expected.
Cocaine:	None expected.
Foods:	None expected.
Marijuana:	None expected.
Tobacco:	None expected.

*See Glossary

VITAMIN E

BRAND NAMES

Aquasol E
Chew-E
Eprolin
Epsilan-M
Pheryl-E
Viterra E

Numerous brands of single vitamin and multivitamin combinations are available.

BASIC INFORMATION

Habit forming? No
Prescription needed? No
Available as generic? Yes
Drug class: Vitamin supplement

USES

- Dietary supplement to promote normal growth, development and health.
- Treatment and prevention of vitamin E deficiency, especially in premature or low-birth-weight infants.
- Treatment for fibrocystic disease of the breast.
- Treatment for circulatory problems to the lower extremities.
- Treatment for sickle-cell anemia.
- Treatment for lung toxicity from air pollution.

OVERDOSE

SYMPTOMS:
Unknown.
WHAT TO DO:
If person uses much larger amount than prescribed or if accidentally swallowed, call poison control center 1-800-222-1222 for instructions or dial 911 (emergency) for help.

DOSAGE & USAGE INFORMATION

How to take:

- Tablet or capsule—Swallow with liquid.
- Extended-release tablet or extended-release capsule—Swallow whole with liquid. Do not crush or chew.
- Chewable tablet—Chew well, then swallow.
- Effervescent tablet—Follow label instructions. Dissolve tablet in water. Drink entire mixture.
- Suspension, solution or drops—Use a dropper or dose-measuring spoon to measure doses.
- Syrup—Follow label instructions. Use a dose-measuring spoon to measure dosage and then swallow.
- Lozenge—Let dissolve completely in mouth.
- Powder or crystal—Measure and mix as directed. Drink all the mixture right away.

When to take:
At the same times each day.

If you forget a dose:
Take as soon as you remember. If it is almost time for the next dose, wait for that dose (don't double this dose) and resume regular schedule.

What drug does:

- Promotes normal growth and development.
- Prevents oxidation in body.

Time lapse before drug works:
Will vary depending on the disorder being treated.

Don't take with:
Any other medicine or any dietary supplement without consulting your doctor or pharmacist.

POSSIBLE ADVERSE REACTIONS OR SIDE EFFECTS

SYMPTOMS	WHAT TO DO
Life-threatening: None expected.	
Common: None expected with normal doses.	
Infrequent: None expected with normal doses.	
Rare: With high doses, symptoms may occur that cause concern.	Discontinue. Call doctor when convenient.

WARNINGS & PRECAUTIONS

Don't take if:
You are allergic to vitamin E. Allergic reactions are rare, but may occur.

Before you start, consult your doctor if:
- You have had blood clots in leg veins (thrombophlebitis).
- You have liver disease.

Over age 60:
No problems expected. Avoid excessive doses.

Pregnancy:
Consult doctor. Use is generally considered safe. Risk category A (see page xviii).

Breastfeeding; Lactation; Nursing Mothers:
This drug is generally acceptable with breastfeeding. Discuss risks and benefits with your doctor.

Infants & children up to age 18:
Read nonprescription product's label to be sure it is approved for your child's age. Follow label instructions on dosage. Consult doctor or pharmacist if unsure.

Prolonged use:
Toxic accumulation of vitamin E. Don't exceed recommended dose.

Skin & sunlight:
No problems expected.

Driving, piloting or hazardous work:
No problems expected.

Discontinuing:
No problems expected.

Others:
A balanced diet should provide all the vitamin E a healthy person needs and make supplements unnecessary. Best sources are vegetable oils, whole-grain cereals, liver.

POSSIBLE INTERACTION WITH OTHER DRUGS

GENERIC NAME OR DRUG CLASS	COMBINED EFFECT
Anticoagulants,* oral	Increased anticoagulant effect.
Cholestyramine	Decreased vitamin E absorption.
Colestipol	Decreased vitamin E absorption.
Iron supplements*	Possible decreased effect of iron supplement in patients with iron-deficiency anemia. Decreased vitamin E effect in healthy persons.
Mineral oil	Decreased vitamin E effect.
Neomycin	Decreased vitamin E absorption.
Vitamin A	Recommended dose of vitamin E—Increased benefit and decreased toxicity of vitamin A. Excess dose of vitamin E—Vitamin A depletion.

POSSIBLE INTERACTION WITH OTHER SUBSTANCES

INTERACTS WITH	COMBINED EFFECT
Alcohol:	None expected.
Beverages:	None expected.
Cocaine:	Unknown. Avoid.
Foods:	None expected.
Marijuana:	Consult doctor.
Tobacco:	None expected.

*See Glossary

VITAMIN K

GENERIC AND BRAND NAMES

MENADIOL
Synkayvite

PHYTONADIONE
Mephyton

Numerous brands of single vitamin and multivitamin combinations may be available.

BASIC INFORMATION

Habit forming? No
Prescription needed? No
Available as generic? Yes
Drug class: Vitamin supplement

USES

- Dietary supplement.
- Treatment for bleeding disorders and malabsorption diseases due to vitamin K deficiency.
- Treatment for hemorrhagic disease of the newborn.
- Treatment for bleeding due to overdose of oral anticoagulants.

DOSAGE & USAGE INFORMATION

How to take:
- May be given by injection in hospital or doctor's office.
- Tablet or capsule—Swallow with liquid.
- Extended-release tablet or extended-release capsule—Swallow whole with liquid. Do not crush or chew.
- Chewable tablet—Chew well, then swallow.
- Effervescent tablet—Follow label instructions. Let tablet dissolve in water and drink entire mixture.
- Suspension, solution or drops—Use a dropper or dose-measuring spoon to measure doses.
- Syrup—Follow label instructions. Use a dose-measuring spoon to measure dosage and then swallow.
- Lozenge—Let dissolve completely in mouth.
- Powder or crystal—Measure and mix as directed. Drink all the mixture right away.

Continued next column

OVERDOSE

SYMPTOMS:
Unknown.
WHAT TO DO:
If person uses much larger amount than prescribed or if accidentally swallowed, call poison control center 1-800-222-1222 for instructions or dial 911 (emergency) for help.

When to take:
At the same time each day.

If you forget a dose:
Take as soon as you remember. If it is almost time for the next dose, wait for that dose (don't double this dose) and resume regular schedule.

What drug does:
- Promotes growth, development and good health.
- Supplies a necessary ingredient for blood clotting.

Time lapse before drug works:
15 to 30 minutes to support blood clotting. For some problems, it may take weeks or months.

Don't take with:
Any other medicine or any dietary supplement without consulting your doctor or pharmacist.

POSSIBLE ADVERSE REACTIONS OR SIDE EFFECTS

SYMPTOMS	WHAT TO DO
Life-threatening:	
None expected.	
Common:	
None expected with normal doses.	
Infrequent:	
None expected with normal doses.	
Rare:	
With high doses, symptoms may occur that cause concern.	Discontinue. Call. doctor when convenient.

WARNINGS & PRECAUTIONS

Don't take if:
- You are allergic to vitamin K. Allergic reactions are rare, but may occur.
- You have G6PD* deficiency.
- You have liver disease.

Before you start, consult your doctor if:
You are pregnant.

Over age 60:
No problems expected.

Pregnancy:
Decide with your doctor if drug benefits justify risk to unborn baby. Risk category C (see page xviii).

Breastfeeding; Lactation; Nursing Mothers:
Drugs in this group are generally acceptable with breastfeeding. Discuss risks and benefits with your doctor.

Infants & children up to age 18:
Read nonprescription product's label to be sure it is approved for your child's age. Follow label instructions on dosage. Consult doctor or pharmacist if unsure.

Prolonged use:
Talk to your doctor about the need for follow-up medical examinations or laboratory studies to check prothrombin time.

Skin & sunlight:
No problems expected.

Driving, piloting or hazardous work:
No problems expected.

Discontinuing:
No problems expected.

Others:
- Advise any doctor, dentist or pharmacist you consult that you take this medicine.
- Don't exceed dose. Too much over a long time may be harmful.
- A balanced diet should provide all the vitamin K a healthy person needs and make supplements unnecessary. Best sources are green, leafy vegetables, meat or dairy products.

POSSIBLE INTERACTION WITH OTHER DRUGS

GENERIC NAME OR DRUG CLASS	COMBINED EFFECT
Anticoagulants,* oral	Decreased anticoagulant effect.
Cholestyramine	Decreased vitamin K effect.
Colestipol	Decreased vitamin K absorption.
Dapsone	Increased risk of adverse effect on blood cells.
Mineral oil (long-term)	Vitamin K deficiency.
Neomycin	Decreased vitamin K absorption.
Sulfa drugs*	Vitamin K deficiency.

POSSIBLE INTERACTION WITH OTHER SUBSTANCES

INTERACTS WITH	COMBINED EFFECT
Alcohol:	None expected.
Beverages:	None expected.
Cocaine:	Unknown. Avoid.
Foods:	None expected.
Marijuana:	Consult doctor.
Tobacco:	None expected.

***See Glossary**

VITAMINS & FLUORIDE

GENERIC AND BRAND NAMES

Adeflor
Cari-Tab
Mulvidren-F
Poly-Vi-Flor
Tri-Vi-Flor
Vi-Daylin/F
Vi-Penta F

Also brands are available in the forms of multiple vitamins & fluoride; vitamins A, D & C & fluoride.

BASIC INFORMATION

Habit forming? No
Prescription needed? Yes
Available as generic? No
Drug class: Vitamins, minerals

USES

- Reduces incidence of tooth cavities (fluoride). Children who need supplements should take until age 16.
- Prevents deficiencies of vitamin included in formula (some contain multiple vitamins whose content varies among products; others contain only vitamins A, D and C).

DOSAGE & USAGE INFORMATION

How to take:
- Chewable tablet—Chew or crush before swallowing.
- Oral liquid—Measure with specially marked dropper. May mix with food, fruit juice, cereal.

When to take:
- Bedtime or with or just after meals.
- If at bedtime, brush teeth first.

If you forget a dose:
Take as soon as you remember. If it is almost time for the next dose, wait for that dose (don't double this dose) and resume regular schedule.

Continued next column

OVERDOSE

SYMPTOMS:
Unknown.
WHAT TO DO:
If person uses much larger amount than prescribed or if accidentally swallowed, call poison control center 1-800-222-1222 for instructions or dial 911 (emergency) for help.

What drug does:
Provides supplemental fluoride to combat tooth decay.

Time lapse before drug works:
8 weeks to provide maximum benefit.

Don't take with:
- Other medicine at the same time.
- Any other medicine or any dietary supplement without consulting your doctor or pharmacist.

POSSIBLE ADVERSE REACTIONS OR SIDE EFFECTS

SYMPTOMS	WHAT TO DO
Life-threatening: None expected.	
Common: None expected with normal doses.	
Infrequent: None expected with normal doses.	
Rare: With high doses, symptoms may occur that cause concern.	Discontinue. Call doctor when convenient.

WARNINGS & PRECAUTIONS

Don't take if:
- Your water supply contains 0.7 parts fluoride per million. Too much fluoride stains teeth permanently.
- You are allergic to any fluoride-containing product. Allergic reactions are rare, but may occur.
- You have underactive thyroid.

Before you start, consult your doctor or dentist:
For proper dosage.

Over age 60:
No problems expected.

Pregnancy:
Risk factor determined by length of pregnancy and dosage amount. See category list on page xviii and consult doctor.

Breastfeeding; Lactation; Nursing Mothers:
This drug is generally acceptable with breastfeeding. Discuss risks and benefits with your doctor.

Infants & children up to age 18:
Read nonprescription product's label to be sure it is approved for your child's age. Follow label instructions on dosage. Consult doctor or pharmacist if unsure.

Prolonged use:
Excess may cause discolored teeth and decreased calcium in blood.

Skin & sunlight:
No problems expected.

Driving, piloting or hazardous work:
No problems expected.

Discontinuing:
No problems expected.

Others:
- Store in original plastic container. Fluoride decomposes glass.
- Check with dentist once or twice a year to keep cavities at a minimum. Topical applications of fluoride may also be helpful.
- Fluoride probably not necessary if water contains about 1 part per million of fluoride or more. Check with health department.
- Don't freeze.
- Don't keep outdated medicine.

POSSIBLE INTERACTION WITH OTHER DRUGS

GENERIC NAME OR DRUG CLASS	COMBINED EFFECT
Anticoagulants*	Decreased effect of anticoagulant.
Iron supplements*	Decreased effect of any vitamin if iron is present in multivitamin product.
Vitamin A	May lead to vitamin A toxicity if vitamin A is in combination.
Vitamin D	May lead to vitamin D toxicity if vitamin D is in combination.

POSSIBLE INTERACTION WITH OTHER SUBSTANCES

INTERACTS WITH	COMBINED EFFECT
Alcohol:	None expected.
Beverages: Milk.	Prevents absorption of fluoride. Space dose 2 hours before or after milk.
Cocaine:	None expected.
Foods:	None expected.
Marijuana:	None expected.
Tobacco:	None expected.

*See Glossary

VORAPAXAR

BRAND NAMES

Zontivity

BASIC INFORMATION

Habit forming? No
Prescription needed? Yes
Available as generic? No
Drug class: Antiplatelet agent

USES

- Used for the prevention of blood clots that can lead to heart attack or stroke.
- Other uses as recommended by your doctor.

DOSAGE & USAGE INFORMATION

How to take:
Tablet—Swallow whole with water. If you can't swallow it whole, ask doctor or pharmacist for advice. May be taken with or without food.

When to take:
Once a day at the same time each day or as directed by your doctor.

If you forget a dose:
Take as you remember. If it is almost time for the next dose, wait for next scheduled dose (don't double this dose).

What drug does:
It prevents blood cells (platelets) from clumping together and forming blood clots in the arteries.

Time lapse before drug works:
Starts working in one to two hours, but may take a week or more for maximum benefit.

Don't take with:
Any other medicine or any dietary supplement without consulting your doctor or pharmacist.

OVERDOSE

SYMPTOMS:
Bleeding problems that may be severe (e.g., vomiting blood, nosebleed, bright-red blood in stool).
WHAT TO DO:
- **Dial 911 for all medical emergencies or call poison control center 1-800-222-1222 for instructions.**
- **See emergency information on last 3 pages of this book.**

POSSIBLE ADVERSE REACTIONS OR SIDE EFFECTS

SYMPTOMS	WHAT TO DO
Life-threatening: In case of overdose, see previous column.	
Common:	
Mild bruising or bleeding, feeling depressed, skin rash.	Continue. Call doctor when convenient.
Infrequent:	
• Bleeding from gums, frequent nosebleeds, vaginal bleeding, heavy menstrual period, pink or brown urine, red or black or tarry stools, coughing up blood, vomiting blood or coffee ground-like material.	Discontinue. Call doctor right away or seek emergency care.
• Anemia (pale skin, tiredness, faintness, rapid heart rate, shortness of breath, dizziness).	Discontinue. Call doctor right away.
Rare:	
• Signs of stroke (face drooping, arm weakness, speech difficulty).	Seek emergency care.
• Changes in vision.	Continue. Call doctor when convenient.

WARNINGS & PRECAUTIONS

Don't take if:
- You are allergic to vorapaxar. Allergic reactions are rare, but may occur.
- You have a history of transient ischemic attack (TIA), stroke, intracranial hemorrhage (ICH or you have an active bleeding problem (e.g., an ulcer).

Before you start, consult your doctor if:
- You have liver problems.
- You have had any bleeding problems.
- You are allergic to any food, medicine or other substance.

Over age 60:
Risk of stroke and bleeding increases with age. Consult doctor about your risks.

Pregnancy:
Decide with your doctor if drug benefits justify risk to unborn baby. Risk category B (see page xviii).

Breastfeeding; Lactation; Nursing Mothers:
This drug is generally acceptable with breastfeeding. Discuss risks and benefits with your doctor.

Infants & children up to age 18:
Safety and effectiveness in this age group have not been established. Consult doctor.

Prolonged use:
Consult with your doctor on a regular basis while taking this drug to monitor your progress and check for any unwanted effects.

Skin & sunlight:
No problems expected.

Driving, piloting or hazardous work:
No problems expected.

Discontinuing:
- Do not stop taking drug without first talking with your doctor.
- The risk of bleeding can continue for several weeks after stopping the drug.

Others:
- Use of this drug may cause bleeding that can be life-threatening or fatal.
- Advise any doctor, dentist or pharmacist whom you consult that you take this drug.
- Your doctor may have you stop taking drug for a short time if you plan to have surgery or a medical or a dental procedure. You will be advised when to stop and when to resume taking the drug.
- Drug lessens the ability of your blood to clot. You may bruise more easily and it may take longer for any bleeding to stop.

POSSIBLE INTERACTION WITH OTHER DRUGS

GENERIC NAME OR DRUG CLASS	COMBINED EFFECT
Enzyme inducers*	Decreased vorapaxar effect. Consult doctor.
Enzyme inhibitors*	Increased vorapaxar effect. Consult doctor.

POSSIBLE INTERACTION WITH OTHER SUBSTANCES

INTERACTS WITH	COMBINED EFFECT
Alcohol:	None expected.
Beverages:	None expected.
Cocaine:	Unknown. Avoid.
Foods:	None expected.
Marijuana:	Consult doctor.
Tobacco:	No interaction, but smoking is a risk factor for blood clots. Avoid.

*See Glossary

VORTIOXETINE

BRAND NAMES

Brintellix

BASIC INFORMATION

Habit forming? No
Prescription needed? Yes
Available as generic? No
Drug class: Antidepressant, atypical

USES

Treatment for major depressive disorder (MDD) (also referred to as clinical depression).

DOSAGE & USAGE INFORMATION

How to take:
Tablet—Swallow whole with liquid. May be taken with or without food. If you can't swallow tablet whole, ask doctor or pharmacist for advice.

When to take:
Once a day at the same time each day (can be with a meal or not). Dosage may be gradually increased by your doctor.

If you forget a dose:
Take as soon as you remember. If it is almost time for the next dose, wait for that dose (don't double this dose) and resume regular schedule.

What drug does:
It increases the amount of serotonin in the brain. Serotonin is a brain chemical (neurotransmitter) that plays a role in emotional and psychological disturbances.

Time lapse before drug works:
Symptoms may improve after 2 weeks of use, but full benefit may take about a month.

Don't take with:
Any other medicine or any dietary supplement without consulting your doctor or pharmacist.

OVERDOSE

SYMPTOMS:
May experience symptoms similar to side effects—nausea, dizziness, diarrhea, itching and flushing, abdominal discomfort.
WHAT TO DO:
If person takes much larger amount than prescribed or if accidentally swallowed, call doctor or poison control center 1-800-222-1222 for help.

POSSIBLE ADVERSE REACTIONS OR SIDE EFFECTS

SYMPTOMS	WHAT TO DO
Life-threatening: None expected.	
Common: Nausea, vomiting, constipation or diarrhea, dry mouth.	Continue. Call doctor when convenient.
Infrequent: Itching, bloating or passing gas, unusual dreams, sexual problems, dizziness.	Continue. Call doctor when convenient.
Rare: Unusual changes in mood or feelings or behaviors, being agitated or anxious or irritable, suicidal thoughts or actions, increased energy, racing thoughts, abnormal bleeding (e.g., bloody nose, blood in stools, easy bruising), low sodium in body (headache, weakness, confusion, memory problems), serotonin syndrome.*	Discontinue. Call doctor right away.

WARNINGS & PRECAUTIONS

Don't take if:
You are allergic to vortioxetine. Allergic reactions are rare, but may occur.

Before you start, consult your doctor if:
- You have liver problems.
- You have (or have a family history of) bipolar disorder, mania or hypomania.
- You have or have had convulsions or seizures.
- You have hyponatremia (low salt [sodium] levels) in your blood.
- You have or have had bleeding problems.
- You have or have had thoughts about suicide.
- You drink excess amounts of alcohol.

Over age 60:
Adverse reactions and side effects may be more frequent and severe than in younger persons.

Pregnancy:
Decide with your doctor if drug benefits justify risk to unborn baby. Risk category C (see page xviii).

Breastfeeding; Lactation; Nursing Mothers:
This drug is generally not recommended with breastfeeding. Discuss risks and benefits with your doctor.

Infants & children up to age 18:
Safety and effectiveness in this age group have not been established. If prescribed, carefully read information provided with prescription. Contact doctor right away if depression symptoms get worse or there is any talk of suicide or suicide behaviors. Also, read information under Others.

Prolonged use:
Consult with your doctor on a regular basis while taking this drug to check for adverse effects and determine the need for continued treatment.

Skin & sunlight:
No special problems expected.

Driving, piloting or hazardous work:
Avoid if you experience dizziness. Otherwise, no problems expected.

Discontinuing:
- It is recommended that you consult doctor before stopping the drug. In some cases, the dosage may be gradually reduced before stopping.
- If any new or unusual symptoms develop (e.g., headache or muscle tension) after you discontinue the drug, consult your doctor.

Others:
- Advise any doctor, dentist or pharmacist whom you consult that you take this drug.
- Adults and children taking antidepressants may experience a worsening of depression symptoms and may have increased suicidal thoughts or behaviors. Call doctor right away if these symptoms or behaviors occur.
- Take drug as directed. Do not increase or reduce dosage without doctor's approval.

POSSIBLE INTERACTION WITH OTHER DRUGS

GENERIC NAME OR DRUG CLASS	COMBINED EFFECT
Anticoagulants*	Increased risk of abnormal bleeding.
Anti-inflammatory drugs, nonsteroidal (NSAIDs)*	Increased risk of abnormal bleeding.
Aspirin	Increased risk of abnormal bleeding.
Enzyme inducers*	Reduced effect of vortioxetine.
Enzyme inhibitors*	Increased effect of vortioxetine.
Monoamine oxidase (MAO) inhibitors*	Severe adverse reactions. Allow 14 to 21 days between use.
Serotonergics*	Increased risk of serotonin syndrome.

POSSIBLE INTERACTION WITH OTHER SUBSTANCES

INTERACTS WITH	COMBINED EFFECT
Alcohol:	May increase risk of certain side effects. Avoid or limit use.
Beverages:	None expected.
Cocaine:	Unknown. Avoid.
Foods:	None expected.
Marijuana:	Unknown. Avoid.
Tobacco:	None expected.

*See Glossary

XYLOMETAZOLINE

BRAND NAMES

Chlorohist-LA
Inspire
Neo-Synephrine II Long Acting Nasal Spray Adult Strength
Neo-Synephrine II Long Acting Nose Drops Adult Strength
Otrivin Decongestant Nose Drops
Otrivin Nasal Drops
Otrivin Nasal Spray
Otrivin Pediatric Decongestant Nose Drops
Otrivin Pediatric Nasal Drops
Otrivin Pediatric Nasal Spray
Otrivin with M-D Pump
Triaminic Decongestant Spray Nasal & Sinus Congestion

BASIC INFORMATION

Habit forming? No
Prescription needed? No
Available as generic? Yes
Drug class: Sympathomimetic

USES

Relieves congestion of nose, sinuses and throat from allergies and infections.

DOSAGE & USAGE INFORMATION

How to take:
Nasal solution, nasal spray—Use as directed on label. Avoid contamination. Don't use same container for more than 1 person.

When to take:
When needed, no more often than every 4 hours.

Continued next column

OVERDOSE

SYMPTOMS:
Rare occurrence with systemic absorption—sweating, severe tiredness, dizziness, slow heartbeat, coma.
WHAT TO DO:
- **Dial 911 for all medical emergencies or call poison control center 1-800-222-1222 for instructions.**
- **See emergency information on last 3 pages of this book.**

If you forget a dose:
Take as soon as you remember. If it is almost time for the next dose, wait for that dose (don't double this dose) and resume regular schedule.

What drug does:
Constricts walls of small arteries in nose, sinuses and eustachian tubes.

Time lapse before drug works:
5 to 30 minutes.

Don't take with:
- Nonprescription drugs for allergy, cough or cold without consulting doctor.
- Any other medicine or any dietary supplement without consulting your doctor or pharmacist.

POSSIBLE ADVERSE REACTIONS OR SIDE EFFECTS

SYMPTOMS	WHAT TO DO
Life-threatening:	
Rare, severe allergic reaction (anaphylaxis)—difficulty breathing, hives, itching, swelling (throat, face), vomiting, dizziness.	Seek emergency treatment immediately.
Common:	
None expected.	
Infrequent:	
Nasal burning or stinging, runny nose.	Discontinue if symptoms persist.
Rare:	
None expected.	

WARNINGS & PRECAUTIONS

Don't take if:
You are allergic to any sympathomimetic nasal spray.

Before you start, consult your doctor if:
- You have heart disease or high blood pressure.
- You have diabetes.
- You have overactive thyroid.
- You have taken a monoamine oxidase (MAO) inhibitor* in past 2 weeks.
- You have difficulty urinating.

Over age 60:
Adverse reactions and side effects may be more frequent and severe than in younger persons.

Pregnancy:
Decide with your doctor if drug benefits justify risk to unborn baby. Risk category C (see page xviii).

Breastfeeding; Lactation; Nursing Mothers:
This drug is generally acceptable with breastfeeding. Discuss risks and benefits with your doctor.

Infants & children up to age 18:
Read nonprescription product's label to be sure it is approved for your child's age. Follow label instructions on dosage. Consult doctor or pharmacist if unsure.

Prolonged use:
Drug may lose effectiveness and cause a return of congestion.

Skin & sunlight:
No problems expected.

Driving, piloting or hazardous work:
No problems expected.

Discontinuing:
May be unnecessary to finish medicine. Follow doctor's instructions.

Others:
No problems expected.

POSSIBLE INTERACTION WITH OTHER DRUGS

GENERIC NAME OR DRUG CLASS	COMBINED EFFECT
Antidepressants, tricyclic*	Possible increased blood pressure.
Maprotiline	Possible increased blood pressure.
Monoamine oxidase (MAO) inhibitors*	Possible increased blood pressure.

POSSIBLE INTERACTION WITH OTHER SUBSTANCES

INTERACTS WITH	COMBINED EFFECT
Alcohol:	None expected.
Beverages:	
Caffeine drinks.	Nervousness or insomnia.
Cocaine:	Risk of heartbeat irregularities and high blood pressure. Avoid.
Foods:	None expected.
Marijuana:	Consult doctor.
Tobacco:	None expected.

*See Glossary

ZALEPLON

BRAND NAMES

Sonata

BASIC INFORMATION

Habit forming? Yes
Prescription needed? Yes
Available as generic? Yes
Drug class: Anti-insomnia, hypnotic, sedative

USES

Short-term treatment for insomnia (trouble sleeping).

DOSAGE & USAGE INFORMATION

How to take:
Capsule—Swallow with liquid. If you can't swallow whole, open capsule and take with liquid or food.

When to take:
Take immediately before bedtime. Ensure that you can get at least 4 hours of rest after taking your medication. Zaleplon may be taken with or without food; however, if taken after a heavy or fatty meal, it may not work as fast as it should.

If you forget a dose:
Take as soon as you remember. If it is almost time for the next dose, wait for that dose (don't double this dose) and resume regular schedule.

What drug does:
Acts as a central nervous system depressant, decreasing sleep problems such as trouble falling asleep, waking up too often during the night and waking up too early in the morning.

Continued next column

OVERDOSE

SYMPTOMS:
Unsteadiness, confusion, tiredness, troubled breathing, sluggishness, low blood pressure, rarely coma.
WHAT TO DO:
- **Dial 911 for all medical emergencies or call poison control center 1-800-222-1222 for instructions.**
- **See emergency information on last 3 pages of this book.**

Time lapse before drug works:
Within 2 hours.

Don't take with:
Any other medicine or any dietary supplement without consulting your doctor or pharmacist.

POSSIBLE ADVERSE REACTIONS OR SIDE EFFECTS

SYMPTOMS	WHAT TO DO
Life-threatening:	
Rare, severe allergic reaction (anaphylaxis)—difficulty breathing, hives, itching, swelling (throat, face), vomiting, dizziness.	Seek emergency treatment immediately.
Common:	
Dizziness, headache, muscle pain, nausea.	Continue. Call doctor when convenient.
Infrequent:	
• Anxiety, vision problems, not feeling like oneself.	Discontinue. Call doctor right away.
• Abdominal pain, burning or prickling or tingling, cough, constipation, dry mouth, eye pain, fever, indigestion, joint stiffness or pain, skin rash, menstrual pain, nervousness, tightness in chest, trembling or shaking, unusual tiredness or weakness, wheezing, memory loss.	Continue. Call doctor if symptoms persist.
Rare:	
• Nosebleed, hallucinations, sleep-induced behaviors.*	Discontinue. Call doctor right away.
• Loss of appetite, back pain, chest pain, ear pain, general feeling of discomfort, sense of smell difficulty, swelling, rapid weight gain, sensitivity of skin and eyes to sunlight, redness, burning, sunburn.	Continue. Call doctor if symptoms persist.

WARNINGS & PRECAUTIONS

Don't take if:
You have had an allergic reaction to zaleplon.

Before you start, consult your doctor if:
- You have a history of alcohol or drug abuse.
- You have impaired kidney or liver function.
- You are pregnant or nursing.
- You have been diagnosed with clinical depression.

Over age 60:
Adverse reactions and side effects may be more frequent and severe than in younger persons.

Pregnancy:
Decide with your doctor if drug benefits justify risk to unborn baby. Risk category C (see page xviii).

Breastfeeding; Lactation; Nursing Mothers:
This drug is generally acceptable with breastfeeding. Discuss risks and benefits with your doctor.

Infants & children up to age 18:
Safety and effectiveness in this age group have not been established. Consult doctor.

Prolonged use:
Not intended for long term use.

Skin & sunlight:
No problems expected.

Driving, piloting or hazardous work:
Don't drive or pilot aircraft, work around dangerous machinery, climb ladders or work in high places for at least 4 hours after taking this medication. Danger increases if you drink alcohol or take medicine affecting alertness and reflexes, such as antihistamines, tranquilizers, sedatives, pain medicine, narcotics and mind-altering drugs.

Discontinuing:
Dose may require gradual reduction. If drug has been taken for a long time, consult doctor before discontinuing. You may have trouble sleeping for the first few nights after you stop taking zaleplon.

Others:
Advise any doctor, dentist or pharmacist whom you consult that you take this medicine.

POSSIBLE INTERACTION WITH OTHER DRUGS

GENERIC NAME OR DRUG CLASS	COMBINED EFFECT
Antidepressants, tricyclic*	Increased effect of either drug. Avoid.
Central nervous system (CNS) depressants*	May increase effect of depressant.
Enzyme inducers*	Decreased zaleplon effect.
Enzyme inhibitors*	Increased zaleplon effect.

POSSIBLE INTERACTION WITH OTHER SUBSTANCES

INTERACTS WITH	COMBINED EFFECT
Alcohol:	Increased sedation. Avoid.
Beverages:	None expected.
Cocaine:	Increased sedation. Avoid.
Foods:	None expected.
Marijuana:	Increased sedation. Avoid.
Tobacco:	None expected.

***See Glossary**

ZINC SUPPLEMENTS

GENERIC AND BRAND NAMES

ZINC ACETATE	ZINC SULFATE
Galzin	Egozinc
ZINC GLUCONATE	PMS Egozinc
Orazinc	Verazinc
	Zinc-220
	Zincate

Many other multivitamins/mineral products.

BASIC INFORMATION

Habit forming? No
Prescription needed? No
Available as generic? Yes
Drug class: Nutritional supplement (mineral)

USES

- Treats zinc deficiency that may lead to growth retardation, appetite loss, changes in taste or smell, skin eruptions, slow wound healing, decreased immune function, diarrhea or impaired night vision.
- In absence of a deficiency, is used to treat burns, eating disorders, liver disorders, prematurity in infants, intestinal diseases, parasitism, kidney disorders, skin disorders and stress.
- May be useful as a supplement for those who are breastfeeding or pregnant (under a doctor's supervision).
- Zinc acetate is used for treatment of Wilson's disease.

OVERDOSE

SYMPTOMS:
Nausea, vomiting, loss of appetite, stomach cramps, diarrhea, headache; other symptoms may result from chronic use of excess zinc.
WHAT TO DO:
If person uses much larger amount than prescribed or if accidentally swallowed, call poison control center 1-800-222-1222 for instructions or dial 911 (emergency) for help.

DOSAGE & USAGE INFORMATION

How to take:
Tablet or capsule—Swallow with liquid. If you can't swallow whole, crumble tablet or open capsule and take with liquid or food.

When to take:
At the same time each day, according to a doctor's instructions or the package label.

If you forget a dose:
Take as soon as you remember. If it is almost time for the next dose, wait for that dose (don't double this dose) and resume regular schedule.

What drug does:
Required by the body for the utilization of many enzymes, nucleic acids and proteins and for cell growth.

Time lapse before drug works:
2 hours.

Don't take with:
Any other medicine or any dietary supplement without consulting your doctor or pharmacist.

POSSIBLE ADVERSE REACTIONS OR SIDE EFFECTS

SYMPTOMS	WHAT TO DO
Life-threatening: None expected.	
Common: None expected.	
Infrequent: None expected.	
Rare:	
• With large doses—fever, chills, sore throat, sores in throat or mouth, unusual tiredness or weakness.	Discontinue. Call doctor when convenient.
• With large doses—Indigestion, heartburn nausea.	Discontinue. Call doctor if symptoms persist.

WARNINGS & PRECAUTIONS

Don't take if:
You are allergic to zinc. Allergic reactions are rare, but may occur.

Before you start, consult your doctor if:
You have low blood levels of copper.

Over age 60:
No special problems expected. Nutritional supplements may be helpful if the diet is restricted in any way.

Pregnancy:
Decide with your doctor if drug benefits justify risk to unborn baby. Risk category not always designated (see page xviii).

Breastfeeding; Lactation; Nursing Mothers:
This drug is generally acceptable with breastfeeding. Discuss risks and benefits with your doctor.

Infants & children up to age 18:
Read nonprescription product's label to be sure it is approved for your child's age. Follow label instructions on dosage. Consult doctor or pharmacist if unsure.

Prolonged use:
No special problems expected.

Skin & sunlight:
No special problems expected.

Driving, piloting or hazardous work:
No special problems expected.

Discontinuing:
No special problems expected.

Others:
The best natural sources of zinc are red meats, oysters, herring, peas and beans.

POSSIBLE INTERACTION WITH OTHER DRUGS

GENERIC NAME OR DRUG CLASS	COMBINED EFFECT
Copper supplements	Inhibited absorption of copper.
Diuretics, thiazide*	Increased need for zinc.
Folic acid	Increased need for zinc.
Iron supplements*	Increased need for zinc.
Tetracyclines*	Decreased absorption of tetracycline if taken within 2 hours of each other.

POSSIBLE INTERACTION WITH OTHER SUBSTANCES

INTERACTS WITH	COMBINED EFFECT
Alcohol:	May increase need for zinc.
Beverages:	None expected.
Cocaine:	Unknown. Avoid.
Foods: High-fiber.	May decrease zinc absorption.
Marijuana:	Consult doctor.
Tobacco:	May increase need for zinc.

***See Glossary**

ZIPRASIDONE

BRAND NAMES

Geodon

BASIC INFORMATION

Habit forming? No
Prescription needed? Yes
Available as generic? Yes
Drug class: Antipsychotic

USES

- Treatment for schizophrenia.
- Treatment for bipolar disorder.

DOSAGE & USAGE INFORMATION

How to take:
Capsule—Swallow with liquid. Should be taken with food. Do not chew capsule.

When to take:
At the same times each day. The prescribed dosage may gradually be increased over the first few days or weeks of use.

If you forget a dose:
Take as soon as you remember. If it is almost time for the next dose, wait for that dose (don't double this dose) and resume regular schedule.

What drug does:
The exact mechanism is unknown. It appears to block certain nerve impulses between nerve cells.

Time lapse before drug works:
One to 7 days. A further increase in the dosage amount may be necessary to relieve symptoms for some patients.

Don't take with:
Any other medicine or any dietary supplement without consulting your doctor or pharmacist.

OVERDOSE

SYMPTOMS:
Sedation, slurring of speech, short-term high blood pressure.
WHAT TO DO:
If person uses much larger amount than prescribed or if accidentally swallowed, call poison control center 1-800-222-1222 for instructions or dial 911 (emergency) for help.

POSSIBLE ADVERSE REACTIONS OR SIDE EFFECTS

SYMPTOMS	WHAT TO DO
Life-threatening: None expected.	
Common: Constipation or diarrhea, indigestion or heartburn, weight gain, rash, belching, stomach pain, nausea, drowsiness, dizziness, restlessness, feeling weak or a loss of strength, lack of muscle and balance control, coordination problems, trouble in speaking, drooling, twisting body movements (of face, neck and back), arms and legs or muscles feel stiff, muscles tremble, shuffling walk.	Continue. Call doctor when convenient.
Infrequent: Appetite loss and weight loss, runny or stuffy nose, dry mouth, sneezing, vision changes, red and itchy skin, dystonia (unable to move eyes, eyelid twitching, eyes blinking more, tongue wants to stick out, trouble in breathing or speaking or swallowing), feeling faint upon standing after sitting or lying down.	Continue, but call doctor right away.
Rare: Faintness, persistent and painful erection, heartbeat irregular or fast or pounding, palpitations, convulsions, high blood sugar (thirstiness, frequent urination, increased hunger, weakness).	Discontinue. Call doctor right away.

WARNINGS & PRECAUTIONS

Don't take if:
You are allergic to ziprasidone. Allergic reactions are rare, but may occur.

Before you start, consult your doctor if:
- You have liver or kidney disease, heart disease, heart rhythm problems, heart failure, QT prolongation or recent heart attack.
- You have a history of seizures.
- The patient has Alzheimer's.
- You have a family history of, or have diabetes.
- You have tardive dyskinesia.
- You have hypokalemia (low potassium) or hypomagnesemia (low magnesium).
- You have neuroleptic malignant syndrome (serious or fatal problems may occur).

Over age 60:
- Adverse reactions and side effects may be more frequent and severe than in younger persons.
- Use of antipsychotic drugs in elderly patients with dementia-related psychosis may increase risk of death. Consult doctor.

Pregnancy:
Decide with your doctor if drug benefits justify risk to unborn baby. Risk category C (see page xviii).

Breastfeeding; Lactation; Nursing Mothers:
This drug is generally acceptable with breastfeeding. Discuss risks and benefits with your doctor.

Infants & children up to age 18:
Safety and effectiveness in this age group have not been established. Consult doctor.

Prolonged use:
Consult with your doctor on a regular basis while taking this drug to check your progress or to discuss any increase or changes in side effects and the need for continued treatment. Also to check blood levels of potassium and magnesium and to monitor you for any heart problems.

Skin & sunlight:
Hot temperatures and exercise, hot baths can increase risk of heatstroke. Drug may affect body's ability to maintain normal temperature.

Driving, piloting or hazardous work:
Don't drive or pilot aircraft until you learn how medicine affects you. Don't work around dangerous machinery. Don't climb ladders or work in high places. Danger increases if you drink alcohol or take medicine affecting alertness and reflexes.

Discontinuing:
Don't discontinue this drug without consulting doctor. Dosage may require a gradual reduction before stopping.

Others:
- Get up slowly from a sitting or lying position to avoid dizziness, faintness or lightheadedness.
- Advise any doctor, dentist or pharmacist whom you consult that you take this medicine.
- Very rarely, drug may cause reaction called DRESS. Seek emergency care if symptoms of rash, fever and swollen lymph nodes occur.
- Take medicine only as directed. Do not increase or reduce dosage without doctor's approval.

POSSIBLE INTERACTION WITH OTHER DRUGS

GENERIC NAME OR DRUG CLASS	COMBINED EFFECT
Antihypertensives*	Increased antihypertensive effect.
Carbamazepine	Decreased effect of ziprasidone.
Central nervous system (CNS) depressants*	Increased sedative effect. Increased effect of ziprasidone.
Central nervous system (CNS) stimulants*	Unknown effect. Avoid.
Dopamine agonists*	Decreased effect of dopamine agonist.
Enzyme inhibitors*	Increased effect of ziprasidone.
Levodopa	May decrease levodopa effect.
QT interval prolongation-causing drugs*	Heart rhythm problems. Avoid.

POSSIBLE INTERACTION WITH OTHER SUBSTANCES

INTERACTS WITH	COMBINED EFFECT
Alcohol:	Increased sedative affect. Avoid.
Beverages: Grapefruit juice.	May increase the effect of ziprasidone.
Cocaine:	Increased risk of side effects. Avoid.
Foods:	None expected.
Marijuana:	Increased risk of side effects. Avoid.
Tobacco:	None expected.

*See Glossary

ZOLPIDEM

BRAND NAMES

Ambien
Ambien CR
Edluar
Intermezzo
ZolpiMist

BASIC INFORMATION

Habit forming? Yes
Prescription needed? Yes
Available as generic? Yes
Drug class: Sedative-hypnotic agent

USES

- Short-term treatment for insomnia.
- May be used for other disorders as determined by your doctor.

DOSAGE & USAGE INFORMATION

How to take:
- Tablet—Swallow with liquid.
- Controlled-release tablet—Swallow tablet whole. Do not crumble, crush or chew tablet.
- Sublingual tablet—Place under tongue and let it dissolve. Do not take with a liquid and do not swallow tablet.
- Oral spray—Follow instructions provided with prescription.

When to take:
- Take immediately before bedtime. For best results, do not take with a meal or immediately after eating a meal.
- Take drug only when you are able to get 7 to 8 hours (4 hours for brand Intermezzo) of sleep before your daily activity begins.

Continued next column

If you forget a dose:
Take as soon as you remember. If it is almost time for the next dose, wait for that dose (don't double this dose) and resume regular schedule.

What drug does:
Acts as a central nervous system depressant, decreasing sleep problems such as trouble falling asleep, waking up too often during the night and waking up too early in the morning.

Time lapse before drug works:
Within 1 to 2 hours.

Don't take with:
Any other medicine or any dietary supplement without consulting your doctor or pharmacist.

OVERDOSE

SYMPTOMS:
Drowsiness, weakness, stupor, coma.
WHAT TO DO:
- **Dial 911 for all medical emergencies or call poison control center 1-800-222-1222 for instructions.**
- **See emergency information on last 3 pages of this book.**

POSSIBLE ADVERSE REACTIONS OR SIDE EFFECTS

SYMPTOMS	WHAT TO DO
Life-threatening:	
Rare, severe allergic reaction (anaphylaxis)—difficulty breathing, hives, itching, swelling (throat, face), vomiting, dizziness.	Seek emergency treatment immediately.
Common:	
Daytime drowsiness, lightheadedness, dizziness, clumsiness, headache, diarrhea, nausea.	Continue. Call doctor when convenient.
Infrequent:	
Dry mouth, muscle aches or pain, tiredness, indigestion, joint pain, memory problems.	Continue. Call doctor when convenient.
Rare:	
Behavioral changes, agitation, confusion, hallucinations, worsening of depression, bloody or cloudy urine, painful or difficult urination, increased urge to urinate, skin rash or hives, itching, sleep-induced behaviors.*	Discontinue. Call doctor right away.

WARNINGS & PRECAUTIONS

Don't take if:
You are allergic to zolpidem.

Before you start, consult your doctor if:
- You have respiratory problems.
- You have kidney or liver disease.
- You suffer from depression.
- You are an active or recovering alcoholic or substance abuser.

Over age 60:
Adverse reactions and side effects may be more frequent and severe than in younger persons.

Pregnancy:
Decide with your doctor if drug benefits justify risk to unborn baby. Risk category B (see page xviii).

Breastfeeding; Lactation; Nursing Mothers:
This drug is generally acceptable with breastfeeding. Discuss risks and benefits with your doctor.

Infants & children up to age 18:
Safety and effectiveness in this age group have not been established. Consult doctor.

Prolonged use:
Not recommended for long-term usage. Don't take for longer than 1 to 2 weeks unless under doctor's supervision.

Skin & sunlight:
No special problems expected.

Driving, piloting or hazardous work:
Don't drive or pilot aircraft until you learn how medicine affects you (including the day after you take the drug). Don't work around dangerous machinery. Don't climb ladders or work in high places. Danger increases if you drink alcohol or take other medicines affecting alertness and reflexes.

Discontinuing:
- Don't discontinue without consulting doctor. Dose may require gradual reduction if you have taken drug for a long time.
- You may have sleeping problems for 1 or 2 nights after stopping drug.

Others:
- Advise any doctor, dentist or pharmacist whom you consult that you take this medicine.
- Don't take drug if you are traveling on an overnight airplane trip of less than 7 or 8 hours. A temporary memory loss may occur (traveler's amnesia).

POSSIBLE INTERACTION WITH OTHER DRUGS

GENERIC NAME OR DRUG CLASS	COMBINED EFFECT
Central nervous system (CNS) depressants*	Increased sedative effect. Avoid.
Chlorpromazine	Increased sedative effect. Avoid.
Imipramine	Increased sedative effect. Avoid.

POSSIBLE INTERACTION WITH OTHER SUBSTANCES

INTERACTS WITH	COMBINED EFFECT
Alcohol:	Increased sedation. Avoid.
Beverages:	None expected.
Cocaine:	Unknown. Avoid.
Foods:	Decreased sedative effect if taken with a meal or right after a meal.
Marijuana:	Consult doctor.
Tobacco:	None expected.

***See Glossary**

ZONISAMIDE

BRAND NAMES

Zonegran

BASIC INFORMATION

Habit forming? No
Prescription needed? Yes
Available as generic? Yes
Drug class: Anticonvulsant, antiepileptic

USES

Treatment for partial (focal) epileptic seizures. May be used alone or in combination with other antiepileptic drugs.

DOSAGE & USAGE INFORMATION

How to take:
Capsule—Swallow with liquid. Do not break or chew capsule. May be taken with or without food and on a full or empty stomach.

When to take:
Your doctor will determine the best schedule. Dosages will be increased rapidly over the first weeks of use. Further increases may be necessary to achieve maximum benefits.

If you forget a dose:
Take as soon as you remember. If it is almost time for the next dose, wait for that dose (don't double this dose) and resume regular schedule.

What drug does:
The exact mechanism is unknown. Studies have suggested different ways in which the drug provides anticonvulsant activity.

Time lapse before drug works:
May take several weeks for full effectiveness.

Don't take with:
Any other medicine or any dietary supplement without consulting your doctor or pharmacist.

OVERDOSE

SYMPTOMS:
Unknown.
WHAT TO DO:

- **Dial 911 for all medical emergencies or call poison control center 1-800-222-1222 for instructions.**
- **See emergency information on last 3 pages of this book.**

POSSIBLE ADVERSE REACTIONS OR SIDE EFFECTS

SYMPTOMS	WHAT TO DO
Life-threatening:	
None expected.	
Common:	
• Unsteady walk, shakiness.	Continue, but call doctor right away.
• Sleepiness, dizziness, anxiety, restlessness, loss of appetite.	Continue. Call doctor when convenient.
Infrequent:	
• Agitation, delusions, hallucinations, bruising of the skin, depression, unusual mood or mental changes, double vision, rash.	Continue, but call doctor right away.
• Constipation or diarrhea, heartburn, dry mouth, flu-like symptoms (chills, fever, headache, aching muscles and joints), problems with speech, difficulty in concentrating, upset stomach, belching, back and forth eye movement, nausea, runny or stuffy nose, tingling or burning sensations.	Continue. Call doctor when convenient.
Rare:	
Other symptoms.	Continue. Call doctor when convenient.

WARNINGS & PRECAUTIONS

Don't take if:
You are allergic to zonisamide or other sulfonamides.* Allergic reactions are rare, but may occur.

Before you start, consult your doctor if:
- You have a history of liver disease.
- You have kidney problems.
- You have depression or thoughts about suicide.
- You are allergic to any medication, food or other substance.
- You have brittle bones or a growth problem.
- You have a history of metabolic acidosis.

Over age 60:
No special problems expected.

Pregnancy:
Decide with your doctor if drug benefits justify risk to unborn baby. Risk category C (see page xviii).

Breastfeeding; Lactation; Nursing Mothers:
This drug is generally not recommended with breastfeeding. Discuss risks and benefits with your doctor.

Infants & children up to age 18:
Safety and effectiveness in this age group have not been established. Consult doctor.

Prolonged use:
No special problems expected. Follow-up laboratory blood studies may be recommended by your doctor.

Skin & sunlight:
No problems expected.

Driving, piloting or hazardous work:
Don't drive or pilot aircraft until you learn how medicine affects you. Don't work around dangerous machinery. Don't climb ladders or work in high places. Danger increases if you drink alcohol or take other medicines affecting alertness and reflexes such as antihistamines, tranquilizers, sedatives, pain medicine, narcotics and mind-altering drugs.

Discontinuing:
Don't discontinue without doctor's approval due to risk of increased seizure activity. The dosage may need to be gradually decreased before stopping the drug completely.

Others:
- Advise any doctor, dentist or pharmacist whom you consult that you take this medicine.
- Zonisamide may be used with other anticonvulsant drugs and additional side effects may also occur. If they do, discuss them with your doctor.
- May alter some laboratory tests.
- Rarely, antiepileptic drugs may lead to suicidal thoughts and behaviors. Call doctor right away if suicidal symptoms or unusual behaviors occur.
- Wear or carry medical identification to show your seizure disorder and the drugs you take.

POSSIBLE INTERACTION WITH OTHER DRUGS

GENERIC NAME OR DRUG CLASS	COMBINED EFFECT
Anticonvulsants,* other	Decreased effect of zonisamide.
Central nervous system (CNS) depressants*	Increased sedative effect.

POSSIBLE INTERACTION WITH OTHER SUBSTANCES

INTERACTS WITH	COMBINED EFFECT
Alcohol:	Increased sedative effect. Decreased effect of zonisamide. Avoid.
Beverages:	None expected.
Cocaine:	Unknown. Avoid.
Foods:	None expected.
Marijuana:	Consult doctor.
Tobacco:	None expected.

*See Glossary

Generic and Brand Name Directory

Generic and Brand Name Directory

How to Read the Lists Below

On some of the drug charts in this book you are referred to this directory to see a listing of the generic and brand names of the drugs. There are just too many names to fit on the chart page itself, so they are listed here for your reference. It is almost impossible to list all brand names available. Those not listed should be considered as effective as those that are listed.

First, look up the drug chart name (printed in large capital letters). ACETAMINOPHEN is the first name listed in this directory. Under that name, you will find the brand names of drugs (in alphabetical order) that contain the generic drug acetaminophen.

The titles on some drug charts are *drug class* names, such as ADRENOCORTICOIDS (Systemic). In this directory, this *drug class* name is followed first by a numbered list of *generic drug names* in that class. Following that list is the *brand name* list. Each *brand name* has a small number at the end that can be used to match up to the number on the *generic name* list.

For example, under ADRENOCORTICOIDS (Systemic), you will find the brand-name drug **Aristocort**[9]. The number [9] means it contains the generic drug 9. TRIAMCINOLONE.

ACETAMINOPHEN

Abenol
Acephen
Aceta
AcetaDrink
Acetaminophen Uniserts
Aclophen
Actamin
Actamin Extra
Actamin Super
Actimol
Advanced Formula Dristan Caplets
Alba-Temp 300
Alka-Seltzer Plus Cold & Cough Effervescent
Alka-Seltzer Plus Cold & Sinus Liquid-Gels
Alka-Seltzer Plus NightTime - Effervescent
Alka-Seltzer Plus Night-Time Liquid-Gels
Alka-Seltzer Plus Original Effervescent
Alka-Seltzer PM
Allerest No-Drowsiness
All-Nite Cold Formula
Aminofen
Aminofen Max
Anexsia
Anolor 300
Anuphen
Apacet Capsules
Apacet Elixir
Apacet Extra Strength Caplets
Apacet Extra Strength Tablets
Apacet Oral Solution
Apacet Regular Strength Tablets
APAP
Apo-Acetaminophen
Arcet
Aspirin Free Anacin Maximum Strength Caplets
Aspirin Free Anacin Maximum Strength Tablets
Aspirin Free Bayer Select Maximum StrengthHeadache Plus Caplets
Aspirin-Free Excedrin Caplets
Atasol Caplets
Atasol Drops
Atasol Forte
Atasol Forte Caplets
Atasol Forte Tablets

Atasol Oral Solution
Atasol Tablets
Banesin
Bayer Select Maximum Strength Pain Relief Formula
Benadryl Allergy/Sinus Headache Caplets
Benylin All-In-One Cold & Flu Caplets
Benylin All-In-One Cold & Flu Night Caplets
Benylin All-In-One Cold & Flu Nightime Syrup
Benylin All-In-One Cold & Flu Syrup
Benylin All-In-One Day & Night Caplets
Benylin Cold & Sinus
Benylin Cold & Sinus Plus
Benylin Cold & Flu With Codeine Narcotic
Benylin DM 12 Hour Nightime Cough Syrup
Bromo-Seltzer
Buffets
Butapap
Campain
Capital with Codeine
Children's Panadol
Children's Rapimed
Coldrine
Colrex Compound
Comtrex Deep Chest Cold
Conacetol
Conar-A
Congespirin
Congespirin for Children ColdTablets
Congespirin for Children Liquid Cold Medicine
Congestant D
Contac Allergy/Sinus Day Caplets
Contac Allergy/Sinus Night Caplets
Contac Cold + Flu
Contac Maximum Strength Sinus Caplets
Contac Night Caplets
Contac Non-Drowsy Formula Sinus Caplets
Contac Severe Cold Formula
Contac Severe Cold Formula Night Strength
Coricidin HBP
Dapa
Datril Extra Strength
Diabetic Tussin Cold & Flu
Diabetic Tussin Night Time Formula
Dolanex
Dolmar
Dristan AF
Dristan AF Plus
Dristan Cold and Flu
Dristan Cold Caplets
Dristan Cold Maximum Strength Caplets
Dristan Cold Multi- Symptom Formula
Dristan Juice Mix-in Cold, Flu, and Cough
Drixoral Cold and Flu
Drixoral Plus
Drixoral Sinus
Duradyne
Esgic Plus
Excedrin Caplets
Excedrin Extra Strength Caplets
Excedrin Back & Body Excedrin Extra Strength Tablets
Excedrin Migraine
Excedrin Tension Headache
Exdol
Exdol Strong
Febridyne
Feverall Children's
Feverall Infants'
Feverall Junior Strength
Feverall Sprinkle Caps
Fioricet
Fioricet w/Codeine
Gelpirin
Gemnisyn
Genapap
Genapap Children's Elixir
Genapap Children's Tablets
Genapap Extra Strength
Genapap Infants'
Genapap Regular StrengthTablets
Gendecon
Gen-D-phen
Genebs
Genebs Extra Strength
Genebs Regular Strength Tablets
Genex
Goody's Extra Strength Tablets
Goody's Headache Powders

Halenol
Halenol Extra Strength
Histagesic Modified
Histosal
Infants' Anacin-3
Infants' Apacet
Infants' Feverall
Infants' Genapap
Infants' Panadol
Infants' Tylenol
Infants' Tylenol Suspension Drops
Kolephrin
Kolephrin/DM Caplets
Liquiprin Children's Elixir
Liquiprin Infants' Drops
Lortab
Mapap Infant Drops
Maximum Strength Tylenol Allergy Sinus Caplets
Maximum StrengthTylenol Flu Gelcaps
Meda Cap
Meda Tab
Myapap Elixir
Mucinex Adult Caplets - Cold & Sinus
Mucinex Adult Caplets - Cold, Flu, & Sore Throat
Mucinex Adult Caplets - Severe Congestion & Cold
Mucinex Fast-Max Cold, Flu & Sore Throat Liquid
Mucinex MaximumStrength Fast Max Cold, Flu & Sore Throat
ND-Gesic
Neocitrin Colds and Flu Calorie Reduced
NeoCitrin Extra Strength Colds and Flu
NeoCitrin Extra Strength Sinus
Neopap
Nighttime Pamprin
NyQuil Liquicaps
Nytcold Medicine
Oraphen-PD
Ornex Maximum Strength Caplets
Ornex No Drowsiness Caplets
Oxycet
Panadol
Panadol Extra Strength
Panadol Junior Strength Caplets
Panadol Maximum Strength Caplets
Panadol Maximum Strength Tablets
Panex
Panex 500
Paracetamol
Parafon Forte
PediaCare Children's Fever Reducer Plus Cough & Runny Nose
PediaCare Children's Fever Reducer Plus Cough & Sore Throat
PediaCare Children'sFever Reducer Plus Flu
PediaCare Fever Reducer Plus Multi-Symptom Cold
PediaCare Infants' Fever Reducer / Pain Reliever
PediaCare Single Dose
Pedric
Percocet
Pertussin All Night PM
Phenapap SinusHeadache & Congestion
Phenaphen
Phrenilin Forte
Phrenilin w/Caffeine & Codeine
Presalin
Redutemp
Refenesen Chest Congestion & Pain Relief PE
Remcol-C
Repan
Rid-A-Pain Compound
Robigesic
Robitussin Honey Flu
Robitussin Night Relief
Robitussin Night Relief Colds Formula Liquid
Rounox
Roxicet
Roxilox
Semcet
Sinubid
Sinus Relief
Slo-Phyllin GG
Snaplets-FR
St. Joseph Aspirin Free Fever Reducer for Children
Stagesic
Sudafed Multi-Symptom Cold & Cough
Summit
Supac
Suppap
Tapanol
Tapanol Extra Strength

Tapar
Tavist Allergy/Sinus/Headache
Tempra
Tempra Caplets
Tempra Chewable Tablets
Tempra Double Strength
Tempra Drops
Tempra D.S.
Tempra Infants
Tempra Syrup
Tenol
Tenol Plus
Theracof Plus Multi-Symptom Cough and Cold Reliever
Theraflu Cold & SoreThroat Hot Liquid
Theraflu Daytime Severe Cold Caplets
Theraflu Flu & ChestCongestion Hot Liquid
Theraflu Flu & Sore Throat Hot Liquid
Theraflu Nighttime Severe Cold Caplets
Theraflu Nighttime Severe Cold Hot Liquid
Theraflu Warming Relief Daytime
Theraflu Warming Relief Nighttime
Triaminic Cough & Sore Throat
Triaminic Flu, Cough & Fever
Triaminic Softchews Cough & Sore Throat
Trigesic
Tylenol
Tylenol 8 Hour
Tylenol Allergy Multi-Symptom
Tylenol Allergy Multi-Symptom Nighttime
Tylenol Arthritis Pain
Tylenol Chest Congestion
Tylenol Cold Head Congestion Daytime
Tylenol Cold Head Congestion Nighttime
Tylenol Cold Head Congestion Severe
Tylenol Concentrated Infants' Drops
Tylenol Cough & Severe Congestion Daytime
Tylenol Cough & Sore Throat Daytime
Tylenol Cough & Sore Throat Nighttime
Tylenol Extra Strength
Tylenol Extra Strength Rapid Release Gels
Tylenol Junior Strength
Tylenol PM
Tylenol Severe Allergy
Tylenol Sinus Congestion & Pain Daytime
Tylenol Sinus Congestion & Pain Nighttime
Tylenol Sinus Congestion & Pain Severe
Tylenol Sinus Severe Congestion Daytime
Tylenol Sore Throat Daytime
Tylenol Sore Throat Nighttime
Tylenol with Codeine No. 3
Tylenol with Codeine No. 4
Ty-Pap
Tylox
Ultracet
Valadol
Valadol Liquid
Valorin
Valorin Extra
Vanquis
Vicks DayQuil Cold/Flu Relief
Vicks DayQuil Cold & Flu Symptom Relief Plus Vitamin C
Vicks DayQuil Sinus LiquiCaps
Vicks Formula 44 Custom Care Body Aches
Vicks Formula 44 Custom Care Cough & Cold PM
Vicks NyQuil Cold & Flu Relief
Vicks NyQuil Cold & Flu Symptom Relief Plus Vitamin C
Vicks NyQuil D
Vicks NyQuil Less Drowsy Cold & Flu Relief Liquid
Vicks NyQuil Sinus LiquiCaps
Women's Tylenol Menstrual Relief Caplets
Xartemis XR
Zydone

ADRENOCORTICOIDS (Systemic)

GENERIC NAMES

1. BETAMETHASONE
2. BUDESONIDE
3. CORTISONE
4. DEXAMETHASONE
5. FLUDROCORTISONE
6. HYDROCORTISONE (Cortisol)
7. METHYLPREDNISOLONE
8. PREDNISOLONE

9. PREDNISONE
10. TRIAMCINOLONE

BRAND NAMES

Apo-Prednisone[9]
Aristocort[10]
Betnelan[1]
Betnesol[1]
Celestone[1]
Cortef[6]
Cortenema[6]
Cortifoam[6]
Cortone[3]
Cortone Acetate[3]
Decadron[4]
Delta-Cortef[7]
Deltasone[9]
Deronil[4]
Dexasone[4]
Dexone 0.5[4]
Dexone 0.75[4]
Dexone 1.5[4]
Dexone 4[4]
Entocort EC[2]
Flo-Pred[8]
Florinef[5]
Hexadrol[4]
Hydeltrasol8[7]
Hydrocortone[6]
Kenacort[10]
Kenacort Diacetate[10]
Medrol[7]
Meticorten[9]
Mymethasone[4]
Nor-Pred-TBA[8]
Oradexon[4]
Orapred[8]
Orapred ODT[8]
Orasone 1[9]
Orasone 5[9]
Orasone 10[9]
Orasone 20[9]
Orasone 50[9]
Pediapred[8]
Prednisone Intensol[9]
Prednicen-M[9]
Prelone[8]
Rayos[9]
Sterapred DS[9]
Solurex[4]
Solurex LA[4]
Uceris[2]
Winpred[9]

ADRENOCORTICOIDS (Topical)

GENERIC NAMES

1. ALCLOMETASONE Topical)
2. AMCINONIDE (Topical)
3. BECLOMETHASONE (Topical)
4. BETAMETHASONE (Topical)
5. CLOBETASOL (Topical)
6. CLOBETASONE (Topical)
7. CLOCORTOLONE (Topical)
8. CORTISOL
9. DESONIDE (Topical)
10. DESOXIMETASONE (Topical)
11. DEXAMETHASONE (Topical)
12. DIFLORASONE (Topical)
13. DIFLUCORTOLONE (Topical)
14. FLUMETHASONE (Topical)
15. FLUOCINOLONE (Topical)
16. FLUOCINONIDE (Topical)
17. FLURANDRENOLIDE (Topical)
18. FLUTICASONE (Topical)
19. HALCINONIDE (Topical)
20. HALOBETASOL
21. HYDROCORTISONE (Dental)
22. HYDROCORTISONE (Topical)
23. MOMETASONE (Topical)
24. PREDNICARBATE
25. TRIAMCINOLONE (Topical)

BRAND NAMES

9-1-1[22]
Aclovate[1]
Acticort-100[22]
Adcortyl[25]
Aeroseb-Dex[11]
Aeroseb-HC[22]
Ala-Cort[22]
Ala-Scalp HP[22]
Allercort[22]
Alphaderm[22]
Alphatrex[4]
Anucort-HC[22]
Aristocort[25]
Aristocort A[25]
Aristocort C[25]
Aristocort D[25]
Aristocort R[25]
Bactine[22]
Barriere-HC[22]

Beben[4]
Beta HC[22]
Betacort Scalp Lotion[4]
Betaderm[4]
Betaderm Scalp Lotion[4]
Betamethacot[4]
Betatrex[4]
Beta-Val[4]
Betnovate[4]
Betnovate 1/2[4]
Bio-Syn[15]
CaldeCORT Anti-Itch[22]
CaldeCORT-Light[22]
Carmol-HC[22]
Celestoderm-V[4]
Celestoderm-V/2[4]
Cetacort[22]
Chloromyceti[22]
Cipro HC[22]
Clobex[5]
Cloderm[7]
Cloderm Pump[7]
Coraz Lotion[22]
Cordran[17]
Cordran SP[17]
Cormax[5]
Cortacet[22]
Cortaid[22]
Cortaid FastStick[22]
Cortate[22]
Cort-Dome[22]
Cortef[22]
Cortef Feminine Itch[22]
Corticreme[22]
Cortifair[22]
Cortisporin Ointment[22]
Cortoderm[22]
Cortril[22]
Cultivate[18]
Cyclocort[2]
Decaderm[11]
Decadron[11]
Decaspray[11]
Delacort[22]
Delta-Tritex[25]
Demarest DriCort[22]
Dermabet[4]
Dermacomb[25]
Dermacort[22]
Dermalleve[22]
DermAtop[24]
DermiCort[22]
Dermovate[5]
Dermovate Scalp Application[5]
Dermtex HC[22]
Desonate[9]
DesOwen[9]
Diprolene[4]
Diprolene AF[4]
Diprosone[4]
Drenison[17]
Drenison-1/4[17]
Ectosone[4]
Ectosone Regular[4]
Ectosone Scalp Lotion[4]
Efcortelan[22]
Elocon[23]
Emo-Cort[22]
Emo-Cort Scalp Solution[22]
Epifoam[8]
Eumovate[6]
Fludroxycortide[17]
Fluocet[15]
Fluocin[16]
Fluoderm[15]
Fluolar[15]
Fluonid[15]
Fluonide[15]
Flurosyn[15]
Flutex[25]
Foille Cort[22]
Gly-Cort[22]
Gynecort[22]
Gynecort 10[22]
Halciderm[19]
Halog[19]
Halog E[19]
Hi-Cor 1.0[22]
Hi-Cor 2.5[22]
Hyderm[22]
Hydro-Tex[22]
Hytone[22]
Kenac[25]
Kenalog[25]
Kenalog in Orabase[25]
Kenalog-H[25]
Kenonel[25]
Lacticare-HC[22]
Lanacort[22]
Lanacort 10[22]
Lemoderm[22]

Licon[16]
LidaMantle HC[22]
Lidemol[16]
Lidex[16]
Lidex-E[16]
Lipsovir[22]
Locacorten[14]
Locoid[22]
Lotrisone[4]
Lyderm[16]
Maxiflor[12]
Maximum Strength Cortaid[22]
Maxivate[4]
Metaderm Mild[4]
Metaderm Regular[4]
Metosyn[16]
Metosyn FAPG[16]
Myco II[25]
Mycogen II[25]
Mycolog II[25]
My Cort[22]
Myco-Triacet II[25]
Mytrex[25]
Nerisone[13]
Nerisone Oily[13]
Novobetamet[4]
Novohydrocort[22]
Nutracort[22]
Olux[5]
Olux-E[5]
Orabase HCA[22]
Oracort[25]
Oralone[25]
Pandel[22]
Penecort[22]
Pentacort[22]
Pharma-Cort[22]
Pramosone E Cream[22]
Prevex B[4]
Prevex HC[4]
Propaderm[3]
Psorcon[12]
Rederm[22]
Rhulicort[22]
Sarna HC[22]
Sential[22]
S-T Cort[22]
Synacort[22]
Synalar[15]
Synalar HP[15]
Synamol[15]
Synemol[15]
Taclonex[4]
Teladar[4]
Temovate[5]
Temovate E[5]
Temovate Emollient[5]
Temovate Gel[5]
Temovate Scalp Application[5]
Texacort[22]
Topicort[10]
Topicort LP[10]
Topicort Mild[10]
Topicort Spray[10]
Topilene[4]
Topisone[4]
Topsyn[16]
Triacet[25]
Triaderm[25]
Trianex[25]
Trianide Mild[25]
Trianide Regular[25]
Tristatin II[25]
Triderm[25]
Tridesilon[9]
Tri-Luma[15]
Trymex[25]
Ultravate[20]
Unicort[22]
Uticort[4]
Valisone[4]
Valisone Reduced Strength[4]
Valisone Scalp Lotion[4]
Valnac[4]
Vanos[16]
Verdeso Foam[9]
Vioform-Hydrocortisone Lotion[22]
Westcort[22]
Xerese Cream[22]
Xyralid RC[22]
Xyralid LP Lotion[22]
Zytopic Cream[25]

ANDROGENS

GENERIC NAMES

1. ETHYLESTRENOL
2. FLUOXYMESTERONE
3. METHYLTESTOSTER-ONE
4. NANDROLONE
5. OXANDROLONE

6. OXYMETHOLONE
7. STANOZOLOL
8. TESTOSTERONE

BRAND NAMES

Anabolin[4]
Anabolin LA 100[4]
Anadrol-50[6]
Anapolon 50[6]
Andro 100[8]
Andro-Cyp 100[8]
Andro-Cyp 200[8]
Androderm[8]
Androgel[8]
Android-10[3]
Android-25[3]
Android-T[8]
Andro-LA 200[8]
Androlone[4]
Andronaq-50[8]
Andronaq-LA[8]
Andronate 100[8]
Andronate 200[8]
Andropository 100[8]
Andryl 200[8]
Axiron[8]
Bio-T-Gel[8]
Deca-Durabolin[4]
Delatest[8]
Delatestryl[8]
Dep Andro 100[8]
Dep Andro 200[8]
Dep-Androgyn[5]
Depotest[8]
Depotestogen[5]
Depo-Testadiol[5]
Depo-Testosterone[8]
Duo-Cyp[5]
Duo-Gen L.A.[5]
Duogex L.A.[5]
Durabolin[4]
Dura-Dumone 90/4[5]
Durabolin-50[4]
Duratest 100[8]
Duratest-200[8]
Duratestin[5]
Durathate 200[8]
Estratest[3]
Estratest H.S.[3]
Everone[8]
Fortesta[8]
Halodrin[4]
Halotestin[2]
Histerone-50[8]
Histerone-100[8]
Hybolin Decanoate[4]
Menoject L.A.[5]
Natesto[8]
Neo-Pause[5]
OB[5]
Premarin with Methyltestosterone[1]
Striant[8]
Teev[5]
Tes Est Cyp[5]
Test-Estro Cypionate[5]
Testim[8]
Tylosterone[2]
Valertest No. 1[5]
Valertest No. 2[5]
Vogelxo[8]

ANESTHETICS (Topical)

GENERIC NAMES

1. BENZOCAINE
2. BENZOCAINE & MENTHOL
3. BUTAMBEN
4. DIBUCAINE
5. LIDOCAINE
6. LIDOCAINE & PRILOCAINE
7. PRAMOXINE
8. TETRACAINE
9. TETRACAINE & MENTHOL
10. TETRACAINE & LIDOCAINE

BRAND NAMES

Americaine[1]
Amercaine Topical Anesthetic First Aid Ointment[1]
Amercaine Topical Anesthetic Spray[1]
Anestafoam[5]
Benzocol[2]
Butesin Picrate[3]
Butyl Aminobenzoate[2]
Cinchocaine[4]
Dermoplast[2]
Emla[6]
Endocaine[1]
Ethyl Aminobenzoate[1]
Lagol[1]
LidaMantle[5]
Lidoderm[5]
Lidopatch[5]

Lignocaine[5]
Nupercainal Cream[4]
Nupercainal Ointment[4]
Pontocaine Cream[8]
Pontocaine Ointment[9]
Pramegel[7]
Pramosone E Cream[7]
Prax[7]
Synera[10]
Tronothane[7]
Unguentine[1]
Unguentine Plus[1]
Unguentine Spray[1]
Xylocaine[5]
Xyralid Cream[5]
Xyralid LP Lotion[5]
Zostrix Neuropathy Cream[5]

ANGIOTENSIN-CONVERTING ENZYME (ACE) INHIBITORS

GENERIC NAMES

1. BENAZEPRIL
2. CAPTOPRIL
3. ENALAPRIL
4. FOSINOPRIL
5. LISINOPRIL
6. MOEXIPRIL
7. PERINDOPRIL
8. QUINAPRIL
9. RAMIPRIL
10. TRANDOLAPRIL

BRAND NAMES

Accupril[8]
Accuretic[8]
Aceon[7]
Altace[9]
Apo-Capto[2]
Capoten[2]
Capozide[2]
Epaned[3]
Lotensin[1]
Lotrel[1]
Mavik[10]
Monopril[4]
Novo-Captoril[2]
Prestalia[7]
Prinivil[5]
Prinzide[5]
Syn-Captopril[2]
Tarka[10]
Teczem[3]
Uniretic[6]
Univasc[6]
Vaseretic[3]
Vasotec[3]
Zestoretic[5]
Zestril[5]

ANTACIDS

GENERIC NAMES

1. ALUMINA & MAGNESIA
2. ALUMINA & MAGNESIUM CARBONATE
3. ALUMINA & MAGNESIUM TRISILICATE
4. ALUMINA, MAGNESIA, & CALCIUM CARBONATE
5. ALUMINA, MAGNESIA, & SIMETHICONE
6. ALUMINA, MAGNESIUM, CALCIUM CARBONATE & SIMETHICONE
7. ALUMINA, MAGNESIUM TRISILICATE, & SODIUM BICARBONATE
8. ALUMINUM CARBONATE, BASIC
9. ALUMINUM HYDROXIDE
10. CALCIUM & MAGNESIUM CARBONATES
11. CALCIUM CARBONATE
12. CALCIUM CARBONATE & MAGNESIA
13. CALCIUM CARBONATE & MAGNESIUM HYDROXIDE
14. CALCIUM CARBONATE & SIMETHICONE
15. CALCIUM CARBONATE, MAGNESIA, & SIMETHICONE
16. MAGALDRATE
17. MAGALDRATE & SIMETHICONE
18. MAGNESIUM HYDROXIDE
19. MAGNESIUM OXIDE

BRAND NAMES

Acid + All[14]
Advanced Formula Di-Gel[15]
Alamag[1]
Algenic Alka[2]
Algenic Alka Improved[2]
Algicon[9]
Alka-Mints[11]
Alkets[11]
Alkets Extra Strength[11]
Almacone[5]

Almacone II[5]
Alma-Mag #4 Improved[5]
Alma-Mag Improved[5]
AlternaGEL[9]
Alu-Cap[9]
Aludrox[5]
Alu-Tab[9]
Amitone[12]
Amphojel[9]
Amphojel 500[1]
Amphojel Plus[4]
AntaGel[5]
AntaGel-II[5]
Basaljel[9]
Calglycine[11]
Camalox[4]
Chooz[11]
Dialume[9]
Di-Gel[5]
Diovol Ex[1]
Diovol Plus[5]
Duracid[9]
Equilet[11]
Foamicon[3]
Gas-X with Maalox[14]
Gaviscon[2]
Gaviscon Extra Strength Relief Formula[9]
Gaviscon-2[7]
Gelusil[5]
Gelusil Extra-Strength[1]
Genalac[11]
Genaton[2]
Genaton Extra Strength[2]
Glycate[11]
Kudrox Double Strength[5]
Losopan[16]
Losopan Plus[17]
Lowsium[16]
Lowsium Plus[17]
Maalox[1]
Maalox HRF[2]
Maalox Plus[5]
Maalox Plus, Extra Strength[5]
Maalox Quick Dissolving Chews[11]
Maalox TC[1]
Magnalox[5]
Magnalox Plus[5]
Mag-Ox 400[19]
Mallamint[11]
Maox[19]
Marblen[10]
Mi-Acid[5]
Mi-Acid Double Strength[5]
Mintox[1]
Mintox Extra Strength[5]
Mygel[5]
Mygel II[5]
Mylagen[5]
Mylagen II[5]
Mylanta[5]
Mylanta Calci Tabs[12]
Mylanta Double Strength[5]
Mylanta Double Strength Plain[5]
Mylanta Gelcaps[10]
Mylanta Maximum Strength[13]
Mylanta Night Time Strength[9]
Mylanta Regular Strength[5]
Mylanta Plain[5]
Mylanta-II[5]
Nephrox[9]
Neutralca-S[1]
Pepcid Complete[13]
Phillips' Milk of Magnesia[18]
Riopan[16]
Riopan Extra Strength[16]
Riopan Plus[17]
Riopan Plus Double Strength[17]
Riopan Plus Extra Strength[17]
Rolaids[12]
Rolaids Antacid Cool[12]
Rolaids Calcium Rich[11]
Rolaids Chewable[11]
Rolaids Extra Strength[12]
Rolaids Sodium Free[11]
Rulox[1]
Rulox No. 1[1]
Rulox No. 2[1]
Rulox Plus[5]
Simaal 2 Gel[5]
Simaal Gel[5]
Tempo[6]
Titralac[11]
Titralac Plus[11]
Triconsil[7]
Tums[11]
Tums Dual Action[13]
Tums E-X[11]
Tums Extra Strength[11]
Tums Kids[11]
Tums Lasting Effects[11]

Tums Liquid ExtraStrength[11]
Tums Liquid Extra Strength with Simethicone[14]
Tums QuikPak[11]
Tums Freshers[11]
Tums Smooth Dissolve[11]
Univol[1]
Uro-Mag[19]
Zygerid Chewable Tablets[18]

ANTIBACTERIALS (Ophthalmic)

GENERIC NAMES

1. AZITHROMYCIN (Ophthalmic)
2. BACITRACIN & POLYMYXIN B
3. BESIFLOXACIN
4. CIPROFLOXACIN (Ophthalmic)
5. ERYTHROMYCIN (Ophthalmic)
6. GATIFLOXACIN (Ophthalmic)
7. GENTAMICIN (Ophthalmic)
8. LEVOFLOXACIN (Ophthalmic)
9. MOXIFLOXACIN (Ophthalmic)
10. NEOMYCIN, POLYMYXIN B
11. NEOMYCIN, POLYMYXIN B & BACITRACIN
12. NEOMYCIN, POLYMYXIN B & GRAMICIDIN
13. OFLOXACIN (Ophthalmic)
14. POLYMYXIN B & TRIMETHOPRIM
15. SULFACETAMIDE (Ophthalmic)
16. TOBRAMYCIN (Ophthalmic)

BRAND NAMES

Ak-Poly-Bac[2]
AzaSite[1]
Besivance[3]
Bleph-10[15]
Cetamide[15]
Ciloxan[4]
Genoptic[7]
Gentak[7]
Iquix[8]
I-Sulfacet[15]
Moxeza[9]
Neo-Polycin[11]
Neo-Polycin HC[11]
Neosporin Ophthalmic Solution[11]
Ocuflox[13]
Polytrim[14]
Quixin[8]
Sulf-10[15]
Sulfamide[15]
Tobradex[16]
Tobradex ST[16]
Tobrex[16]
Vigamox[9]
Zylet[6]
Zymar[16]
Zymaxid[6]

ANTIDEPRESSANTS, TRICYCLIC

GENERIC NAMES

1. AMITRIPTYLINE
2. AMOXAPINE
3. CLOMIPRAMINE
4. DESIPRAMINE
5. DOXEPIN
6. IMIPRAMINE
7. NORTRIPTYLINE
8. PROTRIPTYLINE
9. TRIMIPRAMINE

BRAND NAMES

Adapin[5]
Anafranil[3]
Apo-Amitriptyline[1]
Apo-Imipramine[6]
Apo-Trimip[9]
Asendin[2]
Aventyl[7]
Endep[1]
Etrafon[1]
Etrafon-A[1]
Etrafon-D[1]
Etrafon-F[1]
Etrafon-Forte[1]
Impril[6]
Levate[1]
Norfranil[6]
Norpramin[4]
Novo-Doxepin[5]
Novopramine[6]
Novo-Tripramine[9]
Novotriptyn[1]
Pamelor[7]
PMS Amitriptyline[1]
PMS Impramine[6]
PMS Levazine[8]
Rhotrimine[9]
Silenor[5]
Sinequan[5]
Surmontil[9]

Tipramine[6]
Tofranil[6]
Tofranil-PM[6]
Triadapin[5]
Triavil[8]
Triptil[8]
Vivactil[8]

ANTIDYSKINETICS

GENERIC NAMES

1. BENZTROPINE
2. BIPERIDEN
3. ETHOPROPAZINE
4. PIMOZIDE
5. PROCYCLIDINE
6. TRIHEXYPHENIDYL

BRAND NAMES

Akineton[2]
Apo-Benztropine[1]
Apo-Trihex[6]
Artane[6]
Artane Sequels[6]
Cogentin[1]
Orap[4]
Parsidol[3]
Parsitan[3]
PMS Benztropine[1]
PMS Procyclidine[5]
PMS Trihexyphenidyl[6]
Procyclid[6]
Trihexane[6]
Trihexy[6]

ANTIFUNGALS (Topical)

GENERIC NAMES

1. AMPHOTERICIN B
2. BUTENAFINE
3. CICLOPIROX
4. CLOTRIMAZOLE
5. ECONAZOLE
6. EFINACONAZOLE
7. FLUCONAZOLE
8. HALOPROGIN
9. KETOCONAZOLE (Topical)
10. LULICONOZOLE
11. MICONAZOLE
12. NAFTIFINE
13. NYSTATIN
14. OXICONAZOLE (Topical)
15. SERTACONAZOLE
16. SULCONAZOLE
17. TAVABOROLE
18. TERBINAFINE
19. TOLNAFTATE
20. UNDECYLENIC ACID

BRAND NAMES

Aftate for Athlete's Foot Aerosol Spray Liquid[19]
Aftate for Athlete's Foot Aerosol Spray Powder[19]
Aftate for Athlete's Foot Gel[19]
Aftate for Athlete's Foot Sprinkle Powder[19]
Aftate for Jock Itch Aerosol Spray Powder[19]
Aftate for Jock Itch Gel[19]
Aftate for Jock Itch Sprinkle Powder[19]
Caldesene Medicated Powder[20]
Canesten Cream[4]
Canesten Solution[4]
Conazol[11]
Cruex Aerosol Powder[20]
Cruex Antifungal Cream[20]
Cruex Antifungal Powder[20]
Cruex Antifungal Spray Powder[20]
Cruex Cream[20]
Cruex Powder[20]
Decylenes[20]
Decylenes Powder[20]
Desenex Aerosol Powder[20]
Desenex Antifungal Cream[20]
Desenex Antifungal Liquid[20]
Desenex Antifungal Ointment[20]
Desenex Antifungal Penetrating Foam[20]
Desenex Antifungal Powder[20]
Desenex Antifungal Spray Powder[20]
Desenex Max Cream[20]
Desenex Ointment[20]
Desenex Powder[20]
Desenex Solution[20]
Ecostatin[5]
Ecoza[5]
Ertazo[14]
Exelderm[16]
Extina[9]
Extina Foam[9]
Fungizone[1]
Genaspore Cream[19]
Gordochom Solution[20]
Halotex[8]

Jublia[6]
Kerydin[17]
Lamisil[18]
Lamisil Solution 1%[18]
Loprox[3]
Lotriderm[4]
Lotrimin AF[4]
Lotrimin Cream[4]
Lotrimin Lotion[4]
Lotrimin Ointment[4]
Lotrimin Ultra[2]
Lotrisone[4]
Luzu[10]
Mentax[2]
Mentax TC[2]
Micatin[11]
Monistat-Derm[10]
Mycelex Cream[4]
Mycelex Solution[4]
Myclo Cream[4]
Myclo Solution[4]
Myclo Spray[4]
Mycostatin[13]
Nadostine[13]
Naftin[11]
Nilstat[12]
Nizoral A-D[9]
Nizoral Shampoo[9]
NP-27 Cream[19]
NP-27 Powder[19]
NP-27 Solution[19]
NP-27 Spray Powder[19]
Nyaderm[13]
Nystex[13]
Nystop[13]
Oravig[11]
Oxistat[14]
Penlac[3]
Pitrex Cream[19]
Spectazole[5]
Tinactin Aerosol Liquid[19]
Tinactin Aerosol Powder[19]
Tinactin Antifungal Deodorant Powder Aerosol[19]
Tinactin Cream[19]
Tinactin Jock Itch Aerosol Powder[19]
Tinactin Jock Itch Cream[19]
Tinactin Jock Itch Spray Powder[19]
Tinactin Plus Powder[19]
Tinactin Powder[19]
Tinactin Solution[19]
Ting Antifungal Cream[19]
Ting Antifungal Powder[19]
Ting Antifungal Spray Liquid[19]
Ting Antifungal Spray Powder[19]
Vusion[11]
Xolegel Gel[9]
Zeasorb-AF Powder[11]

ANTIFUNGALS (Vaginal)

GENERIC NAMES

1. BUTOCONAZOLE
2. CLOTRIMAZOLE
3. ECONAZOLE
4. GENTIAN VIOLET
5. MICONAZOLE
6. NYSTATIN
7. TERCONAZOLE
8. TIOCONAZOLE

BRAND NAMES

Canesten[2]
Canesten 1[2]
Canesten 3[2]
Canesten 10%[2]
Ecostatin[3]
FemCare[2]
Femizole Prefil[2]
Femizole-7[2]
Genapax[4]
Gynazole-1[1]
Gyne-Lotrimin[2]
Gyne-Lotrimin 3[2]
Gyno-Trosyd[8]
Monistat[5]
Monistat 1[8]
Monistat 3[5]
Monistat 5[5]
Monistat 7[5]
Mycelex-7[2]
Mycelex-G[2]
Myclo[2]
Mycostatin[6]
Nadostine[6]
Nilstat[6]
Nyaderm[6]
Terazol 3[7]
Terazol 7[7]
Three Day Cream[2]
Vagistat[8]
Vagistat-1[8]

ANTIHISTAMINES

GENERIC NAMES

1. ACRIVASTINE
2. AZATADINE
3. BROMODIPHENHY-DRAMINE
4. BROMPHENIRAMINE
5. CARBINOXAMINE
6. CHLORPHENIRAMINE
7. CLEMASTINE FUMARATE
8. CYPROHEPTADINE
9. DEXBROMPHENIRAMINE
10. DEXCHLORPHENIRAMINE
11. DIMENHYDRINATE
12. DIPHENHYDRAMINE
13. DIPHENYLPYRALINE
14. DOXYLAMINE
15. PHENINDAMINE
16. PHENIRAMINE
17. PHENYLTOLOXAMINE
18. PYRILAMINE
19. TRIPELENNAMINE
20. TRIPROLIDINE

BRAND NAMES

Aclophen[6]
Actacin[20]
Actagen[20]
Actidil[18]
Advil Multi-Symptom Cold[6]
AH-Chew[6]
Alersule[6]
Aleve PM[12]
Alka-Seltzer Plus Cold & Cough Liquid Gels[6]
Alka-Seltzer PM[12]
Allent[5]
Alleract[18]
Aller-Chlor[6]
Allercon[20]
Allerdryl[12]
Allerest Maximum Strength[6]
Allerfrim[20]
AllerMax Caplets[12]
Aller-med[12]
Allerphed[20]
Allert[6]
All-Nite Cold Formula[14]
Ambenyl Cough[3]
Ambophen Expectorant[3]
Anamine[6]
Anamine T. D.[6]
Apo-Dimenhydrinate[11]
Atrohist Pediatric[6]
Atrohist Pediatric Suspension Dye Free[6, 18]
Atrohist Sprinkle[6]
Banophen[12]
Banophen Caplets[12]
Beldin[12]
Belix[12]
Bena-D 10[12]
Bena-D 50[12]
Benadryl Allergy and Sinus Fastmelt[12]
Benadryl Allergy/Sinus Headache Caplets[12]
Benahist 10[12]
Benahist 50[12]
Ben-Allergin 50[12]
Benaphen[12]
Benoject-10[12]
Benoject-50[12]
Benylin All-In-One Cold & Flu Night Caplets[12]
Benylin All-In-One Cold & Flu Nightime Syrup[6]
Benylin Cold & Sinus Plus[6]
Brexin-L.A[6]
Brofed[5]
Bromanyl[3]
Bromatane DX Cough[5]
Bromfed[5]
Bromfed-DM[5]
Bromfed-PD[5]
Bydramine Cough[12]
Calm X[11]
Carbodec[5]
Carbodec DM Drops[5]
Carbodec TR[5]
Cenafed Plus[20]
Cerose-DM[6]
Children's Benadryl Allergy & Cold Fastmelt[12]
Children's Benadryl Perfect Measure[12]
Children's Dramamine[11]
Children's Triaminic Thin Strips Night Time Cold & Cough[12]
Chlo-Amine[6]
Chlor-100[6]
Chlorate[6]
Chlor-Niramine[6]
Chlorphed[3]

Chlorphedrine SR[3]
Chlor-Pro[6]
Chlor-Pro 10[6]
Chlorspan-12[6]
Chlortab-4[6]
Chlortab-8[6]
Chlor-Trimeton 4 Hour Relief[6]
Chlor-Trimeton 12 Hour Relief[6]
Chlor-Tripolon[6]
Codeprex[6]
Codimal DM[6]
Codimal PH[18]
Codimal-A[5]
Codimal-L.A.[6]
Codimal-L.A. Half[6]
Colfed-A[6]
Colrex Compound[6]
Colrex Cough[6]
Coltab Children's[6]
Comhist[6]
Comhist LA[6]
Compoz[12]
Congestant D[6]
Conjec-B[5]
Contac 12-Hour[6]
Contac 12-Hour Allergy[7]
Contac Allergy/Sinus Night Caplets[7]
Contac Night Caplets[12]
Contac Severe Cold Formula[6]
Contac Severe Cold Formula Night Strength[6]
Coricidin HBP[6]
Dallergy[6]
Dallergy Jr[5]
Deconamine[6]
Deconamine SR[6]
Dexaphen SA[9]
Dexchlor[10]
Dexophed[9]
Diabetic Tussin Allergy Relief[6]
Diabetic Tussin Cold & Flu[6]
Diabetic Tussin Night Time Formula[12]
Diamine T.D.[5]
Diclegis[14]
Dihistine[6]
Dimetabs[11]
Dimetane[5]
Dimetapp Plus Caplets[5]
Dimetapp with Codeine[5]
Dimetapp-A[5]
Dimetapp-A Pediatric[5]
Dimetapp-DM[5]
Dimetapp-DM Cough and Cold[5]
Dimetapp-DM Elixir[5]
Dinate[11]
Diphen Cough[12]
Diphenacen-10[12]
Diphenacen-50[12]
Diphenadryl[12]
Disobrom[9]
Disophrol[9]
Disophrol Chronotabs[9]
Dommanate[11]
Donatussin[6]
Donatussin Drops[6]
Dondril[6]
Dormarex 2[12]
Dormin[12]
Dramamine[11]
Dramamine Chewable[11]
Dramamine Liquid[11]
Dramanate[11]
Dramocen[11]
Dramoject[11]
Dristan AF[6]
Dristan Cold and Flu[6]
Dristan Cold Maximum Strength Caplets[5]
Dristan Cold Multi-Symptom Formula[6]
Dristan Formula P[18]
Drixoral[9]
Drixoral Cold and Allergy[9]
Drixoral Cold and Flu[9]
Drixoral Plus[9]
Drixoral Sinus[9]
Drixtab[9]
Dymenate[11]
Ed A-Hist[6]
Father John's Medicine Plus[6]
Fedahist[6]
Fedahist Decongestant[6]
Fedahist Gyrocaps[6]
Fedahist Timecaps[6]
Fenylhist[12]
Fynex[12]
Genahist[12]
GenAllerate[6]
Gendecon[6]
Gen-D-phen[12]
Gravol[11]
Gravol L/A[11]

Hayfebrol[6]
Histagesic Modified[6]
Histaject Modified[5]
Histalet[6]
Histalet-DM[6]
Histatab Plus[6]
Histatan[6]
Histex I/E[5]
Histor-D[6]
Histor-D Timecelles[6]
Hydramine[12]
Hydramine Cough[12]
Hydramyn[12]
Hydrate[11]
Hydril[12]
Hyrexin-50[12]
Insomnal[12]
Karbinal ER[5]
Klerist-D[6]
Kolephrin[6]
Kolephrin/DM Caplets[6]
Kronofed-A[6]
Kronofed-A Jr.[6]
Lodrane D[4]
Lodrane LD[4]
Lodrane 12Hour ER[4]
Lodrane 24D[4]
Marmine[11]
Maximum Strength Tylenol Allergy Sinus Caplets[6]
Meda Syrup Forte[6]
Motion-Aid[12]
Myidil[20]
Nasahist B[5]
Nauseatol[11]
ND Clear T.D.[6]
ND Stat Revised[5]
ND-Gesic[6]
Neocitran A[16]
Neocitran Colds and Flu Calorie Reduced[16]
NeoCitran DM Coughs & Cold[16]
NeoCitran Extra Strength Colds and Flu[16]
Nervine Night-time Sleep-Aid[12]
Nico-Vert[11]
Nidryl[12]
Nisaval[17]
Nolahist[18]
Noradryl[12]
Norafed[18]
Nordryl[10]
Nordryl Cough[10]
Novodimenate[11]
Novopheniram[6]
NyQuil Cough[14]
Nytol Maximum Strength[12]
Nytol with DPH[12]
Optimine[2]
Oraminic II[5]
Palgic[5]
PBZ[19]
PBZ-SR[19]
PediaCare Children's Allergy[12]
PediaCare Children's Allergy & Cold[12]
PediaCare Children's Fever Reducer Plus Cough & Runny Nose[6]
PediaCare Children's Fever Reducer Plus Flu[6]
PediaCare Fever Reducer Plus Multi-Symptom Cold[6]
Pediacof Cough[6]
Pelamine[19]
Periactin[8]
Pertussin All Night PM[14]
Pfeiffer's Allergy[6]
Phenapap Sinus Headache & Congestion[6]
Phendry[12]
Phendry Children's Allergy Medicine[12]
Phenetron[6]
Phenetron Lanacaps[6]
PMS-Dimenhydrinate[11]
Poladex T.D.[10]
Prehist[6]
Prehist Cough Mixture 4[6]
Prehist D[6]
Pseudo-Chlor[6]
Pyribenzamine[19]
Pyrilamine Maleate Tablets[18]
Remcol-C[6]
Rescon-JR[6]
Rhinatate[6]
Rhinosyn[6]
Rhinosyn-DM[6]
Rhinosyn-PD[6]
Robitussin Children's Cough & Cold Long-Acting[6]
Robitussin Cough & Cold Long-Acting[6]
Robitussin Cough & Cold Nighttime[12]

Robitussin with Codeine[16]
Rolatuss Plain[6]
R-Tannamine[6]
R-Tannamine Pediatric[6]
R-Tannate[6]
R-Tannate Pediatric[6]
Ryna[6]
Rynatan[6]
Rynatan Pediatric[6]
Rynatan-S Pediatric[6]
Rynatuss[6]
Rynatuss Pediatric[6]
Scot-Tussin DM[6]
Scot-Tussin Original 5-Action Cold Medicine[14]
Semprex-D[1]
Siladryl[12]
Silphen[12]
Simply Sleep[12]
Sleep-Eze 3[12]
Sominex Formula 2[12]
Sudafed PE Cold & Cough Caplets[12]
Tanoral[6]
Tavist[7]
Tavist-1[7]
Tavist Allergy/Sinus/Headache[7]
Tega-Vert[11]
Telachlor[6]
Teldrin[6]
Theraflu Cold & Cough Hot Liquid[16]
Theraflu Flu & Sore Throat Hot Liquid[16]
Theraflu Nighttime Severe Cold & Cough[12]
Theraflu Nighttime Severe Cold Hot Liquid[12]
Theraflu Thin Strips Multi Symptom[12]
Theraflu Thin Strips Nighttime Severe Cold & Cough[12]
Theraflu Warming Relief Nighttime Severe Cold & Cough[12]
Touro A&H[5]
Travamine[11]
Triafed[20]
Triaminic-D Multi-Symptom Cold[6]
Triaminic Flu, Cough & Fever[6]
Triaminic Night Time Cough & Cold[6]
Triaminic Softchews Cough & Runny Nose[6]
Triaminic Thin Strips Cough & Runny Nose[12]
Triaminic Thin Strips Night Time Cold & Cough[12]
Tricodene Sugar Free[6]
Trimedine Liquid[6]
Tri-Nefrin Extra Strength[6]
Triotann[6]
Triotann Pediatric[6]
Trip-Tone[11]
Tritann Pediatric[6]
Tri-Tannate[6]
Tri-Tannate Plus Pediatric[6]
Trymegen[6]
Tussar DM[6]
TussiCaps[6]
Tussi-12[18]
Tussionex[6]
Tusstat[12]
Twilite[12]
Tylenol Allergy Multi-Symptom[6]
Tylenol Allergy Multi-Symptom Nighttime[12]
Tylenol Cold Head Congestion Nighttime[6]
Tylenol Cold Multi-Symptom Nighttime[14]
Tylenol Cough & Sore Throat Nighttime[14]
Tylenol Severe Allergy[12]
Tylenol Sinus Congestion & Pain Nighttime[6]
Uni-Bent Cough[12]
Unisom Nighttime Sleep Aid[14]
Unisom SleepGels Maximum Strength[12]
Vanex Forte R[6]
Veltane[5]
Vertab[11]
Vicks Children's NyQuil[6]
Vicks Formula 44 Custom Care Cough & Cold PM[6]
Vicks NyQuil Cold & Flu Relief[14]
Vicks NyQuil Cold & Flu Symptom Relief Plus Vitamin C[14]
Vicks NyQuil Cough[14]
Vicks NyQuil D[14]
Vicks NyQuil Less Drowsy Cold & Flu Relief Liquid[6]
Vicks Pediatric Formula 44m Cough & Cold Relief[6]
Viravan DM[18]
Vituz[6]
Wehamine[11]

Wehdryl[12]
Wehdryl-10[12]
Wehdryl-50[12]
Zutripro[6]
ZzzQuil[12]

ANTIHISTAMINES, NONSEDATING

GENERIC NAMES

1. CETIRIZINE
2. DESLORATADINE
3. FEXOFENADINE
4. LEVOCETIRIZINE
5. LORATADINE

BRAND NAMES

Alavert[5]
Alavert D-12[5]
Allegra[3]
Allegra ODT[3]
Allegra Oral Suspension[3]
Allegra-D[3]
Allegra-D 24 Hour[3]
Children's Allegra[3]
Children's Claritin Grape Chewable[5]
Children's Claritin Syrup, Grape[5]
Children's Zyrtec Allergy Bubble Gum Syrup[1]
Children's Zyrtec Perfect Measure[1]
Clarinex D 12 Hour[2]
Clarinex D 24 Hour[2]
Clarinex RediTabs[2]
Clarinex Syrup[2]
Clarinex[2]
Claritin RediTabs for Kids 24 Hour[5]
Claritin 12 Hour[5]
Claritin Extra[5]
Claritin Hives Relief[5]
Claritin Liqui-Gels[5]
Claritin RediTabs 24 Hour[5]
Claritin Syrup[5]
Claritin[5]
Claritin-D[5]
Claritin-D 12 Hour RediTabs for Kids[5]
Claritin-D 12 Hour[5]
Claritin-D 24 Hour[5]
Reactine[1]
Xyzal[4]
Zyrtec[1]
Zyrtec Allergy[1]
Zyrtec Children's Allergy Syrup[1]
Zyrtec Children's Chewable[1]
Zyrtec Children's Hives Relief Syrup[1]
Zyrtec-D[1]
Zyrtec Dissolve Tabs[1]

ANTIHISTAMINES, PHENOTHIAZINE-DERIVATIVE

GENERIC NAMES

1. PROMETHAZINE
2. TRIMEPRAZINE

BRAND NAMES

Anergan 25[1]
Anergan 50[1]
Antinaus 50[1]
Histantil[1]
Panectyl[2]
Pentazine[1]
Phenameth DM[1]
Phenazine 25[1]
Phenazine 50[1]
Phencen-50[1]
Phenergan[1]
Phenergan Fortis[1]
Phenergan Plain[1]
Phenergan VC[1]
Phenergan with Dextromethorphan[1]
Phenoject-50[1]
Pherazine DM[1]
Pherazine VC[1]
PMS Promethazine[1]
Pro-Med 50[1]
Promerhegan[1]
Promet[1]
Prometh VC Plain[1]
Prometh VC with Codeine[1]
Prometh with Dextromethorphan[1]
Prometh-25[1]
Prometh-50[1]
Promethazine DM[1]
Promethazine VC[1]
Prorex-25[1]
Prorex-50[1]
Prothazine[1]
Prothazine Plain[1]
Shogan[1]
TV-Gan-25[1]
V-Gan-50[1]

ANTI-INFLAMMATORY DRUGS, NONSTEROIDAL (NSAIDs)

GENERIC NAMES

1. DICLOFENAC
2. DIFLUNISAL
3. ETODOLAC
4. FENOPROFEN
5. FLOCTAFENINE
6. FLURBIPROFEN
7. IBUPROFEN
8. INDOMETHACIN
9. KETOPROFEN
10. KETOROLAC
11. MECLOFENAMATE
12. MEFENAMIC ACID
13. NABUMETONE
14. NAPROXEN
15. OXAPROZIN
16. PIROXICAM
17. SULINDAC
18. TENOXICAM
19. TOLMETIN

BRAND NAMES

Advil[7]
Advil Allergy and Congestion Relief[7]
Advil Caplets[7]
Advil Chewable Tablets[7]
Advil Cold and Sinus Caplets[7]
Advil Cold and Sinus LiquiGels[7]
Advil Congestion Relief[7]
Advil Film Coated[7]
Advil First[7]
Advil Flu & Body Ache[7]
Advil Liqui-Gel[9]
Advil Migraine[7]
Aleve[14]
Aleve Liquid Gels[14]
Aleve-D Sinus & Cold[14]
Aleve PM[14]
Anaprox[14]
Anaprox DS[14]
Ansaid[6]
Apo-Diclo[1]
Apo-Diflunisal[2]
Apo-Ibuprofen[7]
Apo-Keto[9]
Apo-Naproxen[14]
Arthrotec[1]
Brufen[7]
Cambia[1]
Cataflam[1]
Children's Advil Allergy Sinus[7]
Children's Elixsure[7]
Children's Motrin[7]
Children's Motrin Cold[7]
Clinoril[17]
Daypro[15]
Dolgesic[7]
Duexis[7]
EC-Naprosyn[14]
Feldene[16]
Feldene Melt[16]
Fenopron[4]
Genpril[7]
Genpril Caplets[7]
Haltran[7]
Ibren[7]
Ibudone[7]
Ibu-Tab[7]
Idarac[5]
Ifen[7]
Indocid[8]
Indocin[8]
Junior Strength Motrin[7]
Midol Liquid Gels[7]
Motrin, Children's[7]
Motrin, Infants[7]
Motrin-IB[7]
Motrin Migraine[7]
Nalfon[4]
Naprelan[14]
Naprosyn[14]
Naprosyn-E[14]
Naprosyn-SR[14]
Naxen[14]
Nexcede[9]
Novomethacin[8]
Novopirocam[16]
Novoprofen[7]
Ponstan[12]
Profen[7]
Prevacid NapraPac[14]
Reprexain[7]
Sine-Aid IB[7]
Sprix[10]
Tab-Profen[7]
Tivorbex[8]
Treximet[14]
Vicoprofen[7]
Vimovo[14]

Voltaren XR[1]
Zipsor[1]
Zorvolex[1]

ANTI-INFLAMMATORY DRUGS, STEROIDAL (Ophthalmic)

GENERIC NAMES

1. BETAMETHASONE (Ophthalmic)
2. DEXAMETHASONE (Ophthalmic)
3. FLUOROMETHOLONE
4. HYDROCORTISONE (Ophthalmic)
5. LOTEPREDNOL
6. MEDRYSONE
7. PREDNISOLONE (Ophthalmic)
8. RIMEXOLONE

BRAND NAMES

Ak-Pred[7]
AK-Tate[7]
Alrex[5]
Baldex[2]
Betnesol[1]
Cortamed[4]
Decadron[2]
Dexair[2]
Dexotic[2]
Dexsone[2]
Diodex[2]
Econopred[7]
Econopred Plus[7]
Eflone[3]
Flarex[3]
Fluor-Op[3]
FML Forte[3]
FML Liquifilm[3]
FML S.O.P.[3]
HMS Liquifilm[6]
Inflamase Forte[7]
Inflamase-Mild[7]
Lite-Pred[7]
Lotemax[5]
Maxidex[2]
Neo-Polycin HC[4]
Ocu-Dex[2]
Ocu-Pred[7]
Ocu-Pred Forte[7]
Ocu-Pred-A[7]
PMS-Dexamethasone Sodium Phosphate[2]
Pred Forte[7]
Pred Mild[7]
Predair[7]
Predair Forte[7]
Predair-A[7]
Spersadex[2]
Storz-Dexa[2]
Tobradex[2]
Tobradex ST[2]
Ultra Pred[7]
Vexol[8]
Zylet[5]

ANTISEBORRHEICS (Topical)

GENERIC NAMES

1. CHLOROXINE
2. PYRITHIONE
3. SALICYLIC ACID, SULFUR & COAL TAR
4. SELENIUM SULFIDE

BRAND NAMES

Capitrol[1]
Dan-Gard[2]
DHS Zinc Dandruff Shampoo[2]
Exsel[4]
Glo-Sel[4]
Head & Shoulders[2]
Head & Shoulders Antidandruff Cream Shampoo Normal to Dry Formula[2]
Head & Shoulders Antidandruff Cream Shampoo Normal to Oily Formula[2]
Head & Shoulders Antidandruff Lotion Shampoo 2 in 1 Formula[2]
Head & Shoulders Antidandruff Lotion Shampoo Normal to Dry Formula[2]
Head & Shoulders Antidandruff Lotion Shampoo Normal to Oily Formula[2]
Head & Shoulders Dry Scalp 2 in 1 Formula Lotion Shampoo[2]
Head & Shoulders Dry Scalp Conditioning Formula Lotion Shampoo[2]
Head & Shoulders Dry Scalp Regular Formula Lotion Shampoo[2]
Head & Shoulders Intensive Treatment 2 in 1 Formula Dandruff Lotion Shampoo[4]
Head & Shoulders Intensive Treatment Conditioning Formula Dandruff Lotion Shampoo[4]

Head & Shoulders Intensive Treatment Regular Formula Dandruff Lotion Shampoo[4]
Meted Maximum Strength Anti-Dandruff Shampoo with Conditioners[3]
Sebex-T Tar Shampoo[2]
Sebulex Conditioning Suspension Shampoo[3]
Sebulex Lotion Shampoo[3]
Sebulon[2]
Sebutone[3]
SelRX Shampoo[4]
Selsun[4]
Selsun Blue[4]
Selsun Blue Dry Formula[4]
Selsun Blue Extra Conditioning Formula[4]
Selsun Blue Extra Medicated Formula[4]
Selsun Blue Oily Formula[4]
Selsun Blue Regular Formula[4]
Tersi Foam[4]
Theraplex Z[2]
Vanseb Cream Dandruff Shampoo[3]
Vanseb Lotion Dandruff Shampoo[3]
Vanseb-T[3]
Zincon[2]
ZNP[2]

ASPIRIN

8-Hour Bayer Timed Release
217
217 Strong
Acetylsalicylic Acid
Alka-Seltzer Original
Alka-Seltzer Morning Relief Medicine
Alka-Seltzer Plus Flu Effervescent
Alka-Seltzer PM
Alpha-Phed
Anacin
APAC Improved
APF Arthritic Pain Formula
Arthrinol
Arthrisin
Arthritis Pain Formula
Artria S.R.
A.S.A.
A.S.A. Enseals
Ascriptin
Ascriptin A/D
Aspergum
Astrin
Bayer
Bayer Advanced
Bayer Extra Strength Aspirin
Bayer Quick Release Crystals
Bayer Timed-Release Arthritic Pain Formula
Bayer Women's Caplets Aspirin Plus Calcium
Buffaprin
Bufferin
Buffets II
Buffinol
Cama Arthritis Reliever
Coryphen
Dristan Formula P
Duradyne
Easprin
Ecotrin
Empirin
Endodan
Entrophen
Epromate-M
Equagesic
Equazine-M
Excedrin Back & Body Excedrin Extra Strength Caplets
Excedrin Extra Strength Tablets
Excedrin Migraine
Extra Strength Bayer, Back & Body Pain
Extra Strength Bayer PM
Fasprin
Fiorinal
Fiorinal w/Codeine
Fortabs
Gelpirin
Gemnisyn
Goody's Extra Strength Tablets
Goody's Headache Powders
Halfprin
Headstart
Heptogesic
Laniroif
Lanorinal
Magnaprin
Magnaprin Arthritis Strength
Maprin
Measurin
Meprogesic
Meprogesic Q

Micrainin
Nervine
Night-Time Effervescent Cold
Norwich Aspirin
Novasen
P-A-C Revised Formula
Percodan
Pravigard PAC
Presalin
Riphen
Robaxisal
Sal-Adult
Salatin
Sal-Infant
Salocol
Soma Compound
St. Joseph Adult Chewable Aspirin
St. Joseph Companion Aspirin
Supac
Supasa
Synalgos-DC
Tenol Plus
Therapy Bayer
Triaphen
Trigesic
Ursinus Inlay
Vanquis
Viro-Med
Zorprin

BARBITURATES

GENERIC NAMES

1. BUTALBITAL
2. PHENOBARBITAL
3. SECOBARBITAL

BRAND NAMES

Anolor-300[1]
Arcet[1]
Butace[1]
Butapap[1]
Dolmar[1]
Donnatal[2]
Donnatal Elixir[2]
Donnatal Extentabs[2]
Esgic-Plus[1]
Fioricet[1]
Fioricet w/Codeine[1]
Fiorinal[1]
Fiorinal w/Codeine[1]
Fortabs[1]
Laniroif[1]
Lanorinal[1]
Luminal[2]
Phrenilin Forte[1]
Phrenilin w/Caffeine & Codeine[1]
Repan[1]
Seconal[3]
Solfoton[2]
Theofed[2]

BENZODIAZEPINES

GENERIC NAMES

1. ALPRAZOLAM
2. BROMAZEPAM
3. CHLORDIAZEPOXIDE
4. CLOBAZAM
5. CLONAZEPAM
6. CLORAZEPATE
7. DIAZEPAM
8. ESTAZOLAM
9. FLURAZEPAM
10. HALAZEPAM
11. KETAZOLAM
12. LORAZEPAM
13. MIDAZOLAM
14. NITRAZEPAM
15. OXAZEPAM
16. PRAZEPAM
17. QUAZEPAM
18. TEMAZEPAM

BRAND NAMES

Alprazolam Intensol[1]
Apo-Alpraz[1]
Apo-Chlordiazepoxide[3]
Apo-Clorazepate[6]
Apo-Diazepam[7]
Apo-Flurazepam[9]
Apo-Lorazepam[12]
Apo-Oxazepam[15]
Ativan[18]
Centrax[16]
Clindex[3]
Clinoxide[3]
Dalmane[9]
Diastat[7]
Diazemuls[7]
Diazepam Intensol[7]
Doral[17]
Klonopin[5]
Lectopam[2]

Librax[3]
Libritabs[3]
Librium[3]
Lidoxide[3]
Limbitrol[3]
Limbitrol DS[3]
Lipoxide[3]
Loftran[17]
Lorazepam Intensol[12]
Medilium[3]
Meval[7]
Mogadon[16]
Niravam[1]
Niravam Orally Disintegrating Tablets[1]
Novo-Alprazol[1]
Novoclopate[6]
Novodipam[7]
Novoflupam[9]
Novolorazem[12]
Novopoxide[3]
Novoxapam[15]
Nu-Alpraz[1]
Nu-Loraz[12]
Onfi[4]
Paxipam[10]
PMS Diazepam[7]
Restoril[18]
Rivotril[5]
Serax[15]
Solium[3]
Somnol[9]
T-Quil[7]
Tranxene[7]
Tranxene T-Tab[6]
Tranxene-SD[7]
Valium[7]
Valrelease[7]
Vivol[7]
Xanax[1]
Xanax XR[1]
Zapex[15]
Zebrax[3]
Zetran[7]

BENZOYL PEROXIDE

Acanya
Acetoxyl 2.5 Gel
Acetoxyl 5 Gel
Acetoxyl 10 Gel
Acetoxyl 20 Gel
Acne-5 Lotion
Acne-10 Lotion
Acne-Aid 10 Cream
Acne-Mask
Acnomel B.P. 5 Lotion
Ben-Aqua $2\frac{1}{2}$ Gel
Ben-Aqua $2\frac{1}{2}$ Lotion
Ben-Aqua 5 Gel
Ben-Aqua 5 Lotion
Ben-Aqua 10 Gel
Ben-Aqua 10 Lotion
Ben-Aqua Masque 5
Benoxyl 5 Lotion
Benoxyl 5 Wash
Benoxyl 10 Lotion
Benoxyl 10 Wash
Benoxyl 20 Lotion
Benzac Ac $2\frac{1}{2}$ Gel
Benzac Ac 5 Gel
Benzac Ac 10 Gel
Benzac W $2\frac{1}{2}$ Gel
Benzac W 5 Gel
Benzac W 10 Gel
Benzaclin
Benzagel 5 Acne Lotion
Benzagel 5 Acne Wash
Benzagel 5 Gel
Benzagel 10 Gel
Benzamycin
BenzaShave 5 Cream
BenzaShave 10 Cream
BenzEFoam
Brevoxyl 4 Gel
Buf-Oxal 10
Cleanse and Treat
Clear By Design 2.5 Gel
Clearasil BP Plus 5 Cream
Clearasil BP Plus 5 Lotion
Clearasil Maximum Strength Medicated Anti-Acne 10 Tinted Cream
Clearasil Maximum Strength Medicated Anti-Acne 10 Vanishing Cream
Clearasil Medicated Anti-Acne 10 Vanishing Lotion
Clinac BPO
Cuticura Acne 5 Cream
Del-Aqua-5 Gel
Del-Aqua-10 Gel
Del-Ray
Dermoxyl 2.5 Gel

Dermoxyl 5 Gel
Dermoxyl 10 Gel
Dermoxyl 20 Gel
Dermoxyl Aqua
Desquam-E 2.5 Gel
Desquam-E 5 Gel
Desquam-E 10 Gel
Desquam-X 2.5 Gel
Desquam-X 5 Gel
Desquam-X 5 Wash
Desquam-X 10 Gel
Desquam-X 10 Wash
Dry and Clear 5 Lotion
Dry and Clear Double Strength 10 Cream
Dryox 5 Gel
Dryox 10 Gel
Dryox 20 Gel
Dryox Wash 5
Dryox Wash 10
Duac Topical Gel
Epiduo
Fostex 5 Gel
Fostex 10 Bar
Fostex 10 Cream
Fostex 10 Gel
Fostex 10 Wash
H_2Oxyl 2.5 Gel
H_2Oxyl 5 Gel
H_2Oxyl 10 Gel
H_2Oxyl 20 Gel
Inova 8/2 ACT
Loroxide 5 Lotion with Flesh Tinted Base
Loroxide 5.5 Lotion
NeoBenz Micro
Neutrogena Acne Mask 5
Noxzema Clear-Ups Maximum Strength 10
Noxzema Clear-Ups On-the-Spot 10 Lotion
Onexton
Oxy 5 Tinted Lotion
Oxy 5 Vanishing Formula Lotion
Oxy 5 Vanishing Lotion
Oxy 10
Oxy 10 Daily Face Wash
Oxy 10 Tinted Lotion
Oxy 10 Vanishing Lotion
Oxyderm 5 Lotion
Oxyderm 10 Lotion
Oxyderm 20 Lotion
PanOxyl 5 Bar
PanOxyl 5 Gel
PanOxyl 10 Bar
PanOxyl 10 Gel
PanOxyl 15 Gel
PanOxyl 20 Gel
PanOxyl Acne Creamy Wash
PanOxyl AQ 2½ Gel
PanOxyl AQ 5 Gel
Persa-Gel 5
Persa-Gel 10
Persa-Gel W 5
Persa-Gel W 10
pHisoAc BP 10
Propa P.H. 10 Acne Cover Stick
Propa P.H. 10 Liquid Acne Soap
Stri-Dex Maximum Strength Treatment 10 Cream
Theroxide 5 Lotion
Theroxide 10 Lotion
Theroxide 10 Wash
Topex 5 Lotion
Topex 10 Lotion
Vanoxide 5 Lotion
Xerac BP 5 Gel
Xerac BP 10 Gel
Zeroxin-5 Gel
Zeroxin-10 Gel
Zoderrm
Zoderm Ready Pads

BETA-ADRENERGIC BLOCKING AGENTS

GENERIC NAMES

1. ACEBUTOLOL
2. ATENOLOL
3. BETAXOLOL
4. BISOPROLOL
5. CARTEOLOL
6. CARVEDILOL
7. LABETALOL
8. LEVOBETAXOLOL
9. METOPROLOL
10. NADOLOL
11. NEBIVOLOL
12. OXPRENOLOL
13. PENBUTOLOL
14. PINDOLOL

15. PROPRANOLOL
16. SOTALOL
17. TIMOLOL

BRAND NAMES
Apo-Atenolol[2]
Apo-Metoprolol[9]
Apo-Propranolol[15]
Apo-Timol[17]
Betaloc[9]
Betaxon[8]
Betapace[16]
Bystolic[11]
Blocadren[17]
Cartrol[5]
Coreg[6]
Coreg CR[6]
Corgard[10]
Detensol[15]
Inderal[15]
Inderal LA[15]
Kerlone[3]
Levatol[13]
Lopressor[9]
Lopressor SR[9]
Monitan[1]
Normodyne[7]
Novo-Atenol[2]
Novometoprol[9]
Novo-Pindol[14]
Novopranol[15]
Novo-Timol[17]
NuMetop[9]
Sectral[1]
Slow-Trasicor[12]
Sotacor[16]
Syn-Nadolol[10]
Syn-Pindolol[14]
Tenormin[2]
Toprol[9]
Toprol XL[9]
Toprol XL-XR[9]
Trandate[7]
Trasicor[12]
Visken[14]
Zebeta[4]

BETA-ADRENERGIC BLOCKING AGENTS & THIAZIDE DIURETICS

GENERIC NAMES
1. ATENOLOL & CHLORTHALIDONE
2. BISOPROLOL & HYDROCHLOROTHIAZIDE
3. METOPROLOL & HYDROCHLOROHIAZIDE
4. NADOLOL & BENDROFLUMEHIAZIDE
5. PROPRANOLOL & HYDROCHLOROTHIAZIDE

BRAND NAMES
Corzide[4]
Dutoprol[3]
Inderide[5]
Lopressor HCT[3]
Tenoretic[1]
Ziac[2]

BISPHOSPHONATES

GENERIC NAMES
1. ALENDRONATE
2. ETIDRONATE
3. IBANDRONATE
4. PAMIDRONATE
5. RISEDRONATE
6. TILUDRONATE
7. ZOLEDRONIC ACID

BRAND NAMES
Actonel[5]
Actonel with Calcium[5]
Aredia[4]
Atelvia[5]
Binosto[1]
Boniva[3]
Didronel[2]
Fosamax[1]
Fosamax Plus D[1]
Fosavance[1]
Reclast[7]
Skelid[6]
Zometa[7]

BRONCHODILATORS, ADRENERGIC (Short-Acting)

GENERIC NAMES
1. ALBUTEROL
2. LEVALBUTEROL

3. METAPROTERENOL
4. PIRBUTEROL
5. TERBUTALINE

BRAND NAMES

Accuneb[1]
Combivent[1]
Combivent Respimat[1]
Duoneb[1]
Maxair[4]
Pro-Air[1]
Pro-Air Respiclick[1]
Proventil HFA[1]
Ventolin HFA[1]
Vospire ER[1]
Xopenex[2]
Xopenex HFA[2]

CAFFEINE

Actamin Super
Alka-Seltzer Morning Relief
Anacin
Anolor-300
A.P.C.
Aspirin Free Bayer Select Maximum Strength Headache Pain Relief Caplets
Aspirin-Free Excedrin Caplets
Bayer Quick Release Crystals
Cafergot
Cafergot PB
Cafertine
Cafetrate
Caffedrine
Caffefrine Caplets
Citrated Caffeine
Cotanal 65
Dexitac
Dristan AF
Dristan AF Plus
Dristan Formula P
Enerjets
Ercaf
Ergo-Caff
Esgic-Plus
Excedrin Caplets
Excedrin Extra Strength Caplets
Excedrin Extra Strength Tablets
Excedrin Migraine
Excedrin Tension Headache
Extra Strength Bayer Back & Body Pain
Fioricet
Fioricet w/Codeine
Fiorinal
Fiorinal w/Codeine
Gotamine
Keep Alert
Lucidex
Migergot
P-A-C Revised Formula
Pep-Back
Phrenilin w/Caffeine & Codeine
Quick Pep
Repan
Salatin
Salocol
Scot-tussin Original 5-Action Cold Medicine
Sinapils
Snap Back
Supac
Synalgos-DC
Tirend
Trigesic
Vanquish
Vivarin
Wake-Up
Wigraine

CALCIUM CHANNEL BLOCKERS

GENERIC NAMES

1. AMLODIPINE
2. BEPRIDIL
3. DILTIAZEM
4. FELODIPINE
5. FLUNARIZINE
6. ISRADIPINE
7. NICARDIPINE
8. NIFEDIPINE
9. NISOLDIPINE
10. VERAPAMIL

BRAND NAMES

Adalat[8]
Adalat CC[8]
Adalat FT[8]
Adalat P.A.[8]
Apo-Diltiaz[3]
Apo-Nifed[8]
Apo-Verap[10]
Amturnide[1]
Azor[1]
Bepadin[2]

Caduet[1]
Calan[10]
Calan SR[10]
Cardene[7]
Cardene SR[7]
Cardizem[3]
Cardizem CD[3]
Cardizem LA[3]
Cardizem SR[3]
Cartia XT[3]
Chronovera[10]
Dilacor-XR[3]
Exforge[1]
Exforge HCT[1]
Isoptin[10]
Isoptin SR[10]
Ketorolac[3]
Lotrel[1]
Nifedical[3]
Norvasc[1]
Novo-Diltazem[3]
Novo-Nifedin[8]
Novo-Veramil[10]
Nu-Diltiaz[3]
Nu-Nifed[8]
Nu-Verap[10]
Plendil[4]
Prestalia[1]
Procardia[8]
Procardia XL[8]
Renedil[4]
Sibelium[5]
Sular[9]
Syn-Diltiazem[3]
Tarka[10]
Teczem[3]
Tekamlo[1]
Tiazac[3]
Tribenzor[1]
Twynsta[1]
Vascor[2]
Verelan[10]
Verelan PM[10]

CALCIUM SUPPLEMENTS

GENERIC NAMES

1. CALCIUM CARBONATE
2. CALCIUM CITRATE
3. CALCIUM GLUBIONATE
4. CALCIUM GLUCONATE
5. CALCIUM GLYCEROPHOSPHATE & CALCIUM LACTATE
6. CALCIUM LACTATE
7. DIBASIC CALCIUM PHOSPHATE
8. TRIBASIC CALCIUM PHOSPHATE

BRAND NAMES

Actonel with Calcium[1]
Apo-Cal[1]
Bayer Women's Caplets Plus Calcium[1]
BioCal[1]
Calcarb 600[1]
Calci-Chew[1]
Calciday 667[1]
Calcilac[1]
Calcite 500[1]
Calcium Carbonate/600[1]
Calcium Stanley[4]
Calcium-600[1]
Calcium-Sandoz[3]
Calcium-Sandoz Forte[1,6]
Calglycine[1]
Calphosan[5]
Calsan[1]
Cal-sap[3]
Caltrate[1]
Caltrate-300[1]
Caltrate-600[1]
Caltrate Chewable[1]
Children's Pepto[1]
Chooz[1]
Citracal[2]
Citracal Liquitabs[2]
Gencalc 600[1]
Gramcal[1,6]
Mallamint[1]
Neo-Calglucon[3]
Nephro-Calci[1]
NutraCal[3]
Os-Cal[1]
Os-Cal 500[1]
Os-Cal Chewable[1]
Oysco[1]
Oysco 500 Chewable[1]
Oyst-Cal[1]
Oyst-Cal 500 Chewable[1]
Oystercal 500[1]
Posture[8]
Rolaids-Calcium Rich[1]
Titralac[1]
Tums[1]
Tums E-X[1]

COAL TAR (Topical)

Alphosyl
Aquatar
Balnetar
Balnetar Therapeutic Tar Bath
Cutar Water Dispersible Emollient Tar
Denorex
Denorex Extra Strength Medicated Shampoo
Denorex Extra Strength Medicated Shampoo with Conditioners
Denorex Medicated Shampoo
Denorex Medicated Shampoo and Conditioner
Denorex Mountain Fresh Herbal Scent Medicated Shampoo
DHS Tar Gel Shampoo
DHS Tar Shampoo
Doak Oil
Doak Oil Forte
Doak Oil Forte Therapeutic Bath Treatment
Doak Oil Therapeutic Bath Treatment For All-Over Body Care
Doak Tar Lotion
Doak Tar Shampoo
Doctar
Doctar Hair & Scalp Shampoo & Conditioner
Estar
Fototar
Ionil-T Plus
Lavatar
Liquor Carbonis Detergens
Medotar
Pentrax Extra-Strength Therapeutic Tar Shampoo
Pentrax Tar Shampoo
Psorent
psoriGel
PsoriNail
Scytera Foam
Tar Doak
Taraphilic
Tarbonis
Tarpaste
Tarpaste Doak
T/Derm Tar Emollient
Tegrin Lotion for Psoriasis
Tegrin Medicated Cream Shampoo
Tegrin Medicated Shampoo Concentrated Gel
Tegrin Medicated Shampoo Extra Conditioning Formula
Tegrin Medicated Shampoo Herbal Formula
Tegrin Medicated Shampoo Original Formula
Tegrin Medicated Soap for Psoriasis
Tegrin Skin Cream for Psoriasis
Tersa-Tar Mild Therapeutic Shampoo with Protein and Conditioner
Tersa-Tar Soapless Tar Shampoo
Tersa-Tar Therapeutic Shampoo
T-Gel
T/Gel Therapeutic Conditioner
T/Gel Therapeutic Shampoo
Theraplex T Shampoo
Zetar
Zetar Emulsion
Zetar Medicated Antiseborrheic Shampoo

CONTRACEPTIVES, ORAL & SKIN

GENERIC NAMES

1. DESOGESTREL & ETHINYL ESTRADIOL
2. DROSPIRENONE & ETHINYL ESTRADIOL
3. ESTRADIOL VALERATE & DIENOGEST
4. ETHYNODIOL DIACETATE & ETHINYL ESTRADIOL
5. LEVONORGESTREL & ETHINYL ESTRADIOL
6. NORELGESTROMIN & ETHINYL ESTRADIOL
7. NORETHINDRONE & ETHINYL ESTRADIOL
8. NORETHINDRONE & MESTRANOL
9. NORETHINDRONE ACETATE & ETHINYL ESTRADIOL
10. NORGESTIMATE & ETHINYL ESTRADIOL
11. NORGESTREL & ETHINYL ESTRADIOL

BRAND NAMES

Alesse[5]
Apri[1]
Aviane[5]

Beyaz[2]
Brevicon[7]
Brevicon 0.5/35[7]
Brevicon 1/35[7]
Camrese[5]
Cyclen[10]
Cyclessa[1]
Demulen 1/35[4]
Demulen 1/50[4]
Demulen 30[4]
Demulen 50[4]
Desogen 28[1]
Desogen Ortho-Cept[1]
Emoquette[1]
Estrostep[9]
Estrostep Fe[9]
Femcon Fe[7]
GenCept 0.5/35[7]
GenCept 1/35[7]
GenCept 10/11[7]
Generess Fe[7]
Genora 0.5/35[7]
Genora 1/35[7]
Genora 1/50[8]
Introvale[5]
Jenest-28[7]
Larin Fe[7]
Levlen[5]
Levlite[5]
Levora[5]
Loestrin 1/20[9]
Loestrin 1.5/30[9]
Loestrin 24 Fe[9]
Lo Loestrin Fe[9]
Lo Minastrin Fe[7]
Lo/Ovral[11]
LoSeasonique[5]
Lybrel[5]
Marvelon[1]
Microgestin Fe[7]
Minestrin 1/20[9]
Min-Ovral[5]
Mircette[1]
ModiCon[7]
Myzilra[5]
Natazia[3]
Necon 0.5/35-21[7]
Necon 0.5/35-28[7]
Necon 1/35-21[7]
Necon 1/35-28[7]
Necon 1/50-21[8]
Necon 1/50-28[8]
Necon 10/11-21[7]
Necon 10/11-28[7]
N.E.E. 1/35[7]
N.E.E. 1/50[7]
Nelova 0.5/35E[7]
Nelova 1/35E[7]
Nelova 1/50M[8]
Nelova 10/11[7]
Nelulen 1/35E[4]
Nelulen 1/50E[4]
Nikki[2]
Norcept-E 1/35[9]
Nordette[5]
Norethin 1/35E[7]
Norethin 1/50M[8]
Norinyl 1+35[7]
Norinyl 1+50[8]
Norinyl 1/50[8]
Norlestrin 1/50[99]
Norlestrin 2.5/50[9]
Ocella[2]
Orsythia[5]
Ortho 0.5/35[7]
Ortho 1/35[7]
Ortho 7/7/7[7]
Ortho 10/11[7]
Ortho-Cept[1]
Ortho-Cyclen[10]
Orthro Evra[6]
Ortho-Novum 0.5[7]
Ortho-Novum 1/35[7]
Ortho-Novum 1/50[8]
Ortho-Novum 1/80[8]
Ortho-Novum 2[7]
Ortho-Novum 7/7/7[7]
Ortho-Novum 10/11[7]
Ortho-Tri-Cyclen 21[8]
Ortho-Tri-Cyclen 28[8]
Orthr-Tri-Cyclen Lo[10]
Ovcon-35[7]
Ovcon-50[7]
Ovral[11]
Pimtrea[1]
Previfem[10]
Quartette[5]
Quasense[5]
Safyral[2]
Seasonale[5]
Seasonique[5]
Setlakin[5]

Symphasic[7]
Tri-Cyclen[10]
Tri-Levlen[5]
Tri-Norinyl[7]
Tri-Previfem[10]
Tri-Sprintec[10]
Triphasil[5]
Triquilar[5]
Trivora[5]
Xulane[6]
Yasmin[2]
Yaz[2]
Zovia 1/35E[4]
Zovia 1/50E[4]

CONTRACEPTIVES, VAGINAL

GENERIC NAMES

1. BENZALKONIUM CHLORIDE
2. ETONOGESTREL & ETHINYL ESTRADIOL
3. NONOXYNOL 9
4. OCTOXYNOL 9

BRAND NAMES

Advantage 24[3]
Because[3]
Conceptrol Gel[3]
Conceptrol-Contraceptive Inserts[3]
Delfen[3]
Emko[3]
Encare[3]
Gynol II Extra Strength[3]
Gynol II Original Formula[3]
Koromex Cream[3]
Koromex Crystal Gel[3]
Koromex Foam[3]
Koromex Jelly[3]
K-Y Plus[3]
NuvaRing[2]
Ortho-Creme[3]
Ortho-Gynol[4]
Pharmatex[1]
Pre-Fil[3]
Ramses Contraceptive Foam[3]
Ramses Contraceptive Vaginal Jelly[3]
Ramses Crystal Clear Gel[3]
Semicid[3]
Shur-Seal[3]
VCF[3]

DEXTROMETHORPHAN

2/G-DM Cough
Alka-Seltzer Plus Cold & Cough Effervescent
Alka-Seltzer Plus Cold & Cough Liquid
Alka-Seltzer Plus Cold & Cough Liquid-Gels
Alka-Seltzer Plus Day & Night Effervescent
Alka-Seltzer Plus Day & Night Liquid Gels
Alka-Seltzer Plus Day Cold Liquid
Alka-Seltzer Plus Flu Effervescent
Alka-Seltzer Plus Nighttime Cold Liquid
Alka-Seltzer Plus Night-Time Effervescent
Alka-Seltzer Plus Night-Time Liquid-Gels
All-Nite Cold Formula
Ambenyl-D Decongestant Cough Formula
Anatuss DM
Anti-Tuss DM Expectorant
Balminil DM
Baytussin DM
Benylin All-In-One Cold & Flu Caplets
Benylin All-In-One Cold & Flu Nightime Syrup
Benylin All-In-One Cold & Flu Syrup
Benylin All-In-One Day & Night Caplets
Benylin DM 12 Hour Nightime Cough Syrup
Benylin DM-D
Benylin DM-D for Children Cough & Cold Syrup
Benylin DM-D-E
Benylin DM-E Chest Cough Syrup
Benylin DM-D-E Extra Strength
Benylin DM-E
Benylin DM-E Chest Cough Syrup
Bromfed-DM
Broncho-Grippol-DM
Carbodec DM Drops
Cerose-DM
Cheracol D Cough
Children's Benylin DM-D
Children's Formula Cough
Children's Hold
Children's Tylenol Plus Cold & Cough

Children's Tylenol Plus Cough & Runny Nose
Children's Tylenol Plus Cough & Sore Throat
Children's Tylenol Plus Flu
Children's Tylenol Plus Multi-Symptom Plus Cold
Codimal DM
Codistan No. 1
Colrex Cough
Conar
Conar Expectorant
Conar-A
Concentrin
Congespirin
Contac Night Caplets
Contac Severe Cold Formula
Contac Severe Cold Formula Night Strength
Coricidin HBP
Coricidin HBP Chest Congestion & Cough
Cough X
Creo-Terpin
Delsym
Delsym Grape for Adults
Delsym Grape for Children
Diabetic Tussin Cold & Flu
Diabetic Tussin DM
Diabetic Tussin Night Time Formula
Dimetapp Children's Elixir Cold & Allergy PE
Dimetapp Children's Long Acting Cough Plus Cold
DM Cough
DM Syrup
Donatussin
Dondril
Dristan Cold and Flu
Dristan Juice Mix-in Cold, Flu, & Cough
Drixoral Cough
Efficol Cough Whip (Cough Suppressant/ Decongestant)
Efficol Cough Whip (Cough Suppressant/Expectorant)
Extra Action Cough
Fast-Max DM Adult Liquid
Father John's Medicine Plus
Genatuss DM
Glycotuss-dM
Guiamid D.M. Liquid
Guiatuss-DM
Halotussin-DM Expectorant
Histalet-DM Hold
Humibid DM Sprinkle
Koffex
Kolephrin GG/DM
Kolephrin/DM Caplets
Kophane Cough and Cold Formula
Maximum Strength Mucinex DM
Meda Syrup Forte
Medatussin
Mediquell
Mucinex Adult Caplets - Cold, Flu & Sore Throat
Mucinex Adult Caplets - Severe Congestion & Cold
Mucinex Children's Cough, Expectorant & Suppressant
Mucinex Cough Mini-Melts
Mucinex Fast-Max Cold, Flu & Sore Throat Liquid
Mucinex Maximum Strength Fast Max Cold, Flu & Sore Throat
Mucinex Maximum Strength Fast Max Severe Congestion & Cough
Mytussin DM
Naldecon Senior DX
Naldecon-DX
NeoCitran DM Coughs & Colds
Neo-DM
Nytcold Medicine
Ornex DM 15
Ornex DM 30
Ornex Severe Cold No Drowsiness Caplets
Par Glycerol-DM
PediaCare Children's Cough & Congestion
PediaCare Children's Fever Reducer Plus Cough & Runny Nose
PediaCare Children's Fever Reducer Plus Cough & Sore Throat
PediaCare Children's Fever Reducer Plus Flu
PediaCare Children's Multi-Symptom Cold
PediaCare Fever Reducer Plus Multi-Symptom Cold
Pertussin All Night CS

Pertussin All Night PM Pertussin Cough Suppressant
Pertussin CS
Pertussin ES
Phanatuss
Phenameth DM
Phenergan with Dextromethorphan
Pherazine DM
Prometh with Dextromethorphan
Promethazine DM
Queltuss
Remcol-C
Rhinosyn-DM
Rhinosyn-DMX Expectorant
Rhinosyn-X
Robafen DM
Robidex
Robitussin Children's Cough Long-Acting
Robitussin Children's Cough & Cold Long-Acting
Robitussin Cough & Chest Congestion
Robitussin Cough & Chest Congestion DM
Robitussin Cough & Chest Congestion DM Max
Robitussin Cough & Chest Congestion Sugar Free DM
Robitussin Cough & Cold CF
Robitussin Cough & Cold D
Robitussin Cough & Cold Long-Acting
Robitussin Cough Cold & Flu Nighttime
Robitussin Cough Gels Long Acting
Robitussin Cough Long Acting
Robitussin Maximum Strength Cough + Congestion DM
Ru-Tuss Expectorant
SafeTussin 30
Scot-Tussin DM
Sedatuss
Silexin Cough
Simply Cough
Snaplets-DM
Snaplets-Multi
St. Joseph Cough Suppressant for Children
Sucrets Cough Control
Sudafed Multi-Symptom Cold & Cough
Sudafed PE Cold & Cough Caplets
Suppress Cough with Dextromethorphan
Terphan
Theracof Plus Multi-Symptom Cough and Cold Reliever
Theraflu Cold & Cough Hot Liquid
Theraflu Daytime Severe Cold Caplets
Theraflu Daytime Severe Cold Caplets
Theraflu Nighttime Severe Cold Caplets
Theraflu Thin Strips Daytime Cold & Cough
Theraflu Warming Relief Daytime
Tolu-Sed DM Cough
Touro DM
Triaminic Cough & Sore Throat
Triaminic-D Multi-Symptom Cold
Triaminic Day Time Cold & Cough
Triaminic Flu, Cough & Fever
Triaminic Long Acting Cough
Triaminic Softchews Cough & Runny Nose
Triaminic Softchews Cough & Sore Throat
Triaminic Thin Strips Daytime Cold & Cough
Triaminic Thin Strips Long Acting Cough
Tricodene Sugar Free
Trimedine Liquid
Trocal
Tussar DM
Tuss-DM
Tussi-Bid
Tylenol Cold Head Congestion Nighttime
Tylenol Cold Head Congestion Severe
Tylenol Cold Multi Symptom Daytime
Tylenol Cold Multi Symptom Nighttime
Tylenol Cold Multi Symptom Severe
Tylenol Cough & Severe Congestion Daytime
Tylenol Cough & Sore Throat Daytime
Tylenol Cough & Sore Throat Nighttime
Tylenol Sinus Severe Congestion Daytime
Uni-Tussin DM
Unproco
Vicks Children's NyQuil
Vicks DayQuil Cold/Flu Relief

Vicks DayQuil Cold & Flu Symptom Relief Plus Vitamin C
Vicks DayQuil Cough
Vicks DayQuil Mucus Control DM
Vicks Formula 44 Custom Care Chesty Cough
Vicks Formula 44 Custom Care Congestion
Vicks Formula 44 Custom Care Cough & Cold PM
Vicks Formula 44 Custom Care Dry Cough Suppressant
Vicks NyQuil Cold & Flu `Relief
Vicks NyQuil Cold & Flu Symptom Relief Plus Vitamin C
Vicks NyQuil Cough
Vicks NyQuil D
Vicks NyQuil Less Drowsy Cold & Flu Relief Liquid
Vicks Pediatric Formula 44e Cough & Chest Congestion Relief
Vicks Pediatric Formula 44m Cough & Cold Relief
Viravan DM

DICYCLOMINE

Antispas
A-Spas
Bentyl
Bentylol
Byclomine
Dibent
Di-Cyclonex
Dilomine
Di-Spaz
Forulex
Lomine
Neoquess
Or-Tyl
Protylol
Spasmoban
Spasmoject
Viscerol

DIURETICS, THIAZIDE

GENERIC NAMES

1. BENDROFLUMETHIAZIDE
2. BENZTHIAZIDE
3. CHLOROTHIAZIDE
4. CHLORTHALIDONE
5. CYCLOTHIAZIDE
6. HYDROCHLOROTHIAZIDE
7. HYDROFLUMETHIAZIDE
8. METHYCLOTHIAZIDE
9. METOLAZONE
10. POLYTHIAZIDE
11. QUINETHAZONE
12. TRICHLORMETHIAZIDE

BRAND NAMES

Accuretic[6]
Aldoclor[3]
Aldoril[6]
Anhydron[5]
Apo-Chlorthalidone[4]
Apo-Hydro[6]
Aquatensen[8]
Amturnide[6]
Atacand Plus[6]
Avalide[6]
Benicar HCT[6]
Capozide[6]
Diovan HCT[6]
Diucardin[7]
Diuchlor H[6]
Diulo[9]
Diurese R[12]
Diuril[3]
Duretic[8]
Edarbyclor[4]
Enduron[8]
Exforge HCT[6]
Esidrix[6]
Exna[2]
Hydrex[2]
Hydro-D[6]
Hydrochlor[6]
HydroDIURIL[6]
Hydromox[11]
Hydropine[7]
Hydropine H.P.[7]
Hygroton[4]
Hyzaar[6]
Metahydrin[12]
Micardis HCT[6]
Micardis Plus[6]
Microzide[6]
Minizide[10]
Mykrox[9]
Naqua[12]
Naturetin[1]

Neo-Codema[6]
Novodoparil[6]
Novo-Hydrazide[6]
Novo-Thalidone[4]
Oretic[6]
PMS Dopazide[6]
Prinzide[6]
Regroton[4]
Renese[10]
Renese-R[10]
Saluron[7]
Salutensin[7]
Salutensin-Demi[7]
Supres[3]
Tekturna HCT[6]
Teveten HCT[6]
Thalitone[4]
Tribenzor[6]
Trichlorex[12]
Uniretic[6]
Uridon[4]
Urozide[6]
Vaseretic[6]
Zaroxolyn[9]
Zestoretic[6]

ERYTHROMYCINS

GENERIC NAMES

1. ERYTHROMYCIN ESTOLATE
2. ERYTHROMYCIN ETHYLSUCCINATE
3. ERYTHROMYCIN GLUCEPTATE
4. ERYTHROMYCIN LACTOBIONATE
5. ERYTHROMYCIN STEARATE
6. ERYTHROMYCIN-BASE

BRAND NAMES

Apo-Erythro[6]
Apo-Erythro E-C[6]
Apo-Erythro ES[2]
Apo-Erythro-S[5]
E-Base[6]
E.E.S.[2]
E/Gel[6]
Emgel[6]
E-Mycin[6]
Erybid[6]
ERYC[6]
EryPed[2]
Ery-Tab[6]
Erythraderm[6]
Erythro[2]
Erythrocin[5]
Erythrocot[5]
Erythromid[6]
Ilosone[1]
My-E[5]
Novorythro[5]
PCE Dispersatabs[5]
Pediazole[6]
Sulfimycin[6]
Wintrocin[5]

ESTROGENS

GENERIC NAMES

1. CONJUGATED ESTROGENS
2. DIETHYLSTILBESTROL
3. ESTERIFIED ESTROGENS
4. ESTRADIOL
5. ESTROGEN
6. ESTRONE
7. ESTROPIPATE
8. ETHINYL ESTRADIOL
9. QUINESTROL

BRAND NAMES

Activella[4]
Alora[4]
Angeliq[4]
Cenestin[1]
C.E.S.[1]
Climara[4]
Climara Pro[4]
Clinagen LA 40[4]
CombiPatch[4]
Congest[1]
Deladiol-40[4]
Delestrogen[4]
depGynogen[4]
Depo Estradiol[4]
Depogen[4]
DES[2]
Divigel[4]
Dura-Estrin[4]
Duragen[4]
Duragen-20[4]
Duragen-40[4]
E-Cypionate[4]
Elestrin[4]
Enjuvia[1]
Esclim[4]
Estinyl[8]

Estrace[4]
Estraderm[4]
Estragyn 5[6]
Estragyn LA 5[6]
Estra-L[4]
Estrasorb[4]
Estratab[3]
Estring[4]
Estro-A[6]
Estro-Cyp[4]
Estrofem[4]
Estrogel[4]
Estroject-L.A.[4]
Estro-L.A.[4]
Estro-Span[4]
EvaMist[4]
Femhrt[4]
Femogex[4]
Femring[4]
Femtrace[4]
Gynogen L.A. 20[4]
Gynogen L.A. 40[4]
Honvol [2]
Kestrone-5[6]
Mannest[2]
Menaval-20[4]
Menest[3]
Menostar[4]
Minivelle[4]
Neo-Estrone[3]
Ogen[7]
Ogen 1.25[7]
Ogen 2.5[7]
Ogen 6.25[7]
Ortho-Est[7]
Orthro-Prefest[4]
Premarin[1]
Premarin Vaginal Cream[1]
Premphase[1]
Prempro[1]
SCE-A Vaginal Cream[1]
Stilbestrol[2]
Stilphostrol[2]
Vagifem[4]
Valergen-10[4]
Valergen-20[4]
Valergen-40[4]
Vivelle[4]
Vivelle Dot[4]
Wehgen[6]

GUAIFENESIN

2/G-DM Cough
Ambenyl-D Decongestant Cough Formula
Amonidrin
Anatuss DM
Anatuss LA
Anti-Tuss
Anti-Tuss DM Expectorant
Asbron G
Asbron G Inlay Tablets
Balminil Expectorant
Baytussin DM
Benylin All-In-One Cold & Flu Caplets
Benylin All-In-One Cold & Flu Nightime Syrup
Benylin All-In-One Cold & Flu Syrup
Benylin All-In-One Day & Night Caplets
Benylin DM-D-E
Benylin DM-D-E Extra Strength
Benylin DM-E Chest Cough Syrup
Breonesin
Broncholate
Brontex
Cheracol D Cough
Children's Mucinex Cough
Codimal Expectorant
Codistan No. 1
Colrex Expectorant
Comtrex Deep Chest Cold
Conar Expectorant
Conar-A
Concentrin
Congess JR
Congess SR
Congestac Caplets
Coricidin HBP Chest Congestion and Cough
Deconsal II
Diabetic Tussin DM
Diabetic Tussin EX
Diabetic Tussin Mucus Relief
Donatussin
Donatussin Drops
Duratuss
Efficol Cough Whip (Cough Suppressant/Decongestant)
Entex PSE
Entuss Pediatric Expectorant
Extra Action Cough

Fast-Max DM Adult Liquid
Father John's Medicine Plus
Fedahist Expectorant
Fedahist Expectorant Pediatric Drops
Fenesin
Gee-Gee
Genatuss
Genatuss DM
GG-CEN
Glyate
Glycotuss
Glycotuss-dM
Glytuss
Guaifed
Guaifed-PD
GuaiMAX-D
Guiamid D.M. Liquid
Guiatuss PE
Guiatuss-DM
Halotussin
Halotussin-DM Expectorant
Histalet X
Humibid L.A.
Humibid Sprinkle
Humibid-DM Sprinkle
Hytuss
Hytuss-2X
Kids-EEZE Chest Relief
Kolephrin GG/DM
LiquiBid D
LiquiBid D-R
LiquiBid PD
Malotuss
Maximum Strength Mucinex
Maximum Strength Mucinex D
Maximum Strength Mucinex DM
Meda Syrup Forte
Medatussin
Medatussin Plus
Mucinex
Mucinex Adult Caplets - Cold & Sinus
Mucinex Adult Caplets - Cold, Flu, & Sore Throat
Mucinex Adult Caplets - Severe Congestion & Cold
Mucinex Children's Cough, Expectorant & Suppressant
Mucinex Children's Expectorant
Mucinex Cold Liquid
Mucinex Cough Mini-Melts
Mucinex D
Mucinex DM
Mucinex Fast-Max Cold, Flu & Sore Throat Liquid
Mucinex Junior Strength Expectorant
Mucinex Maximum Strength Fast Max Cold, Flu & Sore Throat
Mucinex Maximum Strength Fast Max Severe Congestion & Cough
Mucinex Mini-Melts
Mytussin DM
Naldecon Senior DX
Naldecon Senior EX
Nasatab LA
NeoCitrin DM Coughs & Colds
Nortussin
Organidin
PediaCare Children's Fever Reducer Plus Flu
Pertussin All Night CS
Phanatuss
Pneumomist
Poly-Histine Expectorant Plain
P-V-Tussin Tablets
Queltuss
Refenesen
Refenesen Chest Congestion & Pain Relief PE
Refenesen PE
Respaire-30
Resyl
Rhinosyn-DMX Expectorant
Rhinosyn-X
Robafen DM
Robafen Syrup
Robitussin Chest Congestion
Robitussin Cough & Chest Congestion
Robitussin Cough & Chest Congestion DM
Robitussin Cough & Chest Congestion DM Max
Robitussin Cough & Chest Congestion Sugar Free DM
Robitussin Cough & Cold CF
Robitussin Cough & Cold D
Robitussin Maximum Strength Cough + Congestion DM
Robitussin with Codeine
Ru-Tuss DE
Ru-Tuss Expectorant

SafeTussin 30
Scot-Tussin
Silexin Cough
Sinumist-SR
Sinupan
Sinutab Non-Drying Liquid Caps
SINUvent PE
Stamoist E
Sudafed Multi-Symptom Cold & Cough
Sudafed Non-Drying Sinus Liquid Caps
Sudafed PE Cold & Cough Caplets
Theolate
Theracof Plus Multi-Symptom Cough and Cold Reliever
Theraflu Flu & Chest Congestion Hot Liquid
Tolu-Sed DM Cough
Touro DM
Touro Ex
Touro LA Caplets
Triaminic Chest & Nasal Congestion
Tuss-DM
Tussi-Bid
Tussi-organidin
Tylenol Chest Congestion
Tylenol Cold Head Congestion Severe
Tylenol Cold Multi Symptom Severe
Tylenol Cough & Severe Congestion Daytime
Tylenol Sinus Congestion & Pain Severe
Uni-Tussin
Uni-Tussin DM
Unproco
Versacaps
Vicks DayQuil Mucus Control DM
Vicks DayQuil Mucus Control Liquid
Vicks Formula 44 Custom Care Chesty Cough
Vicks Pediatric Formula 44e Cough & Chest Congestion Relief
Vicks VapoSyrup Severe Congestion Head & Chest Congestion Relief

INSULIN

Afrezza
Humulin 70/30
Humulin N
Humulin R
Novolin 70/30
Novolin N
Novolin R

IRON SUPPLEMENTS

GENERIC NAMES

1. CARBONYL IRON
2. FERROUS FUMARATE
3. FERROUS GLUCONATE
4. FERROUS SULFATE
5. IRON DEXTRAN
6. IRON POLYSACCHARIDE
7. IRON SORBITOL

BRAND NAMES

Apo-Ferrous Gluconate[3]
Apo-Ferrous Sulfate[4]
Estrostep Fe[2]
Femiron[2]
Feosol[4]
Feosol Caplet[1]
Feostat[2]
Feostat Drops[2]
Fergon[3]
Fer-In-Sol[4]
Fer-In-Sol Drops[4]
Fer-In-Sol Syrup[4]
Fer-Iron[4]
Fero-folic 500[4]
Fero-Grad[4]
Fero-Gradumet[4]
FerraCap[1]
Ferralet[3]
Ferralet 90[1]
Ferralyn[4]
Ferra-TD[4]
Fertinic[4]
Generess Fe[2]
Hemocyte[2]
Hytinic[6]
Icar[1]
Ircon[2]
Jectofer[7]
Larin Fe[2]
Loestrin 24 Fe[2]
Lo Loestrin Fe[2]
Lo Minastrin Fe[2]
Neo-Fer[2]
Novoferrogluc[6]
Novoferrosulfa[3]
Novofumar[4]
Nu-Iron[6]
Nu-Iron 150[6]
Palafer[2]
Palmiron[2]

PMS Ferrous Sulfate[4]
Simiron[3]
Slow Fe[4]
Span-FF[2]

KERATOLYTICS

GENERIC NAMES

1. RESORCINOL
2. RESORCINOL & SULFUR
3. SALICYLIC ACID
4. SALICYLIC ACID & SULFUR
5. SULFUR (Topical)

BRAND NAMES

Acne-Aid Gel[2]
AcnoAcnomel Cake[2]
Acnomel Cream[2]
Acnomel Vanishing Cream[2]
Acnomel-Acne Cream[2]
Acnotex[4]
Adult Acne Clearing Gel[3]
Antinea[3]
Aveeno Acne Bar[4]
Aveeno Cleansing Bar[4]
Bensulfoid Cream[2]
Buf-Puf Acne Cleansing Bar with Vitamin E[3]
Buf-Puf Medicated Maximum Strength Pads[3]
Buf-Puf Medicated Regular Strength Pads[3]
Calicylic[3]
Cleanse and Treat[4]
Clear Away[3]
Clear by Design Medicated Cleansing Pads[3]
Clearasil Adult Care Medicated Blemish Cream[2]
Clearasil Adult Care Medicated Blemish Stick[2]
Clearasil Clearstick Maximum Strength Topical Solution[3]
Clearasil Clearstick Regular Strength Topical Solution[3]
Clearasil Double Textured Pads Maximum Strength[3]
Clearasil Double Textured Pads Regular Strength[3]
Clearasil Medicated Deep Cleanser Topical Solution[3]
Compound W Gel[3]
Compound W Liquid[3]
Creamy SS Shampoo[4]
Cuplex Gel[3]
Cuticura Ointment[5]
Diasporal Cream[4]
Duofilm[3]
Duoplant[3]
Duoplant Topical Solution[3]
Finac[5]
Fostex CM[4]
Fostex Medicated Cleansing Bar[4]
Fostex Medicated Cleansing Cream[4]
Fostex Medicated Cleansing Liquid[4]
Fostex Regular Strength Medicated Cleansing Bar[4]
Fostex Regular Strength Medicated Cleansing Cream[4]
Fostex Regular Strength Medicated Cover-Up[4]
Fostril Cream[5]
Fostril Lotion[5]
Freezone[3]
Gordofilm[3]
Hydrisalic[3]
Inova 8/2 ACT[3]
Ionax Astringent Skin Cleanser Topical Solution[3]
Ionil Plus Shampoo[3]
Ionil Shampoo[3]
Keralyt[3]
Keratex Gel[3]
Lactisol[3]
Listerex Golden Scrub Lotion[3]
Listerex Herbal Scrub Lotion[3]
Lotio Asulfa[3]
Mediplast[3]
Meted Maximum Strength Anti-Dandruff Shampoo with Conditioners[4]
Night Cast R[4]
Night Cast Regular Formula Mask-Lotion[4]
Night Cast Special Formula Mask-Lotion[2]
Noxzema Anti-Acne Gel[3]
Noxzema Anti-Acne Pads Maximum Strength[3]
Noxzema Anti-Acne Pads Regular Strength[3]
Occlusal Topical Solution[3]
Occlusal-HP Topical Solution[3]

Off-Ezy Topical Solution Corn & Callus Removal Kit[3]
Off-Ezy Topical Solution Wart Removal Kit[3]
Oxy Clean Medicated Cleanser[3]
Oxy Clean Medicated Pads Maximum Strength[3]
Oxy Clean Medicated Pads Sensitive Skin[3]
Oxy Clean Regular Strength[3]
Oxy Clean Regular Strength Medicated Cleanser Topical Solution[3]
Oxy Clean Regular Strength Medicated Pads[3]
Oxy Clean Sensitive Skin Cleanser Topical Solution[3]
Oxy Clean Sensitive Skin Pads[3]
Oxy Night Watch Maximum Strength Lotion[3]
Oxy Night Watch Night Time Acne Medication Extra Strength Lotion[3]
Oxy Night Watch Night Time Acne Medication Regular Strength Lotion[3]
Oxy Night Watch Sensitive Skin Lotion[3]
Oxy Sensitive Skin Vanishing Formula Lotion[3]
P&S[3]
Paplex[3]
Paplex Ultra[3]
Pernox Lemon Medicated Scrub Cleanser[4]
Pernox Lotion Lathering Abradant Scrub Cleanser[4]
Pernox Lotion Lathering Scrub Cleanser[4]
Pernox Regular Medicated Scrub Cleanser[4]
Propa pH Medicated Acne Cream Maximum Strength[3]
Propa pH Medicated Cleansing Pads Maximum Strength[3]
Propa pH Medicated Cleansing Pads Sensitive Skin[3]
Propa pH Perfectly Clear Skin Cleanser Topical Solution Oily Skin[3]
Propa pH Perfectly Clear Skin Cleanser Topical Solution Sensitive Skin Formula[3]
R.A.[1]
Rezamid Lotion[2]
Salac[3]
Salacid[3]
Sal-Acid Plaster[3]
Salactic Film Topical Solution[3]
Sal-Clens Plus Shampoo[3]
Sal-Clens Shampoo[3]
Salex Shampoo[3]
Saligel[3]
Salonil[3]
Sal-Plant Gel Topical Solution[3]
Sastid (AL) Scrub[4]
Sastid Plain[4]
Sastid Plain Shampoo and Acne Wash[4]
Sastid Soap[4]
Sebasorb Liquid[4]
Sebex[4]
Sebucare[3]
Sebulex Antiseborrheic Treatment and Conditioning Shampoo[4]
Sebulex Antiseborrheic Treatment Shampoo[4]
Sebulex Conditioning Shampoo[4]
Sebulex Cream Medicated Shampoo[4]
Sebulex Medicated Dandruff Shampoo with Conditioners[4]
Sebulex Medicated Shampoo[4]
Sebulex Regular Medicated Dandruff Shampoo[4]
Sebulex Shampoo[4]
Stri-Dex[3]
Stri-Dex Dual Textured Pads Maximum Strength[3]
Stri-Dex Dual Textured Pads Regular Strength[3]
Stri-Dex Dual Textured Pads Sensitive Skin[3]
Stri-Dex Maximum Strength Pads[3]
Stri-Dex Regular Strength Pads[3]
Stri-Dex Super Scrub Pads[3]
Sulforcin[2]
Sulsal Soap[4]
Tersac Cleansing Gel[3]
Therac Lotion[4]
Trans-Plantar[3]
Trans-Ver-Sal[3]
Vanseb Cream Dandruff Shampoo[4]
Vanseb Lotion Dandruff Shampoo[4]
Verukan Topical Solution[3]
Verukan-HP Topical Solution[3]
Viranol[3]

Viranol Ultra[3]
Wart-Off Topical Solution[3]
X-Seb[3]

LAXATIVES, BULK-FORMING

GENERIC NAMES

1. CALCIUM POLYCARBOPHIL
2. CARBOXYMETHYL-CELLUOSE SODIUM
3. MALT SOUP EXTRACT
4. METHYLCELLULOSE
5. POLYCARBOPHIL
6. PSYLLIUM

BRAND NAMES

Cillium[6]
Citrucel Orange Flavor[4]
Citrucel Sugar-Free Orange Flavor[4]
Cologel[4]
Disolan Forte[2]
Disoplex[2]
Effer-syllium[6]
Equalactin[5]
Fiberall[5]
Fibercon[5]
FiberNorm[5]
Fiberpur[6]
Hydrocil Instant[6]
Karacil[6]
Konsyl[5]
Konsyl Easy Mix Formula[6]
Konsyl-D[6]
Konsyl-Orange[6]
Maalox Daily Fiber Therapy[6]
Maalox Daily Fiber Therapy Citrus Flavor[6]
Maalox Daily Fiber Therapy Orange Flavor[6]
Maalox Sugar Free Citrus Flavor[6]
Maalox Sugar Free Orange Flavor[6]
Maltsupex[3]
Metamucil[6]
Metamucil Apple Crisp Fiber Wafers[6]
Metamucil Cinnamon Spice Fiber Wafers[6]
Metamucil Instant Mix, Orange Flavor[6]
Metamucil Smooth Citrus Flavor[6]
Metamucil Smooth Orange Flavor[6]
Metamucil Smooth, Sugar-Free Citrus Flavor[6]
Metamucil Smooth, Sugar-Free Orange Flavor[6]
Metamucil Smooth, Sugar-Free Regular Flavor[6]
Metamucil Sugar Free[6]
Metamucil Sugar Free Citrus Flavor[6]
Metamucil Sugar Free Lemon-Lime Flavor[6]
Metamucil Sugar Free Orange Flavor[6]
Mitrolan[5]
Modane Bulk[6]
Mylanta Natural Fiber Supplement[6]
Mylanta Sugar Free Natural Fiber Supplement[6]
Naturacil[6]
Natural Source Fibre Laxative[6]
Perdiem[6]
Perdiem Fiber[6]
Perdiem Plain[6]
Prodiem[6]
Prodiem Plain[6]
Prodiem Plus[6]
Pro-Lax[6]
Prompt[6]
Reguloid Natural[6]
Reguloid Orange[6]
Reguloid Orange Sugar Free[6]
SennaPrompt[6]
Serutan[6]
Serutan Toasted Granules[6]
Siblin[6]
Syllact[6]
Syllamalt[3]
Versabran[6]
Vitalax Super Smooth Sugar Free Orange Flavor[6]
Vitalax Unflavored[6]
V-Lax[6]

LAXATIVES, OSMOTIC

GENERIC NAMES

1. GLYCERIN
2. LACTULOSE
3. MAGNESIUM CITRATE
4. MAGNESIUM HYDROXIDE
5. MAGNESIUM OXIDE
6. MAGNESIUM SULFATE
7. MILK OF MAGNESIA
8. MINERAL OIL
9. POLYETHYLENE GLYCOL 3350
10. SODIUM PHOSPHATE

BRAND NAMES

Agarol Plain[7]
Agarol Strawberry[8]
Agarol Vanilla[8]
Bilagog[6]
Cholac[2]
Chronula[2]
Citroma[3]
Citro-Mag[3]
Citro-Nesia[3]
Constilac[2]
Constulose[2]
Duphalac[2]
Evalose[2]
Fleet Pedia Lax Liquid Gels[1]
Generlac[2]
Hayley's M-O[7]
HealthyLax[9]
Heptalac[2]
Kristalose[2]
Lactulax[2]
Magnolax[7]
Mag-Ox 400[5]
Maox[5]
Miralax[9]
Phillips' Chewable[4]
Phillips' Concentrated[4]
Phillips' Magnesia Tablets[4]
Phillips' Milk of Magnesia[4]
Portalac[2]

LAXATIVES, SOFTENER/ LUBRICANT

GENERIC NAMES

1. CASANTHRANOL & DOCUSATE
2. DOCUSATE
3. DOCUSATE CALCIUM
4. DOCUSATE POTASSIUM
5. DOCUSATE SODIUM
6. MINERAL OIL
7. POLOXAMER 188

BRAND NAMES

Afko-Lube[2]
Afko-Lube Lax[2]
Agarol Plain[6]
Agarol Marshmallow[6]
Agarol Raspberry[6]
Agarol Strawberry[6]
Agarol Vanilla[6]
Alaxin[7]
Bilax[2]
Colace[5]
Colace Microenema[5]
Correctol Extra Gentle[2]
Dialose[2]
Diocto[2]
Diocto-C[1]
Diocto-K[2]
Diocto-K Plus[1]
Dioeze[2]
Diosuccin[2]
Dio-Sul[2]
Diothron[1]
Disanthrol[1]
Disolan[2]
Disolan Forte[1]
Disonate[2]
Disoplex[2]
Di-Sosul[2]
Di-Sosul Forte[1]
Docu-K Plus[1]
DOK[2]
DOK Softgels[2]
Doss[2]
Doss Tablets[2]
Doxinate[2]
DSMC Plus[1]
Dulcodos[2]
Duosol[2]
Fleet Enema Mineral Oil[7]
Fleet Pedia-Lax Childrens Liquid Stool Softener[5]
Gentlax-S[2]
Kasof[2]
Kondremul[7]
Kondremul with Cascara[7]
Kondremul Plain[7]
Lansoyl[7]
Laxinate 100[2]
Liqui-Doss[7]
Milkinol[7]
Modane Plus[2]
Modane Soft[2]
Molatoc[2]
Molatoc-CST[2]
Neo-Cultol[7]
Neolax[2]
Nujol[7]
Peri-Colase[1]
Pertrogalar Plain[7]

PMS-Docusate Calcium[2]
PMS-Docusate Sodium[2]
Pro-Cal-Sof[2]
Pro-Sof[1]
Pro-Sof Liquid Concentrate[1]
Pro-Sof Plus[1]
Regulace[2]
Regulax SS[2]
Regulex[2]
Regulex-D[2]
Regutol[2]
Senokot-S[2]
Stulex[2]
Sulfolax[2]
Surfak[2]
Therevac Plus[2]
Therevac-SB[2]
Trilax[2]
Zymenol[7]

LAXATIVES, STIMULANT

GENERIC NAMES

1. ALOE
2. BISACODYL
3. CASANTHRANOL
4. CASCARA
5. CASTOR OIL
6. DEHYDROCHOLIC ACID
7. SENNA
8. SENNOSIDES

BRAND NAMES

Afko-Lube Lax[3]
Alphamul[5]
Aromatic Cascara Fluidextract[4]
Bilax[6]
Bisac-Evac[2]
Bisacolax[2]
Bisco-Lax[2]
Black Draught[3]
Black-Draught Lax-Senna[7]
Caroid Laxative[2]
Carter's Little Pills[4]
Cascara Aromatic Fluidextract[4]
Cascara Sagrada[4]
Cholan-HMB[6]
Dacodyl[2]
Decholin[6]
Deficol[2]
Diocto-C[3]
Diocto-K Plus[1]
Diothron[3]
Disanthrol[3]
Disolan Forte[3]
Di-Sosul Forte[3]
Docu-K Plus[3]
Dosaflex[7]
Dr. Caldwell Senna Laxative[7]
DSMC Plus[3]
Dulcodos[2]
Dulcolax[2]
Dulcolax for Women[2]
Emulsoil[5]
Ex-Lax Gentle Nature[8]
Fleet Bisacodyl[2]
Fleet Bisacodyl Prep[2]
Fleet Flavored Castor Oil[5]
Fleet Laxative[2]
Fleet Pedia Lax Liquid Gels[8]
Fletcher's Castoria[7]
Gentlax S[7]
Gentle Nature[8]
Glysennid[8]
Hepahydrin[6]
Herbal Laxative[8]
Kellogg's Castor Oil[5]
Kondremul with Cascara[4]
Laxit[2]
Molatoc-CST[3]
Mucinum Herbal[8]
Nature's Remedy[4]
Neolax[6]
Neoloid[5]
Nytilax[8]
Perdiem[7]
Peri-Colace[3]
PMS-Bisacodyl[2]
PMS-Sennosides[8]
Prodiem Plus[7]
Prompt[8]
Pro-Sof Plus[3]
Purge[5]
Regulace[3]
Senexon[7]
SennaPrompt[8]
Senokot[7]
Senokot-S[7]
SenokotXTRA[7]
Senolax[7]
Theralax[2]
Trilax[6]
X-Prep Liquid[7]

MEPROBAMATE
Acabamate
Apo-Meprobamate
Epromate-M
Equagesic
Equanil
Equanil Wyseals
Equazine-M
Heptogesic
Medi-Tran
Meprogesic
Meprogesic Q
Meprospan 200
Meprospan 400
Micrainin
Miltown
Neuramate
Novo-Mepro
Pax 400
Probate
Sedabamate
Trancot
Tranmep

METFORMIN
ACTOplus Met
ACTOplus Met XR
Avandamet
Fortamet
Glucophage
Glucophage XR
Glucovance
Glumetza
Invokamet
Janumet
Kazano
Kombiglyze XR
Metaglip
Pazamet
PrandiMet
Riomet
Xigduo XR

NARCOTIC ANALGESICS
GENERIC NAMES
1. BUPRENORPHINE
2. CODEINE
3. FENTANYL
4. HYDROCODONE
5. HYDROCODONE & HOMATROPINE
6. HYDROMORPHONE
7. LEVORPHANOL
8. MEPERIDINE
9. METHADONE
10. MORPHINE
11. OXYCODONE
12. OXYMORPHONE
13. PENTAZOCINE

BRAND NAMES
Abstral[3]
Actifed with Codeine Cough[2]
Actiq[3]
Ambenyl Cough[2]
Ambophen Expectorant[2]
Anexsia[4]
Avinza[10]
Bunavail[1]
Bromanyl[2]
Brontex[2]
Buprenex[1]
Butrans[1]
Calcidrine[2]
Capital with Codeine[2]
Cheracol[2]
Codeprex[2]
Codimal PH[2]
Demerol[8]
Dilaudid[6]
Dilaudid-HP[6]
Duragesic[3]
Embeda[10]
Endodan[11]
Exalgo[6]
Fentora[3]
Fioricet w/Codeine[2]
Fiorinal w/Codeine[2]
Hydromet[5]
Hydrostat IR[6]
Hysingla ER[4]
Ibudone[4]
Kadian[10]
Lazanda[3]
Lortab[4]
Methadose[9]
MS Contin[10]
Opana[12]
Opana ER[12]
Oxecta[11]
Oxycet[11]
Oxycontin[11]

Percocet[11]
Percodan[11]
Phrenilin w/Caffeine & Codeine[2]
Prometh VC with Codeine[2]
Reprexain[4]
Rezira[4]
Robitussin A-C[2]
Robitussin-DAC[2]
Roxicet[11]
Roxilox[11]
Soma Compound w/Codeine[2]
Stagesic[4]
Subsys[3]
Targiniq ER[11]
Temgesic[1]
Triacin C Cough[2]
TussiCaps[4]
Tussigon[5]
Tussionex[4]
Tylenol with Codeine No. 3[2]
Tylenol with Codeine No. 4[2]
Tylox[11]
Vicoprofen[4]
Vituz[4]
Xartemis XR[11]
Zohydro ER[4]
Zubsolv[1]
Zutripro[4]
Zydone[4]

NITRATES

GENERIC NAMES

1. ERYTHRITYL TETRANITRATE
2. ISOSORBIDE DINITRATE
3. ISOSORBIDE MONONITRATE
4. NITROGLYCERIN (GLYCERYL TRINITRATE)
5. PENTAERYTHRITOL TETRANITRATE

BRAND NAMES

Apo-ISDN[2]
BiDil[2]
Cardilate[1]
Cedocard-SR[2]
Coradur[2]
Coronex[2]
Deponit[4]
Dilatrate-SR[2]
Duotrate[5]
Glyceryl Trinitrate[4]
IMDUR[3]
ISMO[3]
Iso-Bid[2]
Isonate[2]
Isorbid[2]
Isordil[2]
Isotrate[2]
Klavikordal[4]
Minitran[4]
Monoket[3]
Niong[4]
Nitro-Bid[4]
Nitrocap[4]
Nitrocap T.D.[4]
Nitrocine[4]
Nitrodisc[4]
Nitro-Dur[4]
Nitro-Dur II[4]
Nitrogard-SR[4]
Nitroglyn[4]
Nitrol[4]
Nitrolin[4]
Nitrolingual[4]
NitroMist[4]
Nitronet[4]
Nitrong[4]
Nitrong SR[4]
Nitrospan[4]
Nitrostat[4]
Novosorbide[2]
NTS[4]
Pentritol[5]
Pentylan[5]
Peritrate[5]
Peritrate Forte[5]
Peritrate SA[5]
P.E.T.N.[5]
Rectiv[4]
Sorbitrate[2]
Sorbitrate SA[2]
Transderm-Nitro[4]
Tridil[4]

OXYMETAZOLINE (Nasal)

4-Way Long Acting Nasal Spray
12-Hour Nostrilla Nasal Decongestant
Afrin 12 Hour Nasal Spray
Afrin 12 Hour Nose Drops
Afrin Cherry Scented Nasal Spray
Afrin Children's Strength 12 Hour Nose Drops

Afrin Children's Strength Nose Drops
Afrin Extra Moisturizing Nasal Decongestant Spray
Afrin Menthol Nasal Spray
Afrin Nasal Spray
Afrin No-Drip Extra Moisturizing
Afrin No-Drip Nasal Decongestant, Severe Congestion with Menthol
Afrin No-Drip Nasal Decongestant Sinus with Vapornase
Afrin No-Drip Sinus
Afrin Nose Drops
Afrin Sinus
Afrin Spray Pump
Allerest 12 Hour Nasal Spray
Cheracol Nasal Spray
Cheracol Nasal Spray Pump Cherry Scented
Coricidin Nasal Mist
Dristan 12-Hour Nasal Spray
Dristan Long Lasting Menthol Nasal Spray
Dristan Long Lasting Nasal Pump Spray
Dristan Long Lasting Nasal Spray
Dristan Long Lasting Nasal Spray 12 Hour Metered Dose Pump
Dristan Mentholated
Drixoral
Duramist Plus Up To 12 Hours Decongestant Nasal Spray
Duration 12 Hour Nasal Spray Pump
Mucinex Full Force Nasal Spray
Mucinex Moisture Smart Nasal Spray
Nasal Decongestant Spray
Nasal Relief
Nasal Spray 12-Hour
Nasal Spray Long Acting
Nasal-12 Hour
Neo-Synephrine 12 Hour Nasal Spray
Neo-Synephrine 12 Hour Nasal Spray Pump
Neo-Synephrine 12 Hour Nose Drops
Neo-Synephrine 12 Hour Vapor Nasal Spray
Nostril Nasal Decongestant Mild
Nostril Nasal Decongestant Regular
NTZ Long Acting Decongestant Nasal Spray
NTZ Long Acting Decongestant Nose Drops
Sinarest 12 Hour Nasal Spray
Vicks Sinex 12-Hour Nasal Spray
Vicks Sinex 12-Hour Ultra Fine Mist
Vicks Sinex Long-Acting 12 Hour Nasal Spray

PENICILLINS

GENERIC NAMES

1. AMOXICILLIN
2. AMPICILLIN
3. BACAMPICILLIN
4. CARBENICILLIN
5. CLOXACILLIN
6. DICLOXACILLIN
7. FLUCLOXACILLIN
8. NAFCILLIN
9. OXACILLIN
10. PENICILLIN G
11. PENICILLIN V
12. PIVAMPICILLIN
13. PIVMECILLINAM

BRAND NAMES

Amoxil[1]
Apo-Amoxi[1]
Apo-Ampi[2]
Apo-Cloxi[5]
Apo-Pen VK[11]
Bactocill[9]
Beepen-VK[11]
Betapen-VK[11]
Cloxapen[5]
DisperMax[1]
Dycill[6]
Dynapen[6]
Fluclox[7]
Geopen Oral[4]
Ledercillin-VK[11]
Megacillin[10]
Moxatag[1]
Nadopen-V[11]
Nadopen-V 200[11]
Nadopen-V 400[11]
Novamoxin[1]
Novo-Ampicillin[2]
Novo-Cloxin[5]
Novo-Pen VK[11]
Nu-Amoxi[11]
Nu-Ampi[2]
Nu-Cloxi[5]
Nu-Pen-VK[11]
Omnipen[2]

Omeclamox-Pak[1]
Orbenin[5]
Pathocil[6]
Pen Vee[11]
Pen Vee K[11]
Penbritin[2]
Penglobe[3]
Pentids[10]
Polycillin[2]
Pondocillin[12]
Principen[2]
Prostaphlin[9]
PVF[11]
PVF K[11]
Selexid[13]
Spectrobid[3]
Tegopen[5]
Totacillin[2]
Trimox[1]
Unipen[8]
V-Cillin K[11]
Veetids[11]
Wymox[1]

PHENOTHIAZINES

GENERIC NAMES

1. ACETOPHENAZINE
2. CHLORPROMAZINE
3. FLUPHENAZINE
4. MESORIDAZINE
5. METHOTRIMEPRAZINE
6. PERICYAZINE
7. PERPHENAZINE
8. PIPOTIAZINE
9. PROCHLORPERAZINE
10. PROMAZINE
11. THIOPROPAZATE
12. THIOPROPERAZINE
13. THIORIDAZINE
14. TRIFLUOPERAZINE
15. TRIFLUPROMAZINE

BRAND NAMES

Apo-Fluphenazine[3]
Apo-Perphenazine[7]
Apo-Thioridazine[13]
Apo-Trifluoperazine[14]
Chlorpromanyl-5[2]
Chlorpromanyl-20[2]
Chlorpromanyl-40[2]
Compazine[9]
Compazine Spansule[9]
Dartal[11]
Duo-Medihaler
Etrafon[7]
Etrafon-A[7]
Etrafon-D[7]
Etrafon-F[7]
Etrafon-Forte[7]
Largactil[2]
Largactil Liquid[2]
Largactil Oral Drops[2]
Levoprome[5]
Majeptil[12]
Mellaril[13]
Mellaril Concentrate[13]
Mellaril-S[13]
Modecate[3]
Modecate Concentrate[3]
Moditen Enanthate[3]
Moditen HCl[3]
Moditen HCl-H.P.[3]
Neuleptil[6]
Novo-Chlorpromazine[2]
Novo-Flurazine[14]
Novo-Ridazine[13]
Nozinan[5]
Nozinan Liquid[5]
Nozinan Oral Drops[5]
Permitil[3]
Permitil Concentrate[3]
Piportil L_4[8]
PMS Levazine[7]
PMS Thioridazine[13]
Prolixin[3]
Prolixin Concentrate[3]
Prolixin Decanoate[3]
Prolixin Enanthate[3]
Prorazin[9]
Prozine [10]
Serentil[4]
Serentil Concentrate[4]
Stemetil Liquid[9]
Suprazine[14]
Terfluzine[14]
Terfluzine Concentrate[14]
Thorazine[2]
Thorazine Concentrate[2]
Thorazine Spansule[2]
Thor-Prom[2]
Tindal[2]

Triavil[7]
Trilafon[7]
Trilafon Concentrate[7]
Ultrazine-10[9]
Vesprin[5]

PHENYLEPHRINE
Aclophen
Advil Allergy and Congestion Relief
Advanced Formula Dristan Caplets
AH-Chew
Alersule
Alka-Seltzer Plus Cold & Sinus Effervescent
Alka-Seltzer Plus Night-Time Effervescent
Alka-Seltzer Plus Original Effervescent
Atrohist Pediatric
Atrohist Pediatric Suspension Dye Free
Atrohist Sprinkle
Benylin All-In-One Cold & Flu Night Caplets
Benylin Cold & Sinus
Benylin Cold & Sinus Plus
Cerose-DM
Codimal DM
Codimal PH
Colrex Compound
Colrex Cough
Coltab Children's
Comhist
Comhist LA
Conar
Conar Expectorant
Conar-A
Congespirin for Children Cold Tablets
Contac Cold + Flu
Dallergy
Dihistine
Doktors
Donatussin
Donatussin Drops
Dondril
Dristan Cold Multi-Symptom Formula
Dristan Formula P
Dristan-AF
Dristan-AF Plus
Ed A-Hist
Father John's Medicine Plus
Gendecon
Histagesic Modified
Histatab Plus
Histatan
Histor-D
Histor-D Timecelles
LiquiBid D
LiquiBid D-R
LiquiBid PD
Meda Syrup Forte
Mucinex Adult Caplets - Cold & Sinus
Mucinex Adult Caplets - Cold, Flu, & Sore Throat
Mucinex Adult Caplets - Severe Congestion & Cold
Mucinex Cold Liquid
Mucinex Fast-Max Cold, Flu & Sore Throat Liquid
Mucinex Maximum Strength Fast Max Cold, Flu & Sore Throat
Mucinex Maximum Strength Fast Max Severe Congestion & Cough
ND-Gesic
Neocitran A
Neocitran Colds & Flu Calorie Reduced
NeoCitran DM Coughs & Colds
NeoCitran Extra Strength Colds & Flu
NeoCitran Extra Strength Sinus
Neo-Synephrine Nasal Drops
Neo-Synephrine Pediatric Nasal Drops
Nostril Spray Pump
PediaCare Children's Allergy
PediaCare Children's Decongestant
PediaCare Children's Fever Reducer Plus Flu
PediaCare Children's Multi-Symptom Cold
PediaCare Fever Reducer Plus Multi-Symptom Cold
Pediacof Cough
Phenameth VC
Pherazine VC
Prehist
Prehist D
Prometh VC Plain
Prometh VC with Codeine
Promethazine VC
Refenesen Chest Congestion & Pain Relief PE
Refenesen PE
Rhinall

Rhinall Children's Flavored Nose Drops
Rhinatate
Robitussin Cough & Cold CF
Robitussin Cough & Cold Nighttime
Robitussin Cough Cold & Flu Nighttime
Rolatuss Plain
R-Tannamine
R-Tannamine Pediatric
R-Tannate
R-Tannate Pediatric
Rynatan
Rynatan Pediatric
Rynatan-S Pediatric
Rynatuss
Rynatuss Pediatric
Scot-Tussin
Scot-Tussin Original 5-Action Cold Medicine
Sinupan
SINUvent PE
Sudafed PE Cold & Cough Caplets
Tanoral
Theracof Plus Multi-Symptom Cough and Cold Reliever
TheraFlu Cold & Cough Hot Liquid
Theraflu Daytime Severe Cold Caplets
Theraflu Daytime Severe Cold Hot Liquid
Theraflu Flu & Sore Throat Hot Liquid
Theraflu Nighttime Severe Cold Caplets
Theraflu Nighttime Severe Cold Hot Liquid
Theraflu Thin Strips Daytime Cold & Cough
Theraflu Thin Strips Nighttime Severe Cold & Cough
Theraflu Warming Relief Daytime
Theraflu Warming Relief Nighttime
Triaminic Chest & Nasal Congestion
Triaminic Cold & Allergy
Triaminic Day Time Cold & Cough
Triaminic Night Time Cough & Cold
Triaminic Thin Strips Cold
Triaminic Thin Strips Daytime Cold & Cough
Triaminic Thin Strips Night Time Cold & Cough
Trimedine Liquid
Triotann
Triotann Pediatric
Tritann Pediatric
Tri-Tannate
Tri-Tannate Plus Pediatric
Tussi-12
Tylenol Allergy Multi-Symptom
Tylenol Allergy Multi-Symptom Nighttime
Tylenol Cold Head Congestion Daytime
Tylenol Cold Head Congestion Nighttime
Tylenol Cold Head Congestion Severe
Tylenol Cold Multi Symptom Daytime
Tylenol Cold Multi Symptom Nighttime
Tylenol Cold Multi Symptom Severe
Tylenol Cough & Severe Congestion Daytime
Tylenol Sinus Congestion & Pain Daytime
Tylenol Sinus Congestion & Pain Nighttime
Tylenol Sinus Congestion & Pain Severe
Vicks DayQuil Cold & Flu Symptom Relief Plus Vitamin C
Vicks DayQuil Sinus LiquiCaps
Vicks Formula 44 Custom Care Congestion
Vicks NyQuil Sinus LiquiCaps
Vicks Sinex Nasal Spray
Vicks VapoSyrup Severe Congestion Head & Chest Congestion Relief
Viravan DM

POTASSIUM SUPPLEMENTS

GENERIC NAMES

1. POTASSIUM ACETATE
2. POTASSIUM BICARBONATE
3. POTASSIUM BICARBONATE & POTASSIUM CHLORIDE
4. POTASSIUM BICARBONATE & POTASSIUM CITRATE
5. POTASSIUM CHLORIDE
6. POTASSIUM GLUCONATE
7. POTASSIUM GLUCONATE & POTASSIUM CHLORIDE
8. POTASSIUM GLUCONATE & POTASSIUM CITRATE
9. POTASSIUM TRIPLEX

BRAND NAMES

Apo-K[5]
Cambia[2]
Cena-K[5]
Effer-K[4]
Gen-K[5]
Glu-K[6]
K+10[5]
K-10[5]
K-8[5]
K+Care[5]
K+Care ET[2]
Kalium Durules[5]
Kaochlor[5]
Kaochlor S-F[5]
Kaochlor-10[5]
Kaochlor-20[5]
Kaochlor-Eff[5]
Kaon[6]
Kaon-Cl[5]
Kaon-Cl 10[5]
Kaon-Cl 20[5]
Kato[5]
Kay Ciel[5]
Kay Ciel Elixir[5]
Kaylixir[6]
KCL[5]
K-Dur[5]
K-Electrolyte[2]
K-G Elixir[6]
K-Lease[5]
K-Long[5]
K-Lor[5]
Klor-Con 8[5]
Klor-Con 10[5]
Klor-Con Powder[5]
Klor-Con/25[5]
Klor-Con/EF[2]
Klorvess[3]
Klorvess 10% Liquid[5]
Klorvess Effervescent Granules[5]
Klotrix[5]
K-Lyte[2]
K-Lyte DS[4]
K-Lyte/Cl[3]
K-Lyte/Cl 50[3]
K-Lyte/CL Powder[5]
K-Med 900[5]
K-Norm[5]
Kolyum[7]
K-Sol[5]
K-Tab[5]
K-Vescent[2]
Micro-K[5]
Micro-K 10[5]
Micro-K LS[5]
Neo-K[3]
Potasalan[5]
Potassium-Rougier[6]
Potassium-Sandoz[3]
Roychlor 10%[5]
Roychlor 20%[5]
Royonate[6]
Rum-K[5]
Slow-K[5]
Ten K[5]
Tri-K[9]
Twin-K[8]

PROGESTINS

GENERIC NAMES

1. DROSPIRENONE
2. ETONOGESTREL
3. HYDROXYPROGESTERONE
4. LEVONORGESTREL
5. MEDROXYPROGESTERONE
6. MEGESTROL
7. NORETHINDRONE
8. NORGESTIMATE
9. NORGESTREL
10. PROGESTERONE

BRAND NAMES

Activella[8]
Amen[5]
Angeliq[1]
Aygestin[7]
Climara Pro[4]
CombiPatch[7]
Crinone[10]
Curretab[5]
Cycrin[5]
Depo-Provera[5]
Duralutin[3]
EContra[4]
Endometrin[10]
Fallback Solo[4]
Femhrt[7]
Gesterol 50[10]
Gesterol L.A.[3]
Hy/Gestrone[3]

Hylutin[3]
Hyprogest[3]
Implanon[3]
Megace[6]
Megace ES[6]
Megace Oral Suspension[6]
Micronor[7]
Next Choice[4]
Norlutate[7]
Nor-Q.D.[7]
Opcicon One-Step[4]
Ortho-Prefest[8]
Ovrette[9]
Plan B One Step[4]
Plan B OTC/Rx[4]
Premphase[5]
Prempro[5]
Prodrox[3]
Prometrium[10]
Provera[5]
ProveraPak[5]

PROTEASE INHIBITORS

GENERIC NAMES

1. ATAZANAVIR
2. DARUNAVIR
3. FOSAMPRENAVIR
4. INDINAVIR
5. LOPINAVIR
6. NELFINAVIR
7. RITONAVIR
8. SAQUINAVIR
9. TIPRANAVIR

BRAND NAMES

Aptivus[9]
Crixivan[4]
Evotaz[1]
Invirase[8]
Kaletra[5&7]
Lexiva[3]
Norvir[7]
Prezcobix[2]
Prezista[2]
Reyataz[1]
Viracept[6]

PSEUDOEPHEDRINE

Actacin
Actagen
Advil Cold and Sinus Caplets
Advil Cold and Sinus LiquiGels
Advil Flu & Body Ache
Advil Multi-Symptom Cold
Alavert D-12
Aleve-D Sinus & Cold
Allegra-D
Allegra-D 24 Hour
Allent
Allercon
Allerest Maximum Strength
Allerest No-Drowsiness
Allerfrim
Allerphed
All-Nite Cold Formula
Ambenyl-D Decongestant Cough Formula
Amdry-D
Anamine
Anamine T.D.
Anatuss DM
Anatuss LA
Balminil Decongestant
Benylin All-In-One Cold & Flu Caplets
Benylin All-In-One Cold & Flu Nightime Syrup
Benylin All-In-One Cold & Flu Syrup
Benylin All-In-One Day & Night Caplets
Benylin DM-D
Benylin DM-D-E
Benylin DM-D-E Extra Strength
Brexin-L.A.
Brofed
Bromatane DX Cough
Bromfed
Bromfed-DM
Bromfed-PD
Carbodec
Carbodec DM Drops
Carbodec TR
Cenafed
Cenafed Plus
Children's Allergy Sinus
Children's Benadryl Allergy & Cold Fastmelt
Children's Motrin Cold
Chlorphedrine SR
Chlor-Trimeton 4 Hour Relief
Chlor-Trimeton 12 Hour Relief
Clarinex D 24 Hour
Claritin-D

Claritin-D 12 Hour
Claritin-D 24 Hour
Codimal-L.A.
Codimal-L.A. Half
Coldrine
Colfed-A
Concentrin
Congess JR
Congess SR
Congestac Caplets
Contac Allergy/Sinus Day Caplets
Contac Allergy/Sinus Night Caplets
Contac Maximum Strength Sinus Caplets
Contac Night Caplets
Contac Non-Drowsy Formula Sinus Caplets
Contac Severe Cold Formula
Contac Severe Cold Formula Night Strength
Cophene-XP
CoTylenol Cold Medication
Dallergy Jr.
Deconamine
Deconamine SR
Deconsal II
Dexaphen SA
Dexophed
Disophrol
Disophrol Chronotabs
Dristan Cold and Flu
Dristan Cold Caplets
Dristan Cold Maximum Strength Caplets
Dristan Juice Mix-in Cold, Flu, & Cough
Drixoral
Drixoral Cold & Allergy
Drixoral Cold & Flu
Drixoral Nasal Decongestant
Drixoral Non-Drowsy Formula
Drixoral Plus
Drixoral Sinus
Drixtab
Eltor-120
Entex PSE
Entuss Pediatric Expectorant
Fedahist
Fedahist Decongestant
Fedahist Expectorant
Fedahist Expectorant Pediatric Drops
Fedahist Gyrocaps
Fedahist Timecaps
Genaphed
Guaifed
Guaifed-PD
GuaiMAX-D
Guiatuss PE
Hayfebrol
Histalet
Histalet X
Histalet-DM
Klerist-D
Kolephrin
Kolephrin/DM Caplets
Kronofed-A
Kronofed-A Jr.
Lodrane D
Lodrane LD
Maxenal
Maximum Strength Mucinex D
Maximum Strength Tylenol Allergy Sinus Caplets
Mucinex D
Mucinex DM
Myfedrine
Nasatab LA
ND Clear T.D.
Nexafed
Nytcold Medicine
Ornex Maximum Strength Caplets
Ornex No Drowsiness Caplets
Ornex Severe Cold No Drowsiness Caplets
Pertussin All Night PM
Phenapap Sinus Headache & Congestion
Phenergan-D
Rescon-JR
Respaire-30
Rezira
Rhinosyn
Rhinosyn-DM
Rhinosyn-PD
Rhinosyn-X
Robafen DM
Robitussin Cough & Cold D
Ru-Tuss DE
Ru-Tuss Expectorant
Ryna
Semprex-D
Simply Stuffy

Sinus Relief
Sinutab Non-Drying Liquid Caps
Stamoist E
Sudafed 12 Hour
Sudafed Multi-Symptom Cold & Cough
Sudafed Non-Drying Sinus Liquid Caps
Sufedrin
Tavist Allergy/Sinus/ Headache
Touro A&H
Touro LA Caplets
Triafed
Triaminic-D Multi-Symptom Cold
Tussar DM
Tussend
Tussend Expectorant
Tussend Liquid
Versacaps
Vicks NyQuil D
Zephrex-D
Zutripro
Zyrtec-D

SALICYLATES

GENERIC NAMES

1. BALSALAZIDE
2. CHOLINE MAGNESIUM SALICYLATES
3. CHOLINE SALICYLATE
4. MAGNESIUM SALICYLATE
5. SALICYLAMIDE
6. SALSALATE
7. SODIUM SALICYLATE

BRAND NAMES

Amigesic[6]
Arthropan[3]
Choline Magnesium Trisalicylate[2]
Colazal[1]
Diagen[6]
Disalcid[6]
Doan's Pills[4]
Dodd's Pills[7]
Giazo[1]
Kolephrin[5]
Magan[4]
Mobidin[4]
Mono-Gesic[6]
Presalin[5]
Rid-A-Pain Compound[5]
Salflex[6]
Salgesic[5]
Salsitab[6]
Scot-tussin Original 5-Action Cold Medicine[7]
Tricosal[2]
Trilisate[2]
Uracel[7]

SCOPOLAMINE (Hyoscine)

Buscopan
Donnatal
Donnatal Elixir
Donnatal Extentabs
Transderm-Scop
Transderm-V

SELECTIVE SEROTONIN REUPTAKE INHIBITORS (SSRIs)

GENERIC NAMES

1. CITALOPRAM
2. ESCITALOPRAM
3. FLUOXETINE
4. FLUVOXAMINE
5. PAROXETINE
6. SERTRALINE

BRAND NAMES

Brisdelle[5]
Celexa[1]
Lexapro[2]
Luvox[4]
Luvox CR[4]
Paxil[5]
Paxil CR[5]
Prozac[3]
Prozac Weekly[3]
Sarafem[3]
Selfemra[3]
Symbyax[3]
Zoloft[6]

SULFONAMIDES

GENERIC NAMES

1. SULFACYTINE
2. SULFADIAZINE
3. SULFAMETHIZOLE
4. SULFAMETHOXAZOLE
5. SULFISOXAZOLE

BRAND NAMES

Apo-Sulfamethoxazole[4]
Apo-Sulfatrim[4]
Apo-Sulfatrim DS[4]

Apo-Sulfisoxazole[5]
Bactrim[4]
Bactrim DS[4]
Cotrim[4]
Cotrim DS[4]
Co-trimoxazole[4]
Gantanol[4]
Novo-Soxazole[5]
Novotrimel[4]
Novotrimel DS[4]
Nu-Cotrimox[4]
Nu-Cotrimox DS[4]
Pediazole[5]
Protrin[4]
Renoquid[1]
Roubac[4]
Septra[4]
Septra DS[4]
SMZ-TMP[4]
Sulfamethoprim[4]
Sulfamethoprim DS[4]
Sulfaprim[4]
Sulfaprim DS[4]
Sulfatrim[4]
Sulfatrim DS[4]
Sulfimycin[5]
Sulfizole[5]
Sulfoxaprim[4]
Sulfoxaprim DS[4]
Sulmeprim[4]
Thiosulfil Forte[3]
Triazole[4]
Triazole DS[4]
Trimeth-Sulfa[4]
Trisulfam[4]
Urobak[4]
Uroplus DS[4]
Uroplus SS[4]

SULFONYLUREAS

GENERIC NAMES

1. ACETOHEXAMIDE
2. CHLORPROPAMIDE
3. GLIMEPIRIDE
4. GLIPIZIDE
5. GLYBURIDE
6. TOLAZAMIDE
7. TOLBUTAMIDE

BRAND NAMES

Albert Glyburide[5]
Amaryl[3]
Apo-Chlorpropamide[2]
Apo-Glyburide[5]
Apo-Tolbutamide[7]
Avandaryl[3]
DiaBeta[5]
Diabinese[2]
Duetact[3]
Euglucon[5]
Gen-Glybe[5]
Glucamide[2]
Glucotrol[4]
Glucotrol XL[4]
Glucovance[5]
Glynase PresTab[6]
Metaglip[4]
Micronase[5]
Novo-Butamide[7]
Novo-Glyburide[4]
Novo-Propamide[2]
Orinase[7]
Tolamide[6]
Tolinase[6]

TETRACYCLINES

GENERIC NAMES

1. DEMECLOCYCLINE
2. DOXYCYCLINE
3. MINOCYCLINE
4. OXYTETRACYCLINE
5. TETRACYCLINE

BRAND NAMES

Acticlate[2]
Adoxa[2]
Apo-Doxy[2]
Apo-Tetra[5]
Arestin[3]
Declomycin[1]
Doryx[2]
Doxy-Caps[2]
Doxycin[2]
Doxy-Tabs[2]
E.P. Mycin[4]
Helidac[5]
Minocin[3]
Monodox[2]
Novodoxlin[2]
Novotetra[5]

Nu-Tetra[5]
Oracea[2]
Panmycin[5]
Periostat[5]
Pylera[5]
Robitet[5]
Solodyn[3]
Sumycin[5]
Terramycin[4]
Tetracyn[5]
Tija[4]
Vibramycin[2]
Ximino[3]

ADDITIONAL DRUG INTERACTIONS

Additional Drug Interactions

The following lists of drugs and their interactions with other drugs are continuations of the POSSIBLE INTERACTIONS WITH OTHER DRUGS section found in the drug charts beginning on page 2. These lists are alphabetized by the drug chart name (shown in large capital letters). Only those lists too long for a particular drug chart are included in this section. For complete information about any generic drug, see the alphabetized charts.

ADRENOCORTICOIDS (Systemic)

GENERIC NAME OR DRUG CLASS	COMBINED EFFECT
Antivirals HIV/AIDS*	Increased adrenocorticoid effect or decreased antiviral effect.
Bupropion	Increased risk of seizures.
Carbamazepine	Decreased adrenocorticoid effect.
Carbonic anhydrase inhibitors	Increased risk of side effects.
Contraceptives, oral*	Increased adrenocorticoid effect or decreased contraceptive effect.
Cyclosporine	May increase effect of adrenocorticoid or cyclosporine.
Digitalis preparations*	Risk of heart rhythm problems. Possible digitalis toxicity.
Diuretics	Decreased levels of potassium.
Ephedrine	Decreased adrenocorticoid effect.
Estrogens*	Increased adrenocorticoid effect.
Insulin	Decreased insulin effect.
Insulin analogs	Decreased insulin analog effect.
Isoniazid	Decreased isoniazid effect.
Mifepristone	Variable. Consult doctor or pharmacist.
Mitotane	Decreased adrenocorticoid effect.
Phenobarbital	Decreased adrenocorticoid effect.
Phenytoin	Decreased adrenocorticoid effect.
Potassium supplements*	Decreased levels of potassium.
Primidone	Decreased adrenocorticoid effect.
Rifamycins	Decreased adrenocorticoid effect.
Salicylates*	Decreased salicylate effect.
Somatropin	Decreased effect of somatropin or adrenocorticoid.
Thyroid hormones*	Effect can vary. Consult doctor or pharmacist.
Thyroid hormones*	Risk of viral infection or decreased vaccine effect.

ANGIOTENSIN-CONVERTING ENZYME (ACE) INHIBITORS

GENERIC NAME OR DRUG CLASS	COMBINED EFFECT
Potassium supplements*	Possible increased potassium in blood.
Sotalol	Increased antihypertensive effects of both drugs. Dosages may require adjustment.

*See Glossary

GENERIC NAME OR DRUG CLASS	COMBINED EFFECT	GENERIC NAME OR DRUG CLASS	COMBINED EFFECT

ANGIOTENSIN-CONVERTING ENZYME (ACE) INHIBITORS continued

Spironolactone	Possible excessive potassium in blood.	**Tiopronin**	Increased risk of toxicity to kidneys.
Terazosin	Decreased effectiveness of terazosin.	**Triamterene**	Possible excessive potassium in blood.

ANTIARRHYTHMICS, BENZOFURAN-TYPE

Cyclosporine	May increase effect of cyclosporine.	**Fentanyl**	May increase risk of low blood pressure.
Digoxin	May increase digoxin effect.	**Flecainide**	May increase effect of flecainide.
Disopyramide	May increase risk of irregular heartbeat.	**HMG-CoA reductase inhibitors**	Risk of muscle injury and kidney failure.
Enzyme inducers*	May decrease effect of benzofuran-type antiarrhythmic and enzyme inducer.	**QT prolongation-causing drugs***	Increased risk of irregular heartbeat. Avoid.
Enzyme inhibitors*	May increase effect and toxicity risk of benzofuran-type antiarrhythmic. May increase effect of enzyme inhibitor.	**Quinidine**	May increase effect of quinidine.

ANTICONVULSANTS, HYDANTOIN

Central nervous system (CNS) depressants*	Oversedation.	**Estrogens***	Increased estrogen effect.
Chloramphenicol	Increased anticonvulsant effect.	**Felbamate**	Increased side effects and adverse reactions.
Cimetidine	Increased anti-convulsant toxicity.	**Furosemide**	Decreased furosemide effect.
Contraceptives, oral*	Increased seizures.	**Gold compounds***	Increased anticon-vulsant blood levels. Hydantoin dose may require adjustment.
Cyclosporine	May decrease cyclosporine effect.	**Griseofulvin**	Increased griseofulvin effect.
Digitalis preparations*	Decreased digitalis effect.	**Hypoglycemics, oral***	Possible decreased hypoglycemic effect.
Disopyramide	Decreased disopyramide effect.	**Hypoglycemics,* other**	Possible decreased hypoglycemic effect.
Disulfiram	Increased anticonvulsant effect.		

*See Glossary

GENERIC NAME OR DRUG CLASS	COMBINED EFFECT

ANTICONVULSANTS, HYDANTOIN continued

GENERIC NAME OR DRUG CLASS	COMBINED EFFECT
Isoniazid	Increased anticonvulsant effect.
Lamotrigine	Decreased lamotrigine effect with phenytoin.
Leucovorin	May counteract the effect of phenytoin or any hydantoin anticonvulsant.
Leukotriene modifiers	Increased phenytoin effect.
Loxapine	Decreased anticonvulsant effect.
Methadone	Decreased methadone effect.
Methotrexate	Increased methotrexate effect.
Methyldopa	Possible decreased methyldopa effect.
Methylphenidate	Increased anticonvulsant effect.
Mifepristone	Decreased effect of mifepristone.
Modafinil	Anticonvulsant dose may need adjustment.
Molindone	Increased phenytoin effect.
Monoamine oxidase (MAO) inhibitors*	Increased polythiazide effect.
Nicardipine	Increased anticonvulsant effect.
Nimodipine	Increased anticonvulsant effect.
Nitrates*	Excessive blood pressure drop.
Nitroimidazoles	Decreased effect of phenytoin.
Nizatidine	Increased effect and toxicity of phenytoin.
Omeprazole	Delayed excretion of phenytoin causing increased amount of phenytoin in blood.
Oxyphenbutazone	Increased anticonvulsant effect.
Para-aminosalicylic acid (PAS)	Increased anticonvulsant effect.
Paroxetine	Decreased anticonvulsant effect.
Phenacemide	Increased risk of paranoid symptoms.
Phenothiazines*	Increased anticonvulsant effect.
Phenylbutazone	Increased anticonvulsant effect.
Potassium supplements*	Decreased potassium effect.
Probenecid	Decreased probenecid effect.
Propafenone	Increased effect of both drugs and increased risk of toxicity.
Propranolol	Increased propranolol effect.
Quetiapine	Decreased quetiapine effect with phenytoin.
Quinidine	Increased quinidine effect.
Rifamycins	Decreased anticonvulsant effect.
Sedatives*	Increased sedative effect.
Sotalol	Decreased sotalol effect.
Sucralfate	Decreased anticonvulsant effect.
Sulfa drugs*	Increased anticonvulsant effect.
Theophylline	Reduced anticonvulsant effect.
Trimethoprim	Increased phenytoin effect.
Valproic acid*	Each drug may need dosage adjusted.
Xanthines*	Decreased effects of both drugs.
Zafirlukast	May increase effect of phenytoin.

*See Glossary

GENERIC NAME OR DRUG CLASS	COMBINED EFFECT	GENERIC NAME OR DRUG CLASS	COMBINED EFFECT

ANTICONVULSANTS, HYDANTOIN continued

GENERIC NAME OR DRUG CLASS	COMBINED EFFECT
Zaleplon	Decreased zaleplon effect.

ANTIDEPRESSANTS, TRICYCLIC

GENERIC NAME OR DRUG CLASS	COMBINED EFFECT
Benzodiazepines*	Increased sedation.
Bupropion	Increased risk of seizures.
Central nervous system (CNS) depressants*	Excessive sedation.
Cimetidine	Possible increased tricyclic anti-depressant effect and toxicity.
Citalopram	Increased tricyclic antidepressant effect and toxicity.
Clonidine	Blood pressure increase. Avoid combination.
Clozapine	Toxic effect on the central nervous system.
Contraceptives, oral*	Increased depression.
Desmopressin	Increased risk of thirstiness which may lead to drinking excess fluids.
Dextrothyroxine	Increased anti-depressant effect. Irregular heartbeat.
Disulfiram	Delirium.
Dofetilide	Increased risk of heart problems.
Ethchlorvynol	Delirium.
Fluoxetine	Increased effect of tricyclic antidepressant. Possible toxicity.
Fluvoxamine	Increased anti-depressant effect.
Furazolidone	Sudden, severe increase in blood pressure.
Guanabenz	Decreased guanabenz effect.
Guanadrel	Decreased guanadrel effect.
Guanethidine	Decreased guanethidine effect.
Leucovorin	High alcohol content of leucovorin may cause adverse effects.
Levodopa	May increase blood pressure. May decrease levodopa effect.
Lithium	Possible decreased seizure threshold.
Methyldopa	Possible decreased methyldopa effect.
Methylphenidate	Possible increased tricyclic anti-depressant effect and toxicity.
Modafinil	Increased anti-depressant effect.
Molindone	Increased molindone effect.
Monoamine oxidase (MAO) inhibitors*	Fever, delirium, convulsions.
Narcotics*	Oversedation.
Nicotine	Increased effect of antidepressant (with imipramine).
Phenothiazines*	Possible increased tricyclic anti-depressant effect and toxicity.
Phenytoin	Decreased phenytoin effect.
Procainamide	Possible irregular heartbeat.

*See Glossary

ANTIDEPRESSANTS, TRICYCLIC continued

GENERIC NAME OR DRUG CLASS	COMBINED EFFECT
Quinidine	Possible irregular heartbeat.
Serotonergics*	Increased risk of serotonin syndrome.*
Sertraline	Increased depressive effects of both drugs.
Sympathomimetics*	Increased sympathomimetic effect.
Thyroid hormones*	Irregular heartbeat.
Tolcapone	May increase incidence of adverse effects of tolcapone.
Zaleplon	Increased effect of either drug. Avoid.
Zolpidem	Increased sedative effect. Avoid.

ANTIDYSKINETICS

GENERIC NAME OR DRUG CLASS	COMBINED EFFECT
Imatinib	Increased effect of pimozide.
Levodopa	Possible increased levodopa effect.
Monoamine oxidase (MAO) inhibitors*	Increased antidyskinetic effect.
Telithromycin	Heart problem risk with pimozide. Avoid.

ANTIFUNGALS, AZOLES

GENERIC NAME OR DRUG CLASS	COMBINED EFFECT
Losartan	Decreased losartan effect.
Methscopolamine	Decreased azole effect.
Methylprednisolone	Increased effect of methylprednisolone.
Mifepristone	Decreased effect of mifepristone.
Nizatidine	Decreased azole effect.
Omeprazole	Decreased azole effect.
Phenytoin	May alter effect of both drugs.
Propantheline	Decreased azole effect.
Proton pump inhibitors	Decreased antifungal effect.
Quetiapine	Increased risk of quetiapine toxicity.
Ranitidine	Decreased azole effect.
Rifamycins	Decreased azole effect.
Ritonavir	Increased ritonavir effect.
Scopolamine	Decreased azole effect.
Sibutramine	Increased effect of sibutramine.
Sildenafil	Effects unknown. Consult doctor.
Sodium bicarbonate	Decreased azole effect.
Warfarin	Increased warfarin effect.

***See Glossary**

GENERIC NAME OR DRUG CLASS	COMBINED EFFECT	GENERIC NAME OR DRUG CLASS	COMBINED EFFECT

ANTIHISTAMINES, PHENOTHIAZINE-DERIVATIVE

GENERIC NAME OR DRUG CLASS	COMBINED EFFECT
Clozapine	Toxic effect on the central nervous system.
Dronabinol	Increased effects of both drugs. Avoid.
Epinephrine	Decreased epinephrine effect.
Ethinamate	Dangerous increased effects of ethinamate. Avoid combining.
Extrapyramidal reaction*-causing medicines	Increased frequency and severity of extra-pyramidal reactions.
Fluoxetine	Increased depressant effects of both drugs.
Guanethidine	Decreased guanethidine effect.
Guanfacine	May increase depressant effects of either medicine.
Leucovorin	High alcohol content of leucovorin may cause adverse effects.
Levodopa	Decreased levodopa effect.
Methyprylon	May increase sedative effect to dangerous level. Avoid.
Metyrosine	Increased likelihood of toxic symptoms of each.
Mind-altering drugs*	Increased effect of mind-altering drugs.
Molindone	Increased sedative and antihistamine effect.
Monoamine oxidase (MAO) inhibitors*	Increased anti-histamine effect.
Nabilone	Greater depression of central nervous system.
Narcotics*	Increased narcotic effect.
Sedatives*	Increased sedative effect.
Sertraline	Increased depressive effects of both drugs.
Sotalol	Increased antihistamine effect.
Tranquilizers*	Increased tranquilizer effect. Avoid.

ANTI-INFLAMMATORY DRUGS, NONSTEROIDAL (NSAIDs)

GENERIC NAME OR DRUG CLASS	COMBINED EFFECT
Cyclosporine	Increased effect of cyclosporine.
Digoxin	Increased digoxin effect.
Diuretics*	May decrease diuretic effect.
Lithium	Increased lithium effect.
Methotrexate	Increased risk of methotrexate side effects.
Phenytoin	Increased phenytoin effect.
Selective serotonin reuptake inhibitors (SSRIs)	Increased risk of bleeding.
Thyroid hormones*	Rapid heartbeat, blood pressure rise.
Tiopronin	Increased risk of toxicity to kidneys.
Triamterene	Reduced triamterene effect.
Aspirin	Increased risk of side effects.

*See Glossary

GENERIC NAME OR DRUG CLASS	COMBINED EFFECT	GENERIC NAME OR DRUG CLASS	COMBINED EFFECT

ASPIRIN

GENERIC NAME OR DRUG CLASS	COMBINED EFFECT
Meloxicam	May increase risk of ulcers and bleeding.
Methotrexate	Increased methotrexate effect.
Phenytoin	Increased phenytoin effect.
Probenecid	Decreased probenecid effect.
Salicylates*	Increased risk of adverse effects.
Selective serotonin reuptake inhibitors (SSRIs)	Risk of bleeding problems.
Spironolactone	Decreased spironolactone effect.
Sulfinpyrazone	Decreased sulfinpyrazone effect.
Ticlopidine	Increased risk of bleeding.
Valproic acid*	May increase valproic acid effect.

ATROPINE, HYOSCYAMINE, METHENAMINE, METHYLENE BLUE, PHENYLSALICYLATE & BENZOIC ACID

GENERIC NAME OR DRUG CLASS	COMBINED EFFECT
Cortisone drugs*	Increased internal eye pressure, increased cortisone effect. Risk of ulcers and stomach bleeding.
Diuretics, thiazide*	Decreased urine acidity.
Furosemide	Possible salicylate toxicity.
Gold compounds*	Increased likelihood of kidney damage.
Haloperidol	Increased internal eye pressure.
Indomethacin	Risk of stomach bleeding and ulcers.
Ketoconazole	Reduced ketoconazole effect.
Meperidine	Increased atropine and hyoscyamine effect.
Methylphenidate	Increased atropine and hyoscyamine effect.
Minoxidil	Decreased minoxidil effect.
Monoamine oxidase (MAO) inhibitors*	Increased belladonna and atropine effect.
Orphenadrine	Increased atropine and hyoscyamine effect.
Oxprenolol	Decreased anti-hypertensive effect of oxprenolol.
Para-aminosalicylic acid (PAS)	Possible salicylate toxicity.
Penicillins*	Increased effect of both drugs.
Phenobarbital	Decreased salicylate effect.
Phenothiazines*	Increased atropine and hyoscyamine effect.
Phenytoin	Increased phenytoin effect.
Pilocarpine	Loss of pilocarpine effect in glaucoma treatment.
Potassium supplements*	Possible intestinal ulcers with oral potassium tablets.

ADDITIONAL DRUG INTERACTIONS

*See Glossary

GENERIC NAME OR DRUG CLASS	COMBINED EFFECT	GENERIC NAME OR DRUG CLASS	COMBINED EFFECT

ATROPINE, HYOSCYAMINE, METHENAMINE, METHYLENE BLUE, PHENYLSALICYLATE & BENZOIC ACID continued

GENERIC NAME OR DRUG CLASS	COMBINED EFFECT
Probenecid	Decreased probenecid effect.
Propranolol	Decreased salicylate effect.
Rauwolfia alkaloids*	Decreased salicylate effect.
Salicylates*	Likely salicylate toxicity.
Sedatives* or central nervous system (CNS) depressants*	Increased sedative effect of both drugs.
Serotonergics*	Serotonin syndrome.*
Sodium bicarbonate	Decreased methenamine effect.
Spironolactone	Decreased spironolactone effect.
Sulfa drugs*	Possible kidney damage.
Sulfinpyrazone	Decreased sulfinpyrazone effect.
Vitamin C (1 to 4 grams per day)	Increased effect of methenamine, contributing to urine acidity; decreased atropine effect; possible salicylate toxicity.

BARBITURATES

GENERIC NAME OR DRUG CLASS	COMBINED EFFECT
Meglitinides	Increased blood level of meglitinides.
Mifepristone	Decreased effect of mifepristone.
Mind-altering drugs*	Dangerous sedation. Avoid.
Modafinil	Increased modafinil effect.
Monoamine oxidase (MAO) inhibitors*	Increased barbiturate effect.
Narcotics*	Dangerous sedation. Avoid.
Nitroimidazoles	Decreased nitroimidazoles effect.
Sertraline	Increased depressive effects of both drugs.
Sotalol	Increased barbiturate effect. Dangerous sedation.
Valproic acid*	Increased barbiturate effect.
Zaleplon	Decreased zaleplon effect.

BENZODIAZEPINES

GENERIC NAME OR DRUG CLASS	COMBINED EFFECT
Levodopa	Decreased levodopa effect.
Narcotics*	Increased sedative effect of both drugs.
Nefazodone	Increased effect of benzodiazepine.
Nicotine	Increased effect of benzodiazepine.
Omeprazole	Increased effect of benzodiazepine.
Probenecid	Increased effect of probenecid and risk of sedation.
Proton pump inhibitors	Increased effect of diazepam.

*See Glossary

GENERIC NAME OR DRUG CLASS	COMBINED EFFECT	GENERIC NAME OR DRUG CLASS	COMBINED EFFECT

BENZODIAZEPINES) continued

GENERIC NAME OR DRUG CLASS	COMBINED EFFECT
Rifamycins	Decreased effect of benzodiazepine.
Telithromycin	Increased effect of benzodiazepine.
Valproic acid*	Increased effect of benzodiazepines.
Zidovudine	Increased effect of zidovudine.

BETA-ADRENERGIC BLOCKING AGENTS

GENERIC NAME OR DRUG CLASS	COMBINED EFFECT
Diazoxide	Additional blood pressure drop.
Estrogens*	May cause blood pressure problems.
Flecainide	Increased effect of toxicity on heart muscle.
Fluvoxamine	Increased beta blocker effect.
Guanabenz	May cause blood pressure problems.
Insulin	Hypoglycemic effects may be prolonged.
Leukotriene modifiers	Increased beta blocker effect.
Meglitinides	Increased risk of low blood sugar.
Miglitol	Decreased effect of propranolol.
Molindone	Increased tranquilizer effect.
Monoamine oxidase (MAO) inhibitors*	High blood pressure following MAO discontinuation.
Nefazodone	Dosages of both drugs may require adjustment.
Nicotine	Increased effect of propranolol.
Nitrates*	Possible excessive blood pressure drop.
Phenothiazines	Increased effect of both drugs.
Phenytoin	Decreased beta blocker effect.
Propafenone	Increased beta blocker effect.
Quinidine	May cause heart problems.
Reserpine	Increased reserpine effect. Excessive sedation and depression. Additional blood pressure drop.
Sympathomimetics*	Decreased effects of both drugs.
Telithromycin	Various effects. Use with caution.
Warfarin	Increased warfarin effect.
Xanthines (aminophylline, theophylline)	Decreased effects of both drugs.

*See Glossary

GENERIC NAME OR DRUG CLASS	COMBINED EFFECT

BETA-ADRENERGIC BLOCKING AGENTS & THIAZIDE DIURETICS

GENERIC NAME OR DRUG CLASS	COMBINED EFFECT
Antidiabetics*	Increased antidiabetic effect.
Antihistamines*	Decreased antihistamine effect.
Antihypertensives*	Increased anti-hypertensive effect.
Anti-inflammatory drugs, nonsteroidal (NSAIDs)*	Decreased antiinflammatory effect.
Barbiturates*	Increased barbiturate effect. Dangerous sedation.
Bumetanide	Increased diuretic effect.
Calcium channel blockers*	Increased anti-hypertensive effect. Dosages of both drugs may require adjustments.
Cholestyramine	Decreased hydro-chlorothiazide effect.
Diclofenac	Decreased anti-hypertensive effect.
Digitalis preparations*	Excessive potassium loss that causes dangerous heart rhythms. Can either increase or decrease heart rate. Improves irregular heartbeat.
Diuretics, thiazide*	Increased effect of other thiazide diuretics.
Ethacrynic acid	Increased diuretic effect.
Furosemide	Increased diuretic effect.
Guanfacine	Increased effect of both drugs.
Hypoglycemics, oral*	Decreased ability to lower blood glucose.
Indapamide	Increased diuretic effect.
Insulin	Decreased ability to lower blood glucose.
Lisinopril	Increased anti-hypertensive effect. Dosage of each may require adjustment.
Metolazone	Increased diuretic effect.
Miglitol	Decreased effect of propranolol.
Monoamine oxidase (MAO) inhibitors*	Increased hydro-chlorothiazide effect.
Narcotics*	Increased narcotic effect. Dangerous sedation.
Nicardipine	Possible irregular heartbeat and congestive heart failure.
Nicotine	Increased beta blocker effect.
Nitrates*	Excessive blood pressure drop.
Phenytoin	Increased beta adrenergic effect.
Potassium supplements*	Decreased potassium effect.
Probenecid	Decreased probenecid effect.
Propafenone	Increased beta blocker effect.
Quinidine	Slows heart excessively.
Reserpine	Increased reserpine effect. Excessive sedation and depression.
Sympathomimetics*	Decreased effectiveness of both.
Theophylline	Decreased effectiveness of both.
Tocainide	May worsen congestive heart failure.
Zinc supplements	Increased need for zinc.

***See Glossary**

GENERIC NAME OR DRUG CLASS	COMBINED EFFECT	GENERIC NAME OR DRUG CLASS	COMBINED EFFECT

BRONCHODILATORS, ADRENERGIC (Long-Acting)

Xanthines*	May decreased levels of potassium.		

CALCIUM CHANNEL BLOCKERS

Lithium	Possible decreased lithium effect.	**Nimodipine**	Dangerous blood pressure drop.
Metformin	Increased metformin effect.	**Nitrates***	Reduced angina attacks.
Nicardipine	Possible increased effect and toxicity of each drug.		

CALCIUM SUPPLEMENTS

Quinidine	Increased quinidine effect.	**Theophylline**	May increase effect and toxicity of theophylline.
Rifamycins	Decreased effect of calcium channel blocker.	**Vitamin A**	Decreased vitamin effect.
Salicylates*	Increased salicylate effect.	**Vitamin D**	Increased vitamin absorption, sometimes excessively; decreased effect of calcium channel blocker.
Sulfa drugs*	Decreased sulfa effect.	**Zafirlukast**	May increase calcium channel blocker effect.
Tetracyclines*	Decreased tetracycline effect.		

CARBAMAZEPINE

Antidepressants, tricyclic*	Confusion. Possible psychosis.	**Bupropion**	Decreased effect of bupropion.
Antifungals, azole	Increased effect of carbamazepine. Decreased effect of antifungal.	**Carbonic anhydrase inhibitors (oral)**	Increased risk of bone loss.
Barbiturates*	Possible increased barbiturate metabolism.	**Central nervous system (CNS) depressants***	Increased sedative effect.
Benzodiazepines	Decreased effect of benzodiazepine.	**Cimetidine**	Increased carbamazepine effect.

***See Glossary**

GENERIC NAME OR DRUG CLASS	COMBINED EFFECT
CARBAMAZEPINE continued	
Citalopram	May lessen effect of citalopram.
Clozapine	Toxic effect on bone marrow and central nervous system.
Contraceptives, oral*	Reduced contraceptive protection. Breakthrough bleeding.
Cyclosporine	Decreased effect of cyclosporine.
Danazol	Increased effect of carbamazepine.
Desmopressin	May increase desmopressin effect.
Digitalis preparations*	Decreased digitalis effect.
Diltiazem	Increased effect of carbamazepine.
Doxycycline	Decreased doxycycline effect.
Estrogens*	Decreased estrogen effect.
Erythromycins*	Increased carbamazepine effect.
Felbamate	Increased side effects and adverse reactions.
Felodipine	Decreased effect of felodipine.
Fluoxetine	Increased carbamazepine effect.
Fluvoxamine	Possible toxicity of carbamazepine.
Guanfacine	May increase depressant effects of either drug.
Haloperidol	Decreased effect of haloperidol.
Isoniazid	Increased effect of carbamazepine.
Lamotrigine	Decreased lamotrigine effect. Increased risk of side effects.
Leucovorin	High alcohol content of leucovorin may cause adverse effects.
Leukotriene modifiers	Increased effect of carbamazepine.
Lithium	Increased risk of side effects.
Meglitinides	Blood sugar problems.
Methylphenidate	Decreased effect of methylphenidate.
Mifepristone	Decreased effect of mifepristone.
Monoamine oxidase (MAO) Inhibitors*	Dangerous overstimulation. Avoid.
Phenobarbital	Decreased carbamazepine effect.
Phenytoin	Decreased effect of both drugs.
Primidone	Decreased carbamazepine effect.
Propoxyphene	Increased toxicity of both. Avoid.
Rifampin	Decreased carbamazepine effect.
Risperidone	Decreased risperidone effect.
Theophylline	Decreases effect of theophylline.
Ticlopidine	Decreased effect of carbamazepine.
Tiopronin	Increased risk of toxicity to bone marrow.
Valproic acid*	Decreased effect of valproic acid.
Vasopressin	Increased effect of vasopressin.
Verapamil	Increased effect of carbamazepine.
Zafirlukast	Increased effect of carbamazepine.
Zaleplon	Decreased zaleplon effect.

*See Glossary

GENERIC NAME OR DRUG CLASS	COMBINED EFFECT	GENERIC NAME OR DRUG CLASS	COMBINED EFFECT

CENTRAL ALPHA AGONISTS

GENERIC NAME OR DRUG CLASS	COMBINED EFFECT
Central nervous system (CNS) depressants*	Increased depressive effects of both drugs.
Monoamine oxidase (MAO) inhibitors*	Dangerous changes in blood pressure. Take at least 14 days apart.

CONTRACEPTIVES, ORAL & SKIN

GENERIC NAME OR DRUG CLASS	COMBINED EFFECT
Dextrothyroxine	Decreased dextrothyroxine effect.
Guanethidine	Decreased guanethidine effect.
Hypoglycemics, oral*	Decreased effect of hypoglycemics.
Insulin	Possibly decreased insulin effect.
Insulin lispro	May need increased dosage of insulin.
Meperidine	Increased meperidine effect.
Meprobamate	Decreased contraceptive effect.
Mineral oil	Decreased contraceptive effect.
Non-nucleoside reverse transcriptase inhibitors	Decreased contraceptive effect. Use alternative birth control method.
Phenothiazines*	Increased phenothiazine effect.
Rifamycins	Decreased contraceptive effect.
Sulfadoxine and pyrimethamine	Reduced reliability of the pill.
Terazosin	Decreases terazosin effect.
Tetracyclines*	Decreased contraceptive effect.
Thiazolidinediones	Decreased contraceptive effect.
Ursodiol	Decreased ursodiol effect.
Vitamin A	Vitamin A excess.
Vitamin C	Possible increased contraceptive effect.

CYCLOSPORINE

GENERIC NAME OR DRUG CLASS	COMBINED EFFECT
Nimodipine	Increased cyclosporine toxicity.
Nitroimidazoles	Increased cyclosporine effect.
Orlistat	Unknown effect. Monitor closely.
Rifamycins	Decreased effect of cyclosporine.
Terbinafine (oral)	Decreased effect of cyclosporine.
Thiazolidinediones	Decreased effect of cyclosporine.
Tiopronin	Increased risk of toxicity to kidneys.
Vancomycin	Increased chance of hearing loss or kidney damage.
Virus vaccines	Increased adverse reactions to vaccine.
Zafirlukast	May increase effect of cyclosporine.

*See Glossary

GENERIC NAME OR DRUG CLASS	COMBINED EFFECT	GENERIC NAME OR DRUG CLASS	COMBINED EFFECT

DIFENOXIN & ATROPINE

GENERIC NAME OR DRUG CLASS	COMBINED EFFECT
Nitrates*	Increased internal eye pressure.
Orphenadrine	Increased atropine effect.
Phenothiazines*	Increased atropine effect.
Pilocarpine	Loss of pilocarpine effect in glaucoma treatment.
Potassium supplements*	Possible intestinal ulcers with oral potassium tablets.
Procainamide	Increased atropine effect.
Sertraline	Increased depressive effects of both drugs.
Vitamin C	Decreased atropine effect. Avoid large doses of vitamin C.

DIGITALIS PREPARATIONS (Digitalis Glycosides)

GENERIC NAME OR DRUG CLASS	COMBINED EFFECT
Laxatives*	Decreased digitalis effect.
Metformin	Increased metformin effect.
Metoclopramide	Decreased digitalis absorption.
Nefazodone	Increased effect of digoxin.
Nicardipine	Increased digitalis effect. May need to reduce dose.
Nizatidine	Increased digitalis effect.
Oxyphenbutazone	Decreased digitalis effect.
Paroxetine	Increased levels of paroxetine in blood.
Phenobarbital	Decreased digitalis effect.
Phenylbutazone	Decreased digitalis effect.
Potassium supplements*	Overdose of either drug may cause severe heartbeat irregularity.
Propafenone	Increased digitalis absorption.
Proton pump inhibitors	Increased effect of digoxin.
Quinidine	Increased digitalis effect.
Ranolazine	Increased effect of digoxin.
Rauwolfia alkaloids*	Increased digitalis effect.
Rifamycins	Possible decreased digitalis effect.
Sotalol	Can either increase or decrease heart rate. Improves irregular heartbeat.
Spironolactone	Increased digitalis effect. May require digitalis dosage reduction.
Sulfasalazine	Decreased digitalis absorption.
Sympathomimetics*	Increased risk of heartbeat irregularities.
Telithromycin	May increase digoxin effect.
Tetracycline	May increase digitalis absorption.
Thyroid hormones*	Digitalis toxicity.
Ticlopidine	Slightly decreased digitalis effect (digoxin only).

***See Glossary**

DIGITALIS PREPARATIONS (Digitalis Glycosides) continued

GENERIC NAME OR DRUG CLASS	COMBINED EFFECT
Trazodone	Possible increased digitalis toxicity.
Triamterene	Possible decreased digitalis effect.
Verapamil	Increased digitalis effect.

DIURETICS, LOOP

GENERIC NAME OR DRUG CLASS	COMBINED EFFECT
Narcotics*	Dangerous low blood pressure. Avoid.
Nephrotoxics*	Increased risk of toxicity.
Nimodipine	Dangerous blood pressure drop.
Nitrates*	Excessive blood pressure drop.
Phenytoin	Decreased diuretic effect.
Potassium supplements*	Decreased potassium effect.
Probenecid	Decreased probenecid effect.
Salicylates* (including aspirin)	Dangerous salicylate retention.
Sedatives*	Increased diuretic effect.

DIURETICS, POTASSIUM-SPARING & HYDROCHLOROTHIAZIDE

GENERIC NAME OR DRUG CLASS	COMBINED EFFECT
Cholestyramine	Decreased diuretic effect. Take 1 hour before diuretic.
Colestipol	Decreased diuretic effect. Take 1 hour before diuretic.
Cyclosporine	Increased potassium levels.
Digitalis preparations*	Increased digitalis effect.
Diuretics,* other	Increased effect of both drugs.
Folic acid	Decreased effect of folic acid.
Lithium	Possible lithium toxicity.
Metformin	Increased metformin effect.
Potassium-containing medications	Increased potassium levels.

DIURETICS, THIAZIDE

GENERIC NAME OR DRUG CLASS	COMBINED EFFECT
Indapamide	Increased diuretic effect.
Indomethacin	Decreased anti-hypertensive effect.
Lithium	Increased effect of lithium.
Meglitinides	Increased blood sugar levels.
Memantine	Increased effect of memantine and hydrochlorothiazide.
Monoamine oxidase (MAO) inhibitors*	Increased anti-hypertensive effect.

ADDITIONAL DRUG INTERACTIONS

GENERIC NAME OR DRUG CLASS	COMBINED EFFECT	GENERIC NAME OR DRUG CLASS	COMBINED EFFECT

DIURETICS, THIAZIDE continued

GENERIC NAME OR DRUG CLASS	COMBINED EFFECT
Nicardipine	Blood pressure drop. Dosages may require adjustment.
Nimodipine	Dangerous blood pressure drop.
Nitrates*	Excessive blood pressure drop.
Opiates*	Dizziness or weakness when standing up after sitting or lying down.
Pentoxifylline	Increased antihypertensive effect.
Potassium supplements*	Decreased potassium effect.
Probenecid	Decreased probenecid effect.
Sotalol	Increased antihypertensive effect.
Terazosin	Decreased terazosin effect.
Toremifene	Possible increased calcium.
Zinc supplements	Increased need for zinc.

FLUOROQUINOLONES

GENERIC NAME OR DRUG CLASS	COMBINED EFFECT
Cyclosporine	Increased cyclosporine effect.
Didanosine	Decreased fluoro-quinolone effect.
Digoxin	Increased digoxin effect.
Iron supplements	Decreased fluoro-quinolone effect.
Oxtriphylline	Increased risk of oxtriphylline toxicity.
Phenytoin	Decreased effect of phenytoin with ciprofloxacin.
Probenecid	Increased effect of fluoroquinolone.
QT interval prolongation causing drugs*	Heart rhythm problems.
Sucralfate	Decreased fluoro-quinolone effect.
Theophylline	Increased risk of theophylline toxicity.
Tizanidine	Dangerous increased tizanidine
Warfarin	Increased warfarin effect.
Zinc supplements	Decreased fluoro-quinolone effect.effect with ciprofloxacin.

GLP-1 RECEPTOR AGONISTS

GENERIC NAME OR DRUG CLASS	COMBINED EFFECT
Warfarin	Increased risk of bleeding (with exenatide).

*See Glossary

GENERIC NAME OR DRUG CLASS	COMBINED EFFECT	GENERIC NAME OR DRUG CLASS	COMBINED EFFECT

HALOPERIDOL

GENERIC NAME OR DRUG CLASS	COMBINED EFFECT
Fluoxetine	Increased depressant effects of both drugs.
Guanethidine	Decreased guanethidine effect.
Guanfacine	May increase depressant effects of either drug.
Leucovorin	High alcohol content of leucovorin may cause adverse effects.
Levodopa	Decreased levodopa effect.
Lithium	Increased toxicity.
Loxapine	May increase toxic effects of both drugs.
Methyldopa	Possible psychosis.
Narcotics*	Excessive sedation.
Nefazodone	Unknown effect. May require dosage adjustment.
Pergolide	Decreased pergolide effect.
Procarbazine	Increased sedation.
QT interval prolongation-causing drugs*	Serious heart rhythm problems.
Sertraline	Increased depressive effects of both drugs.

HISTAMINE H_2 RECEPTOR ANTAGONISTS

GENERIC NAME OR DRUG CLASS	COMBINED EFFECT
Diazepam	Increased effect and toxicity of diazepam.
Digitalis preparations*	Increased digitalis effect.
Dofetilide	Increased risk of heart problems.
Encainide	Increased effect of histamine H_2 receptor antagonist.
Flurazepam	Increased effect and toxicity of flurazepam.
Glipizide	Increased effect and toxicity of glipizide.
Itraconazole	Decreased absorption of itraconazole.
Ketoconazole	Decreased ketoconazole absorption.
Labetalol	Increased antihypertensive effects.
Memantine	Increased effect of either drug.
Metformin	Increased metformin effect.
Methadone	Increased effect and toxicity of methadone.
Metoclopramide	Decreased absorption of histamine H_2 receptor antagonist.
Metoprolol	Increased effect and toxicity of metoprolol.
Miglitol	Decreased effect of ranitidine.
Moricizine	Increased concentration of H_2 receptor antagonist in the blood.
Morphine	Increased effect and toxicity of morphine.
Nicardipine	Possible increased effect and toxicity of nicardipine.
Nimodipine	Possible increased effect and toxicity of nimodipine.
Nitroimidazoles	Increased nitroimidazole effect.

*See Glossary

GENERIC NAME OR DRUG CLASS	COMBINED EFFECT	GENERIC NAME OR DRUG CLASS	COMBINED EFFECT

HISTAMINE H2 RECEPTOR ANTAGONISTS continued

GENERIC NAME OR DRUG CLASS	COMBINED EFFECT
Paroxetine	Increased levels of paroxetine in blood.
Phenytoin	Increased effect and toxicity of phenytoin.
Propafenone	Increased effect of both drugs and increased risk of toxicity.
Propranolol	Possible increased propranolol effect.
Quinidine	Increased quinidine effect.
Tamoxifen	Decreased tamoxifen effect.
Terbinafine (oral)	Increased effect of terbinafine with cimetidine.
Theophylline	Increases theophylline effect.
Triazolam	Increased effect and toxicity of triazolam.
Varenicline	May increase effect of varenicline (with cimetidine).
Venlafaxine	With cimetidine—Increased risk of adverse reactions.
Verapamil	Increased effect and toxicity of verapamil.
Zaleplon	Increases zaleplon effect.

HYDRALAZINE

GENERIC NAME OR DRUG CLASS	COMBINED EFFECT
Lisinopril	Increased anti-hypertensive effect. Dosage of each may require adjustment.
Monoamine oxidase (MAO) inhibitors*	Increased hydralazine effect.
Nicardipine	Blood pressure drop. Dosages may require adjustment.
Nimodipine	Dangerous blood pressure drop.
Sotalol	Increased antihypertensive effect.
Terazosin	Decreased effectiveness of terazosin.

IMMUNOSUPPRESSIVE AGENTS

GENERIC NAME OR DRUG CLASS	COMBINED EFFECT
Probenecid	Increased effect of mycophenolate.
Proton pump inhibitors	Increased effect of tacrolimus.
Telithromycin	Increased effect of tacrolimus.
Vaccinations	Avoid unless doctor approves.

*See Glossary

GENERIC NAME OR DRUG CLASS	COMBINED EFFECT	GENERIC NAME OR DRUG CLASS	COMBINED EFFECT

INDAPAMIDE

GENERIC NAME OR DRUG CLASS	COMBINED EFFECT
Probenecid	Decreased probenecid effect.
Sotalol	Increased antihypertensive effect.
Terazosin	Decreased effectiveness of terazosin.

LITHIUM

GENERIC NAME OR DRUG CLASS	COMBINED EFFECT
Diuretics*	Increased lithium effect or toxicity.
Fluvoxamine	Increased risk of seizure.
Haloperidol	Increased toxicity of both drugs.
Indomethacin	Increased lithium effect.
Iodide salts	Increased lithium effects on thyroid function.
Ketoprofen	May increase lithium in blood.
Meloxicam	Increased lithium effect.
Methyldopa	Increased lithium effect.
Molindone	Brain changes.
Nitroimidazoles	Increased lithium effect.
Oxyphenbutazone	Increased lithium effect.
Phenothiazines*	Decreased lithium effect.
Phenylbutazone	Increased lithium effect.
Phenytoin	Increased lithium effect.
Potassium iodide	Increased potassium iodide effect.
Sodium bicarbonate	Decreased lithium effect.
Sumatriptan	Adverse effects unknown. Avoid.
Theophylline	Decreased lithium effect.
Tiopronin	Increased risk of toxicity to kidneys.

LOXAPINE

GENERIC NAME OR DRUG CLASS	COMBINED EFFECT
Rauwolfia alkaloids*	May increase toxic effects of both drugs.
Sertraline	Increased depressive effects of both drugs.
Thioxanthenes*	May increase toxic effects of both drugs.

ADDITIONAL DRUG INTERACTIONS

***See Glossary**

MAPROTILINE

GENERIC NAME OR DRUG CLASS	COMBINED EFFECT
Clonidine	Decreased clonidine effect.
Clozapine	Toxic effect on the central nervous system.
Disulfiram	Delirium.
Diuretics, thiazide*	Increased maprotiline effect.
Ethchlorvynol	Delirium.
Fluoxetine	Increased depressant effects of both drugs.
Guanethidine	Decreased guanethidine effect.
Guanfacine	May increase depressant effects of either drug.
Leucovorin	High alcohol content of leucovorin may cause adverse effects.
Levodopa	Decreased levodopa effect.
Lithium	Possible decreased seizure threshold.
Methyldopa	Decreased methyldopa effect.
Methylphenidate	Possible increased antidepressant effect and toxicity.
Molindone	Increased tranquilizer effect.
Monoamine oxidase (MAO) inhibitors*	Fever, delirium, convulsions.
Narcotics*	Dangerous oversedation.
Phenothiazines*	Possible increased antidepressant effect and toxicity.
Phenytoin	Decreased phenytoin effect.
Quinidine	Irregular heartbeat.
Selegiline	Fever, delirium, convulsions.
Sertraline	Increased depressive effects of both drugs.
Sympathomimetics*	Increased sympathomimetic effect.
Thyroid hormones*	Irregular heartbeat.

METFORMIN

GENERIC NAME OR DRUG CLASS	COMBINED EFFECT
Ranitidine	Increased metformin effect.
Trimethoprim	Increased metformin effect.
Vancomycin	Increased metformin effect.

METHOTREXATE

GENERIC NAME OR DRUG CLASS	COMBINED EFFECT
Sulfa drugs*	Possible methotrexate toxicity.
Sulfadoxine and pyrimethamine	Increased risk of toxicity.
Tetracyclines*	Possible methotrexate toxicity.
Tiopronin	Increased risk of toxicity to bone marrow and kidneys.
Vaccines, live or killed	Increased risk of toxicity or reduced effectiveness of vaccine.

***See Glossary**

GENERIC NAME OR DRUG CLASS	COMBINED EFFECT	GENERIC NAME OR DRUG CLASS	COMBINED EFFECT

METOCLOPRAMIDE

GENERIC NAME OR DRUG CLASS	COMBINED EFFECT
Phenothiazines*	Increased chance of muscle spasm and trembling.
Sertraline	Increased depressive effects of both drugs.
Tetracyclines*	Slow stomach emptying.
Thioxanthenes*	Increased chance of muscle spasm and trembling.

MONOAMINE OXIDASE (MAO) INHIBITORS

GENERIC NAME OR DRUG CLASS	COMBINED EFFECT
Central nervous system (CNS) depressants*	Excessive depressant action.
Citalopram	Can cause a life-threatening reaction. Avoid.
Clozapine	Toxic effect on the central nervous system.
Doxepin (topical)	Potentially life-threatening. Allow 14 days between use of the 2 drugs.
Dexfenfluramine	Potentially life-threatening. Allow 14 days between use of 2 drugs.
Dextromethorphan	Very high blood pressure.
Diuretics*	Excessively low blood pressure.
Ephedrine	Increased blood pressure.
Fluoxetine	Potentially life-threatening. Avoid.
Fluvoxamine	Potentially life-threatening. Avoid.
Furazolidone	Sudden, severe increase in blood pressure.
Guanadrel	High blood pressure.
Guanethidine	Blood pressure rise.
Guanfacine	May increase depressant effects of either drug.
Indapamide	Increased indapamide effect.
Insulin	Increased hypoglycemic effect.
Leucovorin	High alcohol content of leucovorin may cause adverse effects.
Levodopa	Sudden, severe blood pressure rise.
Maprotiline	Dangerous blood pressure rise.
Meglitinides	Increased risk of low blood sugar.
Methyldopa	Sudden, severe blood pressure rise.
Methylphenidate	Increased blood pressure.
Mirtazapine	Potentially life-threatening. Allow 14 days between use of 2 drugs.
Monoamine oxidase (MAO) inhibitors* (others, when taken together)	High fever, convulsions, death.
Narcotics*	Severe high blood pressure.
Nefazodone	Potentially life-threatening. Allow 14 days between use of the 2 drugs.
Paroxetine	Potentially life-threatening. Avoid.

*See Glossary

GENERIC NAME OR DRUG CLASS	COMBINED EFFECT	GENERIC NAME OR DRUG CLASS	COMBINED EFFECT

MONOAMINE OXIDASE (MAO) INHIBITORS continued

GENERIC NAME OR DRUG CLASS	COMBINED EFFECT
Phenothiazines*	Possible increased phenothiazine toxicity.
Pseudoephedrine	Increased blood pressure.
Serotonergics*	Increased risk of serotonin syndrome.*
Sertraline	Potentially life-threatening. Avoid.
Sympathomimetics*	Blood pressure rise to life-threatening level.
Tolcapone	May reduce effectiveness of MAO inhibitor.
Tramadol	Increased risk of seizures.
Trazodone	Increased risk of mental status changes.
Venlafaxine	Increased risk and severity of side effects. Allow 4 weeks between use of the 2 drugs.

NARCOTIC ANALGESICS

GENERIC NAME OR DRUG CLASS	COMBINED EFFECT
Monoamine oxidase (MAO) inhibitors*	Serious toxicity (including death). Avoid.
Muscle relaxants*	Increased risk of side effects.
Naloxone	Decreased effect of narcotic analgesic.
Naltrexone	Increased risk of withdrawal symptoms of narcotic analgesic.
Narcotics,* other	Increased sedative effect.
P-Glycoprotein inhibitors	Increased effect of narcotic analgesic; increased risk of side effects.

NEFAZODONE

GENERIC NAME OR DRUG CLASS	COMBINED EFFECT
Fluoxetine	Increased risk of side effects.
Haloperidol	Unknown. May need haloperidol dosage adjusted.
Monoamine oxidase (MAO) inhibitors*	Potentially life-threatening. Allow 14 days between use of 2 drugs.
Pentazocine	Increased effect of pentazocine.
Pimozide	Increased effect of pimozide. Avoid.
Propranolol	Unknown effect. May need dosage adjustment of both drugs.
Terfenadine	Increased effect of terfenadine. Avoid.
Triazolam	Increased effect of triazolam.

*See Glossary

GENERIC NAME OR DRUG CLASS	COMBINED EFFECT	GENERIC NAME OR DRUG CLASS	COMBINED EFFECT

NICOTINE

Phenylephrine	Decreased effect of phenylephrine.	**Theophylline**	Increased effect of theophylline.
Propoxyphene	Increased effect of propoxyphene.		

NUCLEOSIDE REVERSE TRANSCRIPTASE INHIBITORS

Probenecid	Increased effect of zidovudine.	**Tetracyclines**	Decreased antibiotic effect.
Rifamycins	Decreased zidovudine effect.	**Valproic acid***	Increased effect of zidovudine.
Tenofovir	Increased effect of didanosine. Take tenofovir 2 hours before or 1 hour after didanosine.		

OLANZAPINE

Eszopiclone	Decreased alertness.	**Levodopa**	May decrease levodopa effect.
Hepatotoxics*	Risk of liver problems.		

PHENOTHIAZINES

Doxepin (topical)	Increased risk of toxicity of both drugs.	**Mind-altering drugs***	Increased effect of mind-altering drug.
Duloxetine	Increased effect of duloxetine.	**Molindone**	Increased tranquilizer effect.
Guanethidine	Increased guanethidine effect.	**Narcotics***	Increased narcotic effect.
Isoniazid	Increased risk of liver damage.	**Procarbazine**	Increased sedation.
Levodopa	Decreased levodopa effect.	**Quetiapine**	Decreased quetiapine effect.
Lithium	Decreased lithium effect.	**Tramadol**	Increased sedation.
		Zolpidem	Increased sedation. Avoid.

GENERIC NAME OR DRUG CLASS	COMBINED EFFECT	GENERIC NAME OR DRUG CLASS	COMBINED EFFECT

POTASSIUM SUPPLEMENTS

GENERIC NAME OR DRUG CLASS	COMBINED EFFECT	GENERIC NAME OR DRUG CLASS	COMBINED EFFECT
Potassium-containing drugs*	Increased potassium levels.	**Vitamin B-12**	Extended-release tablets may decrease vitamin B-12 absorption and increase vitamin B-12 requirements.
Spironolactone	Dangerous rise in blood potassium.		
Triamterene	Dangerous rise in blood potassium.		

PRIMIDONE

GENERIC NAME OR DRUG CLASS	COMBINED EFFECT	GENERIC NAME OR DRUG CLASS	COMBINED EFFECT
Narcotics*	Increased narcotic effect.	**Sedatives***	Increased sedative effect.
Oxyphenbutazone	Decreased oxyphenbutazone effect.	**Sertraline**	Increased depressive effects of both drugs.
Phenylbutazone	Decreased phenylbutazone effect.	**Sleep inducers***	Increased effect of sleep inducer.
Phenytoin	Possible increased primidone toxicity.	**Tranquilizers***	Increased tranquilizer effect.
Rifamycins	Possible decreased primidone effect.	**Valproic acid***	Increased effect of primidone.

PROCARBAZINE

GENERIC NAME OR DRUG CLASS	COMBINED EFFECT	GENERIC NAME OR DRUG CLASS	COMBINED EFFECT
Diuretics*	Excessively low blood pressure.	**Leucovorin**	High alcohol content of leucovorin may cause adverse effects.
Doxapram	Increased blood pressure.	**Levamisole**	Increased risk of bone marrow depression.
Ethinamate	Dangerous increased effects of ethinamate. Avoid combining.	**Levodopa**	Sudden, severe blood pressure rise.
Fluoxetine	Increased depressant effects of both drugs.	**Methyldopa**	Severe high blood pressure.
Guanethidine	Blood pressure rise to life-threatening level.	**Methylphenidate**	Excessive high blood pressure.
Guanfacine	May increase depressant effects of either medicine.	**Methyprylon**	May increase sedative effect to dangerous level. Avoid.
		Monoamine oxidase (MAO) inhibitors,* other	High fever, convulsions, death.

*See Glossary

PROCARBAZINE continued

GENERIC NAME OR DRUG CLASS	COMBINED EFFECT
Nabilone	Greater depression of central nervous system.
Narcotics*	Increased sedation.
Phenothiazines*	Increased sedation.
Rauwolfia alkaloids*	Very high blood pressure.
Reserpine	Increased blood pressure, excitation.
Sertraline	Increased depressive effects of both drugs.
Sumatriptan	Adverse effects unknown. Avoid.
Sympathomimetics*	Heartbeat abnormalities, severe high blood pressure.
Tiopronin	Increased risk of toxicity to bone marrow.

PROTEASE INHIBITORS

GENERIC NAME OR DRUG CLASS	COMBINED EFFECT
Protease inhibitor, other	Increased risk of severe heart problem if both saquinavir and ritonavir taken.
Proton pump inhibitors	May need dosage adjustment of protease inhibitor.
Rifabutin	Decreased effect of protease inhibitor.
Rifampin	Decreased protease inhibitor effect. Don't use with saquinavir.
Trazodone	Increased trazodone effect.
Warfarin	Increased or decreased warfarin effect.

QUETIAPINE

GENERIC NAME OR DRUG CLASS	COMBINED EFFECT
Levodopa or dopamine agonists*	Decreased effect of levodopa or dopamine agonist.
Phenytoin	Decreased quetiapine effect.
Thioridazine	Decreased quetiapine effect.

RIFAMYCINS

GENERIC NAME OR DRUG CLASS	COMBINED EFFECT
Phenytoin	Decreased phenytoin effect.
Probenecid	Possible toxicity to liver.
Protease inhibitors	Decreased protease inhibitor effect.
Quinidine	Decreased effect of both drugs.
Quinine	Decreased quinine effect.
Sildenafil	Decreased sildenafil effect.
Tacrolimus	Decreased tacrolimus effect.
Telithromycin	Decreased effect of rifampin.

*See Glossary

GENERIC NAME OR DRUG CLASS	COMBINED EFFECT	GENERIC NAME OR DRUG CLASS	COMBINED EFFECT

RIFAMYCINS continued

GENERIC NAME OR DRUG CLASS	COMBINED EFFECT
Theophylline	Decreased theophylline effect.
Tocainide	Possible decreased blood cell production in bone marrow.
Trimethoprim	Decreased trimethoprim effect.
Valproic acid*	Decreased effect of valproic acid with rifampin.
Zaleplon	Decreased zaleplon effect.
Zidovudine	Decreased zidovudine effect.

SALICYLATES

GENERIC NAME OR DRUG CLASS	COMBINED EFFECT
Insulin lispro	May need decreased dosage of insulin.
Ketoconazole	Decreased ketoconazole effect with buffered salicylates.
Methotrexate	Increased methotrexate effect and toxicity.
Para-aminosalicylic acid	Possible salicylate toxicity.
Penicillins*	Increased effect of both drugs.
Phenobarbital	Decreased salicylate effect.
Phenytoin	Increased phenytoin effect.
Probenecid	Decreased probenecid effect.
Rauwolfia alkaloids*	Decreased salicylate effect.
Salicylates,* other	Likely salicylate toxicity.
Sotalol	Decreased antihypertensive effect of sotalol.
Spironolactone	Decreased spironolactone effect.
Sulfinpyrazone	Decreased sulfinpyrazone effect.
Terazosin	Decreased effectiveness of terazosin. Causes sodium and fluid retention.
Urinary acidifiers*	Decreased excretion. Increased salicylate effect.
Urinary alkalizers*	Increased excretion. Decreased salicylate effect.
Vitamin C (large doses)	Possible salicylate toxicity.
Zidovudine	increased zidovudine effect.

SCOPOLAMINE (Hyoscine)

GENERIC NAME OR DRUG CLASS	COMBINED EFFECT
Nabilone	Greater depression of central nervous system.
Nitrates*	Increased internal eye pressure.
Nizatidine	Increased nizatidine effect.
Orphenadrine	Increased scopolamine effect.
Phenothiazines*	Increased scopolamine effect.
Pilocarpine	Loss of pilocarpine effect in glaucoma treatment.

***See Glossary**

GENERIC NAME OR DRUG CLASS	COMBINED EFFECT

SCOPOLAMINE (Hyoscine) continued

GENERIC NAME OR DRUG CLASS	COMBINED EFFECT
Potassium supplements*	Possible intestinal ulcers with oral potassium tablets.
Quinidine	Increased scopolamine effect.
Sedatives* or central nervous system (CNS) depressants*	Increased sedative effect of both drugs.
Sertraline	Increased depressive effects of both drugs.
Vitamin C	Decreased scopolamine effect. Avoid large doses of vitamin C.

SELECTIVE SEROTONIN REUPTAKE INHIBITORS (SSRIs)

GENERIC NAME OR DRUG CLASS	COMBINED EFFECT
Benzodiazepines*	Increased benzo-diazepine effect.
Bromocriptine	Increased risk of serotonin syndrome.*
Buspirone	Increased risk of serotonin syndrome.*
Desmopressin	Increased risk of thirstiness which may lead to drinking excess fluids.
Dextromethorphan	Increased risk of serotonin syndrome.*
Digoxin	Increased risk of side effects of both drugs.
Levodopa	Increased risk of serotonin syndrome.*
Lithium	Increased risk of serotonin syndrome.*
Meperidine	Increased risk of serotonin syndrome.*
Moclobemide	Increased risk of side effects and serotonin syndrome.*
Monoamine oxidase (MAO) inhibitors*	Increased risk of adverse effects. May lead to convulsions and hypertensive crisis. Let 14 days elapse between taking the 2 drugs.
Nefazodone	Increased risk of serotonin syndrome.*
Pentazocine	Increased risk of serotonin syndrome.*
Phenytoin	Increased effect of phenytoin.
Propranolol	Increased effect of propranolol.
Ramelteon	Increased effect of ramelteon. Avoid.
Serotonergics*	Increased risk of serotonin syndrome.
Sumatriptan	Increased risk of serotonin syndrome.*
Theophylline	Increased effect of theophylline.
Tizanidine	Dangerous increased tizanidine effect with fluvoxamine.
Vorapaxar	Increased risk of side effects and bleeding.

GENERIC NAME OR DRUG CLASS	COMBINED EFFECT	GENERIC NAME OR DRUG CLASS	COMBINED EFFECT

SEROTONIN & NOREPINEPHRINE REUPTAKE INHIBITORS (SNRIs)

Serotonergics*	Increased risk of serotonin syndrome.*	**Tryptophan**	Increased risk of serotonin syndrome.*
Tramadol	Increased risk of serotonin syndrome.*	**Venlafaxine**	Increased risk of serotonin syndrome.*
Trazodone	Increased risk of serotonin syndrome.*	**Vorapaxar**	Increased risk of side effects and bleeding.

SULFONYLUREAS

Insulin	Increased blood sugar lowering.	**Phenytoin**	Decreased blood sugar lowering.
Insulin lispro	Increased anti-diabetic effect.	**Probenecid**	Increased blood sugar lowering.
Isoniazid	Decreased blood sugar lowering.	**Pyrazinamide**	Decreased blood sugar lowering.
Labetalol	Increased blood sugar lowering, may mask hypoglycemia.	**Ranitidine**	Increased blood sugar lowering.
Leukotriene modifiers	Increased effect of tolbutamide.	**Rifamycins**	Decreased blood sugar lowering.
MAO inhibitors*	Increased blood sugar lowering.	**Sulfa drugs***	Increased blood sugar lowering.
Nicotinic acid	Decreased blood sugar lowering.	**Sulfadoxine and pyrimethamine**	Increased risk of toxicity.
Oxyphenbutazone	Increased blood sugar lowering.	**Sulfaphenazole**	Increased blood sugar lowering.
Phenothiazines*	Decreased blood sugar lowering.	**Thiazolidinediones***	May decrease plasma glucose concentrations.
Phenylbutazone	Increased blood sugar lowering.	**Thyroid hormones***	Decreased blood sugar lowering.
Phenyramidol	Increased blood sugar lowering.	**Zafirlukast**	May increase effect of tolbutamide.

***See Glossary**

GENERIC NAME OR DRUG CLASS	COMBINED EFFECT	GENERIC NAME OR DRUG CLASS	COMBINED EFFECT

TETRACYCLINES

Vitamin A	Increased risk of intracranial hypertension.	**Zinc supplements**	Decreased tetracycline absorption if taken within 2 hours of each other.

THIOTHIXENE

Sertraline	Increased depressive effects of both drugs.	**Tranquilizers***	Increased thiothixene effect. Excessive sedation.

VALPROIC ACID

Salicylates*	Increased effect of valproic acid.	**Zidovudine**	Increased effect of zidovudine.

ADDITIONAL DRUG INTERACTIONS

***See Glossary**

GLOSSARY

Glossary

Many of the following medical terms are found in the drug charts. Where drug names are listed, they indicate the generic or drug class and not the brand names.

A

ACE Inhibitors—See Angiotensin-Converting Enzyme (ACE) Inhibitors.

Acne Preparations—Creams, lotions and liquids applied to the skin to treat acne. These include adapalene; alcohol and acetone; alcohol and sulfur; azelaic acid; benzoyl peroxide; clindamycin; erythromycin; erythromycin and benzoyl peroxide; isotretinoin; meclocycline; resorcinol; resorcinol and sulfur; salicylic acid gel USP; salicylic acid lotion; salicylic acid ointment; salicylic acid pads; salicylic acid soap; salicylic acid and sulfur bar soap; salicylic acid and sulfur cleansing lotion; salicylic acid and sulfur cleansing suspension; salicylic acid and sulfur lotion; sulfurated lime; sulfur bar soap; sulfur cream; sulfur lotion; tetracycline, oral; tetracycline hydrochloride for topical solution; tretinoin.

Acridine Derivatives—Dyes or stains (often yellow or orange) used for some medical tests and as antiseptic agents.

Active Ingredient—An active ingredient is any component that provides pharmacological activity or other direct effect in the diagnosis, cure, mitigation, treatment, or prevention of disease, or to affect the structure or any function of the body of man or animals.

Acute—Having a short and relatively severe course.

Addiction—Psychological or physiological dependence upon a drug.

Addictive Drugs—Any drug that can lead to physiological dependence on the drug. These include alcohol, cocaine, marijuana, nicotine, opium, morphine, codeine, heroin (and other narcotics) and others.

Addison's Disease—Changes in the body caused by a deficiency of hormones manufactured by the adrenal gland. Usually fatal if untreated.

Adrenal Cortex—Center of the adrenal gland.

Adrenal Gland—Gland next to the kidney that produces cortisone and epinephrine (adrenalin).

Agonist—A drug that mimics naturally occurring chemicals that stimulate action in the brain and central nervous system. Examples of agonists include opiates and nicotine.

Agranulocytosis—A symptom complex characterized by (1) a sharply decreased number of granulocytes (one of the types of white blood cells), (2) lesions of the throat and other mucous membranes, (3) lesions of the gastrointestinal tract and (4) lesions of the skin. Sometimes also called granulocytopenia.

Alcoholism—A clinical syndrome that involves heavy alcohol consumption and continued drinking despite severe negative social and physical consequences.

Alkalizers—These drugs neutralize acidic properties of the blood and urine by making them more alkaline (or basic). Systemic alkalizers include potassium citrate and citric acid, sodium bicarbonate, sodium citrate and citric acid, and tricitrates. Urinary alkalizers include potassium citrate, potassium citrate and citric acid, potassium citrate and sodium citrate, sodium citrate and citric acid.

Alkylating Agent—Chemical used to treat malignant diseases.

Allergy—Excessive sensitivity to a substance that is ordinarily harmless. Reactions include sneezing, stuffy nose, hives, itching.

Alpha-Adrenergic Blocking Agents—A group of drugs used to treat hypertension. These drugs include alfuzosin, prazosin, terazosin, doxazosin and labetalol (an alpha-adrenergic and beta-adrenergic combination drug). Also included are other drugs that produce an alpha-adrenergic blocking action such as haloperidol, loxapine, phenothiazines, thioxanthenes.

Amebiasis—Infection with amoebas, one-celled organisms. Causes diarrhea, fever and abdominal cramps.

Amenorrhea—Abnormal absence of menstrual periods.

Aminoglycosides—A family of antibiotics used for serious infections. Their usefulness is limited because of their relative toxicity compared to some other antibiotics. These drugs include amikacin, gentamicin, kanamycin, neomycin, netilmicin, streptomycin, tobramycin.

Amphetamines—A family of drugs that stimulates the central nervous system, prescribed to treat attention-deficit disorders in children and also for narcolepsy. They are habit-forming, are controlled under U.S. law and are no longer prescribed as appetite suppressants. These drugs include amphetamine, dextroamphetamine, lisdexamfetamine, methamphetamine. They may be ingredients of several combination drugs.

ANA Titers—A test to evaluate the immune system and to detect antinuclear antibodies (ANAs), substances that appear in the blood of some patients with autoimmune disease.

Analgesics—Agents that reduce pain without reducing consciousness.

Anaphylaxis—Severe allergic response to a substance. Symptoms are wheezing, itching, hives, nasal congestion, intense burning of hands and feet, collapse, loss of consciousness and cardiac arrest. Symptoms appear within a few seconds or minutes after exposure. Anaphylaxis is a severe medical emergency. Without appropriate treatment, it can cause death. Instructions for home treatment for anaphylaxis are at the end of the book.

Androgens—Male hormones, including fluoxymesterone, methyltestosterone, testosterone, DHEA.

Anemia—Not enough healthy red blood cells in the bloodstream or too little hemoglobin in the red blood cells. Anemia is caused by an imbalance between blood loss and blood production.

Anemia, Aplastic—A form of anemia in which the bone marrow is unable to manufacture adequate numbers of blood cells of all types—red cells, white cells, and platelets.

Anemia, Hemolytic—Anemia caused by a shortened lifespan of red blood cells. The body can't manufacture new cells fast enough to replace old cells.

Anemia, Iron-Deficiency—Anemia caused when iron necessary to manufacture red blood cells is not available.

Anemia, Pernicious—Anemia caused by a vitamin B-12 deficiency. Symptoms include weakness, fatigue, numbness and tingling of the hands or feet and degeneration of the central nervous system.

Anemia, Sickle-Cell—Anemia caused by defective hemoglobin that deprives red blood cells of oxygen, making them sickle-shaped.

Anesthesias, General—Gases that are used in surgery to render patients unconscious and able to withstand the pain of surgical cutting and manipulation. They include alfentanil, amobarbital, butabarbital, butorphanol, chloral hydrate, enflurane, etomidate, fentanyl, halothane, hydroxyzine, isoflurane, ketamine, levorphanol, meperidine, methohexital, methoxyflurane, midazolam, morphine parenteral, nalbuphine, nitrous oxide, oxymorphone, pentazocine, pentobarbital, phenobarbital, promethazine, propiomazine, propofol, scopolamine, secobarbital, sufentanil, thiamylal, thiopental.

Anesthetics—Drugs that eliminate the sensation of pain.

Angina (Angina Pectoris)—Chest pain with a sensation of suffocation and impending death. Caused by a temporary reduction in the amount of oxygen to the heart muscle through diseased coronary arteries. The pain may also occur in the left shoulder, jaw or arm.

Angiotensin-Converting Enzyme (ACE) Inhibitors—A family of drugs used to treat hypertension and congestive heart failure. Inhibitors decrease the rate of conversion of angiotensin I into angiotensin II, which is the normal process for the angiotensin-converting enzyme. These drugs include benazepril, captopril, enalapril, fosinopril, lisinopril, moexipril, perindopril, quinapril, ramipril, trandolapril.

Antacids—A large family of drugs prescribed to treat hyperacidity, peptic ulcer, esophageal reflux and other conditions. These drugs include alumina and magnesia; alumina, magnesia and calcium carbonate; alumina, magnesia and simethicone; alumina and magnesium carbonate; alumina and magnesium trisilicate; alumina, magnesium trisilicate and sodium bicarbonate; aluminum carbonate; aluminum hydroxide; bismuth subsalicylate; calcium carbonate; calcium carbonate and magnesia; calcium carbonate, magnesia and simethicone; calcium and magnesium carbonates; calcium and magnesium carbonates and magnesium oxide; calcium carbonate and simethicone; dihydroxyaluminum aminoacetate; dihydroxyaluminum sodium carbonate; magaldrate; magaldrate and simethicone; magnesium carbonate and sodium bicarbonate; magnesium hydroxide; magnesium oxide; magnesium trisilicate, alumina and magnesia; simethicone, alumina, calcium carbonate and magnesia; simethicone, alumina, magnesium carbonate and magnesia; sodium bicarbonate.

Antacids, Calcium Carbonate—These antacids include calcium carbonate and magnesium, calcium carbonate and simethicone, calcium carbonate and magnesium carbonates.

Antacids, Magnesium-Containing—These antacids include magnesium carbonate, magnesium hydroxide, magnesium oxide and magnesium trisilicate. All these medicines are designed to treat excess stomach acidity. In addition to being an effective antacid, magnesium can sometimes cause unpleasant side effects and drug interactions. Look for the presence of magnesium in nonprescription drugs.

Antagonist—A substance that blocks the effects of another drug by binding with the receptor site for that drug in the brain.

Anthelmintics—A family of drugs used to treat intestinal parasites. Names of these drugs include niclosamide, piperazine, pyrantel, pyrvinium, quinacrine, mebendazole, metronidazole, oxamniquine, praziquantel, thiabendazole.

Antiacne Topical Preparations—See Acne Preparations.

Antiadrenals—Medicines or drugs that prevent the effects of the hormones produced by the adrenal glands.

Antianginals—A group of drugs used to treat angina pectoris (chest pain that comes and goes, caused by coronary artery disease). These drugs include acebutolol, amlodipine, amyl nitrite, atenolol, bepridil, carteolol, diltiazem, felodipine, isosorbide dinitrate, labetalol, metoprolol, nadolol, nicardipine, nifedipine, nitroglycerin, oxprenolol, penbutolol, pindolol, propranolol, sotalol, timolol, verapamil.

Antianxiety Drugs—A group of drugs prescribed to treat anxiety. These drugs include alprazolam, bromazepam, buspirone, chlordiazepoxide, chlorpromazine, clomipramine, clorazepate, diazepam, halazepam, hydroxyzine, imipramine, ketazolam, lorazepam, meprobamate, oxazepam, prazepam, prochlorperazine, thioridazine, trifluoperazine, venlafaxine.

Antiarrhythmics—A group of drugs used to treat heartbeat irregularities (arrhythmias).
These drugs include acebutolol, adenosine, amiodarone, atenolol, atropine, bretylium, deslanoside, digitalis, digitoxin, diltiazem, disopyramide, dofetilide, edrophonium, encainide, esmolol, flecainide, glycopyrrolate, hyoscyamine, lidocaine, methoxamine, metoprolol, mexiletine, nadolol, oxprenolol, phenytoin, propafenone, propranolol, quinidine, scopolamine, sotalol, timolol, verapamil.

Antiasthmatics—Medicines used to treat asthma, which may be tablets, liquids or aerosols (to be inhaled to get directly to the bronchial tubes rather than through the bloodstream). These medicines include adrenocorticoids, glucocorticoid; albuterol; aminophylline; astemizole; beclomethasone; bitolterol; budesonide; cetirizine; corticotropin; cromolyn; dexamethasone; dyphylline; ephedrine; epinephrine; ethylnorepinephrine; fenoterol; flunisolide; fluticasone; ipratropium, isoetharine; isoproterenol; isoproterenol and phenylephrine; loratadine; metaproterenol; oxtriphylline; oxtriphylline and guaifenesin; pirbuterol; racepinephrine; terbutaline; theophylline; theophylline and guaifenesin; triamcinolone.

Antibacterials (Antibiotics)—A group of drugs prescribed to treat infections. These drugs include, amikacin, amoxicillin, amoxicillin and clavulanate, ampicillin, azithromycin, azlocillin, aztreonam, bacampicillin, carbenicillin, cefaclor, cefadroxil, cefamandole, cefazolin, cefonicid, cefoperazone, ceforanide, cefotaxime, cefotetan, cefoxitin, ceftazidime, ceftibuten, ceftizoxime, ceftriaxone, cefuroxime, cephalexin, cephalothin, cephapirin, cephradine, chloramphenicol, cinoxacin, clarithromycin, clindamycin, cloxacillin, cyclacillin, cycloserine, demeclocycline, dicloxacillin, dirithromycin, doxycycline, erythromycin, erythromycin and sulfisoxazole, fidaxomicin, flucloxacillin, fusidic acid, gentamicin, imipenem and cilastatin, kanamycin, lincomycin, methacycline, methenamine, methicillin, metronidazole, mezlocillin, minocycline, moxalactam, nafcillin, nalidixic acid, netilmicin, nitrofurantoin, norfloxacin, oxacillin, oxytetracycline, penicillin G, penicillin V, piperacillin, pivampicillin, rifabutin, rifampin, rifaximin, spectinomycin, streptomycin, sulfacytine, sulfadiazine and trimethoprim, sulfamethoxazole, sulfamethoxazole and trimethoprim, sulfisoxazole, telithromycin, tetracycline, ticarcillin, ticarcillin and clavulanate, tinidazole, tobramycin, trimethoprim, vancomycin.

Antibiotics—Chemicals that inhibit the growth of or kill germs. See Antibacterials.

Anticholinergics—Drugs that work against acetylcholine, a chemical found in many locations in the body, including connections between nerve cells and connections between muscle and nerve cells. Anticholinergic drugs include aclidinium. amantadine, anisotropine, atropine, belladonna, benztropine, biperiden, clidinium, darifenacin, dicyclomine, glycopyrrolate, homatropine, hyoscyamine, ipratropium, isopropamide, mepenzolate, methantheline, methscopolamine, oxybutynin, pirenzepine, propantheline, scopolamine, solifenacin, tolterodine, trihexyphenidyl, trospium.

Anticoagulants—A family of drugs prescribed to slow the rate of blood clotting. These drugs include acenocoumarol, anisindione, apixaban, dabigatran, dicumarol, dihydroergotamine and heparin, heparin, rivaroxaban, vorapaxar, warfarin.

Anticonvulsants—A group of drugs prescribed to treat or prevent seizures (convulsions). These drugs include these families: barbiturates, carbonic anhydrase inhibitors, diones, hydantoins and succinimides. These are the names of the generic drugs in these families: acetazolamide, amobarbital, carbamazepine, carbonic anhydrase inhibitors, clobazam, clonazepam, clorazepate, diazepam, dichlorphenamide, divalproex, ethosuximide, ethotoin, felbamate, fosphenytoin, gabapentin, lacosamide, lamotrigine, levetiracetam lorazepam, mephenytoin, mephobarbital, metharbital, methsuximide, nitrazepam, oxcarbazepine, paraldehyde, phenacemide, paramethadione, pentobarbital, phenobarbital, phenytoin, primidone, secobarbital, tiagabine, topiramate, trimethadione, valproic acid, zonisamide.

Antidepressants—A group of medicines prescribed to treat mental depression. These drugs include amitriptyline, amoxapine, bupropion, citalopram, clomipramine, desipramine, doxepin, duloxetine, escitalopram, fluoxetine, fluvoxamine, imipramine, isocarboxazid, lithium, maprotiline, mirtazapine, moclobemide, nefazodone, nortriptyline, paroxetine, phenelzine, protriptyline, sertraline, tranylcypromine, trazodone, trimipramine, venlafaxine, vortioxetine.

Antidepressants, MAO (Monoamine Oxidase) Inhibitors—A special group of drugs prescribed for mental depression. These are not as popular as in years past because of a relatively high incidence of adverse effects. These drugs include isocarboxazid (Marplan), phenelzine (Nardil), tranylcypromine (Parnate).

Antidepressants, Tricyclic (TCAs)—A group of medicines with similar chemical structure and pharmacologic activity used to treat mental depression. These drugs include amitriptyline, amoxapine, clomipramine, desipramine, doxepin, imipramine, nortriptyline, protriptyline, trimipramine.

Antidiabetic Agents—A group of drugs used in the treatment of diabetes. These medicines all reduce blood sugar. These drugs include acarbose, acetohexamide, canagliflozin, chlorpropamide, dapaglifozin, gliclazide, glimepiride, glipizide, glyburide, insulin, insulin analogs, linagliptin, metformin, nateglinide, pioglitazone, repaglinide, rosiglitazone, saxagliptin, sitagliptin, tolazamide, tolbutamide, troglitazone.

Antidiarrheal Preparations—Medicines that treat diarrhea symptoms. Most do not cure the cause. Oral medicines include aluminum hydroxide; charcoal, activated; kaolin and pectin; loperamide; polycarbophil; psyllium hydrophilic mucilloid. Systemic medicines include carbohydrates; codeine; difenoxin and atropine; diphenoxylate and atropine; glucose and electrolytes; glycopyrrolate;

kaolin, pectin, belladonna alkaloids and opium; kaolin, pectin and paregoric; nitazoxanide, opium tincture; paregoric, rifaximin.

Antidyskinetics—A group of drugs used for treatment of Parkinsonism (paralysis agitans) and drug-induced extrapyramidal reactions (see elsewhere in Glossary). These drugs include amantadine, benztropine, biperiden, bromocriptine, carbidopa and levodopa, diphenhydramine, entacapone, ethopropazine, levodopa, levodopa and benserazide, procyclidine, rasagiline, selegiline, trihexyphenidyl.

Antiemetics—A group of drugs used to treat nausea and vomiting. These drugs include buclizine, cyclizine, chlorpromazine, dimenhydrinate, diphenhydramine, diphenidol, domperidone, dronabinol, haloperidol, hydroxyzine, meclizine, metoclopramide, nabilone, ondansetron, perphenazine, prochlorperazine, promethazine, scopolamine, thiethylperazine, triflupromazine, trimethobenzamide.

Antifibrinolytic Drugs—Drugs that are used to treat serious bleeding. These drugs include aminocaproic acid and tranexamic acid.

Antifungals—A group of drugs used to treat fungus infections. Those listed as systemic are taken orally or given by injection. Those listed as topical are applied directly to the skin and include liquids, powders, creams, ointments and liniments. Those listed as vaginal are used topically inside the vagina and sometimes on the vaginal lips. These drugs include: Systemic—amphotericin B, miconazole, fluconazole, flucytosine, griseofulvin, itraconazole, ketoconazole, potassium iodide, posaconazole, terbinafine. Topical—amphotericin B, carbol-fuchsin, ciclopirox; clioquinol, clotrimazole, econazole, haloprogin, ketoconazole, mafenide, miconazole, naftifine, nystatin, oxiconazole, salicylic acid, silver sulfadiazine, sulconazole, sulfur and coal, terbinafine, tioconazole, tolnaftate, undecylenic acid. Vaginal—butoconazole, clotrimazole, econazole, gentian violet, miconazole, nystatin, terconazole, tioconazole.

Antifungals, Azole—Drugs used to treat certain types of fungal infections. These drugs include fluconazole, itraconazole, ketoconazole, miconazole, posaconazole, sertaconazole, voriconazole.

Antiglaucoma Drugs—Medicines used to treat glaucoma. Those listed as systemic are taken orally or given by injection. Those listed as ophthalmic are for external use. These drugs include: Systemic—acetazolamide, dichlorphenamide, glycerin, mannitol, methazolamide, timolol, urea. Ophthalmic—apraclonidine, betaxolol, bimatoprost, brimonidine, brinzolamide, carbachol ophthalmic solution, carteolol, demecarium, dipivefrin, dorzolamide, echothiophate, epinephrine, epinephrine bitartrate, epinephryl borate, isoflurophate, isopropyl unoprostone, latanoprost, levobetaxolol, levobunolol, metipranolol, physostigmine, pilocarpine, tafluprost, timolol, travoprost, unoprostone.

Antigout Drugs—Drugs to treat the metabolic disease called gout. Gout causes recurrent attacks of joint pain caused by deposits of uric acid in the joints. Antigout drugs include allopurinol, carprofen, colchicine, febuxostat, fenoprofen, ibuprofen, indomethacin, ketoprofen, naproxen, phenylbutazone, piroxicam, probenecid, probenecid and colchicine, sulfinpyrazone, sulindac.

Antihistamines—A family of drugs used to treat allergic conditions, such as hay fever, allergic conjunctivitis, itching, sneezing, runny nose, motion sickness, dizziness, sedation, insomnia and others. These drugs include astemizole, azatadine, brompheniramine, carbinoxamine, cetirizine, chlorpheniramine, clemastine, cyproheptadine, desloratadine, dexchlorpheniramine, dimenhydrinate, diphenhydramine, diphenylpyraline, doxylamine, fexofenadine, hydroxyzine, loratadine, phenindamine, promethazine, pyrilamine, trimeprazine, tripelennamine, triprolidine.

Antihyperammonemias—Medications that decrease the amount of ammonia in the blood. The ones with this pharmacological property that are available in the United States are lactulose, sodium benzoate and sodium phenylacetate.

Antihyperlipidemics—A group of drugs used to treat hyperlipidemia (high levels of lipids in the blood). These include atorvastatin, cerivastatin, cholestyramine, clofibrate, colesevelam, colestipol, ezetimibe, fenofibrate, fluvastatin, gemfibrozil, lovastatin, niacin, pitavastatin, pravastatin, probucol, simvastatin, sodium dichloroacetate.

Antihypertensives—Drugs used to help lower high blood pressure. These medicines can be used singly or in combination with other drugs. They work best if accompanied by a low-salt, low-fat diet plus an active exercise program. These drugs include acebutolol, amiloride, amiloride and hydrochlorothiazide, amlodipine, atenolol, atenolol and chlorthalidone, benazepril, bendroflumethiazide, benzthiazide, betaxolol, bisoprolol, bumetanide, candesartan, captopril, captopril and hydrochlorothiazide, carteolol, carvedilol, chlorothiazide, chlorthalidone, cilazapril, clonidine, clonidine and chlorthalidone, cyclothiazide, debrisoquine, deserpidine, deserpidine and hydrochlorothiazide, deserpidine and methyclothiazide, diazoxide, diltiazem, doxazosin, enalapril, enalapril and hydrochlorothiazide, eplerenone, eprosartan ethacrynic acid, felodipine, fosinopril, furosemide, guanabenz, guanadrel, guanethidine, guanethidine and hydrochlorothiazide, guanfacine, hydralazine, hydralazine and hydrochlorothiazide, hydrochlorothiazide, hydroflumethiazide, indapamide, irbesartan isradipine, labetalol, labetalol and hydrochlorothiazide, lisinopril, lisinopril and hydrochlorothiazide, losartan, losartan and hydrochlorothiazide, mecamylamine, methyclothiazide, methyldopa, methyldopa and chlorothiazide, methyldopa and hydrochlorothiazide, metolazone, metoprolol, metoprolol and hydrochlorothiazide, minoxidil, moexipril, nadolol, nadolol and bendroflumethiazide, nicardipine, nifedipine, nisoldipine, nitroglycerin, nitroprusside, olmesartan, oxprenolol, penbutolol, perindopril, pindolol, pindolol and hydrochlorothiazide, polythiazide, prazosin, prazosin and polythiazide, propranolol, propranolol and hydrochlorothiazide, quinapril, quinethazone, ramipril, rauwolfia serpentina, rauwolfia serpentina and bendroflumethiazide, reserpine, reserpine and chlorothiazide, reserpine and chlorthalidone, reserpine and hydralazine, reserpine, hydralazine and hydrochlorothiazide, reserpine and hydrochlorothiazide, reserpine and hydroflumethiazide, reserpine and methyclothiazide, reserpine and polythiazide, reserpine and quinethazone, reserpine and trichlormethiazide, sotalol, spironolactone, spironolactone and hydrochlorothiazide, telmisartan, terazosin, timolol, timolol and hydrochlorothiazide, torsemide, trandolapril, triamterene, triamterene and hydrochlorothiazide, trichlormethiazide, trimethaphan, valsartan, verapamil.

Anti-Inflammatory Drugs, Nonsteroidal (NSAIDs)—A family of drugs not related to cortisone or other steroids that decrease inflammation wherever it occurs in the body. Used for treatment of pain, fever, arthritis, gout, menstrual cramps and vascular headaches. These drugs include aspirin; aspirin, alumina and magnesia tablets; buffered aspirin; bufexamac; celecoxib; choline salicylate; choline and magnesium salicylates; diclofenac; diflunisal; fenoprofen; flurbiprofen; ibuprofen; indomethacin; ketoprofen; ketorolac, magnesium salicylate; meclofenamate; meloxicam; naproxen; piroxicam; rofecoxib; salsalate; sodium salicylate; sulindac; tolmetin.

Anti-Inflammatory Drugs, Steroidal—A family of drugs with pharmacologic characteristics similar to those of cortisone and cortisone-like drugs. They are used for many purposes to help the body deal with inflammation no matter what the cause. Steroidal drugs may be taken orally or by injection (systemic) or applied locally (topical) for the skin, eyes, ears, bronchial tubes and others. These drugs include: Nasal—beclomethasone, budesonide, dexamethasone, flunisolide, fluticasone, triamcinolone. Ophthalmic (eyes)—betamethasone, dexamethasone, fluorometholone, hydrocortisone, medrysone, prednisolone, rimexolone. Otic (ears)—betamethasone, desonide and acetic acid, dexamethasone, hydrocortisone, hydrocortisone and acetic acid, prednisolone. Systemic—betamethasone, corticotropin, cortisone, dexamethasone, hydrocortisone, methylprednisolone, paramethasone, prednisolone, prednisone, triamcinolone. Topical—alclometasone; amcinonide; beclomethasone; betamethasone; clobetasol; clobetasone; clocortolone; desonide; desoximetasone; dexamethasone; diflorasone; diflucortolone; flumethasone; fluocinolone; fluocinonide; fluocinonide; flurandrenolide; fluticasone; halcinonide; halobetasol; mometasone; procinonide and ciprocinonide; flurandrenolide; halcinonide; hydrocortisone; methylprednisolone; mometasone; triamcinolone.

Antimalarials (also called Antiprotozoals)—A group of drugs used to treat malaria. The choice depends on the precise type of malaria organism and its developmental state. These drugs include amphotericin B, atovaquone and proguanil, clindamycin, chloroquine, dapsone, demeclocycline, doxycycline, halofantrine; hydroxychloroquine, iodoquinol, methacycline, mefloquine, metronidazole, minocycline, oxytetracycline, pentamidine, primaquine, proguanil, pyrimethamine, quinacrine, quinidine, quinine, sulfadoxine and pyrimethamine, sulfamethoxazole, sulfamethoxazole and trimethoprim, sulfisoxazole, tetracycline.

Antimuscarinics—Drugs that relax smooth muscle such as the detrusor muscle in the bladder. They can also decrease the secretion of saliva, sweat and digestive juice. Some dilate the eyes.

Antimyasthenics—Medicines to treat myasthenia gravis, a muscle disorder (especially of the face and head) with increasing fatigue and weakness as muscles tire from use. These medicines include neostigmine, pyridostigmine.

Antineoplastics—Potent drugs used for malignant disease. Some of these are not described in this book, but they are listed here for completeness. These drugs include: Systemic—amifostine, altretamine, aminoglutethimide, amsacrine, anastrazole, antithyroid agents, asparaginase, azathioprine, bicalutamide, bleomycin, busulfan, capecitabine, carboplatin, carmustine, chlorambucil, chloramphenicol, chlorotrianisene, chromic phosphate, cisplatin, colchicine, cyclophosphamide, cyclosporine, cyproterone, cytarabine, dacarbazine, dactinomycin, daunorubicin, deferoxamine, diethylstilbestrol, docetaxel, doxorubicin, dromostanolone, epirubicin, estradiol, estradiol valerate, estramustine, estrogens (conjugated and esterified), estrone, ethinyl estradiol, etoposide, exemestane, floxuridine, flucytosine, fluorouracil, flutamide, fluoxymesterone, gemcitabine, gold compounds, goserelin, hexamethylmelamine, hydroxyprogesterone, hydroxyurea, ifosfamide, interferon alfa-2a and alfa-2b (recombinant), ketoconazole, letrozole, leucovorin, leuprolide, levamisole, levothyroxine, liothyronine, liotrix, lithium, lomustine, masoprocol, mechlorethamine, medroxyprogesterone, megestrol, melphalan, methyltestosterone, mercaptopurine, methotrexate, mitomycin, mitotane, mitoxantrone, nandrolone, paclitaxel, phenpropionate, penicillamine, plicamycin, porfimer, procarbazine, raltitrexed, sodium iodide I 131, sodium phosphate P 32, streptozocin, tamoxifen, temoporfin, teniposide, testolactone, testosterone, thioguanine, thiotepa, thyroglobulin, thyroid, thyrotropin, topotecan, toremifene, trastuzumab, trimetrexate, triptorelin, uracil mustard, valrubicin, vinblastine, vincristine, vindesine, vinorelbine, zidovudine. Topical—fluorouracil, mechlorethamine.

Antiparkinsonism Drugs—Drugs used to treat Parkinson's disease. A disease of the central nervous system in older adults, it is characterized by gradual progressive muscle rigidity, tremors and clumsiness. These drugs include amantadine, benztropine, biperiden, bromocriptine, carbidopa and levodopa, diphenhydramine, entacapone, ethopropazine, levodopa, levodopa and benserazide, orphenadrine, pramipexole, procyclidine, rasagiline, ropinirole, rotigotine, selegiline, trihexyphenidyl.

Antiplatelet Drugs—Drugs used to stop the platelets in the blood from sticking to one another and forming blood clots. These drugs include aspirin, clopidogrel, dipyridamole, prasugrel, rivaroxaban, ticagrelor, ticlopidine.

Antipsychotic Drugs—A group of drugs used to treat the mental disease of psychosis, including such variants as schizophrenia, manic-depressive illness, anxiety states, severe behavior problems and others. These drugs include acetophenazine, aripiprazole, asenapine, carbamazepine, chlorpromazine, chlorprothixene, fluphenazine, flupenthixol, fluspirilene, haloperidol, iloperidone, loxapine, mesoridazine, methotrimeprazine, molindone, paliperidone, pericyazine, perphenazine, pimozide, pipotiazine, prochlorperazine, promazine, risperidone, thioproperazine, thioridazine, thiothixene, trifluoperazine, triflupromazine.

Antitussives—A group of drugs used to suppress coughs. These drugs include benzonatate, chlophedianol, codeine (oral), dextromethorphan, diphenhydramine syrup, hydrocodone, hydromorphone, methadone, morphine.

Antiulcer Drugs—A group of medicines used to treat peptic ulcer in the stomach, duodenum or the lower end of the esophagus. These drugs include amitriptyline, antacids, anticholinergics, antispasmodics, bismuth subsalicylate, cimetidine, doxepin, famotidine, lansoprazole, misoprostol, nizatidine, omeprazole, ranitidine, sucralfate, trimipramine.

Antiurolithics—Medicines that prevent the formation of kidney stones.

Antiviral Drugs—A group of drugs used to treat viral infections. These drugs include: Ophthalmic (eye)—idoxuridine, trifluridine, vidarabine. Systemic—acyclovir, amantadine, didanosine, famciclovir, foscarnet, ganciclovir, oseltamivir, ribavirin, rilpivirine, rimantadine, stavudine, zalcitabine, zanamivir, zidovudine. Topical drugs—acyclovir, docosanol.

Antivirals, HIV/AIDS—A group of drugs used to treat human immunodeficiency virus (HIV) and acquired immune deficiency syndrome (AIDS). They work by suppressing the replication of HIV. These drugs include abacavir, darunavir, delavirdine, didanosine, efavirenz, indinavir, lamivudine, maraviroc, nelfinavir, nevirapine, raltegravir, ritonavir, saquinavir, stavudine, tipranavir, zalcitabine, zidovudine.

Appendicitis—Inflammation or infection of the appendix. Symptoms include loss of appetite, nausea, low-grade fever and tenderness in the lower right of the abdomen.

Appetite Suppressants—A group of drugs used to decrease the appetite as part of an overall treatment for obesity. These drugs include amphetamine and dextroamphetamine, benzphetamine, diethylpropion, fenfluramine; mazindol, phendimetrazine, phentermine, phenylpropanolamine, sibutramine.

Artery—Blood vessel carrying blood away from the heart.

Asthma—Recurrent attacks of breathing difficulty due to spasms and contractions of the bronchial tubes.

Attenuated Virus Vaccines—Liquid products of killed germs used for injections to prevent certain diseases.

B

Bacteria—Microscopic organisms. Some bacteria contribute to health; others (germs) cause disease and infection.

Barbiturates—Powerful drugs used for sedation, to help induce sleep and sometimes to prevent seizures. Except for use in seizures (phenobarbital), barbiturates are being used less and less because there are better, less hazardous drugs that produce the same or better effects. These drugs include amobarbital, aprobarbital, butabarbital, mephobarbital, metharbital, pentobarbital, phenobarbital, secobarbital, secobarbital and amobarbital, talbutal.

Basal Area of Brain—Part of the brain that regulates muscle control and tone.

Benzethonium Chloride—A compound used as a preservative in some drug preparations. It is also used in various concentrations for cleaning cooking and eating utensils and as a disinfectant.

Benzodiazepines—A family of drugs prescribed to treat anxiety and alcohol withdrawal and sometimes prescribed for sedation. These drugs include alprazolam, bromazepam, chlordiazepoxide, clobazam, clonazepam, clorazepate, diazepam, estazolam, flurazepam, halazepam, ketazolam, lorazepam, nitrazepam, oxazepam, prazepam, quazepam, triazolam.

Beta Agonists—A group of drugs that act directly on cells in the body (beta-adrenergic receptors) to relieve spasms of the bronchial tubes and other organs consisting of smooth muscles. These drugs include albuterol, bitolterol, indacaterol, isoetharine, isoproterenol, metaproterenol, terbutaline.

Beta-Adrenergic Blocking Agents—A family of drugs with similar pharmacological actions with some variations. These drugs are prescribed for angina, heartbeat irregularities (arrhythmias), high blood pressure, hypertrophic subaortic stenosis, vascular headaches (as a preventative, not to treat once the pain begins) and others. Timolol is prescribed for treatment of open-angle glaucoma. These drugs include acebutolol, atenolol, betaxolol, bisoprolol, carteolol, labetalol, levobetaxolol, metoprolol, nadolol, oxprenolol, penbutolol, pindolol, propranolol, sotalol, timolol.

Bile Acids—Components of bile that are derived from cholesterol and formed in the liver. Bile acids aid the digestion of fat.

Biologic Response Modifiers—Substances that are produced naturally in the body or manufactured as drugs designed to strengthen, direct, or restore the body's immune response. The drugs are used in treating disease such as rheumatoid arthritis, certain cancers and some infections. These drugs include abatacept, adalimumab, anakinra, atumumab, canakinumab, certolizumab, etanercept, infliximab, golimumab, interferon, interleukin-2, natalizumab, rilonacept, rituximab, tocilizumab, tofacitinib, ustekinumab, various types of colony-stimulating factors.

Blood Count—Laboratory studies to count white blood cells, red blood cells, platelets and other elements of the blood.

Blood Dyscrasia-Causing Medicines—Drugs which cause unpredictable damaging effects to human bone marrow. These effects occur in a small minority of patients and are not dependent upon dosage. These medicines include the following (some of which are not described in this book): ACE inhibitors, acetazolamide, aminopyrine, amodiaquine, anticonvulsants (dione, hydantoin, succinimide), antidepressants (tricyclic), antidiabetic agents (sulfonylurea), anti-inflammatory analgesics, antithyroid agents, bexarotene, captopril, carbamazepine, cephalosporins, chloramphenicol, cisplatin, clopidogrel, clozapine, dapsone, divalproex, felbamate, flecainide acetate, foscarnet, gold compounds, levamisole, loxapine, maprotiline, methicillin, methimazole, methsuximide, metronidazole, mirtazapine, pantoprazole, penicillins (some), penicillamine, pentamidine, phenacemide, phenothiazines, phensuximide, phenytoin, pimozide, primaquine, primidone, propafenone, propylthiouracil, pyrimethamine (large doses), rabeprazole, rifampin, rifapentine, rituximab, sulfamethoxazole and trimethoprim, sulfasalazine, sulfonamides, thioxanthenes, ticlopidine, tiopronin, topiramate, trastuzumab, trimethobenzamide, trimethoprim, valproic acid.

Blood Pressure, Diastolic—Pressure (usually recorded in millimeters of mercury) in the large arteries of the body when the heart muscle is relaxed and filling for the next contraction.

Blood Pressure, Systolic—Pressure (usually recorded in millimeters of mercury) in the large arteries of the body at the instant the heart muscle contracts.

Blood Sugar (Blood Glucose)—Necessary element in the blood to sustain life.

Bone Marrow Depressants—Medicines that affect the bone marrow to depress its normal function of forming blood cells. These medicines include the following (some of which are not described in this book): abacavir, alcohol, aldesleukin, altretamine, amphotericin B (systemic), anticancer drugs, antithyroid drugs, azathioprine, bexarotene, busulfan, capecitabine, carboplatin, carmustine, chlorambucil, chloramphenicol, chromic phosphate, cisplatin, cladribine, clozapine, colchicine, cyclophosphamide, cyproterone, cytarabine, dacarbazine, dactinomycin, daunorubicin, didanosine, docetaxel, doxorubicin, eflornithine, epirubicin, etoposide, floxuridine, flucytosine, fludarabine, fluorouracil, ganciclovir, gemcitabine, hydroxyurea, idarubicin, ifosfamide, imatinib, interferon, irinotecan, lomustine, mechlorethamine, melphalan, mercaptopurine, methotrexate, mitomycin, mitoxantrone, paclitaxel, pentostatin, plicamycin, procarbazine, sirolimus, streptozocin, sulfa drugs, temozolomide, teniposide, thioguanine, thiotepa, topotecan, trimetrexate, uracil mustard, valrubicin, vidarabine (large doses), vinblastine, vincristine, vindesine, vinorelbine, zidovudine, zoledronic acid.

Bone Marrow Depression—Reduction of the blood-producing capacity of human bone marrow. Can be caused by many drugs taken for long periods of time in high doses.

Brain Depressants—Any drug that depresses brain function, such as tranquilizers, narcotics, alcohol and barbiturates.

Brand Name Drug—A brand name drug is a drug marketed under a proprietary, trademark-protected name.

Bronchodilators—A group of drugs used to dilate the bronchial tubes to treat such problems as asthma, emphysema, bronchitis, bronchiectasis, allergies and others. These drugs include albuterol, aminophylline, arformoterol, bitolterol, cromolyn, dyphylline, ephedrine, epinephrine, ethylnorepinephrine, fenoterol, formoterol, indacaterol, ipratropium, isoetharine, isoproterenol, levalbuterol metaproterenol, nedocromil, oxtriphylline, oxtriphylline and guaifenesin, pirbuterol, procaterol, salmeterol, terbutaline, theophylline and guaifenesin.

Bronchodilators, Xanthine-Derivative—Drugs of similar chemical structure and pharmacological activity that are prescribed to dilate bronchial tubes in disorders such as asthma, bronchitis, emphysema and other chronic lung diseases. These drugs include aminophylline, dyphylline, oxtriphylline, theophylline.

BUN—Abbreviation for blood urea nitrogen. A test often used as a measurement of kidney function.

C

Calcium Channel Blockers—A group of drugs used to treat angina and heartbeat irregularities. These drugs include bepridil, diltiazem, felodipine, flunarizine, isradipine, nicardipine, nifedipine, nimodipine, verapamil.

Calcium Supplements—Supplements used to increase the calcium concentration in the blood in an attempt to make bones denser (as in osteoporosis). These supplements include calcium citrate, calcium glubionate, calcium gluconate, calcium glycerophosphate and calcium lactate, calcium lactate, dibasic calcium phosphate, tribasic calcium phosphate.

Carbamates—A group of drugs derived from carbamic acid and used for anxiety or as sedatives. They include meprobamate and ethinamate.

Carbetapentane—An antitussive (cough suppressing) drug similar to dextromethorphan in action. It is an ingredient in some cough and cold remedies.

Carbonic Anhydrase Inhibitors—Drugs used to treat glaucoma and seizures and to prevent high altitude sickness. They include acetazolamide, brinzolamide, dichlorphenamide, dorzolamide, methazolamide.

Cataract—Loss of transparency in the lens of the eye.

Catecholamines—A group of drugs, also found naturally in the body, used to treat low blood pressure or shock. These drugs include dopamine, norepinephrine and epinephrine.

Cationic Drugs—Drugs removed from the body by the kidneys (called renal tubular secretion). If two of these drugs are taken together, one of them may stay in the body longer and increase its effect. These drugs include digoxin, metformin, morphine, pancuronium, tenofovir, trospium, vancomycin.

Cell—Unit of protoplasm, the essential living matter of all plants and animals.

Central Nervous System (CNS) Depressants—These drugs cause sedation or otherwise diminish brain activity and other parts of the nervous system. These drugs include alcohol, aminoglutethimide, anesthetics (general and injection-local), anticonvulsants, antidepressants (MAO inhibitors, tricyclic), antidyskinetics (except amantadine), antihistamines, apomorphine, azelastine, baclofen, barbiturates, benzodiazepines, beta-adrenergic blocking agents, brimonidine, buclizine, carbamazepine, cetirizine, chlophedianol, chloral hydrate, chlorzoxazone, clonidine, clozapine, cyclizine, cytarabine, difenoxin and atropine, diphenoxylate and atropine, disulfiram, donepezil, dronabinol, droperidol, ethchlorvynol, ethinamate, etomidate, fenfluramine, fluoxetine, glutethimide, guanabenz, guanfacine, haloperidol, hydroxyzine, ifosfamide, interferon, loxapine, magnesium sulfate (injection), maprotiline, meclizine, meprobamate, methyldopa, methyprylon, metoclopramide, metyrosine, mirtazapine, mitotane, molindone, nabilone, nefazodone, olanzapine, opioid (narcotic) analgesics, oxcarbazepine, oxybutynin, paliperidone, paraldehyde, paregoric, pargyline, paroxetine, phenothiazines, pimozide, procarbazine, promethazine, propiomazine, propofol, quetiapine, rauwolfia alkaloids, risperidone, scopolamine, sertraline, skeletal muscle relaxants (centrally acting), tapentadol, thalidomide, thioxanthenes, tramadol, trazodone, trimeprazine, trimethobenzamide, zaleplon, zolpidem, zonisamide, zopiclone.

Central Nervous System (CNS) Stimulants—Drugs that cause excitation, anxiety and nervousness or otherwise stimulate the brain and other parts of the central nervous system. These drugs include amantadine, amphetamines, anesthetics (local), appetite suppressants (except fenfluramine), bronchodilators (xanthine- derivative), bupropion, caffeine, chlophedianol, cocaine, dextroamphetamine, diclofenac, doxapram, dronabinol, dyphylline, entacapone, ephedrine (oral), fluoroquinolones, fluoxetine, meropenem, methamphetamine, methylphenidate, moclobemide, modafinil, nabilone, pemoline, rasagiline, selegiline, sertraline, sympathomimetics, topiramate, tranylcypromine, zonisamide.

Cephalosporins—Antibiotics that kill many bacterial germs that penicillin and sulfa drugs can't destroy.

Cholinergics (Parasympathomimetics)—Chemicals that facilitate passage of nerve impulses through the parasympathetic nervous system.

Cholinesterase Inhibitors—Drugs that prevent the action of cholinesterase (an enzyme that breaks down acetylcholine in the body).

Chronic—Long-term, continuing. Chronic illnesses may not be curable, but they can often be prevented from becoming worse. Symptoms usually can be alleviated or controlled.

Cirrhosis—Disease that scars and destroys liver tissue resulting in abnormal function.

Citrates—Medicines taken orally to make urine more acid. Citrates include potassium citrate, potassium citrate and citric acid, potassium citrate and sodium citrate, sodium citrate and acid, tricitrates.

Coal Tar Preparations—Creams, ointments and lotions used on the skin for various skin ailments.

Cobicistst—A drug that in itself does not treat any disorder. It is combined with certain other drugs and acts as a boosting agent to raise the level of the combined drug(s) in the body.

Colitis, Ulcerative—Chronic, recurring ulcers of the colon for unknown reasons.

Collagen—Support tissue of skin, tendon, bone, cartilage and connective tissue.

Colostomy—Surgical opening from the colon, the large intestine, to the outside of the body.

Coma—A sleeplike state from which a person cannot be aroused.

Compliance—The extent to which a person follows medical advice.

Congestive—Characterized by excess accumulation of fluid. In congestive heart failure, congestion occurs in the lungs, liver, kidneys and other parts of the body to cause shortness of breath, swelling of the ankles and feet, rapid heartbeat and other symptoms.

Constriction—Tightness or pressure.

Contraceptives, Hormonal—Any form of contraception (birth control) that contains hormones (such as ethinyl estradiol and others). These forms include oral, injectable, transdermal and implantable.

Contraceptives, Oral (Birth Control Pills)—A group of hormones used to prevent ovulation, therefore preventing pregnancy. These hormones include drospirenone and ethinyl estradiol, ethynodiol diacetate and ethinyl estradiol, ethynodiol diacetate and mestranol, etonogestrel, levonorgestrel and ethinyl estradiol, medroxyprogesterone, norethindrone tablets, norethindrone acetate and ethinyl estradiol, norethindrone and ethinyl estradiol, norethindrone and mestranol, norethynodrel and mestranol, norgestrel, norgestrel and ethinyl estradiol.

Contraceptives, Vaginal—Topical medications or devices applied inside the vagina to prevent pregnancy.

Convulsions—Violent, uncontrollable contractions of the voluntary muscles.

COPD (Chronic Obstructive Pulmonary Disease)—Lung conditions including emphysema and chronic bronchitis.

Corticosteroids (Adrenocorticosteroids)—Steroid hormones produced by the body's adrenal cortex or their synthetic equivalents.

Cortisone (Adrenocorticoids, Glucocorticoids) and Other Adrenal Steroids—Medicines that mimic the action of the steroid hormone cortisone, manufactured in the cortex of the adrenal gland. These drugs decrease the effects of inflammation within the body. They are available for injection, oral use, topical use for the skin, eyes and nose and inhalation for the bronchial tubes. These drugs include alclometasone; amcinonide; beclomethasone; benzyl benzoate; betamethasone; bismuth; ciclesonide; clobetasol; clobetasone 17-butyrate; clocortolone; cortisone; desonide; desoximetasone; desoxycorticosterone; dexamethasone; diflorasone; diflucortolone; fludrocortisone; flumethasone; flunisolide; fluocinonide; fluocinonide, procinonide and ciprocinonide; fluorometholone; fluprednisolone, flurandrenolide; halcinonide; hydrocortisone; medrysone; methylprednisolone; mometasone; paramethasone; Peruvian balsam; prednisolone; prednisone; triamcinolone; zinc oxide.

Cycloplegics—Eye drops that prevent the pupils from accommodating to varying degrees of light.

Cystitis—Inflammation of the urinary bladder.

D

Decongestants—Drugs used to relieve congestion by shrinking swollen membranes. These drugs include:
Cough-suppressing—phenylephrine and dextromethorphan, phenylpropanolamine (phenylpropanolamine products are being discontinued) and caramiphen, phenylpropanolamine and dextromethorphan, phenylpropanolamine and hydrocodone, pseudoephedrine and codeine, pseudoephedrine and dextromethorphan, pseudoephedrine and hydrocodone.
Cough-suppressing and pain-relieving—phenylpropanolamine, dextromethorphan and acetaminophen.
Cough-suppressing and sputum-thinning—phenylephrine, dextromethorphan and guaifenesin; phenylephrine, hydrocodone and guaifenesin; phenylpropanolamine, codeine and guaifenesin; phenylpropanolamine, dextromethorphan and guaifenesin; pseudoephedrine, codeine and guaifenesin; pseudoephedrine, dextromethorphan and guaifenesin; pseudoephedrine, hydrocodone and guaifenesin; phenylephrine, dextromethorphan, guaifenesin and acetaminophen; pseudoephedrine, dextromethorphan, guaifenesin and acetaminophen.
Sputum-thinning—ephedrine and guaifenesin; ephedrine and potassium iodide; phenylephrine, phenylpropanolamine and guaifenesin; phenylpropanolamine and guaifenesin; pseudoephedrine and guaifenesin.
Nasal—ephedrine (oral), phenylpropanolamine, pseudoephedrine.
Ophthalmic (eye)—naphazoline, oxymetazoline, phenylephrine.
Topical—oxymetazoline, phenylephrine, xylometazoline.

Delirium—Temporary mental disturbance characterized by hallucinations, agitation and incoherence.

Detoxification—A process in which the body rids itself of a drug (or its metabolites). During this period, withdrawal symptoms can emerge that may require medical treatment. This is often the first step in drug abuse treatment.

Diabetes—Metabolic disorder in which the body can't use carbohydrates efficiently. This leads to a dangerously high level of glucose (a carbohydrate) in the blood.

Dialysis—Procedure to filter waste products from the bloodstream of patients with kidney failure.

Digitalis Preparations (Digitalis Glycosides)—Important drugs to treat heart disease, such as congestive heart failure, heartbeat irregularities and cardiogenic shock. These drugs include digitoxin, digoxin.

Digoxin—One of the digitalis drugs used to treat heart disease. All digitalis products were originally derived from the foxglove plant.

Dilation—Enlargement.

Disulfiram Reaction—Disulfiram (Antabuse) is a drug to treat alcoholism. When alcohol in the bloodstream interacts with disulfiram, it causes a flushed face, severe headache, chest pains, shortness of breath, nausea, vomiting, sweating and weakness. Severe reactions may cause death. A disulfiram reaction is the interaction of any drug with alcohol or another drug to produce these symptoms.

Diuretics—Drugs that act on the kidneys to prevent reabsorption of electrolytes, especially chlorides. They are used to treat edema, high blood pressure, congestive heart failure, kidney and liver failure and others. These drugs include amiloride, amiloride and hydrochlorothiazide, bendroflumethiazide, benzthiazide, bumetanide, chlorothiazide, chlorthalidone, cyclothiazide, ethacrynic acid, furosemide, glycerin, hydrochlorothiazide, hydroflumethiazide, indapamide, mannitol, methyclothiazide, metolazone, polythiazide, quinethazone, spironolactone, spironolactone and hydrochlorothiazide, triamterene, triamterene and hydrochlorothiazide, trichlormethiazide, urea.

Diuretics, Loop—Drugs that act on the kidneys to prevent reabsorption of electrolytes, especially sodium. They are used to treat edema, high blood pressure, congestive heart failure, kidney and liver failure and others. These drugs include bumetanide, ethacrynic acid, furosemide.

Diuretics, Potassium-Sparing—Drugs that act on the kidneys to prevent reabsorption of electrolytes, especially sodium. They are used to treat edema, high blood pressure, congestive heart failure, kidney and liver failure and others. This particular group of diuretics does not allow the unwanted side effect of low potassium in the blood to occur. These drugs include amiloride, spironolactone, triamterene.

Diuretics, Thiazide—Drugs that act on the kidneys to prevent reabsorption of electrolytes, especially chlorides. They are used to treat edema, high blood pressure, congestive heart failure, kidney and liver failure and others. These drugs include bendroflumethiazide, benzthiazide, chlorothiazide, chlorthalidone, cyclothiazide, hydrochlorothiazide, hydroflumethiazide, methyclothiazide, metolazone, polythiazide, quinethazone, trichlormethiazide.

Dopamine Agonists—Drugs that stimulate activity of dopamine (a brain chemical that helps control movement). These include apomorphine, bromocriptine, cabergoline, pramipexole, quinagolide, ropinirole, rotigotine.

Dopamine Antagonists—Drugs that interfere with dopamine production (brain chemical that helps control movement). These drugs include haloperidol, metoclopramide, phenothiazines, procainamide, thioxanthenes and others.

Dosage Form—A dosage form is the physical form in which a drug is produced and dispensed, such as a tablet, a capsule, an injectable and others.

Duodenum—The first 12 inches of the small intestine.

E

ECG (or EKG)—Abbreviation for electrocardiogram or electrocardiograph. An ECG is a graphic tracing r epresenting the electrical current produced by impulses passing through the heart muscle. This is a useful test in the diagnosis of heart disease, but used alone it usually can't make a complete diagnosis. An ECG is most useful in two areas:
(1) demonstrating heart rhythm disturbances and (2) demonstrating changes when there is a myocardial infarction (heart attack). It will detect enlargement of either heart chamber, but will not establish a diagnosis of heart failure or disease of the heart valves.

Eczema—Disorder of the skin with redness, itching, blisters, weeping and abnormal pigmentation.

EEG—Electroencephalogram or electroencephalograph. An EEG is a graphic recording of electrical activity generated spontaneously from nerve cells in the brain. This test is useful in the diagnosis of brain dysfunction, particularly in studying seizure disorders.

Electrolytes—Substances that can transmit electrical impulses when dissolved in body fluids. These include sodium, potassium, chloride, bicarbonate, and carbon dioxide.

Embolism—Sudden blockage of an artery by a clot or foreign material in the blood.

Emphysema—An irreversible disease in which the lung's air sacs lose elasticity and air accumulates in the lungs.

Endometriosis—Condition in which uterus tissue is found outside the uterus. Can cause pain, abnormal menstruation and infertility.

Enzyme Inducers—Drugs that increase the metabolism of another drug in the liver, resulting in a decrease of that drug's effect. Ask your doctor or pharmacist about this possible interaction. Listed here are the more common used drugs in this category. These drugs include: alcohol (chronic use), barbiturates (especially phenobarbital), carbamazepine, darunavir, dexamethasone, efavirenz, glucocorticoids, glutethimide, griseofulvin, insulin, isoniazid, modafinil, nafcillin, nevirapine, norethindrone, omeprazole, oxcarbazepine, phenylbutazone, phenytoin, pioglitazone, prednisone, primidone, rifabutin, rifampin, rifapentine, saquinavir, secobarbital, St. John's wort, tipranavir, troglitazone. Also included are charbroiled meats, cruciferous vegetables (such as broccoli and cabbage) and smoking. There may be other drugs in this category. Consult doctor or pharmacist.

Enzyme Inducing Antiepileptic Drugs—Drugs used for seizure disorders that increase the metabolism of another drug in the liver, resulting in a decrease of that drug's effect. These drugs include carbamazepine, phenytoin, primidone and phenobarbital.

Enzyme Inhibitors—Drugs that decrease the metabolism of another drug in the liver, resulting in an increase of that drug's effect. Ask your doctor or pharmacist about this possible interaction. Listed here are the more common used drugs in this category. These drugs include: amiodarone, antipsychotics, aprepitant, asenapine, azole antifungals, bicalutamide, bupropion, celecoxib, chloramphenicol, chlorpheniramine, chlorpromazine, cimetidine, cinacalcet, ciprofloxacin, citalopram, clarithromycin, clemastine, clomipramine, cyclosporine, darunavir, delavirdine, diphenhydramine, diltiazem, disulfiram, doxepin, doxorubicin, duloxetine, enoxacin, erythromycins, escitalopram, fenofibrate, felbamate, fluoroquinolones, fluoxetine, fluvoxamine, gemfibrozil, glitazones, halofantrine, histamine H2 receptor antagonists, HIV antivirals (some), hydroxyzine, imatinib, indomethacin, isoniazid, itraconazole, ketoconazole, lansoprazole, levomepromazine, lorcaserin, lovastatin, methadone, methoxsalen, metoclopramide, metronidazole, mibefradil, midodrine, mifepristone, mirabegron, moclobemide, modafinil, montelukast, nefazodone, nelfinavir, norfloxacin, norfluoxetine, omeprazole, oxcarbazepine, pantoprazole, paroxetine, perphenazine, phenothiazines, phenylbutazone, probenecid, protease inhibitors, quercetin, quinidine, rabeprazole, ranitidine, ranolazine, ritonavir, saquinavir, selective serotonin reuptake inhibitors (SSRIs), sertraline, sulfamethoxazole, sulfaphenazole, telithromycin, teniposide, terbinafine, thiotepa, ticagrelor, ticlopidine, topiramate, trimethoprim, tripelennamine, valproic acid, venlafaxine, verapamil, voriconazole, zafirlukast; and grapefruit juice. There may be other drugs in this category. Consult doctor or pharmacist.

Enzymes—Protein chemicals that can accelerate chemical reactions in the body.

Epilepsy—Episodes of brain disturbance that cause convulsions and loss of consciousness.

Erectile Dysfunction Agents—Medicines used to treat male impotence (the inability to develop and sustain an erection). These drugs include: Alprostadil, papaverine, sildenafil citrate, tadalafil, vardenafil, yohimbine.

Ergot Preparations (Alkaloids)—Medicines used to treat migraine and other types of throbbing headaches. Also used after delivery of babies to make the uterus clamp down and reduce excessive bleeding. These drugs include dihydroergotamine, ergoloid mesylates, ergotamine.

Erythromycins—A group of drugs with similar structure used to treat infections. These drugs include erythromycin, erythromycin estolate, erythromycin ethylsuccinate, erythromycin gluceptate, erythromycin lactobionate, erythromycin stearate.

Esophagitis—Inflammation of the lower part of the esophagus, the tube connecting the throat and the stomach.

Estrogens—Female hormones used to replenish the body's stores after the ovaries have been removed or become nonfunctional after menopause. Also used with progesterone in some birth control pills and for other purposes. These drugs include:
Systemic—chlorotrianisene, diethylstilbestrol, estradiol, estrogens (conjugated and esterified), estrone, estropipate, ethinyl estradiol, quinestrol.
Vaginal—dienestrol, estradiol, estrogens (conjugated), estrone, estropipate.

Estrogen agonists/antagonists—Drugs that 1) activate (agonism) estrogenic receptors in certain body tissues to produce an estrogen effect, or 2) block (antagonism) estrogenic effects in others. Drugs in this group include ospemifene, raloxifene, tamoxifen and toremifene.

Eustachian Tube—Small passage from the middle ear to the sinuses and nasal passages.

Extrapyramidal Reactions—Abnormal reactions in the power and coordination of posture and muscular movements. Movements are not under voluntary control. Some drugs associated with producing extrapyramidal reactions include amoxapine, antidepressants (tricyclic), droperidol, haloperidol, loxapine, metoclopramide, metyrosine, moclobemide, molindone, olanzapine, paliperidone, paroxetine, phenothiazines, pimozide, rauwolfia alkaloids, risperidone, tacrine, thioxanthenes.

Extremity—Arm, leg, hand or foot.

F

Fecal Impaction—Condition in which feces become firmly wedged in the rectum.

Fibrocystic Breast Disease—Overgrowth of fibrous tissue in the breast, producing non-malignant cysts.

Fibroid Tumors—Non-malignant tumors of the muscular layer of the uterus.

Flu (Influenza)—A virus infection of the respiratory tract that lasts three to ten days. Symptoms include headache, fever, runny nose, cough, tiredness and muscle aches.

Fluoroquinolones—A class of drugs used to treat bacterial infections, such as urinary tract infections and some types of bronchitis. These drugs include ciprofloxacin, enoxacin, gatifloxacin, gemifloxacin, levofloxacin, lomefloxacin, moxifloxacin, norfloxacin, ofloxacin, sparfloxacin.

Folate Antagonists—Drugs that impair the body's utilization of folic acid, which is necessary for cell growth. These drugs include dione anticonvulsants, hydantoin anticonvulsants, succinimide anticonvulsants, divalproex, methotrexate, oral contraceptives, phenobarbital (long-term use), pyrimethamine, sulfonamides, triamterene, trimethoprim, trimetrexate, valproic acid.

Folliculitis—Inflammation of a follicle.

Functional Dependence—The development of dependence on a drug for a normal body function. The primary example is the use of laxatives for a prolonged period so that there is a dependence on the laxative for normal bowel action.

G

G6PD—Deficiency of glucose 6-phosphate, which is necessary for glucose metabolism.

Ganglionic Blockers—Medicines that block the passage of nerve impulses through a part of the nerve cell called a ganglion. Ganglionic blockers are used to treat urinary retention and other medical problems. Bethanechol is one of the best ganglionic blockers.

Gastritis—Inflammation of the stomach.

Gastrointestinal—Of the stomach and intestinal tract.

Generic Drug—A generic drug is the same as a brand name drug in dosage, safety, strength, how it is taken, quality, performance, and intended use.

Gland—Organ or group of cells that manufactures and excretes materials not required for its own metabolic needs.

Glaucoma—Eye disease in which increased pressure inside the eye damages the optic nerve, causes pain and changes vision.

Glucagon—Injectable drug that immediately elevates blood sugar by mobilizing glycogen from the liver.

Gold Compounds—Medicines which use gold as their base and are usually used to treat joint or arthritic disorders. These medicines include auranofin, aurothioglucose, gold sodium thiomalate.

H

H_2 Antagonists—Antihistamines that work against H_2 histamine. H_2 histamine may be liberated at any point in the body, but most often in the gastrointestinal tract.

Hangover Effect—The same feelings as a "hangover" after too much alcohol consumption. Symptoms include headache, irritability and nausea.

Hematocrit—A blood test that measure how much space in blood is occupied by red blood cells.

Hemochromatosis—Disorder of iron metabolism in which excessive iron is deposited in and damages body tissues, particularly of the liver and pancreas.

Hemoglobin—Pigment in blood that carries oxygen in red blood cells.

Hemolytics—Drugs that can destroy red blood cells and separate hemoglobin from the blood cells. These include acetohydroxamic acid, antidiabetic agents (sulfonylurea), doxapram, furazolidone, mefenamic acid, menadiol, methyldopa, nitrofurans, primaquine, quinidine, quinine, sulfonamides (systemic), sulfones, vitamin K.

Hemorrhage—Heavy bleeding.

Hemorheologic Agents—Medicines to help control bleeding.

Hemosiderosis—Increase of iron deposits in body tissues without tissue damage.

Hepatitis—Inflammation of liver cells, usually accompanied by jaundice.

Hepatotoxics—Medications that can possibly cause toxicity or decreased normal function of the liver. These drugs include the following (some of which are not described in this book): acetaminophen (with long-term use), abacavir, acitretin, alcohol, amiodarone, anabolic steroids, androgens, angiotensin-converting enzyme (ACE) inhibitors, acitretin, anti-inflammatory drugs, nonsteroidal (NSAIDs), antithyroid agents, asparaginase, azlocillin, bexarotene, carbamazepine, carmustine, clindamycin, clofibrate, colestipol, cox 2 inhibitors, cyproterone, cytarabine, danazol, dantrolene, dapsone, daunorubicin, disulfiram, divalproex, dofetilide, epirubicin, erythromycins, estrogens, ethionamide, etretinate, felbamate, fenofibrate, fluconazole, flutamide, gold compounds, halothane, HMG-CoA reductase inhibitors, imatinib, iron (overdose) isoniazid, itraconazole, ketoconazole (oral), labetalol, mercaptopurine, methimazole, methotrexate, methyldopa, metronidazole, naltrexone, nevirapine, niacin (high doses), nilutamide, nitrofurans, pemoline, phenothiazines, phenytoin, piperacillin, plicamycin, posaconazole, pravastatin, probucol, rifampin, rosiglitazone, sulfamethoxazole and trimethoprim, sulfonamides, tacrine, tenofovir, testosterone, tizanidine, tolcapone, toremifene, tretinoin, troglitazone, valproic acid, zidovudine, zidovudine and lamivudine.

Hiatal Hernia—Section of the stomach that protrudes into the chest cavity.

Histamine—Chemical in body tissues that dilates the smallest blood vessels, constricts the smooth muscle surrounding the bronchial tubes and stimulates stomach secretions.

History—Past medical events in a patient's life.

Hives—Elevated patches on the skin that are redder or paler than surrounding skin and often itch severely.

HMG-CoA Reductase Inhibitors—A group of prescription drugs used to lower cholesterol. These include Atorvastatin, fluvastatin, lovastatin, pravastatin, pravastatin & aspirin, and simvastatin.

Hoarseness—Husky, gruff, weak voice.

Hormone Replacement Therapy—A medication (estrogen) or combination of medications (estrogen and progestin or estrogen and androgen) used for treatment of premenopausal and menopausal symptoms and for prevention of diseases that affect women in their later years.

Hormones—Chemical substances produced in the body to regulate other body functions.

Hypercalcemia—Too much calcium in the blood. This happens with some malignancies and in calcium overdose.

Hyperglycemia-Causing Medications—A group of drugs that may contribute to hyperglycemia (high blood sugar). These include oral estrogen-containing contraceptives, corticosteroids, estrogens, isoniazid, nicotinic acid, phenothiazines, phenytoin, sympathomimetics, thyroid hormones, thiazide diuretics.

Hyperkalemia-Causing Medications—Medicines that cause too much potassium in the bloodstream. These include ACE inhibitors; amiloride, anti-inflammatory drugs, nonsteroidal (NSAIDs); cyclosporine; digitalis glycosides; diuretics (potassium-sparing); pentamidine; spironolactone; succinylcholine chloride; tacrolimus; triamterene; trimethoprim; possibly any medicine that is combined with potassium.

Hypersensitivity—Serious reactions to many medications. The effects of hypersensitivity may be characterized by wheezing, shortness of breath, rapid heart rate, severe itching, faintness, unconsciousness and severe drop in blood pressure.

Hypertension—High blood pressure.

Hypervitaminosis—A condition due to an excess of one or more vitamins. Symptoms may include weakness, fatigue, loss of hair and changes in the skin.

Hypnotics—Drugs used to induce a sleeping state. See Barbiturates.

Hypocalcemia—Abnormally low level of calcium in the blood.

Hypoglycemia—Low blood sugar (blood glucose). A critically low blood sugar level will interfere with normal brain function and can damage the brain permanently.

Hypoglycemia-Causing Medications—A group of drugs that may contribute to hypoglycemia (low blood sugar). These include clofibrate, monoamine oxidase (MAO) inhibitors, probenecid, propranolol, rifabutin, rifampicin, salicylates, sulfonamides (long-acting), sulfonylureas.

Hypoglycemics—Drugs that reduce blood sugar. These include acetohexamide, chlorpropamide, gliclazide, glipizide, glyburide, insulin, metformin, tolazamide, tolbutamide.

Hypokalemia-Causing Medications—Medicines that cause a depletion of potassium in the bloodstream. These include adrenocorticoids (systemic), alcohol, amphotericin B (systemic), bronchodilators (adrenergic), capreomycin, carbonic anhydrase inhibitors, cisplatin, diuretics (loop and thiazide), edetate (long-term use), foscarnet, ifosfamide, indapamide, insulin, insulin lispro, laxatives (if dependent on), penicillins (some), salicylates, sirolimus, sodium bicarbonate, urea, vitamin D (overdose of).

Hypomagnesemia-Causing Drugs—Drugs that may increase the loss of magnesium in urine. The loss can lead to low blood levels of magnesium (hypomagnesemia). These drugs include busulfan, cyclosporine, digoxin, foscarnet, lenalidomide, loop diuretics, mycophenolate, nilotinib, proton-pump inhibitors, tacrolimus, thiazide diuretics, valganciclovir, voriconazole and others.

Hypotension—Blood pressure decrease below normal. Symptoms may include weakness, lightheadedness and dizziness.

Hypotension-Causing Drugs—Medications that might cause hypotension (low blood pressure). These include alcohol, alpha adrenergic blocking agents, alprostadil, amantadine, anesthetics (general), angiotensin-converting enzyme inhibitors (ACE inhibitors), angiotensin II receptor antagonists, antidepressants (MAO inhibitors, tricyclic), antihypertensives, benzodiazepines used as preanesthetics, beta-adrenergic blocking agents, brimonidine, bromocriptine, cabergoline, calcium channel-blocking agents, carbidopa and levodopa, clonidine, clozapine, dipyridamole and aspirin, diuretics, docetaxel, droperidol, edetate calcium disodium, edetate disodium, haloperidol, hydralazine, levodopa, lidocaine (systemic), loxapine, magnesium sulfate, maprotiline, mirtazapine, molindone, nabilone (high doses), nefazodone, nitrates, olanzapine, opioid analgesics (including fentanyl, fentanyl and sufentanil), oxcarbazepine, paclitaxel, paliperidone, pentamidine, pentoxifylline, phenothiazines, pimozide, pramipexole, propofol, quetiapine, quinidine, radiopaques (materials used in x-ray studies), ranitidine risperidone, rituximab, ropinirole, sildenafil, thioxanthenes, tizanidine, tolcapone, trazodone, vancomycin, venlafaxine. If you take any of these medications, be sure to tell a dentist, anesthesiologist or anyone else who intends to give you an anesthetic to put you to sleep.

Hypothermia-Causing Medications—Medicines that can cause a significant lowering of body temperature. These drugs include alcohol, alpha-adrenergic blocking agents (dihydroergotamine, ergotamine, labetalol, phenoxybenzamine, phentolamine, prazosin, tolazoline), barbiturates (large amounts), beta-adrenergic blocking agents, clonidine, insulin, minoxidil, narcotic analgesics (with overdose), phenothiazines, vasodilators.

I

Ichthyosis—Skin disorder with dryness, scaling and roughness.

Ileitis—Inflammation of the ileum, the last section of the small intestine.

Ileostomy—Surgical opening from the ileum, the end of the small intestine, to the outside of the body.

Immunosuppressants—Powerful drugs that suppress the immune system. Immuno-suppressants are used in patients who have had organ transplants or severe disease associated with the immune system. These drugs include the following (some of which are not described in this book): azathioprine, basiliximab, betamethasone, chlorambucil, corticotropin, cortisone, cyclophosphamide, cyclosporine, daclixímab, dexamethasone, hydrocortisone, mercaptopurine, methylprednisolone, muromonab, muromonab-CD3, mycophenolate, prednisolone, prednisone, sirolimus, tacrolimus, thalidomide, triamcinolone, ursodiol.

Impotence—Male's inability to achieve or sustain erection of the penis for sexual intercourse.

Insomnia—Sleeplessness.

Interaction—Change in the body's response to one drug when another is taken. Interaction may decrease the effect of one or both drugs, increase the effect of one or both drugs or cause toxicity.

Iron Supplements—Products that contain iron in a form that can be absorbed from the intestinal tract. Supplements include ferrous fumarate, ferrous gluconate, ferrous sulfate, iron dextran, iron-polysaccharide.

J

Jaundice—Symptoms of liver damage, bile obstruction or destruction of red blood cells. Symptoms include yellowed whites of the eyes, yellow skin, dark urine and light stool.

K

Keratosis—Growth that is an accumulation of cells from the outer skin layers.

Kidney Stones—Small, solid stones made from calcium, cholesterol, cysteine and other body chemicals.

L

Label—The approved label is the official description of a drug product which includes indication (what the drug is used for); who should take it; adverse events (side effects); instructions for uses in pregnancy, children, and other populations; and safety information for the patient. Labels are often found inside drug product packaging.

Laxatives—Medicines prescribed to treat constipation. These medicines include bisacodyl; bisacodyl and docusate; casanthranol; casanthranol and docusate; cascara sagrada; cascara sagrada and aloe; cascara sagrada and phenolphthalein; castor oil; danthron; danthron and docusate; danthron and poloxamer 188; dehydrocholic acid; dehydrocholic acid and docusate; docusate; docusate and phenolphthalein; docusate and mineral oil; docusate and phenolphthalein; docusate, carboxymethylcellulose and casanthranol; glycerin; lactulose; magnesium citrate; magnesium hydroxide; magnesium hydroxide and mineral oil; magnesium oxide; magnesium sulfate; malt soup extract; malt soup extract and psyllium; methylcellulose; mineral oil; mineral oil and cascara sagrada; mineral oil and phenolphthalein; mineral oil, glycerin and phenolphthalein; phenolphthalein; poloxamer; polycarbophil; potassium bitartrate and sodium bicarbonate; psyllium; psyllium and senna; psyllium hydrophilic mucilloid; psyllium hydrophilic mucilloid and carboxymethyl-cellulose; psyllium hydrophilic mucilloid and sennosides; psyllium hydrophilic mucilloid and senna; senna; senna and docusate; sennosides; sodium phosphate.

LDH—Abbreviation for lactate dehydrogenase. It is a measurement of cardiac enzymes used to confirm some heart conditions.

Lincomycins—A family of antibiotics used to treat certain infections.

Low-Purine Diet—A diet that avoids high-purine foods, such as liver, sweetbreads, kidneys, sardines, oysters and others. If you need a low-purine diet, request instructions from your doctor.

Lupus—Serious disorder of connective tissue that primarily affects women. Varies in severity with skin eruptions, joint inflammation, low white blood cell count and damage to internal organs, especially the kidneys.

Lymph Glands—Glands in the lymph vessels throughout the body that trap foreign and infectious matter and protect the bloodstream from infection.

M

Macrolides—A class of antibiotic (antibacterial) drugs. They include azithromycin, clarithromycin, dirithromycin, erythromycin, fidaxomicin, and telithromycin.

Male Hormones—Chemical substances secreted by the testicles, ovaries and adrenal glands in humans. Some male hormones used by humans are derived synthetically. Male hormones include testosterone cypionate and estradiol cypionate, testosterone enanthate and estradiol valerate.

Mania—A mood disturbance characterized by euphoria, agitation, elation, irritability, rapid and confused speech and excessive activity. Mania usually occurs as part of bipolar (manic-depressive) disorder.

Manic-Depressive Illness—Psychosis with alternating cycles of excessive enthusiasm and depression.

MAO Inhibitors—See Monoamine Oxidase (MAO) Inhibitors.

Mast Cell—Connective tissue cell.

Medication Guide—A medication guide contains information for patients on how to safely use a drug product.

Meglitinides—Drugs that stimulate the pancreas to produce insulin. Used to treat Type II (non-insulin dependent) diabetes. These drugs include repaglinide and nateglinide.

Menopause—The end of menstruation in the female, often accompanied by irritability, hot flashes, changes in the skin and bones and vaginal dryness.

Metabolism—Process of using nutrients and energy to build and break down wastes.

Migraine Headaches—Periodic headaches caused by constriction of arteries to the skull. Symptoms include severe pain, vision disturbances, nausea, vomiting and sensitivity to light.

Mind-Altering Drugs—Any drugs that decrease alertness, perception, concentration, contact with reality or muscular coordination.

Mineral Supplements—Mineral substances added to the diet to treat or prevent mineral deficiencies. They include iron, copper, magnesium, calcium, etc.

Monoamine Oxidase (MAO) Inhibitors—Drugs that prevent the activity of the enzyme monoamine oxidase (MAO) in brain tissue. MAO inhibitors include drugs that treat depression, Parkinson's and other conditions. MAOs can cause dangerous interactions with certain foods, beverages and other drugs. Drugs in this class include furazolidone, isocarboxazid, linezolid, methylene blue, phenelzine, procarbazine, rasagiline, selegiline, St. John's wort (acts similar to MAO inhibitors), tranylcypromine.

Muscle Blockers—Same as muscle relaxants or skeletal muscle relaxants.

Muscle Relaxants—Medicines used to lessen painful contractions and spasms of muscles. These include atracurium, carisoprodol, chlorphenesin, chlorzoxazone, cyclobenzaprine, metaxalone, methocarbamol, metocurine, orphenadrine citrate, orphenadrine hydrochloride, pancuronium, phenytoin, succinylcholine, tubocurarine, vecuronium.

Myasthenia Gravis—Disease of the muscles characterized by fatigue and progressive paralysis. It is usually confined to muscles of the face, lips, tongue and neck.

Mydriatics—Eye drops that cause the pupils to dilate (become larger) to a marked degree.

N

Narcotics—A group of habit-forming, addicting drugs used for treatment of pain, diarrhea, cough, acute pulmonary edema and others. Most are derived from opium, a milky exudate in capsules of papaver somniferum. Law requires licensed physicians to dispense by prescription. These drugs include alfentanil, buprenorphine, butorphanol, codeine, fentanyl, hydrocodone, hydromorphone, levorphanol, meperidine, methadone, morphine, nalbuphine, opium, oxycodone, oxymorphone, paregoric, pentazocine, propoxyphene, sufentanil.

Nephrotoxics—Under some circumstances, these medicines can be toxic to the kidneys. These medicines include acetaminophen (in high doses); acyclovir (injection of); aminoglycosides; amphotericin B (given internally); analgesic combinations containing acetaminophen and aspirin or other salicylates (with chronic high-dose use); anti-inflammatory analgesics (nonsteroidal); azapropazan; bacitracin (injection of); bezafibrate; capreomycin; carmustine; chlorpropamide; cidofovir; cimetidine; ciprofloxacin; cisplatin; colchicine; cox 2 inhibitors; cyclosporine; deferoxamine (long-term use); diclofenac; edetate calcium disodium (with high doses); edetate disodium (with high dose); fenofibrate; foscarnet; gentamycin; gold compounds; ifosfamide; imipenem; ketoconazole; lithium; mephalan; methicillin, methotrexate (with high dose therapy); methoxyflurane; nafcillin; neomycin (oral); pamidronate; penicillamine; pentamidine; pentostatin, phenacetin; plicamycin; polymyxins (injection of); radiopaques (materials used for special x-ray examinations); rifampin; streptozocin; sulfonamides; sulindac; tacrolimus; tetracyclines (except doxycycline and minocycline); tiopronin; tobramycin; tretinoin; trimethoprim; vancomycin (injection of).

Neuroleptic Malignant Syndrome—Ceaseless involuntary, jerky movements of the tongue, facial muscles and hands.

Neuromuscular Blocking Agents—A group of drugs prescribed to relax skeletal muscles. They are all given by injection, and descriptions are not included in this book. These drugs include atracurium, edrophonium, gallamine, neostigmine, metocurine, pancuronium, pyridostigmine, succinylcholine, tubocurarine, vecuronium.

Neurotoxic Medications—Medicines that cause toxicity to the nerve tissues in the body. These drugs include alcohol (chronic use), allopurinol, altretamine, amantadine; amiodarone; anticonvulsants (hydantoin), capreomycin, carbamazepine, carboplatin, chloramphenicol (oral), chloroquine, cilastatin, ciprofloxacin, cisplatin, cycloserine, cyclosporine, cytarabine, didanosine, disulfiram, docetaxel, ethambutol, ethionamide, fludarabine, hydroxychloroquine, imipenem, interferon, isoniazid, lincomycins, lindane (topical), lithium, meperidine, methotrexate, metronidazole, mexiletine, nitrofurantoin, oxcarbazepine, paclitaxel, pemoline, pentostatin, pyridoxine (large amounts), quinacrine, quinidine, quinine, stavudine, tacrolimus, tetracyclines, thalidomide, vinblastine, vincristine, vindesine, zalcitabine.

Nitrates—Medicines made from a chemical with a nitrogen base. Nitrates include erythrityl tetranitrate, isosorbide dinitrate, nitroglycerin, pentaerythritol tetranitrate.

Nonsteroidal Anti-Inflammatory Drugs (NSAIDs)—See Anti-Inflammatory Drugs, Nonsteroidal.

Nutritional Supplements—Substances used to treat and prevent deficiencies when the body is unable to absorb them by eating a well-balanced, nutritional diet. These supplements include: Vitamins—ascorbic acid, ascorbic acid and sodium ascorbate, calcifediol, calcitriol, calcium pantothenate, cyanocobalamin, dihydrotachysterol, ergocalciferol, folate sodium, folic acid, hydroxocobalamin, niacin, niacinamide, pantothenic, pyridoxine, riboflavin, sodium ascorbate, thiamine, vitamin A, vitamin E. Minerals—calcium carbonate, calcium citrate, calcium glubionate, calcium gluconate, calcium lactate, calcium phosphate (dibasic and tribasic), sodium fluoride. Other—levocarnitine, omega-3 polyunsaturated fatty acids.

O

Opiates—See Narcotics.

Orthostatic Hypotension—Excess drop in blood pressure when arising from a sitting or lying position.

Osteoporosis—Softening of bones caused by a loss of calcium usually found in bone. Bones become brittle and fracture easily.

Ototoxic Medications—These medicines may possibly cause hearing damage. They include aminoglycosides, 4-aminoquinolines, anti- inflammatory analgesics (nonsteroidal), bumetanide (injected), capreomycin, carboplatin, chloroquine, cisplatin, deferoxamine, erythromycins, ethacrynic acid, furosemide, hydroxychloroquine, quinidine, quinine, salicylates, vancomycin (injected).

Ovary—Female sexual gland where eggs mature and ripen for fertilization.

Over-the-Counter Drugs (OTC)—Drugs that are considered safe and effective for use by the general public without a doctor's prescription. Same as nonprescription drugs.

P

Pain Relievers—Non-narcotic medicines used to treat pain.

Palpitations—Rapid, forceful or throbbing heartbeat noticeable to the patient.

Pancreatitis—Serious inflammation or infection of the pancreas that causes upper abdominal pain.

Pancreatitis-associated Drugs—Medications associated with the development of pancreatitis. These include alcohol, asparaginase, azathioprine, didanosine, estrogens, furosemide, methyldopa, nitrofurantoin, sulfonamides, tetracyclines, thiazide diuretics, valproic acid.

Parkinson's Disease or Parkinson's Syndrome—Disease of the central nervous system. Characteristics are a fixed, emotionless expression of the face, tremor, slower muscle movements, weakness, changed gait and a peculiar posture.

Pellagra—Disease caused by a deficiency of the water-soluble vitamin thiamine (vitamin B-1). Symptoms include brain disturbance, diarrhea and skin inflammation.

Penicillin—Chemical substance (antibiotic) originally discovered as a product of mold, that can kill some bacterial germs.

Peripheral Neuropathy (Peripheral Neuritis)—Inflammation and degeneration of the nerve endings or of the terminal nerves. It most often occurs in the nerve tissue of the muscles of the extremities (arms and legs). Symptoms include pain of varying intensity and sensations of numbness, tingling and burning in the hands and feet. It can be caused by certain medications or chemicals, infections, chronic inflammation or nutritive disease.

Peripheral-neuropathy Associated Drugs—Medications that are associated with the development of peripheral neuropathy. These include chloramphenicol, cisplatin, dapsone, didanosine, ethambutol, ethionamide, fluoroquinolones, hydralazine, isoniazid, lithium, metronidazole, nitrofurantoin, nitrous oxide, phenytoin, stavudine, vincristine, zalcitabine.

P-glycoprotein Inducers—Drugs or supplements that can decrease the amount of another drug in the body's cells. These drugs and supplements include dexamethasone, morphine, phenobarbital, rifampin, St. John's wort, trazodone and others. Consult doctor or pharmacist about interaction.

P-glycoprotein Inhibitors—Drugs that can increase the amount of another drug in the body's cells. The interaction may be helpful by increasing a drug's effectiveness. The interaction may be harmful by increasing the risk of a drug's side effects or toxicity. These drugs include amiodarone, amprenavir, clarithromycin, colchicine, cyclosporine A, daunorubicin, digoxin, diltiazem, dronedarone, erythromycin, indinavir, itraconazole, ketoconazole, loperamide, nelfinavir, nicardipine, omeprazole, paroxetine, propafenone, propranolol, quinidine, saquinavir, sertraline, tacrolimus, tamoxifen, ticagrelor, valspodar, verapamil, vinblastine and others. Consult doctor or pharmacist about interaction.

Phenothiazines—Drugs used to treat mental, nervous and emotional conditions. These drugs include acetophenazine, chlorpromazine, fluphenazine, mesoridazine, methotrimeprazine, pericyazine, perphenazine, prochlorperazine, promazine, thiopropazate, thioproperazine, thioridazine, trifluoperazine, triflupromazine.

Pheochromocytoma—A tumor of the adrenal gland that produces chemicals that cause high blood pressure, headache, nervousness and other symptoms.

Phlegm—Thick mucus secreted by glands in the respiratory tract.

Photophobia—Increased sensitivity to light as perceived by the human eye. Drugs that can cause photophobia include antidiabetic drugs, atropine, belladonna, bromides, chloroquine, ciprofloxacin, chlordiazepoxide, clidinium, clomiphene, dicyclomine, digitalis drugs, doxepin, ethambutol, ethionamide, ethosuximide, etretinate, glycopyrrolate, hydroxychloroquine, hydroxyzine, hyoscyamine, isopropamide, mephenytoin, methenamine, methsuximide, monoamine oxidase (MAO) inhibitors, nalidixic acid, norfloxacin, oral contraceptives, orphenadrine, paramethadione, phenothiazines, propantheline, quinidine, quinine, scopolamine, tetracyclines, tridihexethyl, trimethadione.

Photosensitizing Medications—Medicines that can cause abnormally heightened skin reactions to the effects of sunlight and ultraviolet light. These medicines include acetazolamide, acetohexamide, alprazolam, amantadine, amiloride, amiodarone, amitriptyline, amoxapine, antidiabetic agents (oral), barbiturates, bendroflumethiazide, benzocaine, benzoyl peroxide, benzthiazide, captopril, carbamazepine, chlordiazepoxide, chloroquine, chlorothiazide, chlorpromazine, chlorpropamide, chlortetracycline, chlorthalidone, ciprofloxacin, clindamycin, clofazimine, clofibrate, clomipramine, coal tar, contraceptives (estrogen-containing), cyproheptadine, dacarbazine, dapsone, demeclocycline, desipramine, desoximetasone, diethylstilbestrol, diflunisal, diltiazem, diphenhydramine, disopyramide, doxepin, doxycycline, enoxacin, estrogens, etretinate, flucytosine, fluorescein, fluorouracil, fluphenazine, flutamide, furosemide, glipizide, glyburide, gold preparations, griseofulvin, haloperidol, hexachlorophene, hydrochlorothiazide, hydroflumethiazide, ibuprofen, imipramine, indomethacin, isotretinoin, ketoprofen, lincomycin, lomefloxacin, maprotiline, mesoridazine, methacycline, methotrexate, methoxsalen, methyclothiazide, methyldopa, metolazone, minocycline, minoxidil, nabumetone, nalidixic acid, naproxen, nifedipine, norfloxacin, nortriptyline, ofloxacin, oral contraceptives, oxyphenbutazone, oxytetracycline, perphenazine, phenelzine, phenobarbital, phenylbutazone, phenytoin, piroxicam, polythiazide, prochlorperazine, promazine, promethazine, protriptyline, pyrazinamide, quinidine, quinine, sulfonamides, sulindac, tetracycline, thiabendazole, thioridazine, thiothixene, tolazamide, tolbutamide, tranylcypromine, trazodone, tretinoin, triamterene, trichlormethiazide, trifluoperazine, triflupromazine, trimeprazine, trimethoprim, trimipramine, triprolidine, vinblastine.

Pinworms—Common intestinal parasites that cause rectal itching and irritation.

Pituitary Gland—Gland at the base of the brain that secretes hormones to stimulate growth and other glands to produce hormones.

Platelet—Disc-shaped element of the blood, smaller than a red or white blood cell, necessary for blood clotting.

Polymyxins—A family of antibiotics that kill bacteria.

Polyp—Growth on a mucous membrane.

Porphyria—Inherited metabolic disorder characterized by changes in the nervous system and kidneys.

Post-Partum—Following delivery of a baby.

Potassium—Important chemical found in body cells.

Potassium Foods—Foods high in potassium content, including dried apricots and peaches, lentils, raisins, citrus and whole-grain cereals.

Potassium Supplements—Medicines needed by people who don't have enough potassium in their diets or by those who develop a deficiency due to illness or taking diuretics and other medicines. These supplements include chloride; potassium acetate; potassium bicarbonate; potassium bicarbonate and potassium chloride; potassium bicarbonate and potassium citrate; potassium chloride; potassium chloride, potassium bicarbonate and potassium citrate; potassium gluconate; potassium gluconate and potassium chloride; potassium gluconate and potassium citrate; potassium gluconate, potassium citrate and ammonium; trikates.

Prescription Drug Abuse—The use of a medication without a prescription; in a way other than as prescribed; or for the experience or feeling elicited. This term is used interchangeably with "nonmedical" use, a term employed by many of the national surveys.

Premenstrual Dysphoric Disorder (PMDD)—A severe form of premenstrual syndrome which is characterized by severe monthly mood swings as well as physical symptoms that interfere with everyday life, especially a woman's relationships with her family and friends.

Progesterone—A female steroid sex hormone that is responsible for preparing the uterus for pregnancy.

Progestin—A synthetic hormone that is designed to mimic the actions of progesterone. These include etonogestrel, hydroxy-progesterone, medroxyprogesterone, megestrol, norethindrone, norgestrel and progesterone.

Prostaglandins—A group of drugs used for a variety of therapeutic purposes. These drugs include alprostadil (treats newborns with congenital heart disease), carboprost and dinoprost (both used to induce labor) and dinoprostone (used to induce labor or to induce a late abortion).

Prostate—Gland in the male that surrounds the neck of the bladder and the urethra.

Protease Inhibitors—A class of anti-HIV drugs which inhibit the protease enzyme and stop virus replication. These drugs include: abacavir, atazanavir, darunavir, fosamprenavir, indinavir, lopinavir and ritonavir; nelfinavir, nifedipine, ritonavir, saquinavir and tipranavir.

Protein Bound Drugs—This protein is not to be confused with the protein in your food. Drugs that are highly protein bound could interact and cause an increased amount of one drug in the blood. This can increase the risk of adverse effects. These drugs include clofibrate, diazepam, diazoxide, fluoxetine, ibuprofen, indomethacin, naproxen, nonsteroida anti-inflammatory drugs (NSAIDS), raloxifene, sulfonylureas, sulfonamides, tiagabine, valproate, vilazodone, warfarin.

Prothrombin—Blood substance essential in clotting.

Prothrombin Time (Pro Time)—Laboratory study used to follow prothrombin activity and keep coagulation safe.

Psoriasis—Chronic inherited skin disease. Symptoms are lesions with silvery scales on the edges.

Psychosis—Mental disorder characterized by deranged personality, loss of contact with reality and possible delusions, hallucinations or illusions.

Purine Foods—Foods that are metabolized into uric acid. Foods high in purines include anchovies, liver, brains, sweetbreads, sardines, kidneys, oysters, gravy and meat extracts.

Q

QT Interval Prolongation-Causing Drugs—A group of drugs that can cause serious heart rhythm problems. These drugs include amiodarone, aripiprazole, asenapine, azithromycin, azole antifungals, calcium channel blockers (especially bepridil), chloroquine, chlorpromazine, cisapride, clarithromycin, clozapine, dirithromycin, disopyramide, dofetilide, domperidone, dronedarone, droperidol, erlotinib, erythromycins, flecainide, fluoroquinolones, halofantrine, haloperidol, ibutilide, iloperidone, imipremine, macrolide antibiotics, maprotiline, mesoridazine, methadone, olanzapine, pentamidine, phenothiazines, piliperidone, pimozide, procainamide, quetiapine, quinidine, ranolazine, risperidone, sotalol, sparfloxacin, thioridazine, toremifene, tricyclic antidepressants, tyrosine kinase inhibitors, zisprasidone and others.

R

Rauwolfia Alkaloids—Drugs that belong to the family of antihypertensives (drugs that lower blood pressure). Rauwolfia alkaloids are not used as extensively as in years past. They include alseroxylon, deserpidine, rauwolfia serpentina, reserpine.

RDA—Recommended daily allowance of a vitamin or mineral.

Rebound Effect—The worsening or return of symptoms when a drug (such as a decongestant) is discontinued or a patient no longer responds to it.

Renal—Pertaining to the kidney.

Retina—Innermost covering of the eyeball on which the image is formed.

Retinoids—A group of drugs that are synthetic vitamin A-like compounds used to treat skin conditions. These drugs include etretinate, isotretinoin and retinoic acid.

Retroperitoneal Imaging—Special x-rays or CT scans of the organs attached to the abdominal wall behind the peritoneum (the covering of the intestinal tract and lining of the walls of the abdominal and pelvic cavities).

Reye's Syndrome—Rare, sometimes fatal, disease of children that causes brain and liver damage.

Rickets—Bone disease caused by vitamin D deficiency. Bones become bent and distorted during infancy or childhood.

S

Salicylates—Medicines to relieve pain and reduce fever. These include aspirin, aspirin and caffeine, balsalazide, buffered aspirin, choline salicylate, choline and magnesium salicylates, magnesium salicylate, salicylamide, salsalate, sodium salicylate.

Sedatives—Drugs that reduce excitement or anxiety. They are used to produce sedation (calmness). These include alprazolam, amobarbital, aprobarbital, bromazepam, butalbital, chloral hydrate, clonazepam, clorazepate, chlordiazepoxide, diazepam, diphenhydramine, doxylamine, estazolam, ethchlorvynol, ethinamate, eszopiclone, flurazepam, glutethimide, halazepam, hydroxyzine, ketazolam, lorazepam, methotrimeprazine, midazolam, nitrazepam, oxazepam, pentobarbital, phenobarbital, prazepam, promethazine, propiomazine, propofol, quazepam, secobarbital, temazepam, triazolam, trimeprazine, zaleplon, zolpidem, zopiclone.

Seizure—A sudden attack of epilepsy or some other disease can cause changes of consciousness or convulsions.

Seizure threshold lowering drugs—Seizure threshold refers to the minimal conditions required to trigger a seizure. A number of drugs can lower the seizure threshold in susceptible patients and increase the risk for a seizure. These drugs include certain antibiotics, antiasthmatics, antidepressants, antipsychotics, phenothiazines, psychostimulants, hormones, local anesthetics, immunosuppressants, narcotics and others. Herbal remedies may also lower seizure threshold. Consult your doctor about your risks.

Selective Serotonin Reuptake Inhibitors (SSRIs)—Medications used for treatment of depression that work by increasing the serotonin levels in the brain. Serotonin is a neuro- transmitter (brain chemical) having to do with mood and behavior. These drugs include fluoxetine, fluvoxamine, paroxetine, sertraline. More information can be found on the individual drug chart for each drug.

Serotonergics—Drugs that increase the levels of serotonin (a brain chemical). Excess serotonin can lead to serotonin syndrome (a possibly life-threatening condition). These drugs include almotriptan, amitriptyline, amphetamines, bromocriptine, buspirone, citalopram, clomipramine, desvenlafaxine, dextromethorphan, duloxetine, eletriptan, escitalopram, fenfluramine, fluoxetine, fluvoxamine, frovatriptan, imipramine, levodopa, linezolid, lithium, meperidine, moclobemide, monoamine oxidase (MAO) inhibitors, naratriptan, nefazodone, paroxetine, pentazocine, rizatriptan, sertraline, sibutramine, St. John's Wort, sumatriptan, tapentadol, tramadol, trazodone, tricyclic antidepressants, tryptophan, valproic acid, venlafaxine, vilazodone, zolmitriptan and some drugs of abuse (e.g., cocaine, LSD, ecstasy, marijuana and others). Other drugs or herbal supplements may be risk factors also. Ask your doctor or pharmacist about the safety of any drugs you are prescribed and advised them of any herbal supplements you take.

Serotonin Syndrome—A potentially very serious and life-threatening condition caused by drugs that increase serotonin levels in the brain. The syndrome can be caused by an overdose, interaction with other drugs, or rarely, with doses used in treatment. It is more likely to occur when starting a drug or increasing the dosage of a drug. Symptoms often come on quickly and progress rapidly. Symptoms may include confusion, agitation, headache, diarrhea, irritability, muscle rigidity, high body temperature, fast heart beat, rapid change in blood pressure, nausea or vomiting, poor coordination, dilated pupils, restlessness, hallucinations, overactive reflexes, sweating, shivering, tremor, and muscle twitching. More severe symptoms can include seizures, delirium, shock loss of consciousness and other major medical problems.

SGOT—Abbreviation for serum glutamic-oxaloacetic transaminase. Measuring the level in the blood helps demonstrate liver disorders and diagnose recent heart damage.

SGPT—Abbreviation for a laboratory study measuring the blood level of serum glutamic-pyruvic transaminase. Deviations from a normal level may indicate liver disease.

Sick Sinus Syndrome—A complicated, serious heartbeat rhythm disturbance characterized by a slow heart rate alternating with a fast or slow heart rate with heart block.

Sinusitis—Inflammation or infection of the sinus cavities in the skull.

Skeletal Muscle Relaxants (same as Skeletal Muscle Blockers)—A group of drugs prescribed to treat spasms of the skeletal muscles. These drugs include carisoprodol, chlorphenesin, chlorzoxazone, cyclobenzaprine, diazepam, lorazepam, metaxalone, methocarbamol, orphenadrine, phenytoin.

Sleep Inducers—Night-time sedatives to aid in falling asleep.

Sleep-Related Behaviors—Certain behaviors that can occur with the use of sedative-hypnotic drugs. The behaviors include: cooking and eating, using the telephone, having sex and sleep-driving (driving while not fully awake). Typically, the person has no memory of these actions.

Statins—See HMG-CoA Reductase Inhibitors.

Streptococci—A bacteria that can cause infections in the throat, respiratory system and skin. Improperly treated, can lead to disease in the heart, joints and kidneys.

Stroke—Sudden, severe attack, usually sudden paralysis, from injury to the brain or spinal cord caused by a blood clot or hemorrhage in the brain.

Stupor—Near unconsciousness.

Sublingual—Under the tongue. Some drugs are absorbed almost as quickly this way as by injection.

Sulfa Drugs—Shorthand for sulfonamide drugs, which are used to treat infections.

Sulfonamides—Sulfa drugs prescribed to treat infections. They include sulfacytine, sulfamethoxazole, sulfamethoxazole and trimethoprim, sulfasalazine, sulfisoxazole.

Sulfonylureas—A family of drugs that lower blood sugar (hypoglycemic agents). Used in the treatment of some forms of diabetes.

Sympatholytics—A group of drugs that block the action of the sympathetic nervous system. These drugs include beta-blockers, guanethidine, hydralazine and prazosin.

Sympathomimetics—A large group of drugs that mimic the effects of stimulation of the sympathetic part of the autonomic nervous system. These drugs include albuterol, amphetamine, benzphetamine, bitolterol, clonidine, cocaine, dextroamphetamine, diethylpropion, dipivefrin, dobutamine, dopamine, ephedrine, epinephrine, indacaterol, ipratropium, isoproterenol, isoetharine, mazindol, mephentermine, metaproterenol, metaraminol, methoxamine, midodrine, naphazoline, norepinephrine, oxymetazoline, phendimetrazine, phentermine, phenylephrine, pirbuterol, pseudoephedrine, ritodrine, salmeterol, terbutaline, tetrahydrozoline, xylometazoline.

T

Tardive Dyskinesia—Slow, involuntary movements of the jaw, lips and tongue caused by an unpredictable drug reaction. Drugs that can cause this include haloperidol, phenothiazines, thiothixene.

Tartrazine Dye—A dye used in foods and medicine preparations that may cause an allergic reaction in some people.

Tetracyclines—A group of medicines with similar chemical structure used to treat infections. These drugs include demeclocycline, doxycycline, methacycline, minocycline, oxytetracycline, tetracycline.

Thiazides—A group of chemicals that cause diuresis (loss of water through the kidneys). Frequently used to treat high blood pressure and congestive heart failure. Thiazides include bendroflumethiazide, benzthiazide, chlorothiazide, chlorthalidone, cyclothiazide, hydrochlorothiazide, hydroflumethiazide, methyclothiazide, metolazone, polythiazide, quinethazone, trichlormethiazide.

Thiothixines—See Thioxanthenes.

Thioxanthenes—Drugs used to treat emotional, mental and nervous conditions. These drugs include chlorprothixene, flupenthixol, thiothixene.

Thrombocytopenias—Diseases characterized by inadequate numbers of blood platelets circulating in the bloodstream.

Thrombolytic Agents—Drugs that help to dissolve blood clots. They include alteplase, anistreplase, streptokinase, urokinase.

Thrombophlebitis—Inflammation of a vein caused by a blood clot in the vein.

Thyroid—Gland in the neck that manufactures and secretes several hormones.

Thyroid Hormones—Medications that mimic the action of the thyroid hormone made in the thyroid gland. They include dextrothyroxine, levothyroxine, liothyronine, liotrix, thyroglobulin, thyroid.

Tic Douloureux—Painful condition caused by inflammation of a nerve in the face.

Tolerance—A decreasing response to repeated constant doses of a drug or a need to increase doses to produce the same physical or mental response.

Toxicity—Poisonous reaction to a drug that impairs body functions or damages cells.

Tranquilizers—Drugs that calm a person without clouding consciousness.

Transdermal Patches—Medicated patches that stick to the skin. There are more and more medications in this form. This method produces a prolonged systemic effect. If you are using this form, follow these instructions: Choose an area of skin without cuts, scars or hair, such as the upper arm, chest or behind the ear. Thoroughly clean area where patch is to be applied. If patch gets wet and loose, cover with an additional piece of plastic. Apply a fresh patch if the first one falls off. Apply each dose to a different area of skin if possible.

Tremor—Involuntary trembling.

Trichomoniasis—Infestation of the vagina by trichomonas, an infectious organism. The infection causes itching, vaginal discharge and irritation.

Triglyceride—Fatty chemical manufactured from carbohydrates for storage in fat cells.

Tyramine—Normal chemical component of the body that helps sustain blood pressure. Can rise to fatal levels in combination with some drugs. Tyramine is found in many foods:

Beverages—Alcohol beverages, especially Chianti or robust red wines, vermouth, ale, beer.

Breads—Homemade bread with a lot of yeast and breads or crackers containing cheese.

Fats—Sour cream.

Fruits—Bananas, red plums, avocados, figs, raisins, raspberries.

Meats and meat substitutes—Aged game, liver (if not fresh), canned meats, salami, sausage, aged cheese, salted dried fish, pickled herring, meat tenderizers.

Vegetables—Italian broad beans, green bean pods, eggplant.

Miscellaneous—Yeast concentrates or extracts, marmite, soup cubes, commercial gravy, soy sauce, any protein food that has been stored improperly or is spoiled.

U

Ulcer, Peptic—Open sore on the mucous membrane of the esophagus, stomach or duodenum caused by stomach acid.

Urethra—Hollow tube through which urine (and semen in men) is discharged.

Urethritis—Inflammation or infection of the urethra.

Uricosurics—A group of drugs that promotes excretion of uric acid in the urine. These drugs include probenecid and sulfinpyrazone.

Urinary Acidifiers—Medications that cause urine to become acid. These include ascorbic acid, potassium phosphate, potassium and sodium phosphates, racemethionine.

Urinary Alkalizers—Medications that cause urine to become alkaline. These include potassium citrate, potassium citrate and citric acid, potassium citrate and sodium citrate, sodium bicarbonate, sodium citrate and citric acid, tricitrate.

Uterus—Also called the womb. A hollow muscular organ in the female in which the embryo develops into a fetus.

V

Valproic Acid Drugs—Anticonvulsant drugs that are used to prevent seizures in epilepsy, and as treatment for migraine and bipolar disorder. They include oral drugs divalproex and valproic acid and the injectable drug valproate.

Vascular—Pertaining to blood vessels.

Vascular Headache Preventatives—Medicines prescribed to prevent the occurrence of or reduce the frequency and severity of vascular headaches such as migraines. These drugs include atenolol; clonidine; ergotamine, belladonna alkaloids and phenobarbital; fenoprofen; flunarizine; ibuprofen; indomethacin; isocarboxazid, lithium; mefenamic acid; methysergide; metoprolol; nadolol; naproxen; phenelzine; pizotyline; propranolol; timolol, tranylcypromine, verapamil.

Vascular Headache Treatment—Medicine prescribed to treat vascular headaches such as migraines. These drugs include butalbital (combined with acetaminophen, aspirin, caffeine or codeine); cyproheptadine; diclofenac; diflunisal; dihydroergotamine; ergotamine; ergotamine and caffeine; ergotamine, caffeine, belladonna alkaloids and pentobarbital; etodolac; fenoprofen; ibuprofen; indomethacin (capsules, oral suspension, rectal); isometheptene, dichloralphenazone and acetaminophen; ketoprofen; meclofenamate; mefenamic acid; metoclopramide; naproxen; phenobarbital; triptans.

Vasoconstrictor—Any agent that causes a narrowing of the blood vessels.

Vasodilator—Any agent that causes a widening of the blood vessels.

Vertigo—A sensation of motion, usually dizziness or whirling either of oneself or one's surroundings.

Virus—Infectious organism that reproduces in the cells of the infected host. Viruses cause many diseases in humans including the common cold.

W

Withdrawal—Symptoms that occur after chronic use of a drug is reduced abruptly or stopped.

X

Xanthines—Substances that stimulate muscle tissue, especially that of the heart. Types of xanthines include aminophylline, caffeine, dyphylline, oxtriphylline, theophylline.

Y

Yeast—A single-cell organism that can cause infections of the mouth, vagina, skin and parts of the gastrointestinal system.

INDEX

INDEX

INDEX

How to Find a Drug Name in the Index

Look it up by its brand name, such as Tylenol, or its generic name, such as acetaminophen (which is the generic name for Tylenol).

The drug names in the index appear in two different formats:

1. Drug brand (or trade) names. These names appear in ***bold italic***, and are followed by the drug chart name and a page number. (The drug chart name may be a single generic drug name or a drug family name.)

Example: ***Bayer*** - See ASPIRIN 152

Many brand name drugs contain two or more generic ingredients. These brand names will be followed by two or more drug chart names. Refer to all the drug charts listed for complete information on that particular brand name.

Example: ***Allent*** - See
ANTIHISTAMINES 104
PSEUDOEPHEDRINE 666

2. Drug generic names or drug family names (a group of similar generic drugs such as antihistamines). These names appear in all capital letters and are followed by their chart page number. An underlined name in the index indicates that it is a title of a drug chart in the book.

Example: HMG-CoA REDUCTASE INHIBITORS 410
Example: ATORVASTATIN - See HMG-CoA REDUCTASE INHIBITORS 410

Drug class names are also listed. The drug class is usually a treatment (therapeutic) name (e.g., antihypertensive). These names appear in regular type with capital and lowercase letters. Each drug class name is followed by drug chart name(s) that are in that class:

Example: Antifungal - See
ANTIFUNGALS, AZOLES 84
CLOTRIMAZOLE (Oral-Local) 258
GRISEOFULVIN 402
NYSTATIN 584

INDEX

INDEX

B

C

INDEX

D

E

F

G

J

K

L

INDEX

M

N

INDEX

O

P

Q

R

S

T

INDEX

U

V

Emergency

Emergency Guide for Overdose Victims

These are basic steps in recognizing and treating immediate effects of drug overdose.

Take just 5 to 10 seconds to check for a response from the victim. In an adult, check for *normal* breathing. In a child, check for the *presence* or *absence* of breathing (a child may be breathing in a pattern that is not normal, but is adequate).

IF VICTIM IS NOT BREATHING:

Adults and children over age 1:

1. Call 911. The emergency dispatcher can help coach you through CPR (cardiopulmonary resuscitation) until experienced help arrives. If there are other persons available, have one call 911 and one or more persons perform CPR .
2. Begin CPR. Open the victim's airway—tilt the head back and lift chin. Pinch the victim's nose and prepare to give two rescue breaths. Give the first rescue breath (last one second) into his or her mouth. The chest should rise. If it does, give the second breath. If the chest does not rise after the first breath, reopen the airway.
3. Then place your hands on the breastbone in the middle of the chest. Use the heel of both hands (stacked) and push down on the chest 1½ to 2 inches 30 times right between the nipples. Pump hard and fast at the rate of 100/minute (faster than once per second). Let the chest rebound between pumps. For a child ages 1 to 8, use 1 hand or 2 hands as needed to compress the chest one third to one half its depth.
4. Continue with 2 breaths and 30 pumps until victim is breathing or medical help arrives. This ratio is the same for one-person and two-person CPR. In two-person CPR the person pumping the chest stops while the other gives mouth-to-mouth breathing.

Infants under age 1 (excluding newborns):

1. If there are other persons present, have one call 911 and one or more persons begin CPR. If you are alone, perform about 5 cycles or 2 minutes of CPR and then call 911.
2. Begin CPR. Place infant on his or her back. Open the airway—tilt head back and lift chin. Do not tilt the head too far back. Cover the infant's mouth and nose with your mouth. Give 2 small gentle breaths. Each breath should be 1 second long. The infant's chest should rise with each breath.
3. Then place two fingers in the center of the chest just below the nipples. Give 30 gentle chest compressions at the rate of 100 per minute. Press down about one-third the depth of the chest.
4. Continue with 2 breaths and 30 pumps until victim is breathing or medical help arrives. This ratio is the same for one-person and two-person CPR. In two-person CPR the person pumping the chest stops while the other gives mouth-to-mouth breathing.

Note: For all victims, don't try to make the victim vomit. If vomiting does occur, turn the victim's head to the side and try to sweep out or wipe off the vomit. Continue with CPR. Save vomit to take to emergency room for analysis. Take medicine or empty pill bottles with you to emergency room.

IF VICTIM IS UNCONSCIOUS AND BREATHING:

1. Call 911 for medical help.
2. If you are in a location where you can't get emergency help immediately, take victim to the nearest emergency room. Monitor the victim's breathing.
3. Don't try to make victim vomit. If vomiting occurs, save vomit to take to emergency room for analysis.
4. Take medicine or empty pill bottles with you to emergency room.

IF VICTIM IS DROWSY:

1. Call 911 for medical help.
2. If you are in a location where you can't get emergency help immediately, take victim to the nearest emergency room. Monitor the victim's breathing.
3. Don't try to make victim vomit. If vomiting occurs, save vomit to take to emergency room for analysis.
4. Take medicine or empty pill bottles with you to emergency room.

IF VICTIM IS ALERT:

1. Call 911 for medical help or call Poison Control Center–1-800-222-1222 (in the United States) for instructions.
2. If you can't get medical help quickly, take victim to nearest emergency room.
3. Take medicine or empty pill bottles with you to emergency room.

IF VICTIM HAS NO SYMPTOMS, BUT YOU SUSPECT AN OVERDOSE:

1. Call Poison Control Center 1-800-222-1222 (in the United States).
2. Describe the suspect drug with as much information as you can quickly gather. The center will give emergency instructions.
3. Consider calling the victim's doctor or your doctor for instructions.
4. If you have no telephone, take victim to the nearest emergency room.
5. Take medicine or empty pill bottles with you to emergency room.

A note about CPR training: Just about everyone can benefit by taking a certified course in first aid and learn to perform cardiopulmonary resuscitation (CPR). Courses taught by certified instructors are the best way to learn CPR. To find a course in your area, contact:

American Heart Association: 800-242-8721 or check website at www.americanheart.org.

American Red Cross: call the local office or check website at www.redcross.org.

Also, classes are offered in almost every city and town, at YMCAs, civic centers, schools, doctors' offices, and elsewhere.

Emergency Guide for Anaphylaxis Victims

The following are basic steps in recognizing and treating immediate effects of a severe allergic reaction, which is called *anaphylaxis*.

Some people may be highly sensitive to drugs. An anaphylactic reaction to a drug can be life-threatening! Persons suffering these allergic symptoms need immediate emergency treatment!

Symptoms of anaphylaxis:

- **Itching**
- **Rash**
- **Hives**
- **Runny nose**
- **Wheezing**
- **Paleness**
- **Cold sweats**
- **Low blood pressure**
- **Coma**
- **Cardiac arrest**

IF VICTIM IS UNCONSCIOUS AND NOT BREATHING:

1. Call 911. The emergency dispatcher may be able to coach you through CPR until experienced help arrives. If there are other persons available, have one call 911 and one or more persons perform CPR (cardiopulmonary resuscitation).
2. Check to see if the person is carrying special medication (e.g., Epi-Pen or allergy kit) to inject to counter the effects of the allergic attack. If so, inject it if you know how to do so. Avoid oral medication if the person is not breathing.
3. Begin CPR. See instructions 2 through 4 under Emergency Guide for Overdose Victims on page 1088 (2nd previous page).

IF VICTIM IS UNCONSCIOUS AND BREATHING:

1. Call 911 for medical help.
2. Check to see if the person is carrying special medication to inject to counter the effects of the allergic attack. If so, inject it if you know how to do so.
3. Take medicine or empty bottles with you to emergency room for analysis.

IF VICTIM IS CONSCIOUS:

1. Ask the person if he or she is carrying special medication to inhale, swallow or inject to counter the effects of the allergic attack. Help the person use the medication.
2. Give the person an antihistamine pill, if it is available.
3. Encourage the person to breathe slowly and deeply. Stay calm and be reassuring.

DISCLAIMER: The information provided in Emergency Guide for Overdose Victims and Emergency Guide for Anaphylaxis Victims is of a general nature and should not be used as a substitute for professional medical advice, emergency treatment instructions, or formal first aid training. If you think you or a family member may suffer from an allergic or other disease that requires attention, you should discuss it with your health care provider.